# Seas at the Millennium: An Environmental Evaluation

# Seas at the Millennium: An Environmental Evaluation

*Edited by*

**Charles R.C. Sheppard**
*Department of Biological Sciences,*
*University of Warwick,*
*Coventry, U.K.*

## Volume III
## Global Issues and Processes



2000
**PERGAMON**
*An imprint of Elsevier Science*

AMSTERDAM – LAUSANNE – NEW YORK – OXFORD – SHANNON – SINGAPORE – TOKYO

ELSEVIER SCIENCE Ltd
The Boulevard, Langford Lane
Kidlington, Oxford OX5 1GB, UK



First edition 2000

*Library of Congress Cataloging in Publication Data*
Seas at the millennium: an environmental evaluation / edited by Charles Sheppard. -- 1st ed.
p. cm.
Includes bibliographical references.
ISBN 0-08-043207-7 (alk. paper)
1. Marine ecology. I. Sheppard, Charles (Charles R.C.)
QH541.5.S3 S35 2000
577.7--dc21 00-034738

*British Library Cataloguing in Publication Data*
Seas at the millennium: an environmental evaluation
1. Marine ecology 2. Coastal ecology 3. Marine resources conservation
I. Sheppard, Charles R.C.
577.7

ISBN: 0-08-043207-7

(∞) The paper used in this publication meets the requirements of ANSI/NISO Z39.48-1992 (Permanence of Paper).

Printed in The Netherlands.

# CONTENTS


*List of Authors* . . . xi

*Acknowledgements* . . . xxi

**Volume I. Regional Chapters: Europe, The Americas and West Africa**

INTRODUCTION. . . . 1
*Charles Sheppard*
Chapter 1. THE SEAS AROUND GREENLAND . . . 5
*Frank Riget, Poul Johansen, Henning Dahlgaard, Anders Mosbech, Rune Dietz and Gert Asmund*
Chapter 2. THE NORWEGIAN COAST . . . 17
*Jens Skei, Torgeir Bakke and Jarle Molvaer*
Chapter 3. THE FAROE ISLANDS . . . 31
*Maria Dam, Grethe Bruntse, Andrias Reinert and Jacob Pauli Joensen*
Chapter 4. THE NORTH SEA . . . 43
*Jean-Paul Ducrotoy, Mike Elliott and Victor N. de Jonge*
Chapter 5. THE ENGLISH CHANNEL . . . 65
*Alan D. Tappin and P. Chris Reid*
Chapter 6. THE IRISH SEA . . . 83
*Richard G. Hartnoll*
Chapter 7. THE BALTIC SEA, ESPECIALLY SOUTHERN AND EASTERN REGIONS . . . 99
*Jerzy Falandysz, Anna Trzosinska, Piotr Szefer, Jan Warzocha and Bohdan Draganik*
Chapter 8. THE BALTIC SEA, INCLUDING BOTHNIAN SEA AND BOTHNIAN BAY. . . . 121
*Lena Kautsky and Nils Kautsky*
Chapter 9. THE NORTH COAST OF SPAIN . . . 135
*Isabel Díez, Antonio Secilla, Alberto Santolaria and José María Gorostiaga*
Chapter 10. SOUTHERN PORTUGAL: THE TAGUS AND SADO ESTUARIES . . . 151
*Graça Cabeçadas, Maria José Brogueira and Leonor Cabeçadas*
Chapter 11. THE ATLANTIC COAST OF SOUTHERN SPAIN . . . 167
*Carlos J. Luque, Jesús M. Castillo and M. Enrique Figueroa*
Chapter 12. THE CANARY ISLANDS . . . 185
*Francisco García Montelongo, Carlos Díaz Romero and Ricardo Corbella Tena*
Chapter 13. THE AZORES . . . 201
*Brian Morton and Joseph C. Britton*
Chapter 14. THE SARGASSO SEA AND BERMUDA . . . 221
*Anthony H. Knap, Douglas P. Connelly and James N. Butler*
Chapter 15. THE AEGEAN SEA . . . 233
*Manos Dassenakis, Kostas Kapiris and Alexandra Pavlidou*
Chapter 16. THE COAST OF ISRAEL, SOUTHEAST MEDITERRANEAN . . . 253
*Barak Herut and Bella Galil*
Chapter 17. THE ADRIATIC SEA AND THE TYRRHENIAN SEA . . . 267
*Giuseppe Cognetti, Claudio Lardicci, Marco Abbiati and Alberto Castelli*
Chapter 18. THE BLACK SEA . . . 285
*Gülfem Bakan and Hanife Büyükgüngör*
Chapter 19. THE GULF OF MAINE AND GEORGES BANK. . . . 307
*Jack B. Pearce*
Chapter 20. THE NEW YORK BIGHT . . . 321
*Jack B. Pearce*

Chapter 21. CHESAPEAKE BAY: THE UNITED STATES' LARGEST ESTUARINE SYSTEM . . . . . . . . . . . 335
*Kent Mountford*
Chapter 22. NORTH AND SOUTH CAROLINA COASTS . . . . . . . . . . . . . . . . . . . . . . . . . . . . 351
*Michael A. Mallin, JoAnn M. Burkholder, Lawrence B. Cahoon and Martin H. Posey*
Chapter 23. THE GULF OF ALASKA . . . . . . . . . . . . . . . . . . . . . . . . . . . . . . . . . . . . . . 373
*Bruce A. Wright, Jeffrey W. Short, Tom J. Weingartner and Paul J. Anderson*
Chapter 24. SOUTHERN CALIFORNIA . . . . . . . . . . . . . . . . . . . . . . . . . . . . . . . . . . . . . 385
*Kenneth C. Schiff, M. James Allen, Eddy Y. Zeng and Steven M. Bay*
Chapter 25. FLORIDA KEYS . . . . . . . . . . . . . . . . . . . . . . . . . . . . . . . . . . . . . . . . . . 405
*Phillip Dustan*
Chapter 26. THE BAHAMAS . . . . . . . . . . . . . . . . . . . . . . . . . . . . . . . . . . . . . . . . . . 415
*Kenneth C. Buchan*
Chapter 27. THE NORTHERN GULF OF MEXICO . . . . . . . . . . . . . . . . . . . . . . . . . . . . . . . 435
*Mark E. Pattillo and David M. Nelson*
Chapter 28. COASTAL MANAGEMENT IN LATIN AMERICA . . . . . . . . . . . . . . . . . . . . . . . . . . 457
*Alejandro Yáñez-Arancibia*
Chapter 29. SOUTHERN GULF OF MEXICO . . . . . . . . . . . . . . . . . . . . . . . . . . . . . . . . . . 467
*Felipe Vázquez, Ricardo Rangel, Arturo Mendoza Quintero-Marmol, Jorge Fernández, Eduardo Aguayo, E.A. Palacio and Virender K. Sharma*
Chapter 30. THE PACIFIC COAST OF MEXICO . . . . . . . . . . . . . . . . . . . . . . . . . . . . . . . . 483
*Alfonso V. Botello, Alejandro O. Toledo, Guadalupe de la Lanza-Espino and Susana Villanueva-Fragoso*
Chapter 31. BELIZE . . . . . . . . . . . . . . . . . . . . . . . . . . . . . . . . . . . . . . . . . . . . . . 501
*Alastair R. Harborne, Melanie D. McField and E. Kate Delaney*
Chapter 32. NICARAGUA: CARIBBEAN COAST . . . . . . . . . . . . . . . . . . . . . . . . . . . . . . . . 517
*Stephen C. Jameson, Lamarr B. Trott, Michael J. Marshall and Michael J. Childress*
Chapter 33. NICARAGUA: PACIFIC COAST . . . . . . . . . . . . . . . . . . . . . . . . . . . . . . . . . . 531
*Stephen C. Jameson, Vincent F. Gallucci and José A. Robleto*
Chapter 34. EL SALVADOR . . . . . . . . . . . . . . . . . . . . . . . . . . . . . . . . . . . . . . . . . . 545
*Linos Cotsapas, Scott A. Zengel and Enrique J. Barraza*
Chapter 35. JAMAICA . . . . . . . . . . . . . . . . . . . . . . . . . . . . . . . . . . . . . . . . . . . . . 559
*Marjo Vierros*
Chapter 36. PUERTO RICO . . . . . . . . . . . . . . . . . . . . . . . . . . . . . . . . . . . . . . . . . . 575
*Jack Morelock, Jorge Capella, Jorge Garcia and Maritza Barreto*
Chapter 37. THE TURKS AND CAICOS ISLANDS . . . . . . . . . . . . . . . . . . . . . . . . . . . . . . . 587
*Gudrun Gaudian and Paul Medley*
Chapter 38. THE DUTCH ANTILLES . . . . . . . . . . . . . . . . . . . . . . . . . . . . . . . . . . . . . . 595
*Adolphe O. Debrot and Jeffrey Sybesma*
Chapter 39. UK OVERSEAS TERRITORIES IN THE NORTHEAST CARIBBEAN: ANGUILLA, BRITISH VIRGIN ISLANDS, MONTSERRAT . . . . . . . . . . . . . . . . . . . . . . . . . . . . . . . . . . . . . . . . . 615
*Fiona Gell and Maggie Watson*
Chapter 40. THE LESSER ANTILLES, TRINIDAD AND TOBAGO . . . . . . . . . . . . . . . . . . . . . . . . 627
*John B.R. Agard and Judith F. Gobin*
Chapter 41. VENEZUELA . . . . . . . . . . . . . . . . . . . . . . . . . . . . . . . . . . . . . . . . . . . 643
*Pablo E. Penchaszadeh, César A. Leon, Haymara Alvarez, David Bone, P. Castellano, María M. Castillo, Yusbelly Diaz, María P. Garcia, Mairin Lemus, Freddy Losada, Alberto Martin, Patricia Miloslavich, Claudio Paredes, Daisy Perez, Miradys Sebastiani, Dennise Stecconi, Victoriano Roa and Alicia Villamizar*
Chapter 42. THE CARIBBEAN COAST OF COLOMBIA . . . . . . . . . . . . . . . . . . . . . . . . . . . . . . 663
*Leonor Botero and Ricardo Alvarez-León*
Chapter 43. THE PACIFIC COAST OF COLOMBIA . . . . . . . . . . . . . . . . . . . . . . . . . . . . . . . 677
*Alonso J. Marrugo-González, Roberto Fernández-Maestre and Anders A. Alm*
Chapter 44. PERU . . . . . . . . . . . . . . . . . . . . . . . . . . . . . . . . . . . . . . . . . . . . . . 687
*Guadalupe Sanchez*
Chapter 45. THE CHILEAN COAST . . . . . . . . . . . . . . . . . . . . . . . . . . . . . . . . . . . . . . 699
*Ramón B. Ahumada, Luis A. Pinto and Patricio A. Camus*

Chapter 46. TROPICAL COAST OF BRAZIL . . . . . . 719
*Zelinda M.A.N. Leão and José M.L. Dominguez*
Chapter 47. SOUTHERN BRAZIL . . . . . . 731
*Eliete Zanardi Lamardo, Márcia Caruso Bícego, Belmiro Mendes de Castro Filho, Luiz Bruner de Miranda and Valéria Aparecida Prósperi*
Chapter 48. THE ARGENTINE SEA: THE SOUTHEAST SOUTH AMERICAN SHELF MARINE ECOSYSTEM . 749
*José L. Esteves, Nestor F. Ciocco, Juan C. Colombo, Hugo Freije, Guillermo Harris, Oscar Iribarne, Ignacio Isla, Paulina Nabel, Marcela S. Pascual, Pablo E. Penchaszadeh, Andrés L. Rivas and Norma Santinelli*
Chapter 49. THE GULF OF GUINEA LARGE MARINE ECOSYSTEM . . . . . . 773
*Nicholas J. Hardman-Mountford, Kwame A. Koranteng and Andrew R.G. Price*
Chapter 50. GUINEA. . . . . . 797
*Ibrahima Cisse, Idrissa Lamine Bamy, Amadou Bah, Sékou Balta Camara and Mamba Kourouma*
Chapter 51. CÔTE D'IVOIRE . . . . . . 805
*Ama Antoinette Adingra , Robert Arfi and Aka Marcel Kouassi*
Chapter 52. SOUTHWESTERN AFRICA: NORTHERN BENGUELA CURRENT REGION . . . . . . 821
*David Boyer, James Cole and Christopher Bartholomae*

*Index* . . . . . . 841

## Volume II. Regional Chapters: The Indian Ocean to The Pacific

Chapter 53. THE ARABIAN GULF . . . . . . 1
*D.V. Subba Rao and Faiza Al-Yamani*
Chapter 54. NORTHWEST ARABIAN SEA AND GULF OF OMAN. . . . . . 17
*Simon C. Wilson*
Chapter 55. THE RED SEA . . . . . . 35
*Charles R.C. Sheppard*
Chapter 56. THE GULF OF ADEN . . . . . . 47
*Simon C. Wilson and Rebecca Klaus*
Chapter 57. THE INDIAN OCEAN COAST OF SOMALIA . . . . . . 63
*Federico Carbone and Giovanni Accordi*
Chapter 58. TANZANIA . . . . . . 83
*Martin Guard, Aviti J. Mmochi and Chris Horrill*
Chapter 59. MOZAMBIQUE. . . . . . 99
*Michael Myers and Mark Whittington*
Chapter 60. MADAGASCAR . . . . . . 113
*Andrew Cooke, Onésime Ratomahenina, Eulalie Ranaivoson and Haja Razafindrainibe*
Chapter 61. SOUTH AFRICA. . . . . . 133
*Michael H. Schleyer, Lynnath E. Beckley, Sean T. Fennessy, Peter J. Fielding, Anesh Govender, Bruce Q. Mann, Wendy D. Robertson, Bruce J. Tomalin and Rudolph P. van der Elst*
Chapter 62. THE NORTHWEST COAST OF THE BAY OF BENGAL AND DELTAIC SUNDARBANS . . . . . . 145
*Abhijit Mitra*
Chapter 63. SOUTHEAST INDIA. . . . . . 161
*Sundararajan Ramachandran*
Chapter 64. SRI LANKA . . . . . . 175
*Arjan Rajasuriya and Anil Premaratne*
Chapter 65. THE ANDAMAN, NICOBAR AND LAKSHADWEEP ISLANDS . . . . . . 189
*Sundararajan Ramachandran*
Chapter 66. THE MALDIVES . . . . . . 199
*Andrew R.G. Price and Susan Clark*
Chapter 67. THE CHAGOS ARCHIPELAGO, CENTRAL INDIAN OCEAN . . . . . . 221
*Charles R.C. Sheppard*
Chapter 68. THE SEYCHELLES . . . . . . 233
*Miles Gabriel, Suzanne Marshall and Simon Jennings*

Chapter 69. THE COMOROS ARCHIPELAGO . . . . . 243
*Jean-Pascal Quod, Odile Naim and Fouad Abdourazi*
Chapter 70. THE MASCARENE REGION . . . . . 253
*John Turner, Colin Jago, Deolall Daby and Rebecca Klaus*
Chapter 71. THE BAY OF BENGAL . . . . . 269
*D.V. Subba Rao*
Chapter 72. BANGLADESH . . . . . 285
*Dihider Shahriar Kabir, Syed Mazharul Islam, Md. Giasuddin Khan, Md. Ekram Ullah and Dulal C. Halder*
Chapter 73. THE GULF OF THAILAND . . . . . 297
*Manuwadi Hungspreugs, Wilaiwan Utoomprurkporn and Charoen Nitithamyong*
Chapter 74. THE MALACCA STRAITS . . . . . 309
*Chua Thia-Eng, Ingrid R.L. Gorre, S. Adrian Ross, Stella Regina Bernad, Bresilda Gervacio and Ma. Corazon Ebarvia*
Chapter 75. MALACCA STRAIT INCLUDING SINGAPORE AND JOHORE STRAITS . . . . . 331
*Poh Poh Wong*
Chapter 76. EAST COAST OF PENINSULAR MALAYSIA . . . . . 345
*Zelina Z. Ibrahim, Aziz Arshad, Lee Say Chong, Japar Sidik Bujang, Law Ah Theem, Nik Mustapha Raja Abdullah and Maged Mahmoud Marghany*
Chapter 77. BORNEO . . . . . 361
*Steve Oakley, Nicolas Pilcher and Elizabeth Wood*
Chapter 78. CONTINENTAL SEAS OF WESTERN INDONESIA . . . . . 381
*Evan Edinger and David R. Browne*
Chapter 79. THE PHILIPPINES . . . . . 405
*Gil S. Jacinto, Porfirio M. Aliño, Cesar L. Villanoy, Liana Talaue-McManus and Edgardo D. Gomez*
Chapter 80. THE CORAL, SOLOMON AND BISMARCK SEAS REGION . . . . . 425
*Michael E. Huber and Graham B.K. Baines*
Chapter 81. ASIAN DEVELOPING REGIONS: PERSISTENT ORGANIC POLLUTANTS IN THE SEAS . . . . . 447
*Shinsuke Tanabe*
Chapter 82. SEA OF OKHOTSK . . . . . 463
*Victor V. Lapko and Vladimir I. Radchenko*
Chapter 83. SEA OF JAPAN . . . . . 473
*Anatoly N. Kachur and Alexander V. Tkalin*
Chapter 84. THE YELLOW SEA . . . . . 487
*Suam Kim and Sung-Hyun Kahng*
Chapter 85. TAIWAN STRAIT . . . . . 499
*Woei-Lih Jeng, Chang-Feng Dai and Kuang-Lung Fan*
Chapter 86. XIAMEN REGION, CHINA . . . . . 513
*Chua Thia-Eng and Ingrid Rosalie L. Gorre*
Chapter 87. HONG KONG . . . . . 535
*Bruce J. Richardson, Paul K.S. Lam and Rudolf S.S. Wu*
Chapter 88. SOUTHERN CHINA, VIETNAM TO HONG KONG . . . . . 549
*Zhang Gan, Zou Shicun and Yan Wen*
Chapter 89. VIETNAM AND ADJACENT BIEN DONG (SOUTH CHINA SEA) . . . . . 561
*Dang Duc Nhan, Nguyen Xuan Duc, Do Hoai Duong, Nguyen The Tiep and Bui Cong Que*
Chapter 90. CAMBODIAN SEA . . . . . 569
*Touch Seang Tana*
Chapter 91. THE AUSTRALIAN REGION: AN OVERVIEW . . . . . 579
*Leon P. Zann*
Chapter 92. TORRES STRAIT AND THE GULF OF PAPUA . . . . . 593
*Michael E. Huber*
Chapter 93. NORTHEASTERN AUSTRALIA: THE GREAT BARRIER REEF REGION . . . . . 611
*Leon P. Zann*
Chapter 94. THE EASTERN AUSTRALIAN REGION: A DYNAMIC TROPICAL/TEMPERATE BIOTONE . . . . 629
*Leon P. Zann*

Chapter 95. THE TASMANIAN REGION . . . . . . . . . . 647
*Christine M. Crawford, Graham J. Edgar and George Cresswell*
Chapter 96. VICTORIA PROVINCE, AUSTRALIA . . . . . . . . . . 661
*Tim D. O'Hara*
Chapter 97. THE GREAT AUSTRALIAN BIGHT . . . . . . . . . . 673
*Karen Edyvane*
Chapter 98. THE WESTERN AUSTRALIAN REGION . . . . . . . . . . 691
*Diana I. Walker*
Chapter 99. THE SOUTH WESTERN PACIFIC ISLANDS REGION . . . . . . . . . . 705
*Leon P. Zann and Veikila Vuki*
Chapter 100. NEW CALEDONIA . . . . . . . . . . 723
*Pierre Labrosse, Renaud Fichez, Richard Farman and Tim Adams*
Chapter 101. VANUATU . . . . . . . . . . 737
*Veikila C. Vuki, Subashni Appana, Milika R. Naqasima and Maika Vuki*
Chapter 102. THE FIJI ISLANDS . . . . . . . . . . 751
*Veikila C. Vuki, Leon P. Zann, Milika Naqasima and Maika Vuki*
Chapter 103. THE CENTRAL SOUTH PACIFIC OCEAN (AMERICAN SAMOA) . . . . . . . . . . 765
*Peter Craig, Suesan Saucerman and Sheila Wiegman*
Chapter 104. THE MARSHALL ISLANDS . . . . . . . . . . 773
*Andrew R.G. Price and James E. Maragos*
Chapter 105. HAWAIIAN ISLANDS (U.S.A.) . . . . . . . . . . 791
*James E. Maragos*
Chapter 106. FRENCH POLYNESIA . . . . . . . . . . 813
*Pat Hutchings and Bernard Salvat*

*Index* . . . . . . . . . . 827

## Volume III. Global Issues and Processes

Chapter 107. GLOBAL STATUS OF SEAGRASSES . . . . . . . . . . 1
*Ronald C. Phillips and Michael J. Durako*
Chapter 108. MANGROVES . . . . . . . . . . 17
*Colin D. Field*
Chapter 109. CORAL REEFS: ENDANGERED, BIODIVERSE, GENETIC RESOURCES . . . . . . . . . . 33
*Walter H. Adey, Ted A. McConnaughey, Allegra M. Small and Don M. Spoon*
Chapter 110. WORLD-WIDE CORAL REEF BLEACHING AND MORTALITY DURING 1998: A GLOBAL CLIMATE CHANGE WARNING FOR THE NEW MILLENNIUM? . . . . . . . . . . 43
*Clive R. Wilkinson*
Chapter 111. SEA TURTLES . . . . . . . . . . 59
*Jeanne A. Mortimer, Marydele Donnelly and Pamela T. Plotkin*
Chapter 112. WHALES AND WHALING . . . . . . . . . . 73
*Sidney Holt*
Chapter 113. SMALL CETACEANS: SMALL WHALES, DOLPHINS AND PORPOISES . . . . . . . . . . 89
*Kieran Mulvaney and Bruce McKay*
Chapter 114. SEABIRDS . . . . . . . . . . 105
*W.R.P. Bourne and C.J. Camphuysen*
Chapter 115. FISHERIES EFFECTS ON ECOSYSTEMS . . . . . . . . . . 117
*Raquel Goñi*
Chapter 116. BY-CATCH: PROBLEMS AND SOLUTIONS . . . . . . . . . . 135
*Martin A. Hall, Dayton L. Alverson and Kaija I. Metuzals*
Chapter 117. FISHERIES MANAGEMENT AS A SOCIAL PROBLEM . . . . . . . . . . 153
*Douglas C. Wilson*
Chapter 118. FARMING OF AQUATIC ORGANISMS, PARTICULARLY THE CHINESE AND THAI EXPERIENCE . . . . . . . . . . 165
*Krishen J. Rana and Anton J. Immink*

Chapter 119. CLIMATIC CHANGES: GULF OF ALASKA . . . . . 179
*Howard Freeland and Frank Whitney*
Chapter 120. EFFECTS OF CLIMATE CHANGE AND SEA LEVEL ON COASTAL SYSTEMS . . . . . 187
*Shiao-Kung Liu*
Chapter 121. PARTICLE DRY DEPOSITION TO WATER SURFACES: PROCESSES AND CONSEQUENCES . . 197
*Sara C. Pryor and Rebecca J. Barthelmie*
Chapter 122. MARINE ECOSYSTEM HEALTH AS AN EXPRESSION OF MORBIDITY, MORTALITY AND DISEASE EVENTS . . . . . 211
*Benjamin H. Sherman*
Chapter 123. EFFECT OF MINE TAILINGS ON THE BIODIVERSITY OF THE SEABED: EXAMPLE OF THE ISLAND COPPER MINE, CANADA . . . . . 235
*Derek V. Ellis*
Chapter 124. MARINE ANTIFOULANTS . . . . . 247
*Stewart M. Evans*
Chapter 125. EUTROPHICATION OF MARINE WATERS: EFFECTS ON BENTHIC MICROBIAL COMMUNITIES . . . . . 257
*Lutz-Arend Meyer-Reil and Marion Köster*
Chapter 126. PERSISTENCE OF SPILLED OIL ON SHORES AND ITS EFFECTS ON BIOTA . . . . . 267
*Gail V. Irvine*
Chapter 127. REMOTE SENSING OF TROPICAL COASTAL RESOURCES: PROGRESS AND FRESH CHALLENGES FOR THE NEW MILLENNIUM . . . . . 283
*Peter J. Mumby*
Chapter 128. SATELLITE REMOTE SENSING OF THE COASTAL OCEAN: WATER QUALITY AND ALGAE BLOOMS . . . . . 293
*Bertil Håkansson*
Chapter 129. ENERGY FROM THE OCEANS: WIND, WAVE AND TIDAL . . . . . 303
*Rebecca J. Barthelmie, Ian Bryden, Jan P. Coelingh and Sara C. Pryor*
Chapter 130. MULTINATIONAL TRAINING PROGRAMMES IN MARINE ENVIRONMENTAL SCIENCE . . . 323
*G. Robin South*
Chapter 131. GLOBAL LEGAL INSTRUMENTS ON THE MARINE ENVIRONMENT AT THE YEAR 2000 . . . 331
*Milen F. Dyoulgerov*
Chapter 132. COASTAL MANAGEMENT IN THE FUTURE . . . . . 349
*Derek J. McGlashan*
Chapter 133. SUSTAINABILITY OF HUMAN ACTIVITIES ON MARINE ECOSYSTEMS. . . . . 359
*Paul Johnston, David Santillo, Julie Ashton and Ruth Stringer*
Chapter 134. MARINE RESERVES AND RESOURCE MANAGEMENT . . . . . 375
*Michael J. Fogarty, James A. Bohnsack and Paul K. Dayton*
Chapter 135. THE ECOLOGICAL, ECONOMIC, AND SOCIAL IMPORTANCE OF THE OCEANS . . . . . 393
*Robert Costanza*

*Index* . . . . . 405

# Contributing Authors


**Marco Abbiati**
*Dipartimento di Biologia Evoluzionistica, Università di Bologna, Via Tombesi dell'Ova 55, I-48100 Ravenna, Italy*

**Fouad Abdourazi**
*AIDE, Minizi, Mavouna, Moroni, Comoros*

**Nik Mustapha Raja Abdullah**
*Universiti Putra Malaysia, 43400 UPM Serdang, Malaysia*

**Giovanni Accordi**
*Centro di Studio per il Quaternario e l'Evoluzione Ambientale, C.N.R., Dipartimento di Scienze della Terra, Università degli Studi "La Sapienza", P. Aldo Moro, 5, 00185 Roma, Italy*

**Tim Adams**
*Secretariat of the Pacific Community (SPC), B.P. D5, 98848 Nouméa Cedex, New Caledonia*

**Walter H. Adey**
*Marine Systems Laboratory, Smithsonian Institution, NHB E-117, MRC 164, Washington, DC 20560, U.S.A.*

**Ama Antoinette Adingra**
*Centre de Recherches Océanologiques, BP V18, Abidjan, Côte d'Ivoire*

**John B.R. Agard**
*Department of Life Sciences, The University of the West Indies, St. Augustine, Trinidad and Tobago*

**Eduardo Aguayo**
*Instituto de Ciencias del Mar y Limnología, UNAM, Cd. Universitaria, A.P. 70-305, Mexico City, C.P. 04510 Mexico*

**Ramón B. Ahumada**
*Facultad de Ciencias, Universidad Católica de la Santísima Concepción, Campus San Andrés, Paicaví 3000, Casilla 297, Concepción, Chile*

**Faiza Al-Yamani**
*Kuwait Institute of Scientific Research, P.O. Box 1638, 22017 Salmiya, Kuwait*

**Porfirio M. Aliño**
*Marine Science Institute, University of the Philippines, 1101 Diliman, Quezon City, Philippines*

**M. James Allen**
*Southern California Coastal Water Research Project, 7171 Fenwick Lane, Westminster, CA 92683, U.S.A.*

**Anders A. Alm**
*Universidad de Cartagena, Facultad de Ciencias Químicas y Farmacéuticas, Zaragocilla, AA 1661, Cartagena, Colombia*

**Ricardo Alvarez-León**
*Universidad de la Sabana, Depto. Ciencias de la Vida, Campus Universitario Puente del Común, Edif. E-2, of. 232, Chía, Cundi., Colombia*

**Haymara Alvarez**
*Universidad Simon Bolivar, Apdo 89000, Caracas 1080-A, Venezuela*

**Dayton L. Alverson**
*Natural Resources Consultants, 1900 West Nickerson St., Suite 207, Seattle, WA 98119, U.S.A.*

**Paul J. Anderson**
*Kodiak Laboratory, National Marine Fisheries Service, National Oceanic and Atmospheric Administration, P.O. Box 1638, Kodiak, AK 99615, U.S.A.*

**Subashni Appana**
*Marine Studies Programme, University of the South Pacific, P.O. Box 1168, Suva, Fiji*

**Robert Arfi**
*Centre de Recherches Océanologiques, BP V18, Abidjan, Côte d'Ivoire*

**Aziz Arshad**
*Universiti Putra Malaysia, 43400 UPM Serdang, Malaysia*

**Julie Ashton**
*Greenpeace Research Laboratories, University of Exeter, Exeter EX4 4PS, U.K.*

**Gert Asmund**
*National Environmental Research Institute, Department of Arctic Environment, P.O. Box 358, DK-4000 Roskilde, Denmark*

**Amadou Bah**
*The National Center of Halieutic Sciences of Boussoura, B.P. 3060, Conakry, Republic of Guinea*

**Graham B.K. Baines**
*Environment Pacific, 3 Pindari St., The Gap, Brisbane, QLD 4061, Australia*

**Gülfem Bakan**
*Ondokuz Mayis University, Faculty of Engineering, Department of Environmental Engineering, 55139 Kurupelit, Samsun, Turkey*

**Torgeir Bakke**
*Norwegian Institute for Water Research (NIVA), P.O. Box 173 Kjelsaas, 0411 Oslo, Norway*

**Idrissa Lamine Bamy**
*The National Center of Halieutic Sciences of Boussoura, B.P. 3060, Conakry, Republic of Guinea*

**Enrique J. Barraza**
*Ministry of the Environment, San Salvador, El Salvador*

**Maritza Barreto**
*University of Puerto Rico Rio Piedras, Geography Department, Rio Piedras, Puerto Rico*

**Rebecca J. Barthelmie**
*Department of Wind Energy and Atmospheric Physics, Risø National Laboratory, DK-4000 Roskilde, Denmark*

**Christopher Bartholomae**
*National Marine Information and Research Centre, P.O. Box 912, Swakopmund, Namibia*

**Steven M. Bay**
*Southern California Coastal Water Research Project, 7171 Fenwick Lane, Westminster, CA 92683, U.S.A.*

**Lynnath E. Beckley**
*SAAMBR, P.O. Box 10712, Marine Parade, KwaZulu-Natal 4056, South Africa*

**Stella Regina Bernad**
*GEF/UNDP/IMO Regional Programme on Partnerships in Environmental Management for the Seas of East Asia (PEMSEA), P.O. Box 2502, Quezon City, 1165 Metro Manila, Philippines*
**Márcia Caruso Bícego**
*Universidade de São Paulo, Dept. Oceanografia Física do Instituto Oceanográfico, Pca do Oceanográfico, 191, Cidade Universitária, SP, 05508-900, Brazil*
**James A. Bohnsack**
*National Marine Fisheries Service, Southeast Fisheries Science Center, 75 Virginia Beach Drive, Miami, FL 33149, U.S.A.*
**David Bone**
*Universidad Simon Bolivar, Apartado 89000, Caracas 1080-A, Venezuela*
**Alfonso V. Botello**
*Institute for Marine and Limnology Sciences, National Autonomous University of Mexico, Marine Pollution Laboratory, P.O. Box 70305, México City 04510 D.F., Mexico*
**Leonor Botero**
*COLCIENCIAS, Trans. 9A # 133-28, Santafé de Bogotá, Colombia*
**W.R.P. Bourne**
*Department of Zoology, Aberdeen University, Tillydrone Avenue, Aberdeen AB24 2TZ, Scotland*
**David Boyer**
*National Marine Information and Research Centre, P.O. Box 912, Swakopmund, Namibia*
**Joseph C. Britton**
*Department of Biology, Texas Christian University, Fort Worth, Texas 76129, U.S.A.*
**Maria José Brogueira**
*Instituto de Investigação das Pescas e do Mar (IPIMAR), DAA, Av. Brasilia, 1400 Lisboa, Portugal.*
**David R. Browne**
*Department of Biology, McGill University, 1205 Docteur Penfield Avenue, Montreal, PQ, H3A 1B1 Canada*
**Grethe Bruntse**
*Kaldbak Marine Biological Laboratory, FO-180 Kaldbak, Faroe Islands*
**Ian Bryden**
*The Robert Gordon University, School of Mechanical and Offshore Engineering, Aberdeen, AB10 1FR, Scotland*
**Kenneth C. Buchan**
*Bahamian Field Station, San Salvador, Bahamas*
**Japar Sidik Bujang**
*Universiti Putra Malaysia, 43400 UPM Serdang, Malaysia*
**JoAnn M. Burkholder**
*Department of Botany, North Carolina State University, Raleigh, NC 27695-7612, U.S.A.*
**James N. Butler**
*Harvard University, Cambridge, Massachussetts, U.S.A.*
**Hanife Büyükgüngör**
*Ondokuz Mayis University, Faculty of Engineering, Department of Environmental Engineering, 55139 Kurupelit, Samsun, Turkey*
**Graça Cabeçadas**
*Instituto de Investigação das Pescas e do Mar (IPIMAR), DAA, Av. Brasilia, 1400 Lisboa, Portugal.*
**Leonor Cabeçadas**
*Direcção Geral do Ambiente (D.G.A.), R. da Murgueira, Zambujal, 2720 Amadora, Portugal*
**Lawrence B. Cahoon**
*Department of Biological Sciences, University of North Carolina-Wilmington, Wilmington, NC 28403, U.S.A.*
**Sékou Balta Camara**
*The National Center of Halieutic Sciences of Boussoura, B.P. 3060, Conakry, Republic of Guinea*
**C.J. Camphuysen**
*CSR Consultancy, Ankerstraat 20, 1794 BJ Oosterend, Texel, The Netherlands*
**Patricio A. Camus**
*Facultad de Ciencias, Universidad Católica de la Santísima Concepción, Campus San Andrés, Paicaví 3000, Casilla 297, Concepción, Chile*
**Jorge Capella**
*University of Puerto Rico R.U.M., Department of Marine Sciences, Mayagüez, Puerto Rico*
**Federico Carbone**
*Centro di Studio per il Quaternario e l'Evoluzione Ambientale, C.N.R., Dipartimento di Scienze della Terra, Università degli Studi "La Sapienza", P. Aldo Moro, 5, 00185 Roma, Italy*
**P. Castellano**
*Centro de Procesamiento de Imagenes Digitales (CPDI), Sartenejas, Caracas, Venezuela*
**Alberto Castelli**
*Dipartimento di Zoologia ed Antropologia, Corso Margherita di Savoia 15, I-17100 Sassari, Italy*
**Jesús M. Castillo**
*Departamento de Biología Vegetal y Ecología, Facultad de Biología, Universidad de Sevilla, Apdo 1095, 41080 Sevilla, Spain*
**María M. Castillo**
*Universidad Simon Bolivar, Apdo 89000, Caracas 1080-A, Venezuela*
**Michael J. Childress**
*Department of Biological Sciences, Idaho State University, Pocatello, ID 83209-8007, U.S.A.*
**Lee Say Chong**
*National Hydraulic Research Institute of Malaysia, Km 7 Jalan Ampang, 68000 Ampang, Malaysia*
**Nestor F. Ciocco**
*CENPAT-CONICET, Bv. Brown 3000, (9120) Puerto Madryn, Chubut, Argentina*
**Ibrahima Cisse**
*The National Center of Halieutic Sciences of Boussoura, B.P. 3060, Conakry, Republic of Guinea*
**Susan Clark**
*Department of Marine Sciences and Coastal Management, University of Newcastle upon Tyne, Newcastle upon Tyne, U.K.*
**Jan P. Coelingh**
*ECOFYS Energy and Environment, P.O. Box 8408, NL-3503 RK Utrecht, The Netherlands.*
**Giuseppe Cognetti**
*Dipartimento di Scienze dell'Uomo e dell'Ambiente, Università di Pisa, Via Volta 6, I-56124 Pisa, Italy*
**James Cole**
*2 Dolphin Cottage, 31 Penny St., Portsmouth PO1 2NH, U.K.*
**Juan C. Colombo**
*Química Ambiental y Bioquímica, Facultad de Ciencias Naturales y Museo, Universidad Nacional de La Plata, Paseo del Bosque s/n, (1900) La Plata, Argentina*

**Douglas P. Connelly**
*Bermuda Biological Station for Research, 17 Biological Station Lane, St. Georges, Bermuda, GE 01*
**Andrew Cooke**
*Cellule Environnement Marin et Côtier, Office National pour l'Environnement, B.P. 822, Antananarivo, Madagascar*
**Robert Costanza**
*Center for Environmental Science and Biology Department, and Institute for Ecological Economics, University of Maryland, Box 38, Solomons, MD 20688-0038, U.S.A.*
**Linos Cotsapas**
*Research Planning, Inc., 1121 Park St., Columbia, SC 29201, U.S.A.*
**Peter Craig**
*National Park of American Samoa, Pago Pago, American Samoa 96799, U.S.A.*
**Christine M. Crawford**
*Tasmanian Aquaculture and Fisheries Institute, University of Tasmania, Nubeena Crescent, Taroona, Tasmania 7053, Australia*
**George Cresswell**
*CSIRO Marine Research, Castray Esplanade, Hobart, Tasmania 7000, Australia*
**John Croxall**
*British Antarctic Survey, Natural Environment Research Council, High Cross, Madingley Road, Cambridge CB3 OET, U.K.*
**Deolall Daby**
*Faculty of Science, University of Mauritius, Reduit, Mauritius*
**Henning Dahlgaard**
*Risø National Laboratory, DK-4000 Roskilde, Denmark*
**Chang-Feng Dai**
*Institute of Oceanography, National Taiwan University, Taipei, Taiwan, Republic of China*
**Maria Dam**
*Food and Environmental Agency, Debesartrøð, FO-100 Tórshavn, Faroe Islands*
**Manos Dassenakis**
*University of Athens, Department of Chemistry, Division III, Inorganic and Environmental Chemistry, Panepistimiopolis, Kouponia, Athens 15771, Greece*
**Paul K. Dayton**
*Scripps Institution of Oceanography, 9500 Gilman Dr., La Jolla, CA 92093, U.S.A.*
**Adolpe O. Debrot**
*Carmabi Foundation, Piscaderabaai, P.O. Box 2090, Curaçao, Netherlands Antilles*
**E. Kate Delaney**
*Department of Geography, University of Southampton, Southampton, SO17 1BJ, U.K.*
**Tom Dahmer**
*Hyder Consulting Ltd, Hong Kong; Ecosystems Ltd, Hong Kong*
**Yusbelly Diaz**
*Universidad Simon Bolivar, Apdo. 89000, Caracas 1080-A, Venezuela*
**Rune Dietz**
*National Environmental Research Institute, Department of Arctic Environment, P.O. Box 358, DK-4000 Roskilde, Denmark*
**Isabel Díez**
*Departamento de Biología Vegetal y Ecología, Facultad de Ciencias, Universidad del País Vasco, Apdo. 644, Bilbao 48080, Spain*
**José M.L. Dominguez**
*Laboratório de Estudos Costeiros, Centro de Pesquisa em Geofísica e Geologia, Universidade Federal da Bahia, Rua Caetano Moura 123, Federação, Salvador, 40210-340, Bahia, Brazil*
**Marydele Donnelly**
*IUCN/SSC Marine Turtle Specialist Group, 1725 DeSales St. NW #600, Washington, DC 20036, U.S.A.*
**Bohdan Draganik**
*Sea Fisheries Institute, 1 Kollataja Str., PL 81-332 Gdynia, Poland*
**Nguyen Xuan Duc**
*Institute of Ecology and Biological Resources, Nghia Do, Hanoi, Vietnam*
**Jean-Paul Ducrotoy**
*University College Scarborough, CERCI, Scarborough YO11 3AZ, U.K.*
**Do Hoai Duong**
*Institute of Hydrometeology, Lang Thuong, Dong Da, Hanoi, Vietnam*
**Michael J. Durako**
*Center for Marine Science, The University of North Carolina at Wilmington, Wilmington, NC 28403, U.S.A.*
**Phillip Dustan**
*Department of Biology, University of Charleston, Charleston, SC 29424, U.S.A.*
**Milen F. Dyoulgerov**
*International Program Office, National Ocean Service, National Oceanic and Atmospheric Administration, 1305 East West Highway, Silver Spring, MD 20910, U.S.A.*
**Ma. Corazon Ebarvia**
*GEF/UNDP/IMO Regional Programme on Partnerships in Environmental Management for the Seas of East Asia (PEMSEA), P.O. Box 2502, Quezon City, 1165 Metro Manila, Philippines*
**Graham J. Edgar**
*Zoology Department, Tasmanian Aquaculture and Fisheries Institute, University of Tasmania, GPO Box 252-05, Hobart, Tasmania 7011, Australia*
**Evan Edinger**
*Department of Geology, St. Francis Xavier University, P.O. Box 5000, Antigonish, Nova Scotia B2G 2W5, Canada. Present address: Department of Earth Sciences, Laurentian University, Ramsey Lake Road, Sudbury, Ontario, P3E 2C6, Canada*
**Karen Edyvane**
*SA Research and Development Institute, P.O. Box 120, Henley Beach, South Australia 5022, Australia*
**Mike Elliott**
*IECS, University of Hull, Hull HU6 7RX, U.K.*
**Derek V. Ellis**
*Biology Department, University of Victoria, P.O. Box 3020, Victoria, B.C., V8W 3N5, Canada*
**Rudolph P. van der Elst**
*SAAMBR, P.O. Box 10712, Marine Parade, KwaZulu-Natal 4056, South Africa*

**Paul R. Epstein**
*Center for Health and the Global Environment, Harvard Medical School, Boston MA 02115, U.S.A*

**Mark V. Erdmann**
*Dept. of Integrative Biology, University of California, Berkeley, Berkeley, CA 94720, U.S.A.*

**José L. Esteves**
*CENPAT-CONICET, Bv. Brown 3000, (9120) Puerto Madryn, Chubut, Argentina*

**S.M. Evans**
*Dove Marine Laboratory (Department of Marine Sciences and Coastal Management, Newcastle University), Cullercoats, Tyne & Wear, NE30 4PZ, UK*

**Jerzy Falandysz**
*University of Gdansk, 18 Sobieskiego Str., PL 80-952 Gdansk, Poland*

**Kuang-Lung Fan**
*Institute of Oceanography, National Taiwan University, Taipei, Taiwan, Republic of China*

**Richard Farman**
*Southern Province, Department of Natural Resources, B.P. 3718, 98846 Nouméa Cedex, New Caledonia*

**Sean T. Fennessy**
*SAAMBR, P.O. Box 10712, Marine Parade, KwaZulu-Natal 4056, South Africa*

**Roberto Fernández-Maestre**
*Universidad de Cartagena, Facultad de Ciencias Químicas y Farmacéuticas, Zaragocilla, AA 1661, Cartagena, Colombia*

**Jorge Fernández**
*PEMEX-Exploración Producción Región Marina Suroeste, Calle 33 S/N, Edif. Cantarell, Cd. del Carmen, Campeche, C.P. 24170 Mexico*

**Renaud Fichez**
*IRD (Institute of Research for Development), B.P. A5, 98848 Nouméa Cedex, New Caledonia*

**Colin D. Field**
*Faculty of Science (Gore Hill), University of Technology, Sydney, P.O. Box 123, Broadway NSW 2007, Australia*

**Peter J. Fielding**
*SAAMBR, P.O. Box 10712, Marine Parade, KwaZulu-Natal 4056, South Africa*

**M. Enrique Figueroa**
*Departamento de Biología Vegetal y Ecología, Facultad de Biología, Universidad de Sevilla, Apdo. 1095, 41080 Sevilla, Spain*

**Belmiro Mendes de Castro Filho**
*Universidade de São Paulo, Dept. Oceanografia Física do Instituto Oceanográfico, Pca do Oceanográfico, 191, Cidade Universitária, SP, 05508-900, Brazil*

**William S. Fisher**
*U.S. Environmental Protection Agency, National Health and Environmental Effects Laboratory, Gulf Ecology Division, One Sabine Island Drive, Gulf Breeze, FL 32561, U.S.A*

**Michael J. Fogarty**
*University of Maryland Center for Environmental Science, Chesapeake Biological Lab., Solomons, MD, U.S.A. Present address: National Marine Fisheries Service, Northeast Fisheries Science Center, 166 Water St., Woods Hole, MA 02543, U.S.A.*

**Mark S. Fonseca**
*NOAA/National Ocean Service, Center for Coastal Fisheries and Habitat Research, 101 Pivers Island Road, Beaufort, NC 28516-9722, U.S.A.*

**Howard Freeland**
*Institute of Ocean Sciences, P.O. Box 6000, Sidney, B.C., V8L 4B2, Canada*

**Hugo Freije**
*Universidad Nacional del Sur, Química Ambiental, Av. Alem 1253, (8000) Bahía Blanca, Argentina*

**Miles Gabriel**
*Inter-consult Nambia (Pty) Ltd., P.O. Box 20690, Windhoek, Namibia*

**Bella Galil**
*Israel Oceanographic and Limnological Research, National Institute of Oceanography, P.O.Box 8030, Haifa 31080, Israel*

**Vincent F. Gallucci**
*University of Washington, School of Fisheries, Seattle, WA 98195, U.S.A.*

**Zhang Gan**
*Guangzhou Institute of Geochemistry, Chinese Academy of Sciences, Guangzhou 510640, People's Republic of China*

**María P. Garcia**
*Universidad Simon Bolivar, Apdo. 89000, Caracas 1080-A, Venezuela*

**Jorge Garcia**
*University of Puerto Rico R.U.M., Department of Marine Sciences, Mayagüez, Puerto Rico*

**Gudrun Gaudian**
*Sunny View, Main Street, Alne, N. Yorks, YO61 1RT, U.K.*

**Fiona Gell**
*ICLARM Caribbean and Eastern Pacific Office, PMB 158, Inland Messenger Service, Road Town, Tortola, British Virgin Islands*

**Bresilda Gervacio**
*GEF/UNDP/IMO Regional Programme on Partnerships in Environmental Management for the Seas of East Asia (PEMSEA), P.O. Box 2502, Quezon City, 1165 Metro Manila, Philippines*

**Ed Gmitrowicz**
*Hyder Consulting Ltd, Hong Kong; Ecosystems Ltd, Hong Kong*

**Judith F. Gobin**
*Institute of Marine Affairs, Hilltop Lane, Chaguaramas, Port of Spain, Trinidad and Tobago*

**Edgardo D. Gomez**
*Marine Science Institute, University of the Philippines, 1101 Diliman, Quezon City, Philippines*

**Raquel Goñi**
*Centro Oceanografico de Baleares Muelle de Poniente s/n, Apdo. 291, 07080 Palma de Mallorca, Spain*

**José María Gorostiaga**
*Departamento de Biología Vegetal y Ecología, Facultad de Ciencias, Universidad del País Vasco, Apdo. 644, Bilbao 48080, Spain*

**Ingrid Rosalie L. Gorre**
*GEF/UNDP/IMO Regional Programme on Partnerships in Environmental Management for Seas of East Asia (PEMSEA), P.O. Box 2502, Quezon City, 1165 Metro Manila, Philippines*

**Anesh Govender**
*SAAMBR, P.O. Box 10712, Marine Parade, KwaZulu-Natal 4056, South Africa*

**Stephen L. Granger**
*University of Rhode Island, Graduate School of Oceanography, Narragansett, Rhode Island 02882, USA*

**Martin Guard**
*Zoology Department, University of Aberdeen, Tillydrone Road, Aberdeen AB24 2TZ, U.K. and Department of Zoology and Marine Biology, University of Dar es Salaam, P.O. Box 35091, Dar es Salaam, Tanzania*

**Bertil Håkansson**
*Swedish Meteorological and Hydrological Institute, S-60176 Norrköping, Sweden*

**Dulal C. Halder**
*Independent University, Bangladesh (IUB), Plot 3 & 8, Road 10, Baridhara, Dhaka-1212, Bangladesh*

**Martin A. Hall**
*Inter-American Tropical Tuna Commission, 8604 La Jolla Shores Dr., La Jolla, CA 92037, U.S.A.*

**Alastair R. Harborne**
*Coral Cay Conservation, 154 Clapham Park Road, London, SW4 7DE, U.K.*

**Nicholas J. Hardman-Mountford**
*Centre for Coastal and Marine Sciences, Plymouth Marine Laboratory, Plymouth, U.K.*

**Guillermo Harris**
*Fundación Patagonia Natural, Marcos A. Zar 760, (9120) Puerto Madryn, Chubut, Argentina*

**Richard G. Hartnoll**
*Port Erin Marine Laboratory, University of Liverpool, Port Erin, Isle of Man IM9 6JA, British Isles*

**Frank Hawkins**
*Projet ZICOMA, BirdLife International, B.P. 1074, Antananarivo, Madagascar*

**Barak Herut**
*Israel Oceanographic and Limnological Research, National Institute of Oceanography, P.O.Box 8030, Haifa 31080, Israel*

**Sidney Holt**
*Hornbeam House, 4 Upper House Farm, Crickhowell, Powys, NP8 1BP, U.K.*

**Chris Horrill**
*Tanga Coastal Zone Conservation and Development Project, P.O. Box 5036, Tanga, Tanzania*

**Vicki Howe**
*Department of Maritime Studies and International Transport, University of Cardiff, Cardiff, Wales, U.K.*

**Michael E. Huber**
*Global Coastal Strategies, P.O. Box 606, Wynnum, QLD 4178, Australia*

**Manuwadi Hungspreugs**
*Department of Marine Science, Chulalongkorn University, Bangkok 10330, Thailand*

**George L. Hunt Jr.**
*Department of Ecology and Evolutionary Biology, University of California, Irvine, Irvine, CA 92697, U.S.A.*

**Pat Hutchings**
*The Australian Museum, Sydney, NSW 2010, Australia*

**Zelina Z. Ibrahim**
*National Hydraulic Research Institute of Malaysia, Km 7 Jalan Ampang, 68000 Ampang, Malaysia*

**Anton J. Immink**
*Fisheries Department, FAO, Rome, Italy*

**Oscar Iribarne**
*Universidad Nacional de Mar del Plata, Biologia, CC 573, Correo Central, (7600) Mar del Plata, Argentina*

**Gail V. Irvine**
*U.S. Geological Survey, Alaska Biological Science Center, 1011 E. Tudor Rd., Anchorage, AK 99503, U.S.A.*

**Ignacio Isla**
*Universidad Nacional de Mar del Plata, Centro de Geología de Costas*

**Syed Mazharul Islam**
*School of Liberal Arts and Science, Independent University, Bangladesh (IUB), Plot # 3 & 8, Road 10, Baridhara, Dhaka-1212, Bangladesh*

**Gil S. Jacinto**
*Marine Science Institute, University of the Philippines, 1101 Diliman, Quezon City, Philippines*

**Colin Jago**
*School of Ocean Sciences, University of Wales Bangor, LL59 5EY, U.K.*

**Stephen C. Jameson**
*Coral Seas Inc. – Integrated Coastal Zone Management, 4254 Hungry Run Road, The Plains, VA 20198-1715, U.S.A.*

**Woei-Lih Jeng**
*Institute of Oceanography, National Taiwan University, Taipei, Taiwan, Republic of China*

**Simon Jennings**
*CEFAS, Fisheries Laboratory, Lowestoft, NR33 OHT, U.K.*

**Jacob Pauli Joensen**
*Food and Environmental Agency, Debesartrøð, FO-100 Tórshavn, Faroe Islands*

**Poul Johansen**
*National Environmental Research Institute, Department of Arctic Environment, P.O. Box 358, DK-4000 Roskilde, Denmark*

**Paul Johnston**
*Greenpeace Research Laboratories, University of Exeter, Exeter EX4 4PS, U.K.*

**Victor N. de Jonge**
*National Institute for Coastal and Marine Management, Rijkswaterstaat, Ministry of Transport, Public Works and Water Management, P.O. Box 207, 9750 AE Haren, The Netherlands*

**Dihider Shahriar Kabir**
*School of Environmental Science and Management, Independent University, Bangladesh (IUB), Plot # 3 & 8, Road 10, Baridhara, Dhaka-1212, Bangladesh*

**Anatoly N. Kachur**
*Pacific Geographical Institute, Far East Branch, Russian Academy of Sciences, Vladivostok 690022, Russia*

**Sung-Hyun Kahng**
*Korea Ocean Research and Development Institute, Ansan P.O. Box 29, Seoul, 425-600, Korea*

**Kostas Kapiris**
*University of Athens, Department of Biology, Division of Zoology and Marine Biology, Panepistimiopolis, Kouponia, Athens 15784, Greece*

**Lena Kautsky**
*Department of Botany, Stockholm University, S-106 91 Stockholm, Sweden*

**Nils Kautsky**
*Department of Systems Ecology, Stockholm University, S-106 91 Stockholm, Sweden*

**W. Judson Kenworthy**
*NOAA/National Ocean Service, Center for Coastal Fisheries and Habitat Research, 101 Pivers Island Road, Beaufort, NC 28516-9722, U.S.A.*

**Oleg Khalimonov**
*International Maritime Organization, London, UK*

**Md. Giasuddin Khan**
*Centre for Environment and Geographical Information Systems Support, House 49, Road 27, Banani, Dhaka-1212, Bangladesh*

**Ruy K.P. Kikuchi**
*Departamento de Ciências Exatas, Universidade Estadual de Feira de Santana, BR-116, Campus Universitário, Feira de Santana, 44031-160, Bahia, Brazil*

**Suam Kim**
*Korea Ocean Research and Development Institute, Seoul, Korea. Present address: Dept. of Marine Biology, Pukyong National University, 599-1 Daeyeon 3-Dong, Nam-Gu, Pusan, 608-737, Korea*

**Rebecca Klaus**
*Department of Biological Sciences, University of Warwick, Coventry CV4 7RU, U.K.*

**Anthony H. Knap**
*Bermuda Biological Station for Research, 17 Biological Station Lane, St. Georges, Bermuda*

**Kwame A. Koranteng**
*Marine Fisheries Research Division, Ministry of Food and Agriculture, Ghana*

**Marion Köster**
*Institut für Ökologie der Ernst-Moritz-Arndt-Universität Greifswald, Schwedenhagen 6, D-18565 Kloster/Hiddensee, Germany*

**Aka Marcel Kouassi**
*Centre de Recherches Océanologiques, BP V18, Abidjan, Côte d'Ivoire*

**Mamba Kourouma**
*The National Center of Halieutic Sciences of Boussoura, B.P. 3060, Conakry, Republic of Guinea*

**Andreas Kunzmann**
*ZMT Bremen, Germany*

**Pierre Labrosse**
*Secretariat of the Pacific Community (SPC), B.P. D5, 98848 Nouméa Cedex, New Caledonia*

**Paul K.S. Lam**
*Department of Biology and Chemistry, City University of Hong Kong, 83 Tat Chee Avenue, Kowloon, Hong Kong*

**Eliete Zanardi Lamardo**
*University of Miami – RSMAS, Dept. Marine and Atmospheric Chemistry, 4600 Rickenbacker Causeway, Miami, FL 33149, U.S.A.*

**Guadalupe de la Lanza-Espino**
*Institute for Biological Sciences, National Autonomous University of Mexico, Marine Ecology Laboratory, P.O. Box 70233, México City 04515 D.F., México*

**Victor V. Lapko**
*Pacific Fisheries Research Centre, TINRO Centre, Vladivostok, Russia*

**Claudio Lardicci**
*Dipartimento di Scienze dell'Uomo e dell'Ambiente, Università di Pisa, Via Volta 6, I-56124 Pisa, Italy*

**Zelinda M.A.N. Leão**
*Laboratório de Estudos Costeiros, Centro de Pesquisa em Geofísica e Geologia, Universidade Federal da Bahia, Rua Caetano Moura 123, Federação, Salvador, 40210-340, Bahia, Brazil*

**Mairin Lemus**
*Instituto Oceanografico de Venezuela, Universidad de Oriente, Cumana, Venezuela*

**César A. Leon**
*Universidad Simon Bolivar, Apdo. 89000, Caracas 1080-A, Venezuela*

**Shiao-Kung Liu**
*Systems Research Institute, 3706 Ocean Hill Way, Malibu, CA, 90265, U.S.A.*

**Freddy Losada**
*Universidad Simon Bolivar, Apdo. 89000, Caracas 1080-A, Venezuela*

**Carlos J. Luque**
*Departamento de Biología Vegetal y Ecología, Facultad de Biología, Universidad de Sevilla, Apdo. 1095, 41080 Sevilla, Spain*

**Anmarie J. Mah**
*Vancouver Aquarium, P.O. Box 3232, Vancouver, British Columbia, Canada V6B 3X8*

**Michael A. Mallin**
*Center for Marine Science Research, University of North Carolina-Wilmington, Wilmington, NC 28403, U.S.A.*

**Bruce Q. Mann**
*SAAMBR, P.O. Box 10712, Marine Parade, KwaZulu-Natal 4056, South Africa*

**James E. Maragos**
*U.S. Fish and Wildlife Service, Pacific Islands Ecoregion, 300 Ala Moana Blvd., Box 50167, Honolulu, HI 96850, U.S.A.*

**Maged Mahmoud Marghany**
*Universiti Putra Malaysia, 43400 UPM Serdang, Malaysia*

**Alonso J. Marrugo-González**
*Universidad de Cartagena, Facultad de Ciencias Químicas y Farmacéuticas, Zaragocilla, AA 1661, Cartagena, Colombia*

**Michael J. Marshall**
*Coastal Seas Consortium, 5503 40th Avenue East, Bradenton, FL 34208, U.S.A.*

**Suzanne Marshall**
*Environment Agency, Kings Meadow Road, Reading, RG1 8DQ, U.K.*

**Alberto Martin**
*Universidad Simon Bolivar, Apdo. 89000, Caracas 1080-A, Venezuela*

**Ted A. McConnaughey**
*Marine Systems Laboratory, Smithsonian Institution, NHB E-117, MRC 164, Washington, DC 20560, U.S.A.*

**Melanie D. McField**
*Department of Marine Science, University of South Florida, 140 Seventh Ave South, St. Petersburg, FL 33701, U.S.A. and P.O. Box 512, Belize City, Belize*

**Derek J. McGlashan**
*Graduate School of Environmental Studies, Wolfson Centre, 106 Rottenrow East, University of Strathclyde, Glasgow, G4 0NW, U.K.*

**Bruce McKay**
*4058 Rue Dorion, Montreal, PQ H2K 4B9, Canada*

**Paul Medley**
*Sunny View, Main Street, Alne, N. Yorks, YO61 1RT, U.K.*

**Kaija I. Metuzals**
*Biological Sciences, University of Warwick, Coventry, U.K.*

**Lutz-Arend Meyer-Reil**
*Institut für Ökologie der Ernst-Moritz-Arndt-Universität Greifswald, Schwedenhagen 6, D-18565 Kloster/Hiddensee, Germany*

**Patricia Miloslavich**
*Universidad Simon Bolivar, Apartado 89000, Caracas 1080-A, Venezuela*

**Luiz Bruner de Miranda**
*Universidade de São Paulo, Dept. Oceanografia Física do Instituto Oceanográfico, Pca do Oceanográfico, 191, Cidade Universitária, SP, 05508-900, Brazil*

**Abhijit Mitra**
*Department of Marine Science, University of Calcutta 35, B.C Road, Calcutta 700 019, West Bengal, India.*

**Aviti J. Mmochi**
*Marine Environmental Chemistry, Institute of Marine Sciences, P.O. Box 668, Zanzibar, Tanzania*

**Jarle Molvaer**
*Norwegian Institute for Water Research (NIVA), P.O. Box 173 Kjelsaas, 0411 Oslo, Norway*

**Francisco García Montelongo**
*Department of Analytical Chemistry, Nutrition and Food Sciences, University of La Laguna, 38071 La Laguna, Spain*

**Jack Morelock**
*University of Puerto Rico R.U.M., Department of Marine Sciences, P.O. Box 3200, Lajas, Puerto Rico 00667*

**Jeanne A. Mortimer**
*Department of Zoology, University of Florida, Gainesville, FL 32611-8525, U.S.A. and Marine Conservation Society of Seychelles, P.O. Box 445, Victoria, Mahe, Seychelles*

**Brian Morton**
*The Swire Institute of Marine Science and Department of Ecology and Biodiversity, The University of Hong Kong, Hong Kong*

**Anders Mosbech**
*National Environmental Research Institute, Department of Arctic Environment, P.O. Box 358, DK-4000 Roskilde, Denmark*

**Kent Mountford**
*US Environmental Protection Agency, Chesapeake Bay Program, 410 Severn Ave., Suite 109, Annapolis, MD 21403, U.S.A.*

**Kieran Mulvaney**
*1219 W. 6th Avenue, Anchorage, AK 99501, U.S.A.*

**Peter J. Mumby**
*Centre for Tropical Coastal Management Studies, Department of Marine Sciences & Coastal Management, Ridley Building, The University, Newcastle upon Tyne, NE1 7RU, U.K.*

**Michael Myers**
*TCMC, University of Newcastle upon Tyne, Newcastle upon Tyne NE1 7RU, U.K.*

**Paulina Nabel**
*Museo Argentino de Ciencias Naturales-CONICET, Av. A. Gallardo 470, (1405) Buenos Aires, Argentina*

**Odile Naim**
*Laboratoire d'Ecologie Marine, Université de la Réunion, BP 9151,Saint-Denis messag 9, Réunion,France*

**Milika Naqasima**
*Marine Studies Programme, University of the South Pacific, P.O. Box 1168, Suva, Fiji*

**Milika Naqasima**
*Marine Studies Programme, University of the South Pacific, P.O. Box 1168, Suva, Fiji*

**David M. Nelson**
*National Ocean Service, 1305 East-West Highway, Silver Spring, MD 20910, U.S.A*

**Dang Duc Nhan**
*Institute of Nuclear Sciences and Techniques, P.O. Box 5T-160, Hoang Quoc Viet, Hanoi, Vietnam*

**Charoen Nitithamyong**
*Department of Marine Science, Chulalongkorn University, Bangkok 10330, Thailand*

**Scott W. Nixon**
*University of Rhode Island, Graduate School of Oceanography, Narragansett, Rhode Island 02882, USA*

**Steve Oakley**
*Institute of Biodiversity and Environmental Conservation, University of Malaysia, Kota Samarahan 93400, Sarawak, Malaysia*

**Tim D. O'Hara**
*Zoology Department, University of Melbourne, Parkville, Vic., Australia*

**E.A. Palacio**
*Instituto de Ingeniería, UNAM, Cd. Universitaria, Mexico City, C.P. 04510 México*

**Claudio Paredes**
*Universidad Simon Bolivar, Apartado 89000, Caracas 1080-A, Venezuela*

**Marcela S. Pascual**
*Instituto de Biología Marina y Pesquera "A. Storni", (8520) San Antonio Oeste, Río Negro, Argentina*

**Mark E. Pattillo**
*U.S. Army Corps of Engineers, Galveston District, P.O. Box 1229, Galveston, TX 77551-1229, U.S.A.*

**Alexandra Pavlidou**
*University of Athens, Department of Chemistry, Division III, Inorganic and Environmental Chemistry, Panepistimiopolis, Kouponia, Athens 15771, Greece*

**Jack B. Pearce**
*NMFS/NOAA, NE Fisheries Center, Woods Hole, MA 02543, U.S.A.*

**Pablo E. Penchaszadeh**
*Universidad Simon Bolivar, Apartado 89000, Caracas 1080-A, Venezuela*

**Daisy Perez**
*Universidad Simon Bolivar, Apartado 89000, Caracas 1080-A, Venezuela*

**Ronald C. Phillips**
*Commission of Environmental Research, Emirates Heritage Club, Abu Dhabi, United Arab Emirates. Correspondence: 1597 Meadow View Drive, Hermiston, OR 97838, U.S.A.*

**Niphon Phongsuwan**
*Phuket Marine Biological Center, P.O. Box 60, Phuket, 83000, Thailand*

**Nicolas Pilcher**
*Institute of Biodiversity and Environmental Conservation, University of Malaysia, Kota Samarahan 93400, Sarawak, Malaysia*

**Luis A. Pinto**
*Facultad de Ciencias, Universidad Católica de la Santísima Concepción, Campus San Andrés, Paicaví 3000, Casilla 297, Concepción, Chile*
**Pamela T. Plotkin**
*Center for Marine Conservation, 1725 DeSales St. NW #600, Washington, DC 20036, U.S.A.*
**Martin H. Posey**
*Center for Marine Science Research, University of North Carolina-Wilmington Wilmington, NC 28403, U.S.A.*
**Anil Premaratne**
*Coast Conservation Department, Sri Lanka*
**Andrew R.G. Price**
*Ecology and Epidemiology Group, Department of Biological Sciences, University of Warwick, Coventry, U.K.*
**Valéria Aparecida Prósperi**
*CETESB – Companhia de Tecnologia de Saneamento Ambiental, Setor de Ictiologia e Bioensaios com organismos aquáticos, Av. Prof. Frederico Hermann Jr., 345, Alto de Pinheiros, SP 05489-900, Brazil*
**Sara C. Pryor**
*Atmospheric Science Program, Department of Geography, Indiana University, Bloomington, IN 47405, U.S.A.*
**Bui Cong Que**
*Institute of Oceanology, Hoang Quoc Viet, Nghia Do, Hanoi, Vietnam*
**Arturo Mendoza Quintero-Marmol**
*PEMEX-Exploracíon-Produccion-Región Marina Noreste, Calle 31, Esq. Periferica, Cd. del Carmen, Campeche, C.P. 24170 Mexico*
**Jean-Pascal Quod**
*ARVAM, 14, Rue du stade de l'Est, 97490 Réunion, France*
**Vladimir I. Radchenko**
*Pacific Fisheries Research Centre, TINRO Centre, Vladivostok, Russia*
**Arjan Rajasuriya**
*National Aquatic Resources Research and Development Agency, Colombo, Sri Lanka*
**Sundararajan Ramachandran**
*Institute for Ocean Management, Anna University, Chennai 600 025, India*
**Krishen J. Rana**
*Fisheries Department, FAO, Rome, Italy*
**Eulalie Ranaivoson**
*Institut Halieutique et des Sciences Marines, B.P. 141, Toliara, Madagascar*
**Bemahafaly J. de D. Randriamanantsoa**
*Cellule des Océanographes de l'Université de Toliara (COUT), IHSM, B.P. 141, Toliara, Madagascar*
**Ricardo Rangel**
*Instituto de Ciencias del Mar y Limnología, UNAM, Cd. Universitaria, A.P. 70-305, Mexico City, C.P. 04510 Mexico*
**Onésime Ratomahenina**
*Ministère de la Recherche Scientifique, Direction Générale de la Recherche, B.P. 4258, Antananarivo, Madagascar*
**Haja Razafindrainibe**
*Cellule Environnement Marin et Côtier, Office National pour l'Environnement, B.P. 822, Antananarivo, Madagascar*
**P. Chris Reid**
*Sir Alister Hardy Foundation for Ocean Science, 1 Walker Terrace, The Hoe, Plymouth, PL1 3BN, U.K.*
**Andrias Reinert**
*Aquaculture Research Station of the Faroes, við Áir, FO-430 Hvalvík, Faroe Islands*
**Louise Richards**
*Hyder Consulting Ltd, Hong Kong; Ecosystems Ltd, Hong Kong*
**Bruce J. Richardson**
*Department of Biology and Chemistry, City University of Hong Kong, 83 Tat Chee Avenue, Kowloon, Hong Kong*
**Frank Riget**
*National Environmental Research Institute, Department of Arctic Environment, P.O. Box 358, DK-4000 Roskilde, Denmark*
**Andrés L. Rivas**
*CENPAT-CONICET, Bv. Brown 3000, (9120) Puerto Madryn, Chubut, Argentina*
**Victoriano Roa**
*Universidad Simon Bolivar, Apartado 89000 Caracas 1080-A, Venezuela*
**Wendy D. Robertson**
*SAAMBR, P.O. Box 10712, Marine Parade, KwaZulu-Natal 4056, South Africa*
**José A. Robleto**
*University of Mobile, Latin American Campus, San Marcos, Carazo, Nicaragua*
**Carlos Díaz Romero**
*Department of Analytical Chemistry, Nutrition and Food Sciences, University of La Laguna, 38071 La Laguna, Spain*
**S. Adrian Ross**
*GEF/UNDP/IMO Regional Programme on Partnerships in Environmental Management for the Seas of East Asia (PEMSEA), P.O. Box 2502, Quezon City, 1165 Metro Manila, Philippines*
**Bernard Salvat**
*Ecole Pratique des Hautes Etudes, URA CNRS 1453, Université de Perpignan, France, and Centre de Recherches Insulaires et Observatoire de l'Environnement, BP 1013, Moorea, Polynésia Française*
**Guadalupe Sanchez**
*Instituto del Mar del Peru, Callao, Peru*
**David Santillo**
*Greenpeace Research Laboratories, University of Exeter, Exeter EX4 4PS, U.K.*
**Norma Santinelli**
*Universidad Nacional de la Patagonia, Belgrano 504, (9100) Trelew, Chubut, Argentina*
**Alberto Santolaria**
*Departamento de Biología Vegetal y Ecología, Facultad de Ciencias, Universidad del País Vasco, Apdo. 644, Bilbao 48080, Spain*
**Suesan Saucerman**
*Environmental Protection Agency, EPA Region IX – WTR-5, 75 Hawthorne St., San Francisco, CA 94105-3901, U.S.A.*
**Kenneth C. Schiff**
*Southern California Coastal Water Research Project, 7171 Fenwick Lane, Westminster, CA 92683, U.S.A.*
**Michael H. Schleyer**
*SAAMBR, P.O. Box 10712, Marine Parade, KwaZulu-Natal 4056, South Africa*
**E.A. Schreiber**
*National Museum of Natural History, Smithsonian Institution, NHB MRC 116, Washington D.C. 20560, U.S.A.*

**Miradys Sebastiani**
*Universidad Simon Bolivar, Apdo. 89000, Caracas 1080-A, Venezuela*
**Antonio Secilla**
*Departamento de Biología Vegetal y Ecología, Facultad de Ciencias, Universidad del País Vasco, Apdo. 644, Bilbao 48080, Spain*
**Virender K. Sharma**
*Chemistry Department, Florida Tech., 150 West University Blvd., Melbourne, FL 32901-6975, U.S.A.*
**Charles R.C. Sheppard**
*Department of Biological Sciences, University of Warwick, Coventry CV4 7AL, U.K.*
**Benjamin H. Sherman**
*Climate Change Research Center, Institute for the Study of Earth Oceans and Space, OSP HEED MMED program, 206 Nesmith Hall, University of New Hampshire, Durham, NH 03824, U.S.A.*
**Zou Shicun**
*School of Chemistry and Chemical Engineering, Zhangshan University, Guangzhou 510301, People's Republic of China*
**Frederick T. Short**
*Jackson Estuarine Laboratory, University of New Hampshire, 85 Adams Point Road, Durham, NH 03824, U.S.A.*
**Jeffrey W. Short**
*Auke Bay Laboratory, National Marine Fisheries Service, National Oceanic and Atmospheric Administration, 11305 Glacier Highway, Juneau, AK 99801, U.S.A.*
**Jens Skei**
*Norwegian Institute for Water Research (NIVA), P.O. Box 173 Kjelsaas, 0411 Oslo, Norway*
**Allegra M. Small**
*Marine Systems Laboratory, Smithsonian Institution, NHB E-117, MRC 164, Washington, DC 20560, U.S.A.*
**G. Robin South**
*International Ocean Institute – Pacific Islands, The University of the South Pacific, P.O. Box 1168, Suva, Republic of the Fiji Islands*
**Don M. Spoon**
*Marine Systems Laboratory, Smithsonian Institution, NHB E-117, MRC 164, Washington, DC 20560, U.S.A.*
**Dennise Stecconi**
*Centro de Procesamiento de Imagenes Digitales (CPDI), Sartenejas, Caracas. Venezuela*
**Ruth Stringer**
*Greenpeace Research Laboratories, University of Exeter, Exeter EX4 4PS, U.K.*
**D.V. Subba Rao**
*Mariculture and Fisheries Department, Kuwait Institute For Scientific Research, P.O. Box 1638, Salmiya 22017, Kuwait*
**Jeffrey Sybesma**
*University of the Netherlands Antilles, Jan Noorduynweg 111, P.O. Box 3059, Curaçao, Netherlands Antilles*
**Piotr Szefer**
*Medical University of Gdansk, 107 Gen. Hallera Ave., 80-416 Gdansk, Poland*
**Liana Talaue-McManus**
*Marine Science Institute, University of the Philippines, 1101 Diliman, Quezon City, Philippines*
**Touch Seang Tana**
*Department of Fisheries, 186 Norodom Blvd., P.O. Box 582, Phnom Penh, Cambodia*
**Shinsuke Tanabe**
*Center for Marine Environmental Studies, Ehime University, Tarumi 3-5-7, Matsuyama 790-8566, Japan*
**Alan D. Tappin**
*Centre for Coastal and Marine Science, Plymouth Marine Laboratory, Prospect Place, Plymouth PL1 3DH, U.K.*
**Ricardo Corbella Tena**
*Department of Analytical Chemistry, Nutrition and Food Sciences, University of La Laguna, 38071 La Laguna, Spain*
**Gordon W. Thayer**
*NOAA/National Ocean Service, Center for Coastal Fisheries and Habitat Research, 101 Pivers Island Road, Beaufort, NC 28516-9722, U.S.A.*
**Law Ah Theem**
*Universiti Kolej Terengganu, Universiti Putra Malaysia, 21030 Kuala Terengganu, Malaysia*
**Chua Thia-Eng**
*GEF/UNDP/IMO Regional Programme on Partnerships in Environmental Management for the Seas of East Asia (PEMSEA), P.O. Box 2502, Quezon City, 1165 Metro Manila, Philippines*
**Nguyen The Tiep**
*Institute of Oceanology, Hoang Quoc Viet, Nghia Do, Hanoi, Vietnam*
**Alexander V. Tkalin**
*Far Eastern Regional Hydrometeorological Research Institute (FERHRI), Russian Academy of Sciences, 24 Fontannaya Street, Vladivostok 690600, Russia*
**Alejandro O. Toledo**
*Institute for Marine and Limnology Sciences, National Autonomous University of Mexico, Marine Pollution Laboratory, P.O. Box 70305, México City 04510 D.F., México*
**Bruce J. Tomalin**
*SAAMBR, P.O. Box 10712, Marine Parade, KwaZulu-Natal 4056, South Africa*
**Tomas Tomascik**
*Parks Canada – WCSC, 300-300 West Georgia St., Vancouver, British Columbia, Canada V6B 6B4*
**Michael S. Traber**
*University of Rhode Island, Graduate School of Oceanography, Narragansett, RI 02882, USA*
**Lamarr B. Trott**
*National Oceanic and Atmospheric Administration, National Marine Fisheries Service, 1315 East West Highway, Silver Spring, MD 20910, U.S.A.*
**Anna Trzosinska**
*Institute of Meteorology and Water Management, 42 Waszyngtona Str., PL 81-342 Gdynia, Poland*
**Caroline Turnbull**
*Department of Biological Sciences, University of Warwick, Coventry, UK*
**John Turner**
*School of Ocean Sciences, University of Wales Bangor, Marine Science Laboratories, Anglesey, Gwynedd LL59 5EY, U.K.*
**Md. Ekram Ullah**
*Environment Section, Water Resources and Planning Organisation (WARPO), House 4 A, Road 22, Gulshan-1, Dhaka-1212, Bangladesh*

**Wilaiwan Utoomprurkporn**
*Department of Marine Science, Chulalongkorn University, Bangkok 10330, Thailand*
**Marieke M. van Katwijk**
*University of Nijmegen, The Netherlands*
**Felipe Vázquez**
*Instituto de Ciencias del Mar y Limnología, UNAM, Cd. Universitaria, A.P. 70-305, Mexico City, C.P. 04510 Mexico*
**Marjo Vierros**
*UNEP-CAR/RCU, 14–20 Port Royal St., Kingston, Jamaica, Correspondence: Rosentiel School of Marine and Atmospheric Science, University of Miami, Dept. of Marine Geology & Geophysics, 4600 Rickenbacker Causeway, Miami, FL 33149-1098, U.S.A.*
**Alicia Villamizar**
*Universidad Simon Bolivar, Apartado 89000 Caracas 1080-A, Venezuela*
**Cesar L. Villanoy**
*Marine Science Institute, University of the Philippines, 1101 Diliman, Quezon City, Philippines*
**Susana Villanueva-Fragoso**
*Institute for Marine and Limnology Sciences, National Autonomous University of Mexico, Marine Pollution Laboratory, P.O. Box 70305, México City 04510 D.F., México*
**Maika Vuki**
*Chemistry Department, University of the South Pacific, P.O. Box 1168, Suva, Fiji*
**Veikila C. Vuki**
*Marine Studies Programme, University of the South Pacific, P.O. Box 1168, Suva, Fiji*
**Greg Wagner**
*Department of Zoology and Marine Biology, University of Dar es Salaam, Dar es Salaam, Tanzania*
**Diana I. Walker**
*Department of Botany, The University of Western Australia, Perth, WA 6907, Australia*
**Jan Warzocha**
*Sea Fisheries Institute, 1 Kollataja Str., PL 81-332 Gdynia, Poland*
**Maggie Watson**
*ICLARM Caribbean and Eastern Pacific Office, PMB 158, Inland Messenger Service, Road Town, Tortola, British Virgin Islands*
**Tom S. Weingartner**
*University of Alaska Fairbanks, Institute of Marine Science, School of Fisheries and Ocean Sciences, Fairbanks, AK 99775-7220, U.S.A.*
**Yan Wen**
*South China Sea Institute of Oceanology, Chinese Academy of Sciences, Guangzhou 510301, People's Republic of China*
**Frank Whitney**
*Institute of Ocean Sciences, P.O. Box 6000, Sidney, B.C., V8L 4B2, Canada*
**Mark Whittington**
*Frontier International, Leonard St., London EC2A 4QS, U.K.*
**Sheila Wiegman**
*American Samoa Environmental Protection Agency, Pago Pago, American Samoa 96799, U.S.A.*
**Clive R. Wilkinson**
*Australian Institute of Marine Science, PMB No. 3, Townsville MC 4810, Australia*
**Simon C. Wilson**
*Department of Biological Sciences, Warwick University, Coventry, CV4 7RU, U.K. and P.O. Box 2531, CPO 111, Seeb, Oman*
**Douglas C. Wilson**
*Institute for Fisheries Management and Coastal Community Development, P.O. Box 104, DK-9850 Hirtshals, Denmark*
**Poh-Poh Wong**
*Department of Geography, National University of Singapore, Singapore 119260*
**Elizabeth Wood**
*Marine Conservation Society, 9 Gloucester Road, Ross on Wye HR9 5BU, U.K.*
**Bruce A. Wright**
*Alaska Region, National Marine Fisheries Service, National Oceanic and Atmospheric Administration, 11305 Glacier Highway, Juneau, AK 99801, U.S.A.*
**Rudolf S.S. Wu**
*Department of Biology and Chemistry, City University of Hong Kong, 83 Tat Chee Avenue, Kowloon, Hong Kong*
**Sandy Wyllie-Echeverria**
*School of Marine Affairs, University of Washington, Seattle, WA 98105, U.S.A.*
**Alejandro Yáñez-Arancibia**
*Department of Coastal Resources, Institute of Ecology A.C., Km 2.5 Antigua Carretera Coatepec, P.O. Box 63, Xalapa 91000, Veracruz, México*
**Leon P. Zann**
*School of Resource Science and Management, Southern Cross University, P.O. Box 57, Lismore, NSW 2480, Australia*
**Eddy Y. Zeng**
*Southern California Coastal Water Research Project, 7171 Fenwick Lane, Westminster, CA 92683, U.S.A.*
**Scott A. Zengel**
*Research Planning, Inc., 1121 Park St., Columbia, SC 29201, U.S.A.*

# ACKNOWLEDGEMENTS

Several people greatly facilitated the logistical and editorial work of this series of 136 chapters. I am very grateful to Professor Leon Zann, in Australia, and Dr Jack Pearce in the USA who greatly assisted in the process of identifying sensible and manageable regions in their own respective parts of the world, and who helped to identify excellent people or groups of people to write about them. Both of them also contributed more than one excellent chapter themselves. Jack Pearce was also the co-editor of a more-or-less random collection of 16 of these chapters for a special issue of *Marine Pollution Bulletin* published simultaneously in 2000.

Bathymetry in the figures for these volumes was taken from 'GEBCO-97: The 1997 Edition of the GEBCO Digital Atlas'. This excellent product is published on behalf of the Intergovernmental Oceanographic Commission (of UNESCO) and the International Hydrographic Organisation as part of the General Bathymetric Chart of the Oceans (GEBCO) by the British Oceanographic Data Centre, Birkenhead. Further details are available at www.bodc.ac.uk. I wish to thank those who produced this invaluable digital data set for granting permission to use it. Coastlines, political boundaries and other cartographic details including some of the place names were taken from Europa Technologies 'Map elements: International & global map data components'. These GIS products were used in most 'Figure 1' sketch maps, and in many others. Early guidelines to authors bravely said that we would prepare 'standard' style maps from rough materials supplied by authors. The phrase 'rough materials' was taken rather literally by many, and for preparing the maps I am especially grateful to Anne Sheppard, whose interpretations of alleged draft maps often required prolonged diligence and clairvoyance. I thank Rebecca Klaus, who set up the GIS system, and I am also grateful to Olivia Langmead and Sheila O'Sullivan who provided editorial help with several draft chapters by converging numerous and variable dialects of the English language towards a common format. In the whole production process, Justinia Seaman in Elsevier provided masterful co-ordination of the huge project, and finally, when I thought that the editing process was complete, I had cause to be most grateful to Pam Birtles, whose production and copy editing skills added far more than just a final polish and a checking of references.

The final product is, of course, a credit to more than 350 authors. My emphasis throughout this series, particularly in the first two volumes, was on the lesser developed countries, precisely those areas with the least available information and commonly with the most pressing environmental difficulties. The breadth of material asked for in each chapter is considerable, and many large areas have very few resident marine scientists, who are in any case very over-stretched. Some had the data but insufficient time or resources to easily compile a review for a 'foreign language book'. Several had difficulty persuading their governments to allow them to do so, or to present data which might be embarrassing to their own employers. Several chapters have a long and interesting tale behind their gestation. That so many people did write is gratifying: an 'ordinary' chapter from some areas is anything but an ordinary achievement, when a range of obstacles conspired to prevent it. From personal experience, I know that in several countries these obstacles can include considerable censure and risk when describing, for example, environmental problems; which is understandable when it is realised that continuing aid may depend on the government pretending to comply with imposed 'sustainable use' measures. We should all be grateful to these authors, and I am also grateful for their subsequent acceptance of my sometimes drastic editorial changes done in the interests of brevity (mostly), format (usually) and language (sometimes). I hope that one of the benefits of this series will lie in its provision of information, so that one place may learn from the problems of another before repeating the same mistakes, and thus avoid impoverishing yet another bit of the world's coast and its people and, hopefully, reversing many of the problems.

Charles Sheppard

Chapter 107

# GLOBAL STATUS OF SEAGRASSES

Ronald C. Phillips and Michael J. Durako

Seagrasses are a dominant component of many of the world's estuaries and shallow, coastal waters (Phillips and Menez, 1988). The shallow distribution of seagrasses places them in close proximity to the land/sea interface, a region of the globe experiencing explosive growth of human populations. This coastal distribution also places seagrass communities at the end of the watershed; thus, their status reflects not only direct coastal influences, but larger, landscape and regional influences as well. Because most seagrasses are benthic-perennial plants, they are continuously subject to stresses and disturbances that are associated with changes in water quality along the land/sea interface. Seagrasses act as integrators of net changes in water quality variables which tend to exhibit rapid and wide fluctuations when measured directly. For these reasons, seagrasses may be one of the best indicators of changes in coastal hydroscapes.

Seagrasses are found in shallow estuarine and coastal areas of the world with suitable substrates. Their upper limit of growth is determined by their ability to cope with heat and desiccation at low tides. Their lower limits of growth are largely determined by the penetration of a suitable amount of light.

By and large, seagrasses are found along most coastlines of all continents. They appear to be excluded generally north of the Arctic Circle and south of the Antarctic Circle. *Zostera marina* L. (eelgrass) extends north of the Arctic Circle only in northern Russia, presumably due to the warming influence of the Gulf Stream (Phillips and Menez, 1988). Few, if any, seagrasses are found along the west coast of Africa, along the west coast of South America except for a one km$^2$ patch of *Heterozostera* known to occur near Coquimbo in northern Chile, or south of Sao Paulo, Brazil, on the east coast of South America (Phillips and Menez, 1988). It is the role of seagrasses as fisheries habitats that largely drives man's interest in these plants (Zieman, 1982; Thayer et al., 1984; Zieman and Zieman, 1989). Loss or deterioration of seagrass habitat has generally been linked with loss of fisheries and a decline in habitat quality (Orth and Moore, 1983; Lombardo and Lewis, 1985; Dennison et al., 1993). Man's increasing development and use of coastal and estuarine systems has resulted in dramatic alterations of many seagrass beds (Orth and Moore, 1983; Lewis et al., 1985). Consequently, there has been an increasing concern regarding planet-wide losses of seagrasses over the past several decades (Short and Wyllie-Echeverria, 1996). Management actions have recently been directed toward instituting monitoring programs to assess the status and trends of seagrasses and linking these to changes in water quality, and even more recently, in developing strategies to stop or reverse the losses of seagrasses (Neckles, 1994).

*Seas at The Millennium: An Environmental Evaluation (Edited by C. Sheppard)*

## ECOSYSTEM SERVICES

Seagrasses are the only plants that grow totally submerged and are rooted in their substrate in estuarine and coastal marine environments. As such, they represent a distinct and unique ecosystem. In addition, as a result of their spatial position, they form a link between inshore marsh and mangrove ecosystems, and offshore coral, algal, benthic soft-bottom, and planktonic ecosystems. The functions and processes of the seagrass ecosystem can be described as ecosystem services, since they directly relate to not only the health, stability, and wellbeing of the physical environment in which they live, but also to that of human populations as well.

Basically, there are three major functional components in the seagrass ecosystem (Phillips and Menez, 1988): (1) the rate of energy flow through the system, including primary and secondary production, growth, and respiration; (2) the rate of material or nutrient cycling within the system, including decomposition; and, (3) the degree of biological or ecological regulation in the ecosystem, including biological and physical feedback loops. There are many interactions between these three components which involve the diversity of the biota in the system and their individual dynamics.

### Sediment Accumulation and Stabilization

An intact seagrass ecosystem provides sediment stabilization because of the presence of an interlacing rhizome/root mat. Seagrass beds have been documented to remain intact through high winds and wave action during hurricanes in the Caribbean (Thomas et al., 1961). Hurricane Donna in 1960 passed over *Thalassia testudinum* Banks ex Konig (turtlegrass) in Biscayne Bay with winds exceeding 87 km for 24 hours. Windrows of leaves were cast on shore, but only light damage was observed to the seagrass beds. On the other hand, in Salcombe Harbour, U.K., sand banks were lowered 30 cm overnight during a storm after *Zostera marina* L. (eelgrass) disappeared in 1931 (Wilson, 1949). Up to 20 cm of sediment eroded from unvegetated sand banks following a single storm in Chesapeake Bay, while little, if any, sediment disappeared from within an eelgrass meadow (Orth, 1977).

Many studies have documented the trapping and settling of senescent sloughed off leaves, which then decompose and release dissolved and particulate (detritus) organic matter and nutrients to the substrate and water mass. The decomposing leaf mass becomes the substrate for a bacterial flora which forms the basis of diverse complex food chains. The leaf mass becomes, in itself, a shelter for a large animal community which adds to the food chain. In the final stages, the remaining leaf mass is incorporated into the existing substrate, adding a large store of organic matter to the substrate.

### Nutrient Effects and Water Quality

The production of detritus and promotion of sedimentation by the seagrass leaves provide organic matter for the sediments and maintain an active environment for nutrient cycling (Zieman, 1982). Epiphytic algae on the seagrass leaves fix nitrogen, thus adding to the nutrient pool of the region (Zieman, 1982). Seagrasses assimilate nutrients from the sediments, transporting them through the plant and releasing them into the water column through the leaves, thus acting as a nutrient pump from the sediment (Zieman, 1982). Seagrass blades, their epiphytes and macroalgae also pick up water column nutrients (Fonseca et al., 1998).

In an intact, relatively dense seagrass bed, the leaves act as a baffle for wave and current action. This baffling effect effectively reduces water motion within the leaf canopy, allowing incoming and resident particulate matter to settle to the bottom and remain there. This allows water within the seagrass bed to be clearer than water over unvegetated sediments, thus improving water quality for resident plants and animals (Zieman, 1982; Fonseca et al., 1998).

### Primary and Secondary Productivity and Growth

The physical stability, reduced mixing of different water masses, and shelter provided by the complex seagrass structure provides the basis for a highly productive ecosystem (Wood et al., 1969; Fonseca et al., 1998).

The ability of seagrasses to exert a major influence on the marine seascape is due in large part to their extremely rapid growth and high net productivity. Leaves grow at rates typically 5 mm/day, but growth rates of over 10 mm/day are not uncommon under favourable circumstances (Zieman, 1982; Brouns, 1987; Pollard and Greenway, 1993).

Primary production is the most essential function of the seagrass ecosystem (Phillips and Menez, 1988). Production rates are remarkably high. Seagrass primary production rates in Australia vary from 120 g C $m^2$ $yr^{-1}$ to 690 g C $m^2$ $yr^{-1}$, depending on the species (Hillman et al., 1989). Elsewhere, rates for tropical species vary from 280 g C $m^2$ $yr^{-1}$ (*Halodule wrightii* Aschers) to 825 g C $m^2$ $yr^{-1}$ (turtlegrass), while the temperate eelgrass species varies between 190 g C $m^2$ $yr^{-1}$ (Thorne-Millar and Harlin, 1984) to 400–800 g C $m^2$ $yr^{-1}$ (Sand-Jensen and Borum, 1983).

These rates compare with the maximum values for terrestrial plants, but are somewhat lower than the very high rates of the highest agricultural crops or of mangroves (3000 g C $m^2$ $yr^{-1}$ for sugar cane and sorghum and ~2000 g C $m^2$ $yr^{-1}$ for mangroves (Hillman et al., 1989).

Few data exist on the primary production of seagrass epiphytes and the benthic macrophyte and microphyte components with the seagrass ecosystem. Production of seagrass epiphytes can reach 50% of the seagrass production (Jones, 1968; Penhale, 1977; Borum and Wium-Andersen, 1980; Morgan and Kitting, 1984; Brouns and

Heijs, 1986). Loose-lying and attached macroalgae within the seagrass ecosystem may account for 2% to 39% of the total above-ground seagrass biomass at times and various places, ranging to almost 75% (Dawes, 1987). Thus, the total primary production of any given seagrass ecosystem may be extremely high.

### Shelter

Seagrass beds serve as nursery grounds, places of both food and shelter, for juveniles of a variety of finfish and shellfish of commercial and sportfishing importance (Zieman, 1982).

A great variety of commercially and recreationally important animals feed directly on the plants, e.g., manatees in Florida, dugongs in the Indo-Pacific, a wide variety of waterfowl, green turtles, and queen conchs in the Caribbean (McRoy and Helfferich, 1980). In the Arabian Gulf, the valuable pearl oyster finds shelter and food in the extensive seagrass beds (Basson et al., 1977).

Seagrass beds are spatially positioned between the inshore mangroves and marshes and the offshore coral reefs and/or open waters. Thus, a great variety of animals are permanent residents or find temporary shelter within the leaf canopy at some stage of their life cycle. In the tropics, a number of fish and sea urchin species undertake a diel migration from their daylight abode on the coral reef to a night-time feeding residence on the adjacent seagrasses. There are seasonal finfish and shellfish that shelter and feed within the seagrasses during their migratory movements. Finally, there are a great number of juvenile fish species which migrate offshore from the mangrove/marsh habitats, and find refuge in the seagrass meadows before migrating offshore (Thayer et al., 1978; Phillips, 1984).

Kikuchi (1980) enumerated the reasons for this sheltering function: (1) seagrasses form a dense submerged meadow and increase the available substrate surface for epiphytic algae and fauna; (2) dense vegetation reduces the water movement resulting from currents and waves and offers calm underwater space within it; (3) because of the less disturbed water, mineral and organic particles sink to the bottom, creating relatively clear water; and, (4) the leaf mass reduces excessive illumination in the daytime which facilitates the refuge effect for the prey species. Fonseca et al. (1998) summarized the sheltering function as being the result of the high productivity of the seagrasses coupled with the physical stability (perennial plants, tenacious interlacing rhizome/root mat which holds the sediment, a leaf canopy which creates relatively calm water within the leaf canopy) of the plants.

### Food and Feeding Pathways and Biodiversity

In a holistic sense, at least one major ecosystem property emerges when the system is intact, viz., the nursery function of the seagrass meadow. The interplay of structural and functional characteristics results in a dense, stable environment that forms refuge and shelter, as well as food for a myriad of organisms. Some organisms spend their lifetime in the meadow, while many spend only their juvenile life in it, or merely feed in it during a portion of the day, to pass on to an adjacent system to complete their life cycles (Phillips and Menez, 1988). The photosynthetically fixed energy from the seagrasses may follow two general pathways: direct grazing by organisms on the living plant material or utilization of detritus from decaying seagrass material, primarily leaves. The export of seagrass material, both living and detrital, to a location some distance from the seagrass bed allows for further distribution of energy away from its original source (Zieman, 1982).

Of the two pathways, the detrital pathway is overwhelmingly the most important within the seagrass ecosystem. In this pathway, bacteria form the basis of the food web. When senescing, the leaves release both particulate and dissolved carbon and organic matter, which the bacteria assimilate and transform into bacterial matter. Bacteria also aid in leaf decay. The bacteria are consumed by a host of small organisms, which are then eaten by another larger group of organisms, and so on. Detrital food webs are long and complex, and lead to the commercially and recreationally important food animals. Direct food webs are shorter, with perhaps two to three links; they may lead to animals harvested by humans.

There exists a very large literature documenting the food webs and relationships which organisms have in the seagrass ecosystems. Kikuchi and Peres (1973) subdivided the animal community in seagrasses into several components by microhabitat structure and mode of life:

(1) biota on the leaves, e.g., epiphytic flora and meiofauna on the leaves, mobile epifauna, and swimming fauna which often rest on the leaves;
(2) biota attached to the stems and rhizomes, e.g. nest-building polychaeta and amphipoda;
(3) highly mobile animals swimming through and under the leaf canopy, e.g., fishes, cephalopods, decapods; and,
(4) the biota living on or in the sediment (Kikuchi, 1980).

Species composition affects the biological regulation within ecosystems. In the seagrass ecosystem, the cyanobacteria on or in the plants or substrate fix nitrogen for seagrass or epiphyte use. Owing to the high rate of use, nitrogen is considered a rate-limiting factor in the seagrass ecosystem. Seagrass density and biomass variations in space and time are reflections of the nitrogen pool (Short, 1981). These parameters in turn affect sediment accretion and stabilization, water clarity (which affects primary production), nutrient availability (which affects primary production), and further nutrient cycling. Features of the abiotic environment, viz., daily and annual ranges in temperature and salinity, wave activity, and tidal currents, regulate species composition and productivity values.

# Propagation of *Zostera marina* L. from Seed

Stephen L. Granger, Michael S. Traber and Scott W. Nixon

*University of Rhode Island, Graduate School of Oceanography, Narragansett, Rhode Island 02882, USA*

The recent loss of seagrass habitat worldwide, including in estuaries along the East Coast of the United States, has been well documented. It seems clear that the usual technique of transplanting whole plants from donor beds to restoration sites has been unable to keep pace with the loss. With target goals of hundreds to thousands of acres of restored seagrass habitat set in estuaries along the East Coast alone, there is a strong incentive to develop additional approaches to seagrass propagation. Preliminary work of raising seagrass plants from seed may provide a restoration technique that can be used in conjunction with, or as an alternative to, transplanting mature, healthy plants. Experimental results presented here on the shores of Narragansett Bay may be generally applicable to eelgrass, *Zostera marina* L., throughout much of its range.

Ample supplies of seeds become available each year as seagrass plants flower and develop seeds. Flowering plants are easily harvested with low or no impact to the donor bed. With only a modest effort (approximately 55 person-hours), some 500,000 seeds can be routinely harvested. To increase genetic diversity of the seed stock (and ultimately the restored habitat) seed-bearing spathe were collected from a number of field sites and placed in flowing seawater tanks for several weeks until they released the majority of their seeds (Fig. 1). After a series of screening and winnowing processes, the separated seeds are easily transported, require little storage space, and can be held for months. Seeds harvested during summer and stored in flowing seawater tanks at ambient temperature display a constant rate of germination and are virtually completely germinated by the following spring. Seed germination may be hastened through manipulations of temperature and salinity but seeds treated in this manner do not develop into healthier or larger seedlings.

Significantly higher rates of germination and seedling success occur when seeds are planted in the first few centimetres of sediment compared to those simply cast upon the sediment surface (Fig. 2). However, placing seeds deeper than about two centimetres results in lower germination. After hypocotyl emergence and development of true leaves, seedlings display an exponential rate of growth with rising water

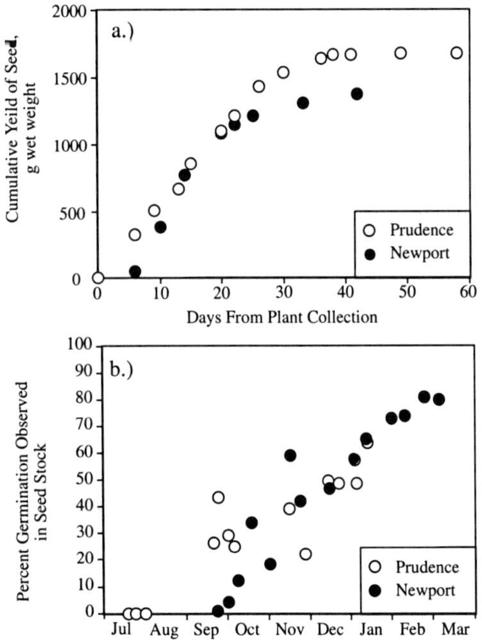

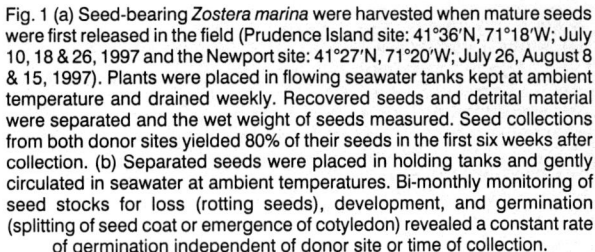

Fig. 1 (a) Seed-bearing *Zostera marina* were harvested when mature seeds were first released in the field (Prudence Island site: 41°36'N, 71°18'W; July 10, 18 & 26, 1997 and the Newport site: 41°27'N, 71°20'W; July 26, August 8 & 15, 1997). Plants were placed in flowing seawater tanks kept at ambient temperature and drained weekly. Recovered seeds and detrital material were separated and the wet weight of seeds measured. Seed collections from both donor sites yielded 80% of their seeds in the first six weeks after collection. (b) Separated seeds were placed in holding tanks and gently circulated in seawater at ambient temperatures. Bi-monthly monitoring of seed stocks for loss (rotting seeds), development, and germination (splitting of seed coat or emergence of cotyledon) revealed a constant rate of germination independent of donor site or time of collection.

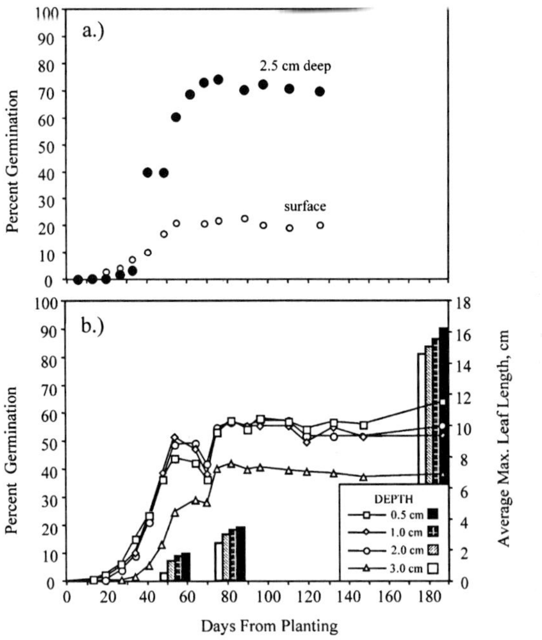


Fig. 2. (a) Germination success of seeds planted Jan 1997 at a depth of 2.5 cm and on the surface of marine sediments. Data points represent a mean of two seed stocks collected from Prudence Island and Newport, R.I. and two sediment sources (Hope and Prudence Islands). Germination success was independent of seed or sediment source. (b) Success of plants raised from seeds planted Jan 1997 at 0.5, 1, 2, and 3 cm depths in marine sediments, at ambient temperature in flowing seawater. Seeds placed in the first 2 cm of sediment displayed similar success. A lower rate of germination but similar growth was observed in seeds planted at a depth of 3 cm. Seedlings were considered successful if they developed leaves, increased in length or leaf number through the study. Mean leaf length was determined to the nearest millimetre on three occasions (shown as bar graph).

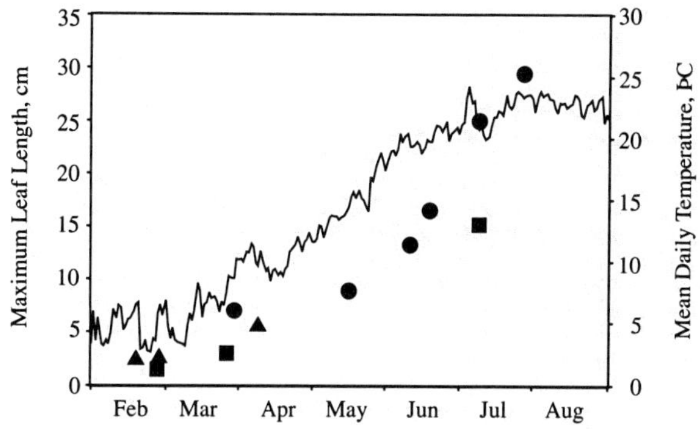


Fig. 3. Canopy height expressed as mean maximum leaf length, and observed during three separate experiments (1997, 1998 and 1999). Seed and sediment collection sites were identical for the experiments. Seeds were planted on 12/12/96 (black triangles), 1/13/97 (black squares) and 12/6/98 (black circles). To demonstrate the close correlation of plant growth to rising water temperatures we have included mean daily seawater temperatures of the test aquaria during the 1998 experiment.

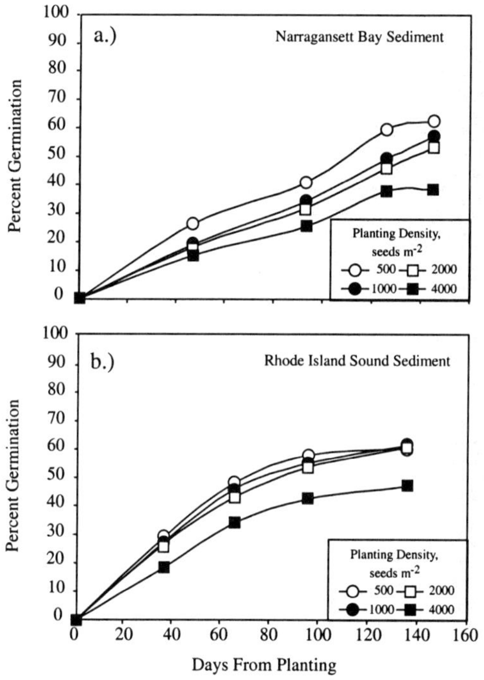


Fig. 4. Seed germination as a function of seeding density and sediment type in flowing seawater aquaria measuring 2.2×6 m, with 1 m water depth lined with 15 cm of sediment from two sites. Triplicate plots were seeded at a depth of 2 cm to achieve final plant densities of 250, 500, 1000 and 2000 shoots $m^{-2}$ (assuming a 60% germination rate). Plots displayed similar germination success in sediment with a low organic content, but there was an inverse relationship between seeding density and germination success in the organically rich sediment collected from Narragansett Bay (panel a).

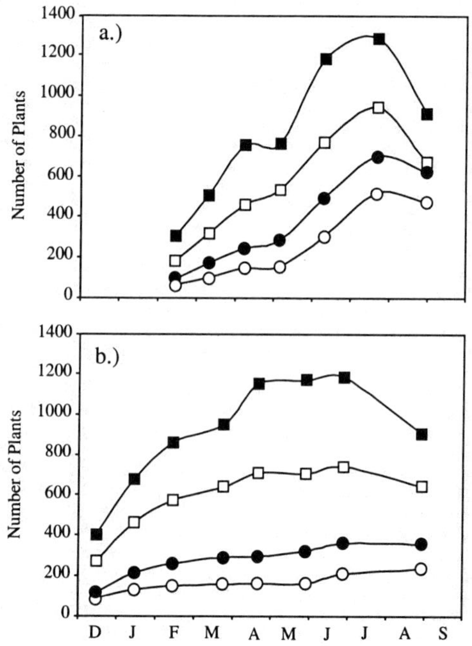


Fig. 5. Monthly counts of healthy seedlings in test plots seeded to achieve plant densities of 250, 500, 1000 and 2000 shoots $m^{-2}$. (a) Narragansett Bay, (b) Rhode Island Sound. Each point represents a mean of three replicates. Young seedlings had begun to produce lateral shoots accounting for the increase in plant numbers from April through July. Plots seeded at lower densities produced more lateral shoots per plant indicating an inverse relationship between the production of lateral shoots per plant and plant density. However, plots seeded at 2000 shoots $m^{-2}$ produced the greatest number of lateral shoots. For legend, see Fig. 4.

temperature and reach a canopy height of 30 centimetres by the following August (Fig. 3). Plants propagated from seeds during the fall do not flower or bear seeds the following summer, though a number of seedlings do develop seeds during their second year of growth. There do not appear to be significant differences in germination success or seedling development with seeds collected from different sites in Narragansett Bay or Rhode Island Sound.

Sowing at high densities resulted in significantly lower germination than lower rates (Fig. 4). This result was observed in both high and low organic sediment. Monthly measurements of plant density indicated that seed germination ended in April (Fig. 5). This was followed by a renewed increase in plant numbers due to the production of lateral shoots from the seedlings. While plots seeded at the highest densities produced more lateral shoots simply as a function of plant numbers, plots seeded at lower densities produced more lateral shoots per plant. As a result, it is not clear that planting more seeds will necessarily produce a greater standing stock of seagrass. Seeds planted directly in field locations within Narragansett Bay have developed into healthy seedlings that have persisted for several years. However, we are still working toward transferring the high success of germination and plant development we observe in our aquaria experiments to the field.

# *Zostera Marina* and The Wadden Sea

Marieke M. van Katwijk

*University of Nijmegen, The Netherlands*

The Wadden Sea is one of the world's largest international marine wetland reserves (approximately 6000 km²), bordering the coasts of The Netherlands, Germany, and Denmark. Before the 1930s, The Dutch Wadden Sea contained large beds of subtidal and low-intertidal *Zostera marina* L. covering an area between 65 and 150 km² (Fig. 1).

These seagrass beds were of great economic importance as the seagrass was used as roofing, furnishing and insulation material. Before 1857, it was used to build dikes. Considering the importance of dykes to The Netherlands, it is no wonder that in the past a proverb was used to describe the harvest (good hay grass, good sea grass), a special prayer day was held to invoke a bumper crop, and lyrical descriptions and poems about seagrass were written during the 18th and 19th centuries. Less is known about the past German and Danish beds which had little or no economic value.

During the 1930s, this seagrass cover was largely lost, and the beds never recovered. At present, *Z. marina* occurs only in the mid-littoral, with approximately 2 km² of *Z. marina* in The Netherlands. In the German Wadden Sea, *Z. noltii* and *Z. marina* together cover approximately 170 km², and in the Danish part there is 30 km². The large-scale decline of *Z. marina* coincided with: (1) the outbreak of wasting disease caused by the slime-mould *Labyrinthula zosterae*; (2) increased dyking and damming activities; and (3) two subsequent years with a sunlight deficit. There is no consensus about which of these events (or combination of events) caused the decline. The Dutch government is currently attempting to return seagrass to the Wadden Sea, in order to 'restore natural values'.

The presence of potential seagrass habitats is the first condition for successful restoration. In the Wadden Sea, a distinction can be made in a higher and a lower zone of potential habitats along the tidal gradient, each suitable for differing morphotypes of *Z. marina* (Fig. 2). The higher zone is inhabited by mostly annual plants. When emerged, the plants lie flat on the moist sediment, in this way protected from desiccation. The lower zone (which disappeared during the thirties) was inhabited by perennial plants, whose stiff sheaths are vulnerable to desiccation during low tide, but which are more resistant to high water dynamics than the former morphotype. Between the two seagrass zones, a bare zone exists, where the habitat is too dynamic for the high morphotype, and the periods of emergence last too long for the low morphotype. The upper limit for *Z. marina* growth in

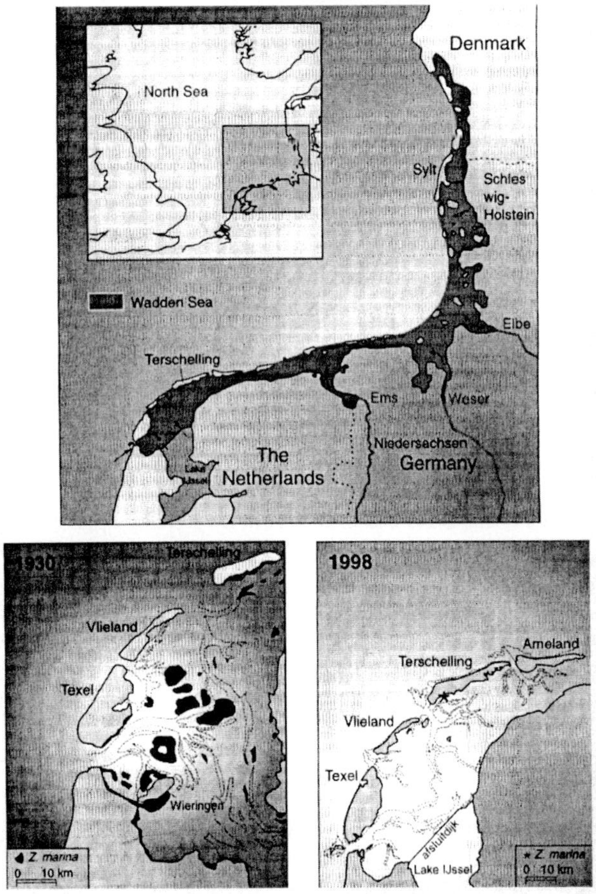


Fig. 1. The Wadden Sea, and the distribution of *Zostera marina* in the 1930s and in 1998. At present, one mid intertidal bed of this seagrass of 1.6 ha remains. Arrows show major freshwater inputs.

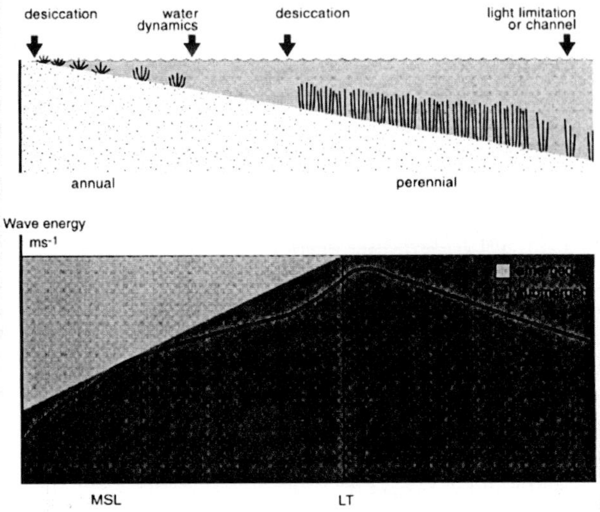


Fig. 2. Zonation of *Zostera marina* in the Wadden Sea along a tidal gradient in relation to wave energy (orbital velocity at the sediment surface) and emergence time.

the high zone is delineated by the degree of desiccation, whereas the low zone is limited by light availability and/or strong currents due to the presence of channels (Fig. 2).

Suitability of potential *Z. marina* habitats is influenced by two main groups of factors:

(1) Water and sediment dynamics, grain size, turbidity, and degree of desiccation (see Fig. 2). Dynamics may increase as a consequence of construction activities, disappearance of subtidal seagrass and increased fishery activities, including shell fisheries. The latter may additionally increase the degree of desiccation when the disappearance of oyster and mussel beds results in an increased superficial drainage.

(2) Interactive effect of nutrients and salinity. High nutrient loads negatively affect *Z. marina* and high salinity stresses the plants which will aggravate the negative effects of high nutrient loads. Nutrients stimulate algal growth which subsequently limits light. Finally, increased nutrient loads in the water causes an increased shoot:root ratio which makes the plants more vulnerable to exposure.

In the Wadden Sea, dynamics, disturbance, and nutrient loads have increased during the 20th century, whereas the overall salinity has remained stable. As a result, the area suitable for re-establishment of *Z. marina* has decreased. However, since the end of the 1980s, turbidity levels in the Wadden Sea have decreased, nutrient levels have decreased or stabilized, and shell fisheries are prohibited in some areas. Both active and passive restoration of *Z. marina* seems now more feasible, although further decreases of these factors are desirable (so as to increase the area of potential *Z. marina* habitats). Until then, potential *Z. marina* habitats will be confined to undisturbed sheltered locations and locations with freshwater influence.

## WORLDWIDE DECLINE OF SEAGRASSES

Following World War II, the need for housing, clothing, and food in many parts of the northern hemisphere, particularly in Europe and Japan, was very great. By necessity, industrialization to satisfy these needs and to provide employment was rapid and intense. During the reconstruction period, economic programs became international in scope, with the development of international shipping and airline transportation systems. The economic boom that developed was accompanied by an escalation in the use of an abundant supply of cheap fossil fuels in every sphere of human activity. Forests were levelled for shelter, paper, and cardboard packing materials; land was cleared for agriculture and housing; the size of cattle, sheep, and goat herds increased; raw materials were acquired by the western nations to aid the reconstruction (Milne and Milne, 1951; Phillips, 1978; Thayer et al., 1975).

Gradually, shallow coastal zones received the impacts of these rapidly growing activities. As development and industrial activity intensified in the more prosperous nations, near the centres of population where the labour force was located, an increasing number of people moved from the farm to urban areas. Thus, demographic changes occurred that turned predominantly rural economies in the 1930s and 1940s to urban and international economies in the 1950s. With fewer people left on the farms, farming became highly mechanized and widespread use of chemical fertilizers, pesticides and herbicides developed. This mechanization, along with the development of the interstate highway system, led to a vast escalation of the automobile industry, and a voracious need for petroleum products. Increasingly, people who had moved to the cities began to travel to resort locations along the coast during their increasing leisure time, then to buy second homes along the coast as their affluence grew. Finally new cities developed and old ones enlarged along the coastal areas.

Human-related impacts on coastal areas in the US were not readily observed until the mid-1950s. Extensive logging in the northeast and northwest USA occurred as early as 1900, resulting in an increase in silt flow into estuaries, but there is no indication whether this was a major problem to submerged aquatic vegetation. It was in the mid-1950s, following the sudden increase in coastal development, that seagrass, as well as coral reef, marsh, and mangrove systems began to show signs of stress and decline.

Dredging in shallow bays in Florida, which contained vast, luxuriant, seagrass meadows, disrupted the ecology of the meadows and resulted in erosion, siltation, and turbid water far removed from the construction activity (Taylor and Salomon, 1968; Phillips, 1974). Maintenance dredging in shallow estuaries was required to allow increasing ship traffic to carry commercial products to ports in the coastal cities. With the population increase came the problem of where and how to dispose of the human and industrial waste. This waste, if not directly toxic, has a fertilizer-effect on estuarine plant systems. Even if treated, it may stimulate phytoplankton growth and noxious benthic algae, leading to a decline in benthic vegetation.

In Long Island, New York, eelgrass was cleared from a large area as it was deemed a nuisance to swimmers at a newly-created beach (Burkholder and Doheny, 1968). Dredging and sewage disposal were the major factors causing seagrass declines in Christiansted Harbor, U.S. Virgin Islands, in the 1950s when the area developed into a major tourist resort (Dong et al., 1972; Zieman, 1975a,b). In Puget Sound, Washington, eelgrass stocks declined near the City of Seattle, following large influxes of people in the early 1950s.

In the Dutch Wadenzee, profound changes in the abundance and location of eelgrass beds were recorded between 1869–1930, but these were regarded as normal long-term fluctuations with a dynamic equilibrium. In 1932, the "wasting disease" decimated subtidal populations (den Hartog, 1996). After 1965, a general decline began of remaining stocks, which is still in progress. While the exact cause is not determined, the decline appears to be related to the amount of increasing pollution (silt, toxic materials, viz., heavy

## Guidelines for the Conservation and Restoration of Seagrasses in the United States and Adjacent Waters: A Synopsis

Mark S. Fonseca, W. Judson Kenworthy and Gordon W. Thayer

*NOAA/National Ocean Service, Center for Coastal Fisheries and Habitat Research, 101 Pivers Island Road, Beaufort, NC 28516-9722, USA*

Seagrass ecosystems are protected in the U.S. under the federal "no-let-loss" policy for wetlands. Greater awareness and public education, however, is essential for conservation of this resource. Tremendous losses of this habitat have occurred as a result of development within the coastal zone. Disturbances usually kill seagrasses rapidly, and recovery is often comparatively slow.

Mitigation to compensate for destruction of existing habitat usually follows when the agent of loss and responsible party are known. Compensation assumes that ecosystems can be made to order and, in essence, trades existing functional habitat for the promise of replacement habitat. While planting seagrass is not technically complex, there is no easy way to meet the goal of maintaining or increasing seagrass acreage. Rather, the entire process of planning, planting and monitoring requires attention to detail and does not lend itself to oversimplification.

The success rate of mitigation projects remains low overall, but this appears to result from failures in the planning process as much as from any other cause. To prevent continued loss of habitat under compensatory mitigation, decisive action must be taken by placing emphasis on improving site selection, compliance, generating desired acreage, and maintaining a true baseline.

Seagrass planting is no longer experimental, but planting will not succeed unless managers appreciate and emphasize the extreme importance of site selection, care in planting, and incorporation of plant demography into the planning and planting processes. Seagrass beds can be restored, but preservation is the most cost-effective course of action to sustain seagrass resources. Although techniques and protocols exist that produce persistent seagrass beds, they are often applied inconsistently, which has resulted in large-scale failures that have biased managers against seagrass restoration.

A logical and ecologically defensible goal is to attain replacement of the lost seagrass species with an area of bottom coverage that compensates for interim lost resource services and a comparable shoot density. Seagrass plantings that persist and generate the target acreage have been shown to quickly provide many of the functional attributes of natural beds. When destruction of the impact site requires planting in another location (ie. off-site) it is often difficult to find a site elsewhere with suitable biological and physical parameters required for seagrass growth and persistence.

As more information is made available to managers regarding the function of seagrass ecosystems and the difficulties involved in mitigating for their loss, fewer permitted impacts are occurring. Recent successful prosecutions in U.S. Federal Court of parties found to be responsible for destruction of seagrass have also contributed to public awareness, and emphasis placed on impact avoidance and minimization.

---

metals, pesticides, PCB, detergents) carried by the river Rhine, which empties into the Waddenzee.

Seagrass systems are also affected by natural environmental impacts. These include periodic and aperiodic disturbances from storms. Population explosions of sea urchins in the northeast Gulf of Mexico ravaged mixed beds of *Thalassia* and *Syringodium* over a length of 26 km and width of 5.9–9.2 km. The numbers of the urchin, *Lytechinus variegatus*, averaged 5.6/m$^2$, with as many as 63.5/m$^2$ at the leading edge. How the numbers became so high so quickly is not known (Camp et al., 1973). Fish, rays, skates, and crabs disturb seagrass vegetation by resting on it, and foraging for food in it. Rays and crabs create holes in the meadow by digging. In Chesapeake Bay, several hectares of eelgrass were removed by rays which scoured out the sediment while removing clams (Orth, 1975). Sirenians, such as the dugong, in the southwest Pacific, graze extensively on seagrass. They use their snouts to shovel out strips of seagrasses, leaving open patches in the meadow (Domning, 1981; pers. obs., 1979). In Izembek Lagoon, Alaska, black brant geese consume 4% of the standing crop of eelgrass with no adverse impacts on the system. These birds crop only the leaves and thus leave the rhizome/root mat in the substrate (McRoy, 1966).

It is human activity, where impacts can be avoided or diminished or mitigated, that constitutes the greatest concern for the health and survival of the seagrass ecosystems. The list of human-related activities and impacts is very long. These include dredging projects for channel construction and maintenance and real estate development (Zieman, 1975, 1982). Other dredging-type impacts are caused by fishermen who drag nets and rakes to collect oysters, scallops, and clams (Thayer and Stuart, 1974). In several parts of the United States, hydraulic dredges have been used to collect clams; the dredges project a stream of water onto the bottom, blasting trenches in the seagrass meadow up to 0.5 m deep and one meter wide (Godcharles, 1971). In Humboldt Bay, California, oyster dredging in eelgrass beds led to a 70% reduction in shoot density, as compared to a 33% reduction from other causes (siltation and turbid water from extensive upland logging, road building, and agricultural practices; Waddell, 1964). Boat propellers do much damage to seagrass meadows, especially in shallow tropical waters (Durako et al., 1992). Recently, Sargent et al. (1995) documented that 6.4% of seagrasses in Florida are scarred due to boat propellers. This scarring results in direct losses of seagrass habitat and may lead to further loss due to erosion and resuspension of sediments.

Dredging activities are so damaging because they not only physically remove the entire plant system, but they also remove the sediment and change the redox potential of the remaining sediments. Seagrasses can only recolonize the remaining sediments after a long period of time when the pre-impact physical–chemical conditions have returned. Plant removal and dredging also create turbid water, hindering plant growth. Heated water discharges from power generators have been released onto seagrass meadows. In Biscayne Bay, Florida, seagrasses died when the bay waters were heated to 5°C above ambient, while up to 60% of the stocks died when the waters were elevated to 4°C above ambient (Roessler and Zieman, 1969). Such heated waters may exceed the adaptive tolerances of species and disrupt their reproductive cycles. The release of sewage and agri-chemicals by industries and farms fertilize and affect the functioning of the flora and fauna. A severe decline in eelgrass and other submerged vegetation was documented in Chesapeake Bay, coincident with the use of atrizine for upland maize cultivation (Correll and Wu, 1982). Flatfish in Elliott Bay, Seattle, Washington, developed liver tumours and were no longer fit for consumption, a fact directly attributable to the release of PCB from industry along the Duwamish River. These chemical additions may result in declining water clarity and plant density, biomass, and production. Even in remote areas, oil drilling and increasing ship traffic related to oil bring the spectre of extensive negative impacts. In these cases, seagrass ecosystems may be greatly affected and in no case may these systems be termed "natural" any longer (Phillips and Menez, 1988).

Recently, Short and Wyllie-Echevarria (1996) published a synopsis of the natural and human-induced disturbances of seagrasses, including world-wide declines in seagrass stocks. These losses occurred in highly industrialized and developing countries. The reasons for an inability to precisely quantify the amount of global seagrass loss are: (1) the lack of work in some areas such as the Arabian Gulf (where there are large quantities of seagrass), South America, and the west coast of Africa, where seagrasses are poorly known or not known at all; and (2) monitoring in most places where research is being performed is not done on a regular basis. Until remote sensing is feasible and is performed regularly on a global basis, it will not be possible to obtain and keep an accurate account of the seagrass stocks on the globe. Until this can be done, we are resigned to monitoring small areas and to attempting to equate changes with a possible cause. Den Hartog (1996) summarized what is known about the "wasting disease" of the 1930s in the North Atlantic. He followed with a report on worldwide sudden declines in seagrasses, and with an in-depth analysis of the biological effects of *Labyrinthula zosterae* Porter & Muehlstein. Under global warming, sea levels would rise and salinities would probably decline. This would produce an inestimable decline in the global distribution and abundance of seagrasses.

## PLANNING, MANAGEMENT, POLICY, GOALS

As seagrasses are subjected to an ever-increasing onslaught of natural and anthropogenic disturbances, it is timely to consider management directions before these communities become so rare that they lack the genetic diversity to cope with change (McRoy, 1966). Unlike many other marine ecosystems, seagrasses may be amenable to management, if management practices are based on knowledge of plant biology and ecosystem interactions (McRoy, 1996). McRoy (1996) made a salient point when he asked, "Are we pursuing the appropriate questions to acquire such knowledge?".

Livingston (1987) listed four action items arising from a continuous, long-term multidisciplinary research program on the Apalachicola Estuary in northern Florida, a program conducted to develop a management plan for the estuary and the watershed: (1) purchase ecologically sensitive lands; (2) translate scientific data into layman's language (an atlas was developed); (3) using scientific data, work with legislative bodies at all levels to protect ecologically sensitive areas; and (4) use these same bodies to provide designations of the sensitive areas (e.g., Class II Waters, Aquatic Preserve, Outstanding Florida Water, Experimental Ecological Preserve, National Estuarine Sanctuary, International Preserve, Area of Critical State Concern).

Fonseca et al. (1998) assembled a lexicon of guidelines for the conservation, restoration, and management of seagrasses in the United States and adjacent waters. No one now doubts that seagrasses create and stabilize structural and functional coastal and estuarine habitats of inestimable value. Going beyond the four criteria listed earlier by Livingston (1987) for preserving seagrass habitat, it is imperative to create and enforce legislation which will guarantee the conservation and restoration of seagrasses on a global basis. A large number of factors should be considered in all management plans, i.e.: (1) pre-project planning such as identification of project goals, assessment of interim losses, planting site selection, minimum size criteria, nutrient and light requirements, and wave, current, and sediment properties, etc.; (2) planting procedures, i.e., the proper technique to use, spacing of the planting units; and (3) monitoring and success evaluation, i.e., monitoring specifications, survival criteria, and monitoring frequency (Fonseca et al., 1998). No one can control the plethora of disturbances to seagrass stocks and ecosystems arising from

# Global Seagrass Declines and Effects of Climate Change

Frederick T. Short and Sandy Wyllie-Echeverria

*Jackson Estuarine Laboratory, University of New Hampshire, 85 Adams Point Road, Durham, NH 03824, U.S.A.*
*School of Marine Affairs, University of Washington, Seattle, WA 98105, U.S.A.*

Seagrasses, an assemblage of true flowering plant species that grow underwater in the world's coastal oceans, are experiencing losses due to human activity as well as natural causes (Short and Wyllie-Echeverria, 1996). Presently, a realistic estimate of seagrass loss worldwide is not possible, and the few available quantitative reports are certainly an underestimate of overall loss. Over the last decade, 290,000 ha of seagrass loss have been documented. Projecting this rate to undocumented parts of the world, over 1.2 million ha of seagrass have likely been lost.

In Australia, Walker and McComb (1992) estimated losses at over 45,000 ha, predominantly from human-induced causes. In the USA, the following reports of large-scale seagrass losses have been documented: Tampa Bay, Florida, lost 25,220 ha of seagrass due to poor water quality from dredge and fill operations, industrial discharge, and sewage discharge (Lewis and Phillips, 1980; Lewis et al., 1985); Laguna Madre, Texas, lost 14,000 ha due to turbidity from continuous maintenance dredging (Quammen and Onuf, 1993); Chesapeake Bay lost 1,746 ha in 1971–74 alone (Orth, 1976) due to poor water quality conditions (Kemp et al., 1983); Florida Bay has a documented loss of 4,000 ha due to as yet unknown causes (Robblee et al., 1991), but recent estimates of loss reach 100,000 ha (Durako, pers. com., 1995).

Additionally, smaller US losses have been documented: Indian River Lagoon, Florida, lost 621 ha of seagrass between 1951 and 1984 due to decreased water quality (Livingston, 1987) and Waquoit Bay, Massachusetts lost 55 ha over a five-year period (90% of its seagrass) due to increased housing development and groundwater nitrogen loading (Short and Burdick, 1996). In Alaska, the documented seagrass loss from the *Exxon Valdez* oil spill was not quantified (Juday and Foster, 1990). In Europe, several studies have discussed recent seagrass losses, suggesting that they result primarily from human pollution, although other factors are also postulated. Loss of both intertidal and subtidal seagrass in the Dutch Wadden Sea has attracted much attention, but the cause has not been clearly identified (den Hartog and Polderman, 1975; Giesen et al., 1990; Philippart et al., 1992; De Jonge and De Jong, 1992). Mechanical damage worldwide is also clearly a factor in seagrass loss, though the overall size of this impact is often considered small. However, some researchers suggest that mechanical damage may cause fragmentation of habitat which may initiate a large-scale decline (Walker et al., 1989; Burdick and Short, 1999).

Natural events have also been identified as causes of seagrass loss, most notably the historic wasting disease of the 1930s, as well as a recurrence of that disease in the USA in the 1980s and to a limited extent in Europe. Animal activity has also been found to cause limited seagrass loss. Obviously, the kinds of natural disturbance that can result in seagrass loss are numerous and often overlapping, making determination of both cause and areal extent of habitat loss problematic. Natural disturbances may be exacerbated by interactions with human-induced impacts, so that exact causal factors are difficult to ascertain. A large-scale loss of seagrass has been reported in Queensland, Australia, where 100,000 ha of several *Halophila* species were lost after two major floods and a cyclone in 1992 (Preen et al., 1995). The die-off resulted from reduced water clarity and mechanical uprooting, although poor catchment management and intensive shrimp trawling likely contributed to this extensive seagrass loss.

## GLOBAL CLIMATE CHANGE IMPACTS ON SEAGRASSES

Global temperature increase will affect seagrasses (see review, Short and Neckles, 1999). A primary effect will be the alteration of growth rates and other physiological functions of the plants themselves. The distribution of seagrasses will shift as a result of increased temperature stress, and changes in the patterns of sexual reproduction. Climate change also produces increased frequency and intensity of extreme weather events which can uproot seagrasses, cause erosion, and increase turbidity; periods of excessive rainfall or drought can alter salinity conditions, which can stress the plants or trigger disease outbreaks.

The direct effects of sea-level rise will be to increase water depths, change tidal variation, alter water movement, and increase seawater intrusion into estuaries. This will redistribute existing habitats. The intrusion of ocean water into formerly fresh or brackish water areas will directly affect estuarine plant distribution by changing conditions at specific locations, causing some plants to relocate to stay within their tolerance zones and allowing others to expand their distribution inland. These distribution changes may result from the effects of salinity change on seed germination, propagule formation, photosynthesis, growth and biomass. Also, some plant communities may decline or be eliminated as a result of increased disease under more highly saline conditions. Worldwide, the reduction in coastal water clarity is one of the major limits on seagrasses and an increase in water depth will exacerbate this problem.

Increasing dissolved carbon dioxide in areas where seagrasses are carbon limited may increase primary production. Although consequences are uncertain this will likely affect species distribution by altering the competition between seagrass species as well as between seagrass and algal populations.

Both mutagenic and physiological damage to plants may result from the assumed increase in UV-B radiation. The reaction of seagrasses may range from inhibition of photosynthetic activity, as seen for terrestrial plants and marine algae, to the increased metabolic cost of producing UV-B blocking compounds within plant tissue. These effects will likely be strongest in the tropics and in southern oceans.

There is every reason to believe that, along with the predicted terrestrial effects of global climate change, the alteration of seagrass habitats will be great, although specific changes to a given seagrass habitat are difficult to predict.

What is very clear is the vast amount of disturbance that human activity imposes on seagrass habitats around the world. Our assessment of recent seagrass declines shows that human-induced damage, namely the impact from human population expansion and associated pollution and upland development, is the primary cause of recent seagrass losses worldwide. However, there is a lack of basic information on seagrass distribution and demography, including trends in habitat change. In order to protect and preserve these critical marine resources, periodic mapping and monitoring of seagrass habitats must be initiated globally. Beyond the gathering of information, there must be a commitment on the part of coastal populations, resource managers, and politicians to preserve and restore seagrass habitats. The size and number of cases of reported seagrass loss attributed to human disturbance should clearly alert scientists and the public to better care for our coastal oceans.

**REFERENCES**

Burdick, D.M. and Short, F.T. (1999) The effects of boat docks on eelgrass beds in coastal waters of Massachusetts. *Environmental Management* **23**, 231–240.

De Jong, V.N. and De Jong, D.J. (1992) Role of tide, light and fisheries in the decline of *Zostera marina* L. in the Dutch Wadden Sea. *Netherlands Institute for Sea Research Publication* **20**, 161–176.

Giesen, W.B.J.T., van Katwyk, M.M. and den Hartog, C. (1990) Temperature, salinity, insolation and wasting disease of eelgrass (*Zostera marina* L.) in the Dutch Wadden Sea in the 1930s. *Netherlands Journal of Sea Research* **25**, 395–404.

Hartog, C. den and Polderman, P.J.G. (1975) Changes in the seagrass populations of the Dutch Wadden Zee. *Aquatic Botany* **1**, 141–147.

Juday, G.P. and Foster, N.R. (1990) A preliminary look at effects of the *Exxon Valdez* oil spill on Green Island Research Natural Area.

Kemp, W.M., Boynton, W.R., Twilley, R.R., Stevenson, J.C. and Means, J.C. (1983) The decline of submerged vascular plants in Upper Chesapeake Bay: summary of results concerning possible causes. *Mar. Soc. Tech. Journal* **17**, 78–89.

Lewis, R.R. and Phillips, R.C. (1980) Seagrass mapping project. Hillsborough County, Florida. Rept. No. 1, Tampa Bay Cooperative Seagrass Project. 15 pp.

Lewis, R.R., Durako, M.J., Moffler, M.D. and Phillips, R.C. (1985) Seagrass meadows of Tampa Bay: a review. In, S.F. Treat, J.L. Simon, R.R. Lewis, and R.L. Whitman (eds.). Proceedings, Tampa Bay Scientific Information Symposium. Florida Sea Grant publ. 65. Burgess Publ. Co., Minneapolis, MN, pp. 210–246.

Livingston, R.J. (1987) Historic trends of human impacts on seagrass meadows in Florida. In *Proceedings of the Symposium on Subtropical–Tropical Seagrasses of the Southeastern United States*, eds. M.J. Durako, R.C. Phillips, and R.R. Lewis. Fla. Mar. Res. Publ. No. 42, pp. 139–151.

Orth, R.J. (1976) The demise and recovery of eelgrass, *Zostera marina*, in the Chesapeake Bay, Virginia. *Aquatic Botany* **2**, 141–159.

Philippart, C.J.M., Dijkema, K.S. and Van Der Meer, J. (1992) Wadden Sea seagrasses: where and why? *Netherlands Institute of Sea Research Publ.* **20**, 177–191.

Preen, A.R., Lee Long, W.J. and Coles, R.G. (1995) Flood and cyclone related loss, and partial recovery, of more than 100 $km^2$ of seagrass in Hervey Bay, Queensland, Australia. *Aquatic Botany* **52**, 3–17.

Quammen, M.L. and Onuf, W.A. (1993) Laguna Madre: seagrass changes continue decades after salinity reduction. *Estuaries* **16**, 302–309.

Robblee, M.B., Barber, T.R., Carlson, P.R., Durako, M.J., Fourqurean, J.W., Muehlstein, L.K., Porter, D., Yarbro, L.A., Zieman, R.T. and Zieman, J.C. (1991) Mass mortality of the tropical seagrass *Thalassia testudinum* in Florida Bay (USA). *Marine Ecology Progress Series* **71**, 297–299.

Short, F.T. and Burdick, D.M. (1996) Quantifying eelgrass habitat loss in relation to housing development and nitrogen loading in Waquoit Bay, Massachusetts. *Estuaries* **19**, 730–739.

Short, F.T. and Neckles, H. (1999) The effects of global climate change on seagrasses. *Aquatic Botany* **63**, 169–196.

Short, F.T. and Wyllie-Echeverria, S. (1996) Natural and human-induced disturbance of seagrasses. *Environmental Conservation* **23**, 17–27.

Walker, D.I. and McComb, A.J. (1992) Seagrass degradation in Australian coastal waters. *Marine Pollution Bulletin* **25**, 191–195.

Walker, D.I., Lukatelich, R.J., Bastyan, G. and McComb, A.J. (1989) Effect of boat moorings on seagrass beds near Perth, Western Australia. *Aquatic Botany* **36**, 69–77.

natural events such as global warming, bottom subsidence, storms and outbreaks of organisms which reduce seagrass stocks. However, we must not only be aware of the human-related activities which lead to the decline of seagrasses and their habitat, but we must stop these activities and make changes which will allow us to preserve our natural heritage and its resources.

To do this will necessitate a reorientation of our morals, goals, and value systems. We will continue to need and practice the best science-based information possible. We will need to expand our base of research. We need seagrass scientists who are willing to leave the laboratory occasionally and plunge into the public arena to exchange information and themselves with elected officials from all levels of government, with policy-makers, with legal staff, with economists, and with sociologists.

A change in human value systems is needed. For years, waterfront property has been valued as the ultimate in living conditions. Government and private interests have viewed this property as private domain, a commodity to be bartered and sold. Few in the past have recognised that waterfront property is also a vital buffer that screens and filters contaminants before runoff reaches the water, that stabilizes that shoreline so that both the upland and water remain healthy. If this trend of shoreline development continues, treated sewage systems must be installed to ensure clean water at the landscape and watershed levels. This value system must include the cost of maintaining our natural ecosystems. Real estate development must factor in the cost of ecosystem services, lest the privileged few impair the rights of all to clean water, clean air, recreation, and food.

## RESEARCH PRIORITIES

Foremost among the various items that require implementation immediately is the need to conduct periodic global monitoring of seagrass resources. There are still a few places on earth for which we know little to nothing of seagrass presence. Even in the Arabian Gulf, where there are three species of seagrass and where the expanse of seagrass beds is potentially vast, we know little about their distribution, abundance, and dynamics.

A system of remote sensing and subsequent mapping which can link presence with density and productivity on a global basis is needed. These efforts would be extremely expensive, but the seagrass resources on earth cannot be effectively monitored by boat or by photography from fixed-wing aircraft. Until we can periodically assess these resources from satellite imagery, we will never begin to focus attention on addressing the critical needs in places which most need them.

A second critical item requiring research is to determine how anthropogenic actions may isolate small seagrass populations and erode genetic diversity. Until relatively recently, we did not know the role of genetics relating to the vegetative plants when transplanted nor the impacts to the donor beds when the transplant units were collected, and equally recently, work in assessing genetic diversity in seagrass populations, using DNA fingerprinting (Alberte et al., 1994; Fain et al., 1992).

More effective and comprehensive coastal and landscape management plans are needed in the next millennium. These plans should be based on the newest and most rigorous scientific research information possible, such as the population genetics data that are becoming available for many species and populations. Fonseca et al. (1998) listed the various criteria needed to make conservation and restoration of seagrasses most effective (Manager's Summary, pp. 139–149).

## REFERENCES


Alberte, R.S., Suba, G.K., Procaccini, G., Zimmerman, R.C. and Fain, S.R. (1994) Assessment of genetic diversity of seagrass populations using DNA fingerprinting: implications for population stability and management. *Proc. Natl. Acad. Sci.* **91**, 1049–1053.

Basson, P.W., Burchard, J.E., Hardy, J.T. and Price, A.R.G. (1977) Biotopes of the Western Arabian Gulf: Marine Life and Environments of Saudi Arabia. Aramco Dept. of Loss Prevention and Environmental Affairs. Dhahran, Saudi Arabia.

Borum, J. and Wium-Andersen, S. (1980) Biomass and production of epiphytes on eelgrass (*Zostera marina* L.) in the Oresund, Denmark. *Ophelia* **1**, 57–64.

Brouns, J.J.W.M. (1987) Growth patterns of some tropical Indo-West Pacific seagrasses. *Aquatic Botany* **28**, 39–61.

Brouns, J.J.W.M. and Heijs, F.M.L. (1986) Production and biomass of the seagrass *Enhalus acoroides* (L.F.) Royle and its epiphytes. *Aquatic Botany* **25**, 21–45.

Burkholder, P.E. and Doheny, T.E. (1968) *The Biology of Eelgrass.* Contr. From the Lamont Geol. Observatory, 1227. Palisades, N.Y. 120 pp.

Camp, D.K., Cobb, S.P. and van Breedveld, J.F. (1973) Overgrazing of seagrasses by a regular urchin, Lytechinus variegatus. *Bioscience* **23**, 37–38.

Correll, D.L. and Wu, T.L. (1982). Atrazine toxicity to submersed vascular plants in simulated estuarine microcosms. *Aquatic Botany* **14**, 151–158.

Dawes, C.J. (1987) The dynamic seagrasses of the Gulf of Mexico and Florida coasts. In *Proceedings of the Symposium on Subtropical-Tropical Seagrasses of the Southeastern United States*, eds. M.J. Durako, R.C. Phillips and R.R. Lewis. Fla. Mar. Res. Publications, No. 42, pp. 25–30.

Dennison, W.C., Orth, R.J., Moore, K.A., Stevenson, J.C., Carter, V., Kollar, S., Bergstrom, P.W. and Batiuk, R.A. (1993) Assessing water quality with submersed vegetation. *BioScience* **43**, 86–94.

Domning, D.P. (1981) Sea cows and sea grasses. *Paleobiology* **7**, 417–420.

Dong, M., Rosenfeld, J., Redmann, G., Elliott, M., Bayazy, J., Poole, B., Ronnholm, K., Kenisberg, D., Novak, P., Cunningham, C. and Karnow, C. (1972) The role of man-induced stresses in the ecology of Long Reef and Christiansted Harbor, St. Croix, U.S. Virgin Islands. St. Croix: West Indies Lab, Fairleigh Dickinson University (Special Publications). 125 pp.

Durako, M.J., Hall, M.O., Sargent, F. and Peck, S. (1992) Propeller scars in seagrass: an assessment and experimental study of recolonization. In Proceedings of the Nineteenth Annual Conference on Wetlands Restoration and Creation, F.J. Webb (ed.). Hillsborough Community College, Tampa, Florida, pp. 42–53.

Durako, M.J. and K.M. Kuss. 1994. Effects of *Labyrinthula* infection on the photosynthetic capacity of *Thalassia testudinum. Bulletin of Marine Science* **54**, 727–732.

Fain, S.R., DeTomaso, A. and Alberte, R.S. (1992) Characterization of disjunct populations of *Zostera marina* (eelgrass) from California: genetic differences resolved by restriction-fragment length polymorphisms. *Marine Biology* **112**, 683–689.

Fonseca, M.S., Kenworthy, W.J. and Thayer, G.W. (1998) Guidelines for the conservation and restoration of seagrasses in the United States and adjacent waters. NOAA Coastal Ocean Program Decision Analysis Series No. 12. NOAA Coastal Ocean Office, Silver Spring, Maryland. 222 pp.

Godcharles, M.F. (1971) A study of the effects of a commercial hydraulic clam dredge on benthic communities in estuarine areas. Fla. Dept. of Nat. Resources. Tech. Ser 64. 51 pp.

Hartog, C. den (1996) Sudden declines of seagrass beds: "Wasting Disease" and other disasters. In *Seagrass Biology: Proceedings of an International Workshop, Jan. 25–29, 1996*, eds. J. Kuo, R.C. Phillips, D.I. Walker and H. Kirkman, pp. 307–314.

Hillman, K., Walker, D.I., Larkum, A.W.D. and McComb, A.J. (1989) Productivity and nutrient limitation. In *Biology of Seagrasses*, eds. A.W.D. Larkum, A.J. McComb and S.A. Shepherd. Elsevier, Amsterdam, pp. 635–685.

Jones, J.A. (1968) Primary productivity by the tropical marine turtle grass *Thalassia testudinum* Konig and its epiphytes. Doctoral Dissertation, Univ. of Miami, Miami, Florida. 196 pp.

Kikuchi, T. (1980) Faunal relationships in temperate seagrass beds. In *Handbook of Seagrass Biology*, eds. R.C. Phillips and C.P. McRoy. Garland STPM Press. New York, pp. 153–172.

Kikuchi, T. and Peres, J.M. (1973) Animal communities in the seagrass beds: a review. Review paper for Consumer Ecology Group, submitted to the Intern. Seagrass Workshop. Leiden. Oct. 1973. 26 pp.

Lewis, R.R., Durako, M.J., Moffler, M.D. and Phillips, R.C. (1985) Seagrass meadows of Tampa Bay: a review. In *Proceedings, Tampa Bay Scientific Information Symposium*, eds. S.F. Treat, J.L. Simon, R.R. Lewis and R.L. Whitman. Florida Sea Grant Publ. 65. Burgess Publ. Co., Minneapolis, MN, pp. 210–246.

Livingston, R.J. (1987) Historic trends of human impacts on seagrass meadows in Florida. In *Proceedings of the Symposium on Subtropi-*

*cal–Tropical Seagrasses of the Southeastern United States*, eds. M.J. Durako, R.C. Phillips and R.R. Lewis. Fla. Mar. Res. Publ. No. 42, pp. 139–151.

Lombardo, R. and Lewis, R.R. (1985) Commercial fisheries data: Tampa Bay. In *Proceedings, Tampa Bay Area Scientific Information Symposium*, eds. S.F. Treat, J.L. Simon, R.R. Lewis and R.L. Whitman. Florida Sea Grant publ. 65. Burgess Publ. Co., Minneapolis, MN, pp. 614–634.

McRoy, C.P. (1966) The standing stock and ecology of eelgrass, *Zostera marina*, Izembek Lagoon, Alaska. M.S. Thesis, Univ. of Washington, Seattle. 135 pp.

McRoy, C.P. (1996) The global seagrass initiative continues. In *Seagrass Biology: Proceedings of an International Workshop. Jan. 25–29, 1996*, eds. J. Kuo, R.C. Phillips, D.I. Walker and H. Kirkman, pp. 3–6.

McRoy, C.P. and C. Helfferich (1980) Applied aspects of seagrasses. In *Handbook of Seagrass Biology*, eds. R.C. Phillips and C.P. McRoy (eds.). Garland STPM Press. New York, pp. 297–343.

Milne, L.J. and Milne, M.J. (1951) The eelgrass catastrophe. *Scientific American* **184**, 52–55.

Morgan, M.D. and Kitting, C.L. (1984) Productivity and utilization of the seagrass *Halodule wrightii* and its attached epiphytes. *Limnology and Oceanography* **29**, 1066–1076.

Neckles, H.A. (1994) Indicator development: seagrass monitoring and research in the Gulf of Mexico. U.S. Environmental Protection Agency, EPA/620/R-94/029, Gulf Breeze, FL. 62 pp.

Orth, R.J. (1975) Destruction of eelgrass, *Zostera marina*, by the cownose ray, *Rhinoptera bonasus*, in the Chesapeake Bay. *Chesapeake Science* **16**, 205–208.

Orth, R.J. (1977) The importance of sediment stability in seagrass communities. In *Ecology of Marine Benthos*, ed. B.C. Coull. Univ. South Carolina Press, Columbia, pp. 122–138.

Orth, R.J. and Moore, K.A. (1983) Chesapeake Bay: an unprecedented decline in submerged aquatic vegetation. *Science* **222**, 51–53.

Penhale, P.A. (1977) Macrophyte-epiphyte biomass and productivity in an eelgrass (*Zostera marina* L.) community. *Journal of Experimental Marine Biology and Ecology* **26**, 211–224.

Phillips, R.C. (1974) Transplantation of seagrasses, with special emphasis on eelgrass, *Zostera marina* L. *Aquaculture* **4**, 1–16.

Phillips, R.C. (1978) Seagrasses and the coastal marine environment. *Oceanus* **21**, 30–40.

Phillips, R.C. (1984) The ecology of eelgrass meadows in the Pacific Northwest: a Community Profile. U.S. Fish and Wildlife Service FWS/OBS-84/24. 85 pp.

Phillips, R.C. and Menez, E.G. (1988) *Seagrasses*. Smithsonian Contr. Mar. Sci. No. 14. 104 pp.

Pollard, P.C. and M. Greenway. 1993. Photosynthetic characteristics of seagrasses, *Cymodocea serrulata, Thalassia hemprichii*, and *Zostera capricorni*, in a low-light environment, with a comparison of leaf-marking and lacunal-gas measurements of productivity. *Australian Journal of Marine and Freshwater Research* **44**, 127–139.

Roessler, M.J. and Zieman, J.C. (1969) The effects of thermal additions on the biota in Southern Biscayne Bay, Florida. In *Proceedings of the Gulf Caribbean Fisheries Institute*, 22nd. Ann. Session, pp. 136–145.

Sand-Jensen, K. and Borum, J. (1983) Regulation of growth in eelgrass (*Zostera marina* L.) in Danish coastal waters. *Marine Tech. Soc. J.* **17**, 15–21.

Sargent, F.J., Leary, T.J., Crevz, D.W. and Kruer, C.R. (1995) Scarring of Florida's seagrasses: assessment and management options. Fla. Dept. of Env. Protection, St. Petersburg, FL. FMRI Tech. Rept. TR-1. 46 pp.

Short, F.T. (1981) Nitrogen resource analysis and modelling of an eel-grass (*Zostera marina* L.) meadow in Izembek Lagoon, Alaska. Ph.D. Diss., Univ. of Alaska, Fairbanks. 173 pp.

Short, F.T. and Neckles, H. (1999) The effects of global climate change on seagrasses. *Aquatic Botany* **63**, 169–196.

Short, F.T. and Wyllie-Echeverria, S. (1996) Natural and human-induced disturbance of seagrasses. *Environmental Conservation* **23**, 17–27.

Taylor, J.L. and Salomon, C.H. (1968) Some effects of hydraulic dredging and coastal development in Boca Ciega Bay, Florida. *Fisheries Bulletin* **67**, 213–241.

Thayer, G.W. and Stuart, H.H. (1974) The bay scallop makes its bed of eelgrass. *Marine Fish. Review* **36**, 27–39.

Thayer, G.W., Wolfe, D.A. and Williams, R.B. (1975) The impact of man on seagrass systems. *American Scientist* **63**, 288–296.

Thayer, G.W., Kenworthy, W.J. and Fonseca, M.S. (1984) The ecology of eelgrass meadows of the Atlantic coast: a community profile. U.S. Fish Wildlife, FWS/OBS-84/02, Washington, D.C. 147 pp.

Thayer, G.W., Stuart, H.H., Kenworthy, W.J., Ustach, J.F. and Hall, A.B. (1978) Habitat values of salt marshes, mangroves, and seagrasses for aquatic organisms. In *Wetland Functions and Values: The State of Our Understanding*, eds. P.E. Greeson, J.R. Clark, and J.E. Clark (eds.). Am. Water Res. Assoc. Minneapolis, MN, pp. 235–247.

Thomas, L.P., Moore, D.R. and Work, R.C. (1961) Effects of Hurricane Donna on the turtlegrass beds of Biscayne Bay, Florida. *Bulletin of Marine Science* **11** (2), 191–197.

Thorne-Miller, B. and Harlin, M.M. (1984) The production of *Zostera marina* L. and other submerged macrophytes in a coastal lagoon in Rhode Island, U.S.A. *Botanica Marina* **27**, 539–546.

Waddell, J.E. (1964) The effect of oyster culture on eelgrass (*Zostera marina* L.) growth. M.S. thesis, Humboldt State University, Arcata, CA, 48 pp.

Wilson, D.P. (1949) The decline of *Zostera marina* L. at Salcombe and its effects on the shore. *Journal of the Marine Biological Association U.K.* **28**, 295–412.

Wood, E.J.F., Odum, W.E. and Zieman, J.C. (1969) Influence of sea grasses on the productivity of coastal lagoons. In *Coastal Lagoons*, eds. A. Ayala Castanares and F.B. Phelger. Universidad Nacional Autonoma de Mexico Ciudad Universitaria, Mexico, D.F. pp. 495–502.

Zieman, J.C. (1975a) Tropical seagrass ecosystems and pollution. In *Tropical Marine Pollution*, eds. E.J.F. Wood and R.E. Johannes. Elsevier Publ. Co., Amsterdam, pp. 63–74.

Zieman, J.C. (1975b) Quantitative and dynamic aspects of the ecology of turtle grass, *Thalassia testudinum*. In *Estuarine Research*, ed. L.E. Cronin. Academic Press, New York, pp. 541–562.

Zieman, J.C. (1982) The ecology of the seagrasses of south Florida: a community profile. U.S. Fish Wildlife, FWS/OBS-82/25, Washington, D.C. 123 pp.

Zieman, J.C. and Zieman, R.T. (1989) The ecology of the seagrass of west Florida: a community profile. U.S. Fish Wildlife, Biological Rept. 85(7.25), Washington, D.C. 155 pp.

## THE AUTHORS

**Ronald C. Phillips**
*Commission of Environmental Research,*
*Emirates Heritage Club,*
*Abu Dhabi, United Arab Emirates*
*Correspondence address: 1597 Meadow View Drive,*
*Hermiston, OR 97838, U.S.A.*

**Michael J. Durako**
*Center for Marine Science,*
*The University of North Carolina at Wilmington,*
*Wilmington, NC 28403, U.S.A.*

# Seagrass Die-off in Florida Bay: A Trend for Coastal Waters in the New Millennium?

Michael J. Durako

*Center for Marine Science, The University of North Carolina at Wilmington, Wilmington, North Carolina 28403, U.S.A.*

Seagrasses, dominated by *Thalassia testudinum* Banks ex König (turtle grass), are the dominant biological community in Florida Bay, historically covering over 90% of the 180,000 ha of subtidal mudbanks and basins within the Bay (Zieman et al., 1989). By comparison, mangrove islands cover only about 7% of the Bay. Because of the shallow nature of Florida Bay (mean depth <2 m, Schomer and Drew, 1982), seagrasses are also the dominant physical feature, and their presence greatly affects physical, chemical, geological, as well as biological processes in this system. Seagrass communities are important to the economy of South Florida, because they provide food and shelter to numerous fish and invertebrate species, many of commercial importance within the region (Tilmant, 1989; Chester and Thayer, 1990). Seagrass abundance to a large extent determines public perception regarding the "health" of the coastal waters of Florida (Goerte, 1994). Thus, the recent changes in the distribution and abundance of seagrasses within Florida Bay have been perceived as a change in the health of the Bay.

Widespread die-off of seagrasses within Florida Bay was first observed in 1987 (Robblee et al., 1991; Durako 1994, 1995). Extensive areas of *Thalassia testudinum* began dying rapidly in central and western basins, and by 1990, 4000 ha were completely lost and 24,000 ha were affected by the die-off (Fig. 1) (Robblee et al., 1991). Factors such as elevated water temperature, prolonged hypersalinity, and excessive seagrass biomass (due to lack of recent hurricanes), leading to increased respiratory demands, hypoxia and sulfide toxicity are some of the physiological stressors thought to have contributed to *T. testudinum* die-off (Robblee et al., 1991; Carlson et al., 1994; Durako, 1994). In addition, die-off has been associated with disease infection of *T. testudinum* by an undescribed species of *Labyrinthula*, a marine slime mould (Robblee et al., 1991); a related species, *Labyrinthula zosterae* Muehlstein, is known to cause wasting disease in eelgrass, *Zostera marina* L. (Muehlstein et al., 1988). Infection by *Labyrinthula* significantly reduced photosynthetic rates in *Thalassia* (Durako and Kuss, 1994), resulting in a negative carbon balance. In addition to reduced photosynthetic capacity and increased respiration, infected *Thalassia* short-shoots also exhibited reduced conductance of oxygen which makes them more susceptible to hypoxia and sulphide toxicity (Carlson et al., 1994). However, the causative mechanisms responsible for initiating the die-off in Florida Bay remain incompletely understood.

The patterns of loss of seagrasses in Florida Bay have undergone at least two phases: the first being the initial die-off which occurred during the relatively dry and clear period of 1987 to early 1991 (Fig. 1) (Robblee et al., 1991). Several years after the initiation of the seagrass die-off, which appeared to affect only *T. testudinum*, Florida Bay began exhibiting widespread and chronic turbidity (Phlips et al.,

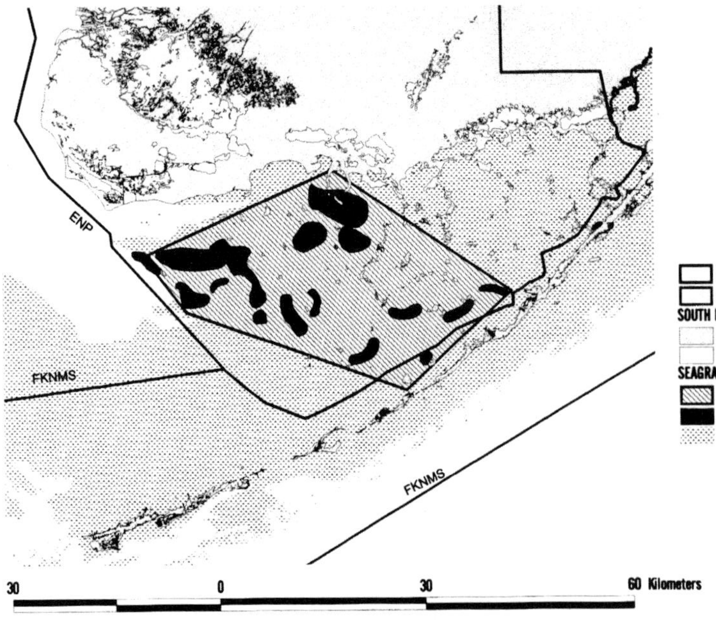


Fig. 1. Seagrass die-off in Florida Bay in 1989–90 based on SPOT satellite imagery (from Robblee et al., 1991). The isolated die-off area adjacent to the Florida Keys is next to the Everglades National Park Key Largo Ranger Station where research vessels are moored. This occurrence suggests the presence of a transmissible agent in the die-off.

1995; Stumpf et al., 1999). The increase in turbidity, which began during the fall of 1991, is principally due to microalgal blooms and resuspended sediments associated with the loss of seagrasses on the western banks, and it has been most severe in the western and central Bay (Phlips and Badylak, 1996). The microalgal blooms, consisting largely of cyanobacteria (possibly *Synechococcus elongatus*; Phlips and Badylak, 1996), may have been initiated by the nutrients liberated from the die-off of seagrasses (Butler et al., 1995). Sponge mortality, changes in juvenile lobster population dynamics (Butler et al., 1995) and indications of cascading effects on plant and animal communities in adjacent systems (e.g., sea urchin population explosions and unbalanced growth of *Syringodium filiforme* in the waters of the Florida Keys National Marine Sanctuary southwest of Florida Bay; Rose et al., 1999; Kenworthy et al., 1998) have also been associated with the continued degradation of seagrass communities within Florida Bay.

Loss of seagrass cover is a major factor in recent increases in sediment resuspension in the Bay (Prager, 1998). This results in a negative feedback loop in which loss of seagrass cover from die-off has led to exposed, easily resuspended sediments, and more, widespread loss of seagrasses due to turbidity—as more seagrass cover is lost, turbidity increases. Recent comparisons of seagrass distributions in Florida Bay between 1984 and 1994 (Hall et al., 1999) and between 1995 and 1998 (Durako et al., 1999) indicate that the chronically turbid regions have exhibited the most significant losses of *T. testudinum* (Fig. 2). In contrast, the lower-light adapted pioneer/disturbance species, *Halodule wrightii* Ascherson (shoal grass), has doubled in abundance, at the bay-scale, and has increased up to 400% in the western basins from 1995 to 1998 (Fig. 3). Recruitment and spread of another low-light-adapted and small-bodied seagrass, *Halophila engelmannii* Ascherson (star grass), in the western basins indicates that a major community response resulting from die-off is a change to a shade-adapted seagrass community.

Establishing the relative contribution of die-off versus light-stress-induced mortality to the current losses of *Thalassia* in western Florida Bay, is clearly problematic. There is a high spatial coincidence among the distribution patterns of seagrass loss, *Labyrinthula* abundance (Blakesley et al., 1998) and turbidity (Phlips et al., 1995; Stumpf et al., 1999). Increases in *Halodule* in the Bay may reflect relatively lower light requirements for this species (Dunton and Tomasko, 1994), ability to rapidly spread into areas where the *Thalassia* canopy has been removed (Thayer et al., 1994), or resistance

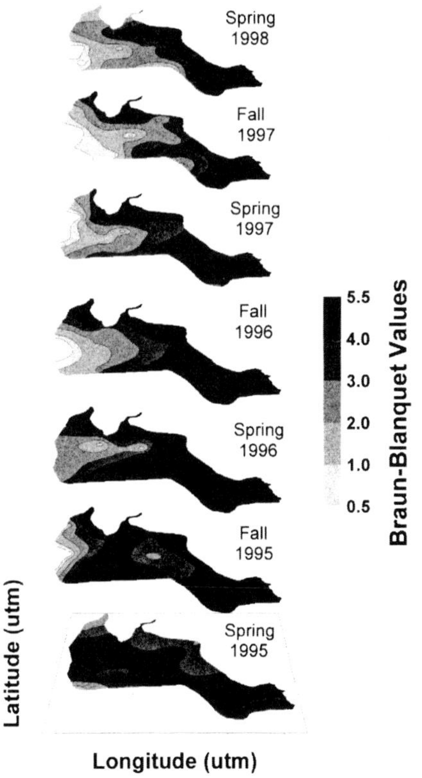


Fig. 2. Changes in distribution and abundance of *Thalassia testudinum* from Spring 1995 to Spring 1998 in Rabbit Key Bay in western Florida Bay. Note the dramatic decrease in cover along the western side of the basin. Axis ticks are 1 km intervals. Braun–Blanquet cover/abundance values: 0 = absent, 0.5 = few individual ramets but <5% cover, 1 = many individual ramets but <5% cover, 2 = 5–25% cover, 3 = 25–50% cover, 4 = 50–75% cover, 5 = 75–100% cover (from Durako et al., in press).

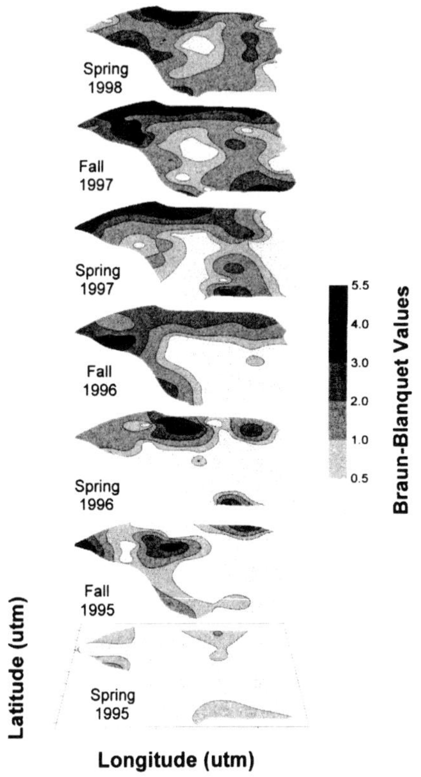


Fig. 3. Changes in distribution and abundance of *Halodule wrightii* from Spring 1995 to Spring 1998 in Johnson Key Basin in western Florida Bay. Recruitment of this seagrass seemed to originate from banks surrounding this basin. Scales as in Fig. 2.

to disease. There has been little change in *Thalassia* and *Halodule* abundances in basins which are periodically subjected to low salinities. This could reflect the existence of a low-salinity refugia from disease since *Labyrinthula* has never been found in Florida Bay at salinities below 15 ppt (Blakesley et al., 1998).

The spatial patterns of change from 1995 to 1998 suggest that, presently, the most perturbed environment is along the western Bay margin, bordering the open waters of the Gulf of Mexico. In contrast, much of the focus of management and restoration efforts in South Florida has been directed toward landscape-scale modifications to an extensive flood-control system to increase the quantity of freshwater delivered to northeast Florida Bay and along the Everglades/Florida Bay land/sea interface. Although the initial die-off is generally thought to have been initiated in the interior basins of the Bay (Robblee et al., 1991), the greatest changes in the present system are occurring far from this land/sea interface. In addition, the waters of western Florida Bay form a hydrodynamic link between the Everglades and the coastal waters of the southwestern Florida peninsula/eastern Gulf of Mexico, to the north, and the Florida Keys reef tract and the Atlantic Ocean to the south (Schomer and Drew, 1982). The seagrass communities of this region form an important buffer by intercepting the flow of water along this region and reducing nutrient and particulate loads in the waters reaching the reef tract (Kenworthy et al., 1998). Continued losses of seagrasses along this margin, coupled with the proposed increase in water flows out of the Everglades could result in greater fluxes of material out of Florida Bay and onto the reef tract (Kenworthy et al., 1998). In the next millennium, resource managers will need to consider actions that might aid in the re-establishment of continuous seagrass cover along western Florida Bay. This would be an important step in reducing sediment-resuspension induced turbidity along this boundary, which could reverse the cascading declines that characterize the present system.

## REFERENCES


Blakesley, B.A., Landsberg, J.H., Ackerman, B.B., Roose, R.O., Styer, J.R., Obordo, C.O. and Lucas-Black, S.E. (1998) Slime mold, salinity, and statistics: implications from laboratory experiments with turtlegrass facilitate interpretation of field results from Florida Bay. Florida Bay Science Conference, Univ. Miami, May 12–14, 1998.

Butler, M.J., Hunt, J.H., Herrnkind, W.F., Childress, M.J., Bertelsen, R., Sharp, W., Matthews, T., Field, J.M. and Marshall, H.G. (1995) Cascading disturbances in Florida Bay, USA: cyanobacterial blooms, sponge mortality, and implications for juvenile spiny lobsters *Panulirus argus*. *Marine Ecology Progress Series* **129**, 119–125.

Carlson, P.R., Yarbro, L.A. and Barber, T.R. (1994) Relationship of sediment sulfide to mortality of *Thalassia testudinum* in Florida Bay. *Bulletin of Marine Science* **54**, 733–746.

Chester, A.J. and Thayer, G.W. (1990) Distribution of spotted seatrout (*Cynoscion nebulosus*) and gray snapper (*Lutjanus griseus*) juveniles in seagrass habitats in western Florida Bay. *Bulletin of Marine Science* **46**, 345–357.

Dunton, K.H. and Tomasko, D.A. (1994) *In situ* photosynthesis in the seagrass *Halodule wrightii* in a hypersaline subtropical lagoon. *Marine Ecology Progress Series* **107**, 281–293.

Durako, M.J. and Kuss, K.M. (1994) Effects of *Labyrinthula* on the photosynthetic capacity of *Thalassia testudinum*. *Bulletin of Marine Science* **54**, 727–732.

Durako, M.J. (1994) Seagrass die-off in Florida Bay (USA): changes in shoot demography and population dynamics. *Marine Ecology Progress Series* **110**, 59–66.

Durako, M.J. (1995) Indicators of seagrass ecological condition: an assessment based on spatial and temporal changes associated with the mass mortality of the tropical seagrass *Thalassia testudinum*. In *Changes in Fluxes in Estuaries: Implications for Science to Management*, eds. K.R. Dyer and R.J. Orth. Olsen and Olsen, Fredensborg, Denmark, pp. 261–266.

Durako, M.J., Hall, M.O. and Merello, M. (In press) Patterns of change in the seagrass-dominated Florida Bay hydroscape. In *Linkages Between Ecosystems in the South Florida Hydroscape: The River of Grass Continues*, eds. J.W. Porter and K.G. Porter. CRC Press, Boca Raton, Florida.

Goerte, R.W. (1994) The Florida Bay economy and changing environmental conditions. U.S. Library of Congress, Congressional Research Service Rept. No. 94-435 ENR. Washington, D.C. 19 pp.

Hall, M.O., Durako, M.J., Fourquerean, J.W. and Zieman, J.C. (1999) Decadal-scale changes in seagrass distribution and abundance in Florida Bay. *Estuaries* **22**, 445–459.

Kenworthy, W.J., Swartzschild, A.C., Fonseca, M.S., Woodruff, D., Durako, M.J. and Hall, M.O. (1998) Ecological and optical characteristics of a large *Syringodium filiforme* meadow in the southeastern Gulf of Mexico. Florida Bay Science Conference, Univ. Miami, May 12–14, 1998.

Muehlstein, L.K., Porter, D. and Short, F.T. (1988) *Labyrinthula* sp., a marine slime mould producing symptoms of wasting disease in eelgrass, *Zostera marina*. *Marine Biology* **99**, 465–472.

Phlips, E.J. and Badylak, S. (1996) Spatial variability in phytoplankton standing crop and composition in a shallow inner-shelf lagoon, Florida Bay. *Bulletin of Marine Science* **58**, 203–216.

Phlips, E.J., Lynch, T.C. and Badylak, S. (1995) Chlorophyll a, tripton, color, and light availability in a shallow tropical inner-shelf lagoon, Florida Bay, USA. *Marine Ecology Progress Series* **127**, 223–234.

Prager, E. (1998) Sediment resuspension in Florida Bay. Proc. Florida Bay Science Conference, Univ. Miami, May 12–14, 1998.

Robblee, M.B., Barber, T.R., Carlson, P.R., Durako, M.J., Fourquerean, J.W., Muehlstein, L.K., Porter, D., Yarbro, L.A., Zieman, R.T. and Zieman, J.C. (1991) Mass mortality of the tropical seagrass Thalassia testudinum in Florida Bay (USA). *Marine Ecology Progress Series* **71**, 297–299.

Rose, C.D., Sharp, W.C., Kenworthy, W.J., Hunt, J.H., Lyons, W.G., Prager, E.J., Valentine, J.F., Hall, M.O., Whitfield, P. and Fourquerean, J.W. (1999) Sea urchin overgrazing of a large seagrass bed in Outer Florida Bay. *Marine Ecology Progress Series* **190**, 211–222.

Schomer, N.S. and Drew, R.D. (1982). An ecological characterization of the lower Everglades, Florida Bay, and the Florida Keys. U.S. Fish Wildlife, FWS/OBS- 82/58.1, Washington, D.C. 246 pp.

Stumpf, R.P., Frayer, M.L., Durako, M.J. and Brock, J.C. (1999) Variations in water clarity and bottom albedo in Florida Bay from 1985 to 1997. *Estuaries* **22**, 431–444.

Thayer, G.W., Murphy, P.L. and LaCroix, M.W. (1994) Responses of plant communities in western Florida Bay to the die-off of seagrasses. *Bulletin of Marine Science* **54**, 718–726.

Tilmant, J.T. (1989) A history and an overview of recent trends in the fisheries of Florida Bay. *Bulletin of Marine Science* **44**, 3–22.

Zieman, J.C., Fourquerean, J.W. and Iverson, R.L. (1989) Distribution, abundance and productivity of seagrasses and macroalgae in Florida Bay. *Bulletin of Marine Science* **44**, 292–311.

Chapter 108

# MANGROVES

Colin D. Field

Mangroves form coastal forests largely confined between 30° north and south of the equator. This range is determined mainly by low and extreme temperatures and, to a lesser extent, by rainfall. They are thus abundant in many lesser developed countries with fast rising populations which exert strong development pressure on this ecosystem.

In recent years, pressures of increasing population, food production and development have led to a significant proportion of the world's mangrove resource being destroyed at a rate faster than they are being regenerated. Much of their elimination has been to create land for aquaculture, particularly shrimp ponds, but in many instances shrimp ponds are quickly abandoned as a result of falling production, leaving highly degraded land on which mangroves do not naturally re-establish themselves. It is estimated that shrimp ponds in Asia rarely last for more than five to ten years, leaving irreversibly degraded environments. The scale of mangrove clearance may be huge: in the Philippines, for example 70% were lost in a period of 60 years, and similar or even greater clearance rates have been seen in 'New World' areas too.

A popular view of mangrove forests is that they are very productive, and under some conditions they compare well with terrestrial forests. The qualitative importance of mangroves as habitat, nursery and source of food for both commercial fisheries species and other non-commercial fauna is generally accepted, as is the fact that a large number of juvenile fish use mangroves as nursery habitats. However, there is a lack of well established quantified relationships between fish yields and area of mangrove, though several studies support the hypothesis that coastal fish resources are closely linked to estuaries and mangroves, even if controversy remains about their degree of dependence. Apart from being a productive shoreline ecosystem, mangroves can help stabilise dynamic coastlines.

There is now an upsurge in the number of rehabilitation projects, but of the 90 or so countries around the world that possess mangroves only some 20 have attempted any form of mangrove replanting, and only nine have planted more than 10 $km^2$ since 1970. Mangrove rehabilitation can bring into conflict the aims of maximising biodiversity and providing maximum benefit to the human communities that rely on them. Often, however, the objective of rehabilitation is sustainable production, and the knowledge that now exists about the extent, structure and function of mangrove ecosystems is substantial and growing; mangroves are fashionable at the end of the 20th century. There is an urgent need to use this to improve their sustainable use for the benefit of the people who rely upon them, and future work must focus on efficient ways to use and rehabilitate mangrove ecosystems for sustainable purposes.

Despite the problems, enormous progress has been made on understanding the structure, function and use of mangroves during this century. Above all there has been a cathartic shift in their understanding, importance and usefulness, much of which has been achieved by education. It is on this basis of understanding that the sustainable use of mangrove ecosystems must be further developed.

*Seas at The Millennium: An Environmental Evaluation (Edited by C. Sheppard)*

## INTRODUCTION

Mangrove trees and shrubs form a special form of vegetation existing at the boundary of two environments: the land and the sea. The species belong to a variety of plant families, and mangroves are ancient plants that evolved soon after the appearance of flowering plants in the late Cretaceous to early Palaeocene. There is a floristic divergence between the 'old' and 'new' world mangroves which can only be explained by historical events and must be seen in the context of plate tectonics and continental drift. It is not easy to define a mangrove, so the terms 'true' mangrove and 'mangrove associate' have been used. The common characteristic that all possess is tolerance to salt and brackish waters. Some 69 species of mangrove plants are recognised from various regions of the world (Duke, 1992), with the highest concentration of species in South East Asia. They are predominantly found in the tropics (Fig. 1).

As mangroves grow in the intertidal region, they often interact intimately with other ecosystems such as seagrasses and coral reefs, and the health of the mangrove forest can be a critical factor in the functioning of a variety of these other ecosystems. The states of oceans and mangroves are inextricably linked. A mangrove forest often possesses a strange and convoluted beauty and it flourishes in conditions of heat, salinity and oxygen-starved mud that would overwhelm other terrestrial plants. To cope with this hostile environment mangroves have undergone selective changes. As time has passed they have adapted and emerged as the most successful coloniser of tropical coastal wetlands. They are often ephemeral but aggressively opportunistic.

Aerial roots are the most noticeable adaptation (Fig. 2). These come with various forms of architecture such as hoop stilts, buttresses or single unbranched structures rising from the mud to the sky and form breathing roots known as pneumatophores. Other adaptations include glands on the leaves for excreting salt, a tendency in some species towards succulence and roots that have an ability to exclude salt. The seeds are often viviparous. They are frequently buoyant, easily dispersed by tides and shaped so that they anchor in the mud.

As a result of these adaptations a fringe ecosystem has developed on tropical shores and estuaries (Fig. 3). The biology of the swamp forest is complex, and although in recent times such ecosystems have attracted a lot of scientific attention, the dynamics and interrelationships of living organisms within the structure are still poorly understood. This ecosystem is important for timber and fish production and as habitat for many forms of wildlife. It also has a role in stabilising river banks and coastlines.

In regions with continuous high temperatures and prolific rainfall, such as the west coast of Malaysia, mangroves have prospered. Slowly, huge productive forests have evolved and the resource has been widely used by coastal people of the tropics for hundreds of years. A balance existed between the complex biological system, represented by the tidal forests, and the swamp dwellers who learnt to exploit the system without destroying it. Historically, the human pressure on the mangrove swamps was limited as, except for some subsistence populations, such places were seen as inhospitable, unhealthy and dangerous. They were not easy places to penetrate, except by small boat, and few communities of people actually lived within the mangrove forests.

In recent years the pressures of increasing population, food production and industrial and urban development have led to a significant proportion of the world's mangrove resource being threatened with destruction. Rapidly rising affluence in many developing countries has meant that areas previously covered with mangrove forests have been converted to domestic, industrial and tourist facilities. This intense exploitation is relatively recent. It is still common for mangroves to be considered as waste lands with

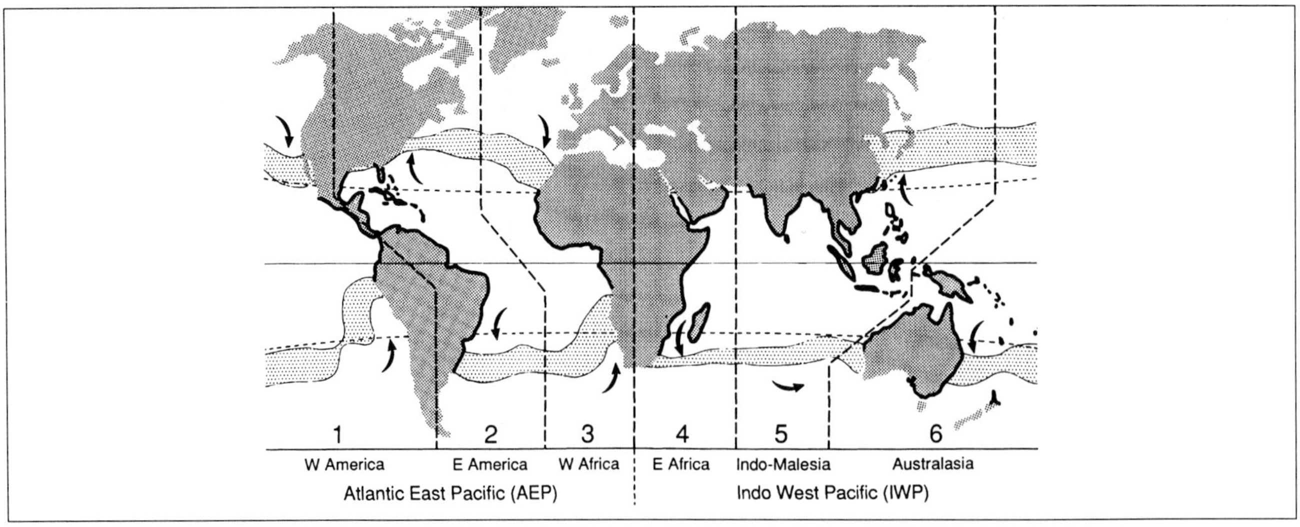


Fig. 1. World distribution of mangroves, showing range of 20°C isotherms in January and July, major ocean currents influencing latitudinal range around the tropic line and six biogeographic regions (Duke, 1992).

Fig 2. Pneumatophores of the common and very widespread Indo-Pacific mangrove *Avicennia marina* (photo C. Sheppard).

Fig. 3. Tangle of roots at low tide of the mangrove *Rhizophora apiculata*, which occurs from India to the islands of the Pacific (photo M. Spalding).

little intrinsic value and their destruction to be encouraged by governments and planners. The usual approach is to dredge and drain the mangrove swamps and then fill them, converting the natural habitat to dry land on which no mangrove can survive. The construction of dams and other engineering projects which divert fresh water from higher ground is another adverse influence on mangrove communities. Such attitudes and activities can be seen in the construction of the Volta dam in Ghana (Rubin et al., 1998), the diversion of water from the Cienaga Grande de Santa Marta (Perdomo et al., 1998), damming the Selangor river in Malaysia (Primavera, pers. comm., 1999); a proposal for 500 ha of new fish ponds in the Santa Cruz Channel, Brazil (Barros, pers. com., 1999) and the destruction of mangroves by oil pollution in the Niger Delta, Nigeria (Mangrove Action Project, 1999). Perhaps the most substantial promises to be the conversion of large parts of the Rufiji delta in Tanzania to aquaculture.

However, in the early 1970s, scientific interest in mangroves began to shift from long-established scholarly biological studies of these curious salt-adapted collections of plants and animals to the more immediate problem of the disappearance, at an alarming rate, of this highly productive and important biological system. There is an impressive literature on mangroves and mangrove ecosystems. Rollet (1981) produced a bibliography of mangrove research from 1600 to 1975. Since the mid-seventies, the number of books, refereed papers and reports has risen dramatically and there are now hundreds of references to mangroves appearing in the literature each year. This trend is continuing to accelerate. An initial impetus to the recent interest in mangroves started with an international symposium on mangroves in Hawaii in 1974 (Walsh et al., 1975) and a UNESCO-UNDP project on mangrove ecosystems of Asia and the Pacific that began in 1983. A number of recent texts give an overview of mangrove distribution, mangrove research, mangrove ecology and mangrove management such as: IUCN (1983), Tomlinson (1986), Hutchings and Saenger (1987), Robertson and Alongi (1992), FAO (1994), Field (1995, 1996), Spalding et al. (1997) and Streever (1999).

Scientists and land managers continue to express concern that natural mangrove forests are disappearing at a faster rate than they are being regenerated. Mangrove forests face all of the problems and pressures that are being experienced by other forest communities around the world. Indeed, because of their unusual nature, prime location and perceptions of their lack of importance, mangroves are subject to more than usual dangers. As the 21st century approaches, it is opportune to review the present extent and rate of loss of global mangroves, their ecological values, their patterns of use and socio-economic values, the steps that are being taken to rehabilitate or afforest degraded coastal land with mangroves and the challenges that the future centuries present.

## THE PRESENT EXTENT AND LOSS OF GLOBAL MANGROVES

Mangroves are largely confined to the regions between 30° north and south of the equator, with notable extensions beyond this to the north in Bermuda (32°20′N) and Japan (31°22′N), and to the south in Australia (38°45′S), New Zealand (38°03′S) and South Africa (32°59′S) (see Fig. 1). Within these confines they are widely distributed, although their latitudinal development is restricted along the western coasts of the Americas and Africa, as compared to the equivalent eastern coasts. In the Pacific Ocean natural mangrove communities are limited to western areas, and they are absent from many Pacific islands.

There are two main centres of diversity for mangrove communities which have been termed the western and eastern groups (Tomlinson, 1986). The eastern group broadly corresponds with the Indo-Pacific and is bound to the east by the limits to natural mangrove occurrence in the west and central Pacific, and to the west by the southern tip of Africa. The western group fringes the coast of the Atlantic

Ocean, the Caribbean Sea and the Gulf of Mexico. These two regions have quite different floristic inventories, and the eastern region has approximately five times the number of species that are found in the western region.

The distribution patterns of mangroves are the result of a wide range of historical and contemporary factors. Perhaps the most obvious distribution pattern, the latitudinal limits, is largely set by low temperatures, both sea surface temperatures and air temperatures, and particularly by extremes of temperature. Rainfall also has a strong influence over mangrove distribution, largely through the reduction of salinity in an otherwise highly saline environment. Although mangroves are adapted to saline or brackish environments, the high salinities associated with intertidal areas, particularly in arid countries, frequently restrict growth. In areas with low, irregular or limited seasonal rainfall the number of mangrove species which can survive is limited. This is clearly one of the major factors leading to sparse mangrove development over wide areas of coast, such as around the Arabian peninsula. Historical and tectonic factors are probably responsible for the easterly limit to mangrove development in the Pacific, although exact mechanisms for these limits are unclear. It may be that mangroves were once more widely distributed in this ocean and have undergone range constrictions, alternatively the current distributions could represent the eastern limits to dispersal from a western centre of origin. The centre of origin of mangroves and their subsequent dispersal routes around the various continents remains a matter of dispute (Saenger, 1998). At the national and local level many other factors influence the distribution of mangroves, including soils, tides, geomorphology, mineral availability, soil aeration, winds, currents and wave action. The influence of man is now considerable and is affecting mangrove distribution patterns at all scales.

It is not easy to identify accurately the areal extent of mangroves in the world (Blasco et al., 1998). Good and current vegetation maps are only available from a few countries and it is rare for such maps to define what they have called mangroves. Correct identification of mangrove species is also a problem. Some areas of mangrove have been mapped using satellite imaging but the application of such techniques covers only a small fraction of the global mangroves. Even when such advanced techniques are used there are still problems with spatial accuracy.

According to Spalding et al. (1997) the total area of mangroves in the world is some 182,000 km$^2$. This figure differs from the figure provided in IUCN (1983). The latter document does not cover all countries, notably in the Red Sea and Arabian Gulf areas and parts of the Americas: Florida, Bahamas and Lesser Antilles. The overlap in the material used in each of these references is minimal, so the figures can be taken as being effectively independent.

The areal statistics in Table 1 give a reasonable assessment of the total area of mangroves in the world but there are likely to be considerable margins of error. Due to differences in definition, age, scale and accuracy of different national sources, the use of global composite statistics as a base-line for monitoring changes in global mangrove area should be employed with extreme caution. Any future composite figures derived in a similar way will probably be subject to similar errors. There is an urgent need for more accurate mapping of mangrove areas at a much higher level of resolution. A compilation of data from more accurate maps would lead to a more reliable base-line for measuring change.

Table 1

Estimates of mangrove areas

| Region | Mangrove area (km$^2$) Spalding et al. (1997) | Mangrove area (km$^2$) IUCN (1983) | Mangrove area (km$^2$) Fisher and Spalding (1993) |
|---|---|---|---|
| South and Southeast Asia | 75,688 | 51,766 | 76,226 |
| Australasia | 18,789 | 16,980 | 15,145 |
| The Americas | 49,485 | 67,446 | 51,286 |
| West Africa | 27,995 | 27,110 | 49,500 |
| East Africa and the Middle East | 10,348 | 5,508 | 6,661 |
| Total Area | 182,305 | 168,810 | 196,825 |

A few countries dominate the areal statistics. Notable among these are Indonesia (42,550 km$^2$), Australia (11,500 km$^2$), Brazil (13,800 km$^2$) and Nigeria (10,515 km$^2$). In total, these countries have some 43% of the world's mangroves and each has between 25% and 50% of the mangroves in their respective regions. Indonesia alone has 23% of the world's mangroves. Aside from the general geographical interest of this situation, it is clear that these four countries have a considerable heritage. Political and management decisions relating to mangroves in each of these countries will have a significant effect on the global status of mangrove ecosystems into the future.

It is difficult to get accurate figures for the global loss of mangroves, though there are some well documented case studies for particular countries. However, there is no doubt that the major threats to mangroves are the increase in global population, with the concomitant increased demand for food and infrastructure, and the expectations of improved standards of living amongst the coast dwellers in the tropics.

The world population reached 6 billion in late 1999 (United Nations Population Division, 1999). Between 1995 and 2000 the world population grew at 1.33% per year, adding an average of 78 million persons each year. In the mid 21st century the world population will be in the range of 7.3 to 10.7 billion, depending on assumed future fertility trends, but 97% of the world population increase will take place in less developed regions. It is in many of these countries that mangroves adorn the coastlines.

Table 2

Causes of loss of mangrove forests

| **Direct human action** | **Indirect human action** |
|---|---|
| Drainage for agriculture crops | Sediment diversion from dams and flooding |
| Drainage for mosquito control | Change in water ways by construction of canals and roads |
| Dredging and flood protection | Increased soil salinity by changes to fresh water run off |
| Wood chipping | |
| Industrial and road development | |
| Conversion to shrimp and fish ponds | |
| Construction of sea walls and dikes | **Natural causes** |
| Discharge and spraying of pesticides | Sea-level rise |
| Mining activities | Drought |
| Oil pollution | Typhoons and other tropical storms |
| Maintenance of salt pans | Soil erosion |
| Construction of tourist facilities | |

Table 2 summarises the causes of loss of mangroves globally. Figures showing mangrove loss are not available for most countries, but where these do exist there is an indication that significant decreases in the mangrove area have occurred. In Southeast Asia, for example, the loss figures for four countries are: Malaysia: 5050 $km^2$ in 1980 to 4460 $km^2$ in 1990 (Chan et al., 1993); the Philippines: 4500 $km^2$ in 1920 to 133 $km^2$ in 1990. (Primavera, 1995); Thailand: 3679 $km^2$ in 1961 to 1905 $km^2$ in 1996. (Platong, 1998); and Vietnam: 4000 $km^2$ in 1943 to 1560 $km^2$ in 1999 (Hong, 1999). The four countries concerned have suffered significant mangrove loss but they are not alone. Ong (1995) considers that the approximate 1% loss of mangrove area per year in Malaysia is a conservative estimate of destruction of mangroves in the Asia-Pacific region. Some figures for Latin America are: Peru: 139 $km^2$ in 1943 to 45 $km^2$ in 1992 (Echevarria and Sarabia, 1993); and Ecuador: 2037 $km^2$ in 1969 to 1618 $km^2$ in 1991 (Bodero, 1993). The total figure of some 9684 $km^2$ of mangroves lost in the second half of this century in the six countries cited represents over 5% of the current global total.

Spalding et al. (1997) state that it is extremely difficult to put a figure to the annual rate of loss of mangroves at a global scale. Indeed, the knowledge of the current global extent of mangroves is, at best, rudimentary. This is partly due to the absence of good vegetation maps and partly because there is a lack of definition of what is considered to be a mangrove. The availability of modern technology such as satellite imaging and Geographical Information Systems (GIS) will improve the situation but there are inherent problems with these techniques and accurate ground truthing is still required. The use of such techniques is also complex and not readily available to many countries.

A case study from Thailand (Platong, 1998) shows that there has been a 48% decline in the areal extent of mangroves in 35 years. Figure 4 shows that, as the areal extent of mangroves has declined, the shrimp production has risen dramatically as more and more mangroves are converted to shrimp ponds. The decline in the area of mangroves is not due solely to conversion to shrimp ponds but is also due to local inhabitants using mangrove wood as a source of fire-wood and charcoal. Other causes of mangrove loss are mining, urbanisation, coastal development and agriculture.

Uncontrolled conversion of mangroves to shrimp ponds has been accompanied by intensive farming methods that require the addition of feed and chemicals to maintain an adequate level of water quality. The result is widespread pollution as the discharge from the ponds spreads into the adjacent environment. In many instances the shrimp ponds have been abandoned as a result of falling production, leaving highly degraded land with high acid sulphate levels. The mangroves do not naturally re-establish themselves on such degraded land and the productive mangrove ecosystem is often destroyed permanently. The Government of Thailand began to show concern about the destruction of mangrove as early as 1981 and in 1987 passed legislation to control the use of mangroves. In the same period, an extensive mangrove planting programme was commenced by the Royal Thai Forestry Department. In 1990 the control of conversion of mangroves to shrimp ponds became tightly controlled. The slight rise in the mangrove area in Thailand after 1991 (Fig. 4) can be accounted for by the vigorous mangrove rehabilitation programme.

In another case study from the Philippines (Primavera, 1995), 70% of the mangrove area was lost in a period of sixty years. As in Thailand, the decline in the area of the mangroves mirrored the increase in the area of fishponds (Fig. 5). About 50% of the mangrove loss can be traced to brackish water pond construction. The fishpond industry in the

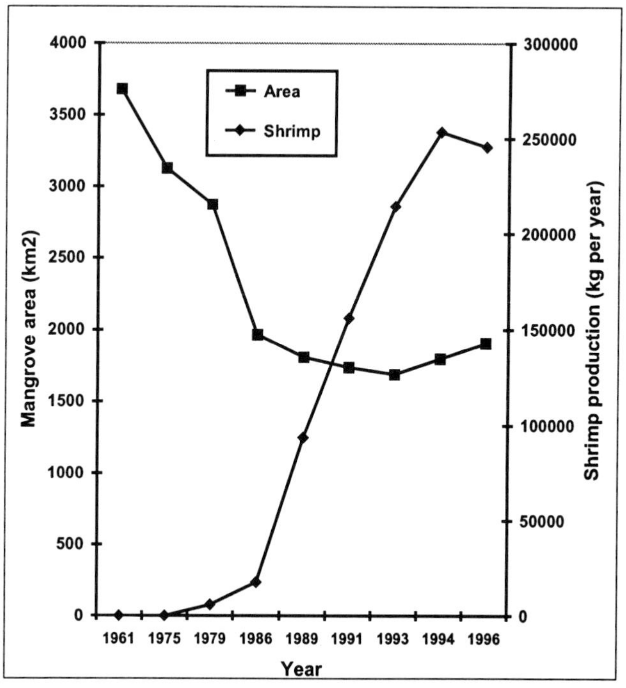


Fig. 4. Comparison between mangrove area ($km^2$) and shrimp production (kg per year) in Thailand (Platong, 1998).

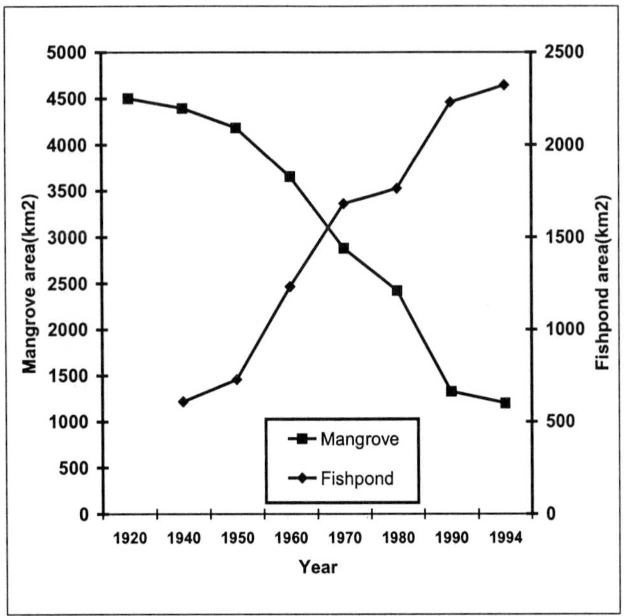


Fig. 5. Comparison between mangrove area (km$^2$) and fish pond area (km$^2$) in the Philippines (Primavera, 1995).

Philippines boomed in the 1950s and 1960s as a result of government sponsorship. There was a slow-down in the 1970s when concerns about the destruction of mangroves began to be voiced. Then in the 1980s there was a massive increase in the area of fishponds; the Asian Development Bank made $US 21.8 million available for the development of hatcheries and ponds at this time. This same institution is now supporting mangrove rehabilitation projects.

The enigma is that fish ponds represent a significant component of the Philippines' economy with 267,00 million tonnes of fish and shrimps being produced in 1990 with a worth of 6.5 million pesos per annum. The same situation pertains in Vietnam, where, notwithstanding poor natural seed supply, shrimp yields and farm incomes are attractive (AIMS, 1999). Unfortunately, the outcome in both countries is very similar to that of Thailand, that is, the life of the shrimp and fish farms is short, with accompanying peripheral pollution. The short-term gain is achieved at the expense of the long-term loss of the naturally productive mangroves. Primavera (1997) points out that, though the Philippines Government has enacted numerous laws concerning the utilisation of mangrove forests since 1975, the less than optimal management of mangrove resources in the Philippines may be traced to overlapping bureaucracy and legislative ambiguities.

The lack of knowledge of the true current global areal extent of mangroves referred to previously means that it is even more difficult to gauge the global rate of loss of mangroves. The rate of loss of mangroves in certain areas is well documented but this does not provide information at a global level. There is an urgent need to refine the base-line data on the current areal extent of mangroves and then to up-date this information on a regular basis, say at ten-yearly intervals, so that the global rate of loss of mangroves can be estimated with some degree of accuracy. There is no indication that such a programme is likely at a global level. The lack of good quantitative data will impede good decision-making on the management of mangroves.

There is considerable interest in the effects that climate change may have on mangrove forest communities (100 references can be found at *http://possum.murdoch.edu.au/~mangrove*). This often involves predicting whether mangrove distribution will be altered by changes in global temperature and carbon dioxide levels and whether any significant changes to sea level may involve coastal abandonment and inland retreat by mangrove communities, due to flooding, saline intrusion and changes to precipitation and catchment run-off (Snedaker, 1995). It has also been postulated that some of the earliest changes to sea level might be indicated by changes to mangrove forests (UNEP, 1994). There is little consensus as to the magnitude of the effects that may occur but it is likely that the dynamic nature of mangrove communities and the natural variability of the environment in which mangroves occur will mask any variation due to man-induced global climate change for some time.

## ECOLOGICAL VALUES

A great deal has been written about the ecology of mangrove ecosystems; some general references are Odum and Heald (1972), Lugo and Snedaker (1974), Chapman (1976), Clough (1982), Tomlinson (1986), Hutchings and Saenger (1987), Robertson and Alongi (1992), Twilly et al. (1996), and Ball (1996). There are also a number of specific country accounts of mangrove ecosystems. These include India (Blasco, 1975), China (Peng, 1988), Africa (Hughes and Hughes, 1992), Vietnam (Hong and San, 1993), Thailand (Aksornkoae, 1993), Africa (Saenger and Bellan, 1995) and Latin America (Kjerfve et al., 1997). A notable research and training programme on mangrove ecology was mounted by UNESCO in Thailand (Macintosh et al., 1991).

In considering the status of mangroves at the end of the 20th century, it is important to identify the value and importance of mangrove ecosystems. In recent years, there have been detailed studies of the fauna, flora, ecology, hydrology, physiology and productivity of many different mangrove ecosystems, most of them pristine. This information, though of intrinsic value, is difficult to synthesize and interpret because of the varying techniques and conditions employed. The discrete nature of much of the scientific data makes identification of the principal factors controlling overall productivity of the mangroves hard to identify. This situation is, perhaps, not surprising given the complexity and heterogeneity of mangrove ecosystems.

A popular view of mangrove forests is that they are very productive (Dugan, 1993) but this statement has to be approached with caution. The problem is that it is difficult to measure 'net primary productivity' accurately as it involves measuring gas exchange between various parts of the plant and the atmosphere. It is also difficult to measure 'biomass' accumulation because of the sheer magnitude of the task. In contrast, it is relatively easy to measure litter fall, though a common mistake is to equate litter fall with 'net primary production' (Clough, 1992) as there is no reason to believe that the amount of material falling from the trees is in any way connected to the total amount of plant material being produced in the mangrove forest. Estimates of 'biomass' accumulation, which do not usually include below--ground biomass, are highly variable and depend on the species of mangrove and a whole range of geological and environmental factors. However, some values have been obtained for *Rhizophora* sp. which range from close to zero to 45 tonnes ha year$^{-1}$ (Clough, 1992). It is clear that the productivity of mangroves is very site and species specific and that there is a lack of information on the contribution that mangroves might make to carbon entering the food chain.

The productivity of a mangrove forest can be very small if environmental conditions are adverse, but under some conditions can compare well with terrestrial forests. Even in well studied mangrove forests the amount and availability of carbon that is produced by the trees and shrubs is not well understood and critical factors controlling productivity have still to be convincingly identified. However, it is generally agreed that where the amount of mangrove forest area is high compared to the area of open water, then mangrove production is the major source of carbon entering the food chain. It is probably only in large lagoons where there are thin strips of mangrove at the edges of the water that the production of carbon by other sources, such as phytoplankton, is greater than that produced by the mangroves.

The qualitative importance of mangroves as habitat, nursery and source of food for both commercial fisheries species and other non-commercial fauna is generally accepted (Fig. 6). Studies carried out thirty years ago (Odum and Heald, 1972) stressed the value that mangroves contributed to regions in which they grow. The importance of mangrove forests in supplying part of the food chain to support fish populations of the coastline was recognised. The importance of mangrove vegetation is directly related to the belief that the carbon and energy present in the leaves and other parts of the mangrove trees and shrubs is a vital link in the supply of food to animal communities in and near mangrove wetlands. The apparent simplicity of that proposition is very appealing and it has been used to justify the conservation of mangroves around the world. The problem is that the various steps in the food chain have been difficult to quantify and the nature of the food chain in various areas has proved to be highly variable and very complex (Twilley et al., 1996).

A great deal of effort has gone into determining what happens to the mangrove litter when it enters the water. Findings are very variable and highly dependent on the mangrove species and location. Despite early indications of the importance of material exports from mangrove forests to adjacent waters, there have been few thorough and quantitative estimates (Robertson et al., 1992), and the level of export of organic material from mangrove habitats has long been a matter of argument. As an example, results obtained from different Caribbean forests estimate that 2.5 kg of carbon material is exported from every hectare of forest each day (Lugo and Snedaker, 1974); a later finding for some Florida mangroves gave a figure of 0.4 kg (Twilley, 1985), while an estimate from a tropical mangrove system in Australia gave a figure of 9.1 kg (Robertson, 1986). Even if significant amounts of organic matter are washed out of the mangroves it is still difficult to know how important this is to the feeding habits of marine animals out at sea.

Fig 6. Roots of the mangrove *Rhizophora* at high tide, showing sheltered habitat for small fish (photo M. Spalding).

It is generally accepted that a large number of juvenile fish use mangroves as nursery habitats but Baran and Hambrey (1998) state that there is a lack of well established quantified relationships between fish yields and area of mangrove. However, they go on to say that several studies support the hypothesis that coastal fish resources are closely linked to estuaries and mangroves, even if controversy remains about their degree of dependence. There is clearly a need to know more about how below-ground mangrove biomass and sedimentation rates (Ellison, 1998) contribute to carbon fluxes to and from mangrove forests and how trophic interactions between plants and animals contribute to plant population processes in mangrove ecosystems (Robertson, 1991). There is also a need to have better quantitative information on how fish production depends on the area of mangroves. A consensus of the evidence must conclude that mangrove ecosystems are often very productive and that they provide a rich source of carbon and other nutrients for numerous secondary consumer organisms across all trophic levels. However, the details of these relationships still need more elaboration.

Apart from being a productive shoreline ecosystem, mangroves can help stabilise dynamic coastlines. Mangroves are opportunistic colonisers of suitable land, usually in protected parts of the coastline, lagoons and estuaries. The occurrence and extent of mangrove forests is the result of geomorphic and hydrodynamic forces that create river deltas, estuaries and lagoons (Woodroffe, 1992). Once mangroves become established they tend to accumulate sediment and so modify tendencies towards erosion (Wolanski et al., 1990). The extent to which mangroves can stabilise land exposed to ocean currents and substantial freshwater inputs varies. If the forces are too great the mangroves can become ephemeral, only to reappear at some other location when an opportunity presents itself. This phenomenon is manifest on the west coast of Africa. In parts of Asia, it has been long recognised that a substantial mangrove fringe can mitigate the impact of severe tropical storms. The theory is that the mangrove prop root system acts as a baffle that dissipates the energy of the surging waves and so protects embankments against damage and erosion. Mangroves are often planted in Bangladesh, China and Vietnam with the sole purpose of preventing storm damage.

An additional ecological value of mangroves is the preservation of biodiversity in the intertidal zone and beyond. Mangroves are considered to be amongst the most species-poor forest ecosystems in the tropics (Lugo, 1998) as the conditions under which they prosper are extremely difficult for non-halophytic plants to survive. However, mangrove forests do support extensive populations of birds, fish, crustacea, meiofauna, microbes and fungi, as well as reptiles and mammals. In the South East Asian mangroves, the Milky Stork, *Mycteria cineres,* the crab-eating frog, *Rana cancrivora,* and leaf monkey, *Presbytis cristata,* are all considered endangered along with many other species (Talaue-McManus, 1999), as is the tiger in the world's largest contiguous mangrove area in the Sundabans of Bangladesh and India. It is clear that extensive loss of mangrove forests in some countries is placing pressure on the unique biodiversity that they support.

In terms of their ecological value, each mangrove ecosystem must be considered on its merits and, though guidance from work on similar systems elsewhere may prove useful, there is no substitute for carrying out measurements at a specific site. It is dangerous to generalise about the performance of mangrove ecosystems. Certainly it is true that some mangrove ecosystems are highly productive and that they support a most valuable terrestrial and aquatic resource but not all mangrove ecosystems fall into this category. The state of knowledge of the processes occurring in mangrove ecosystems is that some of the important things that should be measured have been identified, but what causes their magnitude can vary from one site to another.

It is important not to view mangroves in isolation. They are known to interact with nearby ecosystems and in many instances the functioning of one system has important implications for the others. Saltmarsh is often found adjacent to mangrove communities which, like the mangroves, are considered to be a major source of organic material in the food chains (Adam, 1993). Seagrass meadows and coral reefs likewise interact closely with mangroves (Hutchings and Saenger, 1987).

There are risks in extrapolating findings from one mangrove forest to another and there is much work still to be done, but there are dangers that large quantities of data will be produced without a proper focus on the key things that have to be known. At the present time there are numerous scientists working on a disparate collection of mangrove ecosystems around the world. They are often studying one small aspect of the overall process and there is no real attempt to synthesise their findings into some coherent whole even at the level of the individual site. However, recent attempts to model mangrove ecosystems (Chen and Twilley, 1998; Twilley et al., 1998) may prove a useful approach, as at the very least it will pose problems that need to be answered.

## PATTERNS OF MANGROVE USE

Living with mangroves can be difficult. The constant intrusion by tides, the consequent deep muddy landscape and the phalanx of tangled roots make the mangrove forest a place that is generally inhospitable to man. There are places where human communities have been established deep within the mangroves, isolated from modern ways of communication, but these are exceptional. More often, traditional communities have been established close to the mangrove forests on land above high tide level but with easy access by boat to the produce supplied by the natural processes of the ecosystem. Such communities can operate at a subsistence level which is sustainable but as the size of the population increases there is mounting pressure to modify the environment to support larger numbers of people. Several recent publications have catalogued the benefits that mangroves can bestow and the conservation issues that arise (ISME, 1993a,b,c; Field, 1995; Farnsworth and Ellison, 1997; Ewel et al., 1998). The consensus is that mangroves contribute a plethora of benefits but that they have to be managed carefully. Mangroves often inhabit areas that are prime sites for coastal development such as mining, ports, housing, and tourism. In many ways, the problems have to be dealt with at different levels of organisation. The problem of the indigenous subsistence dweller on the fringe of the mangroves, seeking to maximise the productivity of the mangrove ecosystem for the benefit of a community, can only be addressed at the local level. The problem of development or conservation of mangroves has to be addressed at a national level The economic and social advantages have to be weighed against the ecological benefits. Clearly the twin dilemmas of local progress and state priorities merge as the impact of national development is felt in the local communities. The solutions to these problems are complex and will be inescapably political in the end.

For whatever reason, there is no doubt that the area of mangroves in the world is decreasing. If this trend is to be stopped or reversed there must be compelling reasons why the preservation of mangrove forests should be

Table 3

Benefits from mangroves

| Local communities | National interests | Global interests |
|---|---|---|
| Shelter | Timber production | Conservation |
| Construction timber | Charcoal production | Education |
| Firewood | Fishing industry | Indication of climate change |
| Food | Shrimp and crab industries | Preservation of biodiversity |
| Income from fishing, shrimp culture and wood gathering | Water quality management | |
| Income through cottage industries | Mangrove silviculture | |
| Medicine | Mixed shrimp farms-mangrove forestry enterprises | |
| Fodder for animals | Wetland habitat creation | |
| Protection from storm damage and river bank erosion | Recreation | |
| | Tourism | |
| | Education | |
| | Coastal and estuary protection | |

incorporated into the plans of developing and developed countries. There are many benefits that mangroves can bestow on individual communities, individual countries and the world (Table 3).

Coastal villages face the twin problems of a degrading environment, as they try to extract more wealth from an already stretched resource, and a drift of young people away to the cities. Villagers often realise the long-term dangers of over-exploiting because the tradition of living with the mangroves is strong. In order to improve the wealth and the standard of living of a mangrove village community there are strong pressures to adopt commercial practices. Sometimes these involve the expansion of activities at the village level so that excess produce can be sold in local markets. At other times commercial exploitation of the mangroves can be by outside interests and there is a massive disruption of the mangrove ecosystem.

An example of a popular and profitable cottage industry that relies on the mangrove forest is the exploitation of wood products. This can take the form of producing poles for construction purposes and charcoal wood.

On the west coast of Malaysia there is a vast managed mangrove forest of some 35,000 hectares, known as Matang, that is used extensively for its wood products. The dominant mangroves are *Rhizophora apiculata* and *Rhizophora mucronata*. Species of *Bruguiera, Avicennia* and *Sonneratia* also occur, and the Matang mangrove forest is an interesting example of successful sustainable utilisation, where the management approach today differs little from that employed by Watson (1928). The operation provides significant employment for the local population, and the use of mangrove timber for charcoal and pole production makes a significant contribution to the economy of the west coast of Peninsular Malaysia (Chan, 1996). Matang also provides protection against erosion of the coastal region, breeding grounds for high protein sea-foods, cheap firewood, fishing stakes and structural materials. This area encourages forest research and training in mangrove forestry.

An alternative source of wealth in the mangroves is the exploitation of the fish, molluscs and crustaceans that abound in the mangrove areas. It is common in many mangrove villages to cultivate fish and crabs within the mangrove waters (Lee, 1992; Davie and Sumardja, 1997) by building small fish ponds which are restocked by opening them to the tide from time to time. Traditionally the ponds are often quite small and operated by individual families. The effect of this type of cultivation on the mangrove ecosystem is usually minimal and there is little likelihood of any permanent damage being done, though there is a slight cause for concern that such operations may encroach on natural wildlife habitat. The expansion of this system for commercial purposes is often accompanied by disastrous effects. In the last decade shrimp aquaculture production has grown by a factor of seven and it is considered to have a global value exceeding $6 billion annually (Naylor et al., 1998). It is therefore not surprising that this resource is exploited for commercial purposes as well as for local community consumption. As the demand for wealth in many developing countries increases there is a trend to expand the areas of mangrove land given over to rearing shrimps.

An example of such a development and its impact on local communities can be seen along the coast of Vietnam, though such developments are not limited to that country (Ajiki, 1994; Rivera-Monroy et al., 1999). Since the beginning of 1980, shrimp (*Penaeus merguiensis*) farming for export has been encouraged by the government and is now widespread (Hong, 1994). The early promise of quick profits has encouraged local people to expand their area of fish ponds and has also encouraged mass migration of people from the inland regions in search of a livelihood. Thousands of hectares of mangroves have been cleared to provide space for shrimp ponds, and policies of indiscriminate land allocation have led to widespread deterioration of forest cover. In Minh Hai Province in southern Vietnam more than half the mangrove forest that existed in 1982 has been destroyed to accommodate fish ponds and there is now likely to be a shortage of firewood in the area. The subsequent failure of the shrimp ponds has meant that there are now hundreds of hectares of abandoned unproductive land. This in turn has renewed the threat of malaria in these areas, as the mosquitoes breed in the brackish stagnant pools that are no longer washed by the tides.

Pond aquaculture gives high yields for the first few crops because of the productive natural environment. However, a number of problems rapidly appear. The cleared mangrove land that forms the basis of the ponds reacts with oxygen to release large quantities of sulphuric acid into the ponds so that the water becomes highly acidic. Crop failures appear to be caused by poor environmental management of the ponds. Chemical and biological pollution by shrimp farms results from disposal in coastal waters of pond effluents and sludge, salinization of soil and water, and the introduction of diseases. The result is that both ponds and the nearby environment become degraded. It is estimated that shrimp ponds in Asia rarely last for more than five to ten years (Naylor et al., 1998) and the intensive shrimp and fish farming degrades the environment often in an irreversible fashion (Stevenson, 1997). Nevertheless, the economic return to the operators remains high if the cost to the environment is not included. There is an urgent need for better regulation of the clearing of mangrove forests for the purpose of aquaculture, the introduction of more appropriate technology, such as improved pond design, and the adoption of superior management practices, such as improved water usage, disease prevention and pollution avoidance (Boyd and Clay, 1998; AIMS, 1999).

A relatively recent commercial use of mangroves is for recreation and ecotourism. In theory, at least, this is a non-destructive use which can yield financial benefits for the local people. It has been estimated that there are currently more than 400 million international tourists who create over 70 million jobs and produce more than 200 billion US dollars in revenue. Even a small part of this trade being funnelled through the mangroves in the form of adventure travel would have a significant impact on the economy of the local people. There are problems as well as benefits with the ecotourism trade. There will be no trade, of course, if the environment is degraded, so there is a significant incentive to preserve the mangrove ecosystem in its natural state. A second advantage is that local communities gain employment and an enhanced sense of pride in their own environment. A third benefit is that there is an educational value for the people who visit such usually inaccessible places. A major disadvantage is that a frequent stream of people visiting the same location will inevitably cause some degradation itself, and will also have an indelible impact on the local inhabitants.

It is difficult to quantify all the losses and gains involved in the various options for using mangrove lands but it is clear that the irresponsible use of this resource, without proper consideration of all the options, is something that has to change. It is also clear that the options vary depending on the state of economic development of the country involved. As the ecological values are often perceived as "free" they tend to be ignored in the economic calculations that frequently determine whether mangrove lands should be conserved or developed. Private owners and corporate enterprises frequently decide to drain mangrove swamps or convert them to shrimp ponds because they expect to earn more from growing such crops than from having them in their natural condition. A long-term view of the situation is rarely taken. To compound the problem, many governments and international agencies give only lip-service to the sustainable use of mangroves and well developed management plans. There are also many examples of agencies like the European Community, FAO and the World Bank supporting shrimp pond development, and the dredging and draining of mangroves for crops, in the name of enhanced production. The results of some of these projects can be seen as waste land in Asia and Africa.

There have been a few attempts to assess the economic value of mangrove conservation (Lal, 1990; Grasso, 1998; Spurgeon, 1998). The main problem with all these approaches is that it is difficult to quantify non-use values of the mangrove ecosystem. There are many complex uncertainties inherent in the natural environment that affect the assessment of potential costs and benefits of mangrove rehabilitation. On the assumptions used in the Lal study (1990) the reclamation of mangrove land for sugar cane and shrimp production in Fiji was considered not viable under the existing conditions and the inclusion of figures for loss of fish and timber by removing the mangroves compounded the problem. There is little evidence that such studies are affecting the actual use of mangrove land.

There is much to be learned from the ways in which mangrove lands have been managed for centuries. Many traditional ways were developed under tightly controlled conditions imposed by the village community. These controls are loosening as populations rise and expectancy of increased living standards emerges, with a consequent demand for more agricultural land. At the same time, commercial pressures beyond the control of the villages demand more and more access to the land now occupied by mangroves. Many governments have chosen to invest heavily in converting natural mangrove forests for intensive agriculture because this is perceived as the most effective way of increasing food production. These actions are now under open criticism and are frequently cited for their negative impact.

## REHABILITATION OF MANGROVE ECOSYSTEMS

The realisation that in some parts of the world mangrove ecosystems are being destroyed, with a consequent loss of valuable productivity, has prompted an upsurge in the number of projects aimed at rehabilitation, regarded as the return of degraded mangrove land to a fully functional mangrove ecosystem, regardless of the original state of the degraded land (Brown and Lugo, 1994).

Mangrove ecosystems are very dynamic and their growth and decline often reflect the changing conditions of the coastal environment in which they grow. Any attempt to restore the structure and function of mangrove forests to anything like their perceived original state may prove elusive and impractical (Lugo, 1997). The need for rehabilitation often arises because of intensive use of the land, excessive extraction of materials, chemical contamination and changes to the hydrology of the site; though at times it can arise because of natural events such as sea-level changes, hurricanes, frost, lightning and fires.

Before a rehabilitation project is undertaken it is essential that the goals of the project be defined as a first step in the rehabilitation process. Goals determine the rehabilitation process and help identify the elements which must be included to provide the project with a clear framework for operation and implementation. There are many reasons for mangrove ecosystem rehabilitation (Clough, 1999) but these can be summarised in three main categories:

a) conservation and landscaping (improved biodiversity, education, beautification, and wetland habitat creation);
b) sustainable utilisation (silviculture, aquaculture, wastewater treatment);
c) coastal protection (erosion and sedimentation modification, shelter from tropical storms).

The characteristics of each category are discussed in Field (1998a). There are four main criteria used to measure the success of a mangrove rehabilitation project. They are: effectiveness of the planting, which can be considered as the closeness to which the growth of the new mangrove ecosystem meets the original objectives of the rehabilitation programme; rate of recruitment of flora and fauna, which can be considered a measure of how quickly the rehabilitated site recovers its integrity; efficiency of rehabilitation, which can be measured in terms of the amount of labour, resources and material used; and the long-term sustainability of the new mangrove ecosystem. Effectiveness and efficiency are only sometimes quantified, the recruitment of flora and fauna rarely quantified and the long term sustainability awaits determination.

There are five main practical considerations (Field, 1998a) when contemplating the rehabilitation of a mangrove ecosystem. They are: identifying causes of site degradation, assessing site selection criteria, sourcing of seedlings and planting (natural, field transplant or nursery), monitoring the outcomes and maintaining the resulting mangrove ecosystem. There are well established general criteria in the literature for these activities (JAM, 1994; Lewis, 1994; Field, 1996; JAM, 1997; Turner and Lewis, 1997; Clough, 1999), however local environmental conditions must dominate all other considerations. It is generally agreed that the hydrologic regime is the single most important overall site condition governing the survival and subsequent growth of the mangrove seedlings.

Once a mangrove rehabilitation programme has been completed, it is essential to monitor progress and to maintain the site (Field, 1998a). These activities are similar to those that would be normally undertaken in any forestry programme. Three to five years is often specified as the monitoring period in small-scale rehabilitation programmes but more realistically programmes should be monitored for ten years. For large afforestation projects up to thirty years may be necessary.

The number of mangrove rehabilitation programmes world-wide is extensive (Field, 1996; Field, 1998b). Of the 90 or so countries around the world that possess mangrove vegetation, only some 20 have attempted any form of mangrove replanting. Only nine of these twenty countries have planted more than 10 km$^2$ since 1970. Bangladesh (>1,200 km$^2$), Indonesia(>400 km$^2$), The Philippines(>400 km$^2$), and Vietnam(>1,100 km$^2$) stand out as the countries that have put the most effort into the rehabilitation of mangrove ecosystems. In the case of Bangladesh most of the planting has been in the form of afforestation on newly accreted land (Saenger and Siddiqi, 1993). In Indonesia and The Philippines the plantings have been on degraded areas caused by clear-felling, shrimp ponds and population pressure. In Vietnam the causes are similar but have been compounded by the devastating effects of the recent wars. Erftemeijer and Lewis (1999) and Clough (1999) point out that not all attempts to rehabilitate mangroves have been successful and that there are lessons to be learned from such failures.

Apart from national governments, numerous international organisations have, or are supporting, mangrove replanting programmes. Included in this list are the European Union, the World Bank, the Asian Development Bank, the World Wide Fund for Nature (WWF), the International Union for the Conservation of Nature (IUCN), the Food and Agricultural Organisation (FAO), UNESCO, UNEP, Wetlands International, the International Tropical Timber Organisation (ITTO), the Save the Children Fund and the Australian Centre for International Agricultural Research (ACIAR).

One of the challenges is to gauge how successful rehabilitation projects have been and what lessons have been learnt from failures. It is clearly impossible to carry out such a critical review without access to the myriad of reports that must be hidden in the archives of the many sponsoring agencies. This information is very difficult to obtain.

## BIODIVERSITY AND HUMAN COMMUNITIES IN THE MANGROVES

The rehabilitation of mangrove ecosystems can bring into conflict the aims of maximising biodiversity and providing maximum benefit to the human communities that rely on the mangroves. The conservation of biological diversity is central to the dogma of the international conservation community. It is perceived as pivotal to nature conservation because species extinction threatens not only the idealised Western view of nature but depletes the genetic resources that are essential for continued human prosperity (Davie and Hynes, 1997). This view can be challenged if the objective of the rehabilitation is sustainable production. There is a necessity to rehabilitate mangrove lands to support human communities. This may involve a landscape mosaic composed of mature forests, logged forests, aquaculture and agriculture. The use of such a mosaic can be sustainable.

As a means of maintaining global biodiversity, increasing attention is being given to rehabilitating and creating various coastal habitats. This activity is expensive and probably involves more mangrove research dollars than any other research activity. However, it is interesting to note that the relationship between changes in biodiversity and ecosystem function is not easily quantified in mangrove ecosystems despite the extensive pool of information (Twilley et al., 1996). In addition, the value of focusing the purpose of nature conservation on biodiversity can be queried. Davie and Hynes (1997) question why biodiversity should be paramount in the thinking of conservationists and argue that conservation should be more inclusive of community participation. They believe that nature conservation practice should stand within a context of multiple land-use. They argue that the preservation of ecological processes to maintain habitat, water quality and soil fertility may better integrate nature conservation into other land uses. Perhaps of even more importance is the relationship between biodiversity and productivity. There is some evidence that productivity and biodiversity have a reciprocal relationship (Johnson et al., 1996). However, biodiversity and productivity both contribute to system integrity, which is pivotal to management. The question that needs to be asked with respect to mangrove ecosystems is how much loss of biodiversity can be tolerated in order to maximise productivity and ensure system integrity (Freckman et al., 1997). Grime (1997) points out that it is naïve to assume that species-poor ecosystems always malfunction. There are plenty of highly managed, relatively stable, highly simplified, 'natural' environments that support human populations. However, land conversion and intensive cultivation alter the biotic interactions and resource availability in ecosystems (Matson et al., 1997). It is important that ecological management strategies are employed to ensure the sustainable use of mangrove ecosystems.

Human factors critically influence the outcomes of mangrove rehabilitation projects in populated areas. It is recognised that human involvement in causing degradation is as important as the natural determinants of the structure and function of the mangrove ecosystem in determining the success of a rehabilitation project. It follows that for a mangrove rehabilitation project to be effective, the human population must be viewed as an inherent part of the mangrove ecosystem.

Examples of concern for the human factor in mangrove rehabilitation projects are: the Bais Bay, Philippines (Walters, 1997); Pattani Bay, Thailand (Wetlands International Asia-Pacific, 1997); Tamil Nadu, India (Swaminathan Research Foundation, 1998); Indus Delta, Pakistan (M. Quereshi, 1998, pers. comm.); East Java, Indonesia (Davie and Sumardja, 1997); Mekong Delta, Vietnam (MERC, 1994; AIMS, 1999); and Mida Creek, Kenya (Dahdouh-Guebas et al., 1999). People working on large-scale mangrove rehabilitation projects in poor, highly populated tropical environments consider the most important requirement is to factor in the involvement of the human community. Only by managing the rehabilitation project with the support and involvement of the local people and by using innovative approaches will the project be successful.

There is a consensus that the act of rehabilitation must benefit the local people by offering them economic benefits, using their knowledge and skills and by effectively employing their existing social and political structures. The rehabilitation must also be compatible with local resource use and land tenure. Erftemeijer and Bualuang (1998) while discussing the Pattanu Bay project make the telling comment that 'local ownership of the project and effective community participation are considered crucial to achieve sustainable impacts'.

## THE FUTURE

As the next millennium dawns the knowledge that exists about the extent, structure and function of mangrove ecosystems is substantial and growing at an impressive rate. There are more people studying mangrove ecosystems today than at any time in the past. Mangroves are fashionable at the end of the 20th century, and every facet of their existence is under scrutiny.

In reviewing the literature one is left with an uncomfortable feeling that previous knowledge has been ignored and that more and more obscure aspects of mangrove ecosystems are being subjected to examination. This, perhaps, does not matter if as Arthur Koestler (1963) said: 'The philosophy of nature evolves by occasional leaps and bounds alternating with delusional pursuits, *cul-de-sacs*, regressions, periods of blindness and amnesia'. However, there is an urgent need for better access to existing knowledge and more synthesis of the knowledge that exists.

To illustrate this proposition, it is useful to point out that there is currently no single comprehensive database that can be accessed for reference to mangrove studies. There are, of course, the usual scientific data bases that give access to refereed papers but a large number of references to mangroves exist in the 'grey' literature of reports.

Reference has been made previously to the large number of organisations that fund some form of mangrove investigation. It is almost impossible to get useful information on the results of the work that has been carried out. As an example, the World Bank's web site lists in excess of fifty projects involving mangrove studies with the total cost being in the millions of dollars. As a result of a direct request for information, the World Bank reluctantly supplied two reports. It appears that the World Bank has no overall picture of the money that it expends on mangrove projects and makes no attempt to analyse or synthesise the results from such studies. It appears that officials of the World Bank are unaware of the outcome or importance of many of the mangrove projects that they support. The World Bank is not alone in this neglect. It is tragic that millions of dollars are being spent on studying and rehabilitating mangroves but that the information being collected is not being shared. An opportunity is being missed to construct the studies in such a way that useful research information could be obtained. It is not that the money being used is necessarily wasted but so much more could be achieved.

On the same theme, there is a need to have a better knowledge of the areal extent of mangroves at a global level. Spalding et al. (1997) attempted to catalogue the areal extent of mangrove forests at a gobal level using verifiable data. It is clear that such data was far from perfect and that many countries need a much better inventory of their mangrove vegetation. It is important to establish a reliable base-line for the global areal extent of mangroves, which can be revised every decade. It is only when a reliable base-line has been established that accurate statements can be made as to the rate of loss or gain of mangrove forests at a global level. At this time there is no indication that such work will be undertaken.

The World Commission on Environment and Development (1987) said: 'Sustainable development requires meeting the basic needs of all and extending to all the opportunity to satisfy their aspirations for a better life'. In 1992, Agenda 21 stated: 'Human beings are at the centre of concerns for sustainable development. They are entitled to a healthy and productive life in harmony with nature.' As the new millennium approaches and the global population continues to rise sharply these worthy sentiments sound increasingly hollow.

In the case of mangrove ecosystems there is an urgent need to improve their sustainable use for the benefit of the people who rely upon them. The focus of future work must be on the most efficient ways to use and rehabilitate mangrove ecosystems for sustainable purposes. This may involve using existing information, generating land-use information using technology such as GIS, employing advances in biotechnology and being aware of the important role that social and economic information can play in the processes. If such a process produces useful information for land-use managers, then ecology and management will begin to work together. Underwood (1995) points out that ecologists should not be too eager to confine their efforts solely to the provision of sound ecological advice but should be prepared to have more say in the way the advice and data are used.

There is a need for applied ecological research aimed at testing the decisions made when using and rehabilitating mangroves. This seldom happens. Indeed, much mangrove research is done in isolation from the needs of the managers of the projects. There is an urgent need to study the failures of mangrove ecosystem management. Likewise, there is a need to do research that will enhance mangrove ecosystem rehabilitation.

There is a significant ethical challenge in this work for all people interested in the future of mangrove ecosystems. It is unacceptable for international organisations, national governments, learned societies and individual scientists not to be alert to the implications and practical use of their work. It is important that everyone, whether they operate at a global level or at the village level, keep in touch with the aspirations and wishes of the local people. Unfortunately, the machinations of international and national organisations often mean that little attention is paid to the impact of their grandiose programmes on local communities. On the other hand, the myopia of many ecologists means that they show little interest in the application of their work. It is also important that advances in science be evaluated in a rational manner. As an example, what is the future for many mangrove ecosystems if biotechnologists develop a rice crop, or some other highly productive agricultural crop, that grows well in sea water? Such a development would produce heated arguments about the importance of such things as biodiversity, food chains, conservation, habitat and a myriad of other matters. The critical issue is that the debate be rational and not emotional.

There is a great waste of opportunity, money and intellectual effort in the study and utilisation of mangrove ecosystems. The challenge for the 21st century is to harness all this endeavour and to focus it on the twin problems of how to ensure the well-being of the subsistence dwellers that depend on the mangroves and of how to ensure, whenever possible, that mangrove ecosystems continue to grace the shorelines and estuaries of the tropical world.

It would be remiss to leave the twentieth century with a negative impression concerning the status of mangroves. There has been enormous progress made on the understanding of the structure, function and use of mangroves during this century. Above all there has been a cathartic shift in the understanding of the importance and usefulness of mangrove ecosystems. Much of this release has been achieved by education. It is on this basis of understanding that the sustainable use of mangrove ecosystems must be further developed.

## REFERENCES


Adam, P. (1993) *Saltmarsh Ecology.* Cambridge University Press, 461 pp.

Agenda 21 (1993) *Earth Summit.* United Nations Publication.

AIMS (1999) Mixed shrimp farming-mangrove forestry models in the Mekong Delta. Report of consulting project sponsored by ACIAR. Clough, B. (project leader) (in press).

Ajiki, K. (1994) The clearance of mangrove forests and expansion of fish ponds in Southeast Asian countries. In *Proceedings of the National Workshop: Reforestation and Afforestation of Mangroves in Vietnam.* Mangrove Ecosystem Research Centre, Hanoi. pp. 56–59.

Aksornkoae, S. (1993) *Ecology and Management of Mangroves.* IUCN, Bangkok, Thailand. 176 PP.

Ball, M.C. (1996) Comparative ecophysiology of mangrove forest and tropical lowland moist rainforest. In *Tropical Forest Plant Ecophysiology,* eds. S.S. Mulkey, R.L. Chazdon and A.P. Smith. Chapman and Hall, New York, pp. 461–496.

Baran, E. and Hambery, J. (1998) Mangrove Conservation and Coastal Management in Southeast Asia: what impact on fishery resources. *Marine Pollution Bulletin* **37** (8–12), 431–440.

Blasco, F. (1975) *Les Mangroves De L'Inde.* Institut Francais de Pontichery. Travaux de la Section Scientific et Technique, 175 pp.

Blasco, F., Gauquelin, T., Rasolofoharinoro, M., Denis, J., Aizpuru, M. and Caldairou, V. (1998) Recent advances in mangrove studies using remote sensing data. *Marine and Freshwater Research* **49**, 287–296.

Bodero, A. (1993) Mangrove ecosystems of Ecuador.In *Conservation and Sustainable Utilization of Mangrove Forests in Latin America and African Regions. Part 1: Latin America,* ed. L.D. Lacerda, Mangrove Ecosystems Technical Reports Vol. 2. ISME, Okinawa, Japan. pp. 55–73.

Boyd, C.E. and Clay, J.W. (1998) Shrimp aquaculture and the environment. *Scientific American*, June, 59–65.

Brown. S. and Lugo, A.E. (1994) Rehabilitation of tropical lands: a key to sustaining development. *Restoration Ecology* **2** (2), 97–111.

Chan, H.T., Ong, J.E., Gong, W.K. and Sasekumar, A. (1993) The socio-economic, ecological and environmental values of mangrove ecosystems in Malaysia and their present rate of conservation. In *The Economic and Environmental Values of Mangrove Forests and their present state of Conservation in the Southeast Asia and Pacific Region,* ed. B. Clough. Mangrove Ecosystems Technical Reports Vol. 1. ISME, Okinawa, Japan, pp. 41–77.

Chan, H.T. (1996). Mangrove restoration in Peninsular Malaysia. In *Restoration of Mangrove Ecosystems,* ed. C.D. Field. International Society for Mangrove Ecosystems, Okinawa, Japan. pp. 64–75.

Chapman, V.J. (1976) *Mangrove Vegetation.* L. Cramer, Vaduz. Germany.

Chen, R. and Twilley, R.R. (1998) A gap dynamic model of mangrove forest development along gradients of soil salinity and nutrient resources. *Journal of Ecology* **86**, 1–12.

Clough, B.F. (1982) *Mangrove Ecosystems of Australia.* Australian National University Press, Canberra, 302 PP.

Clough, B.F. (1992) Primary productivity and growth of mangrove forests. In *Tropical Mangrove Ecosystems,* eds. A.I. Robertson and D.M. Alongi, Coastal and Estuarine Studies 41. American Geophysical Union, Washington, DC, pp. 225–249.

Clough, B.F. (1999) Evaluating the success or failure of mangrove afforestation/rehabilitation projects (in press).

Davie, J. and Hynes, R. (1997) Integrating nature conservation and sustainable rural management. *Australian Biologist* **10**(4), 185–199.

Davie, J. and Sumardja, E. (1997) The mangroves of East Java: an analysis of the impact of pond aquaculture on biodiversity and coastal ecological processes. *Tropical Biodiversity* **4** (1),1–33.

Dahdouh-Guebas, F., Mathenge, C., Kairo, J.G. and Koedam, N. (1999) Use of mangrove woodproducts around Mida Creek (Kenya) from a subsistence perspective. *Economic Botany* (in review).

Dugan, P. (ed.) (1993) *Wetlands in Danger.* IUCN. Mitchell Beazley. London, 187 pp.

Duke, N.C. (1992) Mangrove floristics and biogeography. In *Tropical Mangrove Ecosystems,* eds. A.I. Robertson and D.M. Alongi, Coastal and Estuarine Studies 41. American Geophysical Union, Washington, DC, pp. 63–100.

Echevarria, J. and Sarabia, J. (1993) Mangroves of Peru. In *Conservation and sustainable utilization of Mangrove Forests in Latin America and African Regions. Part 1: Latin America,* ed. L.D. Lacerda. Mangrove Ecosystems Technical Reports Vol. 2. ISME, Okinawa, Japan, pp. 43–53.

Erftemeijer, P.L.A. and Bualuang, A. (1998) Participation of local communities in mangrove forest rehabilitation in Pattani Bay, Thailand: learning from successes and failures. *2nd International Conference on Wetlands and Development.* Dakar, Senegal.

Erftemeijer, P.L.A. and Lewis, R.R. (1999) Planting mangroves on intertidal mudflats: habitat restoration or habitat conversion? *Ecotone-V111 seminar 'Enhancing Coastal Ecosystem Restoration for the 21st Century'*, Ranong and Phuket May (in press).

Ellison, C.E. (1998) Impacts of sediment burial on mangroves. *Marine Pollution Bulletin* **37** (8–12), 420–426.

Ewel, K.C., Twilley, R.R. and Ong, J.E. (1998) Different kinds of mangrove forests provide different goods and services. *Global Ecology and Biogeography Letters* **7**, 83–94

Farnsworth, E.J. and Ellison, A.M. (1997) The global conservation status of mangroves. *Ambio* **26**(6), 328–334

FAO (1994) *Mangrove Forest Management Guidelines.* FAO, Rome, 319 pp.

Field, C.D. (1995) *Journey Amongst Mangroves.* International Society for Mangrove Ecosystems, Okinawa, Japan, 140 pp.

Field, C.D. (ed.) (1996) *Restoration of Mangrove Ecosystems.* International Society for Mangrove Ecosystems, Okinawa, Japan, 250 pp.

Field, C.D. (1998a) Rationales and practices of mangrove afforestation. *Marine and Freshwater Research* **49**, 353–358.

Field, C.D. (1998b) Rehabilitation of mangrove ecosystems: an overview. *Marine Pollution Bulletin* **37** (8), 383–392.

Fisher, P. and Spalding, M.D. (1993) Protected areas with mangrove habitat. Draft report. World Conservation Monitoring Centre. UK, 60 pp.

Freckman, D.W., Blackburn, T.H., Brussaard, L., Hutchings, P., Palmer, M.A. and Snelgrove, P.V.R. (1997) Linking biodiversity and ecosystem functioning of soils and sediments. *Ambio* **26** (8), 556–562.

Grasso, M. (1998) Ecological-economic model for optimal mangrove trade off between forestry and fish production: comparing a dynamic optimization and simulation model. *Ecological Modelling* **112**, 131–150.

Grime, J.P. (1997) Biodiversity and ecosystem function: the debate deepens. *Science* **277**, 1260–1261.

Hong, P.N. and San, H.T. (1993) *Mangroves of Vietnam.* IUCN, Bangkok, Thailand, 173 pp.

Hong, P.N. (1994) Causes and effects of the deterioration in the mangrove resources and environment in Vietnam. *Proceedings of the National Workshop: Reforestation and Afforestation of Mangroves in Vietnam.* Mangrove Ecosystem Research Centre, Hanoi. pp. 24–39.

Hong, P.N. (1999) Mangrove Forest Restoration for Enhancing Coastal Ecosystems in Vietnam (in press).

Hughes, R.H. and Hughes, J.S. (1992) *A Directory of African Wetlands.* IUCN Switzerland, UNEP, Nairobi and WCMC, Cambridge, 820 pp.

Hutchings, P. and Saenger, P. (1987) *Ecology of Mangroves.* University of Queensland Press. Brisbane, Australia, 388 pp.

ISME (1993a) *The Economic and Environmental Values of Mangrove Forests and their present state of Conservation in the Southeast Asia and Pacific Region.* Mangrove Ecosystems Technical Reports Vol. 1. ed. B.F. Clough. ISME, Okinawa, Japan, 202 pp.

ISME (1993b) *Conservation and Sustainable Utilization of Mangrove Forests in Latin America and Africa Regions. Part 1: Latin America.* Mangrove Ecosystems Technical Reports Vol. 2, ed. L.D.Lacerda, ISME, Okinawa, Japan, 272 pp.

ISME (1993c) *Conservation and Sustainable Utilization of Mangrove Forests in Latin America and Africa Regions. Part 2: Africa.* Mangrove Ecosystems Technical Reports Vol. 3, ed. E.S. Diop. ISME, Okinawa, Japan, p.262.

IUCN (1983) *Global Status of Mangrove Ecosystems.* Commission on Ecology Papers No. 3. eds. P. Saenger, E.J. Hegerl and J.D.S. Davie. International Union for Conservation of Nature and Natural Resources, Gland, Switzerland, 88 pp.

JAM (Japan Association for Mangrove) (1994) *Development and Dissemination of Re-afforestation Techniques of Mangrove Forests.* Publication of JAM, Tokyo, Japan, 216 pp.

JAM (Japan Association for Mangrove (1997) *Final report of the ITTO project on Development and Dissemination of Re-afforestation Techniques of Mangrove Forests.* Publication of JAM, Tokyo, Japan, 104 pp.

Johnson, K., Vogt, K.A., Clark, H.J., Schmitz, O.J. and Vogt, D.J. (1996) Biodiversity and the productivity and stability of ecosystems.

*Trends in Ecology and Evolution* **11**, 372–377.
Koestler, A. (1963) *The Sleepwalkers.* Macmillan, New York, 624 pp.
Kjerfve, K., Lacerda, L.D., and Diop, H.S. (eds.) (1997) *Mangrove Ecosystem Studies in Latin America and Africa.* UNESCO. Paris, France, 349 pp.
Lal, P.N. (1990) Conservation or conversion of mangroves in Fiji. *Occasional papers of the East–West Environment and Policy Institute.* Paper No. 11. EWEPI, Honolulu.
Lee, S.Y. (1992) The management of traditional tidal ponds for aquaculture and wildlife conservation in Southeast Asia: problems and prospects. *Biological Conservation* **63**, 113–118.
Lewis, R.R. (1994) Enhancement, restoration and creation of coastal wetlands. In *Applied Wetlands Science and Technology,* ed. D.M. Kent, CRC Press. pp. 167–191.
Lugo, A.E. and Snedaker, S.C. (1974) The Ecology of Mangroves. *Annual Review of Ecology and Systematics* **5**, 39–64.
Lugo, A.E. (1997) Old growth mangrove forests in the United States. *Conservation Biology* **11** (1), 11–21.
Lugo, A.E. (1998) Mangrove forests: a tough system to invade but an easy one to rehabilitate. *Marine Pollution Bulletin* **37** (8–12), 427–430.
Macintosh, D.J., Arsornkoae, S., Vannucci, M., Field, C.D., Clough, B.F., Kjerfve,B., Paphavasit, N. and Wattayakorn, G (eds.) (1991) Final report of the integrated multidisciplinary survey and research programme of the Ranong mangrove ecosystem. UNESCO, Paris, 183 pp.
Matson, P.A., Parton, W.J., Power, A.G. and Swift, M.J. (1997) Agricultural intensification and ecosystem properties. *Science* **277**, 504–509.
MERC (1994) *Proceedings of the National Workshop: Reforestation and Afforestation of Mangroves in Vietnam.* Mangrove Ecosystem Research Centre (MERC), Hanoi, 187 pp.
Naylor, R.L., Goldburg, R.J., Mooney, H., Beveridge, M., Clay, J., Folke, C., Kautsky, N., Lubchenco, J., Primavera, L. and Williams, M. (1998) Nature's subsidies to shrimp and salmon farming. *Science* **282**, 83–84.
Odum, W.E. and Heald, E.J. (1972) Trophic analysis of an estuarine mangrove community. *Bulletin Marine Science* **22**, 671–738.
Ong Jin-Eong (1995) The ecology of mangrove conservation and management. *Hydrobiologia* **295**, 343–341.
Peng, L. (1988) *Mangrove Vegetation.* China Ocean Press. Beijing, China, 74 pp.
Perdomo, L., Ensminger, L., Espinosa, F., Elster, C., Wallner-Kersanach, M. and Schnetter, M.L. 1998. The mangrove ecosystem of the Cienaga Grande de Santa Marta (Colombia): observations on regeneration and trace metals in sediment. *Marine Pollution Bulletin* **37** (8), 8–12.
Platong, J. (1998) *Status of Mangrove Forests in Southern Thailand.* Wetlands International. PSU, Publication No. 5, 128 pp.
Primavera, J.H. (1995) Mangroves and brackishwater pond culture in the Philippines. *Hydrobiologia* **295**, 303–309.
Primavera, J.H. (1997) Development and conservation of Philippine mangroves: institution issues. *4th Workshop of the Global Wetlands Economics Network,* Beijer Institute, Stockholm. November 1997.
Rivera-Monroy, V.H., Torres, L.A., Bahamon, N., Newmark, F. and Twilley, R.R. (1999) The potential use of mangrove forests as nitrogen sinks of shrimp effluents: the role of denitrification. *Journal of the World Aquaculture Society* **30**(1),12–25.
Robertson, A.I. (1986) Leaf burying crabs: their influence on energy flow and export from mixed mangrove forests (*Rhizophora* spp.) in North Eastern Australia. *Journal of Experimental Marine Biology and Ecology* **102**, 237–248
Robertson, A.I. (1991) Plant-animal interactions and the structure and function of mangrove forest ecosystems. *Australian Journal of Ecology* **16**, 433–443.
Robertson, A.I. and Alongi, D.M. (eds.) (1992) *Tropical Mangrove Ecosystems.* Coastal and Estuarine Studies 41. American Geophysical Union, Washington, DC, p. 329.
Robertson, A.I., Alongi, D.M. and Boto, K.G. (1992). Food chains and carbon fluxes. In *Tropical Mangrove Ecosystems,* eds. A.I. Robertson and D.M. Alongi. Coastal and Estuarine Studies 41. American Geophysical Union, Washington, DC. pp. 293–326.
Rollet, B. (1981) *Bibliography of Mangrove Research 1600–1975.* UNESCO, Rome. 79 pp.
Rubin, J.A., Gordon, C. and Amatekpor (1998) Causes and consequences of mangrove deforestation in the Volta Estuary, Ghana: some recommendations for ecosystem rehabilitation. *Marine Pollution Bulletin* **37** (8), 441–449.
Saenger, P. (1998) Mangrove vegetation: an evolutionary perspective. *Marine and Freshwater Research* **49**(4), 277–286.
Saenger, P. and Bellan, M.F. (1995) *The mangrove vegetation of the Atlantic coast of Africa.* A review. Université de Toulouse Press., Toulouse, 96 pp.
Saenger, P. and Siddiqi, N.A. (1993) Land from the sea: the mangrove afforestation programme for Bangladesh. *Ocean and Coastal Management* **20**, 23–39.
Snedaker, S.C. (1995) Mangroves and climate change in the Florida and Caribbean region: scenarios and hypotheses. *Hydrobiologia* **295**, 43–49.
Spalding, M.D., Blasco, F. and Field, C.D. (eds.) (1997) *World Mangrove Atlas.* International Society for Mangrove Ecosystems, Okinawa, Japan, 178 pp.
Spurgeon, J. (1998) The socio-economic costs and benefits of coastal habitat rehabilitation and creation. *Marine Pollution Bulletin* **37**(8), 373–383.
Streever, W. (ed.) (1999) *An International Perspective on Wetland Rehabilitation.* Kluwer Academic Publishers, Dordrecht, The Netherlands.
Stevenson, N.J. (1997) Disused shrimp ponds: options for redevelopment of mangrove. *Coastal Management* **25** (4), 423–425.
Swaminathan Research Foundation (1998) *Coastal Wetlands: Mangrove Conservation and Management.* Annual Reports 1, 2, 3. M.S. Swaminathan Research Foundation. Madras, India.
Talaue-McManus, L. (1999) *Transboundary Diagnostic Analysis for the South China Sea.* UNEP, p. 84.
The World Commission on Environment and Development (1987) *Our Common Future.* Oxford University Press. Oxford.
Tomlinson, P.B. (1986) *The Botany of Mangroves.* Cambridge University Press, Cambridge, 413 pp.
Turner, R.E. and Lewis III, R.R. (1997) Hydrologic restoration of coastal wetlands. Special issue: Hydrologic Restoration of Coastal Wetlands. *Wetlands Ecology and Management* **4**(2), 65–72.
Twilley, R.R. (1985) The exchange of organic carbon in basin mangrove forests in a southwest Florida estuary. *Estuarine, Coastal and Science* **20**, 543–557.
Twilley, R.R., Snedaker, S.C., Yanez-Arancibia, A. and Medina, E. (1996) Biodiversity and ecosystem processes in tropical estuaries: perspectives from mangrove ecosystems. In *Biodiversity and Ecosystem Functions: a Global Perspective,* eds. H. Mooney, H. Cushman and E. Mednia. John Wiley and Sons, New York. pp. 327–370.
Twilley, R.R., Rivera-Monroy, V.H., Chen, R. and Botero, L. (1998) Adapting an ecological mangrove model to simulate trajectories in restoration ecology. *Marine Pollution Bulletin* **37** (8–12), 404–419.
Underwood, A.J. (1995) Ecological research and (and research into) environmental management. *Ecological Applications* **5**(1), 232–247.
UNEP (1994) Assessment and monitoring of climatic change impacts on mangrove ecosystems. *UNEP Regional Seas Reports and Studies* No 154, 61 pp.
United Nations Population Division (1999) *World Population Prospects: the 1998 Revision.* United Nations Publication (ST/ESA/SER.A/176) Sales Number E99.X111.6. United Nations.
Walsh, G., Snedaker, S. and Teas, H. (eds.) (1975) *Proceedings of the International Symposium on Biology and Management of Mangroves.*

Vols. 1 and 2. Institute of Food and Agricultural Sciences. University of Florida, Florida. 846 pp.

Walters, B.B. (1997) Human ecological questions for tropical restoration: experiences from planting native upland trees and mangroves in the Philippines. *Forest Ecology and Management* **99**, 275–290.

Watson, J.G. (1928) Mangrove Forests of the Malay Peninsular. *Malayan Forest.* Rec. No 6. Fraser and Neave, Singapore, 276 pp.

Wetlands International Asia-Pacific (1997) Community Participation in Mangrove Forest Management and Rehabilitation in Southern Thailand. Second Interim Report (ECU funded project), 26 pp.

Wolanski, E., Mazda, Y., King, B. and Gray, S. (1990) Dynamics, flushing and trapping in Hinchinbrook Channel, a giant mangrove swamp, Australia. *Estuarine, Coastal and Shelf Science* **31**, 555–580.

Woodroffe, C. (1992) Mangrove sediments and geomorphology. In *Tropical Mangrove Ecosystems,* eds. A.I. Robertson and D.M. Alongi. Coastal and Estuarine Studies 41. American Geophysical Union, Washington, DC, pp. 7–41.

## THE AUTHOR

**Colin D. Field**

*Faculty of Science (Gore Hill),*
*University of Technology, Sydney,*
*P.O. Box 123, Broadway NSW 2007,*
*Australia*

Chapter 109

# CORAL REEFS: ENDANGERED, BIODIVERSE, GENETIC RESOURCES

Walter H. Adey, Ted A. McConnaughey, Allegra M. Small and Don M. Spoon

Coral reefs have the highest overall biocomplexity and species density of any set of ecosystems (biomes) and may even have the greatest number of species. In the past several decades, many serious anthropogenic degrading factors relating to coral reefs have been identified. However the physiology of carbonate construction leaves reefs particularly prone to degradation by large-scale eutrophication and global climatic change. Coral reef ecosystems are especially rich in species specializing in chemical warfare and defence, and as such, have the potential to provide numerous pharmaceutical compounds.

The laboratory culture of coral reef ecosystems (microcosms) and their constituent organisms has now been brought to pilot scale. Thus, the joining together of pharmaceutical research and development with coral reef conservation can be mutually beneficial and may help prevent the extinction of many valuable species.

*Seas at The Millennium: An Environmental Evaluation (Edited by C. Sheppard)*

## CORAL REEFS, PINNACLES OF GLOBAL BIODIVERSITY

Coral reefs have long been known as the "rainforests of the sea" as a result of their high species diversity (Stone, 1995; Maragos et al., 1996). A recent published estimate proposed that coral reefs have approximately 1 million species, considerably higher than previous estimates, with only about 10% described (Reaka-Kudla, 1997). However, based on the diversity of a well-established Caribbean coral reef microcosm, Small et al. (1998) calculated global coral reef biodiversity to be, conservatively, 3 million species. In a similarly conservative accounting, it has been estimated that rainforests, the most diverse terrestrial biome, contain over 2 million species (Reaka-Kudla, 1997). Since rainforest area is 20 times larger than that of coral reefs (Reaka-Kudla, 1997), these shallow marine ecosystems have the highest species density of any set of ecosystems on earth (Small et al., 1998). Even by the simplest biodiversity measure—the number of species—coral reefs may be the most diverse of the earth's biomes.

The great diversity of rainforests lies in the chlorophytic higher plants and the insects of the Phylum Arthropoda (Broombridge, 1992). By contrast, the diversity of coral reefs is distributed equivalently across over 40 phyla (30 animal phyla) and numerous taxonomic subgroups (Paulay, 1997; Small et al., 1998) (Table 1). Consequently, coral reefs have a higher genetic diversity (the most basic level of biological diversity) than that of tropical forests (Gray, 1997). Species-dense ecosystems such as coral reefs, provide for intensive interspecific interaction, which in turn leads to defensive and offensive "chemical warfare." In coral reefs, the chemical information from which this warfare is conducted, being multi-phyletic, is broader than it is in the rainforests. The extremely high species diversity, species density, phyletic diversity, and genetic diversity of coral reef ecosystems creates a base of genetic information that could be valuable to the discovery of new pharmaceuticals and therefore, critical to the future health of humanity.

In this chapter we will discuss how, in this important marine habitat, the calcium carbonate-created spatial heterogeneity of coral reefs—a major factor in high biodiversity—is particularly endangered by global change, especially eutrophication. We also point out that major advances in pilot-scale coral reef culture systems have been developed in the past several decades. Thus a practical basis for the bioprospecting and harvesting of coral reefs for their very large pharmaceutical value, without disturbance of the wild ecosystems, now exists. Since such harvesting can now occur at the level of genetic information, rather than biomass removal, this process can provide a major incentive for international-level conservation of coral reef ecosystems.

Table 1

A comparison of the dominant phyla of plants, animals and protists that occur in coral reefs and rainforests. Bacteria are not included. Minor to absent phyla in both ecosystems (e.g. brachiopods in modern reefs and gymnosperms in rainforests) are not included. Total phyla included: 25 for coral reefs, 14 for rainforests. Dominant phyla: 20 for coral reefs, 7 for rainforests. Exclusive phyla: 13 for coral reefs, 4 for rainforests. Note that 10 dominant reef phyla evolved during the Pre-Cambrian. Only a single rainforest phylum (true fungi) evolved during the Pre-Cambrian, while the most important element of rainforests (flowering plants) is Cenozoic in origin. Insects (the second most important element in rainforests, and a class of the Phylum Arthropoda) evolved in the late Paleozoic but did not explode in biodiversity until a complex of relationships with flowering plants developed in the Cenozoic.

| Phylum (or equivalent) | Coral Reefs | Rain Forests |
|---|---|---|
| **Protista** | | |
| Cyanophycota (blue-green algae, cyanobacteria) | PC | ✔ |
| Phaeophycota (brown algae) | PC | – |
| Baccilariophycota (diatoms) | PC | – |
| Pyrrophycota (dinoflagellates) | PC | – |
| Rhodophycota (red algae) | PC | – |
| Oomycetae (aquatic fungi) | PC | ✔ |
| Eumycetae (true fungi) | ✔ | PC |
| Sarcomastigophora (flagellated & amoeboid protozoa) | PC | ✔ |
| Ciliophora (ciliated protozoa) | PC | ✔ |
| **Green Plants** | | |
| Chlorophycota (green algae) | PC | ✔ |
| Ulvophyta | PAL | – |
| Marchantiomorpha (liverworts & hornworts) | – | PAL |
| Bryophyta (mosses) | – | PAL |
| Filicophyta (ferns) | – | PAL |
| Magnoliophyta (flowering plants) | ✔ | M/C |
| **Animalia** | | |
| Porifera (sponges) | PC | – |
| Cnidaria (coelenterata) | PAL | – |
| Sipunculida | PAL | – |
| Mollusca (molluscs) | PAL | ✔ |
| Annellida (segmented worms) | PAL | ✔ |
| Arthropoda: Crustacea | PAL | – |
| Insecta | – | PAL |
| Arachnida | ✔ | PAL |
| Bryozoa (moss animals) | PAL | – |
| Platyhelminthes (flat worms) | PAL | ✔ |
| Nemertea (ribbon worms) | PAL | – |
| Nematodea (round worms) | PAL | ✔ |
| Echinodermata | PAL | – |
| Tunicata | PAL | – |
| Vertebrata | PAL | PAL |

– = Absent; ✔ = present, but minor; PC = evolved in the Precambrian; PAL = evolved in the Paleozoic; M/C = evolved in the Mesozoic/Cenozoic.

## CORAL REEF BIODIVERSITY IS BASED IN CALCIUM FRAMEWORK BUILDING

Many factors lead to elevated levels of biodiversity in ecosystems. Chief among these are moderate levels of disturbance and primary production (or energy availability) (Connell, 1978; Huston, 1994). High levels of disturbance (e.g., elevated temperatures, extreme wave action, overgrazing) limit the number of species to those few capable of handling the extremes. Low levels of primary productivity reduce the food supply required to deliver energy to a complex trophic structure or food web. Under what would appear to be optimal conditions, low disturbance and high productivity, a few species that specialize in very rapid growth and reproduction can totally dominate a habitat to the exclusion of many other species. Coral reefs have moderately high levels of primary production (Adey 1998), as compared to many other ecosystems, and when best developed are characterized by moderate wave energy disturbance (strong seasonal trade winds and periodic cyclonic storms).

Elevated spatial heterogeneity is key to expanding the number of niches available for many species. A major factor in the development of the high biodiversity of rainforests, for example, lies in complex woody and leafy structures, with myriad surfaces and spaces, developed by forest trees. Likewise, well-developed coral reef habitats are characterized by extensive calcium carbonate structures with a superabundance of surfaces and spaces, many of them internal or cryptic. Although extensive cellulosic structures can develop in the sea, these are ephemeral (e.g., even the large brown algae of "kelp forests" are quickly delivered to shorelines during storms; also overgrazing by urchins periodically reduces these same kelp forests to coralline algae and urchin "barrens"). The calcium carbonate frameworks of stony corals and coralline algae by contrast, can effectively build open ocean breakwaters and develop significant geologic structures that can last for hundreds of millions of years. Thus, like rainforests on land, coral reefs in the sea have the same basic requisites for an elevated biodiversity, in part driven internally by the primary organic structuring elements. However, calcium carbonate is much denser than cellulose; its rapid growth, seen only in coral reefs, has some serious limitations. In the sea, the branch tips of many non-calcified algae can grow at rates of centimetres per day, whereas the fastest growing branch tips of stony corals are measured as centimetres per month.

Many aquatic animals and a considerable number of algae build significant hard structures of calcium carbonate for protection (shells) and support (skeleta), and the basic physiological and structural conditions for calcification (carbonate formation) by organisms have been extensively studied (Simkiss and Wilbur, 1989). While carbonate accumulations as shells and skeleta in the biosphere and in geological time are well-known (and are the basis for much palaeontology), rates of accumulation are an order of magnitude or more below that of coral reef construction (Milliman, 1993). Also, even though most surface waters of open tropical seas are saturated with calcium carbonate, chemical precipitation rates are far below those on coral reefs, where organisms are physiologically directly involved in causing precipitation (Milliman, 1993).

## THE PHYSIOLOGY OF CALCIFICATION

Many organisms, including algae and invertebrate species from numerous phyla, calcify in coral reefs, and some reefs are built mostly by algae such as red corallines and green *Halimeda*. However, the principal structure creating (bioherms), and spatial heterogeneity creating, calcium carbonate formation is accomplished by scleractinian (stony) corals. It is generally accepted that in most cases stony corals build frameworks, with the infill provided by other species.

The most complete model for biological calcification has emerged from geochemical studies along with physiological studies of corals and calcareous algae (reviewed by McConnaughey and Whelan, 1997). These studies determined that molecular $CO_2$ (at least initially) provides most of the carbon for $CaCO_3$ precipitation, and that proton removal from the site of calcification probably occurs in exchange for calcium. This is consistent with the scenario illustrated in Fig. 1a, in which the cell membrane-associated ion pump $Ca^{2+}$/ATPase expels $Ca^{2+}$ from the cytoplasm in exchange for protons (Niggli et al., 1982). The proton translocation in effect creates a carbon trap. $CO_2$ diffuses across the phospholipid cell membranes and ionizes in the alkaline extracellular (but internal) fluid to produce $CO_3^{2-}$, which cannot diffuse back across the membrane. The accumulation of $CO_3^{2-}$ largely drives $CaCO_3$ to supersaturation and thus precipitation. This basic biochemical system in other organisms has been adapted to precipitate calcium phosphate (as is present in our bones) by adding an anion exchange protein to the cell membrane (McConnaughey, 1989). In the case of stony corals, the pumping of protons away from the skeleton into polyp coelenterons brings the protons into contact with the abundant bicarbonate ($HCO_3^-$) of sea water. As shown in Fig. 1a, the neutralization of $HCO_3^-$ results in the production of excess $CO_2$. The zooxanthellae symbiotic within coral tissues provide a "pull" component (by removing accumulating $CO_2$) to the basic "push" of the ATP-ase ion pumping process. As microcosm experiments have shown (Small and Adey, 2000), free-living algal turfs abundantly present on "dead" reef carbonate significantly assist the "pull" component, further driving calcification.

## CALCIFICATION AND NUTRIENT UPTAKE

Calcification obviously helps organisms to support and defend themselves, but its benefits apparently extend well beyond skeletogenesis. As we have discussed, the hydrogen ions from calcification can be used to convert

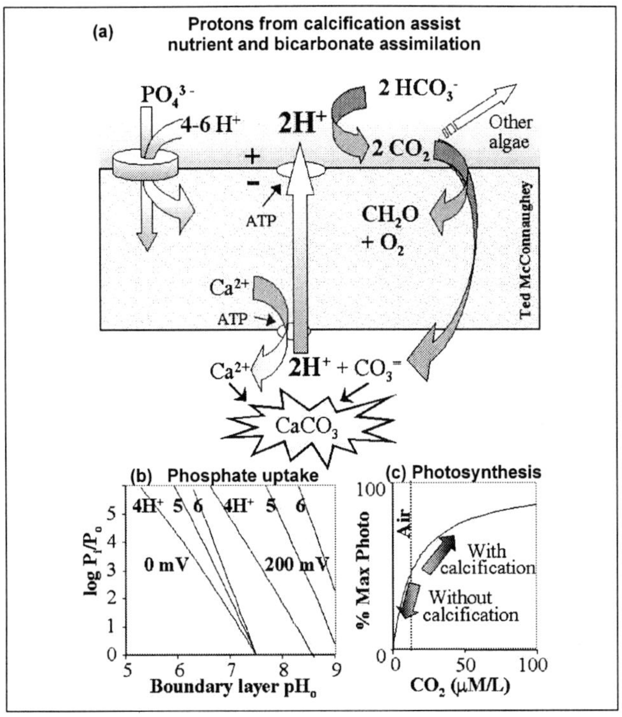


Fig. 1. Calcification physiology. (a) Calcification mechanisms. $Ca^{2+}$ expulsion from the cell in exchange for protons creates an alkaline fluid that absorbs $CO_2$ and precipitates $CaCO_3$. The protons from calcification are expelled at a separate surface, where they protonate $HCO_3^-$ to produce $CO_2$, and participate in nutrient uptake processes. (b) Proton–nutrient co-transport energetics. Calculations of potential phosphate concentration factors for membrane potentials of 0 and 200 mV, assuming a cytoplasmic pH of 7.5. (c) $CO_2$ levels largely determine photosynthetic efficiency. Elevating $CO_2$ by protonating $HCO_3$ potentially stimulates photosynthesis considerably.

bicarbonate to carbon dioxide, the carbon form most readily assimilated by photosynthesis. This can be important because $CO_2$ levels largely determine photosynthetic efficiency (Fig. 1c). However, proton secretion also plays key roles in nutrient uptake. Simply put, the chemical work required to import a molecule across the cell membrane is $RT \ln a_i/a_o$, where $R$ is the gas constant, $T$ is Kelvin temperature, and $a_i$ and $a_o$ are the activities of the chemical inside and outside of the cell. The electrical work is $ZFV$, where Z is the charge on the molecule, $F$ is the Faraday constant, and $V$ is the cell's membrane electrical potential. A passive ion transporter that imports $n$ protons with a nutrient ion can theoretically achieve a nutrient concentration ratio of:

$$\log(C_i/C_o = n(pH_i - pH_o) + (n + Z)FV/2 \cdot RT$$

$V$ is represented as a positive voltage here so that the two terms on the right are additive, and ion activity coefficients are assumed to be similar inside and outside of the cell. Figure 1a,b illustrates the use of this equation for phosphate, which appears to be taken up in high-affinity co-transport with 4–6 protons (Ullrich-Eberius et al., 1984; Sakano, 1990; Schachtman et al., 1998). These calculations indicate how cells can maintain millimolar internal phosphate levels even when surrounding waters contain less than 50 nanomolar phosphate, as is typical around coral reefs (McConnaughey et al., in press). One indication that nutrient uptake significantly motivates coral calcification is the observation that corals sometimes calcify fastest when nutrients are scarce (Marubini and Davies, 1996).

This relationship of calcification and nutrient uptake is not unique to coral reefs. Many land plants growing under nutrient deficient conditions also appear to use calcification in one form or another. Some hyperaccumulate calcium salts, to the extent that cacti and aloes for example sometimes consist largely of organic calcium salts (Meyer and Popp, 1997; Trachtenberg and Mayer, 1991). Some calcify within the vacuoles of root cortical cells or other tissues, sometimes largely filling these tissues with calcite (Jaillard et al., 1991). Most enter into symbiotic relations with soil fungi, many of which precipitate calcium oxalate (Arnott, 1995). All of these physiologies may be viewed as proton generators in the sense illustrated in Fig. 1a.

Kleypas et al. (1999) suggested that the increase in atmospheric $CO_2$ levels is reducing the degree of $CaCO_3$ supersaturation in the oceans, and that this reduces reef calcification rates. However, most biological calcification occurs within somewhat isolated compartments, as beneath the skeletogenic epithelium of corals. These fluids become highly supersaturated mainly as a result of ion pumping by the calcifying cells, and the counteracting effects of increasing $CO_2$ levels are comparatively minor. Increasing $CO_2$ levels might simply reduce the incentive to calcify, or perhaps more critical, increase the competitiveness of free-living algae on coral reefs (as we discuss below).

Nutrient enrichment (other than $CO_2$) is a more serious threat to coral reefs. Non-calcareous "fleshy" macroalgae are now proliferating along most populated coastlines, while coral and algal turf abundance is simultaneously declining (Adey, 1998). Several interactions appear likely here. Larger, fleshy algae and corals compete for space, and many fleshy algae are quite capable of overgrowing corals if not restricted in their growth potential. Increased algal densities encourage generalist consumers such as sea urchins, which then feed on the corals too. At a more basic ecological level, nutrient enrichment encourages the growth of opportunistic, fast-growing species, that build largely cellulosic structures. Slow-growing stony (calcium carbonate structured) corals, exquisitely adapted for a nutrient-deficient environment, are simply overwhelmed by potentially fast-growing non-calcified macroalgae, freed of their severe nutrient constraints (Hughes, 1994; Adey, 1998).

## CORAL REEFS ARE ENDANGERED

Coral reefs are exceptionally sensitive to both direct and indirect human exploitation and worldwide are showing signs of deterioration (Maragos et al., 1996). Models of the effects of habitat destruction on biodiversity show that, coral reefs display extremely high local species extinction rates with only small levels of destruction, whereas tropical forests do not reach the same levels of extinction until destruction is much more severe (Stone, 1995). Cases of direct destruction of coral reefs, involving channelization, carbonate mining, blasting, etc., are certainly obvious and well known. However, unfortunately for reef conservation, the early stages of most coral reef degradation are subtle, being the result mainly of over-fishing (Roberts, 1995), nutrient increase (accompanied by expansion of area covered by larger algae) (Hallock et al., 1993) and global change temperature increase (Pockley, 1999). These key problems plaguing coral reefs are very different from those involved in rainforest degradation (farming and logging), particularly in that the latter can easily be seen on site or from satellites. Through the popular conservation media and news outlets to the general public, a wide audience has been privy to the serious problems of tropical rainforests, whereas the dangers to coral reefs are considerably less widely known.

A worldwide coral reef initiative (Crosby et al., 1995) was declared to analyze the damage, propose solutions, and initiate appropriate conservation programs. However, coral reefs are also subject to significant natural perturbations, such as hurricanes, which can locally greatly reduce their short-term status as complex ecosystems (Hughes, 1994). While such natural periodic stresses are important to maintaining the broader scale of coral reef biodiversity in the long term, they have sometimes confused conservation issues (just as natural temperature fluctuations have tended to mask broad-scale global warming) and often led to scientific disputes that have prevented effective action. Nevertheless, among reef scientists and managers, there is little dispute today as to the linkage between anthropogenic global change and reef ecosystem health and biodiversity (Maragos et al., 1995; Buddemeier and Lasker, 1999; Kleypas et al., 1999; Pockley, 1999).

Coral reef scientists have been slow to describe the impending endangerment of coral reef ecosystems. While eight International Coral Reef Symposia have been held since 1969, the proportion of papers devoted to degradation and conservation issues stayed low until the present decade. By 1993, it was reported that ten percent of reefs were irretrievably lost to coastal siltation and human eutrophication that allows algal growths to smother and kill the reef building communities (Wilkinson, 1993). Another 30% were deteriorated and endangered to the critical state, with an additional 30% threatened. In addition to these broader-scale degenerative factors, dynamite, electric shock, chemicals (like cyanide and rotenone), fish traps, and over-fishing in general were characterized as locally destroying and depleting the fish and shellfish predator species that are important in maintaining the health and continuation of coral reef communities. Coral reef ecosystems are subject to considerable "top down" control by their fish populations. Significant disruptions can result from selective fishing; however equally important broad-scale over-fishing required to support either an expanding native population or imported tourist populations can be devastating (Roberts, 1995). With 70% of the world's reefs predicted to be completely destroyed in the next 40 years if drastic measures are not taken, we will be losing large numbers of species to extinction (Wilkinson, 1993). While massive global species extinction appears to be inevitable in the 21st century, the case of coral reefs will represent an ecocide at the biome level.

Based on 65 pan-tropical samples covering a period of four years, Connell (1997) showed that coral cover declined and did not recover in 42% of the examples (16% Indo-Pacific; 26% W. Atlantic). Connell (1997) also noted that recovery occurred in 69% of the acute, short-term disturbances, but in only 27% of the chronic, long-term disturbances. Although there is some concern as to the representativeness of the samples, these findings suggest that in 20 years (less than 10 years in the W. Atlantic) greater than 80% of reef sites will have shown a significant decline in coral cover without recovery. Jackson (1997) contends that when a significant level of coral reef research began, roughly in mid-century, major degradation had already occurred. He concludes that it is unlikely that Caribbean coral reefs can be "saved" and that "marine protected areas on the scale of hundreds to thousands of square kilometres are" (i.e., would be) "vital to any hope of conserving Caribbean coral reefs and coral reef species." At a World Bank symposium on "sustainable management" of coral reefs, Knowlton (1997) concluded that Caribbean reefs had "collapsed" and percent coral cover was on a steep decline to virtually zero levels (Fig. 2). In this model, oceanic Pacific reefs (the least stressed) had not yet collapsed but were approaching that collapse threshold. In a report to Greenpeace, Hoegh-Guldberg (see Pockley, 1999) identified global warming and high sea surface temperatures, followed by coral bleaching (loss of zooxanthellae) as a significant factor in coral reef degeneration. Hoegh-Goldberg predicted the elimination of reef ecosystems by 2100, with the world's largest single coral reef ecosystem, The Great Barrier Reef, seriously degraded within two to three decades.

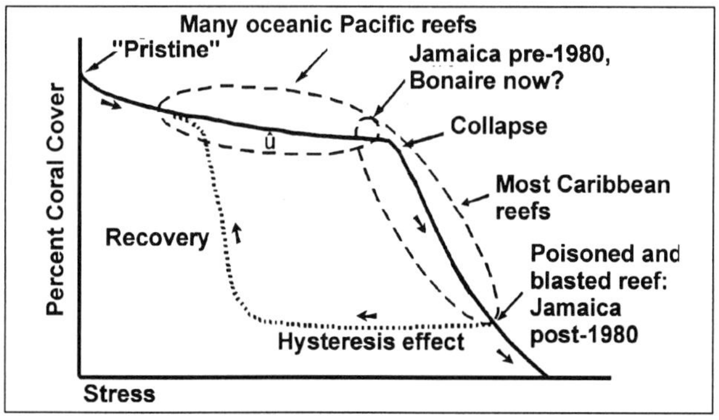


Fig. 2. Threshold effects, from Knowlton (1997). Note, the hysteresis effect suggests that considerable stress removal is required before severely damaged reefs can begin the process of recovery.

## THE ROLE OF NUTRIENTS IN CORAL REEF DEGRADATION

There has been considerable discussion in the scientific literature as to the nutrient (N, P, $CO_2$) sensitivity of reefs (Delgado and Lapointe, 1994; Snidvongs and Kinzie, 1994; Atkinson et al., 1995; Marubini and Davies, 1996; Kleypas et al., 1999). Most of the earlier work assumed a direct relationship between locally derived eutrophication and reef degradation. A long-term study at Kaneohe Bay in Hawaii has shown degradation and then partial recovery with the addition and removal of sewage inputs (Coles and Ruddy, 1995). However, until recently, no widely accepted understanding of the basis for this sensitivity has existed and some dispute as to its existence has been presented (Atkinson et al., 1995). Until the past several years, few scientists have questioned the role of anthropogenic elevated $CO_2$ (e.g., Buddemeier and Smith, 1999), even though oceanic sinks for the so-called "missing" anthropogenic carbon have long been postulated (see e.g. Frankignoulle et al., 1996). More recently, McConnaughey et al. (in press) have demonstrated a linkage between calcification and its related proton release and enhanced nutrient uptake (Fig. 1). The ability of the dominant reef calcifiers to compete with larger macroalgae is in large measure dependent upon the maintenance of extremely low nutrient (oligotrophic) conditions. While $CO_2$ increase may directly enhance the growth of algal competitors to a small degree, increased nutrient concentration (especially of phosphorus and iron) removes the competitive nutrient advantage of calcifiers. As we discussed briefly above, photosynthesis by free-living algal turf communities rich in nitrogen-fixing blue-green cyanobacteria (which occupy a large part of most coral reef surfaces; Adey, 1998), enhances coral calcification. As long as grazing fish are abundant and ambient nutrients are low, larger macro-algae are prevented from achieving dominance over both algal turfs and stony corals.

Many other global anthropogenic factors could also relate to coral animal destruction and consequent reef biodiversity decline, but cause and effect are poorly understood. In the last several decades, widespread coral bleaching (the general and rapid loss of zooxanthellae from corals) frequently resulting in coral death has occurred (Brown, 1997; Pockley, 1999). It has been proposed that zooxanthellae expulsion is an adaptive feature of coral evolution that is taken beyond the physiological limits of the animals (Ware et al., 1996), and is typically due to elevated temperatures (presumably due to $CO_2$-induced global warming) (Jones et al., 1997; Huang et al., 1998). Elevated ultraviolet radiation has also been proposed as a supporting factor (Lesser, 1997). Hoegh-Gulding (see Pockley, 1999) has reviewed the relationship of ultraviolet radiation and temperature, identified the critical change in zooxanthellae chemistry, and pointed out the role of El Niño events in causing mass coral bleaching. Wide-spread cases of coral diseases, such as black band disease, caused by a cyanobacterium, have also been documented (Hallock et al., 1993; Peters, 1997). These are more weakly supported as anthropogenic factors, but may relate to excess $CO_2$ availability, and resulting increased mucous production.

## THE SOCIAL AND ECONOMIC VALUE OF CORAL REEFS

The economic value of coral reefs for tourism, the aquarium and diving hobby/recreation industries and as a source of seafood is critical for many countries, especially small island nations (Broombridge, 1992). At a World Bank conference held in 1997, participants identified ways in which underdeveloped countries could further utilize sustainable economic potential from coral reef ecosystems (Hatziolos et al., 1997). However, many scientists expressed concern as to whether or not it is possible to have adequate "sustainable" management in the case of these very sensitive ecosystems. A lesser known value of coral reefs, though of great importance to the health of human beings, is their potential as a source of pharmaceuticals. If fully understood and made known to a broad public, this aspect of coral reefs could provide the international economic incentive needed to conserve these endangered ecosystems.

## NATURAL PRODUCTS CHEMISTRIES AND PHARMACEUTICALS

Nearly half of the best-selling pharmaceuticals in use in 1991 were from natural products, or chemically derived from them (O'Neill and Lewis, 1993). Most of these came from defensive, secondary metabolites of terrestrial plants, bacteria, and fungi and were discovered by cell-based *in vitro* assays. Many of the drugs derived from terrestrial plants had a long historical foundation, and continued use is based on empirical medical practices (Kinghorn and Balandrin, 1993). Modern, mechanism-based screening for new drugs from natural products, utilizing cellular or biochemical targets in addition to cell-based assays, has additionally yielded many useful therapeutic agents now in clinical use or in trials (Shu, 1998).

Modern methods for drug discovery using combinatorial chemistry and molecular modelling designs using computers have not superseded natural products as a source of new pharmaceuticals. A recent review (Cragg et al., 1996) of anti-cancer and anti-infective agents documents that over 60% of approved drugs and those in late stages of clinical trials are either from natural products, or are chemically derived from, or modeled on natural products (this comparison did not include biologics such as monoclonals and vaccines). A "hit" rate of commercial products from an estimated 12,000 plant species (mainly temperate) from the U.S. National Cancer Institute cell-based anti-cancer assays from 1960 to 1982, was one in 4000 species (Cragg et al., 1998). Although the path of discovery from natural product to viable drug has been shortened in recent years, it

| | |
|---|---|
| | Research continues to improve effectiveness and reduce side effects. |
| 1994 | First total synthesis of Taxol by K. Nicolaou and R. Guy at Scripps Research Institute and R. Holton at Florida State University. |
| 1993 | Bristol-Myers Squibb announces halt to non-renewable wild collection, now using a newer semi-synthesis method of I. Ojima, University of New York at Stony Brook, and R. Holton, Florida State University. |
| 1989 | Bristol-Myers Squibb (BMS) begins large-scale harvesting of Pacific Yew, predicts five-year supply. |
| | Erik Rowinsky, Johns Hopkins, reports high rate of success in shrinking tumours, however bark is in short supply. |
| 1984 | Clinical trials initiated, Dana Farber Cancer Institute, Johns Hopkins Oncology Center, Memorial Sloan-Ketlering Cancer Center. |
| | P. Poitier, National Center Scientic Research, France; A. Greane, Joseph Fourier University at Grenoble – first successful semi-synthesis from European Yew needles (which grow back); 30 active research groups working to fully synthesize |
| 1978 | Susan Horowitz, Albert Einstein College, demonstrates Taxol's attachment to micotubules and the resulting prevention of cell division. |
| 1977 | National Cancer Institute, on suggestions from M. Wall, agrees to further testing of Taxol. |
| | Early testing as anti-cancer drug |
| 1967 | M. Wani and M. Wall, Research Triangle Institute isolate 112 atom active compound from Arthur Barclay's (USDA) Yew bark and call it Taxol. |
| 1962 | Arthur Barclay, USDA, collects bark from Pacific Yew (*Taxus brevifolia*) for testing. |
| Pre-modern | Julius Cesar, 51 BC, describes use of yew bark as a poison by Gallic tribes. Northwest American Indian Tribes use as disinfectant, abortifacient and skin cancer treatment. |

Fig. 3. Time line for the development of the anti-cancer drug Taxol from its natural product source (Nicolaou et al., 1996).

remains a long and tortuous process with no reassurance of ultimate success (Fig. 3). On the other hand, according to one recent estimate, only 5–15% of 250,000 higher plant species have been assayed for bioactive compounds (Balandrin et al., 1993) and in comparison, the assay of marine species has barely begun. Recently, the newly discovered anti-tumour agents from natural products, which includes those from marine species, have been extensively reviewed (Cragg et al., 1997). These authors found natural products to be very valuable in the development of new anti-tumour drugs.

## PHARMACEUTICALS FROM MARINE ENVIRONMENTS AND CORAL REEFS

In recent decades the search for new pharmaceuticals has shifted from our ground-based, terrestrial third of the planet's surface to the vast, phyletically rich marine environment (May, 1994). SCUBA and submersibles have greatly facilitated bioprospecting for bioactive chemicals, especially in tropical and temperate waters. There has been substantial progress toward discovery of novel chemotypes, the development of new pharmaceuticals, skin care products and sunscreens, and the discovery of important molecular probes from marine organisms (Fenical, 1996; Hay and Fenical, 1996). Even an essential infant formula additive has been derived from a marine alga (Radmer, 1996). Marine animals, such as filter-feeding sponges, bryozoans, tunicates, and grazing sea slugs, that have yielded the best results, have a combination of the following traits: sessile, stationary or slow-moving, soft-bodied without structural defences, brightly coloured, light-refracting or with contrasting markings, and possessing endosymbiotic microorganisms that can provide additional toxins. These characteristics indicate species that have put a large part of their energy into constructing chemical as opposed to other forms of defence against predators. Sometimes the defensive toxin is concentrated from the food eaten by the grazer, as in sea slugs, and even modified chemically in their tissues. Alternatively, motile and sessile predators produce toxins to catch their prey. Some of the most powerful toxins ever tested are found in marine macropredators, such as jellyfish, sea snakes, gastropod coneshells, and pufferfish (Meier and White, 1995). Toxins work by disrupting cellular functions, and, for the same reason, can provide the molecules we need to cure human diseases.

In the clear waters of coral reefs are found the world's greatest assemblages of such promising candidates for possessing secondary metabolites with useful bioactivities. Already, promising new drugs have been found in ocean invertebrates and algae, some that are in preclinical development and others beginning clinical trials (Carte, 1996; Fenical, 1996) An example is the discovery (Lindel et al., 1997) from a rare Australian soft-coral, *Eleutherobia*, of the cytotoxin, Eleutherobin, that stabilizes microtubules (like Paclitaxel [Taxol] from the yew tree). Eleutherobin was found to have similar anti-cancer potency and selectivity as Taxol.

In general, the higher the species density of an ecosystem, the greater the likelihood for the development of chemical offences and defences by organisms for survival. The very high species densities of coral reefs provide "breeding grounds" for the evolution of "super K" species—those species which are especially successful in putting a large part of their energy into chemical defences rather than into reproduction and food gathering.

## THE IDENTIFICATION AND EXTRACTION OF NATURAL PRODUCTS FROM CORAL REEFS

The identification and extraction of natural products from coral reefs require major search and collection efforts, many already underway by pharmaceutical companies. Because powerful toxins and defensive compounds are usually found in very small concentrations, even at nM levels, moderate quantities of each organism are needed for testing. Commercial production can require very large quantities. The potential extinction which faced the yew tree a decade ago, due to intensive collecting for extraction of Paclitaxel (Fig. 3), is likely to be repeated many-fold in coral reef environments. While the yew tree was saved from almost certain extinction by the discovery of synthetic processes, this good fortune cannot be counted on for many species.

Methods that allow the culture of the very high biodiversities of coral reefs, at the level of reproductive populations, have been developed (Small et al., 1998; Carlson, 1999), and recently expanded to production scale for the marine aquarium hobby industry (Adey and Loveland, 1998). Thus, with these advancements in aquaculture understanding, supplying the needs of even the most "difficult" of coral reef species to provide adequate and repeatable bulk for the pharmaceutical industry is possible.

The abundant microorganisms that occur in coral reef ecosystems have the advantage of reproducing rapidly and requiring little space for their maintenance. Even a small coral reef microcosm can maintain indefinitely hundreds of species of bacteria, cyanobacteria, protists and micrometazoa, including their various associated symbiotic microorganisms. Marine bacteria, and especially cyanobacteria, have only begun to be studied (Fenical, 1993). These prokaryotes have had billions of years to evolve potent secondary metabolites for their defence. Likewise, some protistan taxa trace their separate lineages to ancient ancestors over a billion years ago. Protists, such as dinoflagellates, have already provided many useful molecular probes. Many protists have effective delivery systems for their offensive and defensive toxins, called extrusomes, where the chemicals are stored and concentrated. Some protists have microorganisms living on or inside them which also possess powerful defence chemicals.

Even though not suitable for therapeutic drug use, toxins produced by dinoflagellates, with their highly specific actions and potency at very low concentrations, serve as invaluable tools for studying basic cellular processes (Carte, 1996; Fenical, 1996). An example is okadaic acid, a cycle polyether fatty acid from a red-tide-causing species of the genus *Prorocentrum*, first discovered in sponges that ingest the plankters and concentrate their toxin. Okadaic acid selectively inhibits protein phosphatase enzymes making it a powerful probe to elucidate phosphorylation–dephosphorylation events and signal transduction pathways. Another example, the neurotoxin saxitoxin, produced by several dinoflagellate genera, is concentrated in shellfish that ingest them, causing paralytic poisoning in humans. Saxitoxin's high potency to specifically block membrane sodium channels has made it very useful in basic neurophysiological studies.

Each year, hundreds of new secondary metabolites are extracted form marine and coral reef organisms. Many of these are shown to be unique and novel molecules, that serve as a resource for new and potentially indispensable molecular probes and pharmaceuticals. Mass culture of essential organisms in a microcosm setting could very well be essential to economic pharmaceutical production of a wide variety of compounds, and as we pointed out above, small-scale production systems already exist (Adey and Loveland, 1998). However, conserving the coral reef ecosystem in the wild is essential for maintaining the genetic base of information needed for the discovery of new pharmaceuticals. Biodiversity is area-based, and while microcosm culture can be used to locate new organisms and their special compounds, and can then be skewed to concentrate desired species, the larger area of the wild is essential to maintaining full biodiversity for the long term.

## CORAL REEF CONSERVATION—AN INTERNATIONAL PRIORITY

Because coral reefs have the highest species density of any biome on earth and may even have the highest number of species, the potential for exotic natural products, and especially increasingly critically-needed pharmaceutical compounds, is particularly large. Less than 10%, and perhaps less than 5%, of this biodiversity has even been described and yet, within a very few decades, the very coral framework that allows this diversity to be maintained appears doomed to serious degradation.

Scientific research of the last decade has identified the primary bases for coral reef degradation. Some of these problems are local, and can perhaps be ameliorated by the countries directly affected, with the help of conservation-directed non-governmental organizations (NGOs) and internationally operated governmental agencies with direct conservation and economic interests. However, a major component of the degrading factors facing coral reefs is global in scope, in large measure linked directly to anthropogenic global change, especially nutrient increase, and perhaps indirectly to temperature increase and ozone depletion.

Scientific research of the last few decades has also indicated the extraordinary economic value of coral reefs, particularly that inherent in the potential for pharmaceutical compounds. Aquacultural research has also shown the way to the development of the pharmaceutical potential of coral reef biodiversity without destruction of reefs themselves. Coral reef ecosystems need to be considered as genetic reserves—most valuable for the chemical information they contain. An extraordinary international conservation effort by governments in association with the pharmaceutical industry is necessary to achieve any significant progress in conserving this biodiversity-based information. "Extinction is forever" and, in this case, human health concerns (which support one of the most economically viable industries of the 20th century) are at stake. Prior to the last decade of the 20th century, it might have been difficult for world governments to accept and act on the economic value of information. This should no longer be an issue.

Only a single, extensive and well developed, large-scale reef system has been subject to a comprehensive reef conservation plan (Australian Great Barrier Reef). Even here, explosions of a coral-eating starfish (Crown-of-Thorns), and more recently extensive bleaching episodes, have led to considerable concerns on the part of many scientists that even this unparalleled local effort at reef conservation may not be enough. Elsewhere, considerable efforts have been made to conserve the most degraded of the world's reefs. Development of marine parks and diversion of pollutants have aided in slowing the decline of many of these devastated reefs. Unfortunately, with the current plan of action, it is mostly those reefs that are showing severe decline that are getting financial support and attention, at the point where biodiversity has already been lost to some degree. Extensive time and effort is being spent "saving" reefs that are already past the point of no return, in large measure because those reefs are tourism destinations. It is time to start from the other end. Only when we realize that our conservation efforts must be concentrated on the remaining healthy reefs, on those capable of revival, and on more global degrading factors, will there be a chance for conserving our world's coral reefs. Viewing coral reefs as endangered resources helps us to realize we need to protect and utilize them wisely. Direct economic value (through the pharmaceutical industry) is presently only a weak driver of conservation efforts relating to coral reefs. Indeed, in some cases today the pharmaceutical connection may be negative, due to excessive bioprospecting and over-harvest. This is particularly true because of the low "hit rate" and the long lag time to a viable product. However, getting the relevant information to the biological community and the general public could provide governmental incentives to action, as it has for the problem of chlorofluorocarbons and ozone depletion.

## REFERENCES


Adey, W.H. (1998) Coral reefs: algal structured and mediated ecosystems in shallow, turbulent, alkaline waters. *Journal of Phycology* **34**, 393–406.

Adey, W.H. and Loveland, K. (1998) *Dynamic Aquaria: Building Living Ecosystems.* 2nd Edition. Academic Press, New York.

Arnott, H.J. (1995) Calcium oxalate in fungi. In *Calcium Oxalate in Biological Systems*, ed. S.R. Khan, pp. 73–112. CRC Press, Boca Raton, FL. ISBN 0-8493-7673-4.

Atkinson, M.J., Carlson, B. and Crow, G.L. (1995) Coral growth in high-nutrient, low-pH seawater: a case study of corals cultured at the Waikiki Aquarium, Honolulu, Hawaii. *Coral Reefs* **14**(4), 215–223.

Balandrin, M.F., Kinghorn, A.D. and Farnsworth, N.R. (1993) Plant-derived natural products in drug discovery and development: An overview. In *Human Medicinal Agents from Plants*, eds. M.F. Balandrin and A.D. Kinghorn, pp. 2–12. American Chemical Society Symposium Series 534. American Chemical Society, Washington, DC.

Broombridge, B. (1992) *Global Biodiversity: Status of the Earth's Living Resources.* Chapman & Hall, London.

Brown, B.E. (1997) Coral bleaching: causes and consequences. *Coral Reefs* **16** (supp), S129–S138.

Buddemeier, R. and Smith, S. (1999) Coral adaptation and acclimatization: a most ingenious paradox. *American Zoologist* **39**, 1–9.

Buddemeier, R. and Lasker, H. (1999) Coral reefs and environmental change-adaptation, acclimation or extinction. *American Zoologist* **39**, 1–183.

Carlson, B. (1999) Organism responses to rapid change: what aquaria tell us about nature. *American Zoologist* **39**, 44–55.

Carte, B.K. (1996) Biomedical potential of marine natural products. *BioScience* **46**, 271–286.

Coles, S.L. and Ruddy, L. (1995) Comparison of water quality and reef coral mortality and growth in southwestern Kaneohe Bay, Oahu, Hawaii, 1990–1992, with conditions before sewage diversion. *Pacific Science* **49** (3), 247–265.

Connell, J.H. (1978) Diversity in tropical rain forests and coral reefs. *Science* 199, 1302–1309.

Connell, J.H. (1997) Disturbance and recovery of coral assemblages. *Proceedings of the 8th International Coral Reef Symposium* **1**, 9–22.

Cragg, G.M., Newman, D.J. and Snader, K.M. (1996) Natural products in drug discovery and development. *Journal of Natural Products* **60** (1), 52–60.

Cragg, G.M., Newman, D.J. and Weiss, R.B. (1997) Coral reefs, forests, and thermal vents: The worldwide exploration of nature for novel antitumor agents. *Seminars in Oncology* **24**, 156–163.

Cragg, G.M., Newman, D.J. and Yang, S.S. (1998) Bioprospecting for drugs. *Nature* **393**, 301.

Crosby, M.P., Drake, S.F., Eakin, C.M., Fanning, N.B., Paterson, A., Taylor, P.R. and Wilson, J. (1995) The United States coral reef initiative: An overview of the first steps. *Coral Reefs* **14** (1), 1–4.

Delgado, O. and Lapointe, B.E. (1994) Nutrient-limited productivity of calcareous versus fleshy macroalgae in a eutrophic, carbonate rich tropical marine environment. *Coral Reefs* **13** (3), 151–159.

Fenical, W. (1993) Chemical studies of marine bacteria: Developing a new resource. *Chemical Reviews* **93**, 1673–1683.

Fenical, W. (1996) Marine biodiversity and the medicine cabinet: The status of new drugs from marine organisms. *Oceanography* **9**, 23–27.

Frankignoulle, M., Gattuso, J.-P., Biondo, R., Bourge, I., Copin-Montegut, G. and Pichon, M. (1996) Carbon fluxes in coral reefs. 2. Eulerian study of inorganic carbon dynamics and measurement of air–sea $CO_2$ exchanges. *Marine Ecology Progress Series* **145** (1–3), 123–132.

Gray, J.S. (1997) Marine biodiversity: Patterns, threats and conservation needs. *Biodiversity and Conservation* **6**, 153–175.

Hallock, P., Muller-Karger, F.E. and Halas, J.C. (1993) Coral reef decline. *Research & Exploration* **9** (3), 358–378.

Hatziolos, M., Hooten, A. and Fodor, M. (eds.) (1997) *Coral Reefs, Challenges and Opportunities for Sustainable Management.* The World Bank, Washington, DC.

Hay, M.E. and Fenical, W. (1996) Chemical ecology and marine biodiversity: Insights and products from the sea. *Oceanography* **9**, 10–20.

Huang, S.P., Lin, K.L. and Fang, L.S. (1998) The involvement of calcium in heat-induced coral bleaching. *Zoological Studies* **37** (2), 89–94.

Hughes, T.P. (1994) Catastrophes, phase shifts, and large-scale degradation of a Caribbean coral reef. *Science* **265**, 1547–1551.

Huston, M. (1994) *Biological Diversity: The Coexistence of Species on Changing Landscapes.* Cambridge University Press, Cambridge.

Jackson, J.B.C. (1997) Reefs since Columbus. *Coral Reefs* **16** (supp), S23–S32.

Jaillard, B., Guyon, A. and Maurin, A.F. (1991) Structure and composition of calcified roots and their identification in calcareous soils. *Geoderma* **50**, 197–210.

Jones, R.J., Berkelmans, R. and Oliver, J.K. (1997) Recurrent bleaching of corals at Magnetic Island (Australia) relative to air and seawater temperature. *Marine Ecology Progress Series* **158**, 289–292.

Kinghorn, A.D. and Balandrin, M.F. (eds.) (1993) *Human Medicinal Agents from Plants.* American Chemical Society Symposium Series 534, Washington, DC.

Kleypas, J., Buddemeier, R., Archer, D., Gattuso, J.P., Langdon, C. and Opdyke, B. (1999) Geochemical consequences of increased atmospheric carbon dioxide on coral reefs. *Science* **284**, 118–120.

Knowlton, N. (1997) Hard decisions and hard science: Research needs for coral reef management. In *Coral Reefs, Challenges and Opportunities for Sustainable Management,* eds. M. Hatziolos, A. Hooten and M. Fodor, pp. 183–187. The World Bank, Washington, DC.

Lesser, M.P. (1997) Oxidative stress causes coral bleaching during exposure to elevated temperatures. *Coral Reefs* **16**(3), 187–192.

Lindel, T., Jensen, P.R., Fenical, W., Long, B.H., Casazza, A.M., Carboni, J. and Fairchild, C.R. (1997) Eleutherobin. A new cytotoxin that mimics Paclitaxel (Taxol) by stabilizing microtubules. *Journal of the American Chemical Society* **119**, 8744–8745.

Maragos, J.E., Crosby, M.P. and McManus, J.W. (1996) Coral reefs and biodiversity: A critical and threatened relationship. *Oceanography* **9** (1), 83–99.

Marubini, F. and Davies, P.S. (1996) Nitrate increases zooxanthellae population density and reduces skeletogenesis in corals. *Marine Biology* **127**, 319–328.

May, R.M. (1994) Biological diversity: Differences between land and sea. *Philosophical Transactions of the Royal Society of London B* **343**, 105–111.

McConnaughey, T.A. and Whelan, J.F. (1997) Calcification generates protons for nutrient and bicarbonate uptake. *Earth-Science Reviews* **42**, 95–117.

McConnaughey, T.A., Adey, W. and Small, A. (in press ) Coral reefs: Fertilization endangers calcareous specialists. *Limnology and Oceanography.*

McConnaughey, T.A. (1989) Biomineralization mechanisms. In *Origin, Evolution, and Modern Aspects of Biomineralization in Plants and Animals,* ed. R.E. Crick, pp. 57–73. Plenum, New York.

Meier, J. and White, J. (eds.) (1995) *Handbook of Clinical Toxicology and Animal Venoms and Poisons.* CRC Press, New York.

Meyer, A.J. and Popp, M. (1997). Free $Ca^{2+}$ in tissue saps of calciotrophic CAM plants as determined with $Ca^{2+}$ selective electrodes. *Journal of Experimental Botany* **48**, 337–344.

Milliman, J.D. (1993) Production and accumulation of calcium carbonate in the ocean: Budget of a non-steady state. *Global Biogeochemical Cycles* **7** (4), 927–957.

Nicolaou, K., Guy, R.K. and Poitier, P. (1996) Taxoids: New weapons against cancer. *Scientific American* (June) 94–98.

Niggli, V., Sigel, E. and Carafoli, E. (1982) The purified Ca++ pump of human erythrocyte membranes catalyzes an electroneural $Ca^{++}$–$H^{+}$ exchange in reconstituted liposomal systems. *Journal of Biological Chemistry* **257**, 2350–2356.

O'Neill, M.J. and Lewis, J.A. (1993) The renaissance of plant research in the pharmaceutical industry. In *Human Medicinal Agents from Plants,* eds. A.D. Kinghorn and M.F. Balandrin, pp. 48–55. American Chemical Society Symposium Series 534. Washington, D.C.

Paulay, G. (1997) Diversity and distribution of reef organisms. In *Life and Death of Coral Reefs,* ed. C. Birkeland, pp. 298–253. Chapman and Hall, New York.

Peters, E. (1997) Diseases of coral reef organisms. In *Life and Death of Coral Reefs,* ed. C. Birkeland, pp. 114–139. Chapman and Hall, New York.

Pockley, P. (1999) Global warming could kill most coral reefs by 2100. A review of Climate Change, Coral Bleaching and the future of the World's Coral Reefs by O. Hoegh-Guldberg. A report to Greenpeace. *Nature* **400**, 98.

Radmer, R.J. (1996) Algal diversity and commercial algal products. *BioScience* **46**, 263–270.

Reaka-Kudla, M.L. (1997) The global biodiversity of coral reefs. A comparison with rain forests. In *Biodiversity II: Understanding and Protecting Our Natural Resources,* ed. M.L. Reaka-Kudla, D.E. Wilson and E.O. Wilson, pp. 83–108. Joseph Henry/National Academy Press, Washington, DC.

Roberts, C. (1995) Effects of fishing on the ecosystem structure of coral reefs. *Conservation Biology* **9**, 65–91.

Sakano, K. (1990) Proton/Phosphate stoichiometry in uptake of inorganic phosphate by cultured cells of *Catharanthus roseus* (L.) G. Don. *Plant Physiology* **93**, 479–483.

Schachtman, D.P., Reid, R.J. and Ayling, S.M. (1998) Phosphorus uptake by plants: from soil to cell. *Plant Physiology* **116**, 447–453.

Shu, Y.-Z. (1998) Recent natural products based drug development: A pharmaceutical industry perspective. *Journal of Natural Products* **61**, 1053–1071.

Simkiss, K. and Wilbur, K.M. (1989) *Biomineralization: Cell Biology and Mineral Deposition.* Academic Press, San Diego.

Small, A.M., Adey, W.H. and Spoon, D.M. (1998) Are current estimates of coral reef biodiversity too low? The view through the window of a microcosm. *Atoll Research Bulletin* **458**, 1–20.

Small A. and Adey, W. (2000) Reef corals, zooxanthellae and free-living algae: a microcosm study demonstrates synergy between calcification and primary production. *Ecological Engineering.* In press.

Snidvongs, A. and Kinzie III, R.A. (1994) Effects of nitrogen and phosphorus enrichment on in vivo symbiotic zooxanthellae of *Pocillopora damicornis. Marine Biology* **118** (4), 705–711.

Stone, L. (1995) Biodiversity and habitat destruction: A comparative study of model forests and coral reef ecosystems. *Proceedings of the Royal Society of London B* **261**, 381–388.

Trachtenberg, S. and Mayer, A.M. (1981) Calcium oxalate crystals in *Opuntia ficus-indica* (L.) Mill. *Protoplasma* **109**, 271–283.

Ullrich-Eberius, C.I., Novacky, A. and van Bel, A.J.E. (1984) Phosphate uptake in Lemna gibba G1: energetics and kinetics. *Planta* **161**, 46–52.

Ware, J.R., Fautin, D.G. and Buddemeier, R.W. (1996) Patterns of coral bleaching: Modelling the adaptive bleaching hypothesis. *Ecological Modelling* **84** (1–3), 199–214.

Wilkinson, C. (1993) Coral reefs are facing widespread extinction's: Can we prevent these through sustainable management practices? In *Proceedings of the 7th International Coral Reef Symposium,* ed. R.H. Richmond, pp. 11–21, University of Guam Press, Guam.

## THE AUTHORS

**Walter H. Adey**
*Marine Systems Laboratory, Smithsonian Institution, NHB E-117, MRC 164, Washington, DC 20560, U.S.A.*

**Ted A. McConnaughey**
*Marine Systems Laboratory, Smithsonian Institution, NHB E-117, MRC 164, Washington, DC 20560, U.S.A.*

**Allegra M. Small**
*Marine Systems Laboratory, Smithsonian Institution, NHB E-117, MRC 164, Washington, DC 20560, U.S.A.*

**Don M. Spoon**
*Marine Systems Laboratory, Smithsonian Institution, NHB E-117, MRC 164, Washington, DC 20560, U.S.A.*

Chapter 110

# WORLD-WIDE CORAL REEF BLEACHING AND MORTALITY DURING 1998: A GLOBAL CLIMATE CHANGE WARNING FOR THE NEW MILLENNIUM?

Clive R. Wilkinson

The most devastating coral bleaching and mortality event coincided with the severe El Niño and La Niña events of 1997–98 and has changed the agenda for assessing the threats to coral reefs around the world. Prior to 1998, the major threats were direct anthropogenic stresses (increased sedimentation and pollution, and over-exploitation). Then the events of 1998, in particular, focused attention on the potential for Global Climate Change to devastate coral reefs in the immediate future. Coral bleaching was associated with both the El Niño and La Niña conditions, and it was more the extent of the change than the nature of the change, that resulted in the massive bleaching and subsequent mortality.

In some areas, bleaching resulted in only minor levels of coral mortality (e.g. Caribbean, Great Barrier Reef), whereas in other areas there were massive levels of mortality and local extinctions of coral species (e.g. Maldives, Chagos, Palau, southern Japan). Many of the severely affected reefs will take decades to possibly hundreds of years to recover to the previous levels of coral cover and community structure, provided that there is no repetition of this severe bleaching event. Unfortunately the 'business as usual' scenario for the release of greenhouse gases points to a very likely probability of repeated and possibly more severe bleaching events in the first decades of the new Millennium.

*Seas at The Millennium: An Environmental Evaluation (Edited by C. Sheppard)*

## THREATS TO CORAL REEFS PRIOR TO 1998

During the 1980s, it became obvious that many coral reefs around the world were in 'poor health' with declines in coral cover, fish populations and incidences of diseases (Salvat, 1992). The first global report was published in three volumes summarising the status of reefs country by country (Wells et al., 1988), however there were large gaps in our knowledge as information on many of the world's reefs was either not published or inaccessible.

The status of reefs was emphasised in two reviews presented at the 7th International Coral Reef Symposium in Guam, 1992, with both focusing on the anthropogenic factors that were causing reef decline (Buddemeier, 1993; Wilkinson, 1993). A second emphasis occurred in 1993 in Miami at a conference that documented reef damage as a series of case studies around the world showing alarming examples of reef degradation (Ginsburg, 1993). These reports stressed that the major effects causing decline in coral reefs were from anthropogenic influences: increased sediment; organic and inorganic pollution; damage from engineering activities and construction; and over-exploitation, particularly that resulting from the use of explosives or poisons to catch fish. These were the major factors considered in the Reefs at Risk analysis (Bryant et al., 1998) which mapped these impacts as diminishing circles of impact spreading away from centres of human population. The critical factor in all these was that damage was almost directly proportional to centres of either large populations or expanding economic growth, and frequently both.

Direct anthropogenic stresses were thus considered to be the major factors prior to 1998. A UN team emphasised these, concluding that the immediate effects of Global Climate Change (GCC) would be minor, but these may cause reef damage in succeeding decades (Wilkinson and Buddemeier, 1994). To quote from that report: 'Climate change by itself is unlikely to eliminate coral reefs, but climate change will create hardships for people dependent on these reefs, because of changes in reef structure, function, distribution and diversity'.

The agenda changed radically in 1998 as a result of the unprecedented and almost simultaneous coral bleaching and mortality in many locations around the world (Wilkinson, 1998; Hoegh-Guldberg, 1999). Many remote reefs were severely damaged, including reefs remote from any direct anthropogenic impacts, such as in the centre of the Indian Ocean (Wilkinson et al., 1999; Sheppard, 1999). Added to this were predictions that increased dissolved $CO_2$ concentrations would cause chemical imbalances in sea water and reduce rates of coral calcification, thereby further threatening reefs (Kleypas et al., 1999). The combination of these events resulted in dire predictions on the immediate future of coral reefs (Buddemeier, 1999; Hoegh-Guldberg, 1999; Wilkinson, 1999).

## MECHANISMS OF CORAL BLEACHING AND MORTALITY

Prior to the bleaching event of 1997–98, there had been an apparent series of increasing bleaching events (1979, 1983, 1987, 1991, 1994, and 1997–98) that often affected large areas, but never at global scales (Hoegh-Guldberg, 1999). These were all apparently associated with higher than normal sea temperatures. In most cases there was a strong correlation between temperatures of at least 1°C higher than the summer maximum and coral bleaching. Some bleaching events involved just parts of a reef, whereas others affected regions (Hoegh-Guldberg, 1999).

Changes in physical and chemical conditions can result in rapid expulsions of most, or all, of the zooxanthellae from coral and other invertebrate hosts (Hoegh-Guldberg, 1999). The physical changes include reduced salinity, increased or decreased light or temperature. Chemical factors include copper ions, cyanide, herbicides, pesticides and biological agents such as bacteria (Rosenberg and Loya, 1999; Toren et al., 1998). Evidence indicates that elevated temperature is the cause of mass bleaching events, with a rapid elevation of 1–3°C being sufficient. Changes to photosynthetically active radiation (PAR) or ultra-violet B (UVB) aggravate the effect. The loss of algal pigments or zooxanthellae turns the corals white, referred to as 'bleaching' (Hoegh-Guldberg, 1999).

The time immediately after bleaching is critical for corals, as they are without their major source of carbon energy. Should the stress be neither too severe nor prolonged, the corals will frequently either re-capture free-living dinoflagellates from surrounding water, or grow residual zooxanthellae in their tissue. However, extended periods of stress or the addition of another one (e.g. a flood) will result in mortality. But even for corals that recover, the consequences are serious for coral reefs as field evidence shows that the production of larvae and new coral recruitment may totally fail for one or two years afterwards.

## THE 1997–98 MASS CORAL BLEACHING EVENT

The 1997–98 bleaching event was the most severe and widespread on record and operated at global scales (Wilkinson, 1998). Bleaching was first observed during the strong El Niño event of 1997 in the far Eastern Pacific, along the reefs of the western coast of Central America (Mexico to Colombia). This pool of warm water increased and spread out to the Galapagos Islands where bleaching was recorded in December 1997.

The extreme El Niño impacted heavily on the Indian Ocean and Southeast Asia causing a cessation of the trade and Indian monsoon winds, which resulted in a marked reduction in ocean currents. Surface waters heated, particularly under the apex of the sun, without the cooling winds and currents. The northerly 'migration' of the 'Hot Spot' can be seen in Fig. 1 as a pool of warm water (yellow) that

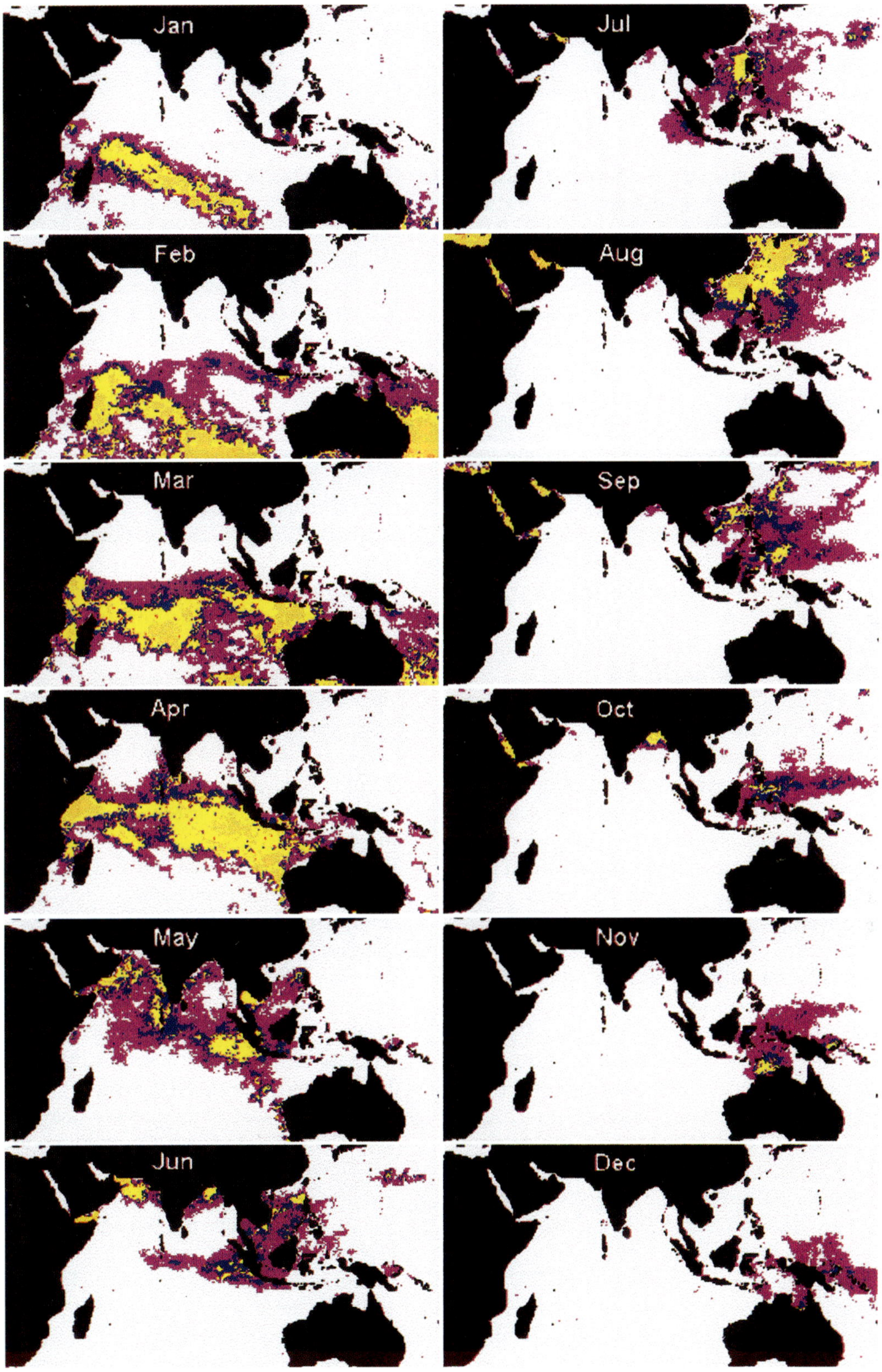


Fig. 1. Sea-surface temperatures (SSTs) in the Indian Ocean and Asia for the 12 months of 1998. Each month is a composite figure of the mean SSTs as they exceeded the maximum summertime climatological values and clearly shows 'Hot Spots'. These 'Hot Spot' anomalies, defined as SSTs equal to or greater than 1°C above the summertime level, are coloured yellow to orange. These were generated from satellite data on sea-surface temperatures collected by the US National Oceanographic and Atmospheric Administration. Region shown is: 35°N to 35°S and 20°E to 165°E and provided by Alan Strong.

matches the vertical position of the sun. There was significant coral bleaching along this path during the first six months of 1998. The 'Hot Spot' reached the northern Indian Ocean and Arabian Sea in May, with resultant bleaching there. There was also bleaching close to the coast of northeastern Australia along the Great Barrier Reef.

The El Niño collapsed in June 1998 and reverted to a La Niña of similar strength. This also caused a cessation of trade winds in Southeast and East Asia. Bleaching also followed the path of the new 'Hot Spot' (Fig. 1). The impacts of the La Niña in the north ceased with the first typhoon in early September. This cooled the waters off Japan and Taiwan, and bleaching immediately ceased. Bleaching, however, continued in Southeast Asia and the far northwestern Pacific, with some reports of the first severe coral bleaching in recorded history. There were also major bleaching reports from the Red and Arabian Seas, and in the wider Caribbean.

Numerous accounts follow in the next pages, summarising details in Wilkinson (1998), and including new observations approximately one year after the bleaching event. Three boxes include other objective examples.

## INTERPRETATION AND CONCLUSIONS

Whether the massive world-wide bleaching in 1997–98 is one of a sequence of ever increasing bleaching events, or whether it is a one in a thousand year event remains to be seen. Clearly, a repetition within the next decade will lend strong evidence to an underlying cause of Global Climate Change. The recent episode far eclipsed any recorded or anecdotal reports of bleaching around the world and resulted in the death of corals over 500 years old e.g. in Australia and Vietnam.

Most episodes coincide with sea surface waters both warming and remaining warm for a period of weeks to months. Bleaching causes stress to the host animal, and temperature rise is the only factor that fits global changes and those impacting on particularly isolated reefs like the Chagos Archipelago. Whether the proximate cause is a bacterial infection (Rosenberg and Loya, 1999; Toren et al., 1998) induced by temperature rise, or changes to photosynthesis of the zooxanthellae resulting in the production of free oxygen radicals (Hoegh-Guldberg, 1999), the initial trigger is a sustained rise in temperature.

There were two principal factors underlying the 1997–98 bleaching event: a rising baseline of sea surface temperatures; and the most extreme and sudden swings in the global climate ever recorded. One of the strongest El Niño events continued from March–April 1987 until April 1998 with Southern Oscillation Index averages of minus 14 to 24. Then in June 1998, there was a rapid switch to a similarly strong La Niña event until April 1999, with SOI averages of +8 to 13 (see Wilkinson et al., 1999). Coral bleaching was associated with all of these changes.

***There was major El Niño associated bleaching in:***

1. The far eastern Pacific between May and December 1997
- Colombia starting May;
- Mexico from July–September;
- Panama in September; and
- Galapagos in December.
2. The Indian Ocean between March and June 1998
- Kenya and Tanzania in March–May;
- Maldives and Sri Lanka in April to May;
- Western Australian reefs from April to June;
- India from May to June;
- Oman and Socotra in May.
3. South East Asia between January and May 1998
- Indonesia from January to April;
- Cambodia, Thailand and East Malaysia during April–May.
4. Eastern Australia and Great Barrier Reef in January and February 1998
5. Southern Atlantic Ocean off Brazil in April 1998.

***There was major La Niña associated bleaching in:***

6. South East and East Asia from July to October 1998
- Philippines from July to September;
- Vietnam in July;
- Japan and Taiwan from July to September;
- Singapore and Sumatra, Indonesia in July.
7. Arabian Gulf and Red Sea from August to October 1998
- Bahrain, Qatar and UAE in August–September;
- Eritrea and Saudi Arabia (Red Sea) in August–September
8. Throughout the Caribbean Sea and Atlantic Ocean from August to October 1998
- Florida from July to September;
- Bahamas, Bonaire, Bermuda in August–September;
- Barbados, BVI, Caymans, Colombia, Honduras, Jamaica and Mexico in September.
9. Far West Pacific from September to November 1998
- Federated States of Micronesia in September;
- Palau in September to November.

Thus, there were almost comparable amounts of bleaching under both El Niño and La Niña conditions throughout the world, signifying two important conclusions: that bleaching may be region-specific; and it is greater with a greater degree of change. Both the El Niño and La Niña events were particularly strong.

Bleaching during the El Niño started when warm waters pooled up in the far east Pacific against the coast of Central America and then expanded out to the Galapagos Islands. The next significant bleaching was in early 1998 when the trade and monsoon winds in the Indian Ocean and Southeast Asia ceased under the influence of this extreme El Niño, resulting in almost doldrum conditions in the Indian Ocean. Without the cooling winds and currents, surface waters warmed from South to North, bleaching reefs in its path. When the El Niño switched to a La Niña, similar conditions prevailed in Southeast and East Asia with a cessation of trade winds until the first typhoon in early September cooled the waters off Japan and Taiwan, resulting in an immediate cessation of bleaching. Bleaching continued South of the typhoon path, affecting the Philippines, and Palau for the first time in its recorded history. The southerly 'migration' of the 'HotSpot' can also be seen in Fig. 1.

Current Global Climate Change (GCC) models predict that the incidence and severity of extreme conditions, like El Niño events, will increase (Timmermann et al., 1999). Similarly, application of a series of GCC models to sea surface temperatures in tropical waters shows a continuing pattern of increasing temperatures near or above the critical limits for coral bleaching (Hoegh-Guldberg, 1999). These were accompanied by predictions that the incidence of bleaching will increase dramatically in frequency and severity in the coming decades, resulting in major losses of coral reefs around the world.

The bleaching in 1997–98 was clearly related to the major climate fluctuations of the El Niño Southern Oscillation with resultant increases in sea surface temperatures (often extending to below 50 m depth). The only remaining question is whether this event was a direct result of GCC, caused by increases in the emission of greenhouse gases. If so, we will witness a marked increase in incidences of such events that will result in many local extinctions of coral species and major losses of coral cover in the short term. Coral reefs will recover in the long term (thousands to millions of years) as they have done to major climate changes in the past. But the sad prediction is that coral reefs will be seriously degraded in the first 50 years of the new millennium.

## SUMMARY OF BLEACHING AND MORTALITY EVENTS

### Arabian Region

#### *Red Sea*

During 1998, bleaching was infrequent except in some localised, severe cases in Saudi Arabia. There were no reports in the northern Red Sea (Egypt, Israel, Jordan, and Sudan). Water temperatures in Eritrea around Massawa and Green Island were up to 40°C in August and September, resulting in bleaching on deep and shallow reefs. Most corals recovered after the temperatures dropped, but some shallow-water corals died. No bleaching was seen on the Assab Islands. The fringing reefs and reef banks off Rabigh (22°33′N to 22°58′N), Saudi Arabia, contained extensive communities of hard and soft corals, and fishes when surveyed in October 1997. When examined in December 1998, the area had suffered wide-scale catastrophic coral mortality, apparently resulting from bleaching during August–September 1998. The fringing reefs were especially affected, with almost all hard (including *Millepora*) and soft corals killed (>95%) from the reef-flat to 15 m depth.*Acropora* species were heavily affected, and even the hardy *Pocillopora verrucosa*, *Stylophora pistillata* and *Porites* were bleached. *Millepora* species that dominated extensive lengths of the reef crest on the coast in 1997 suffered mortality approaching 100%. Mortality of corals was estimated at 70% on reef banks (50 × 20 km) off Rabigh, 25 km offshore, with some corals still bleached three months after the onset. Reefs along the southern mudflat–mangrove dominated coastline of Jizan (16°40′N to 17°10′N), near the boarder with Yemen, were virtually all dead prior to the temperatures rises, probably due to intensive trawling and increased sediment loading. These included large, dead colonies of *Porites* (Yusef Fadlallah, pers. comm.).

#### *Arabian Gulf*

Coral bleaching started in Bahrain, mid-August 1998, when water temperatures rapidly increased from 34°C to 37°C and stayed high for a few weeks (up to 39°C in shallow areas). There was 100% bleaching from Hayr Shutaya (30 km north of Bahrain) south to Fasht Al Adhom, and Fasht Al Dibal (all less than 10 m depth). Coral mortality was 90–95% a few weeks later, and surviving corals were still bleached in October. There had been major bleaching in 1996, when water temperatures were 37.3°C at Fasht Al Dibal and most corals on Fasht Al Adhom (26°N, 50°E) bleached then died. Almost 100% of the remaining corals in less than 4 m died in 1998 when water temperatures exceeded 38°C. By August 1999, a few small *Porites* colonies (1–4 cm diameter) had survived at depths of 4 to 8 m. The best coral area in Bahrain, Abul Thama (26°54′N, 50°58′E, surrounded by 40–50 m deep water) also bleached in 1996. About 50% of the remaining coral bleached in 1998, but there was extensive recovery by mid-1999. Only a small amount of bleaching (less than 5%) was seen in Hayr Shutaya and Fasht Al Dibal in August 1999 with water temperatures 35 to 36°C. No new recruitment of corals has been observed (Roger Uwate, pers. comm.)

Widespread coral bleaching was seen in Qatar with seawater temperatures at 35–36°C in mid-August 1998. Coral mortality was about 95% affecting particularly *Acropora* and other species (especially *Platygyra daedalea*, a common nearshore coral) that had survived bleaching in 1996. Bleaching was minimal on an offshore reef with seawater temperature below 34°C, but severe bleaching in 1996 had killed many of the corals. In Qatar in December 1998, 99% of corals were dead at two sites (of 70–80% cover of predominantly *Acropora* and *Porites* species) following coastal water temperatures of 39°C. Bleaching also impacted the United Arab Emirates in 1996, and it appears that almost all remaining corals died during 1998.

#### *Arabian Sea*

Extensive bleaching was observed at temperatures of 29.5 to 31.5°C around Mirbat, southern Oman, in May 1998.

Between 75 and 95% of abundant *Stylophora* bleached, and 50% of large *Porites* colonies were partially bleached. No bleaching was observed at Sudh, 40 km east of Mirbat, at temperatures between 25 and 25.5°C, nor in the Muscat Area, Gulf of Oman at temperatures around 30.5°C. Upwelling during the southeast monsoon normally drops temperatures to 19°C. No recovery of bleached colonies was seen in mid-October when temperatures increased to 25°C after the summer upwelling. Extensive coral bleaching was seen on the islands of Socotra off the Horn of Africa in May 1998 with high mortality.

## Indian Ocean

Coral bleaching was unprecedented in extent and scale in the Indian Ocean during the first six months of 1998 with massive losses of coral cover in reefs from Africa to Western Australia.

### *Eastern Africa*

There was severe coral bleaching and mortality on the coasts of Kenya and Tanzania from March to May 1998. Bleaching started in Lamu, Kenya (2°S), when temperatures reached 32°C and continued to Mnazi Bay Tanzania (10°S). In Kenya, bleaching varied from 50%, but mostly near 100% on almost all reefs where coral cover had been 20–50%. Bleaching was most severe on shallow reefs, but was 50% at 20 m depth. Now reefs have only 1–10% coral cover. Bleaching along the whole coastline of Tanzania ranged from 15–25% in the South to 25–50% near Zanzibar (6°S) and 25% around Tanga (5°S). Mafia Marine Park, lost 80–100% of corals by October. *Acropora* species were most affected with 80–95% bleached in Chumbe, and 40–70% in other areas. Some *Porites* species bleached whereas others were unaffected. Less than 5% of coral is still alive on Tutia Reef in the South, while there has been massive mortality of *Acropora* in Chole Bay (100%) in the North and in the 'coral gardens' of Kinasi Pass (80–90%). Recovery from bleaching was about 50% in Mnazi Bay, 60–80% in Bawe and Chumbe, and less than 40% in Changuu and Chapwani.

### *Southern Indian Ocean*

There was relatively minor bleaching in Mauritius, with considerable recovery. Bleaching and mortality on the Reunion reef flats and slopes was approximately 30–50%. On Madagascar, there was 30% bleaching of corals at Belo sur Mer (mid-west coast) and similar bleaching at Antananbe, Toliara, Nosy Bé, and Mitsio archipelago. At Mananara-Nord (northeast coast, 15°S) 40–80% of corals bleached with high mortality, and 10–40% of mixed species corals bleached in deeper water. A similar pattern was reported on Mayotte, except on the exposed southern end and the lagoon, which receives cooler water from the north.

### *Central Indian Ocean*

The Seychelles, Chagos, Maldives and reefs off Western Australia were amongst the most severely impacted in the world by the 1998 coral bleaching, with mass mortality of shallow water corals in almost all areas. Some bleaching was observed from the air on two islands of the Comores, but no details are available.

Temperatures in the Seychelles increased to 32°C, and 34°C, followed by extensive bleaching down to 23 m on Aldabra and Providence Group (9°S; 46–51°E), and Alphonse Group (7°S; 53°E) during March–May 1998. Bleaching and mortality affected *Acropora, Pocillopora* and *Millepora,* with 20–55% mortality. Soft corals (85–95% mortality), anemones and giant clams also bleached. In the Seychelles Marine Park there was an average of 75% coral mortality (range 50–95%).

There was severe and rapid bleaching for Maldive coral reefs from April to May 1998. Less than 3% live coral remains on Vaavu (<3%), Ari (<1%) and Haa Dahl (<1%) Atolls. Live coral cover dropped 30–50% to 1–7% on reef tops (2–4 m depth) inside atolls from the far north, the middle and the south. Mortality was around 95% of extensive *Acropora* communities, as well as soft corals and anemones. Most areas reported around 80% of corals wholly or partially bleached on the back reef, with around 45% at 10 m on the reef slope, 30–40% at 20–30 m, and bleaching also at 50 m. The drop in mortality with depth is partly due to a natural drop in cover of the most susceptible species (*Acropora*) at greater depths. Live *Acropora* is now very rare below 20 m. There were signs of partial recovery of *Porites* on South Male and nearby atolls, and some anemones regained colour. Some digitate *Porites* are overgrowing dead skeletons from remnant tissue. However, the net effect was almost total loss of once abundant *Acropora* and *Pocillopora* species and most soft corals, e.g. very few *Acropora* colonies were observed during more than 100 surveys, each 50 m long. New *Acropora* recruits are generally rare, but are common in some locations, which indicates the potential for substantial recovery here, provided the massive bleaching is not repeated in the near future (William Allison, pers. comm.).

Experiments concerning rehabilitation of reef flats devastated by coral mining were nearly obliterated by bleaching (Clark, 1999). Concrete blocks on the reef flats adjacent to Male had high coral recruitment and growth for 3.5 years. Then water up to 3°C above normal summer maxima remained over the site in April and May 1998, resulting in up to 100% coral bleaching and death down to 30 m depth. There was almost 100% mortality of branching corals on the concrete blocks, whereas massive corals suffered much lower mortality. *Porites*, *Favites*, *Pavona* and *Favia* showed initial bleaching, but recovered within 1 to 2 months and dominated the new population. Coral cover declined on adjacent reef flats by 30–60% to 0–5% after massive losses of the previously dominant branching *Acropora* and *Pocillopora.* Prior to 1998, branching corals dominated

## Bleaching in the Chagos Archipelago, Central Indian Ocean

The El Niño of 1998 resulted in a major collapse in coral abundance across the Chagos Archipelago. This appears to be a continuation of a long series of events. Over the last 20 years, there has been a marked decline from 50–75% hard coral cover and 10–20% soft coral and other coelenterates to only 12% live cover on seaward reefs of the Chagos Archipelago in early 1999. These are very remote and largely uninhabited reefs with high biodiversity; about 220 coral species and 770 fish species. Most of the reductions occurred during 1998 when a 'Hotspot' of elevated sea surface temperatures sat over Chagos for two months. Seaward reefs of all six Chagos atolls now are 40% (to about 20 m deep) covered by dead coral, with another 40% covered by dead coral and bare rock. Lagoonal reefs fared better, but still lost 50% of live coral cover. All reefs have large quantities of loose rubble, which will inhibit new recruitment and growth.

Previous El Niño episodes, like those of 1982–83 and 1987–88 are probably the reason for some decline, but these events were not as severe as the 1998 event. Climatic records on these reefs remote from direct anthropogenic influences are a sound indication of global climate change. There has been a 1 to 2°C rise in mean air temperature over 25 years, as well as a fall in mean annual atmospheric pressure, a reduction in cloud cover, and winds have become more variable.

The depleted coral cover and diversity included reductions of tabular *Acropora* sp. in 5–15 m depth, and *A. palifera* in wave-exposed shallows. In February 1999, substantial mortality of corals and soft corals was evident, with seaward reefs most heavily affected with great uniformity of dead coral and bare substrate. Only the large genus *Porites* had more living corals than dead.

Calcareous red algae were also reduced in shallow water along with the associated *Millepora*, and blue *Heliopora* corals, which were both almost completely eliminated in 1999. Soft corals were almost totally eliminated from seaward slopes in 1999, resulting in areas of bare substrate.

Lagoon reefs suffered much lower coral mortality, probably because they experience greater fluctuations in temperature normally and corals may be adapted to this. Massive *Porites* survived far better in lagoons than on seaward reefs. Fish diversity and abundance appeared considerably reduced, especially where coral mortality was nearly total, but no quantitative data were gathered.

*Acropora* was especially vulnerable. After 1996, there were vast areas of dead and still-standing digitate corals, table corals and *Acropora palifera*. The two sites previously found with the greatest *A. palifera*, had been north-western Salomon and western Blenheim; in these areas the dead colonies were mostly still intact, but over 95% were dead and suffering erosion. In early 1999, it was possible to snorkel for 15 minutes in some areas and not see any live coral where there used to be 50–75% cover. In a few sites, up to 10% of visible colonies were alive, but rarely more. The once-common large 'table' forms on reef slopes were all dead with very rare exceptions.

The common digitate forms *Stylophora pistillata* and the two *Pocillopora* species (*P. damicornis* and *P. verrucosa*) were almost always dead. Live colonies existed in lagoons, but many areas contained no living colonies at all. All genera of the large Faviid family suffered very heavy mortality, but some large colonies of the leafy *Echinopora lamellosa* were found in lagoons.

*Porites lutea* and other similar large massive *Porites* species were the best survivors, and covered an average of 8% of all reefs. Dead colonies of *Porites* in contrast covered an average of only 2%, thus this was the only coral with more live colonies than dead skeletons. The blue coral *Heliopora coerulea*, survived well in some areas but not others, and the fire coral, *Millepora*, was mostly dead in all sites. Most living corals contained large areas of recently bare skeleton, with the remaining live tissue mostly around the lower edges and shaded parts of the colonies. These observations illustrate that bleaching and mortality is due to a combination of increased temperature combined with direct illumination, which has been increased over Chagos due to a significant reduction of cloud cover.

Immediate consequences are likely to be a reduction in reef growth, and delayed recovery of corals and the coral community. Erosion of dead coral skeletons by physical and biological processes will result in the production of unstable rubble and sand, which will prevent new larvae from establishing. The supply of new larvae may be remarkably reduced.

The increased surface area covered by algae may result in increases in grazing fish and echinoids, but the reduction in structural complexity may result in fewer habitats for fishes and a drop in fish populations in the future. The massive loss of coral cover, including coral colonies that were 200–300 years old, clearly indicates that partial recovery may need several decades at least, and a few hundred years before the previous widely age-structured community of corals and associated species is re-established.

(From C. Sheppard, 1999)

recruitment, whereas after the bleaching, massive species were dominant (67%), compared to the 2% proportion of massive species recruiting to the blocks earlier. This indicates that a supply of new coral larvae exists either from upstream reefs or from deeper areas of the same reef. Similarly there were new *Acropora* and *Pocillopora* recruits on adjacent reef flats, although apparently much reduced than in previous years. It is predicted that coral recruitment will remain low for the next 5–10 years as there is a scarcity of parent stock, and moreover the new communities will have a greater proportion of slow-growing massive species than before (Susan Clark, pers. comm.).

### Southeast Asia

Coral bleaching in Southeast Asia started in western Indonesia in January and February 1998, and later affected Bali and Lombok in March. During April–May, bleaching developed around Cambodia, Thailand and parts of Sabah, Malaysia. In northern parts and in East Asia, it coincided

## Bleaching and Recovery of Western Australian Coral Reefs

There was highly variable bleaching and mortality on coral reefs off West Australia during the first 6 months of 1998. All of these reefs are either remote from, or receive very minor direct anthropogenic damage. Some isolated northern atoll reefs (Scott and Seringapatam) were devastated, with 80–100% of corals being killed, whereas, other isolated oceanic reefs (Ashmore, Rowley Shoals) suffered only minor bleaching of susceptible species (e.g. Pocilloporids). Unlike the situation on the Great Barrier Reef, the inshore reefs (Dampier Archipelago, Onslow, Ningaloo and Abrolhos reefs) suffered minor or insignificant damage.

Ashmore Reef (S 12°16′ E 123°00′) is the most westerly of these 'atoll' reefs where only minor bleaching within the lagoon was observed, and this was not considered unusual. There are no quantitative data.

Seringapatam Reef (S 13° 39′ E 122°02′) is very isolated, and extensive bleaching occurred in all habitats of this small continental shelf atoll which is surrounded by depths of 400 m. The previous 45% of hard coral cover was reduced to 5%, and 10% cover of soft corals was down to 2% on the outer eastern flank. Many species have become locally extinct. Coral communities in the relatively shallow lagoon suffered almost 100% mortality. Extensive *Acropora* stands have died and only 0.5 m$^2$ of a massive colony of *Pavona minuta* measuring 100 m diameter and 6 m high, is now alive (compared to 90% before the bleaching). Other small patch reefs have suffered between 90 and 100% mortality. There is a six-year data set available on coral and fish communities.

Scott Reef (S 14°11′ E 121°48′) is a continental shelf atoll (50 × 35 km) and was the most seriously affected of the isolated reefs. There was extensive coral bleaching and mortality in all habitats at all depths in April 1998. Most lagoon patch reefs suffered almost 100% mortality with death of extensive mono-specific fields of *Acropora* and no survivors are evident. Reef crest communites (dominated mostly by Acroporids) have been devastated, with few survivors. Coral community composition has changed dramatically and overall coral cover has been substantially reduced. Extensive quantitative data will be made available (from permanent video transects and fish counts) for all habitats and depths for Scott and Seringapatam Reefs.

Almost all benthic organisms with symbiotic algae at 9 m suffered whole or partial bleaching, with highly variable recovery. Branching Acroporids suffered 80 to 100% and there was 100% mortality of *Millepora* at all sites except one. While *Porites* suffered about a 50% reduction in cover, many colonies showed partial mortality with only small areas of live tissue one year later. Pocilloporids, once common, are now extremely rare and the most abundant coral, *Acropora bruggemanni*, is locally extinct at all locations but one. Rapid surveys at 30 m at one location, revealed 80% of corals were bleached, with 100% bleaching of the dominant species of *Diploastrea*.

Mass coral spawning at Scott Reef usually occurs after the full moon in April, but surveys showed a complete recruitment failure of all coral families, at all sites for 1998 and 1999. Previously (1995/96/97) there were between 36 and 135 recruits per terracotta settlement plate, which may indicate that these isolated reefs are likely to be 'self-seeded' and thus may have extremely slow recovery times.

Rowley Shoals including Mermaid Reef (S 17°04′ E 114°38′) Clerke Reef (S 17°17′ E 119°23′) and Imperiuse Reef (S 17°33′ E 118°58′) are on the edge of the continental shelf, surrounded by 300 m deep water. Bleaching in 1998 was minor, with some Pocilloporids and the branching, *Porites cyclindrica* bleached. Lagoon corals bleached more than those on the outer slope, but mortality was minor. No significant reduction in coral cover was seen in permanent video transects monitored since 1994.

Dampier Archipelago (S 20°38′ E 116°39′) consists of 15 islands stretching out from the mainland. The inner fringing reefs of the Dampier Archipelago suffered a severe bleaching in March–April 1998 affecting most benthic organisms with symbionts. However bleaching of species and locations was patchy, with most hard corals suffering some bleaching, while soft corals, anemones and zoanthids suffered varying degrees of bleaching. Recovery was highly variable, with most Pocilloporids and Fungiids dying, while many massive corals (*Turbinia*, Mussids, Pectinids) recovered. Since 1998, coral communities have a reduced cover of Pocilloporids and Fungiids. The outer island areas of the Dampier Archipelago suffered little damage, with quantitative surveys showing only a single species of *Fungia* being affected.

Onslow–Mangrove Islands (S 21°27′ E 115°22′) experienced extensive coral bleaching on the inshore reefs (North of the town of Onslow) in late March 1998, whereas there was little bleaching in the outer areas. Video footage showed that approximately 80% of corals were bleached, but there are no quantitative data on this or extent of recovery or mortality.

Ningaloo (S 22°13′ E 113°49′) is a fringing reef that runs approximately 220 km south along the Western Australian coastline from NorthWest Cape, with the outer slope between 2 and 5 km from the shore and enclosing extensive lagoon coral communities. There were anecdotal reports of minor bleaching in some areas from February to April 1998, but quantitative surveys showed no measurable impact.

Abrolhos Islands (S 28°41′ E 113°50′) much further to the South showed no evidence of coral bleaching in 1998.

(From Luke Smith and Andrew Halford,
*Australian Institute of Marine Science, Western Australia*)

with the onset of La Niña in June and July, affecting Philippines, eastern Vietnam, Taiwan, and Japan. Bleaching spread south to Singapore and the Riau islands off Sumatra in June and July.

Corals around Sihanoukville, Cambodia were moderately to severely bleached in mid-May affecting most species with approximately 80% mortality in some places. In Taiwan, bleaching started in June, around Penghu Islands (Pascadores Is) with about 30–40% of corals bleached in 1–5 m, with water temperatures around 30°C. Extensive coral bleaching was observed in August around Posunotao, offshore southeast Taiwan in the Kuroshio current, with >80% of corals bleached down to 20 m, at water temperatures of 31°C at 20 m and 34°C at 1 m.

### *Indonesia*

In Indonesia it started in the Riau Islands with a warm current from the South China Sea that flowed to Lombok. At Bali Barat National Park (northwest Bali), and Tulamben (eastern Bali), 75–100% of the 25% coral cover bleached, including many soft corals. Bleaching was less at Nusa Penida and Nusa Lembongan. Bleaching in Pulau Seribu off Jakarta, and Karimunjawa Marine National Park (north of Java) started in January–February 1998 and continued to May, with 0–46% bleached at 3 m (mainly *Acropora* and *Galaxea*), and 1–25% at 10 m (*Pachyseries, Hydnopora* and *Galaxea*), with 50–60% mortality. Almost 90% of corals bleached down to 20 m (especially *Acropora*) in March 1998, on the Gili Islands, Lombok Strait. Some massive corals, especially *Porites*, had recovered by August. There was cold water bleaching (60–70% at some depths), in East Kalimantan (Borneo) during January, and significant recovery. Corals were excellent at Pulau Bangka and the Lembeh Straits, but cyanide fishing was in progress (Jonathan Simon, Mark Erdmann, Roy Caldwell, pers. comms.).

Mixed reports came from the Bunaken National Park, Northeast Sulawesi, Indonesia, where there was little bleaching of the 52% coral cover on the reef front and 35% on the outer flat, with dead coral cover at 4–5% (Corrado Villa, pers. comm.). Corals in the Bunaken-Manado Tua Reserve are very healthy, with abundant fish life, although extensive bleaching was reported on the walls down to 30 m depth with virtually 100% of corals, soft corals and hydroids bleached.

Extensive bleaching started in October 1998 in North Sulawesi (Manado Bunaken, Manado Tua, Siladen, Mantehage, and Nain) after earlier bleaching, with rapid recovery. Estimates of the new bleaching were 20–40% of corals on the reef crests of these islands (prior live coral cover on the crests was 50–100%). Most affected were *Montipora digitata, Acropora palifera* and encrusting/platelike forms of *Montipora* spp., and agaricids and pectinids. There was less bleaching for branching *Acropora*, massive *Porites, Hydnophora, Seriatopora* and *Pocillopora*. Up to 90% of soft corals bleached (*Sinularia* and *Nepthea* species), but *Sarcophyton* was unaffected. Deep-water corals to at least 40 m depth were bleached, but this was highly variable. No bleaching occurred in the Banda islands (Maluku province) in October 1998; likewise in the Lembeh Strait (east side of North Sulawesi), while anemones were heavily bleached (up to 50% in some areas). There was a lot of bleaching in the Togian Islands (central Sulawesi), with 30–35% bleaching of the 60–90% live coral cover (Mark Erdmann, pers. comm.).

### *Malaysia and the Philippines*

Coral bleaching was localised and relatively minor in Sabah, Malaysia in mid-May 1998. At Pulau Gaya, there was 30–40% bleaching of all corals in 1–2 m, with water temperatures of 32°C. In Pulau Sakar, 30% of all species bleached with 10% mortality down to 20 m, including 90% of *Acropora* colonies and some giant clams (20% bleached). Less than 5% of corals were bleached in Pulau Baik and there was minor bleaching on Mamutik Island, Turtle Islands Park, Darvel Bay and off Semporna.

In the Philippines massive bleaching started in mid-July, from Bolinao (northwest Luzon), to Puerto Galera, southern Negros Island, central Philippines (Dumaguete, Campomanes Bay, Danjugan Island, El Nido (Bacuit Bay) and Coron Island Palawan), and Pag-asa Island (Spratleys). Temperatures were 33–34°C. Up to 75% of the plating, branching, foliose and faviid corals bleached in some areas, including some massive corals, but large *Porites* resisted bleaching below 5 m, (except on the reef flats of Bolinao and Negros). Mortality, however, appears to be low. Massive bleaching was reported in Danao Bay, near Baliangao, northwest Mindanao in October 1998, affecting branching corals and soft corals, but fire coral (*Millepora*) was not affected.

Significant bleaching at Bolinao (16°20′N; 119°55′E) started in June 1998 and peaked in August with temperatures between 30.7 and 33.2°C. Bleaching varied between 5 and 95% of live coral cover, often more extensive at 20 m depth than shallower. Bleaching like this has never been seen previously in this area (Peter van der Wateren, pers. comm.). There was combined bleaching and crown-of-thorns starfish damage to Cuatro Islas, Leyte (10°32′N, 124°39′E) in 1998. These reefs suffered heavy illegal fishing damage until the late 1980s, but recovered remarkably when effective community management was implemented from 1993 to 1998, including control of starfish outbreaks. Heavy bleaching started in August–September when water temperatures stayed well above 30°C, killing 90–95% of shallow water soft corals, 80% of shallow hard corals, and 50% of deeper corals. This coincided with another starfish outbreak, until November when local people controlled the outbreak. The reefs were severely depleted, negating the establishment of sanctuaries and strict law enforcement efforts (Sabine Schoppe, pers. comm.).

On Apo Island, southern Negros, about 90% of *Galaxea fasicularis* was bleached, with 50–60% of the surface area dead. On nearby Sumilon Island, most of the bleached coral did not recover (Laurie Raymundo, pers. comm.).

Bleaching on Danjugan Island, Negros, started in August 1998 affecting 35–90% of corals down to 25 m depth, including almost total loss of soft corals (*Sacrophyton, Sinularia*). Prior to 1998, there was 10–30% standing dead coral in 0–10 m, mainly branching *Acropora* and *Seriatopora*. By April 1999, bleaching was less than 5% at all sites, with recently dead coral cover at 20–30%. There has been some recovery in deeper sites, mainly massive corals (e.g. *Porites*) with remaining live tissue growing over the dead skeleton. Some new recruitment has been observed. There is anecdotal evidence of bleaching before July 1998, at Cagdanao Island, Palawan. In 1999, the estimated dead coral cover was 30–60% below 10 m depth. Cover of macroalgae is up to 90%

shallower than 10 m on all islands in the Taytay Bay region (Maria Beger, pers. comm.).

More than 92% of corals were dead and overgrown by algae to 8 m depth at Mactan Island, Cebu (10°13′N; 123°57′E), but 33% of massive corals have small surviving patches. However, only single tips remain alive of 8% of colonies of the formerly dominant branching *Acropora* corals (Thomas Heeger, pers. comm.).

There appears to be minimal impact from coral bleaching at Anilao, Batangus, with only a few percent hard corals dead or dying, and the coral community looks healthy (Karl Lung, pers. comm.).

### Singapore, Thailand, Vietnam

The first time that mass bleaching has been reported in Singapore occurred during June and July 1998 when seawater temperatures were 33°C, compared to the normal of 28–30°C. The bleaching affected all hard and soft coral species over the entire depth range, with high mortality.

Coral bleaching started in April 1998 in the Gulf of Thailand from the Malaysian border to that of Cambodia, but there was no bleaching on the Andaman Sea coast. Water temperatures increased from 28–29°C to above 32°C, and as high as 35°C. Bleaching affected 100% of *Acropora*, 80% of *Pocillopora damicornis*, and about 60–70% of massive *Porites*, especially in shallow water in some places. Bleaching affected 30–50% of corals around Chumporn, 50–60% of corals around Sichang and Mun Islands (Rayong), with mortality in about half of these corals. This is the first widespread bleaching report in the Gulf of Thailand.

In Vietnam, extensive bleaching started in mid-July 1998 off Nha Trang (south-central Vietnam), with moderate coral mortality of shallow species, especially *Acropora*. In Con Dao National Park, 70% of corals bleached down to 15 m, including 90% of the dominant table *Acropora*, with total losses by mid-September of about 70–80% of the shallow water coral cover, and 90% mortality of the deeper water dominant massive *Porites* and *Lobophyllia*. There was 60–70% coral cover loss in deeper water including hundreds of 2–3 m diameter *Porites* colonies, including 500-year-old or more 9 m diameter colonies. No bleaching was seen at Hon Mun Island or at Halong Bay.

### East Asia

#### *Japan*

Coral bleaching started on Okinawa Island (26°N) in mid-July 1998, when temperatures rose from 25–28°C to 28–31.5°C, but was less extensive near shore (30°C). All shallow water corals were affected. Bleaching spread to the Japanese mainland (33°N), and below 20 m on Okinawa but stopped with the first typhoon in September. Bleaching was severe on Ishigaki Island (50–70% bleached) and Amakusa, Kyushu. Most species bleached, except for the blue coral *Heliopora*. By mid-October most bleached *Acropora* were dead and covered with algae on Okinawa, however, many *Acropora* colonies in shallow moats survived. At Ie Shima, Okinawa in October 1998 it was estimated that 50% of the coral coverage had bleached. Some anemones were bleached, whereas others were unaffected.

On Akajima Island further to the North, there was a marked reduction in coral fertilization in 1999. There had been 56–97% bleaching on these reefs from July to August, but by March 1999, 49% of bleached corals had recovered. Fertilization success of common *Acropora* species dropped from above 90% to an average of 42%, due mainly to marked reductions in sperm numbers and motility. These reductions are interpreted as being due to bleaching stress. This is a further indication that reef recovery will be delayed following the devastation of 1998 (Makoto Omori, pers. comm.).

No bleaching was seen to 30 m in September off Shikoku Island (33°N), where there is 75% coral cover of plate *Acropora* down to 10 m. Previous bleaching was in 1980 and 1983.

### Northwest Pacific Ocean

#### *Federated States of Micronesia*

About 20% of corals from a wide range of genera bleached to 20 m on the north of Yap, in September 1998. Soft corals (*Sarcophyton, Lobophyton*), anemones (*Heteractis*), and *Heliopora* also bleached. Water temperatures were 30–31°C, but no bleaching occurred in Chuuk.

Major coral bleaching occurred in Palau in September 1998 with water temperatures from 30–32°C. Bleaching was evident at all depths, with a mean of 53.4% of corals at 3–5 m, 68.9% at 10–12 m, and an estimate of 70% bleached down to 30 m (Peleliu, the Blue Corner, the Big Drop-off, Iwayama Bay–Rock Islands). Severe bleaching occurred in Arakabasan and Cemetary island. The major damage was to corals in oceanic waters, with much less bleaching and mortality in shallow areas and those close to the more turbid and stressed habitats near the shore. The least effect was to corals around the sewage outfall near Koror (John Bruno, pers. comm.).

### Southwest Pacific Ocean

#### *Australia*

Although there were alarming press reports of massive coral bleaching along the Great Barrier Reef (GBR) in early 1998, the final evidence was of extensive coral bleaching and mortality on inshore reefs, whereas bleaching was relatively minor on offshore (middle- and outer-shelf) reefs. Sea surface temperatures warmed considerably off eastern Australia during January to March 1998 and aerial surveys of 654 reefs between March and April showed that extensive bleaching occurred along the entire length of the (GBR) from Elford Reef (17°S), to Heron Island (23°S). Overall 87% of inshore reefs (<20 km from the coast) showed at least some bleaching, compared to 28% of offshore reefs. Of the inshore reefs that bleached, 25% had very severe levels of bleaching (>60% of coral affected), a further 30% had very high bleaching (30–60% of corals affected) and another 12% had bleaching between 10–30%. There was no severe bleaching on off-shore reefs and only 5% had more than 30% bleached corals (Berkelmans and Oliver 1999). On Orpheus Island (an inshore reef of the Central GBR), 84–87% of corals bleached, but five weeks later, mortality was 2.5–17%, with *Acropora* species being most affected; *Pocillopora* species were hardly affected. However, there was almost 100% coral mortality 10 km away on Pandora reef, where temperature rises were compounded by extensive freshwater flows.

Experimental studies showed that it requires only five days of a 2–3°C increase over mean summer temperatures of 29°C for bleaching to occur in the common branching coral *Acropora formosa*, and another 1°C more for bleaching to occur in *Pocillopora damicornis* and *A. elseyi* (Berkelmans and Willis, 1999). Moreover there was often a month delay between elevated temperature stress and the onset of bleaching. Recorded temperatures exceeded 31°C for 27 days with a maximum of 32.7°C at Magnetic Island, near the centre of the inner-shelf bleaching. The effects of these raised temperatures were exacerbated by massive flows of rainwater in January in the central GBR where some of the most severe bleaching occurred.

On the GBR, over 100 coral species bleached, including some large *Porites* colonies that were centuries old. Almost all species of soft coral were extensively bleached on inshore reefs with almost 100% bleached and killed in the upper 5 m, and about 20% bleached at 8–12 m depth. Mortality at all depths was high, even for the normally resistant *Sinularia*.

Bleaching also extended south of the GBR on rocky reefs with up to 50% coral cover (Gneering Shoals; 26°S) and northern New South Wales (28°30′S) in March. But on the Flinders Reefs (27°S) which are in oceanic waters, there was no bleaching. Bleaching occurred when water temperatures were 28–30°C, whereas the normal maxima are 26°Cs. *Pocillopora damicornis* and *Stylophora pistillata* were most affected, with 60–70% of these species bleached to 15 m depth.

Analysis of long-term temperature records show that the first 6 months of 1998 were the most extreme for the period since 1982 for the number of days exceeding normal summer maxima and the temperatures reached (Lough, 1999).

*Papua New Guinea*

In Kimbe Bay, in August 1998, there was high coral mortality with 75% of *Acropora* affected, and bleaching in *Porites*, *Platygyra* and *Montipora* down to 50 m. Temperatures below 10 m were 31.5°C, and 33°C on the surface. There was <10% bleaching of *Acropora* on southwest Kimbe Bay. Bleaching was extensive from shallow water to 20 m (almost 100% in some areas, including soft corals and anemones) of large parts of Normanby Island to Cape Vogel, and Tufi (far southeast PNG) in February–March 1998. Reefs north of Normanby and Fergusson Islands did not bleach.

*New Caledonia, Solomon Islands, Vanuatu*

No bleaching was reported in New Caledonia in 1997–98 where reefs remain healthy (Michel Pichon, pers. comm.). There was no evidence of bleaching in the Guadalcanal, Florida Islands, Russell Islands and Marovo Lagoon at near normal temperatures (Charles Delbeek, pers. comm.).

## Central and Eastern Pacific Ocean

Most of the Pacific Ocean escaped bleaching in 1997–98, with the exceptions of the far east and west. Throughout most of the remainder, water temperatures were often colder than normal.

No bleaching was observed in American Samoa (14°S, 170°W), and bleaching in Samoa (Western) of 60–70% of all *Acropora* on the reef flat in February 1998, was linked to extreme low tides and exposure to air. No bleaching was reported in Fiji, Hawaii and Johnston Atoll where temperatures were below normal.

Bleaching in early 1998 was variable among islands and atolls of French Polynesia, e.g. 20% coral cover was reduced to 12% on Takapoto; there was severe bleaching and mortality on Rangiroa and Manihi; but on Moorea, Bora Bora and Tikehau there was minimal bleaching.

## Pacific Coast of the Americas

The first bleaching was in May 1997 in Malpelo Island (3°51′N and 81°35′W) Colombia, although less than 1% of the living tissue was affected. Increasing signs of bleaching were observed in July–August 1997; in September 1997, up to 30% of *Porites lobata* and some *Pocillopora* bleached, whereas similar corals nearby appeared normal.

Bleaching started in the Galapagos in mid-December 1997, when water temperatures reached 28°C. In February the water was 2°C warmer and bleaching continued. Nearly all corals had bleached to some extent by March 1998, especially *Porites* and *Pavona*; *Psammocora*, *Diaseris* and *Cycloseris* bleached on top, but many had pigment around the bases. Corals bleached most in 10–15 m depth, but there was some at 30 m. *Pocillopora*, which was heavily impacted in 1982–83, was largely unaffected this time.

Bleaching on the Pacific Coast of Mexico was observed for the first time in July 1997, peaking in August–September, from the Gulf of California (25°N) to Jalisco (19°N) and the remote Revillagigedo Islands (18°N). No bleaching was observed on Clipperton Atoll. About 25% of total coral cover bleached (water temperatures were 31–34°C), with the most extensive (60% of shallow corals) at Nayarit (20°N). Bleaching was 10–15% in the Revillagigedos, with some mortality. Very minor bleaching was seen at Oaxaca (16°N); all colonies recovered. A sudden *drop* in surface water temperatures (from 30 to 20°C in three days) in mid-September 1998 also caused extensive bleaching in Puerto Angel–Huatulco region (96°33'W–96°10'W and 15°39'N–15°45'N), reducing coral cover from 16% to 2% on 2 reefs: La Tijera and Mazunte. The principal species bleached were *Pocillopora verrucosa, P. damicornis, P. meandrina, P. capitata* and no recovery was observed. *Pavona gigantea, P. varians* and *Porites panamensis* resisted bleaching (Gerardo Leyte, pers. comm.).

All coral species down to 20 m bleached in mid-September 1997 at Uva Island, Gulf of Chiriqui Panama, including the remaining *Millepora intricata* (the most common species remaining after the 1982–1983 El Niño). Bleaching again in mid-April 1998 in the Gulf of Chiriqui region affected almost all coral species, with 50–90% of corals affected. Corals from nearby Gulf of Panama were less affected.

### Caribbean and Atlantic Ocean

Bleaching in the Caribbean did not follow a clear pattern and was regionally variable. Even in areas where it was severe, mortality was generally low, except when a secondary stress like Hurricane Mitch hit the area. Most bleaching was associated with the La Niña event of the latter half of 1998. This contrasts with the Pacific coasts which bleached in late 1997 associated with El Niño conditions. For example, early reports in Belize and Puerto Rico were of massive bleaching, but mortality was eventually low in most places.

#### *Bahamas*

Over 60% of all head corals bleached to 15 m around Walker's Cay, New Providence Island, Little Inagua, Sweetings Cay, Chubb Cay, Little San Salvador, San Salvador and Egg Island, Bahamas in August 1998, and up to 80% between 15–20 m depth. Samana Cay was less effected. *Montastrea cavernosa* was not bleached, and *Acropora palmata* bleached on the upper sides. A large area of the Carribee bank, Barbados apparently bleached in September 1998. Whereas on Bermuda bleaching started in August 1998 and continued into October when surface temperatures rose to 30°C. There was 2–3% bleaching of the 25% coral cover at 8 m on rim reefs, 5–10% bleaching of 40% coral cover at 15 m on offshore terrace reefs, and 10–15% of the 15–20% lagoon coral cover. Mortality was low, perhaps 1–2% of affected colonies.

#### *Netherlands Antilles*

Less than 15% of corals bleached on Bonaire in August–September 1998, with partial bleaching in *Montastrea annularis* between 10 and 20 m and in *Agaricia* below 20 m. Few corals in shallow water bleached. By September, nearly 100% of all *Agaricia* bleached from 8 m to 30 m. Less than 1% of corals bleached during the summer of 1999. In November, a large proportion of the deep *Agaricia lamarcki* and *A. grahamae* bleached to depths in excess of 40 m in Curacao (Rolf Bak, pers. comm.).

#### *Caymans*

Unprecedented bleaching occurred in September 1998 with 90% of all corals heavily affected, with some bleaching of *Acropora palmata* and *Montastrea annularis* at 1–5 m depth, and widespread bleaching and some mortality of the abundant *Millepora*.

#### *South America*

There was mass bleaching on reefs off Bahia State (12°S; 38°W) Brazil in April 1998, with 60% of *Mussismilia hispida* (endemic coral), 80% of *Agaricia agaricites*, and 79% of *Siderastrea stellata* (endemic) bleached. Similar bleaching was reported on the Abrolhos Reefs (18°S; 40°W). By October, all colonies had recovered.

In Colombia, only a few bleached corals were seen at Isla San Andres in September 1998. By early October, there was minor bleaching (5–10%) at Islas del Rosario affecting several corals and some gorgonians. No significant bleaching was seen in the Santa Marta area, in October. San Andrés island had only a few partially bleached corals in September 1998 with a similar situation in September 1999. Chengue bay experienced only a minor bleaching event in November 1998 affecting less than 5% of the living coral cover. The most affected species was *Colpophyllia natans*, with many colonies totally bleached, including large ones, greater than 1 m diameter. Other bleached species were *Montastraea faveolata, M. annularis, Millepora alcicornis* and *Porites astreoides*. Another minor bleaching (less than 5%) event occurred in October 1999 with warm (30°C) and very turbid waters, due to continental run-off during the rainy season. Similarly the first bleaching was observed on large *Colpophyllia natans* and 6% of 440 colonies examined between 15–18 m depth had some bleaching. Other species most affected were *Meandrina meandrites* and *Porites astreoides* along with *Millepora alcicornis, Siderastrea siderea, Montastraea faveolata, M. cavernosa, Colpophyllia natans, Diploria strigosa, Stephanocoenia intersepta* and *Dichocoenia stokesi*. But there was no associated coral mortality (Jaime Garzon-Ferreira, pers. comm.).

Up to 50% of live coral in Honduras bleached from 10 to 25 m around Roatan in mid-September 1998, with most species affected, especially *Agaricia, Montastrea* and some *Diploria*. Some *Acropora* and *Millepora* were slightly affected, and no bleaching was seen around the Bay Islands.

#### *Greater Antilles*

Extensive bleaching occurred near Santiago, Herradura and Varadero Cuba (west and east of Havana respectively) in late August 1998, with *Millepora* being extensively bleached, and some *Montastrea annularis, Porites* and zoanthids

## Belize after the 1998 Bleaching and Hurricane Mitch

First reports from September–October 1998 were of large-scale coral bleaching, varying from very severe to relatively minor. Water temperatures were consistently between 30–32°C, and surface temperatures near some cays were between 36–38°C. Then severe Hurricane Mitch hit these reefs a glancing, but strong, blow in November.

Coral populations on Glovers Atoll were disturbed by severe coral bleaching and Hurricane Mitch in 1998. First measures of bleaching were 76% on the western fore reef (near Baking Swash) at 12–15 m, and 70–80% on the shallow patch reefs in the lagoon and on the eastern fore reef down to at least 25 m. Then the atoll barrier of Middle Caye bore the brunt of Hurricane Mitch. Storm damage was severe on the fore-reef (<12 m) to the formerly abundant populations of *Millepora complanata*, *Agaricia tenuifolia*, *Acropora cervicornis* and *A. palmata*. The tops of leaf-like colonies of *M. complanata* and *A. tenuifolia* were sheared off, leaving only remnant live tissue around the base. The branching *A. cervicornis* and *A. palmata* were reduced to rubble. Below 12 m on the 'Wall' there was no obvious signs of mechanical damage, although bleaching was still prevalent down to 30 m, with minimal recovery four months after onset of bleaching. The massive corals *Diploria* and *Montastrea* were less damaged by the storm except those on the upper fore-reef. Columns of *Dendrogyra cylindrus* were broken in shallow areas.

Bleaching was still widespread four months later on the fore-reef, back-reef and lagoon patch reefs, with signs of both recovery and mortality. Mortality was obvious in *Diploria* and *Montastrea* populations, and in addition some colonies have numerous small patches of black band disease. There was an overall loss of 30–50% living coral cover from 10 m depth on the atoll barrier to 200 m into the lagoon.

Impacts were studied on coral recruits (2–20 mm diameter), at 8–10 m depth on the fore-reef slope. Between 70–90% of adult corals were completely bleached, and 25% of smaller recruits exhibited signs of bleaching at all four study sites. One month after the bleaching, which lasted about 3.5 months, many adult colonies regained usual coloration and only 1% of recruits showed any bleaching. The combined bleaching and hurricane disturbance reduced total recruit densities to 20% of pre-disturbance levels, with considerable local variability (Mumby, 1999).

Offshore Cays, the Barrier Reef and inshore patch reefs were originally sites of severe bleaching, with between 25 and 30% mortality to 14–18 m. Total to high bleaching was prevalent in *Millepora*, *Agaricia*, and *Porites*. High to moderate bleaching affected *Montastrea, Siderastrea*, and *Diploria*. Moderate to low bleaching occurred in *Dendrogyra* and *Acropora*, although *A. palmata* was only moderately, or occasionally totally, bleached. Mortality of bleached corals throughout Belize appeared to be about 20–25%.

When these reefs were re-examined after four months and after Hurricane Mitch, the reports were variable depending on the region and on whether the storm had damaged the already stressed corals. At South Water Caye, incidence of bleaching was very low, even in shallow protected areas. Some colonies of *S. siderea, P. astreoides* and *M. annularis* were partially to completely bleached at all sites, but there was considerable recovery from storm damage. The Ambergris Caye barrier reef tract was severely impacted, with little recovery, although some fragments of *A. palmata* and *A. cervicornis* were re-attaching to the southern cayes. There were many juvenile coral and fish recruits and the South Water and Tobacco Caye reefs were quite healthy.

The Silk Keys, which are about 300 m behind the barrier reef, suffered the most hurricane damage on the exposed shallow side at 1–5 m depth. There are very few living corals amongst the *A. palmata* rubble. There was an almost complete absence of living *Acropora* at all sites, where once all three species were relatively abundant. Corals (mostly *Montastrea* spp.) on the protected sites were largely unaffected. There was severe coral bleaching in the lagoon with most *A. tenuifolia* and *Millepora* colonies dead. Most colonies of *Diploria strigosa* and *D. clivosa*, *P. astreoides*, *Siderastrea siderea*, and *Mycetophyllia* spp., were still totally bleached, but living, 4 months after onset. Colonies of *Porites* and *Montastrea* were healthy with no recently dead colonies. The lagoon reefs of Southern Belize have particularly healthy populations of *Montastrea*. Juvenile corals were absent, including species with high levels of recruitment e.g. *Agaricia agaricites* and *Porites astreoides*.

Hurricane Mitch and coral bleaching mortality at Ambergris Caye included approximately 45–50% of corals on the back reef, especially *Montastrea* spp. at Bacalar Chico and Hol Chan Marine Reserves. This is less than the original estimate of 70% of corals bleached on the barrier and back reef in September to November 1998. The extent of bleaching was similar to the 1995 event.

Virtually all corals on the rhomboid shoals of the central shelf lagoon in Belize bleached in October 1998, when surface temperatures were 28.5–30°C. This included almost 100% of *Agaricia tenuifolia*, formerly the predominant coral for many years, and *Acropora cervicornis,* which was the dominant coral until drastic reduction by white-band disease after 1986.

Spurs on the outer barrier reef, Curlew Bank, south of Carrie Bow Cay were covered with partially bleached corals in October 1998. By January 1999, these spurs were wholly or partially devoid of living *A. tenuifolia* and *Porites porites* because of Hurricane Mitch (wave heights 5–6 m). Damage was less severe in the grooves between the spurs, however, broken branches of *Acropora palmata* had moved around, causing significant damage.

Massive corals around Curlew Bank fared better. Most of the surviving foliose, branching, and massive corals are recovering, whereas recovery of *M. annularis* and *M. faveolata* is variable between colonies. In summary, where either bleaching or the Hurricane effects were severe, there have been major losses of corals (>50%), whereas in many other areas, there has been relatively high recovery.

(Compiled from information received from:
Richard Aronson, Eric Borneman, Thomas Bright,
John Bruno, Dr. Burke, Peter Mumby)

(nearly 30% of all colonies). No bleaching has been reported on southern Cuba.

No mass bleaching was seen by dive operators on Dominican Republic, but severe bleaching was reported in Guadeloupe and Haiti.

Temperatures below 30 m were 29°–30°C at Discovery Bay, Jamaica in September 1998 with 70–75% of all *Montastrea* colonies bleached. Recovery was extensive and coral populations are returning to 12–15% cover after extensive losses in the 1980s, through the arrival of new recruits (Jeremy Woodley, pers. comm.).

There was massive bleaching on coral reefs of Puerto Rico in 1998. Many colonies of corals, hydrocorals, crustose octocorals and zoanthids bleached completely (100% of living tissue) down to 40 m in the southwest and Mona Island. The intensity and spread of the bleaching increased significantly after Hurricane George passed, while water temperatures dropped 0.7°C at all depths and remained low for seven days. Visibility was reduced to almost zero for a week, with high turbidity for three weeks. Temperatures measured during 1998 in several reef localities ranged from 26.6–30.2°C (at 20 m), 27.0–31.3°C (at 10 m) and 26.5–31.8°C at the surface. Many corals and zoanthid colonies remained bleached for four months, but recovered when water temperatures returned to normal, such that coral colony mortality was minimal (<0.1%), with some 5% of colonies showing partial mortality.

Bleaching in September 1999 was moderate in Puerto Rico compared to 1998, with fewer colonies affected; none were 100% bleached after water temperatures reached 29.5°C. About 15% of 386 colonies that were severely bleached in August 1998, bleached again in 1999, with most of the bleaching only partial (from 5 to 15% of total surface area). No coral mortality was observed and 99% had regained normal coloration by late December 1999. Water temperatures in September 1999 were 30.2°C at 20 m, 30.5°C at 10 m, and 31.0°C at the surface, but were below 29°C in October (Ernesto Weil, pers. comm.).

### *Mexico and Gulf of Mexico*

Bleaching occurred in August–October 1998 in Quintana Roo, Mexico with temperatures up to 33°C in the lagoon. Bleaching was variable, with *Agaricia* and *Millepora* more affected than *Montastraea* and *Diploria*, and *Acropora* was not affected. Bleaching was extensive in Xcalak, Mahaual, Sian Ka'an Biosphere Reserve in September 1998, affecting mainly *A. tenuifolia*, with high mortality. About 95% of *Montastrea annularis* and 100% of *Diploria labrynthyformis* were bleached but did not die in shallow water (5–7 m) (Rodrigo Garza, pers. comm.).

Virtually no bleaching occurred on the Flower Garden Banks, Gulf of Mexico, USA, during 1998 (Emma Hickerson, pers. comm.). Scattered bleaching was observed in inshore waters of the Florida Keys USA in August 1997, but there was significant recovery. But in July–September 1998 there was bleaching in the Florida Keys, Dry Tortugas National Park and on the Tortugas Banks, with water temperatures of 30–31°C. In some areas bleaching was up to 90% of *Acropora palmata*, with some mortality, 50–80% of *Montastrea annularis* and *A. cervicornis*, and 40–60% of other corals. Minimal bleaching was observed in *Millepora* colonies. By September there was significant recovery of corals on inshore patch reefs and little mortality. Some bleaching was reported during the summer of 1999.

### *Lesser Antilles*

The worst ever bleaching occurred in Soufriere, St. Lucia, with 100% of *Diploria* and extensive bleaching in other species, including *Monrastrea annularis*, *Porites astreoides*, and *Agaricia*. There were unconfirmed reports of severe bleaching in St. Vincent and the Grenadines. Widespread coral bleaching was seen in U.S. Virgin Islands in September 1998, south of St. Thomas, with at least 50% of colonies affected, including *Montastrea annularis*, *Porites* (branching and massive), *Colpophyllia*, some *Millepora*, agaricids, some *Siderastrea*. Bleaching was patchy and there was little mortality by mid-October. Moderate bleaching affected about 20% of the corals on Virgin Gorda. Bleaching on St. John was the worst seen in 25 years, with large expanses of *Montastrea annularis*, and moderate to extreme bleaching in *Colpophyllia*, *Millepora*, *Agaricia*, *Porites*, *Acropora*, and *Siderastrea* (Bob Steneck, pers. comm.).

Other reef localities that were surveyed earlier in the season (June to August 1999) included Bonaire (Netherland Antilles), Venezuela and Bermuda. Bleaching was minimal in all localities sampled. Relative proportions of bleached colonies ranged from 0.07% to 1.0–1.11% of all coral colonies surveyed in Bonaire, Los Roques–Venezuela, and Bermuda respectively.

## ACKNOWLEDGEMENTS

Particular thanks go to those who financially support coordination of the Global Coral Reef Monitoring Network, the Department of State and the National Oceanographic and Atmospheric Administration of USA, and Australian Institute of Marine Science. Thanks are also due to the GCRMN co-sponsors (Intergovernmental Oceanographic Commission of UNESCO, United Nations Environment Programme, IUCN–World Conservation Union, and the Environment Department of the World Bank), and co-hosts (AIMS and ICLARM). The authors who contributed reports and anecdotes, first reported the Wilkinson (1998) report and contributed since are especially thanked. Most of these responded to a broadcast request over the e-mail 'coral-list', administered by Jim Hendee of NOAA in Florida, USA. This is contribution number 1003 from the Australian Institute of Marine Science, but the opinions expressed here are mine and do not necessarily reflect any position of the Institute.

## REFERENCES


Berkelmans, R. and Oliver, J.K. (1999) Large scale bleaching of corals on the Great Barrier Reef. *Coral Reefs* **18**, 55–60.

Berkelmans, R. and Willis, B.L. (1999) Seasonal and local spatial patterns in the upper thermal limits of corals on the inshore Central Great Barrier Reef. *Coral Reefs* **18**, 219–228.

Buddemeier, R.W. (1993) Corals, climate and conservation. *Proceedings 7th International Coral Reef Symposium, Guam 1992, University of Guam*, Vol. 1, pp. 3–10.

Buddemeier, R.W. (1999) Is it time to give up? Presentation to International Coral Reef Symposium, Ft Lauderdale Florida, USA, 16 April 1999.

Bryant, D., Burke, L., McManus, J., and Spalding, M. (1998) Reefs at Risk: a map-based indicator of the threats to the world's coral reefs. World Resources Institute, International Center for Living Aquatic Resources Management, World Conservation Monitoring Centre, and United Nations Environment Programme, Washington, D.C. 56 pp.

Clark, S. (1999) Rehabilitation of degraded reefs using artificial reef blocks (R4533). Field survey reports on the status of the artificial reef site following a major coral bleaching event in April 1998. Unpublished report to Department for International Development, UK, pp. 21.

Ginsburg, R.N. (1993) *Global Aspects of Coral Reefs: Health Hazards and History.* University of Miami, Miami, FL, 420 pp.

Hoegh-Guldberg, O. (1999) Climate Change, coral bleaching and the future of the world's coral reefs. *Marine and Freshwater Research* **50**, 839–866.

Kleypas, J.A., Buddemeier, R.W., Archer, D., Gattuso, J-P., Langdon, C. and Opdyke, B.N. (1999) Geochemical consequences of increased atmospheric carbon dioxide on coral reefs. *Science* **284**, 118–120.

Lough, J.M. (1999) Sea surface temperatures on the Great Barrier Reef: a contribution to the study of coral bleaching. Research Publication No. 57, Great Barrier Reef Marine Park Authority, Townsville pp. 31.

Mumby, P.J. (1999) Bleaching and hurricane disturbances to populations of coral recruits in Belize. *Marine Ecology Progress Series* **190**, 27–35.

Rosenberg, E. and Loya, Y. (1999) *Vibrio shiloi* is the etiological (causative) agent of *Oculina patagonica* bleaching: general implications. *Reef Encounter* **25**, 8–11.

Salvat, B. (1992) Coral reefs—a challenging ecosystem for human societies. *Global Environmental Change* (March) 12–18.

Sheppard, C.R.C. (1999) Coral decline and weather patterns over 20 years in the Chagos Archipelago, Central Indian Ocean. *Ambio* **28**, 472–478.

Timmermann, A., Latif, M., Bacher, A., Oberhuber, J.M. and Roeckner E. (1999) Increased El Nino frequency in a climate model forced by future greenhouse warming. *Nature* **398**, 694–696

Toren, A., Laundau, L., Kushmaro, A, Loya, Y. and Rosenberg, E. (1998) Effect of temperature on the adhesion of Vibrio AK-1 to *Oculina patagonica* and coral bleaching. *Appl. Environ. Microbiol.* **64**, 1379–1384

Wells, S.M., Sheppard, D. and Jenkins, M.D. (eds.) (1988) *Coral Reefs of the World*, Vols. 1–3, UNEP Regional Seas Directories and Bibliographies. IUCN Publication Services, Gland and Cambridge. UNEP Nairobi.

Wilkinson, C.R. (1993) Coral reefs of the world are facing widespread devastation: can we prevent this through sustainable management practices? *Proceedings 7th International Coral Reef Symposium*, Guam 1992, University of Guam, Vol. 1, pp. 11–21.

Wilkinson, C. (1998) *Status of Coral Reefs of the World: 1998*. Australian Institute of Marine Science and Global Coral Reef Monitoring Network, Townsville, 184 pp.

Wilkinson, C.R. (1999) Global and local threats to coral reef functioning and existence: review and predictions. *Marine and Freshwater Research* **50**, 867–878.

Wilkinson, C.R. and Buddemeier, R.W. (1994) *Global Climate Change and Coral Reefs: Implications for People and Reefs.* Report of the UNEP-IOC-ASPEI-IUCN Global Task Team on Coral Reefs. IUCN, Gland Switzerland, 124 pp.

Wilkinson, C., Lindén, O., Hodgson, H., Cesar, J., Rubens J. and Strong, A. (1999) Ecological and socio-economic impacts of 1998 coral mortality in the Indian Ocean: an ENSO (El Niño/Southern Oscillation) impact and a warning of future change? *Ambio* **28**, 188–196.


## THE AUTHOR


**Clive R. Wilkinson**

*Australian Institute of Marine Science,*
*PMB No. 3,*
*Townsville MC 4810, Australia*

Chapter 111

# SEA TURTLES

Jeanne A. Mortimer, Marydele Donnelly and Pamela T. Plotkin

Modern sea turtles are an important component of a wide range of tropical, temperate, and cold water marine ecosystems. Their inclusion on various lists of endangered species reflects past over-exploitation and the current need for better management. Today, seven or eight species of sea turtles are recognised, in two families and six genera.

Adults typically migrate between resident foraging grounds and natal nesting beaches; females reproduce on one to nine year cycles, laying multiple clutches in a season, while males may breed more frequently. Hatching success is usually 80% or more unless external factors intervene. Immature sea turtles grow slowly and delayed maturity requires that large numbers of eggs, hatchlings, juveniles and subadults be maintained in a population in order to sustain even a relatively small number of reproductively active adults. Some populations of some species may take 30 to 50 years (or more) to reach adulthood.

Sea turtle populations have survived and indeed flourished until very recent times. Today they are classed as Critically Endangered, Endangered or Vulnerable.

Human impacts include both purposeful and accidental take, and loss of nesting beach and foraging habitats. Collection of eggs and breeding adults historically has been the most important factor in their demise, but each year hundreds of thousands also die accidentally in numerous fisheries. Well intentioned, but misguided, management efforts, such as poorly run hatcheries, can undermine their survival further. Marine pollution is also important, as are climate change and global warming. Because all sea turtle populations have distributions comprising multiple range states, priorities for action need to be set regionally as well as nationally. The present status of a population can only be properly evaluated when considered in light of past exploitation, and nesting populations (outwardly) respond only slowly to both over-harvest and protection. Critical habitats need to be identified and protected, and both rookeries and habitats should be monitored. Sources of mortality need to be identified, quantified and mitigated. Community involvement is critical to long-term support for conservation programmes, and there is a growing interest among governments and resource managers to collaborate regionally on marine turtle conservation and management. While numerous agreements to conserve sea turtles and their habitats are being created, their widespread implementation is needed.

*Seas at The Millennium: An Environmental Evaluation (Edited by C. Sheppard)*

## EVOLUTIONARY HISTORY

The earliest sea turtles appeared in the late Jurassic, some 150 million years ago (Gaffney and Meylan, 1988), and the fossil record contains more than 30 genera of extinct sea turtles (Pritchard and Trebbau, 1984). This abundant fossil record is the result of several factors, including the relatively good chances of fossilisation of such heavy-boned animals and the disappearance of entire isolated oceans (Pritchard, 1997). Nevertheless, their evolutionary patterns of diversity parallel those found in other taxa (Gould, 1980), with over-specialised types evolving and then disappearing.

Modern turtles are an important component of a wide range of tropical, subtropical and even cold-water marine ecosystems, and are generally less specialised than their forebears (Zangerl, 1980). They should not be considered a relict group. Their inclusion on lists of threatened or endangered species reflects past over-exploitation by humans and the current need for better management, rather than inherently poor adaptation to post-Pleistocene conditions (Pritchard, 1997).

Today, seven or eight species of sea turtles are recognised, comprising two families and six genera (Bowen and Karl, 1997; Pritchard, 1997; Pritchard and Mortimer, 1999). These include: family Dermochelyidae, represented by a single species, the leatherback (*Dermochelys coriacea*); and family Cheloniidae which comprises the green turtle (*Chelonia mydas*), the black turtle (*Chelonia agassizii*) (possibly only a subspecies of *C. mydas*), the hawksbill (*Eretmochelys imbricata*), the loggerhead (*Caretta caretta*), the olive ridley (*Lepidochelys olivacea*), the Kemp's ridley (*Lepidochelys kempii*), and the flatback (*Natator depressus*).

## SEA TURTLE LIFE CYCLE

Prior to the 1950s, when the writings of Professor Archie Carr first attracted attention to marine turtles, few studies had addressed their natural history. Today, although many unanswered questions remain, much more is known about their biology.

### Reproduction

All modern sea turtle species share similar life cycles (Fig. 1). They spend their entire lives in marine or estuarine habitats, the females emerging into terrestrial habitats only to lay eggs or, in restricted cases, for basking (Musick and Limpus, 1997). Copulation occurs at sea (although at certain Pacific rookeries, copulating green turtles sometimes bask). Mating occurs in the month or two just preceding the first ovipositional cycle of the nesting season (Limpus and Miller, 1993). It is usually observed in the vicinity of the nesting beach (Miller, 1997), but also has been recorded during the adult migration between foraging and nesting habitats (JAM pers. obs., Pitman, 1990). Both sexes are promiscuous (Crowell-Commuzzie and Owens, 1990).

All marine turtles emerge from the sea to lay eggs in warm sandy beaches in tropical or subtropical latitudes. Adults typically migrate between resident foraging grounds and natal nesting beaches to reproduce (Bowen and Karl, 1997). During a nesting season, individual females lay an average of 2 to 6 clutches (depending on species and population) at approximately two week intervals (Miller, 1997; Mortimer and Bresson, 1999; Limpus, unpubl. data), and usually reside in the internesting habitat in the vicinity of the nesting beach. In general, females do not reproduce

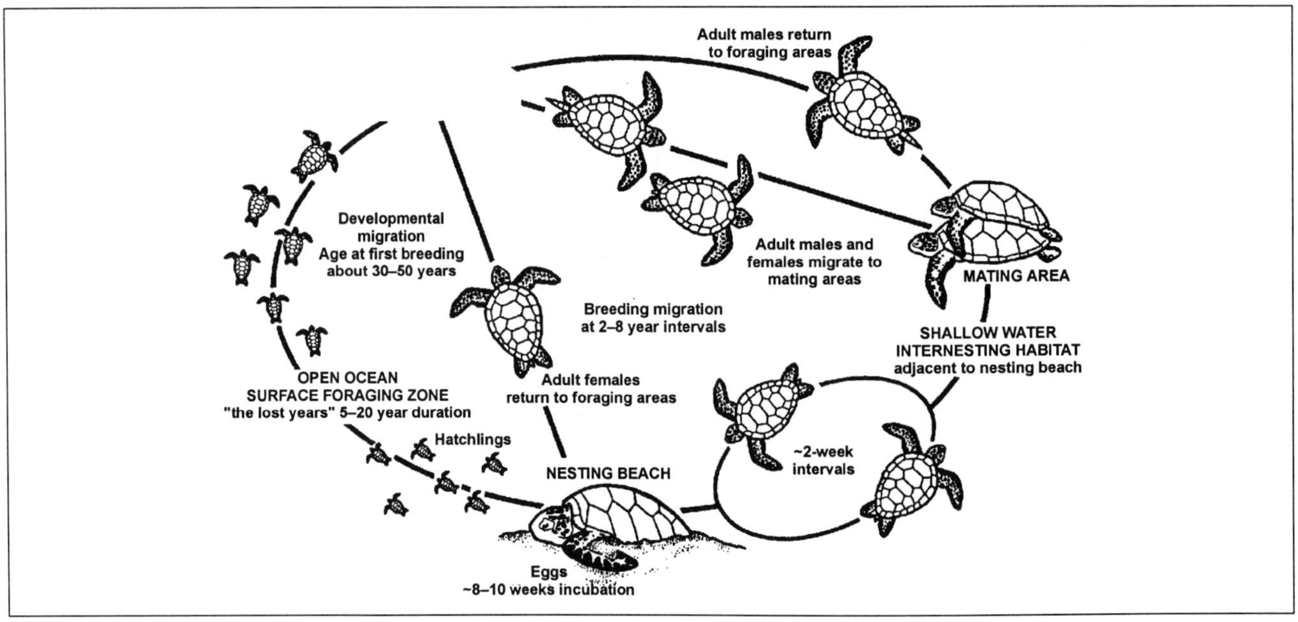


Fig. 1. Generalised life cycle of sea turtles; species vary in the duration of phases. *Dermochelys coriacea* and at least some populations of *Lepidochelys olivacea* remain pelagic foragers throughout their lives. (Reproduced from Miller, 1997.)

every year. The duration of the period between reproductive seasons, defined as the remigration interval, varies among species and populations, ranging from 1 to 9 or more years for females (Miller, 1997). Males may breed at more frequent intervals (Wibbels et al., 1990; Limpus, 1993).

Egg diameter, clutch size and nest depth vary according to species and population. Eggs are approximately spherical in shape, averaging 32 to 55 mm in diameter (Miller, 1997; Pritchard and Mortimer, 1999). Egg clutches averaging some 65 to 180 eggs (Pritchard and Mortimer, 1999) are deposited in nests excavated by the hind flippers of the female to a depth of approximately 60 to 85 cm. After laying eggs, the female covers the nest with sand and returns to the sea (Ehrhart, 1982; Miller, 1997).

The nest environment needs to be within the limits of embryonic tolerance in respect to gas exchange, moisture and temperature. Incubation temperature determines the sex ratio of the offspring, with warmer temperatures producing more females and cooler temperatures more males (Ackerman, 1997). The pivotal temperature (i.e., that which produces an even sex ratio) varies among species and between populations (Limpus et al., 1985). A temperature above 33°C or below 23°C is usually lethal (Ackerman, 1997).

Eggs usually hatch after about eight weeks of incubation (6–13 weeks depending on temperature) (Miller, 1985). Natural hatching success is typically 80% or more unless external factors intervene (e.g. predation, environmental change, microbial infection) (NRC, 1990). Upon emerging from the egg, newly hatched offspring struggle inside the egg chamber, causing the sand above them to be scratched away and sift down through the struggling mass of hatchlings to the bottom of the egg chamber. The empty shells are left in the bottom of the nest and the mass of wriggling hatchlings moves (while maintaining a chamber of air) upward through the sand column (Miller, 1997; Lohmann et al., 1997). If sand temperatures are high near the surface of the sand, the hatchlings become quiescent, usually emerging through the final few centimetres of sand at night or on cool, cloudy or rainy days (Gyuris, 1993).

### Migrations, Growth and Population Structure

Upon reaching the surface of the sand, hatchlings immediately crawl towards the sea, primarily guided by visual cues. They orient towards the brightest point on the horizon, being especially attracted to the shortest wavelengths of visible light (Lohmann et al., 1997), and in some cases they also orient away from elevated silhouettes (Salmon et al., 1992). Upon entering the sea, hatchling loggerheads from Florida at first orient into the waves. By the time they swim beyond the wave refraction zone, they seem to have transferred their initial seaward heading to their magnetic 'compasses', on which they apparently rely to maintain an offshore course long after swimming beyond sight of land (Lohmann et al., 1997).

Most species of post-hatchling turtles (except for the flatback) apparently undertake a passive pelagic migration drifting in oceanic gyres (Muzick and Limpus, 1997) for two to ten or more years depending on the species. The ability of hatchling loggerheads to detect magnetic inclination angle and magnetic field intensity (Lohmann and Lohmann, 1994, 1996) may enable young turtles to stay within the boundaries of gyres or other favourable currents. Adult green turtles at Ascension Island, however, may depend on cues other than magnetic field during navigation (Papi and Mencacci, 1999; Hays et al., 1999). In green turtles, Kemp's ridleys, loggerheads and hawksbills, the larger and older juveniles actively recruit to demersal neritic developmental habitats in the tropical and temperate zones, and eventually move into adult foraging habitats. In some populations, adult habitats are geographically distinct from juvenile developmental habitats, while in others they may overlap or coincide (Musick and Limpus, 1997). Adult turtles show great fidelity to both their nesting (Meylan, 1982) and feeding grounds (Limpus et al., 1992) and migrate between them at appropriate times, even though they may be separated by thousands of km (Mortimer and Carr, 1987).

Mark–recapture studies of immature sea turtles have demonstrated very slow growth rates under natural conditions for those species studied (Bjorndal and Zug, 1995). Data indicate that some populations of green turtles and hawksbills may take 30 to 50 years (or more) to reach adulthood (Chaloupka and Limpus, 1997; Limpus and Chaloupka, 1997) while loggerheads and Kemp's ridley grow more rapidly (Zug, 1990; Frazer and Ehrhart, 1985; Zug et al., 1986). Less is known about growth rates in the more pelagic olive ridleys and leatherbacks. Studies using laparoscopic surgery and hormonal assays demonstrate that temperature-dependent sex determination is capable of producing unbiased, as well as female- and male-biased sex ratios in sea turtle populations (Wibbels et al., 1993). Among populations surveyed to date (including green turtles, loggerheads, and hawksbills), female-biased sex ratios predominate (Owens, 1997).

Delayed maturity requires that a very large number of eggs, hatchlings, juveniles and subadults be maintained in a population in order to sustain even a relatively small number of reproductively active adults. The population model developed by Crouse et al. (1987) for U.S. loggerheads demonstrated that more than 498,000 female eggs and immature turtles would be necessary to maintain a stable adult female population of only 1277 animals (Crouse and Frazer, 1995).

## BIOLOGICAL STATUS OF LIVING SEA TURTLES

### Leatherback (*Dermochelys coriacea*)

This, the largest species of sea turtle, typically reaches a carapace length of 150 to 180 cm and a weight of some 500 kg (Pritchard and Mortimer, 1999). An adult male recovered in

Wales weighed 916 kg. Leatherbacks apparently attain their great size on a diet of jellyfish, salps and other gelatinous organisms (Bjorndal, 1997). Although they nest on tropical beaches, they are inertial endotherms (most efficiently so when >110–120 cm carapace length) and range into cold waters of the high latitudes of both hemispheres (Musick and Limpus, 1997; Hughes et al., 1998). Long-distance tag returns and satellite telemetry show them to be powerful swimmers able to travel 7,000 km in less than four months (Hughes et al., 1998). They forage at the surface and throughout the water column, being able to dive to depths in excess of 1,500 m (Eckert et al., 1989). Although primarily pelagic, they recruit seasonally to temperate and boreal (Musick and Limpus, 1997) as well as tropical coastal habitats (Suarez and Starbird, 1996) to feed on concentrations of jellyfish. Because animals <110 cm carapace length are rarely encountered (Musick and Limpus, 1997) no growth or age data have been obtained from free living animals. Indirect evidence based on skeletochronological analysis and histology of vascular cartilage (Zug and Parham, 1996) suggests rapid growth rates and sexual maturity in as few as nine years, but some investigators place the figure higher (Spotila et al., 1996).

The largest leatherback populations occur in the Pacific and Atlantic Oceans, with only minor rookeries in the Indian Ocean. The species is considered as Endangered by IUCN (1996). Many populations, especially in the Pacific Ocean, are in severe decline (Spotila et al., 1996), largely in response to over-harvest of eggs and incidental capture in fishing gear. The Malaysian population at Terengganu, which comprised some 2000 females nesting annually in the 1960s, is now virtually extinct (Chan and Liew, 1996; Mortimer in press). Several major rookeries along the Pacific coast of Mexico have declined precipitously from over a thousand turtles ten years ago, to only a few hundred in recent years (Sarti et al., 1996). In contrast, the leatherback population of Tongaland, South Africa, has increased from some 20 females nesting annually in the 1960s to over 100 during the 1990s, in response to conservation efforts at the nesting beach (Hughes, 1996).

### Green Turtle (*Chelonia mydas*)

The green turtle, also known as the "edible turtle," is named for the colour of its fat, and may (in some populations) attain a carapace length of 120 cm and weigh up to 230 kg (Hirth, 1997). The systematic status and nomenclature of the "black turtle" or east Pacific green turtle (sometimes referred to as *Chelonia agassizii* or *C. mydas agassizii)* remains uncertain and is the subject of ongoing debate (Pritchard and Mortimer, 1999). Post-hatchling green turtles are believed to be omnivorous and to reside in open ocean pelagic habitats until attaining carapace lengths of 20–25 cm in the Atlantic and about 35 cm in the Indo-Pacific. At that time they enter benthic foraging habitats and shift to an herbivorous diet, feeding primarily on seagrasses and algae, and occasionally on jellyfish, salps and sponges (Bjorndal, 1997). Green turtles are characterised by particularly slow growth, and some populations may take 40 to 60 years to reach sexual maturity (Bjorndal and Zug, 1995). Green turtle distribution is circumglobal throughout tropical seas. Adults travel between feeding pastures and nesting beaches in migrations that may encompass thousands of km (Hirth, 1997).

The green turtle is listed as Endangered (IUCN, 1996). Although relatively abundant compared to most other sea turtle species, green turtle populations are much reduced from historic levels, with some nesting populations now extinct. The apparent abundance of the species is deceptive in that slow growth rates dictate a need for populations to include a relatively large number of immature turtles in order to sustain a much smaller number of breeding adults. The green turtle has been, and continues to be, in great demand for its meat.

### Hawksbill (*Eretmochelys imbricata*)

The hawksbill can attain a carapace length of 90 cm and weigh up to 80 kg in parts of its range, but is significantly smaller in the vicinity of the Arabian Peninsula (Witzell, 1983). Both its nesting and feeding distributions are the most tropical of any sea turtle. Little is known about the ecology of post-hatchling hawksbills, but they appear to be omnivorous and pelagic. At a carapace length of 20–25 cm in the Caribbean and 30–35 cm in the Indo-Pacific they first appear at benthic foraging habitats that include coral reefs, rock outcroppings, seagrass pastures and mangrove-fringed bays (Bjorndal, 1997). Hawksbills subsist primarily on sponge in the Caribbean (Meylan, 1988), but may be more omnivorous in the Indo-Pacific (Bjorndal, 1997; JAM unpubl. data). They grow slowly and, in some populations, may take 30 to 50 years to reach sexual maturity (Chaloupka and Limpus, 1997; JAM, unpubl. data).

Since antiquity, they have been harvested for tortoise shell (i.e., the scales covering their carapace and plastron), long considered a precious material (Parsons, 1972). Today, most populations are declining, depleted, or remnants of larger aggregations. Although some authors have considered the hawksbill to be naturally rare (e.g., Groombridge and Luxmoore, 1989), evidence of their former abundance is provided by high-nesting density at a few remaining sites and by trade statistics reflecting immense past harvests (Meylan and Donnelly, 1999). The exploitation of already depleted populations intensified after World War II in response to the high prices paid by the Japanese market and to more efficient capture techniques (i.e., use of snorkel, SCUBA, spearguns, underwater torches, outboard engines) (Mortimer, 1984). The hawksbill is listed as Critically Endangered (IUCN, 1996), but illegal international trade in tortoiseshell still continues. Throughout much of its range the hawksbill also provides eggs, meat, and shell for the domestic market, and is threatened by habitat destruction—in particular of nesting beaches from coastal development and of foraging habitat from climate change (Meylan and Donnelly, 1999).

### Olive Ridley (*Lepidochelys olivacea*)

The olive ridley is one of the smallest and most abundant sea turtles in the world, growing to 78 cm in carapace length (Pritchard and Plotkin, 1995). These nomadic migrants occupy oceanic habitats in temperate and tropical waters where they swim, feed and bask near the surface (Plotkin, 1994). They appear to be opportunistic feeders of jellyfish, tunicates, crabs, shrimp, and algae (Mortimer, 1982). In regions of greatest population abundance, the species engages in a distinctive nesting behaviour in which hundreds to thousands of breeding females emerge at once to lay eggs on specific beaches in Mexico, Nicaragua, Panama, Costa Rica, India and Surinam (Cornelius, 1986). These annual mass emergences, known by the Spanish term "arribada", may include as many as 500,000 turtles.

Although very large, relatively stable populations of olive ridleys still exist in the Indian and Pacific Oceans, the species is considered to be Endangered (IUCN, 1996). Some nesting populations in Mexico have become extinct as a result of over-harvest of breeding adults by a Mexican turtle fishery (Clifton et al., 1982) that was shut down in 1990 (Aridjis, 1990). Recent downward trends at other nesting beaches have also been reported (Valverde et al., 1998), primarily attributed to incidental capture in shrimp trawl fisheries, but may in part also reflect the movement of turtles to adjacent beaches that have shown an increase in numbers during the same period (Pandav et al., 1994). A small, declining population of olive ridleys occurs in the western Atlantic Ocean along the coasts of Guyana, Surinam and French Guiana (Reichart, 1993). The survival prospects are questionable for this population which has been reduced by past over-harvest of eggs, and incidental capture of juveniles and adults in bottom trawl fisheries in recent years (Pritchard, 1997).

### Kemp's Ridley (*Lepidochelys kempii*)

One of the smallest sea turtles, the Kemp's ridley can reach a carapace length of 80 cm and weigh up to 50 kg (Ross et al., 1989). This nearshore, coastal migrant occurs primarily in temperate and sub-tropical waters of the Gulf of Mexico and western Atlantic Ocean, but juveniles occasionally are recovered in the eastern Atlantic (Weber, 1995). Kemp's ridleys forage in benthic habitats and prey primarily on crabs (Shaver, 1991). Although the Kemp's ridley shares a tendency with its congener the olive ridley to nest in "arribadas", it is restricted to only a single mass nesting site and nests in far fewer numbers. Neither the life span nor the age to maturity have been well documented, but recent estimates suggest that females may reach sexual maturity within 11–12 years (Zug, 1990).

Tens of thousands of Kemp's ridleys once nested at Rancho Nuevo on Mexico's Gulf coast, but the over-harvest of eggs and incidental capture of turtles in shrimp trawl fisheries reduced the population size significantly (Pritchard, 1997). By the 1980s only a few hundred adult females returned to nest at Rancho Nuevo. The species was declared the most endangered of all sea turtles (Pritchard, 1997), and listed as Critically Endangered (IUCN, 1996). A very encouraging upward trend in the number of nests at Rancho Nuevo during the past decade suggests the beginning of a population recovery (Marquez et al., 1996). Conservation efforts, including the mandatory use of turtle excluder devices (TEDs) by shrimp fishing vessels and also protection of adults, eggs and hatchlings at Rancho Nuevo, have apparently decreased mortality and increased recruitment of Kemp's ridleys.

### Loggerhead (*Caretta caretta*)

The broad, massive skull is the most distinguishing characteristic of loggerheads. This large-bodied species can reach a carapace length of 105 cm and weigh up to 180 kg (Dodd, 1988). Loggerheads occur throughout the world, primarily in temperate and sub-tropical (but occasionally in tropical) waters of the Atlantic, Pacific and Indian Oceans. Adult loggerheads primarily occupy nearshore benthic habitats where they feed opportunistically on crustaceans, molluscs, and other invertebrates (Mortimer, 1982). Nesting occurs on beaches throughout its range, but is most dense in the western Atlantic along the U.S. coast, Mediterranean, Australia, southeast Asia and southern Africa (Dodd, 1988). Hatchling loggerheads, upon leaving the natal beaches, undertake developmental migrations that span entire ocean basins (Bolten et al., 1998). Thus they are associated with pelagic habitats far from land, feeding opportunistically on epipelagic invertebrates (Plotkin, 1996). Age to sexual maturity appears to vary among populations; some mature between 24–26 years while others take 30 or more years (Frazer and Ehrhart, 1985; Parnham and Zug, 1987).

Loggerhead populations throughout the world have been depleted due to incidental capture in commercial fisheries, and the species is considered Endangered (IUCN, 1996). Bottom trawl fisheries (NRC, 1990) as well as pelagic longline fisheries (Achaval and Marin, 1998) take thousands each year. Some populations appear to be stable while others are decreasing and have special conservation needs.

### Flatback (*Natator depressus*)

The flatback, also known as the Australian flatback, is a medium sized turtle that reaches a carapace length of 100 cm and weighs up to 90 kg (Limpus et al., 1981). It has a skull similar to the Kemp's ridley turtle (Pritchard, 1997) and an oval shaped carapace similar to the green turtle, but covered with thin, oily epidermal scutes (Zangerl et al., 1988). This species was once believed to be closely related to the green turtle and was classified in the genus *Chelonia*, but a recent taxonomic reassessment based on morphological and biochemical data (Limpus et al., 1988) supported placement of this species in its own genus. In contrast to all other sea turtle species, flatbacks have a very restricted range. They are endemic to the continental shelf of Australia and occur less abundantly in Papua New Guinea and southern Indonesia, apparently residing in shallow water benthic habitats (Limpus et al., 1983). Juvenile flatbacks appear not to undertake long-distance developmental migrations, and are assumed to live sympatrically with adults in Australian waters (Musick and Limpus, 1997). The largest recorded populations of nesting flatbacks occur on Crab Island in the northeastern Gulf of Carpentaria. Nesting is widespread, however, with large populations in the western Torres Strait as well as on Deliverance and Kerr Islands (Limpus et al., 1989). Flatbacks produce the smallest clutch size of any species of sea turtle (averaging 53 eggs), but the largest eggs of any of the Cheloniid sea turtle (Miller, 1997).

The flatback is protected by Australian laws and also listed as Vulnerable (IUCN, 1996). Its status appears to be stable. There are at least two distinct flatback populations in Australian waters (Limpus et al., 1993) and although current population estimates are imprecise several thousand females nest annually. Flatbacks and their eggs are harvested by aboriginal people, but the largest threat to their survival is incidental capture in fisheries. Flatbacks are the most common sea turtle species captured in the prawn trawl fishery (Poiner et al., 1990).

## MODERN RESEARCH NEEDS AND TOOLS

Although a typical adult sea turtle will have spent less than 0.01% of its life in or on nesting beaches as an embryo, a hatchling, or while nesting, approximately 90% of the literature on sea turtle biology is based on nesting beach studies. Arguably, the most vulnerable points in the life cycle of sea turtles are those associated with reproduction and thus warrant thorough study, but the real reason for the bias is that nesting beach studies tend to be less expensive and logistically simpler than in-water studies conducted in turtle foraging habitats.

This focus is shifting, however, in part, due to a realisation of the importance of foraging ground studies, and in part to improvements in the research tools and techniques now available. There is a recognised need (Bjorndal, 1999) to address topics that include: how sea turtles affect the structure and function of ecosystems; the early pelagic stages in the life cycle of sea turtles (virtually unknown for most sea turtle populations); what management units comprise the various sea turtle populations of the world; and their migratory patterns. For all species, the following need to be derived: reliable methods to estimate size of foraging populations; quantitative descriptions of population structure; and measurements of critical demographic parameters so that population models can be developed. Quantitative assessments of human impacts on foraging grounds are needed for effective management and conservation. Beach studies will still be needed. Even basic nesting beach surveys have yet to be conducted in parts of Africa, Asia, and Oceania, and long-term monitoring of known nesting populations remains a useful tool for evaluating population trends (Bjorndal et al., 1999).

Today, complementary techniques are available to achieve recognition of individuals or cohorts and to obtain information about reproductive biology, movements, strandings, residency and growth rates. Tags made of plastic or metal are applied externally to flippers, while passive integrated transponder (PIT) tags are inserted internally (Balazs, 1999). The highest rates of tag retention usually occur with titanium and Inconel tags (available since the late 1970s) and PIT tags. Satellite telemetry can be used to identify migratory routes and corridors (Papi and Luschi, 1996), to locate the sites of foraging habitats (Mortimer and Balazs, in press), and to conduct experiments that test hypotheses about navigational mechanisms (Papi et al., 1997). Interpretation of these data is enhanced with information obtained by remote sensing which can be used to evaluate water temperature, current patterns, and habitat availability (Sheppard, 1999).

Genetic inventories of nesting populations are being used to address evolutionary questions of phylogenetic affiliation and to define management units (Bowen and Karl, 1997). Genetic markers and mixed stock assessments can identify the source of turtles in feeding areas vulnerable to human disturbance, including directed harvests or incidental capture in fishing gear (Bowen and Karl, 1997). They can also elucidate patterns of migration during various stages in the life cycle (Broderick et al., 1994; Mortimer and Broderick, 1999). Genetic markers have application in forensic identification of species and geographic origin for marine turtle products in the market place (Bowen and Karl, 1997).

The sex of juvenile and subadult sea turtles cannot be determined externally, so laparoscopic techniques and hormonal assays are used to assess the structure of foraging populations (Owens, 1997). Laparoscopic surgery also enables an evaluation of reproductive history as well as the stage of follicular development in females and testicular cycles in males (Limpus et al., 1994; Owens, 1997).

### Information Networks

In 1966 IUCN established the Marine Turtle Specialist Group to address the decline of sea turtle populations around the world. A loosely organized network of volunteer specialists, the MTSG has grown to be a dynamic network of several hundred members in 69 countries. It is governed by a Chairperson and an Executive Committee which oversee the work of a full-time Program Officer and various Task Forces. Efforts focus on establishing and strengthening regional networks for sea turtle conservation and promoting the recovery of depleted and declining sea turtle populations. The MTSG has produced global as well as regional Marine Turtle Conservation Strategies and Action Plans, and recently published a manual on Research and Management Techniques for the Conservation of Sea Turtles (Eckert et al., 1999).

A number of resources pertaining to sea turtles are now available on the Internet. These include: the global sea turtle listserve (CTURTLE) and the On-line Sea Turtle Bibliography (both maintained by the Archie Carr Center for Sea Turtle Research at the University of Florida); Spanish and English versions of the quarterly Marine Turtle Newsletter; proceedings of the Annual Symposia on Sea Turtle Biology and Conservation as well as those from various regional meetings; and web sites for a number of organisations devoted to sea turtle conservation.

The Annual Symposium on Sea Turtle Biology and Conservation, conducted each year since 1981, in each of recent years has been attended by more than 700 people from 60 countries. Such international conferences and regional workshops are critical for information exchange and the establishment of networks to promote research, conservation and management. During the last decade regional workshops have been held in the Atlantic, Indian and Pacific Oceans and in the Mediterranean. In 1995 the South Pacific Regional Environment Programme organised a Year of the Sea Turtle campaign to promote recovery of sea turtle resources in the region.

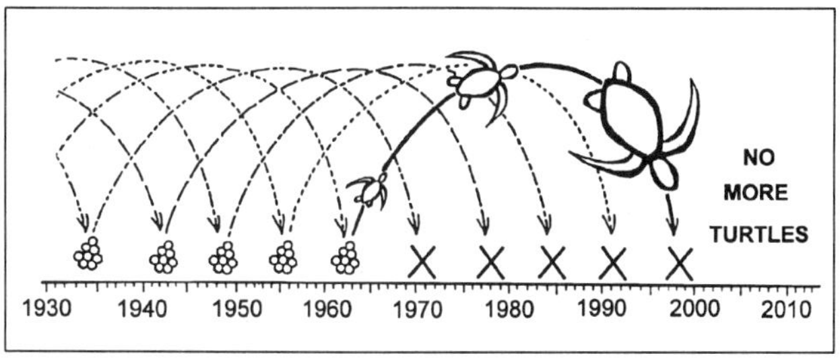


Fig. 2. Schematic diagram showing the response of a nesting population to over-harvest of nesting females. Because many populations of sea turtles take a long time to reach maturity, effects of over-harvest of nesting females may not manifest themselves on the nesting beach until it is too late to save the population. For the purpose of illustration, it is here assumed that the females take 35 years to reach maturity (as is the case for many populations of green turtles and hawksbills). If, each season beginning in 1965, every female is killed on the nesting beach before she can lay her eggs, no serious decline in the annual number of nesting females will be apparent until the year 2000, when the population will have already reached the brink of extinction. (Adapted from Mortimer, 1995.)

## HISTORICAL AND MODERN THREATS TO SPECIES SURVIVAL

Sea turtle populations flourished until very recent times. Undoubtedly, human interference is the cause of their recent collapse. Human activities impact every stage of their life cycle, and range from mortality caused by both purposeful and accidental take, to loss of nesting beach and foraging habitats.

### Direct Harvest

Directed take of eggs and breeding adults, historically, has been the most important factor in the demise of sea turtle populations (Mortimer in press). Using the most rudimentary of implements, small numbers of people on nesting beaches can harvest virtually every egg laid or indeed every female that emerges to lay eggs. Likewise, courting males are easily harpooned from small boats in waters immediately adjacent to the beach. But, for species that take decades to attain sexual maturity, the impact of such intense harvests will not manifest itself on the nesting beach until the population has become dangerously close to extinction. A healthy population that includes only a small number of adults will also comprise relatively large numbers of juvenile and subadults in the 20 to 40 year classes of immature turtles that make up the total population. Thus at least 20 to 40 years can pass after harvesting began before there will be a serious decline in the numbers of females nesting annually (Figs. 2 and 3) (Mortimer, 1995; Bjorndal, 1999).

Aggregations of nesting turtles can produce an illusion of abundance and population stability that does not reflect reality. In theory, a small harvest conducted at a nesting beach could be sustainable. But, in practice, nesting ground harvests have been difficult to control, they interfere with reproduction, and the products derived are often exported to distant markets, either domestic or international. In developing countries, the lure of foreign exchange has provided a strong incentive to maximise exploitation. International trade has been particularly damaging to green turtle and hawksbill populations, and even domestic markets take a toll when eggs, meat, and shell are transported to population centres within the country or when human populations increase in coastal areas. Many populations that have suffered a history of over-harvest are now composed primarily of ageing adults, and their reproductive output needs to be maximised to prevent population collapse (Mortimer, 1992).

Populations that have been well protected for decades at the nesting beach seem able to tolerate some level of harvest at the foraging grounds (Bjorndal et al., 1999). The limits of this population resilience, however, have not been defined. Quantitative descriptions of population structure are needed along with research to determine the regulatory mechanisms that control these demographic parameters (Bjorndal, 1999).

### Mortality Associated with Fisheries

Sea turtles are captured and incidentally drowned or injured in numerous trawl, driftnet, gill net, longline, and other fisheries. The extent of this mortality is difficult to quantify, but hundreds of thousands of turtles die accidentally in fisheries around the world each year (Oravetz, 1999). These interactions are expected to increase as fisheries expand. Mortality-reducing technologies have been developed in some fisheries, but far more research is needed.

Nearly two decades ago scientists identified the capture of sea turtles in shrimp fisheries as the greatest source of incidental mortality (Hillestad et al., 1982). More recently, Oravetz (1999) estimated that as many as 150,000 turtles die in shrimp trawls each year. Turtles captured in trawls drown after long periods of forced submergence or undergo physiological changes that result in death. In the early 1980s gear specialists developed net inserts or Turtle Excluder Devices (TEDs) that allow entrapped turtles to escape. TEDs are required in US waters and other areas of the Western Hemisphere; worldwide use remains a critical but elusive goal.

Indiscriminate and wasteful high seas drift net fisheries for swordfish, billfish, sharks and tuna were abolished nearly a decade ago. Today sea turtles are captured in areas where small-scale or illegal drift net fishing still occurs. Rapidly expanding pelagic longline fisheries have replaced high seas drift net fisheries, and each year billions of hooks are set. Although capture data are very incomplete, longline fleets capture tens of thousands of sea turtles annually (Balazs and Pooley, 1994; Oravetz, 1999). Sea turtles swallow baited hooks, are hooked externally, or become entangled in lines. In some fisheries, such as the Japanese tuna fishery, turtle mortality is observed to be in excess of 40%. In other fisheries most turtles are released alive, but animals that have ingested hooks may face debilitating injuries and are unlikely to survive. Longline fisheries have enormous potential to adversely impact sea turtle populations everywhere. To date, these fisheries in the Pacific have been implicated in the demise of Pacific leatherback populations. Closures by area and season as well as changes to gear and bait will be needed to reduce incidental capture.

Other sources of mortality in fisheries include gill nets, pound nets, purse seines, and trammel nets set for a variety of species, such as coastal sharks and sturgeon. These fisheries capture and drown a significant number of turtles. In some areas, such as Brazil, gill nets are a greater threat to sea turtles than longlines (Oravetz, 1999). Sea turtles also become entangled in the lines of crab and lobster pots and submerged fish traps. The largest leatherback ever recorded, a giant dead male weighing 916 kg, was found entangled in whelk fishing lines off the coast of Wales.

Before the advent of TEDs, incidental shrimp fishery mortality often masked the effects of other fisheries, but in areas where TEDs are required, managers can identify problematic fisheries. Fisheries affect sea turtles of different age classes in different ways. Nets set off nesting beaches, for example, impact breeding adults while the majority of turtles caught in shrimp fisheries are juveniles and subadults. Some fisheries are restricted to certain areas or certain times of the year, such as during migration. The total extent of incidental sea turtle mortality in fisheries remains unquantified because reporting programs are either non-existent or inadequate.

#### Other Mortality Factors

Feral animals (pigs, dogs, cats, fox, mongoose, etc.) or wild ones (such as raccoons) that occur in unnaturally high numbers due to year-round access to human refuse can in some situations cause clutch mortality approaching 100%. Other documented sources of turtle mortality include: boat strikes; dredging operations; and underwater explosions detonated to remove unwanted oil and gas platforms (Lutcavage et al., 1997). Well intentioned, but misguided management efforts, such as poorly run hatcheries (Mortimer, 1992), can also undermine the survival of sea turtle populations.

#### Damage to Nesting and Foraging Habitat

Humans can inadvertently damage sea turtle populations by modifying or eliminating their habitats (Lutcavage et al., 1997). Unregulated coastal development can destroy nesting habitat. Because most sand beaches are dynamic, coastal buildings placed too near the sea are typically protected from erosion by concrete sea walls, wooden walls, rock revetments, sandbag structures, or by constructing groins and jetties to control longshore sand movement. Although any of these structures can impede the access of a turtle to nesting habitat, they pose a greater threat in their potential to completely eliminate nesting habitat (Lutcavage et al., 1997). By creating a scouring effect or by impeding sand flow, such structures often exacerbate erosion. An alternative solution to beach armouring is beach nourishment, whereby sand is mechanically dumped or pumped onto the beach. The suitability of the new habitat for sea turtle nesting will depend on the quality of the fill material. Sand mining operations sometimes remove large quantities of sand from beaches for construction activities. Destruction of dune vegetation at the nesting beach can increase incubation temperatures and thus feminise the offspring produced.

Artificial light visible on a beach can discourage nesting by gravid females and disrupt the nocturnal sea-finding behaviour of both hatchlings and adult turtles. Short wavelength light (UV, violet, blue and green) is more disruptive to turtles than longer wavelengths (yellow and red) (Lutcavage et al., 1997). The presence of people, dogs and mechanised watercraft on and near the nesting beach can discourage female turtles from emerging to lay eggs. Traffic, including pedestrian traffic, has the potential to damage buried eggs and pre-emergent hatchlings. In addition, tyre ruts can "entrap" hatchlings, allowing them to become exhausted or taken by predators (Lutcavage et al., 1997). Egg clutches can be impaled, hatchlings injured or trapped, and nesting emergences of female turtles obstructed by the placement of beach furniture. A wide variety of human rubbish and debris may occur in such density that it can impede or entangle nesting females and emerging hatchlings. Although some objects are dumped directly onto the beach, most material is carried ashore from ships or river mouths (JAM pers. obs.). Tar balls are found on most of the world's beaches, and oil slicks a metre deep have been recorded at rookeries in the Arabian Gulf (Lutcavage et al., 1997).

There is direct evidence that sea turtles are seriously harmed by oil spills at sea and near nesting beaches (Lutcavage et al., 1997). In ocean convergent zones and drift lines, where pelagic juveniles are found, floating tar also accumulates. Sea turtles will pursue and swallow tar balls indiscriminately, and though they appear unable to detect and avoid oil slicks, they are biologically very sensitive to the effects of oil. In juvenile loggerheads, many major physiological systems were adversely affected by short exposures to weathered crude oil. At sea, a turtle maintains prolonged physical contact with floating oil when resurfacing through an oil slick to breathe, inhaling petroleum vapour into its lungs, and ingesting contaminated food or tar balls. Young turtles can starve to death when beak and esophagus become blocked.

Plastics and other non-biodegradable debris are prevalent, and ingestion of and entanglement by such debris has been responsible for high levels of mortality in all species and sizes of sea turtle (Lutcavage et al., 1997). Leatherbacks may not be able to distinguish between their gelatinous prey and transparent plastic. In other species, the list of materials found in turtle digestive tracks is extraordinary, and suggests that hungry sea turtles will swallow almost any material of a suitable size and consistency. Although ingestion of debris can cause a fatal stoppage of the gut, the effects are more often sub-lethal. Entanglement in derelict fishing gear, especially monofilament line, is a particularly serious problem for sea turtles.

Other forms of pollution affecting sea turtles at their feeding grounds include industrial and urban wastes, and agricultural run-off. Chronic pollution from industrial or agricultural sources has been linked with immune suppression in some species, and high tissue levels of organochlorine pollutants have been found (Rybitski et al., 1995). Sea turtle fibropapilloma disease, a major epizootic disease affecting a variety of sea turtle species, is reaching epidemic proportions in some localities. The disease is currently associated with a viral infection, but its expression may be mediated by a compromised immune system (Herbst, 1994).

Climate change and global warming have the potential to impact sea turtles at both the nesting beach and the foraging grounds. Rising temperatures at the nesting beach would feminise the offspring produced. An increase in sea water temperature has already produced a massive die-off of coral reef habitat in the Indian Ocean (Sheppard, 1999). Exactly how such changes may affect sea turtle populations in the long term is yet unknown.

## SEA TURTLE CONSERVATION PRIORITIES

A successful sea turtle conservation programme needs to be multifaceted and to approach problems from a broad perspective, both geographically and temporally (Eckert, 1999). All sea turtle populations have distributions comprising multiple range states, so priorities for action need to be set by thinking regionally as well as nationally. The present status and health of a sea turtle population can only be properly evaluated when considered in light of past exploitation. Where the goal is sustained recovery of depleted stocks, long-term commitment is necessary. Sea turtle nesting populations (outwardly) respond slowly to both over-harvest and protection. With foresight some problems may be circumvented.

### National Level

The status of nesting and feeding populations needs to be evaluated in terms of species, numbers of turtles, and whether the populations are stable, increasing or decreasing. Index rookeries and habitats should be monitored over time. Sources of mortality (including direct harvest, incidental capture, and entanglement in debris) need to be identified and quantified. Mitigating solutions need to be designed and implemented, and incorporated into national legislation as appropriate.

Critical habitats need to be identified and protected from both existing and anticipated threats. Important nesting beaches and feeding habitats should be protected to the highest degree possible. This is usually easiest to achieve before the area is directly threatened by shoreline development. A case in point is the sea turtle nesting habitat now being purchased for the Archie Carr Wildlife Refuge in Florida USA at prices far in excess of what the land would have cost 30 years ago.

Strategies for the protection of important habitats should be fully incorporated into integrated coastal zone management initiatives. Set-back lines need to be established. Lighting ordinances can be adopted to minimise visible light on nesting beaches. Complete darkness is best, but light sources can be regulated in number and wattage, shielded, redirected, lowered, recessed, or repositioned behind objects so light does not reach the beach (Lutcavage et al., 1997).

Awareness campaigns are needed for the general public, and particularly policy makers, fishermen, coastal residents, enforcement personnel, educators, and children. Because slow growth and delayed maturation can mask declines in turtle populations, many people become strong advocates of sea turtle conservation only after they are made aware of the vulnerability of the populations (Figs. 2 and 3) (Mortimer, 1995). Community involvement is critical to long-term support for conservation programmes, and is most effective when stakeholders can realise economic benefits, as for example, in programmes associated with ecotourism.

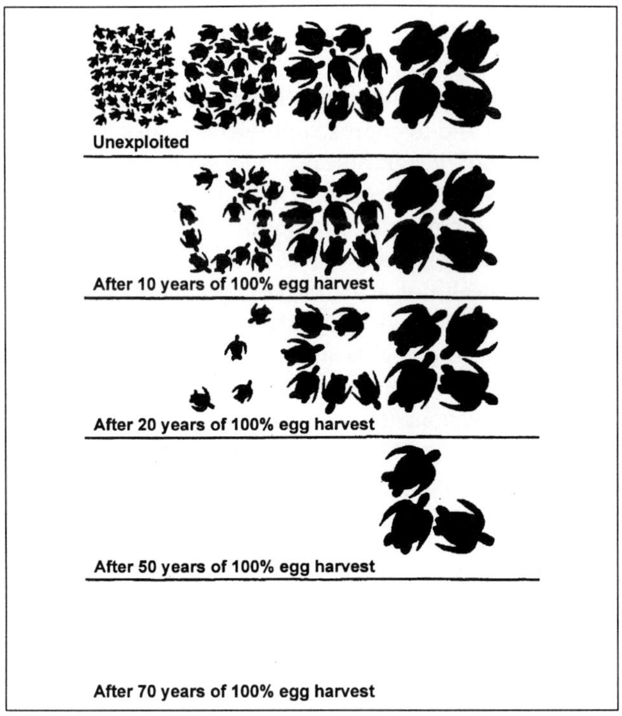


Fig. 3. Diagram illustrating the destruction of a nesting population through over-harvest of eggs. The population is destroyed "from the bottom up" because hatchling turtles are not allowed to enter the population. Because green turtles can take 20 to 50 years to reach adulthood, more than 20 years will pass before the egg collectors (on the nesting beach) will notice a decrease in the numbers of breeding females. By the time such a decline manifests itself it may be too late to save the population. (Adapted from Mortimer, 1995.)

### International and Regional Levels

On a global level, the most important conservation treaties for sea turtles include CITES, and the Convention on Migratory Species (CMS). After CITES came into force in 1975, the enormous, unregulated trade in sea turtle products was significantly reduced. Only Japan continued to import shell, leather, and stuffed specimens under a reservation it maintained until 1994. CITES does not restrict domestic trade, but encourages its Parties to pass complementary, implementing legislation at the national level. Although sea turtle products are still sold in substantial quantities within some countries, international trade is currently banned. Sea turtles are a priority for CMS, which obligates Parties to protect these species and their habitats, and provides a mechanism for regional conservation agreements. In 1999, a number of CMS Parties in West Africa adopted a Memorandum of Understanding to halt the decline of sea turtle populations in the Eastern Atlantic and promote their conservation.

In addition to these global conventions, there is a growing interest among governments and resource managers to collaborate regionally on marine turtle conservation and management. Since the early 1990s, strategic planning workshops and training sessions organized by non-governmental organizations as well as governments have been held in Central America, the wider Caribbean, West Africa, East Africa, North Africa, the Mediterranean, the western Indian Ocean, the northern Indian Ocean, Southeast Asia, the wider Indian Ocean, and the South Pacific. Efforts are under way to promote comprehensive regional programs in the Western Hemisphere. There, two agreements are expected to come into force in 2000: the SPAW (Specially Protected Areas and Wildlife) Protocol of the Cartagena Convention, and the Inter American Convention for the Protection and Conservation of Sea Turtles, the world's first treaty dedicated entirely to sea turtles and their habitats. In 1999, 23 nations in the Indian Ocean participated in a preliminary intergovernmental meeting in Australia to develop a Regional Agreement on the Conservation and Management of Marine Turtles and Their Habitats in the Indian Ocean and Southeast Asian Region. Negotiations for this agreement are scheduled to continue in 2000. While development of these agreements to conserve sea turtles and their habitats will create a mosaic of similar agreements around the world, all range states will need to implement conservation recommendations if these agreements are to be effective.

### Sea Turtles as Flagship Species for Conservation

Because of their charismatic nature and intriguing life cycle, sea turtles are ideal for educational and research activities and are also model flagship species for both local and international conservation (Frazier, 1999). Conserving nesting habitats requires protecting not only the beach, but also vegetation behind the beach, and the marine habitat adjacent to it. The neritic feeding grounds of sea turtles include seagrass and algal pastures, coral reefs, oyster reefs, worm reefs, sand flats, mud flats, estuaries, and mangroves. The pelagic habitats of post-hatchling turtles include the surface waters of convergence zones and major gyre systems throughout the tropical and temperate oceans of the world, while the foraging grounds of leatherbacks include the deep waters of tropical, temperate and even polar seas. In the words of Frazier (1999): "by conserving these animals and their habitats, vast areas of the planet have to be taken into consideration, and managed adequately. ... Conserving sea turtles means protecting the seas and coastal areas, which in turn means protecting a complex, interconnected world on which human societies depend."

### REFERENCES

Achaval, F. and Marin, Y.H. (1998) Incidental capture of sea turtles in the swordfish long line fishery. 17th Annual Symposium on Sea Turtle Biology and Conservation. U.S. Dept. Commerce NOAA Tech. Memo. NMFS-SEFSC-415.

Ackerman, R.A. (1997) The nest environment and the embryonic development of sea turtles, In *The Biology of Sea Turtles*, eds. P.L. Lutz, and J.A. Musick, pp. 83–106. CRC Marine Science Series, CRC Press, Inc., Boca Raton, Florida.

Aridjis, H. (1990) Mexico proclaims total ban on harvest of turtles and eggs. *Marine Turtle Newsletter* **50**, 1–3.

Balazs, G.H. (1999) Factors to consider in the tagging of sea turtles. In *Research and Management Techniques for the Conservation of Sea Turtles*, eds. K.L. Eckert, K.A. Bjorndal, F.A. Abreu-Grobois and M. Donnelly, pp. 101–109. IUCN/SSC Marine Turtle Specialist Group Publication No. 4.

Balazs, G.H. and Pooley, S.G. (1994) Research plan to assess marine turtle hooking mortality: results of an expert workshop held in Honolulu, Hawaii November 16–18 1993. NOAA Technical Memorandum NMFS-SWFSC-201.

Bjorndal, K.A. (1997) Foraging ecology and nutrition of sea turtles. In *The Biology of Sea Turtles*, eds. P.L. Lutz and J.A. Musick, pp. 199–231. CRC Marine Science Series, CRC Press, Inc., Boca Raton, Florida.

Bjorndal, K.A. (1999) Priorities for research in foraging habitats. In *Research and Management Techniques for the Conservation of Sea Turtles*, eds. K.L. Eckert, K.A. Bjorndal, F.A. Abreu-Grobois and M. Donnelly, pp. 12–14. IUCN/SSC Marine Turtle Specialist Group Publication No. 4.

Bjorndal, K.A., Wetherall, J.A., Bolten, A.B. and Mortimer, J.A. (1999) Twenty-six years of green turtle nesting at Tortuguero, Costa Rica: an encouraging trend. *Conservation Biology* 13(1), 126–134.

Bjorndal, K.A. and Zug, G.R. (1995) Growth and age in sea turtles. In *Biology and Conservation of Sea Turtles*, Rev. edn., ed. K.A. Bjorndal, pp. 599–600. Smithsonian Institution Press, Washington, D.C.

Bolten, A.B., Bjorndal, K.A., Martins, H., Dellinger, T., Biscoito, M.J., Encalada, S.E. and Bowen, B.W. (1998) Transatlantic developmental migrations of loggerhead sea turtles. *Ecological Applications* 8(1), 1–7.

Bowen, B.W. and Karl, S.A. (1997) Population genetics, phylogeography, and molecular evolution. In *The Biology of Sea Turtles*, eds. P.L. Lutz and J.A. Musick, pp. 29–50. CRC Marine Science Series, CRC Press, Inc., Boca Raton, Florida.

Broderick, D., Moritz, C., Miller, J.D., Guinea, J.D., Prince, R.I.T. and Limpus, C.J. (1994) Genetic studies of the hawksbill turtle, *Eretmochelys imbricata*: evidence for multiple stocks in Australian waters. *Pacific Conservation Biology* **1**, 123–131.

Chaloupka, M. and Limpus, C.J. (1997) Robust statistical modelling of hawksbill sea turtle growth rates (southern Great Barrier Reef). *Marine Ecology Progress Series* **146**, 1–8.

Chan, E.H. and Liew H.C. (1996) Decline of the leatherback population in Terengganu, Malaysia, 1956–1995. *Chelonian Conservation and Biology* 2(2), 196–203.

Clifton, K. Cornejo, D.O. and Felger R.S. (1982) Sea turtles of the Pacific coast of Mexico. In *Biology and Conservation of Sea Turtles*, ed. K.A. Bjorndal, pp. 199–209. Smithsonian Institution Press, Washington, D.C.

Cornelius, S.E. (1986) *The Sea Turtles of Santa Rosa National Park.* Fundacion de Parques Nacionales, San Jose.

Crouse, D.T.,Crowder, L.B. and Caswell, H. (1987) A stage-based population model for loggerhead sea turtles and implications for conservation. *Ecology* **68**, 1412–1423.

Crouse, D.T. and Frazer, N.B. (1995) Population models and structure. In *Biology and Conservation of Sea Turtles*, Rev. edn., ed. K.A. Bjorndal, pp. 601–603. Smithsonian Institution Press, Washington, D.C.

Crowell-Commuzzie, D.K. and Owens, D.W. (1990) A quantitative analysis of courtship behavior in captive green sea turtles (*Chelonia mydas*). *Herpetologica* 46, 195–202.

Dodd, C.K. (1988) Synopsis of the biological data on the loggerhead sea turtle *Caretta caretta* (Linnaeus 1758). U.S. Fish Wildl. Serv. Biol. Rep. 88 (14).

Eckert, K.L. (1999) Designing a conservation program. In *Research and Management Techniques for the Conservation of Sea Turtles*, eds. K.L. Eckert, K.A. Bjorndal, F.A. Abreu-Grobois and M. Donnelly, pp. 6–8. IUCN/SSC Marine Turtle Specialist Group Publication No. 4.

Eckert, K.L., Bjorndal, K.A.., Abreu-Grobois, F.A. and Donnelly, M. (eds.). (1999) *Research and Management Techniques for the Conservation of Sea Turtles.* IUCN/SSC Marine Turtle Specialist Group Publication No. 4.

Eckert, S.A., Eckert, K.L., Ponganis, P. and Kooyman, G.L. (1989) Diving and foraging behavior of leatherback sea turtles (*Dermochelys coriacea*). *Canadian Journal of Zoology* **67**, 2834–2840.

Ehrhart, L.M. (1982) A review of sea turtle reproduction. In *Biology and Conservation of Sea Turtles*, ed. K.A. Bjorndal, pp. 29–38. Smithsonian Institution Press, Washington, D.C.

Frazer, N.B. and Ehrhart, L.M. (1985) Preliminary growth models for green, *Chelonia mydas*, and loggerhead, *Caretta caretta*, turtles in the wild. *Copeia* 1985, 73–79.

Frazier, J.G. (1999) Community-based conservation. In *Research and Management Techniques for the Conservation of Sea Turtles*, eds. K.L. Eckert, K.A. Bjorndal, F.A. Abreu-Grobois and M. Donnelly, pp. 15–18. IUCN/SSC Marine Turtle Specialist Group Publication No. 4.

Gaffney, E.S. and Meylan, P.A. 1988. A phylogeny of turtles. In *The Phylogeny and Classification of Tertrapods*, ed. M.J. Benton, pp. 157–219. Clarendon Press, Oxford.

Gould, S.J. (1980) An early start. In *The Panda's Thumb. More Reflections in Natural History.* Norton, New York. pp. 217–226.

Groombridge, B. and Luxmoore, R. (1989) The green turtle and hawksbill (Reptilia: Cheloniidae): World status, exploitation, and trade. CITES Secretariat, Lausanne Switzerland. 601 pp.

Gyuris, E. (1993) Factors that control the emergence of green turtle hatchlings from the nest. *Wildlife Research* **20**, 345–353.

Hays, G.C., Luschi, P., Papi, F., Del Seppia, C. and Marsh, R. (1999) Changes in behavior during the inter-nesting period and post-nesting migration for Ascension Island green turtles. *Marine Ecology Progress Series* **189**, 263–273.

Herbst, L.H. (1994) Fibropapillomatosis of marine turtles. *Annual Reviews of Fish. Diseases* **4**, 389–425.

Hillestad, H.O., Richardson, J.I., McVea, Jr., C. and Watson, Jr., J.M. (1982) Worldwide incidental capture of sea turtles. In *Biology and Conservation of Sea Turtles*, ed. K.A. Bjorndal, pp. 489–502. Smithsonian Institution Press, Washington, D.C.

Hirth, H.F. (1997) Synopsis of the biological data on the green turtle, *Chelonia mydas* (Linnaeus 1758). Biological Report 97(1). Fish and Wildlife Service, Washington DC.

Hughes, G.R. (1996) Nesting of the leatherback turtle *(Dermochelys coriacea)* in Tongaland, KwaZulu-Natal, South Africa, 1963–1995. *Chelonian Conservation and Biology* **2** (2), 153–158.

Hughes, G.R., Luschi, P., Mencacci, R. and Papi, F. (1998) The 7000-km oceanic journey of a leatherback turtle tracked by satellite. *Journal of Experimental Marine Biology and Ecology* **229**, 209–217.

IUCN (1996) *1996 IUCN Red List of Threatened Animals.* IUCN, Gland, Switzerland.

Limpus, C.J. (1993) The green turtle, *Chelonia mydas*, in Queensland: breeding males in the southern Great Barrier Reef. *Wildlife Research* **10**, 513–523.

Limpus, C.J. and Chaloupka, M. (1997) Nonparametric regression modelling of green sea turtle growth rates (southern Great Barrier Reef). *Marine Ecology Progress Series* **149**, 23–34.

Limpus, C.J., Couper, P.J. and Read, M.A. (1994) The loggerhead turtle, *Caretta caretta*, in Queensland: population structure in a warm temperate feeding area. *Memoirs of the Queensland Museum* **37**, 195–204.

Limpus, C.J., Couper, P.J. and Couper, K.L.D. (1993) Crab Island revisited: reassessment of the world's largest flatback turtle rookery after twelve years. *Memoirs of the Queensland Museum* **33** (1), 277–289.

Limpus, C.J., Gyuris, E., and Miller, J.D. (1988) Reassessment of the taxonomic status of the sea turtle genus *Natator* McCullough, 1908, with redescription of the genus and species. *Transactions of the Royal Society of S. Australia* **112**, 1–10.

Limpus, C.J. and Miller, J.D. (1993) Family Cheloniidae. In *Fauna of Australia, Vol. 2A. Amphibia and Reptilia*, eds. C.J. Glasby, G.J.B. Ross, and P.L. Beesley. Australian Government Publishing Service, Canberra, Australia.

Limpus, C.J., Miller, J.D., Parmenter, C.J., Reimer, D., McLachlan, N. and Webb, R. (1992) Migration of green (*Chelonia mydas*) and loggerhead (*Caretta caretta*) turtles to and from eastern Australian rookeries. *Wildlife Research* **19**, 347–358.

Limpus, C.J., Parmenter, C.J., Baker, V. and Fleay, A. (1983) The flatback turtle, *Chelonia depressa*, in Queensland: Post-nesting migration and feeding round distribution. *Aust. Wildlife Research* **10**, 557–561.

Limpus, C.J., Parmenter, C.J., Parker, R. and Ford, N. (1981) The flatback turtle *Chelonia depressa* in Queensland: The Peak Island rookery. *Herpetofauna* **13**(1), 15–19.

Limpus, C.J., Reed, P., Miller, J.D. 1985. Temperature dependent sex determination in Queensland sea turtles: Intraspecific variation in *Caretta caretta*. In *Biology of Australasian Frogs and Reptiles*, eds. G. Grigg, R. Shine and H. Ehmann, pp. 343–351. Surrey Beatty and Sons, Sidney, Australia.

Limpus, C.J., Zeller, D., Kwan, D. and Macfarlane, W. (1989) The sea turtle rookeries of north western Torres Strait. *Aust. Wildlife Research* **16**, 517–525.

Lohmann, K.J. and Lohmann, C.M.F. (1994) Acquisition of magnetic directional preference in hatchling loggerhead sea turtles. *Journal of Experimental Biology* **190**, 1–8.

Lohmann, K.J. and Lohmann, C.M.F. (1996) Orientation and open-sea navigation in sea turtles. *Journal of Experimental Biology* **199**, 73–81.

Lohmann, K.J., Witherington, B.E., Lohmann, C.M.F. and Salmon, M. (1997) Orientation, navigation, and natal beach homing in sea turtles. In *The Biology of Sea Turtles*, eds. P.L. Lutz and J.A. Musick, pp. 107–135. CRC Marine Science Series, CRC Press, Inc., Boca Raton, Florida.

Lutcavage, M.E., Plotkin, P., Witherington, B. and Lutz, P.L. (1997) Human impacts on sea turtle survival. In *The Biology of Sea Turtles*, eds. P.L. Lutz and J.A. Musick, pp. 387–409. CRC Marine Science Series, CRC Press, Inc., Boca Raton, Florida.

Marquez, R.M., Byles, R.A., Burchfield, P., Sanchez, M.P., Diaz, J.F, (1996) Good news! Rising numbers of Kemp's ridleys nest at Rancho Nuevo, Tamaulipas, Mexico. *Marine Turtle Newsletter* **76**, 2–5.

Meylan, A.B. (1982) Sea turtle migration — evidence from tag returns. In *Biology and Conservation of Sea Turtles*, ed. K.A. Bjorndal, 91–100. Smithsonian Institution Press, Washington D.C.

Meylan, A.B. (1988) Spongivory in hawksbill turtles: a diet of glass. *Science* **239**, 393.

Meylan, A.B. and Donnelly, M. (1999) Status justification for listing the hawksbill turtle (*Eretmochelys imbricata*) as Critically Endangered on the 1996 IUCN Red List of Threatened Animals. *Chelonian Conservation and Biology* **3** (2), 200–224.

Miller, J.D. (1985) Embryology of marine turtles. In C. Gans, F. Billett and P.F.A. Maderson, pp. 269–328. *Biology of the Reptilia*, Vol. 14A. Wiley-Interscience, New York.

Miller, J.D. (1997) Reproduction in sea turtles. In *The Biology of Sea Turtles*, eds. P.L. Lutz and J.A. Musick, pp. 51–82. CRC Marine Science Series, CRC Press, Inc., Boca Raton, Florida.

Mortimer, J.A. (1982) Feeding ecology of sea turtles. In *Biology and Conservation of Sea Turtles*, ed. K.A. Bjorndal, pp. 103–109. Smithsonian Institution Press, Washington, D.C.

Mortimer, J.A. (1984) Marine turtles in the Republic of the Seychelles: Status and Management. IUCN, Gland Switzerland. 80 pp.

Mortimer, J.A. (1992) Marine turtle conservation in Malaysia. *Malayan Nature Journal* **45** (1–4), 353–361.

Mortimer, J.A. (1995) Teaching critical concepts for the conservation of sea turtles. *Marine Turtle Newsletter* **71**, 1–4.

Mortimer, J.A. (in press) Sea turtle conservation programs: factors determining success or failure. In *Marine and Coastal Protected Areas. A Guide for Planners and Managers*, eds. R.V. Salm and J.R. Clark. IUCN, Gland, Switzerland.

Mortimer, J.A. and Balazs, G.H. (in press) Post-nesting migrations of hawksbill turtles in the granitic Seychelles and implications for conservation. Proceedings of the 19th Annual Symposium on Sea Turtle Conservation and Biology. South Padre Island, Texas, USA.

Mortimer, J.A. and Bresson, R. (1999) Temporal distribution and periodicity in hawksbill turtles *(Eretmochelys imbricata)* nesting at Cousin Island, Republic of Seychelles, 1971–1997. *Chelonian Conservation and Biology* **3**(2), 318–325.

Mortimer, J.A. and Broderick, D. (1999) Population genetic structure and developmental migrations of sea turtles in the Chagos Archipelago and adjacent regions inferred from mtDNA sequence variation. In *Ecology of the Chagos Archipelago*, eds. C.R.C. Sheppard and M.R.D. Seaward, pp. 185–194. Linnaean Society Occasional Publications 2. Westbury Press.

Mortimer, J.A. and Carr, A. (1987) Reproduction and migrations of the Ascension Island green turtle *(Chelonia mydas)*. *Copeia* 1987(1), 103–113.

Musick, J.A. and Limpus, C.J. (1997) Habitat utilization and migration in juvenile sea turtles. In *The Biology of Sea Turtles*, eds. P.L. Lutz and J.A. Musick, pp. 137–164. CRC Marine Science Series, CRC Press, Inc., Boca Raton, Florida.

NRC (National Research Council) (1990) *The Decline of Sea Turtles*. National Academy of Science Press, Washington, D.C. 259 pp.

Oravetz, C. (1999) Reducing incidental catch in fisheries. In *Research and Management Techniques for the Conservation of Sea Turtles*, eds. K.L. Eckert, K.A. Bjorndal, F.A. Abreu-Grobois and M. Donnelly, pp. 189–193. IUCN/SSC Marine Turtle Specialist Group Publication No. 4.

Owens, D.W. (1997) Hormones in the life history of sea turtles. In *The Biology of Sea Turtles*, eds. P.L. Lutz and J.A. Musick, pp. 315–341. CRC Marine Science Series, CRC Press, Inc., Boca Raton, Florida.

Pandav, B., Choudhury, B.C. and Kar, C.S. (1994) A status survey of olive ridley sea turtle *(Lepidochelys olivacea)* and its nesting habitats along the Orissa coast, India. Wildlife Institute Of India, Dehradun. 40 pp. + illustrations.

Papi, F. and Luschi, P. (1996) Pinpointing 'Isla Meta': the case of sea turtles and albatrosses. *Journal of Experimental Biology* **199**, 65–71.

Papi, F., Luschi, E., Crosio, E. and Hughes, G.R. (1997) Satellite tracking experiments on the navigation ability and migratory behaviour of the loggerhead turtle *Caretta caretta*. *Marine Biology* **129**, 215–220.

Papi, F. and Mencacci, R. (1999) The green turtles of Ascension Island: a paradigm of long-distance navigational ability. *Rend. Fis. Acc. Lincei* s 9, **10**, 109–119.

Parnham, J.F. and Zug, G.R. (1997) Age and growth of loggerhead sea turtles (*Caretta caretta*) of coastal Georgia: An assessment of skeletochronological age-estimates. *Bulletin of Marine Science* **61** (2), 287–304.

Parsons, J.J. (1972) The hawksbill turtle and the tortoise shell trade. In *Etudes de Geographie Tropicale Offertes a Pierre Gourou*, pp. 45–60. Mouton, Paris.

Pitman, R.L. (1990) Sea turtle associations with flotsam in the eastern tropical Pacific Ocean. In M. Salmon and J. Wyneken (eds.) Proceedings of the 11th Annual Workshop on Sea Turtle Biology and Conservation, p. 94. NOAA-TM-NMFS-SEFSC 302.

Plotkin, P.T. (1994) Migratory and reproductive behavior of the olive ridley turtle, *Lepidochelys olivacea* (Eschscholtz 1829), in the eastern Pacific Ocean. Ph.D. dissertation, Texas A&M University, College Station, TX.

Plotkin, P.T. (1996) Occurrence and diet of juvenile loggerhead sea turtles, *Caretta caretta*, in the northwestern Gulf of Mexico. *Chelonian Conservation and Biology* **2**(1), 78–80.

Poiner, I.R., Buckworth, R.C. and Harris, A.N.M. (1990) Incidental capture and mortality of sea turtles in Australia's northern prawn fishery. *Australian Journal of Marine and Freshwater Research* **41**, 97–110.

Pritchard, P.C.H. (1997) Evolution, phylogeny, and current status. In *The Biology of Sea Turtles*, eds. P.L. Lutz and J.A. Musick, pp. 1–28. CRC Marine Science Series, CRC Press, Inc., Boca Raton, Florida.

Pritchard, P.C.H. and Mortimer, J.A. (1999) Taxonomy, external morphology, and species identification. In *Research and Management Techniques for the Conservation of Sea Turtles*, eds. K.L. Eckert, K.A. Bjorndal, F.A. Abreu-Grobois and M. Donnelly, pp. 21–38. IUCN/SSC Marine Turtle Specialist Group Publication No. 4.

Pritchard, P.C.H. and Plotkin, P.T. (1995) Status of the olive ridley sea turtle *(Lepidochelys olivacea)*. In: P.T. Plotkin (ed.), National Marine Fisheries Service and U.S. Fish and Wildlife Service Status Reviews for Sea Turtles Listed under the Endangered Species Act of 1973. National Marine Fisheries Service, Silver Spring, MD.

Pritchard, P.C.H. and Trebbau, P. 1984. *The Turtles of Venezuela*. Society for Study of Amphibians and Reptiles, Oxford, Ohio. 403 pp.

Reichart, H.A. (1993) Synopsis of biological data on the olive ridley sea turtle *Lepidochelys olivacea* (Escholtz 1829) in the western Atlantic. NOAA Technical Memorandum, NMFS-SEFSC-336, Miami, Florida. 78 pp.

Ross, J.P., Beavers, S., Mundell, D. and Airth-Kindree, M. (1989) The status of Kemp's ridley. Center for Marine Conservation, Washington, D.C. 51 pp.

Rybitski, M.J., Hale, R.C. and Musick, J.A. (1995) Distribution of organochlorine pollutants in Atlantic sea turtles. *Copeia* 1995(2), 379–390.

Salmon, M., Wyneken, J., Fritz, E. and Lucas, M. (1992) Seafinding by hatchling sea turtles: role of brightness, silhouette and beach slope as orientation cues. *Behaviour* **122** (1–2), 56–77.

Sarti, M., Eckert, S.A., Garcia T. and Barragan, A.R. (1996) Decline of the world's largest nesting assemblage of leatherback turtles. *Marine Turtle Newsletter* **74**, 2–5.

Shaver, D.J. (1991) Feeding ecology of wild and headstarted Kemp's ridley sea turtles in south Texas waters. *Journal of Herpetology* 25(3), 327–334.

Sheppard, C.R.C. (1999) Coral decline and weather patterns over 20 years in the Chagos Archipelago, Central Indian Ocean. *Ambio* 28(6), 472–478.

Spotila, J.R., Dunham, A.E., Leslie, A.J., Steyermark, A.C., Plotkin, P.T. and Palodino, F.V. (1996) Worldwide population decline of *Dermochelys coriacea*: are leatherback turtles going extinct? *Chelonian Conservation and Biology* **2**(2), 209–222.

Suarez, A. and Starbird, C.H. (1996) Subsistence hunting of leatherback turtles, *Dermochelys coriacea*, in the Kai Islands, Indonesia. *Chelonian Conservation and Biology* **2**(2), 190–195.

Valverde, R.A., Cornelius, S.E. and Mo, C.L. (1998) Decline of the olive ridley sea turtle *(Lepidochelys olivacea)* nesting assemblage at Nancite Beach, Santa Rosa National Park, Costa Rica. *Chelonian Conservation and Biology* **3**(1), 58–63

Weber, M. (1995) Status of the Kemp's ridley sea turtle, *Lepidochelys kempii*. In: P.T. Plotkin (ed.), National Marine Fisheries Service and U.S. Fish and Wildlife Service Status Reviews for Sea Turtles Listed under the Endangered Species Act of 1973. National Marine Fisheries Service, Silver Spring, MD.

Wibbels, T., Balazs, G.H., Owens, D.W. and Amoss, M.S. (1993) Sex ratio of immature green turtles inhabiting the Hawaiian Archipelago. *Journal of Herpetology* **27**(3), 327–329.

Wibbels, T., Owens, D.W., Limpus, C.J., Reed, P.C. and Amoss, M.S., Jr. (1990) Seasonal changes in serum gonadal steroids associated with migration, mating and nesting in the loggerhead sea turtle *(Caretta caretta)*. *Gen. Comp. Endocrinol.* **79**(1), 154–164.

Witzell, W.N. (1983) Synopsis of biological data on the hawksbill turtle, *Eretmochelys imbricata* (Linnaeus, 1766). FAO Fish. Synop. (137), 78 p.

Zangerl, R. (1980) Patterns of phylogenetic differentiation in the toxochelyid and cheloniid sea turtles. *American Zoologist* **20**, 585–596.

Zangerl, R., Hendrickson, L.P. and Hendrickson, J.R. (1988) A redescription of the Australian flatback sea turtle, *Natator depressus*. *Bishop Museum Bulletins in Zoology, Honolulu* **1**, 1–69.

Zug, G.R. (1990) Estimates of age and growth in *Lepidochelys kempii* from skeletochronological data, pp. 285–286. In Richardson, T.H., Richardson, J.I., Donnelly, M. (Compilers). Proceedings of the tenth annual workshop on sea turtle biology and conservation. NOAA Tech.

Zug, G.R. and Parham, J.F. (1996) Age and growth in leatherback turtles, *Dermochelys coriacea* (Testudines: Dermochelyidae): a skeletochronological analysis. *Chelonian Conservation and Biology* **2** (2), 244–249.

Zug, G.R., Wynn, A.H. and Ruckdeschel, C.A. (1986) Age determination of loggerhead sea turtles, *Caretta caretta*, by incremental growth marks in the skeleton. Smithsonian Contributions in Zoology, No. 427.

## THE AUTHORS

**Jeanne A. Mortimer**
*Department of Zoology, University of Florida, Gainesville, FL 32611-8525, U.S.A.*
*and*
*Marine Conservation Society of Seychelles, P.O. Box 445, Victoria, Mahe, Seychelles*

**Marydele Donnelly**
*IUCN/SSC Marine Turtle Specialist Group, 1725 DeSales St. NW #600, Washington, DC 20036, U.S.A.*

**Pamela T. Plotkin**
*Center for Marine Conservation, 1725 DeSales St. NW #600, Washington, DC 20036, U.S.A.*

Chapter 112

# WHALES AND WHALING

Sidney Holt

Until recently most of what we know about whales came from the whaling industry, and the things we thought we knew, but really did not, also came from that source. Much of whales' natural history is now familiar, but there are still misunderstandings. Their economic history is not so well known, even by many of those who are concerned about their fate.

There are nearly one hundred surviving species of cetaceans, in two sub-groups distinguished essentially by their sizes. The bigger are by far the biggest animals that have ever lived. A few species venture up rivers, and a few others live entirely—and now survive precariously—in tropical freshwaters. Most of the larger species are toothless (Mysticetes), but a few are odontocetes (toothed), notably the sperm whale and four species of bottlenose whales, pilot whales, the white whale (beluga), and the narwhal.

A brief history of whaling provides a background to the present condition of the world's depleted cetacean stocks. The development of methods which attempted to determine allowable catch rates and to assess population levels is described; most are shown to have been flawed, often obviously and severely so, with the result that today we are largely still ignorant of what population sizes existed historically, and how big population sizes are now, for most species. Current controversies remain over levels of catch, 'scientific whaling', illegal and pirate whaling, aboriginal catches and sanctuaries, but it is clear that there now is a global shift away from the practices which dominated most of the 20th century.

*Seas at The Millennium: An Environmental Evaluation (Edited by C. Sheppard)*

## HISTORICAL PERSPECTIVE

For millennia, humans have occasionally eaten whales. Equally, cetaceans have attracted human attention as awesome creatures. Orcas, for example, are reputed to have driven fish ashore for the benefit of Australian Aboriginals, while the now much persecuted minke whale was known to Norwegian fishers as the whale that drove herring into the bays. They have been common objects of prose and poetry, of painting, sculpture and music (McIntyre, 1974).

It seems that the Basques of Northwestern Spain invented commercial pelagic whaling soon after the turn of the second millennium. At first they hunted the Biscay/Greenland right whale close to home, but as these were depleted they ventured further afield and eventually reached North American shores. In that process of expansion they appear to have invented what later came to be called the try works, an apparatus for processing whales aboard their vessels. Later, having depleted the right whales on the western side of the north Atlantic they turned to cod fishing, and for a couple of centuries they kept secret the location of their far offshore fishing and whaling operations (Kurlansky, 1999; Ellis, 1992).

We know little about the next few centuries. In a period known as "old whaling", techniques differed little though operations expanded thanks to larger sailing vessels on which whales were processed, and the introduction of the ships' chronometer. The British, Dutch and, later, North Americans (East coast "Yankees") were principally engaged in this industry. Between them they nearly exterminated the right whales everywhere in the northern hemisphere, and the Pacific grey whales that migrated between Alaska and Mexico. Their main target then became, in the 19th century, the sperm whale, oil from which illuminated the rising cities of North America and Europe. This industry collapsed, it seems, not so much from a dearth of whales as from the development of mineral oil resources and the industrial production of gas.

One legacy from old sperm whaling days is a form of so-called aboriginal subsistence whaling. Members of American and European whalers' crews from time to time jumped ship in faraway places such as Tonga in the tropical south Pacific, St Vincent and Grenadines in the Caribbean, and the Azores. In some of those places they introduced harpoon whaling from small boats. Until a decade ago a Tongan family named Cook (!) were still catching humpback whales, and a similar hunt of humpbacks continues in St Vincent and Grenadines. Sperm whaling from open boats continued in the Azores well into the second half of the 20th century. In Japan—now the main whaling nation—catching whales from shore in nets rather than with harpoons has continued unbroken for a millennium.

Modern whaling came into being in the latter half of the 19th century. The Norwegian historians J.N. Tønnessen and A.O. Johnsen asked "why the Norwegians were its pioneers and why the county of Vestfold in southeastern Norway was to be its cradle"? Tønsberg in Vestfold was the birthplace of Svend Foyn, the inventor, in mid-century, of the harpoon fired from a cannon and carrying an explosive grenade, which revolutionised commercial whaling. This, combined with the steamship, was the key. Although Sandefjord later became the capital of modern pelagic whaling, the Foyn method began its world-wide growth in bases in Arctic Finnmark. In the southeast were capital and potential crewmen, in the far north were the marine mammals.

Tønnessen and Johnsen assert that "no whaling was carried out around the middle of the 19th century from the shores of Norway", giving the lie to current claims that whaling in Norway is an unbroken tradition extending back to the Norsemen. "The Americans were the dominant whaling nation, and both they and the British were far ahead of the Norwegians in technical development. Both had...gone so far as to combine the harpoon and the grenade in a single shot."

The late 1860s signalled the emergence of Japan from a feudal community to a modern industrial power. Many went to the West to learn technologies. One was Juro Oka who went to Vestfold, where he ordered equipment, then to Finnmark to see it being used, then to the Azores to inspect sperm whaling, and then on to Newfoundland where Norwegian-style catching had been implanted. Reporting back to Tokyo, Oka found fertile ground for his ideas.

The sale of whaling technology to the Japanese, and others, caused a huge reaction in Norway, since this technology was seen as the principal element of the whaling economy, and so began the moves by which the Norwegians sought, but ultimately failed to hold, a monopoly in large-scale whaling. Towards the end of the century a Norwegian newspaper wrote: "Once the Japanese have appeared on the scene in any whaling ground, then the Norwegians will soon be banished from it". This indeed happened but the banishment took another half century to be completed.

Steam-powered whaling vessels and the Foyn method permitted the capture of fast-swimming rorquals. Initially they were hunted close to shore and towed back to shore stations for processing ("flensing"). Steam-power, of course, also facilitated towing; even in cold waters the commercial quality of the carcass deteriorates rapidly, so speed has always been important. The available power also provided the compressed air used to keep the heavier-than-water rorquals afloat.

As rorquals in the Norwegian Sea became rapidly depleted, whalers began to look further afield, to the west, north and east coasts of Iceland. The Norwegians sent their largest purpose-built steam whale catcher there, but whales still had to be processed on land.

The next big effort was in waters between Spitzbergen, Jan Mayen and the east coast of Greenland, which were known to have been extremely abundant in right whales, which were reduced to near extinction by "old whalers". The new entrepreneur was the shipbuilder Christen

Christensen, of Sandefjord, who developed the first two steam-powered floating factories, designed to operate mainly when moored. In the last years of the century he sent some of his vessels to the Antarctic, where they found rorquals in unimagined abundance.

Meanwhile the Russians caught whales in summer between Hajdamak and Sakhalin, and in the winter in the Sea of Japan, their meat and un-reduced blubber then being sold in Nagasaki, in a country in which "the market for meat for human consumption was insatiable". At the turn of the century the Russians had acquired the world's first "mobile whale factory", but the enterprise collapsed with the outbreak of the Russo-Japanese war in 1904. An important feature was that the Japanese whalers began to extract oil, so when they ventured into pelagic whaling in the Antarctic, the production of edible oil became of immense strategic importance.

If the Norwegian Sea was the cradle of modern whaling, that around Antarctica was later to prove its grave. As the century opened, occasional sealers were killing a few whales; the first was a humpback killed in the Straits of Magellan in 1903 by a Sandefjord man, A.A. Andresen. By 1905 Andresen had brought a new whale catcher from Norway, and with Chilean partners brought a floating factory which operated around the South Shetlands; Deception Island became the first whaling centre of the Antarctic. In 1905 the British Governor of the Falklands offered South Georgia as a whaling centre, and shepherds from the Falklands set sail for South Georgia to work it. On their arrival they were astonished to find an Argentine whaling company already operating there; the Norwegian whaler C.A. Larsen had called at Grytvigen in 1903, discovering it to be ideal.

One difficulty with the use of the first mobile factories was that the whales had to be flensed while still in the water. Various techniques were tried, sometimes with near catastrophic consequences, eventually resulting in the now familiar stern slipway. It was not until 1921 that the first purpose-built pelagic factory ship operated in the Antarctic (Tønnessen and Johnsen, 1982). From 1928 to 1930, 24 floating factories became operative, 110 new whale catchers were built and 23 new whaling companies were started. This was in the years of global depression when pelagic whaling itself practically ceased.

Fleets expanded again in 1936 to 1938. The Antarctic industry was dominated by Norway and Britain. Germany was by far the biggest market for oil for margarine production. Unilever, the British-based corporation, was formed in 1929. It dealt in most Norwegian and British whale oil production, most of which ended up in Germany. The First World War had shown that supply of edible fats was Germany's Achilles' Heel, so to shake loose from this dependence, Nazi Germany itself then embarked on pelagic whaling. This had devastating effects on the Norwegian whaling economy and consequences for the failures during the 1930s of international attempts to regulate whaling (Gatenby, 1983).

## CONSERVATION

When one species after another became rarer, questions would be asked about whether there were limits to which "stocks" would be bound. Mørch, a Norwegian engineer, was a key figure, with 13 patents relating to whaling; he dreamt of an integrated, rational industry. In 1908 he called for the licensing of all whaling operations, for what we would now term the "full utilization of the whale carcass", and for an expansion of research. He thought—at a time when whale migrations were not understood—that the humpbacks he had seen off Africa in winter were the same as those seen at South Georgia in the summer. In 1910 Mørch appealed to the Natural History Museum in London to help ensure the keeping of logbooks on whaling vessels recording details of every whale caught, as well as environmental factors such as the presence of plankton, water temperatures and the like; his specifications were almost exactly what were collected first by the Bureau of International Whaling Statistics, in Sandefjord, and later by the IWC. Second, he called for research to be funded. This led to the establishment of The Discovery Committee, a British institution largely funded by the Norwegian whaling industry.

Mørch's memorandum coincided with a report on an International Fur Seal Conference. That Conference had also heard a proposal for ten years' protection of both species of right whale, and another to prohibit the use of floating factories. In 1912 the British proposed a conference on whaling, directed at the decline of whales in the South Atlantic, and soon after that Germany made a similar proposal. The war intervened, but not before the British had set up an Interdepartmental Committee on Whaling and the Protection of Whales to consider and report to Parliament on "the question of the protection of whales and the regulation of the whaling industry". The Norwegian Whaling Association was asked for its cooperation, which led to Professor Johan Hjort, of Oslo, working with the Natural History Museum.

The emerging science of population dynamics at this time was of great interest to people concerned about fishing. The International Council for the Exploration of the Sea (ICES) was set up. Hjort, who was one of its guiding lights, applied the S-curve, the logistic, to data—not from fishing but from whaling (Hjort et al., 1933; Graham, 1939). He realised the dramatic implications: the steepest part of this curve was at about its middle, at its inflexion, where the population was increasing at its fastest. So, if a whale population could be held at roughly one half of its original "virgin" or "pristine" number, a maximum catch could be taken from it, year after year, without further depletion. Almost exterminated whales should be left to recover, but only to about half their original number. But, by the same logic, a previously unexploited "stock" could be reduced—perhaps *should* be reduced—to that level in order to provide maximal sustainable catches.

## The Whales

Most cetaceans are far-ranging, predatory, and exhibit intelligent behaviour. They use complex vocal means for communication and navigation, take several years to mature, and reproduce slowly.

There are six major species of fast-swimming baleen whales, called rorquals, five in the genus *Balaenoptera*, in descending order of size: blue, fin, sei, Bryde's and minke. This is also the order in which their populations have been depleted. The sixth rorqual is the humpback whale, *Megaptera*, noted for recordings of its wonderful singing. All rorquals are migratory, feeding in polar waters in summer and breeding in tropical and sub-tropical waters. Females are larger than males, which is exceptional among mammals. There appear to be some small self-contained, non-migratory populations of certain species. Southern and northern populations do not intermingle, and their migrations are half-a-year out of phase.

There are two large, slow-swimming baleen species, the right whales (Balaenids), so-called because they float when killed. The bowhead (*Balaena mysticetus*) is restricted to the Arctic and is still hunted. Before it was nearly exterminated in the Atlantic sector it was known as the Greenland whale. The great right whale, Biscay or Basque whale (*B. glacialis*), has populations in cold waters around both poles and in sub-tropical waters. A third smaller balaenid, the pygmy right whale (*Caperea*) is much rarer, with a south-circumpolar distribution. It has not been subject to whaling. The grey whale, *Eschrichtius*, is primarily a bottom feeder, unlike the others. It survives only in the North Pacific.

The sperm whale, or cachelot, is the largest odontocete. It has a global distribution, is highly migratory, with distinct northern and southern populations. They feed mainly on large prey, especially squids. Sperm whale meat is unpalatable except to groups in Indonesia and eastern Japan, but its oil was widely used as fuel and lubricating oil. Sperm whaling largely ended in the 1980s due to the reduction of demand for the oils, the decline of whaling generally, and the tightening of IWC regulations. A pygmy relative, *Kogia breviceps*, inhabits tropical and sub-tropical latitudes. It is occasionally caught in Indonesia and Japan, but there is no systematic fishery for it. Catches are not regulated by the IWC. An even smaller species, the dwarf sperm whale (*Kogia simus*) has a similar distribution.

There are four bottlenose whales. Baird's beaked whale (*Berardius bairdii*), is a North Pacific species, hunted by Japanese whalers. Classed at present by the IWC as a "small cetacean" its hunt is not regulated. It is hunted in the same operations as for minke whales, so when catches of minke whales were reduced, whalers simply increased beaked whale catches. The smaller Northern Bottlenose (*Hyperoodon ampullatus*) lives in the north Atlantic. This species was depleted by Norwegian whalers, mainly to feed the British petfood market and to produce oil. The other two bottlenose whales, *Berardius arnuxii* and *Hyperoodon planifrons*, occur in southern polar waters, though *H. planifrons* extends further northward also.

The narwhal (*Monodon monoceros*) and beluga (*Delphinapterus leucas*) are both circum-arctic species. They have long been exploited by indigenous peoples for meat, and in the case of the narwhal for its spectacular "unicorn" tooth. Their catching is not regulated by the IWC, and as recently as September 1999 Russian whalers killed many belugas in the Northwest Pacific.

There are two pilot whale species: shortfin (*Globicephala macrorhynchus*) and longfin (*G. melaena*). The longfin is common in the North Atlantic, Mediterranean and Black Seas, and throughout the southern hemisphere in temperate and sub-tropical waters. It is still caught in large numbers in the Faroe Islands and to a lesser extent in Newfoundland. The shortfin lives in the tropics and sub-tropics of all oceans, from Northern Australia to southern Japan, and from Namibia to Gibraltar. It is hunted in substantial numbers in Okinawa, the Izu Peninsula of Japan, and in St Vincent and Grenadines in the Caribbean. The regulation of pilot whale fisheries is highly controversial; the IWC occasionally passes non-binding resolutions, which are generally ignored.

Orca, the "killer whale" (*Orcinus orca*) is usually classed as the largest of the dolphins. It is widely distributed and is now protected from whaling. Orca eats other mammals such as seals and even blue whale calves, as well as penguins.

Growing anti-whaling sentiment was fuelled by rising concerns about cruelty, by revelations of their extraordinary biology, and in part by the realisation that the whaling industry was out of control. The question of the IWC's legal competence with respect to "smaller" cetaceans has for years been controversial, but under UNCLOS, nearly all marine cetaceans have a special status. All species of cetaceans are listed in Appendix II of CITES except that all large whales are listed in Appendix I. The Governments of Japan and Norway are seeking the downlisting of the minke whale and the grey whale to Appendix II.

### REFERENCE

Watson, L. (1981) *Sea Guide to Whales of the World*. Hutchinson, London.

Baranov came to essentially the same conclusion, but his work was initially lost. Hjort's exposition was nurtured and elaborated, and for many years provided the basis for practically all attempts to manage international marine fisheries.

With this scientific background, ICES, in 1926, launched moves towards the drafting of an international agreement on the regulation of whaling. In 1925 the League of Nations called for action to protect whales. ICES met in April 1927, and passed back to the League a request to convene a meeting of experts, and in 1931 the Report was adopted by 26 governments. This Geneva Convention provided the basis for all subsequent whaling agreements, including the International Convention for the Regulation of Whaling (ICRW) of 1946. It prohibited the catching of right whales, of calves, sexually immature animals and lactating mothers, and set whalers' bonuses according to the size, species and oil yield of the whales killed.

The setting of any "quotas" was highly controversial: Unilever maintained that the market was the best and only needed a regulator (Burgmans, 1998). But a production agreement signed in 1932 limited the durations of Antarctic whaling seasons and set quotas in terms of oil-based "Blue Whale Units" (BWU) which had been agreed many years previously as a standard production measure for rorquals: 1 blue whale = 2 fin whales = 2.5 humpback whales = 6 sei whales, and so on.

The League's involvement had one very special interest. During the negotiations a legal expert from Argentina suggested that a part of the Antarctic waters, in the Pacific sector, should be established as a Sanctuary for whales that might be fleeing from whalers. At the renewed negotiations, just before the outbreak of the Second World War, the diplomatic conference asked delegates to go home and seek authority for governments to make such a declaration. They did, and in 1946 it was written specifically and generally into the new convention that created the IWC.

Despite the good expressed intentions to protect whales, the reality of the 1930s was constant bickering among whaling countries over economics and related politics. The leading zoological organisations of the United States set up a council for the preservation of whales in 1929, which paid unreserved tribute to Norway for leading the way, its Whaling Act of 1929 being described as "the most constructive legislation ever drawn up to save the waning animal life. Norway, by Royal Decree, has accomplished more than all the nations of the world." Norway repeatedly accused the British of "hypocrisy" for not being helpful about holding down oil production. British intransigence and procrastination delayed for years any relief to the whale populations other than was given temporarily by the great Depression. However, Norway, while demanding that Britain accede to reduced quotas, was busily making agreement more difficult by transferring companies to other flags and ownerships, by attempting vigorously to maintain a monopoly of skills and equipment and to control foreign operations.

Further international agreements in 1937 and 1938 were apparently flavoured by the rise in influence of the diplomats and scientists, over that of the whaling industries. But the 1938 agreement was weak because compromises had to be made to get any movement at all from Japan and France. Still, important new conservation measures were agreed: protection of humpbacks in most places; temporary restrictions on the deployment of pelagic expeditions in some ocean areas, specifically between 40°S and 20°N, but including a "Sanctuary" in the South Pacific south of 40°S. As was said at the time, however, "it is easy to bar pelagic whaling from oceans where there is hardly any pelagic whaling!" No progress was made in the crucial matter of limiting the number of expeditions, and to get Japan's adherence it was agreed that they would be permitted to catch smaller individual whales from land stations.

## 'THE ORDERLY DEVELOPMENT OF THE WHALING INDUSTRY'

"To provide for the proper conservation of whale stocks and thus make possible the orderly development of the whaling industry" is the core mandate given by the 1946 Conference to the fledgling IWC (Wallace, 1994–1997). Those words contain the seeds of intense disagreements between those who think the whaling industry simply cannot be brought to "order" and therefore should pause or even cease, and those who wish to continue to kill whales. The problem of trust underlies this conflict, as well as differences of opinion about mistreatment of animals. The pre-war and the 1946 conferences eventually provided for national inspectors on every floating factory and at land stations too. This reflected the problem of confidence between government and industry, and was intended to enable governments to meet in a dignified way and claim that specific agreements on rules were being adhered to. But of course the governments did not trust each other. This difficulty was sharpened when, in the 1960s, there was an attempt not merely to agree overall limits, but also on how to share these among the five participating countries. That required the appointment of international inspectors. The idea of the IWC employing such inspectors was rejected (purportedly on grounds of cost) and, instead, arrangements were made for inspectors to be exchanged among the different countries.

By the time this came into effect there were only two countries—Japan and the USSR—still engaged in pelagic whaling in the Antarctic. In the early 1990s hitherto secret files became available showing that the Soviets had caught immensely more whales than had been reported, including protected species such as the blue whale. Furthermore, since nearly all the meat produced by the Soviets was destined for the Japanese market, and some of it was transferred at sea or in foreign ports to transport vessels, Japanese collusion seemed clear.

This was the worst of several incidents that convinced an already sceptical public that control and surveillance provisions must be absolutely watertight and transparent. This is the backdrop to the present hesitant negotiations about the resumption of commercial whaling following the moratorium declared in 1982.

After the War, Japan resumed Antarctic whaling, with the help of the US occupiers. The Soviet Union also built up its whaling fleet, at first to feed people, but soon to sell products to Japan for hard currency (it happens that a good part of that currency landed in private bank accounts in Oslo—of all places!—and elsewhere held by high officials in the Soviet Ministry of Fisheries).

The next big technical change was the installation of freezers on vessels, and all the pelagic operators turned from mainly oil production to the much more profitable meat production. The British tried to sell whale meat to their public but failed dismally. Norway had a limited domestic market for whale meat which was soon saturated. The Japanese demand was virtually unlimited.

In its first years the IWC set BWU limits for Antarctic catches only, and so one species of rorqual after another declined steadily, and by 1959 the IWC and much of the whaling industry was in deep crisis. It was agreed that whale scientists needed help, so three mathematically inclined "experts" from non-whaling countries were asked to examine the data with "modern" methods and to make proposals for sustainable catch levels. (I was one of them, British but working for the UN, so my nationality was deemed not to count.) Their proposals were for catches so low that the whalers choked on them. In particular the two countries that had just invested huge sums in new fleets—Japan and USSR—were resistant. The 'Three Wise Men' also said that catch limits should be set by species, not BWUs. The whalers didn't much like that, either.

While these negotiations over mainly blue, fin and humpback whales were going on, the industry turned to the smaller sei whale. By the time limits tailored to blue and fin whales were agreed, sei whales had also been catastrophically depleted. And so it went on for the rest of the decade. By 1972, the world seemed to be ready for a change, and a call by the UN for the IWC to declare a ten-year moratorium was adopted overwhelmingly. Norway supported it, Japan abstained, and the USSR did not participate. On a suggestion from Australia—at the time a whaling country—the IWC adopted a New Management Procedure (NMP) which came into effect in 1976. It was described by proponents as "a modified moratorium" in that, if a whale stock had been shown to have been depleted to below a somewhat arbitrary level of 54% of its pristine number, the catch limits would be set to zero; otherwise catches would be set at 90% of the estimated maximum sustainable yield.

Now that all the fuss was being made about the dire state of the sei as well as of the blues and fins, Antarctic whalers had quietly turned their attention to the little (nine tonnes!) minke whale. It had now been accepted that this species was, after all, a real whale, not merely a "small cetacean", but there were insufficient data on which to base quota decisions. The efforts by scientists to squeeze something out of nothing in those years are painful to look back at (Holt, 1995).

So the eighth decade went by. By the end of it sei and fin whales were so obviously depleted that they were protected anyway. As meat supplies dwindled, Japanese operators set up new or revived shore operations in Brazil, Chile, Peru, Philippines and Canada, and made deals with South Africa, South Korea and China. "Pirate" whaling continued, notoriously, under flags of convenience, hunting especially Bryde's whales which had hitherto barely been exploited. The pirates were mini-factory catcher boats. They hauled whales over the side, took the best meat and dumped the rest. They could handle Bryde's and sei whales. But one caught a fin whale and capsized trying to haul it over; its Norwegian skipper was last seen on the bridge, it is said, with a tankard of beer in his fist.

In 1979 the Australian scientist, Dr William de la Mare, an engineer and relative of the English poet, demonstrated by elegant computer simulation that the NMP could never have worked anyway; its application was a recipe for continued depletion (de la Mare, 1986). The scientists were about to be sent back to their drawing boards. This paved the way for yet one more proposal for a moratorium at last to be accepted, put forward in 1982 by the Seychelles. The requisite majority was obtained by virtue of four convergent trends. First, several new Indian Ocean states had joined in 1979 because they wanted to have the Indian Ocean declared as a whale sanctuary. Second, countries such as Australia and Chile had recently ceased whaling and changed their policies. Third, some Caribbean countries had joined for reasons best known even now to themselves, wielding anti-whaling votes. Fourth, some countries harbouring Japanese-managed shore stations, such as Brazil, were becoming "unreliable". The decisive event was the offer by Spain to cease its coastal whaling if the onset of the moratorium could be delayed for three seasons. That was agreed.

The indefinite moratorium is still in place. So is the NMP, but it is of course dormant, being overridden by the moratorium decision. Japan, Norway, USSR, Peru lodged "objections" to the 1982 decision, which had the effect of exempting them from it. Norway and USSR (now Russia) maintain their objection, but Russia and Ukraine ceased commercial whaling nevertheless, and scrapped their fleets. Norway stopped for a few years but resumed on the legal basis of its objection (Birnie, 1985). Peru stopped whaling and left the IWC. Iceland also stopped, because its Parliament over-rode its Government by one vote in a close internal battle. For a few years that country engaged in some commercial catching of fin whales under the guise of taking scientific samples. Iceland was induced to stop by threats of consumer boycotts of its economic staple—fish.

It is suspected that Icelandic scientific whaling was dreamed up by Japanese meat dealers—a trial balloon. Japan had objected to the moratorium and was thus free to continue, without the IWC being able even to determine quotas. However, Japan was induced to withdraw its objection before 1986 by US pressure, since it wanted to continue to be permitted to catch fish for a few years in US waters; soon after that the US Government decided to build up the US-based fishing industry instead, so lost its hold over Japan. "Scientific whaling" was the answer, and Japan has continued to this day to kill hundreds of minke whales annually. These "samples" are of course brought back and sold on the open market. (Some have labelled the cannon "sampling devices", the flensing deck of the factory ship "the operating table" and the below-decks processing machinery "the dissecting gear"!)

### A NEW WAVE

During the decade after the moratorium, IWC scientists had a breathing space: they no longer had to try to calculate plausible catch limits. Following de la Mare's bombshell, he and colleagues set to work to develop a new approach to managing commercial whaling. They worked competitively. Several came up with what looked like feasible systems of calculating "safe"catch limits for baleen whales—no one can even contemplate the much more difficult problem of regulating any resumed sperm whaling. In a close race they selected an "algorithm" for a Revised Management Procedure (RMP) developed by Dr. Justin Cooke. This was accepted by the IWC itself, but is not yet implemented; it awaits the assembly of a Revised Management System (RMS) which would also incorporate other all-important elements for any safe resumption of whaling: effective international inspection; strict and monitored controls over international trade; means of ensuring that no whaling can take place outside IWC rules; and adequate monitoring of whale stocks (Cooke, 1995). Some Governments also insist that the issue of cruelty be taken more seriously.

The development of the RMP has revolutionised ideas about how to manage whaling in the third millennium. This revolution is spreading to fisheries management generally. It introduces a quantified "precautionary approach". Apart from erring on the safe side in case of scientific uncertainty, an attempt is made to shift the burden of proof in favour of the whales rather than of the whalers—by contrast, in the NMP, all the burden was on demonstrating that depletion had occurred. The RMP dispenses with the traditional mathematical models of the dynamics of whale populations, which were mostly derived from the ideas behind the Volterra–Hjort formulation, as variants of the logistic curve (Hjort et al., 1933; Volterra, 1938; Graham, 1939; Beverton and Holt, 1993). There is a model buried deeply in the algorithm, but the results are not critically dependent on that. The basic idea is to think of a formula, try it out in a one-hundred year computer simulation of a regulated whaling operation, and see whether it ever causes accidental depletion of the virtual whale stock, whether it nevertheless yields fairly good total catches over the century (though not necessarily near to a theoretical maximum) and with limited year-to-year variation. Then keep fiddling with the algorithm until you have something that performs acceptably.

This route to a paradise fit both for whalers and their opponents was strewn with discarded "science". First, most previous data had come from whaling operations. Some were suspect just for that reason. But it came to be realised that these data could not provide the acceptable estimates of the numbers of whales at a given time and place that the RMP called for. Such estimates had previously been obtained in two ways. One was by shooting darts into living whales and later recovering some of them from the blubber boilers in the factories or, later, finding them in the meat using metal detectors and the like. This whale tagging was not planned scientifically; it was not systematic and results could not be unambiguously interpreted in terms of abundance estimates. At best, information was obtained about migrations, but even that was hard to interpret: the marks were only recovered where and when commercial whaling was permitted or wished to go.

The second method was to look at the apparent decline of an exploited stock, and compare that with the time-series of catch statistics. Making various plausible assumptions, these figures could be manipulated to provide a corresponding time series of estimates of the numbers of live whales left. Unfortunately the only way of finding the rate of decline was, in most situations, to look at the catch per unit whaling effort, such as the days of work by a "standard" catcher boat. An enormous effort has been put over the years into refining such "CPUE" indices. At best, however, they serve to reveal large changes in numbers of whales, and to conceal slower changes, and are now reckoned to be generally unreliable.

Whales can be counted where they are regularly and easily seen. For decades the grey whales migrating annually up and down the coasts off Alaska, California and Mexico have been subject to census: what is actually seen depends on the weather, how far they are offshore each year and there is also the complicated business of turning sightings into absolute figures. After many years of theoretical work (related to the military task of spotting submarines and such) and field studies and experiments, some scientists are now confident that they can make the necessary corrections to raw sightings data.

Of course, such confidence in methodology has been misplaced more than once before. In the 1960s it was believed that the ages of baleen whales could reliably be determined by counting grooves on baleen plugs and rings in their waxy ear plugs—but it turned out that whales were twice as old as had been thought! In Alaska, scientists' estimates did not match with what the Inuit thought was happening: native hunters said whales went far under the ice; the scientists were only counting them in open water. The Inuit were right.

Visual counting thus has unexpected hazards. Using recordings made by the US military, sounds made by some types of whales can give an idea of their distributions and rough abundances. These data are being worked on now. Acoustic listening and echo-ranging methods may be useful for counting sperm whales, but whaling countries are not at present much interested in applying these on a large scale. Most attention is being given to counting minke whales at sea, visually. It is the minke whales that Japanese and Norwegian whalers want to continue catching, and they would prefer, for political reasons, that it be done in future on a scientifically acceptable basis. Minke counting has now been done systematically, under IWC auspices, in the Antarctic for several decades, with promising results.

Similar surveys have been made of minke whales in the North Atlantic, by Norwegians, Icelanders and others, for fewer years, less frequently and under national auspices. Similar surveys have been undertaken, mainly by Japanese scientists, in the Northwestern Pacific.

An enormous effort has been put into tackling the problem of estimating numbers from sightings. Results are promising, but there are unresolved questions as to whether minke whales move away from survey ships, swim towards them or ignore them. There is the more important question as to whether a critical feature—the estimated distance of the sighted whale from the ship, and its bearing—is without bias. These estimates are judged by eye, added to which, minke whales are notoriously difficult to spot anyway.

The RMP calls not only for good historical catch statistics and estimates of the current number of whales; it requires knowledge of the statistical errors of the latter estimates. A "better" estimate, having low error, will provide a higher catch limit, and vice versa. Analytical methods so far devised do provide values for the statistical error, just as do opinion polls. These are based mainly on the density of the primary observations, and most surveys to date give figures with a statistical error of around ±50%. If a whale population changes by a few percent a year this means that such changes are impossible to detect except perhaps over decades or centuries. But, in addition, there may be other types of error which make that 50% a minimum value, given the economically feasible density of the survey. Maybe the whales do not appear at quite the same place every year? Such possibilities, called technically "process error", are just beginning to be looked into.

A last problem with the otherwise admirable RMP approach is that it should be applied to discrete populations of whales. But for each species, in each ocean region, there may be two or more morphologically indistinguishable, partially mixing, sub-populations, or "biological stocks". Very little indeed is yet known about such divisions. For some of the minke whales even fundamental migration routes are unknown. Some southern hemisphere minke whales breed off the equatorial coast of Brazil; we know because the Brazilian/Japanese operation caught them there. But although feeding whales being hunted by the Norwegians in the Barents and Norwegian Seas migrate south in the winter, we do not know where they go.

The RMP gurus have invented a cunning way of applying the algorithm when practically nothing is known about sub-populations and where whaling is confined to the feeding grounds. It depends on making fail-safe style assumptions about possible sub-divisions. The less you are prepared to assume, the lower will be the calculated overall catch limit. There is thus a strong incentive for those who wish urgently to resume regulated commercial whaling to pretend we know much more about this matter than we really do.

The identity and dynamics of sub-populations may eventually be revealed by DNA "fingerprinting". Biopsy can tell us other useful things, too, such as the content in the fat of POPs—persistent organic contaminants such as ubiquitous endocrine mimics. Proposals have been made for ocean wide systematic surveys by this method. Meanwhile, DNA fingerprinting has found another, more immediate use. Analyses of samples from meat on sale in shops and restaurants can tell us the species of whale it is from, and the location in which it was killed. Such analyses have already revealed sales of meat from protected species, and of minke whales from areas in which catches have not been reported. This is, politically speaking, a very tricky business. The arguments about who should take samples, who should carry out such analysis, where the data files will be held and who has access to them, look to slow down the general application of this most powerful technique to management.

Identification of individual whales began with tagging, but of course when the tag is re-observed, the whale is dead. Attempts have been made to fix radio tags to live whales, with some success. This can give information about movements, diving frequency and so on, but not about population dynamics. Individual whales of some species, such as humpbacks and right whales can be recognised visually, and catalogues of their photographs have been assembled. By this means the number of individuals in some small populations have been estimated. Even individual sperm whales and orcas can be recognised and much learned about their natural history from years of close observation, especially when combined with recognition of individual voices. But these cases are exceptional (Watson, 1981).

An important consequence of this type of research is psychological. Humans cannot resist the temptation to give these identified whales names. Name-giving establishes an identity of a different order. It has contributed to the human consciousness of whales, and has also contributed significantly to funds for researches of kinds not usually funded by official sources, as well as for campaigning for the whales' protection.

## HOW MANY WHALES ARE THERE, AND WHAT IS HAPPENING?

At last I am ready to try to answer the apparently straightforward questions put to me by the editor of this series. Most statements about the numbers of whales left in the sea, and whether or not they are recovering from the depredations of the twentieth and previous centuries should be set aside. They include statements like: there are a million sperm whales left; there were once two million. These should be forgotten. Such numbers were either plucked out of the air or given by methods now discredited. We have, to continue the example, practically no idea how many sperm whales there are now or once were (Dufault et al., 1999).

So we can concentrate on the species and populations of which we now have acceptable knowledge. These are the grey whale (Reilly, 1984) the bowhead (Zeh et al., 1995), the minke whale in the Northeast Atlantic and the Northwest Pacific, minke whales in the Southern Ocean, as well as the blue, fin and sei whales seen and counted during decades of minke whale surveys there (Butterworth et al., 1993). There are also some useful estimates of the numbers of tiny residual populations, some of which seem to be recovering at various rates under protection, such as right whales in the North and South Atlantic (Best, 1993), and humpback whales in some locations in both hemispheres. Here I try to summarise what we now think we know (Reports of the IWC).

### Grey Whales

Data about the grey whale were last comprehensively reviewed by IWC scientists in 1990, and they have examined some new data each year since. There are thought to be very roughly 25,000 grey whales alive. They were increasing at about 3% annually over the period 1967/68 to 1987/88, despite Soviet catches of about 1% annually. These numbers have the usual wide statistical error ranges: current numbers could be between 15,000 and 50,000 with 95% probability, and the increase rate could be between 2 and 5 or 6%. Scientists have been unable to agree how close this number might be to an eventual maximum number but now think that is more likely to be over 60% of the way to full recovery than below that magic number—the level assumed to be that for MSY. It is also unclear whether the carrying capacity has been changing during the observed period of partial recovery. Calculation of the entire population trajectory since intensive commercial whaling began in 1846 is hampered by uncertainties about the magnitudes of those catches as well as of the aboriginal catches since the 17th century.

Unexpected difficulties have been encountered in aligning sightings data on which these estimates are based, with the "classical" model of how such a population would be expected to behave. Because of the incompleteness of the data it seems unlikely that these anomalies will ever be resolved. Nevertheless, the increase, which is thought to be continuing but at a slowing pace, is within the wide range of possible values assumed for baleen whales during the development of the RMP: 1 to 6% when the population is in its mid-range, close to the presumed MSY level.

### Bowhead Whales

A number of distinct stocks of bowhead whales are recognised by biologists: three in the Atlantic (Davis Strait/Baffin Bay; Hudson Bay/Foxe Basin; Spitzbergen) and two in the Pacific sector—Okhotsk Sea, and Bering–Chukchi–Beaufort Seas (B-C-B). Of these only the last has been extensively and continuously studied. It is by far the most numerous and is the target of aboriginal subsistence whaling by the Inuit.

There have in recent years been small unauthorised catches from the Okhotsk Sea stock by indigenous people, presumably supplementing their grey whale catches. But this stock has been identified as one of the world's most endangered whales (perhaps fewer than 100 remain); it is thought to be the most vulnerable to climate change, and oil and gas development is imminent in the region. The Davis Strait stock is also still very small—perhaps 250–350 animals. A few have been killed in the last few years, from Greenland. In 1996 Canadian aboriginals, by permission of the Government of Canada (which is not a member of IWC), resumed killing occasional animals from the Hudson Bay/Foxe Basin stock, which also numbers only perhaps 100–300 individuals. In 1998 Canada also authorised killing of bowheads from the Baffin Bay/Davis Strait stock which might number 300–400.

The most recent comprehensive assessment of the B-C-B stock was in 1991, and IWC scientists looked at these more recently as well. A survey was planned for 1999, but at the time of writing results have not been analysed. Some people believe this stock is threatened by oil exploration and extraction in the Beaufort Sea (Zeh et al., 1995).

Although the results of sightings surveys are well-established, much controversy surrounds the methods and interpretation of the dynamics of this much studied whale. The present belief is that there were about 7000–8000 in 1993 (with an error range of less than ±20–30%) and they appear to have been increasing, since 1978 at least, at just over 2% annually—but this rate has an error range from 1 to 3.5%. This is despite an aboriginal subsistence kill in recent decades of 10 to 40 annually. These catches are tiny compared with the commercial catches of up to 2000–3000 in the mid-19th century, and several hundred every year until their collapse in the first decade of this century. Not very reliable estimates, by different methods, of the number at the onset of commercial pelagic whaling in 1848 are 13,000 and 18,000, but the former number has a statistical error range from 8000 to 30,000. Generally speaking, therefore, this stock of bowhead whales may be about half-way to full recovery and close to its putative level for MSY.

Dr. Peter Best, of South Africa, has published estimates of the rates of apparent recovery of several small, severely depleted stocks of whales, including the right whales off South Africa and off Argentina (Best, 1993). These were about 7% annually, though there are uncertainties associated with the possibility that the whales might have been moving in to the study areas during the research periods. Nevertheless, since theory suggests that the expected rate of recovery of very depleted populations is about twice that of stocks near to their MSY levels, the North Pacific bowhead and the southern right whale observations are not incompatible.

Best also cites substantially higher recovery rates for humpback whales in the Northwest Atlantic, and off Australia (Pacific and Indian Oceans) of about 10%. Blue whales were reported to be recovering at 5% annually in the North Atlantic, but this figure has been challenged because it was deduced from sightings made by operating whaling vessels off Iceland which shifted their whaling grounds through the period of observation.

### Minke Whales

The minke whale is the species to which most attention has been given in the past three decades. Populations in the Northeast Atlantic and in the Southern Ocean are being targeted by Norwegian and Japanese whalers respectively, and those which feed in the summer in the Antarctic have been the subject of continuous surveys, some under the auspices of the IWC during the first two International Decades of Cetacean Research (IDCR) launched in the 1970s, others as part of the Japanese Whale Research Programme under Special Permit (JARPA). This latter is the umbrella under which Japanese whalers have, very controversially, been authorised by their Government to kill large numbers of whales every year for scientific purposes, but also to market the products from them commercially.

IDCR surveys covered the entire Antarctic latitudes southward of about 55°S twice in sixteen years, in each year covering one of six sectors. It was estimated that there were 600,000–700,000 minke whales, with an error range of about ±50%. These numbers were very much higher than had been estimated in the period 1973 to 1979 by now discredited methods. No official calculations have been made of what would be the annual catch limit from these populations if the RMP were to be applied, but it is commonly believed that it might have been between 1000 and 2000. Such calculations are not authorised while the commercial moratorium is in place, and since 1994 are precluded by the sanctuary designation.

There was no significant difference between the totals in each of the six-year periods, nor in each sector. Thus we do not know whether these populations have been increasing, declining or remained stable. Any expected rates of change between, say, –10% and +10% annually would scarcely have been detectable given the limited accuracy of the surveys and the short time over which they were conducted.

Interestingly, the JARPA surveys from 1991/92 to 1995/96 gave very different results in the sectors in which they were carried out. Thus in the combined sector from 70°E to 170°W (counting eastward) the IWC-sponsored surveys estimated more than 200,000 minke whales while JARPA gave less than 100,000. So despite many years of research and considerable expenditures, doubts remain as to the abundance of the southern hemisphere minke whale.

We still have little idea about the existence of subpopulations. Since whaling began only in 1972/73, and fewer than 100,000 whales were killed before the moratorium was implemented in 1985, stocks as a whole were not depleted, though any distinct stocks in the South Atlantic and Indian Ocean sectors might have been more severely affected by whaling, which was concentrated in those regions.

While there have been well planned surveys of minke whales in the western North Pacific—where Japanese whalers are anxious to resume minke hunting—it is too early to appraise these. Partial surveys in the 1990s indicate a total of about 25,000 animals. It has been difficult to assess their status because of doubts about previous and even current, kills—many are entangled in shore nets, both in Japan and Korea, and end up on dinner plates—because it seems that there is more than one biologically distinct population.

Minke whales in the Northeast Atlantic have been given almost as much attention as those in the southern hemisphere, but estimates of numbers remain even more controversial. Divergencies between different recent estimates are similar to those between the IWC/IDCR and JARPA figures in the south. For the moment the IWC scientists accept numbers given by Schweder et al. (1997) from Norwegian surveys, though the numbers obtained in international surveys were different, and lower.

Norwegian estimates for 1995 were in the range 91,000 to 137,000 (95% confidence limits; the "best" estimate was 112,000). The "best" value for 1989 was 65,000, with a likely range of 44,000 to 94,000. The two surveys were not precisely comparable but there is an indication that the number of whales in this region may have increased during the six years between the surveys, though, if so, at a rate that cannot yet be determined. Some increase could be expected, despite continuing commercial whaling under the Norwegian objection to the moratorium, since these animals were extensively depleted throughout the period 1931 to the mid-1980s.

### Baleen Whales

During the IDCR surveys for minke whales, sightings of blue, fin, sei and humpback were also recorded. The first circumpolar series, 1978/79–1983/84, covered only two-thirds of the ice-free areas south of 60°S, as did the second series, 1985/86–1990/91. The IDCR cruises did not fully cover the geographical ranges of the species because they were aimed at the minke which, like the blue whale, feeds close to or even within the ice boundary; the sei whale especially feeds further to the north, and the fin whale has a wider latitudinal spread. So Butterworth et al. (1993) supplemented the IDCR results with data from confidential information made available to them from Japanese whaling vessels used for scouting. They arrived at figures for the entire area south of 30°S, but the results especially for sei, humpback, and to a lesser extent fin whales, have big question marks over them. Nevertheless, they are instructive.

There were no significant differences between the numbers from the first and second set of surveys; such could hardly be expected since in any case the total number of sightings was very small indeed, especially of blues and sei. The IDCR results—essentially south of 60°S—suggest about 500 blue whales, 2000 fin, 1200 sei and 6000 humpbacks. The numbers surmised to be in the wider band, south of 30°S were, by comparison, 6500 blue, 12,000 fin, 17,000 sei and 20,000 humpback. These last are much less reliable than the first series, but an explanation is due for the difference in the blue whale numbers. It is now known that there exists a pygmy sub-species or race of blue whale, living northward of the better-known "normal" blue whale feeding range. For some time these were presumed to be simply young blue whales. Butterworth et al. (1993) reckoned that 5600 of the total of 6500 were pygmies, which does not leave many big blues from the roughly 250,000 thought to have been there originally, as estimated from the history of Antarctic catching. A similar rough estimate of the pristine number of fin whales is 500,000.

By the now discredited methods of assessment using CPUE data it had been thought that there were several thousand blue whales remaining, and several tens of thousands of fin whales from the original half million. Thus it was a shock even to the more pessimistic biologists to learn that these species populations had been brought much closer to extinction than had been supposed. Illegal excess catching by the Soviet fleets in the period 1950–1970 no doubt contributed to that situation.

This review should be closed with a sombre warning. The numbers quoted are derived from a very, very few actual sightings. In 12 years of survey effort, fewer than 40 sightings were made of blue whales and fewer than 50 of fin whales. It is not unlikely that in the two "passes" of the entire Antarctic some of those whales were seen twice or, since they are known to move in different seasons from one sector to another, perhaps more often.

During the IDCR sightings surveys from 1978/79 to 1990/91, 1100 schools, 1500 individual and 1100 schools (groups of two or more individuals) of beaked whales were seen (Kasamatsu and Joyce, 1992). Of these, 900 individuals, 400 schools were southern bottlenose whales (*Hyperoodon planifrons*) and nearly all the rest were unidentified beaked whales, surely including among them a high percentage (more than 90%) of southern bottlenose. These observations permit a very rough estimate, for the first time, of the abundance of this species; it was said to be in the range 170,000 to 320,000 in the waters south of 60°S. The distribution of sightings was circumpolar, but this species was more abundant in the south Atlantic and Indian Ocean sectors than in the south Pacific, as were the large baleen whales in the early days of Antarctic pelagic whaling. Individuals weigh about half as much as a minke whale. Like the northern bottlenose and sperm whale, the southern bottlenose makes prolonged hunting dives, and deeply; this makes estimation from sightings difficult. However, unlike those other species it is not accepted as being under the regulatory jurisdiction of the IWC. It is also unclear whether it would be covered by the Southern Ocean sanctuary, but in present circumstances probably not.

There were a few sightings of the fourth bottlenose whale, Arnoux's (*Berardius arnuxii*), and even fewer of incompletely identified beaked whales of the Genus *Mesoplodon*, which contains many species.

## SUBSISTENCE, AND INDIGENOUS RIGHTS

Near the end of the 1946 negotiating conference, the Soviet delegation asked for the floor in order to raise an unscheduled item of Any Other Business. They said the native people of the far eastern Siberian Arctic needed meat of grey whales for their survival. No one else present had ever heard of this. It seems that they formerly killed right whales; those were now gone, but the grey whales were too dangerous to kill by hand harpoon from small boats. This might have been true, since the Americans hunting grey whales along the West Coast called them "devil fish" after some boats had been stoved in and sailors drowned by enraged animals. So, the Russians were hunting them, and handing out the meat to the local people. The conference duly inserted new words in the Schedule to the new convention, which forms an integral part of the convention itself: "It is forbidden to take or kill grey whales or right whales except when the meat or products of such whales are to be used exclusively for local consumption by the aborigines." There was no substantial discussion, no other reference to "aborigines", no consideration of quotas or criteria, and no reference anywhere to "subsistence".

Until the early 1970s there was no further consideration of this matter; catch limits were being set only for Antarctic baleen whaling. However, the 1972 UN Conference, and IWC debates from 1973 to 1975–6, led to agreement that limits would be set for all species (except as scientific samples). This included in particular the hunting of bowheads by Alaskan native people—"aborigines" in IWC terminology. In succeeding years, *ad hoc* catch limits were set for the bowhead hunt and also for minke and fin whales hunted by Greenlanders, with similar conditions of "only for local consumption". There continued much uncertainty about the meaning of "local". It later emerged that whale meat was on sale in Copenhagen for the benefit of Greenlanders visiting or working there!

In 1982, the year the commercial whaling moratorium was adopted, the IWC also adopted an aboriginal subsistence whaling management procedure. This is still essentially the same as the NMP except that it introduced the idea of satisfying aboriginal subsistence need and, instead of protecting stocks below 54% of their pristine level, it permitted killing of animals from stocks in such a depleted condition but above a certain minimum level, the permissible catches being such that they will allow stocks to move (recover) to the presumed MSY level of 60% of pristine. Again, this was adopted on the spur of the moment, with no discussion, and without reference to the IWC's Scientific Committee. A footnote to this decision asks the Committee to advise on the minimum stock level below which whales shall not be taken, and to specify a desirable rate of allowable increase for each stock. The scientists have been unable to provide such advice.

Subsequently, the Commission ignored its own criterion of "aboriginal" when it set a quota for the catching of humpbacks in St Vincent and the Grenadines, by whalers who are certainly not Carib Indians.

The IWC's treatment of the subsistence whaling issue has caused nothing but trouble and bad feeling for 17 years, yet it has failed consistently to confront it. Each year catch limits are set with no scientific basis except for the bowhead and the continuing Russian catches of grey whales, which were in reality always used to feed captive fur animals as well as local people. Submissions of cases to justify "need" are not rigorously examined. It has been argued that there is little substantive difference between most Greenlandic whaling and the smaller vessels in the Norwegian and Icelandic commercial minke whaling. Japanese officials have for years consistently claimed that their "small-type" land-based catching of Baird's and minke whales has most of the characteristics of aboriginal subsistence whaling. Words like "traditional" with no clear, unambiguous meaning are regularly bandied about, essentially as political epithets.

The reality is that there is now only one truly subsistence and traditional whaling operation in the world by "aborigines". That is the sperm whale hunt from the village of Lamalera on the island of Lembata, Indonesia. The meat is mostly consumed in that extremely poor village, and some is traded for other foodstuffs with neighbouring inland villages. In the 1970s the FAO made a misguided and disastrous attempt to introduce on Lembata a motor vessel and a harpoon cannon; perhaps it was a coincidence that the instigator of this effort was at the time on secondment from the Japan Fisheries Agency (which oversees Japanese whaling) and the skipper in the field was a Norwegian fisherman.

Evidence from Portuguese and Dutch sources of the 16th and 18th centuries respectively show that the Indonesian whale fishery was in existence long before American and European sperm whalers arrived in the area. Elsewhere, from the late 18th century, they left their marks with open boat whaling with hand-held harpoons: in the Caribbean, Tonga, Azores. None of these operations is traditional and, except in Tonga, they were not conducted by aboriginals.

In 1995 the IWC asked its scientists to devise a new aboriginal subsistence whaling procedure analogous to the RMP. The basic idea is to permit limited whaling on stocks for which the application of the RMP would give zero quotas—which is probably most of them. The difficulty now is that the RMP calls for regular surveys of a kind far beyond the financial and technical capacities of those administering such whaling, except in the USA—and Canada if that country were to rejoin the IWC. Furthermore, it must be asked, if a relaxed RMP could be safely applied to subsistence whaling, why not make such a relaxation for commercial whaling?

## SHARING RESOURCES

The IWC does not authorise the allocation of slices of an overall catch limit to different nations, vessels or operations. When agreement on shares became economically and politically necessary in the 1960s, the shares were negotiated among the whalers themselves, outside the IWC but simultaneously with its meetings. This process caused serious problems in reaching agreements on overall catch limits, and led to near breakdown of the organisation. It could happen again. The device that the scientists have adopted is to divide regions up into smaller areas containing separate putative stocks of a species, in such a way that only one country's land adjoins each. That can work for coastal whaling in many places but it cannot work if pelagic commercial whaling is ever legitimised on the high seas. That might even be a reason for declaring all the high seas as a Sanctuary!

## SANCTUARY

The sanctuary established in the South Pacific sector of the Southern Ocean in the original 1946 agreement was merely symbolic, and applied only to pelagic catching of baleen whales. It was abolished in 1955 when the big whales were few everywhere else; whalers wanted to get at what there was (which was not much) and the specious argument for this was to take the pressure off elsewhere.

When the Republic of Seychelles joined the IWC in 1979 it launched a three-pronged initiative: to get the catching of Baird's beaked whale under IWC control; to secure a global moratorium on sperm whaling; and to close the Indian Ocean to commercial whaling. With the help of nearly all Indian Ocean coastal states it succeeded in this last. The agreement was a compromise on the original intention, which was to include the Antarctic sector; that could not have secured the requisite three-fourths vote, so the sanctuary boundary was re-drawn to allow continued pelagic minke whaling by Japan and USSR south of 55°S. Nevertheless the decision had an effect: Soviet expeditions stopped killing sperm whales on their way to and from the Antarctic; on the southward journey they were accustomed to doing target practice as well. But, more than anything, it established confidence that small non-whaling countries, especially when allied in regional groupings, could make waves.

In 1992 France proposed that the entire Southern Ocean be declared a sanctuary in which all commercial whaling would pause indefinitely. This time, after a faltering start, the southern hemisphere countries, none of them now engaged in whaling—and especially the Valdivia Group comprising South Africa, Australia, New Zealand, Chile and Argentina—all declared their support. The measure was adopted in 1994 with only Japan voting against and subsequently lodging an objection. This decision had a more significant consequence: whatever happened about the moratorium, the IWC would not calculate catch limits under the RMP for Antarctic whaling. This was entirely in accord with global sentiment (except in Japan) regarding the Antarctic region as a whole; France had been one of the prime movers in establishing a 50-year moratorium on minerals exploration on the continent.

Five additional sanctuaries have now been mooted. Two are specific, though tentative, proposals by IWC member states: Australia and New Zealand regarding the tropical and sub-tropical south Pacific, and Brazil concerning the south Atlantic. New Zealand hinted that it sees its proposal as a step towards the entire southern hemisphere becoming a sanctuary, as it would, if combined with a successful Brazilian bid.

Mediterranean coastal states in the early 1990s declared their wish that the entire Mediterranean should be a sanctuary. A first step was made in 1999 with the trilateral declaration by Italy, France and Monaco of the waters of the Ligurian Sea as a sanctuary. However, there are no broad zones of national jurisdiction such as EEZs in the Mediterranean beyond the territorial seas; to consolidate and extend their decision, the three states, with Spain, the other Mediterranean IWC member, might move in the IWC. The interests of the Mediterranean countries, now they are aware that there are significant numbers of large whales in their adjacent waters, is not so much in preventing whaling but rather in regulating the growing industry of whale watching and securing the safety of the whales from mercantile shipping and tourist vessels and some types of fishing operations.

Under a special protocol to the Cartagena Convention there have been moves to ensure that whales are protected throughout the Wider Caribbean region; the IWC is not so far involved with this.

As one element in a package of four items aimed at unblocking the political impasse in the IWC, where neither the pro-whaling group nor the pro-moratorium group is now in a position to ensure a three-fourths majority vote on anything controversial, the Commissioner for Ireland has suggested that all remaining high seas be designated as a sanctuary.

I shall not comment here on the pros and cons of these, or the likelihood that any of them will be enacted. However, they reveal the way the world is moving in relation to the conservation of marine mammals. Japan continues to maintain that the 1994 decision for the Southern Ocean was illegal, and has brought forward foreign legal "experts" to argue this. However, it made no such objection in principle to the Indian Ocean decision, and no other country has seriously supported its assertions.

Lastly it should be mentioned that Greenpeace and other NGOs have, in 1999, been circulating on the World Wide Web a petition calling for the designation of the entire ocean as a "global sanctuary". Administratively speaking it is difficult to see how this would differ from the existing moratorium; both would be indefinite and require a three-fourths contrary majority to be over-turned. However, in the eyes of some, the idea of a sanctuary rather than an indefinite pause in commercial whaling may be weightier (Mangel at al., 1996).

## JUST LOOKIN'

The IWC has no way of resolving conflicts between different consumptive and non-consumptive users of whales. It is not clear whether both whaling and whale watching can be conducted simultaneously on the same whale populations; there are issues of physical interference, and of possible effects on the behaviour of the whales. And also on the behaviour of people; tourists recently visiting Iceland to watch whales have said that if Iceland resumed whaling they would not return as tourists.

In 1983 eight NGOs and the IWC co-sponsored a Global Conference on the Non-Consumptive Utilisation of Cetacean resources (IFAW, 1996–1999). This reported to the IWC on a series of issues: whale watching, as a rapidly growing activity; holding cetaceans in captivity; educational, cultural, scientific uses and values of cetaceans; the value of protected areas; ecological values of cetaceans; benign (non-lethal) research; conflicts among different uses; and the legal aspects of all these (Birnie, 1985).

The IWC eventually agreed that it could, within its mandate, consider matters relating to whale watching, and make recommendations about them, though it had no power or indeed any wish to regulate that activity; that was up to the countries involved, though they might on occasions cooperate with each other and also exchange information through the IWC. A number of resolutions have since been passed reinforcing these positions and meanwhile, NGOs, particularly IFAW, have continued to examine the issues.

As the industry has expanded it has become clear that whale watching changes the attitude of many participants to whaling (Connecticut Cetacean Society, 1983). For a few years some IWC whaling countries opposed the organisation's consideration of the question, and Japan still drags its feet. But the opposition has changed tack. It now accepts the vitality of this activity, but begins to argue that it is perfectly compatible with commercial whaling.

## COMPETITION

When it was found that there were up to one million minke whales in the southern hemisphere but only a few hundred blue whales left, Japanese government scientists began to say that the minke whales, which they claimed must have been increasing rapidly because they were no longer being hunted on a large scale, were, by consuming the krill, evidently preventing the recovery of the blue whale. We were told that "the balance of nature" should be restored: that is, we should reduce the minke whales to restore nature's balance!

This was before it was discovered that Soviet whalers had been secretly killing blue whales anyway. But there are almost as many other flaws in this argument as holes in a fisherman's net. First, there is no evidence that the blue whales are not increasing. We have no idea what has been happening to them over the past 40 years—apart from being opportunistically counted or harpooned. Second, although the diets of the largest and smallest species of Balaenoptera are similar, their feeding locations and times and habits and preferences are not identical. Third, many other Antarctic animals eat krill—birds, seals, squids—in vast quantities; no simple direct interaction between minke and blue should be expected in such circumstances. Fourth, if there is a long-term threat to the availability of krill, it is more likely to arise from inadequately regulated expansion of fisheries for that species. Fifth, there is no evidence that minke whales have been or are increasing.

A claim to be taken more seriously is that whales eat fish that belong to humans, so they should be reduced, prevented from increasing, and generally regarded as our enemies. Such claims, about seabirds and seals as well as cetaceans as enemies, are being voiced in more places, more stridently, more determinedly, every year. Some of the claims may be plausible. But even when it is true that marine mammals are consuming large quantities of fishes of commercial interest—and it often isn't—the complexities of the biological system involved (for example marine mammals eat both commercial and non-commercial prey; other predators eat those species, too; fishes often eat each other; the mammals often eat fish of different ages from those caught, and so on and on) and the details of the feeding behaviour of the predatory mammal mean that both the direction of any effect, and its scale, are highly uncertain and probably not scientifically predictable. A UNEP study takes the view that an assumption that if whales, seals or dolphins eat fish then this must be detrimental to our economic interests, is not a sound basis for policy, and that the burden of proof should rest with those promoting "culling" (Planning and Consultative Committee, 1999).

## WHALES IN A CHANGING WORLD

In Volterra's logistic model for population growth, the upper asymptote, later called the carrying capacity, was a steady state reached after the passage of infinite time, when the rate of change would be zero. The point of steepest change, at the inflexion of the S-curve, later identified as the point of maximum sustainable yield of an exploited population of wild animals, was, in the calculus, another zero and thus a potential equilibrium of sorts: the point at which the rate of acceleration of change was zero, changing from positive to negative.

From these same ideas we have inherited, with a suitable time-lag, that of the "balance of nature", when applied not to single populations of one species but to complex ecosystems. As we have seen, some approaches both to conservation and to "scientifically managed" exploitation call for restoration of this intangible balance. They seem to hark back to a mythical paradise. We see this in, for example, current arguments about whether we should now try to eradicate "alien" species introduced intentionally or accidentally into ecosystems by human agency. Proponents of the balance of nature tend to use the metaphors of "virginity" and "pristine" states.

Late 20th century physics and mathematics focuses more on discontinuities in change, and on non-linear processes leading to chaos. Poincaré understood this, but nearly another century passed before computers made exploration of the phenomenon practicable. In the language of evolutionary biology, the image of smooth sequences of tiny changes has, in the same period, been replaced by concepts such as that of "punctuated equilibrium".

Most of us still tend to think of whales as having lived for millions of years in an unchanging ocean. They had, as far as we know, no enemies except the occasional orca—until humans arrived in boats. We have looked at what is still the main threat to the survival of whales: the global markets for meat, oil and fish. The rest is speculation, but speculation is the spice of scientific life. But dramatic changes are now being made to the oceans by humans. Some of these, such as pollution by nutrients, silt from deforestation and coastal construction, rubbish from consumptive urban living, even spilt mineral oil, are chiefly local, and probably short-term in their deleterious effects.

Accidental introductions of alien species are another matter. We cannot possibly control most introductions in the ocean, but some are important. A case in point is the infestation of the Black Sea by a foreign jellyfish, a voracious predator on, among other things, the larvae of the once abundant fishes there, such as anchovy. Thus the main fishery of the region is now gone, but anchovies were also a staple food of the once abundant dolphins in the region. Other factors have surely contributed to the Black Sea disaster for both humans and dolphins, but the jellyfish contribution was utterly unsuspected.

Another potentially harmful effect may be uncontrolled and largely unstudied introductions of persistent synthetic molecules that affect hormone systems: endocrine mimics. Many are soluble in fats and oils so accumulate naturally in long-lived species, and especially in the blubber of cetaceans. We do not yet know what effects these may have, but they are unlikely to be beneficial (WWF, 1998).

It is also possible that global warming will affect the distributions of whales and their biological productivity through changes in their food supplies. Perhaps the most likely regional change might be a postulated reversal of the so-called "Atlantic Conveyor", the great ocean system which transports water that originated in the deep Southern Ocean to the northern hemisphere, and from the North Atlantic back southward. This would, among other things, shift the track of the Gulf Stream. However, such events have occurred naturally in the history of this and other ocean regions; whales have repeatedly survived those changes and seem likely to survive further repetitions of them, rather more easily than will less mobile marine and most terrestrial fauna. I don't think we should worry for the whales in this regard.

Smaller, local habitat changes could in some special cases have dramatic effects on whales. The most imminent such change is deterioration of preferred breeding locations of the grey whale in coastal lagoons of western Mexico. These whales used to breed in and, we now think, just outside, lagoons in Baja California and Southern California where they were slaughtered by American whalers. The Southern California lagoons are no more, and those in Mexico are under threat. In particular a giant Japanese corporation is pressing to be allowed to turn one of them into a vast salt pan. Salt is, of course, an absolute necessity for humans and their domestic animals. What is interesting is why Mitsubishi has decided that greatly increased production is now necessary. It seems that it expects growing demand. World-wide, the hardening of fresh waters is happening, and in coastal regions seawater percolates in. In some places, distillation of seawater is possible, but this is enormously energy-intensive. Elsewhere softeners using ion-exchange principles will be better and cheaper. They use salt—hence the likely hard times for the grey whales. This threat to a protected animal, which has only recently recovered from near extermination, has energised a wide range of people to question priorities, and they now have the ear of the Mexican Government. It is too early to say if they will succeed in blocking Mitsubishi. A common slogan in those parts is "Share the Planet"—with animals and trees as well as with other humans. Good luck to them.

## REFERENCES


Best, P.B. (1993) Increase rates in severely depleted stocks of baleen whales. *ICES Journal of Marine Science* **50** (3), 169–186.

Beverton, R.J.H. and Holt, S.J. (1993) On the Dynamics of Exploited Fish Populations. Rev. repr. of 1957 edn. Chapman and Hall, London,

Birnie, P.W. (1985) *International Regulation of Whaling: From Conservation of Whaling to Conservation of Whales and Regulation of Whale Watching.* 2 volumes. Oceana Publications, London.

Burgmans, A. (1998) Securing the future of the seafood industry: A Unilever perspective. Unilever Corporate Relations, Rotterdam, Netherlands.

Butterworth, D.S., Borchers, D.L. and Chalis, S. (1993) Updates of abundance estimates for southern hemisphere blue, fin, sei and humpback whales incorporating data from the second circumpolar set of IDCR cruises. IWC Document SC/44/SHB 19.

Connecticut Cetacean Society (1983) Whales Alive: Report of Global Conference on the non-consumptive utilisation of cetacean resources. Cetacean Society International, P.O. Box 9145, Wethersfield, CT 06109, USA; or The Animal Welfare Institute, P.O. Box 3650, Washington DC 20007, USA.

Cooke, J.G. (1995) The International Whaling Commission's Revised Management Procedure as an example of a new approach to fishery management. In *Whales, Seals, Fish and Man*, eds. A.S. Blix, L. Walløe and Ø. Ulltang. Elsevier, Oxford and Amsterdam.

de la Mare, W.K. (1986) Simulation studies on management procedures. *Rep. Int. Whaling Commission* **36**, 429–450.

Dufault, S., Whitehead, H. and Dillon, M. (1999) An examination of the current knowledge on the stock structure of sperm whales (*Physeter macrocephalus*) worldwide. *Journal of Cetacean Research and Management* **1** (1), 1–10.

Ellis, R. (1992) *Men and Whales.* Robert Hale, London.

Gatenby, G. (1983) *Whales: A Celebration.* Little, Brown and Co., Toronto.

Graham, M. (1939) The sigmoid curve and the overfishing problem. *Cons. Int. Explor. Mer, Rapp. et Proc. Verb.* **1** (10), 15–20.

Hjort, J., Jahn, G. and Otterstad, P. (1933) The optimum catch: Essays on population. *Hvalradets Skrifter* **7**, 92–127.

Holt, S.J. (1995) Viewpoint: Creating confidence. *Marine Pollution Bulletin* **30** (9), 583–585.

Holt, S.J. (1999) Whaling and international law and order. *Marine Pollution Bulletin* **38** (7), 531–534.

IFAW (1996–1999). Reports of six international workshops on whale watching. International Fund for Animal Welfare (1996, 1997, 1999), Old Chapel, Fairview Drive, Bristol, UK.

Kasamatsu, F. and Joyce, G.G. (1992) Abundance of beaked whales in the Antarctic. IWC Document SC/43/O 12.

Kurlansky, M. (1999) *The Basque History of the World.* Alfred A. Knopf, Canada, 387 pp.

Mangel, M. et al. (1996) Principles for the conservation of wild living resources. *Ecological Applications* **6** (2), 338–362.

McIntyre, J. (1974) *Mind in the Waters: A Book to Celebrate the Consciousness of Whales and Dolphins.* Scribners, New York; Sierra Club Books, San Francisco, USA.

Planning and Consultative Committee for the Marine Mammals Action Plan (1999) Protocol for the scientific evaluation of proposals to cull marine mammals. Report of the Scientific Advisory Committee (SAC) of the PCC of UNEP. SAC, at Route d'Amonines 15, 6987 Rendeux, Belgium.

Reilly, S.B. (1984) Observed and maximum rates of increase in gray whales, *Eschrichtius robustus. Report of the International Whaling Commission* (Special issue 6), 389–399.

Reports of the International Whaling Commission. These large books are published annually and are available for purchase from the IWC, The Red House, Station Road, Histon, Cambridge, UK. The volumes for 1975 (1976), 1979 (1980), 1982 (1983), and 1994 (1995) are especially important from the perspective of this chapter. They contain the relevant scientific papers as well as reports of IWC debates and decisions. From April 1999, the scientific papers are published separately in the new *Journal of Cetacean Research and Management.*

Schweder, T. et al. (1997) Abundances of northeastern Atlantic minke whales, estimates for 1987 and 1995. *Report of the International Whaling Commission* **47**, 453–484.

Tønnesen, J.N. and Johnsen, A.O. (1982) *The History of Modern Whaling.* C. Hurst, London; Australian National University Press, Canberra.

Volterra, V. (1938) Population growth, equilibria and extinction under specified breeding conditions: A development and extension of the logistic curve. *Human Biology* **10** (1), 1–11.

Wallace, R.L. (compiler) (1994, 1997) *Compendium of Selected Treaties, International Agreements, and Other Relevant Documents on Marine Resources, Wildlife and the Environment.* 3 volumes. Marine Mammal Commission, Washington, USA.

Watson, L. (1981) *Sea Guide to Whales of the World.* Hutchinson, London.

WWF (World Wide Fund for Nature) (1998) *Creating a Sea Change; Resolving the Global Fisheries Crisis.* Report of an international conference sponsored by WWF's Endangered Seas and Living Planet Campaigns, held in September 1998, Lisbon, Portugal. WWF International, Gland, Switzerland.

Zeh, E.J., George, J.C. and Suydam, R. (1995) Population size and rate of increase, 1978–1993, of bowhead whales, *Balaena mysticetus. Rep. of the International Whaling Commission* **45**, 339–344.

## THE AUTHOR

**Sidney Holt**

*Hornbeam House, 4 Upper House Farm, Crickhowell, Powys, NP8 1BP, U.K.*

Chapter 113

# SMALL CETACEANS: SMALL WHALES, DOLPHINS AND PORPOISES

Kieran Mulvaney and Bruce McKay

The term small cetacean has generally been used to describe those species which do not have their names appended to the International Convention for the Regulation of Whaling (ICRW). This 'Annex of Nomenclature' was drawn up at the 1946 meeting at which the ICRW was completed, and from which the International Whaling Commission (IWC) was born. Unlisted species came to be referred to as "small cetaceans", and many of the IWC's Member States (primarily, but not exclusively, countries which exploit these species) argue that their non-listing means they are beyond the Commission's purview.

Threats to small cetaceans come from a wide range of events, including directed hunting and indirect catches, as well as from rising pollutant loads and possibly from changes to stocks of their prey and other, less direct changes to the marine food webs. This chapter briefly summarises the taxonomic groups of small cetaceans, and then describes in more detail the impacts affecting these mammals and assesses their current status.

*Seas at The Millennium: An Environmental Evaluation (Edited by C. Sheppard)*

## CLASSIFICATION OF SMALL CETACEANS

The mammalian order of *Cetacea* is divided into two sub-orders: the *mysticeti*, or baleen whales, and *odontoceti*, or toothed whales. The baleen whales include the blue whale, (*Balaenoptera musculus*), the largest animal to have ever lived on Earth. The toothed whales include all species of dolphins and porpoises, including the smallest and possibly rarest cetacean, the vaquita or Gulf of California porpoise (*Phocoena sinus*). All the "small cetaceans" are odontoceti. With the exception of the sperm whale and its smaller relatives, the dwarf sperm whale and pygmy sperm whale, this chapter includes all odontocetes and excludes the mysticetes. This serves the purpose of including all the dolphins and porpoises, which, notwithstanding the large size of the orca or killer whale (*Orcinus orca*), legitimately qualify as small cetaceans.

The following taxonomic arrangement is a broadly accepted breakdown (after Castello, 1996; Jefferson et al., 1993):

**Order: Cetacea**
**Sub-Order: Odontoceti**

| | |
|---|---|
| Family Platanistidae | |
| *Platanista gangetica* | Ganges river dolphin |
| *Platanista minor* | Indus river dolphin |
| Family Pontoporiidae | |
| *Lipotes vexillifer* | Baiji/Chinese river dolphin |
| *Pontoporia blainvillei* | Franciscana |
| Family Iniidae | |
| *Inia geoffrensis* | Boto/Amazon river dolphin |

These are mostly freshwater species. The exception is the franciscana, or La Plata river dolphin, which is primarily a marine and estuarine species, found in coastal waters of South America from Sao Paulo, Brazil, south to Peninsula Valdes, Argentina, and including the estuary of the La Plata river (Evans, 1987). Their bodies are small (approx. 2–2.5 m in most species; 1.5–1.7 m for the franciscana), and the beaks are long and narrow.

None are particularly common, but the status of the Indus river dolphin and the baiji give cause for concern. The baiji, having formerly occurred along nearly 2,000 km of the Yangtze river, is now restricted to small groups in limited stretches of the lower and middle parts of the river (Chen and Hua, 1989). Its continued survival is considered to be unlikely in the long term (Perrin, 1999). Threats include habitat loss from dam and barrage construction and flood control measures; environmental pollution; disturbance; conflicts with fisheries, notably gear entanglement; and directed hunts (Smith, 1996).

| | |
|---|---|
| Family Monodontidae | |
| *Monodon monoceros* | Narwhal |
| *Delphinapterus leucas* | Beluga/white whale |

Restricted to Arctic and sub-arctic waters, both species are highly distinctive. Narwhals measure approximately 4–5 m in length, and have a mottled greyish body. Males, which are larger, and females have just one pair of teeth in the upper jaw; in the male, the left tooth erupts as a long (1.5–3 m) spiralling tusk. With the exception of this tusk in males, none of the narwhal's teeth erupt, and so the species can be considered functionally toothless. Adult belugas are white all over, and measure about 3–5 m.

The Scientific Committee of the IWC recently recognized 29 stocks of belugas throughout the Arctic and sub-arctic. Of these, eight were strongly or tentatively considered to be stable, although three of those were already considered depleted. An additional eleven were considered depleted; of those, one in Cook Inlet, Alaska is continuing to decline; one in West Greenland has been reduced by approximately 60% between 1981 and 1994 and a third, in Ungava Bay in the Canadian Arctic, is close to extirpation (IWC, 1999).

Although estimates do exist for the abundance of narwhal populations, the IWC has been "unable to make a meaningful assessment of any stocks" (IWC, 1999).

| | |
|---|---|
| Family Phocoenidae | |
| *Phocoena phocoena* | Harbor porpoise |
| *Phocoena spinipinnis* | Burmeister's porpoise |
| *Phocoena sinus* | Vaquita |
| *Neophocoena phocaenoides* | Finless porpoise |
| *Australophocoena dioptrica* | Spectacled porpoise |
| *Phocoenoides dalli* | Dall's porpoise |

The "true" porpoises, easily distinguishable from other cetaceans by their rounded bodies and snub noses, are relatively small. Harbor porpoises may measure between approximately 1.5 and 2 m in length, depending on location (Read, 1998); based on relatively few recorded samples, the vaquita, the smallest of all cetaceans, has a length range of 90.3–143.5 cm (Brownell et al., 1987; Vidal et al., 1998).

With the exception of the oceanic Dall's porpoise (Houck and Jefferson, 1998), most species are primarily coastal. As a result, porpoises frequently conflict with human activities.

Harbor porpoises and Dall's porpoises are generally abundant, although there are concerns for the status of some stocks as a result of direct and indirect catches (IWC, 1991). There is little information on abundance or status of the finless, spectacled and Burmeister's porpoises (IWC, 1991). The vaquita is apparently restricted in range to the upper Gulf of California and is critically endangered. On the basis of several research surveys, its population was estimated in a 1997 review as being between 224 and 885 individuals (Barlow et al., 1997) and in 1999 at 567 animals, with a 95% confidence interval from 177 to 1073 (Jaramillo-Legorreta et al., 1999).

| | |
|---|---|
| Family Delphinidae | |
| *Steno bredanensis* | Rough-toothed dolphin |
| *Sousa chinensis* | Indo-Pacific humpbacked dolphin |
| *Sousa teuszii* | Atlantic humpbacked dolphin |

| | |
|---|---|
| *Sotalia fluviatilis* | Tucuxi |
| *Lagenorhynchus albirostris* | White-beaked dolphin |
| *Lagenorhynchus obliquidens* | Pacific white-sided dolphin |
| *Lagenorhynchus acutus* | Atlantic white-sided dolphin |
| *Lagenorhynchus obscurus* | Dusky dolphin |
| *Lagenorhynchus cruciger* | Hourglass dolphin |
| *Lagenorhynchus australis* | Peale's dolphin |
| *Grampus griseus* | Risso's dolphin |
| *Tursiops truncatus* | Bottlenose dolphin |
| *Stenella frontalis* | Atlantic spotted dolphin |
| *Stenella attenuata* | Pantropical spotted dolphin |
| *Stenella longirostris* | Spinner dolphin |
| *Stenella clymene* | Clymene dolphin |
| *Stenella coeruleoalba* | Striped dolphin |
| *Delphinus delphis* | Common dolphin |
| *Lagenodelphis hosei* | Fraser's dolphin |
| *Lagenodelphis borealis* | Northern right whale dolphin |
| *Lagenodelphis peronii* | Southern right whale dolphin |
| *Cephalorhynchus commersonii* | Commerson's dolphin |
| *Cephalorhynchus eutropia* | Black dolphin |
| *Cephalorhynchus heavisidii* | Heaviside's dolphin |
| *Cephalorhynhus hectori* | Hector's dolphin |
| *Peponocephala electra* | Melon-headed whale |
| *Feresa attenuata* | Pygmy killer whale |
| *Pseudorca crassidens* | False killer whale |
| *Orcinus orca* | Orca/killer whale |
| *Globicephala melas* | Long-finned pilot whale |
| *Globicephala macrorhynchus* | Short-finned pilot whale |
| *Orcaella brevirostris* | Irrawaddy dolphin |

The delphinidae is the most speciose of all marine mammal families (Ridgway and Harrison, 1998). The bottlenose dolphin is the most common species in dolphin shows, and the orca or killer whale, similarly also frequently seen performing in captivity, is unmistakable, its black-and-white markings and considerable size, up to 7.7 m in females, 9 m in males (Dahlheim and Heyning, 1998), causing it to stand out.

Appearances vary greatly among members of this family. The bottlenose, common and, to some extent, rough-toothed dolphins, and the *Lagenorhynchus* and *Stenella* species, perhaps most conform to the commonly-held stereotype of a "typical" dolphin: streamlined, with dorsal fins and prominent beaks. The pilot whales, in contrast, have a much more robust body shape, with a large bulbous melon and sickle-shaped pectoral fins. They are also among the larger delphinids, males reaching 7.2 m in the western North Pacific and 6.17 m in the western North Atlantic (Bernard and Reilly, 1998). The smallest of the delphinidae are the tucuxi, with a maximum length of about 1.9 m., and the *Cephalorhynchus* species, which range from approximately 1.4 m to 1.7 m in length (Leatherwood and Reeves, 1983).

Several of these species have wide ranges; indeed, the orca is the most widely-distributed mammal in the world (Dahlheim and Heyning, 1998). The bottlenose dolphin, common dolphin, Risso's dolphin, false killer whale and both species of pilot whales are found in all the world's temperate and tropical oceans (Bernard and Reilly, 1998; Evans, 1987; Kruse et al., 1998; Leatherwood and Reeves, 1983; May, 1990; Odell and McClune, 1998; Wells and Scott, 1998). Some others are more localized in distribution. The *Cephalorhynchus* species, for example, are all restricted to coastal waters of the southern hemisphere: the Commerson's dolphin is found in the western South Atlantic, near the tip of South America and in the vicinity of Kerguelen Island in the Indian Ocean; the black dolphin is restricted to coastal waters of Chile, the Hector's dolphin to the waters of New Zealand, and the Heaviside's dolphin to the coastal waters of southwestern Africa (Leatherwood and Reeves, 1983). Similarly, the Irrawaddy dolphin is found only in the tropical Indo-Pacific, primarily near shore and in some large rivers, while the tucuxi is limited to rivers and near-shore marine waters of northeastern South America and eastern Central America (Leatherwood and Reeves, 1983).

Several populations of delphinidae have been impacted by human activities. Two of them, the northern offshore form of the spotted dolphin, and the eastern form of spinner dolphin, accounted for 80% of the total mortality in the Eastern Tropical Pacific purse-seine fishery for yellowfin tuna. It is likely that these stocks were significantly reduced in the early years of the fishery, although more recent analysis suggests that, while probably depleted, the populations are now apparently stable (IWC, 1992, Perrin, 1999). The northern stock of common dolphins, also targeted by the fishery, has, however, shown signs of significant recent decline (IWC, 1993).

The IWC Scientific Committee has on several occasions expressed concern over the status of the striped dolphin in Japanese waters as a result of directed hunts there (IWC, 1982, 1993; Perrin, 1999). Dusky dolphin numbers may be being heavily impacted by directed takes off Peru; recent evidence suggests that duskys now constitute an ever-smaller proportion of the catch, and the average body size of those being caught is also decreasing, both signs of a possible decline in numbers (IWC, 1997). Peale's dolphins had become extremely rare in certain parts of the Magellan region by the late 1980s, presumably due to hunting for use as crab bait in Chile; however, indications now are that they are returning to the region (IWC, 1997). Species with limited range may be especially vulnerable to human impacts: Hector's dolphins, for example, are frequently caught in fishing nets in New Zealand waters, apparently causing at least two of three populations to decline (Martien et al., 1999). Some populations of several species, including the northern right whale dolphin, striped dolphin, common dolphin, and Pacific white-sided dolphin, may have been significantly depleted by high levels of incidental mortality in high seas drift net fisheries (IWC, 1992). The Irrawaddy dolphin may be in decline in parts of its range, possibly as a result of a wide combination of factors, including incidental mortality in fishing gear, shooting by soldiers and villagers, the use of explosives to catch fish, and general habitat deterioration as a result of the Vietnam war (IWC, 1994).

Family Ziphiidae

| | |
|---|---|
| *Tasmacetus shepherdi* | Shepherd's beaked whale |
| *Berardius bairdii* | Baird's beaked whale |
| *Berardius arnuxii* | Arnoux's beaked whale |
| *Mesoplodon pacificus* | Longman's beaked whale |
| *Mesoplodon bidens* | Sowerby's beaked whale |
| *Mesoplodon densirostris* | Blainville's beaked whale |
| *Mesoplodon europaeus* | Gervais' beaked whale |
| *Mesoplodon layardii* | Strap-toothed whale |
| *Mesoplodon hectori* | Hector's beaked whale |
| *Mesoplodon grayi* | Gray's beaked whale |
| *Mesoplodon stejnegeri* | Stejneger's beaked whale |
| *Mesoplodon bowdoini* | Andrews' beaked whale |
| *Mesoplodon mirus* | True's beaked whale |
| *Mesoplodon ginkgodens* | Ginko-toothed beaked whale |
| *Mesoplodon carlhubbsi* | Hubbs' beaked whale |
| *Mesoplodon peruvianus* | Pygmy beaked whale |
| Mesoplodon sp. A | Unnamed beaked whale |
| *Ziphius cavirostris* | Cuvier's beaked whale |
| *Hyperoodon ampullatus* | Northern bottlenose whale |
| *Hyperoodon planifrons* | Southern bottlenose whale |

Most beaked and bottlenosed whales are little-known and relatively rarely seen. Some species are known only from isolated sightings, stranded animals or skulls. Not all species have even been fully described.

The mesopolodonts range in size from roughly 4.5 m to about 6.5 m in length. The northern and southern bottlenose whales reach lengths of approximately 9.8 m, while the Baird's and Arnoux's beaked whales may reach almost 13 m (Leatherwood and Reeves, 1983).

Abundance and distribution are not well known for many species, as few are commonly seen, but the fact that some species, such as the relatively widespread Cuvier's beaked whale, strand with some frequency suggests that at least some species may be more common than the relative paucity of sightings indicates.

Baird's beaked whales are hunted by a coastal operation off Japan; 54 whales are taken annually. The northern bottlenose whale was hunted intensively by Norwegian and, to a lesser extent, Scottish and Canadian whalers earlier this century; as a result, it is possible that some stocks may be depleted (Leatherwood and Reeves, 1983).

## HUMAN IMPACTS

### Directed and Indirect Catches

There have been numerous reviews of small cetacean exploitation, including directed kills and conflicts with fisheries (e.g. Beddington et al., 1985; Currey et al., 1990; Holliday et al., 1996; IWC, 1991; Mitchell, 1975; Mulvaney, 1996; Northridge, 1984; Northridge and Pilleri, 1986; Northridge and Hofman, 1999; Perrin, 1999; Read, 1996). The purpose of this chapter is to focus on those principal hunts, and major cases of incidental catches as a result of fishing operations, known to be occurring in the final few years of the twentieth century.

#### *Beluga and Narwhal in the Arctic*

Hunts for beluga, or white whale, and narwhal have long been conducted by native peoples in the Arctic and subarctic, providing blubber oil for lighting and cooking and food in the form of *muktuk*—the name given to the whales' skin and adhering blubber—and meat, which is also fed to sled dogs (Mitchell, 1975). In addition, male narwhals are considered commercially valuable for their long tusks (Holliday et al., 1996).

Although there is considerable variance from year to year, annual takes of belugas in recent years in Alaska have averaged approximately 400, from five recognized populations (IWC, 1999). In Greenland, it is estimated that approximately 600 whales are taken annually. Although catch figures are not widely available for the number of belugas caught in Canadian waters (IWC, 1999), efforts are ongoing to compile accurate catch statistics (R. Reeves, pers. comm.). The IWC (1992) noted a recorded total of 703 narwhals killed in Greenland in 1987—an increase from 278 in 1975. Catch figures in Canada over the same period showed no particular sign of increase or decrease, with highs of 406, 404 and 350 in 1981, 1982 and 1980 respectively, and lows of 152, 181 and 255 in 1974, 1987 and 1977. The same review noted, however, increasing difficulties in gathering accurate and complete data; by 1999, the Commission was reduced to noting that "directed takes are known to be continuing in Canada and Greenland, presumably at a similar scale to what they have been for at least the past decade." (IWC, 1999).

#### *Pilot Whales in the Faroe Islands*

In the Faroe Islands, North Atlantic, pilot whales have been the targets of a regular hunt for several hundred years (Bourne, 1965; Mitchell, 1975). In a process known as the *grindadrap*, pods of pilot whales are herded toward shore by small boats, then killed in shallows or on the shore with spears and knives. The meat is distributed free of charge to local inhabitants.

Hunting statistics extend back to 1584, and largely unbroken records exist from 1709 to the present (Bloch et al., 1990). Between 1709 and 1996, 239,413 whales were recorded as being killed (IWC, 1992, 1998, 1999).

#### *Use of Small Cetaceans as Fish Bait in Chile*

Small cetaceans have been taken deliberately since the mid-1970s in the Magellan region of Chile for use as bait in traps set for southern king crab or centolla (*Lithodes santolla*) and false king crab or centollon (*Paralomis granulosa*). Despite prohibition on the catch, transportation and commercialization, possession or processing of small cetaceans since 1977, Peale's dolphins, black dolphins and Commerson's dolphins continue to be taken (IWC, 1995). Marine mammal meat has been preferred to fish because it can stay in the water for three days before deteriorating and it is free; fish deteriorates within 24 hours and costs money. One dead dolphin can bait 350 traps (Stone et al., 1987).

In recent years, the catch seems to have declined as a result of a decrease in fishing effort, lessening the demand for bait; the fact that, by 1992, wastes from slaughterhouses and industrial fisheries were providing most of the bait needs; and an apparent decline in abundance of the dolphins in fishing areas. In 1994, the IWC estimated that "the take of small cetaceans [did] not exceed 10% (45T) of the total demand of bait, or an equivalent of 600 dolphins per year" (IWC, 1995).

*Directed and Indirect Catches in Peru*

Catches of small cetaceans in Peruvian waters, both deliberate and incidental, had been relatively low until the mid-1970s, when they suddenly increased, possibly as a consequence of the decline in the region's anchoveta fishery (Read et al., 1988; Van Waerebeek and Reyes, 1990). Cetaceans are caught directly and indirectly in drift and set gill-nets, are harpooned or are caught by purse-seiners (and often landed alive) operating in the fishmeal industry. The cetaceans are used primarily for human consumption. Principal species include dusky dolphins, long-beaked common dolphins, Burmeister's porpoises, and bottlenose dolphins (IWC, 1995).

In 1988, Read et al. estimated the total annual catch of small cetaceans in Peruvian waters at about 10,000 animals. In November 1990, the Peruvian government passed a national ban on cetacean utilization and exploitation. The effect on the hunt seems to have been minimal at best. Studies by researcher Koen van Waerebeek tentatively suggest that, even since the ban, the annual catch is now roughly 17,600 (IWC, 1995).

*Directed and Indirect Catches in Sri Lanka*

Large numbers of small cetaceans are caught in gill-nets, incidentally and deliberately, in Sri Lanka as a direct result of a fisheries modernization program initiated by the Food and Agriculture Organization of the United Nations (FAO). Small cetaceans had long been caught occasionally by coastal gill-net fisheries in Sri Lanka, but the nets were made of natural fibres such as jute or cotton, and most dolphins could either detect them with their sonar or break free if they did become entangled. However, under the FAO program, nets were supplied that were made of stronger materials, which the dolphins can neither detect nor break. Initially, incidentally-caught dolphins were probably discarded by most fishermen, but as uses for the dolphins were identified and the market grew, gillnets came to be set intentionally to catch dolphins for food and bait (Leatherwood and Reeves, 1989).

Estimates of the number of small cetaceans killed annually in this way vary from 8,042–11,821 to 12,950 (IWC, 1992; Leatherwood, 1995). At least seventeen species of small cetaceans are involved; catches also include a few large cetaceans (Leatherwood and Reeves, 1989).

*Directed Hunts of Small Cetaceans in Japan*

Of the 21 or so species of small cetaceans occurring in Japanese waters, 16 are, or have recently been, subject to significant directed hunts (Miyazaki, 1983).

The Baird's beaked whale, which, as already noted, on the grounds of size alone barely qualifies for consideration as a "small" cetacean, is the target of a coastal whaling operation which uses small coastal vessels equipped with harpoon guns armed with explosive harpoons. The hunt is not regulated by the IWC; Japan presently assigns itself an annual quota of 54 whales (Balcomb and Goebel, 1977; IWC, 1992, 1999; Kasuya and Ohsumi, 1984; Mitchell, 1975). A similar operation also conducts hunts for short-finned pilot whales and Risso's dolphins out of Taiji on the Kii Peninsula, with quotas of up to 50 and 30 respectively (IWC, 1993).

The majority of species taken in Japanese waters are killed in so-called drive fisheries. When a herd of dolphins is spotted out to sea, a number of small boats drive the dolphins toward shore and into a bay or inlet. The dolphins may be corralled by means of a net drawn across the mouth of the bay; they may then be left there overnight and killed the next morning (May, 1990).

These drive fisheries have operated in western Japan since the late 17th century. Historically, there were drive fisheries in at least four villages on the Sea of Japan, nine villages on the Izu coast, one in Iwate on the Pacific coast of northern Japan and islands off northern Kyushu and one in the Ryukyus. Today, there are two groups still operating: at Futo on the Izu coast, and at Taiji (IWC, 1993). Eleven species are taken, of which the most frequently caught are striped dolphins, short-finned pilot whales, bottlenose dolphins, false killer whales, Risso's dolphins and spotted dolphins (IWC, 1993). Some hunting in some locations is conducted with hand harpoons or crossbows rather than via drive fisheries (IWC, 1993).

By far the most catches have been of striped dolphins, although takes of the species at both Izu and Taiji have declined in recent years. Prior to 1963, annual catches were as high as 10,000–20,000, although catch records are incomplete. After 1963, the catch dropped to a mean average of 7,350, until 1980 when, following a peak catch year of 16,237, the mean over the following decade dropped to 2,390. 449 were reported killed in 1998 (IWC, 1993, 1999).

Two species, the Pacific white-sided dolphin and Dall's porpoise, are killed with hand-harpoons, the former because it is difficult to herd and the latter because it does not form in large enough groups to be caught in a drive fishery (May, 1990).

Dall's porpoises have been hunted in Japanese waters since at least the 1940s, and the species has generally been the most heavily exploited by fishermen (Miyazaki, 1983). Prior to 1988, the annual take was approximately 10,000 (IWC, 1999); however, in 1988, the recorded take soared to 40,367, before declining again to 29,048 in 1989 and 21,804 in 1990 (IWC, 1990, 1991, 1992). Following criticism of the increased hunt by environmental organizations and by the IWC Scientific Committee, the Government of Japan established an annual quota of 17,700. The actual reported take in 1998 was 11,385 (IWC, 1999).

### *Mortality Associated With Yellowfin Tuna Purse-seine Fisheries in the Eastern Tropical Pacific*

For reasons which still are not entirely clear, herds of certain dolphin species in the Eastern Tropical Pacific (ETP) swim in association with schools of large yellowfin tuna. Since at least the 1920s, fishermen who have known of this association have taken advantage of it to catch fish. Initially, however, this was done without harming the cetaceans: the surface disturbances created by the dolphins were used to locate the schools of tuna, at which point fishermen would throw live bait overboard, sending the tuna into a feeding frenzy; the tuna would then bite at everything, including the hooks lowered to catch them. The dolphins could avoid the hooks with their sonar and were "rewarded" by taking some of the bait (Allen, 1985)

However, beginning in the 1950s, the nature of the fishery changed from small-scale to highly commercial with the development of the "power block," a hydraulic pulley which allowed the rapid retrieval of huge lengths of purse-seine net. Instead of hauling tuna aboard one fish at a time, it became possible to deploy a net around an entire herd of dolphins and trap both tuna and dolphins (Gosliner, 1999). Many dolphins were drowned or were crushed in the nets; others were crushed after they passed through the winch. Annual mortality was estimated at between 200,000 and 500,000 for the period 1959–1972; altogether, over seven million dolphins are believed to have been killed this way over the course of around four decades (Allen, 1985; Gosliner, 1999; May, 1990; Scheele and Wilkinson, 1988).

Following intense and prolonged campaigning by environmental organizations, an international agreement, negotiated in 1992 by the Inter-American Tropical Tuna Commission (IATTC), set annual quotas of dolphin by-catch for IATTC member states and cooperating nations. Having already declined to just over 15,000 in 1992, reported catches continued to decrease markedly over subsequent years, to 3,274 in 1995, 2,547 in 1996, and 3,005 in 1997 (IWC, 1999).

### *Incidental Catch of Vaquita in the Gulf of California*

In comparison with some of the other directed and indirect catches noted, the absolute number of vaquitas killed as a result of human activities is low. Between March 1985 and January 1994, 76 vaquitas were recorded as being caught in gill nets set for the endangered totoaba fish (*Totoaba macdonaldi*) in the upper Gulf of California (Vidal, 1995). However, this is certainly an underestimate of the total number caught; in addition, vaquitas have also been reported entangled in nets set for other fisheries in the region (Vidal et al., 1998).

Given that, as already observed, the vaquita population is almost certainly at a critically low level, it is considered essential that even these levels of by-catch are ended if the species is to have a chance of recovery (IWC, 1995, 1999).

### *By-Catch of Harbor Porpoises in the North Atlantic*

Largely as a result of their habitat in productive coastal waters, harbor porpoises are captured incidentally in commercial fisheries throughout their range (Gaskin, 1984; Read, 1998). Although several different gear types are involved, the majority of fisheries interactions are as a result of bottom-set gill nets (Read, 1998). The most notable by-catches occur in the waters of the Bay of Fundy and Gulf of Maine, Newfoundland and Labrador, Britain and Ireland, Denmark, the Baltic, western Greenland, Iceland, and Sweden and Norway (Hutchinson, 1996; Read, 1998). Read (1998) compiled annual by-catch estimates in several locations from different sources: 1200–2900 in the Gulf of Maine; 101–424 in the Bay of Fundy; 134–1531 for West Greenland; 17 for the Skaggerak; 4,629 for the (Danish) North Sea; and 2,049 for the Celtic shelf. Fisheries closures have resulted in declines in incidental catches in some areas, but by-catch levels have remained high elsewhere.

### *Other Directed and Indirect Takes*

Among some of the other instances over which concern has been expressed at various times are: possible continued illegal hunting of dolphins in the Black Sea; incidental catches of Indus and Ganges river dolphins and the baiji, or Yangtze river dolphin; gillnet fisheries off the coast of California; entanglement in shark nets off South Africa and Australia; incidental mortality of Hector's dolphins off New Zealand; high levels of incidental mortality of franciscana in Uruguay and Brazil and, indeed, of delphinid species throughout the Americas. Concerns persist, also, over occasional continued illegal use of drift nets.

The above list is not comprehensive. The cited examples are relatively well-known simply because of the intensive efforts to investigate by-catches and direct kills in, for example, American and European waters. It is safe to say that, wherever coastal gillnets are in operation, there are likely to be very high by-catches of small cetaceans (R. Reeves, pers. comm.). Many are in less developed countries, with limited documentation, and where priority is on meeting day-to-day nutritional and economic needs (see, for example, Mulvaney 1996). In addition, as has been the case in, for example, Peru and Sri Lanka, these situations can lead easily from indirect to directed catches and the establishment of a market for small cetaceans. Such situations are likely to remain by far the largest source of mortality (for reviews see, for example, Holliday et al., 1996; IWC, 1995).

## Pollution

Perhaps surprisingly, there is little evidence of the extent to which pollutants may affect individual cetaceans or impact populations. Nonetheless, concern has been expressed, to varying levels, about the possible consequences of contamination from, among others, heavy metals, organochlorines, oil, and radionuclides.

### *Heavy Metals*

Concern over possible effects of metals results at least partly from their observed impacts on other species. For example, human symptoms following ingestion of methylmercury include loss of coordination, loss of vision and hearing and mental deterioration (Law, 1995). However, marine mammals have some physiological adaptations enabling them to neutralize some heavy metals. For example, in dolphin bodies mercury can sometimes be found as granules in the liver, where it arrives as a result of a process which effectively takes it out of general circulation (Simmonds et al., 1999). Nonetheless, there is evidence that there is a limit to the level of mercury contamination which cetaceans can safely accumulate. One study found an apparent correlation between liver disease and mercury concentrations in dolphins; disease symptoms appeared to be initiated at liver concentrations between 50 and 61 ppm. This is significantly greater than levels that have generally been recorded in stranded dolphins, although higher levels have been found by some researchers (Gaskin et al., 1974; Honda et al., 1983; Simmonds et al., 1999a). Furthermore, Bennet et al. (1999) showed that the mean liver concentrations of mercury, selenium and zinc in porpoises that died of infectious diseases were higher than those which died of physical trauma.

It is also interesting to note that, despite criticisms about the possible impacts on pilot whale populations in the North Atlantic of the Faroese *grindadrap*, high levels of contaminants, including mercury, in pilot whale blubber could ultimately spell the end to the hunt as the meat is increasingly considered safe to eat only in the smallest quantities (Simmonds et al., 1994).

### *Organic Chemicals*

Many organic chemical pollutants are persistent and readily accumulate in body tissue, and are transferred by the female to their offspring during gestation and via their milk.

The number of chemicals entering the environment is growing, perhaps by more than 1,000 each year (Caroli et al., 1996; National Research Council, 1984). Today, as many as 70,000 may have common application, and adequate information to assess overall toxicity risk is available for approximately 2%. Little is known of the effects of most in many living organisms (McCarthy and Shugart, 1990).

Of particular concern to marine mammal populations are the halogenated hydrocarbons (HHCs) such as the PCBs, DDT, chlordane, dioxins and furans, and the chlorinated and brominated diphenyl ethers. Other chemical groups of concern include organometals such as tributyltin, and polycyclic aromatic hydrocarbons (PAHs) (O'Shea, 1999).

Ethical and logistical considerations have made it difficult accurately to determine impacts on marine mammals. Overall, however, there is general scientific agreement that a number of health-related components are likely being compromised in at least some marine mammal populations and that, ultimately, their survival and reproduction are being affected.

The few experimental studies which have attempted to discern the possible impacts of contaminants on marine mammals have generally focused on pinnipeds. Harbor seals fed fish from contaminated European waters showed a range of immune system impacts (DeSwart, 1994; Ross et al., 1995) as well as reproductive failure (Reijnders, 1986). Elevated DDT and PCB concentrations have been implicated in premature pupping and still births in California sea lions (DeLong et al., 1973), although a cause-and-effect relationship was not established. Similarly, gray and ringed seals in the Baltic Sea have shown impaired reproductive ability and a variety of lesions in kidneys, intestines, arteries, adrenal glands and skull bones that have been associated with environmental contaminants (see, e.g.: Bergman et al., 1992; Bergman and Olsson, 1985; Helle, 1980).

Contaminant-induced immunosuppression has been suggested as the reason for the high incidence and severity of bacteria-related lesions found in dead beluga whales from the St. Lawrence River and estuary in eastern Canada (DeGuise et al., 1995). Subsequent research in which beluga whale immune cells were exposed to environmentally-relevant doses of mercury, cadmium, DDT and various PCBs, showed immune-system-related impacts, such as reduced proliferation of lymphocytes (DeGuise et al., 1996, 1998). Lahvis et al. (1995) noticed reduced immune system response in bottlenose dolphins from the U.S. Gulf of Mexico that was associated with increased blood concentrations of PCBs and DDT. The small and geographically-isolated population of St. Lawrence belugas has also been discovered with a high incidence of other ailments and abnormalities, including cancer, gastric ulcers, and hermaphroditism (DeGuise et al., 1994, 1994a,b; DeGuise et al., 1995; Lair et al., 1997) as well as having abnormally low levels of recruitment relative to arctic populations (Beland et al., 1993).

Ross et al. (in press) have found high levels of PCBs in three populations of orcas in the Puget Sound region, on the western U.S./Canada border. Mean levels in male transient orcas were 251 ppm, and 151 ppm in males from the southern resident population. Ross et al. consider the orcas to be "among the most contaminated marine mammals in the world," noting that mean PCB concentrations in the transient males are four to five times higher than those for the St. Lawrence belugas. The southern resident population has declined from 96 to 82 whales over a four-year timespan, although Ross et al. note that this is still an increase since the end of a live-capture fishery in the region. The Canadian government has declared the orcas as threatened, citing diminished food supply and increased boat traffic, as well as contaminants, as possible threats.

High levels of, notably, organochlorines were recorded in dolphins involved in mass die-offs in U.S. waters and from the Mediterranean Sea (Aguilar and Borrell, 1994; Kuehl et al., 1991, 1994). Organotins were found in dolphins from the U.S. event (Kannan et al., 1997). The extent of the role of contaminants in these die-offs, if any, is uncertain, although much speculation has focused on the possibility that they may have played a role in compromising the animals' immune systems, thus rendering them more vulnerable to an infectious agent.

### *Oil*

There are very few studies of the effect of either chronic or acute oil pollution on dolphins and porpoises. There are indications that, except in situations where oil on the water surface is so thick that it forms mousse, small cetaceans do not generally avoid it (Geraci, 1990; Harvey and Dahlheim, 1994; Simmonds and Hutchinson, 1992). This apparent lack of avoidance has led to concerns about cetaceans inhaling volatile hydrocarbons, or oil adhering to their skin or eyes or contaminating their prey (Harvey and Dahlheim, 1994). Based on observations of other species, among the possible impacts of such contact are inflammation of lung membranes, lung congestion and pneumonia (Simmonds and Hutchinson, 1992).

The *Exxon Valdez* oil spill, in which approximately 11 million gallons of crude oil were spilled into Alaska's Prince William Sound on March 24, 1989, apparently resulted in the deaths and/or disappearance of 14 killer whales from one pod, although the exact causes of this could not be ascertained (Dahlheim and Matkin, 1994). This pod has still not recovered, although overall numbers of orcas in the Prince William Sound area remain roughly the same or perhaps slightly higher than before the spill (Matkin et al., 1994; Oil Spill Trustee Council, 1999).

### *Radionuclides*

Ionizing radiation can result in a wide range of effects in mammals, including changes in behaviour, growth and development, and effects such as mutations and carcinomas (Eisler, 1994; O'Shea, 1999). Anthropogenic radionuclides that contaminate today's ecosystems derive primarily from fallout from atmospheric nuclear weapons testing, the 1986 Chernobyl accident, nuclear reactor operations, nuclear fuel processing and disposal, and applications in medicine, industry, agriculture, and research (O'Shea, 1999).

There have been only a few studies of radionuclide contamination in marine mammals, and none has identified any associated effects. Calmet et al. (1992) found negligible concentrations in muscle and liver tissues from spotted, spinner and common dolphins from the eastern Pacific. A study of milk and tissues of gray seals collected from the North Sea and North Atlantic in 1987 revealed low levels of cesium-137, about 70% of which was ascribed to the nuclear reprocessing industry in England, and the rest to Chernobyl (Anderson et al., 1990). Analysis of liver and muscle tissue from harbor seals, gray seals and harbor porpoises found stranded along the UK coast revealed that radionuclide contamination decreased with distance from the reprocessing plant at Sellafield. The maximum radiation dose to the marine mammals from radiocaesium was higher than doses previously assessed for critical groups of humans living near Sellafield, while the maximum dose from plutonium was comparable to the doses for humans (Watson et al., 1999).

## Small Cetaceans and Environmental Change

Cetaceans potentially face an additional range of consequences from broader environmental change. This includes such factors as habitat change and associated disturbance, eutrophication and the spread of "harmful" algal blooms and toxins, depletion of fish stocks, reduction in the stratospheric ozone layer, and global climate change.

### *Habitat Degradation and Change*

Numerous human activities degrade or alter the physical marine environment, primarily in coastal areas. In addition to those issues already addressed, these activities include coastal development, the introduction of exotic species and the damming of rivers (see, for example: IWCO, 1998; McKay et al., 1997; Thorne-Miller, 1999). However, although there have been numerous studies on the impacts of these activities, few have considered their effects on small cetaceans.

It has been suggested that most oceanic cetacean species may be relatively immune to changes in habitat as a result of human activity, and that such changes are most likely to affect species with largely inshore habitats or specific populations with restricted ranges that are close to human activity (Kemp, 1996). For example, the Chinese white dolphin, a subspecies or possibly close relation, of the Indo-Pacific humpbacked dolphin, has apparently been adversely affected by habitat loss due to construction and increased shipping traffic, among other factors; its numbers have reportedly dropped from around 200–400 in 1989 to 80 in 1995 (Kemp, 1996). There are widespread concerns that construction of the Three Gorges Dam on the Yangtze River will result in a number of fundamental changes to the riverine ecosystem that may seriously impact the already-depleted baiji, or Yangtze river dolphin (Chen and Hua, 1989a; Topping, 1995). Populations of the Ganges and Indus river dolphins have already been fragmented and diminished by dam construction and the alteration of habitat through irrigation and flood-control measures (Smith, 1996).

Of rising importance are the noise and disturbances associated with a wide range of coastal and off-shore human-related activities including shipping, fishing, recreational boating and whale-watching, mariculture (acoustic harassment or warning devices), seismic exploration, geophysical surveys, dredging, minerals mining, oil and gas drilling, military activities (low-frequency active sonar) and even climate monitoring programs (Acoustic Thermometry of Ocean Climate, or ATOC). Ambient noise, mainly due to shipping, has likely increased by some 15 dB between 1950 and 2000 (Ross, 1987) though, of course, levels will vary. Wells and Scott (1997) note, for example, that there are more than 700,000 registered boats in Florida and express concern over these vessels' impact on resident bottlenose dolphin populations.

The effects of noise on the behaviour, distribution and physiological condition of marine mammals are extremely difficult to quantify (Richardson and Wursig, 1997; Gisiner, 1998), though odontocetes are expected to sustain the least impacts because of their relatively poor sensitivity to low-frequency sound (Richardson et al., 1995). Differences in responses from species to species can be considerable: where various odontocetes actively approach ships, others such as beluga in the Canadian Arctic show noticeable avoidance behaviour as much as 50 km from the sound source (Finley et al., 1990). Low-frequency active sonar was suggested (Frantzis, 1998) as the likely cause of the atypical and fatal strandings of 12 Cuvier's beaked whales off Greece, an event that closely coincided temporally and spatially with a NATO submarine detection testing program.

The sensitivity of marine animals, including odontocetes, to intense sound is not known (Gisiner, 1998), although hearing damage could reduce or prevent an animal's ability to detect prey, avoid predation or boats, communicate, or care for young.

### *Impacts of Fisheries*

In addition to direct impacts, commercial fishing operations have the potential to affect small cetacean populations by reducing their prey or by scattering the required aggregations for effective feeding. Furthermore, community interactions may be altered in such a way that nutrient-deficient prey may come to proliferate.

The possibility of such impacts is highlighted by the extent to which commercial fishing operations have affected fish stocks worldwide. According to FAO, of the world's fish stocks for which assessment data are available, one in four is classified as over-exploited, depleted, or recovering from depletion. A further 44% are fully- or heavily-exploited (FAO, 1994).

Although no links have been established between declines in small cetacean populations and fisheries-induced forage reduction, it is still reasonable to infer (Jennings and Kaiser, 1998), that reductions in marine mammal abundance will be the likely outcome of a decline in their prey. Various observations suggest a correlation between the behaviour, abundance and distribution of other cetacean and marine mammal populations and the status of their food sources. Perhaps most notable was the drowning of over 80,000 starving harp seals (*Pagophilus groenlandica*) in fishing nets along the coast of Norway in the late 1980s, as they moved out of the Barents Sea in search of food following the collapse of the region's capelin fishery (Haug et al., 1991). Other examples include a shift in the distribution of humpback whales (*Megaptera novaeangliae*) in the NE Atlantic during the late 1980s after the collapse of the capelin stock (Christensen, 1990); and, during the late 1970s, the movement of humpbacks off the northeastern United States in apparent response to the overfishing of herring and mackerel (Payne et al., 1990; Weinrich et al., 1997). Thompson et al. (1997) noted anemia in harbor seals in NE Scotland that correlated with switching from a clupeid-based diet (e.g. herring, sprat) to one that was gadoid-based (e.g. whiting, sandeel), although it was unclear if this was related to changes in the foraging behaviour required for the alternative prey or due to the composition of the prey itself.

The Norwegian "harp seal invasion" led to calls by many fishermen for a seal cull, apparently in the belief that the increased visibility of the seals was the result of an actual increase in the seal population, and that this was the cause of the fish stock collapse. This response is not atypical in instances of perceived conflicts between fisheries and marine mammals, including small cetaceans. Most famously, between 1976 and 1982, fishermen on Iki Island, Japan, killed a recorded total of 4,147 bottlenose dolphins, 466 Pacific white-sided dolphins, 953 false killer whales, and 525 Risso's dolphins, in the belief that the dolphins were responsible for declines in their catches of yellowtail (Kasuya, 1985). Although there are at present no such large-scale "culls" of small cetaceans, the argument is frequently made that the cetaceans and other marine mammals need to be "controlled" to limit their impact on fish stocks.

The rationale behind such culls is scientifically suspect. The complexity of marine ecosystems is such that it is, at best, difficult to draw a direct correlation between two individual elements of such systems. As several observers have pointed out, it is equally probable that reducing marine mammal populations could result in fewer fish for a commercial fishery, as cetaceans or pinnipeds frequently prey on fish which are themselves predators of the commercial fish species (D.M. Lavigne, pers. comm.).

### *Harmful Algal Blooms and Toxins*

Of the 3400–4100 species of microalgae in the oceans (Sournia, 1995), a number undergo "blooms", and there is compelling evidence to suggest that an increase in bloom frequency, intensity and geographic extent has been occurring recently (Hallegraeff, 1993, Smayda, 1990). Species hitherto undetected or previously characterized as benign have emerged as significant problems (see e.g., Burkholder et al., 1992; Todd, 1993). Eutrophication, a condition plaguing increasingly large sections of coastline (Nixon, 1995), is frequently invoked as an important factor (e.g. Paerl and Whitall, 1999; Smayda, 1990), while dams or water diversions may alter nutrient ratios (Humborg et al., 1997). The geographic spread of some toxic species has increasingly occurred through ballast water transport (Carlton and Geller, 1993; McMinn et al., 1997).

Of particular relevance to marine mammals are species, primarily dinoflagellates, which produce toxins. These can be sequestered into their food where sub-lethal or lethal doses can be ingested (Geraci et al., 1989). Algal toxins have been implicated in a variety of mortality events including humpback whales, manatees, monk seals, and California sea lions (see Anderson and White, 1989) and of approximately 2000 bottlenose dolphins along the east coast of the United States in 1987–88. A government-appointed investigation argued that the dolphins had been poisoned after eating fish which had previously fed on the toxic dinoflagellate *Gymnodinium breve* (Geraci et al., 1989); however, this conclusion remains controversial in the face of the potential roles of, for example, environmental contaminants (Kannan et al., 1997; Kuehl et al., 1991,1994) and morbillivirus-related diseases (Schulman et al., 1997). Dead dolphins have been discovered in the same time and general area of *G. breve* blooms in the U.S. portion of the Gulf of Mexico (Gunter et al., 1948; Marine Mammal Commission, 1997).

### *Atmospheric Impacts: Ozone Depletion and Global Climate Change*

Numerous studies have pointed to the potential impact of increased UV-B radiation as a result of depletion in the stratospheric ozone layer on marine ecosystems, particularly by causing increased mortality in fish eggs and larvae and marine phytoplankton (e.g. Cullen and Neale, 1994; Hunter et al., 1979; Karentz et al., 1994; Smith and Cullen, 1995; Vincent and Roy, 1993). The most likely impact on cetaceans is as a result of reduced productivity and ecosystem disruption, an impact likely to be felt most acutely by the baleen whales of the Southern Ocean. However, the development of an "ozone hole" in Arctic regions has also led to some concerns that small cetacean species such as belugas and narwhals may be susceptible to the prolonged effects of reduced productivity (IWC, 1997; Perry and Trent, undated). However, tremendous uncertainties on the severity of effects of ozone depletion and increased UV-B fluxes remain (e.g., Roberts, 1989; Voytek, 1990). Even under the most optimistic scenarios of reduction in ozone-destroying chemicals, to increased levels of UV-B radiation will continue for at least 50 years.

According to the IPCC, among the potential impacts of global warming on the oceans are: increases in sea surface temperature; increased pollution in coastal and marine waters as a result of increased precipitation and run-off; decreases in average primary productivity; and possible changes in ocean upwellings and currents (Ittekkot, 1996). Additionally, sea-level rise could also adversely affect coastal ecosystems such as mangroves, which provide important nursery areas for many marine species (Bijlsma, 1996). Drawing linkages from these predictions to the potential impacts on cetaceans is difficult. However, MacGarvin and Simmonds (1996) note that matters other than overall reduction in productivity and possible distribution of their prey need to be considered, as the apparent rate of climate change is likely to be well outside the evolutionary experience of existing cetacean species.

## CONCLUSION

At the dawn of the twenty-first century, it seems safe to say that the great majority of small cetacean populations are in a healthier state than those of their larger brethren, of which a number of populations remain highly endangered as a result of decades of commercial whaling (Clapham et al., 1999). Nonetheless, concerns remain. A number of small cetacean populations have been or are being reduced by human activities, primarily direct hunting and mortality as a result of fishing operations. In addition, uncertainty remains over the status of many populations in parts of the world where relatively little research has been conducted. There is a genuine danger that at least two species, the baiji and the vaquita, will, unless adequate conservation measures are taken, become extinct by the time a future author conducts a review similar to that in this chapter one hundred years from now.

Furthermore, cetaceans' position as apex predators potentially makes them especially vulnerable to the bioaccumulative effects of harmful toxins and anthropogenic toxic subtances, and to possible ecosystem changes as a result of, for example, increased UV-B radiation and global climate change. Most importantly, they are particularly susceptible to the cumulative impacts of these and other pressures.

A classic example of the consequences of such cumulative and cascading impacts is provided by the changes underway in the Bering Sea ecosystem. These changes have been highlighted by dramatic declines in populations of Steller sea lion (*Eumetopias jubatus*) in the Bering Sea and Gulf of Alaska. This decline apparently began in the eastern

Aleutian Islands in the early 1970s (National Research Council, 1996), and in most other areas by the early 1980s (Trites and Larkin, 1992). By 1989, the range-wide population estimate of 116,000 was only about 39 to 48 per cent of that estimated 30 years before (Loughlin et al., 1992). Populations of other wildlife species, including harbor seals and sea ducks, are also experiencing similar declines (National Research Council, 1996).

One apparent consequence of the declines in pinnipeds has been recent recorded decreases in sea otters (*Enhydra lutris*) in the Aleutian Islands. According to Estes et al. (1998), the most likely cause is increased predation by orcas; they speculate that, given the lack of nutritional value in sea otters, and the absence of any records of previous eating of sea otters by orcas in the region, the development is a direct result of declines in seals and sea lions, the cetaceans' usual marine mammal prey.

The causes of the declines in Steller sea lions and harbor seals, and other associated changes in the Bering Sea ecosystem, are apparently complex. A 1996 review by the National Research Council of the National Academy of Sciences in the United States proposed a "cascade hypothesis", in which natural climatic fluctuations, combined with human exploitation of predators such as whales and fish, resulted in a change in the ecosystem from one dominated by clupeids to one dominated by gadoids, specifically, pollock. As a result, some forage fishes that have higher nutritional value than pollock may have become less available to some marine mammals and birds, leading to their decline. In addition, concentrated and intensive commercial fishing of pollock, particularly in the region of sea lion rookeries and haulouts and at times when young animals most need nutrition, may be limiting the amount of pollock available (National Research Council, 1996). There is also growing evidence that further climatic changes are leading to continuing impacts on the coastal and marine ecosystems of the Bering and Chukchi Seas (Gibson and Schullinger, 1998; Weller and Anderson, 1998).

Similar complexities surround issues concerning chemical pollutants. The toxicity of the vast majority of chemicals now released into the environment is poorly known, and even less so for the interactive effects of complex mixtures. Little is known about the cumulative effects of contaminants in conjunction with the variety of other stresses facing marine mammals, or on the indirect effects of contaminant-induced foodweb changes.

It is easier to demonstrate the effects of continued takes, directed and indirect, on small cetacean populations, and because hunts and incidental takes provide powerful images around which public sentiment can be mobilized, issues of, say, the impacts of eutrophication on cetacean populations may tend to be ignored.

Many observers have argued the need for an international body to govern such catches, with the IWC frequently cited as the ideal forum (see Currey et al., 1990). Although it has been argued that the IWC does not have the authority to regulate small cetaceans, it is, as noted earlier, widely agreed that the only reason most small cetacean species were not listed on the "Annex of Nomenclature" at the signing of the ICRW is because it simply did not occur to those drawing up the list to include them (Holt, 1985). In addition, international law allows the IWC to decide for itself what it is and is not competent to regulate (Gambell, 1999; Mulvaney, 1987). Indeed, as is the case with the orca, the IWC has in the past chosen to express jurisdiction over species not explicitly covered in 1946 (Gambell, 1999; Mulvaney, 1996). The IWC has consistently demonstrated its unique and invaluable qualities with regard to the conservation of small cetaceans, for example, through the work of the IWC Scientific Committee, the organizing of such conferences as the 1990 Workshop on Cetacean Mortality in Passive Fishing Nets and Traps, and such resolutions as those passed on Japan's Dall's porpoise catches, which did much to bring about a reduction (albeit probably insufficient) in the size of that hunt. It is likely to remain the most widely-used, and most suitable, vehicle for addressing small cetacean hunts for the foreseeable future.

However, most of the countries where directed and indirect catches of small cetaceans take place are not represented in the IWC, and there is little incentive for them to be. Countries which are members cannot agree on the issue of competence, as evidenced by the ongoing debate over whether or not it can or should regulate the Japanese Baird's beaked whale hunt. Most importantly, whereas the IWC was formed to regulate and manage a very specific industry, most catches of small cetaceans have relatively little in common, other than that they involve small cetaceans. Although commercial whaling, for example, is a clear-cut case of an industry dependent on the exploitation of cetaceans for its existence, directed kills of small cetaceans tend to result principally from broader issues, such as over-fishing. This is, clearly, even more the case when dealing with indirect catches.

In order adequately to address the issues facing small cetaceans, it would be advisable for those concerned with the species' conservation not to look at such issues purely in terms of the species involved. In the same way that reducing contaminant burdens in cetaceans can only be achieved by addressing the sources of those contaminants so, in many cases, protection for small cetacean populations which are affected by directed catches can best be attained by looking beyond the fact that small cetaceans are being killed, and addressing the social and environmental issues behind these kills.

In this sense, the wide range of seemingly disparate and unrelated human impacts affecting small cetacean issues worldwide can be seen to have one common thread. Small cetaceans are, in a sense, the canaries in the coal mine, harbingers of broader environmental change, or illustrations of wider economic or social pressures. For that reason, it can be argued that addressing the threats facing small cetaceans is necessary, not only for the cetacean populations themselves, but also for what it can tell us about the problems affecting marine ecosystems as a whole.

## ACKNOWLEDGMENTS

Some information for this review was provided by Stella Duff, International Whaling Commission; Jennifer Lonsdale, Environmental Investigation Agency; and Mark Simmonds, Whale and Dolphin Conservation Society. Portions of this chapter were reviewed by Randall Reeves. All mistakes and omissions remain the responsibility of the authors.

## REFERENCES

Aguilar, A. and Borrell, A. (1994) Abnormally high polychlorinated biphenyl levels in striped dolphins (*Stenella coeruleoalba*) affected by the 1990–1992 Mediterranean epizootic. *The Science of the Total Environment* **154**, 237–247.

Allen, R.L. (1985) Dolphins and the purse-seine fishery for yellowfin tuna. In *Marine Mammals and Fisheries*, eds. J.H. Beddington, R.J.H. Beverton and D.M. Lavigne. George Allen & Unwin, London, pp. 236–252.

Anderson, D.M. and White, A.W. (eds.) (1989) Toxic Dinoflagellates and Marine Mammal Mortalities. Proceedings of an Expert Consultation Held at the Woods Hole Oceanographic Institution. Technical Report, Woods Hole Oceanographic Institution, Woods Hole.

Anderson, S.S., Livens, F.R. and Singleton, D.L. (1990) Radionuclides in gray seals. *Marine Pollution Bulletin* **21**, 343–345.

Balcomb, K. and Goebel, C.A. (1977) Some information on a *Berardius hairdii* fishery in Japan. *Report of the International Whaling Commission* **27**, 485–486

Barlow, J., Gerrodette, T. and Silber, G. (1997) First estimates of vaquita abundance. *Mar. Mamm. Sci.* **9**, 89–94.

Beddington, J.R., Beverton, R.J.H., and Lavigne, D.M. (eds.) (1985) *Marine Mammals and Fisheries.* George Allen & Unwin, London.

Beland, P., DeGuise, S., Girard, C., Lagace, A., Martineau, D., Michaud, R., Muir, D.C.G., Norstrom, R.J., Pelletier, E., Ray, S. and Shugart, L.R. 1993. Toxic compounds and health and reproductive effects in St. Lawrence beluga whales. *Journal of Great Lakes Research* **19**, 766–775.

Bennet, P.M., Jepson, P.D., Law, R.J., Jones, B.R., Kuiken, T., Baker, J.R., Rogan, E. and Kirkwood, J.K. 1999. Infectious disease mortality is associated with exposure to mercury in harbour porpoises from England and Wales. Abstract, European Cetacean Society 13th Annual Conference, Valencia, Spain.

Bergman A. and Olsson, M. (1985) Pathology of Baltic grey seal and ringed seal females with special reference to adrenocortical hyperplasia: Is environmental pollution the cause of a widely distributed disease syndrome? *Finnish Game Research* **44**, 47–62.

Bergman, A., Olsson, M. and Reiland, S. (1992) Skull-bone lesions in the Baltic grey seal (*Halichoerus grypus*). *Ambio* **21**, 517–519.

Bernard, H.J. and Reilly, S.B. (1998) Pilot whales *Globicephala* Lesson, 1828. In: *Handbook of Marine Mammals. Volume 6: The Second Book of Dolphins and the Porpoises*, eds. S.H. Ridgway and R. Harrison. Academic Press, London, pp. 245–279.

Bijlsma, L. (1996) Coastal zones and small islands. In: *Climate Change 1995: Impacts, Adaptations and Mitigation of Climate Change: Scientific-Technical Analyses. Contribution of Working Group II to the Second Assessment Report of the Intergovernmental Panel on Climate Change*, eds. R.T. Watson, M.C. Zinyowera and R.H. Moss. Cambridge University Press, Cambridge, pp. 289–324.

Bloch, D., Hoydal, K., Joensen, J.S. and Zachariassen, P. (1990) The Faroese catch of the long-finned pilot whale. Bias shown of the 280 year time series. *North Atlantic Studies* **2**(1&2), 45–46.

Bourne, A.G. (1965) Exploitation of the small whales in the North Atlantic. *Oryx* **8**, 185–193.

Brownell, R.L., Findley, L.T., Vidal, O., Robles, A. and Manzanilla, N.S. 1987. External morphology and pigmentation of the vaquita, *Phocoena sinus* (Cetacea: Mammalia). *Mar. Mamm. Sci.* **3**, 22–30.

Burkholder, J.M., Noga, E.J., Hobbs, C.W., Glasgow, H.B. and Smith, S.A. (1992) New "phantom" dinoflagellate is the causative agent of major estuarine fish kills. *Nature* **358**, 407–410.

Calmet, D., Woodhead, D. and André, J.M. (1992) $^{210}$Pb, $^{137}$Cs and $^{40}$K in three species of porpoises caught in the Eastern Tropical Pacific Ocean. *Journal of Environmental Radioactivity* **15**, 153–169.

Carlton, J.T. and Geller, J.B. (1993) Ecological roulette: the global transport of nonindigenous marine organisms. *Science* **261**, 78–82.

Caroli, S., Menditto, A. and Chiodo, F. (1996) The International Register of Potentially Toxic Chemicals—Challenges of data collection in the field of toxicology. *Environmental Science and Pollution Research* **3**(2), 104–107.

Castello, H.P. (1996) An introduction to the whales and dolphins. In *The Conservation of Whales and Dolphins: Science and Practice*, eds. M.P. Simmonds and J.D. Hutchinson. John Wiley & Sons, Chichester, pp. 1–22.

Chen, P. and Hua, Y. (1989) Distribution, population size and protection of *Lipotes vexillifer. Occasional papers of the Species Survival Commission (SSC)* **3** ,81–85.

Chen, P. and Hua, Y. (1989a) Projected impacts of the Three Gorges Dam on the baiji *Lipotes vexillifer*, and the eneds for the conservation of the species. Southwest Fisheries Center Administrative Report LJ-89-23.

Christensen, I. (1990) A review of the distribution, migrations, food, reproduction, exploitation and present abundance of humpback whales in the northeast Atlantic. ICES, Committee Meeting 1990/ N:10.

Clapham, P.J., Young, S.B. and Brownell, R.L. (1999) Baleen whales. conservation issues and the status of the most endangered populations. *Mammal Rev.* **29**(1), 35–60.

Cullen, J.J. and Neale P.J. (1994) Ultraviolet radiation, ozone depletion, and marine photosynthesis. *Photosynthesis Research* **39**, 303–320.

Currey, D., Lonsdale, J., Thornton, A. and Reeve, R. (1990) *The Global War Against Small Cetaceans.* Environmental Investigation Agency, London.

Dahlheim, M.E. and Heyning, J.E. (1998) Killer whale *Orcinus orca* (Linnaeus, 1758). In *Handbook of Marine Mammals. Volume 6: The Second Book of Dolphins and the Porpoises*, eds. S.H. Ridgway and R. Harrison, Academic Press, London.

Dahlheim, M.E. and Matkin, C.O. (1994) Assessment of injuries to Prince William Sound killer whales. In *Marine Mammals and the* Exxon Valdez, ed. T.R. Loughlin. Academic Press, San Diego, pp. 163–171.

DeGuise, S., Bernier, J., Martineaur, D., Beland, P. and Fournier, M. (1996) Effects of *in vitro* exposure of beluga whale splenocytes and thymocytes to heavy metals. *Environmental Toxicology and Chemistry* **15**, 1357–1364.

DeGuise, S., Lagace, A. and Beland, P. (1994) True hermaphroditism in a St. Lawrence beluga whale. *Journal of Wildlife Diseases* **30**, 287–290.

DeGuise, S., Lagace, A. and Beland, P. (1994a) Gastric papillomas in eight St. Lawrence beluga whales. *Journal of Veterinary Diagnostics and Investigation* **6**, 385–388.

DeGuise, S., Lagace, A. and Beland, P. (1994b) Tumors in St. Lawrence beluga whales. *Veterinary Pathology* **31**, 444–449.

DeGuise, S., Lagace, A., Beland, P., Girard, C. and Higgins, R. (1995) Non-neoplastic lesions in beluga whales (*Delphinapterus leucas*) and other marine mammals from the St. Lawrence estuary. *Journal of Comparative Pathology* **112**, 257–271.

DeGuise, S., Bernier, J., Martineau, D., Beland, P. and Fournier, M. (1996) In vitro exposure of beluga whale lymphocytes to selected heavy metals. *Environmental Toxicology and Chemistry* **15**, 1357–1364.

DeGuise, S., Martineau, D., Beland, P. and Fournier, M. (1998) Effects of *in vitro* exposure of beluga whale leukocytes to selected organochlorines. *Journal of Toxicology and Environmental Health* Pt. A **55**, 479–493.

DeLong, R.L., Gilmartin, W.G. and Simpson, J.G. (1973) Premature births in California sea lions: association with high organochlorine residue levels. *Science* **181**, 1168–1170.

DeSwart, R.L., Ross, P.S., Vedder, L.J., Timmerman, H.H., Heisterkamp, S.H., van Loveren, H., Vos, J.G., Reijnders, P.J.H. and Osterhaus, A.D.M.E. (1994) Impairment of immune function in harbor seals (*Phoca vitulina*) feeding on fish from polluted waters. *Ambio* **23**, 155–159.

Eisler, R. (1994) *Radioactive Hazards to Fish, Wildlife and Invertebrates: A Synoptic Review.* Biological Report 26. National Biological Service, Washington, D.C.

Estes, J.A., Tinker, M.T., Williams, T.M. and Doak, D.F. (1998). Killer whale predation on sea otters linking oceanic and nearshore ecosystems. *Science* **282** (5388), 473–476.

Evans, P.G.H. (1987) *The Natural History of Whales and Dolphins.* Christopher Helm, London.

FAO (1994) *Review of the State of World Marine Fishery Resources.* FAO Fisheries Technical Paper 335, Food and Agricultural Organization of the United Nations, Rome.

Finley, K.J., Miller, G.W., Davis, R.A. and Greene, C.R. (1990) Reactions of belugas, *Delphinapterus leucas*, and narwhals, *Monodon monoceros*, to ice-breaking ships in the Canadian high arctic. *Canadian Bulletin of Fisheries and Aquatic Sciences* **224**, 97–117.

Frantzis, A. (1998) Does military acoustic testing strand whales? *Nature* **392**, 29.

Gambell, R. (1999) The International Whaling Commission and the contemporary whaling debate. In *Conservation and Management of Marine Mammals*, eds. J.R. Twiss and R.R. Reeves. Smithsonian Institution Press, Washington, D.C., pp. 179–198.

Gaskin, D. 1984. The harbour porpoise *Phocoena phocoena* (L.): regional populations, status, and information on direct and indirect catches. *Report of the International Whaling Commission* **34**, 569–586.

Gaskin, D.E., Smith, G.J.D., Arnold, P.W. and Louisy, M.V. 1974. Mercury, DDT, dieldrin and PCB in two species of odontoceti from St. Lucia, Lesser Antilles. *Journal of the Fisheries Research Board of Canada* **31**, 1235–1239

Geraci, J.R. (1990) Physiologic and toxic effects on cetaceans. In *Sea Mammals and Oil: Confronting the Risks*, eds. J.R. Geraci and D.J. St. Aubin. Academic Press, San Diego, pp. 167–197.

Geraci, J.R., Anderson, D.M., Timperi, R.J., St.Aubin, D.J., Early, G.A., Prescott, J.H. and Mayo, C.A. (1989) Humpback whales fatally poisoned by dinoflagellate toxin. *Canadian Journal of Fisheries and Aquatic Sciences* **46**, 1895–1898.

Gibson, M.A. and Schullinger, S.B. (1998) *Answers from the Ice Edge: The Consequences of Climate Change on Life in the Bering and Chukchi Seas.* Greenpeace/Arctic Network, Anchorage.

Gisiner, R.C. (1998) Proceedings: Workshop on the Effects of Anthropogenic Noise in the Marine Environment, 10–12 February 1998. Office of Naval Research.

Gosliner, M. (1999) The tuna–dolphin controversy. In *Conservation and Management of Marine Mammals*, eds. J.R. Twiss and R.R. Reeves. Smithsonian Institution Press, Washington, D.C., pp. 120–155.

Gunter, G., Williams, R., Davis, C.C. and Smith, F.G.W. (1948) Catastrophic mass mortality of marine animals and coincident phytoplankton bloom on the west coast of Florida, Nov. 1946–Aug. 1947. *Ecological Monographs* **18**, 309–324.

Hallegraeff, G.M. (1993) A review of harmful algal blooms and their apparent global increase. *Phycologia* **32**, 79–99.

Harvey, J.T. and Dahlheim, M. E. (1994) Cetaceans in oil. In *Marine Mammals and the* Exxon Valdez, ed. T.R. Loughlin. Academic Press, San Diego, pp. 257–264.

Haug, T., Krøyer, A.B., Nilssen, K.T., Ugland, K.I. and Aspholm, P.E. 1991. Harp seal (*Phoca groenlandica*) invasions in Norwegian coastal waters: age composition and feeding habits. *ICES Journal of Marine Science* **48**, 363–371.

Helle, E. 1980. Lowered reproductive capacity in female ringed seals (*Pusa hispida*) in the Bothnian Nay, northern Baltic Sea, with special reference to uterine occlusions. *Annales Zoologici Fennici* **17**, 147–158.

Holliday, S., Johns, R., Ryle, J. and Thornton, A. (1996) *The Continuing Global War Against Small Cetaceans.* Environmental Investigation Agency, London.

Holt, S. (1985) Let's all go whaling. *The Ecologist* **15**(3), 113–124.

Honda, K., Tatsukawa, R., Itano, K., Miyazaki, N. and Fujiyama, T. 1983. Heavy metal concentrations in muscle, liver and kidney tissue of striped dolphin, *Stenella coeruleoalba*, and their variations with body length, weight, age and sex. *Agricultural and Biological Chemistry* **47**,1219–1228.

Houck, W.J. and Jefferson, T.A. (1998) Dall's porpoise *Phocoena dalli* (True, 1885). In *Handbook of Marine Mammals. Volume 6: The Second Book of Dolphins and the Porpoises*, eds. S.H. Ridgway and R. Harrison. Academic Press, London, pp. 443–472.

Humborg, C., Ittekkot, V., Cociasu, A. and v. Bodungen, B. (1997) Effect of Danube River dam on Black Sea biogeochemistry and ecosystem structure. *Nature* **386**, 385–388.

Hunter, J.R., Taylor, J.H. and Moser, H.G. 1979. Effect of ultraviolet radiation on eggs and larvae of the northern anchovy, *Engraulis mordax*, and the Pacific mackerel, *Scomber japonicus*, during the embryonic stage. *Photochemistry and Photobiology* **29**, 325–338.

Hutchinson, J.D. (1996) Fisheries interactions: the harbour porpoise—a review. In *The Conservation of Whales and Dolphins: Science and Practice*, eds. M.P. Simmonds and J.D. Hutchinson. John Wiley & Sons, Chichester, pp. 129–165.

Ittekkott, V. (1996) Oceans. In *Climate Change 1995: Impacts, Adaptations and Mitigation of Climate Change: Scientific-Technical Analyses. Contribution of Working Group II to the Second Assessment Report of the Intergovernmental Panel on Climate Change*, eds. R.T. Watson, M.C. Zinyowera and R.H. Moss. Cambridge University Press, Cambridge, pp. 267–288.

IWC (1977) *Report of the International Whaling Commission* **27**, 480

IWC (1982) Report of the Sub-Committee on Small Cetaceans. *Report of the International Whaling Commission* **32**, 113–126.

IWC (1991) Report of the Sub-Committee on Small Cetaceans. *Report of the International Whaling Commission* **41**, 172–190.

IWC (1992) Report of the Sub-Committee on Small Cetaceans. *Report of the International Whaling Commission* **42**, 178–234.

IWC (1993) Report of the Sub-Committee on Small Cetaceans. *Report of the International Whaling Commission* **43**, 130–145.

IWC (1994) Report of the Sub-Committee on Small Cetaceans. *Report of the International Whaling Commission* **44**, 108–119.

IWC (1995) Report of the Sub-Committee on Small Cetaceans. *Report of the International Whaling Commission* **45**, 165–179.

IWC (1997) Report of the IWC Workshop on Climate Change and Cetaceans. *Report of the International Whaling Commission* **47**, 293–318

IWC (1998) Report of the Sub-Committee on Small Cetaceans. *Report of the International Whaling Commission* **48**, 258–280.

IWC (1999) Report of the Sub-Committee on Small Cetaceans. *IWC/51/4*, Annex I.

IWCO (1998) *One Ocean, Our Future: The Report of the Independent World Commission on the Oceans.* Cambridge University Press, Cambridge.

Jaramillo-Legorreta, A.M., Rojas-Bracho, L. and Gerrodette, T. (1999) A new abundance estimate for vaquitas: first step for recovery. *Marine Mammal Science* **15** (4), 957–973.

Jefferson, T.A., Leatherwood, S. and Webber, M.P. (1993) *Marine Mammals of the World.* FAO Species Identification Guide. UNEP-FAO, Rome.

Jennings, S. and Kaiser, M.J. (1998) The effects of fishing on marine ecosystems. *Advances in Marine Biology* **34**, 201–352.

Kannan, K., Senthilkumar, K., Loganathan, B.G., Takahashi, S., Odell, D.K. and Tanabe, S. (1997) Elevated accumulation of tributyltin and its breakdown products in bottlenose dolphins (*Tursiops truncatus*) found stranded along the US Atlantic and Gulf coasts. *Environmental Science and Toxicology* **31**, 296–301.

Karentz, D., Bothwell, M.L., Coffin, R.B., Hanson, A., Herndl, G.J., Kilham, S.S., Lesser, M.P., Lindell, M., Moeller, R.E., Morris, D.P., Neale, P.J., Sanders, R.W., Weiler, C.S. and Wetzel, R.G. (1994) Impact of UV-B radiation on pelagic freshwater ecosystems: Report of working group on bacteria and phytoplankton. *Arch. Hydrobiol. Beih.* **43**, 31–69.

Kasuya, T. (1985) Fishery–dolphin conflict in the Iki Island area of Japan. In *Marine Mammals and Fisheries*, eds. J.R. Beddington, R.J.H. Beverton and D.M. Lavigne. George Allen & Unwin, London, pp. 253–272.

Kasuya, T. and Ohsumi, S. (1984) Further analysis of the Baird's beaked whale stock in the western North Pacific. *Report of the International Whaling Commission* **34**, 587–595

Kemp, N.J. (1996) Habitat loss and degradation. In *The Conservation of Whales and Dolphins: Science and Practice*, eds. M.J. Simmonds and J.D. Hutchinson. John Wiley & Sons, Chichester, pp. 263–280.

Kruse, S., Caldwell, D.K. and Caldwell, M.C. (1998) Risso's dolphin *Grampus griseus* (G. Cuvier, 1812). In *Handbook of Marine Mammals. Volume 6: The Second Book of Dolphins and the Porpoises*, eds. S.H. Ridgway and R. Harrison. Academic Press, London, pp. 183–212.

Kuehl, D., Haebler, R. and Potter, C. (1991) Chemical residues in dolphins from the U.S. Atlantic coast including Atlantic bottlenose obtained during the 1987/88 mass mortality. *Chemosphere* **22**, 1071–1084.

Kuehl, D., Haebler, R and Potter, C. (1994) Coplanar PCB and metal residues in dolphins from the U.S. Atlantic coast including bottlenose obtained during the 1987/88 mass mortality. *Chemosphere* **28**,1245–1253.

Lahvis, G.P., Wells, R.S., Kuehl, D.W., Stewart, J.L., Rhinehart, H.L. and Via, C.S. (1995) Decreased lymphocyte response in free-ranging bottlenose dolphins (*Tursiops truncatus*) are associated with increased concentrations of PCBs and DDT in peripheral blood. *Environmental Health Perspectives* **103** (Suppl. 6), 67–72.

Lair, S., Beland, P., DeGuise, S. and Martineau, D. (1997) Adrenal hyperplastic and degenerative changes in beluga whales. *Journal of Wildlife Diseases* **33**, 430–437.

Law, R.J. (1995) Metals in marine mammals. In *Environmental Contaminants in Wildlife: Interpreting Tissue Concentrations*, eds. W.N. Beyer, G.H. Heinz and A. Redmon-Norwood. CRC Press, Boca Raton.

Leatherwood, S. (1995) Re-estimation of incidental catches in Sri Lanka. In Report of the Workshop on Mortality of Cetaceans in Passive Fishing Nets. *Report of the International Whaling Commission* Special Issue 16.

Leatherwood, S. and R.R. Reeves (1983) *The Sierra Club Handbook of Whales and Dolphins*. Sierra Club Books, San Francisco.

Leatherwood, S. and Reeves, R.R. (1989) *Marine Mammal Research and Conservation in Sri Lanka 1985–1986*. UNEP Marine Mammal Technical Report 1.

Loughlin, T.R., Perlov, A.S. and Vladimirov, V.V. (1992) Range-wide survey and estimation of total number of Steller sea lions in 1989. *Marine Mammal Science* **8**, 220–239.

McCarthy, J.F. and Shugart, L.R. (eds.) (1990) *Biomarkers of Environmental Contamination*. Lewis Publishers, Boca Raton.

MacGarvin, M. and Simmonds, M. (1996) Whales and climate change. In *Conservation of Whales and Dolphins: Science and Practice*, eds. M.P. Simmonds and J.D. Hutchinson. John Wiley & Sons, Chichester, pp. 321–332.

McKay, B., Mulvaney, K. and Thorne-Miller, B. (1997) *Danger at Sea: Our Changing Ocean*. SeaWeb, Washington, D.C.

McMinn, A., Hallegraeff, G.M., Thomson, P., Jenkinson, A.V. and Heijnis, H. (1997) Cyst and radionucleotide evidence for the recent introduction of the toxic dinoflagellate *Gymnodinium catenatum* into Tasmanian waters. *Marine Ecology Progress Series* **161**, 165–172.

Marine Mammal Commission (1997) *Annual Report to Congress, 1996*. Marine Mammal Commission, Bethesda.

Martien, K.K., Taylor, B. L., Slooten, E. and Dawson, S. (1999) A sensitivity analysis to guide research and management for Hector's dolphin. *Biological Conservation* **90**(3), 183–191.

Matkin, C.O., Ellis, G.M., Dahlheim, M.E. and Zeh, J. (1994) Status of killer whales in Prince William Sound, 1985–1992. In *Marine Mammals and the* Exxon Valdez, ed. T.R. Loughlin. Academic Press, San Diego, pp. 141–162

May, J. (ed.) (1990) *The Greenpeace Book of Dolphins*. Century Editions, London.

Mitchell, E. (1975) *Porpoise, Dolphin and Small Whale Fisheries of the World: Status and Problems*. IUCN, Gland.

Miyazaki, N. (1983) Catch statistics of small cetaceans taken in Japanese waters. *Report of the International Whaling Commission* **33**,621–631

Mulvaney, K. (1987) A time of transition. *International Whale Bulletin* **1**, 3

Mulvaney, K. (1996) Directed kills of small cetaceans worldwide. In *The Conservation of Whales and Dolphins: Science and Practice*, eds. M.P. Simmonds and J.D. Hutchinson. John Wiley & Sons, Chichester, pp. 89–108.

National Research Council (1984) *Toxicity Testing: Strategies to Determine Needs and Priorities*. National Academy Press, Washington, D.C.

National Research Council (1996) *The Bering Sea Ecosystem*. National Academy Press, Washington, D.C.

Nixon, S.W. (1995) Coastal eutrophication: a definition, social causes and future concerns. *Ophelia* **41**, 199–219.

Northridge, S.P. (1984) *World Review of Interactions Between Marine Mammals and Fisheries*. Food and Agriculture Organization Fisheries Technical Paper 251.

Northridge, S. and Pilleri, G. (1986) A review of human impact on small cetaceans. *Investigations on Cetacea* **XVIII**, 222–261.

Northridge, S.P. and Hofman, R.J. (1999) Marine mammal interactions with fisheries. In *Conservation and Management of Marine Mammals*, eds. J.R. Twiss and R.R. Reeves. Smithsonian Institution Press, Washington, D.C., pp. 99–119.

O'Shea, T.J. (1999) Environmental contaminants and marine mammals. In *Biology of Marine Mammals*, eds. J.E. Reynolds and S.A. Rommel. Smithsonian Institution Press, Washington, D.C., pp. 485–564.

Odell, D.K. and McClune, K.M. (1998) False killer whale *Pseudorca crassidens* (Owen, 1846). In *Handbook of Marine Mammals. Volume 6: The Second Book of Dolphins and the Porpoises*, eds. S.H. Ridgway and R. Harrison. Academic Press, London, pp. 213–244.

Oil Spill Trustee Council (1999) Exxon Valdez *Oil Spill Restoration Plan: Update on Injured Resources and Services. Exxon Valdez* Oil Spill Trustee Council, Anchorage.

Paerl, H.W. and Whitall, D.R. (1999) Anthropogenically-derived atmospheric nitrogen deposition, marine eutrophication and harmful algal bloom expansion: Is there a link? *Ambio* **28**, 307–311.

Payne, P.M., Wiley, D.N., Young, S.B., Pittman, S., Clapham, P.J. and Jossi, J.W. (1990) Recent fluctuations in the abundance of baleen whales in the southern Gulf of Maine in relation to changes in selected prey. *Fishery Bulletin* **88**, 687–695.

Perrin, W.F. (1999) Selected examples of small cetaceans at risk. In *Conservation and Management of Marine Mammals*, eds. J.R. Twiss and R.R. Reeves. Smithsonian Institution Press, Washington, D.C., pp. 296–310.

Perry, C. and Trent, S. (undated). *Storm Warning: The Environmental Threats to Whales, Dolphins and Porpoises*. Environmental Investi-

gation Agency, London.
Read, A.J. (1996) Incidental catches of small cetaceans. In *The Conservation of Whales and Dolphins: Science and Practice*, eds. M.P. Simmonds and J.D. Hutchinson. John Wiley & Sons, Chichester, pp. 109–128.
Read, A.J. (1998) Harbour porpoise *Phocoena phocoena* (Linnaeus, 1758). In *Handbook of Marine Mammals. Volume 6: The Second Book of Dolphins and the Porpoises*, eds. S.H. Ridgway and R. Harrison. Academic Press, London, pp. 323–355.
Read, A.J., van Waerebeek, K., Reyes, J.C., McKinnon, J.S. and Lehman, L.C. (1988) The exploitation of small cetaceans in coastal Peru. *Biological Conservation* **46,** 53–70.
Reijnders, P.J.H. (1986) Reproductive failure in common seals feeding on fish from polluted coastal waters. *Nature* **324,** 456–457.
Richardson, W.J. and Wursig, B. (1997) Man-made noise and cetacean behaviour. *Marine and Freshwater Physiology and Behaviour* **29,** 183–209.
Richardson, W.J., Green, C.R., Malme, C.I. and Thomson, D.H. (1995) *Marine Mammals and Noise.* Academic Press, New York.
Ridgway, S.H. and Harrison, R. (eds.) (1998) *Handbook of Marine Mammals. Volume 6: The Second Book of Dolphins and the Porpoises.* Academic Press, London.
Roberts, L. (1989) Does the ozone hole threaten Antarctic life? *Science* **244,** 288–289.
Ross, D. 1987. *Mechanics of Underwater Noise.* Peninsula Publishing Co, Los Altos.
Ross, P.S., DeSwart, R.L., Reijnders, P.J.H., Van Loveren, H.V., Vos, J.G. and Osterhaus, A.D.M.E. (1995) Contaminant-related suppression of delayed-type hypersensitivity and antibody responses in harbor seals fed herring from the Baltic Sea. *Environmental Health Perspectives* **103,** 162–167.
Ross, P.S., Ellis, G.M., Ikonomou, M.G., Barret-Lennard, L.G. and Addison, R.F. (In press) High PCB concentrations in free-ranging Pacific killer whales, *Orcinus orca*: effects of age, sex and dietary preference. *Marine Pollution Bulletin.*
Scheele, L. and Wilkinson, D. (1988) Background paper on the Eastern Tropical Pacific tuna/dolphin issue. Greenpeace, Washington, D.C.
Schulman, F.Y., Lipscomb, T.P., Moffett, D., Krafft, A.E., Lichy, J.H., Tsai, M.M., Taubenberger, J.K. and Kennedy, S. (1997) Histologic, immunohistochemical, and polymerase chain reaction studies of bottlenose dolphins from the 1987–1988 United States Atlantic coast epizootic. *Veterinary Pathology* **34,** 288–295.
Simmonds, M. and Hutchinson, J. (1992) The dying seas. *International Whale Bulletin* **8,** 1–4
Simmonds, M.P., Johnston, P.A., French, M.C., Reeve, R. and Hutchinson, J.D. (1994) Organochlorines and mercury in pilot whale blubber consumed by Faroe Islanders. *The Science of the Total Environment* **149,** 97–111.
Simmonds, M., Dolman, S. and Perry, C. (1999) Recent important developments in the cetacean environment. IWC/SC/51/E14. Paper presented to the International Whaling Commission Scientific Committee, Grenada.
Simmonds, M.P., Hanly, K. and Dolman, S.J. (1999a) Cetacean contaminant burdens: regional examples. IWC/SC/51/E13. Paper presented to the International Whaling Commission Scientific Committee, Grenada
Smayda, T.J. (1990) Novel and nuisance phytoplankton blooms in the sea: Evidence for a global epidemic. In *Toxic Marine Phytoplankton*, eds. E. Graneli, B. Sundstrom, L. Edler and D. Anderson. Elsevier, New York, pp. 29–40
Smith, A. (1996) The river dolphins: the road to extinction. In *The Conservation of Whales and Dolphins: Science and Practice*, eds. M.P. Simmonds and J.D. Hutchinson. John Wiley & Sons, Chichester, pp. 355–390.
Smith, R.C. and Cullen J.J. (1995) Effects of UV radiation on phytoplankton. *Reviews of Geophysics* **33** (Supplement), 1211–1223.
Sournia, A. (1995) Red tide and toxic marine phytoplankton of the world ocean: An inquiry into biodiversity. In *Harmful Marine Algal Blooms*, eds. P. Lassus, G. Arzul, E. Erad, P. Gentien and C. Marcaillou. Lavoisier, Paris, pp. 103–112.
Stone, G., Cardenas, J.C. and Stutzin, M. (1987) Notes on cetacean conservation: issues in Chile. *Whalewatcher* Summer, 17–18.
Thompson, P.M., Tollit, D.J., Corpe, H.M., Reid, R.J. and Ross, H.M. (1997) Changes in haematological parameters in relation to prey switching in a wild population of harbour seals. *Functional Ecology* **11,** 743–750.
Thorne-Miller, B. (1999) *The Living Ocean: Understanding and Protecting Marine Biodiversity.* Island Press, Washington, D.C.
Todd, E.C.D. (1993) Domoic acid and amnesic shellfish poisoning—A review. *Journal of Food Protection* **56,** 69–83.
Topping, A.R. (1995) Ecological roulette: Damming the Yangtze. *Foreign Affairs* **74,** 132–146.
Trites, A.W. and Larkin, P.A. (1992) *The Status of Steller Sea Lion Populations and the Development of Fisheries in the Gulf of Alaska and Aleutian Islands.* Fisheries Center, University of British Columbia, Vancouver.
Van Waerebeek, K. and Reyes, J.C. (1990) Catch of small cetaceans at Pucusana port, central Peru, during 1987. *Biological Conservation* **51** (1), 15–22.
Vidal, O. (1995) Population biology and incidental mortality of the vaquita, *Phocoena sinus. Report of the International Whaling Commission* Spec. Issue No. 16.
Vidal, O., Brownell, R.L. and Findley, L.T. (1998) Vaquita *Phocoena sinus.* In *Handbook of Marine Mammals. Volume 6: The Second Book of Dolphins and the Porpoises*, eds. S.H. Ridgway and R. Harrison. Academic Press, London, pp. 357–378.
Vincent, W.F. and Roy, S. (1993) Solar ultra-violet-B radiation and aquatic primary production damage, protection and recovery. *Environmental Reviews* **1,** 1–12.
Voytek, M.A. (1990) Addressing the biological effects of decreased ozone on the Antarctic environment. *Ambio* **19**(2), 52–61.
Watson, W.S., Sumner, D.J., Baker, J.R., Kennedy, S., Reid, R. and Robinson, I. (1999) Radionuclides in seals and porpoises in the coastal waters around the UK. *The Science of the Total Environment* **234**(1–3), 1–13.
Weinrich, M., Martin, M., Griffiths, R., Bove, J. and Schilling, M. (1997) A shift in distribution of humpback whales, *Megaptera novaeangliae,* in response to prey in the southern Gulf of Maine. *Fishery Bulletin* **95,** 826–836.
Weller, G. and Anderson, P.A. (1998) *Implications of Global Change in Alaska and the Bering Sea Regions: Proceedings of a Workshop, University of Alaska Fairbanks, June 1997.* Center for Global Change and Arctic System Research, Fairbanks.
Wells, R.S. and Scott, M.D. (1997) Seasonal incidence of boat strikes on bottlenose dolphins near Sarasota, Florida. *Marine Mammal Science* **13**(3), 475–480.
Wells, R.S. and Scott, M.D. (1998) Bottlenose dolphin *Tursiops truncatus* (Montagu, 1821). In *Handbook of Marine Mammals. Volume 6: The Second Book of Dolphins and the Porpoises,* eds. S.H. Ridgway and R. Harrison. Academic Press, London, pp. 137–182.

## THE AUTHORS

**Kieran Mulvaney**
*1219 W. 6th Avenue,*
*Anchorage, AK 99501, U.S.A.*

**Bruce McKay**
*4058 Rue Dorion,*
*Montreal, PQ H2K 4B9, Canada*

Chapter 114

# SEABIRDS

W.R.P. Bourne and C.J. Camphuysen

---

Seabirds once formed an important source of meat, fat, eggs, feathers and fertiliser for primitive human communities, and in some areas still do so. Wildfowl are also still hunted widely for sport, and other species may sometimes cause problems as sources of pollution, carriers of disease, hazards to aircraft and, debatably, predators upon fisheries. But while in general seabirds are now seldom of much commercial importance, they are now attracting growing attention and concern. They are the marine animals which are seen most often above the water surface, are often of interesting or attractive appearance, and have many allies which also attract attention on land. Their bodies float and wash up on the shore, so that they may also provide the most conspicuous indication of adverse events such as bad weather, oceanic fluctuations, failure of food supply, by-kills in fisheries or pollution at sea. In consequence they have given rise to a vast popular and specialised literature, which unfortunately often shows little connection with work on other branches of marine biology.

---

*Seas at The Millennium: An Environmental Evaluation (Edited by C. Sheppard)*

## EVOLUTION AND HISTORY OF SEABIRDS

The fossil record of birds is reviewed by Olson (1985). That for seabirds is patchy, with more information for the northern hemisphere, whereas it seems likely that much of their evolution occurred in the south, while much of the existing material has apparently not been adequately described yet. Birds appear to be derived from small, carnivorous, feathered and therefore presumably warm-blooded, theropod dinosaurs (Sloan, 1999), which have survived because they acquired the ability to fly. Since flight limits their size, they have later repeatedly also given rise to large flightless forms in the sea as well as on land, which usually failed to survive. Thus at least two early groups of toothed birds started to exploit the sea in the Mesozoic, the large, aquatic, flightless, foot-propelled Hesperornithiformes, and the smaller, aerial Ichthyornithiformes. These apparently disappeared before the final cataclysm, possibly owing to the development of acanthopterygian fish.

Some often controversial intermediate forms have been found in the late Mesozoic and early Tertiary, and the development of modern seabirds apparently reached a climax about the Miocene, when there were more groups and species than there are now. Their numbers may then have declined following the appearance of the marine mammals, the break-up of the northern Tethys Ocean, and the subsequent deterioration in climate. They included a widespread group of gigantic, long-winged, very lightly built, gliding Pelecaniformes with jagged bills, the pseudodontorns, and a northern group of flightless, wing-propelled divers derived from the Sulids, the Plotopteridae. In the south there were larger, more predatory penguins (which now feed largely on plankton), and in the north, in addition to the recently lost North Atlantic Great Auk *Pinguinus impennis*, other flightless auks culminating in the genus *Mancalla* in the North Pacific.

At the height of their development in the Tertiary there appear to have been more rich and varied cool-current seabird communities of the sort still found off western South America (Murphy, 1936) along both coasts of North America, northwest Europe and southwest Africa (the record is deficient elsewhere), and pelagic, largely cephalopod-eating species such as the albatrosses, and fish-eating forms such as the aerial sulids and aquatic

Table 1

The main groups of modern seabirds. Regions of occurrence: C = cosmopolitan, N = Northern, S = Southern hemisphere, P = polar, T = temperate, S = subtropical, E = tropical. Habitats and breeding sites: W = inland waters, C = coast and inshore waters, O = offshore waters over the continental shelf, P = pelagic, and I = islands. Food: A = animal, V = vegetable, F = fish, S = cephalopods, I = other invertebrates, W = scavenger, P = may parasitise other species. In the case of clutch sizes with "+" more than one bird may lay in the same nest.

| | Family | No. Species | Mass (kg) | Region | Habitat | Food | Breeds | Clutch |
|---|---|---|---|---|---|---|---|---|
| Divers/Loons | Gaviidae | 4–5 | 1.0–5.8 | NPT | C | FI | W | 1–3 |
| Grebes | Podicipedidae | 22 | 0.1–1.5 | TSE | C | FI | W | 1–10 |
| Penguins | Spheniscidae | 15–17 | 0.7–38 | SP–E | COP | FSI | CI | 1–2 |
| Albatrosses | Diomedeidae | 13–25 | 2.0–12 | C | OP | SFIW | I | 1 |
| Petrels | Procellariidae | 56–80 | 0.1–6 | C | COP | SFIW | IC | 1 |
| Diving petrels | Pelecanoididae | 4 | 0.1–0.5 | SPTS | COP | IF | I | 1 |
| Storm petrels | Hydrobatidae | 21–23 | 0.02–0.1 | C | COP | IW | IC | 1 |
| Tropicbirds | Phaethontidae | 3 | 0.2–1.1 | SE | OP | FS | IC | 1 |
| Pelicans | Pelecanidae | 8 | 4–13 | TSE | CO | F | CW | 1–6 |
| Gannets and Boobies | Sulidae | 9 | 0.7–3.6 | TSE | COP | FS | IC | 1–4 |
| Cormorants | Phalacrocoracidae | 33–40 | 0.4–3.6 | C | CO | FSI | CIW | 2–7 |
| Frigatebirds | Fregatidae | 5 | 0.7–1.6 | SE | COP | PFSI | CI | 1 |
| Herons etc. | Ardeidae etc. | 80? | 0.1–2.6 | TSE | CW | A | CW | 1–10 |
| Flamingoes | Phoenicopteridae | 5 | 1.9–4.1 | TSE | WC | IV | W | 1–2 |
| Ducks, Geese, Swans | Anatidae | 147? | 0.2–14 | C | CW | AV | CI | 3–12+ |
| Rails and Crakes | Rallidae | 133? | 0.03–1.2 | C | WCI | VA | WCI | 1–0+ |
| Oystercatchers | Haemotopodidae | 10–11 | 0.4–0.7 | C | CW | I | CW | 1–4 |
| Avocets | Recurvirostridae | 7 | 0.1–0.4 | C | CW | I | CW | 3–4 |
| Plovers | Charadriidae etc. | 66 | 0.1–0.7 | C | CWI | I | CW | 1–4+ |
| Sandpipers | Scolopacidae | 88 | 0.1–0.5 | C | CWI | I | CW | 1–4+ |
| Phalaropes | Phalaropopidae | 3 | 0.02–0.1 | C | W–P | I | W | 3–4 |
| Sheathbills | Chionidae | 2 | (0.5?) | SPT | C | AVW | C | 1–4 |
| Skuas/Jaegers | Stercorariidae | 4–7 | 0.2–1.8 | C | W–P | PAVW | IC | 1–2 |
| Gulls | Laridae | 48–55 | 0.1–2.3 | C | W–P | AVWP | ICW | 1–3+ |
| Terns | Sternidae | 43–45 | 0.04–0.8 | C | W–P | A | WCI | 1–3+ |
| Skimmers | Rhynchopidae | 3 | 0.1–0.2 | TSE | CW | A | WC | 2–4+ |
| Auks | Alcidae | 22–23 | 0.1–1.3 | NPTS | COP | A | CI | 1–2 |

shearwaters and alcids were more widespread throughout the northern Tethys Ocean, in the way still found with seabirds in the southern hemisphere. Since the closure of the Strait of Panama in the late Pliocene the first are now confined to the Pacific, the second to the Atlantic, and other groups are now often represented by different forms in these oceans. The break-up of the Tethys Ocean in the late Tertiary may also have led to a further proliferation of coastal fish-, shellfish- and vegetable-feeding cormorants, charadriiformes and wildfowl.

## MODERN SEABIRDS

Modern marine birds include two elements: wildfowl and waders, widespread inland, which also feed along the shore, and true seabirds, with a distinct biology summarised by Ashmole (1971), Nelson (1980) and Furness and Monaghan (1987). Basically, to facilitate free flight they have rigidly retained the reptilian means of reproduction by eggs, so they have to carry out two contrasting activities, dispersal to feed at sea, and concentration to breed on land. Both require great powers of endurance which implies that while their adult survival in a comparatively predator-free environment may be high, with occasional disasters, they are slow to mature and usually have a low reproductive rate (Wynne-Edwards, 1955).

There is currently much regrettable instability in bird classification and nomenclature, owing in part to the introduction of molecular techniques (Sibley and Ahlquist, 1990), still in an active phase of development, partly to attempts to harmonise variations in usage in different parts of the world, and partly to cyclical fluctuations in taxonomic fashion between the lumping and splitting of related groups. A compromise is adopted here (Bourne and Casement, 1996) (Table 1). Bird identification and distribution is usually summarised in local field guides, and for the more marine species by Harrison (1991), shorebirds by Hayman et al. (1986), and wildfowl by Madge and Burn (1988).

The main groups are more fully described and the information for all species summarised by del Hoyo et al. (1992, 1996), and individual species reviewed in more detail for the western Palaearctic by Cramp et al. (1977–85), for North America by Poole and Gill (1992 onwards), for South America by Murphy (1936), for South Africa by Brown et al. (1982) and Urban et al. (1986) and for Australasia and Antarctica by Marchant and Higgins (1990 onwards), who between them cover most species. There are also major monographs on the penguins by Williams (1995), the albatrosses by Tickell (in press), the petrels by Warham (1990, 1996), the sulids by Nelson (1978), the shorebirds by Burger and Olla (1984a,b), the skuas by Furness (1987), and the auks by Gaston and Jones (1998), among many others.

In general many waterbirds, such as the divers, grebes, pelicans, cormorants, wildfowl (ducks, geese and swans), gulls, terns, and waders in the broad American sense (including the herons, storks, ibises, flamingoes and cranes, as well as the plovers and sandpipers, known as shorebirds in America and waders in a narrower sense in Europe), often disperse to breed in fertile areas inland at the most favourable season, usually the summer in high latitudes and aftermath of the rains in the tropics, where they often have comparatively high reproductive and mortality rates. They may then go to estuaries and the coast, often in lower latitudes or the opposite hemisphere, to mature, moult and survive the unfavourable season. Some, such as the pelicans, cormorants, wildfowl, gulls, terns, waders and molluscivorous wildfowl, may also breed in force along the more fertile coasts, but except for the parasitic and predatory skuas, some gulls and terns, and the planktivorous phalaropes, few go far out to sea.

The more specialised seabirds, including the auks, penguins, tropicbirds, frigatebirds, sulids, and petrels in the broad sense (albatrosses, fulmars, shearwaters, storm- and diving-petrels), normally stay out at sea, and are clumsy on land. Owing to a shortage of breeding places, they tend to be more highly colonial, leading to competition for a limited supply of nest-sites and food, so that they may have to travel long distances to breed and feed, and may also moult, mature and pass any adverse season elsewhere. In consequence they tend to have longer and sometimes more irregular life- and breeding-cycles, with lower reproductive rates, leading to prolonged dispute about the regulation of their numbers (e.g. Wynne-Edwards, 1962; Ashmole, 1963; Lack, 1967).

## MOVEMENTS AND MEASUREMENTS

Bird habitats used to be divided between land and water. Murphy (1936) introduced ornithologists to the developing concepts of oceanography, including the roles of upwelling and biological concentration. He paid particular attention to the southern convergences and Humboldt Current; the tropical current-systems, eddy formation on the lee sides of the oceans, and front-formation in areas of turbulence, now also appear important (Bourne, 1980; Hunt and Schneider, 1987; Stone et al., 1995). Most seabird distributions can be related to major ecosystems, which have seldom been defined precisely, with the addition that owing to their great mobility birds may also be able to move seasonally between different latitudes and habitats to feed, mature, moult and breed.

Seabirds tend to disperse most widely on fledging, and may not return to breed for some years. The more marine species tend to concentrate to breed at suitable sites in fertile areas, often in higher latitudes in the summer, but sometimes, as with the North Pacific albatrosses and associated species, in the lower latitudes in the winter, and then disperse at other times. Some may undertake long migrations, sometimes to the opposite hemisphere, or perform more complex movements, to suitable places to moult after breeding, during droughts in the summer, hard weather in the winter, or when blown astray by gales.

# Antarctic Birds

John Croxall

*British Antarctic Survey, Natural Environment Research Council, High Cross, Madingley Road, Cambridge CB3 OET, U.K.*

The Antarctic Region includes the Continent of Antarctica and the Antarctic Peninsula, with surrounding waters north to the Antarctic Polar Front, the circumpolar boundary where north-flowing Antarctic surface water sinks beneath warmer south-flowing sub-Antarctic water. The rich supply of nutrients generated by upwelling here sustains a high spring/summer production.

The region bounded by the northern limit of pack ice contains an assemblage of high latitude, ice-associated species typified by the Emperor *Aptenodytes forsteri* and Adelie *Pygoscelis adeliae* Penguins and Antarctic *Thalassoica antarctica* and Snow *Pagodroma nivea* Petrels (Croxall, 1984). The Emperor Penguin breeds throughout the Antarctic winter, but all others breed in the summer, several remaining closely associated with ice.

A very different avifauna is associated with the peri-Antarctic islands. These have thick grassland with deep, peaty soils providing cover for numerous burrowing petrel species, and substrate for vast colonies of several species of albatross and penguin. Further north, at islands such as Tristan and Gough, Amsterdam and St Paul, and the groups south of New Zealand, there are other, closely related, albatross, petrel and penguin taxa. In the Atlantic sector the sub-Antarctic islands and the Antarctic Continent are linked by the Antarctic Peninsula and its associated islands.

The seabirds tend to occur in latitudinal zones, with distinct populations at the island groups. There are many examples of species which doubtless evolved in adjacent zones and now show largely non-overlapping circumpolar distributions with varying degrees of co-existence. A few species/populations, such as the Adelie Penguin, winter in association with the marginal ice zone, but in winter most breeding species migrate north, and a few are transequatorial migrants. The northern Arctic Tern *Sterna paradisaea* is the only non-breeding summer visitor to the region, undertaking one of the world's longest migrations to achieve this.

The key prey species is the Antarctic Krill *Euphausia superba* whose stock is perhaps 100 million tonnes. Competition for it is avoided mainly by differences in foraging range and the timing of peak demand. In years of krill scarcity some species can switch, at least in part, to other prey (amphipods, copepods), but most suffer a drastic reduction in breeding success. These years of low krill availability are linked both to sea-ice cycles (through the effect of these on krill reproduction, development and recruitment) and to the ENSO cycle (Croxall, 1992; Croxall et al., 1999).

Some seabird species specialize, at least in part, on copepods (e.g. storm-petrels, diving-petrels, Antarctic Prions), myctophid fish (e.g. King Penguins), squid (e.g. Wandering and Grey-headed Albatrosses) and carrion (e.g. giant petrels). Only the Blue-eyed Shag specializes on fish. Several of the squid and carrion-eating species, together with the Black-browed Albatross *Diomedea melanophris* and White-chinned Petrel *Procellaria aequinoctialis*, associate with fishing vessels. With the development of long-line fisheries this has led to a mortality reaching 100,000 birds in several recent years, with considerable impact on populations.

Our recent ability to delineate foraging ranges and define feeding areas has shown that many species- and season-specific variations exist within apparently homogenous groups like albatrosses and penguins. Thus Wandering Albatrosses are distinctive in relying on large scale, basin-wide, largely random searches, often alternating short-distance trips to provision offspring with longer ones to refuel themselves. By contrast, Black-browed Albatrosses are essentially confined to the shelf and shelf-slope throughout the year, and Grey-headed Albatrosses to frontal zones, repeatedly returning to relatively restricted areas to target different prey in different domains (Prince et al., 1999). Similarly, tracking and dive-recording of polar penguins has shown marked species-specific differences in habitats and ranges exploited (Kooyman et al., 1999). Combination of such information with data on feeding rates and meal-sizes explains aspects of the relationship between seabirds and fisheries, both in terms of potential competition (e.g. with krill fisheries, particularly for penguins around the Antarctic Peninsula), and the use of seasonal and area closures of long-line fisheries to minimise mortality.

## REFERENCES

Croxall, J.P. (1984) Seabirds. In *Antarctic Ecology, 2*, ed. R.M. Laws, pp. 533–619. Academic Press, London.

Croxall, J.P. (1992) Southern Ocean environmental change: effects on seabird, seal and whale populations. *Philosophical Transactions of the Royal Society of London* **B338**, 319–328.

Croxall, J.P., Reid, K. and Prince, P.A. (1999) Diet, provisioning and productivity responses of predators to differences in availability of Antarctic krill. *Marine Ecology Progress Series* **177**: 115–131.

Kooyman, G., Hull, C., Olsson, O., Robertson, G., Croxall, J. and Davis, L. (1999) Foraging patterns of polar penguins. In Adams, N.J. and Slotow, R.H. (eds.) 1999. Proceedings of the 22nd International Ornithological Congress, 16–22 August 1998, Durban: Making Rain for African Ornithology. 1 CD-ROM. Johannesburg: Bird Life South Africa, pp. 2021–2039.

Prince, P., Weimerskirch, H., Wood, A.G. and Croxall, J. (1999) Areas and scales of interactions between albatrosses and the marine environment: Species, populations and sexes. In Adams, N.J. and Slotow, R.H. (eds.). Proceedings of the 22nd International Ornithological Congress, 16–22 August 1998, Durban: Making Rain for African Ornithology. 1 CD-ROM. Johannesburg: Bird Life South Africa, pp. 2001–2020.

Birds occurring along the coast often have a fluctuating distribution, influenced by age, season, time of day, tidal cycle, location of good feeding-places, and weather. While immature birds may remain there throughout the year, adults may move inland to breed. Some, such as the cormorants, may disperse to feed by day and assemble to roost at night, and others, such as the wildfowl and shorebirds, disperse to feed on a rising or a falling tide, and gather to roost at other times. Others, such as the gulls, may feed out at sea in fine weather, but along the shore or inland when it deteriorates. Since these birds tend to concentrate at times into large, comparatively accessible groups, it may be easier to define their movements and food, and count, catch, mark and follow them on the coast than when they disperse over larger and often more inaccessible areas in high latitudes inland to breed.

The first attempt to quantify birds at sea was made by Jesperson (1930), who plotted those seen per day in 10 degree rectangles during Atlantic research cruises, and found that numbers were proportional to the density of macroplankton. Wynne-Edwards (1935) then investigated their seasonal fluctuations by recording the number seen per hour in ten 5 degree sectors during eight voyages between Britain and Canada. Following experience during the International Indian Ocean Expedition in the 1960s (Bailey, 1966), Bailey and Bourne (1972) listed difficulties with such estimates, including variations in the conspicuousness and density of the birds and their reaction to ships, and suggested that the most convenient measure of bird distribution might be multiple sample counts of the number seen in ten minutes, and this was developed in the first comprehensive seabird atlas by Brown et al. (1975).

Subsequent observers have used various additional techniques (Bourne, 1976b), notably aerial surveys, which give a better view of irregular bird distributions, and have used counts of the birds seen within 300 m of a moving ship to calculate their density per $km^2$ (Tasker et al., 1984; Stone et al., 1995), and a number of such atlases have now been produced for different parts of the oceans. There are numerous difficulties with such methods (Haney, 1985), including, in addition to variations in the density and conspicuousness of the birds, the measurement of the census area when a margin of error of 50 m in assessing its outer border involves an area at least a third as large as that within this zone, the proportion of the birds detected within this distance (Briggs et al., 1985), the way birds move about (Gaston et al., 1987), and variations between observers (Van der Meer and Camphuysen, 1996).

Although it is normally considered desirable to record the weather, and to stop counting birds when it deteriorates, there has been little study of its effect on both seabird conspicuousness and distribution (Blomqvist and Peterz, 1984), involving both the thoroughness of observations, and the extent to which birds may move around with strong winds and rough seas.

The interpretation of observations at breeding sites may also be difficult. Most larger seabirds have a long period of immaturity and disperse most widely at this time. When they start to mature they may visit a number of breeding sites for increasing periods before settling down at one, usually close to where they were hatched. Once they settle, young birds may spend more years courting before attempting to breed, often with little success at first, so that, for example, the largest albatrosses may be ten years old before they are successful.

As with other marine animals, the productivity and mortality of many seabirds appears erratic, with for example occasional incidents and sometimes long periods of either good or bad breeding success or adult survival. These may be due to a variety of factors, such as fluctuations in weather, oceanography, food-supply, predation, or disease. Thus the generations are not necessarily renewed annually, but by good year-classes. In consequence young birds may survive better and breed sooner when the numbers of a species are low after some disaster, and there are many bereaved adults looking for new partners, but suffer higher mortality and postpone breeding longer, or move elsewhere, when the area becomes congested. It may be difficult to assess such situations since only part of the population, usually the breeding adults, may be easily countable, though much may also often be learnt from the study of marked birds.

Counting breeding birds may also present difficulties. With social species it may be difficult to see the whole colony, so that it is necessary to make sample counts and infer the total. It may also be difficult to distinguish between casual visitors, whose numbers may fluctuate, prospecting birds which are not breeding yet and may eventually go elsewhere, breeding birds which fail at an early stage before the eggs are recorded, and established breeders, a proportion of which may be away feeding. If the birds are nesting in holes they may not all be occupied, or there may be more than one species present, so that it becomes necessary to determine the proportion of each species, which may vary. The most useful figure is usually considered the "apparently occupied nest" (AON), which is liable to include prospecting and failed as well as successful breeders, though it may be necessary to make do with counts of individuals.

The birds may follow complex annual cycles. Thus for example while the large southern molluscivorous wildfowl, the South American steamer ducks *Tachyeres* sp., are sedentary, and in some cases flightless, some of their northern counterparts, the scoters *Melanitta* sp., breed inland, and others, the Common Eider *Somateria mollissima* and its allies, breed on the coast. Both then go to feed on the outer marine shellfish beds when they become flightless during the moult later in the summer, and then move to more sheltered inshore beds in the winter. In contrast, the more marine, larger auks may exploit a series of food fish, breeding on outer islands when sand-eels *Ammodytes* are feeding

## The Ill Winds of El Niño

E.A. Schreiber

*National Museum of Natural History, Smithsonian Institution, NHB MRC 116, Washington D.C. 20560, U.S.A.*

The major El Niño event of 1982–83 affected seabirds worldwide (Schreiber and Schreiber, 1984, 1989; Duffy, 1990). An El Niño, or ENSO event, depresses the upwelling in the eastern Pacific, and similar events occur in the Atlantic and other oceans. This causes major changes to rainfall, wind and current patterns. Some seabirds, such as those on Christmas Island (central Pacific), respond to a coming ENSO event before oceanographers can detect it; fewer birds return to nest, and the chicks of those that are nesting grow more slowly, beginning 2–3 months before an event is recognised (Schreiber and Schreiber, 1989; Schreiber, 1994).

The effects of ENSO events on birds vary with the strength and timing of the event and distance from the tropical Pacific. Responses include: (1) mortality of adults, (2) desertion of nests, followed by loss of eggs and chicks, (3) reduced numbers of birds breeding, (4) delayed breeding, (5) chicks take longer to fledge, or die of starvation, (6) extralimital occurrences of birds, and (7) failure to breed because floods or droughts make the habitat unsuitable. Marine birds often feed at upwellings, so changes in water temperature, upwelling patterns and currents also cause changes in fish distribution and increased mortality of their food.

Many reported "massive breeding failures" of marine birds in the past occurred during ENSO events (Schreiber and Schreiber, 1989), and events severe enough to cause 84–90% adult mortality (Peru: Duffy 1990; Christmas Island: Schreiber and Schreiber 1989) could exert selection pressure. Long-lived marine birds, with a reproductive lifetime of 20–40 years, could experience 4 to 10 ENSOs in a lifetime. Thus a Red-footed Booby hatched in 1960 would have experienced 11 years of ENSO events, with 40–45% of its reproductive years affected by them.

Two traits may have evolved as a result: (1) flexible growth rates in chicks, and (2) juveniles inhabiting separate oceanic areas from adults while they are maturing. In several species, young raised during ENSOs are able to survive on a lower daily food delivery by reducing their growth rate and staying in the nest longer (see Schreiber, 1994). Once fledged, young birds spend two to eight years maturing before returning to the colony to nest. During this time, the juveniles of some species are known to inhabit different areas from the adults. During a severe ENSO, which causes adult mortality, these young may be the individuals that survive and return to nest. Unfortunately, our lack of extensive populations of banded/ringed birds followed over long periods has hampered our ability to determine these demographic effects.

**REFERENCES**

Duffy, D.C. (1990) Seabirds and the 1982–84 Southern Oscillation. In *Global Ecological Consequences of the 1982–83 El Niño-Southern Oscillation*, ed. P.W. Glynn, pp. 395–415. Elsevier, Amsterdam.

Schreiber, E.A. (1994) El Niño-Southern Oscillation effects on provisioning and growth of Red-tailed Tropicbirds. *Colonial Waterbirds* **17**, 105–119.

Schreiber, E.A. and Schreiber, R.W. (1989) Insights into seabird ecology from a global natural experiment. *National Geographic Research* **5**, 64–81.

Schreiber, R.W. and Schreiber, E.A. (1984) Central Pacific seabirds and the El Niño-Southern Oscillation: 1982 to 1983 perspectives. *Science* **225**, 713–716.

on an outburst of plankton along frontal areas out at sea in the early summer, then taking their growing young to feed partly on small gadoids along sheltered lee coasts when they also become flightless during the moult later in the summer, and finally following the growing clupeids south in the winter.

In contrast to the almost universal spring seabird breeding season in the western Palaearctic, the larger seabirds occurring in the Persian/Arabian Gulf have an aberrant autumn breeding season, and the reason appears to be as follows. Unidentified fish-eggs were found along the coast of Oman in the autumn, and wintering nothern gulls, terns and phalaropes later gathered to feed on small marine organisms presumably including the resulting larval fish when the north-east monsoon led to increased upwelling and biological activity over the shelfbreak offshore during the winter. The birds appeared to follow the fry as they were swept by an inflowing current into the Persian/Arabian Gulf as it warmed up in the spring, where vast numbers of breeding terns fed on the growing fish in the summer. Then the larger late-breeding seabirds including innumerable endemic Socotra Cormorants *Phalacrocorax nigrogularis* and fewer Red-billed Tropicbirds *Phaethon aethereus* and Ospreys *Pandion haliaetus* fed on shoals of mature fish moving south to spawn in the autumn (Bourne, 1991).

Other waterbirds may make limited movements to the coast and lower latitudes in the winter, and many shorebirds in particular much more extensive ones, breeding in the Arctic, and moulting in some good feeding area during a return trip to winter quarters extending to southern Africa, Patagonia or New Zealand. Marine species may disperse equally far; thus some breeding in the Tristan-Gough group in the Southeast Atlantic were first recorded far to the west off Argentina, and the Great Shearwater *Puffinus gravis* in its winter quarters in the North Atlantic. Immature Northwest European seabirds may similarly reach the Grand Banks of Newfoundland and the Davis Strait; petrels from the eastern North Atlantic archipelagoes reach United States waters, young Sooty Terns *Sterna fuscata* from the Dry Tortugas off Florida reach the 'dome' at the end of the equatorial counter-current in the Gulf of Guinea; and young Lesser Frigates *Fregata ariel* from the central Pacific archipelagoes reach much of the west Pacific.

Some petrels, skuas, small gulls and terns migrate between the northern and southern temperate zones, and the younger ones in particular may perform figure-of-eight movements with the prevailing winds around the anticyclones in the centres of the oceans, and a clockwise movement with the SW Monsoon around the Indian Ocean, though the older birds may follow more direct routes, and that followed by the Sooty Shearwater *Puffinus griseus* around the Pacific may vary with El Niño (Spear and Ainley, 1999). The Arctic Tern *Sterna paradisaea* breeds on the most northerly exposed ground and moults in the Antarctic pack-ice, and young South Polar Skuas *Stercorarius maccormicki* may carry out a similar movement in the opposite direction.

## FOOD AND FISHERIES

It seems obvious from the variety of bird feeding techniques that birds take a wide variety of marine foods, often with much overlap (Ashmole and Ashmole, 1967; Bourne, 1982). Apparently it is often more a question of each species being adapted to take whatever food is available in different ways, than a different food, though most seem to have preferences. A pioneer attempt in which Lockley (1953) attempted to correlate the summer moult migration of the Balearic Shearwater *Puffinus (puffinus) mauretanicus* from the western Mediterranean to the Bay of Biscay with the movements of pilchards, and feeding movements of breeding Manx Shearwaters *P. (p.) puffinus* south from the Welsh islands shows the difficulties accompanying such exercises, since the Balearic Shearwater now appears to be a distinct species, while breeding Manx Shearwaters appear to feed in the Irish Sea.

Human fisheries might have various effects in such situations (Bourne, 1972). In the first place, they might remove bird food, causing a decline in their numbers, as with the guano birds of Peru following the development of an anchovy fishery (Schaefer, 1970). Secondly, they might remove large predatory fish that compete with birds for food, also providing extra food in the form of spilt fish, vomited stomach contents, discards and offal, as suggested by Fisher (1952) to explain the recent increase of the Fulmar in the North Atlantic. The fisheries might also have incidental adverse effects through the loss of birds on hooks, in nets, or through pollution, discussed later.

It was suggested in the nineteenth century that seabirds might have a harmful effect on fisheries, but it soon emerged that owing to such factors as losses in nets it was probably the other way round (McIntosh, 1903). The best-studied seabird populations in the North Sea (Dunnet et al., 1990; Furness and Tasker, 1999) have dramatically increased since then, with little further evidence for damage either way. While there have recently been some fluctuations in the distribution, breeding success and numbers of some species, especially the Arctic Tern in Orkney and Shetland, which have been attributed to the recent development of an industrial fishery for sandeels, similar fluctuations have been known to occur for nearly a century (Bourne and Saunders, 1992), and their numbers are still much larger than any recorded before the 1960s.

While it is questionable whether birds have much impact compared to man on fish stocks, it is also alleged that they may have a more serious impact on shellfish, as at one of the major winter quarters of the Oystercatcher *Haematopus ostralegus* in the Burry Inlet in South Wales (Davidson, 1968).

## EXPLOITATION AND CONSERVATION

Seabird remains are commonly present in both island and coastal archaeological deposits throughout the world, sometimes including the remains of extinct species such as the Great Auk, which have been found all round the North Atlantic and into the Mediterranean. Unfortunately they have seldom been studied in detail, but it would appear that seabird flesh and eggs must have formed an important resource, together with fish, for primitive peoples, especially when the birds came to land to breed when other foods were scarce in the spring in high latitudes, when some seabirds were appointed honorary fish by the mediaeval Church so that they could be eaten when other flesh was forbidden during Lent. They were also exploited for fat, feathers, fish-bait and fertiliser, including both dead bodies and guano. Precise routines were often adopted for this exploitation, with the different people concerned having recognised rights, and careful regulation of the numbers taken.

Outside the Arctic this traditional exploitation of an important resource has tended to break down in recent times when not specially maintained, for example by widespread game preservation and the regulation of wildfowling in developed countries, and where Hebrideans still harvest young Gannets or 'gugas' from Sula Skeir in Scotland, and the New Zealand Maoris young petrels or 'muttonbirds' from offshore islands. In consequence there then appears to have been a severe but poorly documented loss of seabirds to the growing human population in many parts of the world, which reached a climax in western Scotland and Ireland with the failure of the potato crop through blight in the late 1840s. Later many of the people affected emigrated to North America and Australasia, and apparently continued to take seabirds there, so that for example the gannets were nearly exterminated at this time as well as the Great Auk.

Following the development of the railways and breech-loading firearms, new, more wasteful practices developed in Britain of running excursions to shoot seabirds for "sport", and a more widespread fashion for the use of feathers for adornment. The plumage trade nearly led to the extermination of the North Pacific albatrosses and severe

# Seabirds and Fisheries in the North Sea

C.J. Camphuysen

Most North Sea seabirds have expanded their breeding range and have drastically increased in numbers over the last century (Cramp et al., 1974; Lloyd et al., 1991; Spaans, 1998). Over the last 15 years, the growth in some of these populations has ceased, but seabird numbers in the North Sea are generally at an historically high level (Hunt and Furness, 1996). While the relaxation of persecution, egging and exploitation of seabirds are probably the most important factors, it has often been suggested that commercial fisheries have influenced these trends (Camphuysen and Garthe, 1999). Seabirds and fisheries compete for fish, and seabirds might reduce commercial catches by eating fish that would otherwise be available to man. Conversely, commercial fisheries over-exploit fish stocks, depleting food resources of seabirds. Fisheries probably always have greater effects on seabirds than vice versa (Hunt and Furness, 1996).

There is little doubt that discards and offal in commercial fisheries are of great significance for some species, especially scavengers (Camphuysen et al., 1995; Garthe et al., 1996). The amount currently discharged in commercial fisheries in the North Sea could potentially support over 6 million seabirds. In all carefully studied species, however, it appeared that non-breeders were most frequently using discards and offal, while nesting adults preferred more natural resources. Several species experienced poor breeding success when proportionally large amounts of discards occurred in chick diets in years when natural prey was relatively scarce.

Most piscivorous seabirds specialise on small shoaling fish. Sandeels are particularly important prey in the breeding season, whereas clupeoids and gadoids gain importance in the winter period. The ICES working group on Seabird Ecology estimated the annual consumption of 18 North Sea seabirds at 600,000 tonnes per annum (including nearly 200,000 tonnes of sandeel, 30,000 tonnes of sprat and young herring, 22,000 tonnes of gadoid fish, 13,000 tonnes of mackerel, 13,000 tonnes of large herring, 71,000 tonnes of offal, 109,000 tonnes of discards and 146,000 tonnes of 'other prey'). It has been suggested that the gross overfishing of large predatory fish over the last century has led to increases in the survival and stocks of young fish. (Andersen and Ursin, 1977; Daan, 1980; Furness, 1982; Furness and Monaghan, 1987; Daan et al., 1990). Similarly, congruent shifts in sand eel abundance in the western and eastern North Atlantic ecosystems were explained by the relative scarcity of large predatory fish as a result of overfishing (Sherman et al., 1981). While these shifts in age structure, species composition and length distribution of North Sea fish are largely beyond doubt, the suggestion that seabirds have benefited is largely speculative. Although the increase in the amount of small fish must have resulted in a larger resource, the availability of prey may not have changed that dramatically. Clear negative examples of the impact of commercial fisheries on seabirds are common, and one of the best examples is the very poor reproduction of Puffins nesting on the Lofoten islands over a 20-year period after the collapse of herring stocks in these waters (Barth, 1978; Lid, 1981).

Catches by industrial fisheries (of sandeel, sprat, herring, Norway pout) have increased dramatically over the last 40 years. A major crash of sandeel stocks around Shetland in the late 1980s could not be attributed to industrial fisheries with certainty, but there is still substantial controversy over the impact of the Shetland fishery. If industrial fisheries in the North Sea continue to increase their catches, stock collapses may be expected. Areas at risk from industrial fisheries within the North Sea are the west and northwest, where most seabirds breed and where sandeels are the staple foods in summer.

While fish stocks, seabird numbers, the energetic requirements of seabirds and the distribution of birds at sea are now generally well known, the foraging strategies of seabirds have so far received insufficient attention (Camphuysen and Garthe, 1999). There is mounting evidence that inter- and intraspecific interactions of foraging seabirds are essential factors influencing levels of prey availability and the potential for successful (joint) exploitation of prey resources (Camphuysen and Webb, 1999). As a result, fluctuations in the availability of prey are poorly understood and any switch in fish stock, whether or not impacted by shifts in fishing effort or changes in the fishery policy can have unexpected side-effects on seabirds.

## REFERENCES


Andersen, K.P. and Ursin, E. (1977) A multispecies extension to the Beverton and Holt theory of fishing, with accounts of phosphorous circulation and primary production. *Meddelelser Danmarks Fisk og Havundersokelser* **7**, 319–435.

Barth, E.K. (1978) Lundetragedien på Røst. *Fauna (Oslo)* **31**, 273–274.

Camphuysen, C.J., Calvo, B., Durinck, J., Ensor, K., Follestad, A., Furness, R.W., Garthe, S., Leaper, G., Skov, H., Tasker, M.L. and Winter, C.J.N. (1995) Consumption of discards by seabirds in the North Sea. Final report to the European Commission, study contract BIOECO/93/10, NIOZ-Report 1995-5, Netherlands Institute for Sea Research, Texel, 202 + lviii pp.

Camphuysen C.J. and Garthe S. (1999) Seabirds and commercial fisheries: population trends of piscivorous seabirds explained? In *Effects of Fishing on Non-target Species and Habitats: Biological, Conservation and Socio-Economic Issues*, eds. M.J. Kaiser and S.J. de Groot, Chap. 11, pp. 163–184. Blackwell Science, Oxford.

Camphuysen, C.J. and Webb, A. (1999) Multi-species feeding associations in North Sea seabirds: jointly exploiting a patchy environment. *Ardea* 87(2).

Cramp, S., Bourne, W.R.P. and Saunders, D. (1974) *The Seabirds of Britain and Ireland*. Collins, London, 287 pp.

Daan, N. (1980) A review of replacement of depleted stocks by other species and the mechanisms underlying such replacement. *Rapp. P.-v. Réun. Cons. int. Explor. Mer* **177**, 405–421.

Daan, N., Bromley, P.J., Hislop, J.R.G. and Nielsen, N.A. (1990) Ecology of North Sea fish. *Netherlands Journal for Sea Research* **26**, 343–386.

Furness, R.W. (1982) Competition between fisheries and seabird communities. *Advances in Marine Biology* **20**, 225–307.

Furness, R.W. and Monaghan, P. (1987) *Seabird Ecology*. Blackie, Glasgow.

Garthe, S., Camphuysen, C.J. and Furness, R.W. (1996) Amounts of discards in commercial fisheries and their significance as food for seabirds in the North Sea. *Marine Ecology Progress Series* **136**, 1–11.

Hunt, G.L. and Furness, R.W. (eds) (1996) Seabird/fish interactions, with particular reference to seabirds in the North Sea. ICES Cooperative Research report No. 216, International Council for the Exploration of the Sea, Copenhagen, 87 pp.

Lid, G. (1981) Reproduction of the Puffin on Røst in the Lofoten Islands in 1964–1980. *Fauna Norvegica Serie C, Cinclus* **4**, 30–39.

Lloyd, C., Tasker, M.L. and Partridge, K. (1991) *The Status of Seabirds in Britain and Ireland*. T. and A.D. Poyser, London, 355 pp.

Sherman, K., Jones, C., Sullivan, L., Smith, W., Berrien, P. and Ejsymont, L. (1981) Congruent shifts in sand eel abundance in western and eastern North Atlantic ecosystems. *Nature* **291**, 487–489.

Spaans, A.L. (1998) Booming gulls in the Low Countries during the 20th century. *Sula* **12**, 121–126.

# The Relation between Ocean Physics and Foraging Seabirds in the North Pacific

George L. Hunt Jr.

*Department of Ecology and Evolutionary Biology, University of California, Irvine, Irvine, CA 92697, U.S.A.*

Physical processes in the ocean influence prey density and availability by two different pathways. Prey abundance may be enhanced where upwelling occurs, or may be concentrated in zones of convergence and water mass sinking. Where prey abundance is enhanced, there may be a regional-scale increase in the density of prey, but not necessarily more small-scale aggregations. Under these circumstances, marine birds may show a diffuse, region-wide increase in abundance. In the North Pacific Ocean, more seabirds are found in productive areas such as in the upwelling systems of the eastern boundary currents. Here aggregations of foraging birds occur in response to prey patches formed both by physical processes and by the schooling and swarming of the prey. If the physical processes are influenced by bathymetric features (sills, reefs, island passes: Wolanski and Hamner, 1988; Franks, 1992), then concentrations of prey may be predictable in space and time. In contrast, prey patches formed by mating behaviour, or the tendency for social aggregation, may not be predictable.

The Northern Hemisphere, and in particular the North Pacific Ocean, supports a populous and speciose assemblage of auks, birds that have compromised their flying abilities through the use of their wings to pursue prey under water to depths of 200 m or more. These species, in particular the smaller auklets that forage for copepods and euphausiids, often exploit prey concentrations resulting from the interaction of currents with bathymetry. Thus in the Aleutian Islands and near St Lawrence Island in the Bering Sea, Crested Auklets *Aethia cristatella* forage where the bathymetry causes upcurrents to force euphausiids towards the surface in the daytime. In contrast, surface convergences at the edges of currents, or downstream from sills over which the tide has passed, are preferred by Least Auklets *A. pusilla* taking near-surface aggregations of copepods. Least Auklets also prefer to forage in stratified water rather than in well-mixed water, possibly because in well-mixed water turbulence is more likely to have disrupted the patch structure of their prey (Hunt et al., 1998; 1999).

The dependence of marine birds on high prey concentrations increases their vulnerability to perturbations in the marine environments. In the North Pacific, warming of surface waters and a deepening in the warm upper mixed layers off the west coast of North America has had a strong impact on the productivity of the California Current system and the inshore shelf waters. The result has been a decrease in zooplankton abundance and the numbers of shearwaters migrating through the region (Veit et al., 1997). Likewise, climate-related changes in the Bering Sea have coincided with decreases of certain prey species, and a reduction of the populations of several seabird species nesting on the Pribilof Islands.

At smaller spatial scales, the attraction of seabirds to fishing vessels, which provide artificial patches of easily-obtained food, is also a concern. While the availability of supplemental forage enhances over-winter survival, for example of albatrosses, the incidental by-catch of birds in longline fisheries can have a negative impact. While the losses of individuals of species with large populations may have little influence on the demographic process, the loss of Short-tailed Albatrosses *Diomedea albatrus*, with a total world population of less than 2000 birds, is of concern. Fortunately, the fishing industry is attempting to develop fishing techniques that will minimise albatross by-catch.

Oil spills remain of concern. Floating oil and near-surface aggregations of seabird prey may be similarly concentrated in zones of surface convergence. Seabird deaths have occurred in a number of oil spills, but the greatest mortality by far occurred in the *Exxon Valdez* incident. Losses of seabirds were greatest inshore and in the vicinity of island passes. Although few data were gathered on the circumstances under which birds became oiled, it seems likely they may have been there because of the foraging opportunities that these inshore habitats offer. Although the spill occurred in early spring, before breeding would have commenced, prebreeding assemblages of birds off colonies may in some instances have exacerbated their vulnerability. Although hundreds of thousands of birds were killed by this spill, the long-term population-level consequences of this mortality are still to be determined. In the vicinity of breeding colonies, the timing of a spill will have a major influence on its potential effects.

In the continental shelf regions of the world ocean, and in particular over the continental shelves of the northern hemisphere where auks are abundant, seabirds aggregate to forage at sites where currents interact with bathymetry to produce predictable concentrations of prey. The resulting concentrations of birds may contain a large proportion of local populations, and can be particularly vulnerable to events that reduce local prey availability, or which permit pollutants to contact seabirds. Protection of these regions of local concentration is important for seabird conservation.

## REFERENCES

Franks, P.J.S. (1992) Sink or swim: accumulation of biomass at fronts. *Marine Ecology Progress Series* **82**, 1–12.

Hunt, G.L. Jr., Mehlum, F., Russell, R.W., Irons, D., Decker, M.B. and Becker, P.H. (1999) Physical processes, prey abundance and the foraging ecology of seabirds. *Proceedings of the International Ornithology Congress* **22**, 2040–2056.

Hunt, G.L. Jr., Russell, R.W., Coyle, K.O. and Weingartner, T. (1998) Comparative foraging ecology of planktivorous auklets in relation to ocean physics and prey availability. *Marine Ecology Progress Series* **167**, 241–259.

Veit, R.R., McGowan, J.A., Ainley, D.G., Wahl, T.R. and Pyle, P. (1997) Apex marine predator declines 90% in association with changing oceanic climate. *Global Change Biology* **3**, 23–28.

Wolanski, E. and Hamner, W.M. (1988) Topographically controlled fronts in the ocean and their biological influences. *Science* **241**, 177–181.

losses of the herons, gulls and terns all round the northern hemisphere later in the last century, leading to some of the first bird protection legislation on both sides of the North Atlantic. When in consequence the Northern Gannet, among other species, began to increase again, there were complaints that the birds might be harmful to fisheries, and various fish-eating species, such as the Great Cormorant *Phalacrocorax carbo* and saw-billed ducks *Mergus* sp., are still sometimes excluded from protection for this reason.

In addition to the direct effects of human activity on seabirds there are numerous indirect effects. Normally the first human colonists to arrive on innumerable islands soon introduced a variety of domestic animals and commensal species, including large herbivores which destroyed the vegetation and trampled bird burrows, hogs which also feed on the birds and their eggs and young, rodents, and cats and other carnivores to control the rodents, which also prey on birds. Problems have also been caused by the fur trade. At the end of the last century Arctic Foxes *Alopex lagopus* were widely introduced to northern Pacific islands to live on the seabirds, and more recently captive North American Mink *Mustela vison* have also been fed on fish offal all round the North Atlantic, and have regularly escaped, both on the mainland and islands such as the Hebrides and Iceland. The disastrous impact of introduced mammals on the native wildlife of such places as New Zealand is only now being corrected (Croxall et al., 1984; Moors, 1985; Croxall, 1991; Nettleship et al., 1994).

While human exploitation and introduced animals appear to have led to the devastation of many, still inadequately studied, insular seabird populations, with the development of firearms and motor-boats the situation has become increasingly serious. Many seabirds are still taken for food and fish-bait in some undeveloped areas, and large parts of the tropics have now lost their larger waterbirds. A large trade in guano for fertiliser also led to the disturbance of many seabird colonies (Hutchinson, 1950), which is now usually carefully regulated in the more productive areas, such as southwest Africa and Peru, since continued production depends on retention of the birds, though two important endemic species, the formerly widespread Abbott's Booby *Papasula abbotti* and Christmas Frigatebird *Fregata andrewsi*, remain at risk on the Indian Ocean Christmas Island.

Seabirds have also begun to suffer increasingly over the last century from pollution (Bourne, 1976a). This first attracted attention after ships were converted to use fuel oil during World War I, and ornithologists have led the agitation for the control of oil pollution, which is now at last beginning to have some effect in NW Europe (Camphuysen, 1998). The study of bird bodies washed up on beaches also revealed a number of other problems. These included pollution by toxic metals and organochlorines, now fortunately usually also declining, except possibly where Glaucous Gulls *Larus hyperboreus* feed on other birds' polluted eggs and young in the Arctic (Gabrielsen et al., 1995). Persistent problems are also caused by the ingestion of, or entanglement in, waste and especially plastic materials, poisoning by toxic micro-organisms, and epidemic disease. Fortunately most of these problems have only become serious during temporary local incidents, such as those due to the repeated wrecks of large oil tankers, and the leakage of DDT from a factory in California and cyclodiene insecticides from one in NW Europe, whose control requires perpetual vigilance.

## REFERENCES


Ashmole, N.P. (1963) The regulation of numbers of tropical oceanic birds. *Ibis* **103b**, 458–473.

Ashmole, N.P. (1971) Seabird ecology and the marine environment. In *Avian Biology 1*, eds. D.S. Farner, J.T. King and K.C. Parkes, pp. 223–287. Academic Press, New York.

Ashmole, N.P. and Ashmole, M.J. (1967) Comparative feeding ecology of sea birds of a tropical oceanic island. *Peabody Museum of Natural History, Yale University, Bulletin* **24**, 1–131.

Bailey, R.S. (1966) The sea-birds of the southeast coast of Arabia. *Ibis* **108**, 224–264.

Bailey, R.S. and Bourne, W.R.P. (1972) Counting birds at sea. *Ardea* **60**, 124–127.

Blomqvist, S. and Peterz, M. (1984) Cyclones and pelagic seabird movements. *Marine Ecology Progress Series* **20**, 85–92.

Bourne, W.R.P. (1972) Threats to seabirds. ICBP Bull. 11: 200–218.

Bourne, W.R.P. (1976a) Seabirds and pollution. In *Marine Pollution*, ed. R. Johnston, pp. 403–502. Academic Press, London.

Bourne, W.R.P. (1976b) Birds of the North Atlantic Ocean. *Proc. Int. Ornithol. Congr.* **16**, 705–715.

Bourne, W.R.P. (1980) The habitats, distribution and numbers of northern seabirds. *Transactions of the Linnean Society, New York* **9**, 1–14.

Bourne, W.R.P. (1982) The distribution of Scottish seabirds vulnerable to oil pollution. *Marine Pollution Bulletin* **13**, 2760–273.

Bourne, W.R.P. (1991) The Seabirds of Arabia. *Sea Swallow* **40**, 4–12.

Bourne, W.R.P. and Casement, M.B. (1996) Checklist of Seabirds. *Sea Swallow* **45** (suppl.), 12.

Bourne, W.R.P. and Saunders, D. (1992) Operation seafarer and Arctic terns. *Scottish Birds* **16**, 205–210.

Briggs, K.T., Tyler, W.B. and Lewis, D.B. (1985) Aerial surveys for seabirds: methodological experiments. *Journal of Wildlife Management* **49**, 412–417.

Brown, L.H. et al. (1982) *Birds of Africa. Vol. 1.* Academic Press, London.

Brown, R.G.B., Nettleship, D.N., Germain, P., Tull, C.E., and Davis, T. (1975) *Atlas of Eastern Canadian Seabirds.* Canadian Wildlife Service, Bedford Institute for Oceanography, Dartmouth, 220 pp.

Burger, J. and Olla, B.L. (eds.) (1984a) Shorebirds: Breeding Behavior and Populations. *Behavior of Marine Animals, Vol. 5.* Plenum Press, New York.

Burger, J. and Olla, B.L. (eds.) (1984b) Shorebirds: Migration and Foraging Behavior. *Behavior of Marine Animals, Vol. 6.* Plenum Press, New York.

Camphuysen, C.J. (1997) Oil pollution and oiled seabirds in the Netherlands 1969–97: signals of a cleaner sea. *Sula* **11** (2), 43–156. (In Dutch with English summary).

Camphuysen, C.J. (1998) Beached bird surveys indicate decline in chronic oil pollution in the North Sea. *Marine Pollution Bulletin* **36**, 519–526.

Camphuysen, C.J., Calvo, B., Durinck, J., Ensor, K., Follestad, A., Furness, R.W., Garthe, S., Leaper, G., Skov, H., Tasker, M.L. and Winter, C.J.N. (1995) Consumption of discards by seabirds in the North Sea. Final report to the European Comm., study cont.

BIOECO/93/10, NIOZ-Report 1995-5, Netherlands Institute for Sea Research, Texel, 202+ lviii pp.
Cramp, S. et al. (eds.) (1997–1985) *Birds of Europe, the Middle East and North Africa. Vols. 1, 3, 4.* Oxford University Press, Oxford.
Croxall, J.P. (ed.) (1991) Seabird Status and Conservation: A Supplement. ICBP Tech. Publ. 11, Cambridge.
Croxall, J.P., Evans, P.G.H. and Schreiber, R.W. (eds.) (1984) Status and conservation of the world's seabirds. ICBP Tech. Publ. 2, Cambridge.
Davidson, P.E. (1968) The Oystercatcher—a pest of shellfisheries. In *The Problems of Birds as Pests*, eds. in R.K. Murton and E.N. Wright, pp. 141–169. Academic Press, London.
del Hoyo, J. et al. (eds.) (1992, 1996). *Handbook of the Birds of the World, Vols. 1, 3.* Lynx Edicions, Barcelona.
Dunnet, G.M., Furness, R.W., Tasker, M.L. and Becker, P.H. (1990) Seabird ecology in the North Sea. *Netherlands Journal for Sea Research* **26**, 387–425.
Fisher, J. (1952) *The Fulmar.* Collins, London.
Furness, R.W. (1987) *The Skuas.* Poyser, Calton.
Furness, R.W. and Monaghan, P. (1987) *Seabird Ecology.* Blackie, Glasgow.
Furness, R.W. and Tasker, M.L. (eds.) (1999) Diets of seabirds and consequences of changes in food supply. ICES Cooperative Research Report No. 232, International Council for the Exploration of the Sea, Copenhagen, 65 pp.
Gabrielsen, G.W., Skaare, J.U., Polder, A. and Bakken, V. (1995) Chlorinated hydrocarbons in glaucous gulls (*Larus hyperboreus*) in the southern part of Svalbard. *The Science of the Total Environment* **160/161**, 337–346.
Garthe, S., Camphuysen, C.J. and Furness, R.W. (1996) Amounts of discards in commercial fisheries and their significance as food for seabirds in the North Sea. *Marine Ecology Progress Series* **136**, 1–11.
Gaston, A.J., Collins, B.L. and Diamond, A.W. (1987) The "snapshot" count for estimating densities of flying seabirds during boat transects: a cautionary comment. *Auk* **104**, 336–338.
Gaston, A.J. and Jones, I.L. (1998) *The Auks.* Oxford University Press, Oxford.
Harrison, P. (1991) *Seabirds—An Identification Guide.* Revised edition. Christopher Helm, London.
Haney, J.C. (1985) Counting seabirds at sea from ships: comments on interstudy comparisons and methodological standardization. *Auk* **102**, 897–900.
Hayman, P., Marchant, J. and Prater, T. (1986) *Shorebirds—An Identification Guide to the Waders of the World.* Christopher Helm, London.
Hunt, G.L. and Furness, R.W. (eds.) (1996) Seabird/fish interactions, with particular reference to seabirds in the North Sea. ICES Cooperative Research Report No. 216, International Council for the Exploration of the Sea, Copenhagen, 87 pp.
Hunt, G.L. Jr. and Schneider, D.C. (1987) Scale-dependent processes in the physical and biological environment of marine birds. In *Seabirds: Feeding Biology and Role in Marine Ecosystem*, ed. J.P. Croxall, pp. 7–41. Cambridge University Press, Cambridge.
Hutchinson, G.E. (1950) The biogeochemistry of vertebrate excretion. *Bulletin of the American Museum of Natural History* **96**, 1–554.
Jesperson, P. (1930) Ornithological observations on the North Atlantic Ocean. *Danish Dana Exp. Rep. 1920–22, Oceanogr. Rep.* **7**, 1–36.
Lack, D. (1967) Interrelationships in breeding adaptations as shown by marine birds. *Proceedings of the International Ornithology Congress* **14**, 3–42.
Lockley, R.M. (1953) On the movements of the Manx Shearwater at sea during the breeding season. *British Birds* **46**, Suppl.
Madge, S. and Burn, H. (1988) *Wildfowl—An Identification Guide to the Ducks, Geese and Swans of the World.* Christopher Helm, London.
Marchant, S. and Higgins, P.J. (eds.) (1990) *Handbook of Australian, New Zealand and Antarctic Birds.* Oxford University Press, Melbourne.
McIntosh, W.C. (1903) The effects of marine piscatorial birds on the food fishes. *Annals and Magazine of Natural History* **7** (11), 551–553.
Moors, P.J. (ed.) (1985) *Conservation of Island Birds.* ICBP Tech. Publ. 3, Cambridge.
Murphy, R.C. (1936) *The Oceanic Birds of South America.* 2 vols, American Museum of Natural History, New York.
Nelson, J.B. (1978) *The Sulidae: Gannets and Boobies.* Oxford University Press, Oxford.
Nelson, J.B. (1980) *Seabirds—Their Biology and Ecology.* Hamlyn, London.
Nettleship, D.N., Burger, L. and Gochfeld, M. (eds.) (1994) *Seabirds on Islands.* Birdlife Conservation Series 1, Birdlife International, Cambridge.
Olson, S.L. (1985) The Fossil Record of Birds. In *Avian Biology 8*, eds. D.S. Farner, J.R. King and K.C. Parkes, pp. 80–238. Academic Press, Orlando.
Poole, A. and Gill, F. (eds.) (1992 onwards) *Birds of North America.* American Ornithologists' Union, Washington D.C.
Schaefer, M.B. (1970) Men, birds and anchovies in the Peru Current: dynamic interactions. *Transactions of the American Fisheries Society* **99**, 461–467.
Sibley, C.G. and Ahlquist J.E. (1990) *Phylogeny and Classification of Birds. A study in molecular evolution.* Yale University Press, New Haven.
Sloan, C.P. (1999) Feathers for T. Rex? New birdlike fossils are missing links in dinosaur evolution. *National Geographic* **196** (5), 98–107.
Spear, L. and Ainley, D. (1999) Migration routes of Sooty Shearwaters in the Pacific Ocean. *Pacific Seabirds* **26**, 45.
Stone, C.J., Webb, A., Barton, C., Ratcliffe, N., Reed, T.C., Tasker, M.L. Camphuysen, C.J. and Pienkowski, M.W. (1995) *An Atlas of Seabird Distribution in North-west European Waters.* Joint Nature Conservation Committee, Peterborough.
Tasker, M.L., Jones, P.H., Dixon, T.J. and Blake, B.F. (1984) Counting seabirds from ships: a review of methods employed and a suggestion for a standardized approach. *Auk* **101**, 567–577.
Tickell, L. (in press). *Albatrosses.* Pica Press, Sussex.
Urban, E.K. (1986) *Birds of Africa.* Academic Press, London.
Van der Meer, J. and Camphuysen, C.J. (1996) Effects of observer differences on abundance estimates of seabirds from ship-based strip transect surveys. *Ibis* **138**, 433–437.
Warham, J. (1990) *The Petrels: Their Ecology and Breeding Seasons.* Academic Press, London.
Warham, J. (1996) *The Behaviour, Population Biology, and Physiology of the Petrels.* Academic Press, London.
Williams, T.D. (1995) *The Penguins.* Oxford University Press, Oxford.
Wynne-Edwards, V.C. (1935) On the habits and distribution of birds on the North Atlantic. *Proceedings of the Boston Society of Natural History* **40**, 233–346.
Wynne-Edwards, V.C. (1955) Low reproductive rates in birds, especially sea birds. *Proceedings of the International Ornithology Congress* **11**, 540–547.
Wynne-Edwards, V.C. (1962) *Animal Dispersion in Relation to Social Behaviour.* Oliver and Boyd, Edinburgh.

## THE AUTHORS

**W.R.P. Bourne**
*Department of Zoology, Aberdeen University, Tillydrone Avenue, Aberdeen AB24 2TZ, Scotland*

**C.J. Camphuysen**
*CSR Consultancy, Ankerstraat 20, 1794 BJ Oosterend, Texel, The Netherlands.*

Chapter 115

# FISHERIES EFFECTS ON ECOSYSTEMS

Raquel Goñi

The relentless growth of the global population imposes increasing pressure on the earth's natural resources. Unlike minerals, fish, like forests, are renewable natural resources. This means that they can be exploited endlessly as long as exploitation rates do not exceed renewal rates. Simply put, for marine organisms mortality from all causes—natural deaths from predation, disease or starvation, and deaths caused by fishing—should not surpass the arrival of new individuals to the exploited population through recruitment and growth. Fishery managers and the scientific community have for long acknowledged that excess fishing mortality—overfishing—has negative effects on the populations of marine organisms that are the targets of exploitation. But the unintended effects of fishing activities on other species and on the exploited marine ecosystems have only been recognised recently and, in a rapid transition from oblivion to fashion, they are now among the main concerns of fishery managers and scientists. The unintended nature of these effects implies that they are not controlled and that they are indiscriminate in their impacts.

*Seas at The Millennium: An Environmental Evaluation (Edited by C. Sheppard)*

## INTRODUCTION

Fishing activities interact with marine ecosystems in many different ways and through interconnected paths. Figure 1 is a conceptual model of these interactions. The model indicates the links between fishing and the various sources of mortality of marine organisms that it causes directly or indirectly. In a simplified manner, fishing can cause mortality of marine organisms through three main pathways: (a) intended fishing mortality (F) directed to target organisms; (b) fishing mortality (F) which is neither intended nor directed and that indiscriminately affects both target and non-target organisms; and (c) changes in levels of natural mortality (M) of target and non-target organisms that indirectly result from the above sources of mortality or from the fishing activities themselves. While the first two sources of mortality are first order effects of fishing, the third one encompasses higher order effects that are mediated by competitive interactions, predator–prey relationships, habitat degradation or other processes. Only the fishing mortality directed to target species is intended and, to a large but varying extent, is accounted for by managers and scientists in their decision-making processes and assessment models. In contrast, the other sources of fishing mortality affect indiscriminately both target and non-target organisms when they are either discarded dead at sea or killed by passing or lost gear. Although more familiar every day, these sources of mortality are generally ignored in fishery statistics and stock assessments. Moving down in Fig. 1 it becomes clear that understanding and evaluating second and higher order effects of fishing in marine ecosystems is a daunting and complex task. Impacts of fishing mediated by biological interactions or habitat degradation translate into changes in the rates of natural mortality of the species affected, favouring some (e.g., opportunistic, fast-growing species) in detriment of others (e.g., slow-growing species). These higher order effects of fishing are not negligible as predator removal or habitat degradation by their insidious action may lead to ecosystem changes of substantial proportions.

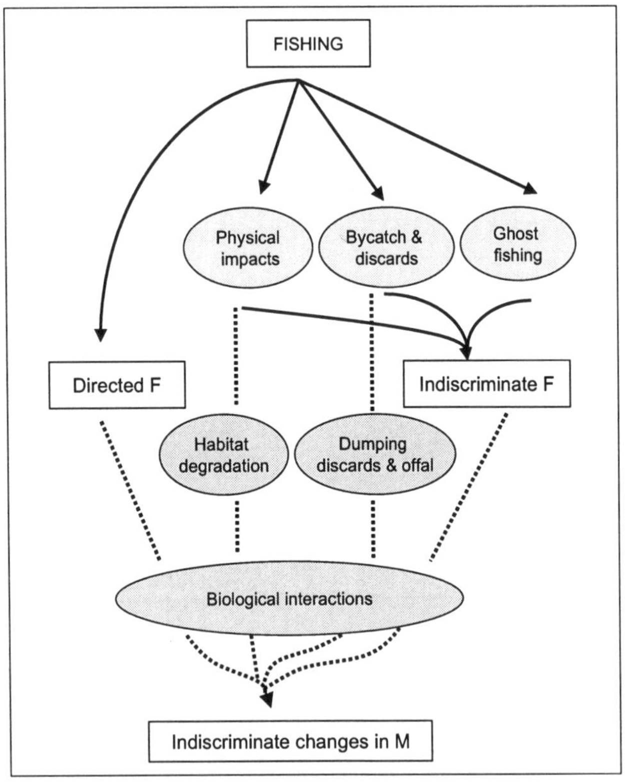


Fig. 1. Conceptual model of fisheries–ecosystem interactions. The model shows the links between fishing and the different sources of mortality. F: fishing mortality. M: natural mortality. Continuous lines are first order effects. Broken lines are second and higher order effects.

This chapter reviews the effects of fishing activities on marine ecosystems. The review is structured using the model in Fig. 1 as a backbone, for it provides a simple but realistic conceptualisation of the interactions between fishing and marine ecosystems. For a more in-depth study of the effects of fishing on marine ecosystems the reader is referred to the comprehensive reviews of Jennings and Kaiser (1998) and Hall (1999) published recently. Also recommended but more succinct reviews are those of Jones (1992), Gislason (1994), Dayton et al. (1995) and Goñi (1998).

## DIRECTED FISHING MORTALITY OF TARGET ORGANISMS

### Population Size

In newly developed fisheries the initial effects of fishing mortality translate into declines in abundance of the species targeted by the fishery. In general the exploitation strategy consists of augmenting fishing effort up to the level which produces the maximum yield (MSY) (Fig. 2). The obvious problem is that to determine the level of fishing effort that corresponds to the maximum yield, it is necessary to pass the inflection point of the curve effort/yield, and then, when a decline in yield is detected, the effort must be reduced. But once the fishery has developed and the financial investment has been made, it becomes nearly impossible to reduce fishing effort before the commercial extinction of the target species occurs (Hilborn and Walters, 1992; Ludwig et al., 1993). Furthermore, commercial extinction often does not provide a strict limit to exploitation as government subsidies contribute to the economic viability of fisheries where biological productivity has been severely reduced. This is illustrated in Fig. 2 by the possibility of displacing downwards the Total Cost line, thus shifting the equilibrium point of an open-access fishery to lower levels of production and higher levels of fishing effort. Pitcher (1998) argues that the risk of driving exploited species to extinction has never been considered seriously by the fisheries community because of this perception that declining abundance should cause fishing to diminish for economic

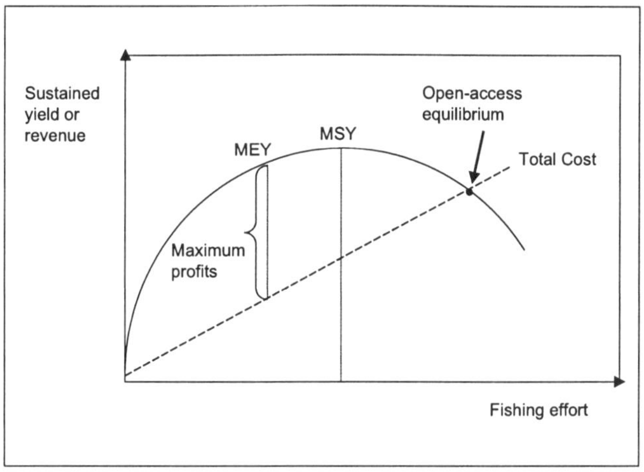


Fig. 2. Model functions relating fishery yield (or revenue) to fishing effort and total cost to fishing effort. MSY: Maximum sustainable yield. MEY: Maximum economic yield. The open-access equilibrium point is located at the intersection of the yield (or revenue) curve and the total cost line. Redrawn with permission from Fig. 8 of Caddy and Mahon (1995).

reasons. But experience shows that it does not because, with the help of government subsidies, fishers improve gear, go further afield and switch to fishing species further down the food web (Pitcher, 1998).

## Overfishing

The effects of overfishing are evident in many exploited marine populations of the world which no longer sustain historical catch levels. About 60% of the world's exploited fish and shellfish species are being either fished to capacity, overfished or recovering from overfishing (FAO, 1997). In well developed fisheries such as those of the Mediterranean (Caddy and Griffiths, 1990), North Sea (NSTF, 1993; Cook et al., 1997), the Northeast Pacific (Goñi et al., 1993) or the Northwest Atlantic (Anthony, 1993), up to 30 to 60% of the biomass of the main exploited species is extracted each year. Figure 3 illustrates the intensification of marine fisheries since 1950. The proportion of marine populations which

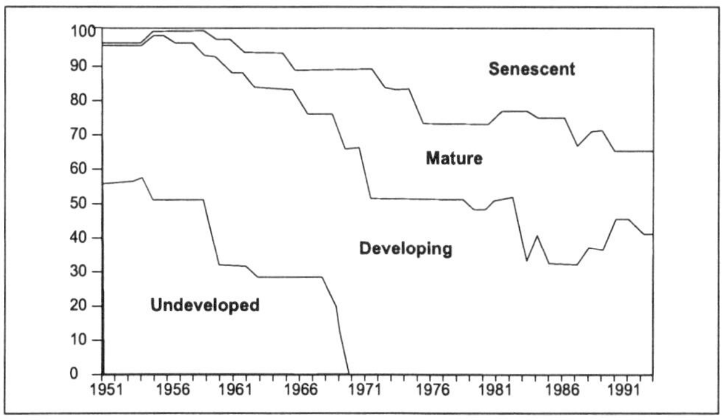


Fig. 3. History of development of the world marine fisheries: Evolution of the proportion of marine fisheries in each of the four states of development defined by FAO. Redrawn from Fig. 26 of Grainger and García (1996).

have declined in productivity—classified as 'senescent' by the Food and Agriculture Organisation (FAO)—is increasing, while the number of fully developed but not yet overfished–('fully mature') fisheries and fisheries in expansion have decreased at an alarming rate (Grainger and García, 1996). 'Sequential overfishing' (see below) underlies the stabilisation of the world marine capture fisheries production that peaked in 1994 at about 90 million tonnes (FAO, 1997).

## Life History Traits

The impact that a certain level of fishing mortality will have on the size of an exploited population and on its ability to survive depends on the species' life history traits (Gislason, 1994). Species characterised by short life spans, rapid population growth and high reproductive output (*r*-selected species) respond rapidly to fishing and may sustain relatively high levels of mortality at young ages. Typical marine examples of this species-type are the small pelagic fishes like sardines or anchovies. Conversely, species with low natural mortality which allocate more energy to individual growth through competitive fitness than to reproduction (*k*-selected species), will support relatively low rates of fishing mortality and at older ages (Gislason, 1994; Jennings and Kaiser, 1998). In this category are included especially vulnerable organisms such as marine mammals and some species of elasmobranchs, among others (see Box).

## Population Collapses

There are a number of well-documented cases of collapses of exploited populations linked to excess fishing pressure. But documenting the effects of excess fishing mortality on the size of the target populations is not easy due to the difficulty in separating the effects of fishing from those of natural variability and to the lack of population assessment studies prior to the onset of exploitation. Thus, where a direct link between stock collapse and excess fishing has been established, natural changes (such as unusual hydrographic conditions) have also been brought into play (e.g., Hillborn and Walters, 1992; Wooster, 1992a; Hutchings and Myers, 1994).

Some of the best known examples of this problem of confounding are provided by fisheries on small pelagic fishes in upwelling systems, such as the sardine (*Sardinops sagax*) in the California Current ecosystem and the Peruvian anchoveta (*Engraulis ringens*) in the Humboldt Current ecosystem (Sherman, 1994). The collapse of the Peruvian anchoveta is a well documented case where high exploitation rates together with productivity fluctuations associated with a sequence of El Niño events led to the commercial extinction of what in the 1970s was the largest fishery in the world (Caviedes and Fik, 1992). Although these populations later recovered (FAO, 1997), in other cases, such as those of the Alaska king crab (*Paralithodes*

## Fishing Vulnerable Species

Many exploited deep-water fish exhibit extreme k-selected life history strategies with low growth rates, late maturation, large egg size and low fecundity. The basis for this strategy is the relative environmental stability of the deep sea. As a consequence fisheries for deep-water species follow characteristic "boom and bust" cycles: the accumulated biomass of previously unfished populations is typically fished down to the point of commercial extinction or to very low sustainable levels (e.g., 1–2% of virgin biomass) within 5–10 years (Koslow et al., 1999). Pacific Ocean Perch (POP, *Sebastes alutus*) is a slow-growing, long-lived teleost fish that attains a maximum age of 90 years (Archibald et al., 1981). Before fishing began in the 1960s, in the Gulf of Alaska the virgin POP population consisted of large number of age classes, from 0 to 90, but by 1991 only 12% of the individuals were older than 15 years (Heiffetz and Clausen, 1992). Some fishery scientists have likened fishing on long-lived rockfish more to strip-mining than to harvesting a renewable resource (Francis, 1985).

The same preoccupation has been expressed for some species of elasmobranchs. Most condrichthyan species are of low productivity relative to teleost species. For example the shark *Galeorhinus galeus* produces 2–3 offspring every two years. This is reflected in the poor record of sustainability of target shark fisheries. A large proportion of the world catches of condrichthyans is taken as by-catch in teleost fisheries and thus, they are subject to high fishing mortality directed to the more productive teleost species. As a result, some skates and deep-water sharks have been virtually extirpated from large regions. Other species that have more *r*-selected life history characteristics and are comparatively resilient to fishing, having expanded in some intensively fished ecosystems (Stevens et al., 1999). Evidence of the virtual extinction of some large, bottom-living elasmobranchs exists for some Northeast Atlantic fishing grounds that are believed to be associated with overfishing. In the Irish Sea, the skate *Raja batis* was last seen in the 1960s (Brander, 1981). Similarly, the shark *Echinorhinus brucus*, the angelfish *Squatina squatina* and several *Raja* species have been absent from the catches of the shelf and basin around the Bay of Arcachon for the last 15 years (Quero and Cendrero, 1996). By 1980, intensive fishing in the Gulf of Thailand reduced the biomass of rays and shark species 110 and 50 times respectively. These species groups were the ones exhibiting the largest reductions (Christensen, 1998).

### REFERENCES

Archibald, C.P., Shaw, W. and Leaman, B.M. (1981) Growth and mortality estimates of Rockfishes (Scorpenidae)from British Columbia coastal waters, 1977–1979. Canadian Technical Report of Fisheries and Aquatic Sciences, No. 1048: 57 pp.

Brander, K. (1981) Disappearance of the common skate Raia *batis* from the Irish Sea. *Nature* **290**, 48–49.

Christensen, V. (1998) Fishery-induced changes in a marine ecosystem; insight from models of the Gulf of Thailand. *Journal of Fish Biology* **53** (Suppl. A), 128–142.

Francis, R.C. (1985) Fisheries research and its application to west coast groundfish management. In Proc. Symp. Fish. Mngt: Issues & Options, Alaska Sea Grant Report: 85-2.

Heiffetz, J. and Clausen, D.M. (1992) Slope rockfish. In: Stock Assessment and Fishery Evaluation Document for groundfish resources in the Gulf of Alaska for 1993. North Pacific Fishery Management Council, Anchorage, Alaska, November.

Koslow, A., Boehlert, G., Gordon, J.D.M., Haedrich, R.L., Lorance, P. and Parin, N. (1999) The impact of fishing on continental slope and deep-sea ecosystems. Contribution presented at the ICES/SCOR Symposium on Ecosystem Effects of Fishing, Montpellier, France, 16–19 March 1999.

Quero, J.C. and Cendredo, O. (1996) Incidence de la pêche sur la biodiversité ichthyologique marine: le Basin d'Arcachon et le plateau continental Sud Gascogne. *Cybium* **204**, 323–356.

Stevens, J., Walker, P., Bonfil, R. and Dulvy, N. (1999) The effects of fishing on condricthyans, and the implications for marine ecosystems. Contribution presented in the ICES/SCOR Symposium on Ecosystem Effects of Fishing, Montpellier, France, 16–19 March 1999.

*camtschatica*) (Wooster, 1992b) or the herring (*Cuplea harengus*) in the Barents Sea (Hamre, 1994), the combination of overfishing and unfavourable oceanographic conditions has had long-lasting effects.

Cases of demersal fish populations that have suffered sustained high exploitation levels and that have been depleted to the point of commercial extinction are also common. It has been argued that these extinctions are the culmination of a series of local extinctions (Pitcher, 1998) that went unnoticed by the innovative and shifting behaviour of fishing operations. The Gulf of Thailand has often been used as a model case of how fishery resources are depleted through human intervention (Christensen, 1998). In this case, between 1963 and 1980 large fish virtually disappeared from the catches and the trophic level of the whole fishery combined declined. The fishery then became more and more dependent on 'trash fish' caught for animal feed and other uses (Christensen, 1998). The most timely illustration of the collapse of a groundfish population is the Atlantic cod (*Gadus morhua*) fishery off Newfoundland where, after more than 500 years of exploitation, unsustainable harvest rates in recent decades led to the closure of the fishery and more than 20,000 fishermen were put out of work (Hilborn and Walters, 1992).

However, the closure of a fishery is not a common event because when productivity is severely reduced fishery operations switch target species or areas, leading to 'sequential overfishing'. The groundfish fishery of the Gulf of Alaska exhibited the characteristic pattern of overfishing virgin stocks, with quick rises and subsequent sharp declines in fish catches; when abundance of the preferred target species declined, the fleet switched to other species (Megrey and Wespestad, 1990; Goñi, 1998). During the heavy fishing years in the 1960s, an estimated 60–99% of the virgin biomass of the long-lived Pacific ocean perch (*Sebastes alutus*) was removed (Ito, 1982). Catches were subsequently dominated first by sabelfish (*Anoploma fimbria*) and pollock (*Theragra chalcogramma*), and later by flatfishes and Pacific cod (*Gadus macrocephalus*) (Goñi, 1998). Similarly, during the 1960s, groundfish fishing fleets in the north-eastern US continental shelf moved sequentially from one species to another, overfishing haddock (*Melanogrammus aeglefinus*), silver hake (*Merluccius bilinearis*), redhake (*Urophycis chuss*), herring (*Clupea harengus*) and mackerel (*Scomber scombrus*) (Anthony, 1993).

### Density-dependent Responses

Reductions in the size of exploited populations are expected to be compensated by density-dependent responses as a result of greater per-capita food supply in populations that have been thinned (Whale, 1997). In fact, the majority of fish populations show evidence of increased survival at lower population levels (Myers et al., 1995). Compensatory responses such as increased growth rates—e.g. cod, *Gadus morhua* (Whale, 1997); plaice, *Pleuronectes platessa* (Jennings and Lock, 1996); thornyback ray, *Raja clavata* (Walker and Hensen, 1996) or relative fecundity (lobster, *Palinurus marginatus* (DeMartini et al., 1993)—have been observed in a variety of species groups (ICES, 1996a). In spite of these density-dependent compensatory processes, other processes that are generally overlooked help to accelerate the process of population decline. Among these are the possibility of depensation at low population density (Allee effect) due to, for example, reduced mating or fertilisation success (Myers et al., 1995), reductions on the average size of the individuals and ensuing loss of reproductive potential, and the loss of genetic diversity in the exploited populations (Whale, 1997). However, the complex and poorly understood relationships between the genetic components of the growth rate, the size and the age of first reproduction, as well as the non-genetic responses of these traits to changes in population density and to environmental parameters, makes it difficult to separate the genetic and non-genetic impacts of fishing on natural populations (Smith, 1996).

### Size-selective Fishing

Large fishes are more desirable because they provide a better economic return. Thus fishing tends to deplete selectively larger and older fish, changing the size and age structure of exploited populations (Ricker, 1963; Bonhsack, 1992; Roberts and Polunin, 1991). Despite the compensatory responses described above, the continued reduction in abundance of the larger individuals eventually leads to a decline in the reproductive potential of the population. The magnitude of this effect depends on the biology of each species. Thus, species with low fecundity and high age of maturation will suffer the largest reductions in spawning potential for a given level of fishing mortality. Examples of significant shifts in size and age structure caused by exploitation abound. The size structure of ray species in the North Sea is currently truncated at 60–65 cm, while individuals of up to 100 cm used to be common (Walker, 1996). Similarly, the mean adult body weight of pink salmon (*Oncorhynchus gorbuscha*) of British Columbia has decreased significantly since 1950 (McAllister and Paterman, 1992).

Of particular concern are reef fish, where many species are hermaphrodites that mature with one sex and change sex at sizes above their size of first capture (Bonhsack, 1992; Jennings and Lock, 1996). In these species, the catchability of one of the sexes is markedly higher and fishing can strongly affect the sex composition of the population and as a result its reproductive potential. Roberts and Polunin (1991) report on a study by Beinssen (1989) that demonstrated a reduction of 5 cm in mean size of a population of the reef fish *Plectropomus leopardus* (Serranidae) after 18 months fishing. By altering size composition and sex ratios, selective fishing may also interfere with mating success. In some crustaceans, such as the spiny lobsters *Panulirus argus* and *Jasus edwardsii* or the Dungeness crab *Cancer magister*, the size of the males available for mating appears to be the prime determinant of reproductive success (Smith and Jamieson, 1991; Whale, 1997). This is important since, in many lobster and crab fisheries, males are preferentially targeted because of their larger size while females are partially protected by minimum landing size regulations.

Evidence of changes in the size structure at the community level has come from the analyses of long-term groundfish survey data. In the North Sea these studies show that large fish have become considerably less abundant since the 1970s (ICES, 1996a; Rijnsdorp et al., 1996a). Conversely, it is well established that the abundance and average size of exploited fish and invertebrate species increase rapidly in areas closed to fishing (Fig. 4) (see also e.g., McClanahan, 1989; García-Rubies and Zabala 1990; Bennet and Attwood,

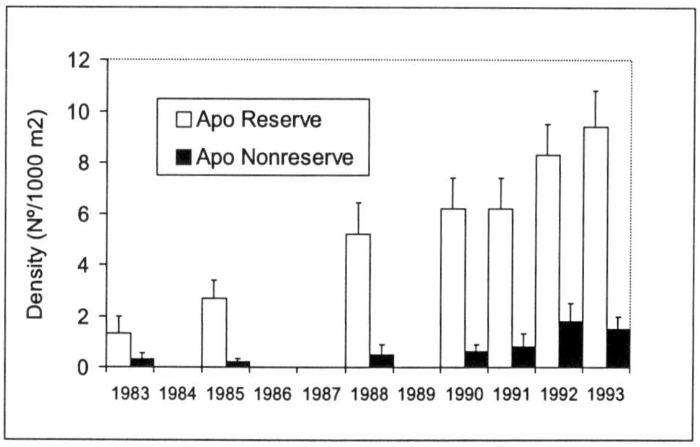


Fig. 4. Temporal evolution of mean (+1 SEM) density of large predatory reef fish (families Serranidae, Lutjanidae, Lethrinidae, and Carangidae as a group) at one reserve and at an open-fishing site in Apo (Philippines). Fishing ceased in Apo reserve in 1982. Redrawn with permission from Fig. 3 of Russ and Alcala (1996).

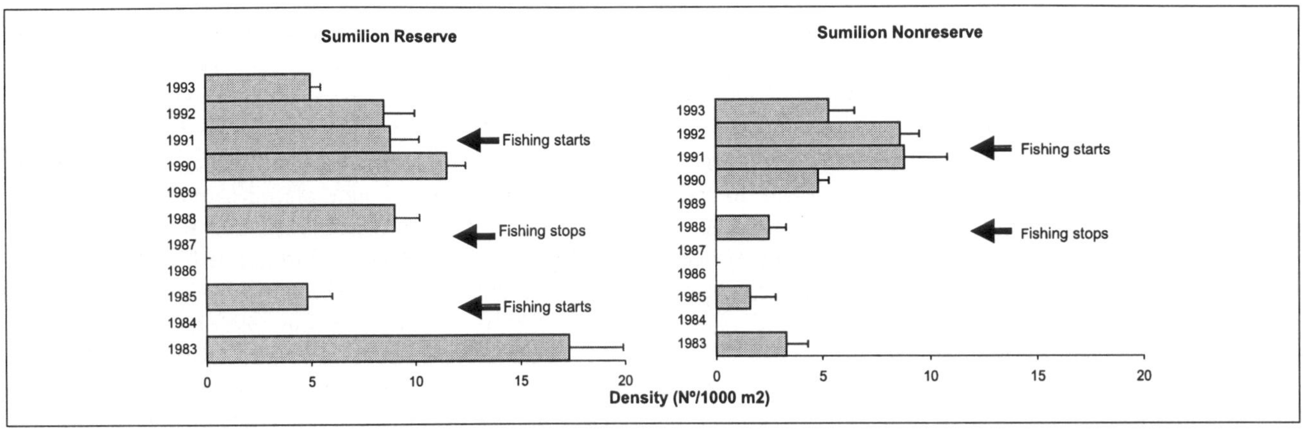


Fig. 5. Temporal evolution of mean (+1 SEM) density of large predatory reef fish (families Serranidae, Lutjanidae, Lethrinidae, and Carangidae as a group) at two sites in Sumilion (Philippines). Arrows indicate years in which fishing began or ceased. Redrawn with permission from Fig. 3 of Russ and Alcala (1996).

1991; Roberts and Polunin, 1991; Bayle and Ramos, 1993; Harmelin et al., 1995, Russ and Alcala, 1996; Reñones et al., 2000; Goñi et al., 1999). Similarly, where protected populations have been exposed to fishing, changes in population density and demographic structure with fewer large individuals are soon observed (Russ and Alcala, 1996) (Fig. 5).

### Loss of Genetic Diversity

In natural populations fishing is an important source of mortality that is not random with respect to the size and the age of the individuals (Smith, 1996). Hence, size-selective fishing, as well as other sources of stress associated with fishing, can impose new selection pressures on the affected populations or may alter the existing selection forces (Thorpe et al., 1981). This has implications for the survival of the exploited population at two levels. On the one hand, size-selective fishing mortality tends to select against rapid growth (Thorpe et al., 1981; Anon, 1994) and, again, reproduction at smaller sizes in general means lower reproductive output. On the other hand, since fishing activities have the potential to induce directional genetic changes, such as alterations of size and age structures or sex ratios which have genetic components, fishing may reduce the genetic diversity in exploited populations (Thorpe et al., 1981). Genetic diversity provides the variation that is the raw material of evolution (May, 1994) and the impoverishment of the genetic make-up of the exploited populations may reduce their homeostatic properties (i.e., their ability to withstand fluctuations in their environment).

## INDISCRIMINATE FISHING MORTALITY

In addition to the directed removal of commercially exploited species, other features of fishing cause additional indiscriminate fishing mortality. Three main causes are: (1) physical impacts of mobile gears and of destructive fishing practices on organisms that are not caught; (2) the mortality of organisms that are caught by fishing gears and returned dead to the sea (discards); and (3) the fishing mortality caused by lost or abandoned gear (ghost fishing).

### Fishing Mortality Caused by Physical Impacts and Destructive Fishing Practices

Of all physical impacts of fishing gears, those caused by towed gears on benthic organisms are the most widespread and the best known. A large number of studies have described and sometimes quantified direct effects after the passage a towed gear (reviews in e.g. Jones, 1992; Goñi, 1998; Jennings and Kaiser, 1998; Hall, 1999). In many shelf sea areas, trawl-fishing intensity is very high and many grounds are impacted at least once a year (Jennings and Kaiser, 1988). It has been estimated that heavily fished areas of the North Sea (14%) are trawled four or more times a year (Jennings et al., 1999), the average penetration depth of a beam trawl being 4–7 cm (de Groot, 1995). On a commercial beam trawl, the tickler chains from the groundrope are changed on average every six weeks, indicating the extent of the abrasion by the chains on the seabed (Rijnsdorp, 1988). As a result, a beam trawl with heavy tickler chains potentially kills about 10 times more benthic organisms than a ground trawl without ticklers (de Groot, 1984).

Bottom towed gears are widely used in all the continental shelves of the world while dredges are employed to scrape the seabed in many nearshore and estuarine ecosystems. The vulnerability of the organisms to the physical destruction caused by towed gears depends on their size and fragility (Jones, 1992; Rijnsdorp et al., 1996b). For species that form reef-like aggregates, such as coral or beds of calcareous algae, vulnerability to bottom trawling or dredging is particularly high. Species such as the tube worm *Sabellaria*, the coralline algae (Maerl), and coral-like bryozoa fall into this category (Hall, 1999). Assemblages of some echinoderms (e.g. *Asterias rubens, Echinocardium conchilega*) and tube-building polychaetes suffer reductions of up to 60% after triple experimental beam-trawling (BEON, 1990;

Bergman and Hup, 1992). Studies of injuries in starfishes (*Asteria rubens, Astropecten irregularis*) such as damaged or regenerated arms (Kaiser, 1996) and of scars in clam shells—*Arctica islandica* (Witbaard and Klein, 1994) and *Ensis siliqua* (Gaspar et al., 1994)—have been carried out to assess the intensity of dredging or trawling and the potential magnitude of this source of fishing mortality.

Most assessments of mortality rates of benthic species due to trawling have been short-term experiments and indicate that mortality is highly dependent on the substrate and the hydrodynamic environment in which trawling takes place (Hall, 1999). One of the most comprehensive studies showed that in trawling for flatfish a relatively high proportion of the fauna in the path of the beam trawl is killed (BEON, 1990). In high-energy environments, such as sandy-subtidal areas, scallop dredging was found to affect significantly the epifauna, but little the infaunal organisms. This has been attributed to adaptations that make the latter well suited to natural disturbances (Eleftheriou and Robertson, 1992). Hall (1999) reviewed the results of most long-term studies of trawling effects in the north-east Atlantic and found that although the evidence for these effects exists, fishing impacts were often confounded with other effects (e.g., changes in phytoplankton dynamics). Nevertheless, Collie et al. (1997) demonstrated that gravel habitats are very sensitive to physical disturbance and that the main fisheries impact is the removal of emergent epifauna which changes the nature of the communities present on those grounds. Other evidence comes from a study of experimental trawling on a soft sediment study site off Scotland that led to a disproportionate increase in the abundance of a few opportunistic dominant species. While the physical effects were not distinguishable after 18 months, community-level effects were more long lasting (Hall, 1999). In general it is observed that, when subjected to intense fishing disturbance, communities dominated by long-lived suspension feeders are likely to be replaced by mobile and opportunistic epifauna (Jennings and Kaiser, 1998; Hall, 1999).

Blast fishing, poison, and drive netting commonly used by small-scale fishers in tropical reefs are also unselective and destructive fishing techniques. The use of explosives is potentially the most destructive because the reef may be reduced to rubble (Munro et al., 1987). Although blast fishing is usually carried out with the objective of catching valuable pelagic species, depending on the depth, the pressure wave from the explosion may affect the coral reefs and the reef-fishes below (Saila et al., 1993). Swim bladders of fish explode when fish are within the lethal zone of the blast, the radius of this zone being in the order of 10–20 m (Saila et al., 1993). However, a typical charge may kill most marine organisms within a radius of up to 77 m (Jennings and Kaiser, 1998).

While subsistence fishermen have used vegetal poisons for thousands of years without known adverse effects, the use of modern synthetic chemicals such as sodium cyanide and chlorine is another matter. Since the early 1960s, sodium cyanide has been used as an asphyxiant in the collection of aquarium fish in the Philippines (the source of 75–80% of tropical marine fish worldwide) (Rubec, 1986). In the simplest form, sodium cyanide is dissolved in seawater and squirted in the direction of the fish. Affected fish may suffer mortalities as high as 75% at the point of collection (Rubec, 1986). Given that cyanide is an inhibitor of mitochondrial respiration, it also causes great damage to corals and non-target organisms. Although cyanide fishing has been banned in many Southeast-Asian countries, its use appears to be expanding to supply human consumption of live fish in that region (Jones and Steven, 1997). Simulation models in conditions of moderate blast fishing, poison fishing and anchor damage indicate that both fish diversity and coral area cover may be expected to decline for 25 years before any recovery occurs. Recovery to the initial state, where 50% of live coral was assumed to be destroyed, is predicted to take about 60 years (Saila et al., 1993). This is in line with other studies reporting no coral recovery 16 years after coral mining had ceased (Saila et al., 1993).

Other methods that involve the use of stones, chains or poles to break up corals and drive fish out of crevices and into nets are also very destructive (Hall, 1999). These drive-netting techniques are extensively used on coral reefs. These include the 'muro-ami' technique in which boulders are bounced over the reefs by swimmers in order to drive fish out and towards a bag net (Munro et al., 1987). These operations may involve from four or five fishers to hundreds of divers targeting offshore reefs in the Philippines and South China Sea (Jennings and Kaiser, 1998). Up to 6% of a 1-ha reef may be damaged during a single 'muro-ami' operation involving 50 fishers each striking the bottom 50 times with a 4 kg weighted scareline (Jennings and Kaiser, 1998).

Although more geographically restricted than bottom towed gears these fishing practices can cause highly indiscriminate mortality and habitat destruction in tropical reefs (Jennings and Kaiser, 1998). These destructive fishing practices are common in tropical reef regions where 'poor fishermen faced with declining catches and lacking other alternatives resort to wholesale destruction in their effort to maintain catches'; Pauly et al. (1989) termed this process 'Malthusian overfishing'.

### By-catch and Discard Mortality

One of the ecological costs of fishing is the incidental catch of organisms which are not the target of the fisheries (Hall, 1998). Many current fishing methods are highly indiscriminate and result in the catch of non-target species or of undesirable sizes of target species (referred to as "by-catch") which are then discarded. Major reasons for discarding include regulations (e.g., prohibited species, season closures, minimum sizes, etc.) and market demand. With some exceptions, in most fisheries and for most species, discarded organisms are dead when returned to the sea (Saila, 1983; Van Beek et al., 1990). Among the non-target organisms, mammals and birds have received the greatest public attention. Recently, however, other aspects of the problem of discard mortality are moving into the managers' agendas. Since most by-catch is discarded, this source of fishing mortality is unaccounted, and fishing mortality is underestimated. Concern over wasteful fishing practices is also growing as discard mortality in many fisheries has been shown to exceed landed catch mortality (Alverson et al., 1994).

Different fishing techniques lead to distinct types and rates of by-catch. Bottom trawling is a fairly unselective technique that captures any organism encountered in the path of the trawl not fast enough to escape. Minimum mesh size regulations, oriented to allow escapement of immature individuals, have limited effectiveness because meshes tend to become clogged after some time towing the trawl and because of the relatively high mortality of organisms escaping through the meshes (Sangster et al., 1996; Suronen et al., 1996). In contrast, pelagic trawls are more selective with regard to the target species. Nevertheless, given their large dimensions (the mouth of the gear can be larger than a football field) and relatively high towing speed, the incidental catch of non-target species is not rare (Northridge and di Natale, 1991; Fritz et al., 1995). For example, sea lion (*Eumetopias jubatus*) by-catch mortality attributed to Alaskan groundfish fisheries (to a large extent mid-water trawling) averaged 1600 individuals per year in 1968, declined to 600 by 1985 and continued to drop along with the size of the sea lion populations (Fritz et al., 1995).

In fisheries using purse seine for small pelagic fishes, incidental catches are almost non-existent. Conversely, the tuna purse-seine fisheries in the Central-eastern Pacific, whereby fishermen set their seines around groups of dolphins congregated over tuna schools, generated high dolphin mortality (CEC, 1994; Hall, 1998). In 1989, dolphin mortality in these fisheries was estimated at around 100,000 animals (Alverson et al., 1994) but has subsided since then (2547 animals in 1996; Hall, 1998) due to changes introduced in the fishing operations. However, while sets on dolphins are decreasing, sets on floating objects—logs and fish-aggregating devices placed by fishermen—have increased dramatically. Tuna sets on logs or on floating objects generate the highest by-catch rates of non-target species (some endangered, e.g. sea turtles, others vulnerable, e.g. sharks) as well as the highest discard mortality of juvenile tuna of all purse seine tuna fishing methods (Hall, 1998) (Fig. 6).

The by-catch in fixed gill net fisheries is largely dependent on the fishing area and the time of the year. Set-net fisheries in the Baltic and North Sea have been shown to produce significant by-catch of harbour porpoise (*Phocoena phocoena*) (e.g., 7000 per year in the Danish gill net fisheries for cod and turbot) (Kock and Benke, 1996). Rates of entanglement of California sea lions (*Zalophus californianus*) of up to 7.9%, mainly juvenile, have been recorded in colonies along the Gulf of California (Zavala-Gonzalez and Mellink, 1997). Dayton et al. (1995) cite coastal gill nets as the main agent for the near extinction of the Gulf of California harbour porpoise.

Drifting gill nets or driftnets, extensively used to catch squid, salmon, tuna and sword fish, are problematic due to their large size and may cause high incidental catches of non target species. As an illustration, during 1990 an estimated 2 to 2.4 million elasmobranchs, mainly blue shark (*Prionace glauca*), were incidentally caught in the Pacific driftnet flying squid fisheries alone (Bonfil, 1994). Estimates of the cetaceans incidentally caught in the Italian driftnet fishery each year range between 3000 and 7000 individuals (Northridge and di Natale, 1991). Also, the highest seabird

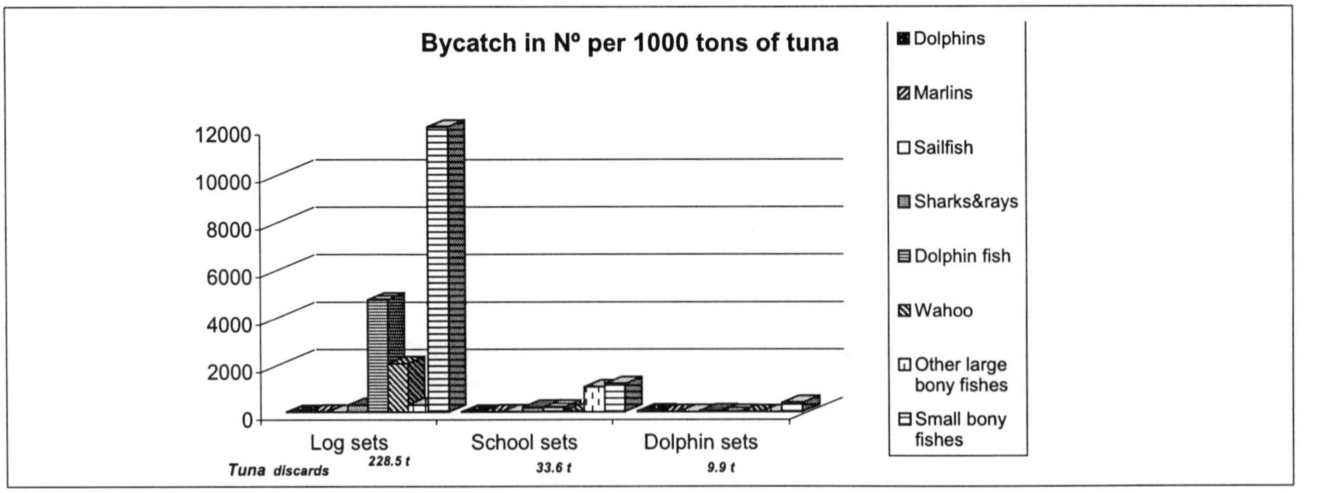


Fig. 6. By-catch per 1000 tons of tuna landed in number of individuals of different species groups from each of three purse-seine fishing methods: sets on logs or floating objects, sets on tuna schools, and sets on dolphins. Figures below horizontal axis are discards of undersized tuna (tons) produced by each purse seining method. Turtle by-catch (0.1) not shown. Source: Hall (1998).

by-catch is produced in gillnets and there are indications that this type of mortality is rising and may at present be the largest cause of mortality in certain seabird populations (Gislason, 1994).

Longlines, although in principle moderately selective, can cause important by-catch of non target and protected species. In fact, longline fisheries are the most important source of shark incidental catch in the high seas, being responsible for an estimated by-catch of more than 8 million fish annually (Bonfil, 1994). In the Western Mediterranean the incidental catch of the loggerhead turtle, *Caretta caretta*, by the Spanish longline sword-fish fleet ranged between a minimum of 1953 individuals in 1993 and a maximum of 23,886 individuals in 1990 (Camiñas, 1996). This estimate does not include other longline fleets operating in the Western Mediterranean. Clearly, the world estimate of turtle annual incidental catch in longline fisheries of about 40,000 individuals per year (with a mortality of 45%), reported by Alverson et al. (1994), underestimates the true world longline mortality of turtles.

Except for isolated examples (e.g., Hall, 1998; Wickens and Sims, 1994) little is known about the incidental catch of fish, invertebrates, reptiles, seabirds or marine mammals in relation to the size of their populations and therefore it is not possible to quantify the impact of this source of unintended fishing mortality. For fish and invertebrates detailed quantification of discards is often difficult and the extent of this practice is unknown. The extent to which the non-target catch (or by-catch) is retained or thrown back to the water largely depends on the economics that drive each particular fishery. Thus, small-scale artisanal fisheries in developing countries will produce few, if any, discards while in many large-scale fisheries discards of low-value species will be common. In 1994, Alverson et al. estimated annual global discards in commercial operations at about 27 million metric tons. This corresponds to a global discard rate of 20% and amounts to nearly one third of the world landings of about 90 million tons reported by FAO (1997). In their study Alverson et al. found that prawn and shrimp trawl fisheries generate more discards than any other fishery type. For example, up to 50,000 tons of fish are discarded annually by prawn trawlers in the Gulf of Papua and Torres Strait (South Pacific) alone (Kan et al., 1995). Hill and Wassemberg (1990) estimated that offshore Australian prawn trawlers on average discard 3000 t of by-catch while catching 500 tons of prawns. However, by-catch to catch ratios vary from tropical versus temperate and inshore versus offshore prawn trawl fisheries and can be as low as 0.1:1 and as high as 10:1 (Liggins et al., 1996; Liggins and Kenelly, 1996).

High by-catch mortality rates are also generated in other fisheries. Sole trawl fisheries in the Northeast Atlantic have been reported to discard 5 to 10 times the weight of sole in fish and invertebrates (work by various authors cited in ICES, 1994). Recent studies of the Spanish bottom trawl fisheries in the North Atlantic and in the Mediterranean resulted in somewhat lower discard rates, 35-59% and 25% respectively (Pérez et al., 1995; Carbonell et al., 1997). In the North Atlantic, gillnets produced the second highest discard rates (25%), followed by long-line (8–18%) and purse seine (1–30%) (Perez et al., 1995). In general terms, the lowest discard rates in weight correspond to pelagic trawl, purse-seine for small pelagic fish and some high seas driftnet fisheries (Alverson et al., 1994).

## Fishing Mortality Caused by Lost Gear: Ghost Fishing

Fishing operations often result in the accidental loss of gear or in the dumping of dismissed gear which may continue to fish and entangle seabirds or mammals for some time after being discarded or lost (ICES, 1994; Gislason, 1994). The scale of the impacts of ghost fishing mortality is unknown but there are indications that its extent is not negligible. The problem is primarily produced by gill nets and traps (ICES, 1995) and to some extent by trawl net fragments (Fowler, 1987). In the case of gill nets the amount of time that they continue to fish depends on factors such as depth, habitat type, current speed and amount of algae and floating material entangled, which at some point force the net to collapse and stop fishing (Kaiser et al., 1996; Erzini et al., 1997). Also, in the long term, the nets accumulate a layer of material and organisms which make them visible to fish and other organisms (ICES, 1992). In a study done in Norway it was found that nets lost in 1983 continued fishing in 1990 (ICES, 1992). In Canada a study was carried out in response to complaints by fishermen on the occurrence of ghost fishing in Georges Bank. In the study 8% of the experimental tows recovered lost gillnets (ICES, 1992). In experiments conducted with monofilament set nets on rocky bottoms the fishing lifetime of lost nets ranged from over four months (Erzini et al., 1997) up to the nine months that the study lasted (Kaiser et al., 1996). However, given the longevity of the monofilament, a net may fish for over two years (Kaiser et al., 1996). Traps also continue fishing once lost unless they are built with a piece of biodegradable material. Breen (1987) estimated an annual rate of trap loss of 10.7% in the Dungeness crab (*Cancer magister*) fishery in the Fraser River Estuary (British Columbia, NE Pacific). Given an estimated life-span of lost traps of 2.2 years, Breen calculated that at any one time about 3336 traps were ghost fishing in the area causing the mortality of over 31,000 crabs annually (or 7% of the catch) in that fishery alone.

Entanglement of mammals and seabirds also occurs (ICES, 1992; Zavala-Gonzalez and Mellink, 1997). The incidence of entanglement of marine mammals in floating synthetic debris in the Bering Sea has been related to the growth in fishing effort and to the use of plastic materials for trawl netting and packing bands (Fowler, 1987). To grasp the extent of the problem of mammal entanglement related mortality, a northern fur seal (*Callorhinus ursinus*) is expected to encounter on average 3–25 pieces of net debris along the 8000 k distance along the NE Pacific coast travelled each year. In his study, Fowler estimates a 15% debris-related mortality of youngsters of the fur seal population of the Pribilof Islands.

Jennings and Kaiser (1998) estimate that the fishing mortality caused by ghost fishing is probably low compared with the fishing mortality caused by attended gears. However, its effects may be important locally (e.g. Fraser River Dungeness crab fishery).

The magnitude of the fishing mortality inflicted by towed gears, by lost traps and nets and by destructive fishing practices on organisms that are never caught is nearly impossible to estimate. Even discards, which could be reported, are regularly under-reported or not reported at all (Alverson et al., 1994). As a result, the total fishing mortality of target and non-target marine organisms is always underestimated. Systematic underestimation of fishing mortality in populations of vulnerable species or in heavily exploited populations increases the risk of population depletion. While the levels of fishing mortality of target exploited populations are underestimated, no estimates of fishing mortality are available for non-target species. Thus the impacts of fishing on those species are considered disputable or non-existent.

## INDISCRIMINATE CHANGES IN NATURAL MORTALITY: IMPACTS ON THE STRUCTURE OF COMMUNITIES

In exploited ecosystems, fishing may indirectly affect the rates of natural mortality of marine species to a substantial extent and these changes are potentially more important than those of fishing mortality itself (Bostford et al., 1997). There are a number of ways in which this can occur. Habitat degradation and the destruction of the physical support for living communities generated by destructive fishing practices is an important mechanism mediating these changes (ICES, 1996a). Also, the dumping of discards and offal from fish processing at sea when concentrated in time and space also disturbs the structure of communities, favouring scavenger species to the detriment of others. Additionally, species that are not directly exploited by a fishery may be affected by the removal of a substantial proportion of their prey, predator or competitor (Hall, 1999). All of these impacts result in changes in the structure of the communities in exploited ecosystems.

### Changes Mediated by Habitat Degradation

Trawls, dredges, and blast and poison fishing may produce substantial alterations in benthic habitats. Physical disturbance of the substrate can result in the loss of biological organisation. In particular biogenic structures (e.g. corals, bryozoans, tube-building polychaetes, seagrasses, algal beds) are easily destroyed by towed fishing gears (e.g. Bradstock and Gordon, 1983; Collie et al., 1997). The ICES study group on the effects of bottom trawling recognised that this activity had both long and short-term effects on benthic communities (ICES, 1988) as recovery times vary. In general it is expected that the effects of physical disturbance on habitat and community structure will be short-lived in communities adapted to frequent natural perturbations and longer-lived in habitats exposed to fewer disturbances (Jennings and Kaiser, 1998).

Based on 60 years of observations, a study in the Wadden Sea (NE Atlantic) revealed long-term changes in the species composition of benthic communities as a result of continued trawling (Rijnsdorp, 1988). Slow-growing epibenthic species had been replaced by fast-growing species, the total number of individuals had grown, and the diversity of species of molluscs and crustaceans had decreased, while that of polychaetes had increased. A further example is the destruction of the oyster habitat by the *Cassostrea virginica* fishery of Chesapeake Bay (NW Atlantic) after a century of exploitation first with dredges and later with hydraulic-powered gear (Rothschild et al., 1994). Collie et al. (1997) found higher diversity indices at the less intensively dredged sites of Georges Bank which was attributed to the larger number of organisms, such as shrimp, brittle stars, mussels, polychaetes and small fishes that were associated with biogenic fauna. Similarly, scallop dredging over beds of the calcareous algae *Lithothamnion,* which can take hundreds of years to accumulate, causes long-term changes in the composition of the associated epifauna (Jennings and Kaiser, 1998).

The removal or destruction of erect benthic fauna, such as corals and sponges, and of flora, such as seagrass and kelp beds, constitutes a negative impact of fishing gears whose consequences go well beyond the immediate physical disturbance (Bostford et al., 1997). In reef ecosystems, declines of the abundance of large, substrate providing epifaunal organisms (coral, sponges, alcionarians, gorgonians) as a result of trawl, blast or poison fishing leads to reduced abundance of fishes which shelter and feed among the large epibenthos and to concomitant increases in abundance of fishes which occur mostly over open sand (Sainsbury, 1988). Bradstock and Gordon (1983) report on the destruction of coralline grounds in Tasman Bay with specially designed trawling nets geared with chains, sledges and rolling bobbins. With the loss of shelter and of food organisms, a reduction of the number of juveniles of the target fish species followed. Seagrasses in temperate coastal ecosystems perform important functions in the same manner as do corals in tropical environments; they provide primary production, habitat for invertebrates and fishes, stabilise sediments and protect the coast from wave exposure. The destruction of suitable habitat for the settlement of juveniles and for refuge and food of juvenile and adult phases of many marine organisms has been identified as the primary cause of the decline of their populations (Langton et al., 1996).

### Changes Mediated by the Dumping of Offal, Discards and Other Food Subsidies

The dumping of large amounts of organic material resulting from the processing of fish at sea and from discards may cause changes in the structure and diversity of marine communities, in principle favouring the proliferation of scavenger and decomposer species (Wassemberg and Hill, 1987; Hill and Wassemberg, 1990; Olaso et al., 1996; ICES, 1996b). This is in addition to the benthic animals damaged or disturbed by trawling (e.g., sea urchins, scallops) which in some areas constitute a significant food subsidy for opportunistic species such as dogfish (*Scyliorhinus* spp.) and gunnards (*Trigla* spp.) (Kaiser and Spencer, 1994).

Assessment of these effects requires knowledge about the fate of discards and offal which until recently have been largely neglected in studies of fishery–ecosystem interactions. Also, there may be the perception that the problem can be ignored because nutrient and energy fluxes will not be altered by the introduction of fishery-related organic matter. However, this perception overlooks the concentrated area- and time-specific distribution of fishery operations and of its associated discarding and dumping. In the Northeast Pacific shelf, where amounts of fish processed by factory trawlers are vast, e.g., 1.4 million t in 1990, input of detritus from offal by catcher-processor ships is potentially high since most fishery products extract less than 50% of the wet weight biomass of processed fish (Hartmann et al., 1992). In the North Sea fisheries, between 6.5 and 12.5% of the groundfish processed for market is dumped as offal at sea (ICES, 1992).

All these sources of material are important in maintaining populations of major scavengers locally, such as rays (ICES, 1994; ICES, 1996a), sharks and dolphins (Hill and Wassemberg, 1990), crabs (Wassemberg and Hill, 1987) and seabirds (Hill and Wassemberg, 1990; ICES, 1994, 1996a). These food subsidies to some organisms change the competitive equilibrium among different species in the ecosystems. But the evaluation of their ecological implications is difficult since reliable estimates of the intake of this food source relative to the total consumption are scarce (Hudson and Furness, 1988; Berghahn and Rosner, 1992). Most work has been done on seabird populations (ICES, 1996b). Time series data since the beginning of the century show large increases in the populations of several seabird species indicating that fishery-induced mortality, such as entanglement in gillnets (Gislason, 1994), has not hampered the increases (NSTF, 1993). In a study of the shrimp fishery in the Wadden Sea, Berghahn and Rosner (1992) estimated at 68–90% the proportion of roundfish discards taken by seabirds. Similar estimates exist for the whole North Sea fisheries (Hudson and Furnace, 1988; Garthe and Huppop, 1994). Considering that 30–40% of the biomass of the North Sea commercially exploited fish species (about 5.85 million t, 1985 figures) is caught each year (discards not included) and that annual discard rates in the various fisheries of the North Sea ranges from 27–56% (NSTF, 1993), it is clear that some 0.55–1.15 million t of fish were discarded in the North Sea in 1985 (NSTF, 1993). To this must be added the estimated annual offal production in the North Sea fisheries of some 79,000 t (ICES, 1994). The expansion of scavenging seabird populations in that region since the beginning of the century has been attributed to these large inputs of discards and offal coupled with the introduction of protective measures (ICES, 1992; NSTF, 1993; Garthe and Huppop, 1994). In the small nature reserve of the archipelago of Columbretes (NW Mediterranean) the seagull (*Larus audouinii*) has established colonies which are wholly dependent on discards from the groundfish trawl fleets operating in the area. Interestingly, managers of the nature reserve are concerned with the survival of the colonies because the recent timing of the annual two-month trawl closure coincides with the seagull-breeding season.

Studies indicate that a certain amount of the offal and discards sink beyond the reach of seabirds and become available to benthic scavengers (ICES, 1996a). Benthic organisms are associated in specific ways with the sediment in which they live and dumping of excess organic matter may change the patchiness of the benthic habitat favouring highly mobile scavengers, such as fish or crabs, relative to slow-moving ones, such as starfishes (Dayton et al., 1995). The populations of dogfish (*Scyliorhinus canicula*) in the Cantabrian Sea (Olaso et al., 1996) and of *Raja radiata* in the Greenland Sea (ICES, 1996a) have expanded apparently as a result of increased food availability from discards of the whiting and rockfish fisheries respectively. Also there is some evidence that the boom of some flatfishes in trawled areas in the North Sea and in the Gulf of Thailand is related to the food subsidies made available to them by trawling (Hall, 1999).

### Changes Mediated by Biological Interactions: Competition and Predation

Fishery managers are becoming aware that the impacts of fishing can cascade through the food chain by affecting competitive and predatory interactions (Sherman, 1991; Payne et al., 1990). According to the trophic cascade model, the effects of environmental conditions climb upward through the system (bottom-up control) to meet the effects of fishing cascading down from the top (top-down control) (Daan, 1989; McQueen et al., 1986). For this reason, it is very difficult to separate natural and man-induced causes of the changes observed at different levels of the ecosystem.

The few documented cases of top-down controls in community structure involve the loss of a predator such as sea otters, lobsters, gastropods, or carnivorous fish with the ensuing domination of its principal prey (examples in Bostford et al., 1997), although predator control is often not demonstrated (e.g., Jennings and Polunin, 1997). In some littoral ecosystems, cascading effects of overfishing of predators has clearly altered the patterns of abundance of

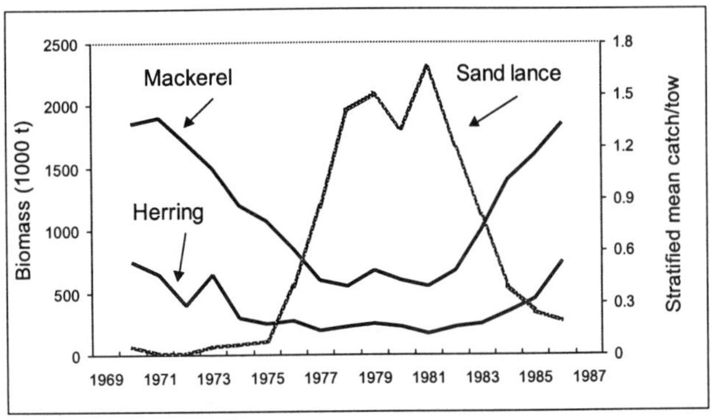


Fig. 7. Trends in biomass of mackerel (age 1+) and of herring (age 3+) (estimated using virtual population analysis) and of the relative abundance of sand lance (research survey data) off the northeastern United States. Redrawn with permission from Fig. 1 of Fogarty et al. (1991).

benthic organisms, such as kelps, corals and sea urchins (e.g., Moreno et al., 1986; Durán and Castilla, 1989; Roberts and Polunin, 1991; Sala and Zabala, 1996). In a Kenyan reef, fish predator removal through fishing resulted in the ecological release of the sea urchins which apparently led to a decrease in live coral cover and to loss of topographic complexity, species diversity and fish biomass (McClanahan and Muthiga, 1988). As a further example of presumable predator control, Fig. 7 illustrates the decline of the mackerel (*Scomber scombrus*) and herring (*Cuplea harengus*) populations in the NW Atlantic continental shelf and the release of the sand lance (*Ammondytes* spp.) population during the period of low abundance of these predators (Fogarty et al., 1991).

Examples of community level changes in continental shelf groundfish communities in association with fishing abound. In the Yellow Sea changes in dominance associated with heavy fishing between the 1950s and the 1980s resulted in a shift from demersal to small planktophagous fish species which were their prey (Jin and Tang, 1996). As assessed by resource surveys, a shift occurred in the groundfish species composition of the Eastern Bering Sea since 1979: the relative abundance of the main gadoid species (*Theragra chalcogramma*) decreased (40%) while that of flatfish and skates increased dramatically (450% and 600% respectively) (Fig. 8). This trend was also observed in the Gulf of Alaska surveys of 1984 and 1990 (Goñi *et al.*, 1993).

These important shifts in the structure of the fish and benthic communities in the north-eastern Pacific (see also marine mammals and birds below) during the last 40 years have also been attributed to changes in ocean temperature, circulation or upwelling events (Fritz et al., 1995). Enlargements of flatfish stocks, apparently indicating an increased energy flow to the shelf bottom, have also been associated with replacements of collapsed fish stocks in the Gulf of Thailand (Ursin, 1982). From trawl survey data of the Bay of Arcachon (NE Atlantic) Quero and Cendrero (1996) report dramatic changes in the species composition of the groundfish community since 1924, seemingly associated with fishing. In this case, however, the abundance of rays and *Triglidae* fish had undergone the most dramatic declines.

Recent studies of evolution of the species diversity and the structure of North Sea groundfish assemblages during the period 1925 to 1996 showed that species diversity decreased in areas where fishing intensity was highest (Jennings and Cotter, 1999). An example is the reduction of ray species richness since the period 1929–1939 reported by Walker (1996). However, in the North Sea there was no evidence of major species replacement cycles like those reported above and fishing effects were largely confined to reductions in the size of the populations of 'vulnerable' species. For the North Sea it is believed that because intensive fishing has been going on for more than one century, the most dramatic changes in species composition of the exploited communities occurred well before these studies were undertaken (Jennings and Kaiser, 1998). The rapidity with which changes arise in marine populations at the onset of exploitation (e.g. Russ and Alcala, 1996) (Figs. 4 and 5) supports this belief.

The above cases show that the observed modifications in community structure differ both in direction and intensity, suggesting that the ways in which fishing affects the relative survival of the various organisms in communities affected by exploitation are the result of complex interactions and feedback mechanisms.

Fisheries exploit many species of fish and shellfish that are preyed upon by seabirds and marine mammals. Because most marine mammals and birds depend on an abundant supply of local food for their survival (NSTF, 1993), fishing may negatively affect their survival by reducing the availability of prey or by inducing its dispersal

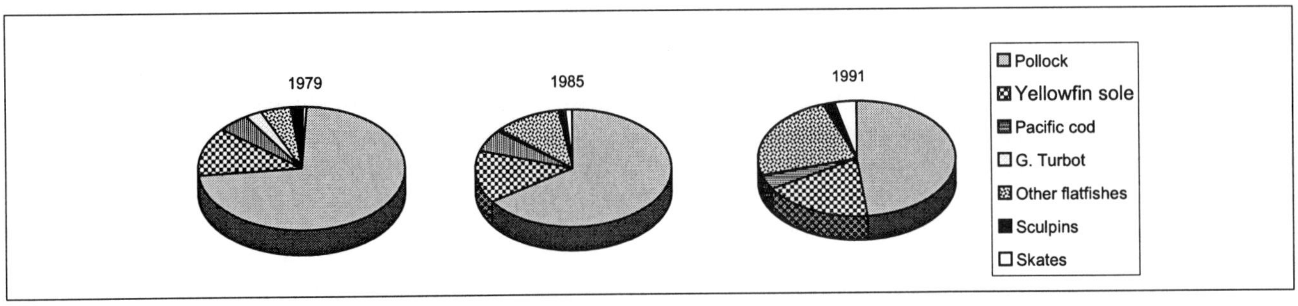


Fig. 8. Changes in the species composition (selected species: only species for which biomass estimates are available over time) of the groundfish assemblage of the Eastern Bering Sea as estimated by NMFS resource surveys. Source: NPFMC (1992).

(Anon, 1994). This common utilisation has led to concern over potential competition between fisheries and these predators. For seabirds feeding on small pelagic fish, a number of correspondences between collapses of fish stocks and breeding failures or population declines have been established (ICES, 1996b). Examples are the collapse of the large stocks of Peruvian anchoveta that was followed by that of the guano birds (Lalli and Parsons, 1994) or the decline of kittiwakes (*Rissa tridactila*) after collapses of capelin stocks in the Barents Sea (Barret and Krasnov, 1996). Conversely in fisheries targeting large demersal fishes, indirect effects on seabird populations may be positive by reducing predatory pressure on small prey fish (ICES, 1996b). An example cited by Dayton et al. (1995) is the removal of southern ocean baleen whales that may have resulted in the release and reallocation of their prey (krill) to seals and seabirds. In all the cases, proper assessments of these linkages are hampered by the lack of independent estimations of fish abundance at the small scales relevant for mammal or seabird foraging (ICES, 1996b; Barret and Krasnov, 1996).

The Bering Sea sustains some of the largest populations of marine mammals and seabirds in the world (Bakkala et al., 1979). In the 1970s, the total fish consumption by marine mammals and birds was estimated to be about 3 million t (Laevastu and Favorite, 1981). During the expansion of the pollock fishery in the past decades, there have been concurrent declines in the populations of several marine mammals: sea lions (*Eumetopias jubatus*, 76% decline since 1975) and seals (*Callorhinus ursinus*, 60% since 1950, and *Phoca vitulina*, 85% since 1970) (Fritz et al., 1995; Springer, 1992; Pitcher, 1990). In all these species, pollock makes up an important portion of the diet, from 21% to 90% depending on the species, age, location and season (Lowry et al., 1989). Concurrently, populations of piscivorous seabirds (*Urea algae, U. lomvia, Rissa brevirostris, R. tridactyla*) underwent declines which have been associated with inadequate food availability and have paralleled the reduction of young pollock abundance since the late 1970s (Springer and Byrd, 1989). Pollock comprises 30–90% of their diet depending on species and location (Hunt et al., 1981; Springer and Byrd, 1989). Causes of the marine mammal and bird declines other than reduced prey availability (legal direct harvest, subsistence harvests, shooting, and incidental takes) have been considered and ruled out as being not significant. Conversely, the role that possible alterations in the oceanographic regimes may have played in the declines has been the subject of continuous debate (Wooster, 1992a; Fritz et al., 1995).

In the Mediterranean, the sustained decline of the populations of monk seal (*Monachus monachus*) has been attributed to the low physical condition of the animals which would make them more vulnerable to diseases and the effects of pollution (Anon, 1994). This seal has been chased by fishermen due to its habit of feeding on the nets, apparently as a consequence of the reduction of their main prey fish populations (Northridge and di Natale, 1991).

In contrast, and paralleling the stability of the groundfish communities in the North Sea indicated earlier, there is no evidence of similar general declines in the populations of marine mammals and seabirds in the North Sea, despite the intense fishing activity having taken place there (except for some reductions in sightings of harbour porpoise in near-shore areas of the Wadden Sea) (NSTF, 1993).

As these examples amply demonstrate, indirect effects mediated by changes in biological interactions are difficult to ascertain. The complexity of the biological relationships between the species affected and the influence of variability in the natural environment compound these difficulties. Nevertheless, biological interactions must be considered when management decisions are intended to protect depleted populations. For example, the recovery of the Northwest and Southwest Atlantic seal populations may put increasing pressure on exploited species which are their prey (Shelton, 1992). In view of the slow progress in developing an understanding of multispecies interactions and incorporating them into management procedures, it is prudent to consider management schemes that are robust to uncertainty (Shelton, 1992).

## DISCUSSION

There has been increasing interest in recent years in developing new fishery management schemes based on knowledge of ecosystem dynamics instead of individual fish populations. This is because in some marine systems commercial species are depleted or have collapsed despite the use of best traditional management practices. Simultaneously, non-commercial ecosystem components, such as macro-invertebrates, non-target fish, birds, and marine mammals have undergone significant, yet largely unexplained, changes in abundance. Commercial and in some instances recreational fishing are obvious contributing factors, yet external environmental factors and interactions among species may also be important determinants.

The effects here described provide a qualitative overview of the impacts of fishing on marine ecosystems, separating those that are directed and controlled from those that are indiscriminate and uncontrolled. Despite the wide recognition of their importance, often on the basis of common sense alone, measuring many of these effects in nature is not possible. While estimating changes in levels of natural mortality in components of exploited communities is hardly possible, most sources of fishing mortality can be assessed experimentally. However, on their own, these estimates would be of limited value because it is necessary to know the effects that the changes in mortality rates have at the population level (Hall, 1999).

At present many efforts are underway to assess the extent to which changes in marine ecosystems have occurred in association with fishing activities, but the evidence is often equivocal. Nevertheless, caution must be exercised because the absence of evidence does not imply that these changes have not taken place. In most instances impacts are not documented or a causal relationship cannot be established because the research has not been specifically designed and carried out with that purpose and data collected for other objectives is proven inadequate (Goñi et al., 1993; Dayton et al., 1995). Also, as has been repeatedly noted, efforts to quantify ecosystem effects of fishing are bound to fail because the most sensitive components of marine ecosystems have been for long impacted by fishing and, as a consequence, there are few meaningful controls. As Dayton (1998) correctly points out, 'fishing has been going on for so many years and is reaching every ecosystem of the world, that each successive generation of marine biologists has a different perception of what is natural because they study increasingly altered ecosystems'. In many marine regions, areas closed to fishing may be the only way to evaluate impacts of fishing on benthic and resident pelagic species (Roberts, 1998).

Despite these difficulties, the ample information base accumulated during recent years points to some inescapable facts: many fisheries are overcapitalised, overfishing is rampant and many fishing techniques have undesirable and indiscriminate effects on species which are not the targets of exploitation. Habitat degradation caused by destructive fishing methods further exacerbates these problems and reduces the capability of decimated populations to recover. The urgent challenge for the next millennium is to halt and then to reverse this trend. Pro-active, precautionary management now and, as information becomes available, adaptive, ecosystem management, must replace single-species, reactive management which is the state of the art in fisheries management today. Having recognised all the complexities and uncertainties inherent to fisheries–ecosystem interactions, there is no question about what must be the immediate course of action: effective fishing effort must be reduced in nearly every fishery of the world and indiscriminate and destructive fishing practices should be eliminated across the range.

## REFERENCES


Alverson, D.L., Freeberg, M.H., Murawski, S.A. and Pope, J.G. (1994) A global assessment of fisheries bycatch and discards. FAO Fisheries Technical Paper 339. 235 pp. FAO, Rome.

Anon. (1994) Report on the meeting on the data base for evaluation of biological impact of fisheries. Commission staff working paper. Commission of the European Communities. SEC(94) 1453.

Anthony, V.C. (1993) The state of groundfish resources off the coast of the north-eastern United States. *Fisheries* **18** (3), 12–17.

Bakkala, R.G., Hirschberger, W. and King, K. (1979) The groundfish resources of the eastern Bering Sea and Aleutian Islands regions. *Marine Fish Review*, November, 24 pp.

Barret, R.T. and Krasnov, Y.V. (1996) Recent responses to changes in stocks of prey species by seabirds breeding in the Southern Barents Sea. *ICES Journal of Marine Science* **53**, 713–722.

Bayle, J.T. and Ramos, A.A. (1993) Some population parameters as bioloindicators to assess the "reserve effect" on the fish assemblage. In *Qualité du Milieu Marin. Indiacateurs Biologiques et Phycochimiques*, eds. C.F. Boudouresque, M. Avon and C. Pergent-Martini. GIS Posidonie, pp. 189–214.

Beinssen, K. (1989) Boult reef revisited. Reeflections. Newsletter Great Barrier Reef Marine Park Authority, Townsville, 61 pp.

Bennet, B.A. and Attwood, C.G. (1991) Evidence for the recovery of a surf-zone fish assemblage following the establishment of a marine reserve on the southern coast of South Africa. *Marine Ecology Progress Series* **75**, 173–181.

BEON (1990) Effects of beam trawl fishery on the bottom fauna in the North Sea. Netherlands Institute for Sea research, BEON Report 8.

Berghahn, R. and Rosner, H.U. (1992) A method to quantify feeding of seabirds on discard from the shrimp fishery in the North Sea. *Netherlands Journal of Sea Research* **28** (4), 347–350.

Bergman, M.J.N. and Hup, M. (1992) Direct effects of beam trawling on macrofauna in a sandy sediment in the Southern North Sea. *ICES Journal of Marine Science* **49**, 5–11.

Bonfil, R. (1994) Overview of world elasmobranch fisheries. FAO Fisheries Technical Paper 341, 119 pp.

Bonhsack, J.A. (1992) Reef resource habitat protection: The forgotten factor. In Stemming the tide of fish habitat loss, ed. R.H. Stroud (ed). *Marine Recreational Fisheries* **14**, 114–129.

Bostford, L.W., Castilla, J.C. and Paterson, C.H. (1997) The management of fisheries and of marine ecosystems. *Science* **277**, 509–515.

Bradstock, M. and Gordon, D.P. (1983) Coral-like bryozoan growths in Tasman Bay and their protection to conserve commercial fish stocks. *New Zealand Journal of Marine Freshwater Research* **17**, 159–163.

Breen, P.A. (1987) Mortality of Dungeness crabs caused by lost traps in the Fraser River Estuary, British Columbia. *North American Journal of Fisheries Management* 7, 429–435.

Caddy, J.F. and Mahon (1995) Reference points for fisheries management. FAO Fisheries Technical Paper No. 1347. FAO, Rome, 83 pp.

Caddy, J.F. and Griffiths, R.C. (1990) Tendencias recientes en las pesquerias y el mediambiente en la zona del Consejo General de Pesca del Mediterraneo (CGPM). Analisis y Estudios Consejo General de Pesca del Mediterraneo, No. 63, FAO, Rome, 83 pp.

Camiñas, J.A. (1996) Capturas accidentales de tortuga boba (*Caretta caretta* L., 1758) en el Mediterráneo occidental en la pesquería de palangre de superficie de pez espada. 25th Anniversary of ICCAT. Symposium on Tunids. SYMP/031. 18 pp.

Carbonell, A., Martin, P. and Ranieri, S. (1997) Discards of the Western Mediterranean trawl fleets. Rapport Commission International Exploration de la Mer Mediterranée. Vol. 35, 392–393.

Caviedes, C.N. and Fik, T.J. (1992) The Peru–Chile eastern Pacific fisheries and climatic oscillation. In: *Climate Variability, Climate Change and Fisheries*, ed. M.H. Glantz. Cambridge University Press, pp. 255–416.

CEC (Commission of the European Communities) (1994) The use of large driftnets under the Common Fisheries Policy. COM(94) 50 final. Brussels, 08.04.1994.

Cook, R.M., Sinclair, A. and Stefansson, G. (1997) Potential collapse of North Sea cod stocks. *Nature* 385, 521–522.

Collie, J.S., Escvanero, G.A. and Valentine, P.C. (1997) Effects of bottom fishing on the benthic megafauna of Georges Bank. *Marine Ecology Progress Series* **155**, 159–172.

Christensen, V. (1998) Fishery-induced changes in a marine ecosystem; insight from models of the Gulf of Thailand. *Journal of Fish Biology* **53** (Suppl. A), 128–142.

Daan, N. (1989) The ecological setting of the North Sea fisheries. *Dana* **8**, 17–31.

Dayton, P.K. (1998) Reversal of the burden of proof in fisheries management. *Science* **279**, 821–822.

Dayton, P.K., Trush, S.F., Agardy, M.I. and Hofman, R.J. (1995) Environmental effects of marine fishing. Aquatic conservation. *Marine and Freshwater Ecosystems* **5**, 205–232.

Durán, L.R. and Castilla, J.C. (1989) Variation and persistence of the middle rocky intertidal community of Central Chile, with and without human harvesting. *Marine Biology* **103**, 555–562.
de Groot, S.J. (1984) The impact of bottom trawling on the benthic fauna of the North Sea. *Ocean Management* **9**, 177–190.
de Groot, S.J. (1995) On the penetration of the beam trawl in the seabed. ICES CM/B:36, 10 pp.
DeMartini, E.E., Ellis, D.M. and Honda, V.A. (1993) Comparisons of spiny lobster, *Panulirus marginatus*, fecundity, egg size and spawning frequency before and after exploitation. *Fishery Bulletin U.S.* **91**, 1–7.
Eleftheriou, A. and Robertson, M.R. (1992) The effects of experimental scallop dredging on the fauna and physical environment of a shallow sandy community. *Netherlands Journal of Sea Research* **30**, 289–299.
Erzini, K., Monteiro, C.C., Ribero, J., Santos, M.N., Gaspar, M., Monteiro, P. and Borges, T.C. (1997) An experimental study of gill net and trammel net 'ghost fishing' of the Algarve (southern Portugal). *Marine Ecology Progress Series* **158**, 257–265.
FAO (Food and Agricultural Organization) (1997) Review of the State of World Fishery Resources: Marine Fisheries. FAO Fisheries Circular No. 920, Rome, FAO. 173 pp.
Fogarty, M.J., Cohen, E.B., Michaels, W.L. and Morse, W.W. (1991) Predation and the regulation of sand lance populations: an exploratory analysis. *ICES Marine Science Symposium* **193**, 120–124.
Fowler, Ch.W. (1987) Marine debris and northern fur seals: a case study. *Marine Pollution Bulletin* **18** (63), 326–335.
Fritz, L.W., Ferrero, R.C., Berg, R.J. (1995) The threatened status of Steller sea lions, *Eumetopias jubatus*, under the Endangered Species Act: Effects on Alaska groundfish fisheries management. *Marine Fisheries Review* **57** (2), 14–27.
García-Rubies, A. and M. Zabala (1990) Effects of total fishing prohibition on the rocky fish assemblages of Medes Islands marine reserve (NW Mediterranean). *Scientia Marina* **54**, 317–328.
Garthe, S. and Huppop, U. (1994) Distribution of ship-following seabirds and their utilisation of discards in the North Sea in summer. *Marine Ecology Progress Series* **106**, 1–9.
Gaspar, M.B., Richardson, C.A. and Monteiro, C.C. (1994) The effects of dredging on shell formation in the razor clam *Ensis siliqua* from Barrinha, Southern Portugal. *Journal of Marine Biology Assessment* **74**, 927–938.
Gislason, H. (1994) Ecosystem effects of fishing activities in the North Sea. *Marine Pollution Bulletin* **29** (6–12), 520–527.
Goñi, R. (1998) Ecosystem effects of marine fisheries: An overview. *Ocean & Coastal Management* **40**, 37–64.
Goñi, R., Hartmann, H. and Mathews, K. (1993) Groundfish fisheries and dynamics of the Northeastern Pacific Ecosystems. Report to Greenpeace International. 56 pp.
Goñi, R., Reñones. O., Quetglas, A. and Mas, J. (1999) Effects of protection on the abundance and distribution of red lobster (*Palinurus elephas*, Fabricius 1767) in the marine reserve of Columbretes Islands (Western Mediterranean) and surrounding area. Contribution presented at the 1st International Workshop on Marine Reserves. Murcia (Spain), 1999.
Grainger, R.J.R. and García, S.M. (1996) Chronicles of marine fishery landings (1950–1994): Trend analysis and fisheries potential. FAO Fisheries Technical Paper No. 359. FAO, Rome. 51 pp.
Hall, M.A. (1998) An ecological view of the tuna–dolphin problem: impacts and trade-offs. *Reviews in Fish Biology and Fisheries* **8**, 1–34.
Hall, S.J. (1999) *The Effects of Fishing on Marine Ecosystems and Communities.* Fish Biology and Aquatic Resources Series I. Blackwell Science. 274 pp.
Hamre, J. (1994) Biodiversity and exploitation of the main fish stocks in the Norwegian–Barents Sea ecosystem. *Biodiversity and Conservation* **3**, 274–492.
Harmelin, J.G., Bachet, F. and García, F. (1995) Mediterranean Marine Reserves: Fish indices as Tests of Protection efficiency. *P.S.Z.N.I.: Marine Ecology* **16** (3), 233–250.
Hartmann, H.J., Lehner, J. and Briclayer, E. (1992) Estimation of discarded catch by commercial fishing in Alaska waters in 1990 and potential ecosystem impacts. Contribution to World Fisheries Congress, Athens, Greece, May 1992.
Hill, B.J. and Wassemberg, I.J. (1990) Fate of discards from prawn trawlers in Torres Strait (Australia). *Australian Journal of Marine and Freshwater Research* **41**(1), 53–64.
Hilborn, R. and Walters, C. (1992) *Quantitative Fisheries Stock Assessment: Choice, Dynamics and Uncertainty.* Chapman & Hall, New York, 570 pp.
Hudson, A.V. and Furness, R.W. (1988) Utilization of discarded fish by scavenging seabirds behind whitefish trawlers in Shetland. *Journal of Zoology, London* **215**, 151–166.
Hunt, G.L., Burgesson, B. and Sanger, G.A. (1981) Feeding ecology of sea birds in the eastern Bering Sea. In *The Eastern Bering Sea Shelf: Oceanography and Resources. Vol. 1*, eds. D.W. Wood and J.A. Calder, pp. 629–648. University of Washington Press, Seattle, Washington.
Hutchings, J.A. and Myers, R.A. (1994) What can be learned from the collapse of a renewable resource? Atlantic cod, *Gadus morhua*, of Newfoundland and Labrador. *Canadian Journal of Fisheries and Aquatic Science* **5**, 2126–2146.
ICES, (1988) Report of the study groups on the effects of bottom trawling. ICES C.M. 1988/B:56
ICES, (1992) Report of the study group on the ecosystem effects of fishing activities. ICES C.M. 1992/G:11.
ICES, (1994) Report of the working group on ecosystem effects of fishing activities. ICES C.M. 1994/ASSESS/ENV:1.
ICES, (1995) Report of the study group on unaccounted mortality in fisheries. ICES CM 1995/B:1.
ICES, (1996a) Report of the working group on ecosystem effects of fishing activities. ICES CM 1966/ASSESS/ENV:1.
ICES, (1996b) Seabird/fish interactions with particular reference to seabirds in the North Sea. ICES Cooperative Research Report No. 216. G.L. Hunt and E.W. Furness (eds.). 87 pp.
Ito, D.H. (1982) A cohort analysis of Pacific ocean perch stocks from the Gulf of Alaska and Bering Sea regions. NWAFC Processed Rep., 82-15, 157 pp. Northwest and Alaska Fishery Center, NOAA 7600 Sandpoint Way N.E. Bin C15700, Seattle, WA, 98115.
Jennings, S., Lock, J. (1996) Population and ecosystem effects of fishing. In *Reef Fisheries*, eds. N.V.C: Polunin and C.M. Roberts. Chapman and Hall, London, pp. 193–218.
Jennings, S. and Polunin, N.V.C. (1997) Impacts of predator depletion by fishing on the biomass and diversity of non-target reef fish communities. *Coral Reefs* **16**, 71–82.
Jennings, S. and Kaiser, M.J. (1998) The effects of fishing on marine ecosystems. *Advances in Marine Biology* **34**, 201–351.
Jennings, S. and Cotter, A.J.R. (1999) Fishing effects in the Northeast Atlantic shelf seas: patterns in fishing effort, diversity and community structure. I. Introduction. *Fisheries Research* **40**, 104–106.
Jennings, S., Alvsvag, J., Cotter, A.J.R., Ehrich, S., Greenstreet, S.P.R., Jarre-Teichmann, A., Mergardt, N., Rijndorp, A.D. and Smedstad. O. (1999) Fishing effects in northeast Atlantic shelf seas: patterns in fishing effort, diversity and community structure. III. International trawling effort in the North Sea: an analysis of spatial and temporal trends. *Fisheries Research* **40**, 125–134.
Jin, X. and Tang, Q. (1996) Changes in fish species diversity and dominant species composition in the Yellow Sea. *Fisheries Research* **26**, 337–352.
Jones, J.B. (1992) Environmental impact of trawling on the seabed: a review. *New Zealand Journal of Marine and Freshwater Research* **26**, 59–67.
Jones, R.J. and Steven, A.L. (1997) Effects of cyanide on corals in relation to cyanide fishing on reefs. *Marine and Freshwater Research* **48**, 517–522.
Kaiser, M.J. (1996) Starfish damage as an indication of trawling inten-

sity. *Marine Ecology Progress Series* **134**, 303–307.

Kaiser, M.J. and Spencer, B.E. (1994) Fish scavenging behavior in recently trawled areas. *Marine Ecology Progress Series* **112**, 41–49.

Kaiser, M.J., Bullimore, B., Newman, P., Lock, K. and Gilbert, S. (1996) Catches in 'ghost fishing' set nets. *Marine Ecology Progress Series* **145**, 11–16.

Kan, T.T., Aitsi, J.B., Kasu, J.E., Matsuoka, T. and Nagaleta, H.L. (1995) Temporal changes in a tropical nekton assemblage and performance of prawn selective gear. *Marine Fisheries Review* **57** (3–4), 21–34.

Kock, K-H., Benke, H. (1996) On the bycatch of harbour porpoise (*Phocoena phocoena*) in German fisheries in the Baltic and North Sea. *Arch. Fish. Mar. Research* **44** (1/2), 95–114.

Laevastu, T. and Favorite, F. (1981) Ecosystem Dynamics in the Eastern Bering Sea. In *The Eastern Bering Sea Shelf: Oceanography and Resources. Vol. 1*, eds. D.W. Wood and J.H. Calder, pp. 611–625. US Government Printing Office, Washington, DC.

Lalli, C.M. and Parsons, T.R. (1994) *Biological Oceanography: An Introduction.* Pergamon, New York, 301 pp.

Langton, R.W., Steneck, R.S., Gotceitas, V., Juanes, F. and Lawton, P. (1996) The interface between fisheries research and habitat management. *North American Journal of Fisheries Management* **16**, 1–7.

Liggins, G.W. and Kenelly, S.J. (1996) Bycatch from prawn trawling in the Clarence River Estuary, New South Wales, Australia. *Fisheries Research* **25**, 347–367.

Liggins, G.W., Kenelly, S.J. and Broadhurst, M.K. (1996) Observer based survey of bycatch from prawn trawling in Botany Bay and Port Jackson, New South Wales. *Marine and Freshwater Research* **47**, 877–888.

Lowry, L.F., Frost, K.J. and Loughlin, T.R. (1989) Importance of walleye pollock in the diets of marine mammals in the Gulf of Alaska and Bering Sea, and implications for management. Proceedings International Symposium on Biology and Management of Walleye Pollock, November 1989, Alaska Sea Grant Report, No. 89-1, University of Alaska.

Ludwig, D., Hilborn, R. and Walters, C. (1993) Uncertainty, resource exploitation and conservation: lessons from history. *Science* **260** (17) 36.

May, R. (1994) Biological diversity: Differences between land and sea. *Philosophical Transactions of the Royal Society of London B* **334**, 105–111.

McAllister, M.K. and Paterman, R.M. (1992) Decision analysis of a large-scale fishing experiment designed to test for a genetic effect of size-selective fishing on British Columbia pink salmon (*Oncorhynchus gorbuscha*). *Canadian Journal of Fisheries and Aquatic Science* **49**, 1305–1314.

McClanahan, T.R. (1989) Kenyan coral reef associated gastropod fauna: a comparison between protected and unprotected reefs. *Marine Ecology Progress Series* **53**, 11–20.

McClanahan, T.R. and Muthiga, N.A. (1988) Changes in Kenyan coral reef community structure and function due to exploitation. *Hydrobiologia* **166** (3), 269–276.

McQueen, D.J., Post, J.R. and Mills, E.L. (1986) Trophic relationship in freshwater pelagic ecosystems. *Canadian Journal of Fishes and Aquatic Science* **43**, 1571–1581.

Megrey, B.A. and Wespestad, V.G. (1990) Alaskan groundfish resources: 10 years of management under the Magnuson Fishery Conservation and Management Act. *North-American Journal of Fisheries Management* **10** (2), 125–143.

Moreno, C.A., Lunecke, K.M. and Lépez, M.I. (1986) The response of an intertidal *Concholepas concholepas* (Gastropods) populations to protection in southern Chile and effects on sessile assemblages. *Oikos* **46**, 359–364.

Myers, R.A., Barrowman, N.J., Hutchins, J.A. and Rosemberg, A.A. (1995) Population dynamics of exploited fish stocks at low population levels. *Science* **269**, 1106–1108.

Munro, J.L., Parrish, J.D. and Talbot, F.H. (1987) The biological effects of intensive fishing upon coral reef communities. In *Human Impacts on Coral Reefs; Facts and Recommendations*, ed. B. Salvat. Antenne Museum E.P.H.E. French Polynesia, pp. 41–49.

NPFMC (1992) Stock assessment and fishery evaluation document for groundfish resources in the Bering Sea and Aleutian Islands region for 1993. North Pacific Fisheries Management Council, NPFMC, Anchorage, Alaska, November 1992.

Northridge, S. and di Natale, A. (1991) The environmental effects of fisheries in the Mediterranean. Report to the European Commission's Directorate General of Environment, Nuclear Safety and Civil Protection. 48 pp.

NSTF (North Sea Task Force) (1993) North Sea Quality Status Report 1993. Oslo and Paris Commissions, London. Olsen & Olsen, Fredensborg, Denmark.

Olaso, I., Velasco, F., Pereda, P. and Perez, N. (1996) Importance of blue whiting (*Micromesistius poutassou*) discarded in the diet of lesser-spotted dog-fish (*Scyliorhinus canicula*) in the Cantabrian Sea. ICES CM 1996/Mini: 2, 18 pp.

Pérez, P., Pereda, P., Uriarte, A., Trujillo, V., Olaso, Y. and Lens, S. (1995) Discards of the Spanish fleet in ICES Divisions. Study contract EU-DG XIV. Ref. No. PEM/93/005.

Pauly, D., Silverstre, G. and Smith, I.R. (1989) On development, fisheries and dynamite: a brief review of tropical fisheries management. *Natural Resource Modelling* **3**, 307–329.

Pauly, D., Christensen, V., Dalsgaard, J., Forese, R. and Torres, F. (1998) Fishing down marine foodwebs. *Science* **279**, 860–63.

Payne, P.M., Wiley, D.N., Young, S.B., Pittman, S., Clapman, P.J. and Jossi, J.W. (1990) Recent fluctuations in the abundance of baleen whales in the Southern Gulf of Maine in relation to changes in selected prey: *Fishery Bulletin* **88**, 687–696.

Pitcher, K.W. (1990) Major declines in number of harbor seals (*Phoca vitulina*) on Tugidak Island, Gulf of Alaska. *Marine Mammal Science* **6** (2), 121–131.

Pitcher, T.J. (1998) A cover story: fisheries may drive stocks to extinction. *Reviews in Fish Biology and Fisheries* **8**, 367–370.

Quero, J.C. and Cendredo, O. (1996) Incidence de la pêche sur la biodiversité ichthyologique marine: le Basin d'Arcachon et le plateau continental Sud Gascogne. *Cybium* **204**, 323–356.

Reñones, O., Goñi, R., Pozo, M., Deudero, S. and Moranta, J. (2000) Effects of protection on the demographic structure and abundance of *Epinephilus marginatus* (Lowe, 1834). Evidence from the Cabrera Archipelago National Park (West-Central Mediterranean). *Marine Life*, in press.

Ricker, W.E. (1963) Big effects from small causes: two examples from fish population dynamics. *Journal of the Fisheries Board of Canada* **20**, 257–284.

Rijnsdorp, A.D. (1988) Report of the study group on the effects of bottom trawling. ICES CM 1988/b:56. ICES, Copenhagen.

Rijnsdorp, A.D., van Leeuwen, P.I., Daan, N., Hessen, H.J.L. (1996a) Changes in abundance of demersal fish species in the North Sea between 1906–1909 and 1990–1995. *ICES Journal of Marine Science* **53**, 1054–1062.

Rijnsdorp, A.D., Brijs, A.M., Storbeck, F. and Visser, E. (1996b) Micro-scale distribution of beam trawl effort in the Southern North Sea between 1993 and 1996 in relation to trawling frequency of the sea bed and the impact on benthic organisms. ICES CM/Mini: 11: 31 pp.

Roberts, C.M. (1998) Marine reserves as a strategic tool. *EC Fisheries Cooperation Bulletin* **11** (3–4), 10–12.

Roberts, C.M. and Polunin, N.V.C. (1991) Are marine reserves effective in management of reef fisheries? *Reviews in Fish Biology and Fisheries* **1**, 65–91.

Rothschild, B.J., Ault, J.S., Goulletquer, P. and Héral, M. (1994) Decline of the Chesapeake Bay oyster populations: a century of habitat destruction and overfishing. *Marine Ecology Progress Series* **111**, 29–39.

Rubec, P.J. (1986) The effects of sodium cyanide on coral reefs and marine fish in the Philippines. In *The first Asian Fisheries Forum*, eds. J.L. Maclean, L.B. Dizon and L.V. Hosillos. Asian Fisheries Society, Manila, Philippines, pp. 297–302.

Russ, G.R. and Alcala, A.C. (1996) Marine reserves: rates and patterns of recovery and decline of large predatory fish. *Ecological Applications* **6** (3), 947–961.

Saila, S.B. (1983) Importance and assessment of discards in commercial fisheries, FAO, Fisheries Circular No. 765. FAO, Rome. 62 pp.

Saila, S.B., Kocic, V.L.J. and McManus, J.W. (1993) Modelling the effects of destructive fishing practices on tropical coral reefs. *Marine Ecology Progress Series* **94**, 51–60.

Sainsbury, K.J. (1988) The ecological basis of multispecies fisheries and management of a demersal fishery in tropical Australia. In *Fish Populations Dynamics*, ed. J.A. Gulland. John Wiley & Sons, pp. 349–382.

Sala, E. and Zabala, M. (1996) Fish predation and the structure of the sea urchin *Paracentrotus lividus* population in the NW Mediterranean. *Marine Ecology Progress Series* **140**, 71–81.

Sangster, G.I., Lehman, K. and Breen, M. (1996) Commercial fishing experiments to assess the survival of haddock and whiting after escape from four sizes of diamond mesh cod-ends. *Fisheries Research* **25**, 323–345.

Shelton, P.A. (1992) Detecting and incorporating multispecies effects into fisheries management in the Northwest and Southeast Atlantic. *South-Africa Journal of Marine Science* **12**, 723–737.

Sherman, K. (1991) The large marine ecosystem concept: research and management strategy for living marine resources. *Ecological Applications* **1** (4), 349–360.

Sherman, K. (1994) Sustainability, biomass yields, and health of coastal ecosystems: an ecological perspective. *Marine Ecology Progress Series* **112**, 277–301.

Smith, P.J. (1996) La diversidad genetica de los recursos pesqueros marinos: posibles repercusiones de la pesca. FAO Fisheries Technical Document No. 344. FAO, Rome. 59 pp.

Smith, B.D. and Jamieson, G.S. (1991) Possible consequences of intensive fishing for males on the mating opportunities of Dungeness Crabs. *Transactions of the American Fisheries Society* **120**, 650–653.

Springer, A.M. (1992) A review: Walleye pollock in the North Pacific: How much difference do they really make? *Fish Oceanography* **1** (1).

Springer, A.M. and Byrd, G.V. (1989) Seabird dependence on walleye pollock in the Southeastern Bering Sea. Proc. Int. Symp. Mgmt. Walleye Pollock, Alaska Sea Grant Report, No. 89-1. November 1988.

Suronen, P., Erickson, D.L. and Orrensalo, A. (1996) Mortality of herring escaping from pelagic trawl codends. *Fisheries Research* **25**, 305–321.

Thorpe, J.E., Koonce, J.F., Borgeson, D., Henderson, B., Lamsa, A., Maitland, P.S., Ross, M.A., Simon, R.C. and Walters, C. (1981) Assessing and managing Man's impact in fish genetic resources. *Canadian Journal of Fisheries and Aquatic Sciences* **38**, 1899–1907.

Ursin, E. (1982) Stability and variability in marine ecosystems. *Dana* **2**, 51–67.

Van Beek, F.A., Van Leewen, P.I. and Rinjsdorp, A.D. (1990) On the survival of plaice and sole discards in the otter trawl and beam trawl fisheries in the North Sea. *Netherlands Journal of Sea Research* **26** (1), 151–160.

Walker, P.A. (1996) Sensitive skates or resilient rays? Spatial and temporal shifts in ray species composition in the Central and Northwestern North Sea. ICES CM 1996/Mini:4.

Walker, P.A., Hessen, H.J.L. (1996) Long-term changes in ray populations. *ICES Journal of Marine Science* **53**, 1085–1093.

Wassemberg, T.J. and Hill, B. (1987) Feeding by the sand crab (*Portunus pelagicus*) on material discarded from prawn trawlers in Moreton Bay, Australia. *Marine Biology* **95**, 387–393.

Whale, R.A. (1997) Consequences of fishing with regard to lobster fisheries: report from a workshop. *Marine Freshwater Research* **48**, 1115–1119

Wickens, P.A. and Sims, P.F. (1994) Trawling operations and South Africa for seals, *Arctocephalus pusillus pusillus*. *Marine Fisheries Review* **56** (3), 1–2.

Witbaard, R. and Klein, R. (1994) Long-term trends on the effects of the southern North Sea fishery of the bivalve mollusc *Arctica islandica* L. (Mollusca, Bivalva). *ICES Journal of Marine Science* **51**, 99–105.

Wooster, W.S. (1992a) Report of the Is it food? Workshop, March 9–14, 1991, Fairbanks, Alaska. Unpublished manuscript. School of Marine Affairs, University of Washington, Seattle.

Wooster, W.S. (1992b) King Crab dethroned. In Climate Variability, Climate Change and Fisheries, ed. M.H. Glantz. Cambridge University Press, pp. 15–29.

Zavala-Gonzalez, A. and Mellink, E. (1997) Entanglement of California sea lions, *Zalophus californianus californianus*, in fishing gear in the central-northern part of the Gulf of California, Mexico. *Fishery Bulletin* **95**, 180–184.

## THE AUTHOR

**Raquel Goñi**

*Centro Oceanografico de Baleares Muelle de Poniente s/n,*
*Aptdo 291,*
*07080 Palma de Mallorca, Spain*

Chapter 116

# By-catch: Problems and Solutions

Martin A. Hall, Dayton L. Alverson and Kaija I. Metuzals

By-catch is one of the most significant issues affecting fisheries management today. For some, incidental mortality of species which are long-lived and have low reproductive rates is a conservation problem affecting marine mammals, sea birds, sea turtles, sharks, etc. By-catches can also affect biodiversity through their impacts on top predators, through the removal of individuals from many species, or through the elimination of prey. For others, the by-catch issue is one of waste; the millions of tons of protein dumped in the ocean, and the waste of animal lives is condemned on moral grounds. For the economist, it generates additional costs without affecting the revenues, and may hinder profitability. For the fishers, it causes conflicts among fisheries, it gives fishers a bad public image, it generates regulations and limitations on the use of resources, and frequently has negative effects on the resources harvested through the mortality of juvenile and undersized individuals of the target species before they reach their optimal size from the point of view of future yield.

Several examples of major by-catch issues are described, focusing also on the solutions to the problems which have been developed by scientists, fisheries managers and members of the fishing industry itself. By-catch is an extremely complex set of scientific issues, not only an economic, political, or moral one. Although only a few fisheries include by-catches of the target species in their stock assessment (e.g. Pacific halibut), it is clear that by-catch management will be an integral part of future ecosystem management schemes. These considerations, together with the introduction of environmental variability and a better handling of scientific uncertainty, should lead to more intelligent ways to harvest our resources.

*Seas at The Millennium: An Environmental Evaluation (Edited by C. Sheppard)*

## INTRODUCTION

The ecologist has never been asked before how to harvest an ecosystem optimally. In the mid 1950s, population dynamicists developed the quantitative models that have been the basis for 50 years of management. They were single-species models and, with many, the control mechanisms were simply the selection of optimum or minimum size to be taken. One of the early problems for gear technologists was to develop nets to fulfil these requirements. Selective fishing as a concept meant catching the desirable sizes of the target species.

More recently, the concept of selective fishing has changed, and its new meaning includes avoiding not only undesirable sizes of target species, but also avoiding forbidden species or those without economic value. If ecologists, rather than population dynamicists, had recommended a way to harvest an ecosystem, it would probably have resulted in a very different scheme. To concentrate all the impact of the harvest on a narrow range of sizes of one species seems, intuitively, a very unlikely way to preserve ecosystem structure and function (Hall, 1996). Economics and technology rather than ecological principles, have determined the way an ecosystem is exploited. Very selective fishing may be desirable in some cases but, from the ecological point of view, there is no experimental or theoretical evidence showing that this is the best, or the least harmful, way to extract a sustainable harvest from an ecosystem. The complexities of handling and processing the mixture of species and sizes, together with the lack of markets for many of the products, prevent the alternative ("non-selective") strategy of exploitation from happening. But this does not mean that things will never change. Sooner or later, the human species may have to give ecological principles a higher priority when choosing a foraging strategy.

Another area where ecologists and markets collide is in the choice of target species. If exploitation is targeted on a few species, the ecologist may recommend taking those at lower levels of the food chain on the grounds that fewer transfers between trophic levels should allow a much larger biomass harvest. However, markets in developed nations demand and pay a high price for swordfishes, tunas, sharks, and other top predators. Targeting these large species requires gear that frequently causes by-catches of other large and frequently long-lived species such as marine mammals, sea turtles, sea birds, etc. Purse-seines, gillnets, longlines and trawls take a toll of these groups (Northridge, 1984, 1991a; Magnuson et al., 1990; Brothers, 1991; Andrew and Pepperell, 1992; Stevens, 1992; Alverson et al., 1994; Bonfil, 1994; IWC, 1994; Jefferson and Curry, 1994; Wickens, 1995; Williams and Capdeville, 1996, Dayton et al., 1995; Alexander et al., 1997; Hall, 1998; Vinther, 1999).

By-catches (Fig. 1) can create a conservation problem when endangered species are affected (i.e. the vaquita porpoise; Rojas-Bracho and Taylor, 1999), or when the level of the take is not sustainable for the non-target species. They

Fig. 1. Crew shovelling by-catch from the vessel deck (photo D. Bratten).

can affect biodiversity, and can disturb the ecosystem by transferring biomass between water layers (i.e. discards of bottom trawls on surface waters; Hill and Wassenberg, 1990), or by causing accumulations of biomass that affect the normal flow of nutrients and matter, may cause anoxia, or have other impacts on the benthos (ICES, 1995; Dayton et al., 1995). They may become a subsidy to those species that learn to utilise the fishing operations to find feeding opportunities at the expense of their competitors (Wickens, 1995; Ramsay et al., 1997; Laptikhovsky and Fetisov, 1999). On longer time scales, biomass discarded in deep water may not be recycled vertically, but may enter the circulation patterns of bottom water masses, that may take centuries to return its components to the ecosystem of origin.

For decades, by-catches were mostly ignored by scientists working on stock assessment, by fisheries managers, and by environmentalists. There were several reasons for this neglect:

1. It was not visible. With most fisheries data being collected in ports, events happening at sea were not witnessed by scientists. There was ignorance on the existence or level of the problem. The by-catch issue became highly visible when the public found out about cases involving charismatic species such as dolphins (Perrin, 1968), or endangered species such as sea turtles (Magnuson et al., 1990).
2. It was probably smaller in magnitude. The increase in scale of industrial fisheries has resulted in evolution of gear that covers huge volumes of water, moving in some cases at high speeds, and is much less selective. Examples of this are the transition of tuna fisheries from pole-and-line to purse seining, or from small coastal gillnets to large pelagic driftnets.
3. It was less significant for stock assessment. With resources in earlier stages of exploitation, the waste of some of the target species was not perceived as a major factor affecting fishing mortality estimates. With fisheries closer to their upper limit, by-catches may make a difference.

4. The interference among fisheries was less intense. One of the main constituencies of the by-catch issue today is fishers affected by the waste of some gears or fleets. The more diversified the fisheries, and the higher the level of effort, the more intense is competition for the resources.
5. The ecosystem concerns were not a management priority. The emphasis on single-species management models and schemes did not leave much room for consideration of by-catches.

But all this has changed, and although often difficult to assess, the by-catch issue is now globally important, and a source of conflicts. By-catches of marine mammals, sea birds and sea turtles are a very significant if not dominant factor in the present management of some fisheries (Alverson et al., 1994; Alverson and Hughes, 1996; Hall, 1996, 1998). By-catches of fishes (i.e. halibut; Clark and Hare, 1998) or crustaceans (i.e. crabs) in one fishery may cause or accelerate the closure of another.

Alverson et al. (1994) have provided a global assessment of fish discards of 27.0 million Mt with a range of 17.9 to 39.5 million Mt. This means that a significant proportion of the world catch, estimated at around 100 million Mt, is discarded. After examining over 800 articles concerning by-catch and discards in the world fisheries, the authors estimated that the region with the highest discard level is the northwest Pacific Ocean. Shrimp trawl fisheries were found to have higher by-catch/catch ratios in weight than any other gear type and accounted for over one third of the global total. On a weight basis, 14 of the highest 20 by-catch/catch ratios were associated with shrimp trawls. This is clearly a significant quantity in a planet with an increasing human population, and this review emphasised the wasteful nature of some exploitations. An update (Alverson, 1998) reviews many of those figures in view of changes in gear or fishing practices in the different regions and of additional data provided by several researchers.

Discards and by-catch are neither new issues nor new problems. Many authors (Saila, 1983; Alverson et al., 1994) point out that by-catches have existed since fishing first began. Programmes and techniques designed to reduce capture of non-target species or undersized target species are not just the product of recent fishery managers. Attempts to deal with problems caused by the use of non-selective fishing gears have been tried many years ago. Regulations to reduce the catch of undersized target species and to limit catch of non-target species constitute long-accepted fisheries management measures (Alverson and Hughes, 1996).

## DEFINITIONS

McCaughran (1992) defined by-catch as that 'portion of the catch returned to the sea as a result of economic, legal or personal considerations plus the retained catch of non-targeted species'. This definition can be misleading, and it lumps together a waste product with an additional source of income. Sometimes it is difficult to establish which is the target of a fishery. The definition used here is that of Hall (1996): "it is that part of the capture that is discarded at sea, dead (or injured to an extent that death is the result)". *Capture*, in turn, means all that is taken in the gear. The capture can be divided into three components: (a) the portion retained because it has economic value (*catch*), (b) the portion discarded at sea dead (*by-catch*), and (c) the portion released alive (*release*). In this sense, the term by-catch has a clear negative connotation for fishers or environmentalists, and programs and actions to "reduce by-catch" can be considered as ways to improve the fishery, without being detrimental to the fishers.

Catch could be subdivided further into two main components: *target catch* and *non-target catch*, the latter including other species caught incidentally but retained because of their economic value. By-catch and release have the same components. If necessary, one could distinguish primary and secondary target species.

But not all the catch loaded in the vessel reaches consumers. Once the catch reaches port, buyers or processors may reject some because of size or condition. This proportion of the catch is the *rejects*, the rest is the *marketable catch*. While the latter is being prepared for sale or processed, another fraction, the *processing waste*, is lost; what remains is the *yield*. In very efficient fisheries, the yield should be a high proportion of the *capture*, or more accurately of the difference *capture – release*.

Special cases of by-catch exist also: *Prohibited species*: any species which must by law be returned to the sea, and *High Grading*: the discard of a marketable species in order to retain the same species at a larger size and price, or to retain another species of higher value.

## REASONS FOR DISCARDING

Some of the above definitions have already clarified the main reasons for discarding fish species. A fisher is a business-person, and with the best of intentions intends to make a living. It is likely that any incidental or extra catch is purely accidental, and he then makes a decision whether or not to land it or to risk searching for higher priced fish. Clucas (1997) summarised the main reasons:

- Fish caught are of the wrong species, size or sex, or the fish are damaged
- Fish are incompatible with the rest of the catch (from the point of view of storage)
- Fish are poisonous
- Fish spoil rapidly (i.e. before it is brought on board)
- Lack of space on board
- High grading
- Quotas reached
- The catch was of prohibited species, in prohibited season or fishing ground, or with prohibited gear

For some gears, most of the fish discarded will be dead. In other cases, even if the fish are alive when returned to the water, their survival rate is low.

Besides these discards, there is another type of loss of fish that also adds to mortality. Sometimes the net, or other type of gear, is ripped apart, or breaks under the stress of its load because of malfunctions, defective materials, etc. This is not strictly a by-catch, because there is no intent to discard, but the complete or partial loss follows. This source of mortality is probably negligible in some fisheries, but it is never accounted for in the estimation of fishing mortality because there are no data available on its frequency or the magnitude of its impact.

## REGULATIONS AND GUIDELINES

The first and most obvious set of regulations and guidelines are the UN FAO Code of Conduct and the Kyoto Convention.

### FAO Code of Conduct for Responsible Fisheries

The FAO code of conduct for responsible fisheries (FAO, 1995) encourages nations to establish principles and criteria for the elaboration and implementation of national policies for responsible conservation of fisheries resources and fisheries management and development, and states precisely that discarding should be discouraged. But besides its obvious good intentions, its implementation faces many challenges. The fisheries sector in many countries constitutes powerful lobbies, or groups large numbers of participants (i.e. artisanal small-scale fishers). As many restrictions concerning by-catch affect the productivity of the fishery, at least initially, there is a strong resistance to the constraints that should be imposed. The economic costs of gear modifications or replacements add to the costs of the fisheries, and unless major incentives are offered or significant outside pressures exerted, changes will not happen.

In some of the relevant articles, the Code states:

8.4.5 States, with relevant groups from industry, should encourage the development and implementation of technologies and operational methods that reduce discards. The use of fishing gear and practices that lead to the discarding of catch should be discouraged and the use of fishing gear and practices that increase survival rates of escaping fish should be promoted.

8.4.6 States should co-operate to develop and apply technologies, materials and operational methods that minimise the loss of fishing gear and the ghost fishing effects of lost and abandoned fishing gear.

8.4.8 Research on the environmental and social impacts of fishing gear and in particular, on the impact of such gear on biodiversity and coastal fishing communities should be promoted.

11.3.3 States should simplify their laws, regulations and administrative procedures applicable to trade in fish and fishery products without jeopardising their effectiveness.

11.1.8 States should encourage those involved in fish processing, distribution and marketing to: a) Reduce post harvest losses and waste, b) Improve the use of by-catch to the extent that this is consistent with responsible fisheries management practices.

12.4 States should collect reliable and accurate data, which are required to assess the status of fisheries and ecosystems, including data on by-catch, discards and waste. Where appropriate, this data should be provided, at an appropriate time and level of aggregation, to relevant State and sub regional, regional and global fisheries organisations.

12.10 States should carry out studies on selectivity of fishing gear, the environmental impact of fishing gear on target species and on the behaviour of target and non-target species in relation to such fishing gear as an aid for management decisions and with a view to minimising non-utilised catches as well as safeguarding the biodiversity of ecosystems and the aquatic habitat.

12.12 States should investigate and document traditional fisheries knowledge and technologies; in particular those applied to small-scale fisheries, in order to assess their application to sustainable fisheries conservation, management and development.

Some individual nations are developing their own versions of a Code of Conduct. In some cases, the industry has taken the initiative (i.e. Canadian Code of Conduct for Responsible Fishing Operations, Consensus Code 1998).

Of particular interest is the case of Norway, which has adopted a policy of "No discards." Fishers are not allowed to discard anything caught in the net, and that forces them to fish selectively by avoiding periods, areas or times of the day with high by-catches, and by developing technology that contributes to that goal. Norway is the only country that has prohibited discards by law and fishermen are obliged to bring all their catch ashore (Olsen, 1995; Isaksen, 1997). Fishermen also have to keep logbooks with detailed records of their operations. This is controlled by frequent inspections, but the success of efforts like this depends on the good faith of the fishers, or on a very extensive and costly monitoring system based on on-board observers. Without being too pessimistic about human nature, the need for monitoring stands as a clear pre-requisite to the implementation of this type of program.

These programs (a) encourage research on by-catch reduction gear and techniques with a clear economic disincentive, which is to fill the boat with low-value fish; (b) encourage behavioural changes in fishers with regard to avoiding areas and seasons of high by-catches; (c) help reduce the waste of life and protein caused by the fishery, by forcing the utilisation of what was already harvested. However, they are costly; and may result in the development of markets for undersized fish, juveniles, etc.

### The Kyoto Declaration and Plan of Action

The States that met in Kyoto for the International Conference on the sustainability contribution of fisheries to Food security in December 1995 endorsed the provisions of the FAO Code of Conduct and in Declaration 15 stated that "they would promote fisheries through research and development and use of selective, environmentally safe and cost effective fishing gear and techniques". This resulted in the following being included in the plan of action (Clucas, 1997):

- To increase efforts to estimate the quantity of fish, marine mammals, sea birds, sea turtles and other sea life which are incidentally caught and discarded in fishing operations;
- To assess the effect on the populations or species;
- To take action to minimise waste and discards through measures including, to the extent practicable, the development and use of selective, environmentally safe and cost effective fishing gear and techniques, and,
- To exchange information on methods and technologies to minimise waste and discards.

## CLASSIFICATION OF BY-CATCH: WHY IS IT USEFUL?

By-catch is not a phenomenon that exists by itself; it is simply the result of deficiencies in our ability to select what we harvest from the ocean. As such, the label covers a wide variety of situations. By-catches happen for many different reasons, and have widely different characteristics, so it helps to analyze the problem by classifying by-catches to illustrate how they happen, their ecological or economic origin, and their significance. The classifications proposed by one of us are based on eight different criteria that highlight some special characteristics of the problem, and in many cases point to likely approaches for its solution (Hall, 1996):

1. by the spatial pattern of by-catch rates (more or less aggregated in space);
2. by the temporal stratification (more or less "aggregated in time");
3. by the level of control (controllable or uncontrollable by the fishers);
4. by the frequency of occurrences (rare or common);
5. by the degree of predictability (predictable or unpredictable);
6. by the ecological origin of the by-catch (species associated with the target or random encounters);
7. by the level and type of impact;
8. by legal or economic considerations.

Criteria 1 and 2 show the cases where closures (spatial or temporal) can be effective in reducing by-catch. This is not the only piece of information necessary to make an intelligent decision on the matter. The spatial and temporal distribution of the catches is also relevant. If the distribution of catches and by-catches is similar, then the cost in catch losses of the closures will be high. The key variable is the ratio by-catch to catch in all spatial or temporal strata considered. When it is high, the potential value of closures is high. The models proposed by Hall (1996) allow an approximate assessment of the costs incurred when applying a closure system to control by-catches. If reasonable assumptions can be used to re-allocate effort from the closed areas or periods to alternative ones then the model can be modified to introduce the new effort distribution; otherwise, the basic model will produce a worst-case scenario.

If a by-catch is controllable to some extent (criterion 3), then the fishers may be required to meet performance standards. The incidental dolphin mortality in the eastern Pacific is a clear example; the more capable skippers, with the best trained crews have much lower mortality rates than others. It is possible then to demand minimum performance standards by those participating in the fishery on dolphins. Some countries implemented maximum acceptable values of average dolphin mortality per set, and skippers exceeding those were removed from the fleet. The whole international fleet is managed with a system of individual vessel limits that has been steadily declining; this decline is a reflection of the fact that the fleet can modify its behaviour. But it is not sensible to set standards for the fisher that got a whale entangled in a lobster trap; there is no modification of the fishing technique that could have avoided that incident. Between these extremes, there is a continuum of levels of control. A gill-netter may affect his by-catch rates by selecting fishing areas known to him to have lower by-catches, or by moving away from areas where the results were negative.

Very infrequent events (criterion 4) make planning for prevention impossible. Patterns cannot be established, methods cannot be tested, but still the impact may be important if the event is catastrophic (i.e. a massive entanglement due to special circumstances).

Predictability (criterion 5) is a requirement for the effectiveness of closures. Consistent migrations, such as sea turtle "arribadas" are easy targets for management controls, and there are many other situations where predictability can be used to mitigate impacts. Species that are known to be active at night, and not during the day, or that are especially active when there is a full moon, etc., provide the manager with an opportunity.

When by-catch species are ecologically associated with the target species (criterion 4), a difficult choice is offered. A selective fishery may result in an imbalance in the ecosystem. The harvest of a predator but not the prey, or vice versa, or of one of a guild of competitors, is likely to result in changes in community structure, but unfortunately, for most ocean basins, there is very little baseline data to compare with. From the point of view of ecosystem utilisation, a diversified harvest may be a better way to preserve its structure and function, than a very selective one (Hall, 1996).

The level of the by-catch (criterion 7) gives a good idea of the priority it deserves. Unsustainable by-catches, especially those of species in danger of extinction, will generate a need for actions that could be drastic and with high economic and social impacts (i.e. coho salmon in British Columbia). It is these types of by-catches, together with those of charismatic species, that have brought the issue to the forefront of fisheries management. Today, by-catch considerations play a significant role in the regulation of several major fisheries.

The last criterion (criterion 8) simply separates by-catches that are imposed by regulation from those that are generated by economic considerations (low price, no market etc.) This second type can be reduced through marketing campaigns, changes in food processing, or other ways to increase the value of the discarded products.

## BY-CATCH AS A COMPONENT OF FISHING MORTALITY

A basic need of fisheries management is to quantify the mortality caused to the resource (Chopin et al., 1996). This value, together with natural mortality, constitutes the total loss of individuals from the target population. The traditional formula for total mortality ($Z$) is the sum of the natural mortality ($M$) as well as mortality due to fishing ($F$). That is

$$Z = M + F.$$

As total mortality can be estimated from catches by age or other techniques, and natural mortality has proven very difficult to estimate independently, many thought that a good approximation to $F$ could be obtained from fisheries data. But omissions, incomplete or biased nature of the data have been a problem. An ICES study group (review in Chopin et al., 1996), tried to explore all possible sources of fishing mortality, and of other uncertainties in the data. The total impact of fishing was quantified as an aggregate of all catch mortalities including discards, illegal fishing and misreported mortalities.

The complexity of the $F$ term is described by decomposing it into a series of components that identify the potential sources of fishery-induced mortality and complete the picture by adding non- or under-reporting. The equation to specify all the components is the following, where $F$ is the sum of all direct and indirect fishery-induced mortalities:

$$F = (F_{cl} + F_{rl} + F_{sl}) + F_i + F_d + F_o + F_a + F_e + F_g + F_p + F_h$$

where

$F_{cl}$ = commercial landing mortalities;

$F_{rl}$ = recreational landing mortalities (these are mortalities associated with sport fishing). In the UK for example there are many anglers but there is no licensing system for sea fishing. This mortality is therefore largely unknown despite the fact that over 1 million people fish at the sea);

$F_{sl}$ = subsistence landing mortalities (this is a mortality associated with fishing by indigenous people);

$F_i$ = illegal and misreported landing mortality;

$F_d$ = discard mortality;

$F_o$ = drop-out mortality;

$F_a$ = mortality resulting from fish or shellfish that avoid gear but die from stress or injuries;

$F_e$ = mortality resulting from fish or shellfish contacting but escaping gear that subsequently die;

$F_g$ = mortality resulting from fish or shellfish that are caught and die in ghost fishing gears;

$F_p$ = mortality resulting from predation of fish or shellfish escaping from or stressed by fishing gear that would otherwise live;

$F_h$ = mortality due to gear impacts on the habitat (i.e. increased predation because of loss of shelters or disturbances, etc.).

This equation helps point out the need for research plans aimed at understanding and quantifying each of these components, and eventually to devising programs to reduce waste. It also serves to emphasize the little data available on most of these sources of mortality in the majority of our fisheries.

## BY-CATCH REDUCTION PROGRAMS

As Alverson and Hughes (1996) point out, the emergence of by-catch as a major management issue can be traced to the rapid growth of world fisheries and their increasing competition, the rise of environmental groups and the resulting efforts to protect populations of marine mammals, birds and turtles affected by commercial fisheries. Facing the task of reducing by-catch, it quickly became apparent that there were only two levers that could be moved to achieve reductions (Hall, 1996). Either the level of effort is reduced or the average by-catch caused by each unit of effort is reduced (Table 1).

### Reductions in Effort

Reduction in the level of effort amounts to a reduction in the fishery, and it is frequently a costly solution. The United Nations Driftnet ban is an example (Burke et al., 1994; Bache and Evans, 1999). A reduction in effort is often imposed indirectly when a total by-catch limit is established as has been done in the tuna–dolphin program (Hall, 1998).

Table 1

Basic by-catch equation ("two lever system") and ways to reduce both terms

| Total by-catch = | "effort" | × | by-catch-per-unit effort (BPUE) |
|---|---|---|---|
| | Regulatory bans | | Technological changes |
| | Regulatory limits | | Training |
| | Trade sanctions | | Regulations |
| | Consumer boycotts | | |
| | Gear changes | | |

Temporal or spatial closures may result in effort limitations if the fishers cannot increase the level of effort in open areas or seasons to compensate for the closures. Policies such as the "dolphin-safe" one, that try to eliminate the market for the catch produced by a specific gear or type of effort, have the target of reducing or eliminating it (Scott, 1996).

### Reductions in BPUE

#### *Technological Changes in Gear and Other Equipment*

The options to reduce BPUE are many. Technological changes in equipment may be used. This has proved very successful in many fisheries, provided that some conditions are met during the experimental stage (Pikitch, 1992). Traditionally, changes in mesh size or type have been used to improve selectivity; hook size and bait type have been used in longline and hook-and-line fisheries; addition of Turtle-Excluder Devices (TEDs) to trawls has reduced sea turtle mortality dramatically (Magnuson et al., 1990); sorting grids, mesh changes and other modifications are also used (Larsen and Isaksen, 1993; Stone and Bublitz, 1996; Kennelly, 1995; Kennelly and Broadhurst, 1996). The Medina Panel and many other changes of the basic purse-seine gear have helped reduce dolphin mortality (Coe et al., 1984; Francis et al., 1992). Sorting grids are also in use or have been tested in many purse seine fisheries (Misund and Beltestad, 1994; Beltestad and Misund, 1996) and bird-scaring lines have been successful in reducing bird by-catches in longlines (Alexander et al., 1997). Other examples are provided by Alverson (1998).

#### *Deployment and Retrieval Changes*

Sometimes changes in the procedures used to deploy or retrieve the gear are sufficient to alter its selectivity, and it is not necessary to modify it. The backdown procedure applied to purse-seining on dolphins has been a major contribution to the reduction in dolphin mortality from 133,000 in 1986 to under 2,000 in 1998 (Hall, 1998; Anon., 2000). Deploying longlines at night, or bringing the point of release to 1.5 to 2 meters below the surface reduces seabird by-catches (Alexander et al., 1997; Løkkeborg, 1998). Gillnets and longlines can be deployed at different levels in the water column. The speed, depth or duration of a trawl haul can also affect selectivity.

#### *Training*

Training of fishers may include providing them with information that helps them avoid conditions that lead to high by-catches, and specific training in the use of the gear and manoeuvres (Bratten and Hall, 1996).

#### *Management Actions*

Many options are available, including programs and approaches pursued in different countries and fisheries (Olsen 1996; Duthie, 1997; Everett, 1997; Isaksen, 1997; Witzig, 1997; Alverson, 1998). Considering the goal is to reduce the average BPUE, any change which switches effort from areas or time periods with high BPUE to those with lower values will result in an average decrease. Closures can be effective if the areas or periods closed have higher averages. When by-catches are controllable, selection of skipper and crews that meet some performance standards could also lower averages. But even more effective has been the setting of individual vessel by-catch limits. In the tuna purse seine fishery of the eastern Pacific, the total dolphin mortality limit set by the participating nations in the international program is divided by the number of vessels requesting a limit, and each of them is allowed to make sets on dolphin-associated tuna until its individual limit is reached (Hall, 1998; Gosliner, 1999).

Management actions which improve performance include: selective licensing; economic advantages so that those with the best performance get licences for the best areas, or for longer periods, or for preferred species; total retention of by-catch; acknowledging fishers with outstanding records or, as is done in some fleets, publishing lists of worst offenders; use of labels identifying responsible fishers that meet required standards, to allow consumers to use their buying power to provide an incentive.

## HISTORY OF THE BY-CATCH ISSUE: SOME EARLY EXAMPLES

The emergence of the by-catch issue was a result of a few highly visible cases. From the typical, early responses to the by-catch of charismatic species, the more ecologically-minded groups moved on to face effects of by-catches at the ecosystem level. The former had economic successes, while the latter were less glamorous but addressed the very significant issues of effects on ecosystems.

### The Tuna–Dolphin Problem

Incidental mortality of dolphins in the tuna purse-seine fishery in the eastern Pacific Ocean during the 1960s was the first by-catch problem that generated intense public attention. Initially tuna was fished with pole and line, using live bait. In the late 1950s, the U.S. fleet switched to purse-seine gear, following several technological developments. This fishery had much higher catch rates. The tunas were detected in three ways: association with floating objects, association with herds of dolphins, and as free-swimming schools visible at the surface. When tunas were associated with dolphins, the net encircled both, many of the latter being killed. When environmental groups became aware of this (Perrin, 1968), the reaction was intense, leading to the passing of the Marine Mammal Protection Act (MMPA) in 1972 (Gosliner, 1999). Dolphin mortality during the 1960s was estimated to be several hundreds of

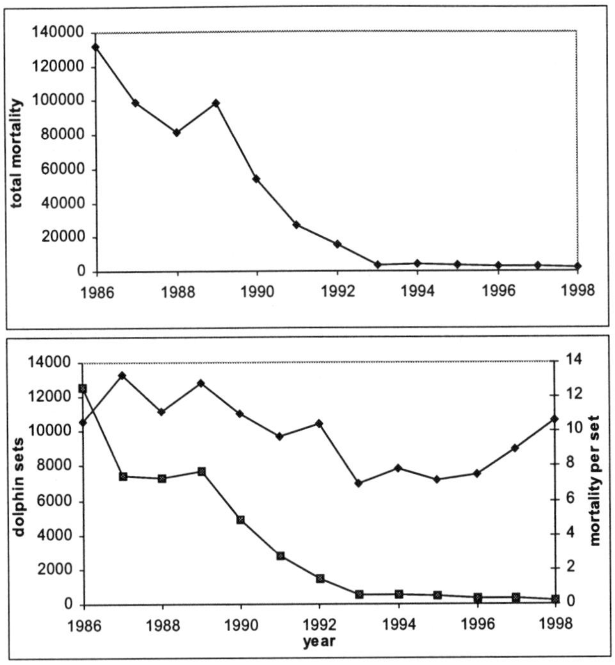


Fig. 2. Evolution of the total by-catch, and the components of the by-catch equation for the tuna–dolphin problem (1986–1998).

thousands of animals per year, but the estimates are not solid (Francis et al., 1992). Most dolphin populations declined into the early 1970s, and several stocks were declared depleted in the early 1990s because their levels were much lower than in earlier assessments (Gosliner, 1999). The association of yellowfin tuna and dolphins has also been observed in other oceans but the frequency of setting in the eastern Pacific is much greater (Hall, 1998). The National Marine Fisheries Service and, since 1979, the Inter-American Tropical Tuna Commission (IATTC) ran

Table 2

Estimates of population abundance (pooled for 1986–1990), of incidental mortality in 1998, and of relative mortality

| Stock | Population abundance | Incidental mortality | Relative mortality | |
|---|---|---|---|---|
| | | | Estimate | 95% Cl |
| Northeastern spotted | 730,900 | 298 | 0.04% | (0.031,0.061) |
| West/Southern spotted | 1,298,400 | 341 | 0.03% | (0.020,0.037) |
| Eastern spinner | 631,000 | 422 | 0.07% | (0.042,0.101) |
| Whitebelly spinner | 1,019,300 | 249 | 0.025 | (0.015,0.032) |
| Northern common | 476,300 | 261 | 0.05% | (0.030,0.118) |
| Central common | 406,100 | 172 | 0.04% | (0.022,0.083) |
| Southern common | 2,210,900 | 33 | <0.01% | (0.001,0.004) |
| Other dolphins | 2,802,300 | 101 | <0.01% | (0.002,0.004 |
| All | 9,576,000 | 1,877 | 0.02% | (0.017,0.022) |

observer programs to produce mortality estimates and other data (Fig. 2).

Mortalities of dolphins through fishing which were estimated at 133,000 in 1986 were only 1,877 in 1998 (Table 2). Recent levels of mortality are not significant from the population point of view; mortality levels for all stocks are less than 0.1%, much lower than the 2% value used as a conservative estimate of net recruitment (Hall, 1998). All dolphin stocks have estimated population sizes between 400,000 and 2,200,000, and the most recent survey shows much higher numbers for the depleted stocks. As more surveys are planned for the coming years, a more solid estimate will be available at the end of these studies. But in any case, it is quite safe to state that the declines in these populations have been halted, and that their recovery is under way.

What brought the mortality down? Public awareness and pressure were critical, and kept the process moving, while campaigns of environmental organisations were effective. When it became clear that a large sector of the public would not accept the high levels of mortality, the industry abandoned a policy of fighting restrictions. It was clear that they were fighting for their survival, which provided the motivation. Most of the change originated in technological improvements, and on the performance in releasing dolphins. Most of the solutions came from the fishers themselves; the role of the scientists was to facilitate communication, identify the causes of high dolphin mortality, promote the testing of new ideas, and validate statistically the experiments performed. Several changes in the gear (additions of fine mesh panel, rescue platforms, etc.) and in the procedures used (backdown, hand rescue, etc.) were instrumental for the solution of the problem. The training of skippers and crews run by the Tuna–Dolphin Program of the IATTC (Bratten and Hall, 1996) played a role in the quick diffusion of new ideas, and communicated to all crews the standards that were expected (Fig. 3).

Management actions covered a broad spectrum, going from global quotas for dolphin mortality, to more detailed approaches such as prohibitions of night sets, mandatory use of equipment, and gear of specified characteristics, etc. Probably the most successful approach proposed by the Tuna–Dolphin Program of the IATTC was the establishment of individual vessel dolphin mortality limits. A total limit was divided by the number of participating vessels, and each vessel received an individual limit. This was fair to individual vessels, and was convenient because it avoided conflicts over resource allocation. Those capable of staying within the limit were allowed to continue fishing on dolphins; those reaching the limit had to cease fishing on dolphins, which entailed a considerable economic impact. This set in action what can be called the "Darwinian selection of fishers". Less-skilled captains became a liability and were replaced. Good captains became valuable and were sought after. The individual limits decreased year after year, which forced the continued improvement. In the early years of the fishery, the levels of mortality made it a

Fig. 3. Crewman rescuing dolphin from the net (photo D. Bratten).

conservation problem, but by the early 1990s, the by-catch had become sustainable and dolphin populations were no longer declining. Public pressure required that the reductions continued, and they did so until reaching the current level of under 2,000 individuals taken from populations probably numbering over 10 million individuals.

Many lessons were learned from this first major battle against a by-catch problem (Hall, 1996, 1998), and the solution came from a combination of factors (Hall, 1996):

Fishers began to avoid large herds, or herds of those dolphin stocks whose behaviour made them more vulnerable. They also changed deployment conditions, avoiding sets in areas with strong currents, or at night. The major improvements came in the release process; the backdown manoeuvre was adopted by all fleet and refined. Additional changes in the net such as a fine mesh panel to reduce entanglements, and a highly motivated hand rescue effort based on rafts, swimmers, and divers, completed the process.

The solution came from a combination of actions and changes. Change was gradual, but between 1986 and 1998 mortality declined by 98%. The fishing industry proved very effective in setting standards, and the fishers themselves produced most of the solutions. The process was fairly transparent; environmental and industry NGOs were deeply involved, and had direct access to the basic information on compliance, etc.

### The Shrimp–Sea Turtle Problem

Trawls are used in most countries of the world to harvest valuable resources of shrimps and prawns. Sea turtles are one of the most critically endangered groups of species to be taken incidentally in some of these trawl fisheries. With the possible exception of the olive ridley (*Lepidochelys olivacea*) which number in the hundreds of thousands, numbers of other species are low, and some populations have declined markedly. The urgency to reduce the by-catch reflected a clear conservation issue.

Again technology played a role in mitigating the problem. Turtle-Excluder Devices (TEDs) were installed in nets to let go sea turtles that entered. TEDs consist of rigid or soft structures that direct out of the net species larger than the spacing of vertical bars that create a grid. They proved extremely efficient in releasing sea turtles, and became mandatory after some years (Magnuson et al., 1990). In addition, some closures reduce captures when high numbers of sea turtles are present, especially during peak nesting periods. In some cases captured turtles can be released alive after a "resuscitation" procedure. The major difficulty has been the spread of this technology to all nations that trawl for shrimp in areas with sea turtles present. The USA has used its very large market share to threaten embargoes on countries not using TEDs. There are other threats for the sea turtles—by-catches in other fisheries, habitat destruction, excessive harvest of individuals or eggs, etc.—but the solution to the by-catch in shrimp and prawn trawls is available.

### Discards in Shrimp and Prawn Trawls

Sea turtles are not the only problem caused by the shrimp and prawn trawls. Alverson et al. (1994) show that these trawls have the highest discard/catch ratios of all fisheries considered. Of the 16 worse offenders in this category, 14 are shrimp or prawn trawls. The ratios go from 3:1 to around 15:1 kg discarded per kg landed. These fisheries are perceived by the public as extremely wasteful, and environmentally harmful, not only by the removal of biomass and diversity, but also by the potential impact on habitat, and the impacts of the discards on pelagic and bottom communities. In some fisheries, the lack of markets results in the discard of most of the species caught with the shrimps or prawns; by-catches are high, and the moral condemnation of the fishery follows these wasteful practices. But in many others, most of the biomass captured is utilised for human consumption or animal feeds (Clucas, 1997). In developing nations, the utilisation of the capture is almost 100%; there is no by-catch. The fact that non-target species are utilised may remove the moral problem of the protein waste, but it does not remove the problem of the poor utilisation of most of those species. They are still taking individuals at sizes that are much smaller than desirable. The solution reached in many countries, that of utilisation of the by-catch, is not the most satisfactory way to utilise the resources (Table 3).

Although the number of studies on the fate of discards is very limited, they are a clear subsidy to some components of the pelagic and benthic communities (scavengers, etc.), and cause changes in structure and function of these communities (Hill and Wassenberg, 1990; Kennelly, 1995; Dayton et al., 1995; Wickens, 1995; Ramsay et al., 1997). Of particular significance in the Gulf of Mexico is the incidental mortality of juvenile red snappers (Gutherz and Pellegrin, 1988;

Table 3

By-catches in numbers of individuals per 1000 tons of tuna loaded for the different types of sets. Combined data for 1993–1998.

| | Dolphin sets ($n$=33,927) | School sets ($n$=19,210) | Log sets ($n$=21,567) |
|---|---|---|---|
| **Dolphins** | 19.47 | 0.15 | 0.40 |
| **Billfishes** | | | |
| Sailfish | 4.02 | 6.86 | 0.42 |
| Black marlin | 0.40 | 1.28 | 4.60 |
| Striped marlin | 0.53 | 1.39 | 1.12 |
| Blue marlin | 0.41 | 1.55 | 6.01 |
| Unidentified marlin | 0.22 | 0.36 | 0.83 |
| Swordfish | 0.07 | 0.19 | 0.10 |
| Shortbill spearfish | 0.06 | 0.02 | 0.08 |
| Unidentified billfish | 0.67 | 0.15 | 0.43 |
| **Large Bony Fishes** | | | |
| Mahi-Mahi | 2.13 | 182.12 | 4288.51 |
| Wahoo | 3.67 | 23.54 | 2307.38 |
| Rainbow runner | 0.05 | 36.82 | 383.77 |
| Yellowtail | 14.12 | 448.22 | 330.24 |
| Other large bony fish | 0.27 | 367.69 | 206.09 |
| **Sharks** | | | |
| Blacktip shark | 22.08 | 98.78 | 163.07 |
| Whitetip shark | 2.06 | 3.25 | 31.52 |
| Silky shark | 3.57 | 16.48 | 73.97 |
| Hammerhead shark | 0.99 | 8.41 | 6.72 |
| Other sharks | 3.02 | 9.57 | 29.78 |
| Unidentified sharks | 5.70 | 11.83 | 43.53 |
| **Rays** | | | |
| Mantaray | 3.77 | 29.84 | 0.93 |
| Stingray | 2.19 | 8.72 | 1.02 |
| **Sea Turtles** | | | |
| Olive ridley | 0.18 | 0.41 | 0.42 |
| Green-black | 0.01 | 0.08 | 0.07 |
| Loggerhead | 0.01 | 0.03 | 0.01 |
| Unidentified turtles | 0.07 | 0.14 | 0.14 |

Graham, 1996) that is the cause of friction between different groups of fishers. Even though the use of by-catch-reduction devices has helped mitigate the problem, some authors believe that they are not as effective as needed for the recovery of the red snapper stock (Gallaway and Cole, 1999).

By-catch-reduction devices (BRDs) are being developed to deal with the issue of finfish by-catches. Differences in escape responses, size and shape between fishes and the target crustaceans are being utilised for these new designs (Kennelly, 1995; Kennelly and Broadhurst, 1996; Broadhurst et al., 1997). One of the concepts used is the so-called "fish-eye", which is an opening, or group of openings, strategically placed in the trawl to allow the escape of the finfish. But rigid grids such as the Nordmøre are the most effective solution available for some fisheries (Broadhurst et al., 1997).

Another way to reduce by-catch from these trawls is to improve the handling and survival rates of the individuals captured (Ross and Hokenson, 1997). Some species have relatively high survival rates after capture, while others are fragile (Alverson et al., 1994).

The impact of trawls is clearly a priority issue. These fisheries produce roughly 2% of the world catch of fish in weight, but result in more than one third of the by-catch (Alverson et al., 1994), although there are some changes to these figures in a recent update of that work (Alverson, 1998). However, their economic and social significance is also out of proportion to their landings weight. Continued research on the technology may produce more efficient solutions, and there are experiments in progress testing several modifications of the by-catch reduction devices (i.e. Rogers et al., 1997).

### Gillnets and Cetaceans

Cetaceans of many species are killed incidentally in gillnets. These nets can be deployed at different levels in the water column according to the target, and because of their low cost and effectiveness, they are used by a large number of small inshore boats throughout the world. A major review of the mortality of cetaceans in gillnets was produced a few years ago (Perrin et al., 1994), and some of the proposed solutions are considered there.

Probably the best studied interaction of cetaceans–gillnets is in the Bay of Fundy and Gulf of Maine fisheries. Demersal gillnets, which target cod, pollock and hake, have by-catches of several dolphin species, long-finned pilot whales, humpback, minke, fin and right whales (IWC, 1994), as well as harbour porpoises. Many large whales survive entanglement, although they may carry off portions of

**Measuring the impact of by-catch on the target species: Target utilization efficiency.**

Target utilization efficiency

TUE = Yield target species/(Catch target species + By-catch target species)

How many tons of the resource are used (catch + by-catch) to bring one ton to the consumer.
This measure includes the impact of by-catch plus other losses due to transportation, storage and production losses

**Measuring the impact of by-catch on the ecosystem: Biomass transfer efficiency.**

Biomass transfer efficiency

BTE = Yield of all species/(Catch all species + By-catch all species)

How many tons of biomass (catches + by-catches **of all species**) are needed to produce one ton of target species to the consumer.

gear. A rescue group has been put together to try to help in the release of entangled individuals (Lien et al., 1994).

Entanglement is almost always fatal for small cetaceans. Harbour porpoises *Phocoena phocoena* are the most frequently killed cetaceans in these nets. In the North Sea annual mortality estimates can range to several thousand per year (Vinther, 1999). A number of studies (Northridge, 1984, 1991a, 1991b; Palka, 1995; Read and Gaskin, 1988; Lien, 1994; Jefferson and Curry, 1994) have attempted to assess the by-catch levels in different regions and fisheries. In the Gulf of California, the vaquita, *Phocoena sinus* is in danger of extinction, and the major risk factor is the incidental mortality caused by gillnets (Rojas-Bracho and Taylor, 1999).

In order to minimise the impact of these entanglements, special devices have been developed (Lien et al., 1994). These 'pingers' or sound-emitting alarms on nets have decreased by-catch considerably (Kraus et al., 1997). Some experimental studies, notably in the Bay of Fundy (Trippel et al. 1999), and in the California coast (Cameron, 1999) have shown a reduced by-catch of small cetaceans by 77% and 70% respectively for nets with pingers. The reasons why the entanglements are reduced are still poorly known (Nachtigall et al., eds. 1995). The pingers may affect the distribution of the prey items of the small cetaceans, and indirectly affect their distribution, or they may make the net "visible" to the cetaceans.

## Trawls and Cetaceans

Midwater trawls have a much greater potential to capture cetaceans than bottom trawls (Read, 1994). The nets can be towed much faster, and their targets are often fish or squid that are important prey for marine mammals. Thus dolphins and small whales may be captured while feeding on schools of these species, or they may learn to associate the presence of midwater trawls with concentrations of food in a patchy environment.

Couperus (1997) reviewed the interactions between cetaceans and trawling and also reported target fish species in the stomachs of by-caught common and bottlenose dolphins, but these were absent from stomachs of white-sided dolphins. Descriptions of cetaceans feeding associated with trawls were first reviewed by Fertl and Leatherwood (1997). At least 16 cetacean species all over the world are known to feed in association with fishing trawlers.

A recent study to quantify midwater trawl by-catch in the eastern Atlantic showed that most dolphin by-catch occurred during night or close to dawn (Morizur et al., 1999). Some factors which may be important in the cetacean–trawl interaction include the target fish species, time of day, tow duration, level of tow, size of net opening, haul back speed and gear design. A better understanding of these factors should help find the solutions to the problem.

Other studies such as Tregenza et al. (1997) in the Celtic Sea, Tregenza and Collet (1998) in the northeast Atlantic, Dans et al. (1997) and Crespo et al. (1997) in Patagonia, and Lens (1997) southeast of Newfoundland, show that the incidental takes happen in many trawl fisheries. Without observers on the vessels in most cases, a solid quantification of these impacts will not be feasible. Some technological approaches to a solution have been explored (de Haan, 1997). These include acoustic deterrents (i.e. pingers) in the trawl, and excluder panels. But the solution is not yet available, and given the widespread use of trawls, even low BPUE values may still yield important impacts.

## High Seas Driftnets

Pelagic driftnets came to the public's attention in the mid 1980s. This type of fishing had developed a few years earlier, and was based on the deployment of nets that were miles long. Although some European nations had fisheries of this type in the mid 1970s, the Pacific Ocean was the first to be extensively exploited with this technique. The technique was effective, but by-catches in some fishing grounds were high, and involved charismatic species (Northridge, 1991b; Hobbs and Jones, 1992). NGOs started campaigns to ban the "walls of death." Only a few countries participated in this fishery in the Pacific (Japan, Korea, Taiwan), but the level of effort they deployed at its peak was very high. Some coastal nations from the Pacific region, especially island nations, felt threatened because of the potential impact of this fishery on their artisanal fisheries. The Mediterranean Sea is another area where ecological impacts of driftnets were significant (Silvani et al., 1999). The issue was brought to the UN by a coalition of nations and NGOs, and a moratorium was imposed on the utilisation of nets of more than 2.5 km long. Some authors questioned the scientific basis for these actions, because the analyses performed had been incomplete (Burke et al., 1994), or simply stated that the support for them by some nations had been politically motivated rather than science-based (Bache and Evans, 1999.) In any case, with only a few nations participating in the fishery it was easy for the coalition opposing it to pass a moratorium. The impacts on the countries participating in the fishery were high; employment losses alone were estimated at more than 15,000 jobs (Huppert and Mittleman, 1993). This is one example of a fishery where the by-catches were eliminated by reducing the effort to zero, or close to zero.

## Longlines and Sea Birds

Longlines are used in many ocean areas of the world to catch a variety of species including tunas, swordfishes, sharks, toothfishes, etc. Bait and offal discarded from the boats attract seabirds, which become hooked and drown. When longlines are set on the surface, the probability of hooking persists during the set; when they are set on the bottom, hooking can occur while deploying or retrieving the lines. Two of the fisheries where these interactions play a major role in their management are the albacore fisheries off Australia (Brothers, 1991; Murray et al., 1993; Gales et al., 1998), and the North Pacific longline fishery for Pacific cod, halibut, and blackcod. There is special concern with the by-catches of several species of albatross, and in some cases declines in population numbers have been observed (Weimerskirch and Jouventin, 1987; Croxall et al., 1990; Brothers, 1991; Murray et al., 1993). A variety of solutions have been proposed, and their implementation is resulting in major reductions in by-catch levels. Some solutions are very simple: an object is towed or an additional line carrying streamers is placed to scare the birds; weighting lines or thawing the bait to increase sink rates of the line; changes in the method of setting (Løkkeborg, 1998) or in the time of deployment (i.e. at night to reduce the visibility of the baited hooks). Other solutions are more expensive: a lining pipe to take longlines to 1–2 m deep to "hide" the baited hooks from the birds (see Matsen 1997 for illustrations). Also, offal can be discharged away from the place where the line is being set. Many of these modifications also reduce the loss of bait. The developments from the Australian zone fisheries were quickly adopted in the North Pacific. The FAO has set in motion an International Plan of Action for reducing incidental catch of seabirds in longline fisheries (see www.fao.org). An excellent review, also produced by FAO, covers the problem and techniques available for mitigation (Brothers et al., 1999). Many nations have developed policies to deal with the issue (Bergin, 1997, Bache and Evans, 1999). An excellent review of the longlines–Albatrosses problem can be found in Alexander et al., (1997), and a manual produced for fishers by CCAMLR (1996) is a good example of how to communicate with fishers to promote awareness.

### Coastal Gillnets and Sea Birds

Because of the very large number of coastal gillnets deployed, this type of interaction is probably much more common than suggested by the literature. Diving birds, mostly Alcids, are entangled in gill nets while diving for prey, or when trying to steal the catch from the net. The best known example of a by-catch reduction program is that of the salmon drift gillnet fishery of Puget Sound, Washington State, USA (Melvin, 1995; Melvin and Conquest, 1996; Melvin et al., 1997, 1999). Again, options were found that would allow a substantial reduction of the by-catch without a significant impact on the catches. The solution is not simple, but is a combination of technological modifications, changes in deployment, and management of seasonal closures. The technological modifications include the addition of a visual alert (a strip of highly visible netting in the upper part of the net), and an acoustic alert (a 'pinger' (Lien, 1996). The deployment time, (i.e. eliminating dawn sets) would contribute to the reduction in by-catch. Finally, managing the opening and closing of the fishery to coincide with periods of high target catch rates and low by-catch rates provides an additional mechanism to increase selectivity. The implementation of these measures is held up by legal issues (Melvin et al., 1999), but the key to the solution is available, without significant losses in catches.

### Longlines and Sea Turtles

Unlike many other fisheries, the incidental capture of sea turtles in longlines is one with no clear solution (Balazs, 1982; Bolten et al., 1996). Besides some obvious options, such as closing areas where high densities of sea turtles aggregate near nesting beaches, it is difficult to find technological answers because of the heterogeneity of the problem. Sea turtles are taken in longlines when trying to take the bait (especially piscivorous sea turtles, i.e. loggerheads), or they get snagged by a hook, or become entangled with the fishing line (Witzell, 1984). Given that sea turtles can survive for long periods of time underwater, there has been considerable effort in establishing the conditions that will reduce the mortality rate of those captured. As the turtles dive quite deeply, changes in the depth of gear deployment may not be very effective.

After the sea turtles have taken a hook, it is still possible to release them, so a good deal of research has gone into the mortality caused by the hooks (Balazs and Pooley, 1994). If the hook has been ingested, mortality may follow, but few cases are fully documented (Skillman and Balasz, 1992), and there are no statistics showing what proportion recovers fully. Further research may produce hooks or baits that reduce the incidental captures. The effect of injuries and evidence that a release will result in survival, still needs to be examined.

### Northeast Pacific Groundfisheries

By-catch became an issue in the Bering Sea during the late 1950s following the development of the Japanese high seas salmon fisheries. The issue was first brought to the attention of fishery managers when it was found that birds and marine mammals were taken in the high seas gillnet fishery. However, fisheries managers of the region (Alaska) became greatly concerned following the development of the extensive trawl and line fisheries by the Japanese and Soviets during the 1960s and later, by Korean, Polish and other foreign nationals. Fishery scientists pointed out that there were extensive by-catches of halibut and crab taken in these fisheries as well as salmon. These concerns led to the development of a foreign observer program which documented by-catch of "preheated species", species being fully utilized by US and Canadian fishermen, of great economic and social interest to the coastal fishermen of the regions.

Following unilateral extension of fisheries jurisdiction by the US and Canada during the late 1970s and the development of US domestic fisheries for bottom fish off Alaska, particularly during the 1980s, the National Marine Fisheries

Table 4

Measuring the impact of by-catch on the ecosystem: BPUE is the by-catch caused per-unit of effort. Example is by-catch per set for different set types, Eastern Pacific tuna fishery 1993–1998.

| | Dolphin sets ($n$=33,927) | School sets ($n$=19,210) | Log sets ($n$=21,567) |
|---|---|---|---|
| **Dolphins** | 0.371 | 0.002 | 0.001 |
| **Billfishes** | | | |
| Sailfish | 0.065 | 0.093 | 0.012 |
| Black marlin | 0.007 | 0.017 | 1.129 |
| Striped marlin | 0.009 | 0.019 | 0.031 |
| Blue marlin | 0.007 | 0.021 | 0.168 |
| Unidentified marlin | 0.004 | 0.005 | 0.023 |
| Swordfish | 0.001 | 0.003 | 0.003 |
| Shortbill spearfish | 0.001 | 0.000 | 0.002 |
| Unidentified billfish | 0.011 | 0.002 | 0.012 |
| **Large Bony Fishes** | | | |
| Mahi-Mahi | 0.035 | 2.472 | 120.202 |
| Wahoo | 0.060 | 0.319 | 64.673 |
| Rainbow runner | 0.001 | 0.500 | 10.757 |
| Yellowtail | 0.230 | 6.084 | 9.256 |
| Other large bony fish | 0.004 | 4.991 | 5.777 |
| **Sharks** | | | |
| Blacktip shark | 0.359 | 1.341 | 4.571 |
| Whitetip shark | 0.033 | 0.044 | 0.884 |
| Silky shark | 0.058 | 0.224 | 2.073 |
| Hammerhead shark | 0.016 | 0.114 | 0.188 |
| Other sharks | 0.049 | 0.130 | 0.835 |
| Unidentified sharks | 0.093 | 0.161 | 1.220 |
| **Rays** | | | |
| Mantaray | 0.061 | 0.405 | 0.026 |
| Stingray | 0.036 | 0.118 | 0.029 |
| **Sea Turtles** | | | |
| Olive ridley | 0.003 | 0.006 | 0.012 |
| Green-black | 0.000 | 0.001 | 0.002 |
| Unidentified turtles | 0.001 | 0.002 | 0.004 |

Service (NMFS) in response to management actions taken by the North Pacific Fisheries Management Council (NPFMC) implemented a 100% observer program on vessels over 125 ft and 30% for vessels between 65 and 125 ft. This program led to one of the most comprehensive documentations of by-catch rates in the world.

Regulation of by-catch of the fisheries of the region was ultimately based on control of the level of prohibited species taken (quota and area management), closed areas for trawling and other fishing gears, required landing of by-catch taken in certain fisheries having bottom fish quotas, and seasonal quotas and closed periods.

The overall by-catch and discard rate in the Alaska groundfishery is not exceptional compared with other fisheries of the world (about 16%) and is much lower in the large pollock fishery of the region (Alverson et al., 1994). However, the 2–3 million tonne fishery is so immense that the total by-catch volume, and forgone opportunities they may represent, have driven efforts to further reduce by-catch levels. For example the 1995 mid-water pollock fishery landed a total of 1.1 million tonnes of which 46,000 tonnes was by-catch of undersized pollock—a small percentage, but still a high quantity of discards. However, some of the region's trawl fisheries have relatively high discard rates, e.g. rock sole and flat head sole (NMFS, 1998), and further regulation is expected.

## CONCLUSIONS: INTO THE 21st CENTURY

This list of case studies is very incomplete. Sections on shark by-catches in longlines, pinnipeds in trawls, and many others could have been included. Presenting these case studies in a simplistic form (one by-catch group of species × one type of gear) has the merit of focusing attention on a specific problem, but readers should keep in mind that by-catch problems cannot be isolated or generalised. The same longlines that affect sea turtles may be taking sharks and billfishes. Shrimp and prawn trawls have high by-catch/catch ratios in some fisheries, but they are considerably lower in others. Because of the nature of the problem, solutions have to include a complete analysis of the consequences of the current practices, and of proposed changes. They also have to be specific to the circumstances of each fishery. Switching gear or modes of fishing to avoid one problem may create or exacerbate others (Hall, 1998). Closing an area or season to reduce by-catch of some species may concentrate the effort in other areas or periods with other problems (Table 4).

The case studies synthesised above do not cover all by-catch issues, but they provide a broad sketch of the direction in which programs are moving. In most of them, attempts are being made to keep a viable fishery. The fishers usually play a major role in developing and testing the modifications proposed. Most solutions involve relatively minor changes in gear and procedures, and usually a combination of technology and management achieves the desired improvements. But not all the problems listed above yet have a satisfactory solution, and it is necessary to think of the fisheries as dynamic systems, where evolution is taking place, and changes should be expected.

Understanding the behavioural and ecological differences between target and non-target species, and their responses to fishing gear pays off, showing any opportunities available. But given the multitude of problems we are facing, the research needs are massive, and only with vigorous research programs and with the active involvement of the fishers will progress be made. But many of the battles are being won. Whenever the challenge is faced, the solutions appear. The first step towards the solution is the commitment to finding one.

The fishers of the world are evolving, and must continue doing so. The process that has started is a process of natural selection. Only the fishers that can adapt to the new conditions will "survive"; those that can produce catches with the lowest ecological costs, with the least waste, with the least impact on the habitat, will inherit the fisheries. Adaptation happens through education and technological change; the ways of fishing that have become maladaptive must be replaced by others that are sustainable.

Technology played a role in the generation of the problems, but it is also a major contributor to the solutions (Prado, 1997; Alverson, 1998). An area of especial interest is the development of instruments that can help fishers make better-informed choices before setting gear. If fishers know in advance the species and size composition of the potential targets in an area, they could make better decisions concerning whether to make the set or not, modify the deployment conditions, etc. In this way, sets with high by-catches, or with high numbers of endangered species, would be reduced or avoided. This will also allow a more sophisticated management system. Acoustic, optical, laser, or other devices could provide this information.

Scientists need to quantify the impacts of by-catches on the target species and others, and to incorporate them into management schemes. But even more important is to understand the effects of the discard process on the ecosystem (Kennelly, 1995; Hall, 1999). The tonnage discarded is so large that we cannot continue ignoring its impact on the marine communities and ecosystems involved (Garthe et al., 1996). Research on this subject, hampered in many cases by technological difficulties and high costs, has been incredibly limited.

Finally, scientists and managers need to maintain a close cooperation with fishers, to develop practical solutions and regulations that allow the process of change to continue, without unnecessarily strict regulations. The natural selection process develops because there is variability, and that variability needs to be preserved.

## REFERENCES


Alexander, K., Robertson, G. and Gales, R. (1997) The incidental mortality of albatrosses in longline fisheries. Rept. Workshop First International Conference on the Biology and Conservation of Albatrosses. Australian Antarctic Division, Hobart, Australia, September 1995. 44 pp.

Alverson, D.L., Freeberg, M.H., Murawski, S.A. and Pope, J.G. (1994) A global assessment of fisheries bycatch and discards. FAO Fisheries Tech. Pap. 339, 235 pp.

Alverson, D.L. and Hughes, S.E. (1996) Bycatch: from emotion to effective natural resource management. In Alaska Sea Grant College Program (Eds.) "Solving bycatch: considerations for today and tomorrow." Alaska Sea Grant College Program Report Nr. 96-03, Univ. of Alaska, Fairbanks, pp. 13–28.

Alverson, D.L. (1998) Discarding practices and unobserved fishing mortality in marine fisheries: an update. Washington Sea Grant 98-06.

Andrew, N.L. and Pepperell, J.G. (1992) The by-catch of shrimp trawl fisheries. In *Oceanographic and Marine Biology Annual Review, Vol 30*, eds. M. Barnes, A.D. Ansell and R.N. Gibson. UCL Press, UK, pp. 527–565.

Anon. (2000). Annual Report for 1998. Inter-American Tropical Tuna Commission, La Jolla, CA.

Bache, S.J. and Evans, N. (1999) Dolphin, albatross and commercial fishing:Australia's response to an unpalatable mix. *Marine Policy* **23** (3), 259–270.

Balasz, G.H. (1982) Annotated bibliography of sea turtles taken by longline fishing gear. NOAA, NMFS, Southwest Fisheries Center, Honolulu Lab. Unpub. Rept., 4 pp.

Balazs, G.H. and Pooley, S.G. (1994) Research plan to assess marine turtle hooking mortality: Results of an expert workshop held in Honolulu, Hawaii, November 16–18, 1993. NOAA Tech. Memo. NOAA-TM-NMFS-SWFSC-201, 166 pp.

Beltestad, A. and Misund, O. 1996. Size selection in purse seines. In Alaska Sea Grant College Program (Eds.) Solving Bycatch: Considerations for Today and Tomorrow. Alaska Sea Grant College Program Report Nr. 96-03, Univ. of Alaska, Fairbanks, 227–233.

Bergin, A. (1997) Albatross and longlining—managing seabird bycatch. *Marine Policy* **21** (1), 63–72.

Bolten, A.B., Weatherall, J.A., Balazs, G.H. and Pooley, S.G. (comps.) (1996) Status of marine turtles in the Pacific Ocean relevant to incidental take in the Hawaii-based pelagic longline fishery. NOAA Tech. Memo. NOAA-TM-NMFS-SWFSC-230, 167 pp.

Bonfil, R. (1994) Overview of world elasmobranch fisheries. FAO Fish. Tech. Pap. 341, 119 pp.

Bratten, D. and Hall, M.A. (1996) Working with fishers to reduce bycatch: the tuna–dolphin problem in the eastern Pacific Ocean. In Alaska Sea Grant College Program (Eds.) Fisheries Bycatch: Consequences and Management. Proc. of the meeting held at Dearborn, Michigan, August 27–28, 1996. Alaska Sea Grant Rpt. 97-02, 97–100.

Broadhurst, M.K., Kennelly, S.J., Watson, J.W. and Workman, I.K. (1997) Evaluations of the Nordmøre grid and secondary bycatch reducing devices (BRD's) in the Hunter River prawn-trawl fishery, Australia. *Fisheries Bulletin* **95** (2), 209–218.

Brothers, N. (1991) Albatross mortality and associated bait loss in the Japanese longline fishery in the Southern Ocean. *Biological Conservation* **55**, 255–268.

Brothers, N.P., Cooper, J. and Løkkegorg, S. (1999) The incidental catch of seabirds by longline fisheries: worldwide review and technical guidelines for mitigation. FAO Fish. Circ. No. 937, 100 pp.

Burke, W., Freeberg, M. and Miles, E. (1994) United Nations resolution on driftnet fishing: an unsustainable precedent for high seas and coastal fisheries management. *Ocean Development and International Law*, **25**, 127–186.

Cameron, G. (1999) Report on the effect of acoustic warning devices (pingers) on cetacean and pinniped bycatch in the California drift gillnet fishery. NMFS, SWFSC, Admin. Rpt. LJ-99-08C, 71 pp.

CCAMLR (1996) Fish the sea not the sky: how to avoid by-catch of seabirds when fishing with bottom longlines. Commission for the Conservation of Atlantic Marine Living Resources, 23 Old Wharf, Hobart, Tasmania, 7000 Australia. 46 pp.

Chopin, F., Inoue, Y., Matsushita, Y. and Arimoto, T. (1996) Sources of accounted and unaccounted fishing mortality. In Alaska Sea Grant College Program (Eds.) Solving Bycatch: Considerations for Today and Tomorrow. Alaska Sea Grant College Program Report Nr. 96-03, Univ. of Alaska, Fairbanks, 41–47.

Clark, W.G. and Hare, S.R. (1998) Accounting for bycatch in management of the Pacific halibut fishery. *North American Journal of Fisheries Management* **18**(4), 809–821.

Clucas, I.J. (1997) The utilization of bycatch/discards. In I.J. Clucas and D.G. James (eds.) Papers presented at the Technical consultation on reduction of wastage in fisheries, Tokyo, Japan, October 28–November 1, 1996. FAO Fisheries Rpt. 547. Suppl., pp. 25–44; 338 pp.

Coe, J.M., Holts, D.B. and Butler, R.W. (1984) The "tuna–porpoise" problem: NMFS dolphin mortality reduction research, 1970–1981. *Marine Fisheries Review* **46**(3), 18–33.

Couperus, A. (1997) Interactions between Dutch midwater trawls and Atlantic white-sided dolphins (*Lagenorhynchus acutus*) southwest of Ireland. *Journal of Northwest Atlantic Fisheries Science* **22**, 209–218.

Crespo, E.A., Pedraza, S.N., Dans, S.L. Koen Alonso, M., Reyes, L.M. Garcia, N.A., Coscarella, M and Schiavini, A.C.M. (1997) Direct and indirect effects of the high seas fisheries on the marine mammal populations in the northern and central Patagonian coast. *Journal of Northwest Atlantic Fisheries Science* **22**, 189–207.

Croxall, J.P., Rothery, P., Pickering, S.P.C., and Prince, P.A. (1990) Reproductive performance, recruitment, and survival of Wandering Albatrosses, *Diomedea exulans*, at Bird Island, South Georgia. *Journal of Animal Ecology* **59**, 773–794.

Dans, S.L., Crespo, E.A., Garcia, N.A., Reyes, L.M., Pedraza, S.N. and Koen Alonso, M. (1997) Incidental mortality of Patagonian dusky dolphins in mid-water trawling: retrospective effects from the early 1980s. Reports of the International Whaling Commission **47**, 699–704.

Dayton, P.K., Thrush, S.F., Agardy, M.T. and Hofman, R.J. (1995) Environmental effects of marine fishing. *Aquatic Conservation: Marine and Freshwater Ecosystems* **5**, 205–232.

de Haan, D. (1997) Prevention of by-catch cetaceans in pelagic trawls by technical means. AIR-III Project CT94-2423. CETASEL, Consolidated Interim Rpt. No. 2, RIVO-DLO

Duthie, A. (1997) Canadian efforts in responsible fishing operations: the impact in the Canadian northern shrimp fishery, and the success of bycatch reduction devices in Canadian fisheries. In: Technical consultation on reduction of wastage in fisheries. FAO, Fisheries Rpt. No. 547 (Suppl.), pp. 99–116. Tokyo, Japan.

Everett, G.V. (1997) Actions to reduce wastage through fisheries management. In: Technical consultation on reduction of wastage in fisheries. FAO, Fisheries Rpt. No. 547 (Suppl.), pp. 45–58. Tokyo, Japan.

FAO (1995) Code of Conduct for Responsible Fisheries. Food and Agriculture Organization of the United Nations, Rome, 41 pp.

Fertl, D. and S. Leatherwood. (1997) Cetacean Interactions with trawls: a preliminary review. *Journal of Northwest Atlantic Fisheries Science* **22**, 219–248.

Francis, R.C., Awbrey, F.T., Goudey, C.L., Hall, M.A., King, D.M., Medina, H., Norris, K.S., Orbach, M.K., Payne, R. and Pikitch, E. (1992) *Dolphins and the Tuna Industry*. National Research Council, National Academy Press, Washington D.C.

Furness, R.W., Hudson, A.V. and Ensor, K. (1988) Interactions between scavenging seabirds and commercial fisheries around the British Isles. In *Seabirds and Other Marine Vertebrates: Competition, Predation and Other Interactions*, ed. J. Burger, pp. 232–260. Columbia Univ. Press, New York.

Gales, R., Brothers, N. and Reid, T. (1998) Seabird mortality in the Japanese tuna longline fishery around Australia, 1988–1995. *Biological Conservation* **86**, 37–56.

Gallaway, B.J. and Cole, J.G. (1999) Reduction of juvenile red snapper bycatch in the U.S. Gulf of Mexico shrimp trawl fishery. *North American Journal of Fisheries Management* **19** (2), 342–355.

Garthe, S., Camphuysen, C.J. and Furness, R.W. (1996) Amounts discarded by commercial fisheries and their significance as food for seabirds in the North Sea. *Marine Ecology Progress Series* **136**, 1–11.

Gosliner, M.L. (1999) The tuna–dolphin controversy. In *Conservation and Management of Marine Mammals*, eds. J.R. Twiss and R.R. Reeves. Smithsonian Inst. Press, Washington and London, pp. 120–155.

Graham, G.L. (1996) Finfish bycatch from the southeastern shrimp fishery. In Alaska Sea Grant College Program (Eds.) Solving Bycatch: Considerations for Today and Tomorrow. Alaska Sea Grant College Program Report Nr. 96-03, Univ. of Alaska, Fairbanks, pp. 115–119.

Gutherz, E.J. and Pellegrin, G.J. (1988) Estimate of the catch of red snapper, *Lutjanus campechanus*, by shrimp trawlers in the U.S. Gulf of Mexico. *Fisheries Review* **50** (1), 17–25.

Hall, M.A. (1996) On bycatches. *Review of Fish Biology and Fisheries*, **6**, 319–352.

Hall, M.A. (1998) An ecological view of the tuna–dolphin problem: impacts and trade-offs. *Review of Fish Biology and Fisheries*, **8**, 1–34.

Hall, S.J. (1999) *The Effects of Fishing on Marine Ecosystems and Communities*. Blackwell Science, 274 pp.

Hill, B.J. and Wassenberg, T.J. (1990) Fate of discards from prawn trawlers in Torres Strait. *Australian Journal of Marine and Freshwater Research* **41**, 53–64.

Hobbs, R.C. and Jones, L.L. (1992) Impacts of high seas driftnet fisheries on marine mammal populations in the North Pacific. In *Symposium on Biology, Distribution and Stock Assessment of Species Caught in the High Seas Driftnet Fisheries in the North Pacific Ocean*, eds. J. Ito, W. Shaw and R.L. Burgner. International North Pacific Fisheries Commission Bulletin, 53 (III), pp. 409–434.

Huppert, D.D. and Mittleman, T.W. (1993) Economic effects of the United Nations moratorium on high seas driftnet fishing. NOAA Tech. Memo. NOAA-TM-NMFS-SWFSC-194, 59 pp.

ICES (1995) Report of the study group on ecosystem effects of fishing activities. ICES Cooperative Res. Rpt. Nr. 200, 120 pp.

IWC (International Whaling Commission) (1994) Report of the Workshop on Mortality of Cetaceans in passive fishing nets and traps. *Reports of the International Whaling Commission* (Special Issue) **15**, 1–71.

Isaksen, B. 1997. The Norwegian approach to reduce bycatch and avoid discards. In I. Clucas and D. James (eds.) Technical Consultation of Reduction of Wastage in Fisheries. FAO, Fish. Rpt. Nr. 547, Supplement, FAO, Rome, pp. 89–93.

Jefferson, T. and Curry, B. (1994) A global review of porpoise (Cetacea:Phocoenidae) mortality in gillnets. *Biological Conservation* **67**, 167–183.

Kennelly, S.J. (1995) The issue of bycatch in Australia's demersal trawl fisheries. *Review of Fish Biology and Fisheries* **5**, 213–234.

Kennelly, S.J. and Broadhurst, M.K. (1996) Fishermen and scientists solving bycatch problems: examples from Australia and possibilities for the northeastern United States. In Alaska Sea Grant College Program (Eds.) Solving Bycatch: Considerations for Today and Tomorrow. Alaska Sea Grant College Program Report Nr. 96-03, Univ. of Alaska, Fairbanks, pp. 121–128.

Kraus, S., Read, A., Solow, A., Balwin, K., Spradlin, T., Anderson, E. and Williamson, J. (1997) Acoustic alarms reduce porpoise mortality. *Nature* **388** (7 Aug.), 525.

Laptikhovsky, V. and Fetisov A. (1999) Scavenging by fish of discards from the Patagonian squid fishery. *Fisheries Research* **41**, 93–97.

Larsen, R.B. and Isaksen, B. (1993) Size selectivity of rigid sorting grids in bottom trawls for Atlantic cod (*Gadus morhua*) and haddock (*Melanogrammus aeglefinus*). *ICES Marine Science Symp.* **196**, 178–182.

Lens, S. (1997) Interactions between marine mammals and deep water trawlers in the NAFO regulatory area. International Council for the Exploration of the Sea, CM. 1997/Q:08, 10 pp.

Lien, J. (1994) Entrapments of large cetaceans in passive inshore fishing gear in Newfoundland and Labrador (1979–1990). In Perrin, W.F., Donovan, G. and Barlow, J. (eds.), Gillnets and Cetaceans. *Reports of the International Whaling Commission, Special Issue* **15**, 149–158.

Lien, J., Stenson, G.B., Carver, S. and Chardine, J. (1994) How many did you catch? The effects of methodology on bycatch reports obtained from fishermen. In Perrin, W.F., Donovan, G.P. and Barlow, J. (eds.) Gillnets and Cetaceans. *Reports of the International Whaling Commission*, Special Issue **15**, 535–545.

Lien, J. 1996. Conservation aspects of fishing gear: cetaceans and gillnets. In Alaska Sea Grant College Program (Eds.) Solving Bycatch: Considerations for Today and Tomorrow. Alaska Sea

Grant College Program Report Nr. 96-03, Univ. of Alaska, Fairbanks, pp. 219–224.

Løkkeborg, S. (1998) Seabird by-catch and bait loss in long-lining using different setting methods. *ICES, Journal of Marine Science* **55**, 145–149.

Magnuson, J.J., Bjorndal, K.A., DuPaul, W.D., Graham, G.L., Owens, D.W., Peterson, C.H., Pritchard, P.C.H., Richardson, J.I., Saul, G.E. and West, C.W. (1990) *Decline of the Sea Turtles: Causes and Prevention.* National Research Council, National Academy of Sciences, Washington D.C., 259 pp.

Matsen, B. (1997) For the birds. *National Fisherman*, Jan., 20–23.

McCaughran, D.A. (1992) Standardized nomenclature and methods of defining bycatch levels and implications. In *Proceedings of the National Industry Bycatch Workshop, Feb. 4–6, Newport, Oregon*, eds. R.W. Schoning, R.W. Jacobson, D.L. Alverson, T.H. Gentle and J. Auyong. Natural Resources Consultants, Inc., Seattle, Washington, USA.

Melvin, E.F. (1995) Reduction of seabird entanglements in Puget Sound drift gillnet fisheries through gear modification. WSG-95-01. Washington Sea Grant Program, Seattle.

Melvin, E.F. and Conquest, L.L. (1996) Reduction of seabird bycatch in salmon driftnet fisheries: 1995 sockeye/pink salmon fishery final report. Project A/FP-2(a). WSG-AS 96-01. Washington Sea Grant Program, Seattle.

Melvin, E.F., Conquest, L.L. and Parrish, J.K. (1997) Seabird bycatch reduction: new tools for Puget Sound drift gillnet salmon fisheries: 1996 sockeye and 1995 chum non-treaty salmon test fisheries final report. Project A/FP-7, WSG-AS 97-01. Washington Sea Grant Program.

Melvin, E.F., Parrish, J.K. and Conquest, L.L. (1999) Novel tools to reduce seabird bycatch in coastal gillnet fisheries. *Conservation Biology* **13** (6), 1–12.

Misund, O.A. and Beltestad, A.K. (1994) Size selection of mackerel and saithe in purse seines. ICES C.M. 1994/B:28, Ref. G, H. 12 pp.

Morizur, Y., Berrow, S., Tregenza, N., Couperus, A. and Pouvreau, A. (1999) Incidental catches of marine mammals in the pelagic trawl fisheries of the northeast Atlantic. *Fisheries Research* **41**, 297–307.

Murray, T.E., Bartle, J.A., Kalish, S.R. and Taylor, P.R. (1993) Incidental capture of seabirds by Japanese Southern bluefin tuna longline vessels in New Zealand waters, 1988–1992. *Bird Conservation International* **3**, 181–210.

Nachtigall, P.E., Lien, J., Au, W.W.L. and Read, A.J. (eds.) (1995) *Harbor Porpoises, Laboratory Studies to Reduce Bycatch.* De Spil Publishers, Woerden, The Netherlands.

NMFS (1998) Managing the nation's bycatch. U.S. Dept. of Commerce, NOAA, Washington DC. 174 pp.

Northridge, S.P. (1984) World review of interactions between marine mammals and fisheries. FAO, Fish. Tech. Pap. 251, 190 pp.

Northridge, S.P. (1991a) An updated world review of interactions between marine mammals and fisheries. FAO, Fish. Tech. Pap. 251, Suppl. 1. I–VI, 58 pp.

Northridge, S.P. (1991b) Driftnet fisheries and their impact on non-target species: a world-wide review. FAO Fisheries Tech. Pap. 320, 115 pp.

Olsen, V.J. (1995) Ways in which Norway is solving the bycatch problem. In Alaska Sea Grant College Program (eds.) Solving Bycatch: Considerations for Today and Tomorrow. Alaska Sea Grant College Program Report Nr. 96-03, Univ. of Alaska, Fairbanks, 289–291.

Palka, D.L. (1995) Abundance estimates of Gulf of Maine harbor porpoise. *Reports of the International Whaling Commission* Special Issue **16**, 27–50.

Perrin, W.F. (1968) The porpoise and the tuna. *Sea Frontiers* **14** (3), 166–174.

Perrin, W.F., Donovan, G.P. and Barlow, J. (eds.) 1994. Gillnets and cetaceans. *Reports of the International Whaling Commission* Special Issue **15**, 629 pp.

Pikitch, E.K. (1992) Potential for gear solutions to bycatch problems. In *Proc. of the National Industry Bycatch Workshop*, eds. R.W. Schoning, R.W. Jacobsen, D.L. Gentle and J. Auyong. Natural Resources Consultants, Seattle, Washington, USA, pp. 128–138.

Prado, J. (1997) Technical measures for bycatch reduction. In Clucas, I.J, and James, D.G. (eds.) Papers presented at the Technical consultation on reduction of wastage in fisheries, Tokyo, Japan, October 28–November 1, 1996. FAO Fisheries Rpt. 547. Suppl., pp. 25–44; 338 pp.

Ramsay, K., Kaiser, M.J. and Moore, P.G. (1997) Consumption of fisheries discards by benthic scavengers: utilization of energy subsidies in different marine habitats. *Journal of Animal Ecology* **66** (6), 884–896.

Read, A.J. and Gaskin, D. (1988) Incidental catch of harbor porpoises by gill nets. *Journal of Wildlife Management* **52**, 517–523.

Read, A.J., Krause, S., Bisack, K. and Palka, D. (1993) Harbour porpoises and gill nets in the Gulf of Maine. *Conservation Biology* **7** (1), 189–193.

Read, A. (1994) Interactions between cetaceans and gillnet and trap fisheries in the Northwest Atlantic. In Perrin, W.F., Donovan, G. and Barlow, J. (eds.), Gillnets and Cetaceans. *Reports of the International Whaling Commission* Special Issue **15**, 133–148.

Rogers, D.R., Rogers, B.D., de Silva, J.A. and Wright, V.L. (1997) Effectiveness of four industry-developed bycatch reduction devices in Louisiana's inshore waters. *Fisheries Bulletin* **95** (3), 552–565.

Rojas-Bracho, L. and Taylor, B.L. (1999) Risk factors affecting the vaquita (*Phocoena sinus*). *Marine Mammal Science* **15** (4), 974–989.

Ross, M.R. and Hokenson, S.R. (1997) Short-term mortality of discarded finfish bycatch in the Gulf of Maine fishery for northern shrimp *Pandalus borealis. North American Journal of Fisheries Management* **17**(4), 902–909.

Saila, S. (1983) Importance and assessment of discards in commercial fisheries. FAO Circular 765, FAO, Rome, 62 pp.

Scott, M.D. (1996) The tuna–dolphin controversy. *Whalewatcher* **30** (1), 16–20.

Silvani, L., Gazo, M. and Aguilar, A. (1999) Spanish driftnet fishing and incidental catches in the western Mediterranean. *Biological Conservation* **90**, 79–85.

Skillman, R.A. and Balasz, G.H. (1992) Leatherback turtle captured by ingestion of squid bait on swordfish longline. *Fisheries Bulletin* **90**, 807–808.

Stevens, J.D. (1992) Blue and mako shark by-catch in the Japanese longline fishery off southeastern Australia. *Australian Journal of Marine and Freshwater Research* **43** (1), 227–236.

Stone, M. and Bublitz, C.G. (1996) Cod trawl separator panel: potential for reducing halibut bycatch. In Alaska Sea Grant College Program (Eds.) Solving Bycatch: Considerations for Today and Tomorrow. Alaska Sea Grant College Program Report Nr. 96-03, Univ. of Alaska, Fairbanks, 71–78.

Tregenza, N.J.C., Berrow, S.D., Leaper, R. and Hammond, P.S. (1997) Harbour porpoise *Phocoena phocoena* bycatch in set gillnets in the Celtic Sea. *ICES, Journal of Marine Science* **54**, 896–904.

Tregenza, N.J.C. and Collet, A. (1998) Common dolphin *Delphinus delphis* bycatch in pelagic trawl and other fisheries in the northeast Atlantic. *Reports of the International Whaling Commission* **48**, 453–459.

Trippel, E., Strong, M.B., Terhune, J.M. and Conway, J. (1999) Mitigation of harbour porpoise (*Phocoena phocoena*) bycatch in the gillnet fishery in the lower Bay of Fundy. *Canadian Journal of Aquatic Science* **56**, 113–123.

Vinther, M. 1999. Bycatches of harbour porpoises (*Phocoena phocoena* L.) in Danish set-net fisheries. *J. Cetacean Research Management* **1**(2), 123–135.

Wade, P.R. and Gerrodette, T. (1993) Estimates of cetacean abundance and distribution in the eastern tropical Pacific. *Reports of the International Whaling Commission* **43**, 477–493.

Weimerskirch, H. and Jouventin, P. (1987) Population dynamics of the wandering Albatross, *Diomedea exulans*, of the Crozet Islands: causes and consequences of the population decline. *Oikos* **49**, 315–322.

Wickens, P.A. (1995) A review of operational interactions between pinnipeds and fisheries. FAO Fish. Tech. Pap. 346, 85 pp.

Williams, R. and Capdeville, D. (1996) Seabird interactions with trawl and longline fisheries for *Dissostichus eleginoides* and *Champsocephalus gunnari*. *CCAMLR Science* **3**, 93–99.

Witzell, W.N. (1984) The incidental capture of sea turtles in the Atlantic U.S. Fishery Conservation Zone by the Japanese tuna longline fleet, 1978–1981. *Marine Fisheries Review* **46**(3), 56–58.

Witzig, J.F. (1997) Development of a plan for managing bycatch in U.S. fisheries. In: Technical consultation on reduction of wastage in fisheries. FAO, Fisheries Rpt. No. 547 (Suppl.), pp. 117–135. Tokyo, Japan.

## THE AUTHORS

**Martin A. Hall**
*Inter-American Tropical Tuna Commission,*
*8604 La Jolla Shores Dr.,*
*La Jolla, CA 92037, U.S.A.*

**Dayton L. Alverson**
*Natural Resources Consultants,*
*1900 West Nickerson St., Suite 207,*
*Seattle, WA 98119, U.S.A.*

**Kaija I. Metuzals**
*Biological Sciences, University of Warwick,*
*Coventry, U.K.*

Chapter 117

# FISHERIES MANAGEMENT AS A SOCIAL PROBLEM

Douglas C. Wilson

This chapter looks at fisheries management from the perspective of social science. It reviews a number of issues around the question of how we can create fisheries management institutions that are socially as well as biologically sustainable. The first section looks at factors which influence the legitimacy, or acceptability, of management measures. This section begins by examining the economic incentive structures that measures create, their basis in accurate science, and their fairness. Next the current state of the treaties and negotiations that make up international fisheries law is reviewed. Finally, the issue of who participates in creating management measures and how they do so is touched upon briefly. The next two sections review first the surveillance and monitoring of fishing activities, and then issues in international law related to effective enforcement. The chapter concludes with a discussion of the social implications of particular types of management measures, with particular attention to the idea of individual transferable quotas.

*Seas at The Millennium: An Environmental Evaluation (Edited by C. Sheppard)*

## INTRODUCTION

It is doubtful that anyone reading this book would dispute the need for effective fisheries management in nearly every corner of the global seas (see Goñi, Chapter 115 in this volume). This chapter does not approach fisheries management as a biological and technical problem—that task should be done by fisheries scientists—rather it looks at fisheries management as a social and political problem. It addresses two complex themes in what can only be, in a single chapter, a brief, schematic outline: what it takes to modify fishing behaviour so that fishing becomes sustainable, and where we are in the effort to make such a change happen.

Fisheries management is not a new problem. Bogue (1993) found the same kinds of arguments in her study of international disputes between Canada and the United States in 1892 as we find in the myriad failed attempts to create effective international management regimes today. However, the size of the global fishing fleet, its ability to find fish, and its capacity to harvest fish is on a scale far beyond what we have faced before. The world's fishing capacity has doubled since 1970 (NRDC, 1997). Can human institutions adapt in the face of a challenge of this magnitude? Thus far the record has not been good, but there is reason for hope. After the failure at international cooperation between Canada and the US in the 1890s it took two more failed attempts and an invasion of sea lamprey before the Great Lakes Fisheries Commission was formed in 1955. In 1999 this same commission received a conservation award from the American Fisheries Society. We should not fool ourselves that we know how to solve this problem, but the history of fisheries management has been a history of much more than failure.

The chapter begins with a short discussion of how social scientists have approached fisheries management. The bulk of the chapter reviews the most important current debates and developments in fisheries management institutions in terms of three factors—legitimacy, surveillance and enforcement—that determine the effectiveness of any social institution. The section on legitimacy examines several questions that are almost always raised about the acceptability of fisheries management measures: rational incentive structures; scientific realism; equity and fairness; the basis of measures in law; and who should participate in management decisions. The section on surveillance considers the current state of fisheries monitoring and record keeping. The section on enforcement focuses on international efforts to improve the enforcement of fisheries law. The last section discusses social issues that arise around particular types of fisheries management measures. The reader should be cautioned that the treatment of the topic's substantive broadness and geographical scope is necessarily distorted, especially in the choice of illustrations, by limits on this author's experience with marine fisheries beyond North America.

## FISHERIES MANAGEMENT FROM A SOCIAL SCIENCE PERSPECTIVE

Social scientists have taken three basic approaches to the problem of fisheries management. The best-known begins with the concept of the "tragedy of the commons". Fisheries are common pool resources and it is difficult to establish property rights to the fish while they remain *in situ*. As a result, in situations where access to the fishery is open to a sufficiently large group of potential fishers, fisheries are exploited to a point where the wealth produced is dissipated. This is because fishers receive the entire benefit of their own exploitation while the long-term costs of overfishing are shared by all users (Gordon, 1954). The response is to create rational institutions which prevent fishers from benefitting while "externalizing" the costs of their behaviour onto the other users. Institutions, following Scott (1995), are "cognitive, normative, and regulative structures and activities that provide stability and meaning to social behaviour". Markets and their property rules, regulations and their enforcement mechanisms, and fishing communities and their social networks, are all examples of institutions. Effective fisheries management institutions must either force fishers to bear all the costs of their actions, often through mimicking property rights, or exert some sort of direct control on fishing activities. The goal of the "tragedy of the commons" approach is to learn how to design these rational institutions. It has produced several commonly recognized ideas for improving management such as "fisheries co-management" and "individual transferable quotas".

The second basic social science approach to fisheries management applies the techniques of political economy to fisheries management. Such studies chart how management outcomes reflect struggles over profits between different groups such as nations (Peterson, 1993), ethnic groups (Bailey, 1986), and fishing sectors (Duff and Harrison, 1997). It is impossible to understand real-world fisheries management without looking at its politics. Some (Shipman, 1996) have pointed out the benefits of this active contestation, such as management institutions in which many parties have a stake. Others have emphasized the down-side: management measures constantly postponed when compromise cannot be reached; management measures chosen primarily because of their allocative implications (Hanna and Smith, 1993); or, incompatible management policies couched in fuzzy language that hides the contradictions (Hersoug, 1996). Many fisheries professionals, particularly frustrated experts with a good technical grasp of what should be done, often talk about management as being "distorted" by politics. But management is not distorted by politics, it is politics. Politics is how societies make decisions.

Both of these approaches are necessary for understanding management as a social process, and the third basic social science approach consists of attempts to bring them together. Some authors (Kooiman et al., 1999) suggest the term "governance" to describe such a synthetic approach because they want to understand the process of creating management institutions in the real world of competing user groups. A critical concept in this approach is "institutional embeddedness" (McCay and Jentoft, 1998). Studies that look at embeddedness draw on well known theories from anthropology (Geertz, 1971), sociology (Giddens, 1984; Granovetter, 1985) and political science (Dryzek, 1987) to investigate how larger social contexts, such as cultural understandings, social networks, and polity structures, influence the creation of management institutions. This chapter's use of the concepts of legitimacy, surveillance and enforcement reflects this governance approach.

## ISSUES RELATED TO THE LEGITIMACY OF FISHERIES MANAGEMENT

The legitimacy of an institution is the degree to which people are willing to allow it to influence their actions. It is a cultural phenomenon, i.e., it has to do with the meanings that people attach to things. Many anthropological studies have demonstrated the importance of cultural meanings in forming and maintaining institutions (Douglas, 1985). McCay (1984), for example, investigated the causes of a lack of compliance with management in a New Jersey fishing community. She found that illegal fishing was used to maintain a diverse and opportunistic approach to resource use. It was a "patterned deviant behaviour" that was accepted and organized by the community and even acted to increase the local sense of community. Economists have linked legitimacy to economic behaviours. Kuperan and Sutinen (1998) found that survey-based measures of legitimacy improved the fit of economic models that predicted fishing behaviour in the Philippines. While inherent subjectivity always makes it suspect to attach numbers to meanings, these results suggest a quantitative, fishing-related confirmation of the many anthropological observations.

It is helpful to distinguish between substantive and process legitimacy (Jentoft, 1993). Substantive legitimacy is how acceptable management measures are because of their own characteristics. Process legitimacy is the acceptability of the process that creates the measures. People often obey laws they think are silly because they want to live in a society with a legal system that works. Similarly, fishers may or may not like a particular management measure but they may want management to work. This means that the effectiveness of individual measures is reduced when the overall management system that produces it is seen, for various reasons, as being of little value.

This section discusses three kinds of questions that are commonly raised about the substantive legitimacy of fisheries management: the economic incentive structure that measures create; the science on which the measures are based; and, the fairness of the measures toward those affected. Then we turn to issues of process legitimacy, which includes the legal basis of management and important instruments of international fisheries law will be discussed. The other key question for process legitimacy is who participates in creating the measures, and how they do so.

## INCENTIVE STRUCTURES

Most fishing is done through a business and most fish are caught by businesses. The laws of economics apply; boat mortgages and crew have to be paid and profits have to be made if families are to be fed. Beyond simple economics, many fishers, whether commercial, recreational or subsistence, exist within cultures that value catching many large fish and honour those "high-liners" who succeed in doing so. Management measures must rationally reflect the operational and market realities of this competitive fishing environment.

Derby fisheries are a great example. The halibut fishery in the United States Pacific Northwest had a fishing fleet of 2900 vessels in 1981 which grew to about 4400 vessels in 1991. The fishing season, meanwhile, had shrunk from 120 days to 48 hours (Baker, 1996). Four thousand boats all trying to catch halibut in two days is a huge investment being used for a very short time, basically to drive the price of halibut through the floor. This was a patently absurd way to manage a fishery and everyone involved knew it. In 1993 the fishery was shifted to an individual quota system. A management measure that fails to account for the competitive nature of fishing and creates inefficiencies in asset use and price will quickly undermine its own legitimacy.

Subsidies are a common attempt by governments to create legitimacy that have had devastating long-term effects on fisheries resources. One of the implications of the tragedy of the commons is that investment in fishing boats and gear will soon create more capacity to catch fish than a sustainable fish harvest can support. The World Wildlife Fund estimates that the global fishing fleet now has two and one half times more fishing equipment than is needed to sustainably harvest the oceans (WWF, 1999). This problem is exacerbated by governments providing support to the fishing industry. This support is not trivial; it amounts to an estimated 22% of the value of world's estimated commercial landings. The subsidy leaders are Japan at an estimated $750 million dollars in yearly subsidies and the European Union at $500 million dollars (FFITF, 1999).

Subsidies are a complex issue and many subsidies are necessary and justifiable. Some provide needed support to marginal coastal communities and others, fishing vessel buy-back programs for example, are needed to help the fishing industry reduce capacity. If reducing fishing capacity is needed to protect a public resource it is fair that the public help pay for that reduction. Nevertheless, many subsidies create a perverse incentive structure by increasing potential profits from fishing. This may work in the short term but increased profit potential leads to increased investment, greater excess fishing capacity, more overfishing, and the eventual loss of both fishing profits and fish (FFITF, 1999).

For many fishers the worst example of a perverse incentive structure is regulatory discards. Regulatory discards happen when fishers at sea catch fish that they cannot legally sell. With many common fishing methods the fish are sorted when they are already dead or dying and throwing them back in the water does nothing for conservation. Many management measures create regulatory discards. In some cases small "bycatch quotas" allow fishers to sell the otherwise illegal fish, but there is always a concern that such quotas will become a targeted fishery. The waste that regulatory discards create angers both fishers and the general public. Corey and Williams (1995) quote one fisher who calls these fish "government bycatch". "Bycatch of any kind is not desired," he says, "but watching cod, fluke, swordfish, or other such fish go overboard dead is a crime. It's certainly not conservation." Fishers in the Northeast United States have put the issue of regulatory discards to very effective use in soliciting public support for their resistance to management measures (Wojtas, 1996).

## SCIENTIFIC REALISM

Fisheries management's legitimacy depends on it being based on accurate science. Fisheries science is an example of what sociologists of science call "mandated science", i.e. science that is produced under pressure from legal mandates. Indeed, it was the demand for fisheries management that gave birth to fisheries science. The formation of national fisheries research organizations in the late nineteenth century, and the International Committee for the Exploration of the Seas in 1902, resulted from politicians looking for ways to resolve disputes between fishers. These organizations brought together biologists from various backgrounds who formed the core of what later became a scientific discipline and community in its own right.

Regardless of specialty, mandated science exhibits a number of patterns. It differs from other areas of science in that it is usually about evaluating a situation rather than discovering new knowledge (Salter, 1988). To achieve its objective of legitimating policies, mandated science must be intelligible to non-scientific audiences and facilitate clear choices. It must be seen as value-free, methodical, and susceptible to open debate, anonymous peer review, and academic publication. The problem, as Salter (1988) argues, is that mandated science meets none of these ideals. It often makes moral dilemmas more open rather than hiding them. It produces results quickly, often uses data from private sources, and deals with problems of little academic interest. It is very often seen as a corrupt process even when it is not, because interest groups use both the knowledge produced and gaps in that knowledge to define issues in terms of their own social objectives (Jasanoff, 1990). Mandated science makes its last stand in the court room where the legal process, not the scientist, defines the factual question that must be answered (Jasanoff, 1995). In many areas of science, being a scientific witness has become an advocacy-oriented specialty of its own that operates independently of how science is done elsewhere (Wynne, 1989). Indeed, many scientists refuse to play the role at all (Jasanoff, 1995).

Fisheries science has done a good job of facing these challenges while still maintaining its scientific integrity. Leading fisheries scientists are confident in the basic approach used in fisheries stock assessments. Rosenberg et al. (1993) argue that there is a sound theoretical and empirical basis for sustainable fishing based on the established principle of density-dependent population regulation. They point out that overfishing has historically not been the result of inadequate scientific advice as much as ignoring that advice. Fisheries science is meeting the challenge of the inevitable uncertainty of stock assessments through developing better and better probabilistic approaches.

Fisheries stock assessments routinely come under fire in the course of setting management policy. Many of these attacks would not stand up in a fair and open debate. Most cite the commonly inadequate data used to make the assessments. This inadequacy is readily acknowledged by those performing the assessments who, in fact, would love to have better data. Nothing does as much damage to the legitimacy and effectiveness of management than the all too common decision not to adequately fund the gathering of fisheries data, particularly survey data gathered separately from fishing activities.

Two general lines of criticism of the general approach to stock assessments, however, enjoy fairly wide support among fisheries scientists and other interested parties. The first is the issue of the relationship between present stocks and future stocks. For many species the assumption, basic to stock assessment models, that more fish now will mean more fish later is perfectly reasonable. For other species, however, effects on recruitment from environmental factors may be so large as to overwhelm the effect of the number of spawning fish (Peterson, 1993). This problem has been the basis, for instance, of a long-standing debate between state and federal scientists, and fishers, over assessments of lobster (*Homarus americanus*) stocks in northeastern United States.

The second line of criticism has been over the validity of assessing single fish stocks in multi-species ecosystems. This problem was more of a concern among fisheries scientists during the 1970s than it is today (Peterson, 1993). While there is still a great deal of interest in how we might be able to do "ecosystem management", few scientists still believe that predator–prey interactions and the like undermine the validity of stock assessments for single species. What does draw concern about single-species models is not so much their accuracy as the use to which they are put. A precautionary approach to managing each species is often interpreted (and in the United States at least this interpretation is legally required) to mean establishing a spawning stock biomass (SSB) that achieves maximum sustainable yield (MSY). Basically because the ocean is a large place, models for most species put such an SSB at quite a large number of fish, often near the historically high levels of abundance. However, as fishers are very quick to point out, in multi-species ecosystems historically high levels of abundance do not occur for all species of fish at the same time. Trying to manage many species at the same time using this precautionary approach results in a contradiction.

This contradiction, and the fact that fishers are usually among the first to point it out, points at a deeper problem in the relationship between fisheries science and the legitimacy of fisheries management. Unlike most scientific specialties where mandated science is done, for example the regulation of toxic chemicals, in fisheries science the interested public, i.e. the fishers, also have a broad knowledge of the subject. Fishers's insights are often described, even by fishers, as "anecdotal", in contrast to the systematic data gathered through statistically valid population sampling procedures. Fishers's insights, however, are not anecdotes in the sense of random stories. They are the result of observations that they make within the context of long-term experience with what is and is not an important indicator of the health of fish stocks. Some authors suggest that managers' too heavy reliance on population models, and failure to heed the warnings of inshore fishers, contributed to the collapse of cod stocks in Atlantic Canada (Grafton and Silva-Echenique, 1997). Studies suggest that fishers view the sea in ways that are alien to the assumptions of stock assessment models. These models treat fisheries as mathematically tractable systems while fishers see them as more chaotic (Smith, 1990). Fishers see the ocean as a textured and differentiated place, while population models treat it as uniform and homogeneous. Fishers emphasize the importance of habitat considerations, while population models are not able to directly consider habitat (Pinkerton, 1989). Finally, fishers view the resource in small temporal and spatial scales while models consider large ones (Smith, 1995).

This is not meant to support a romantic vision of wise old salts putting the college boys in their place; too many wise old salts have publicly denounced people doing transect-based population surveys for being too stupid to sample where the fish are. It does imply two things about the way science supports the legitimacy of fisheries management. One is that fishers have information that is potentially valuable for management and this information needs to be considered in making decisions. One of the key elements in process legitimacy is that all points of view are perceived to get a hearing rather than being ignored (Wilson and McCay, 1998a). The second is that the cultural gap between fishers, fisheries scientists and fisheries managers needs to be considered and understood in the creation and the implementation of management measures.

## EQUITY AND FAIRNESS

Fisheries management measures are designed for conservation of fish, but almost all of them also affect the allocation of fish between different user groups. As a result, an important aspect of the legitimacy is the degree to which these allocative effects are seen as fair. In their discussion of fairness and equity in management, Loomis and Ditton (1993) draw on two social psychological theories of perceptions of fairness. According to equity theory people evaluate fairness using principles of equity such as (1) the proportionality of inputs to returns to effort, i.e. those who put the most in should get the most out, (2) the equality with which people are treated, and (3) whether or not people's relative needs are considered. Relative deprivation theory, on the other hand, suggests that people's judgements about fairness come from comparisons with others. If one group is seen as sacrificing then another group is more likely to agree to sacrifice.

In fisheries management, one important equity issue is "inter-generational" equity, the idea that we are being unfair to those in the future if we fail to conserve fish. This notion of equity is at the heart of management legislation and is accepted, in principle, by everyone from fishers to environmentalists. It is also the moral basis of the "precautionary principle" that says that when science is uncertain we must err on the side of conservation. The precautionary principle, however, raises its own equity issues because the costs of caution are born by today's fishing communities, while its benefits may well accrue to others. A fishing family that loses its boat because a fishery is closed bears the costs of conservation and gains none of the benefits. Hence, application of equity principles often face the difficult problem of who exactly the winners and losers are.

This problem takes us to the issue of relative deprivation, but from the angle of how principles of equity can be used to define membership in the groups that the management system is creating. One common principle used to address this problem has been "historical participation". The longer a fisher has been fishing for a particular species the more "right" that fisher has to future access to that species. Historical participation derives more from the principle of proportionality mentioned above than it does from any actual legal claim, an important exception being the treaty rights of indigenous peoples. Historical participation is usually the basis of the allocation of both individual and area-based fishing quotas. Although very often, for reasons of either ideology or convenience, these allocations redefine "fisher" as "fishing boat owner", which can undermine perceptions of fairness.

Equity is also part of the link between process and legitimacy. When fishers see management as corrupted by the undue influence of others it is undermined. Therefore, transparency in the decision-making process is critical (Jentoft, 1989).

## EMERGING GLOBAL FISHERIES LAW

The current legal form of many fisheries management regimes dates from the mid-1970s when the United Nations Convention of the Law of the Sea (UNCLOS) was in progress. The convention itself came formally into force in 1982, but it was during the 1970s that most nations created their 200 mile Exclusive Economic Zones (EEZ). This change forced countries to reconsider and renegotiate how they were going to manage their fisheries. This effected both the way they approached the management of their own coastal stocks, which were now, of course, much more extensive. It also effected a large number of bilateral and multilateral agreements and conventions. Several ceased to exist and others had substantial changes in their terms of reference. The eight regional commissions governing shared fisheries among southern countries changed their mandates to include management in addition to development (Peterson, 1993).

Regional agreements form the bulk of international legal instruments related to fishing. There are many of them. FAO (1996) lists 18 fishing-related international treaties but describes this list as only a set of examples. Peterson (1993) lists 23 international commissions that have been set up to manage a shared stock or region and his list is also not exhaustive. Some of these, the International Pacific Halibut Commission, set up between the United States and Canada, for example, have been relatively successful. After a dip in the 1970s, Pacific halibut landings have remained fairly steady. Others have been much less successful. The International Convention for the Conservation of Atlantic Tuna (ICCAT) has remained essentially deadlocked and impotent in the face of an 80% drop in the breeding population of bluefin tuna in the last two decades.

Peterson (1993) argues that four patterns can be seen in how domestic patterns created by this new situation influenced the way nations approach international fisheries law. The first are nations with large, privately owned long-distance fishing fleets that exert influence on their governments. Japan is the best example but the same basic pattern applies to Britain, South Korea, Norway, Spain, Taiwan and to some extent the United States. While still very important, distant water fleets are generally in rapid decline. The Japanese fleet, for example, declined by half between 1990 and 1993 (Jonsson, 1996). A second, similar pattern can be seen in former Soviet Block countries with the difference that these fleets, in even more rapid decline, are owned by governments. Peterson's (1993) third pattern is developing countries who are intent upon developing their own fisheries. The fourth is those countries, Iceland being the prime example, where coastal fleets influence their government to help them against long-distance competitors. Canada, some European countries that do not have long distance fleets, and for some fisheries the United States, also fit this pattern.

The key variable influencing the behaviour of nations in negotiating international fisheries law seems to be the size of the fleets involved (Peterson, 1993; Sutherland, 1989). The larger their fleets the more they support the status quo and the greater their reluctance to make the needed changes. For example, larger nations have been slower to ratify one critical theory (Hyvarinen et al., 1998), the Agreement for the Implementation of the Provisions of the United Nations Convention on the Law of the Sea of 10 December 1982 Relating to the Conservation and Management of Straddling Fish Stocks and Highly Migratory Fish Stocks (The Straddling Stocks Agreement). As of this writing, ratifications have come from 24 of the 30 nations required for the agreement to enter into force. Basically reflecting Peterson's (1993) patterns, Canada, Iceland, Norway, Russia and the United States have acceded, Australia, the European Community, Japan, New Zealand, Spain and the United Kingdom have not (UN, 1999).

UNCLOS provided the basis for international fisheries law but it had few details on how it was to be applied to fisheries (Hyvarinen et al., 1998). This gap has begun to be filled in the 1990s. Important impetus was given by the UN Conference on Environment and Development (UNCED) in Rio de Janeiro in 1992. The UNCED product, Agenda 21, recognized that the right to fish is conditional and is accompanied by a duty to manage and conserve stocks (Hyvarinen et al., 1998). The 1993 Agreement to Promote Compliance with International Conservation and Management Measures by Fishing Vessels on the High Seas (The Reflagging Agreement) began the job of fleshing out the fishing-related details of the UNCLOS principle that nations are responsible for vessels that fly their flag. These included requirements pertaining to vessel authorization to fish, the provision of data about authorized vessels to a central data base administered by the Food and Agriculture Organization of the United Nations (FAO), and for internationally recognized fishing vessel markings. States should also only flag vessels over which they can effectively exert control. This treaty has not yet entered into force.

In 1995, the Straddling Stocks Agreement, although it does not apply to nations' own stocks, incorporated a number of important principles. It was the first international commitment to the use the precautionary principle in setting management policy and it requires setting specific precautionary parameters as both management targets and management triggers (Juda, 1997). Furthermore, it mandates the use of scientific advice and review and requires that other species in the relevant ecosystems be considered, not just the species for which the plan is developed (Juda, 1997). It calls on states to minimize waste, discards and by-catch, and to prevent and eliminate excessive fishing capacity. It also establishes the principle that bilateral and multi-lateral fishing agreements and regional management bodies must meet international standards and that nations that are not members of regional bodies must still cooperate in the protection of highly migratory and straddling stocks (Juda, 1997). Regional fisheries management bodies have been described as obstacles to the implementation of the Straddling Stocks Agreement because they have been slow bring their own operations into line with it (Hyvarinen et al., 1998).

Finally, the 1995 FAO Code of Conduct for Responsible fishing, while non-binding in the sense that its provisions are not enforceable on countries, is an international agreement that establishes that the precautionary principle applies to the coastal fisheries resources of nations. The technical guidelines (FAO, 1996) for the implementation of the Code draw on the other agreements, particularly the Reflagging Agreement, and emphasize the effective monitoring and control of fishing vessels.

Trade agreements are another type of international agreement that directly affect fisheries management. The World Trade Organization (WTO) has been a hindrance to both national and international environmental protection and this has proven true in fisheries as well. The best known fisheries-related dispute was a challenge by Thailand, Malaysia, Pakistan and India to a United States law banning the importation of shrimp caught by vessels failing to use turtle exclusion devices. This measure brought imported shrimp into line with requirements on the US domestic industry. WTO ruled against the United States after famously refusing to consider amicus briefs provided by environmental organizations. The relationship between international trade and regional fisheries agreements is also an unresolved issue. Several such agreements, ICCAT for example, relies on trade restrictions for enforcement of provisions. The question of whether or not such restrictions would violate free trade agreements has not yet been tested.

## PARTICIPATION IN FISHERIES MANAGEMENT

A crucial potential source of legitimacy is the various forms of participation by fishers and other "stake holders" in making management decisions. When the focus is on participation by fishers a commonly used term is "co-management". McCay and Jentoft (1996) describe co-management as cooperative fisheries management where "the basic principle is self-governance, but within a legal framework established by government, and power is shared between user groups and the government". They oppose co-management to "consultative" management in which the advice of the fishers is solicited but there is no real participation in decision making. Co-management is argued to mobilize several assets to aid effective management. One is facilitated access to information (Pinkerton, 1989). Others are increased legitimacy through increased transparency in decision making (Jentoft, 1989), greater accountability for officials (Magrath, 1989), and increased respect for indigenous perspectives (Pomeroy and Carlos, 1997).

The relationship between participation and management, however, is more complicated than simply the possibility of increased legitimacy though programs such as co-management. There is the critical question of who legitimately participates (Wilson and McCay, 1998a). Even the choice of the words "stake holder" to describe participants is fraught with the politics of participation. The alternative terms "fishers" and "user groups" both draw the line narrowly excluding groups, such as environmentalists, who may not actually be making use of the resource. Stake holder is perhaps the most neutral term because it includes anyone who sees themselves as having a "stake" or an interest in marine resources. Stake holders may be too broad, however, because it invites participation from anyone from anywhere who defines themselves as a stake holder.

Representation is a key issue in any but the smallest management regime. The US Federal management system uses a complex combination of co-management and consultative management approaches. In the co-management aspects, i.e. the participation of industry representatives on Regional Management Councils with some decision-making power, the industry representatives are appointed by state governments. Participation in the many consultative aspects, i.e. advisory panels, public hearings, etc. depends either on who knows whom or on who shows up. Who is representing whom in the process is usually entirely unclear (Wilson and McCay, 1998a), which is particularly unsettling in light of Hanna's (1996) documentation of the critical importance for success of clear representation in three case studies of management plans by the Pacific Fisheries Management Council.

## ISSUES RELATED TO SURVEILLANCE IN FISHERIES MANAGEMENT

Surveillance straddles legitimacy and enforcement. That surveillance of fishing behaviour is necessary for the enforcement of management measures is obvious. What might be less obvious is the importance it plays in legitimacy. People need to know about the behaviour of other people in order to decide their own behaviour (Ostrom, 1987) and they contribute to collective goods at higher rates when they have information about others' contributions and they know that others have that information as well (Sell and Wilson, 1991). Seeing others, and being seen by others, conforming to an institution is an independent and essential part of maintaining that institution. The National Marine Fisheries Service (NMFS) in the United States is taking advantage of this in a very clever way: they are publishing bycatch statistics for fishers so that the public, most importantly the other fishers, can see how they are doing in trying to maintain low bycatch rates.

The management and use of information is the key to surveillance. The Reflagging Agreement, discussed above, requires flag states to maintain records of their vessels operating in international waters and report such information to FAO (Buck, 1996). Ocean regions, notably the South Pacific with its combination of vast sea area and relatively resource-poor management agencies, have based their regional management cooperation in part on compiling and comparing landings data within the regions (Lodge, 1997; Moore, 1993). In the Pacific, vessel information is collated in a regional register and this has made it possible to monitor the number and size of fishing vessels operating in the region. The global data base administered by FAO has established a system to maintain records of high seas fishing vessels. Data maintained on vessels is extensive and includes such items as its name, previous names, registration number and port, size, engine capacity, as well as the species it is licensed for, the quotas it holds, and its compliance records (Fitzpatrick, 1996). The major source of this data is the flag state's mandated process of authorizing vessels to fish.

A central issue for any fisheries management measure is whether or not compliance can be monitored from land. Some measures, such as the banning of certain gears, can be fairly easily monitored in port. In most northern countries at least, the amount and size of fish that a boat sells can also be reasonably well verified. Other measures, such as closed areas or restrictions on high grading (keeping only the best of the legal fish) have to be monitored at sea.

Observer programs, where government officers or private contractors are stationed on fishing vessels, are a critical component of surveillance that are now used by many northern and southern nations organized at both national and regional levels. In many cases observers not only monitor compliance with regulations but have some biological training and gather information useful for research. Perhaps the largest observer program is the North Pacific Groundfish Observer Program (NPGOP) who report training and placing 370 observers on 400 vessels in 1996 (NOAA, 1999). The observer program in the Pacific has become an industry in itself and observers have organized a professional association. Observing is difficult work, sometimes conducted in hostile or corrupt environments, and this can limit its effectiveness. Observers are generally not well paid, even NPGOP contractors pay relatively low wages for a job that requires a bachelors degree in biology. Roberts (1997) reports that West African observer programs are oversupplied with poorly paid and poorly qualified observers.

Areal surveillance is becoming more and more important in fisheries management. The 1984 FAO World Congress on Fisheries Management and Development gave initial impetus to standardize markings of fishing vessels that would make areal surveillance possible and the Reflagging Agreement requires such markings. Areal surveillance is very expensive even for wealthy countries. The United States and Canada often piggyback areal surveillance for fisheries management on marine safety and other programs. The South Pacific Forum Fisheries Agency coordinates regional area surveillance in cooperation with the Australian and New Zealand armed forces (Lodge, 1997).

More comprehensive, and less expensive, are satellite-based vessel monitoring systems (VMS). Australia, New Zealand, Canada, the European Union and the United States have all introduced remote sensing for determining the position of fishing vessels. The EU has been scheduled to require VMS for all vessels in the Atlantic greater than 15 meters in 1999 (Fitzpatrick, 1996), but strong opposition from fishers has emerged. It is very likely that these VMS systems will eventually be linked to the FAO database on fishing vessels. Fitzpatrick (1996) describes how satellites can observe more than simply vessel position. In addition to heading and speed, data on the performance of main machinery, deck machinery, steering, gyro compass, and information generated by electronic equipment can be monitored. Catch reports can easily be added to these systems.

The traditional emphasis in fisheries monitoring at sea has been on physical inspections and the use of surface craft. This is time consuming, expensive and difficult to do effectively. These advances in new programs and technologies promise to change this emphasis. All of them are fairly expensive, even VMS involves installing tamper-proof equipment on each vessel. Surveillance programs are generally charged one way or another to the vessels being observed and this makes them economical only in fisheries using high capacity vessels. Most fishing vessels are still small-scale. For reasons of both rural economics and sustainability this is probably a good thing, and in any event it will remain true for the foreseeable future. It remains a serious challenge, however, as we try to find ways to monitor small-scale fisheries and this issue once again points us in the direction of co-management.

## ISSUES RELATED TO ENFORCEMENT IN FISHERIES MANAGEMENT

The various penalties related to fisheries management measures are currently a hodge-podge of widely varying sanctions in different jurisdictions. Slow improvement is being made, however, on the enforcement front. Regional efforts at the harmonization of fisheries policy are being attempted around the world and these include nations sanctioning violators of other nation's rules (Cirelli and Van Houtte, 1997; Sutherland, 1989). Areal and VMS programs have been linked to enforcement through both blacklisting of individual vessels and putting pressure on flag-states to enforce their own vessels compliance (Moore, 1993). Blacklisting, linked to a regional register, has worked well in the Pacific where the threat of blacklisting has often proven to be enough to bring a wanted vessel to dock (Moore, 1993;

Sutherland, 1989). The regional enforcement provisions in the Pacific also authorize hot pursuit of fishing vessels into foreign jurisdictions.

The UN and FAO's approach to fisheries enforcement has centred on flag state responsibility and the World Conference on Fisheries Management and Development in 1984 adopted it as a guideline in its strategy for Fisheries Management and Development. It has historically received more lip service than real compliance (Moore, 1993). Buck (1996) argues that the Reflagging Agreement is the potential key to the effectiveness of other international fisheries treaty because it set in place the mechanisms that will make enforcement possible. It addresses the central issue of vessels in states that acceded to international fisheries agreements evading restrictions by reflagging in non-member states. The Reflagging Agreement is based on the UNCLOS requirement that a genuine link exists between nations and the vessels they register. The 1993 agreement strengthens this link through the authorization and reporting requirements but it still lacks clear enforcement guidelines (Buck, 1996). The agreement has not yet entered into force and only 13 of the required 25 nations have ratified it (FAO, 1999).

## SOCIAL ASPECTS OF MANAGEMENT MEASURES

Fisheries management measures are as varied as the commercial, recreational and subsistence fisheries they manage. Traditionally, fisheries management has focused on measures that control the fishing for specific species. Now the most rapidly expanding regulatory areas are for the control of bycatch and the protection of the marine environment. Bycatch has been a driving issue in management, and the development of fishing technology, for some years now. Habitat and the marine environment have now begun to move to the forefront, requirements to address habitat are now part of United States fisheries law, for example. The issue of the effects of mobile fishing gear on the sea floor is rapidly emerging. This has profound implications for how fishing is done and threatens to be the next major area of fisheries management debate.

Another emerging issue is the management of recreational fisheries. Not only is recreational fishing already generating large economic gains from fisheries, $25 billion dollars a year in the United States (NMFS, 1997), but they are becoming increasingly important as alternatives for commercial fishers displaced by the loss of fish stocks. The fact that the product of the recreational industry is an experience rather than actual fish raises many questions for management that we still do not know how to address. Maximum sustainable yield is replaced by maximum visitor attractiveness. In recreational fisheries one fish can be very different from another in terms of value. The single fish that attracts the foreign world record seeker or wins the big tournament can be worth tens of thousands of dollars to fishing communities (Wilson and McCay, 1998b).

Traditional management measures involve controls on the techniques used in fishing, controls on fishing effort (the number, time and/or place of boats or gear), and controls on the size and amount of fish that can be taken. Controls on techniques often have the great political advantage of having fewer implications for the allocation of catch: it is much easier to tell all the fishers that they have to use a larger mesh size than it is to try to fairly divide a quota or close an area or season with equal effects on all. Restrictions on techniques, however, are economically inefficient, especially compared to individual quotas, because they place an external constraint on the fishers' ability to bring home as much catch for as little effort as possible. Furthermore, restrictions on technique, while they are likely to be effective against growth overfishing (in short, catching fish that would be more profitably allowed to grow), are much less effective against recruitment overfishing (reducing a stock's ability to reproduce itself) (Peterson, 1993). Because of the strength of both these economic and technical arguments, there is a general consensus among fisheries scientists and other technically expert observers that controls on techniques are nearly always inadequate by themselves.

Restrictions on effort have a long history. Closed or restricted areas and seasons are found in many traditional "folk" approaches to management (Acheson, 1975), which have a longer track record of success than modern management. Marine protected areas (MPAs) are getting a great deal of attention, initially in tropical areas as protection for reefs, but now in temperate areas as well. While they have many technical advantages for protecting fish, their habitat, and biodiversity, MPAs are politically difficult measures. Their very nature means that the burden of their creation will be born by some fishing communities, and so by those communities' elected officials, more than by others. Their effectiveness contributes to their political volatility. Recent area closures off the northeast US were put in place to protect groundfish. The scallop populations in these areas increased rapidly leading to intense pressure from scallop fishers, who were being badly cut back in other areas, and from Senator Teddy Kennedy to open the areas to scallop dredging.

Direct controls on the size and amount of fish caught are at the heart of modern fisheries management. Taken alone, controls on size are no more effective than controls on techniques. Quotas systems have serious problems with bycatch, regulatory discards and high grading, but they do have a track record of bringing back fisheries that have been seriously damaged by recruitment overfishing. Quotas are a set amount of fish that is allowed to be caught by someone, they are usually derived from a total allowable catch (TAC) that is, in turn, derived from a stock assessment model. The most common form of quota management specifies an aggregate quota for some population of fishers. When the quota is caught by the fishing fleet then the fishery is closed. This is almost always done in conjunction with other management measures because of the perverse incentives that quotas create. Fishers rush to get as much of the quota as possible for themselves and in the process they drive down the price of the fish. Controls on effort, such as trip limits, are commonly found in conjunction with quotas in order to slow the rush down, but they can never get rid of it entirely.

Therefore, it has become clear to many well informed people that because of the promise of the biological effectiveness of quotas for many fisheries, and the economic perversity of aggregate quotas, that many fisheries would be best managed by individual quotas. Each fisher would have his or her own quota of fish that he or she could fish for in the time and manner of his or her choosing. These decisions would tend to maximize prices while minimizing costs. Because not all fishers are equally skilled at maximizing prices and minimizing costs, the best of all possible economic worlds would be to have individual transferable quotas (ITQs) that fishers could sell to one another. This way the more skilled fishers, because they make more money with less effort, would buy the quota of the less skilled and the fishery would end up being fished in a manner that is simultaneously biologically sustainable and economically efficient.

It all adds up. The only problem is that if one were to set out to design a fisheries management measure to be as politically difficult as possible one would probably invent the ITQ. Most existing ITQ systems specifically state that an ITQ is a right of access and not a property right, but it seems inevitable that they will become more and more property-like. People buy, sell and base investment decisions on ITQs. Courts may well feel it necessary to give considerable weight to the rights of those who purchase ITQs to maintain control over them. To be allocated an ITQ is often a major windfall for a fishers. Hence, the first hurdle is the initial allocation of quota. The principle that is usually invoked to guide the allocation is historical participation, but with this kind of prize a great many people are going to by trying to get designated an historical participant. To make the situation even more difficult, ITQ systems are often introduced into already managed fisheries where "historical" participation patterns have been heavily influenced by regulations (McCay et al., 1996). Initial allocation is not just a huge political problem among the fishers who want quota, it is a legal problem for the society at large. In December 1998 the Icelandic Supreme Court found that the ITQ system was unconstitutional. What they found unconstitutional was precisely the use of historical participation to allocate the ITQs because, they said, this was discriminatory (Pálsson, 1999).

Initial allocation, however, is far from the only problem with ITQs. Many, Greenpeace is a prominent example (King and Townsend, 1996), object to the idea of essentially privatizing common resources that have heretofore been held as a public trust. The doctrine of public trust is an increasingly important legal tool for environmentalism in general (Johnson and Galloway, 1996). Mace (1993) points out that with ITQ systems the harvesters discount rates on the value of the resource may well be higher than those of the public or those with conservationist interests. ITQ systems will almost certainly lead to a concentration of access rights to the fishery as more and more people sell out. Indeed, getting excess people and harvest capacity out of the fishery is part of the idea and ITQ systems often begin by allocating less quota to fishers than can sustain a fishing operation (McCay et al., 1996). While in theory it is the more efficient operators who stay, in practice this means larger operators, sometimes with company fleets replacing owner-operators. Some ITQ systems put caps on how much quota a single fisher can own, but others do not.

The Icelandic system, now twenty years old, is the longest running ITQ system and perhaps the most comprehensive in terms of stocks covered. The results have been mixed and very complex. The Icelandic fishing industry now has the highest productivity in the North Atlantic and some stocks, particularly herring which has been under ITQ management for the longest time, have increased (Hannesson, 1996). Pálsson and Helgasson (1996) report the system's social effects. The initial pattern was for small holders to lease their ITQ to the big companies. Then, as the TAC was reduced, the small holders began to go under and sell out to large companies. Now these companies are the lessors and the lease contracts have resulted in lower profits and smaller wages for the crew, as leasing prices are taken out as an overall expense before the catch is shared (Eythorsson, 1996). As a result, there were national strikes in 1994 and 1995. Public discontent with this concentration of ownership has increasingly been articulated with feudal metaphors such as "tenancy", "lords of the sea", and "quota kings" (Pálsson and Helgasson, 1996). ITQs have reduced excess harvesting capacity in the quota fisheries (Hannesson, 1996), but between 1984 and 1993 the overall Icelandic fleet expanded 9% and engine power increased 17% (Eythorsson, 1996).

ITQs have a real promise as a tool of fisheries management, but political realities make them a much more complex tool than the technical models suggested. We will eventually figure out how to design ITQ systems well and under what environmental and social conditions they work best. The time is passed, however, when ITQ were something people "believed" in rather than one more item in the management tool kit.

## CONCLUSION

Fisheries management is a challenging social problem. The effectiveness of management is frustrated by the difficulties of decision making at every level. It has taken three international agreements, two not yet in force, just to agree upon a system for keeping track of the identity of large fishing vessels. When such a basic step is so difficult is there any reason to believe we will ever achieve sustainable fishing?

One hopeful sign is the emergence of an international "epistemic community" in fisheries management. This term became well known through an article by Haas (1989) in which he documented how an agreement to control pollution in the Mediterranean was created. "Epistemic" comes from the word epistemology, the philosophy of how to ask questions. Epistemic communities are international networks of people who agree about what the important questions are. The emergence and acceptance of the precautionary principle is a sign of the fisheries epistemic community at work. Around the world, people working in fisheries have learned a great deal about how to talk to one another and they are placing a lot of creative energy into looking for solutions.

I believe we will eventually learn how to live sustainably with ocean resources. We are never going to reach the point where fisheries are not depleted and have to be rebuilt. We are never even going to reach the point where we do not have periodic crises in fisheries. In fact, I wonder if we really want to rationalize fisheries management to the point that we have such absolute control. Part of the value of fishing is its risk-taking, independent, life style. Hopefully, we can use this life style, along with our technology, science, economic knowledge and, above all, ability to listen and negotiate with one another, to create ever better management institutions.

## REFERENCES


Acheson, J.M. (1975) The lobster fiefs: economic and ecological effects of territoriality in the Maine lobster industry. *Human Ecology* **3**, 183–207.

Baker, D.J. (1996) The World Around Us: A Science Perspective. Paper presented at the Columbia River Inter-Tribal Fish Commission Commissioners' Retreat Welches, Oregon 24 July.

Bailey, C. (1986) Government Protection of Traditional Resource Use Rights—The Case of Indonesian Fisheries" Chapter 21, pp. 292–308. In *Community Management: Asian Experience and Perspectives*, ed. D.C. Korten. Kumarian Press, West Hartford, CT.

Bogue, M.B. (1993) To save the fish: Canada, the United States, the Great Lakes, and the Joint Commission of 1892. *The Journal of American History* **79** (4), 1429–1454.

Buck, E.H. (1996) Agreements to Promote Fishery Conservation and Management in International Waters. Congressional Research Service Report for Congress, Washington DC.

Cirelli, M.T. and Van Houtte, A. (1997) Legal Aspects of Cooperation in Monitoring, Control and Surveillance in the Southwestern Indian Ocean, pp. 77–110 in Report of a Regional Workshop on Fisheries Monitoring Control and Surveillance. Albion, Mauritius 16–20 December 1996. FAO, Rome.

Corey, T. and Williams E. (1995) Bycatch: Whose issue is it, anyway? *Nor'Easter* Fall/Winter.

Delbos, G. and Prémel, G. (1996) The Breton Fishing Crisis in the 1990s: Local Society in the Throes of Enforced Change, pp. 129–140 In *Fisheries Management in Crisis*, eds. K. Crean and D. Symes. Fishing News Books, Oxford.

Douglas, M. (1985) *Risk Acceptability According to the Social Sciences*. Social Research Perspectives 11. Russell Sage Foundation, New York.

Dryzek, J. (1987) *Rational Ecology: Environment and Political Economy*. Basil Blackwell, Oxford.

Duff, J.A. and Harrison W.C. (1997) The law, policy, and politics of gillnet restrictions in the state waters of the Gulf of Mexico. *St. Thomas Law Review* **9** (Winter), 389–417.

Eythorsson, E. (1996) Coastal communities and ITQ management. The case of Icelandic fisheries. *Sociologica Ruralis* **36** (2), 212–223.

FAO (1996) FAO Technical Guidelines for Responsible Fisheries. No. 1. FAO, Rome, 26 pp. 6 annexes.

FAO (1999) The FAO Legal Office. [Online] www.fao.org/Legal/home.htm.

FFITF (1999) Federal Fisheries Investment Task Force Report to Congress. Washington: National Oceanic and Atmospheric Administration.

Fitzpatrick, J. (1996) Satellite Data Communications Systems, Remote Sensing and Other Techniques as an Aid to Monitoring, Control, Surveillance and Enforcement. In Report of a Regional Workshop on Fisheries Monitoring, Control and Surveillance, Albion, Mauritius, 16–20 December. GCP/INT/606/NOR Field Report 97/37.

Geertz, C. (1971) *The Interpretation of Cultures*. Basic Books, New York.

Giddens, A. (1984) *The Constitution of Society*. The University of California Press, Berkeley.

Grafton, R.Q. and Silva-Echenique, J. (1997) How to manage nature? Strategies, predator–prey models and chaos. *Marine Resource Economics* **12**, 127–143.

Granovetter, M. (1985) Economic action and social structure: The problem of embeddedness. *American Journal of Sociology* **91**, 481–510.

Gordon, H.S. (1954) The economic theory of a common-property resource: the fishery. *Journal of Political Economy* **62**, 124–142.

Haas, P.M. (1989) Do regimes matter? Epistemic communities and Mediterranean pollution.

Hanna, S.S. and C.L. Smith (1993) Resolving allocation conflicts in fishery management. *Society and Natural Resources* **6**, 55–69.

Hanna, S.S. (1996) User participation and fishery management performance within the Pacific Fishery Management Council. *Ocean and Coastal Management* **28** (1–3), 23–44.

Hannesson, R. (1996) Exclusive rights to fish: Toward a rational fisheries policy. *Geojournal* **39** (2), 179–184.

Hersoug, B. (1996) Social considerations in fisheries planning and management—real objectives or a defense of the status quo, pp. 19–24. In *Fisheries Management in Crisis*, eds. K. Crean and D. Symes. Fishing News Books. Oxford.

Hyvarinen, J., Wall, E. and Lutchman, I. (1998) *The United Nations and Fisheries in 1998*. Ocean Development and International Law 29, pp. 323–338.

Jasanoff, S. (1990) *The Fifth Branch: Science Advisors as Policy Makers*. Harvard University Press, Cambridge.

Jasanoff, S. (1995) *Science at the Bar*. Harvard University Press, Cambridge.

Jentoft, S. (1989) Fisheries co-management. *Marine Policy* **13**, 137–154.

Jentoft, Svein (1993) *Dangling Lines: The Fisheries Crisis and the Future of Coastal Communities—the Norwegian Experience*. Institute of Social and Economic Research, Memorial University of Newfoundland, Social and Economic Studies No. 50.

Johnson R.W. and Galloway, W.C. (1996) Can the public trust doctrine prevent extinctions? pp. 157–164. In *Biodiversity and the Law: Challenges and Opportunities*, ed. W.J Snape. Island Press, Washington.

Jonsson, O.D. (1996) The geopolitics of fish: the case of the North Atlantic, pp. 92–98. In *Fisheries Management in Crisis*, eds. K. Crean and D. Symes. Fishing News Books, Oxford.

Juda, L. (1997) The 1995 United Nations Agreement on straddling fish stocks and highly migratory fish stocks: a critique. *Ocean Development and International Law* **28**, 147–166.

King, R. and Townsend, J. (1996) Sinking Fast: How Factory Trawlers are Destroying U.S. Fisheries and Marine Ecosystems. A Greenpeace Report. [Online] www.greenpeace.org.

Kooiman, J., van Vliet, M. and Jentoft, S. (eds.) (1999) *Creative Gover-*

*nance: Opportunities for Fisheries in Europe.* Ashgate, Aldershot, UK.

Kuperan, K. and Sutinen, J.G. (1998) Blue water crime: Deterrence, legitimacy and compliance in fisheries. *Law and Society Review* **32** (2), 309–337.

Lodge, M.W. (1997) The South Pacific Forum Fisheries Agency and Legal Aspects of Fisheries Monitoring, Control and Surveillance, pp. 153–170 in Report of a Regional Workshop on Fisheries Monitoring Control and Surveillance. Albion, Mauritius 16–20 December 1996 Rome: FAO GCP/INT/606/NOR-Field Report 97/37.

Loomis, D.K. and Ditton, R.B. (1993) Distributive justice in fisheries management. *Fisheries* **18** (2), 14–18.

Mace, P.M. (1993) Will private owners practice prudent resource management. *Fisheries* **18** (9), 29–31.

Magrath, W. (1989) The Challenge of the Commons: the Allocation of Nonexclusive Resources. World Bank: Environment Department Working Paper #14.

McCay, B.J. (1984) The pirates of piscary: ethnohistory of illegal fishing in New Jersey. *Ethnohistory* **31** (1), 17–37.

McCay, B.J., Creed, C.F., Finlayson, A.C., Apostle, R. and Mikalsen, K. (1996) Individual transferable quotas (ITQs) in Canadian and US fisheries. *Ocean and Coastal Management* **26** (1–3), 85–116.

McCay, B.J. and Jentoft, S. (1996) From the bottom up: Participatory issues in fisheries management. *Society and Natural Resources* **9**, 237–250.

McCay, B.J. and Jentoft, S. (1998) Market or community failure? Critical perspectives on common property research. *Human Organization* **57** (1), 21–29.

Moore, G. (1993) Enforcement without force: New techniques in compliance control for foreign fishing operations based on regional cooperation. *Ocean Development and International Law* **24**, 197–204.

NDRC (1997) Hook Line and Sinking: the Crisis in Marine Fisheries. Natural Resources Defense Council, Washington.

NMFS (1997) Accomplishment Report under the Recreational Fishery Resources Conservation Plan. National Marine Fisheries Service, Silver Spring MD.

NOAA (1999) The North Pacific Groundfish Observer Program. The National Oceanic and Atmospheric Administration (NOAA) [Online] www.refm.noaa.gov.

Ostrom, E. (1987) Institutional arrangements for resolving the commons dilemma, pp. 250–265. In *The Question of the Commons: The Culture and Ecology of Communal Resources*, eds. B.J. McCay and J.M. Acheson. The University of Arizona Press, Tucson.

Pálsson, G. and Helgason, A. (1996) Figuring fish and measuring men: the individual transferable quota system in the Icelandic cod fishery. *Ocean and Coastal Management* **26** (1–3), 117–146.

Pálsson, G. (1999) Individual transferable quotas: unconstitutional regimes? *Common Property Resource Digest* **48**, 1–5.

Peterson, M.J. (1993) International fisheries management, pp. 249–305. In Institutions for the Earth: Sources of Effective International Environmental Protection, eds. P.M. Haas, R.O. Keohane and M.A. Levy. MIT Press, Cambridge.

Pinkerton, E. (1989) Introduction: Attaining better fisheries management through co-management—prospects, problems and propositions, pp. 3–36. In *Co-Operative Management of Local Fisheries*, ed. E. Pinkerton. University of British Columbia Press, Vancouver.

Pomeroy, R.S. and Carlos, M.B. (1997) Community-based coastal resource management in the Philippines: A review and evaluation of programs and projects 1984–1994. *Marine Policy* **21** (5), 445–464.

Roberts, K. (1997) Regional Cooperation in MCS: The Experience of the Member States of the Sub-Regional Fisheries Commission (West Africa), pp. 181–191. In Report of a Regional Workshop on Fisheries Monitoring Control and Surveillance. Albion, Mauritius 16-20 December 1996 Rome: FAO GCP/INT/606/NOR-Field Report 97/37.

Rosenberg, A.A., Fogarty, M.J., Sissenwine, M.P., Beddington, J.R. and Shepherd, J.G. (1993) Achieving sustainable use of renewable resources. *Science* **262**, 828–829.

Salter, L. (1988) *Mandated Science: Science and Scientists in the Making of Standards.* Kluwer Academic Publisher, Dordrecht, Netherlands.

Scott, R. (1995) Institutions and Organizations. Sage Publications, Thousand Oaks, CA.

Sell, J. and Wilson, R.K. (1991) Levels of Information and Contributions to Public Goods. *Social Forces* **70**, 107–124.

Shipman, S. (1996) Do commercial users influence marine fisheries management? *Fisheries* **21** (12), 22–23.

Smith, M.E. (1990) Chaos in fisheries management. *Marine Anthropological Studies* **3** (2), 1–13.

Smith, M.E. (1995) Chaos, consensus and common sense. *The Ecologist* **25**, 80–85.

Sutherland, W.M. (1989) Policy, Law, and Management in Pacific Island Fisheries. Chapter 3, pp. 33–46. In *Development and Social Change in the Pacific Islands*, ed. A.D. Couper. Routledge, New York.

UN (1999) Oceans and the Law of the Sea. United Nations. [Online] www.un.org/depts/los.

Wilson, D.C. and McCay, B.J. (1998a) How the participants talk about participation in mid-Atlantic fisheries management. *Ocean and Coastal Management* **41**, 41–61.

Wilson, D.C. and McCay, B.J. (1998b) A Social and Cultural Impact Assessment of the Highly Migratory Species Fisheries Management Plan and the Amendment to the Atlantic Billfish Fisheries Management Plan. for the National Marine Fisheries Service, Highly Migratory Species Office. 174 pp.

Wojtas, J. (1996) Prosecutor drops fluke quota charges, saying fishermen 'trying to make a point'. *New London Day*, Dec. 17th.

Wynne, B. (1989) Establishing the rules of laws: constructing expert authority, pp 23–55. In *Expert Evidence: Interpreting Science in the Law*, eds. R. Smith and B. Wynne. Routledge, London.

WWF (1999) Underwriting Overfishing. Washington: World Wildlife Fund. Pamphlet.

## THE AUTHOR

**Douglas C. Wilson**

*Institute for Fisheries Management and Coastal Community Development,*
*P.O.Box 104,*
*DK-9850 Hirtshals, Denmark*

Chapter 118

# FARMING OF AQUATIC ORGANISMS, PARTICULARLY THE CHINESE AND THAI EXPERIENCE

Krishen J. Rana and Anton J. Immink

For centuries people have reaped the benefits that the sea provides, believing that this zero-input treasure trove was boundless. A very different picture was painted in the 20th century and humankind has had to rethink its strategy. As catches from the world's major fish stocks dwindled, diversification became the norm, and it became apparent that, as had happened on land, there were not enough natural resources to provide humans with a bottomless larder. There was only one solution remaining—the sea would have to be farmed.

Towards the end of the 20th century fisherfolk were forced to abandon some traditional fisheries and to diversify areas and species caught. Fishing moved further offshore, aided by technological advances such as vessel sonar and ice storage. Even the new stocks and species became overexploited and, by the end of the last millennium, fish, crustacean and mollusc catches from the sea were constant at around 85 million mt (metric tonnes). In the 1980s and more noticeably in the 1990s, aquaculture production in both inland and marine areas increased rapidly, with aquaculture production (including plants) from the seas reaching more than 18 million mt in 1997.

This production, however, is not evenly distributed throughout the waters of the world. Indeed, most aquaculture production is recorded from the developing countries. Of the 18 million mt of aquaculture products that originate from the seas of the world, 87% comes from the Pacific Ocean, where growth has been strongest since 1990, reflecting and directly influencing the major global trend. Although total production is only one tenth of that in the Pacific, the Atlantic, which includes the Mediterranean and Caribbean, has shown a similar growth trend since 1993. Fish and mollusc production in European waters are important factors in this growth.

This chapter provides an overview of the status of aquaculture, with particular reference to mariculture and the important producers, China and Thailand.

*Seas at The Millennium: An Environmental Evaluation (Edited by C. Sheppard)*

## OVERVIEW OF GLOBAL AQUACULTURE

As we move into the next millennium, the domestic and international requirement for both high and low valued aquatic animals and plant species for direct and indirect consumption is likely to increase due to a combination of rising populations, living standards and disposable incomes. While the predictions of supply of fish required to meet future needs varies, it is widely acknowledged that the yield from traditional marine and inland capture fisheries, which reached 93 million tonnes (excluding plants) in 1997, is unlikely to increase substantially and that expectations from aquaculture will probably increase (see Fig. 1). For aquaculture, therefore, the challenges are how to: (i) sustain and increase the current mean annual global growth rate and (ii) strengthen and promote aquaculture of finfish, shellfish and plants as a legitimate long-term farming activity. In addressing these challenges it should be appreciated that issues faced by aquaculture have not changed greatly, but their prioritisation has changed, and that these priorities vary among nations depending on the state of aquaculture development.

The potential of aquaculture in securing food supply, generating employment and foreign exchange is clearly demonstrated by the rapid expansion of this sector. Since 1990 global aquaculture has expanded at an average annual rate of almost 13%, compared with only 3% for livestock and meat and 1.6% for capture fisheries production. The global geographical distribution of aquacultural output continues to be highly skewed towards Asia, and in particular, China. In 1997 Asia accounted for 91% of the world's reported tonnage and China alone accounted for a staggering 67% of world output (new adjusted values) followed by India (4.9%) and Japan (3.7%).

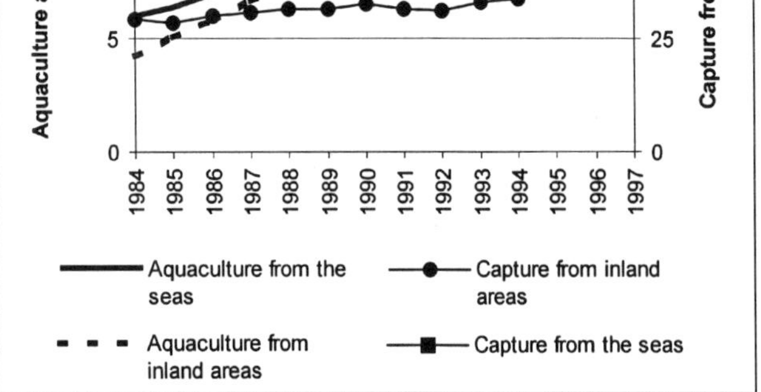


Fig. 1. Production from the waters of the world (all data sourced, unless otherwise stated, from FAO/FIDI, 1999).

Table 1
Expansion rate of aquaculture within the sub-Asian regions

| | 92–94 | 95–97 | 96–97 |
|---|---|---|---|
| Asia | 16.6 | 9.2 | 6.0 |
| China | 23.6 | 12.1 | 8.2 |
| South Asia | 9.0 | 7.5 | 2.8 |
| Southeast Asia | 7.9 | 7.0 | 3.8 |
| East Asia | 2.7 | –4.8 | -5.3 |

### Significance of Asian Aquaculture

The overall global average, however, masks the huge variation in the rate at which regional and national aquaculture is developing. For example, between 1990 and 1997 the change in expansion rate of production reported by individual countries in East, Southeast and South Asia ranged between –6 and +61% per year and highlights the huge heterogeneity in aquaculture development and achievements between countries and sub-regions.

Although the overall expansion rate of production of the sector in Asia averaged 12%/year between 1990 and 1997, it ranged from –0.7%/year for East Asia to 17%/year for China. These differences may reflect recent changes, which include shifts in the patterns of national and international demand for aquaculture products, improvements in sectoral and farm-level management, and marketing, availability of input resources and production losses due to sub-optimal management and changes in reporting of production.

Closer evaluation of the data shows that the rate at which aquaculture production is expanding is slowing in the region (see Fig. 10) (Table 1). Output from East Asia between 1995 and 1997 declined by 5.3%/year. This decline was largely attributed to the drastic decline of seaweeds in Korea and Japan. Similarly, in Southeast Asia the rate of development decelerated considerably from 10.3%/year for the biennium 1993–95 to just 3.8%/year for 1996–97. In the case of Southeast Asia, this was largely due to the decline in the reporting of *Penaeid* spp. and blood-cockle production from Thailand, Philippines and Malaysia. Despite this recent regional decline in rate of expansion, global output of cultured aquatic products from the seas is still increasing.

## OVERVIEW OF MARINE AQUACULTURE

Marine and coastal areas of the Atlantic and Pacific Oceans are most extensively used for aquaculture and the rate of utilisation has increased sharply since 1990 (Fig. 2). Within the Pacific Ocean area the leading producers are China, Japan, the Republic of Korea, the Philippines and Indonesia (Fig. 3). China has the greatest influence on trends in aquaculture production, with most of the production being for

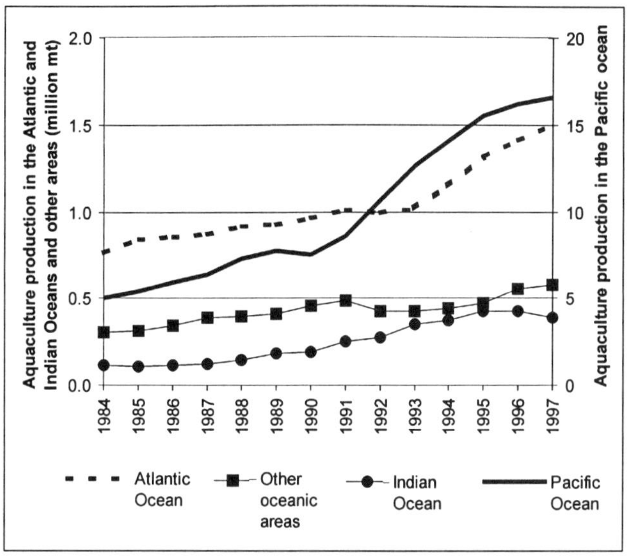


Fig. 2. Aquaculture production in marine and brackish water environments in and around the oceans.

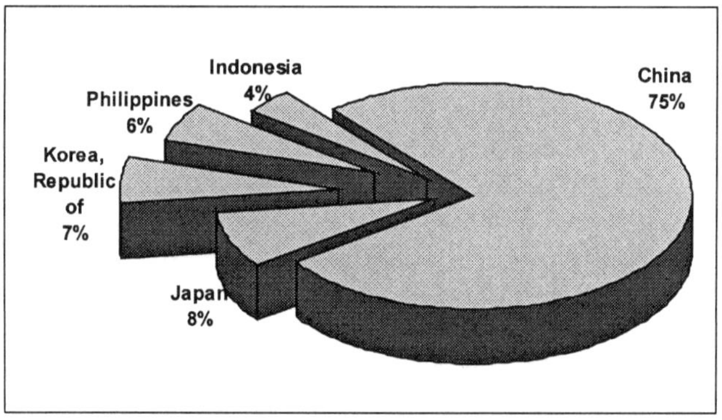


Fig. 3. Top five countries utilising the coastal environment in the Pacific area for mariculture production in 1997. Mariculture production in the coastal areas of these five countries totalled over 15.4 million tonnes alone.

local consumption. Asian countries are the largest utilisers of the marine and brackish water environments (mariculture), with production largely consisting of aquatic plants and molluscs. Europe produces a high percentage of marine fish, but is also well represented with molluscs. Aquaculture production in South America is dominated by salmon—in 1997 Chile produced 240,000 mt of salmonids in the sea.

On a global scale, the near-shore areas of the oceans have been predominantly used to produce molluscs and aquatic plants (Table 2). Fish and crustaceans are only produced in comparatively small amounts, with only one species each in the top ten. Of the major cultured groups, the plants are natural consumers of dissolved organic matter and the molluscs filter feed from the waters of the oceans to remove particulate matter. Essentially these aquaculture products contribute towards reducing the eutrophication of the environment (United Nations, 1992).

## Mollusc Farming

Mussels and oysters have been traditionally taken from the sea for free, and they have long been the common person's food. Today, after over-exploitation and disease problems, oysters are considered a delicacy and fetch a higher market price. The demand-driven higher price means that culture of the product was justified. There have been setbacks and in some areas the species cultured has changed completely (Elzière-Papayanni, 1993; Menzel, 1991). Although there is still artisanal collection and limited commercial harvest of oysters and mussels, most of the world's production now comes from culture in near-shore areas (Fig. 4).

## Aquatic Plant Farming

There is little scope for farming plants in inland waters. Freshwater plants usually grow slowly and are less palatable than seaweeds. Seaweeds have been farmed in the seas off Japan, China and Korea for decades, and seaweed harvesting was a traditional livelihood in northern Europe and Canada (Perez et al., 1992). Now the costs of harvesting seaweeds from the wild in Europe and Canada are higher than the market price and the industry has effectively died. In 1997, four of the top ten cultured species were seaweeds (Table 2).

Many species of seaweeds are farmed for direct human consumption locally (usually as dried product) in countries

Table 2

Top ten species produced in the brackish and marine water of the world

| Species | Quantity (tonnes) | Value (US$ million) | Value ranking | Trophic level |
|---|---|---|---|---|
| *Laminaria japonica* (Japanese kelp) | 4,401,931 | 2874 | 3 | P |
| *Crassostrea gigas* (Pacific cupped oyster) | 2,968,266 | 3164 | 2 | FF |
| *Ruditapes philippinarum* (Japanese carpet shell) | 1,275,104 | 1670 | 6 | FF |
| *Pecten yessoensis* (Yesso scallop) | 1,256,162 | 1700 | 5 | FF |
| Mollusca (Marine molluscs nei) | 1,130,652 | 539 | 14 | FF |
| *Porphyra tenera* (Laver = Nori) | 861,231 | 1335 | 7 | P |
| *Salmo salar* (Atlantic salmon) | 638,639 | 2112 | 4 | C |
| *Eucheuma cottoni* (*Eucheuma cottoni*) | 589,263 | 50 | 50 | P |
| *Undaria pinnatifida* (Wakame) | 535,357 | 169 | 29 | P |
| *Penaeus monodon* (Giant tiger prawn) | 490,188 | 3501 | 1 | C/FF |

Key: P = plant/primary producer; FF = filter feeder; C = carnivore.

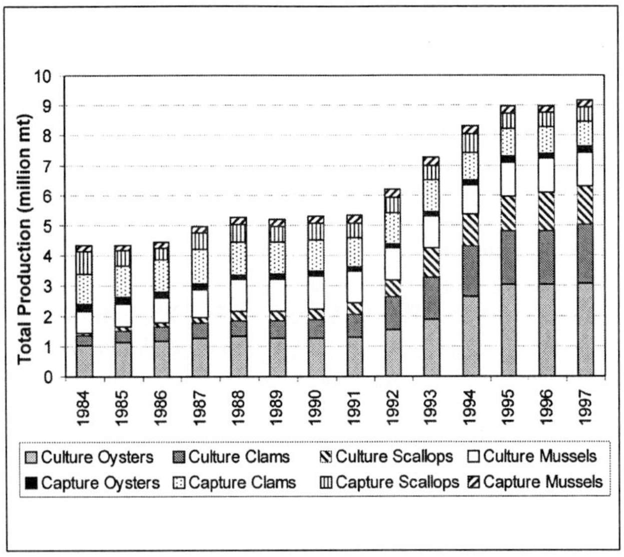


Fig. 4. Capture and culture of four main mollusc groups.

Table 3

Top five exporting and importing countries of traded seaweed products (dry weight)

| Country | Trade flow (mt) | 1994 | 1995 | 1996 | 1997 |
|---|---|---|---|---|---|
| Japan | Import | 66,478 | 65,716 | 63,833 | 71,475 |
| USA | Import | 39,125 | 55,380 | 45,333 | 53,486 |
| UK | Import | 17,134 | 18,006 | 10,339 | 18,493 |
| Denmark | Import | 21,575 | 20,799 | 17,574 | 16,502 |
| China | Import | 6,668 | 8,448 | 9,573 | 15,844 |
| China | Export | 36,135 | 38,325 | 42,815 | 51,719 |
| Chile | Export | 34,483 | 39,073 | 41,375 | 47,858 |
| Mexico | Export | 26,470 | 39,413 | 35,000 | 32,738 |
| Korea, Rep. of | Export | 28,551 | 24,578 | 21,249 | 21,729 |
| Indonesia | Export | 19,517 | 25,888 | 23,291 | 13,335 |

such as China, Korea, Philippines and Japan, although the majority of traded seaweeds (Table 3) are used to produce alginates for the food and pharmaceutical industries (De Silva, 1998).

## The Development of Marine Fish Aquaculture

By the time farming the sea had become important, the techniques and technology to successfully farm in freshwaters already existed. Trout was cultured in many temperate countries across the world: China had been farming fish for centuries, Egypt had a strong history of aquaculture and all across Asia aquaculture was common practice, from backyard culture to large-scale operations. The necessity for food forced many countries with expanding populations to make the maximum use of this resource, and most rural smallholdings used their local water-storage areas for this purpose.

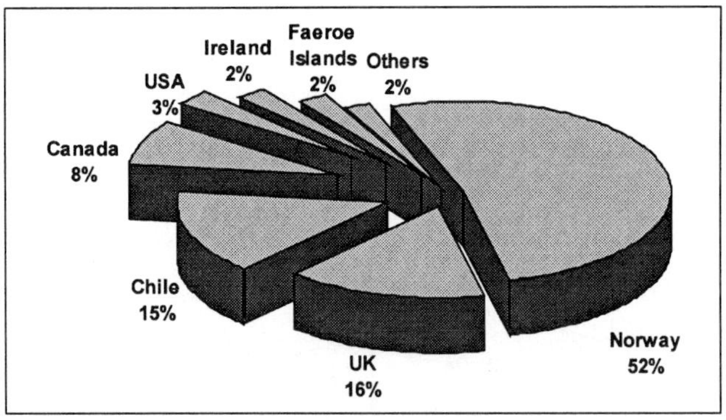


Fig. 5. Breakdown of global Atlantic salmon (*Salmo salar*) production in 1997 (total = 638,600 mt)

The first major success with a marine fish was the salmon, now well documented (Heen et al., 1993, for an overview). This success was largely due to the fact that the young are raised in freshwater. Their eggs are large, the juveniles are relatively easy to feed because they are a reasonable size at the stage of first feeding, and the basic technology was there because of the experiences with trout. Probably the single most important issue for the spread of Atlantic salmon (*Salmo salar*) across the globe is the fact that fertilised eggs are hardy enough to be shipped. This meant that millions of eggs could be transported with high viability for very little cost. For other species, e.g. carps, this was not possible and their spread was much slower. The high market price for salmon also meant that the investment in technology was economically justified, and Atlantic salmon is now produced all over the world (Fig. 5), although Norway remains the largest producer of this species.

Interestingly, 125,800 mt of Rainbow trout (*Oncorhynchus mykiss*)—considered to be a freshwater fish—were produced in the sea in 1997.

### *Closing the Production Cycle*

For most marine fish species the facilities used for the culture of adult fish are much the same as for salmon—usually conducted in cages. However, the hurdle that has to be overcome is the supply of juveniles, and the challenge for aquaculture is to provide feed of the correct size and quality to match the requirements at an economic level. For shrimp, however, the constraints differ, and the art of captive reproduction is still difficult (McVey, 1993). These technological and cost constraints have meant that the farming of most marine organisms started with juveniles collected from the wild. If unregulated, this practice may lead to environmental consequences including loss of species diversity, habitat destruction and loss of the target species from its original areas. Most aquaculturists, however, would prefer to have the cycle 'closed' (every stage of production conducted on-farm), because this gives greater flexibility and the ability to plan production. If the cycle is not closed, production is usually erratic, as demonstrated with grouper in Fig. 6.

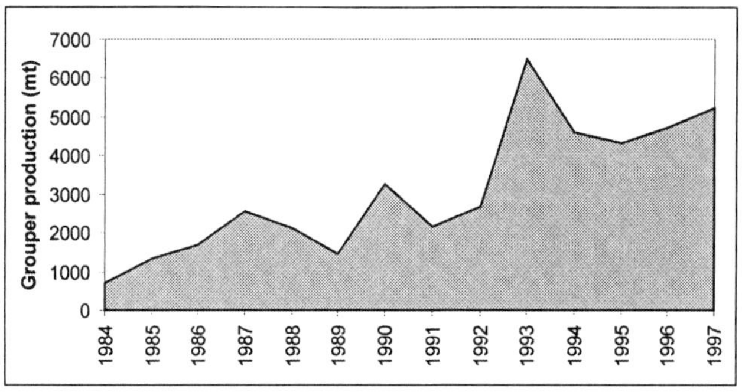


Fig. 6. Global growth in the production of groupers—a high-value species for which the production cycle is still not closed. At present, this fish is important for rural incomes because the grow-out technology is relatively simple and the market price is high.

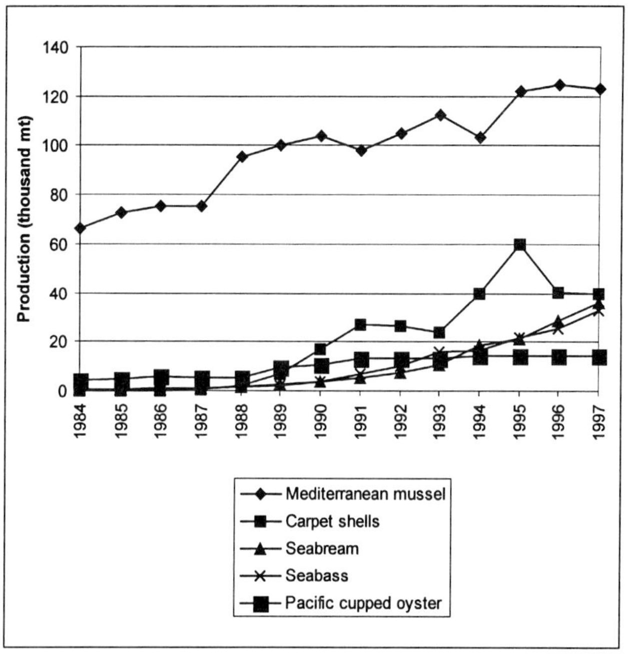


Fig. 7. Aquaculture production in the Mediterranean.

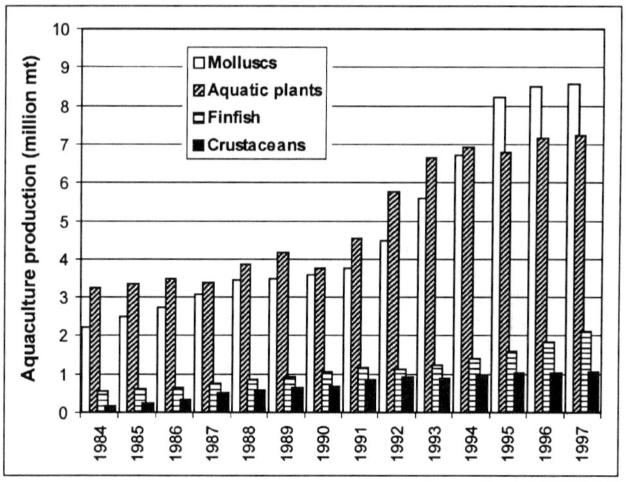


Fig. 8. Production of the major groups in marine and brackish environments.

The economic commitment required to overcome these technological constraints has meant that marine fish farming has concentrated on traditionally high-value species, but once the technology has been stabilised for the production of these species, the quantity produced increases and the market price drops. With the continued decline in catches of traditional marine fish, there will always be new species reaching a market price high enough to justify their inclusion in aquaculture production.

New species that have recently entered into commercial production, or are at an advanced experimental phase, include cod, haddock, halibut, turbot, rabbitfishes and a more diverse range of basses and breams. Eventually, production will include marine species of lower price, contributing to food security and sustainable rural livelihoods in coastal areas.

### *The Mediterranean Example*

One region where aquaculture development in the marine environment has been very significant is the Mediterranean, which traditionally produced molluscs, but has seen the rapid development of seabass (*Dicentrachus labrax*) and seabream (*Sparus auratus*) production (Stephanis and Divanach, 1993). The reproduction cycle, including breeding and feeding was closed around 1990 and since that time seabass and seabream production has increased nine-fold, from 7800 tonnes to 69,000 tonnes in 1997 (Fig. 7). Most of this production comes from Greece, where new finfish species, mostly other bass and bream type fish, are also being trialed. Mussel production in the Mediterranean, which has increased by almost 50% since 1984, is centred in Italy, which produced 84% of the total 123,000 tonnes in 1997.

## Crustacean Farming

Shrimp farming has had a bad press, but, as is shown in Fig. 8, crustacean farming makes up a very small percentage of aquaculture practices. Although shrimp production was allowed to develop largely unregulated, economic and technological pressure has now meant that most shrimp are produced with more consideration for the environment. This has followed the introduction of better farming practices and coordination of production for all farmers within certain areas. Overall, 66 countries produced just over one million tonnes of shrimps and crabs in 1997. Thailand is the largest penaeid shrimp producer in the world, but other Asian countries and Latin American countries (notably Ecuador) are important. In recent years disease problems and a related drop in market price has meant that production has decreased (Fig. 9), but with better management practices the industry should stabilise at a sustainable level.

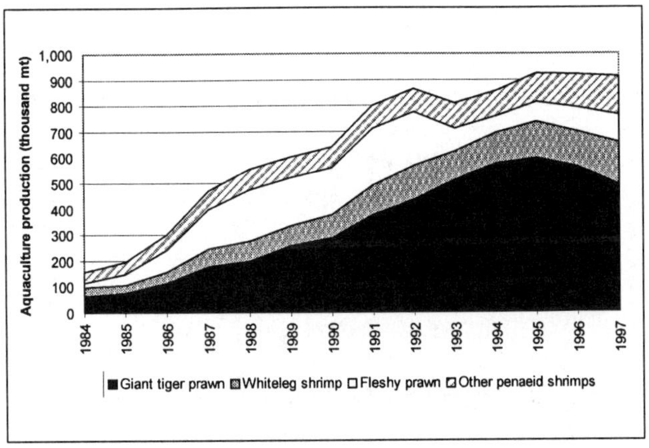


Fig. 9. The development of shrimp farming, showing major species.

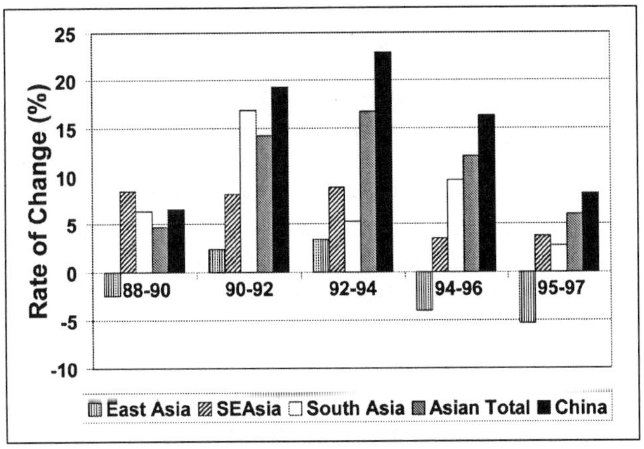


Fig. 10. Recent changes in the rate of expansion of aquaculture in the sub-Asian region.

## Culture Facilities

Within the marine and brackish water environments, the increasing diversity of species is cultured in/on a diversity of facilities. The first culture of organisms in saline water involved growing of fish and crustaceans in ponds flooded by the tide at high water, and the farming of fish, aquatic plants or molluscs in enclosed or semi-enclosed shallow sea areas or lagoons. For mollusc culture the utilisation of the near-shore area has brought about the design and construction of trays, nets, raft, poles, bags and suspended ropes. Aquatic plants are cultured using similar techniques, as well as in ponds. One of the main design criteria is the exclusion of predators. Crustacean farming is largely carried out in ponds. It is becoming more common to integrate this with some form of fish, mollusc or aquatic plant farming in the inflow and outflow waters to remove either predators or wastes. The notable finfish species cultured in marine or brackish water ponds are the milkfish (*Chanos chanos*) and mullets (Mugilidae). Most other marine fish are cultured largely, although not exclusively, in cages. The materials used for the supporting structures of cages range from wooden planks floating on polystyrene (styrofoam) blocks to submersible metal platforms designed for the open sea (Beveridge, 1996). The competition between aquaculture and other users of coastal resources means that aquaculture is likely to move offshore. As witnessed with traditional agricultural land, the owner can make more money using it for residential or commercial property than for primary production. This move away from near-shore areas for some types of culture may also be driven by the need for clean water.

## The Culture Environment

Aquaculture is often accused of polluting the environment, but in fact successful aquatic farming requires clean water. There have been some cases of poor management, but most producers are now aware that, in order for aquaculture to be sustainable, there is a strong need for effective planning and management. What appears to be omitted from reports is that effluents are localised and largely consist of biodegradable material. For example, under salmon cages there can be a build-up of organic waste products. When the problem becomes too great, the salmon themselves are affected, and it would not be economically viable for aquaculturists to ignore these problems. Any long-term effects are likely to be minimised because the waste is broken down and utilised by organisms naturally present in the environment.

The use of theraputants is limited by economic constraints. Their use is often carefully controlled and, on completion of treatment, the dilution factor makes them almost undetectable. The problem arises when many farms are located within a confined area. In recent years, the need for these products has been reduced by better planning of water flow through farms, vaccines and overall better management. However, where they have been necessary, farms are encouraged to coordinate their health management programmes for their benefits and that of the environment. As farming moves offshore it is envisaged that there will be less of a need to use theraputants and less risk of accumulation.

## The Need for Integrated Coastal Area Management

Beyond the need to bridge the gap between falling fish capture and rising demand, the necessity for aquaculture in the marine environment has come about because of a change in the way our coastal resources are used. More people now live in coastal areas, often occupying whole stretches of coast that used to be wild. Private ownership of coastal areas for commercial or private developments has altered access to areas that used to be used for artisanal collection of coastal products. The need for clean water has meant that food production has been squeezed into areas with little development and little conflict of interest. This has brought employment to remote areas where rural communities were in decline and has provided food security for remote communities with little easy access to other sources of food.

In general, however, there are now too many users for all resources, and so countries are having to implement plans for the integrated use of both coastal and inland areas.

## SUSTAINABLE AQUACULTURE: CHINESE AND THAI EXPERIENCE

China and Thailand are used here as examples, as the challenges have been addressed in these countries within an evolving national aquaculture framework since the 1980s, aimed at increasing and sustaining the efficiency of output. The conceptual premise and approaches aimed at developing a stable aquaculture sector together with changes in living standards in some of the key markets will be discussed. Between 1990 and 1997, China and Thailand expanded aquaculture production at an average annual rate of 20 and 10%, respectively, and diversified production. Since the 1980s China's open-door policy together with the decision to gradually decentralise management, have played important roles in transforming aquaculture from a centrally planned to a market-based activity. Available indicators suggest that growth in Chinese, as well as Thai aquaculture production, were attributed to an increase in land area used for aquaculture, as well as an increase in national average production efficiencies from culture systems.

### Selected National Perspectives on Issues and Challenges

The dynamic flux in national and regional aquaculture output illustrated above could be used to illustrate two extreme scenarios for considering issues and challenges. These scenarios could be used to provide a context within which aquaculture can be evaluated and developed in a stable or sustainable manner. An overall evaluation of the production of species or species groups suggests a sigmoid relationship between production (tonnage) and time (see Fig. 16 for grouping by feeding types). Such developments are usually defined by an initial lag or development/pioneering phase, a second rapid-growth phase as technological impediments are better understood, overcome and solutions disseminated, and a tertiary mature phase governed by market supply and demand and required natural and managerial inputs. The precise form of the relationship is species-specific. Moreover, such a general relationship, is also representative of the macro-sectoral development, particularly since, in many Asian and non-Asian countries, few species or species groups dominate output. For example, in China, Chinese and common carps, which are largely cultured for local consumption, accounted for 55% of national fin and shell fish tonnage; and in Thailand, the giant tiger shrimp, *Penaeus monodon*, and the tilapia, *Oreochromis niloticus*, which are produced for export and national consumption, account for 37% and 18% of national cultured fin and shell fish production, respectively. Similarly, in the US, catfish accounts for 54% of aquaculture output, while in Chile, Norway and the UK, salmonids account for 90, 99.7, and 91%, respectively, of total fin and shellfish output.

Clearly, countries like China and Thailand have recognised that to maintain and increase the current production levels during and beyond the mature development phase of the species or the sector, strategic national institutions and frameworks need to be developed and strengthened to identify and address challenges and to provide new opportunities in aquaculture. These challenges will have to accommodate national demographic changes and changes in living standards in both rural and urban populations. In

## Scenarios for Opportunities to Secure and Increase Aquaculture Production

- *Increasing growth rate and National Aquaculture Framework = Opportunities for Stable Expansion.*
- *Stagnating growth rate and National Aquaculture Framework = Opportunities for Stable Diversification and Innovation.*

The rapid growth phase is largely driven by entrepreneurship, availability of seed and nutritional inputs of suitable quality and in sufficient quantity, widespread dissemination of and access to relevant technology, high profit margin (even at low production efficiencies), acceptable mortalities and access to required natural, human and financial resources. To sustain high growth, the sector would need to be regulated through a national aquaculture framework to stabilize both the internal and external environment on aquaculture, in particular those elements that can negatively impact on the sector.

The stagnating phase is typically due to resource limiting factors, market saturation and falling prices, unacceptable seed quality and quantity and production losses, higher operating costs and reducing profit margins, and potential social conflict. Under these scenarios, the national aquacultural framework would need to be biased and focused towards applied research and dissemination for improving production efficiency, developing innovative technologies, promoting practical and enforceable regulations governing the availability of and equitable access to natural resources, conflict resolution, movement of aquatic organisms and market development. To assist in the upturn of the expansion, national institutions will be required to address, amongst others, the internal and external environments, untapped national (and international) markets, the need to diversify cultured products and post-harvest products and their presentation to meet diverging and emerging niche markets. To reach distant intra- and international markets in acceptable quality, infrastructure and the shelf-life of products also needs to be improved.

instances where the sector is growing, appropriate regulation of the sector—particularly in the expansion phase—is desirable to rationalise development and access to resources.

## National Support For Aquaculture Development (Statutory Protection of Waters for Aquaculture)

In 1997 total world aquaculture production reached 36 million tonnes and was valued at US$50.37 billion, confirming the fundamental premise for many countries that aquaculture can make a significant contribution to the well-being of its people, and to both the local and national economies. Consequently, to protect and foster this sector, governments (e.g. China, India, Thailand, etc.) have recognized that one key challenge will be to effectively regulate those activities external to the aquaculture sector (i.e. the external environment) in order to minimize their impact on the culture medium. Water quality deterioration, arising from domestic and industrial sewage discharge, and the reducing availability of freshwater arising from increasing water abstraction rates and diversion for domestic, agricultural and industrial purposes are key challenges faced by several countries. Concurrently, national institutions and aquafarmers alike have also recognized (through a process of learning) that activities within the aquaculture sector (i.e. the internal environment) require similar planning to collectively support the sector in order to attain and maintain a wholesome environment, a prerequisite for individual farms as well the aquafarming community. The holistic objective in regulating the external and internal environments is to minimize any negative impact on both environments, whilst improving production efficiency during and beyond the economic life of the operation, thereby ensuring stability, return on investment, equity, etc.

As in any other food production sector, sustainable development of aquaculture will depend on the extent to which each of the several inter-dependant components develop and, more importantly, the extent to which each successfully co-ordinates, contributes and interacts to support the other (Fig. 11).

The greater the overlap (i.e. better understanding and uptake of each other's objectives) the higher the probability of sustainable development and growth. In practice, whilst the perceived need for developing an interactive frame in the early phases of aquaculture development is low, there is often a growing realization that as output and competition increases, readily available markets saturate and input resources become limiting (i.e. the mature phase), common needs between the components of the above aquaculture frame increases. For example, a greater effort will be placed between the enabling environment and technological and social development on issues such as equitable access to and allocation of required resources and the need to increase and sustain production efficiencies. Similarly, between marketing and production there is likely to be

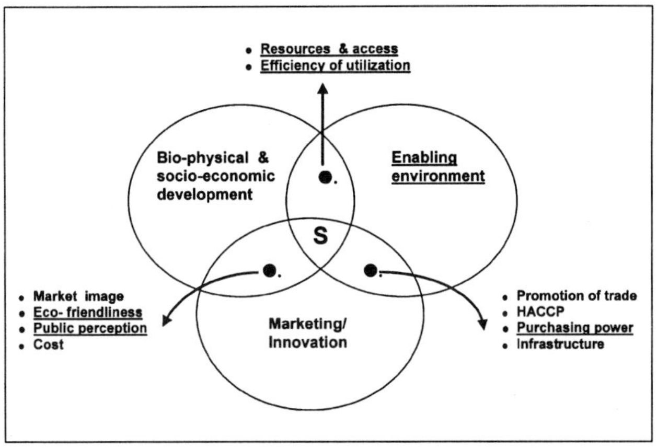


Fig. 11. Aquaculture frame for sustainable development (S): interaction and contribution of key components.

greater interaction to promote local and international trade, develop capacity to improve product quality and infrastructure to reach distant markets. To maintain and expand national and international markets in the light of competition of fish in the market place with other animal products, in particular poultry, marketing institutions, private enterprises and producers alike will have to place more emphasis in addressing public perceptions of environmental concerns and this would require greater interaction to develop strategies to develop identifiable products, perhaps with brand or trade names, and generally promote the sector.

Many of these challenges are being addressed on an ongoing basis in countries including China and Thailand.

## Changing Scenarios in Aquaculture Development in China

The Chinese population is predicted to rise from the present 1.2 to 1.6 billion by 2026, reducing further the *per capita* share of land resources for food production. *Per capita* agricultural land has steadily decreased from 0.19 hectares (ha) in 1949 to 0.09 ha in 1995 (Wu, 1996). These considerations and the rapid changes in population structure, and rising living standards, have presented the Chinese with several challenges and opportunities to meet the rising demand for low and high quality animal products, in particular aquatic products. Between 1991 and 2020 the national *per capita* consumption of fish is projected to increase annually by 5.6% (Huang et al., 1997). The acknowledgement of stagnating wild fish stocks has focused Chinese fishery development policies on expanding inland, brackish and in particular marine aquaculture as a key strategy for meeting changing national demand and consumer patterns.

## Recent Changes in Chinese Living Standards and Consumer Preferences

The gradual transformation of the Chinese economy from a centrally based to a market-based economy and the decentralisation of the economy to minimise urban migration has had a significant impact on the living standards of Chinese urban and rural populations. Between 1990 and 1996 net *per capita* income of rural population rose by nearly 19%/year from 650 in 1990 to 4200 yuan in 1996. The increases in living standards of the rural sector derived through diversifying and industrialising the rural economy, were even higher and rose by 21%/year from 350 to 1700 yuan in the same period (Fig. 12).

The accompanying increases in purchase power and the growing affluence of urbanised and rural populations has influenced the consumption patterns of main commodities (Fig. 13).

Since the reform policies in the 1980s, Chinese household surveys suggest a gradual shift from a predominately vegetarian (grain and vegetable) to a meat diet. In both urban and rural areas the *per capita* purchases of grain and vegetables have declined and that of red meat and aquatic products increased (Fig. 13). Urban populations with their higher disposable income, however, consume more aquatic products than red meat and between 1985 and 1996 the *per capita* purchases of aquatic products increased by 2.6%/year from 7 to 9.3 kg/year. In contrast the *per capita* purchases of red meat was higher (11–12 kg/year) in the rural sector and showed little change in the same time period. For aquatic products (fish and shrimps), however, *per capita* purchases increased at 7%/yr and more than doubled from 1.6 to 3.4 kg/year.

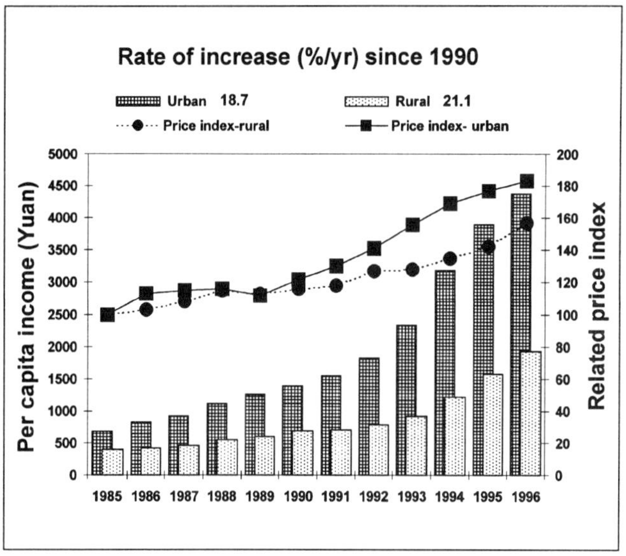


Fig. 12. Recent increases in net per capita annual income in rural and urban areas and Related Price Index. 1985 = 100 (data adapted from Chinese Statistical Year Book, 1997).

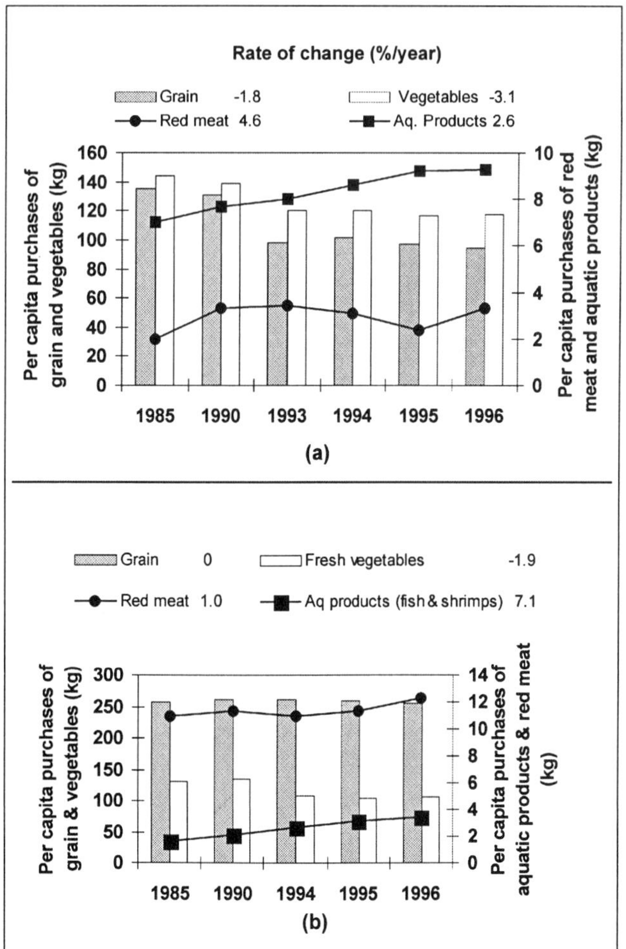


Fig. 13. Consumption pattern of major commodities in (a) urban and (b) rural populations in mainland China. (Data adapted from Chinese Statistical Year Book, 1997.)

## Addressing the Challenges

To meet these rising demands China has formulated and refined its aquaculture development policies targeting specific national, provincial and farm-level issues aimed at transforming the aquaculture sector from a centrally based to a market-based activity. At the national level, the development of inland aquaculture production was part of the strategy for rural industrial development. Freshwater aquaculture expanded from the traditional southern aquafarming provinces south of the Huai river, into northeastern, western, and northern regions of China. At the local level, China aimed to provide the necessary incentives for individuals, collectives, and state-run farms to increase production (Tables 4 and 5).

To increase fish production and employment in the Provinces, the area allocated for culture was increased and the types of water bodies approved for aquaculture broadened, attracting hitherto uninterested households, state-owned farms and water conservation departments in many villages and towns into taking up aquaculture as an additional viable economic activity. By 1990, over two million people were attracted to aquaculture from other industries, taking the number engaged in aquafarming to over six million (Qian, 1994). Total fishery (aquaculture and capture) labour in 1996 totalled 12 million. The number of people who were employed full-time in aquaculture in 1996 reached three million, an increase of 76% over 1990 (Zhao,

Table 4

Reported utility rate of land and water resources for aquaculture in China in 1996

| Total usable area | Area (× 1000 ha) | % Used |
|---|---|---|
| Marine | 9360 | 61 |
| Mud flats | 797 | 67 |
| Bays | 180 | 97 |
| Shallow seas | 1622 | 7 |
| Inland | 6760 | 72 |
| Ponds | 1960 | 100 |
| Lakes | 2150 | 41 |
| Reservoirs | 1884 | 81 |
| Waterways | 756 | 48 |
| Paddy fields | 6867 | 18 |

Adapted from Zhao (1997).

Table 5

Producer and local level incentives for developing the aquaculture sector

- Management decentralized to county-level administration and producers
- Target production in co-operatives linked with sole responsibility on profitability
- Support on credit, materials, processing and marketing
- Around 3350 large and 2200 cold-storage facilities to foster handling and accommodate increased production.
- Deregulation of price controls, mainly for high valued species. (Some control over low valued species to ensure their affordability to most Chinese.)

1997). This increased opportunity, particularly in freshwater production, played an important role in alleviating rural poverty and increasing the income of farmers engaged in capture fisheries and aquaculture.

## Allocation and Utilisation of Natural Resources

Aquaculture development in China focused on increasing production from inland as well as marine waters. Since 1984, the area and intensity of production from ponds, and the utilisation of open waters such as lakes and reservoirs, rivers and rice paddies for freshwater aquaculture have steadily increased. In 1996, 100% of available ponds and 81% of reservoirs were utilised and opportunities still exist to further develop shallow seas and rice paddy fields.

Ponds and reservoirs are the principal types of water bodies utilised for freshwater aquatic production, accounting for 72% of inland area (excluding paddy fields) used for aquaculture in 1996 (Zhao, 1997). Ponds accounted for 40% or 1.96 million ha of inland cultured area. Unlike reservoirs, pond area utilised for aquaculture continued to increase at an average annual rate of 5.6% between 1990 and 1996. Similar increases (5.7%) in the utilisation of marine areas were also reported in the same period (Fig. 14).

In addition to area, production yields have also increased at an annual average of 9.4%/year with improved and more focused technological development. For ponds, national average yields have increased from 2385 kg/ha in 1990 to around 4100 kg/ha in 1996. Higher increases in fish production yield were achieved with marine fishes. Between 1990 and 1996, national yields increased at an average of 20.1%/year (Fig. 14).

## Production Diversification

Although the culture of the traditional carp and tilapia species, which are farmed across China, continues to dominate production, in recent years there have been greater efforts to diversify production to other carps and into higher-valued freshwater species such as mandarin fish, freshwater crabs and prawns, soft-shelled turtle and eels. In 1996 these new groups were valued at US$ 1.4 billion and represent 6.5% of total aquaculture value (Fig. 15).

China has also demonstrated that relatively high-value species are not necessarily carnivorous species. In matching resources and market opportunities, China has also focused

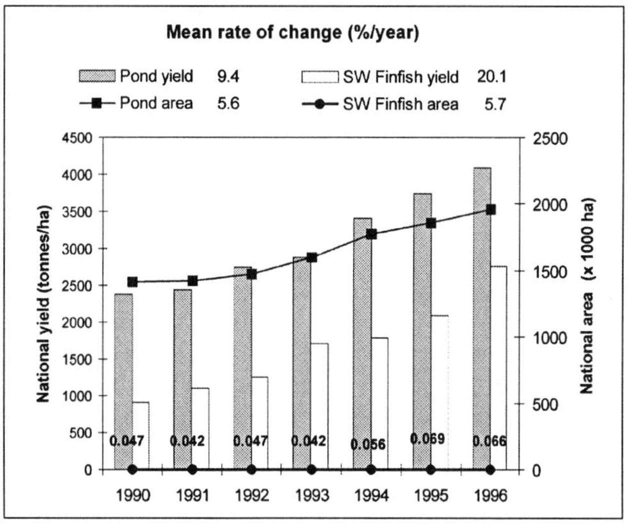


Fig. 14. Changes in land and water use, and production efficiencies in China.

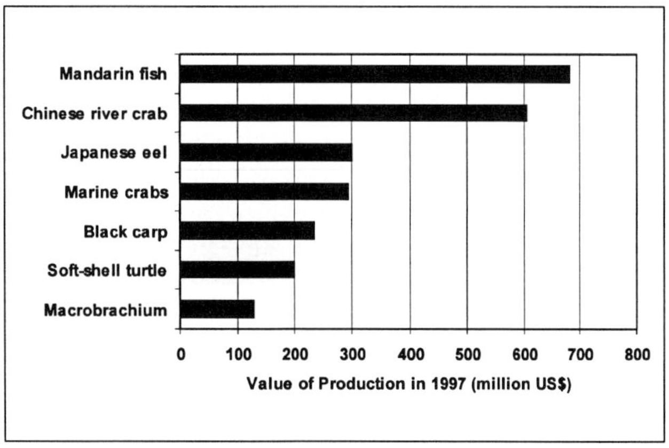


Fig. 15. Total value of recently targeted species in China in 1997.

Table 6

Regulations strengthening the internal/external environment in China

| |
|---|
| Fisheries law of the P.R. of China (1986) |
| - Article 10—Best use of suitable water and land for promoting aquaculture |
| - Article 12—Establish mangrove nature reserve to protect spawning and nursery grounds |
| The law of Marine Environmental Protection (1986) |
| The law of Water Pollution Prevention (1986) |

on developing relatively high value species which feed low in the food chain such as filter-feeding invertebrates e.g. oysters, scallops, razor shells, cockles, mussels, etc. Between 1990 and 1995 production of invertebrate filter feeders increased at 24%/year. In the same period non-filter feeding invertebrates (e.g. crabs, etc.) increased at 49% annually. Demand for marine finfish has mainly contributed to an annual rise of 33% in the same period.

In the last decade, the ongoing transformation from a centrally planned economy to a market economy has highlighted several shortcomings for the aquaculture sector. These include: (a) competition with other industrial sectors for land and aquatic resources, particularly in coastal regions; (b) degradation of water quality and culture environments through efforts of urbanisation, industrialisation and uncontrolled intensification of aquaculture; (c) limited processing capacity of aquaculture products; (d) slow or, in some cases, no implementation of market-oriented policies on price de-regulation; (e) unpredictable fluctuations in the quantity and quality of seed, particularly of high-value fresh and marine finfish and shellfish species; and (f) poorly maintained culture facilities. China has recognised these issues and is addressing them.

To address key issues such as pollution the government has introduced legislation for controlling water quality in order to protect aquaculture and capture fisheries. Since 1979, over 500 laws and regulations have been issued by the State Council (Zhao, 1997). Recent regulations are shown in

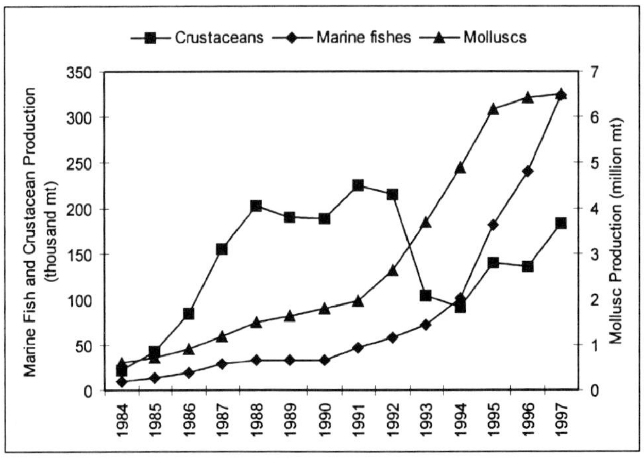


Fig. 16. Growth in Chinese mariculture production by feeding types.

Table 6. For farmers producing high-value species, including small shrimp, eel, mandarin fish etc., fluctuations in fry cost, supply and quality, increased feed and medication and other inputs costs, price fluctuations of end products and high quality standards for export products, have all increased investment risk. To promote sustained production of high-valued species, the state is promoting private investment and the formation of joint-ventures with foreign companies, which should continue to improve technology transfer and reduce some of the investment risk.

## The Thai Shrimp Sector

In Thailand, cultured *Penaeid* spp. account for around 6% of total fish and shell fish output, making Thailand the world's largest producer of *Penaeid* shrimps, valued at over US$1.5 billion in 1997. Given the economic importance of aquaculture and competition for coastal land and water resources, one key challenge prioritised in Thailand is to monitor and analyse the structure of this sub-sector. Clearly, issues such as source, area and access of land used, number, size and distribution of sites, farming practices and output efficiency and diversity and the socio-economic benefits of the activity are all crucial elements for developing and strengthening the aquaculture framework. These issues, which are evaluated below for Thailand, have recently become the focus of national and international attention primarily because of use of mangroves and disease outbreaks.

## Source of Land for Shrimp Production

In Thailand, mangroves have traditionally been viewed as a multipurpose natural resource and mangrove areas are used for roads, ports, salt pans, agriculture, aquaculture, etc. Since the 1960s, but notably since the late 1970s, one of these uses has been *Penaeid* culture. In 1993, according to the Royal Forestry and Land Development Departments and the National Research Council of Thailand, only 17.5% or 65,000 ha of the Thai mangrove resources have been used for aquaculture (Fig. 17). Around 36% have been

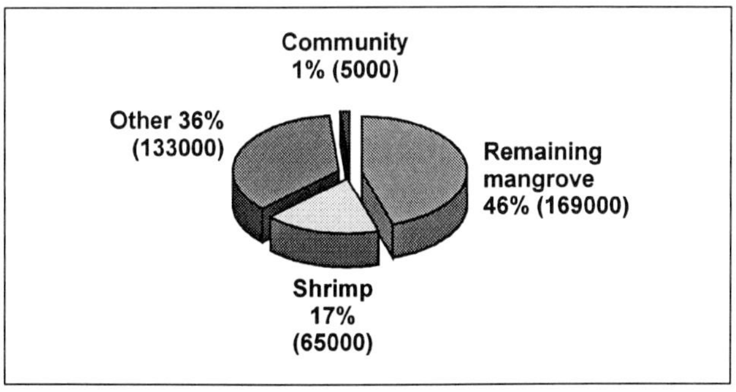


Fig. 17. National utilisation of mangroves by various stakeholders in 1993). Other uses include agriculture, road development, ports, salt farms, fuel etc. (adapted from Menasveta, 1997).

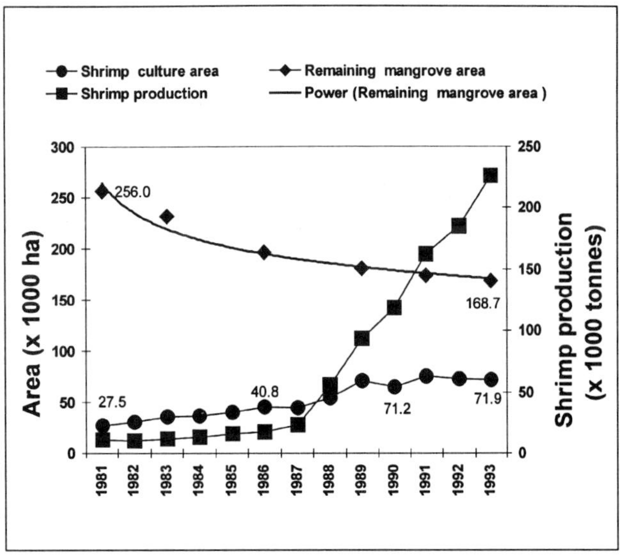


Fig. 18. Land utilisation: changes in total mangrove area in relation to development of shrimp farming (data adapted from Mesasveta, 1997).

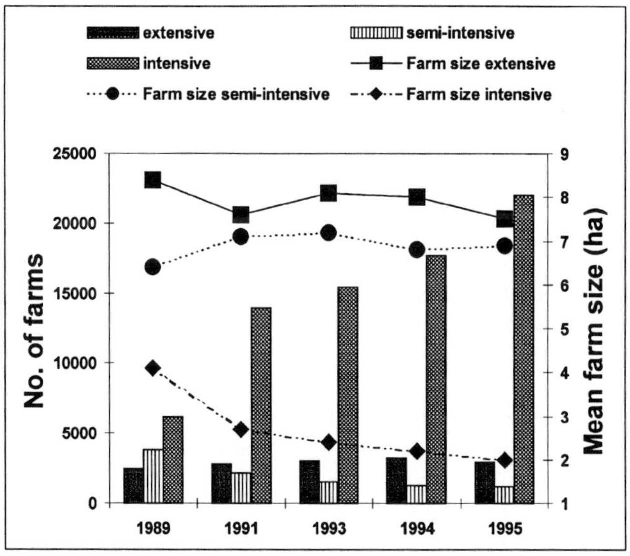


Fig. 19. Recent changes in the number and mean size of extensive, semi-intensive and intensive shrimp farms in Thailand (data adapted from Statistics of Shrimp Culture, 1993–1997).

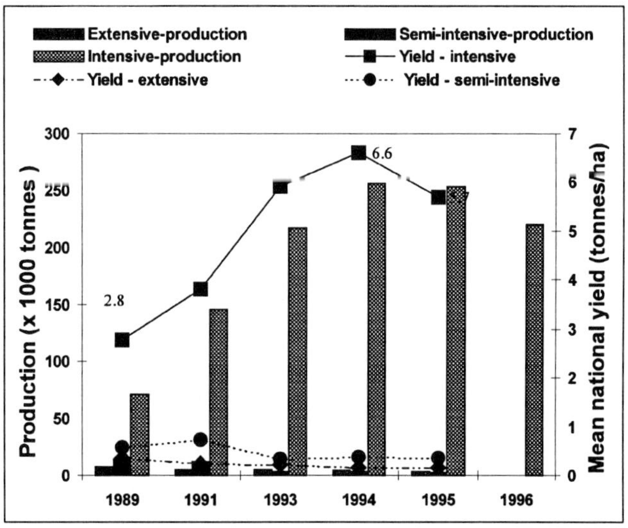


Fig. 20. Recent changes in annual national Thai shrimp production and mean yields under varying culture practices (data adapted from Statistics of Shrimp Culture, 1993–1997 and FAO/FIDI).

utilised for other purposes and just under half remains unused.

The utilisation of mangrove areas in relation to recent shrimp production and total shrimp farming area is shown in Fig. 18. Between 1989 and 1993 production increased by 141%. During the same period there was only a decline of 1.6%/year in mangrove areas resulting from total uses. Data suggest that much of the expansion in the shrimp sub-sector has originated from non-mangrove areas. In contrast to earlier and larger extensive shrimp farms, these increases are attributable to intensive shrimp culture practices on considerably smaller farms (see later) utilising higher ground that is more suitable to pond drainage and construction.

The shift from larger, less productive (in terms of yield) extensive farms to more intensive practices using aeration and pelletised feeds has been evident in recent years, concomitant with a sharp increase in the number of smaller intensive farms (Fig. 19). Between 1991 and 1995, intensive farms increased at a rate of 12% annually, in contrast with a sharp decline in average farm size which more than halved in the same period. The compactness of smaller farms, which allows for the better management of waste, feed, production, handling, harvesting, inventory management, security, etc., was also accompanied by improved yields when compared with larger extensive and semi-intensive farms (Fig. 20). Since 1989, the average yield from farms increased from 2.8 tonnes/year and reached a maximum of 6.6 tonnes/year in 1994 before declining to 5.7 tonnes/year in 1995 (Fig. 20). During the same period yields from extensive and semi-intensive farms ranged from 0.16 to 0.58 tonnes/ha and decreased by 13 and 7.8%/year, respectively. The plateauing of yield from intensive farms may indicate the onset of sub-optimal farm management manifested as production losses resulting from disease and reduced growth performance of shrimps. This decline is also reflected in the stagnation and subsequent decline in total shrimp tonnage since 1994, despite the increase in the number of intensive farms from 17,000 to 22,000 (see Fig. 19). The socio-economic impact of these production losses may be considerable, particularly as the majority of these farms mainly employ local people, and capital investment and working capital in these small farms is high. In 1995, 70% of those employed on farms were from the local area. Moreover, 73% of farms are owned by small farmers, the majority (65%) of which are less than 1.6 ha. In 1995, only 8% of all shrimp farms in Thailand were larger than 8 ha (Waiyaslip, 1997).

Table 7

Regulations guiding sustainable shrimp farming in Thailand

| | |
|---|---|
| 1. | Area allocated to shrimp farming limited to 76,000 ha |
| 2. | To assist monitoring all shrimp farms will be registered |
| 3. | All farms greater than 8 ha must be equipped with water treatment facilities. |
| 4. | BOD limit of discharge effluent must be to <10 mg/l |
| 5. | Discharged seawater should not enter agricultural lands or freshwater |
| 6. | Sludge or pond sediments cannot be discharged into canals or public areas. |

To foster and sustain shrimp farming the Department of Fisheries in Thailand has been promoting policies to guide the sub-sector to ensure that shrimp farming continues to be of benefit to the farmers and the country (Table 7).

## CONCLUSIONS

Overall, reported global aquaculture production suggests that the sector is likely to expand, though the economic downturn in a few key import markets may reduce demand for luxury products. Demand in producer countries, however, may increase. In China, for example, real rises in disposable income have increased fish consumption of low and high value species in rural and urban areas. In Thailand, species produced for local consumption is likely to increase. As markets for high-valued species saturate, farmers are likely to improve production efficiency, use facilities for other species and diversify product base to exploit niche markets. To ensure the long-term development of aquaculture, national aquaculture frameworks will need to be strengthened and their development placed in context with national food production strategies and development.

## REFERENCES


Anon. (1997) *China Statistical Yearbook No. 16.* China Statistical Publishing House, Beijing, China. 851 pp.

Beveridge M.C.M. (1996) *Cage Aquaculture* (2nd edition). Fishing News Books. Oxford, UK. 346 pp.

De Silva, S.S. (ed.) (1998) *Tropical Mariculture.* Academic Press. London. 487 pp.

Elzière-Papayanni, P. (ed.) (1993) *Coquillages.* Informations Techniques des Services Vétérinaires Français. Paris. 522 pp.

FAO/FIDI (FAO Fisheries Department, Fisheries Information, Data and Statistical Unit) (1999) Aquaculture, Capture, Global Production and Trade Databases (included in FISHSTAT Plus: universal software for fishery statistical time series). FAO, Rome.

Heen, K., Monahan, R.L. and Utter, F. (eds.) (1993) *Salmon Aquaculture.* Fishing News Books. Oxford, UK. 278 pp.

Huang, J., Rozelle, S. and Rosegrant, M.W. (1997) China food economy to the twenty-first century: Supply, demand and trade. International Food Research Institute, Washington, DC. 18 pp.

McVey, J.P. (ed.) (1993) *CRC Handbook of Mariculture: Crustacean Aquaculture.* CRC Press. Florida, USA. 526p.

Menasveta, P. (1997) Mangrove destruction and shrimp culture systems. *World Aquaculture,* Dec., 36–42.

Menzel, W. (ed.) (1991) *Estuarine and Marine Bivalve Mollusk Culture.* CRC Press. Florida, USA. 362 pp.

Perez, R. et al. (1992) *La Culture des Algues Marines dans le Monde.* IFREMER. Plouzane, France. 614 pp.

Stephanis, J. and Divanach, P. (1993) Farming of Mediterranean finfish species: present status and potentials. In *From Discovery to Commercialisation,* eds. M. Carrillo et al. European Aquaculture Society. Oostende, Belgium. 290–291 pp.

United Nations (1992) Earth Summit Agenda 21. The UN Programme of Action from Rio. United Nations, New York.

Waiyaslip, M. (1997) Dramatic changes in shrimp culture in Thailand. Fisheries Economics Division, Department of fisheries, Bangkok, Thailand. 13 pp.

Wu, Y. (1996) Pollution threatens fisheries. *China Daily,* 25 August 1996, p. 8.

Zhao, W. (1997) Research on the sustainable development of aquaculture in China. Paper prepared for the first meeting of the FAO/APFIC Aquaculture and Inland Fisheries Committee meeting. 22 pp.


## THE AUTHORS


**Krishen J. Rana**
*Fisheries Department, FAO,*
*Rome, Italy*

**Anton J. Immink**
*Fisheries Department, FAO,*
*Rome, Italy*

Chapter 119

# CLIMATIC CHANGES: GULF OF ALASKA

Howard Freeland and Frank Whitney

The Gulf of Alaska forms a distinct oceanographic and ecological system within the northern region of the Pacific Ocean. It is bounded to the north and east by continental margins, and to the south by the subarctic front. The western boundary has no clear definition and merges smoothly into the waters of the northwestern Pacific, but for convenience we take the international dateline as the western limit, although it has neither ecological nor topographical significance. Between 35° and 45°N the flows across the western boundary are inbound to the Gulf of Alaska. These flows are the Subarctic and the North Pacific currents, and both have their origins in the warm water currents extending eastwards from the Asian coast. At some point west of Vancouver Island these currents split. The part that heads south is known as the California Current, and the part that heads northwards is the Alaska Current. This current turns northwards, following the coast of the American continent, then flows eastwards along the coast of Alaska, and ultimately along the Aleutian Island chain, at which point it is referred to as the Alaskan Stream. The large anti-clockwise circulating body of water within the Gulf of Alaska is known as the Alaska Gyre. The sense of rotation implies, through geostrophy, that sea level must rise as we approach the coasts and, to compensate, the main pycnocline must dome towards the centre of the gyre. This puts more saline and nutrient-rich waters significantly closer to the surface at the centre of the gyre than at the coasts, where it is relatively easily accessible by mid-winter mixing events. However, nutrients are also injected at the coasts of the Gulf of Alaska. The west coast of North America is well known to come under the influence of wind systems that induce strong coastal upwelling events during the summer. The effect is strongest off southern California, decreases in intensity as we progress northwards and is virtually absent at the border between British Columbia and Alaska.

This chapter discusses some of the important properties of the Gulf of Alaska that must be understood before we look at the subsequent changes and evolution of the sub-arctic Pacific. Then it shows examples of important changes now taking place within the Gulf, and discusses the importance of these changes to its marine life.

*Seas at The Millennium: An Environmental Evaluation (Edited by C. Sheppard)*

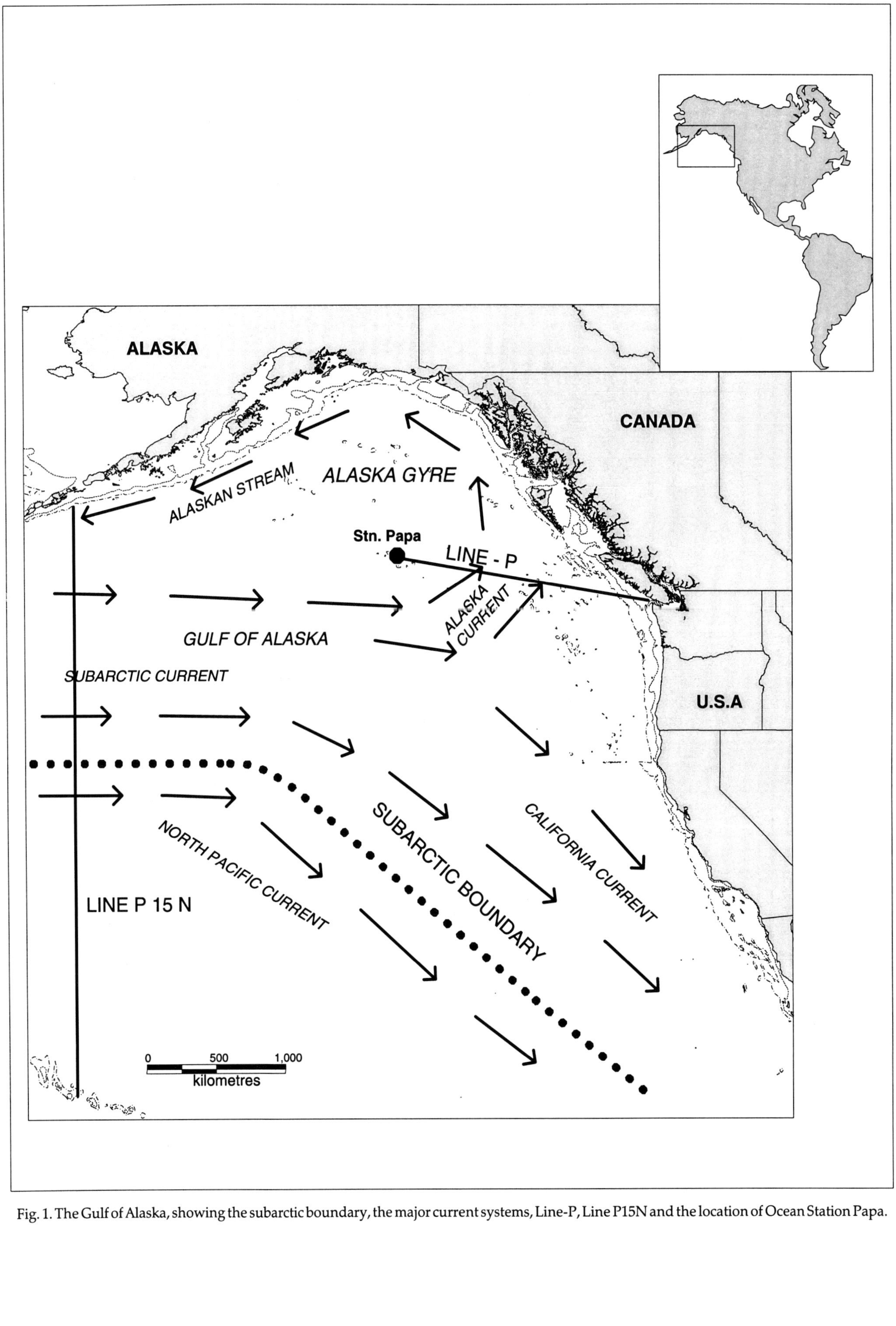


Fig. 1. The Gulf of Alaska, showing the subarctic boundary, the major current systems, Line-P, Line P15N and the location of Ocean Station Papa.

## PHYSICAL PROPERTIES

The distinctiveness of ocean properties in this region results from changes in the nature of the stratification of the Pacific Ocean near the subarctic boundary. The density of seawater is determined by both temperature and salinity. However, the northern regions of the north Pacific are unique in that the surface layer is distinctly fresh and winter cooling does not drive deep convection. The stability of a water column can be written as the density gradient, viz.

$$\frac{d\sigma_t}{dz} = \frac{\partial \sigma_t}{\partial T}\frac{\partial T}{\partial z} + \frac{\partial \sigma_t}{\partial S}\frac{\partial S}{\partial z}$$

showing that there is both a salinity and a temperature contribution to the vertical stratification. A stratification ratio may be written as the ratio of those two contributions

$$R = \left|\frac{\partial \sigma_t}{\partial S}\frac{\partial S}{\partial z}\right| / \left|\frac{\partial \sigma_t}{\partial T}\frac{\partial T}{\partial z}\right|$$

which may be computed along a north–south section using data collected along the WOCE Line P15N (survey completed in 1994) (Figs. 1 and 2).

The white areas in Fig. 2 correspond to areas where the vertical temperature gradient approaches zero. However, it is evident that an abrupt change occurs in the vicinity of the subarctic front. South of the front, stratification is dominated by temperature effects, and, north of the front, stratification is dominated by salinity. In Fig. 2 properties are plotted only to a pressure of 200 dbars, because the pycnocline in the northern North Pacific lies between 100 and 200 dbars, making a rather strong contrast with processes in the North Atlantic Ocean. Thus, the subarctic front represents not just a boundary between waters that ultimately head northwards and those that head southwards, but also the front represents a fundamental change in the nature of the stratification of the north Pacific Ocean. This change has many implications. It was pointed out by Freeland et al. (1997) that this implies that XBT surveys cannot be used for estimation of the depth of stratification north of the front. However, a more fundamental result is that the mode of response of the Pacific Ocean to global climate change is likely to be very different north of the front, compared with the subtropical response. Rainfall is known to be large north of the front and evaporation relatively small. It is likely that the Gulf of Alaska will be far more responsive to changes in evaporation or precipitation, resulting from global climate change, than to temperature changes. That being the case, we must also note that changes in temperature can and do affect the stability of the water column and that any such changes will themselves be reflected in changes in the salinity of the upper ocean. The ocean is intrinsically non-linear and it is very hard to tease out cause and effect in the upper ocean system.

The change in the nature of the stratification alters the depth of penetration of deep winter mixing and so alters the rate at which nutrients are supplied. It is striking how clearly, as shown in the relationship between Figs. 2 and 3, the nutrient supply in the North Pacific is determined by position relative to the subarctic front. It is well known, of course, that the subarctic Pacific is a high nitrate, low chlorophyll region (HNLC), meaning that the supply of the macro-nutrients is in excess of what the phytoplankton community can utilize, and some other effect or parameter keeps biomass (chlorophyll) low (Boyd et al., 1995). It is presently believed that in the central Pacific primary production is limited by supplies of a micro-nutrient, iron, and this hypothesis (see Miller et al., 1991) remains an active field of research.

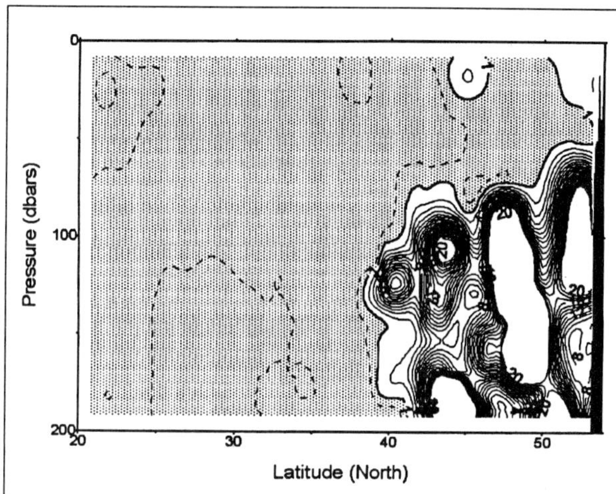


Fig. 2. The ratio of the salinity and temperature contributions to the stratification in the northern Pacific Ocean varying with latitude along WOCE line P15N. (The dashed line indicates a value of $R = 0.25$ and shading areas of $R < 1$, temperature dominant, regions.)

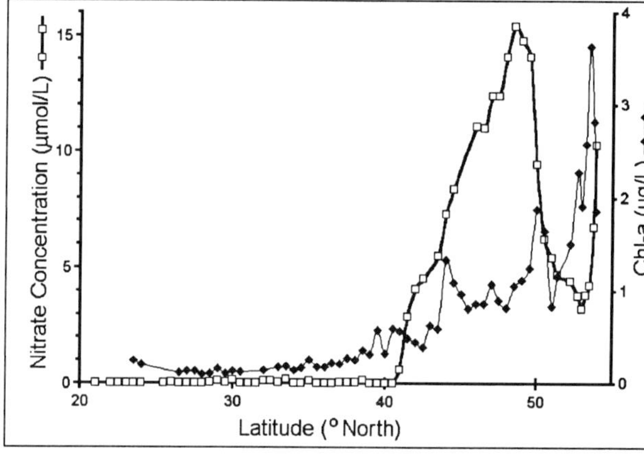


Fig. 3. The concentration of dissolved nitrate (open squares) and Chl-a (filled diamonds) in the surface waters of the N. Pacific along the same line as in Fig. 2.

## CHANGES IN PHYSICAL PROPERTIES

It is well known that the Earth is warming in response to global climate change (e.g. Houghton et al., 1995). The trends in air temperatures have been very widely reported. It is also known that ocean temperatures are rising in most

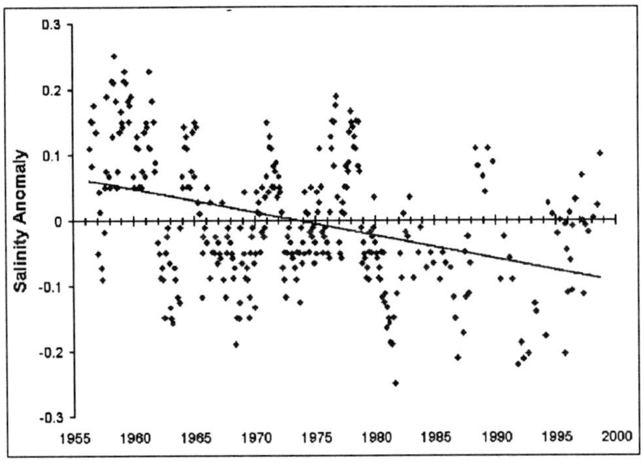


Fig. 4. Monthly averaged salinity anomalies at Ocean Station Papa.

parts of the world. However, any systematic changes in ocean temperatures will inevitably alter evaporation rates and so ultimately change the distribution of precipitation around the world. Precipitation is notoriously hard to predict and the forecasts from the current generation of global climate models should be approached cautiously, but we should reasonably expect that there will be changes in the global hydrology. Recently, Wong et al. (1999) show evidence of changes in the salinity at the deep salinity minimum in the North Pacific, which is evidence of a progressive change in the surface hydrology of the northern North Pacific, where the Pacific Intermediate Water mass outcrops. Figure 4 shows the deviation from normal salinity as seen at Ocean Station Papa, at 50°N, 145°W. It is clear that, though there is a large amount of scatter, the salinity is declining. Similar trends have been reported by Royer (personal communication) from a time series station off the coast of Alaska. The trend is large, amounting to the equivalent of 0.35/century.

The source of the salinity trend is something of a mystery. The obvious candidate would be global warming, which should induce changes in the distribution of precipitation. But it was pointed out by both Freeland et al. (1997) and Wong et al. (1999), that the database on precipitation is inadequate to examine this hypothesis. Ebbesmeyer (private communication) has suggested that the changes are driven by changes in the seasonality of freshwater supply from major rivers. However, we note that the salinity at Ocean Station Papa averaged over the top 250 metres shows no evidence of any significant trend. Does this imply that there is no net freshwater input to the surface of the waters of the Gulf of Alaska? Or, alternatively, is increased freshwater input altering the circulation in such a way that saltier water is being advected in from other regions of the Pacific below the mixed layer? If the former explanation is correct, then could this change in salinity actually be a response to changes in the depth of mixing in the central Pacific? Irrespective, large changes in surface salinity are occurring and this has large implications for the dynamics of the oceanic mixed layer.

In the central Gulf of Alaska the surface temperature is increasing as surface salinity is decreasing. Both of these conspire to decrease the density of the surface waters of the Northeast Pacific. We see little change in deep densities (Whitney and Freeland, 1999) so the net result is to change the energetics governing the formation of mid-ocean mixed layers. The problem is as follows. During the winter there is an increase in general storminess and general agitation of the surface ocean. This generates mixing that deepens the oceanic mixed layer. To do this, we start with a piece of ocean with a shallow low-density layer overlying higher-density water. Wind mixes the sub-surface dense waters into the near-surface waters, raising its density. Thus the centre-of-mass of the water column is raised, and its potential increased. There is a balance between the work done on the surface by wind action, generating turbulence, and the change in potential energy. If the rate of working by wind action does not change, then the change in potential energy of the water column cannot change. However, if the density contrast in the upper ocean has been increased, as observed, then to keep the potential energy change constant, the depth of mixing must decrease. This elementary physics has surprising consequences. At Ocean Station Papa we can survey the archive of density observations and calculate the mid-winter mixed layer depth. This is shown in Fig. 5, which is an extension, to include more recent observations, of a plot shown previously by Freeland et al. (1997). The trend line (a linear regression) is shown on the plot and represents a shallowing trend of 47 m/century. This trend is significant at the 95% confidence level.

Although, as discussed earlier, salinity contributes more to density structure than does temperature in this part of the ocean, we can see events in Fig. 5 that are responses to temperature changes. The very shallow mixed layer depths marked on Fig. 5 with the labels A and B occurred during the winters of 1982/83 and 1997/98, respectively, i.e. during winters associated with extreme warm conditions during El Niño events. The surface waters of the Gulf of Alaska cooled to the long-term (1956–91) average during the

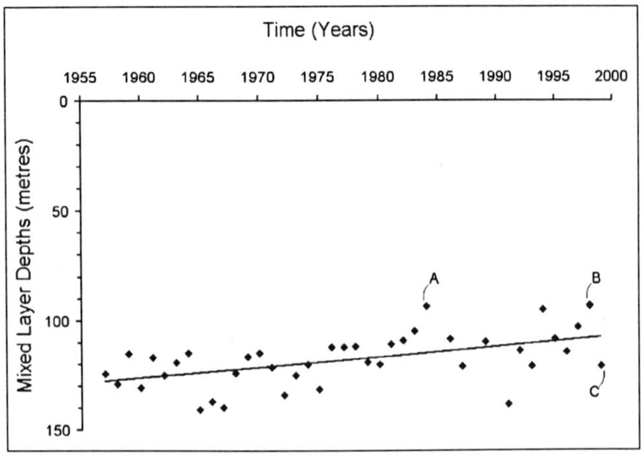


Fig. 5. Mid-winter mixed layer depth at Ocean Station Papa.

winter of 1998/99 (label C) but were cooler than observed between 1992 and 1998 by 1.5°C, resulting in a relatively deep winter mixed layer. It is striking how effective the salinity changes are; the ocean was exceptionally cold during that winter, which only managed to return the mid-winter mixed layer depths back to what one might call "normal conditions". As of writing (September 1999) conditions remain cool in the central Pacific and we anticipate that the winter of 1999/2000 will also result in a relatively *normal* winter mixed-layer depth.

## IMPACT OF EL NIÑO EVENTS

So far, the focus has been on the so-called secular changes, changes that might be attributed to global warming or the greenhouse effect. But the Pacific ocean is also subject periodically to very large changes which have a large impact on the physics, chemistry and biology of the Gulf of Alaska. El Niño events originate in the tropical Pacific and have global influence. However, the influence of these events is particularly large and striking around the coastal regions of the eastern Pacific.

The effect of the 1997/98 El Niño is shown in Fig. 6. The panels comprising this figure show the distribution of the sea surface temperature anomaly across the Gulf of Alaska with a line drawn showing the location of Line-P, and section plots of temperature and salinity anomaly observed along Line-P during September 1997 and February 1998. Before the El Niño event started, sea surface temperature anomalies were weak throughout the Gulf of Alaska. By the autumn of 1997 (see the first row of panels in Fig. 6), a major warming event has clearly taken place throughout the Northeast Pacific Ocean. However, even as late as the autumn of 1997, the warming is apparent only at the surface of the ocean, with very little penetration of heat showing in the thermal anomalies along Line-P. In the salinity field significant anomalies do show up as a linear feature near a depth of 100 dbars; this feature owes its existence to the shallowness of the mixed layer compared with the long-term normal.

In February 1998 the situation evidently changed in dramatic fashion. In some respects the peak of the El Niño appeared to have passed by before 1998 began. In the first column it can be seen that the area of the Gulf of Alaska affected by the large positive sea-surface temperature anomalies has in fact decreased markedly, and the largest temperature anomalies have decreased. But inspection of the section plots shows that Line-P is strongly affected by a sub-surface temperature anomaly just seaward of the coast and with peak anomalies at a pressure of around 200 dbars. The salinity field remains influenced by the elongate feature, due to the shallower mixed layer, but close into the coast is a strong negative salinity anomaly. The properties of these water masses strongly suggest a more southerly origin and so appear to result from advection of water along the coast of the Americas.

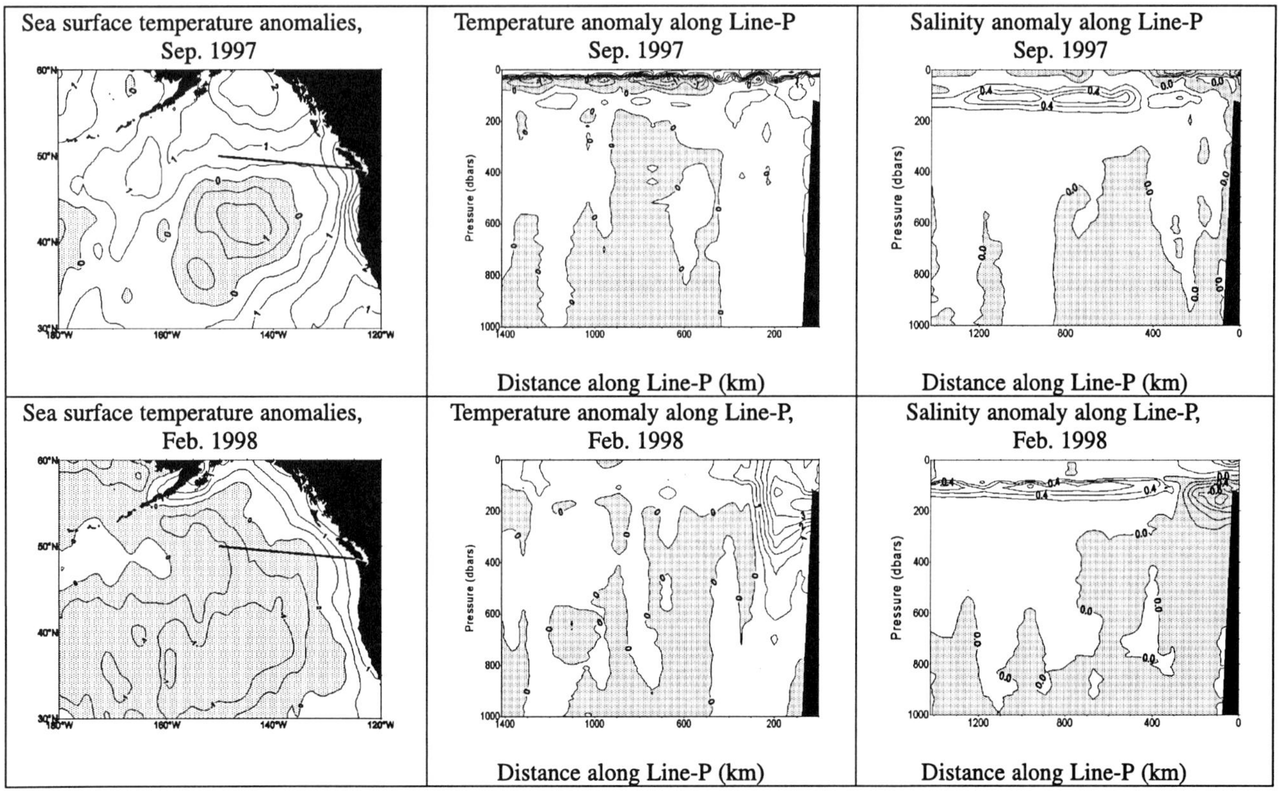


Fig. 6. Sea surface temperature anomalies in the Gulf of Alaska (1st column) and temperature and salinity anomalies along Line-P, 2nd and third columns for September 1997 and February 1998. In all cases, areas with negative anomalies are shaded.

Coincident with the occurrence of the large sub-surface warm anomaly, sea level at the coast moved abruptly upwards, reaching record heights of 40 cm above normal. This sea-level anomaly lasted for 3–4 months. Observations of sea level using the Topex-Poseidon satellite indicated that the sea level anomalies associated with the 1997/98 El Niño were largest at the coast and decayed rapidly offshore. The shape of the distribution of sea level was suggestive of Kelvin wave dynamics (Crawford, personal communication). We suggest that the initial superficial warming was due to an oceanic response to changes in the atmosphere, associated with the El Niño, but that the large anomalies of February and March 1998 were due to the effects of a propagating Kelvin wave. Some support for this view is given by Melsom et al. (1999) who report that interannual variability in the upper ocean coastal circulation in the Gulf of Alaska is linked to the El Niño–Southern Oscillation phenomenon in the tropical Pacific Ocean, via coastal Kelvin waves and atmospheric teleconnections. Irrespective of whether that event was in fact some kind of wave motion, the decay in surface elevation offshore does indicate that a strong anomalous along-shore flow must have occurred, and, from geostrophic considerations alone, this can be estimated at about 10 cm/s, lasting for three months. This is a large velocity anomaly and suggests that a substantial amount of heat was advected northwards during this period. In fact, we can do much better than that. We can compute the temperature anomaly along Line-P during February 1998 (Fig. 6). Further, we can estimate the geostrophic currents relative to the sea surface from any Line-P survey, and then correct those using Topex data to create an estimate of the absolute currents across Line-P. If we look at the difference between estimates in February 1998, compared with estimates before the El Niño arrived, we find a difference field that we can call the velocity anomaly. Multiplying the velocity and temperature anomaly fields, and then multiplying by the specific heat of seawater, yields a plot of the distribution of heat flux across Line-P (Fig. 7). The oceanic contribution to the heat transport implied by Fig. 7 is large and suggests that the oceanic wave guide does have a major contribution to make to the warming associated with El Niño events.

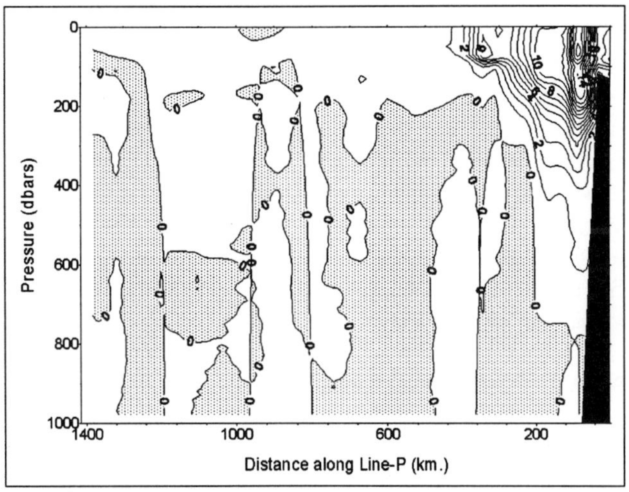


Fig. 7. Heat flux in units of cal $s^{-1}$ $cm^{-2}$. Positive values imply northward transport, and as before, negative contributions are shaded.

## IMPLICATIONS OF OBSERVED CHANGES

The tendency in recent years towards shallow mixed layers leads inevitably to a reduced nutrient supply. This is a simple consequence of the normal cycle of nutrient dynamics in the open ocean. Normal spring and summer primary production consumes 7 $\mu$mol/L nitrate in the mixed layer along Line-P that is replenished by deep mixing in the winter. A shallower mid-winter mixed layer implies a weaker supply of nutrients (Whitney et al., 1998). Despite the lower supply, the nutrient supply at the beginning of spring is ample to drive spring production, but a reduced initial supply increases the probability that, at some stage during the summer, nitrates will be wholly consumed over a broader area.

Normally Fisheries and Oceans Canada acquires data along Line-P only 3 times per year, in February, May/June and September. However, observations were enhanced during 1997/98 to ensure a thorough view of the passage of the El Niño signal. Thus it was possible to create the picture shown in Fig. 8. This shows the near-surface nutrient concentrations through 1998 (triangle symbols) compared with a long-term average. The change from normal is startling. Through the 1990s we have observed a steady decline in nutrient concentrations along Line-P, associated with the steady decline in supply implied by Fig. 5. However, during the winter of 1997/98 conditions were greatly exacerbated by the 1997 El Niño. The warming during this event was profound and created the largest density contrast in the Gulf of Alaska ever observed along Line-P. This produced

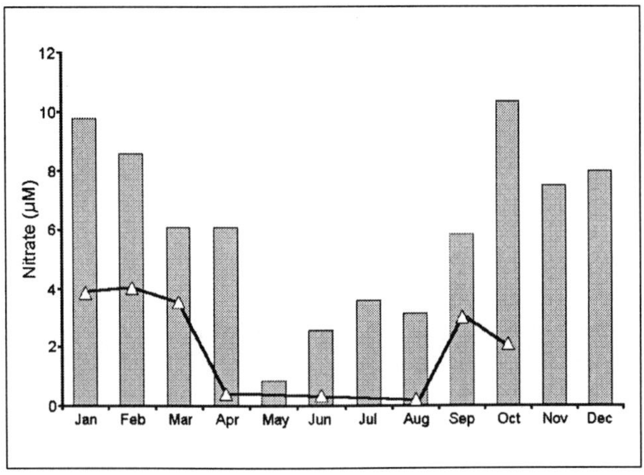


Fig. 8. Nutrients at inshore stations along Line-P during 1998. The bars show long-term averages along Line-P and the triangle symbols show the observations of 1998.

the shallowest mid-winter mixed layer ever observed, and so, as shown above, the lowest supply of nutrients. The primary production of early spring allowed a rapid depletion of nutrients that reached undetectable levels by April 1998. This anomalous depletion of nutrients is primarily the result of progressive changes in the northern Northeast Pacific which are made more evident by the occurrence of the El Niño events. Ancillary evidence (Whitney et al., 1998) suggests that new nutrient depletion events are occurring over a large area of the Gulf of Alaska.

In late summer of 1992, 95 and 96 experiments were carried out (Whitney et al., 1998) to compare primary production between areas that are replete with nitrates and areas that are depleted. The results indicated carbon fixation rates of $920 \pm 160$ mg $m^{-2}$ $d^{-1}$ in waters containing nitrate compared with $580 \pm 180$ mg $m^{-2}$ $d^{-1}$ in depleted waters, a reduction of more than 40%. Apparently the reduction in nutrient supply is having a direct impact on primary production.

Another consequence of a thinner mixed layer, outlined in Freeland et al. (1997) is an apparent increase in the availability of light in early spring. This could advance early spring growth and create a spring demand for nutrients earlier in spring than previously. This could in turn stimulate a tendency towards the earlier development of zooplankton. In fact, just such a shift in the timing of zooplankton growth has been observed (Mackas et al., 1998).

## POSSIBLE BIOLOGICAL EFFECTS: SALMON AS AN EXAMPLE

Changes such as these can have a profound impact on the relationships among all elements of the food web in the Northeast Pacific, and probably are doing so. The fact that Pacific salmon do respond to changes in ocean temperature is well known to Canadian fisheries managers. Forecasts are made annually of the fraction of returning sockeye salmon that will choose a northern route around Vancouver Island as opposed to the normal southerly route through the Strait of Juan de Fuca. These forecasts are based entirely on temperature observations. However, it is interesting to go one step further and speculate on how a reduction in food supply will impact the Pacific salmon.

The association of ocean dynamics and decadal-scale changes in the marine habitat of commercially important fishes has recently become an important issue in fisheries management. In the 1970s and 1980s there was a tendency to assume that the impact of marine habitat changes could be considered to be random and it was fishing that most affected the long-term stability of fish populations. Assessment models generally considered that the dynamics of a population could be defined using data obtained from a fishery. The impacts of the environment and associated species were seldom incorporated into the assessment. This was particularly true for Pacific salmon which were thought to be strongly influenced by the fishery, the numbers that escaped the fishery to spawn, and the quality of the freshwater habitat. Harvest rates for the various species of salmon were high, frequently exceeding 50% and sometimes exceeding 70% (Groot and Margolis, 1991). With such high fishing mortality it appeared logical that the abundance in the ocean would not reach a level in which there was some degree of self regulation or, to use fisheries terminology, there was density-dependent mortality. Consequently, the limits of abundance were believed to be related directly to the number of juveniles entering the ocean. Because the fishery removed potential spawning females, there was a trade-off in the management of salmon between having the number of juveniles that would produce the maximum return possible and having a sustainable fishery that provided food and employment.

It is now necessary to suggest a revision of some of these views. At the time of writing this chapter we do not have firm conclusions that would require a secure model of the North Pacific ecosystem including all relationships between climate fluctuations and changes in salmon mortality, which would be a tall order! However, consider the following:

- We have witnessed profound changes in the physical and chemical structure of the Gulf of Alaska over the past 45 years.
- We appear to be witnessing decreases in the supply of phytoplankton and zooplankton in the surface layers of the Gulf of Alaska.
- The salmon fishery is being managed conservatively, with the result that juvenile salmon are heading to sea in healthy, if not near record, numbers.
- Mature salmon are returning to the fisheries of British Columbia in near record low numbers.

The last two points taken together suggest that a change has occurred in the survival of salmon in the open Pacific; could this change result from simple starvation? We do not know the answer to that question. One obvious symptom of "starvation" would be a decrease in the fat content in the flesh of returning salmon. Unfortunately there has never been a program to monitor the fat content of fish, however, there is some anecdotal information (fishermen reporting that returning sockeye salmon are in poor physical condition) that suggest that this might well be the case.

In fact, many of these ideas are not entirely new, and indeed the concept that a reduction in food supply might lead to lean times for fish is probably not a surprise to anyone. Further, a food web model developed by Iverson (1990) suggests that a 40% reduction in nitrate utilization should cause a 40% decrease in carnivorous fish growth. Elsewhere in the North Pacific there are other, closely related, scenarios being developed. McGowan et al. (1998) have shown that warming and reduced nutrient supply in California coastal waters have resulted in a 70% drop in zooplankton biomass. So the scenario being postulated in the Gulf of Alaska should not come as a great surprise.

In conclusion, as scientists, we feel excited that as we enter the third millennium we appear to be making enormous strides in our understanding of the dynamics of the Gulf of Alaska ecosystem. However, this excitement must be tempered by a genuine concern that the ecosystem that we hand over to our descendants is changing dramatically.

## REFERENCES


Boyd, P.W., Whitney, F.A., Harrison, P.J. and Wong, C.S. (1995) The N.E. Pacific in winter: II. Biological rate processes. *Marine Ecology Progress Series* **128**, 25–34.

Freeland, H.J., Denman, K.L., Wong, C.S., Whitney, F. and Jacques, R. (1997) Evidence of change in the N.E. Pacific Ocean. *Deep-Sea Research* **44** (12), 2117–2129.

Groot, C. and Margolis, L. (1991) *Pacific Salmon Life Histories.* UBC Press. Vancouver, BC.

Houghton, J.T., Meira-Filho, L.G., Callender, B.A., Harris, N., Kattenberg, A. and Maskell, K. (eds.) (1995) *Climate Change 1995: The Science of Climate Change,* Second Assessment Report of the Intergovernmental Panel on Climate Change, Working Group 1, Cambridge University Press.

Iverson, R.L. (1990) Control of marine fish production. *Limnology and Oceanography* **35**, 1593–1604

Mackas, D.L., Goldblatt, R. and Lewis, A.G. (1998) Interdecadal variation in developmental timing of Neocalanus plumchrus populations at Ocean Station P in the subarctic North Pacific. *Canadian Journal of Fish. Aquatic Science* **55**, 1–16.

McGowan, J.A., Cayan, D.R. and Dorman, L.M. (1998) Climate-ocean variability and ecosystem response in the northeast Pacific. *Science* **281**, 210–217.

Melsom, A., Meyers, S.D., Hurlburt, H.E., Metzger, E.J. and O'Brien, J.J. (1999) ENSO effects of Gulf of Alaska eddies. *Earth International* **3**, paper 001.

Miller, C.B., Frost, B.W., Wheeler, P.A., Landry, M.R., Welschmeyer, N. and Powell, T.M. (1991) Ecological dynamics in the subarctic Pacific, a possible iron-limited ecosystem. *Limnology and Oceanography* **36**, 1600–1615.

Whitney, F.A, Wong, C.S. and Boyd, P.W. (1998) Interannual variability in nitrate supply to surface waters of the Northeast Pacific Ocean. *Marine Ecology Progress Series* **170**, 15–23.

Whitney, F.A. and Freeland, H.J. (1999) Variability in upper-ocean water properties in the NE Pacific. *Deep-Sea Research, Part II*, **46**, 2351–2370.

Wong, A.P.S., Bindoff, N.L. and Church, J.A. (1999) Large-scale freshening of intermediate waters in the Pacific and Indian Oceans. *Nature* **400**, 440–443.


## THE AUTHORS


**Howard Freeland**
*Institute of Ocean Sciences,*
*P.O. Box 6000,*
*Sidney, B.C., V8L 4B2, Canada*

**Frank Whitney**
*Institute of Ocean Sciences,*
*P.O. Box 6000,*
*Sidney, B.C., V8L 4B2, Canada*

Chapter 120

# EFFECTS OF CLIMATE CHANGE AND SEA LEVEL ON COASTAL SYSTEMS

Shiao-Kung Liu

Strong expansion of human activity in coastal regions of the world requires engineers and policy makers responsible for resources development, to examine the potential impacts of changes to the coastal region. One important example is sea-level rise. Sea-level rise may induce long-term changes in near-shore tidal circulation patterns resulting from the alteration of residual currents. Rise in sea level would also cause more areas to be flooded due to storm surge. Existing sea walls and dikes designed for 100-year storms would be exposed to higher risks if faced with rising mean sea levels. Sea-level rise may also cause alteration of long-term coastal sediment transport direction, and may induce possible impacts on ecosystems due to changes in the coastal circulation patterns. Presently, coastal numerical models, after adjustment and verification for a particular coastal region, can be used for predicting potential impacts for a region. This is shown here with several examples, assuming a 1 m sea-level rise.

*Seas at The Millennium: An Environmental Evaluation (Edited by C. Sheppard)*

## CLIMATE CHANGE AND SEA LEVEL

Global mean surface temperature has increased by more than half a degree Celsius during the last century. This warming trend is thought to be at least partly the consequence of human activities. Solar energy drives the earth's weather and climate system. It also heats the earth's surface; in turn, the earth radiates energy back into space. Atmospheric greenhouse gases (carbon dioxide, methane, nitrous oxide, chlorofluorocarbons, and water vapour) absorb and re-emit infrared radiation, which includes the wavelengths of radiation emitted by atmospheric gases and clouds, and by the earth's land and oceans. Several scientists have concluded that atmospheric concentrations of greenhouse gases have increased significantly above pre-industrial levels, and that the increase is due to anthropogenic activities.

Based on observed atmospheric concentrations of greenhouse gases and samples of air trapped in ice cores, it has been concluded (Barron, 1995) that the concentration of carbon dioxide is approximately 30% above the pre-industrial level, the observed atmospheric concentration of methane is 100% above pre-industrial level, and the concentration of nitrous oxide has increased by about 10%. The observed atmospheric concentrations of chlorofluorocarbon (CFC) are due solely to anthropogenic sources. Due to the heat-trapping property of these gases, global temperatures are thus gradually rising. Because of the increasing rate of global population growth and the similarly increasing rate of industrial/economic development, the global average temperature is expected to continue increasing, by an additional 1.0–4.5°C by the year 2100 (Barron, 1995; Karl et al., 1997).

Over the past century, as a result of global warming, the snow cover in the Northern Hemisphere and the floating ice in the Arctic Ocean have decreased. Globally, the sea level has risen 10–30 cm. By 2100, sea level is likely to rise about 0.75 m along most of the U.S. coast. Estimates of climate change for specific areas are much less than global ones.

As a result of global warming, more evaporation is induced, leading to more precipitation. Warmer sea surface temperatures would then induce more frequent and more intense tropical cyclones in the next century.

For estimating the long-term climate change induced by the increase of atmospheric anthropogenic greenhouse gases, climate models are used together with observed records and estimates of the response of the climate to various forcings in the geological past.

## COASTAL EFFECTS

As the global sea level rises, its effects on the regional coastal waters differ according to location (Liu, 1997). Because bottom frictional effects are much more pronounced over the continental shelf than they are in the deep ocean, effects of sea-level rise on the coast will be higher. Hydrodynamic processes in the coastal zone are complex. Sea-level rise may cause changes in shoreline configuration, near-shore depth distribution, and changes in various estuarine/river interactions. It may also induce gradual changes in tidal propagation as well as changed coastal responses to passing storms. From a hydrodynamic point of view, rising sea level alters the near-shore momentum transfer process in the water column. A depth change causes a redistribution of turbulent energy density, as this is generated from surface/bottom frictional stress, which decays over the water column. These are nonlinear processes.

When tides or long waves (e.g. storm surges, tsunamis) enter a coastal sea, they will not only change in amplitude but will also transfer turbulent energy from one wave frequency to another. This energy dispersion process generates higher-order tides and thus induces tidal residual circulation through nonlinear advection and transport dynamics (Liu, 1997). Bottom friction generates odd harmonics of the basic tidal frequencies while depth and bathymetry generate even harmonics. In a three-dimensional case, the rise of sea level may also cause changes in baroclinic circulation through density structure. In the Arctic Ocean, it may change the location of the density fronts and the area of the marginal ice zone (Liu and Leendertse, 1981, 1982, 1984, 1987a, 1987b). This may in turn induce significant impact on marine ecology and coastal fisheries.

## EFFECTS ON TIDES AND TIDAL CURRENTS

From a physical point of view, the propagation of tides in the coastal area is dominated by the bathymetry and to a certain extent by the vertical density structure. The density structure in the vertical is as important as the horizontal distribution for its effects on the residual circulation and dispersion process. Bathymetry and bottom friction generate residual circulation through nonlinear advective mechanisms in the equations of motion. These higher-order mechanisms not only modify tidal levels along the coastline, but, more important, they create transport responsible for carrying floating and soluble substances for longer time periods, which is of particular importance to the impact of sea-level rise. The tidal residual in homogeneous water is different from the tidal residual of stratified water. In stratified water the vertical momentum transfer is suppressed, the vertical velocity profiles are different, and different nonlinear advection and dispersion patterns generate a different residual. When deep-water ocean tides enter a shallow shelf area, their amplitude and phase changes depend on the coastal bathymetric distribution and shoreline configuration.

Generally, the magnitude of residual tidal circulation is inversely proportional to the local depth. Consequently, the effect of sea-level rise is more pronounced in shallower coastal areas. One good example is the Gulf of Alaska (Fig. 1) which covers the area from Vancouver Island to Unimak

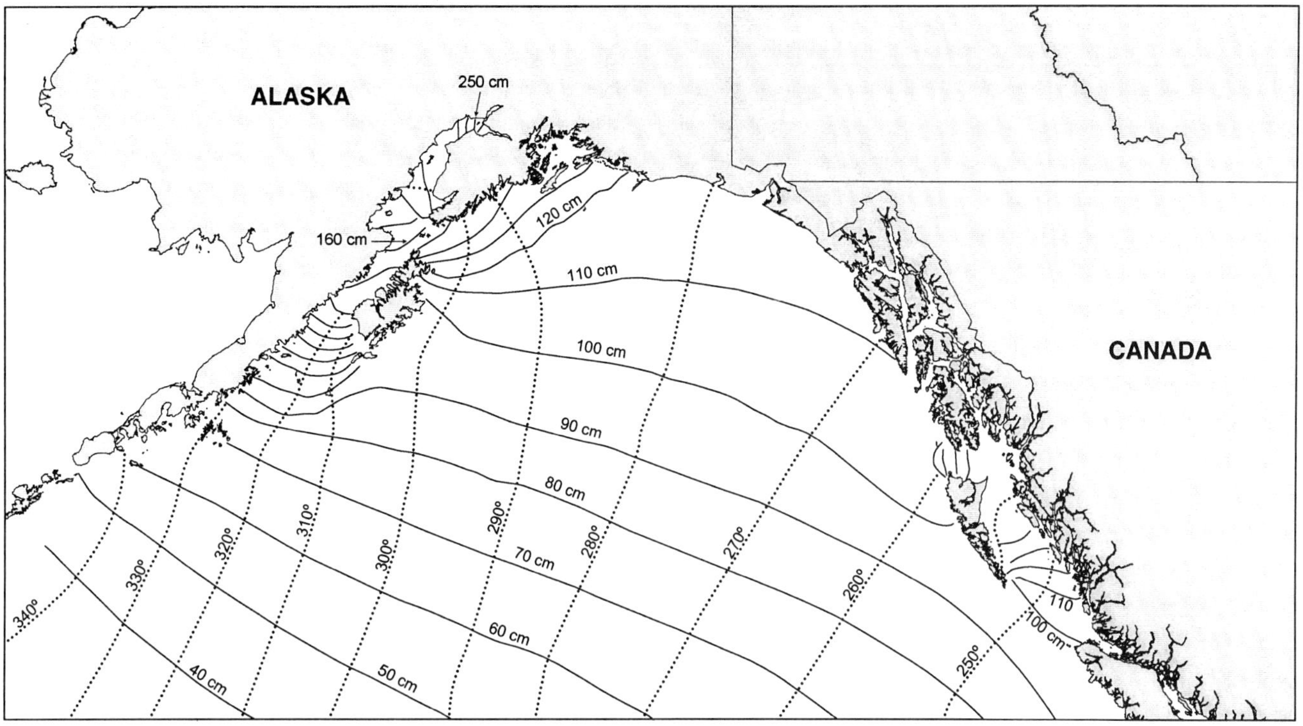


Fig. 1. Computed co-tidal chart for the semidiurnal component (primarily M2). Each 10° in phase represents approximately 20 minutes lag relative to the Greenwich mean phase. The maximum tidal amplitudes, reaching 250 cm, are found in the Cook Inlet.

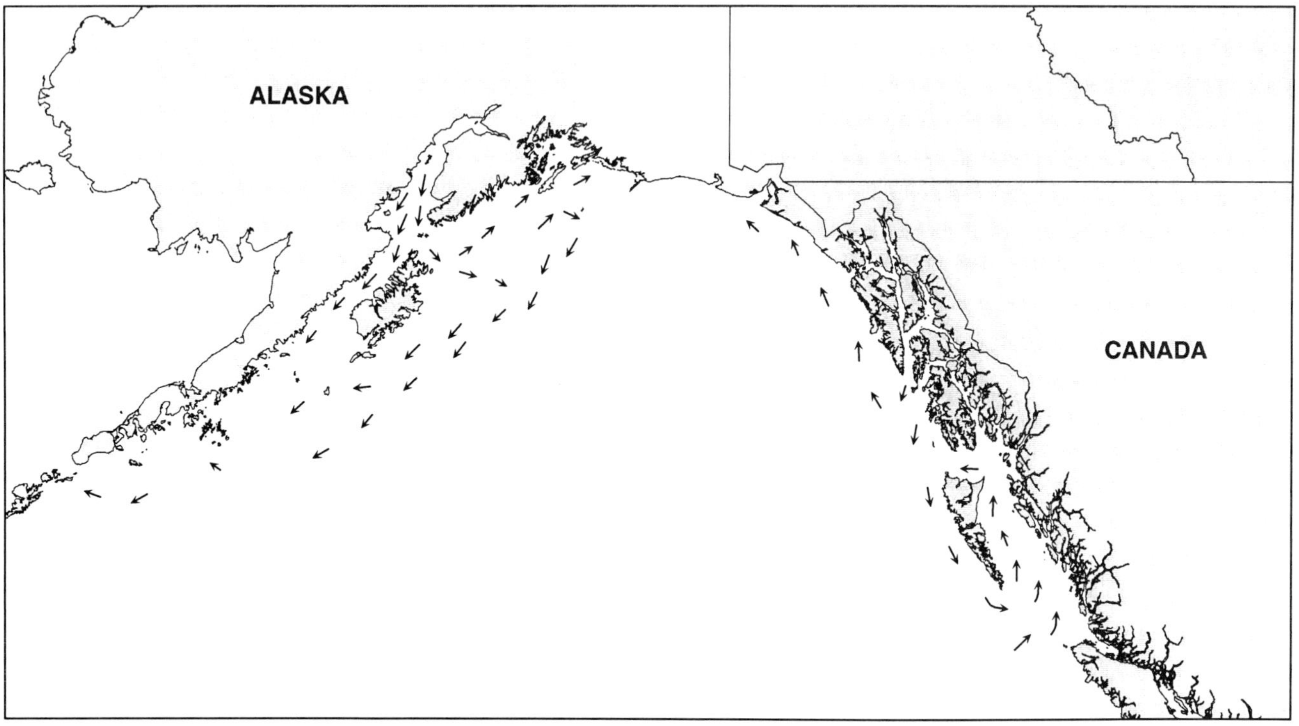


Fig. 2. Computed tidal residual circulation (at 5 m level) in the Gulf of Alaska. Maximum residual current in the middle Cook Inlet can reach a speed of 7.5 cm/sec, which is approximately 5.5% of the local maximum tidal current. Over the shelf break and in the Shalikof Strait, residual currents flow primarily toward the southwest. For more detailed residual current distribution and the computational grid, see Liu and Leendertse (1987b).

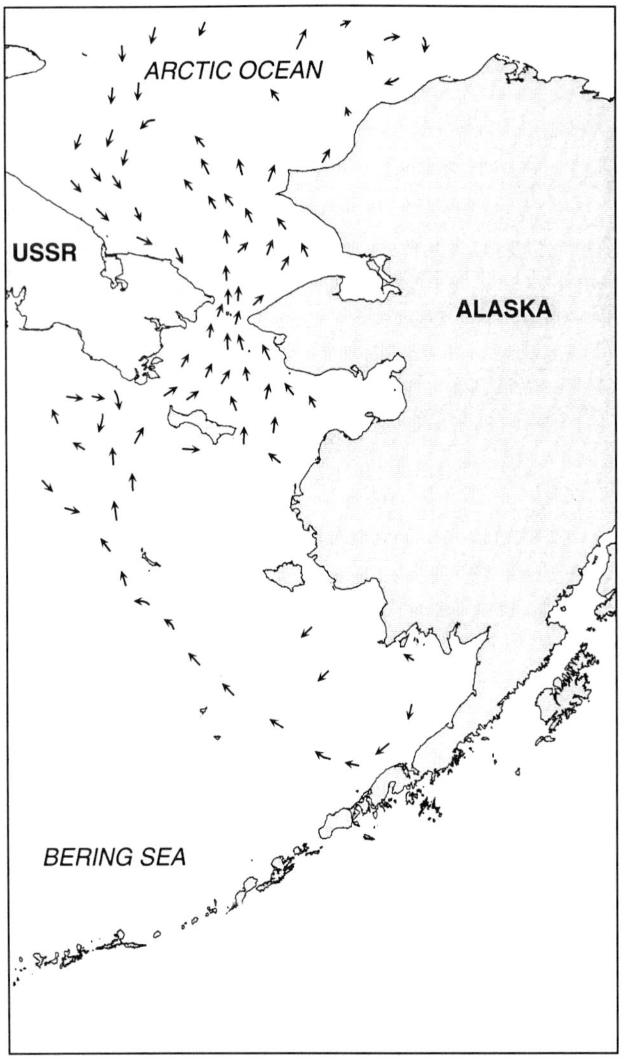


Fig. 3. Baroclinically induced residual currents at 5 m level after the initial 5-day spin-up. To reflect the realistic energy level over the continental shelf, astronomical tides are included in the computation and subsequently filtered with a two-dimensional numerical tidal eliminator. For detailed current distribution and computational grid, see Liu and Leendertse (1987b).

Island of the Aleutians. As the tide propagates from the northern Pacific toward the Gulf of Alaska, its amplitude increases. The maximum tidal amplitudes, reaching 250 cm (i.e. 500 cm tidal range), are found in the Cook Inlet. Maximum tidal excursions are found in the middle of Cook Inlet where the tidal currents can reach 140 cm/s in either direction. Currents over the shelf break near Kodiak are of elliptical rotary-type and can reach a maximum speed of 70 cm/s. As shown in Fig. 2, the maximum in the middle Cook Inlet can reach a speed of 7.5 cm/s, which is approximately 5.5% of the local current. In the Bering and Chukchi Seas, only in the shallow shelf areas are there substantial residual currents (Fig. 3). Figure 4 presents the distribution of ice concentration and thickness as compiled from existing data for the summer period over the Beaufort Sea area of the Arctic Ocean, while in Fig. 5 can be seen the residual circulation in the surface layer of the Beaufort Sea.

One way to study the effects of sea-level rise is to use these numerical models of a given area once they are adjusted and verified (Liu, 1991). However, in order to evaluate possible coastal flooding due to sea-level rise, the model should also have moving boundaries as part of the computational scheme. For analyzing the effect of sea-level rise, we use a section of an operational three-dimensional model of the China Sea, (Liu, 1988, 1993) as an example.

In the model (Fig. 6), mean sea level was increased by 1 m to make comparisons with the present condition. Two cases were considered. Under the normal condition, the model was driven by astronomical tides only. The other condition involves a typical typhoon (920 mbar) with its track moving directly toward the coast. To make a meaningful comparison, the three-dimensional model with density field and moving boundaries was used. For easier comparison, only horizontal displacements are considered, integrated from the differences of tidal velocity components under two conditions over time. These two conditions are the present one and that of a 1 m sea-level rise. Figure 7 shows the cumulative displacements in the north–south (along-shore) direction. The diagram implies that rise in sea level induces a northward net drift in the bottom layer while the increase in the surface layer is less significant. In the onshore/offshore direction, the net changes are only in the short-term tidal frequencies (Fig. 8). The net changes are mainly in the bottom layers, indicating the pronounced bottom frictional effects.

## EFFECTS ON COASTAL STORM SURGES AND ESTUARINE FLOOD RISK

When a hurricane or typhoon passes a coastal area, the inverse barometric effect and the onshore wind stress cause the sea level to rise above the ordinary tide level. Storm surge will cause further flood damage if mean sea level has risen. One reason for this is that higher sea level brings the breaking waves closer to the shoreline, which will thus increase the breaking wave forces on the shore and on any coastal dikes. For an unprotected coast, a rising sea level tends to bring breaking waves further inland, causing more severe beach erosion problems to the existing coastal communities. Additional dike protection would therefore be needed.

To study the effect of sea-level rise on storm surge, a moving typhoon passing perpendicularly across a coastline was modelled. The model typhoon has a central pressure of 920 mbar and a forward speed of 5 m/s. The momentum transfer between the typhoon and the sea surface is computed considering the vertical instability induced by the air–sea temperature difference between the typhoon model and the three-dimensional ocean model.

Under the present condition, with astronomical tidal forcing, the time history of water level, including storm surge, is presented in Fig. 9. The peak water level under the

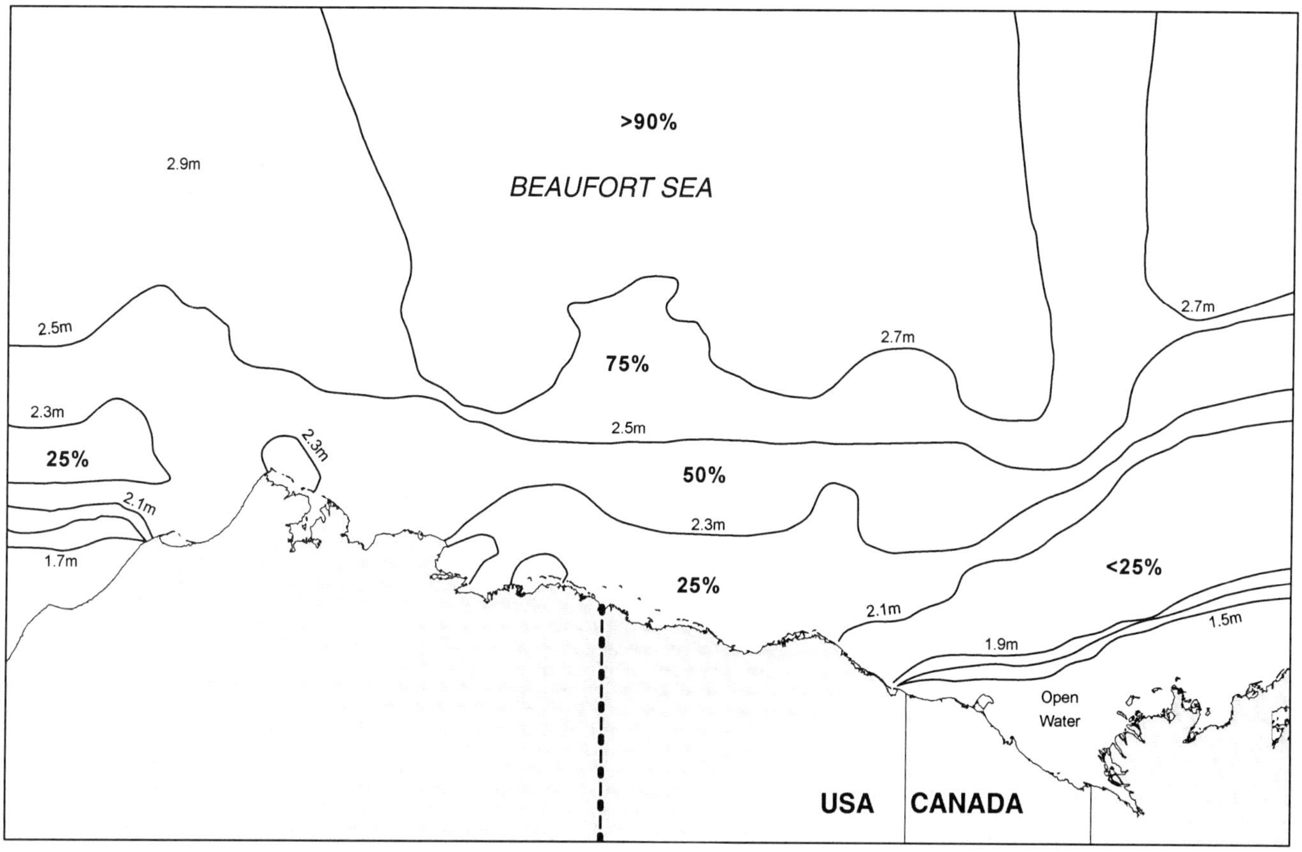


Fig. 4. Spatial distribution of ice concentration and thickness as compiled from existing data for the summer period.

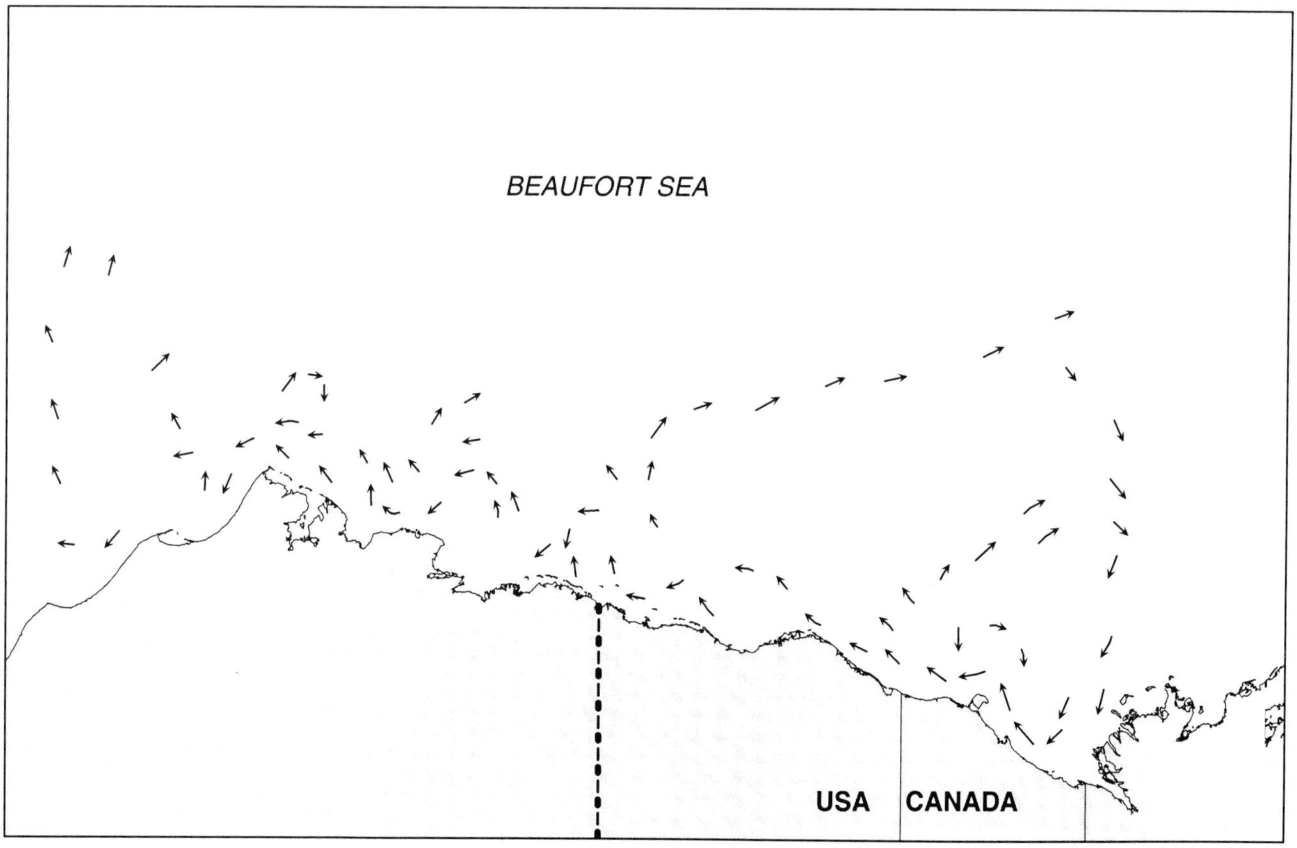


Fig. 5. Computed residual circulation in the surface layer with the plotting scale of 2 cm/sec per grid spacing. For detailed current distribution and computational grid, see Liu and Leendertse (1987b).

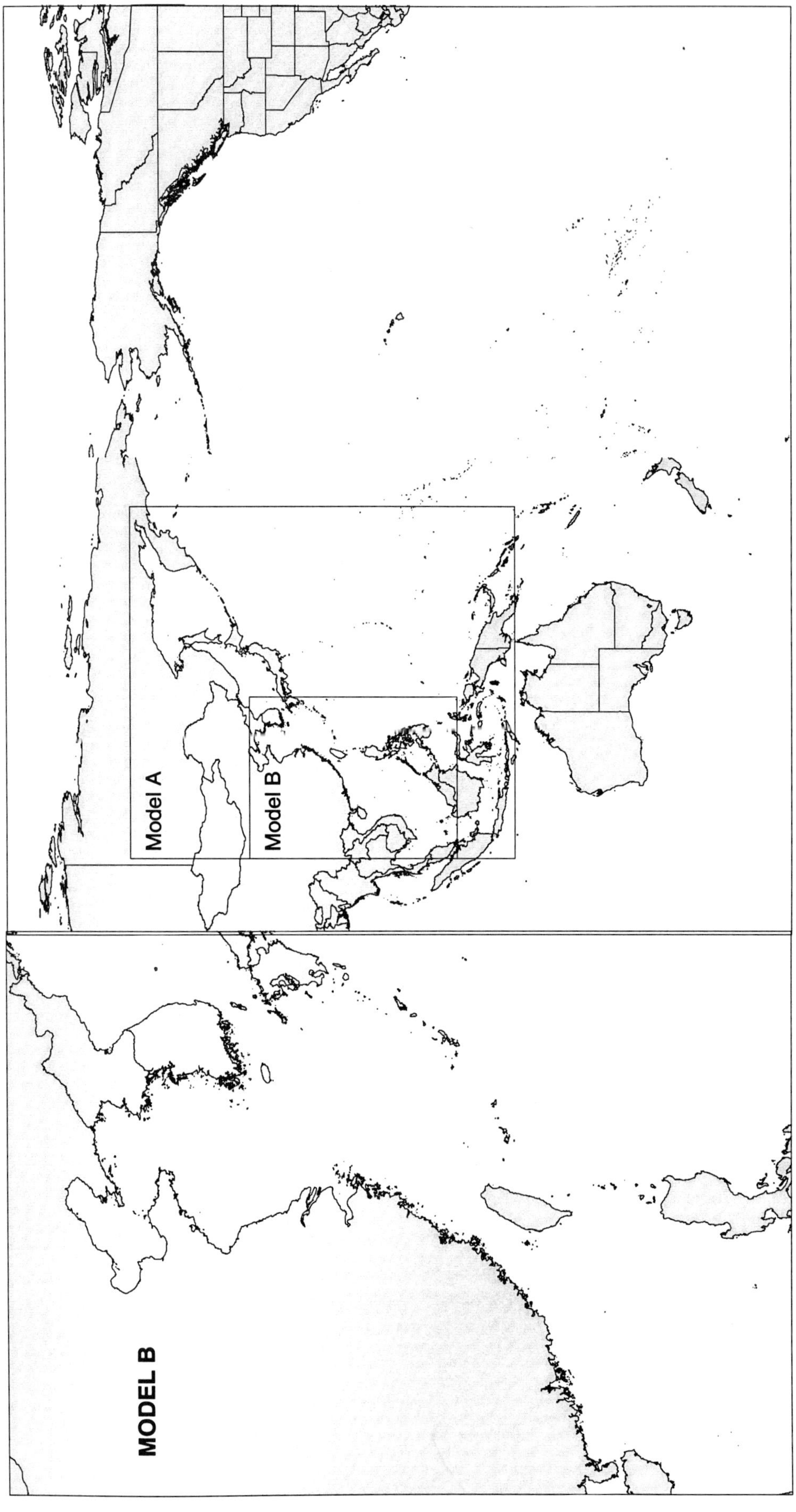


Fig. 6. Three-dimensional model and the nested sub-models of the China Sea. For detailed computational grid, see Liu (1988, 1993, 1995a,b).

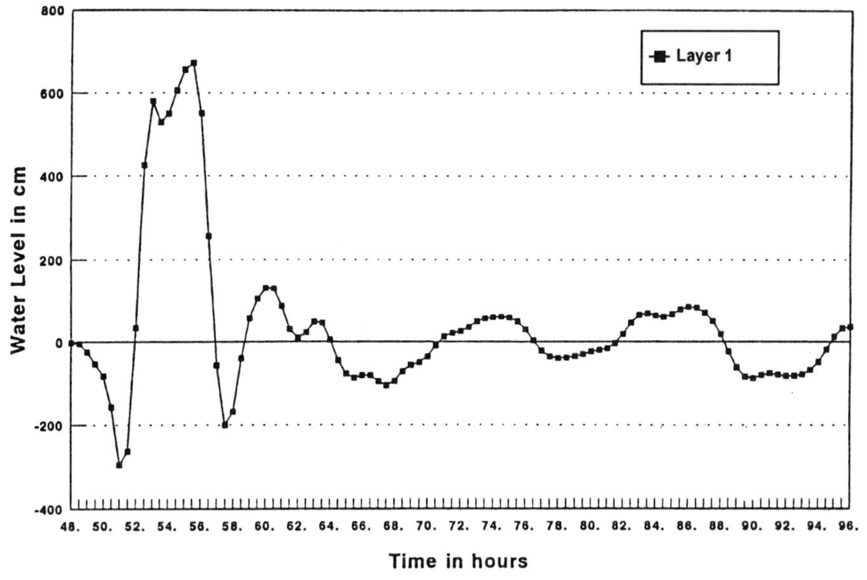


Fig. 7. Effects of sea-level rise. Typhoon surge height.

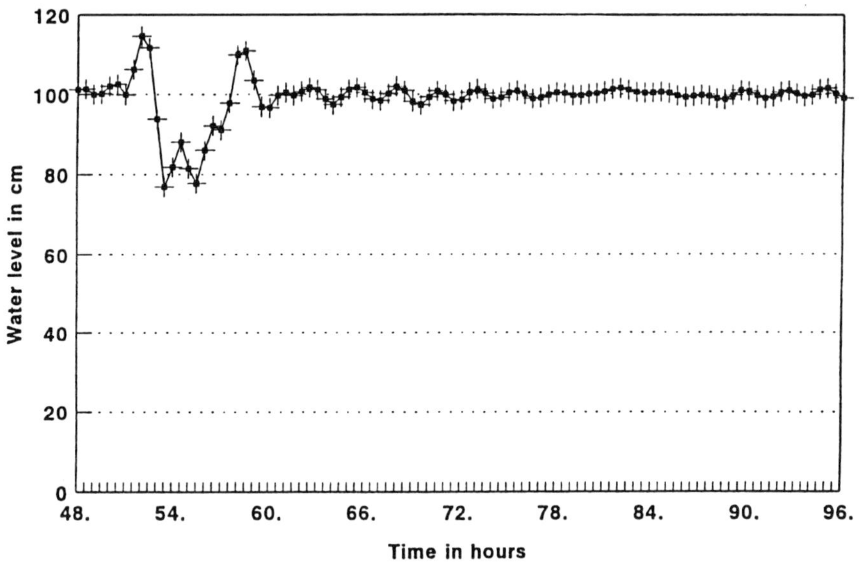


Fig. 8. Effects of sea-level rise. Difference in typhoon surge height.

combined force of tide and typhoon is approximately 6.8 m. After the passage of the typhoon, the water level gradually recovers to the normal astronomical tidal level. The simulation was carried out again with a 1 m sea-level rise. The time history of the water level difference is illustrated in Fig. 10.

As the figure indicates, the basic water level increased generally by about 1 m. The effect of sea-level rise on the surge height is about 20% during the passage of a storm at the same geographic location. At the front and the back of the typhoon, storm surge heights increased. The value of the maximum surge is somewhat reduced. The location of maximum surge is moved further inland in the shallower area. The spatial transition depends on the surrounding shoreline configuration and bathymetric distribution.

When combined with past storm statistics, model analyses may be used for coastal vulnerability assessments and flood risk management. For example, in a China Sea model, 24 stations were selected in the coastal regions (Liu, 1995a). Numerical simulations were made using past typhoon data. Storm surge risk diagrams have been developed for the coastal regions of the China Sea.

Figures 11 and 12 illustrate the relationship between storm surge height versus return period for two stations located near the southwest and northeast coastal regions of Taiwan. For example, in Fig. 11, the return period for a 5 m storm surge is approximately 75 years. If the mean sea level were 1 m higher, the return period for a 4 m surge would be only 19 years.

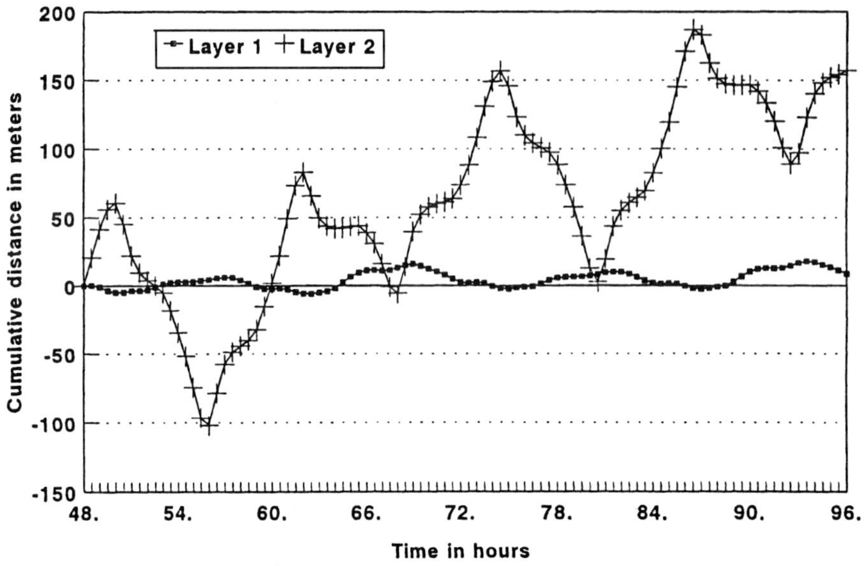


Fig. 9. Effects of sea-level rise. Cumulative displacement in the north–south tidal currents.

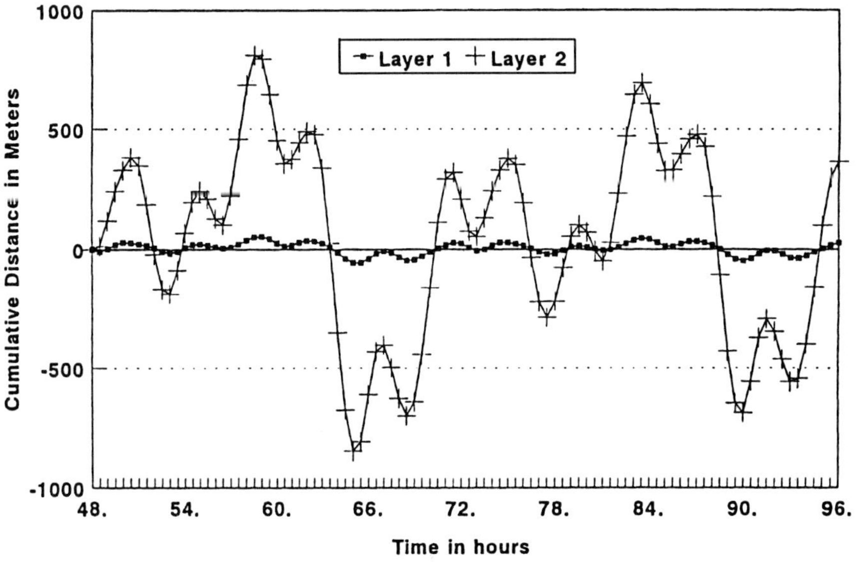


Fig. 10. Effects of sea-level rise. Cumulative displacement in 3D east–west tidal currents.

Tropical and extra-tropical cyclones are often accompanied by heavy rainfall. Enhanced greenhouse effects suggested more frequent and more intense tropical cyclones. Because global warming will lead to increased ocean temperature, it will thus provide a greater supply of moisture. For those metropolitan areas located near estuaries (e.g., New York, London, Shanghai, Taipei), the storm-induced flood wave from the upstream area will meet the storm surge from the downstream area at the estuaries. If the downstream mean sea level were higher due to the sea-level rise, the backwater effect would cause the local flood level to increase. Sea-level rise would also back-up a storm sewer drainage system such as New York City's combined sewer system (Liu and Leendertse, 1978). In another example, the newly completed flood control system of Taipei is designed for 200-year floods. If the downstream mean sea level were 1 m higher due to sea-level rise, the backwater effect would reduce the 200-year flood protection to approximately 40-year flood, which constitutes a much higher risk (Liu, 1995b).

## OTHER EFFECTS

Sea-level rise may induce other effects in coastal areas. Hydrodynamic and ecological processes associated with each coastal region are complex, particularly when they interact with human activities such as future urban

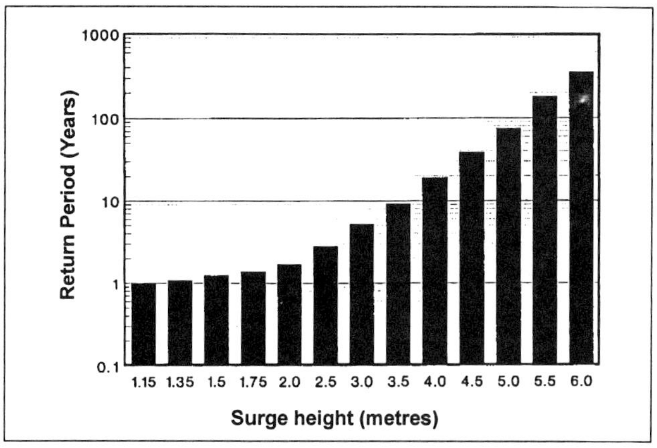


Fig. 11. Typhoon surge height and the estimated return period for the northeast Taiwan coast.

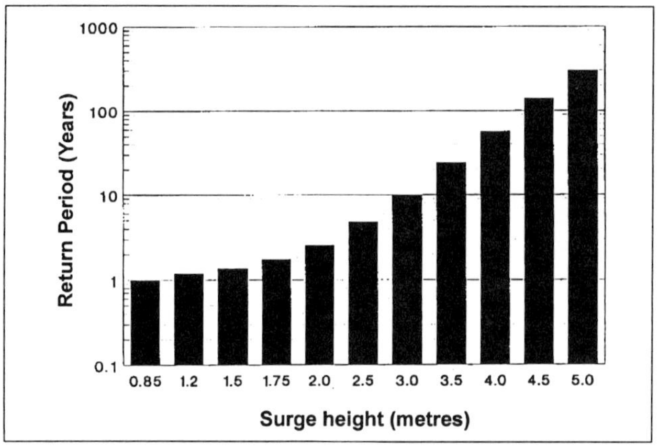


Fig. 12. Typhoon surge height and the estimated return period for the southwest Taiwan coast.

developments. Coastal models can be used to study the following areas associated with sea-level rise:

1. changes in the propagation of coastal tides;
2. changes in coastal wave heights, location of the breaker zone, and the along-shore current direction;
3. changes in aspects of coastal/estuarine interaction such as salinity distribution, pollution dispersion pattern, and the marine ecosystem in the inter-tidal zone;
4. changes in near-shore current patterns and potential impacts on fishery practices;
5. modifications of coastal engineering design criteria, flood protection, and dike design parameters;
6. assistance in coastal urban development planning for long-term morphological change and shoreline variation.

## SUMMARY

Based on data and climate models, scientists are confident that the global average temperature is expected to continue increasing by an additional 1.0–3.5°C by the year 2100. As a result of global warming, sea level is likely to rise along most coastal areas. Warmer sea-surface temperatures would induce more frequent and more intense tropical cyclones, causing a higher coastal flood risk. In shallower near-shore areas, sea-level rise also changes the depth distribution. Because the propagation of tides in the coastal area is dominated by the bathymetry, tidal residual circulation generated by the nonlinear dynamics would alter the near-shore net transport mechanism.

The mechanisms described above may have long-term effects on pollution movement, the direction of sediment transport, and near-shore marine ecology. On a typical straight coast, sea-level rise would have more significant impact on net tidal residual currents in the long-shore direction, in the lower layers. In the onshore direction, the change of net transport would be less significant.

Sea-level rise would also cause storm flooding further inland. The maximum storm surge height offshore would be slightly less, due to increased depth. Storm surge always coexists with high astronomical tides. The dynamic effects through the momentum transfer process change nonlinearly as a function of depth. Therefore, to study the effects of sea-level rise on storm surge, astronomical tidal forcing has to be included in the model.

Because ocean–atmosphere interactions are connected with climate change, salinity/temperature variables have to be included in the formulation of coastal models for it to be effective in studying the baroclinic circulation changes due to sea-level rise.

## REFERENCES


Barron, E.J. (1995) Global change researchers assess projections of climate change. *American Geophysical Union, EOS* 76 (18), 185, 189–190.

Karl, T.R., Nicholls, K.N. and Gregory, J. (1997) The coming climate. *Scientific American*, May.

Liu, S.K. (1991) Verification of hydrodynamic/water quality models—methods and techniques. RAND 5389/DOC–NOAA.

Liu, S.K. (1988) A three-dimensional model of the China Seas. Science and Technology Advisory Group, The Executive Yuan, Taiwan, 1988.

Liu, S.K. (1993) Three-dimensional modeling of tides and wind-wave dynamics. *National Conference on Hydraulic Engineering*, American Society of Civil Engineers, San Francisco, 1993.

Liu, S.K. (1995a) Numerical prediction of tides and typhoon surge using a three-dimensional of the China Sea. Central Weather Bureau Report No. 84-30-01, Taiwan, 1995.

Liu, S.K. (1995b) Numerical modeling of urban hydrology, flood control and basin management for the Taipei metropolitan area. Ministry of Economical Affairs Report No. 84EC2A371005, Taiwan.

Liu, S.K. (1997) Using coastal models to estimate effects of sea level rise. *Ocean and Coastal Management* **37** (1), 85–94.

Liu, S.K. and Leendertse, J.J. (1978) *Multidimensional Numerical Modeling of Estuaries and Coastal Seas*. Advances in Hydrosciences, Academic Press, New York.

Liu, S.K. and Leendertse, J.J. (1981) A three-dimensional model of Norton Sound under ice cover. *Sixth Intern. Conf. Port and Ocean Eng. Under Arctic Conditions*. POAC, Quebec, Canada, pp. 433–443.

Liu, S.K. and Leendertse (1982) Three-dimensional model of Bering and Chukchi Sea. *Coastal Engineering*, Vol. 18, American Society of Civil Engineers, New York, 1982, pp. 598–616.

Liu, S.K. and Leendertse, J.J. (1984) A three-dimensional model of Beaufort Sea. *Coastal Engineering*, Vol. 19, American Society of Civil Engineers, New York.

Liu, S.K. and Leendertse, J.J. (1987a) A three-dimensional model of the Gulf of Alaska. *Coastal Engineering*, American Society of Civil Engineers, New York.

Liu, S.K. and Leendertse, J.J. (1987b) Modeling the Alaskan Continental Shelf Waters. R-3567-NOAA/RC, RAND Corp. for National Oceanic and Atmospheric Administration.

## THE AUTHOR

**Shiao-Kung Liu**
*Systems Research Institute*
*3706 Ocean Hill Way*
*Malibu, CA, 90265, U.S.A.*

Chapter 121

# PARTICLE DRY DEPOSITION TO WATER SURFACES: PROCESSES AND CONSEQUENCES

Sara C. Pryor and Rebecca J. Barthelmie

Algal blooms (increased abundance of phytoplankton) are an increasingly common phenomenon which has been causally linked to increased fluxes of nutrient (particularly nitrogenous) compounds to aquatic ecosystems. These blooms have implications for water quality and human health in addition to ecosystem productivity, health and ecological diversity. Anthropogenic emissions of nitrogen to the atmosphere are estimated to be comparable to, or greater than, biogenic emissions but are considerably more concentrated in space. Although fluvial pathways typically dominate the annually averaged nitrogen flux to coastal waters, atmosphere–surface exchange represents a significant component of the total flux and may be particularly critical during the summertime when both the riverine input and ambient nutrient concentrations are often at a minimum. In this chapter we present an overview of the physical and chemical processes which dictate the quantity (and direction) of atmosphere–surface fluxes of trace chemicals to (and above) water surfaces with particular emphasis on the role of particles.

Dry deposition (transfer to the surface in the absence of precipitation) of particles is determined by meteorological conditions, atmospheric concentrations, surface type/condition and the specific chemical and physical properties of the particle. Dry deposition can be conceptualised as a three-step process: (1) the gas or particle is moved toward the surface by thermally or mechanically driven eddies; (2) it is transferred by diffusion across a thin layer close to the surface where turbulence is absent; and (3) the gas or particle is captured by the surface. In the case of larger particles a second parallel pathway exists; particles are drawn towards the surface by gravity. Atmospheric particles determine dry deposition fluxes not only by serving as a conduit for transfer but also due to their action as sources or sinks of trace gases. The example given here is the transfer of nitric acid to sea salt particles as a result of heterogeneous chemistry acting as a competing sink to surface removal. To illustrate the importance of current uncertainties in our understanding of dry deposition processes and to highlight the role of some of the key parameters in determining the transfer rate (the deposition velocity) a simple model of particle dry deposition is presented. The model describes the calculation of the rate at which a particle of a given size and chemical composition will be moved towards the surface under given environmental conditions. Observational and experimental techniques for measuring dry deposition fluxes are also reviewed. The techniques used for gases are largely reliant on use of

*Seas at The Millennium: An Environmental Evaluation (Edited by C. Sheppard)*

highly temporally resolved sampling (e.g. concentrations sampled 10 times per second) or highly accurate and precise measurements of concentrations, either in the vertical to resolve the gradient to or from the surface or conditionally sampled by the direction of transfer (to or from the surface). These stringent measurement requirements represent significant barriers to application to measurement of particle dry deposition fluxes although, as discussed, innovative solutions are now becoming available.

In the final section we examine meteorological controls on deposition to the coastal zone. This region of the world's oceans and seas is most significantly impacted by human activities. More than half of the world's population lives within 100 km of a coast and hence the overwhelming majority of anthropogenic fluxes to aquatic systems occur in the coastal zone. We discuss the particular challenges that arise from efforts to simulate and measure fluxes close to the coastline. These arise in part from the complexity of atmospheric flow in this region where energy and chemical fluxes are highly inhomogeneous in space and time and thermally generated atmospheric circulations are commonplace.

---

## INTRODUCTION

The exchange of momentum, energy and mass at the Earth's surface plays a key role in climate, ecology and human health. However, our understanding of the chemistry of our atmosphere and biosphere–atmosphere feedbacks is limited by uncertainties in spatial and temporal patterns of the emission and deposition of trace species. In order to understand biogeochemical cycles, anthropogenic impacts on those cycles, and potential system feedbacks as a result of modification of those cycles, it is necessary to understand and quantify surface–atmosphere exchanges.

## NUTRIENT FLUXES AND AQUATIC ECOSYSTEM RESPONSES

Phytoplankton (surface water dwelling, single-celled algae capable of photosynthesis) play a key role in both food chains in the oceans and uptake of carbon dioxide by the oceans (Paerl, 1995). The use of carbon dioxide by phytoplankton in photosynthesis to create carbohydrates and other organic compounds creates an air–sea gradient which in turn leads to greater dissolution of carbon dioxide from the atmosphere (Bigg, 1996).

Most oceanic primary production by phytoplankton is not limited by sunlight (or carbon dioxide availability) but rather the availability of nutrients particularly nitrogen (N) and phosphorus (P). Nitrogen represents a fundamental building block of plant cells (Owens et al., 1992) and P is a critical component of DNA (Graneli et al., 1986). Although P may be the limiting nutrient for open waters (Jickells et al., 1990), there is some evidence that N is of primary concern in estuaries and coastal waters where it is more commonly the limiting nutrient, particularly during the summer (Fisher and Oppenheimer, 1991), and that these waters may be of primary interest in terms of global primary productivity. Indeed Paerl (1995) suggests that 'nitrogen-limited estuaries, shallow coastal and continental shelf waters account for nearly half the global oceanic primary production.' Typically the N is not supplied by the organisms' fixing of N gas from the atmosphere but rather absorption of nitrate ($NO_3^-$), nitrite ($NO_2^-$) and ammonium ($NH_4^+$) from ocean waters.

Different species of phytoplankton preferentially utilise different wavelengths of light for photosynthesis and hence are grouped into classes which are named by the colour of light least used by the algae (e.g. green algae tend not to absorb the green portion of the visible spectrum).

The term 'algal bloom' is typically applied to rapid increases in the concentration of phytoplankton over a short period of time and an extensive geographic area. Plankton growth typically shows a strong seasonal cycle determined by sunlight and nutrient availability (Parsons et al., 1984) culminating in spring 'blooms' during which phytoplankton growth rapidly increases resulting in a yellow-green coloration of the water. This 'bloom' declines due to depletion of nutrient concentrations and consumption by zooplankton. Some of the nutrients are recycled into the water column when bacteria decompose the remains of phytoplankton or other organisms in the food chain. However, due to stratification of the water column, nutrients may not reach a depth where they can be re-used by other phytoplankton. Subsequent blooms may occur in the summer and autumn which are dominated by dinoflagellates which tend to cause a red-brown tint to the water and in extreme situations lead to 'red tides' (Haumann, 1989).

Rivers and lakes with clear water and low biological productivity are described as oligotrophic, while those water systems rich in organisms and organic materials are eutrophic. In the broadest sense eutrophication describes increased nutrient levels and resulting biological productivity. Human activity can greatly accelerate eutrophication, and the resulting high biological productivity, often characterised by blooms of algae, increases in sediment accumulation and bacterial populations and hence water opacity, and in some cases depletion of dissolved oxygen during algal decomposition causing fish and shellfish to be suffocated (Paerl, 1995; Barth and Nielsen, 1989). Red tides have also been associated with loss of human life due to bioaccumulation of toxins from the algae when shellfish ingest the algae and are subsequently eaten by humans. Although not all algal blooms are associated with toxic algae, and not all cause discoloration of water, studies have shown that the number of red-tide events has increased during recent decades particularly in the more populated coastal regions (Anderson, 1994). It is also clear that addition of N to surface waters, by fluvial or atmospheric pathways, can greatly influence phytoplankton productivity and ultimately the carbon cycle, water quality and aquatic ecosystem health are also affected.

Nitrogen emissions to the atmosphere may have doubled relative to the 'natural background' due to anthropogenic activity (Vitousek et al., 1997), and are predicted to increase by a further 25% to over 50% in less developed regions by 2020 (Galloway et al., 1994). For example, ApSimon et al. (1987) report that ammonia emissions rose by more than 50% over Europe between 1950 and 1977 and more than doubled in the Netherlands. It should be noted that although much research to date has focused on inorganic N fluxes to aquatic systems, dissolved organic N (DON) is also a major component of the total N flux to water bodies. Cornell et al. (1995) suggested that DON represents over half that input of dissolved fixed N to oceans via fluvial pathways. DON has also been found to be a ubiquitous component of rain (even at remote locations) and is present in particle samples. For example, Scudlark et al. (1998) determined that over one-fifth of N in rainwater is present as DON. The origin of the DON is extremely varied and can contain contributions from urea, amino acids and proteins (Mopper and Zika, 1987; Cornell et al., 1995). The impact of DON on phytoplankton growth may be considerable, particularly in light of research that suggests that the majority of DON may be bioavailable (Cornell et al., 1995).

Many coastal environments are already exhibiting signs of eutrophication, and with projected increases in N emissions to the atmosphere there is great potential for an enhanced role of atmospheric deposition in eutrophication problems in the 21st century. Indeed some researchers have suggested that atmospheric inputs of fixed N have shifted surface ocean waters in parts of the north Atlantic from N- to P-limited, although this remains a subject of considerable debate (Jickells et al., 1990). Here we shall focus on N deposition in terms of nutrient loads because P has no significant atmospheric sources.

Much of this chapter is focused on atmospheric deposition, and particularly particle deposition in the context of nutrient supply to aquatic ecosystems. However, it is important to note that the atmosphere supplies not only nutrients such as N compounds to water systems, but also a myriad of other chemicals from heavy metals (Mason et al., 1997; Sweet et al., 1998) to persistent organo-chlorines and polynuclear aromatic hydrocarbons (Hoff et al., 1996). Many of these chemicals are also highly likely to be transferred in association with particles.

## ATMOSPHERIC PARTICLES

Particles are a ubiquitous component of the Earth's troposphere. They are of interest in a number of current environmental issues because they can directly and indirectly (via cloud processes) modify climate and visibility by their interactions with radiation, have been linked to detrimental health impacts (including increased mortality associated with elevated respiration of fine particles) and, as is discussed here, they act as carriers of nutrients and toxins to ecosystems.

The term 'atmospheric particles' is used to describe fine solid or liquid particles suspended in air, which range in diameter between approximately 0.002 and 100 $\mu$m. Traditionally, standards for particle concentrations to protect health have focused on three classes of particle, defined based on particle diameter. TSP (total suspended particles) encompasses particles of all sizes and hence reflects the total atmospheric burden. $PM_{10}$ (particles of less than 10 $\mu$m in diameter) and $PM_{2.5}$ (particles of diameters $\leq$ 2.5 $\mu$m) dominate visibility impairment and may be the most active portion of the total particle burden in terms of human health impacts (Spengler et al., 1990). Atmospheric particles originate from a plethora of sources: some natural, others anthropogenic. On a global basis natural sources of particle emissions greatly exceed anthropogenic emissions. The IPCC (1996) estimates the ratio of natural to anthropogenic particle emissions to be approximately nine to one. However, anthropogenic emissions are frequently greatly concentrated in both space and time. The ambient particle mass may be broadly ascribed to two classes based on the origin of the particles. Primary particles are emitted directly into the atmosphere and are frequently mechanically derived (e.g. wind blown dust from an agricultural field). Primary particles are the largest contributor to coarse mass in the form of soil dust (dominated by salts of silicon, aluminium, iron and calcium), sea spray (sodium and chloride), and plant particles (organics). Other sources of coarse aerosols include products of reactions of gaseous nitric acid ($HNO_3$) on soil and sea salt particles (Pakkanen, 1996). Secondary particles form within the atmosphere from primary gaseous emissions (e.g. sulphate formed from sulphur dioxide emissions).

The actual size distribution of atmospheric particles varies greatly in space and time. However, observations conform to either: a tri-modal (three peaked) distribution with an Aitken (or nucleation) mode characterised by particles with diameters less than 0.1 $\mu$m; an accumulation mode characterised by particles with diameters greater than 0.1 $\mu$m and less than 1 $\mu$m, and a coarse mode characterised by particles with diameters greater than 1 $\mu$m; or a bi-modal (two peaked) distribution where the Aitken mode is absent. As implied by this statement, the atmospheric particle size distribution can change as a result of physical and chemical processes, for example:

- *Nucleation.* Mass transfer from the gas to aerosol phase (gas to particle conversion) may occur via condensation (heterogeneous nucleation) thus adding mass to existing particles or by homogeneous nucleation (new particle production). Partitioning between these two processes is a complex function of environmental conditions and the compounds present (e.g. particle number concentration), and is critical to determining the size distribution of different species and hence their deposition.
- *Growth/shrinkage due to condensation/evaporation.* A key property of atmospheric particles, in terms of determining both their environmental effects and removal processes, is hygroscopicity. A hygroscopic particle is one which will readily take up and retain moisture and hence can, for example, act as a cloud condensation nucleus. The particle size distribution varies as a function of relative humidity (RH) because of the presence of hygroscopic materials. As a dry particle is exposed to increasing relative humidity, the particle diameter initially remains essentially unchanged but after a critical RH is passed, will suddenly increase as the particle changes phase, as the RH continues to increase the radius continues to increase and the solution becomes more dilute (thus changing the chemical environment within the drop). If the RH is subsequently reduced, a hysteresis phenomenon will be observed, where the change of phase occurs at a lower RH than under conditions of increasing RH. For some compounds gas-particle partitioning is also a function of temperature, RH and other environmental conditions.
- *Coagulation.* Coagulation occurs as the result of collisions due to Brownian motion. It changes the size distribution by increasing the size of particles and decreasing the number, but the particle volume is conserved. Coagulation is an important process in determining the particle number and size distribution, for example fine nuclei particles tend to preferentially coagulate with accumulation mode particles and this process has been used to explain observations of a number minimum at around 0.1 μm in diameter (Raes et al., 1995).
- *Entrainment of particle (and ozone) enriched air from residual layers during daytime growth of the boundary layer.* The phenomenon of elevated pollutant layers stored in stable layers aloft and vented to the surface during convective growth of the boundary layer has been observed in many environments (e.g. Wakimoto and McElroy, 1986; Hoff et al. 1997) and plays a key role in determining near surface concentrations, composition and subsequent chemical reactions (for example by changing available aerosol surface area).
- Cloud processing (see below).
- Deposition (wet and dry) (see below).

## DEPOSITION PROCESSES

Chemical constituents of the atmosphere may be removed by three processes: dry deposition, wet deposition and occult deposition. The latter two refer to incorporation of the chemical of interest into precipitation or cloud/fog water, respectively. Wet deposition may remove chemicals from the atmosphere either by in-cloud scavenging, where gases and particles are incorporated into cloud droplets and below-cloud scavenging where gases and particles are removed by falling rain and snow.

Solubility of gases in water (typically described by a Henry's law coefficient) determines, to a large extent, the rate of removal of gases by wet deposition. Particles may also be removed by wet deposition where the role of hygroscopic species as condensation nuclei is key for in-cloud removal. Removal of particles in below-cloud scavenging is not very efficient. The scavenging rate for particles normalised by rainfall rate varies with particle size, it is highest for small and large particles and has a minimum at approximately 0.2–1 μm diameter. The surface may also intercept cloud and fog droplets directly (occult deposition). Hill cloud and fog droplets typically have a diameter of 3–50 μm and are efficiently captured by vegetation (Erisman and Draijers, 1995).

The rate of deposition in the absence of wet or occult processes is often specified as a dry deposition velocity ($v_d$) given by:

$$v_d(z) = \frac{F}{C(z) - C(0)}$$

where: $C(z)$ is the concentration at measurement height $z$, $C(0)$ is the concentration at the surface (which may or may not be zero, depending on surface uptake), and $F$ is the flux.

Deposition velocities ($v_d$) may be computed as the inverse of the sum of a number of resistances (Davidson and Wu, 1990):

$$v_d = \frac{1}{r_a + r_b + r_c} + v_g$$

where:

$r_a$ = aerodynamic resistance (which is a function of surface roughness, height, stability, wind speed). The aerodynamic resistance refers to the resistance to transport of air to a thin viscous sub-layer just above the surface. Vertical transport in the free atmosphere is dominated by turbulence transferring chemicals down the concentration gradient towards the surface caused by surface removal.

$r_b$ = viscous sub-layer resistance (which is a function of surface roughness, wind speed and stability). Very close to the surface a laminar boundary layer forms which is essentially free of turbulence. Hence transport across this layer (again down a concentration gradient) is largely the result of Brownian diffusion. The viscous sub-layer resistance for particles is typically larger than for gases because they have lower diffusivities—except very small particles (with particle diameters $D_p < 0.1\ \mu m$)—and hence do not diffuse across this layer very efficiently. However, for larger particles ($D_p > 1\ \mu m$) transport is more efficient because interception and impaction are important (Davidson and Wu, 1990). For submicron particles impaction rates are low and this combined with low diffusion rates results in a minimum for $v_d$ of particles in the accumulation mode ($D_p \sim 0.3$–$0.5\ \mu m$).

$r_c$ = surface or canopy resistance (which is a function of surface type and conditions, chemical species). For highly reactive or soluble gases the surface resistance is small. For less soluble or reactive gases $r_c$ may dominate $v_d$ and over vegetated surfaces $r_c$ is highly dependent on transfer through the stomata and hence may show large diurnal and seasonal variability. In the case of particle dry deposition to surfaces not covered by water there is the possibility for particles to bounce off the surface, however in the case of deposition to a water surface it is assumed that surface capture is 100% efficient.

$v_g$ = gravitational settling velocity. Transport by sedimentation reflects the role of gravity. This transport process is negligible for gases and fine particles, but is significant for particles in the coarse mode. For these particles the downward gravitational force exceeds the drag force due to the viscosity of air.

Additional processes which are of importance to the dry deposition of particles include interception and inertial forces which may transport the more massive particles across the viscous sub-layer. The sub-layer typically has a depth of only a few to tens of micron and so larger particles may collide with the surface as they are transported in the flow above the viscous sub-layer. Equally a massive particle may deviate from streamlines due to inertial forces and hence impact the surface. Other processes which may enhance particle deposition include electrical migration due to surface charge and thermophoresis caused by temperature gradients. Of these effects, diffusiophoresis which is caused by a gradient of gas concentrations may be particularly important over water surfaces. Evaporation from a water surface causes a water vapour concentration gradient away from the surface which must result in a downward motion of air molecules to replace the upward moving water vapour molecules. A particle above the surface is thus affected by collisions with upward motion of water vapour molecules and downward moving air molecules. Because air molecules are more massive than water vapour molecules this will result in downward motion of the particle (Seinfeld and Pandis, 1998).

## THE IMPORTANCE OF DRY DEPOSITION PROCESSES IN NUTRIENT FLUXES

Deposition of atmospheric N to water bodies occurs via deposition of organic and inorganic N compounds. The dominant inorganic compounds and pathways are as follows; wet deposition of ammonium ($NH_4^+$) and nitrate ($NO_3^-$) ions, and dry deposition of N gases—principally ammonia ($NH_3$), oxides of nitrogen ($NO_x$) and nitric acid ($HNO_3$), although other N species may also contribute (e.g. PAN)—and particles containing $NH_4^+$ and $NO_3^-$. While wet deposition dominates in terms of mass flux in many environments, dry deposition may also represent a significant, but highly uncertain, contribution.

Paerl (1995) reports a number of studies which indicate that from 10 to over 50% of coastal N loading is attributable to atmospheric fluxes. Model calculations of atmospheric deposition and observations of fluvial sources of annually averaged nutrient addition to the seas around Denmark indicate 1/3 of the total N comes from the atmosphere (Asman et al., 1995; Hertel, 1995) and approximately 3/5 of the atmospheric deposition derives from particulate matter (Asman et al., 1995). Hertel and Frohn (1997) present estimates of total N input to Danish waters from streams, direct discharge and the atmosphere, based on a mix of modelling and measurements. Their results indicate an average atmospheric contribution of 2/3 to open waters with uncertainties of 40% attributed to the modelled atmospheric deposition estimates. It has further been suggested that atmospheric input to the seas of Northern Europe may be largest in the late spring and summer months when nutrient concentrations are generally at a minimum and hence that the net impact of this flux may be large (Markeger et al., 1998). For example, Rendell et al. (1993) made an assessment of atmospheric N deposition to the North Sea and found that wet and dry deposition contributed approximately equal amounts to the total atmospheric N flux (of approximately $228 \times 10^3$ t N $yr^{-1}$) and estimated that the atmospheric pathway represents at least one-quarter of the total anthropogenic flux of N to the North Sea. Additionally they postulated that during the growing season, when primary production is nutrient-limited, the atmosphere may represent the dominant source of N at non-coastal locations.

Atmospheric N deposition has also been shown to be a significant source of N to watersheds in the eastern USA. Scudlark et al. (1998) estimate that the atmospheric pathway may contribute 43% of the total N influx to the Chesapeake Bay. Winchester et al. (1995) used ten years' of data from the National Acid Deposition Program (NADP) (which under-estimates dry deposition fluxes due to the sampling methodology) and twenty years of riverine samples from the US Geological Survey to show that atmospheric deposition is the principal source of N to a dozen watersheds in northern Florida.

## DRY DEPOSITION TO WATER SURFACES

Measurement and modelling of particle dry deposition is confounded by the myriad of processes that determine the deposition velocity. Dry deposition of gases is a function of meteorological conditions, atmospheric concentrations, surface type (and condition) and species specific chemical properties. Dry deposition of particles is additionally related to particle size (due to diffusion, impaction/interception and settling effects), which may change in response to hygroscopic growth close to the surface (Slinn and Slinn, 1980).

## DRY DEPOSITION: MODELLING FRAMEWORKS AND ALGORITHMS

### Illustration of Mathematical Treatment of Physical Processes

Pryor et al. (1999) recently presented a modified version of a model for the deposition velocity for particles to a water surface developed by Williams (1982) and adapted by Hummelshøj et al. (1992). The form of the model is as follows:

$$v_d = \frac{(v_h + v_{g,d})(v_\delta + v_{g,w})}{v_h + v_\delta + v_{g,d}}$$

where:

$v_h$ = transfer velocity in the layer dominated by turbulent transfer

$v_{g,d}$ and $v_{g,w}$ = the transfer velocity due to gravitational settling (the subscripts d and w indicate dry and wet particles)

$v_\delta$ = transfer velocity across the laminar surface layer

The details of the model are provided in Pryor et al. (1999) and so will not be reiterated here (formulae for calculating the components of the deposition velocity are provided in the Appendix). Rather the model can be used to illustrate the functional dependencies of deposition on particle, atmospheric and surface characteristics. Turbulent transport is a function of atmospheric stratification/mixing and enhancement or suppression of transport due to mechanically generated turbulence (due, for example, to passage of air over a rough surface) and thermally generated turbulence (due, for example, to surface heating). Variation of transport of the chemical of interest from a reference height to the viscous sub-layer due to changing atmospheric stability (or stratification) is described using Monin–Obukhov theory (stability corrections are applied based on the value of the Monin–Obukhov length which is the ratio of mechanically to thermally produced turbulence). Transfer velocity across the viscous sub-layer is treated as having two components. The first is diffusional transfer which is a function of particle size and the depth of the layer, and is slower for large particles due to their lower diffusivity. The second is the increase in downward movement of particles due to bubble burst activity associated with white-cap production. As bubbles of gas burst at the sea surface they inject film and jet drops (of sea water) into the atmosphere (Gong et al., 1997). The jet drops are formed by the hydraulic recovery of the water surface and are relatively large and hence fall back to the water surface quickly.

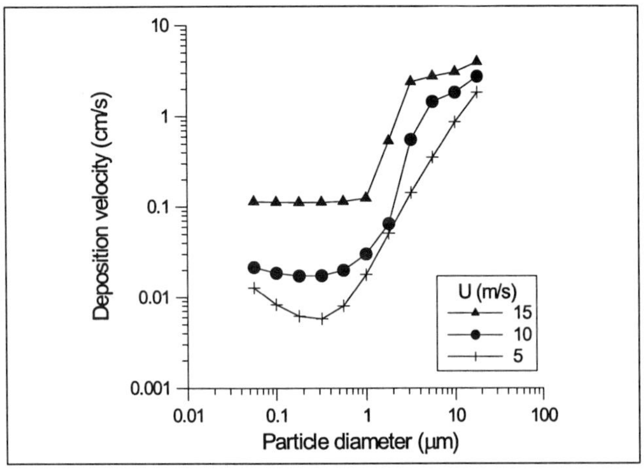


Fig. 1. Particle deposition velocities as a function of wind speed and particle diameter as calculated using the model presented in Pryor et al. (1999).

Film droplets are much smaller and hence have longer atmospheric lifetimes. The ejection and deposition of these droplets is speculated to cause enhancement to particle deposition via two mechanisms. First, this mechanism induces turbulence in the laminar surface layer which is proposed to enhance down gradient transport and secondly, as the jet droplets fall they may sweep *in situ* gases and particles towards the surface. In addition to these processes, as particles are transported across this near-surface layer, they encounter elevated humidity levels and hence hygroscopic particles will absorb water and increase in diameter (thus changing their diffusivity).

Figure 1 shows the particle dry deposition velocities calculated using the model described above for a particle size distribution comprised entirely of ammonium nitrate ($NH_4NO_3$), for near-neutral stability and wind speeds of 5, 10 and 15 $ms^{-1}$. This figure illustrates the role of meteorological conditions in determining deposition fluxes and the critical importance of particle size information in determining deposition.

### Illustration of Current Modelling Uncertainties

Resistance to particle transfer across the viscous sub-layer is large, and initial studies postulated that bubble burst by disrupting the layer may increase the deposition velocity by perhaps orders of magnitude. However, Larsen et al. (1995) reported results from wind–water tunnel experiments that indicate 'a change in simulated white-cap cover from 0 to 25% results in less than 30% change in the average particle deposition velocity'. This may indicate that the magnitude of this effect was initially over-estimated. Figure 2 shows the difference in $v_d$ calculated using the model described above for a simulation without white cap activity and associated bubble burst enhancement of $v_d$ and a base scenario with these processes included (again the particle is assumed to be pure $NH_4NO_3$). This simulation illustrates the size-dependent nature of the importance of white-cap

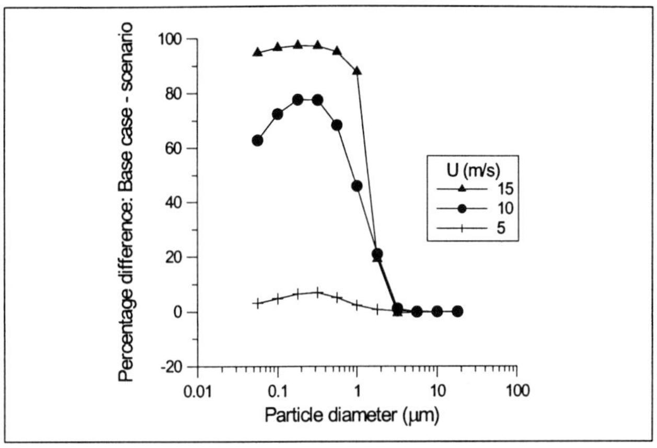


Fig. 2. Change in deposition velocity due to bubble burst activity. The results are represented as a percentage difference: (base case – scenario) / base case in deposition velocity calculated by particle diameter where the scenario run excludes the effect of bubble burst on particle deposition.

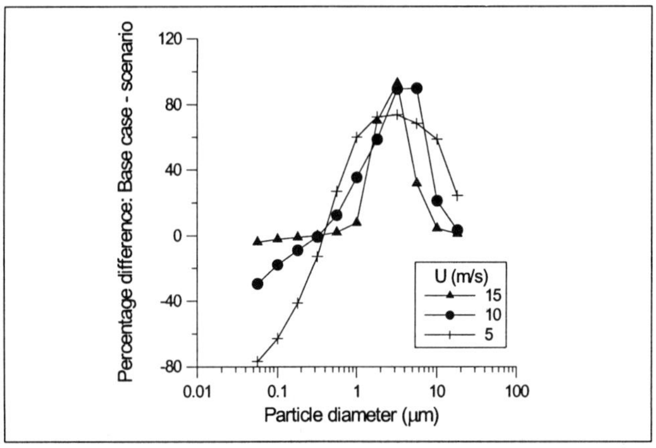


Fig. 3. Change in deposition velocity due to hygroscopic growth. The results are represented as a percentage difference: (base case – scenario) / base case in deposition velocity calculated by particle diameter where the scenario run excludes the growth of particles due to addition of water.

production and bubble burst activity in determining particle deposition. As shown bubble burst is highly dependent on wind speed and so, for a wind speed of 5 m/s, the results are in accord with the results of Larsen et al. (1995) with no particle diameter showing greater than 10% enhancement of deposition due to bubble burst activity. However, for wind speeds of 10 and 15 m/s, the modelled enhancement for particles with diameter $\leq 1\,\mu m$ is clearly in excess of 30%.

Hygroscopic growth of particles as they approach the surface (where relative humidity may approach 98.3%) is also key to determining deposition velocities. In order to make treatment of this process computationally feasible it is assumed that absorption of water is a function of fraction of hygroscopic material. This neglects the possibility that heterogeneous particles may exhibit non-ideal behaviour because of, for example, coasting of hygroscopic particles by compounds which are not water absorbing. To bound the magnitude of this effect, a simulation was performed for identical conditions to those portrayed in Fig. 2 except hygroscopic growth of the particles was excluded. The results are shown in Fig. 3 relative to the base scenario and clearly indicate that hygroscopic growth enhances deposition of larger particles, while suppressing deposition of finer particles due to a reduction in particle diffusivity.

Other uncertainties in model formulation and constants are discussed and quantified in Pryor et al. (1999). These uncertainties include, but are not limited to, the actual number of film and jet drops generated by white-cap activity (Spiel, 1994), the effectiveness with which particles are collected by the spray drops, the dependence of white-cap production on wind speed (Wu, 1988) and applicability of stability corrections for heat, acting as surrogates for stability corrections for particles.

## DRY DEPOSITION: MEASUREMENT TECHNIQUES

Dry deposition fluxes are most commonly measured using the following micrometeorological techniques (see review in Dabberdt et al. (1993)):

- Gradient techniques use samples taken at multiple heights to derive a gradient a few metres above the surface. In this technique the flux is given by:

  $$F = -K_z \frac{dc}{dz}$$

  where: dc/dz is measured vertical concentration gradient and $K_z$ is the transfer coefficient (or eddy diffusivity) of the contaminant of interest. Aside from issues involving determination of $K_z$ particularly over rough surfaces the most stringent measurement requirements focus on precise determination of the concentration gradient.
- Eddy correlation/covariance techniques are based on simultaneous measurement of vertical wind velocity (w-component) and concentration of the chemical of interest. From the covariance (mutual variability) of the velocity and concentration, the flux can be determined:

  $$F = \overline{w'c'}$$

  Measurement requirements include fast response instruments to accurately measure fluctuations of vertical velocity and trace component (usually at 10 Hz or higher) (Leuning and King, 1997).
- Eddy accumulation techniques use conditional sampling based on the sign of the vertical velocity. A fast response sensor and sophisticated electronics are configured such that air is differentially sampled based on the flow direction. During updrafts air is drawn into one sampling canister and under downdraft conditions air is drawn into a separate canister. At the end of the sampling period the accumulated concentration in each container is measured, and the difference in the concentration of the entity of interest in the two containers is calculated. The mean vertical flux is then estimated as being proportional to the concentration difference:

$$F = \beta\sigma_w(\overline{A}^+ - \overline{A}^-)$$

where: $\beta$ = an experimental coefficient (Schery et al., 1998) determined by the probability distribution of w (Dabberdt et al., 1993) and sampling height, $\overline{A}^{+/-}$ = average concentrations from the canisters filled when w is positive and negative respectively, and $\sigma_w$ = standard deviation of w.

Relaxed Eddy Accumulation (REA) is a variant of eddy accumulation in which sampling is only initiated when vertical velocities exceed a threshold specified as a function of the mean velocity. The primary requirements for accurate determination of fluxes using this technique are highly accurate measurements of the mean concentration and fast response electronics and system control.

Whilst eddy correlation and eddy accumulation techniques are theoretically superior methods for determining dry deposition fluxes, due to the high demands on instrument technology it is still common to use gradient techniques for some gases and particles.

Each of the techniques outlined above is dependent on the constant flux layer assumption (i.e. that the measurement or determined flux at a given height above the surface is equal to the flux at the surface). However, this assumption may not be applicable due to advection (due to horizontal concentration gradients) or storage effects (due to changes in the concentration within the air column), which are approximately proportional to the measurement height, and chemical reactions between the measurement height and the ground. If chemical reactions are rapid in comparison with atmospheric transport times (Kramm and Dlugi, 1994) it may be necessary to estimate chemical loss/production in order to correct the measured flux.

Atmosphere–surface fluxes may also be determined using surface analysis methods. These techniques rely on accumulating material over time on natural (e.g. foliage) (Schaefer and Reiners, 1990) or surrogate surfaces (e.g. artificial collectors such as petri-dishes) (Zufall et al., 1998). The surface is exposed for a known period of time and the flux determined from the available surface area and the accumulated mass. The major uncertainties in this type of analysis result from differences in deposition as a result of the character of the surface and the stability of the deposit. The former arises from the problems focused on the role of the physical, chemical and biological nature of the surface in determining the flux; the depth of the viscous sub-layer, existence of roughness elements on which impaction might be important, the uptake efficiency (efficiency with which the particles are incorporated by the surface) and so forth. The latter refers to the likelihood of particle bounce or volatilisation from the surface due to changing sampling conditions. Particle bounce may be minimised by coating of the measurement substrate with an adhesive and is usually small for moist particles. Volatilisation losses are substantially more difficult to overcome and are critical for particles such as ammonium nitrate ($NH_4NO_3$) which may volatise to produce $HNO_3$ and $NH_3$ gas if the concentrations of $HNO_3$ and $NH_3$, temperature or humidity change during the sampling period.

$$NH_3 + HNO_3 \Leftrightarrow NH_4NO_3$$

For these and other reasons deposition to the surrogate surface may not be a direct analogue to the deposition to the surface of interest.

### Wind Tunnel Experiments

Largely as a result of the difficulties in undertaking detailed field measurements of particle dry deposition with sufficient accuracy to be able to determine the magnitude of individual parameters or processes, studies have been undertaken in laboratory (wind tunnel) settings. Wind tunnel experiments offer the potential to examine processes which determine particle (and gas) deposition in a controlled environment. Although many of the wind tunnel experiments focused on particle deposition have concentrated on analogies to vegetated surfaces (Davidson and Wu, 1990), some have examined deposition to water surfaces. Larsen et al. (1995) used a wind/water tunnel apparatus and introduced a measured distribution of hydrophobic MgO particles into the flow above the water surface. Bubble burst activity was simulated using 2600 aquarium airing frits which produce a known spray particle spectrum. They examined a range of conditions with up to 25% sea cover by white-cap activity but report results that appear to show little evidence for significant enhancement of particle deposition due to bubble-burst processes. They state for particle diameters between 0.2 and 1.3 $\mu$m, and wind speeds of 1.6–9.0 $ms^{-1}$ "a change in simulated white-cap cover from 0 to 25% results in less than 30% change in the average particle deposition velocity". As the authors acknowledge, there is also experimental uncertainty and since it is impossible to change the bubble burst activity without also changing surface roughness it is not possible within this experimental set-up to examine solely the role of bubble burst on the transfer through the viscous sub-layer. Nevertheless this research has been critical to illuminating potential uncertainties regarding the role of bubble burst in deposition processes which may have been over-estimated in numerical models of deposition processes.

### Use of Surrogate Surfaces

Deposition of metals associated with coarse particle deposition to water surfaces has been examined by researchers using surrogate surfaces deployed in the field. The study by Zufall et al. (1998) focused on southern Lake Michigan and collected samples on surrogate surfaces (airfoils covered by coated Teflon filters). Individual particles from the samples were analysed for size and elemental composition using scanning electron microscopy. Dry deposition velocities

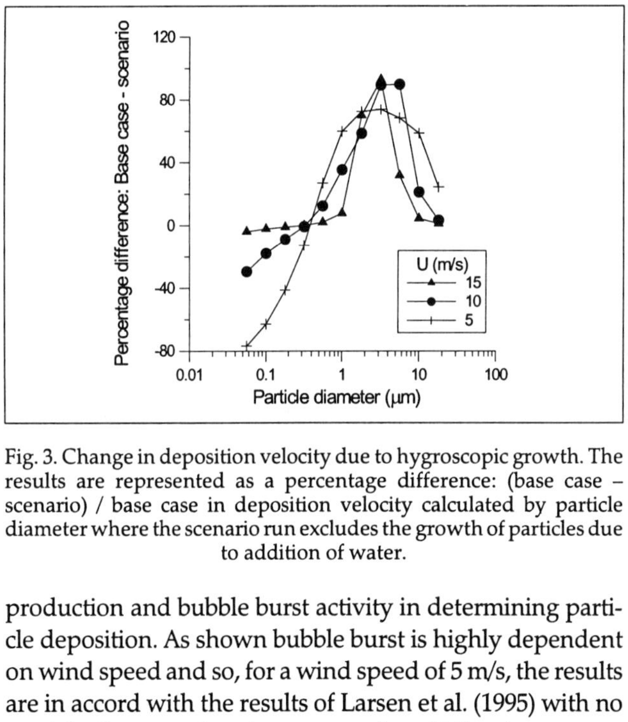

were determined by the ratio of particle mass on the airfoil filters and a filter system deployed to measure the ambient concentration. This type of analysis is highly useful for providing elemental information on stable (solid) particles, but does not provide information on the volatile components of the aerosol mass (such as ammonium nitrate and organics), is subject to uncertainties regarding the role of surface character on deposition, and requires significant time investment for the individual analysis of particles.

### Field Measurement of Particle Dry Deposition: Current Status

Studies designed to offer comprehensive assessment of N fluxes are confounded by both the range of N compounds which may contribute to the flux and the difficulties inherent in both determining the ambient concentrations of certain N-compounds and more critically in determining gradients/fluxes of gas phase compounds (and violation of the assumptions in application of gradient theory) and describing size-resolved particle concentrations and assigning appropriate deposition velocities. Hence, typically most studies do not measure the dry deposition flux of particles (or some gases) but rather collect bulk particle concentrations on a filter pack and either assign an average deposition velocity or assign a theoretical size distribution to the bulk sample $NH_4^+$ and $NO_3^-$ concentrations and calculate deposition velocities for each size class using a model such as that described above. Alternatively, and increasingly commonly, impactors (which provide size resolved deposits for chemical analysis) are used to provide size resolved composition measurements which can be combined with a numerical model to quantify deposition fluxes.

A limited number of studies have reported particle deposition velocities based on eddy correlation (Buzorius et al., 1998; Gallagher et al., 1997) but the technology for particle measurement severely limits the potential for very fast response measurements of highly resolved particle distributions even neglecting chemical composition. Hence most highly temporally resolved measurements are conducted at the expense of size resolution. For example, Buzorius et al. (1998) present spectra of total bulk particle counts obtained using a condensation particle counter (CPC) and Fontan et al. (1997) report measurements for which "80% of the detected particles have a size between 0.05 and 0.5 $\mu$m". Additionally, low frequency fluctuations due to advection of heterogeneities in the distribution of sources and sinks are more important for particles than some gases (Fontan et al., 1997) and complicate the application of eddy correlation.

Recently, a number of prototype devices have been designed for application of relaxed eddy accumulation (REA) to measurement of particles. For example, Schery et al. (1998) report the design and testing of a REA system for measurement of the flux of ultrafine particles (0.001 $\mu$m diameter) using radon decay products.

## GAS DEPOSITION AND THE ROLE OF PARTICLES

Particles are of importance to dry deposition processes not only as a pathway for deposition but also because some highly reactive gases exhibit flux divergence (non-conservative behaviour) due to heterogeneous reactions on particle surfaces or changes in particle equilibrium. The constant flux layer assumption (that a flux measured above the surface is equal to the flux at the surface) is only applicable where the response time of chemical processes ($\tau_c$) is greater than the response time ($\tau_d$) of the diffusive transfer (Kramm and Dlugi, 1994). These time scales may be approximated using:

$$\tau_c = \frac{1}{3D\int_0^\infty \left(1+\frac{\lambda}{\alpha RP}\right) Rp^2 \rho_p}$$

from Wexler and Seinfeld (1992) and

$$\tau_d = \frac{\kappa(z+z_0)}{1.56u_*}$$

from De Arellano and Duynkerke (1992), where $D$ = molecular diffusivity, $\alpha$ = accommodation coefficient, $m(Rp)$ = mass distribution, $Rp$ = particle radius, $\lambda$ = mean free path of molecules, $\rho_p$ = particle density, $\kappa$ = von Karman constant, $z$ = measurement height, $z_0$ = roughness length, $u_*$ = friction velocity.

Using these algorithms it can be shown that for reactions of $HNO_3$ on sea salt the constant flux layer assumption may be violated (Geernaert et al., 1998) for commonly observed conditions (sea salt size distributions, and meteorological/surface conditions) in the coastal zone (see Fig. 4). The flux divergence occurs due to the following reaction:

$$HNO_3 + NaCl \rightarrow HCl + NaNO_3$$

which leads to transfer of nitric acid from the gas phase to the aerosol phase rather than to the sea surface and release of hydrochloric acid (see Pryor and Sørensen, 2000).

## ATMOSPHERIC DEPOSITION TO THE COASTAL ZONE

### Impacts in the Coastal Zone

Eutrophication has been observed in a diverse range of aquatic environments from estuaries to open waters. Examples include; Chesapeake Bay and the Neuse River Estuary in the eastern USA (Russell et al., 1998; Pinckney et al., 1998), seas such as the Baltic and Mediterranean Seas in Europe (Paerl, 1995), open oceans (Owens et al., 1992). The coastal zone and enclosed seas are particularly vulnerable to human-induced eutrophication due to the proximity to massive sources of inorganic and organic emissions combined with reduced mixing and hence dilution of fluxes

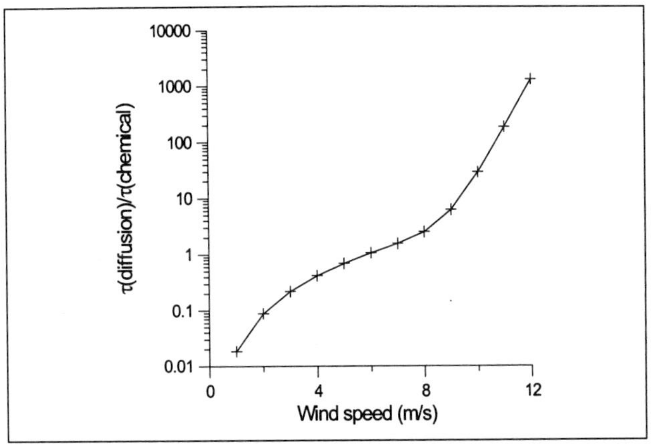


Fig. 4. The ratio of chemical to diffusive time-scales for nitric acid flux over a sea surface for a hypothetical sea salt size distribution as a function of wind speed. Where $\tau_d/\tau_c << 1$, the constant flux layer assumption is not significantly violated due to chemical reactions. Where $\tau_d/\tau_c >> 1$ chemical time scales are shorter than the transport time scale and hence chemical reactions significantly modify the transfer of nitric acid to the sea surface. For this example the generation of sea-salt droplets per unit area of sea surface (by bubble burst and spume), per increment of droplet radius ($r$), per second is determined from:

$$\text{Bubble burst: } \frac{dF_0}{dr} = 2.373 U_{10}^{3.41} r^{-3} (1 + 0.057 r^{1.05}) \times 10^{1.19 e^{-B^2}}$$

$$B = \frac{0.380 - \log r}{0.65}$$

$$\text{Spume } (r > 10\,\mu\text{m}): \frac{dF_1}{dr} = 8.6 \times 10^{-6} e^{2.08 U_{10}} r^{-2}$$

where: $r$ = radius, $F$ = flux (due to bubble burst (subscript 0), or spume (subscript 1)), $U_{10}$ = wind speed at 10 m. Algorithms from Gong et al. (1997). Here the total flux is assumed to enter the bottom meter of the atmosphere and to produce an equilibrium size distribution over a 10 second period. Although transport of nitric acid towards the sea surface is increased by higher wind speeds, due to the highly non-linear dependence of sea spray production on wind speed, in this scenario the higher wind speeds are associated with greater potential for flux divergence.

(both fluvial and atmospheric). In a number of coastal environments increased N loading has led to toxic red tides (blooms of dinoflagellates) and resulted in documented fish kills in both European and North American coastal environments (Paerl, 1995) and changes in ecosystem composition (Pinckney et al., 1998). The losses associated with coastal eutrophication are not only environmental but also economic. For example, Paerl (1995) reports that the fall outbreak of the Florida red tide dinoflagellate along the North Carolina coast in 1987 caused damages amounting to $40 million due to losses in fishing and recreation.

The Chesapeake Bay estuary is perhaps the region most synonymous with eutrophication problems and remediation efforts. It is the largest eastern US coastal estuary and has shown evidence of severe affliction due to nutrient enrichment. Fisher and Oppenheimer (1991) report that water quality in the bay has been declining since the industrial revolution but this decline accelerated rapidly during the 1960s, 1970s and 1980s, resulting in increasingly frequent blooms of blue-green algae. The late summer period when phytoplankton growth is strongly dependent on N availability is also the season most strongly associated with algal blooms and anoxia. Hence, increased N input during this season, which has a large atmospheric component, has been the particular focus of research. Fisher and Oppenheimer (1991) estimate that on an annual basis during the 1980s 25% of the anthropogenic N flux to the bay is in the form of atmospheric $NO_3^-$ deposition and a further 14% of the total is due to atmospheric flux of $NH_4^+$. It can also be postulated that, as measures are implemented to reduce loss from agricultural and industrial sources to water systems which lead to the bay, it is possible that the importance of atmospheric sources may increase.

## Atmospheric Flow in the Coastal Zone and Atmosphere–Surface Exchange

The coastal zone is typically close to sources of pollution and receives relatively large pollutant fluxes via atmospheric and fluvial pathways. In addition to the effects of proximity to nutrient and pollutant sources, concentrations of trace chemicals may also be enhanced by the specific atmospheric and oceanic conditions. Much of this discussion is beyond the scope of this chapter. For a description of coastal oceanography the reader is directed to Pickard and Emery (1986) and other chapters in this volume. Flux divergence in the coastal zone due to horizontal inhomogeneities is also beyond the scope of this review; for further information the interested reader is directed to Geernaert and Geernaert (1998). Here we shall only briefly consider an overview of the characteristics of atmospheric flow in the coastal zone and how these might influence deposition and dispersion.

Pollutants released into the atmosphere are transported, dispersed and diluted at rates primarily determined by the wind speed conditions and atmospheric stratification. Dispersion and deposition processes in the coastal zone are complicated by the change in boundary-layer flow which occurs as air flows from land surfaces over sea surfaces (and *vice versa*). To describe flow adjustment moving across a surface discontinuity (e.g. moving offshore) the concept of the internal boundary layer is used. At the discontinuity, an internal boundary layer develops below which the atmosphere responds to the properties of the new surface. Above this layer, the atmosphere continues to respond to the old surface. In-between these two layers an adjustment layer exists. Hence, in the coastal zone the vertical profiles of momentum, temperature and humidity (as well as pollutant concentrations) are highly complex (see Ayra, 1988).

Internal boundary layers develop in response to changes in the surface thermal fluxes (heat and water vapour) and also differences in the momentum flux. First let us consider thermal effects and flow moving offshore. Due to the greater heat capacity of water, the sea surface has a much lower temperature range than a typical land surface, thus mean sea surface temperatures show little diurnal variation and mean temperatures typically lag those of adjacent land surfaces by about one month (in northern Europe). If the sea surface is cool relative to the air flowing over it from land a stable internal boundary layer develops which inhibits mixing and hence transport of pollutants towards the surface. Elevated plumes will remain aloft from the surface and will show little vertical transport or dilution (Stull, 1988). Thus, chemicals within the plume can be transported great distances prior to removal. These stable layers are frequently found during daytime offshore and in the spring and can propagate for up to 200 km offshore (Garratt, 1987). However, if the sea surface is warm relative to the air flowing over it from land, an unstable layer develops in which mixing is enhanced. Under these circumstances convection can fumigate pollutants held aloft to the water surface and thus greatly enhance deposition close to the shoreline. These unstable layers can occur during the night-time or autumn when sea surface temperatures are frequently warmer than land temperatures. In addition to modification of thermally induced turbulence, the transition from a land to water surface also modifies mechanically generated turbulence. Although sea surface roughness increases with increasing wind speed, sea surfaces are generally two orders of magnitude less rough than typical land surfaces; thus mechanically generated turbulence is reduced over water surfaces. Hence the atmosphere over sea surfaces is typically subject to reduced mechanical and convective mixing and thus increased aerodynamic resistance and reduced vertical mixing of pollutant species aloft from the surface.

Coastal areas are also subject to mesoscale flows such as the well-known sea breeze (Simpson, 1994). Differences between land and sea surface temperatures cause flows during the day to be directed onshore (when the land is warmer than the sea surface and so a relative surface low pressure forms) and over night to be directed offshore (when the converse is true). The sea breeze is a three-dimensional circulation with inflow at the surface balanced by rising air, return flow aloft and return of air to the surface. The third dimension is induced as an along-coast component in the circulation, due to the action of the coriolis effect. Due to the diurnal cycle between land and sea-breezes and the nature of the sea-breeze circulation, there is the potential for recirculation of pollutants that can become trapped in narrow elevated layers offshore, re-advected over the coastline and ultimately fumigated back to the surface at/near the coast on the following day (see Hsu, 1988).

## SUMMARY

The atmospheric pathway represents a significant source of pollutant and nutrient fluxes to aquatic ecosystems. The atmospheric N deposition may represent 10–50% of the anthropogenic N flux to water surfaces. For example, in Chesapeake Bay 25% of the anthropogenic N flux to the bay is in the form of atmospheric $NO_3^-$ deposition and a further 14% of the total is due to atmospheric flux of $NH_4^+$. The role of particle dry deposition in this exchange is relatively poorly quantified but is substantial in the limited number of studies available. In this chapter the physical processes responsible for atmosphere–surface exchange of particles have been outlined along with the major uncertainties in the measurement and modelling of particle fluxes.

## APPENDIX

Outline of parameters used in the model presented by Pryor et al. (1999).

Transfer via turbulent processes is given by:

$$v_h = \frac{1}{\frac{1}{\kappa u_*}\left[\ln\left(\frac{z_r}{z_{0m}}\right) - \psi_h\left(\frac{z_r}{L}\right) + \psi_h\left(\frac{z_{0m}}{L}\right)\right]}$$

where: $u_*$ = friction velocity, $z_{0m}$ = roughness length for momentum, $z_r$ = reference height, $\kappa$ = von Karman constant (0.4), $L$ = Monin–Obukhov length, $\psi_h$ = stability correction.

The transfer velocity due to gravitational settling is given by:

$$v_g = \left[\frac{g}{18\mu}\left(\frac{\rho_p}{\rho_{air}}\right)d^2\right]C$$

where: $\rho_p$ = particle density, $\rho_{air}$ = density of air (temperature corrected density), $g$ = acceleration due to gravity, $\mu$ = dynamic viscosity of air, $C$ = Cunningham slip correction factor:

$$C = 1 + \frac{\lambda}{d}\left[2.514 + 0.8\exp\left(-0.55\frac{d}{\lambda}\right)\right]$$

where: $\lambda$ = mean free path of air, $d$ = particle diameter.

Transfer velocity across the laminar surface layer is given by:

$$v_\delta = (1-\alpha_{bob})(cu_* Sc^{-0.5}\,\mathrm{Re}^{-0.5} + u_* 10^{-3/St}) + \alpha_{bob}\left(\frac{u_*^2}{U_{10}} + Eff(2\pi r_{drop}^2)(2z_{drop})(2z_{drop})\frac{q_{drop}}{\alpha}\right)$$

The two component parts of this equation pertain to diffusional transfer and to the increase in downward movement of particles due to bubble burst activity.

$$c = \frac{1}{c_1\sqrt{c_2}},$$

$c_1 = f$ (the thickness of the molecular diffusion layer), $c_2 = f$ (ratio of the height of roughness elements to aerodynamic roughness length).

*Sc* = Schmidt number (*f*(Diffusivity (*D*))), $\alpha_{bob}$ = the relative area with burst bubbles (defined below), *d* = particle diameter. Because *D* is being calculated for the surface laminar layer, the diameter of the particle in equilibrium with the higher near-surface humidity is used (i.e. $d = d_w$).

$$Re = \text{Reynolds number: } Re = \frac{u_* z_0}{\nu}$$

$$St = \text{Stokes number: } St = \frac{u_*^2 v_{g,w}}{v_g}$$

*Eff* = effectiveness with which particles are collected, $z_{drop}$ = average height the spray drops reach, $r_{drop}$ = average radius of the spray drops, $q_{drop}$ = the flux of spray drops from the surface, $\alpha$ = area of the sea surface covered by whitecaps, $\alpha_{bob}$ = area with bursting bubbles.

## ACKNOWLEDGEMENTS

The authors would like to thank Gary Geernaert for his comments on a draft of this chapter. They also gratefully acknowledge financial support from the following agencies: Environment Canada, the Midwestern Regional Center of the National Institute for Global Environmental Change, the National Science Foundation, the American–Scandinavian Foundation, Nordic Council of Ministers, Indiana University, and the National Environment Research Institute of Denmark.

## REFERENCES

Anderson, D.M. (1994) Red Tides. *Scientific American* August, 62–68.
ApSimon, H., Kruse, M. and Bell, J.N. (1987) Ammonia emissions and their role in acid deposition. *Atmospheric Environment* **21**, 1939–1946.
Asman, W.A.H. et al. (1995) Atmospheric nitrogen input to the Kattegat. *Ophelia* **42**, 5–28.
Ayra, S.P. (1988) *Introduction to Micrometeorology*. Academic Press, San Diego.
Barth, H. and Nielsen A. (1989) The occurrence of *Chrysochromulina polyepis* in the Skagerrak and Kattegat in May/June 1988: An analysis of extent, effects and causes. Commission of the European Communities, Water pollution research report #10. EUR 12069.
Bigg, G.R. (1996) *The Oceans and Climate*. Cambridge University Press, Cambridge, 266 pp.
Buzorius, G., Rannik, U., Hakela, J.M., Vesala, T. and Kulmala, M. (1998) Vertical aerosol particle fluxes measured by eddy covariance technique using condensational particle counter. *Journal of Aerosol Science* **29**, 157–171.
Cornell, S., Rendell, A. and Jickells, T. (1995) Atmospheric inputs of dissolved organic nitrogen to the oceans. *Nature* **376**, 243–246.
Dabberdt, W. et al. (1993) Atmosphere–surface exchange measurements. *Science* **260**, 1472–1481.
Davidson, C.L. and Wu, Y.L. (1990) Dry deposition of particle and vapors. In *Acidic Precipitation. Volume 3. Sources, Deposition and Canopy Interaction*, eds. S.E. Linberg, A.L. Page and S.A. Norton. Springer-Verlag, New York, pp. 103–209.
De Arellano, J. and Duynkerke, P. (1992) The influence of chemistry on the flux-gradient relationships in the $NO\text{–}O_3\text{–}NO_2$ system. *Boundary-Layer Meteorology* **61**, 375.
Erisman, J. and Draijers, G.P.J. (1995) *Atmospheric Deposition in Relation to Acidification and Eutrophication*. Studies in Environmental Sciences 63. Elsevier, Amsterdam, 405 pp.
Fisher, D. and Oppenheimer, M. (1991) Atmospheric nitrogen deposition and the Chesapeake Bay estuary. *Ambio* **20**, 102–108.
Fontan, J., Lopez, A., Lamaud, E. and Druilhet, A. (1997) Exchange of particles: Vertical flux measurements of submicronic aerosol particles and parameterization of the dry deposition velocity. In *Biosphere–Atmosphere Exchange of Pollutants and Trace Substances. Transport and Chemical Transformation of Pollutants in the Troposphere*, ed. S. Slanina. Springer, Berlin, pp. 382–390.
Gallagher, M. et al. (1997) Measurements of aerosol fluxes to Speulder forest using a micrometeorological technique. *Atmospheric Environment* **31**, 359–373.
Galloway, J., Levy , H. and Kasibhatla, P. (1994) Year 2020: consequences of population growth and development on deposition of oxidized nitrogen. *Ambio* **23**(2), 120–123.
Garratt, J.R. (1987) The stably stratified internal boundary layer for steady and diurnally varying offshore flow. *Boundary-Layer Meteorology* **38** (4), 369–394.
Geernaert, G.L. and Geernaert, L.L. (1998) Air–sea exchange of nutrients: Governing processes and models. In *Sea–Air Exchange: Processes and Modelling*, eds. J. Pacyna, D. Broman and E. Lipiatou. ISBN 92-828-2577-9. European Commission, Luxemborg.
Geernaert, L.L.S., Geernaert, G.L., Granby, K. and Asman, W.A. (1998) Fluxes of soluble gases in the marine atmosphere surface layer. *Tellus* **50B**, 111–127.
Gong, S.L., Barrie, L.A. and Blancher, J.P. (1997) Modelling sea-salt aerosols in the atmosphere. 1. Model development. *Journal of Geophysical Research* **102**(D3), 3805–3818.
Graneli, E., Graneli, W. and Rydberg, L. (1986) Nutrient limitation at the ecosystem and phytoplankton community level in the Laholm Bay, south-east Kattegat. *Ophelia* **26**, 181–194.
Haumann, L. (1989) Algal blooms. In: Barth and Nielsen, The occurrence of *Chrysochromulina polyepis* in the Skagerrak and Kattegat in May/June 1988: An analysis of extent, effects and causes. Commission of the European Communities, Water pollution research report #10. EUR 12069.
Hertel, O. (1995) Transformation and deposition of sulphur and nitrogen compounds in the marine boundary layer. Available from; Danish Ministry of Environment and Energy, National Environmental Research Institute, Roskilde, Denmark Thesis, Norway, Bergen, Norway, 215 pp.
Hertel, O. and Frohn, L. (1997) Nitrogen deposition to Danish waters 1989 to 1995. Estimation of the contribution from Danish sources. 215, National Environmental Research Institute, Denmark.
Hoff, R. et al. (1997) Use of airborne LIDAR to determine aerosol sources and movement in the Lower Fraser Valley, BC. *Atmospheric Environment* **31**, 2123–2134.
Hoff, R.M. et al. (1996) Atmospheric deposition of toxic chemicals to the Great Lakes: A review of data through 1994. *Atmospheric Environment* **30** (20), 3505–3527.
Hsu, S.A. (1988) *Coastal Meteorology*. Academic Press, London.
Hummelshøj, P., Jensen, N.O. and Larsen, S.E. (1992) *Particle Dry Deposition to a Sea Surface, 5th International Conference on Precipitation Scavenging and Atmosphere–Surface Exchange Processes*. AMS, Richland, Washington, USA, pp. 12 pp.
IPCC (1996) Intergovernmental Panel on Climate Change. Climate Change 1995. Cambridge University Press, 572 pp.
Jickells, T., Knap, A.H., Sherriff-Dow, R. and Galloway, J. (1990) No ecosystem shift. *Nature* **347**, 25–26.
Kramm, G. and Dlugi, R. (1994) Modelling the vertical fluxes of nitric acid, ammonia and ammonium nitrate. *Journal of Atmospheric Chemistry* **18**, 319–357.
Larsen, S.E. et al. (1995) Dry deposition of particles to ocean surfaces. *Ophelia* **42**, 193–204.
Leuning, R. and King, K.M. (1997) Comparison of eddy covariance

measurements of $CO_2$ fluxes by open and closed path $CO_2$ analyzers. *Boundary-Layer Meteorology* **85**, 293–307.

Markeger, S., Conley, D., Hertel, O. and Skov, H. (1998) Sådan påvirker luftbaren kvælstof havet. *Ingineer* **31**, 4 (In Danish).

Mason, R.P., Lawson, N.L. and Sullivan, K.A. (1997) Atmospheric deposition to the Chesapeake Bay watershed—Regional and local sources. *Atmospheric Environment* **31**(21), 3531–3540.

Mopper, K. and Zika, R.G. (1987) Free amino acids in marine rains: evidence for oxidation and potential role in nitrogen cycling. *Nature* **325**, 246–249.

Owens, N.J.P., Galloway, J.N. and Duce, R.A. (1992) Episodic atmospheric nitrogen deposition to oligotrophic oceans. *Nature* **357**, 397–399.

Paerl, H.W. (1995) Coastal eutrophication in relation to atmospheric nitrogen deposition: current perspectives. *Ophelia* **41**, 237–259.

Pakkanen, T.A. (1996) Study of formation of coarse particle nitrate aerosol. *Atmospheric Environment* **30**, 2475–2482.

Parsons, T.R., Takahashi, M. and Hargrave, B. (1984) *Biological Oceanographic Processes*. Pergamon, Oxford, 330 pp.

Pickard, G.L. and Emery, W.J. (1986) *Descriptive Physical Oceanography: An Introduction*. Pergamon Press, Oxford, 249 pp.

Pinckney, J., Paerl, H., Harrington, M. and Howe, K. (1998) Annual cycles of phytoplankton community-structure and bloom dynamics in Neuse River Estuary, North Carolina. *Marine Biology* **131**, 371–381.

Pryor, S.C. and Sørensen, L.L. (2000) Nitric acid–sea salt reactions: implications for nitrogen deposition to water surfaces. *Journal of applied Meteorology* (in press).

Pryor, S.C., Barthelmie, R.J., Geernaert, L.L.S., Ellermann, T. and Perry, K. (1999) Speciated particle dry deposition to the sea surface: Results from ASEPS'97. *Atmospheric Environment* **33**(13), 2045–2058.

Raes, F.J.W. and Van Dingenen, R. (1995) Aerosol dynamics and its implication for the global aerosol climatology. In *Aerosol Forcing of Climate: Report of Dalem Workshop*, eds. R. Charlson and J. Heintzenberg. Wiley and Sons, London.

Rendell, A.R., Ottley, C.J., Jickells, T.D. and Harrison, R.M. (1993) The atmospheric input of nitrogen species to the North Sea. *Tellus* **45B**, 53–63.

Russell, K.M., Galloway, J.N., Macko, S.A., Moody, J.L. and Scudlark, J.R. (1998) Sources of nitrogen deposition to Chesapeake Bay Region. *Atmospheric Environment* **32**(14), 2453–2465.

Schaefer, D.A. and Reiners, W.A. (1990) Throughfall chemistry and canopy processing mechanisms. In *Acidic Precipitation. Volume 3. Sources, Deposition and Canopy Interaction*, eds. S. Lindberg, A. Page and S. Norton. Springer-Verlag, New York.

Schery, S., Wasiolek, P., Nemetz, B., Yarger, F. and S.W. (1998) Relaxed eddy accumulator for flux measurement of nanometer-size particles. *Aerosol Science and Technology* **28**, 159–172.

Scudlark, J.R., Russell, K.M., Galloway, J.N., Church, T.M. and Keene, W.C. (1998) Organic nitrogen in precipitation at the mid-Atlantic U.S. coast—methods evaluation and preliminary measurements. *Atmospheric Environment* **32**, 1719–1728.

Seinfeld, J. and Pandis, S. (1998) *Atmospheric Chemistry and Physics: from Air Pollution to Climate Change*. Wiley-Interscience, 1326 pp.

Simpson, J.E. (1994) *Sea Breeze and Local Winds*. Cambridge University Press, Cambridge, 234 pp.

Slinn, S. and Slinn, W. (1980) Predictions for particle deposition on natural waters. *Atmospheric Environment* **14**, 1013–1016.

Spengler, J., Brauer, M. and Koutrakis, P. (1990) Acid air and health. *Environmental Science and Technology* **24**, 946–956.

Spiel, D.E. (1994) The sizes of the jet drops produced by air bubbles bursting on sea- and fresh-water surfaces. *Tellus* **46B**, 325–338.

Stull, R.B. (1988) *An Introduction to Boundary Layer Meteorology*. Kluwer, Dordrecht, 666 pp.

Sweet, C.D., Weiss, A. and Vermette, S. (1998) Atmospheric deposition of trace metals at three sites near the great Lakes. *Water, Air and Soil Pollution* **103**, 423–439.

Vitousek, P. et al. (1997) Human alterations of the global nitrogen cycle: causes and consequences. *Issues in Ecology* (Spring 1997), 1–16.

Wakimoto, R. and McElroy, J. (1986) Lidar observation of elevated pollution layers over Los Angeles. *Journal of Applied Meteorology* **25**, 1583–1599.

Wexler, A. and Seinfeld, J. (1992) Analysis of aerosol ammonium nitrate: departures from equilibrium during SCAQS. *Atmospheric Environment* **26** (4), 579–591.

Williams, R.M. (1982) A model for the dry deposition of particles to natural surface waters. *Atmospheric Environment* **16**, 1933–1938.

Winchester, J.W., Escalona, L., Fu, J. and Furbish, D.J. (1995) Atmospheric deposition and hydrogeological flow of nitrogen in northern Florida watersheds. *Geochimica et Cosmochimica Acta* **11**, 2215–2222.

Wu, J. (1988) Variations of whitecap coverage with wind stress and water temperature. *Journal of Physical Oceanography* **18**, 1488–1453.

Zufall, M.J., Davidson, C.I., Caffrey, P.F. and Ondov, J.M. (1998) Airborne concentrations and dry deposition fluxes of particulate species to surrogate surfaces deployed in southern Lake Michigan. *Environmental Science and Technology* **32**, 1623–1628.

## THE AUTHORS

**Sara C. Pryor**
*Atmospheric Science Program,*
*Department of Geography,*
*Indiana University,*
*Bloomington, IN 47405, U.S.A.*

**Rebecca J. Barthelmie**
*Department of Wind Energy and Atmospheric Physics,*
*Risø National Laboratory,*
*DK-4000 Roskilde, Denmark*

Chapter 122

# MARINE ECOSYSTEM HEALTH AS AN EXPRESSION OF MORBIDITY, MORTALITY AND DISEASE EVENTS

Benjamin H. Sherman

Over the last fifty years, national, international and private stewardship and conservation organizations have spent billions of dollars collecting marine ecosystem information. The information remains divided among many custodians, scattered among thousands of published sources and, from a global perspective, is fragmentary in nature. It is argued that new resource management questions regarding coastal ecosystem health can be addressed through the recovery and "data mining" of this previously collected and often discarded information. A retrospective marine epidemiological approach was developed to demonstrate that marine morbidity, mortality and disease information is recoverable by keyword searching of academic journals and through the retrieval of publicly available digital and print-media information. Observational records compiled from disturbances occurring within the Northwestern Atlantic, Gulf of Mexico and Caribbean Sea confirm that anomalous marine morbidity and mortality events have increased in number and frequency during the last 30 years. A global approach is summarized for systematically reconstructing spatial and temporal disturbance indicator time series using data mining and data reduction techniques.

*Seas at The Millennium: An Environmental Evaluation (Edited by C. Sheppard)*

*"Errors of Nature, Sports and Monsters correct the understanding in regard to ordinary things, and reveal general forms. For whoever knows the ways of Nature will more easily notice her deviations; and, on the other hand, whoever knows her deviations will more accurately describe her ways."* Francis Bacon (1620)

## INTRODUCTION

Red tides, bleaching of coral reefs, the stranding of marine mammals and mass mortality of fish and invertebrates have been observed for millennia. Aristotle, Homer (Iliad), Tacitus, and early European navigators were astute observers of these episodic phenomena (Baden et al., 1995). This century, the frequency, extent and duration of disturbances appear to be increasing (HEED, 1998; Harvell et al., 1999), but reports on these phenomena are fragmented.

Several groups have begun to characterize the health of entire coastal marine ecosystems (e.g. GESAMP, 1990; NOAA, 1993; NRC, 1999) and to understand the pressures upon them, compare management techniques and guide better studies. Institutions are now designing new monitoring programs to collect ecosystem health information. National and international "ecosystem health" stewardship and conservation organizations have spent billions of dollars funding research and data collections relevant to ecosystem health questions, but global-scale marine ecosystem health information has not yet been consolidated (NRC, 1995). Thousands of researchers have, however, published studies describing anomalies among various taxonomic and indicator responses to a wide range of global disturbances, so the potential exists to reassemble and characterize disturbance and impact information from published accounts and to reconstruct time series at a variety of scales to find common response patterns among multiple indicators.

Information technologies have recently allowed "mining" of bibliographic archives, geographic information systems and Internet databases to re-collect scattered information. From observational accounts, patterns of some disturbance indicators appear to serve as proxies for each other in time. Because collecting long time-series data for any single indicator is problematic (Magnuson, 1990), reconstruction of historical time series by scavenging available indicator and/or proxy data provides a new information base to evaluate large-scale and long-term disturbance patterns. The purpose of recovering disturbance records is to match impact "symptoms" with exogenous factors. Epidemiologists use similar retrospective techniques to track public health problems. A marine epidemiological (marine ecosystem health) approach incorporates multiple techniques and the methods of many disciplines (Epstein, 1994).

The methods summarized here are a first-approximation data mining–data rescue effort to address both marine ecosystem health and disturbance questions. Methods rely upon a basic knowledge of species' natural histories, as synchronized with, and/or responsive to ecological, climate and oceanographic variability. From such an understanding, a large number of reported anomalous marine morbidity and mortality observations can be placed within an information system and grouped to represent a smaller number of disturbance indicators. Impact-based indices (grouping indicators by co-occurrence in time and space) are used to organize scattered and fragmented species monitoring data (Table 1). These indices measure Multiple Marine Ecological Disturbances (MMEDs) and are used to help define Major impacts (Williams and Bunkley-Williams, 1992). The centralized MMED information provides more complete historical (e.g. time series) information than would be typically available to researchers or managers.

Table 1

Selection of disturbance observations (only occasionally reported) correlated with time-series measurements (regularly monitored) can provide rapid marine health and impact diagnostic information (from HEED, 1998). Fisheries economic and basic meteorological and oceanographic time series for defining impact conditions are required to establish anomalies (near continuous data). By locating and tracking significant "anomalies", determined by observation of threshold conditions and magnitude deviation from a place-specific average, correlative statistics can be run on the combined time series.

| Morbidity, mortality and disease reports. Observational data (only events are reported) | Anomaly time-series report-types. Both events and non-events reported |
|---|---|
| 1. Aquatic birds events | 1. Monitored shellfish poisoning toxins (SPT) |
| 2. Coral and sponge events | 2. Morbidity and mortality reports for public health monitoring (Promed/WHO) |
| 3. Fish events | 3. Sea level pressure (SLP) |
| 4. Harmful algae blooms | 4. Sea surface temperature anomaly (SSTA) |
| 5. Human marine related illness | 5. Precipitation anomaly (PPTA) |
| 6. Mortalities and disease of marine mammals | 6. Wind stress/direction (WSD) |
| 7. Mollusc shellfish and crustacean events | 7. Climate indices (NAO, ENSO) |
| 8. Sea turtle events | 8. Drought indices and severe storms (WMO) |
| 9. Large invertebrate mortality | 9. Fisheries economic statistics (FES) |
| 10. Seagrass disease and loss | 10. Mandatory discharge and dumping reports |

## SCALE SPECIFICATION

Scale is used to optimally aggregate anomaly indicators to construct MMED indices. Point data (*x,y* sample locale), river, estuary, harbour, bay, state, drainage basin, province, nation, large marine ecosystem and their subdivisions are the spatial scales researchers associate with marine indicators. Observers use temporal sampling increments of hours to decades, often reporting conceptual divisions of time

# Marine Ecosystem Health

Paul R. Epstein

*Center for Health and the Global Environment, Harvard Medical School, Boston MA 02115, U.S.A.*

Ecosystem health is a concept circumscribing a wide range of disciplines, for which a single precise definition is problematic (Costanza et al., 1992; Haskell et al., 1992). Common in economics, medicine and biology is the "health" concept pertaining to anomaly tracking and symptom definition. Following Costanza (1992), to be healthy and sustainable an ecosystem must maintain its metabolic activity level, its internal structure and organization, and must be resistant to stress over a wide range of temporal and spatial scales. Methods to assess the health of marine ecosystems are being developed from indicator time series and indices described by several investigators (Costanza, 1992; Costanza and Mageau, 1999; Rapport, 1992; Karr, 1992).

In the epidemiological framework, disease results from a combination of influences: the pathogenic (disease-causing) agent must be present, a host's depressed defences must allow opportunistic infection, and suitable environments are needed for the transmission of the agent (Epstein, 1994). Within each of the overlapping circles (Fig 1), specific information from a variety of disciplines and range of scales is aggregated to find the origin and extent of an emergent disease or epidemic.

In marine ecosystems, stressors and impacts result from physical/oceanographic, biogeochemical, socioeconomic and political/institutional factors. Adapting the health framework to suit marine epidemiology, the "host" may be seen as the ecosystem, with its own set of ecological defences and buffers against disturbance, and the "environment" represents exogenous and variable forcing factors within the Earth's system (e.g., climate and oceanographic). The pressure agent is human influence upon these systems through chemical inputs (e.g. pollutants, climate change, eutrophication) and resource depletion directly and indirectly impacting or exacerbating the influence of climate variability. In a marine system, disease is one of many aggregate disturbances that result from interactions within the circles of influence in Fig. 1.

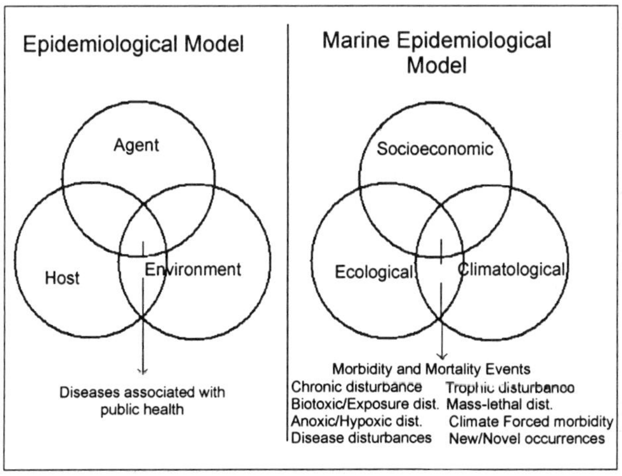


Fig. 1. Analogous health models (adapted from Epstein, 1997).

One of the main components of the marine epidemiological approach involves the tracking of harmful algal blooms (HABs). HABs are good disturbance indicators because: (a) populations of algae may rapidly alter in response to environmental changes; (b) HABs contribute to many marine health impacts; and (c) phytoplankton may serve as reservoirs and vectors for human and animal illness (Colwell, 1996; Epstein, 1997).

## REFERENCES

Colwell, R.R. (1996) Global climate and infectious disease: the cholera paradigm. *Science* **274**, 2025–2031.

Costanza, R. (1992). Toward an operational definition of ecosystem health. In *Ecosystem Health: New Goals for Environmental Management*, eds. R. Costanza, B.G. Norton and B.D. Haskell, pp. 239–256. Island Press, Washington, D.C.

Costanza, R. and Mageau, M. (1999) What is a healthy ecosystem? In *The Gulf of Mexico Large Marine Ecosystem: Assessment, Sustainability, and Management*, eds. H. Kumpf, K. Steidinger and K. Sherman (eds.), pp. 385–415. Blackwell Science Inc., Massachusetts.

Costanza, R., Norton, B. and Haskell, B.D. (eds.) (1992) *Ecosystem Health: New Goals for Environmental Management*. Island Press, Washington, D.C.

Epstein, P.R. (1994). Framework for an integrated assessment of climate change and ecosystem vulnerability, In *Disease in Evolution: Global Changes and Emergence of Infectious Diseases*, eds. M.E. Wilson, R. Levins and A. Spielman, pp. 423–435. New York Academy of Sciences, New York.

Epstein, P.R. (1997) Climate, ecology, and human health. *Consequences* **3**, 3–19.

Haskell, B.D., Norton, B.G. and Costanza, R. (1992) Introduction: what is ecosystem health and why should we worry about it? In *Ecosystem Health—New Guides for Environmental Management*, eds. R. Costanza, B.G. Norton and B.D. Haskell, pp. 3–20. Island Press, Washington, D.C.

Karr, J.R. (1992) Ecological integrity: protecting earth's life support systems. In *Ecosystem Health: New Goals for Environmental Management*, eds. R. Costanza, B.G. Norton and B.D. Haskell, pp. 223–228. Island Press, Washington, D.C.

Rapport D.J. (1992) What is clinical ecology. In *Ecosystem Health: New Goals for Environmental Management*, eds. R. Costanza, B.G. Norton and B.D. Haskell, pp. 223–228. Island Press, Washington, D.C.

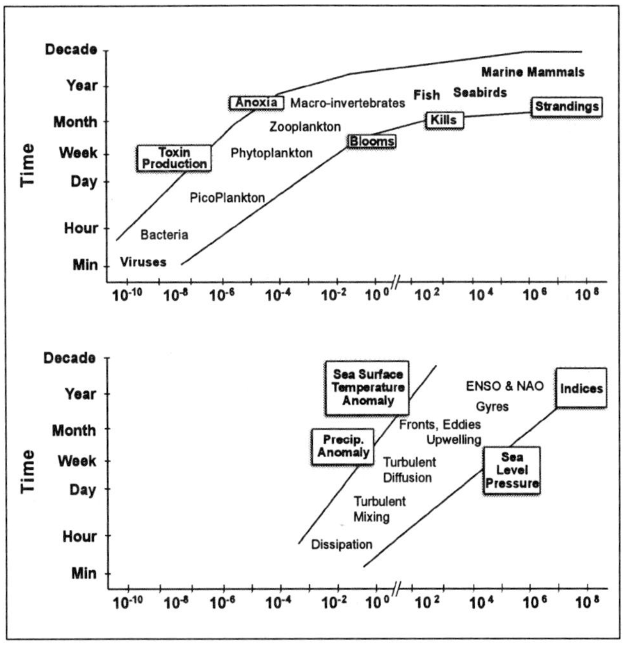


Fig. 1. Some prescribed spatio/temporal scales associated with functional impacts within marine systems. The boxes within this figure are functional groupings of indices detected along a wide range of spatio-temporal scales (sliding along and in between the lines). Aggregate information from Linnaean taxonomic groups (top) or sets of processes (bottom) are the items between the lines. The likelihood of matching oceanographic or climate forcing (bottom log plot) with biological processes (top) is improved because the boxed functional groupings (both plots) are not fixed to a particular scale. Researchers may "float" (Allen and Hoekstra, 1992) or slide the scale with each log-plot until forcing factor anomalies and possible impact category anomalies (from each log-plot) are matched in space–time. The log plot values and between the line elements are adapted from Murphy et al. (1988).

such as peak, duration, phase and seasonality (i.e. timing of phenological changes). Matching appropriately scaled data, many categories of data can be collapsed to fewer, either creating indices (for evenly sampled data time series) or general taxonomic functional classes for episodic observational data.

Episodic events involving the 10 major taxonomic divisions and the 10 regularly reported time-series data types within Table 1, are globally available and provide a first impression of co-occurring anomalies associated with scale relative functional impacts (Fig. 1). If multiple disturbances can be detected in the same time and space, "mining" of distributed sources of archived information will elucidate the relationships among the co-occurring disturbances.

Because anomaly, space and time are common descriptors for event data, co-occurring anomalies (matched in space and time) allow merger of different data types. Multiple events, even if recorded by different observers for different purposes, if found to co-occur may be defined as a distinct disturbance type (by its characteristic behaviour or components). Scale similarity and matching co-occurrence in space and time allows a fuller evaluation and tabulation of cost (to society) for each disturbance or anomaly type.

## HEED APPROACH

A pilot effort to develop a rapid global survey of possible connections and costs associated with marine disturbance types was initiated in 1995. The effort's information system is capable of helping researchers consolidate disturbance reports regionally and nationally, and could serve as a prototype for a global monitoring database.

Marine health assessments like this bring together the expertise of multiple disciplines to: (1) recommend reporting standards for literature-derived morbidity, mortality and disease events, (2) assess the integrity and coverage of monitoring data, and (3) recommend future data collection standards.

## SURVEY METHODS

The first step is to "mine" the literature for relevant categorical information. As an example, a literature search including journals, symposia, workshops, books, government technical reports, semi-popular journals, selected institutional reports, and research communications from 1945 to 1996 was conducted to populate a categorical database. Using keywords, the surveyors were able to extract meta-data (data about data) from 2127 published disturbance accounts. Source material ranged from key primary journal citations to review literature (secondary source citations) to anecdotal accounts of anomalous events (partially verified information). Additional keywords taken from either the title, an author's list of keywords and abstracts of collated reference material were added to the pre-defined list of keywords until 2021 distinct marine disturbance keywords were catalogued to assist in future surveys. Of those keywords, 314 disturbance indicator types were classified and 246 pathogen toxin and disease combinations were derived. Well-described anthropogenic impacts, such as oil/chemical spills, by-catch, dredging and errors in food preparation were excluded unless relevant long-term or recurrent disturbance characteristics were defined. It is accepted that much would be learned by merging an anthropogenic disturbance and marine health impact database.

One of the most important features of literature and data "mining exercises" is to ensure that surveys are thorough and categorical classifications complete. Graphically tracking key concepts within the marine literature, establishing the epistemological evolution for defined disturbance groupings, is one way to evaluate thoroughness. The depictions are created by backtracking citations until primary foundation-works for each disturbance type are identified.

To reduce the number of published disturbance types to a manageable number, spatial and temporal criteria can be applied using data models (Sherman, 1994). Data models help lump categorical data together, based upon similar scale, resolution, quality, and relationships (e.g. Tables 2, 3). Scale groupings integrate presumed functional ecological

## HEED MMED Program

HTTP://HEED.UNH.EDU

A group of investigators and collaborators at Harvard University and associated institutions interested in marine disturbance convened a series of workshops in 1995, 1996, 1997 to determine if "marine disturbances are getting worse as a result of global change," and to determine "the costs of the impacts, if any." This collaboration has become known as the Health Ecological and Economic Dimensions (HEED) of MMEDs Program and is now a multi-institutional consortium of the Center for Health and the Global Environment at Harvard Medical School, the University of New Hampshire's Climate Change Research Center and Office of Sustainability Programs, Tufts University School of Veterinary Medicine and Wildlife Preservation Trust International. Presently over 250 researchers participate.

The program uses workshops to exchange information and obtain consensus on the design and implementation of disturbance information methods. In 1995, participants were asked to provide bibliographic reference material and share their knowledge of the distribution of infectious disease epizootics, mass mortality events, strandings, harmful algal blooms, and anomalous changes in marine species abundance and composition. The material was entered into a database and GIS. Expert reviews determined whether reporting standards could be derived regarding disturbance types from the assembled information.

Systematic data acquisition methods using keywords were used to populate a subset of global maps focusing on the well documented Western Atlantic Ocean, Caribbean Sea, and Gulf of Mexico regions (in 1996). Sources for keywords included journal articles, symposia proceedings, workshops, books, technical reports, semi-popular journals, manuscripts, and institutional reports. The "HOLLIS" library reference system at Harvard University was used to search over 10,000 periodicals covering 1945 to 1996. The "Lexis/Nexis" mass-media database (Dayton, Ohio, USA) was used to search millions of global news sources using a separate, less technical, set of keywords (1995 to 1998). Keywords were added to the search strategy until they were then consolidated into a finite number of event types based upon impact rather than solely taxonomy. Final keywords and event types (functional groupings) were then used to re-search the literature.

The bibliographic information system, GIS and Internet-based search interface developed provides researchers with a unique centralized tool to help promote consistent standards for event reporting and documentation.

The objectives of the database were to construct a time series of disturbance events for the past 40 years, to use these to evaluate ecosystem variability, provide standards for reporting marine disturbance data and to provide a context for marine-related human health and market-related impact data. The resulting database and GIS catalogued species-specific disturbance information, co-occurring physical, chemical and economic disturbance data for several Large Marine Ecosystems (LMEs) along the East Coast of North America, Gulf of Mexico and Caribbean Sea. Anecdotal data were also included.

The HEED system incorporates 5 main elements: (1) Cited information obtained from literature reviews and data archives. (2) This is referenced to particular times and places. (3) Climate anomalies and changes in biophysical characteristics may be joined to the similarly indexed biological anomaly data. (4) Ecosystem data provide a context for co-occurring anomalies by filling data gaps in time series and more fully describing interdependent impacts through modelling and simulation. (5) Enumeration of events is then linked to observed public and economic reactions to disturbances (e.g. number of newspaper reports, supply/demand for fisheries products) to determine societal impact.

Table 2

Pathogen Toxin Disease Model (a relational example)

| Taxonomic indicator, report type | Pathogen keyword | Toxin keyword | Disease keyword |
|---|---|---|---|
| Human: doctors reports, water proximity | *Gymnodinium breve* | Brevatoxin | Respiratory distress |
| Shellfish: mussel harvesting ban | *Alexandrium tamerensis* | Saxitoxin | Paralytic Shellfish Poisoning (PSP) |
| Coral: sea fan decline report | *Aspergillus* spp. | Fungal infection | Gall formation |
| Fish: Menhaden mortality report | *Pfiesteria piscicida* | Biotoxin | Ulcerative mycosis |

categories (e.g. community, population, etc.) and lags or behaviours when disturbed (e.g. fast-acute, intermediate resilient, slow-chronic or cyclical variable and threshold cascade) using insight drawn from idea mapping.

### DATA ASSIMILATION METHODS

Place, time and type of occurrence represent the minimum information necessary for generating an observational report. Additional meta-data information about each event can be obtained from databases held by international, national, regional and local sources. Data may then be extracted for the following data types: sea-surface temperature anomaly (SSTA), sea-level pressure (SLP), wind stress (WS), precipitation (PPT), nutrients and other climate, oceanographic, health and economic data, shellfish toxicity (PSP), fisheries economic variability, El Niño-ENSO and

Table 3

A marine disturbance data model (scale-based coarse aggregate example)

| Some indicators | Spatial types | Temporal types | General parameters |
|---|---|---|---|
| Shellfish toxicity, human illness & harmful algae blooms (HABs) | Harbour, estuary | High frequency Fast/acute | Biotoxins Storm pulses |
| Seagrass disease, shellfish toxicity & HABs & birds & mackerel mass mortality | Bays and sounds | Episodal events Cyclic and variable | Physical, Chemical Biological combinations |
| HABs & pinnipeds | | | |
| Coral Reefs & Herring mass mortality | Gulf, seascape | Translates downscale Threshold cascade | Oceano-graphic forcing |
| HABs & cetaceans & sea turtles | Wide-spread Large Marine Ecosystem | Lowest frequency Slow/chronic | Climate forcing Seasonality shifts |

& = suggested correlation.

North Atlantic Oscillation–NAO indices) and merged with the biological anomaly data for correlative analyses. A variety of natural history reference sources describing the context of each place (e.g. bay, harbour, estuary, river) can be used to identify the resident organisms and habitats anticipated to represent co-occurring biological anomalies. In this regard, the life history of species potentially present and the ramification of impact over time can be better understood even if biological observations are incomplete. To anticipate impacts from climate anomalies for instance, large quantities of data must be assimilated. Fortunately much of the necessary marine climate data is readily available from Internet-accessible world data centres (Blumenthal, 1998).

The purpose of marine health literature surveys, data mining and development of physical, chemical and economic anomaly databases is to retrospectively derive new time series. Single species anomaly reports, and most types of biological monitoring data, when evaluated separately, rarely provide time series of sufficient length to evaluate trend or compare with other time series. However, when co-occurring biological disturbance data are pooled together, a single MMED series representing temporal patterns of morbidity, mortality, and disease for particular places can be more usefully evaluated. Considered as

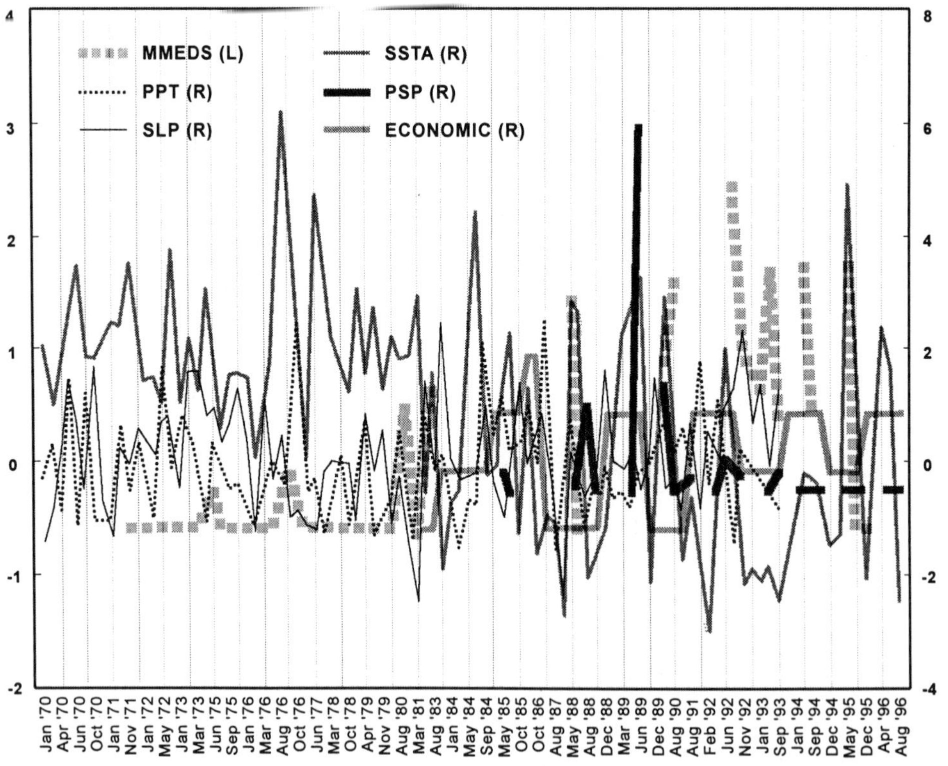


Fig. 2. Combined time-series disturbance regime example. Combined time series help fill in data gaps using a more complete representation of anomalies than available using any single incomplete time series. Depicted here is an example of a time-series combined disturbance regime from 1970–1996 in Passamaquoddy Bay, Maine, USA. Much of the interannual variability is reflected within each series. PSP (paralytic shellfish poisoning) values and MMED events may be suitable proxies for each other in this region because the predominant MMED occurrence type consists of PSP inducing harmful algae blooms.

ecosystem health indices, MMED observational anomaly data, once enumerated, can also be standardized as a time series. Combined anomaly time series, representing place-specific disturbance regimes, provide a natural history of impact and response for specific locations (e.g. Fig. 2).

## DISTURBANCE TYPE DERIVATION

Co-occurring biological anomalies reinforce the importance of particular events, point to periods of time where more intensive data searches ought to be conducted, and typically fill the gaps in incomplete monitoring series. For example, harmful algae bloom incidents, shellfish toxicity data, and reports of human illness from ingestion of shellfish biotoxins are often linked in time (e.g. Fig. 3). In many cases, one of these health indicators can serve as a proxy for the other two within a time series. Cross-correlation analyses are helpful in determining quantitatively which disturbance indicators are related and which may effectively serve as proxies for each other.

Multidimensional data reduction techniques provide another effective means of identifying categorical relationships from large amounts of survey data (e.g. Hoekstra et al. (1991) for methodology). Grouping of disturbance types using principal components analysis (PCA), for example, appears to reflect scaling, common forcing factors, and/or synchronization in response to exogenous disturbance (e.g. Fig. 4). In the HEED database thousands of keyword combinations were reduced to 147 related disturbance types that were further aggregated using PCA to form 8 disturbance types: Anoxic-hypoxic, biotoxin-exposure, disease, keystone-chronic, mass-lethal, new-novel-invasive, physically forced and trophic-magnification disturbances. The eight predominant impact types support the relationships established from cross-correlation analyses, data modelling and classifications generated by an interdisciplinary marine "expert" review group.

## EIGHT CATEGORIES OF DISTURBANCE

A first-approximation global survey of morbidity, mortality and disease using the eight functional disturbance categories is provided. The accompanying maps (Figs. 5 to 9) and lists of occurrence types are not meant as a comprehensive review of the health of the world's marine ecosystems, but rather suggest a starting point for a more complete survey.

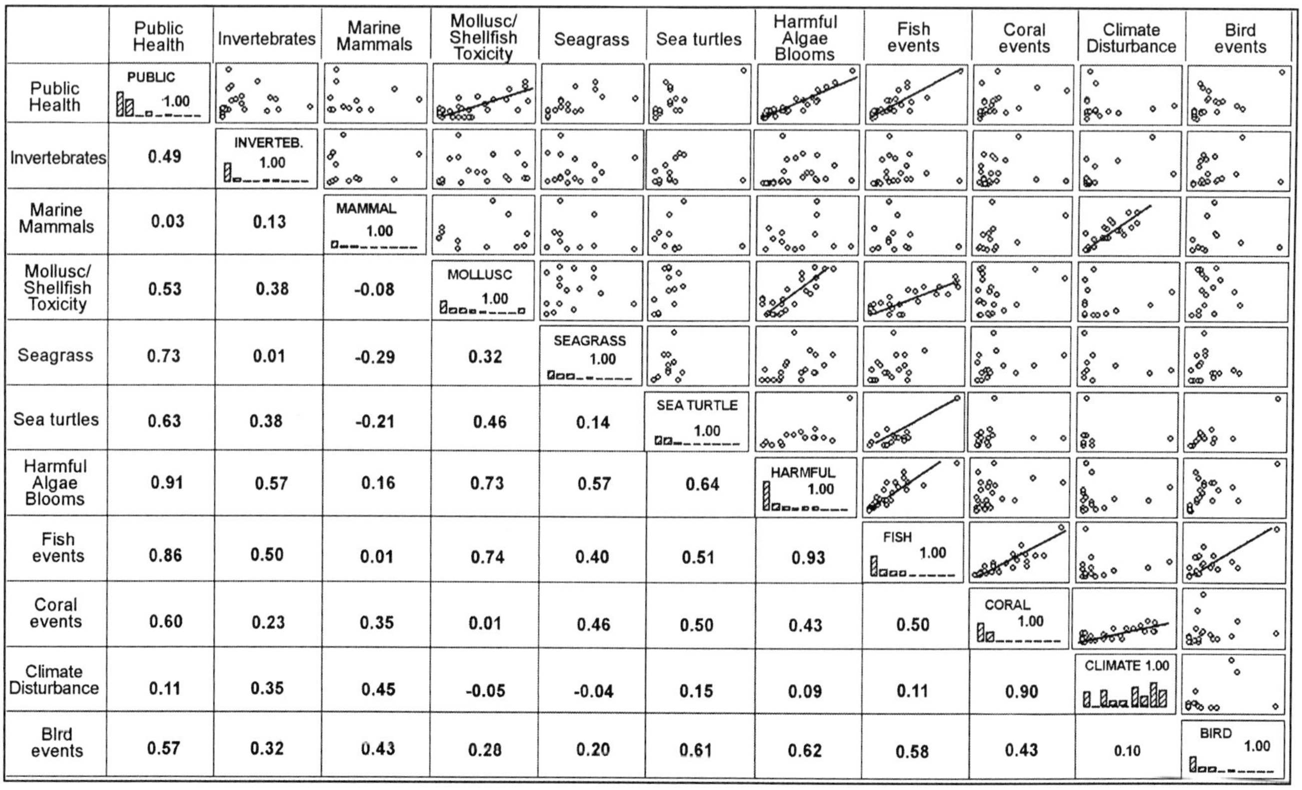

| | Public Health | Invertebrates | Marine Mammals | Mollusc/ Shellfish Toxicity | Seagrass | Sea turtles | Harmful Algae Blooms | Fish events | Coral events | Climate Disturbance | Bird events |
|---|---|---|---|---|---|---|---|---|---|---|---|
| Public Health | PUBLIC 1.00 | | | | | | | | | | |
| Invertebrates | 0.49 | INVERTEB. 1.00 | | | | | | | | | |
| Marine Mammals | 0.03 | 0.13 | MAMMAL 1.00 | | | | | | | | |
| Mollusc/ Shellfish Toxicity | 0.53 | 0.38 | -0.08 | MOLLUSC 1.00 | | | | | | | |
| Seagrass | 0.73 | 0.01 | -0.29 | 0.32 | SEAGRASS 1.00 | | | | | | |
| Sea turtles | 0.63 | 0.38 | -0.21 | 0.46 | 0.14 | SEA TURTLE 1.00 | | | | | |
| Harmful Algae Blooms | 0.91 | 0.57 | 0.16 | 0.73 | 0.57 | 0.64 | HARMFUL 1.00 | | | | |
| Fish events | 0.86 | 0.50 | 0.01 | 0.74 | 0.40 | 0.51 | 0.93 | FISH 1.00 | | | |
| Coral events | 0.60 | 0.23 | 0.35 | 0.01 | 0.46 | 0.50 | 0.43 | 0.50 | CORAL 1.00 | | |
| Climate Disturbance | 0.11 | 0.35 | 0.45 | -0.05 | -0.04 | 0.15 | 0.09 | 0.11 | 0.90 | CLIMATE 1.00 | |
| BIrd events | 0.57 | 0.32 | 0.43 | 0.28 | 0.20 | 0.61 | 0.62 | 0.58 | 0.43 | 0.10 | BIRD 1.00 |

Fig. 3. Categorical Correlation Matrix Using Spearman Rank Order Correlation from the HEED (1998) survey. Represented are Linnaean categorical associations and one climate category. Identity matrices are read like graphs: the intersection of the categories listed at the left and at the top correspond with the axes of the mini-cross correlation graphs (right half of figure) and the correlation coefficient (number values) at the intersection of category labels (bottom left hand side of figure). Frequency histograms depicting number of data points are on the diagonal. The axes for each "graphlet" represent the number of occurrences from 1945 to 1996 at the yearly level of aggregation within the Northwestern Atlantic, Caribbean Sea and Gulf of Mexico.

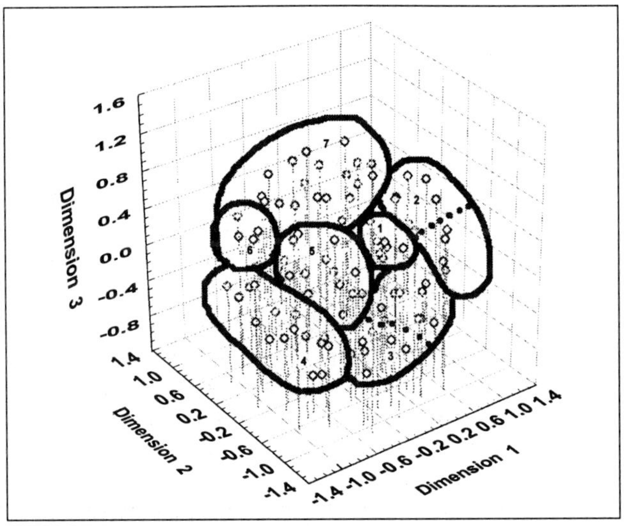


Fig. 4. PCA of 147 occurrence types within the HEED (1998) Database. Numbered labels represent interpreted clusters. Dotted lines represent possible subdivisions within the major clusters, noted as a "/" in the disturbance category title. The eight categories of disturbance are: (1) anoxic-hypoxic, (2) biotoxin/exposure, (3) keystone/chronic, (4) mass-lethal, (5) disease, (6) physically forced, and (7) trophic–magnification. (8) The new/novel or invasive category is comprised of elements from each cluster that, due to temporal uniqueness and rarity within the database (reported only once per location), were given their own class. The eight disturbance categories are not mutually exclusive. A single observation can be part of any number of categories simultaneously. More information regarding the examples and provisional metadata discussed within this chapter may be obtained on the following website: Http://heed.unh.edu.

To illustrate the worldwide extent of available data or lack of survey information, large marine ecosystems (LMEs) are used for data aggregation (Sherman and Alexander, 1989). Detailed data, if available from the HEED (1998) survey, are also provided.

Many LMEs are stressed from the growing depletion of fisheries resources, coastal zone degradation from erosion and over-development, habitat damage, and excessive nutrient loadings and pollution from drainage basin effluents (Duda and Sherman, 1999). Using the eight disturbance types as a guide, the health of the LMEs appears to be further impacted by the growing number of morbidity, mortality and disease events. Considerable effort will be required to quantify the frequency and extent of these impacts on the health of all 50 LMEs.

The eight marine disturbance types represent events that are acute, occurring within a short time-span, (e.g. biotoxin exposure), protracted (evolving over time, as when fungi infect coral reefs and seagrasses), or chronic conditions (e.g., tumour development, eutrophication). Other categories have specific correspondence to forcing factors (e.g. coral bleaching events and sea surface temperatures; population declines from climate variability and altered ecosystem dynamics). The majority of disturbance reports, however, involve harmful algae blooms (HABs) as the indirect cause of morbidity (illness) and mortality. Because HABs can affect a wide variety of living marine resources and dramatically alter ecological relationships among species (see Burkholder (1998) for review), they are a leading indicator of marine disturbance, and thus, ecosystem health.

The documentary evidence systematically provided since 1972 indicates that HABs are increasingly problematic and their duration and numbers are increasing (Smayda and Shimizu, 1993). General explanations of why the frequency and distribution of HABs may have increased during the last three decades are given in Smayda (1990) and Harvell et al. (1999). Harmful algae blooms disturb ecosystems in three primary ways, through direct exposure of wildlife and humans to algal biotoxins, indirect exposure through the food chain and by contributing to hypoxic conditions that may lead to anoxia.

### Biotoxin and Exposure Disturbances

Harmful algae species produce potent toxins (biotoxins) that cause mortalities of whales, dolphins, fish, shellfish and significantly impact commercial fisheries (NRC, 1999). Toxic dinoflagellates are responsible for a majority of major marine mortalities involving HABs. *Alexandrium* species regularly cause mortalities of fish, birds, and mammals (White, 1996). Less frequent toxic diatom blooms (e.g., *Pseudonitzschia*) are of particular concern, because of their impacts upon migratory waterfowl populations (Work et al., 1993) but also because of the severe debilitating illness of humans associated with the diatom biotoxins (Todd, 1993). Diatoms such as *Chaetoceros, Skeletonema, Rhizosolenia* spp., or *dictyophytes* have spines that can become trapped in the gills and soft tissue of fish or shellfish. The resulting mechanical damage and respiratory impairment have led to mortalities in the eastern and western Pacific (Parry et al., 1989; Albright et al., 1993; Tester and Mahoney, 1995). Cyanobacteria (e.g., *Anabeaena* spp.) have been implicated in marine mammal mortalities in Germany (Nehring, 1993; 1995), have been shown to impair the feeding behaviour of zooplankton (Guo and Tester, 1994), damage coral polyps (Endean, 1977), and may be involved in coral bleaching (Coles, 1994). Cyanobacterial blooms have been associated with sponge mortalities and reduction in recruitment of spiny lobsters in Florida Bay (Butler et al., 1995), and are cited as problems in the Latvian portion of the Baltic (Thulin, 2000). Cyanobacteria biotoxins can cause tumours (Falconer and Humpage, 1996) and have been implicated in chronic diseases of freshwater and estuarine animals (Phillips et al., 1985). Cattle, bears, raccoons, otters and birds that spend time in estuaries may also be at risk from cyanobacteria biotoxin exposure (Falconer, 1993).

In the Benguela Current, red tides have resulted in widespread mortalities of near-coastal populations of fish, shellfish, and crustaceans (Pitcher, 1998). Along the Pacific Rim, mortalities of shellfish and finfish species, supporting growing aquaculture industries, have been reported within the Gulf of Thailand subsystem of the South China Sea, where *Trichodesmium erythraeum, Noctiluca scintillans,* and

*Ceratium furca* were the dominant toxic species (Pauly and Christensen, 1993). Mortalities of bivalve and shrimp species supporting mariculture activities along the coast of China, attributed to toxic red tides, have been reported for coastal waters of the Yellow Sea and East China Sea LMEs (Chen and Shen, 1999). Recent information indicates that since the 1970s the frequency and extent of toxin-bearing red tide organisms have increased in the Pacific coastal waters of China (She, 1999).

There is circumstantial evidence for a correlation between human shellfish poisoning and the co-occurrence of tumours, neoplasia and germinomas in bivalves (Landsberg, 1995). Toxic benthic dinoflagellates (*Prorocentrum* spp.) can also produce tumour-promoting agents, like okadaic acid, that have been found within the fibroid tissue of dead green sea turtles (Landsberg et al., 1998). Exposure to okadaic acid can occur through the consumption of macroalgae and seagrasses that provide substrate for *Prorocentrum* growth (Fujiki and Suganuma, 1993).

Direct exposure to biotoxins can arrest reproduction and feeding in shellfish, cause tumours in tropical fish (Landsberg et al., 1998) and turtles, and respiratory irritation (Steidinger 1993) and neurocognitive disease in humans (Grattan et al., 1998). Respiratory irritation due to the aerosolized transport of toxic sea-spray in the vicinity of a bloom is considered a serious public health threat, and has resulted in numerous beach closures along the Florida coast (Landsberg and Steidinger, 1998).

Exposure-related disturbance impacts are not only due to algae blooms. The 1983 United States Environmental Protection Agency report "Health Effects Criteria for Marine Recreational Waters" describes several types of infections that can result from casual exposure to water. Gastroenteritis and hepatitis are the two most common swimming-associated illnesses. Other human impacts include: cellulitis, skin infection, conjunctivitis, outer ear infection, swimmer's itch, seabather's eruption, jelly fish stings, and amoebicencephalitis (fresh water). While the duration of biotoxin and exposure-related impacts is relatively short, long-term ecosystem and use changes can result. Significant reports of toxins and exposure occurrences are usually tied to beach closures, and negative publicity in tourist-related economies.

Biotoxin and exposure impacts occur with greater constancy once species have evolved alleopathic offensive and defensive responses to out-compete rival species (Baden et al., 1995; Baden and Trainer, 1993). Evolutionary mechanisms, such as cyst formation and chemosensitive triggers, allow these same toxic species to more easily spread where disturbances are frequent and/or of sufficient magnitude to permanently carve these opportunists new ecological niches (Burkholder, 1998). Mass mortalities and subsequent species substitutions eventually provide adequate benthic substrate supporting toxic biofilms and eventually displacement of predators and competitors so that once infrequent biotoxin and exposure disturbances become the new stable ecosystem configuration. Early indicators of threshold changes in ecosystem health from the biotoxin-exposure disturbance type can be measured by the temporal periodicity of events and the magnitude of single events (Fig. 5) (Table 4).

## Anoxic/Hypoxic Disturbances

Blooms of non-toxic micro-algae and nuisance macro-algae are defined as harmful because they dramatically reduce sunlight penetration in the water column and, when spent, decompose on the bottom, contributing to hypoxia (Cosper et al., 1989). If blooms are frequent, anoxia can occur. Once a hypoxic disturbance regime is established, an anoxic trajectory is likely to be established. Prolonged anoxia gives sulphur-reducing bacteria a foothold, which also accelerates mortality among benthic organisms. Subsequent deaths of fish and invertebrates provide more substrate for the decomposition cycle (Mahoney and Steimle, 1979).

The magnitude of ecosystem impact from oxygen depletion can be quite significant. In the Black Sea, a region of approximately 35,000 $km^2$ of the northwest shelf is subjected to annual recurring hypoxic conditions leading to mass mortalities of benthos estimated at 60 million tons of bottom-living animals, including 5000 tons of fish between 1973 and 1990 (Zaitsev, 1992). Seasonal mass mortalities are also observed in the southern Baltic Sea (Thulin, 2000) and Gulf of Guinea (Ibe et al., 1998). Reports of massive fish kills including hake, have also been reported as the result of upwelling, blooms, and oxygen depletion in the Benguela Current (O'Toole, 1998). In the Gulf of Alaska, plankton blooms leading to oxygen depletion have impacted fisheries and the economically important king crab (Epstein, 1996) and in Indonesia anoxia has also hurt fisheries (Adnan, 1989).

Seasonal occurrence of the hypoxic trajectory has been observed in the Gulf of Mexico since 1985. The hypoxic zone has grown as large as 18,000 $km^2$ encompassing as much as 80% of the water column (Rabalais et al., 1999). Turbid nutrient-rich waters flowing into the Mississippi River and then the Gulf contain domestic sewage, agricultural and urban run-off from 41% of the United States. Since extreme flooding of the Mississippi river in 1993, the receiving waters within the Gulf of Mexico crossed a hypoxic threshold and the ecosystem has reorganized around a stable 7000 square-mile anoxic "dead zone", off the Louisiana coast. The Gulf seafood industry has been heavily impacted, and that impact appears as anomalies within fisheries economic statistics (Prasad, 1998).

Another mode of anoxic impact is from non-toxic "brown tide" algae (picoplankton), *Aureococcus anophagefferens* and *Aureoumbra lagunensis*, for example (Cosper et al., 1987; Buskey and Stockwell, 1993). In the short term, the large number of small cells within the water column reduces bivalve filter-feeding activity, as brown picoplanktonic cells are poorly processed and of low nutritive

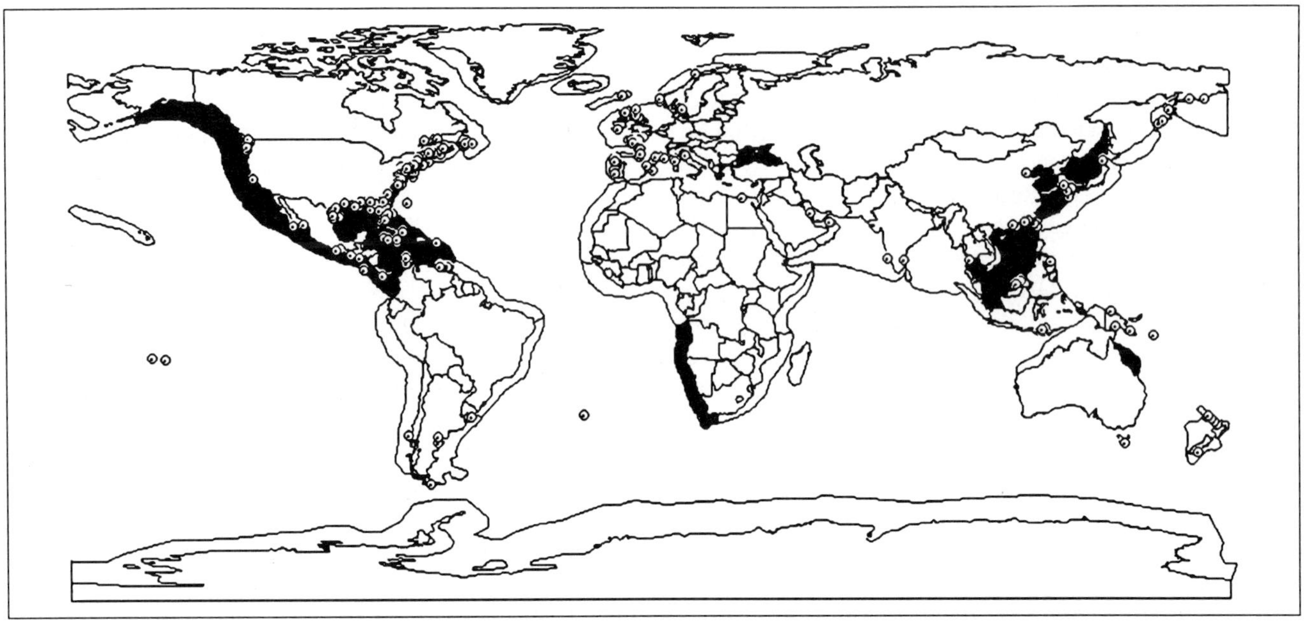

Fig. 5. Biotoxin/exposure disturbance type. The map illustrates the wide distribution of reported occurrences of harmful algae bloom events (dots) from 1945–1998. The 50 Large Marine Ecosystems are outlined and those that are shaded have been identified as having regionally accessible information regarding these data types (vs. insufficient/not surveyed empty polygons). Biotoxin disturbances include algal blooms (red tides), cyanobacteria or diatom blooms, all of varying toxicity that kill fish, invertebrates, mammals and humans. Exposure disturbances include mortality of birds associated with ASP and NSP toxins, lesions on fish or toxicity due to domoic acid, human eye irritation or memory loss or neurological damage or respiratory irritation or skin irritation (including swimmer's itch or seabather's eruption or jellyfish stings). These disturbances are subsetted from a list of 147 disturbance types derived from the HEED (1998) rapid survey.

Table 4

A portion of the 147 sub-disturbance types most associated with biotoxin and exposure occurrences

| Biotoxin disturbances | Exposure disturbance |
|---|---|
| Algal bloom (red tide) | bird mortality ASP |
| Algal bloom (toxic) | bird mortality NSP |
| Algal bloom (toxicity unknown) | fish lesions |
| Algal bloom (unusual red tide) | fish toxicity ASP |
| Algal cyanobacteria bloom | human eye irritation |
| Algal diatom bloom | human memory loss |
| Algal diatom bloom (toxic) | human neurological damage |
| Fish kill (toxic) | human respiratory irritation |
| Fish kill (toxic?) | human seabather's eruption |
| Human ciguatera poisoning | human skin irritation |
| Invertebrate crustacean kill (toxic) | human swimmer's itch |
| Invertebrate crustacean kill (toxic?) | |
| Invertebrate kill (toxic) | |
| Mammal cetacean mortality (toxic?) | |
| Mollusc bivalve kill (toxic) | |
| Mollusc bivalve kill (toxic?) | |
| Mollusc shellfish toxicity | |

content. Persistent brown tides lead to high bivalve mortalities (Tracey, 1988). Low oxygen, combined with reduction of incoming sunlight, stresses aquatic macrophytes arresting further oxygen production that may have offset hypoxic effects from ordinary blooms. Losses of seagrass meadows lead to a decline in egg hatching success of the red and black drum, and declines in zooplankton abundance (Draper et al., 1990). Combined with the high mortalities of shellfish, persistent economic losses in many fisheries can be attributed to the positive feedbacks set into motion by brown tides. Persistent brown tides within the Peconic Bay of Long Island USA, for instance, caused the scallop industry, valued at $ 1.2 million, to net only $ 2400 after four years of recurrent blooms (Grigalunas and Diamantides, 1996).

Fish kills due to hypoxia are the most commonly reported impact within the anoxic/hypoxic disturbance type, and have been used as indirect measures of ecosystem variability (NOAA, 1991). Beyond the regular seasonal oxygen depletion observed in shallow estuaries, mass mortalities of species due to hypoxia have also been used as early indicators of anoxic trajectories (Sindermann, 1994). There are also statistically significant relationships between the temporal co-occurrence of nuisance algae blooms and independently reported invertebrate, crustacean, and mollusc mortality events (HEED, 1998). Of the 147 disturbance types derived from the HEED 1998 survey, algal bloom-anoxic, fish kill-anoxic, invertebrate and or crustacean kill-anoxic and mollusc bivalve kill-anoxic are the classified sub-disturbance types.

Recognizing the importance of eutrophication and the potential for anoxic trajectories, the Ministers of Environment representing North Sea adjacent countries set targets for 50% reduction of nutrient inputs based upon a single status report issued in 1993 (North Sea Status Quality Report, 1993). This proactive approach has become a topic for negotiation within the Oslo and Paris Commissions (OSPAR) (Reid, 1999). The North Sea case is an excellent example of Environmental Ministers, as ecosystem stewards, acting in the best interests of marine ecosystem health, using indicator data.

### Trophic-Magnification Disturbances

A trophic disturbance is one that can be attributed to interspecies food-web relationships. Humans are particularly vulnerable and become sick from the consumption of foods harbouring toxins assimilated and passed up the food chain during HABs (Epstein, 1996). Illness associated with consumption of filter-feeding shellfish and fish having bioaccumulated microalgal toxins are associated with several toxic dinoflagellates (e.g., *Alexandrium, Gymnodinium, Pyrodinium, Dinophysis, Prorocentrum* sp.), diatoms (e.g., *Pseudonitzschia* sp.) and cyanobacteria (e.g., *Anabaena* sp.) (Shumway, 1990; Steidinger 1993). Amnesic shellfish poisoning (ASP) from eating mussels contaminated with domoic acid (from the diatom *Pseudonitzschia* sp.) has caused short term and permanent amnesia due to the loss of brain tissue and, depending upon exposure, death from acute toxicity (Todd, 1995). Illnesses in Canada in 1987 were followed by reports throughout the 1990s in California, Mexico, Spain, Korea and New Zealand. Seabirds have also been poisoned (e.g., Fritz et al., 1992; Sierra-Beltran et al., 1997). In the case of the November and December 1987, Scotian Shelf outbreak of amnesic shellfish poisoning, *Pseudonitzschia delicatissima* var *f. multiseries* bloomed in response to unusually high levels of nitrate-rich run-off complemented by stagnant and anomalous sea surface temperatures (Smith, 1996). The blooming cells remaining within an unusually stable water column were able to produce significant quantities of domoic acid.

More frequent and widespread are paralytic shellfish poisoning (PSP) events due to the ingestion of shellfish contaminated with *Alexandrium* sp. PSP acts on the nervous system (e.g. Price et al., 1991) and though usually not fatal, human mortalities have been recorded. PSP has also been implicated in cases of bioaccumulation with the death of seabirds and marine mammals (Nisbet, 1983; Geraci et al., 1989). PSP has a worldwide distribution co-occurring with the dinoflagellate variants of *Alexandrium*. Significant reports of PSP come from many seas and oceans. *Protogonyaulax tamarense* (a variant of *Alexandrium*), has been reported as the source of PSP from the ingestion of green mussels, *Perna viridis*, in the Gulf of Thailand (Piyakarnchana, 1989; Piyakarnchana and Tamiyavanich, 1979). PSP is caused by *Gymnodinum breve* in the Gulf of Mexico and periodically causes toxicity along the Southeastern U.S. Continental Shelf Ecosystem (Tester et al., 1993). *Gymnodinium catenatum* is associated with PSP in the Caribbean Sea, Iberian coastal LME, Pacific Central American Coastal LME, Southern Benguela Current and New Zealand Shelf. The dinoflagellate *Pyrodinium* has caused PSP in the Flores Sea, Philippines, South China, Coral and Bismarck Seas.

Neurotoxic shellfish poisoning (NSP) affects humans who have consumed shellfish contaminated with biotoxins produced by *Gymnodinium* dinoflagellates. NSP is a significant problem in the Gulf of Mexico and recently, the eastern coast of Florida (Kumpf et al., 1999). Ingestion or inhalation of brevetoxins has also contributed to mass mortalities of manatees (*Trichechus manatus latirostris*) within Florida. Diarrhetic shellfish poisoning (DSP) associated with *Dinophysis* has been reported coincident with eutrophication within the Norwegian Sea, North Sea, Celtic-Biscay Shelf, Mediterranean Sea and the Latvian portion of the Baltic Sea (Thulin, 2000).

In the Caribbean, the toxic dinoflagellates Benthic *Gambierdiscus, Prorocentrum, Ostreopsis,* and *Coolia* sp. are associated with Ciguatoxic fish poisoning (CFP) (De Sylva, 1994). CFP is passed from algae to grazers and eventually to large piscivores within an ecosystem. Tourists have a higher incidence rate of CFP due to their preferential consumption of barracuda and grouper. Carriers of the toxin were thought to be unaffected by CFP (Anderson and Lobel, 1987). However, during the 1980 and 1993/94 reef mortalities and CFP disease outbreaks in Florida (Landsberg, 1995), concurrent immunosuppression was often observed within the lower parts of the food chain associated with ciguatera toxins (Landsberg et al., 1998).

Areas with clear trophic pathways for disturbance can also be conduits for bacteria and viruses. Zooplankton and phytoplankton serve as reservoirs for bacteria such as *Vibrio cholerae* (Colwell and Spira, 1992). Studies in the Gulf of Mexico, Caribbean Sea, Indian Ocean, Humboldt Current and Gulf of Guinea suggest a viable trophodynamic transport mechanism for cholera (Epstein, 1996; Colwell, 1996; Lowenhaupt, 1998). Trophodynamically acquired disease is a significant global issue. In aquaculture, feces from production provide ample nutrients for more frequent toxic blooms, antibiotics make harmful species more resistant (eg. samonellosis) and the combination of wider export, increased crowding and polluted water provide a perfect environment for epidemics (Todd, 1994). Norwalk-type viral diseases, *E. coli*, shigellosis and cholera can all be passed through the food chain and spread from one ecosystem to another (Todd, 1994).

Trophodynamic disturbances affect more than just groups of species, but also the habitat supporting these organisms. Seagrasses and coral reefs, for instance, provide the habitat for extensive marine communities. Due to their highly productive nature, toxin production and bioaccumulation can be rapid given the number of food-web connections. Wetlands and mangroves, also a structural habitat, buffer against trophodynamic disturbances. When harvested, drained or flooded for agriculture and aquaculture, these systems become more vulnerable to disturbance (Pimm and Lawton, 1998) speeding the recurrence of biotoxin-harbouring algae blooms. These

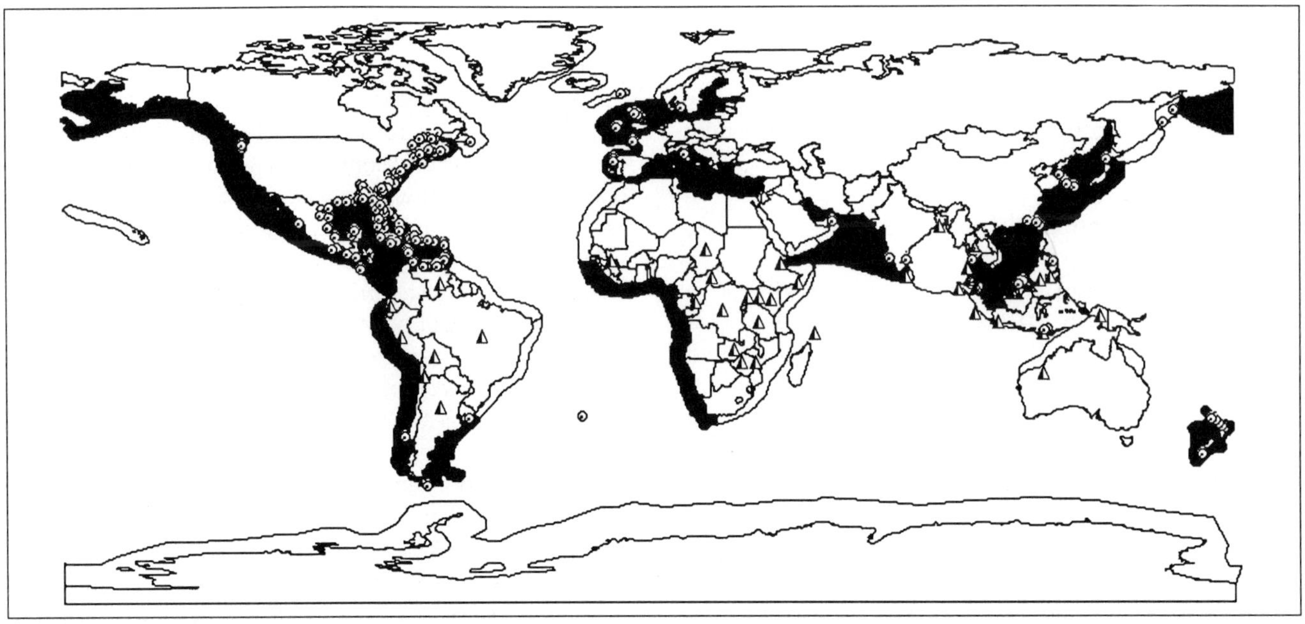

Fig. 6. Trophic-magnification disturbances from 1945–1998. The 50 Large Marine Ecosystems are outlined and those that are shaded have been identified as having regionally accessible information regarding these data types (vs. insufficient/not surveyed empty polygons). Also depicted are locations of 1997–98 Cholera epidemics (triangles) from a rapid global survey of incidences. HABs affecting food chains are depicted as circles. Bird NSP or PSP, fish PSP, human ASP or DSP or NSP or PSP shellfish poisoning or ciguatera fish poisoning, human Hepatitis A, Cholera, gastroenteritis, parahemaliticus or vulnificus, mollusc toxicity or shellfish harvesting bans due to ASP or DSP or NSP or PSP biotoxins are the trophic magnification disturbance types subsetted from 147 occurrence types.

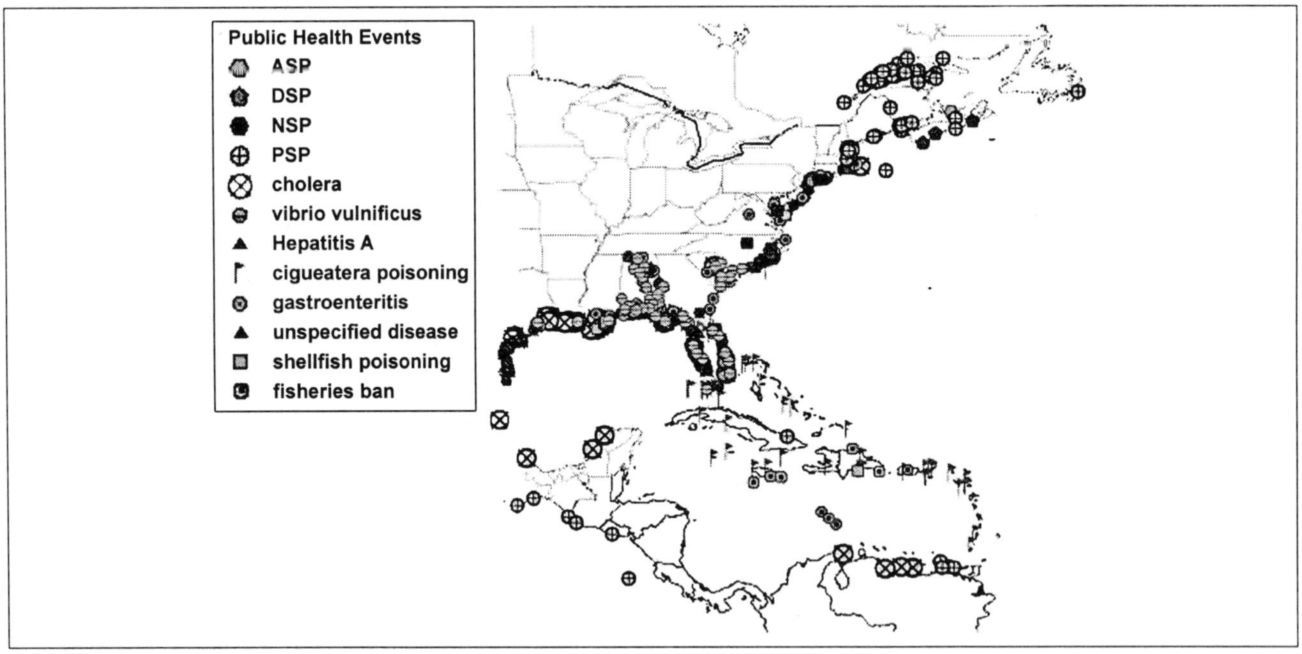


Fig. 7. North American view of public health trophic-magnification events from the HEED (1998) study.

structural habitats are self-perpetuating systems and protect species from trophic disease by providing sufficient co-evolved biodiversity to keep disturbance-causing organisms in check. Often damage to them is the first step in a cascade of disturbances that can undermine ecosystem recovery.

Marine trophodynamic disturbances and habitat loss account for significant portions of lost GDP (Todd, 1995). Indirect costs from marine morbidity, mortality and disease events are not surveyed for many maritime economies and for many impacted LME regions (Figs. 6 and 7) (Table 5) direct costs and losses are not calculated. Losses due to

Table 5

Trophic Magnification Disturbance Type indicators subsetted from a list of 147 disturbance types

| |
|---|
| Bird disease NSP |
| Bird mortality PSP |
| Fish PSP |
| Fish toxicity PSP |
| Human ASP |
| Human DSP |
| Human Hepatitis A |
| Human NSP |
| Human PSP |
| Human Cholera |
| Human gastroenteritis |
| Human shellfish poisoning |
| Human *vibrio vulnificus* |
| Mammal cetacean PSP |
| Mollusc shellfish harvesting ban ASP |
| Mollusc shellfish harvesting ban DSP |
| Mollusc shellfish harvesting ban NSP |
| Mollusc shellfish harvesting ban PSP |
| Mollusc shellfish toxicity ASP |
| Mollusc shellfish toxicity DSP |
| Mollusc shellfish toxicity NSP |
| Mollusc shellfish toxicity PSP |

harvesting bans, hospital costs and monitoring and analytical services calculated for Canada, the Philippines and Japan and the United States during public health emergencies (1988 and 1991 respectively) were over a billion dollars (Todd, 1994).

### Mass Lethal Mortality Disturbances

Mortality disturbance events include groups of occurrence reports clustered in space or time involving a single species or multiple species mortalities not yet attributed to a particular cause. In many cases the causative agents are unidentified harmful blooms. However these massive and widespread mortalities can be due to other factors. For instance, a significant and widespread black sea urchin (*Diadema antillarum*) mortality occurred in the Caribbean in 1983, coincident with El Niño conditions. Careful examination of the mass mortality, however, complicates a certain climate relationship. Hence this event is still classified as a mass mortality. Around the island of Jamaica, during that same time, a number of stresses combined resulting in the collapse of many reef species. Previous over-harvesting of reef fish, cleaners of algae and detritus, led to algae overgrowth upon bleached coral reefs (already inundated by sediment and nutrient run-off). A disease, leading to the mortality of overcrowded sea urchin populations, also feeding on the algae, may have been a natural control on urchin overpopulation. This cycle has twice been observed with Canadian green urchin and kelp populations (Miller and Colodey, 1983; Li et al., 1982). In the urchin case, because the mechanisms are unclear, they are best classified as mortalities until the climate and/or disease etiologies can be demonstrated. Mass mortalities of a species can push an ecosystem past threshold conditions to collapse and reorganize around a new stable state (Holling, 1973). In Jamaica, macroalgal mats now smother the old reefs, and populations of all species, particularly fish, remain depressed. In this example multiple stresses contributed to threshold conditions (coral reef collapse), that subsequently affected the resilience of the entire island ecosystem (Lessios, 1984). The mass mortalities are early indicators of restructuring (Table 6).

Fish are the best monitored of those species apt to succumb. Anchoveta, herring, mullet, and other reef fish within the Caribbean at separate times have inexplicably died in large numbers. Catfish within the Gulf of Mexico and along the Northeastern Brazilian Shelf ecosystem have also mysteriously died in large numbers. Fish kills involving menhaden were prevalent in North Carolina's estuaries since the early 1980s (Noga et al., 1988). Yet it was not until 1991 that a possible causative agent, the extremely toxic dinoflagellate *Pfiesteria piscicida*, was found (Burkholder et al., 1992). *Pfiesteria piscicida* is a small heterotroph with an extremely complex life cycle, and it has been implicated in shellfish mortalities, fish kills, ulcerative lesions, and public health controversy in North Carolina almost every year since its identification. However, it is extremely difficult to detect, leading to confusion over the leading cause of mortality. It is believed that *Pfiesteria* and *Pfiesteria*-like species weaken and cause mass morbidity in fish, allowing opportunistic micro-organisms to secondarily cause the high mortalities.

Mass mortality reports also have described deaths of brown pelicans in the Caribbean, marine mammal strandings in Peru, Chile, along the Caribbean, Gulf of Mexico, the Atlantic and Pacific U.S coastlines, the Azores,

Table 6

Mass-lethal mortality indicators subsetted from a list of 147 disturbance types

| | |
|---|---|
| Algal bloom | Mammal cetacean stranding |
| Bird kill | Mammal kill |
| Bird mortality | Mammal pinniped stranding |
| Coral destruction | Mammal sirenian mass mortality |
| Coral mass mortality | Mammal sirenian mortality |
| Coral mortality | Mollusc bivalve kill |
| Coral ridge mortality | Mollusc bivalve mass mort. |
| Coral sponge destruction | Mollusc bivalve mortality |
| Coral sponge mortality | Mollusc gastropod mortality |
| Fish kill | Sea turtle mortality |
| Fish mass mortality | Seagrass mortality |
| Invertebrate crustacean kill | Mammal cetacean mass mortality |
| Invertebrate kill | |
| Invertebrate mortality | Mammal cetacean mass stranding |
| Invertebrate mass mortality | Mammal cetacean mortality |

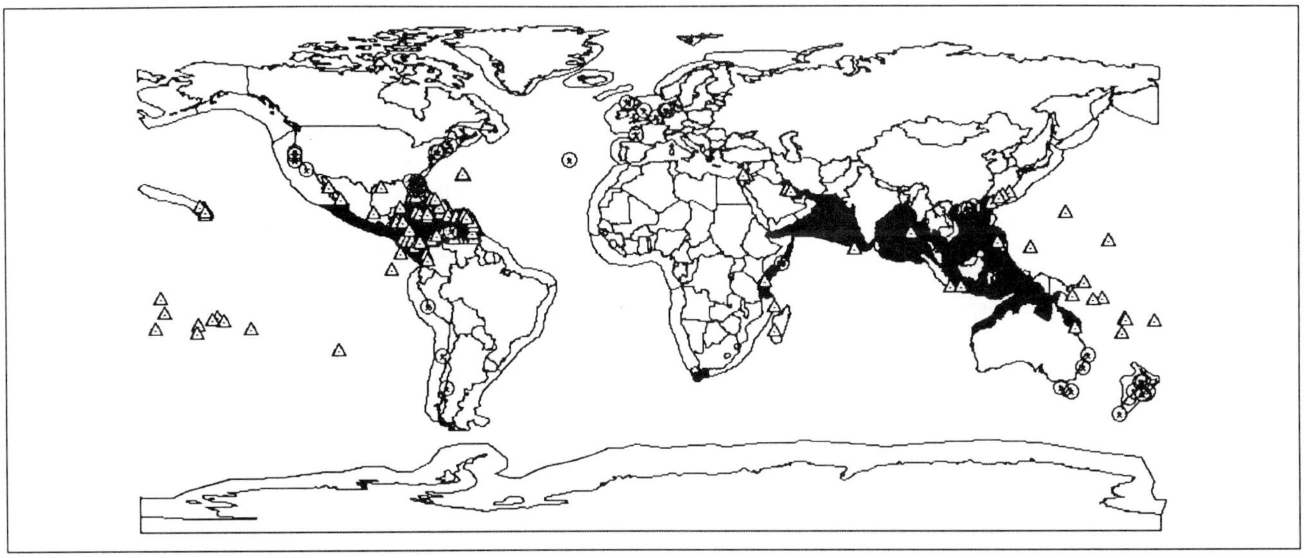

Fig. 8. Physically forced disturbances. The 50 Large Marine Ecosystems are outlined and those that are shaded have been identified as having regionally accessible information regarding these data types (vs. insufficient/not surveyed empty polygons). Also presented are data (triangles) representing coral bleaching and marine mammal strandings (circles) co-occurring with anomalous sea surface temperatures during the 1997–1998 El Niño year. Coral or sponge bleaching, coral shut down reaction, fish or invertebrate or cetacean or sea turtle cold kills (clam bleaching and sea turtle cold stunning and egg destruction also included) are classified as physically forced-climate disturbance indicators (a subset of 147 disturbance types).

North Sea, and continental waters of China, New Zealand and Australia. Many reports do coincide with climate extremes, however, due to the wide diversity of possible causes and uncertainty of the percentage of populations involved, observational reports need to be followed by more in-depth investigation.

## Physically Forced (Climate/Oceanographic) Disturbances

Mortality events caused by shifts in climate or weather may be under-represented within the literature because many of these events are not viewed as anomalous. What distinguishes significant climate disturbances from normal interannual variability is the frequency of extreme events, extraordinary chemical and physical pulses, and subtle shifts in seasonality associated with regional climate change. These features can be viewed as anomalous because populations and communities of organisms adapted to specific tolerance ranges may be ill-suited for the extremes and unable to escape or avoid impact (Bakun, 1993). For threatened or endangered species, the consequences of lost feeding or breeding seasons due to climate irregularities can be significant. Some harmful bloom events have also been linked to changes in the behaviour of water-masses and possibly oceanographic warming (Tester et al., 1993; Fraga and Bakun, 1993).

Of particular concern are coral bleaching events. Bleaching occurs when sea surface temperatures exceed 29°C. This has occurred during recent El Niño years for longer durations than reefs can tolerate. The subsequent mortality of hard and soft corals over large areas of the world has fundamentally altered disturbance regimes within coral reef systems. The most severe bleaching ever reported has occurred on many Indian Ocean islands and continental reefs, affecting as much as 90% or more of the corals and has led to localized species extinction (Hughes and Connell, 1999). Coral bleaching is being reported as a major marine disturbance throughout the world. In the Caribbean, bleaching has been followed by increased incidence of coral disease. The exact etiology of many of the coralline algal polyp mortalities and diseases is not known. Stresses other than water temperature (e.g. increased UV irradiance) appear to act synergistically with bleaching (Harvell et al., 1999), and the resultant mortality has the potential to be costly for both tourism and fishing (Wilkinson et al., 1999).

With dwindling populations of protected species, and the concentration of remaining healthy communities in a few protected marine reserves, unusually severe storms can now have a disproportional impact upon ecosystems. In this century, hurricanes Allen, David, Flora, Frederic, Andrew, Hugo in the Caribbean are responsible for a great deal of habitat destruction and mass mortalities of invertebrates, fish and the loss of seagrass meadows (HEED, 1998).

To document climate variability impacts on ecosystem structure at large scales (hundreds to thousands of km) time-series reconstruction strategies are needed (e.g. Colebrook, 1986; McGowan et al., 1998). A notable example is the detection of a regime shift that affected much of the Pacific basin in the mid-1970s using reconstructed plankton, fish abundance, temperature and sea-level pressure indices. Successful large area studies, like those involving the Peruvian anchovy, can demonstrate that low-frequency trends in ecosystem structure can be correlated with co-occurring indices of atmospheric and oceanic physical structure (e.g. climate change indices) (Polovina et al.,

1995; Hayward, 1999). Hence, indices of weather and climate, once a relationship is understood, can serve as a proxy for species level indicator data and vice versa. This method has allowed researchers to extend and continue time series to better understand the natural history of climate variability and extremes (Guilderson and Schrag, 1998).

Many climate influences are acting against a backdrop of long-term, global-scale environmental changes (Karl et al., 1997). These factors must all be considered in an integrated assessment. Some influences may be anthropogenic while others are "natural" in origin; and the changes occurring may be cyclic or random. If the climate regime is indeed changing, as coral records and shifts in marine ranges now suggest (e.g., Barry et al., 1995), the base-line upon which anomalies are measured will need to be modified. The long-term trends in ocean temperatures, tidal amplitudes, current patterns, ultraviolet radiation and freshwater run-off on marine communities and ecological processes will also require closer scrutiny.

### Disease Disturbances

For ecosystem health assessment, diseases are second only to harmful algae blooms as useful indicators for marine epidemiologists. Diseases are outcome measures reflecting feedbacks or perturbations with ecosystems. New diseases can emerge, and old ones resurge and undergo redistribution when the environment stresses an ecosystem and overwhelms the mechanisms providing resistance to the penetration and spread of pests and pathogens.

A high prevalence of morphological abnormalities (e.g. disease) in particular places and among multiple species is almost universally considered a straightforward indicator that the health of a particular system is suspect (Sindermann, 1996). However, even with adequate time series, it has been difficult to establish specific standards that define the quantitative degradation of particular habitats or regions over time. For instance, tabulation of crab and shellfish spots, pit formation on shells, deformed or missing points of the shell in "one-shot" surveys may tell us a lot about the prevalence of a particular disease, but fail to establish that trend over time (Butler et al., 1995). Even if a time series were available, standard criteria for defining the relationships among multi-species disease incidence are needed.

Significant physical environmental anomalies can render entire populations vulnerable to infection and/or replacement by more hardy competitors. Massive North Sea marine mammal kills in 1988, and dolphin die-offs in the western (1990) and eastern Mediterranean (1992) have been associated with several strains of morbilli (*phocine* distemper) viruses (Harvell et al., 1999). While pollutants (e.g., PCBs) may increase mammal susceptibility to infection, environmental changes particularly associated with El Niño conditions act as concentrators for ill animals, providing narrow migratory and feeding alternatives for stressed individuals and greatly increasing transmission among crowded populations.

The frequency and extent of impact from disease in marine mammals has increased many-fold (Harvell et al., 1999). Mass deaths involving agents such as morbillivirus, or influenza virus, are becoming less "unusual mortalities" by those charged with responding to marine mammal mortalities (e.g. MMC, 1999). In addition to the mass deaths, morbidity events of lesser magnitude are reported at a much greater frequency. The benefit of retrospective time-series reconstruction allows later reclassification once veterinary reports are evaluated and epidemiologists have established an etiology. Smaller scale events are now considered important outside of the veterinary sciences and increasingly, autopsy, necropsy and case reviews are found in a wider range of publications (HEED, 1998).

Diseases among migratory birds, as in their mammalian counterparts, represent similar trends. Outbreaks of disease among birds can occur for a variety of reasons. Migratory birds, often reliant upon small and ever-shrinking oases and feeding grounds, crowd into areas where unhealthy conditions develop and diseases are easily spread (Friend and Pearson, 1973; Friend, 1987). Crowded, commercially-raised species, as evidenced by Hong Kong Avian flu and by several North American Duck Plague outbreaks can subject other species to unique stresses and risks. The 1997/98 ENSO event, altered food distribution patterns and severe weather heavily impacted both mammals and seabirds in several regions of the world (Duffy, 1998). Birds within the Gulf of Mexico, Caribbean and Eastern U.S. seaboard, for instance, are vulnerable to the same stresses that occasionally lead to outbreaks of illness and death in mammal populations (e.g. PCPs, POPs, pesticides). In these populations disease is an ecological opportunist (e.g., Mashima et al., 1998; Daoust et al., 1998). The pathogenesis and epizootiology of duck plague virus and other aquatic bird diseases are complicated and involve multiple strains of disease that vary in virulence but can affect multiple-species across a range of age classes (Spieker et al., 1996). Unexplained outbreaks of a herpesvirus in ducks, geese and swans have been observed across the United States and Canada (Leibovitz, 1970). Better understanding of these complex outbreaks will require an integrated form of reporting that extends the entire length of migratory pathways.

The geographic array of marine pathogens, disease outbreaks, and estimated mortalities assembled recently by Harvell et al. (1999) underscores their global extent and the wide range of taxa involved (Table 7). What distinguishes the significance in these disease accounts is the magnitude of mortalities caused by pathogens, not the increased effort in classifying the disease types. There will always be a background of increased reporting of classifiable diseases, but with impact-specific criteria, trends in patterns of occurrence can be distinguished from improved observation.

Table 7

Disease disturbances subsetted from 147 occurrence types

| | |
|---|---|
| Algal mortality | Invertebrate crustacean shell disease |
| Bird disease (plague) | Invertebrate illness |
| Coral black band disease | Canine distemper in marine mammals |
| Coral rapid wasting disease | Mammal pinniped distemper |
| Coral white band disease | Mammal pinniped influenza |
| Coral white band disease (similar) | Mammal sirenian virus |
| Coral white plague type I | Mollusc bivalve JOD |
| Coral white plague type II | Mollusc bivalve disease (MSX) |
| Coral white pox | Mollusc bivalve epizootic neoplasia |

The global epidemic of coral disease underscores the recent increased prevalence theory. Black-band disease of corals is now found throughout the Caribbean where no incidence had been observed or catalogued in photographs or film footage in prior decades. White-band disease is now found to have spread throughout the Caribbean. In each country or reef system, disease incidences have spread and no prior records of incidence were observed (Goreau and Hayes, 1994; T. Goreau, unpublished). Similar trends have been observed with Red-band and Yellow-Band/Blotch disease in the Caribbean, sea fan *Aspergillus* disease and rapid wasting disease in parts of the Caribbean, and coral neoplasia in Australian waters.

Diseases spread from and among aquaculture facilities have become problematic in recent years (McGladdery, 1998). Mass bivalve mortalities have resulted from transfer of infectious stocks (Fisher, 1988). Shrimp viral diseases, though closely monitored, have also rapidly spread globally (Lightner and Redman, 1998). Taura Syndrome, for instance, moved from a single shrimp farm in Ecuador to sites throughout the Americas. Shrimp-eating birds may be a vector. The Indo-Pacific hypodermal and hematopoietic necrosis virus regularly causes catastrophic epidemics within aquaculture areas and both wild and raised shrimp are susceptible (Lightner and Redman, 1998). It is probably not a coincidence that incidences of disease occur in highly polluted areas (Sindermann, 1996). The fish disease lymphocystis, as an example, appears to become problematic when heavy rains flush toxic chemicals into fisheries areas. Though the causative agent is a virus, the correlation with pollution has been demonstrated in many regions (Sindermann, 1996; Williams, 1992).

## New, Novel Occurrences and Invasive Disturbances

New or novel occurrences are defined as disturbances appearing for the first time in a species or within an area. This includes toxic or newly harmful toxic algae species or known harmful algae entering a new area and unprecedented mass mortalities among populations for which there are no previously recorded instances. The first instance of species invasions and significant changes in seasonality, distribution, and susceptibility of populations are also classified as new or novel occurrences (Fig. 9).

Humans, intentionally through aquaculture or unintentionally through ballast water, carry pathogens and vectors into new areas. Where co-evolved pathogen host population relationships are absent, species are vulnerable to newly introduced disease. Human activity has not only increased the rate of spread of organisms but also changed the meaning of distance. Natural diffusion of genetic material had generally been local and gradual, allowing species

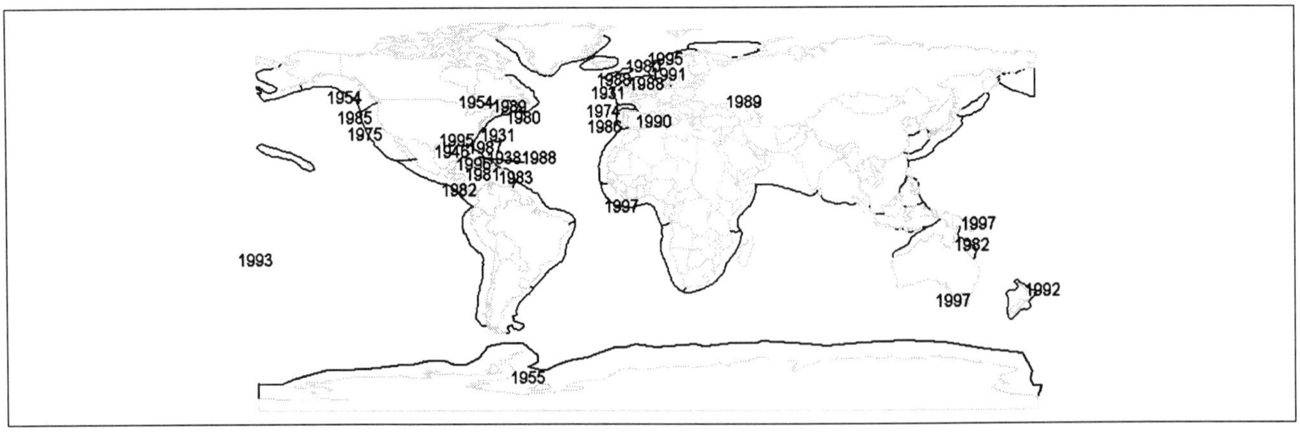


Fig. 9. Selected new/novel/invasive occurrences (adapted from Harvell et al., 1999). 1931: Seagrass/slime mold, N. America, Europe. 1938: Sponges/fungus? N. Caribbean. 1946: Oyster/*Perkinsus*, Gulf Coast, USA. 1954: Herring/pathogen, Gulf St. Lawrence. 1955: Seal/virus, Antarctica. 1974: Flat oyster/*Marteilia*, NW Spain. 1975: starfish/?, Western USA. 1980: Urchin/amoeba, NW Atlantic. 1980: Oyster/*bonamia*, Netherlands. 1981: Coral/bacteria. 1982: Coral/? Central America. 1982: Abalone/*Perkinsus* sp., Australia. 1983: Corals/bacteria? Caribbean-wide. 1983: Scallop/*Perkinsus*, W Canada. 1983: Urchin/bacteria? Caribbean-wide. 1985: Abalone/?, NE Pacific. 1986: Clam/*Perkinsus*, Portugal. 1987: Seagrass/slime mold, Florida, USA. 1988: Scallop/protozoan, N Caribbean. 1988: Seals/virus, NW Europe. 1988: Porpoise/virus, NE Ireland. 1989: Scallop/*Perkinsus*, E Canada. 1989: Seals/virus, Lake Baikal. 1990: Dolphin/virus, W Mediterranean. 1991: Herring/pathogen, W Sweden. 1992: Kelp/?, NE New Zealand. 1993: Coralline algae/bacteria? S Pacific. 1995: Urchin/nematode, Norway. 1995: Corals/fungus, Caribbean-wide. 1995: Coral/bacteria, Florida, USA. 1996: Coral/bacteria, Puerto Rico. 1997: Algae/fungus, Samoa. 1997: Pilchard/virus?-S Australia. 1997: Seal/virus, W Africa.

and ecosystems to adapt. But preferred routes or destinations, often between similar but geographically distant ecosystems, have made the world figuratively smaller. In one survey, 367 different species were identified in ballast water of ships travelling between Japan and Coos Bay, Oregon (Carlton and Geller, 1993). Lockwood (1993) reports that 60 million tons of ballast water are discharged each year into 40 Australian ports. Hallegraeff (1993) was able to reconstruct the arrival of Japanese red tide dinoflagellates from captains' logs of ships anchored in Tasmanian waters. Well known biological invasions, such as the ctenophores (*Mnemiopis leidyi*), in the Black and Azov Seas (Travis, 1993) are only a small part of a greater unmonitored, unregulated, and uncontrolled experiment waged in the coastal zone. Human casualties from *Vibrio cholerae*, believed to be associated with shipping and South American seafood (Epstein, 1999), and the movement of MSX, *Haplosporidium nelsoni*, disease into New England Oyster populations, only hint at the potential economic consequences and vulnerabilities faced by human assisted redistribution.

Tester et al. (1993) refer to a first occurrence of *Gymnodinium breve*, that bloomed during anomalous climatic and oceanographic conditions in North Carolina (USA) as a possible future indicator of algae redistribution due to larger climatic changes. Changes in distribution of organisms and concentrations of weakened hosts in smaller areas can lead to the evolution of new diseases. Duck plague virus first appeared in 1967, for instance, on Long Island, New York, coincident with the development of farm breeding programs (Leibovitz, 1970). New or first occurrences of a pathogenic species test the elasticity and inertia of a marine ecosystem. The ease with which species invade is a measure of the porosity and vulnerability of the system to disturbance. Both anthropogenic facilitation and climatic changes have improved the odds for survival among new, novel and opportunistic pathogens and agents of disturbance.

## Keystone-Endangered and Chronic Cyclical Disturbances

Howarth (1991) describes two general processes typically leading to ecosystem instability and collapse. Following a disturbance, these include (1) structural damage as the first noticeable impact. Subsequent disturbances can eventually impact and (2) change function or ecosystem critical processes. Individual impacts represent a low threshold for collapse due to inertia within the system. However, if disturbances are increasingly measured as rates or trajectories of change, increases or decreases in the amplitude and constancy of disturbance can be viewed as a chronic response to instability within ecosystems. The cycle leading to a chronic recurrence is known as a disturbance regime.

When disturbance becomes a regular feature of an ecosystem, the term anomaly must be redefined. At these periods of transition, a threshold is reached in the composition and function of the ecosystem. Under these circumstances an addition of a single significant perturbation can often lead to long-term significant changes. Ray (1996) uses the example of discontinuity/ threshold in describing the reorganization of the Chesapeake Bay ecosystem. He defines oysters and oyster reefs as the keystone species. With their loss due to the spread of disease, overharvest and anthropogenic stress, the entire system reached a threshold where climatic variability shifted the ecosystem to a new stable state or configuration, from which the system has never recovered.

Stress favours domination by smaller organisms (Rapport et al., 1985). Over time, these opportunistic microorganisms exert selective pressures on the most vulnerable, typically the endangered, rare or threatened, members of an ecosystem. In this scenario, changes in silicate to nitrogen ratios can enable dinoflagellates to prosper at the expense of diatoms (Smayda, 1990). Moreover, an abundance of smaller zooplankton predators feeding on the dinoflagellates can further alter a system by displacing large species. The smaller, more gelatinous, zooplankton can then prosper at the expense of fish larvae and adult fish unable to compete for food or effectively feed within the nekton are doubly impacted. It may be no coincidence that unhealthy systems are dominated by jellyfish (e.g. Black Sea; Mee, 1992).

Chronic disease conditions are particularly important to monitor within endangered populations. Long-term increases in neoplasia (cancer) incidences and other physiological ailments can be classified as chronic disturbances when they affect whole populations. Fibropapilloma incidences (including cutaneous papillomas and fibromas) on green turtles, *Chelonia mydas* (Herbst, 1994), is an example of conditions that are now understood to be population-wide increases. Cutaneous fibropapillomas develop over a long period of time and may play a significant role in turtle mortality (Williams, 1992; Pinto-Rodriguez et al., 1995). These effects, when not fatal themselves, can lower immune system function, exhaust and generally debilitate the carrier. The condition increases susceptibility to entanglement, by-catch, starvation, vessel collision and other diseases (Williams et al., 1994). Prior to the 1980s, occurrence of visceral tumours, the cause of visual obstructions and organ dysfunction among surveyed green turtle populations was considered rare (Herbst, 1994). Fibropapillomas rapidly increased during the 1990s, recognized first in Florida, the Cayman Islands and within Hawaiian Island populations, and eventually the epizootic (wildlife epidemic) was recognized throughout the Caribbean (Williams et al., 1994) and within all known turtle populations (panzootic) (including hawksbill and loggerhead species) (Herbst, 1994; Williams, 1992).

The diversity of functional groups of species, e.g., predators, prey, competitors, recyclers, scavengers and nitrogen fixers, provide the buffers that dampen the impacts of

## Gulf of Mexico Aquatic Mortality Network

William S. Fisher

*U.S. Environmental Protection Agency, National Health and Environmental Effects Laboratory, Gulf Ecology Division, One Sabine Island Drive, Gulf Breeze, FL 32561, U.S.A.*

High-quality data are critical for evaluating environmental influences and interpreting the significance of marine morbidity and mortality reports. Yet reporting efforts vary dramatically from place to place and over time. The lack of consistency undermines the credibility and usefulness of the data, particularly given that times and places of non-events (i.e., when and where events do not occur) is critical for distinguishing between normal and anomalous conditions. And, even though marine disturbance phenomena behave independently of jurisdiction, those who report, document, or respond to mortality events for different state agencies often have differing standards, credentials, and motives. The Gulf of Mexico Aquatic Mortality Network (GMNET) is a collaboration among U.S. state and federal agencies to improve and standardise mortality response approaches, procedures, documentation, and interstate communication and to promote research in forensic pathology across the five U.S. states bordering the Gulf of Mexico. Several attributes of GMNET are instructive with respect to obtaining high-quality data for a global epizootiological approach. Consistency and quality are maintained through the involvement and professional accountability of trained response personnel employed by each state. These resource managers have adopted standard investigative procedures and foster consistency through periodic interagency training exercises.

To encourage state participation in a common regional data repository, GMNET is devising a comprehensive data entry system that eliminates any need for double-entry of data or a separate state database management system. It consists of three data tiers, reflecting essential and non-essential data to be shared regionally and investigation information that is not. The first tier comprises basic information that must be collected for an occurrence to be included in the database. This information describes when, where, what, how many and any circumstantial evidence that may exist to trace probable causes and consequences. Tier 2 data involve the ancillary and specialist observational data including water quality, necropsy samples, weather conditions, methods of analysis. Tier 3 data describe the chain of custody of samples, evidence for litigation, and other privileged information from public health and regulatory databases that are typically unavailable to the public. Only the first two tiers of data are shared in a common database that maintains links to the original custodians of the information, observational data, or samples. The participants maintain control of their own data, yet share standards and update data in the common repository as classification and interpretations change. The three tiers allow participants to use the same formats for shared data while retaining the flexibility and control over unshared data. GMNET is a successful model due to a realization that marine disturbance phenomena behave independently of jurisdiction and adjacent jurisdictions need to coordinate their limited resources efficiently in response to disturbance impact and follow-up studies.

---

stressors. Multiple species performing similar functions and tasks can provide "insurance" in case one or more keystone species decline (Pimm, 1984). Often, the first signs of significant structural and functional damage are the loss of the more stress-sensitive species (Howarth, 1989). The endangered or rare sensitive species are the biological indicators of disturbance, and presence or absence of these species can indicate a level of chronic disturbance within an ecosystem. In the 50 LMEs, those of north and central America, Caribbean, Mediterranean/Black Sea, northeast Australia and Japan have good records and regionally accessible information of the keystone Sensitive Species/Cyclical Chronic disturbance indicators (Table 8).

Whether bio-deposition changes the character of the sea floor through anoxia or biotoxins lead to tumour development in exposed wildlife, chronic disturbances are the most constant of the disturbance types. The aggregated disturbance indicators may be used to rapidly survey marine ecosystems and the eight general types of disturbance may provide a first-approximation of the comparative health of coastal marine ecosystems.

Table 8

Keystone Sensitive Species/Cyclical Chronic disturbance indicators subsetted from the list of 147 disturbance types (from the HEED 1998 Survey)

| | |
|---|---|
| Algae macroalgae overgrowth | Invertebrate decline |
| Algal bloom (shading) | Mollusc bivalve growth suppression |
| Algal bloom (shading/wasting) | Mollusc bivalve mortality (wasting) |
| Algal bloom (wasting) | Mollusc bivalve reproductive failure |
| Bird reproductive failure | Mollusc shellfish contamination |
| Coral disappearance | Mollusc shellfish harvesting ban |
| Coral neoplasia | Mollusc shellfish harvesting ban (contamination) |
| Human beach closing | Sea turtle fibropapillomas |
| Human fisheries ban | Seagrass decline |
| Human river closing | |

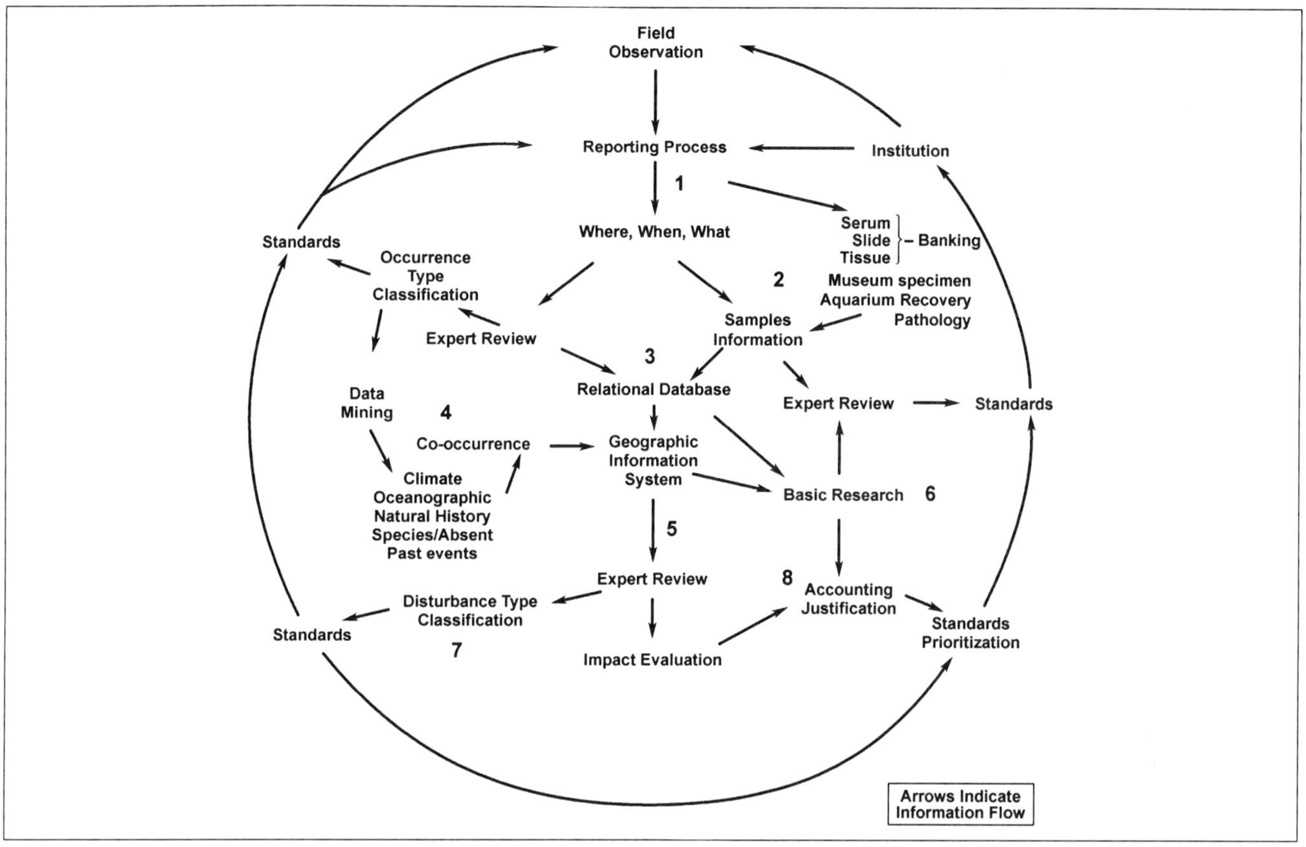


Fig. 10. Information flows in a marine epidemiological information system. Field observations are part of a reporting process that records "where, what, when" information (1). If samples are taken they are sent to the appropriate laboratories and results and/or specimens are archived (2) and information regarding the unique laboratory custodian identification number is returned to a central relational database (3). Queries of the database generate occurrence type classifications, enable comparison of co-occurring anomaly data, and help extract ancillary life history and contextual information (including mass-media accounts if no other information is available (4). Output is provided in GIS format for expert review (5) and basic research exploration and hypothesis generation (6). The expert review process adjusts the disturbance type classification appropriately (7). Research scientists may use new insights gained from participating in quality control and observation to justify their efforts. Further analysis could reveal the cost of particular events that may even further justify the importance of the epidemiological information system approach (8). All steps feed back to the generation of nomenclature and definition standards that ultimately may lead to the better reporting of observations taken in the field or new types of data collection or archiving by institutions.

## NETWORK FOR DEVELOPING STANDARDS AND ACHIEVING CONSENSUS

Determination of impact and ecosystem change is made by closely monitoring biological indicators, the condition of keystone species and co-occurring physical, chemical and ecological anomalies associated with disturbances. Unfortunately, it is rare for multiple monitoring systems spread among regions collecting dissimilar indicators, to provide either consistent temporal resolution or maintain consistency of data collection even within the same ecosystem (Gross et al., 1995). Academic, federal and state personnel working within the northern Gulf of Mexico serve as a notable exception. The Gulf of Mexico aquatic mortality network (GMNET) has developed multi-jurisdictional standards for data reporting and provides participants with regularly observed meta-data (data about data) reports. The shared meta-data and increased research communication have greatly reduced data comparison problems, redundancy in data collection, resolution and quality difficulties (Fisher et al., 1999).

As managers and specialists create standardized methodologies for reporting MMEDs, it becomes more important to provide an equally standardized means for exchanging information. A World Wide Web geographic information system (GIS) developed by the HEED (1998) Program, has allowed researchers interested in hypotheses testing and managers interested in reconstructing disturbance regimes to interactively access data-sets to query anomaly and categorical data by space and time (Fig 10). The HEED system represents a demonstration of how information might be consolidated in a central location to explore standards and provide a common interface to research managers.

Climate, pollution and trophodynamic co-occurrence information have already been of value in assessing collateral impact from known disturbances (HEED, 1998). Counts of the number of disturbance types (variety), their frequency in particular areas and raw numbers of occurrence can be used to better understand the natural history of marine disturbances and establish retrospective time-series reconstruction of past events. At present, the

HEED database and GIS are most often used to describe known events in greater detail and to facilitate communication among researchers and resource managers. In the future, the internet-based marine disturbance data system can be linked to tissue and serum banks, tumour registries, disease image archives, museum specimen collections, and even be available to field veterinary response teams investigating events. Coupling field observation, data analyst recommendations and institutional monitoring provides the necessary components for a comprehensive marine epidemiological research system (Fig. 10). The capacity for real-time event reporting via the Internet is also under investigation.

## BASIS FOR ASSESSING THE HEALTH OF LARGE MARINE ECOSYSTEMS

Chronic illnesses, mass mortality and disease epidemics are being reported across a wide spectrum of marine taxonomic groups. Novel occurrences involving pathogens, invasive species and illnesses affecting humans and wildlife are globally distributed and appear to be increasing. These disturbances impact multiple components of marine ecosystems, disrupting the structural and functional relationships among species and the ability of systems to recover from natural perturbations. Climate change, involving extreme events, combined with human-induced stressors may compromise an open-ecosystem or migratory population's resistance to opportunistic micro-organisms. Fungi, bacteria and viruses are quick to exploit the weak within a disturbed enclosed ecosystem and eventually create suitable environments for continued proliferation. Assessments using a marine epidemiological approach can track these changes in ecosystem health (Epstein, 1996).

Better understanding of the natural history of disturbance regimes requires the inventory and reconstruction of event time series. Time series allow multiple marine ecological disturbances to eventually be classified as Major marine ecological disturbances (MMEDs). The number and frequency of MMEDs in a place can be used as indicators of a decline in ecosystem health and loss of essential services (e.g. economic, nutrient cycling etc.; see Costanza (1992) for review). Reconstructing a natural history of these events and related environmental conditions facilitates a better understanding of the local, regional and global causes of disturbance. MMEDs can be mapped using geographic information systems to define spatial "hotspots" and temporal clusters of events (e.g., during El Niño events). Reconstructed time series have helped researchers identify eight general impact categories based upon the spatial and temporal dynamics of the observed disturbances. Assuming the appropriate observational meta-data information (noting source and quality) is available, resource stewardship agencies and research institutions can work on standards to better define regionally more appropriate or optimal indicators for characterizing each of the sub-disturbance types for ecosystem health assessment.

Data collected using a marine ecosystem health framework depicts an increase in the frequency, severity and geographic spread of MMEDs over the past several decades

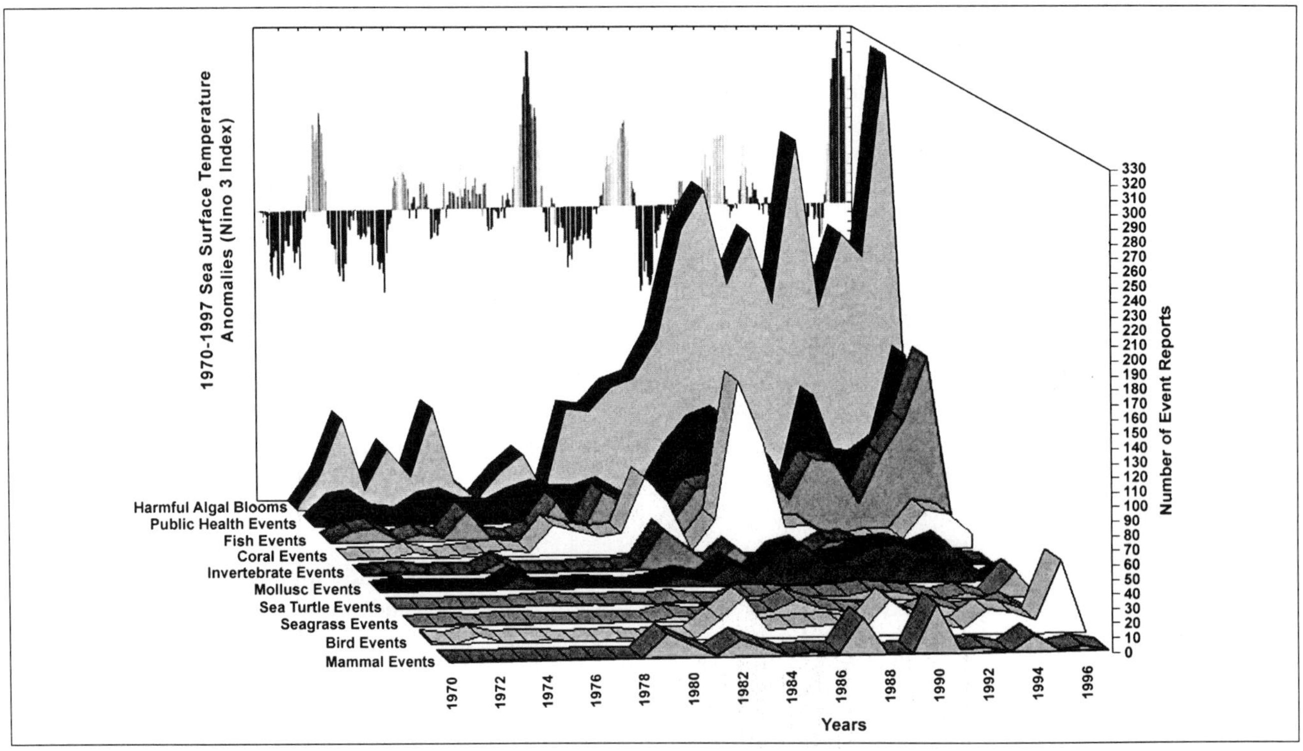


Fig. 11. Reported Marine Ecological Disturbances (1970–1996) within the Northwestern Atlantic, Caribbean Sea and Gulf of Mexico as part of the HEED (1998) rapid survey.

(e.g. Fig. 11). These adverse biological events carry with them significant human health and economic costs. International institutions can use the ecosystem health framework presented in this chapter for assessing the health and consequence of disturbance within large marine ecosystems. Initial assessments could be followed by improved monitoring and surveillance followed by focused basic laboratory and field research. With new insight, early warning systems to enhance response capability to ecosystem health disturbances could be initiated, and coordinated mitigating actions taken, to address the underlying forcing factors leading to disturbance.

The combined HEED retrospective data mining and GMNET integrated multi-jurisdictional monitoring approaches could be applied beyond their present scales to include morbidity, mortality disease and comparative health assessments among the 50 large marine ecosystems.

Standard global characterization is particularly important due to the uncertainties surrounding the cause and consequence of marine disturbance and the unknown origin and distribution of many disturbance types. In many cases complex physical, chemical and ecological processes, for which mechanistic understanding is lacking, mask both human and climate influences. By taking a global, LME-scale perspective, natural, human-induced impacts and the synergy of forcing factors resulting in marine disturbance can be better inventoried, separated and understood. This will remain as an important challenge for marine science in the coming millennium.

## REFERENCES


Adnan, Q. (1989) Red tides due to *Noctiluca scintillans* (MacCartney) Ehrenb. and mass mortality of fish in Jakarta Bay. In *Red Tides, Biology, Environmental Science and Toxicology*, eds. T. Okaichi, D.M. Anderson and T. Nemoto, pp. 53–55. Elsevier, New York.

Albright, L.J., Yang, C.Z. and Johnson, S. (1993) Sub-lethal concentrations of the harmful diatoms, *Chaetocerus concavicornis* and *C. convolutus*, increase mortality rates of penned Pacific salmon. *Aquaculture* **117**, 215–225.

Allen, T.F.H. and Hoekstra, T.W. (1992) *Toward a Unified Ecology*. Columbia University Press, New York.

Anderson, D.M. and Lobel, P.S. (1987) The continuing enigma of ciguatera. *Biological Bulletin* **172**, 89–107.

Bacon, F. (1620) *Novum Organum: With Other Parts of the Great Instauration* (Paul Carus Student Editions, Vol. 3, In P. Urbach and J. Gibson (eds.) (translators 334 pages 1994). Open Court Publishing Company.

Baden, D.G. and Trainer, B.L. (1993) Mode of action of toxins of seafood poisoning. In *Algal Toxins in Seafood and Drinking Water*, ed. I.R. Falconer, pp. 47–49. Academic Press, London.

Baden, D.G., Fleming, L.E. and Bean, J.A. (1995) *Handbook of Clinical Neurology. Vol 21 (65): Intoxications of the Nervous System, Part II*, ed. F.A. de Wolff, Elsevier, Amsterdam.

Bakun, A. (1993) The California Current, Benguela Current and Southwestern Atlantic Shelf ecosystems: a comparative approach to identifying factors regulating biomass yields. In *Large Marine Ecosystems: Stress, Mitigation and Sustainability*, eds. K. Sherman, L. M. Alexander and B.D. Gold, pp. 199–221. AAAS Press, Washington, D.C.

Barry, J.P., Baxter, C.H., Sagarin, R.D. and Gilman, S.E. (1995) Climate-related, long-term faunal changes in a California rocky intertidal community. *Science* **267**, 672–675.

Blumenthal, B. (1998) Guide to the Data. Http:// Ingrid.ldeo.Columbia.edu. Lamont-Doherty Earth Observatory of Columbia University (Online Publication).

Burkholder, J.M., Noga, E.J., Hobbs, C.H., Glasgow Jr., H.B. and Smith, S.A. (1992) New "phantom" dinoflagellate is the causative agent of major estuarine fish kills. *Nature* **358**, 407–410.

Burkholder, J.M. (1998) Implications of harmful microalgae and heterotrophic dinoflagellates in management of sustainable marine fisheries. *Ecological Application* **8**, S37–S62.

Buskey, E.J. and Stockwell, D.A. (1993) Effects of a persistent 'brown tide' on zooplankton populations in the Lagune Madre of South Texas. In *Toxic Phytoplankton Blooms in the Sea*, eds. T.J. Smayda and Y. Shimuzu, pp. 659–666. Elsevier, Amsterdam.

Butler IV, M.J., Hunt, J.J., Hernkind, W.F., Childress, M.J., Bertelsen, R., Sharp, W., Matthews, T., Field, J.M. and Marshall, H.G. (1995) Cascading disturbances in Florida Bay, USA: cyanobacteria blooms, sponge mortality, and implications for juvenile spiny lobsters *Panulirus argus*. *Marine Ecology Progress Series* **129**, 119–125.

Carlton, J.T. and Geller, J.B. (1993) Ecological roulette: the global transport of non-indigenous marine organisms. *Science* **261**, 78–82.

Chen, Y-Q, and Shen, X-Q. (1999) Changes in the biomass of the East China Sea ecosystem. In *Large Marine Ecosystems of the Pacific Rim: Assessment, Sustainability, and Management*, eds. K. Sherman and Q. Tang, pp. 221–239. Blackwell Science Inc., Massachusetts.

Colebrook, J.M. (1986) Environmental influences on long-term variability in marine plankton. *Hydrobiologia* **142**, 309–325.

Coles, S.L. (1994) Extensive coral disease outbreak at Fahl Island, Gulf of Oman, Indian Ocean. *Coral Reefs* **13**, 242.

Colwell, R.R. and Spira, W.M. (1992) The ecology of Vibrio cholera. In *Cholera: Current Topics in Infectious Disease*, eds. D. Barua and W.P. Greenrough III, pp. 107–127. Plenum Medical Book Company, New York.

Colwell, R.R. (1996) Global climate and infectious disease: the cholera paradigm. *Science* **274**, 2025–2031.

Cosper, E.M., Dennison W.C., Carpenter, E.J., Bricelj, V.M., Mitchell, J.G., Kuenstner, S.H., Colflesh, D. and Dewey, M. (1987) Recurrent and persistent brown tide blooms perturb coastal marine ecosystem. *Estuaries* **10**, 284–290.

Cosper, E.M., Dennison, W., Milligan, A., Carpenter, E.J., Lee, C., Holzapfel, J. and Milanse, L. (1989) An examination of the environmental factors important to initiating and sustaining 'brown tide' blooms. In *Novel Phytoplankton Blooms Lecture Notes on Coastal and Estuarine Studies*, eds. E.M. Cosper, E.J. Carpenter and V.M. Biceji, pp. 317–340. Springer-Verlag, Berlin.

Costanza, R. (1992) Toward an operational definition of ecosystem health, In *Ecosystem Health: New Goals for Environmental Management*, eds. R. Costanza, B.G. Norton and B.D. Haskell (eds.), pp. 239–256. Island Press, Washington, D.C.

Daoust, P.Y., Conboy, G., McBurney, S. and Burgess, N. (1998) Interactive mortality factors in common loons from Maritime Canada. *Journal of Wildlife Disease* **34** (3), 524–531.

De Sylva, D.P. (1994) Distribution and ecology of ciguatera fish poisoning in Florida, with emphasis on the Florida Keys. *Bulletin of Marine Science* **54** (3), 944–954.

Draper, C., Gainey, L., Shumway, S. and Shapiro, L. (1990) Effects of *Aureococcus anophagefferens* ("brown tide") on the lateral cilia of 5 species of bivalve molluscs. pp. 128–131. In *Toxic Marine Phytoplankton*, eds. E. Graneli, B. Sundstrom, L. Edler and D.M. Anderson. Elsevier, New York.

Duda, A.M. and Sherman, K. (1999) An ecosystem approach to global assessment and management of coastal waters. *Marine Ecology Progress Series* **190**, 271–287.

Duffy, D. (1998) Website: http://darwin.bio.uci.edu/~sustain/ENSO.html. visited 12 November 1998.

Endean, R. (1977) Destruction and recovery of coral reef communities. In *Biology and Geology of Coral Reefs*, eds. O.A. Jones and R. Endean, pp. 215–254. Academic Press, New York.

Epstein, P.R. (1994). Framework for an integrated assessment of climate change and ecosystem vulnerability, In *Disease in Evolution: Global Changes and Emergence of Infectious Diseases*, eds. M.E. Wilson, R. Levins and A. Spielman, pp. 423–435. New York Academy of Sciences, New York.

Epstein, P.R. (1996) Emergent stressors and public health implications in large marine ecosystems: An overview. In *The Northeast Shelf Ecosystem: Assessment, Sustainability, and Management*, eds. K. Sherman, N.A. Jaworski and T.J. Smayda, pp. 417–438. Blackwell, Massachusetts.

Epstein P.R. (1999) Large marine ecosystem health and human health. In *The Gulf of Mexico Large Marine Ecosystem: Assessment, Sustainability, and Management*, eds. H. Kumpf, K. Steidinger and K. Sherman, pp. 459–475. Blackwell, Massachusetts.

Falconer, I.R. (1993) Mechanism of toxicity of cyclic peptide toxins from blue-green algae. In *Algal Toxins in Seafood and Drinking Water*, ed. I.R. Falconer (ed.), pp. 177–186. Academic Press, London.

Falconer, I.R. and Humpage, A.R. (1996) Tumour promotion by cyanobacterial toxins. *Phycologia* **35**, 74–79.

Fisher, W. (1988) Environmental influence on host response: environmental influence on bivalve hemocyte function. *American Fisheries Society Special Publication* **18**, 225–237.

Fisher, W., Epstein, P.R. and Sherman, B.H. (1999) Overview of health, ecological, and economic dimensions of global change: tracking marine disturbance and disease. In *Marine Mammals and Persistent Ocean Contaminants*: Proceedings of the Marine Mammal Commission Workshop. Keystone, Colorado, 12–15 October 1998, eds. T.J. O'Shea, R.R. Reeves and A.K. Long. Marine Mammal Commission. 4340 E-W Highway, Room 905 Bethesda, Maryland. April 1999.

Fritz, L., Quilliam, M.A. and Wright, J.L.C. (1992) An outbreak of domoic acid poisoning attributed to the pennate diatom *Pseudonitzchia australis*. *Journal of Phycology* **28**, 439–442.

Fraga, S. and Bakun, A. (1993) Global Climate Change and harmful Algal Blooms: The example of *Gymnodinium catenatum* on the Galician Coast. In *Toxic Phytoplankton Blooms in the Sea*, eds. T.J. Smayda and Y. Shimizu. Elsevier, Amsterdam.

Friend, M. (1987) Avian cholera. In *Field Guide to Wildlife Diseases, Vol. 1*, eds. M. Friend and C.J. Laitman, pp. 69–82. U.S. Fish and Wildlife Service Resource Publication. Washington, D.C.

Friend, M. and Pearson, G.L. (1973) Duck plague: The present situation. *Proceedings Western Association State Game and Fish Commission* **53**, 315–325.

Fujiki, H. and Suganuma, M. (1993) Tumour promotion by inhibitors of protein phosphatases 1 and 2A: the okadaic acid class of compounds. *Advances in Cancer Research* **61**, 143–194.

Geraci, J.R., Anderson, D.M., Timperi, R.J., Staubin, D.J., Early, G.A., Prescott, J.H. and Mayo, C.A. (1989) Humpback whales (Megaptera novaeangliae) fatally poisoned by dinoflagellate toxin. *Canadian Journal Fisheries Aquatic Science* **46**, 1895–1898.

GESAMP (Group of Experts on the Scientific Aspects of Marine Pollution) (1990) The state of the marine environment. UNEP Regional Seas Reports and Studies No. 115, Nairobi, Kenya.

Goreau, T.J. and Hayes, R.L. (1994) Coral bleaching and ocean "hot spots." *Ambio* **176**, 176–180.

Grigalunas, T.A. and Diamantides, J. (1996) *The Peconic Estuary System: Perspective on Uses, Sectors, and Economic Impacts*. Economic Analysis Inc., Rhode Island.

Grattan, L.M., Oldach, D., Perl, T.M., Lowitt, M.H., Matuszak, D.L., Dickson, C., Parrott, C., Shoemaker, R.C., Kaufman, C.L., Wassreman, M.P., Hebel, J.R., Charache, P. and Morris Jr., J.G. (1998) Learning and memory difficulties after environmental exposure to water-ways containing toxin-producing *Pfiesteria* or *Pfiesteria*-like dinoflagellates. *Lancet* **352**, 532–539.

Gross, K.L., Pake, C.E., Allen E., Bledsoe, C., Colwell, R., Dayton, P., Detheir, M., Helly, J., Holt, R., Morin, N., Michener, W., Pickett, S.T.A. and S. Stafford. (1995) Final Report of the Ecological Society of America Committee on the Future of Long-term Ecological Data (FLED). Available online: http://sdsc.edu/ecmtext.htm.

Guilderson, T.P. and Schrag, D.P. (1998) Abrupt shift in subsurface temperatures in the tropical Pacific associated with changes in El Niño. *Science* **281**, 5374–240.

Guo, C. and Tester, P.A. (1994) Toxic effect of the bloom-forming *Trichodesmium* sp. (*Cyanophyta*) to the copepod *Acartia tonsa*. *Natural Toxins* **2**, 222–22.

Hallegraeff, G.M. (1993) A review of harmful algal blooms and their apparent global increase. *Phycologia* **32**, 79–99.

Harvell, C.D., Kim, K., Burkholder, J.M., Colwell, R.R., Epstein, P.R., Grimes, D.J., Hofmann, E.E., Lipp, E.K., Osterhaus, A.D.M.E., Overstreet, R.M., Porter, J.W., Smith G.W. and Vasta, G.R. (1999) Emerging marine diseases—Climate links and anthropogenic factors. *Science* **285**, 1505–1510.

Hayward, T.L. (1999) Long term change in the North Pacific Ocean. In *Large Marine Ecosystems of the Pacific Rim*, eds. K. Sherman and Q. Sheng Tang, pp. 56–62. Blackwell Science, Malden, Mass.

HEED (Health Ecological and Economic Dimensions of Global Change Program) (1998) Marine Ecosystems: Emerging Diseases as Indicators of Change. Year of the Ocean Special Report on Health of the Oceans from Labrador to Venezuela. (Eds. P.R. Epstein, B.H. Sherman, E.S. Siegfried, A. Langston, S. Prasad and B. Mckay. The Center for Health and the Global Environment, Harvard Medical School. Massachusetts.

Herbst, L.H. (1994) Fibropapillomatosis of marine turtles. *Annual Review of Fish Diseases* 4, 389–425.

Hoekstra T.W., Allen, T.F.H. and Flather, C.H. (1991) Implicit scaling in ecological research. On when to make studies of mice and men. *BioScience* **41**, 148–154.

Holling C.S. (1973) *Resilience and Stability of Ecological Systems*. Institute of Resource Ecology. University of British Columbia, Vancouver, Canada.

Howarth, R. (1989) Comparative Responses of Aquatic Ecosystem to Toxic Chemical Stress. In *Comparative Analyses of Ecosystems. Patterns, Mechanisms, and Theories*, eds. J. Cole, G. Lovett and S. Findlay. Springer Verlag, New York

Howarth, R.W. (1991) Comparative responses of aquatic ecosystems to toxic chemical stress. In *Comparative Analysis of Ecosystems: Patterns, Mechanisms, and Theories*, eds. J.J. Cole, S. Findlay and G. Lovett, pp. 161–195. Springer-Verlag, New York.

Hughes, T.P. and Connell, J.H. (1999) Multiple stressors on coral reefs: a long-term perspective. *Limnology and Oceanography* **44**, 932–940.

Ibe, C., Oteng-Yeboah A.A., Zabi, S.G. and Afolabi, D. (eds.) (1998) Integrated environmental and living resources management in the Gulf of Guinea: the large marine ecosystem approach. *Proceedings of the 1st Symposium on GEF's LME project for the Gulf of Guinea, Abidjan. 26–30 January 1998*. Institute for Scientific and Technological Information, CSIR, Accra, Nigeria.

Karl, T.R, Nicholls, N. and Gregory, J. (1997) The coming climate. *Scientific American* May, 78–83.

Kumpf, H., Steidinger, K. and Sherman, K. (1999) *The Gulf of Mexico Large Marine Ecosystem. Assessment, Sustainability, and Management*. Blackwell, MA, 688 pp.

Landsberg, J.H. (1995) Tropical reef-fish disease outbreaks and mass mortalities in Florida, USA: what is the role of dietary biological toxins? *Diseases of Aquatic Organisms* **22**, 83–100.

Landsberg, J.H. and Steidinger, K.A. (1998) A historical review of *Gymnodium breve* red tides implicated in mass mortalities of the manatee (*Trichechus manatus latirostris*) in Florida, USA. In *Harmful Algae*, eds. B. Reguera et al., pp. 97–100. Xunta de Galicia and

Intergovernmental Oceanographic Commission of UNESCO.

Landsberg, J.H., Balazs, G.H., Steidinger, K.A., Baden, D.G., Work, T.M. and Russell, D.G. (1998) Preliminary evidence links sea turtle fibropapillomatosis to tumour-promoting microalgae. *Proceedings of the 3rd International Symposium on Aquatic Animal Health.* Building Partnerships for the 21st Century. Baltimore, Maryland.

Leibovitz, L. (1970) Duck Plague (duck virus enteritis). *Journal American Veterinary Medical Association* **156**, 1276–1277.

Lessios, H. (1984) Spread of *Diadema* mass mortality through the Caribbean. *Science* **226**, 335–337.

Li, M.F., Cornick, J.W. and Miller, R.J. (1982) Studies of recent mortalities of the sea urchin (*Stronglyocentrotus droebachiensis*) in Nova Scotia. Int. Council Explor. Mer Cm 1982/L, 46.

Lightner, D.V. and Redman, R.M. (1998) Emerging Crustacean Diseases. *Proceedings of the 3rd International Symposium on Aquatic Animal Health.* Building Partnerships for the 21st Century. Baltimore, Maryland.

Lockwood, A.P.M. (1993) Aliens and interlopers at sea. *Lancet* **342**, 942–943.

Lowenhaupt, E. (1998) *In situ* cholera detection in near coastal lagoons within the Gulf of Guinea. Technical Paper. Harvard University Environmental Sciences and Public Policy Program and the Institute for Scientific and Technological Information. CSIR, Accra.

McGowan, J.A., Cayan, D.R. and Dorman, L.M. (1998) Climate–ocean variability and ecosystem response in the northeast Pacific. *Science* **281**, 21.

Mahoney, J.B. and Steimle Jr., F.W. (1979) A mass mortality of marine animals associated with a bloom of *Ceratium tripos* in the New York Bight. In *Toxic Dinoflagellate Blooms*, eds. D.L. Taylor and H.H. Seliger, pp. 225–230. Elsevier, New York.

Magnuson, J. (1990) Long-term ecological research and the invisible present. *Bioscience* **40**, 495–501.

MMC (Marine Mammal Commission) (1999) Annual Report to U.S. Congress, 31 January 1999. 4340 E-W Highway, Room 905 Bethesda, Maryland 20814.

Mashima, T.Y., Fleming, W.J. and Stoskopf, M.K. (1998) Metal concentrations in oldsquaw (*Clangula hymalis*) during an outbreak of avian cholera, Chesapeake Bay, 1994. *Ecotoxicology* **7** (2), 107–111.

Mcgladdey, S.E. (1998) Emerging Molluscan Disease. *Proceedings of the 3rd International Symposium on Aquatic Animal Health.* Building Partnerships for the 21st Century Baltimore, Maryland.

Mee, L. (1992) The Black Sea in crisis: a need for concerted international action. *Ambio* **21**(4), 1278–1286.

Miller, R.J. and Colodey, A.G. (1983) Widespread mass mortalities of the green sea urchin in Nova Scotia, Canada. *Marine Biology* **73**, 263–267.

Murphy, E.J., Morris, D.J., Watkins, J.L. and Priddle, J. (1988) Scales of interaction between Antarctic Krill and the environment. In *Antarctic Ocean and Resources Variability*, ed. D. Sahrage, pp. 120–303. Springer-Verlag, Berlin.

NOAA (National Oceanic and Atmospheric Administration) (1991) Fish Kills in Coastal Waters 1980–1989. Strategic Environmental Assessments Division, Office of Ocean Resources Conservation and Assessment, National Ocean Service. National Oceanic and Atmospheric Administration. USA. J.A. Lowe, D.R.G. Farrow, A.S. Pait, S.j. Arenstam and E.F. Lavan (ed.) 69 pp.

NOAA (National Oceanic and Atmospheric Administration) (1993) Emerging Theoretical Basis for Monitoring the Changing States (Health) of Large Marine Ecosystems. Summary Report of Two Workshops: 23 April 1992, National Marine Fisheries Service, Narragansett, Rhode Island, and 11–12 July 1992, Cornell University, Ithaca, New York. National Oceanic and Atmospheric Administration Technical Memorandum NMFS-F/NEC-100.

NRC (National Research Council) (1990) *Managing Troubled Waters.* National Research Council. National Academy Press, Washington, DC, USA.

NRC (National Research Council) (1995) *Finding the Forest in the Trees: The Challenge of Combining Diverse Environmental Data.* National Research Council. National Academy Press, Washington, DC, USA.

NRC (National Research Council) (1999) From Monsoons to Microbes—Understanding the Ocean's Role in Human Health. Committee on the ocean's Role in Human Health, Ocean Studies Board, Commission on Geosciences, Environment, and Resources. National Research Council. National Academy Press, Washington, D.C.

Nehring, S. (1993) Mortality of dogs associated with mass development of *Nodularia spumingena* (Cyanophyceae) in a brackish lake of the German North Sea coast. *Journal of Plankton Research* **15**, 867–872.

Nehring, S., Hesse, K.J. and Tillmann, U. (1995) The German Wadden Sea: a problem area for Nuisance blooms: Harmful Marine Algal Blooms. *ACS Symposium Series* **262**, 199–204.

Nisbet, I.C.T. (1983) Paralytic shellfish poisoning effects on breeding terns. *Condor* **85**, 338–345.

Noga E.J., Levine J.F., Dykstra M.J. and Hawkins, J.H. (1988) Pathology of ulcerative mycosis in Atlantic menhaden, *Brevoortoria tyrannus*. *Dis. Aquatic. Org.* **4**, 189–197.

North Sea Status Quality Report (1993) Oslo and Paris Commission. Olsen and Olsen, Fredensborg, Denmark.

O'Toole, M. (1998) Strategic Action Program for the Benguela Large Marine Ecosystem. UNDP, Windhoek, Namibia. 40 pp.

Parry, G.D., Langdon, I.S. and Huisman, J.M. (1989) Toxic effects of a bloom of the diatom *Rhizosolenia chunk* on shellfish in Port Phillip Bay, Southeastern Australia. *Marine Biology* **102**, 25–41.

Pauly, D. and Christensen, V. (1993) Stratified models of large marine ecosystems: a general approach and an application to the South China Sea. In *Large Marine Ecosystems: Stress, Mitigation, and Sustainability*, K. Sherman, L.M. Alexander and B.D. Gold, pp. 148–174. AAAS Press, Washington, D.C.

Phillips, M.J., Roberts, R.J., Stewart, J.A. and Codd, G.A. (1985) The toxicity of the Cyanobacterium *Microcystis aeruginosa* to rainbow trout, *Salmo gairdneri* Richardson. *Journal Fish Disease* **8**, 339–344.

Pimm, S.L. (1984) The complexity and stability of ecosystems. *Nature* **307**, 321–326.

Pimm, S.L. and Lawton, J.H. (1998) Planning for Biodiversity. *Science* **279**, 2068–2069.

Pinto-Rodriguez, B., Mignucci-Giannoni, A.A. and Hall, K.V. (1995) Sea turtle mortality assessment and the newly established Caribbean Stranding Network. Proceedings of the 12th annual workshop on sea turtle biology and conservation. NMFS. SE Fisheries Science Center, pp. 92–93.

Pitcher, T. (1998) Harmful Algal Blooms in the Benguela Current. UNDP/World Bank Symposium Volume, Cape Town South Africa 48 pp.

Piyakarnchana, T. (1989) Yield dynamics as an index of biomass shifts in the Gulf of Thailand ecosystems. In *Biomass Yields and Geography of Large Marine Ecosystems*, eds. K. Sherman and L.M. Alexander, pp. 95–142. AAAS Selected Symposium 111. Westview Press Inc., Boulder.

Piyakarnchana, T. and S. Tamiyavanich. (1979) Frequent occurrence of the red tides in the inner sector of the Gulf of Thailand. *Journal of Aquatic Animal Diseases* **2**, 203–215.

Polovina, J.J., Mitchum, G.T. and Evans, G.T. (1995) Decadal and basin-scale variation in mixed layer depth and the impact on biological production in the central and north Pacific, 1960–88. *Deep-Sea Research* **42**, 1701–1716.

Prasad, S. (1998) Marine Economic Impact Assessment and Evaluation Using Senior Thesis, Environmental Science and Public Policy Program. Harvard University. 130 pp.

Price, D.W., Kizer, K W. and Hansgen, K.H. (1991) California's paralytic shellfish poisoning program. 1927–1989. *Journal of Shellfish Research* **10**, 119–145.

Rabalais, N., Turner, R.E. and Wiseman Jr., W.J. (1999) Hypoxia in the northern Gulf of Mexico: linkages with the Mississippi River. In *The Gulf of Mexico Large Marine Ecosystem: Assessment, Sustainability, and Management*, eds. H. Kumpf, K. Steidinger and K. Sherman, pp. 297–322. Blackwell, Massachusetts.

Rapport, D.J., Regier, H.A. and Hutchinson, T.C. (1985) Ecosystem behaviour under stress. *The American Naturalist* **125**, 617–640.

Ray, G.C. (1996) Coastal-marine discontinuities and synergism: implications for biodiversity conservation. *Biodiversity and Conservation* **5**, 1095–1108.

Reid, P.C. (1999) North Sea ecosystem: status report. In *The Gulf of Mexico Large Marine Ecosystem: Assessment, Sustainability, and Management*, eds. H. Kumpf, K. Steidinger and K. Sherman, pp. 476–489. Blackwell, Massachusetts.

She, J. (1999) Pollution in the Yellow Sea large marine ecosystem: monitoring, research, and ecological effects. In *Large Marine Ecosystems of the Pacific Rim: Assessment, Sustainability, and Management*, eds. K. Sherman and Q. Tang, pp. 419–426. Blackwell, Massachusetts.

Sherman, B.H. (1994) Toward a Unified Environmental Studies. Masters Thesis, Land Resources Program. University of Wisconsin, Madison. 113 pp.

Sherman, K. and Alexander, L.M. (eds.) (1989) *Biomass Yields and Geography of Large Marine Ecosystems*. AAAS Selected Symposium 111. Westview Press Inc., Boulder.

Shumway, S. (1990) A review of the effects of algal blooms on shellfish and aquaculture. *Journal of the World Aquaculture Society* **21**, 65–104.

Sierra-Beltran, A., Palafox-Uribe, M., Grajales-Montiel, J., Cruz-Villacorta, A. and Ochoa, J.L. (1997) Sea bird mortality at Cabo San Lucus, Mexico: Evidence that toxic diatom blooms are spreading. *Toxicology* **35** (3), 447–453.

Sindermann, C.J. (1994) Quantitative Effects of Pollution on Marine and Anadromous Fish Populations. NOAA Technical Memorandum NMFS-F/NEC-104.

Sindermann, C.J. (1996) *Ocean Pollution Effects on Living Resources and Humans*. CRC Press, New York.

Smayda, T.J. and Shimizu, Y. (eds.) (1993) *Toxic Phytoplankton Blooms in the Sea*. Elsevier, London, UK.

Smayda, T.J. (1990) Novel and nuisance phytoplankton in the sea: Evidence for a global epidemic. In *Toxic Marine Phytoplankton*, eds. E. Graneli et al., pp. 29–40. Elsevier, London, UK.

Smith, J.C. (1996) Toxicity and *Pseudonitzschia pungens* in Prince Edward Island, 1987–1992. Harmful Algae News. An IOC newsletter on toxic algae and algal blooms. No. 6. IOC UNESCO Paris. pp. 1, 8.

Sournia, A. (1995) Red tide and toxic marine phytoplankton of the world ocean: an inquiry into biodiversity. In *Harmful Marine Algal Blooms. Technique et Documentation*, eds. P. Lassus, G. Arzal, E. Erard-Le-Denn, P. Gentien and C. Marcaillou-Le-Baut, pp. 103–112. Lavoisier, Paris.

Spieker, J.O., Yuill, T.M. and Burgess, E.C. (1996) Virulence of six strains of duck plague virus in eight waterfowl species. *Journal of Wildlife Diseases* **32** (3), 453.

Steidinger, K.A. (1993) Some taxonomic and biological aspects of toxic dinoflagellates. In *Algal Toxins in Seafood and Drinking Water*, ed. I.R. Falconer, pp. 1–28. Academic Press, New York.

Tester, P.A., Geesey, M.E. and Vukovich, F.M. (1993) *Gymnodinium breve* and global warming: What are the possibilities. In *Toxic Phytoplankton Blooms in the Sea*, eds. T.J. Smayda and Y. Shimizu, pp. 67–72. Elsevier, New York.

Tester, P.A., Fowler, P. K., and J. T. Turner. (1989) Gulf Stream transport of the toxic red tide dinoflagellate *Ptychodiscus brevis* from Florida to North Carolina. pp. 349–358. In Coastal and Estuarine Studies, 35. Novel Phytoplankton Blooms. Causes and impacts of recurrent brown tides and other unusual blooms. Springer Verlag, Germany.

Tester, P.A. and Mahoney, B. (1995) Implication of the diatom, *Chaetocerus convolutus*, in the death of red king crab, *Paralithodes camischafica*, Captains Bay, Unalaska Island, Alaska. In *Harmful Marine Algal Blooms, Technique et Documentation*, eds. P. Lassus, G. Arzul, E. Erard-Le-Denn, P. Gentien and C. Marcaillou-Le-Baut, pp. 95–100. Lavoisier, Paris.

Thulin, J. (2000) Long Term Changes in the Baltic LME in North Atlantic Large Marine Ecosystem: Assessment Sustainability and Management (eds.) K. Sherman and H.R. Skjoldal. Blackwell Science. In press.

Todd, E.C.D. (1993) Domoic acid and Amnesic Shellfish Poisoning—a review. *Journal of Food Protection* **56**(1), 69–83.

Todd, E.C.D. (1994) The Cost of Marine Disease. In *Disease in Evolution: Global Changes and Emergence of Infectious Diseases*, eds. M.E. Wilson, R. Levins and A. Spielman, pp. 423–435. New York Academy of Sciences, New York.

Todd, E.C.D. (1995) Estimated Costs of Paralytic Shellfish, Diarrhetic Shellfish and Ciguatera Poisoning in Canada. Harmful Marine Algal Blooms. In *Harmful Marine Algal Blooms, Technique et Documentation*, eds. P. Lassus, G. Arzul, E. Erard, P. Gentien and C. Marcailou. Lavoisier, Paris.

Tracey, G.A. (1988) Feeding reduction, reproductive failure, and mortality in *Mytilus edulis* during the 1985 'brown tide' in Narragansett Bay, Rhode Island. *Marine Ecology Progress Series* **50**, 73–81.

Travis, J. (1993) Invader threatens Black Azov Seas. *Science* **262**, 1366–1367.

White, A.W. (1996) Biotoxins and the health of living marine resources of the Northeast Shelf Ecosystem. *The Northeast Shelf Ecosystem*. Blackwell Science Inc., Massachusetts.

Wilkinson, C., Linden, O., Cesar, H., Hodgson, G., Rubens, J. and Strong, A.E. (1999) Ecological and socioeconomic impacts of 1998 coral mortality in the Indian Ocean: an ENSO impact and a warning of future change? *Ambio* **28** (2), 188–196.

Williams Jr., E.H. (1992) Mass mortality of marine organisms in the Caribbean. Proceedings of a Workshop on Red Tides and Mortality of Marine Organisms in the Caribbean. Cumana, Venezuela.

Williams Jr., E.H. and Bunkley-Williams, L. (1992) 1989–1991 Worldwide Coral Reef Bleaching. *Proceedings 7th International Coral Reef Symposium*, Guam, p. 108.

Williams Jr., E.H. et al. (1994) An epizootic of cutaneous fibropapillomas in green turtles of the Caribbean: Part of the panzootic. *Journal for Aquatic Animal Health* **6**, 70–78.

Work, T.M., Beale, A.M., Fritz, L., Quilliam, M.A., Silver, M., Buck, K. and Wright, J. (1993) Domoic acid intoxication of brown pelicans and cormorants in Santa Cruz, California. *Toxic Phytoplankton Blooms in the Sea*. Elsevier, Amsterdam and New York, pp. 643–649.

Zaitsev, Y.P. (1992) Recent changes in the trophic structure of the Black Sea Fish. *Oceanography* **1** (2), 180–189.

## THE AUTHOR

**Benjamin H. Sherman**

*Climate Change Research Center,*
*Institute for the Study of Earth Oceans and Space,*
*OSP HEED MMED program, 206 Nesmith Hall,*
*University of New Hampshire, Durham, NH 03824,*
*U.S.A.*

Chapter 123

# EFFECT OF MINE TAILINGS ON THE BIODIVERSITY OF THE SEABED: EXAMPLE OF THE ISLAND COPPER MINE, CANADA

Derek V. Ellis

The state of benthic biodiversity on a sediment seabed subjected to the placement of mine tailings has been monitored over a 29-year period (1970–1998) from before to after operations. Losses of biodiversity were generally restricted to deep areas of relatively thick and quick tailings deposition. Tolerated deposition rates varied, but at times up to ~20 cm/year had little impact. Because the tailings flowed to their deposition target by a meandering, narrow, density current, much of the tailings deposits could stabilize after the density current moved elsewhere. Within 1–2 years of stabilization, biodiversity losses on the tailings silts had recovered to within the range of numbers of species on natural fjord silts nearby, and a sustainable ecological succession had been initiated. The final biodiversity recovery after mine closure followed the same pattern. The sustainable succession was initiated by polychaetes of the widespread genera *Chaetozone, Cossura, Euchone, Tharyx, Lumbrineris* and *Nephtys*, followed by a bivalve mollusc species *Axinopsida*. A crab fishery was maintained in the area during the lifetime of the mine. The presence of organisms too large or too deeply burrowing to be collected by the sampler used was monitored by identifying juveniles. Collected juveniles of some large forms, e.g. cockles, some clams, and holothuria, showed that these could colonize and grow on tailings. A rapid assessment protocol was developed, allowing reporting of basic biodiversity parameters within a few days of sampling. Results from coastal and island mines in other parts of the world, and from dredging operations, suggest that the threshold rate of deposition, and the time and pattern to sustainable succession, at this mine site should have some predictive value elsewhere.

*Seas at The Millennium: An Environmental Evaluation (Edited by C. Sheppard)*

## INTRODUCTION

It is important for countries whose mining industry is economically important to resolve the question of where to place mine tailings with minimal, and reversible, environmental losses. For coastal and island mines, placement deep in the sea may be a viable alternative to disposal on land (Caldwell and Welsh, 1982). On-land placement may create serious risks of unacceptable environmental impact, e.g. where a mine development is in or adjacent to areas that are highly populated, or rich in biological resources, or prone to torrential rains and earthquakes. The 30-year history of submarine placement of mine tailings around the world has shown how this can be engineered (Ellis and Poling, 1995) to provide an alternative to on-land disposal that is socially acceptable as well as physically possible. It is now obvious, for example, that where tailings have been milled to silt-sized particles, a deep silty bottom target area should be selected to receive the tailings, so as not to change the physical nature of the seabed.

This chapter summarizes the details of a unique 29-year marine monitoring program on the biodiversity of a seabed under impact and recovery from mine tailings in a targeted area of a temperate zone fjord in western Canada. This is the longest time-series (1970–1998) of benthic data documenting the seabed's response to deposition of mine tailings, and the processes of biodiversity recovery after a mine has closed.

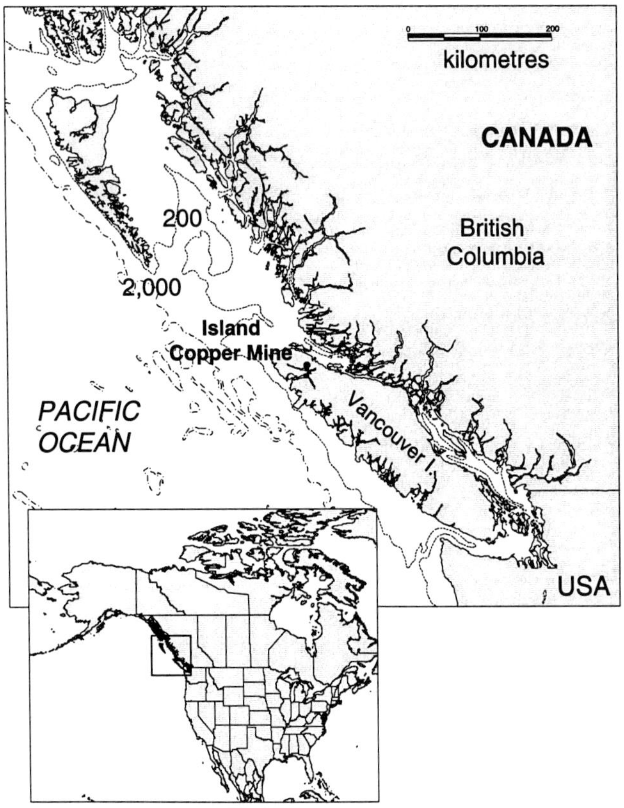


Fig. 1. Location of Island Copper Mine in western Canada.

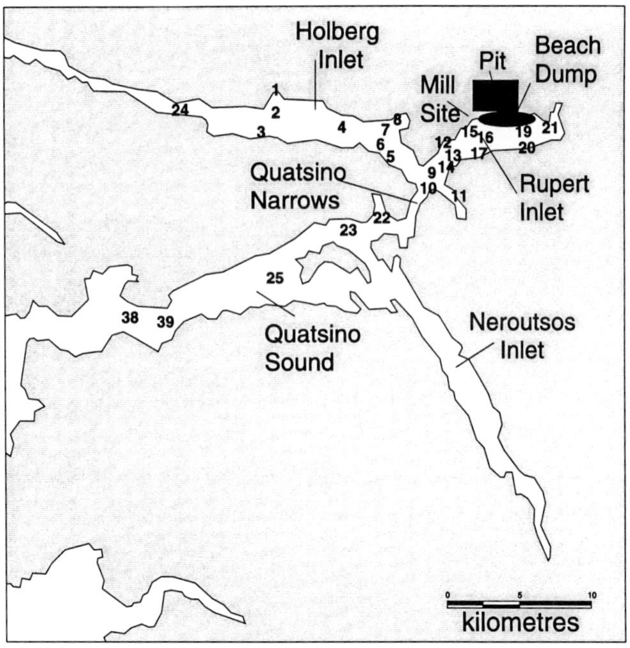


Fig. 2. Location of the benthic sampling stations monitored by Island Copper Mine 1970–1998.

The site is Island Copper Mine (ICM) located on the north shore of Rupert Inlet (Figs. 1 and 2) in northern Vancouver Island. Rupert Inlet is continuous with Holberg Inlet with a basin, originally about 150 m deep, between the two. These two inlets are connected by a narrow, shallow channel of 30 m deep to an outer inlet, Quatsino Sound, which in turn opens to the Pacific Ocean. The mine, which produced copper, molybdenum, gold, silver and rhenium, opened in October 1971 and was in operation until December 1995. During operations ~50,000 metric tonnes of tailings were discharged each day into Rupert Inlet, totalling more than 350 million tonnes over the 25 years. The tailings were discharged at ~50 m depth, below the surface euphotic zone, and flowed by density current to their deposition target. Depth reductions (i.e. tailings thickness) ranged to 50 m in the deep basin, but subsequent tidal scouring after mine closure returned the depth to 140 m by 1998.

The mine undertook a pre-operational environmental impact assessment (January 1970 to September 1971), and during operations was required to maintain an extensive monitoring programme. This programme continued until 1998 (three years after closure), and at the time of writing is being reviewed for further extension. There is much information available about the mine and its monitoring results through its annual environmental reports (ICM, 1972–1999), and published articles (e.g. Ellis et al., 1995). A short description after closure is given by Marcus (1997).

The monitoring results have been checked annually by a group of university scientists, who have also undertaken or supervised some of the necessary research needed for understanding the environmental processes in action (e.g. Jones, 1974; Taylor, 1986). Details of the whole program have been reported annually (ICM, 1972–1999), and details or reviews of the data periodically in the refereed literature (e.g. Ellis et al., 1995; Ellis, 1989). The full data set is expected to be available eventually through a university science information network.

This review adds to the substantial current background knowledge on environmental impact of marine placement of tailings (e.g. Ellis and Poling, 1995), by providing a review of biodiversity recovery after mine closure. It also has some interest and predictive value for any locations where the seabed is subject to smothering amounts of non-contaminating particulate deposits, such as is derived from dredging clean marine shallows for aggregate extraction (e.g. Newell et al., 1998), or where removal is made for navigational purposes.

It should be noted that contamination from bioactivation of trace metals from the tailings has not been a problem at ICM, due to the combination of: (1) insoluble chalcopyrite as the orebody; and (2) tailings placement well below the surface, thus minimizing acid waste generation (Pedersen, 1985; Ellis and Robertson, 1999).

This review, in conjunction with Ellis and Robertson (1999), is also relevant in understanding the benthic environmental issues associated with deep-sea placement of mine tailings with substantial risk from acid-waste generation or from other contamination potential.

The annual reports and prior reviews have shown: (1) the manner in which the tailings flowed by a meandering density current to, and spread through, the targeted area (Fig. 3); (2) the extent of impact on the seabed biodiversity; and (3) the manner in which stabilized areas of the tailings beds (i.e. no longer affected by the density current after it changed location) would start an ecological succession.

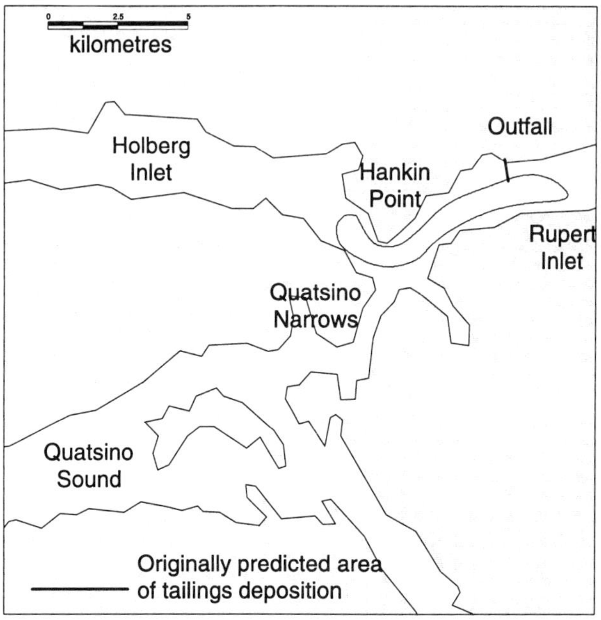


Fig. 3. At Island Copper Mine, the marine target area to receive tailings was primarily below 100 m (300 ft), with the expectation that most tailings would deposit within the marked area (Anon, 1970).

## THE DATA SET

The benthos monitored quantitatively for biodiversity estimates at ICM is the infauna (Thorson, 1957). The infauna and their sediment habitat represent the greater part of the seabed ecosystem (both worldwide and in the subject area), as reefs and underwater cliffs with their epifauna are restricted to almost vertical slopes and ledges in currents or where wave action precludes sediments deposition. The infaunal biodiversity of sedentary small animals can be assessed quantitatively (by a grab sampler) more accurately and expeditiously than the often large and mobile epifauna inhabiting reefs and algal forests. The infaunal biodiversity assessments were made by obtaining samples from a constant surface area (using a 0.05 $m^2$ Ponar grab), and screening through a no. 30 mesh (~0.5 $mm^2$); for methods of infaunal sampling see Holme and McIntyre (1984). From these samples, two basic biodiversity statistics were calculated (Gaston and Spicer, 1998): (1) the number of species per unit area (species richness); and (2) the numbers of organisms (total number, and the number of each species to give statistics for species evenness) per unit area.

In addition, data are presented here on yields from a test trap fishery for crabs conducted alongside the commercial fishery in the fjord. The crab species involved (the Dungeness crab, *Cancer magister*) is fished throughout the fjord, although it spawns in shallow seagrass and algal habitat. The data reviewed here thus span the infaunal macrobenthos and one of their large epifaunal predators.

The basic sampling design for the infaunal benthos of ~20 test and reference stations from deep and shallow water (Fig. 2) was set in 1970 over 1.5 years prior to mine operations. The design was set to monitor: (1) the targeted deposition area (generally >100 m depth as shown in Fig. 3 and described in Anon, 1970); (2) shallow water sediments ringing the targeted deep area; and (3) some remote areas expected to be unaffected, hence able to function as reference stations. Minor changes and additions eventually gave 26 sampling stations. The sampling design was continued until September 1998, over 2.5 years after mine closure in December 1995. The design therefore embodied both monitoring principles of (1) before and after sampling, and (2) test and reference stations. Sampling stations within the main placement area are considered as the test stations (nos. 9, 12–17, 19); reference stations are taken to be the deeper remote stations (nos. 1–3, 24, 23, 25, 38,39). Other stations are either unaffected shallow water stations (nos. 8,10, 21) or stations which showed some but variable impact (4–7, 11, 20, 22).

Usually three replicates were collected at each sampling station. During 1970–1972, one replicate sample at each station was sorted to species, the species identified, and the numbers of each counted. Since 1977, all three replicates at each station have been processed in that way. Details of sampling procedures and statistical summaries each year are available in the annual environmental reports, especially those for 1986 and 1997 (ICM 1997; 1998).

In the early years of monitoring (1972–1976), several changes were made in procedures in attempts to increase time/cost-efficiency, for example by not identifying all specimens to species. These first attempts at time/cost-efficiency in the then new field of monitoring benthic responses to industrial waste discharges were not satisfactory. In particular, procedures tried during the years 1973–1976 when no collected specimens were identified to species, render the data during that period generally unusable. Also, these early specimens were subsequently lost in a fire, though some species data from that time remain (Jones, 1974: Jones and Ellis, 1976; Ellis et al., 1991). By 1977 the definitive procedure that all replicate samples were sorted, identified and counted by species was set. These attempts at time/cost-efficiency in benthic sampling are reviewed in this chapter for the guidance of others in undertaking similar long-term benthic monitoring elsewhere.

This article concentrates on reviewing those results from the enormous data set that demonstrate (1) the level and distribution of environmental impact, (2) environmental recovery, and (3) the natural variation in biodiversity at sampling stations remote from the tailings. The main statistics used are numbers of different species (species richness), their identifications to species, and numbers of the most abundant species (species evenness).

Table 1

Comparison of 1998 and 1991 species richness data; 1998 (three years after mine closure) was the last year for which data has been obtained; 1991 was the last year with widespread impact on species richness.

| Sampling Station | 1998 | | 1991 | |
|---|---|---|---|---|
| | No. species/ 3 replicates (0.15 m$^2$) | No. organisms/m$^2$ | No. Species/ 3 replicates (0.15 m$^2$) | No. organisms/m$^2$ |
| 2 | 52 | 4846 | 31 | 6166 |
| 9 | 81 | 5586 | 8 | 160 |
| 13 | 70 | 5073 | 8 | 306 |
| 15 | 59 | 13420 | 3 | 286 |
| 16 | 58 | 7480 | 15 | 4986 |

Table 2

Biodiversity descriptive scaling developed for the Island Copper Mine tailings placement area (based on 29 years of monitoring data, and the sampling procedures used).

| Biodiversity scaling | No. species/3 replicates (0.15 m$^2$) | No. organisms/m$^2$ |
|---|---|---|
| Impoverished | <10 | <100 |
| Low | 10–19 | 100–999 |
| Moderate | 20–50 | 1000–5000 |
| High | >50 | >5000 |

## SPECIES RICHNESS

Species richness under impact and during recovery is summarized in Table 1, where numbers of species per unit area are shown for five selected sampling stations from 1991, a year during mining operations when impact on species richness was substantial, and from 1998 almost three years after mine closure. Station 2 is a reference station remote from the mine (Fig. 2), whereas the others are within the main tailings placement area. In 1991 there was a typical reduction in species richness on the tailings beds compared to the reference station, but note that the fauna was not totally obliterated. Only twice during the 25 years of mine operation were no species collected, and on each occasion the following year species were restored.

By 1998, species richness at the impacted stations shown (9,13, 15 16) was restored to levels well within the range at the reference stations. A species richness descriptive scale based on the 29 years of data has been developed using the terms Impoverished, Low, Moderate and High (Table 2).

Tables 1 and 2 also show organism abundance data (total number of collected organisms per unit area). The results are similar to those for species richness: there is a reduction in organism abundance under impact, but recovery 2–3 years after tailings placement ceased (in 1995) to levels within the range of unaffected sampling stations.

The validity of the number of different species in these surveys as a biodiversity statistic is enhanced in that sampling procedures were identical from year to year, hence variability in sampling procedures could not affect the numbers of species collected. For example, richness is the number of species per 0.15 m$^2$, i.e. the number of species in three samples each of 0.05 m$^2$. The only major difference in procedures was that the specimen sorters and their supervisors became more proficient with experience, thus there was inevitably some increase in counts of organisms over the years of the surveys. This means that species numbers before and after mining cannot be compared. Also, the generally larger numbers in the final 10 years of sampling cannot be taken to mean a greater biodiversity than premining.

There is considerable merit in using actual counts of species, rather than other diversity indices requiring calculations, as these are subject to errors (Wu, 1982) derived from variable levels of taxonomic identification. Also, the meaning of each of the different kinds of diversity indices is not as intuitively obvious as the actual counts giving species richness. However the species data includes calculations of a number of the conventional biodiversity indices, although there has been no analysis of these to date.

Some other patterns emerge from the overall data:

- *Initial decreases.* Decreases in species richness started within 3–9 months of the mine opening
- *Extent of decreases.* Only twice was the decrease to 0 organisms, i.e. total obliteration: Station 15 in 1982, and Station 16 in 1978. Generally major decreases were to

<10 species, i.e. some organisms could survive the rate of deposition, or could colonise within a year.

- *Temporary recoveries.* Each year that there was a major decrease, the numbers recovered to some extent over the next few years. This appears to be derived from the tailings density current changing position, and leaving areas to stabilize, hence allowing organisms to settle and grow. Only Stations 15 and 16, closest to the discharge point, had species numbers <10 for many years.
- *Final recovery.* Numbers of species were restored to levels (generally >30) within the range of those outside the thick tailings area by September 1996, nine months after mining ceased, except for Station 16 which required a further year. This recovered species richness was then sustained.
- *Initial data interpretation.* Initial data interpretation (1971–1974), in the then new field of monitoring seabed impact from an industrial waste, was substantially incorrect: due in part to the low numbers of specimens reported while personnel were gaining experience with implementing the surveys and processing the samples, and partly due to misinterpretation. As late as 1974, the data from the most impacted stations was still being interpreted as meaning obliteration of the fauna throughout the tailings area (ICM, 1975), while missing the significance of recognizing (as early as 1972) that the low numbers of species present could mean that recolonization was occurring (ICM, 1973). In 1975 this misunderstanding was corrected (Ellis, 1975).

Similarity analyses (SIGTREE software of Nemec and Brinkhurst, 1988) of the species and abundance data each year since 1986 (ICM, 1987–1999) have shown that deep remote sampling stations in Holberg Inlet and Quatsino Sound tended to be more similar to each other than to the impacted stations in Rupert Inlet. The remotest deep stations are 39, 38 and 25 in Quatsino Sound, and Stations 24, 2 and 3 in Holberg Inlet. The numbers of species in all these stations ranged to >100 but even in the last 10 years (of greatest sorting efficiency) numbers of species in these reference stations could nevertheless range down to 20–30. Burd (1999) has shown that variability from year to year in these remote stations can vary from 50–70%.

The remaining deep stations (4, 5, 6, 19 and 23) have had some species decreases, indicating some biodiversity losses, but these were always followed by a recovery within 1–2 years. By 1996 and 1997, these stations at the edge of the thickest tailings deposits had recovered their numbers of species, and the numbers were subsequently sustained.

Analyses for Cu in the sediments indicate that values >~100 ppm dry weight reflect some deposition of tailings (pre-mining ambient levels, and levels at remotest stations 38 and 39 were generally less than 50 ppm). Stations 2, 3 and 25, and possibly 24, show some Cu elevation (to ~200 ppm). Evidently, there is a level of copper increase, which in insoluble chalcopyrite is an expression of tailings deposition, not of Cu leaching potential from the tailings (Pedersen, 1985) which is tolerated by the infauna. This appears to be at least 200 ppm. Burd (1999) also indicates that particulate copper has not adversely affected the organisms. This level of 200 ppm should not be taken as a threshold to toxicity of copper in the mine tailings; it is a surrogate measure of amount of tailings.

This in turn means that there is a rate of tailings deposition which is tolerated by the infauna, either by organisms burrowing up faster than the tailings deposits or by larvae settling rapidly as some organisms are smothered and leave space in the tailings deposits for colonization.

The monitoring design included sampling stations in shallow water to check whether tailings were present in water depths less than the targeted deposition area of 100+ m. These are Stations 1, 7, 8, 10, 11, 12, 20, 21 and 23. Species richness is within the range of the deep reference stations. In general this is to be expected as the shallow water infauna tends to be more diverse than the deep water infauna. Also, Cu values at these stations show some elevations indicating that some tailings were suspended into shallow water (Stations 7 and 8 in Holberg Inlet, Stations 11, 12, 20 in Rupert Inlet, and Station 22 in Quatsino Sound). There has been no detectable impact on the infauna at these Cu levels generally <200 ppm, hence it appears that these shallow water areas also can tolerate some low rate of tailings deposition.

The raised Cu levels in shallow water at Stations 7, 8 and 11 in the Holberg/Rupert complex, and in Station 22 in Quatsino Sound (also in deeper water in Quatsino Sound: Station 23) reflect the resuspension and upwelling of tailings by tidal jets at the junction of Rupert and Holberg Inlets, and redeposition in shallow water with some escape into Quatsino Sound (less than 0.3%: ICM, 1980). There is no detectable impact of these tailings at these sampling stations in water depths of <100 m, and also deeper in Quatsino Sound.

## TAILINGS DEPOSITION LEVELS WHICH AFFECT THE FAUNA

What is the tolerated rate of tailings deposition, according to the species richness statistics? At Station 14 between 1977 and 1978, 46 cm of tailings deposited, and although numbers were already low (16 species) they did not decrease. Also, at Station 19 between 1991 and 1992 36.5 cm of tailings deposited with a drop in numbers of species from 22 to 6. At the stations which received massive deposits of tailings within the first year of mine operations, i.e. more than the 60 cm measurable by the corer used (Stations 9, 13, 15, 16 and 17), there were corresponding drops in species numbers.

At all other stations, e.g. 3, 4 and 5, in contrast, maximum annual changes of ~20 cm produced no obvious decreases in species numbers. It appears that in parts of this area the infaunal benthos can tolerate tailings deposition at rates to ~20 cm (possibly more) in a year. This is not unexpected from data elsewhere with tolerance values (due to upward burrowing by the infauna) between 4 and 30 cm of rapidly deposited sediment (Newell et al., 1998). It is also a conservative estimate of tolerance, since the erratic distribution of maximum deposition increases through the years suggests that deposition is patchy and quick in the remoter parts of the inlets from clouds of tailings, rather than continuing at a constant average rate through a year from a widely spread, layered, plume. If it were constant, 20 cm per year would represent <1 mm of deposition per day.

This estimate of the tailings deposition rate which been tolerated by the benthos, i.e. ~20 cm per year, replaces an earlier, even more conservative, estimate of 1 cm per year (ICM, 1993). This was derived (Ellis et al., 1995) from SIGTREE similarity analyses which showed that sampling stations that had accumulated <~20 cm tailings to 1993 (21 years) were indistinguishable from other sampling stations that had not received tailings. The revised tolerance estimate of ~20 cm/year is closer to values found elsewhere.

## NUMBERS OF EACH SPECIES (SPECIES EVENNESS)

The total number of organisms collected (of all species lumped together) is not a particularly informative statistic after the numbers of different species have been considered (as above). In the early stages of ecological succession, following habitat changes, a few opportunist species can be present in enormous numbers, giving values for total numbers of organisms close to or greater than those of a more diverse community in equilibrium (in climax) with its habitat (Pearson and Rosenberg, 1978).

More significant in the context of the seabed adjacent to Island Copper Mine are the species present and their numbers per unit area. With over 1000 species collected from over 2000 samples (~26×3 samples annually for 29 years), and averaging 300 species per year, it is not possible to show all the species numerical data here. Species identifications from the early years have been updated or corrected for consistency in species evenness counts and biodiversity calculations throughout the 29 years of the surveys. The most recent taxonomic changes may not have been incorporated in the analysis for this chapter, but should have little effect on the calculations.

It has previously been shown (Ellis and Hoover, 1991; Ellis, 1998) that the first species to colonize the tailings (within 1–2 years) and to sustain their populations (primary opportunists) are polychaete worm species such as: *Chaetozone acuta, Cossura pygodactylata, Euchone incolor, Tharyx multifilis, Lumbrineris luti* and *Nephtys cornuta*. The first four are particle-feeding Sedentariate worms, whereas *Lumbrineris luti* and *Nephtys cornuta* are mobile omnivores or carnivores (for which there must be a food supply). Secondary opportunist species appearing a year or so later are particle-feeding bivalve molluscs, especially the small clam *Axinopsida serricata*.

Table 3

The five most abundant species at two mud bottom impacted stations 1998, plus other known opportunist species (numbers/m$^2$ converted from 0.15 m$^2$ sampled area)

| Station 15 | | Station 16 | |
|---|---|---|---|
| Species | No/m$^2$ | Species | No/m$^2$ |
| *Tharyx multifilis* | 5733 | *Tharyx multifilis* | 2547 |
| *Axinopsida serricata* | 2140 | *Axinopsida serricata* | 1013 |
| *Lumbrineris luti* | 1113 | *Lumbrineris luti* | 753 |
| *Chaetozone acuta* | 907 | *Bugula pacifica* | 727 |
| *Cossura pygodactylata* | 673 | *Lafoea* sp. | 500 |
| + *Euchone incolor* | 127 | + *Cossura pygodactylata* | 340 |
| *Ophelina acuminata* | 47 | *Chaetozone acuta* | 247 |
| *Nephtys cornuta* | 13 | *Ophelina acuminata* | 93 |
| | | *Nephtys cornuta* | 73 |

Table 4

The five most abundant species at two reference stations, plus known opportunist species (numbers/m$^2$ converted from 0.15 m$^2$ sampled area)

| Station 2 (1998) | | Station 24 (1997) | |
|---|---|---|---|
| Species | No/m$^2$ | Species | No/m$^2$ |
| *Axinopsida serricata* | 2220 | *Axinopsida serricata* | 2487 |
| *Lafoea* sp. | 933 | *Cossura pygodactylata* | 1473 |
| *Cossura pygodactylata* | 247 | *Paraonella spinifera* | 1040 |
| *Nephtys cornuta* | 193 | *Mediomastus* sp. | 540 |
| *Chaetozone acuta* | 147 | *Minuspio multibranchiata* | 407 |
| + *Lumbrineris luti* | 100 | + *Lumbrineris luti* | 373 |
| *Ophelina acuminata* | 13 | *Nephtys cornuta* | 293 |
| | | *Ophelina acuminata* | 40 |
| | | *Euchone incolor* | 33 |

Table 3 shows the top five species present at impacted stations 15 and 16 in Rupert Inlet in 1998 (three years after mining ceased), plus some other known opportunist species. The above-listed species regularly appeared in these species lists from 1994, one year after major defaunations at Stations 15 and 16, and then remained abundant. The secondary opportunist *Axinopsida serricata* appeared in 1994, and then fluctuated in numbers. Another polychaete *Ophelina acuminata* also appeared in 1994 and then sustained itself, but usually only in low numbers.

Table 4 provides data from reference stations 2 and 24 for comparison with the impacted stations 15 and 16. The primary and secondary opportunists frequently appear in the top five species. It appears that the primary opportunists in this locality are not quickly, if at all, displaced by later colonists.

There is considerable similarity in the most abundant species between test and reference stations 2–3 years after mine closure (Tables 1, 3 and 4). The differences between the recovering and unaffected stations appears to lie in the greater diversity of relatively rare species.

Table 5 provides data for 1998 on the most abundant species in the two impacted stations (Stations 9 and 13) closest to the deep pit at the junction of Rupert and Holberg Inlets, where tidal jets have caused some resuspension and upwelling of tailings (Ellis et al., 1995; ICM, 1999), and where, since mine closure, there has been a gradual habitat change—the depth is increasing (from 120 m in 1995 to 140 m in 1998) and the deposits are both coarsening and becoming harder to penetrate by corer (compacting). Both Stations 9 and 13 show a decrease in fine particles (ICM, 1999) such that both now classify as fine sands rather than as the silts or clays at the remaining impacted stations (15, 16 and 17) and the unaffected stations along the fjord trough (2, 24, etc).

In 1993, at both these two, now sand-bottom, stations, the typical mud-bottom opportunists had dominated the low biodiversity of 16 and 14 species (Table 1). As the biodiversity increased to the high plateau of 50–80 species in 1998, so sand-bottom species appeared and increased in numbers (Table 5). They include the burrowing sea anemone *Halcampa decemtentaculata*, the sand-grain-tube-building polychaete worm *Pectinaria californiensis*, and several bivalve molluscs *Psephidia lordi* and *Yoldia scissurata*.

In summary, the effects in the area where the habitat is changing physically from mud bottom to sand bottom following mine closure (lower Rupert Inlet to the junction with Holberg Inlet) are as follows. The area suffered species reductions during mine operations, but within one year after mine closure the increased biodiversity was showing changes to a sand-bottom fauna. These changes are probably still in progress, but their endpoint in terms of the species making up a sand-bottom equilibrium community, the numbers present, and their variations from year to year and place to place cannot be predicted from the data available. It is probable, however, that species richness and total numbers of organisms will remain at the levels reached by 1998.

Biodiversity recovery has been sustained on stabilized tailings over a sufficient number of years for a set of criteria to be developed for determining whether the ecological succession is being sustained at other locations (Ellis, 1998), with the expectation that it will continue. These are: (1) the number of species and the total number of organisms must fall within or above the ranges at unaffected stations; at ICM the specific values developed for the sampling procedure used are 20 or more species per 0.15 m$^2$, and more than 1000 organism per m$^2$; (2) several opportunist (i.e. quickly colonizing) species must have sustained themselves in large numbers for one or more years. At ICM the values developed are that more than three species have sustained themselves at greater than 100 per m$^2$ for two or more years.

## BIODIVERSITY IN TERMS OF HIGH ORDER TAXA

It became apparent that certain types of organisms normally present in moderate to large numbers on mud (silt/clay) and sand bottom were either absent from the recovering areas, or only present as few species. These include particularly echinoderms, such as brittle stars, sea cucumbers and sea urchins, frequently present on fjord mud bottoms in western Canada (Ellis, 1967; 1968).

In almost all types of organisms (Table 6) the recovering stations are within or close to the range of variation in the two reference stations. The exception is diversity of

Table 5

The five most abundant species at two impacted stations 1998 changing after mine closure from mud to sand bottom, plus other detected opportunist species (numbers/m$^2$ converted from 0.15 m$^2$ sampled area)

| Station 9 | | Station 13 | |
|---|---|---|---|
| Species | No/m$^2$ | Species | No/m$^2$ |
| *Lumbrineris luti* | 1467 | *Lafoea* sp. | 667 |
| *Mediomastus ambiseta* | 747 | *Lumbrineris luti* | 627 |
| *Glycera nana* | 353 | *Mediomastus ambiseta* | 507 |
| *Psephidia lordi* | 287 | *Tharyx multifilis* | 373 |
| *Nitidella gouldii* | 213 | *Ophelina acuminata* | 320 |
| + *Tharyx multifilis* | 100 | + *Chaetozone acuta* | 120 |
| *Axinopsida serricata* | 93 | *Cossura pygodactylata* | 107 |
| *Ophelina acuminata* | 80 | *Euchone incolor* | 80 |
| *Nephtys corniuta* | 80 | *Nephtys cornuta* | 47 |
| *Pectinaria californiensis* | 26 | *Psephida lordi* | 40 |
| *Halcampa decemtentaculata* | 7 | *Pectinaria californiensis* | 20 |
| | | *Yoldia scissurata* | 7 |

Table 6

The biodiversity in 1998 at impacted stations 15, 16 and at reference stations 2 and 24 expressed in terms of high order taxa (1997 data for Station 24). No. species/unit area (0.15 m$^2$)

| | Impacted sampling stations | | Reference stations | |
|---|---|---|---|---|
| Taxonomic unit | Stn 15 | Stn 16 | Stn 2 | Stn 24 |
| Anthozoa | | | | |
| Brachiopoda | | | | |
| Bryozoa | | 1 | 1 | |
| Crustacea | 3 | 4 | 11 | 13 |
| Echinoderms | | 1 | 1 | 1 |
| Hirudinoidea | | | 1 | |
| Hydrozoa | 2 | 1 | 2 | |
| Kinorhyncha | | | | 1 |
| Molluscs: bivalves | 6 | 11 | 7 | 7 |
| Molluscs: gastropods | 4 | 4 | 2 | |
| Nematoda | 1 | | | 1 |
| Nemertea | 3 | 1 | 1 | 2 |
| Phoronida | | | | |
| Polychaetes errantia | 13 | 10 | 9 | 21 |
| Polychaetes sedentaria | 27 | 23 | 17 | 35 |
| Totals | 59 | 58 | 52 | 81 |

crustacea (three and four species at Stations 15 and 16, compared to 11 and 13 at the reference stations 2 and 24).

A surprising finding is the scarcity of echinoderms, which can be abundant (up 100–200/m$^2$) on muddy fjord seabeds in western Canada (Ellis, 1967; 1968). The pre-discharge surveys at ICM support this. Asteroids, Ophiuroids and Holothuroids occurred only at a few stations each survey, and only in small numbers. Heart urchins, genus *Brisaster*, which are common elsewhere in the west coast fjords were not collected at all pre-discharge. Evidently Echinoderms are naturally few in species and numbers throughout the Rupert/Holberg fjord complex, so cannot be used as bioindicators of recovery to the equilibrium. This is unlikely to be an artifact of the early sample processing; echinoderms are very easy to recognize!

In summary, it appears that the mud-bottom impacted stations will sustain their succession to a stage where there is a greater diversity of rare species, particularly crustacea, than at present. This appears to be the only major biodiversity recovery change still in progress on the deep fjord mud seabeds three years after mining operations ceased. The time for numbers of small crustacean species to increase to within their equilibrium range cannot be predicted from the data available.

## SIMILARITY ANALYSES

Similarity analyses were routinely implemented on these surveys from 1986 (reported ICM, 1986–1999) using the Bray–Curtis coefficient and the SIGTREE software (Nemec and Brinkhurst, 1988) for probability analysis. The collated results have been examined in detail by Burd and Ellis (1994) and Burd (1999). Although in some years there were distinct clusters of similar stations, e.g. severely impacted stations 15–17, these were often at low levels of similarity, e.g. <50%. It appears that similarities between and within stations changed greatly from year to year. This is a well-known phenomenon. It was first shown by Petersen and Jensen (1911) and has been expressed quantitatively for the Rupert Inlet area (Burd, 1999) as a normal similarity level of only 50–70%, i.e. at any one station from year to year. Stations on similar habitats, e.g. shallow stations with little or no tailings, e.g. Stations 1, 8, 11, 21, 22, etc., had equally low levels of similarity in any one year.

## LARGE INFAUNAL SPECIES

The small sampling grab used in these surveys sample an area 0.05 m$^2$ to a depth of 5–10 cm, so cannot collect large, deeply burrowed or fast organisms. Fast organisms such as seabed-inhabiting fish tend to be sufficiently mobile that their growth arises from a broad area, hence they cannot be considered as characteristic of the area sampled by grab. In contrast, large sedentary organisms, whether at the surface of the sediments or burrowed, are a product of the sampling site. These include such organisms as large clams, e.g. the commercial geoduck *Panope generosa*, and holothuria of several genera.

Records of large species seen on and off tailings during submersible dives have been documented by Ellis and Heim (1985). About 20 species of fish and invertebrates were seen on the tailings, both mobile and sedentary forms.

During the final years of surveys, particularly from 1993, records were kept of mollusc sizes to determine the largest forms collected by the 0.05 m$^2$ grab. Generally molluscs to ~3 cm in length were collected. Volumes of sediment collected by the grab were also recorded (small grabs were rejected from the collections) to estimate the depth of penetration. In general grabs of 5 litres (10 cm penetration) were available, except on the compacted sands at Stations 9 and 13 during the final years, from which lesser volume samples had to be accepted. Geoducks and some other molluscs, echinoderms and anthozoa can dig deeper than 10 cm. Fortunately, there is a way to estimate the presence of deeply burrowing species: through their growth stages after settling as larvae from the plankton. From 1993 the taxonomic laboratory processing the samples extended their earlier methods of identifying young stages of the infauna, to meet the requirement of determining whether large species were settling.

Young stages of selected large species include the two largest bivalve molluscs known to inhabit the Canadian west coast (*Panope generosa* and *Tresus nuttalli*), two other genera with large species (*Mya* and *Clinocardium*), the heart urchin *Brisaster latifrons* and Holothuria of several genera (*Molpadia, Eupentacta, Cucumaria*, etc.) (Table 7).

Juveniles of several of the larger organisms were collected, but not frequently. The two large bivalve molluscs, *Tresus* and *Panope*, were not collected at all. They may be present in the fjord system, but if so have not spawned with a widespread settlement, neither on nor off tailings from 1993–97. Occasional spawning, i.e. not annually, is not unexpected (Thorson, 1957). *Clinocardium* appears fairly regularly, including at Station 7 with some tailings deposition. *Compsomyax* is interesting in that it has appeared at

Table 7

Stations at which juveniles of selected species of large organisms were collected from 1993–1997

| Species | 1993 | 1994 | 1995 | 1996 | 1997 |
|---|---|---|---|---|---|
| *Panope generosa* | | | | | |
| *Tresus nuttalli* | | | | | |
| *Clinocardium* spp. | 21 | 13, 21, 22 | 12, 21 | 1, 7, 13, 21, 38 | 21 |
| *Mya* spp. | 39 | | | | |
| *Compsomyax subdiaphana* | | | 2, 12, 21 | 17 | 17 |
| *Brisaster latifrons* | | | 39 | | |
| *Holothuroidea* (several species) | 22 | | 13 | 23, 39 | 13, 22, 38, 39 |

Station 17, where there had been severe faunal losses to 1993. Station 13, also with severe losses to 1993, has also received settlements of *Clinocardium* and Holothuria.

It appears that the system of identifying juveniles, at least to genera, can monitor the settlement of organisms that are too large or too deeply burrowing to be collected by a small sampling device.

## TIME- AND COST-EFFECTIVENESS OF THE BENTHOS SURVEYS

Benthic surveys involving the collection of many thousand specimens, each of which is identified to species and numbers of which are accumulated to give total counts, are time-consuming hence expensive. Also, trained identifiers are in short supply; so are experienced identifiers who can quality control the less experienced. The initial surveys at ICM from 1970–1972 attempted to restrain costs by using only one replicate at each sampling station for full identifications, while the other two replicates were identified to high order taxonomic groups only, e.g. molluscs, echinoderms, etc. Biomass measures (weights of organisms) were also made. At the time, this was a reasonable approach for monitoring impact of an industrial discharge, especially in view of the shortage of species identifiers (which remains to the present).

From 1973 there was an attempt to speed analyses so as to provide results to the mine operators within a few weeks of the surveys (Table 8). Also there was a wish to increase time/cost-effectiveness, i.e. to speed obtaining results and to reduce costs while still getting sufficient measures of impact. This was at a time when the complexity of the seabed biodiversity reaction to the tailings discharge was not realised (it was initially concluded—incorrectly—that under tailings the benthos was obliterated and was not recolonizing). From 1973–1977, full species identifications were stopped, and all three replicates were merely sorted and counted to high order taxonomic groups only, thus giving an insufficient number of categories to accurately represent biological diversity. Also 20 large specimens were extracted for a minimal biomass estimate. Part of the objective was to give an indication of biomass changes, and also a count of numbers of molluscs. It had been documented by 1972 (ICM, 1973) that reductions in the numbers of molluscs showed impact of the tailings. Again at the time it seemed a reasonable approach.

In 1977 the problems for station and year comparisons created by these attempts to increase time/cost-effectiveness were realised. From 1977, the traditional approach to identifying all specimens (of all replicate samples) to species was finally implemented. Subsequently in 1987 biomass determinations were stopped, as it was then considered that biomass measures were linked to calculations of benthic productivity, and the objective of monitoring the benthos was to determine standing crop, not productivity. Productivity was undoubtedly affected in some ways

Table 8

Stages in the development of time- and cost-effective benthic surveys at Island Copper Mine (three replicate samples collected at each sampling station)

| Time period | Comment |
|---|---|
| 1970–1972 | Full species sort, identification and counts for 1 replicate only. Two replicates sorted, identified and counted to high order taxonomic groups only, e.g. molluscs, polychaetes, etc. Total weights recorded as a measure of sample biomass |
| 1973–1976 | No samples fully sorted, identified to species and counted. All three replicates sorted, identified and counted to polychaetes, molluscs and others. Additional sorting provided 20 of the largest specimens for a biomass measure. |
| 1977–1994 | All replicates fully sorted, identified to species and counted. Biomass measures stopped. |
| 1994 | Biomass measures restarted, for each species. |
| 1996–1998 | A rapid assessment protocol developed, tested and used exclusively in 1998. |

(Jones and Ellis, 1976) but the biodiversity counts are clearer indicators of impact and recovery. Weighing the specimens in bulk also damaged them making species identification less accurate.

From 1994, biomass determinations were reinstated, by weighing separated species after identification to provide a means, if needed, of assessing the relative importance of large organisms (Burd, 1991).

Finally in 1996, a new attempt to increase time/cost-effectiveness was attempted (Ellis and Macdonald, 1998), and was shown to be effective, provided an experienced species identifier was available to be on site at the time of the surveys. The procedure involves making a sort to apparent species of the most revealing sampling stations, and using these counts for a first analysis. Later, the other samples can be processed if needed, and the apparent species designations can be confirmed or modified. The ICM specimens are available for further study, having been deposited in the taxonomic collections of the Royal British Columbia Museum, Victoria, and the Los Angeles County Natural History Museum. Thus the ~30 years of monitoring at ICM show how the time/cost-effectiveness of benthic surveys can be maximised, while retaining the ability for accurate species separation and counts.

We know now that for benthic monitoring, full species separation and counts are essential, and that the most accurate way to achieve this is by species identification. Time/cost-effectiveness cannot be achieved by sacrificing these procedures, but can be achieved by providing an accurate preliminary sort based on use of experienced identifiers (who are in short supply). The procedure should not be tried without a skilled taxonomist with a broad identification background.

In summary, it was found that the best way to ensure accurate species richness data was to make a preliminary sort to apparent species, then to identify all specimens to species. The species identifications then allow obtaining accurate species evenness data. This procedure has another, logistical, advantage in that skilled species identifiers can be used only for identifications, and specimen sorting can be done by lesser trained personnel.

## THE CRAB FISHERY AND YIELDS

Dungeness crab (*Cancer magister*) was selected for monitoring, as a large epifaunal predatory benthos species that can be collected in a standardized fashion by a trap fishery. The species is also useful as a monitor of trace metal biomagnification up the seabed food-chain. The catch data are surrogates for the catches of the commercial fishery in the inlets, since the commercial crab catch data are not useful for monitoring as they apply to a much broader region than the three inlets. The other main fisheries present in the three inlets are the Pacific salmons (five species of the genus *Oncorhynchus*), but a monitoring fishery could not be conducted for these. As with the crab fishery, the commercial catch statistics of salmon are not useful due to the size of the local catch statistics area. Occasionally during the years of mine operations, a small gill-net fishery on pink salmon (*O. gorbuscha*) was permitted for a few days in any one year.

The crab monitoring fishery was conducted throughout the three inlets, Rupert, Holberg and Quatsino, four times yearly: in March, June, September and December at six sampling stations, two in each inlet. Two standard commercial traps were set for 18 hours (overnight) at each station. They were baited with fish offal, similarly to the commercial fishery. All specimens were identified to species (*C. magister* or *C. productus*), measured (carapace width), weighed and sexed. Three legal-size crabs were retained for trace metal analysis; the remainder were released. The commercial traps used are designed to allow small crabs less than 14 cm carapace width to escape. As a result, size data is biased, and is not considered here. The raw data are available in the annual environmental reports.

The most important results are visible in Fig. 4. In all years there were catches of crabs from the tailings areas in Rupert and Holberg Inlet. The commercial fishery continued throughout the inlets during mine operations, and since then. Rupert Inlet has catches intermediate in value between Holberg Inlet and Quatsino Sound. The catch data fluctuates from year to year, but not in a way that bears a relationship with mine operations, although initially the decline in Rupert Inlet from 1971–73 caused considerable alarm at the time.

The lowest value in Rupert Inlet in 1989 has no obvious cause, and was accompanied by low values throughout all the inlets that year. Similarly the low 1996–97 values are area-wide; hence not mine-related (mine closure was in 1995).

Trace metal monitoring has also been documented (ICM, 1972–1999). At no time was there indication of biomagnification of trace metals through the Dungeness crab pathway. The only clear expression of bioaccumulation in the fjord ecosystem was a doubling of copper and zinc levels in mussels (*Mytilus edulis*) on pilings of the dock for the concentrate freighters. These increases are most likely due to fugitive dust during loading operations.

In summary, the crab fishery was maintained during mine operations, and the mine's monitoring data show that crabs were available throughout all inlets with fluctuations from year to year unrelated to mine operations.

## CONCLUSIONS

The placement of mine tailings in the deep basin of a fjord over 25 years caused biodiversity losses at deposition rates of ~>20 cm per year, but wherever the thick tailings

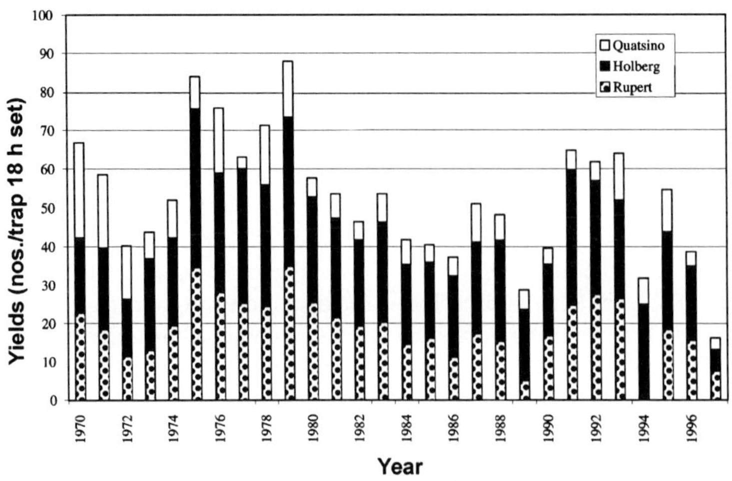


Fig. 4. Crab test fishery catches 1970–1997.

deposits stabilized (through the tailings density current moving elsewhere) a sustainable ecological succession was initiated within 1–2 years. After mine closure, the entire tailings target area had started a similar sustaining ecological succession; also within 1–2 years. The succession was initiated largely by a suite of about six polychaete worms, many of which remained abundant after the succession had progressed to the point where numbers of species were within the range of variation at reference stations. Two years after mine closure (1997) the expected biodiversity changes still to come within the sustaining ecological succession were: (1) an increase in the number of small crustacean species and other rare species in the siltier areas; and (2) increased numbers of sandy bottom species in the sandier areas. Commercial and test fisheries for Dungeness crab, *Cancer magister*, were maintained throughout mine operations and showed erratic fluctuations unrelated to tailings disposal. The presence of other large species, not collected as adults by the sampling grab used, was monitored by developing identification systems for juvenile specimens within the size range collected by the grab. Juveniles of several such large species were collected, some from tailings stations. Others, if present in the area, appeared not to have spawned and settled at either test or reference stations since the juvenile identification system was developed in 1993.

These results add to growing information that placement of mine tailings in a suitable silt-bottom target area deep in the sea can cause few resource losses, and those losses can be recovered within a few years. In principle, submarine tailings placement can be an effective means of tailings management at some coastal and island minesites.

## ACKNOWLEDGEMENTS

I greatly appreciate the co-operation that I have received over almost 30 years by staff of Island Copper Mine, and fellow academics on the universities' environmental review group for the mine; also the work of my graduate students at the University of Victoria, and student assistants at many levels, on whose results I have drawn. Staff and principals of several environmental consultancies have also provided much help over the years. My wife and professional colleague, Katharine Ellis, has provided editorial support and assistance with this and many other ICM reports and projects. Thank you all.

## REFERENCES

Anon. (1970) Public inquiry into the application from the Utah Construction and Mining Company. Rept. of the Inquiry, 2 volumes.

Burd, B.J. (1991) Quantitative and qualitative studies of benthic infaunal communities in British Columbia coastal waters. Ph.D. thesis, University of Victoria.

Burd, B.J. (1999) Post-mining recovery of infaunal benthos in Rupert Inlet. Part 2, Vol. III of ICM 1999.

Burd, B.J. and Ellis, D.V. (1994) ICM Closure Plan: Review of benthic surveys 1970 to 1992 for Rupert/Holberg/Quatsino inlet system. Report to ICM.

Caldwell, J.A. and Welsh, J.D. (1982) Tailings disposal in rugged, high precipitation environments: an overview and comparative assessment. In *Marine Tailings Disposal*, ed. D.V. Ellis, Ch. 1, pp. 50–62. Ann Arbor Science.

Ellis, D.V. (1967) Quantitative benthic investigations. II. Satellite Channel species data, February 1965–May 1967. Fisheries Research Board of Canada, Tech. Rept. No 35.

Ellis, D.V. (1968) Quantitative benthic investigations. V. Species data from selected stations (Straits of Georgia and adjacent inlets), May 1965–May 1966. Fisheries Research Board of Canada, Tech. Rept. No 73.

Ellis, D.V. (1975) Pollution controls on mine discharges to the sea. Proceedings International Conference on Heavy Metals in the Environment, Toronto, Canada, October 27–31, 1975. pp. 677–686.

Ellis, D.V. (1989) *Environments at Risk*. Springer. 329 pp.

Ellis, D.V. (1998) Ecological criteria for determining the sustainability of restoration. Proc. Helping the Land Heal Conference, Victoria, pp. 133–137

Ellis, D.V. and Heim, C. (1985) Submersible surveys of benthos near a turbidity cloud. *Marine Pollution Bulletin* **16** (5), 197–203.

Ellis, D.V. and Hoover, P.M. (1991) Benthos recolonizing tailings beds in British Columbian fjords. *Marine Mining* **9**, 441–457.

Ellis, D.V. and Macdonald, V.I. (1998) Rapid preliminary assessment of seabed biodiversity for the marine and coastal mining industries. *Marine Georesources & Geotechnology* **14** (4), 1–13.

Ellis, D.V., Pedersen, T.F., Poling, G.W., Pelletier, C. and Horne, I. (1995) Review of 23 years of STD: Island Copper Mine, Canada. *Marine Georesources & Geotechnology* **13** (1&2), 59–100.

Ellis, D.V. and Poling, G. (eds.) (1995) Submarine tailings disposal. *Marine Georesources and Geotechnology* **13** (1&2), 233 pp.

Ellis, D.V. and Robertson, J.D.R. (1999) Underwater placement of mine tailings: Case examples and principles. In *Environmental Impact of Mining Activities*, ed. J. Azcue, Ch. 9, pp. 123–141. Springer, 300 pp.

Ellis, D.V., Samoszynski, R. and Jones, A.A. (1991) Re-analysis of species associational data using bootstrap significance tests. *Water, Air and Soil Pollution* **59**, 347–358,

Gaston, K.J. and Spicer, J.I. (1998) *Biodiversity. An Introduction.* Blackwell Science, UK, 113 pp.

ICM (1972–1999) Annual Environmental Reports, 1–4 volumes annually. Reports by Island Copper Mine.

ICM (1980) A study of the suspended sediment movement through Quatsino Narrows. Environmental Department, Island Copper mine.

Jones, A.A. (1974) Effects of mine tailing on benthic infaunal composition in a B.C. inlet; with special reference to sampling, instrumentation and the biology of *Ammotrypane aulogaster* (Polychaetea; Opheliidae). M.Sc. thesis, University of Victoria.

Jones, A.A. and Ellis, D.V. (1976) Sub-obliterative effects of mine-tailing on marine infaunal benthos. *Water, Air and Soil Pollution* **5**, 299–307.

Holme, N.A. and McIntyre, A.D. (1984) *Methods for the Study of Marine Benthos*. IBP Handbook 16, Blackwell, UK, 387 pp.

Marcus, J.J. (1997) Closing BHP's Island Copper Mine. *Engineering & Mining Journal*. February 1997.

Nemec, A.F.L. and Brinkhurst, R.O. (1988) Using the bootstrap to assess statistical significance in the cluster analysis of species abundance data. *Canadian Journal Fisheries and Aquatic Science* **45**, 965–970.

Newell, R.C., Seiderer, L.J. and Hitchcock, D.R. (1998) The impact of dredging works in coastal waters: a review of the sensitivity to disturbance and subsequent recovery of biological resources on the sea bed. *Oceanography and Marine Biology. Annual Review* **36**, 128–178.

Pearson, T.H. and Rosenburg, R. (1978) Macrobenthic succession in relation to organic chemical enrichment and pollution of the marine environment. *Oceanography and Marine Biology. Annual Review* **16**, 229–311.

Pedersen, T.F. (1985) Early diagenesis of copper and molybdenum in mine tailings and natural sediments in Rupert and Holberg Inlets, British Columbia. *Canadian Journal Earth Sciences* **22** (10), 1474–1484.

Petersen, C.J.G. and Jensen, P.B. (1911) Valuation of the Sea. I. Animal life of the sea bottom, its food and quantity. Rept. Danish Biological Station No. 20.

Taylor, L.A. (1986) Marine Macrobenthic Colonization of Mine Tailings in Rupert Inlet, British Columbia. M.Sc. thesis, University of Victoria. 155 pp.

Thorson, G. (1957) Bottom Communities (sublittoral or shallow shelf). *Memoirs Geological Society of America* **67**, 461–534.

Wu, R.S.S. (1982) Effects of taxonomic uncertainty in species diversity indices. *Marine Environmental Research* **6**, 215–225.

## THE AUTHOR

**Derek V. Ellis**

*Biology Department, University of Victoria,*
*P.O. Box 3020, Victoria, B.C., V8W 3N5, Canada*

Chapter 124

# MARINE ANTIFOULANTS

Stewart M. Evans

Effective antifouling brings enormous economic benefits to the shipping industry. There are also indirect environmental benefits. Ships with clean hulls use less fuel, reducing emissions of 'greenhouse' and 'acid rain' gases, and they also provide less opportunity for 'invasive' species to hitch-hike across the world's oceans. The use of tributyltin (TBT), first in free association paints and then in self-polishing copolymers (TBT SPC), were breakthroughs for the paint industry—they are the most efficient antifoulants ever devised. Unfortunately, TBT leaches from the paints into the water column and caused harm to some non-target organisms, particularly during the period when it was used solely in free association paints. The best-documented impacts are on cultivated oysters *Crassostrea gigas* in west France and dogwhelks *Nucella lapillus* in Britain, but similar impacts were described worldwide. The use of TBT-based antifoulants was regulated in several countries during the 1980s and early 1990s. In most cases, their use was prohibited on small vessels (<25 m in length) which were believed to be the main cause of contamination in coastal waters. These regulations have been effective in reducing ambient levels of TBT in the water column, sediments (although less so) and in the tissues of marine organisms. Open waters are now largely free of contamination at biologically harmful levels but commercial harbours, especially those with dry docks, are still hot-spots of contamination. Nevertheless, even in these cases, impacts are localised. The International Maritime Organisation is likely to ban the use of TBT-based paints altogether in the near future but there is a danger that this measure will be introduced prematurely. The ban should not be enforced until alternative antifoulants are available which perform as well as TBT-based coatings on economic and environmental cost–benefit analyses.

*Seas at The Millennium: An Environmental Evaluation (Edited by C. Sheppard)*

## INTRODUCTION

Man-made structures which are immersed in the sea, such as the cages used in fish farms, oil rig supports and hulls of ships, become fouled by growths of marine organisms. These include seaweeds, barnacles, hydroids and molluscs. The shipping industry, in particular, suffers because fouled ships travel more slowly through the water, are less manoeuvrable and use more fuel than those with clean bottoms. The extent of fouling can be enormous. An unprotected vessel hull may gather 150 kg of fouling per square metre in less than six months at sea. This amounts to 6000 tonnes of biological matter on a large ship which has an underwater surface area of 40,000 $m^2$.

Not surprisingly, successful antifouling protection is a matter of high priority. Ship owners must suffer not only higher fuel bills, but also the costs associated with slipping and re-painting. Fouling is also the indirect cause of environmental concerns. Higher fuel consumption of fouled ships results in increased emissions of carbon dioxide and sulphur dioxide, which contribute to 'greenhouse gases' and acid rain, respectively. An additional problem is that fouled hulls present an opportunity for so-called invasive species to hitch-hike their way across the seas. Some of them have disastrous ecological and economic impacts once they become established in 'foreign' habitats. For example, the slipper limpet *Crepidula fornicata*, which was introduced into Britain from North America during the 19th century, is now a serious pest of oyster beds, introduced Zebra mussels have caused enormous ecological and economic harm in the Great Lakes of North America, and an invasive polychaete *Sabella spallanzanii* is smothering cultivated scallops in Port Philip Bay, Australia. It was thought that ballast water was the means by which such organisms, or more usually their larvae, were carried, but it is now evident that fouled hulls are also a means of transport, probably an even more common one (Eno et al., 1997).

In the past, antifouling practices on wooden hulls included coating them with pitch, lead or arsenical compounds, and sheathing them with copper. However, these materials were no longer appropriate on vessels with steel hulls when they were introduced in the late 18th century. Antifouling paints first became available for use on them in the mid-19th century, 60 or so years later. The large majority of these paints incorporate biocides into the matrix which leach slowly, killing organisms which attempt to settle on the ship's hull. Cuprous oxide was used in early formulations but numerous other toxins, including organo-mercury, lead and arsenical compounds, and biocides, such as DDT, were added as boosters. They were needed to ensure that the paint was effective against the full range of fouling organisms. However, these compounds were potentially harmful to the environment and human health and their use was discontinued in the 1960s. They were replaced largely by the synthetic organotin, tributyltin, or TBT, as it is commonly known. Paints incorporating this highly toxic substance are the most effective antifoulants ever devised, and they have dominated antifouling practices for the past three decades. TBT-based antifoulants were adopted by most major shipping companies during the 1970s, eventually taking some 70–80% of the market share for commercial vessels. They were also used widely by small boat owners, including yachtsmen and fishermen. TBT is not a chemical compound in its own right but a moiety which is part of a tributyltin compound. It exists in various guises, such as bis-tributyltin oxide, tributyltin fluoride and the copolymer (poly)tributyltinmethacrylate-methylmethacrylate. TBT, usually bis-tributyltin oxide, was incorporated into so-called free association paints in the first formulations. It was dispersed in a soluble resinous matrix and was released as this dissolved slowly in water. Unfortunately, the release rates from these paints were uncontrolled. The initial rates were high so that paints were effective when they were first applied (Fig. 1). However, the biocide in them was virtually exhausted within 12 or so months and vessels coated with these free association paints needed dry-docking and repainting every one or two years. This was obviously expensive for the ship owners but such high release rates of TBT were also bad for the environment. TBT leached from them into the water column, contaminating coastal waters. The use of free association paints, containing TBT, left a pollution legacy from which the paint and chemical industries are still suffering.

However, there was a breakthrough in paint technology with the introduction of self-polishing copolymer paints in the late 1960s, although it took a decade or so before they were in widespread use. The biocide was bound into an insoluble copolymer resin matrix. In the case of TBT-based paints, this was via an organotin-ester linkage. The biocide was released slowly as this linkage was hydrolysed on contact with water. This occurred at the surface of the paint only and, as a consequence, successive layers of biocide were released in a controlled fashion as the paint surface was progressively worn away (Fig. 2). The advantage was that biocide continued to be released, and therefore acted as an effective antifoulant, until there was no paint left. These self-polishing TBT systems brought huge benefits to the shipping industry. They provided antifouling cover for five or more years, more than doubling the performance of free

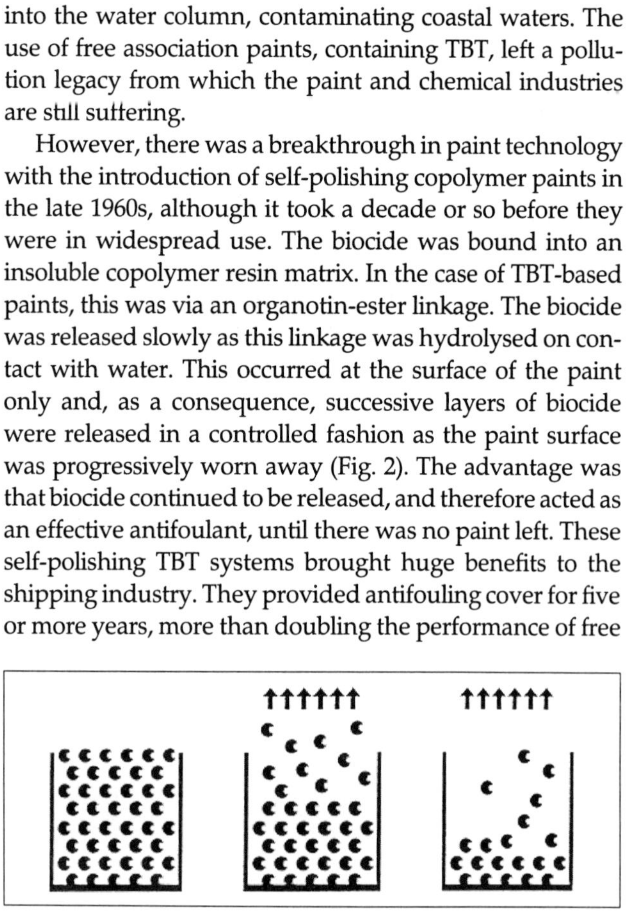

Fig. 1. The mechanism of free association paints. The biocide leaches from the resinous matrix at a rapid and uncontrolled rate. (Bennett, 1996).

## TBT: An Endocrine Disruptor

Some PCBs, pesticides and non-halogenated compounds, which are released into the marine environment as a result of human activities, are known to disrupt endocrine activity. Recent examples include feminization of male fish either by exposure to industrial discharges containing the oestrogen mimic nonylphenol, or by natural or synthetically made oestrogenic hormones in sewage effluents (Harries et al., 1997). Endocrine disruption was also the probable cause of reproductive failure in alligators in Lake Apopka, Florida following a spillage of the insecticide dicofoil in 1980. However, endocrine systems of invertebrates are less well-known and, not surprisingly therefore, unequivocal examples of their disruption have been lacking. TBT-induced masculinization of female gastropods (i.e. the development of imposex) is therefore of particular interest. It is the clearest example of invertebrate endocrine disruption to date (Matthiessen and Gibbs, 1998) and its link to a specific environmental pollutant (TBT) guarantees its place in future generations of marine ecology and physiology textbooks.

The causal link between imposex and TBT has been established in several gastropods but has been particularly well-studied in the dogwhelk *Nucella lapillus*. The first indication that TBT was responsible for the development of imposex in *N. lapillus* came from the observation that it was most advanced in populations from shores which were adjacent to centres of boating or shipping activity, such as marinas or harbours. These were also hot-spots of TBT contamination. Subsequently it was established that there were close dose–response relationships between tissue burdens of TBT and the development of the condition. Field and laboratory experiments confirmed that TBT and imposex were connected. Imposex could be induced by transplanting female dogwhelks from 'clean' shores to TBT-contaminated ones. It could also be induced by exposing dogwhelks to very low doses of these substances in the laboratory. Concentrations of < 1 ng/l were sufficient to trigger the growth of a penis and vas deferens but not enough to impair breeding performance. Higher concentrations of between 5 and 8 ng/l caused sterility in some females.

Spooner et al. (1991) demonstrated that there were changes in the hormone balance in female dogwhelks which were exposed to TBT. Concentrations of the male hormone testosterone became elevated, while those of the female hormones progesterone and 17β-oestradiol remained unchanged. Heightened levels of testosterone were evidently the cause of imposex because the severity of the condition was related to the degree of testosterone elevation. Furthermore, direct injections of testosterone stimulated the development of imposex.

There is strong evidence that TBT causes the build-up of testosterone by inhibiting the cytochrome P450-mediated aromatase systems which is responsible for converting testosterone to 17β-oestradiol in the healthy animal. This hypothesis is supported by the observation that exposure of dogwhelks to a steroid aromatase inhibitor (which acts in the way proposed for TBT) caused the development of imposex. Additional support came for an experiment in which dogwhelks were exposed to a mixture of TBT and the anti-androgen cyproterone acetate. The latter inhibits the build-up of testosterone and it should therefore cancel out the effects of TBT. Significantly, the dogwhelks subjected to the mixture did not develop imposex.

### REFERENCES


Harries, J.E., Sheahan, D.A., Jobling, S., Matthiessen, P., Neall, P., Sumpter, J.P., Tylor, T. and Zaman, N. (1997) Oestrogenic activity in five United Kingdom rivers detected by measurement of vitellogenesis in caged male trout. *Environmental Toxicology and Chemistry* **16**, 534–542.

Matthiessen, P. and Gibbs, P.E. (1998) Critical appraisal of the evidence for tributyltin-mediated endocrine disruption in mollusks. *Environmental Toxicology and Chemistry* **17**, 37–43.

Spooner, N., Gibbs, P.E., Bryan, G.W. and Goad, L.J. (1991) The effect of tributyltin upon steroid titres in the female dogwhelk, *Nucella lapillus* and the development of imposex. *Marine Environmental Research* **32**, 37–48.


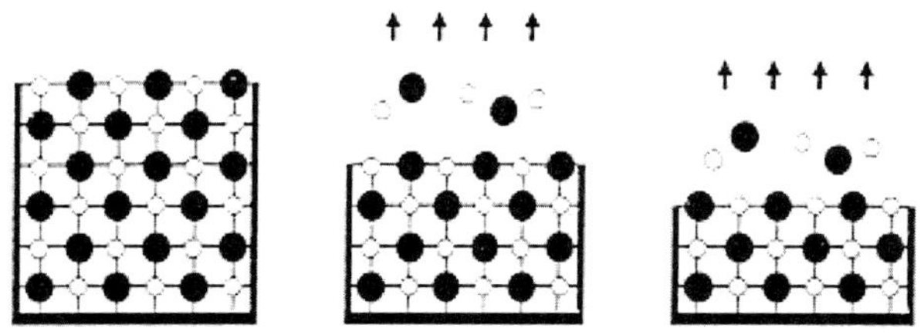

Fig. 2. The mechanism of TBT self-polishing copolymer paints. Seawater hydrolyses the TBT copolymer bond at the surface of the paint. The biocide is released slowly and at a controlled rate. A uniform anti-fouling performance is achieved throughout the life of the paint (Bennett, 1996).

association paints. Their slower TBT release rates meant that they were also less damaging to the environment than their predecessors.

## THE PERSISTENCE OF TBT IN THE ENVIRONMENT

The ideal antifouling biocide is one which is highly toxic at the surface of the ship's hull but degrades immediately on its release into the water column. In fact, TBT performs reasonably well in the water column, where it is degraded first to less toxic dibutyltin (DBT), then monobutyltin (MBT) and finally inorganic tin. Degradation is due predominantly to microbial action and, to a lesser extent, photolysis but the precise rate depends on environmental factors, such as pH, temperature and turbidity. In one study, the estimated half life of TBT in turbid waters in summertime was 5.8 days, and that in non-turbid waters in winter was 127 days.

Unfortunately, TBT becomes adsorbed onto particles in the water, and is much more stable when these aggregate in sediment. Estimates of its half life in sediment vary but it is in the order of a few months in the upper layers of oxic sediment to several years in the deeper anoxic layers. TBT therefore tends to accumulate, and occur at high concentrations, in deep sediments of harbours and marinas, where there are continuous inputs of TBT. Severe contamination was first recorded in these pollution hot-spots, during the 1970s and 1980s when there was no regulatory control of the TBT-based antifoulants.

TBT also accumulates in the tissues of marine organisms. This is because its molecules have both lipophilic and ionic properties, encouraging it to accumulate in lipids and bind to macromolecules, such as glutathione. For instance, it is known to bioaccumulate in bacteria, phytoplankton, molluscs, crustaceans and fish. However, there is no clear pattern of susceptibility; bioconcentration factors vary enormously in different species from <1 to 30,000. Nor is the route of uptake well-established, although greater bioaccumulation in the crab when TBT contamination occurs via its food than via the water column, raises the possibility that biomagnification can occur through the food chain.

TBT is toxic at low concentrations to a wide range of marine organisms. Larvae are particularly sensitive to it. For example, the lethal dose concentration of mussel *Mytilus edulis* larvae in 15-day LC50 tests was 10 ng/l (Beaumont and Budd, 1984), that of oyster larvae *Crassostrea gigas* in 48-h LC50 tests was 16 ng/l (Thain, 1983), and that of sole *Solea solea* larvae in 96-h LC50 tests was 21 ng/l (Thain, 1983). However, sub-lethal effects occur at even lower concentrations. Alzieu (1998) summarised no observable effect levels for invertebrates as follows:

- concentrations of <1 ng/l can cause imposex in female gastropods;
- concentrations >1 ng/l limit cell division in diatoms and reproduction in zooplankton;
- concentrations >2 ng/l are responsible for shell calcification anomalies in oysters;
- concentrations >20 ng/l cause disturbances in the reproduction of molluscs;
- concentrations of <500 $\mu$g/l result in disturbed crustacean exuviation.

There have also been concerns about the accumulation of TBT in vertebrates. There have been a number of studies which have identified TBT in the tissues of fish, seabirds and sea mammals, including porpoises and whales. Kannan et al. (1997) suggested that organotins, together with elevated levels of PCBs, may have contributed to mortalities of dolphins by suppression of the immune system. However, the relationship is controversial and unproven (Green et al., 1997). Furthermore, risk assessments of the oystercatcher, as a representative bird, and the sea otter, as a representative mammal, both of which have diets which would put them at high risk, suggest that current environmental levels of TBT do not pose a threat to top predators in the system (Cardwell et al., in press).

## THE ENVIRONMENTAL IMPACTS OF TBT

It is not surprising that TBT had serious environmental impacts during the period of unregulated use of free association paints. The best documented example is from Arcachon Bay, which is located along the French Atlantic coast halfway between the Gironde estuary and the Spanish border (Alzieu, 1998). The Bay opens to the Atlantic Ocean via a narrow sandy inlet. Shellfish farms occupy about 1000 hectares in the Bay, and are responsible for the production of some 10–15,000 tonnes of oysters in normal years. However, the farms are surrounded by ten marinas which have a mooring capacity of up to about 15,000 boats in summer. The problem first became apparent in the mid-1970s. The oysters were suffering from poor spatfall, and therefore low reproductive success, and surviving individuals showed growth abnormalities. They had thickened shells, which contained characteristic chambers, and their growth was stunted. In the most acute cases, this resulted in misshapen, ball shaped oysters, which were unmarketable. Production dropped to 3000 tonnes in 1981, representing a loss of about 880 million French francs to the industry. Subsequent experimental work confirmed that TBT was to blame. Ambient concentrations of TBT in the water column were mostly above lethal doses for oyster larvae, and the growth abnormalities of adults could be induced by exposure to TBT.

It became apparent a few years later, during the mid-1980s, that TBT was also having a severe impact on populations of dogwhelks *Nucella lapillus* in southwest England. Female dogwhelks there had developed the condition known as imposex in response to TBT poisoning. It is characterised by the development of male characters, including a penis and vas deferens, which become superimposed on the female's genitalia (Fig. 3). In extreme cases, the vas deferens forms a nodule which occludes the genital pore, preventing the release of egg capsules and rendering the female sterile. Imposex is a dose-related response, and has been developed as a biological indicator of TBT contamination (Gibbs et al., 1987). Two indices have been particularly useful: the Relative Penis Size Index (RPSI) which compares the size of the female penis with that of the male as a standard; and the Vas Deferens Sequence Index (VDSI) which assesses the development of imposex on a six-point scale. The condition was advanced in populations of *N. lapillus* in areas of intense boating/shipping activity in the Fal Estuary, Dart Estuary and Plymouth Sound (Bryan et al., 1986) (Fig. 4). The worst affected populations were totally sterile and inevitably died out, although there were,

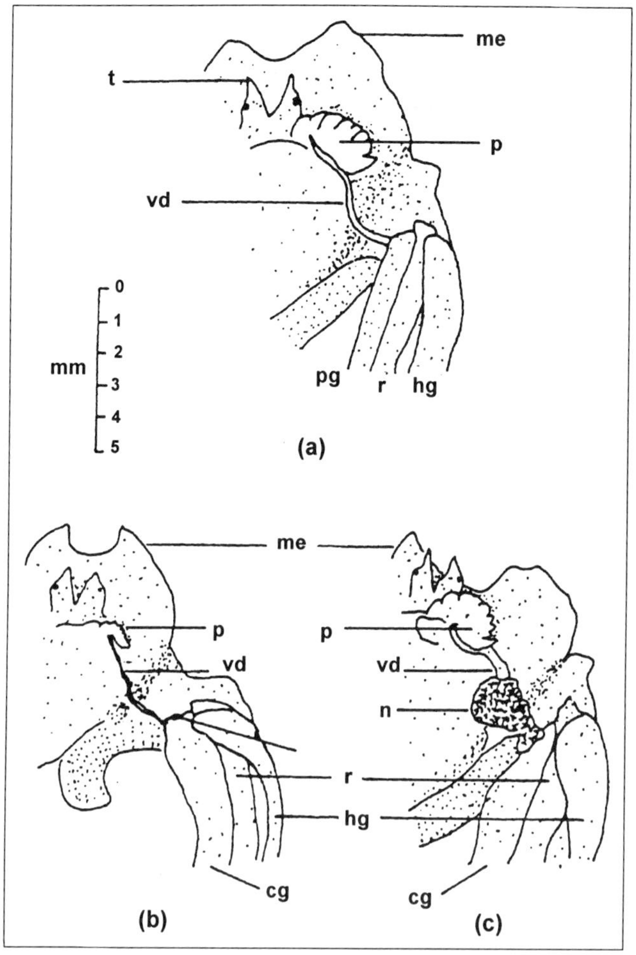


Fig. 3. Imposex in the dogwhelk *Nucella lapillus*. (a) Dorsal view of a dissection of the mantle cavity of a male; (b) female with intermediate imposex; and (c) female with severe imposex. In (c) the vas deferens has formed a nodule blocking the genital pore and rendering the female sterile. Abbreviations are as follows: cg, capsule gland; gp, genital papilla; hg, hypobranchial gland; me, mantle edge; n, nodule; p, penis; pg, prostate gland; r, rectum; t, tentacle; vd, vas deferens. (After Gibbs, P.E. and Bryan, G.W. (1986). Reproductive failure in populations of the dog-whelk, *Nucella lapillus*, caused by imposex induced by tributyltin from antifouling paints. *Journal of the Marine Biological Association of the United Kingdom*, **66**, 767–777.)

and still are, fecund populations along open coasts between them. Again, TBT was confirmed as the culprit. There were high environmental concentrations of TBT in these areas, and the severity of imposex in dogwhelks from them correlated with TBT tissue burdens. Furthermore, it was shown that imposex could be induced in laboratory experiments by exposing dogwhelks to leachates from anti-fouling paints or injecting them with TBT and, in field experiments, by transplanting them from 'clean' to TBT-polluted sites.

Subsequently, it became clear that these abnormalities were having impacts on reproductive performance of molluscs at locations worldwide. Oyster culture, for example, was having problems in southern England, New Zealand and Australia. At the same time, imposex was described in populations of *N. lapillus* throughout Europe and, in other species of whelks, in North America, Malaysia, Singapore, Indonesia, West Africa, Japan, New Zealand and Australia. TBT contamination had become a global problem (Ellis and Pattisina, 1990).

## LEGISLATIVE CONTROL OF TBT-BASED ANTIFOULANTS

There was a clear need to regulate the use of TBT-based antifoulants and the French Government, motivated by the need to protect its oyster industry, was the first to react. Pleasure boats were believed to be the main source of contamination in coastal waters, especially in Arcachon Bay, and restrictions were aimed primarily at them. France initially prohibited the application of organotin-based paints on boats <25 m in length. The majority of industrialised nations, including UK (1987), USA (1988), Canada, New Zealand and Australia (1989) and the European Community (1991), adopted similar regulations soon afterwards. With the exception of Japan, which banned the use of

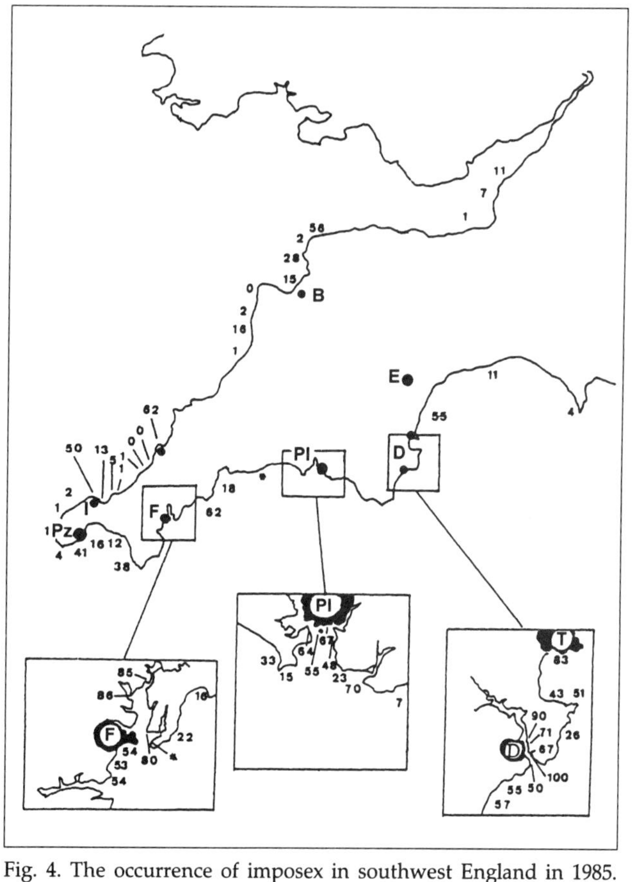


Fig. 4. The occurrence of imposex in southwest England in 1985. Figures indicate the Relative Penis Size Index (RPSI); the higher this index, the more severe the condition. Major cities or ports are indicated as follows: E = Exeter, T = Torquay; D = Dartmouth; Pl = Plymouth; F = Falmouth; Pz = Penzance; I = St. Ives; N = Newquay; B = Bideford. Note the high RPSI scores close to the major ports/harbours. (Data from Bryan, G.W., Gibbs, P.E., Burt, G.R. and Hummerstone, L.G. (1986). The decline of the gastropod *Nucella lapillus* around south-west England: evidence for the effects of tributyltin anti-fouling paints. *Journal of the Marine Biological Association of the United Kingdom*, **66**, 611–640.)

# Genetic Aspects of Imposex: the Dumpton Syndrome

TBT contamination on shores adjacent to pleasure boat centres and commercial harbours was sufficiently high during the late 1980s to cause sterility, and subsequently the local extinction of populations of the dogwhelk *Nucella lapillus*. Recolonisation has occurred on relatively few of these shores, in spite of much reduced levels of contamination during the past decade. This is because *N. lapillus* has direct development and, in the absence of a dispersive larval stage in its life history, invasion of 'new' habitats or recolonisation of 'old' ones, is an unusual occurrence.

The species has become extinct along most of the north coast of Kent, bordering the Thames Estuary (Fig. 1). *N. lapillus* was abundant on suitable shores there in pre-TBT times but it is no longer present, with the exception of the area of shoreline from North Foreland to Dumpton Gap. The species is common at Dumpton and, to a lesser extent, on immediately adjacent shores. The survival of this isolated enclave is an anomaly but it has now been related to a genetic defect, which affects the development of the genital system. It causes carrier males to become sterile, but conversely it inhibits the development of imposex in carrier females so that they remain fertile in areas of high TBT contamination (Gibbs, 1993).

It was noticed that imposex in the Dumpton population was unusual in the sense that there were some females with advanced imposex but others with mild symptoms or no imposex at all (Fig. 2). This contrasted with the typical situation in which all females in a population tended to be at similar stages in the development of the condition. Thus, VDSI stages would occur in clusters so that all females at a mildly contaminated site would be at stages 0 to 3 or 4. Alternatively, those at a highly polluted site would all be at stages 4–6, or even 5–6. A further difference was that some of the Dumpton males were aphallic, and might also have abnormal vas deferens and prostate glands.

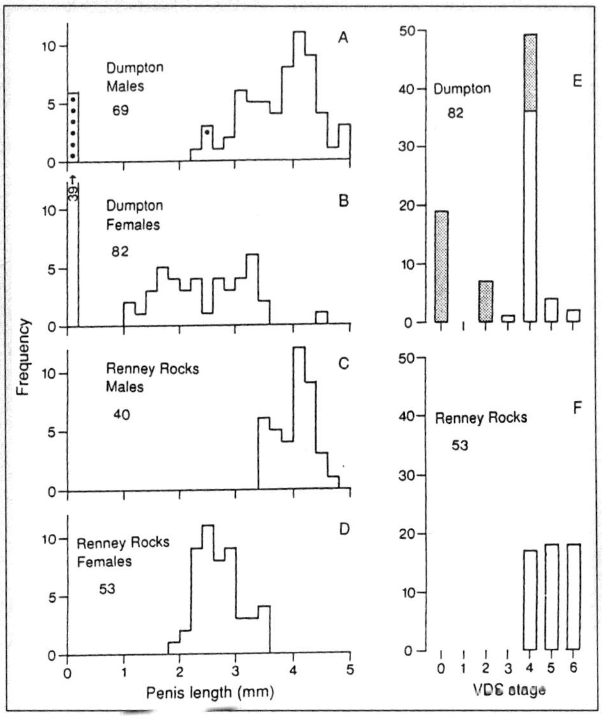


Fig. 2. A comparison of the development of imposex in the populations of dogwhelks *Nucella lapillus* at Dumpton Gap and Renney Rocks (Plymouth Sound). There was relatively high TBT contamination at both locations. The Dumpton Gap population includes aphallic males (i.e. with the genetic defect) and females with imposex but no penis development. There were no aphallic males or females in the Renney Rocks population. Males with split prostate glands are indicated by •. In the comparison of VDSI, hatched areas indicate aphallic females. (Gibbs, 1993.)

The genetic defect, the inability to form a penis, and probably other parts of the male genital system, is now known as Dumpton Syndrome and is a permanent feature of the Dumpton population. It is not caused by environmental conditions at Dumpton Gap (and is therefore genetic) because the same characteristics occur in individuals bred under laboratory conditions. Males carrying the defect are incapable of reproducing. However, there are always some non-carrier fertile males in the natural population. It also includes some fertile females because the development of male genitalia, and therefore imposex, is inhibited in female carriers. These females remain fertile therefore at levels of TBT contamination which have been sufficient to cause sterility in neighbouring populations. The Dumpton population has been able to survive because, even during periods of high TBT pollution, non-carrier males have been able to mate successfully with carrier females.

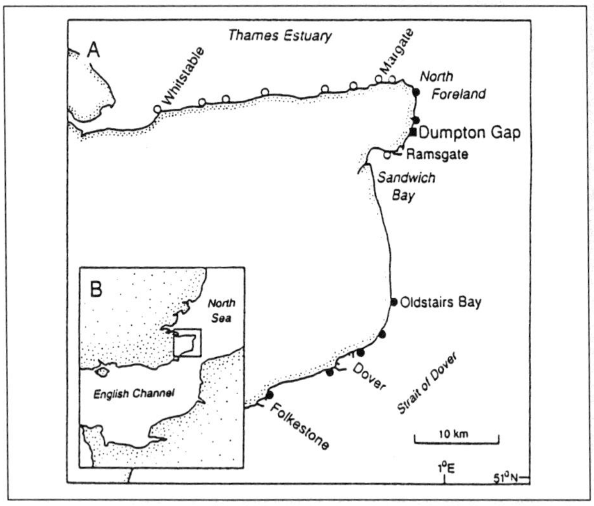


Fig. 1. (A) Map of south-east England indicating the presence or absence of dogwhelks *Nucella lapillus* in 1987/1989 at stations which had been sampled previously before TBT-based paints were widely used (pre-1975). The site of Dumpton Gap is shown by a filled square. (B) Location map of south-east England. (Gibbs, 1993).

## REFERENCES

Gibbs, P.E. (1993) A male genital defect in the dog-whelk, *Nucella lapillus* (Neogastropoda), favouring survival in a TBT-polluted area. *Journal of the Marine Biological Association of the United Kingdom* **73**, 667–678.

TBT-based paints altogether, large commercial vessels, predominantly ocean-going craft, were still permitted to use TBT-based coatings. Nevertheless, some countries introduced refinements to the regulations. These included the permitting the application of paints with slow release rates only (usually 4 $\mu$g TBT/cm$^2$/day), the compulsory registration of antifouling paints, the provision for water quality monitoring and the establishment of environmental quality standards (EQS). The UK, for example, has an EQS of 2 ng/l in the water column, while the USA has one of 10 ng/l.

## THE EFFECTIVENESS OF REGULATIONS: MEASURING AND MONITORING TBT IN THE ENVIRONMENT

There is an obvious need to assess the extent to which TBT contamination has increased or declined since the introduction of regulations. There are, however, difficulties in measuring environmental levels of TBT, and of interpreting the results of monitoring programmes meaningfully (Foale, 1993). This is for at least three reasons. First, TBT can be effective at concentrations which are close to the limits of detection (1–2 ng/l). Analysis at these levels is not only plagued by problems, such as contaminated reagents or sampling jars, but it is also expensive. Second, TBT may be released into the environment in pulses (e.g. from dry docks), and biologically harmful doses may therefore be missed by periodic sampling. Third, the distribution of TBT in the natural environment is by no means uniform. The substance occurs at much higher concentrations in the surface microlayer and sediment than in the water column.

Largely as a result of these difficulties, biological responses of some organisms to TBT have been used widely as indicators of environmental levels of contamination. Imposex has been used, not only in *Nucella lapillus*, but in many of the >100 species of prosobranch mollusc in which the condition has been described. They include: *Nucella canaliculata, Nucella emarginata* and *Nucella lamellosa* (Bright and Ellis, 1989), *Cronia margariticola, Drupella rugosa, Morula musiva* and *Thais luteostoma* (Ellis and Pattisina, 1990), *Thais haemastoma* (Spence et al., 1990), *Lepsiella albo-marginata* and *Lepsiella scobina* (Smith and McVeagh, 1991), *Morula marginata* and *Thais orbita* (Wilson et al., 1993), *Lepsiella vinosa* (Nias et al., 1993), *Thais bronni* and *Thais clavigera* (Horiguchi et al., 1994) and *Buccinum undatum* (Hallers-Tjabbes et al., 1994). Abnormal growth of oysters has also been used as an indicator of TBT contamination. Key et al. (1976) used the cavity/depth ratio of shells as an indicator, Alzieu et al. (1986) developed the shell thickness index and His and Robert (1987) used shell density.

There is overwhelming evidence that the regulations have been effective in reducing TBT contamination. Monitoring programmes worldwide have reported: (i) reduced environmental levels of organotins; and (ii) recovery of populations of marine organisms from the earlier impacts

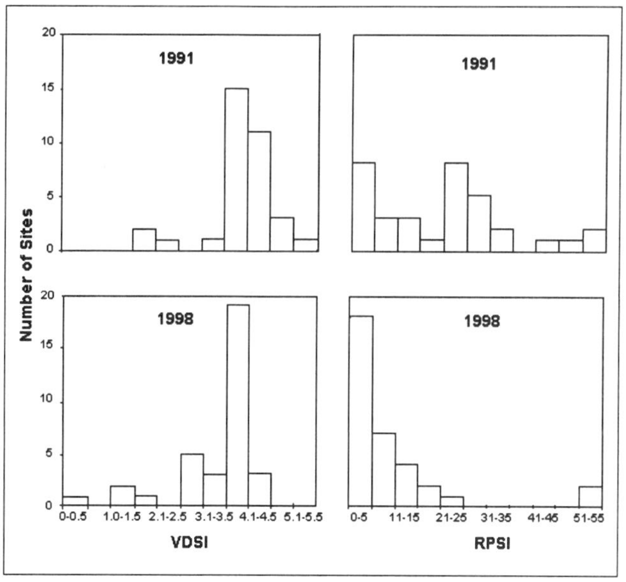


Fig. 5. Recovery of North Sea populations of dogwhelks *Nucella lapillus* from imposex. Scores reported for two indices: the Relative Penis Size Index (RPSI) and the Vas Deferens Sequence Index (VDSI). Data for 1991 are from Harding et al. (1992). Many RPSI were >25, and VDSI >4, indicating impairment of breeding in 1991. Imposex was poorly-developed, and dogwhelks were common or abundant on the same shores when they were re-visited in 1998. RPSI and VDSI measures were significantly lower than they had been in 1991. (Data for 1998 from Evans et al., 2000.)

of TBT. In some cases, the improvements have been almost immediate and dramatic. Oysters recovered from growth defects and harvests in Arcachon Bay returned to former levels of production within two years of the introduction of the 1982 restrictions in France. Similarly, populations of dogwhelks *Nucella lapillus* have shown widespread recovery in the North Sea and elsewhere (Fig. 5). Populations which were described as totally or partially sterile as a result of surveys in the early 1990s, were fecund and common or abundant less than a decade later.

Open seas are largely free of TBT contamination at biologically significant levels. They are sufficiently vast to dilute inputs to below no observable effect concentrations. Evans et al. (1998) found that imposex was mild or absent in *Nucella lapillus* from coastal areas, which were distant from shipping activity, around the rim of the North Atlantic, as it was in other species collected from open areas of the Azores, Pacific coast of North America, west Africa and Indonesia. TBT impacts might have been expected on shores adjacent to busy shipping lanes but Evans et al. (1998) found no evidence of elevated levels of imposex in populations of *N. lapillus* on shores close to lanes where they funnelled through Pentland Firth, the Minches and North Channel in northern Britain. The impacts of contamination was also low on open shores near enclosed bays and harbours in which there were high concentrations of TBT (Fig. 6). Chemical measures of TBT in most water samples of open seas are also low. Yamada et al. (1997) estimated that concentrations of TBT in oceanic waters were between

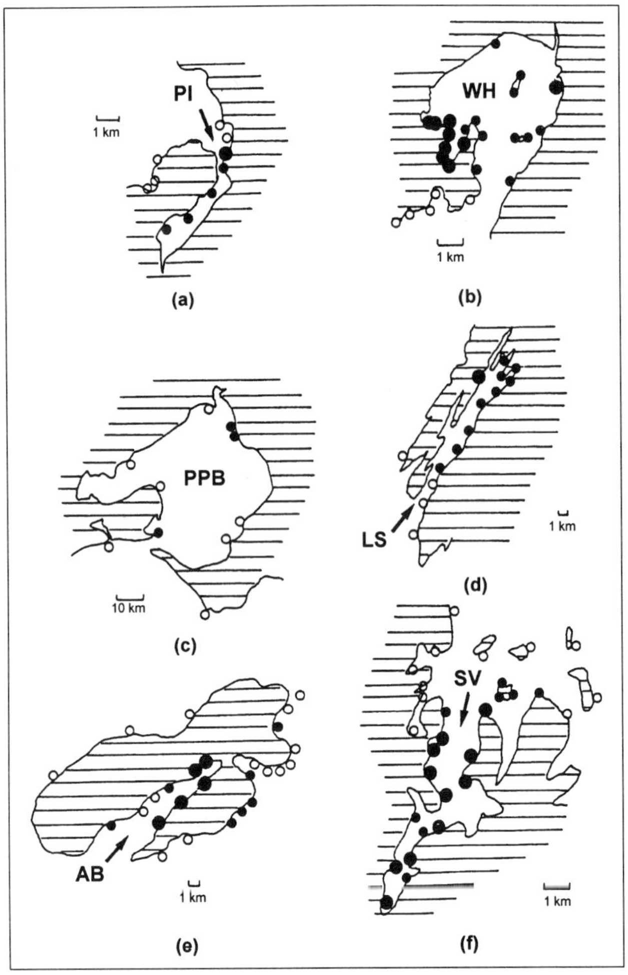


Fig. 6. Imposex in whelk populations in semi-enclosed bays, sealochs and harbours and adjacent open coasts. Data from (a): Porirua Inlet, New Zealand (Smith and McVeagh, 1991); Wellington Harbour (Smith and McVeagh, 1991); (c) Port Philip Bay, Melbourne (Foale 1993); (d) Loch Sween, Scotland (Davies et al., 1987); Ambon Bay, Indonesia (Evans et al., 1995): and Sullom Voe, Shetland Islands (Bailey and Davies, 1988). The measure used is the Relative Penis Size Index (RPSI): in (a), (b), (c), (d) and (f): ○, imposex mild or absent (RPSI < 5); •, RPSI 5–40; ●, RPSI > 40. In (e): ○, no imposex; •; imposex in < 25% of females; ●, either imposex in > 25% of females or females now absent. (Evans, S.M., Dawson, M., Day, J., Frid, C.L.J., Gill, M.E., Porter, J. and Pattisina, L. (1995). Domestic waste and TBT pollution in coastal areas of Ambon Bay (eastern Indonesia). *Marine Pollution Bulletin* **30**, 109–115.)

traces and 0.4 ng/l in the southern hemisphere and between traces and 0.8 ng/l in the northern hemisphere.

Reductions in contamination have not occurred in many commercial harbours. This is undoubtedly due to the fact that large vessels frequenting them continue to use TBT-based antifoulants. However, the impacts of these hot-spots of pollution are surprisingly localised. Ko et al. (1995) found that there were extremely high concentrations of TBT in sediments of two shipyards in Hong Kong but that they had decreased to background levels within 100 m of them. Gradients of diminishing concentrations of TBT, and their biological impacts, have also been described from several other parts of the world including Iceland, Puget Sound (USA) and Sullom Voe (Scotland) (Fig. 7).

## THE BAN ON TBT; FINDING SAFE ALTERNATIVE ANTIFOULANTS

The International Maritime Organisation is almost certain to introduce a total ban on the use of TBT-based antifouling paints within the next decade. This is not controversial in itself. In an ideal world, it is undesirable to introduce any toxic compound into the environment, and it must be the aim of research workers and regulators to eliminate the use of biocides in paints altogether. However, it is important that the ban on TBT is not introduced prematurely, before there are alternative antifoulants which are known to perform at least as well as TBT on economic and environmental grounds. Biocide-free 'non-stick' coatings, based on silicon elastomer technology, provide an exciting development, and a real possibility for the future. They provide smooth, low energy surfaces to which organisms either cannot adhere or from which they are easily dislodged. Unfortunately, there are reports that, while the non-stick paints which are currently available perform well on fast-moving vessels, such as catamarans, patrol boats and fast ferries, they are less efficient on slow-moving craft. This includes most ocean-going vessels. It seems inevitable therefore that the next generation of antifoulants will contain biocides. A range of different products is being developed by paint companies to exploit the market opportunity. Biocides which are incorporated into them include: zinc pyrithione, copper pyrithione, triazine, dichloro-isothiazolone and dichlorophenyl-dimethyl urea. In order to provide cover against the full spectrum of fouling organisms, most of them also include copper compounds as boosters, usually in the form of copper oxide, copper thiocyanate or metallic copper. There is concern among some scientists that the environmental profiles of these new products are not sufficiently well-known to make informed assessments of their environmental impacts (Stewart, 1996). The concern is highlighted by reports of ecological impacts

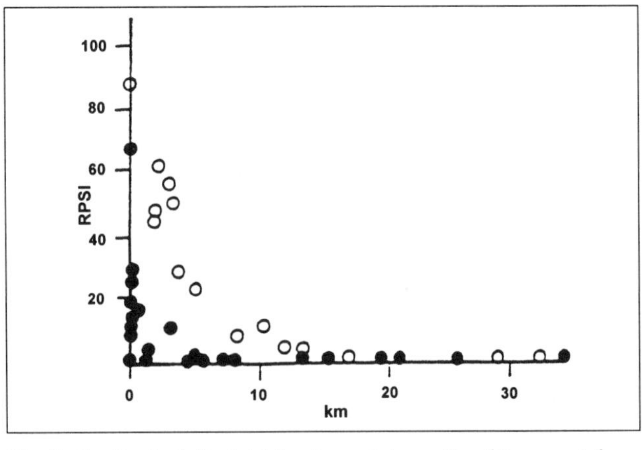


Fig. 7. Gradient of diminishing impact (severity of imposex) from commercial harbours (○) and fishing harbours (●). The measure of imposex is the Relative Penis Size Index (RPSI). (Svavarsson, J. and Skarphedinsdottir, H. (1995). Imposex in the dogwhelk *Nucella lapillus* (L.) in Icelandic waters. *Sarsia* **80**, 35–40.)

of paints containing the herbicide triazine. They have been used on small boats following the regulations prohibiting the application of TBT-based paints. There is increasing evidence that triazine is already at levels in some coastal areas which are sufficient to cause damage to plant communities, such as phytoplankton and seagrass beds. There is also concern about the physical properties of 'new' paints. Some ship owners claim to have had bad experiences with tin-free formulations which have been tried as alternatives to TBT-based paints. They have reported a number of performance failures, including premature fouling, high fuel consumption and poor structural integrity of the paint resulting in cracking and flaking (Bennett, 1996).

## ACKNOWLEDGEMENT

I would like to thank Dr. Judy Foster-Smith for her helpful comments on this manuscript.

## REFERENCES


Alzieu, C. (1998) Tributyltin: case study of a chronic contaminant in the coastal environment. *Ocean and Coastal Management* **40**, 23–26.

Alzieu, C., Sanjuan, D., Detreil, J.P. and Borel, M. (1986) Tin contamination in Arcachon Bay: effects on oyster shell anomalies. *Marine Pollution Bulletin* **17**, 494–498.

Bailey, S.K. and Davies, I.M. (1988) Tributyltin contamination around an oil terminal in Sullom Voe (Shetland). *Environmental Pollution* **55**, 161–172.

Beaumont, A.R. and Budd, M.D. (1984) High mortality of the larvae of the common mussel at low concentrations of tributyltin. *Marine Pollution Bulletin* **15**, 402–425.

Bennett, R.F. (1996) Industrial manufacture and applications of tributyltin compounds. *In Tributyltin: Case Study of an Environmental Contaminant*, ed. S.J. de Mora, Cambridge University Press, Cambridge, pp. 21–61.

Bright, D.A. and Ellis, D.V. (1989) A comparative survey of imposex in north-east Pacific neogastropods (Prosobranchia) related to tributyltin contamination and choice of a suitable bioindicator. *Canadian Journal of Zoology* **68**, 1915–1924.

Bryan, G.W., Gibbs, P.E., Burt, G.R. and Hummerstone, L.G. (1986) The decline of the gastropod *Nucella lapillus* around southwest England: long-term field and laboratory experiments. *Journal of the Marine Biological Association of the United Kingdom* **66**, 611–640.

Cardwell, R.D., Brancato, M.S., Toll, J. and De Forest, D. (in press) Aquatic ecological risk posed by tributyltin in U.S. surface waters: pre-1989–1996 data. *Environmental Toxicology and Chemistry*.

Davies, I.M., Bailey, S.K. and Moore, D.C. (1987) Tributyltin in Scottish sea-lochs, as indicated by the degree of imposex in the dogwhelk, *Nucella lapillus* (L.). *Marine Pollution Bulletin* **18**, 400–404.

Ellis, D.V. and Pattisina, L.A. (1990) Widespread neogastropod imposex: a biological indicator of global contamination. *Marine Pollution Bulletin* **21**, 248–253.

Eno, N.C., Clark, R.A. and Sanderson, G.W. (1997). *Non-native Marine Species in British Waters: a Review and Directory*. Joint Nature Conservation Committee, Peterborough.

Evans, S.M. (1999) Tributyltin pollution: the catastrophe that never happened. *Marine Pollution Bulletin* **38**, 629–637.

Evans, S.M. (1999) TBT or not TBT: that is the question. *Biofouling* **14**, 117–129.

Evans, S.M., Birchenough, A.C. and Fletcher, H. (2000) The value and validity of community-based science: TBT contamination of the North Sea. *Marine Pollution Bulletin* (in press).

Evans, S.M., Leksono, T. and McKinnell, P.D. (1995) Tributyltin pollution: a diminishing problem following legislation limiting the use of TBT-based anti-fouling paints. *Marine Pollution Bulletin* **30**, 14–21.

Evans, S.M., Nicholson, G.J., Browning, C., Hardman, E., Seligman, O. and Smith, R. (1998) An assessment of tributyltin contamination in the North Atlantic using imposex in the dogwhelk *Nucella lapillus* (L.) as a biological indicator of TBT pollution. *Invertebrate Reproduction and Development* **34**, 277–287.

Foale, S. (1993) An evaluation of the potential of gastropod imposex as a bioindicator of tribuyltin in Port Philip Bay, Victoria. *Marine Pollution Bulletin* **26**, 546–552.

Gibbs, P.E., Bryan, G.W., Pascoe, P.L. and Burt, G.R. (1987) The use of the dogwhelk (*Nucella lapillus*) as an indicator of TBT contamination. *Journal of the Marine Biological Association of the United Kingdom* **67**, 507–524.

Green, G.A., Caldwell, R. and Brancato, M.S. (1997) Comment on 'Elevated accumulation of tributyltin and its breakdown products in bottle-nosed dolphins (*Tursiops truncatus*) found stranded along the US Atlantic Gulf coasts'. *Environmental Science and Technology* **31**, 3032–3034.

Hallers-Tjabbes, ten C.C., Kemp, J.F. and Boon, J.P. (1994) Imposex in whelks (*Buccinum undatum*) from the open North Sea: relation to shipping traffic intensities. *Marine Pollution Bulletin* **28**, 311–313.

Harding, M.J.C., Bailey, S.K. and Davies, I.M. (1992) TBT imposex survey of the North Sea. UK Department of the Environment. Contract PECD 7/8/214.

His, E. and Robert, R. (1987) Comparative effects of two antifouling paints on the oyster *Crassostrea gigas*. *Marine Pollution Bulletin* **24**, 408–410.

Horiguchi, T., Shiraishi, H., Shimuzu, M. and Morita, M. (1994) Imposex and organotin compounds in *Thais clavigera* and *T. bronni* in Japan. *Journal of the Marine Biological Association of the United Kingdom* **74**, 651–669.

Kannan, K., Senthilkumar, K., Loganathan, B.G., Takahshi, S., Odell, D.K., and Tanabe, S. (1997) Elevated accumulation of tributyltin and its breakdown in bottle-nosed dolphins (*Tursiops truncatus*) foud stranded along the US Atlantic and Gulf coasts. *Environmental Science and Technology* **32**, 296–301.

Key, D., Nunny, R.S., Davidson, P.E. and Leonard, M.A. (1976) Abnormal shell growth in the Pacific oyster (*Crassostrea gigas*): some preliminary results from experiments undertaken in 1975. *ICES Paper C.M.* 1976/K: 11.

Ko, M.M., Bradley, C.C., Neller, A.H., and Broom, M.J. (1995) Tributyltin contamination of marine sediments of Hong Kong. *Marine Pollution Bulletin*, **31**, 249–253.

Nias, D.J., McKillup, S.C. and Edyvane, K.S. (1993) Imposex in *Lepsiella vinosa* from southern Australia. *Marine Pollution Bulletin* **26**, 380–384.

Nicholson, G.J., Evans, S.M., Palmer, N. and Smith, R. (1998) The value of imposex in the dogwhelk *Nucella lapillus* and the common whelk *Buccinum undatum* as indicators of TBT contamination. *Invertebrate Reproduction and Development* **34**, 289–300.

Smith, P.J. and McVeagh, M. (1991) Widespread organotin pollution in New Zealand coastal waters as indicated by imposex in dogwhelks. *Marine Pollution Bulletin* **22**, 409–413.

Spence, S.K., Hawkins, S.J. and Santos, R.S. (1990) The mollusc *Thais haemastoma* an exhibitor of imposex and potential bioindicator of tributyltin contamination. *Marine Ecology* **11**, 147–156.

Stewart, C. (1996) The efficacy of legislation in controlling tributyltin in the marine environment. In: *Tributyltin: Case Study of an Environmental Contaminant*, ed. S.J. de Mora. Cambridge University Press, Cambridge, pp. 264–297.

Stroben, E., Schulte-Oehlmann, U., Fiorini, P. and Oehlmann, J. (1995) A comparative method for easy assessment of coastal TBT pollution by degree of imposex in prosobranch species. *Haliotis* **24**, 1–12.

Thain, J.E. (1983) The acute toxicity of bis, tributyltin oxide to the adults and larvae of some marine organisms. *ICES Paper C.M.* 1983/E: 13.

Wilson, S.P., Ahsanullah, M. and Thompson, G.B. (1993) Imposex in neogastropods: an indicator of tributyltin contamination in Eastern Australia. *Marine Pollution Bulletin* **26**, 44–48.

Yamada, H., Takayanagi, K., Tateishi, M., Tagata, H. and Ikeda, K. (1997) Organotin compounds and polychlorinated biphenyls of livers of squid collected from coastal waters and open oceans. *Environmental Pollution* **26**, 217–226.

## THE AUTHOR

**S.M. Evans**

*Dove Marine Laboratory (Department of Marine Sciences and Coastal Management, Newcastle University), Cullercoats, Tyne & Wear, NE30 4PZ, UK*

Chapter 125

# EUTROPHICATION OF MARINE WATERS: EFFECTS ON BENTHIC MICROBIAL COMMUNITIES

Lutz-Arend Meyer-Reil and Marion Köster

During the last century organic pollution in coastal areas of the sea has become a serious problem around the world. One of the major stresses comes from the input of excessive macronutrients (nitrogen, phosphorus) resulting in a change of the trophic status of a given body of water, which leads to eutrophication. Although the effects of eutrophication are well known, the mechanisms governing its effects are poorly understood. In particular, effects on microbial processes are key to many aspects of the functioning of the ecosystem, and commonly are inadequately addressed.

The effects of eutrophication on benthic microbial communities are demonstrated using shallow-water coastal inlets in the southern Baltic Sea as an example. These so-called "Bodden" are characterized by pronounced gradients of inorganic and organic nutrients. For the hypertrophic innermost parts of the Bodden, critical points can be identified at which the chronic stress caused by eutrophication could no longer be compensated for by the system. Signs of eutrophication of sediments of the Bodden include increases in inorganic and organic carbon, nitrogen and phosphorus, microbial biomass and enzymatic decomposition potential of substrates, nitrification, denitrification, and nutrient fluxes from the sediments, all of which can be measured.

Above certain carbon concentrations, further increases in organic carbon are not necessarily paralleled by corresponding increases in biological parameters. This might be taken as an indication of a different status of nutrient enrichment. Eutrophication effects became most obvious from changes in the ratios of pelagic to benthic primary production, oxygen to sulphate respiration, and proteolytic to carbohydrate decomposing enzyme activities. The structure and function of microbial biofilms colonizing stones and sediments also reflected the changed trophic status. With increasing eutrophication, the ratio of autotrophic to heterotrophic microbial processes becomes greatly reduced. Drifting filamentous macroalgae, mats of sulphur oxidizing and anaerobic phototrophic bacteria, represent visible signs of eutrophication. Although the external nutrient loads in the example of the Bodden have been greatly reduced during the last decade, the internal loads of the sediments remain a serious problem. Remediation concepts can only support the natural self-purification potential of a marine coastal ecosystem.

*Seas at The Millennium: An Environmental Evaluation (Edited by C. Sheppard)*

## DEFINITION AND SOURCES OF POLLUTION

Human population growth has drastically increased during the last century, and since coastal areas are intensively inhabited, pollution pressure on marine environments has been dramatically enhanced over this time. A broad, widely accepted definition of pollution has been given by the International Advisory Board of the United Joint Group of Experts on the Scientific Aspects of Marine Pollution (GESAMP) as: "the introduction of substances or energy by man, directly or indirectly, into the marine environment (including estuaries) resulting in such deleterious effects as harm to living resources, hazards to human health, hindrance to marine activities including fisheries, impairment of quality for use of seawater and reduction of amenities".

Given this definition, a number of questions remain. Probably the most important problem is how to define (and measure) deleterious effects, which cover a broad spectrum of metabolic responses ranging from repairable disturbances of the normal metabolic pathways to the death of individual organisms or even populations. Besides direct effects on cells, the long-term effects on the behaviour and reproduction of organisms have to be considered as well. These problems demonstrate that a clear distinction between pollution and contamination (level of waste substrates exceeds the background level) is difficult. Taking into consideration the world-wide increase in concentrations of waste substances, what is the background level? Which concentrations of waste substances can be "tolerated" by the biological communities, and which is the level of substances at which damaging effects are manifested? This certainly depends not only on the community structure and the physiology of the organisms, but also on the variety of environmental conditions and their interactions. Another difficult problem is to differentiate between effects on marine biological communities arising from pollution and those from natural changes in hydrographical or meteorological conditions. As has been shown many times, increasing concentrations of waste substances in combination with unusual weather conditions have led to catastrophic effects on marine biological communities.

The release of materials into the marine environment derives from both point and non-point sources. Dumping or discharges of industrial waters and municipal sewage sludges represent primary point-source pollutants. Non-point pollution originates from diffuse sources, such as urban and agricultural run-off, groundwater transport and atmospheric inputs by dry or wet deposition. Non-point pollution is widespread and spatially and temporally variable, occurring in all coastal waters. Although non-point run-off is significant, its contribution to marine pollution is difficult to quantify because of its diffusive character. Commonly occurring pollutants from land-based sources include excessive inorganic and organic nutrients, xenobiotic compounds including radioactive substances, metals, as well as pathogens and other infectious agents. Various aspects of the problems of marine pollution have been addressed by a number of comprehensive overviews (Vollenweider et al., 1992; Sundbäck, 1994; Jørgensen and Richardson, 1996; Caumette et al., 1996; Kennish, 1997; Smith, 1998).

The following focuses on effects of eutrophication on the marine environment. Special emphasis will be paid to the response of benthic microbial communities to excessive nutrients. In a case study, the effects of eutrophication on benthic microbial communities in shallow-water coastal inlets in the southern Baltic Sea will be demonstrated.

## EUTROPHICATION

The discharge of excessive nutrients from municipal and industrial waste waters, urban and agricultural run-off leads to the enrichment of inorganic and organic material in marine waters. These inputs, resulting in eutrophication, presents one of the major stresses to the marine environment. Eutrophication can be defined as the process by which increasing nutrients cause a change of the nutritional status of a given body of water (Richardson and Jørgensen, 1996). Most frequently, the increase in the availability of mineral macronutrients (particularly nitrogen and phosphorus) is meant, although the supply of excessive decomposable organic carbon leads to eutrophication as well. Based on primary production, Nixon (1995) proposed four categories to describe the trophic status of marine coastal and estuarine ecosystems: oligotrophic, mesotrophic, eutrophic, and hypertrophic. The categories of increasing eutrophication are defined by increasing rates of primary production, ranging from less than 100 to more than 500 g carbon $m^{-2}$ $y^{-1}$. Whereas ocean waters can be regarded as oligotrophic, most of the marine coastal systems range from mesotrophic to eutrophic, because of the anthropogenic and riverine inputs.

Eutrophication has become a serious problem in virtually all coastal waters around the world. The problem is enhanced in areas with restricted exchange of water. The observation of large areas of anoxia in bottom waters of nutrient-enriched regions of the Baltic Sea represents well-known examples (Brügmann and Bachor, 1990; Nehring, 1992; Rheinheimer, 1998). Characteristic of this sea is the stratification of the water column in a surface layer of lower salinity and bottom water of higher salinity. Since the halocline restricts the vertical exchange of water, deep basins of the Baltic Sea especially suffer from long periods of anoxia, because the exchange of water occurs only occasionally, depending on the hydrographical conditions via inflow of saline water from the North Sea. As pointed out by Stachowitsch (1992), eutrophication can be regarded as a contributing or enhancing factor, which, together with adverse meteorological and hydrographical conditions, causes depletion of oxygen and subsequent mortality of benthic communities. Although it appears that developed countries have recognized the problem by developing strategies to reduce the nutrient burden, undeveloped countries seem to suppress the problem, almost totally relying on the dilution effect of exchange processes in marine waters. Worldwide, there are many other examples too. Low tidal exchange in parts of the Mediterranean, for example, has led to several embayments in which stratified deeper layers have become anoxic—much of the large Black Sea is a case in point—while in several other areas, man-made restrictions such as solid fill causeways connecting islands of atolls, for example, cause large areas to deteriorate much more rapidly than might otherwise have been the case.

## EFFECTS OF EUTROPHICATION ON MARINE COMMUNITIES

The effects of eutrophication are obvious. With the availability of macronutrients (nitrogen, phosphorus), primary production by algae and macrophytes is stimulated. Subsequent decay of high plant biomass causes an increase in oxygen consumption which may lead to anoxic conditions in bottom waters and sediments, since the biological oxygen consumption exceeds the supply of oxygen by diffusion by orders of magnitude. With increasing eutrophication, the structure of phytoplankton communities changes, resulting in the unpredictable occurrence of novel algae blooms. Increased turbidity of the water column and oxygen depletion greatly influence benthic communities with changes in their distribution, abundance, diversity, and physiological state. Losses of submerged aquatic vegetation causes the sediment to become mobile. Benthic metabolic processes change from aerobic to anaerobic decomposition of organic matter, causing liberation of hydrogen sulphide into the bottom water. End products of the anaerobic decomposition processes cause additional oxygen demand. The problem of oxygen depletion is even enhanced by the stratification of the water column, especially in summer conditions.

Eutrophication leads to an increase in the availability of nitrogen and phosphorus. Since silica does not increase simultaneously, the ratio of nutrients is altered, thus influencing phytoplankton species composition and succession. Species of algae that do not require silica dominate; diatoms requiring silica are discriminated against. Alterations in the ratio of the macronutrients may be responsible for the occurrence of novel and toxic algae blooms. Since diatoms play a key role in pelagic and benthic assemblages—representing the basis of food webs—limited diatom growth has important impacts on the trophic structure and the cycles of nutrients in marine environments (Conley et al., 1993; Sundbäck, 1994). Changes in food web structures of benthic communities by eutrophication were found by Nilsson et al. (1991) and McClelland and Valiela (1998).

Although the effects of eutrophication can be described, the mechanisms governing the effects are poorly understood. An approach towards a better understanding may be derived from a general consideration of the structural and functional attributes of ecosystems. An ecosystem is characterized as a natural area, in which organisms are interacting with each other and with environmental conditions, thus causing an exchange of energy and materials. In this sense, an ecosystem is an entity that is self-sustaining. The use of the term ecosystem refers primarily to processes and functions. Whereas the term function relates to the dynamics of the system, the term process refers to the mechanisms that contribute to overall ecosystem function. For the study of dynamics of ecosystems, the terms persistence, buffer capacity, resistance, and resilience may be used as organizing concepts to describe the degree to which ecosystems vary from steady state (Fig. 1) (Vogt et al., 1997). Persistence defines the length of time an ecosystem remains in a certain status. The buffer capacity is the property of an ecosystem to compensate disturbances through internal regulation mechanisms. Resistance is the capacity to balance impairments after disturbance; the amplitude describes the range in which parameters of an ecosystem vary. Resilience is the time (speed) an ecosystem needs after disturbance to return to the initial status.

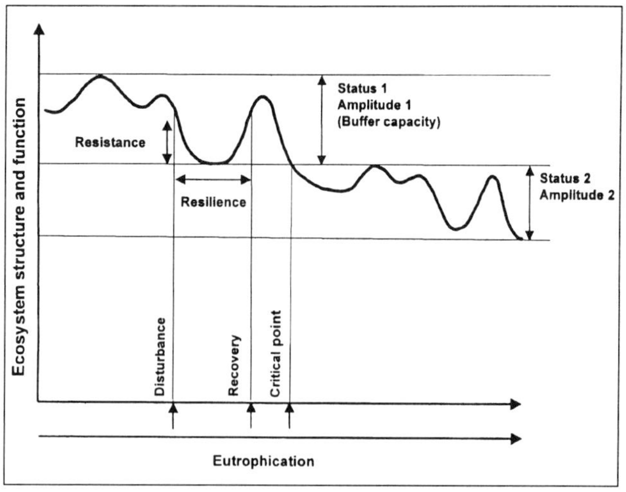

Fig. 1. Schematic diagram of structural and functional attributes of ecosystems (modified from Vogt et al., 1997).

Although in theory useful, it is difficult to apply these terms to the evaluation of variations in natural ecosystems for a number of reasons. Knowledge about only the "normal" ranges of operations of ecosystems requires innumerable measurements of various parameters at different time scales including daily and annual variations. The problem of heterogeneity of natural ecosystems has to be considered as well. Furthermore, which are the "key" parameters that are predicative enough to characterize different status of the ecosystem? Some of the terms listed above are related to time, so that only monitoring programmes carried out over numbers of years may help to identify when the status of a system has changed. Most of the impacts on ecosystems are chronic stresses such as pollution, causing the gradual loss of the integrity of an ecosystem—a process called retrogression (Vogt et al., 1997). Since the stress occurs almost continuously at a low level, changes in the ecosystem are slow and difficult to detect against variations of the ecosystem caused by natural impacts. Frequently, changes become obvious only after fundamental components or functions of the ecosystem are already definitely lost.

## EUTROPHICATION OF COASTAL INLETS IN THE SOUTHERN BALTIC SEA: AN EXAMPLE

### Characteristics of the Baltic Sea and the Bodden

The Baltic Sea is one of the largest brackish seas in the world, covering a total area of 415,000 km$^2$, a water volume of almost 22,000 km$^3$, and a coastline of about 22,000 km. The catchment area comprises 1.7 million km$^2$ with a population of more than 80 million inhabitants. Since the Baltic Sea is almost totally surrounded by land, pollution effects are especially pronounced. Because of the narrow connections with the North Sea, the exchange of water between the Baltic Sea and the North Sea is limited. The salinity reveals a strong gradient throughout the Baltic Sea from about 30 ppt at the connection with the North Sea (Kattegat) to less than 2 ppt at the innermost areas (Gulfs of Bothnia and Finland). Tidal activities are not pronounced; differences in water level of the coastal inlets result from wind-induced water movements. Because of the outflow of water of low salinity from the Baltic and the inflow of water with high salinity from the North Sea, water bodies of different densities are stratified, separated by a halocline.

Important for the formation of the present coastline of the southern Baltic Sea was the Holocene rise in the sea level. During the late stages of the overflow, an island archipelago arose along the southern coast of the Baltic Sea. By the accumulation of sand, barriers (so called "Haken" or "Nehrungen") were formed in a west–east direction along the islands, some of which were subsequently connected, thus separating shallow water inlets from the Baltic Sea. These inlets, called "Bodden", were for the most part surrounded by land. Until the last century, sea passages were present between the Bodden and the Baltic Sea, some of which were ultimately closed by human activities (e.g. construction of roads). Following the embankment, sedimentation significantly increased in the Bodden.

The Bodden along the southwest coast of the Baltic Sea consist of individual basins or chains of linked basins, some of which communicate only via small connections (Fig. 2). Their general features include: polymictic shallow water estuaries with horizontal salinity gradients, strongly influenced by hydrodynamical factors (e.g. freshwater supply, wind-induced water movements, water exchange processes with the Baltic Sea), high anthropogenic inputs of inorganic nutrients causing high phytoplankton productivity, dominance of species that can tolerate the variability of

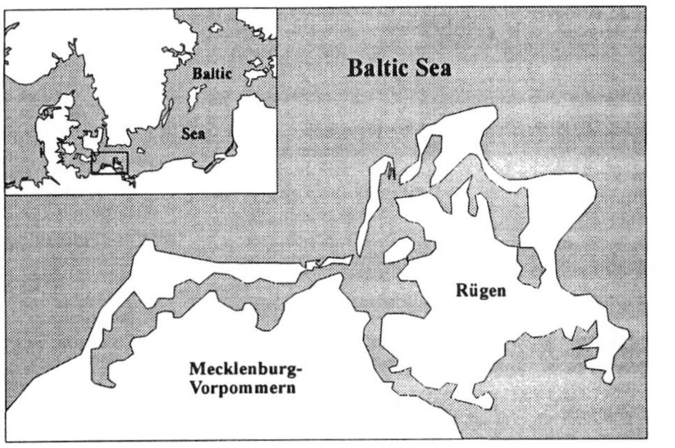


Fig. 2. Map of the Bodden along the southwest coast of Mecklenburg-Vorpommern (Germany). Inset: location of the Bodden in the Baltic Sea.

the environmental conditions, frequent exchange processes between sediment and water (sedimentation, resuspension), intensive cycling of inorganic and organic matter, and pronounced small-scale temporal and spatial variabilities in biological and chemical parameters (e.g. Nausch and Schlungbaum, 1991; Schiewer and Gocke, 1996; Hübel et al., 1998; Meyer-Reil, 1999; as well as literature cited therein). The Bodden serve as drainage basins for anthropogenic and natural inputs between the land and the Baltic Sea. They possess a high filter and buffer capacity, and act as important sinks and sources for pollutants. Today, however, the self-purification potential of the inner parts of the Bodden is overloaded, and the filter and buffer capacity is restricted or even lost.

### Eutrophication of the Bodden

At the time of their origin, the trophic status of the Bodden has to be regarded as oligotrophic. With decreasing exchange with the Baltic Sea (closure of the sea passages), the status changed towards mesotrophic. At the end of the 19th century, the establishment of industries and the population growth with increasing inputs of nutrients led to eutrophication. Caused by high inputs of inorganic and organic matter from agriculture, atmospheric deposition, and municipal sewage, a strong nutrient gradient has built up during the last decades. Concentrations of inorganic and organic nutrients drastically increased from the outer to the inner parts of the Bodden, changing their trophic status from mesotrophic to eutrophic or even hypertrophic in the innermost parts. This led to drastic changes in the structure and function of biological communities inhabiting waters and sediments of the Bodden.

Pronounced signs of the drastic nutrient changes in the Bodden are shifts in population structure and metabolic activities (Messner and von Oertzen, 1991; Hübel et al., 1998; Schiewer, 1998a,b; Meyer-Reil, 1999). Eutrophication impoverished the flora and fauna. Through alterations in the ratio of macronutrients (nitrogen, phosphorus, silica) the dominance of diatoms in the plankton community was lost; instead picoplankton (especially cyanobacteria) dominated. Increasing particle loads in the water with the subsequent reduction of light, restricted the growth of macrophytes in the sediments. This caused the sediments to become mobile, thus leading to a further increase in turbidity of the water column. Changes in the structure went along with changes in nutrient balances. Because of the dominance of microbial food webs, nutrients were no longer fixed in biomass of higher organisms for extended periods of time. Instead remineralisation processes were enhanced. Nitrogen and phosphorus especially were recycled on a short time scale. This caused an almost continuous availability of nutrients. During eutrophication, "critical points" (points of no return; cf. Fig. 1) can be identified, at which the Bodden could no longer compensate the disturbance caused by the chronic stress. The buffer capacity was exhausted; the resistance was exceeded. The Bodden changed their trophic status. One of the critical points was certainly the loss of the dominance of diatoms in the plankton community; another critical point was the disappearance of macrophytes (Schiewer and Gocke, 1996).

## Effects of Eutrophication on Benthic Microbial Communities

Marine sediments are intensively colonized by microorganisms (bacteria, cyanobacteria, fungi, algae; size < 150 $\mu$m). Most are organized in biofilms, complex associations of microbes, immobilized at surfaces and embedded in an extracellular organic matrix, consisting of extracellular polymeric substances (EPS) secreted by the cells. By their organization in biofilms, the organisms create their own microhabitats with pronounced gradients of biological and chemical parameters. Along these gradients they can use substrates and energy effectively (Meyer-Reil, 1994).

Microorganisms are present in high numbers in sediments (about $10^{10}$ cells $g^{-1}$ of dry weight). Their biomass is higher than the biomass of all other benthic organisms. The cell surface of the microbes by far exceeds that of all other organisms. Microbes possess a high surface to volume ratio, indicating their high metabolic activity rates. Dissolved inorganic and organic substrates can be metabolized with high substrate affinity and specificity. Particulate organic matter can be decomposed in close contact with the substrate by hydrolytic enzymes. Beside oxygen, microbes may use alternative electron acceptors (nitrate, manganese, iron, sulphate, carbon dioxide) for the oxidation of organic material. Combined with their logarithmic growth and short generation times (less than one hour), microbes possess a high metabolic potential.

Due to their low morphological differentiation, microorganisms (especially bacteria) cannot be easily identified. Future molecular genetic techniques may help to overcome this problem, but now only a small percentage of the total benthic bacterial community detected by fluorescence microscopy can be cultured and analyzed according to its taxonomy and metabolic potential. Because of these problems, concentration gradients of inorganic and organic substances may be interpreted as the net result of decomposition processes carried out by different physiological groups of microbes in the field.

In the Nordrügensche Bodden, concentrations of inorganic and organic substances increase with increasing eutrophication (Hübel et al., 1998; Meyer-Reil, 1999). Since sediment concentrations undergo strong local and seasonal variations, different metabolic pathways cannot simply be related to eutrophication. From the data of Köster et al. (1997) the general picture becomes obvious. With increasing eutrophication, concentrations of organic carbon and nitrogen, microbial biomass (ATP, phospholipids, chlorophyll, bacterial carbon), microbial decomposition potential of substrates (enzymatic hydrolysis of the model substrate

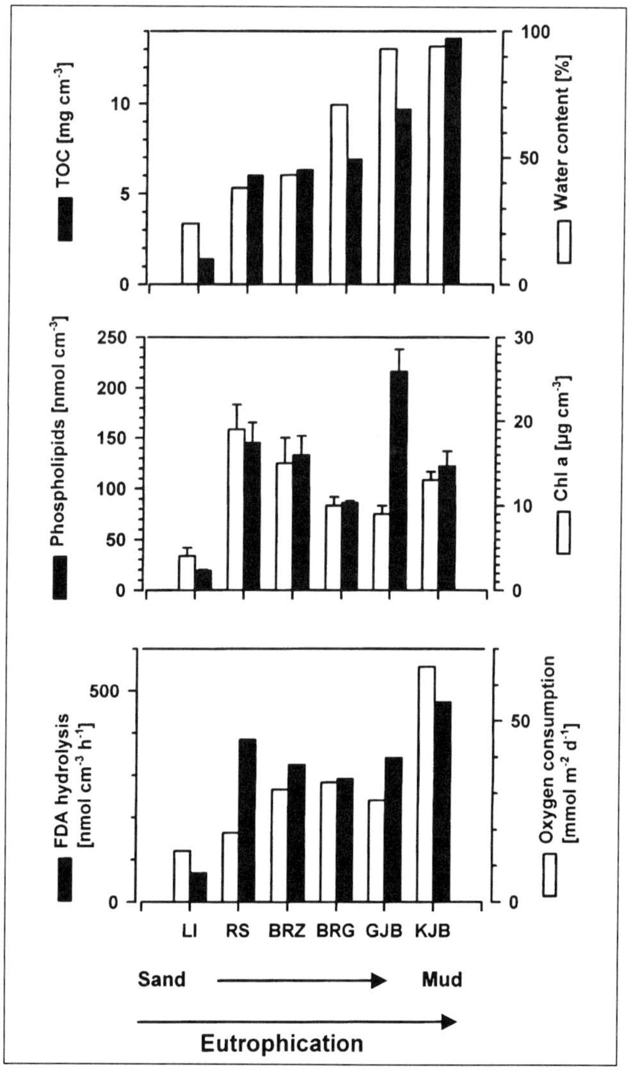


Fig. 3. Variations of chemical and biological parameters at different locations from sandy to muddy sediments along the gradient of eutrophication in the Nordrügensche Bodden (sampling date May 25, 1994; LI: Libben, RS: Rassower Strom, BRZ: Breetzer Bodden, BRG: Breeger Bodden, GJB: Großer Jasmunder Bodden, KJB: Kleiner Jasmunder Bodden). See text for details.

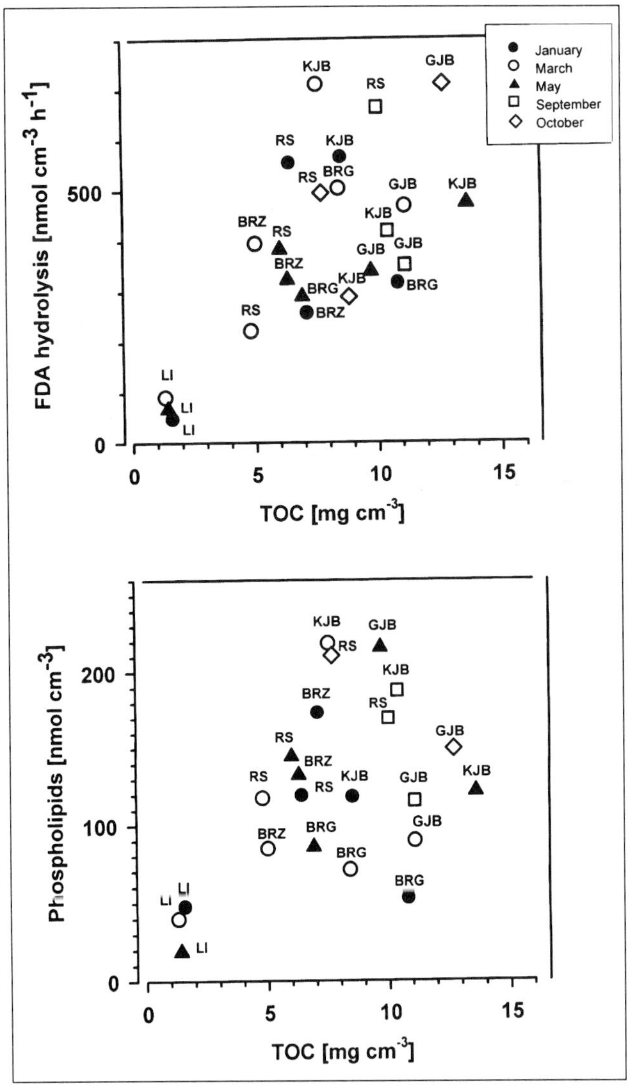


Fig. 4. Relationships between organic carbon and microbial biomass (measured as concentrations of phospholipids), and enzymatic decomposition potential of substrates (measured as enzymatic hydrolysis of fluorescein diacetate) for different locations (cf. Fig. 3) along the gradient of eutrophication in sediments of the Nordrügensche Bodden (adapted from Köster et al., 1997).

fluorescein diacetate) and community oxygen consumption increased (Fig. 3). This coincided with changes from sandy to more muddy sediments (decreasing grain size, increasing water content). The microbial biomass and decomposition potential of substrates were obviously directly correlated. However, the relationships between organic carbon on the one hand and microbial biomass and decomposition of substrates on the other hand were not simply linear with increasing eutrophication (Fig. 4). Up to carbon concentrations of about 10 mg per cm$^3$ sediment, the parameters were directly correlated: increasing carbon concentrations were reflected by corresponding increases in microbial biomass and decomposition potential. However, further increases in carbon concentration were not paralleled by corresponding increases in microbial biomass and decomposition potential. We interpret this observation as a reflection of different states of eutrophication, each of which is characterized by certain relationships between organic carbon and microbial biomass and activities. Above a certain threshold concentration of carbon ("critical point"), the trophic status of the Bodden changed as indicated by a shift in the relationship between carbon and microbial parameters. Furthermore, the change in the trophic status is reflected by changes in the ratios of oxygen respiration to sulphate respiration, total dissolved organic carbon to available organic carbon, and proteolytic to carbohydrate decomposing enzymes. With increasing eutrophication the oxygen penetration depths into the sediments were greatly reduced, sulphate respiration dominated over oxygen respiration, the availability of organic carbon decreased, and proteolytic enzymes dominated over carbohydrate decomposing enzymes. These observations represent impressive examples for the identification of key parameters to characterize shifts in trophic status (Köster et al., 1997; 2000).

There are other interesting observations on the effects of eutrophication on benthic microbial communities. Benthic and pelagic primary production were measured at two locations of different trophic status in the Bodden (Meyercordt and Meyer-Reil, 1999; Meyercordt et al., 1999). With increasing eutrophication, pelagic production was favoured over benthic production because of the shading of the sediments. This applies not only to macrophytes, the disappearance of which was a visible sign of eutrophication during the last decades, but to benthic algae (microphytobenthos) as well. By the liberation of oxygen, the fixation of nutrients released from internal sedimentary nutrient sources, and the stabilization of sediments, macrophytes and benthic microalgae contribute significantly to re-mesotrophication.

## Investigations of the Impact of Eutrophication on the Nitrogen Cycle

Wolff (1999) showed that nitrification increased with enhanced eutrophication. However, the increase was not proportional, obviously caused by the limited availability of oxygen in sediments under hypertrophic conditions. The same applies to denitrification, which did not increase proportionally to the external input of nitrogen (cf. Sloth et al., 1995). During most of the year, denitrification was dependent on nitrate supplied by nitrification (Dahlke, 1990).

Whereas the Bodden represented only a small source for atmospheric nitrous oxide, their contribution to the overall methane emission of the Baltic Sea was significant (Bange et al., 1998). Since both gases are products of microbial decomposition processes in sediments, a link to eutrophication seems to be obvious. With increasing eutrophication, the fluxes of inorganic nutrients (ammonia, phosphate) from the sediments increase (Dahlke, pers. comm.). This applies especially to phosphate, which is released following changes in redox potential from the sediments. The resulting shifts in the ratio of nitrogen to phosphate favour the bloom of cyanobacteria in the water.

During their development, microbial biofilms accumulate the influence of environmental parameters and may serve as indicators for eutrophication. Photoheterotrophic biofilms colonizing stones represent characteristic microbial communities, consisting of bacteria, cyanobacteria, diatoms and green algae. It was shown that the structure and function of the biofilms reflected the trophic status of the Bodden (Meyer-Reil et al., 1998). With increasing eutrophication, the heterotrophic enzymatic decomposition potential of substrates increased and net photosynthesis decreased, due to increasing rates of respiration. At mesotrophic, eutrophic and hypertrophic locations, aerobic respiration amounted to 20, 50, and 70% of the gross primary production. This had important consequences for the oxygen supply of the biofilms. Whereas biofilms from the mesotrophic location were oxic from the surface to the base, biofilms from the hypertrophic location were oxic only at the uppermost surface. These observations are in accordance with the investigations of Borum (1985), who could show that differences in nutrient conditions along a eutrophication gradient in a Danish estuary were more clearly reflected by epiphytes on eelgrass than by phytoplankton.

There are visible signs of eutrophication as well. At certain seasons, drifting filamentous algae can be observed which are caught in plant canopies. Finally, the decaying algae accumulate in deeper regions of the Bodden. If there is no light available, the algae die, thus contributing to local eutrophication. Most obvious are the so-called black and white "spots" at the surface of the Wadden Sea (North Sea) and in shallow water areas of the Baltic Sea, respectively. The black spots arise from iron sulphide, precipitated by hydrogen sulphide, which originates from the intensive microbial decomposition of organic material, oxidized with sulphate as the electron acceptor. In areas of high organic matter accumulation in the Bodden and deeper basins of the Baltic Sea, white spots can be observed at the sediment surface (Fig. 5). These are mats of sulphur-oxidizing bacteria (*Beggiatoa* species), which oxidize at the aerobic/anaerobic interface hydrogen sulphide, originating from microbial sulphate reduction, to elemental sulphur, which is accumulated within the cells (Dahlke and Meyer-Reil, 1996). Frequently, these white spots cover areas in which drifting senescent algae have accumulated. At calm, nutrient-enriched locations in shallow waters of the Bodden, reddish to violet carpets of sulphur-purple bacteria can be observed at the sediment surface, especially in summer. These bacteria carry out anoxigenic photosynthesis using hydrogen sulphide as electron donor which is supplied by microbial sulphate reduction. The above-mentioned microbial communities are visible signs of the rapid exhaustion of oxygen caused by the mineralization of excessive organic material supplied through eutrophication.

Major changes of ecosystem structure and function following eutrophication are compiled in Table 1. Although these were derived from a case study in the Bodden, corresponding effects can be observed in other coastal ecosystems.

Fig. 5. Mats of sulphur-oxidizing bacteria covering areas of decomposing drifting filamentous algae at the sediment surface of the Bodden (from Dahlke and Meyer-Reil, 1996).

## CONCLUSIONS

As documented by a number of investigations, eutrophication is a serious problem of coastal areas around the world. The trophic status of mainly the innermost parts of estuaries and coastal inlets is worrying. Concepts for remediation have to consider the reduction of the external as well as the internal loads. In coastal inlets in the southern Baltic Sea the external loads have been greatly reduced during the last decade. The reduction of intensive livestock farming, semi-liquid manure, and fertilizers in agriculture led to a drastic reduction of the input of nutrients. Further improvements result from the construction of sewage treatment plants, and the restoration of inflows into the Bodden. However, the expected improvement of water quality could not be demonstrated. The reason is the internal load of inorganic and organic nutrients that accumulated over years in sediments. The sediments are easily resuspended by biological activities (bioturbation) and physical impacts (wind driven water movements). The nutrients released into the water led to a stimulation of primary production by phytoplankton which undergoes sedimentation, thus leading again to an enrichment of organic material in sediments.

There is no easy way to get rid of the internal loads. Improvement of the water exchange between the inner and the outer parts of the Bodden only shifts the problem from the heavily into the less eutrophicated Bodden or even into the Baltic Sea. Precipitation by chemicals contributes to the immobilization of macronutrients (phosphorus) in sediments, but nobody can ignore their possible release in the future. The most effective measure would be the removal of the eutrophicated sediments by dredging. However, this is not possible for economic reasons. Furthermore, the

Table 1

Changes in ecosystem structure and function following eutrophication in water and sediments of coastal inlets of the southern Baltic Sea (Bodden)

| | |
|---|---|
| Concentrations, biomass | |
| – increased concentrations of inorganic and organic nutrients (carbon, nitrogen, phosphorus) | Hübel et al. (1998); Schiewer (1998a) |
| – increased autotrophic and heterotrophic biomass | Schiewer (1998b) |
| – shifts in phytoplankton composition from algae to cyanobacteria dominated blooms | Schiewer (1998a,b) |
| Autotrophic and heterotrophic activities | |
| – increased ratio of pelagic to benthic primary production | Meyercordt et al. (1999) |
| – decreased ratio of primary production to respiration in biofilms | Meyer-Reil et al. (1998) |
| – increased ratio of total dissolved organic carbon to microbial available organic carbon | Köster et al. (in prep.) |
| – increased enzymatic decomposition potential of substrates | Köster et al. (1997) |
| – increased ratio of protein to carbohydrate decomposing enzymes | Köster et al. (1997) |
| – increased remineralisation rates | Köster et al. (2000) |
| – decreased ratio of oxygen to sulphate respiration | Köster et al. (2000) |
| – decreased oxygen penetration depths in sediments | Köster et al. (2000) |
| – increased nitrification and denitrification rates | Dahlke (1990); Wolff (1999) |
| – increased fluxes of ammonia and phosphate | Rieling et al. (2000) |
| – increased emission of nitrous oxide and methane | Bange et al. (1998) |
| Visible signs | |
| – decrease in water transparency | Hübel et al. (1998); Schiewer (1998a,b) |
| – black anaerobic sediments | Dahlke and Meyer-Reil (1996) |
| – disappearance of macro- and microphytobenthos | Schiewer and Gocke (1996) |
| – mass abundances of drifting filamentous algae | Dahlke and Meyer-Reil (1996) |
| – occurrence of mats of sulphur-oxidizing bacteria on sediments | Dahlke and Meyer-Reil (1996) |
| – occurrence of mats of anaerobic phototrophic bacteria on sediments | Dahlke and Meyer-Reil (1996) |

question arises of where to deposit the material. There seems to be no other choice than to support the natural self-purification potential of the benthic microbial communities. These processes, however, need time and patience.

## ACKNOWLEDGEMENTS

The authors gratefully acknowledge the excellent technical assistance of T. Brüggmann, M. Gau, S. Kläber, and I. Kreuzer during past research projects funded by the German Umweltbundesamt, Deutsche Forschungsgemeinschaft, and Bundesministerium für Bildung und Forschung. We are especially grateful to H. Ehmke and T. Kreuzer, the crew of the research vessel "Prof. F. Gessner", for their help in taking samples from the Bodden. A number of interested colleagues contributed through their comments to the discussion of the data.

## REFERENCES

Bange, H.W., Dahlke, S., Ramesh, R., Meyer-Reil, L.-A., Rapsomanikis, S. and Andreae, M.O. (1998) Seasonal study of methane and nitrous oxide in the coastal waters of the southern Baltic Sea. *Estuarine, Coastal and Shelf Science* **47**, 807–817.

Borum, J. (1985) Development of epiphytic communities on eelgrass (*Zostera marina*) along a nutrient gradient in a Danish estuary. *Marine Biology* **87**, 211–218.

Brügmann, L. and Bachor, A. (1990) Present state of the Baltic coastal waters off Mecklenburg-Vorpommern, Germany. *GeoJournal* **22.2**, 185–194.

Caumette, P., Castel, J. and Herbert, R. (1996) *Coastal Lagoon Eutrophication and Anaerobic Processes (C.L.E.AN.).* Kluwer Academic Publishers, Dordrecht, 225 pp.

Conley, D.J., Schelske, C.L. and Stoermer, E.F. (1993) Modification of the biogeochemical cycle of silica with eutrophication. *Marine Ecology Progress Series* **101**, 179–192.

Dahlke, S. (1990) Denitrification in sediments of the North Rugian estuaries, preliminary communication. *Limnologica* **20**, 145–148.

Dahlke, S. and Meyer-Reil, L.-A. (1996) Bedeutung der Mikroorganismen. In *Warnsignale aus der Ostsee—Wissenschaftliche Fakten*, eds. J.L. Lozan, R. Lampe, W. Matthäus, E. Rachor, H. Rumohr and H. von Westernhagen. Parey Buchverlag Berlin, pp. 129–134.

Hübel, H., Wolff, C. and Meyer-Reil, L.-A. (1998) Salinity, inorganic nutrients, and primary production in a shallow coastal inlet in the southern Baltic Sea (Nordrügenschen Bodden). Results from long-term observations (1960–1989). *Internationale Revue Hydrobiologie* **83**, 499–519.

Jørgensen, B.B. and Richardson, K. (1996) *Eutrophication in Coastal Marine Ecosystems.* American Geophysical Union, Washington, 272 pp.

Kennish, M.J. (1997) *Pollution Impacts on Marine Biotic Communities.* CRC Press, Boca Raton, New York, 310 pp.

Köster, M., Babenzien, H.-D., Black, H.J., Dahlke, S., Gerbersdorf, S.,

Meyercordt, J., Meyer-Reil, L.-A., Rieling, T., Stodian, I. and Voigt, A. (2000) Significance of aerobic and anaerobic carbon mineralization processes in sediments of a shallow coastal inlet in the southern Baltic Sea (Nordrügensche Bodden). In *Muddy Coasts—Processes and Products*, eds. B.W. Flemming, M.T. Delafontaine and G. Liebezeit. Elsevier Oceanography Book Series, SCOR WG 106. Elsevier, Amsterdam. In press.

Köster, M., Dahlke, S. and Meyer-Reil, L.-A. (1997) Microbiological studies along a gradient of eutrophication in a shallow coastal inlet in the southern Baltic Sea (Nordrügensche Bodden). *Marine Ecology Progress Series* **152**, 27–39.

McClelland, J.W. and Valiela, I. (1998) Changes in food web structure under the influence of increased anthropogenic nitrogen inputs to estuaries. *Marine Ecology Progress Series* **168**, 259–271.

Messner, U. and von Oertzen, J.A. (1991) Long-term changes in the vertical distribution of macrophytobenthic communities in the Greifswalder Bodden. *Acta Ichthyologica et Piscatoria* **21**, 135–143.

Meyercordt, J., Gerbersdorf, S. and Meyer-Reil, L.-A. (1999) Significance of pelagic and benthic primary production in two shallow coastal lagoons of different degrees of eutrophication in the southern Baltic Sea. *Aquatic Microbial Ecology* **20**, 273–284.

Meyercordt, J. and Meyer-Reil, L.-A. (1999) Primary production of benthic microalgae in two shallow coastal lagoons of different trophic status in the southern Baltic Sea. *Marine Ecology Progress Series* **178**, 179–191.

Meyer-Reil, L.-A. (1994) Microbial life in sedimentary biofilms—the challenge to microbial ecologists. *Marine Ecology Progress Series* **112**, 303–311.

Meyer-Reil, L.-A. (1999) The Nordrügensche Boddengewässer—aspects of microbial ecology of an unique environment. *Archive Hydrobiology Special Issues Advances Limnology* **54**, 33–42.

Meyer-Reil, L.-A., Neudörfer, F. and Köster, M. (1998) Struktur und Funktion photoheterotropher Biofilme in den Nordrügenschen Boddengewässern. Final report of the research project "Struktur und Funktionsanalyse natürlicher mikrobieller Lebensgemeinschaften". Deutsche Forschungsgemeinschaft (DFG), 31 pp.

Nausch, G. and Schlungbaum, G. (1991) Eutrophication and restoration measures in the Darß-Zingst Bodden chain. *Internationale Revue der gesamten Hydrobiologie* **76**, 451–463.

Nehring, D. (1992) Eutrophication in the Baltic Sea. In *Marine Coastal Eutrophication*, eds. R.A. Vollenweider, R. Marchetti and R. Viviani. Elsevier, Amsterdam, pp. 673–682.

Nilsson, P., Jönsson, B., Swanberg, I.L. and Sundbäck, K. (1991) Response of a marine shallow-water sediment system to an increased load of inorganic nutrients. *Marine Ecology Progress Series* **71**, 275–290.

Nixon, S.W. (1995) Coastal marine eutrophication: a definition, social causes, and future concerns. *Ophelia* **41**, 199–220.

Rheinheimer, G. (1998) Pollution in the Baltic Sea. *Naturwissenschaften* **85**, 318–329.

Richardson, K., and Jørgensen, B.B. (1996) Eutrophication: definition, history and effects. In *Eutrophication in Coastal Marine Ecosystems*, eds. B.B. Jørgensen, and K. Richardson. American Geophysical Union, Washington, pp. 1–19.

Rieling, T., Gerbersdorf, S., Stodian, I., Black, H.J., Dahlke, S., Köster, M., Meyercordt, J. and Meyer-Reil, L.-A. (2000) Benthic microbial decomposition of organic matter and nutrient fluxes at the sediment-water interface in a shallow coastal inlet of the southern Baltic Sea (Nordrügensche Bodden). In *Muddy Coasts—Processes and Products*, eds. B.W. Flemming, M.T. Delafontaine and G. Liebezeit. Elsevier Oceanography Book Series, SCOR WG 106. Elsevier, Amsterdam. In press.

Schiewer, U. (1998a) 30 years' eutrophication in shallow brackish waters—lessons to be learned. *Hydrobiologia* **363**, 73–93.

Schiewer, U. (1998b) Hypertrophy of a Baltic estuary—changes in structure and function of the planktonic community. *Verh. Internat. Verein. Limnol.* **26**, 1503–1507.

Schiewer, U. and Gocke, K. (1996) Ökologie der Bodden und Förden. In *Meereskunde der Ostsee*, ed. G. Rheinheimer. Springer-Verlag, Berlin, pp. 216–221.

Sloth, N.P., Blackburn, H., Hansen, L.S., Risgaard-Petersen, N. and Lomstein, B.A. (1995) Nitrogen cycling in sediments with different organic loading. *Marine Ecology Progress Series* **116**, 163–170.

Smith, V.H. (1998) Cultural eutrophication of inland, estuarine, and coastal waters. In *Successes, Limitations, and Frontiers in Ecosystem Science*, eds. M.L. Pace and P.M. Groffman. Springer-Verlag, New York, pp. 7–49.

Stachowitsch, M. (1992) Benthic communities: eutrophication's "memory mode". In *Marine Coastal Eutrophication*, eds. R.A. Vollenweider, R. Marchetti and R. Viviani. Elsevier, Amsterdam, pp. 1017–1027.

Sundbäck, K. (1994) The response of shallow-water sediment communities to environmental changes. In *Biostabilization of Sediments*, eds. W.E. Krumbein, D.M. Paterson and L.J. Stal. Bibliotheks und Informationssystem der Carl von Ossietzky Universität Oldenburg, pp. 17–40.

Vogt, K.A., Gordon, J., Wargo, J.P., Vogt, D.J., Asbjornsen, H., Palmiotto, P.A., Clark, H.J. O'Hara, J.L., Keeton, W.S., Patel-Weynand, T. and Witten, E. (1997) *Ecosystems Balancing Science with Management.* Springer-Verlag, New York, 470 pp.

Vollenweider, R.A., Marchetti, R. and Viviani, R. (1992) *Marine Coastal Eutrophication.* Elsevier, Amsterdam, 1310 pp.

Wolff, C. (1999) Die autotrophe Nitrifikation in Küstengewässern unterschiedlichen Trophiegrades. PhD thesis, Ernst-Moritz-Arndt-Universität Greifswald, 150 pp.

## THE AUTHORS

**Lutz-Arend Meyer-Reil**

*Institut für Ökologie der Ernst-Moritz-Arndt-Universität Greifswald,*
*Schwedenhagen 6, D-18565 Kloster/Hiddensee, Germany*

**Marion Köster**

*Institut für Ökologie der Ernst-Moritz-Arndt-Universität Greifswald,*
*Schwedenhagen 6, D-18565 Kloster/Hiddensee, Germany*

Chapter 126

# PERSISTENCE OF SPILLED OIL ON SHORES AND ITS EFFECTS ON BIOTA

Gail V. Irvine

Over two million tonnes of oil are estimated to enter the world's oceans every year. A small percentage, but still a large volume, of this oil strands onshore, where its persistence is governed primarily by the action of physical forces. In some cases, biota influence the persistence of stranded oil or the rate of its weathering. Oil's deleterious effects on biota are frequently related to the persistence and degree of weathering of the oil, with long-lasting effects in low-energy environments such as salt marshes and coastal mangroves, or in higher-energy environments where oil is sequestered. However, an oil spill can have disproportionately large biological effects when it affects key species or processes (e.g., structurally important species, predators, prey, recruitment, or succession). In these cases, the continuing presence of oil is not always a prerequisite for continuing biological effects.

There are relatively few long-term studies of the effects of oil spills; data from these suggest that oil can persist for decades in some environments or situations, and that biological effects can be equally persistent. Broad-based, integrated studies have been the most revealing in terms of the importance of direct and indirect effects, spillover effects between different parts of the environment, and continuing linkages between residual oil and biologic effects. Clean-up and treatment techniques applied to spilled or stranded oil can also have significant, long-lasting effects and need to be carefully evaluated prior to use.

*Seas at The Millennium: An Environmental Evaluation (Edited by C. Sheppard)*
*Published by Elsevier Science Ltd.*

## INTRODUCTION

Ancient carbon, compressed by the earth and transformed into oil, has become an important currency of modern times. Drilling activities to liberate the oil occur globally, but the concentration of oil in large, irregularly distributed reservoirs has led to extensive shipment of crude and refined oils to areas of demand. This large-volume transport occurs most commonly via ship or pipeline, and during such transport spills frequently occur. Oil spills also occur during extraction and use of oil and its products. The largest releases into the marine environment in 1990, the last year that comprehensive figures were available, were from municipal/industrial sources (Table 1). Although the major sources have varied over time, the estimated total volume of inputs since 1973 has reduced significantly, though the statistics can be influenced by the occurrence of large spills (Fig. 1). These have generally been better studied than smaller or chronic releases. Since 1960, at least 55 spills greater than 10 million gallons have affected marine waters and shores. The complex issues surrounding oil spills have been the focus of numerous investigations, some of which are reviewed in this chapter.

Table 1
Estimated annual input of petroleum hydrocarbons into the marine environment by source

| Source | Estimated input (tonnes) | | |
|---|---|---|---|
| | 1973[a] | 1981[b] | 1990[c] |
| Transportation (total) | 2,133,000 | 1,470,000 | 564,000 |
| Bilge/fuel discharges | 500,000 | 300,000 | |
| Tanker operations | 1,080,000 | 700,000 | |
| Tanker accidents | 200,000 | 400,000 | |
| Marine terminals | 3000 | 20,000 | |
| Non-tanker accidents | 100,000 | 20,000 | |
| Dry-docking | 250,000 | 30,000 | |
| Scrapping of ships | N/A | N/A | |
| Municipal and industrial | 2,700,000 | 1,180,000 | 1,175,000 |
| Atmosphere | 600,000 | 300,000 | 305,500 |
| Natural sources | 600,000 | 200,000 | 258,500 |
| Offshore production / exploration | 80,000 | 50,000 | 47,000 |
| Total | 6,113,000 | 3,200,000 | 2,350,000 |

Sources: a = NRC (1975), b = NRC (1985), c = GESAMP (1993).

## OIL PERSISTENCE ON SHORES

Oil is a complex substance with many quickly weathering components. Once spilled, it may be transported rapidly by winds, tides, and currents, changing as it moves and as its components become partitioned into the air and water. When it becomes stranded, its persistence on shores is affected by the type of oil spilled, its weathering history prior to reaching shore, the volume stranded, shoreline

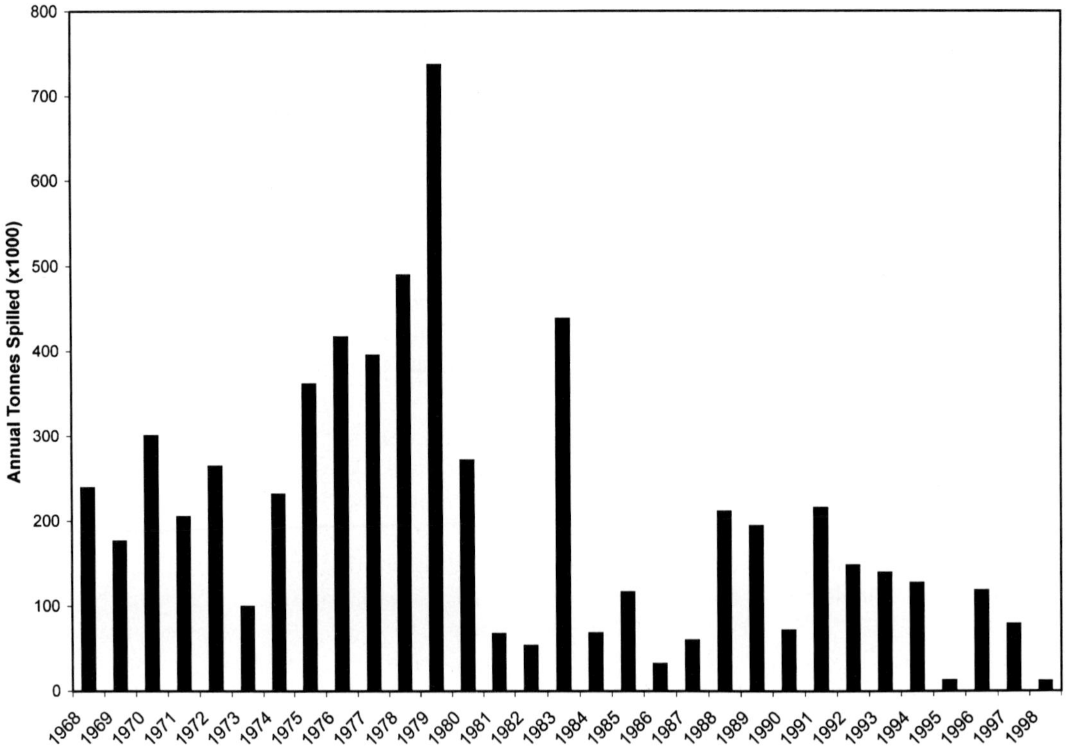


Fig. 1. Amount of oil (tonnes) spilled from tankers, 1968–1998. These data do not include oil spills from other sources that also affect marine waters (e.g., onshore storage tank ruptures: 1991 Gulf War incident and the 1986 Refinería las Minas, Panama spill).

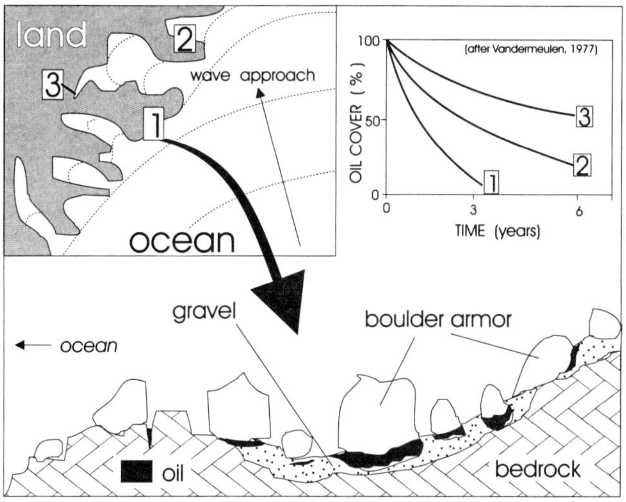


Fig. 2. Surface oil persists longest on beaches with lower wave energies (upper two figures). However, this may not be true for subsurface oil. Where a boulder armour protects gravel substrate from wave erosion, subsurface oil persists for years, despite high wave energies (from Irvine et al., 1999).

geomorphology, and other physical factors. Concerns regarding oil persistence relate to effects from lingering oil, alterations of the physical environment, and cultural issues. Primary among the factors that affect stranded oil persistence is the amount of mechanical energy acting on a shore, which is a function of wind, waves, tides and ice (Owens, 1978; Hayes et al., 1979). Quantifying the degree of exposure of sites allows for more consistent comparison of the relationship between oil persistence and exposure (e.g., Hayes, 1996). Wave fetch, wind duration, and wind velocity have been used to create indices of exposure.

Wave energy acts in concert with shoreline erosion and sediment reworking to affect the persistence of oil. There is an inverse relationship between wave energy and oil persistence (Fig. 2). Exposed rocky shores retain oil for shorter periods of time than do protected ones. Sandy beaches undergoing continual depositional and erosional cycles, are generally subject to greater natural cleansing than are sheltered rocky areas having similar exposure levels, and the timing of oil deposition may therefore also affect its persistence and fate (Hayes et al., 1979). If oil is stranded during a depositional phase, it is likely to be buried and to persist until later reworking of the sediments exposes it. The effectiveness of erosional dynamics in removing this oil depends upon whether the oil has migrated vertically (Long et al., 1981) and upon its recalcitrance to weathering (e.g., asphalt). Sediment-laden oil may alter the stability of a beach locally by altering the erodability of sediment. Shoreline stability and the rate of shoreline change may be more important than wave energy in affecting the potential persistence of oil (e.g, low-energy arctic coasts of the Beaufort Sea; Owens, 1985).

Tidal energy, in combination with winds and waves, affects the height at which oil is deposited, and influences the magnitude of coastal currents. Oil is deposited frequently at one level and then is lifted by a later flooding tide and deposited higher in the intertidal. When oil is deposited above most subsequent tidal heights, it is exposed to less mechanical energy and thus is slower to weather. At these higher levels, it is also subject to slower rates of biodegradation.

The interplay between coastal geomorphology and wave regimes means there can be redundancy in relating oil persistence to geomorphological type and wave exposure, longshore transport, etc. Studies and observations following spills have led to the development of an "Environmental Sensitivity Index" (ESI). This index was originally named the "vulnerability index". It is based primarily on the predicted persistence or retention of oil on different shoreline types, but also includes some biological criteria. Table 2 portrays the version of the index developed after the *Amoco Cadiz* spill (Hayes et al., 1979). Exposed rocky headlands have the lowest vulnerability (environmental sensitivity) and salt marshes (and mangroves) have the highest. In some habitat types, there can be interactions between biota and geomorphology that influence the persistence and long-term potential effects of the oil. ESIs are continually revised as our knowledge of where and why oil persists is refined. A synopsis of various factors controlling oil fate and persistence follows.

## Permeability

The ability of oil to penetrate different substrates affects its persistence. Although permeability effects were incorporated into ESIs starting with the *Amoco Cadiz* spill (Hayes et al., 1979), further understanding of their importance has only been revealed by recent spills. Viscosity and type of oil affect penetration; diesel or other refined oils may easily permeate sediments due to their low viscosity, but they are also more likely to be flushed out. Fluid crude oils penetrate mineral sediments more readily than weathered crudes whose viscosity has increased, or mousse, the water-in-oil emulsions formed by some oils when they are subjected to wind and wave energy.

Sediment grain size is clearly critical in determining permeability to oil. Oil penetrates more readily beaches composed of larger grain sizes. The mobility of the substrate or degree of tidal flushing will also strongly affect oil persistence and weathering. The effect of *armouring* of a beach on oil persistence illustrates the dynamics between these factors. Armouring occurs when wave action selectively removes smaller grain sizes, leaving a remnant "armour" of larger grain sizes (e.g., cobbles or boulders) that is unmoved by the force of typically occurring waves. Oil penetrates between these larger grains, and persists there because this armour and its underlying sediments are seldom disturbed. The degree of oil penetration below the surface armouring depends on the size of the subsurface particles, and oil persistence is related to surface grain size and to wave energy.

Table 2

Oil spill vulnerability index. Higher index values indicate greater long-term damage by the spill (modified from Hayes et al., 1979; comments refer to the *Amoco Cadiz* oil spill). Now known as an Environmental Sensitivity Index (ESI). VI = Vulnerability Index.

| VI | Shoreline type | Comments |
|---|---|---|
| 1 | Exposed rocky headlands | Wave reflection kept most of the oil offshore; no cleanup needed |
| 2 | Eroding wave-cut platforms | Exposed to high wave energy; initial oiling removed within 10 days |
| 3 | Fine-grained sand beaches | All only lightly oil-covered after one month mainly by new oil swashes |
| 4 | Coarse-grained sand beaches | Oil coverage and burial after one month remains at moderate levels |
| 5 | Exposed, compacted tidal flats | No oil remained on the sand flat but did cause enormous mortality of urchins and bivalves |
| 6 | Mixed sand and gravel beaches; no really good example of this beach type | The index value is due to rapid oil burial and penetration; all areas had compacted subsurface which inhibited both actions |
| 7 | Gravel beaches | Oil penetrated deeply (30 cm) into the sediment; cleanup by use of tractors to push gravel into surf zone seemed effective and not damaging to the beach |
| 8 | Sheltered rocky coasts | Thick pools of oil accumulated in these areas of reduced wave action; cleanup by hand and high pressure hoses removed some of the oil (this process is valid in non-biologically active areas) |
| 9 | Sheltered tidal flats | Tidal flats were heavily oiled; cleanup activities removed major oil accumulations but left remaining oil deeply churned into the sediment; biological recovery yet to be determined |
| 10 | Salt marshes | Extremely heavily oiled with up to 15 cm of pooled oil on the marsh surface; cleanup activities removed the thick oil accumulation but also trampled much of the area; biological recovery yet to be determined |

Thus armouring of beaches completely alters the relationship between oil persistence and wave energy (Fig. 2).

The presence of impermeable layers affects the ability of oil to penetrate and persist on beaches. The location of such a layer also can affect the migration of oil, and hence, the natural cleansing pattern of a beach. Mud, sand, and bedrock are examples of impermeable or low permeable substrates. Permeability is affected also by high water content (Little and Scales, 1987). High water content can be caused by particle size (e.g., silt vs gravel) or by groundwater discharge from shorefaces.

### Wetting of Beach; Adhesive Properties of Oil

Poor drainage and resultant high water content of beaches provide hydrological protection of the sediments, inhibiting oil adhesion and penetration (Little and Scales, 1987). The point of the tide cycle when the oil arrives is critical. As the tide falls, coarse-grained beaches "bleed" water stored in the substrates. Freshwater seeping out of the beach also is important, as is the "stickiness" of oil. Stickiness can be affected by chemical treatments (dispersants, surfactants, etc.) and by natural weathering. There is some indication that adhesive qualities change in the days following a spill, which may account for a change from refloating of the oil on successive days after stranding, to its adherence later (Hayes et al., 1979). Possibly the greater dryness of the upper intertidal, where much of the oil ultimately strands, is also an important factor. Further interactions with sediments can cause the oil to become sediment-bound, which may inhibit refloating and result in the oil being deposited in the low intertidal (Hayes et al., 1979) or subtidal.

### Formation of Asphalts

Thick deposits of oil on low-angle surfaces may form asphalts. These stable mixtures of weathered hydrocarbons and sediments form persistent, erosion-resistant deposits (Owens et al., 1986). Although the surface weathers, weathering of the interior of the asphalt mass is usually retarded, and those conditions that favoured asphalt formation (including reduced disturbance rates) promote its persistence. Studies following the *Metula* spill in Chile, where no cleanup occurred, have documented the persistence of asphalt under a variety of conditions (e.g., sand/gravel beaches and marshes). Asphalts were common two years after the spill in 10 different locations, and 12.5 years later, when most of the coast was free of oil, the asphaltic pavements remained relatively unaltered (Owens, 1993). The thick, resistant nature of asphalts means they have the potential to alter the local sediment dynamics of beaches and reduce the rate of erosion. Following the Gulf War spill, the formation of asphalts on muddy beaches trapped more fluid oil that had penetrated the substrates via the burrows of invertebrates (Hayes et al., 1995).

### Oil-contaminated Sandy Beaches

Following the *Amoco Cadiz* spill in Brittany, oil persisted for at least 3 years on several beaches (Long et al., 1981). Oil layers (1–2 cm thick), sometimes multiple layers, migrated downwards within beach sediments, eventually stabilizing deep within the beach at or near the water table. Migration patterns were related to water table movement, porosity of the sediments and presence of an impermeable basement layer. In beaches with deep sand layers and without an impermeable subsurface layer, tidal pumping drove the downward migration of oil, with the rate a function of sediment porosity. When there was an impermeable layer, the downward movement of the oil became stalled and the oil then moved along the basement layer, emerging at the face of the beach in lower intertidal and subtidal zones. Based on these observations, Long et al. (1981) predict that oil can have a multi-year residence time in thick sand beaches. Such beaches can store large volumes of oil; one beach in their study was predicted to contain an estimated 600 metric tons of oil. In a second model, with the presence of an underlying impermeable basement, the residence time of the oil is expected to be relatively short, perhaps less than 1.5 years, because most of the oil exits the beach into the lower intertidal or subtidal zone. This pattern of oil migration is expected to increase contamination of lower intertidal and subtidal sediments (Long et al., 1981).

### Oil Dynamics in Low-energy Environments

Low-energy tidal flats are considered to have high vulnerability to oil, because of long oil retention times. However, some observations (Owens et al., 1993) and experimental studies (Harper et al., 1985) suggest that this is not always true. In experimental studies, sediment size, particularly fines or mud content, appeared to be the most critical factor influencing oil retention, which was inversely proportional to the mud content. The effect may be mediated by the correlative relationships of fine sediments with water retention and permeability. This finding reinforces the notion that sediment water content can afford protection from oil penetration, and hence retention. The studies also found that oil retention was proportional to the time sediments were exposed, suggesting that intertidal sediments with longer emergence times (higher in the intertidal) would have greater retention. Oil retention was also proportional to the loading thickness of oil, although there appeared to be a maximum beyond which there was little increase in retention (Harper et al., 1985).

Another process that may reduce the retention of oil in low-energy, fine sediment environments is *clay–oil flocculation* or *oil–fines interaction*. This physicochemical process involves the binding of fine mineral particles to oil droplets, frequently with the oil droplet(s) being surrounded by the fines. These aggregates form with many types of oil and minerals, provided there is enough water turbulence, although the rate of flocculation or amount of oil incorporated is also affected by oil viscosity (Lee et al., 1998). The oil–mineral aggregates can be quite stable and prevent oil from recoalescing. It has been proposed that the formation of these clay–oil flocculants reduces the adhesion of oil to sediments on the shore. Their low density then results in their dispersion, leading to natural cleansing of sheltered, low-energy beaches (Bragg and Yang, 1995). Rates of biodegradation of oil are also increased with clay–oil flocculation presumably because the surface area available to microorganisms is increased. Although clay–oil flocculation may be important in natural cleansing of some beaches, its significance is unclear due to possible confounding factors related to water turbulence and wave energy. The dispersion of oil into the water column may lead to other biological effects.

## INFLUENCES OF BIOTA ON OIL PERSISTENCE

Biota can influence the persistence of oil by setting up situations that increase oil entrapment and slow its weathering. The structure of mussel beds, which mimics the armouring created by boulders and cobbles, is one such example. The three-dimensional structure of the mussels and their byssal threads traps sediments and protects sequestered oil from weathering (Babcock et al., 1996; 2000). Following the *Exxon Valdez* oil spill, hydrocarbon levels in mussels and their underlying sediments were elevated at some locations for at least 10 years (G. Irvine, unpub. data). The hydrocarbon concentrations have been gradually declining, but most oiled beds are not expected to reach background levels until three decades after the spill (Babcock et al., 2000).

Mangrove trees, with their pneumatophores or prop roots, are likely to increase oil retention for similar reasons. Mangrove sediments may retain oil for at least 20 years (Burns and Knap, 1989; Corredor et al, 1990; Duke et al., 1997). Oil sequestered in sediments within mangrove stands undergoes slow weathering due to anoxic conditions in the sediments.

Burrowing animals can increase the penetration of oil into sediments and may significantly affect persistence and weathering. Following the Gulf War spill, oil penetrated to depths of >40 cm in muddy habitats by flowing down burrows created by infauna such as polychaete worms and crabs (Hayes et al., 1995). In some cases, these muddy habitats were covered by asphalt, which probably severely restricted the weathering processes affecting the subsurface oil (Hayes et al., 1995). Oil penetration also is increased within mangrove sediments by the presence of burrows (e.g., Getter et al., 1984; Duke et al., 1997) and a similar pattern has been observed in marshes, where oil has penetrated deeper through the hollow stalks of dead marsh plants (Baker et al., 1993).

Burrowing animals can, on the other hand, increase weathering of subsurface oil by reworking sediments, bringing contaminated sediments to the aerated surface where weathering processes are more rapid (Gordon et al., 1978).

## EFFECTS OF OIL ON NEARSHORE POPULATIONS AND COMMUNITIES

### General Effects

Species differ widely in their sensitivity to oil, but the actual effects are influenced as well by the amount of contact, the type and volume of oil spilled, weather, sea state, and the timing of the spill. Organisms known to be quite sensitive include certain crustaceans (e.g., ampeliscid amphipods, ostracods), echinoderms, molluscs, diving sea birds, certain marine mammals (e.g. sea otters), marsh vegetation (e.g., *Spartina alterniflora*), and mangroves. Others, such as the lugworm (*Arenicola*), the clam (*Macoma*), priapulids, and some nematodes, appear to be relatively insensitive (Teal and Howarth, 1984). Within a species, some life stages are more sensitive than others, with developing larvae and eggs generally more sensitive than older stages (NRC, 1985).

Acute, lethal effects frequently occur after spills, causing declines in species abundances that can last for years (e.g., Dauvin and Gentil, 1990; Holland-Bartels, 1999). Although effects on individuals may be the most easily noticed, effects occur at all levels of biological organization, from subcellular to ecosystem (e.g., NRC, 1985; Suchanek, 1993). Sublethal effects can be prolonged, and can result in delayed mortality. Many effects are probably indirect, for instance effects on behaviour that increase predation (Teal and Howarth, 1984).

The effects of actual oil spills are highly variable. For example, considerable variation appears to exist in the effect of oil on *Fucus* and closely related genera (e.g., *Ascophyllum*), which are dominant intertidal flora in northern temperate regions. After the *Tsesis* spill, *Fucus vesiculosus*, the predominant intertidal plant along the Swedish Baltic coast, was not measurably affected. In contrast, following the *Arrow* spill (Nova Scotia), *F. vesiculosus* became restricted in its vertical distribution for five years. A congener, *F. spiralis*, which is usually confined to the mean high-tide level, was killed outright, and even after 6 years had not reappeared in the oiled region. Stranded oil released by the *Arrow* was concentrated in the upper two-thirds of the intertidal and was most persistent about the mean high-tide level. After the *Amoco Cadiz* spill (France), the fucoid, *Ascophyllum* sp., was killed, and was replaced by the more resistant *Fucus* sp., as long as there were *Fucus* plants close enough in the vicinity to serve as a source of propagules (Teal and Howarth, 1984). *F. vesiculosus* increased in abundance through indirect effects following the *Torrey Canyon* spill in England (Southward and Southward, 1978, Hawkins and Southward, 1992) and *F. gardneri* decreased following the *Exxon Valdez* oil spill off Alaska (e.g., Houghton et al., 1997), but the effects derived from both oil and clean-up procedures.

Species vulnerability also depends on life history patterns. Many pronounced effects occur in the intertidal zone where large volumes of oil often strand. Obviously, sessile species are vulnerable, but mobile species that seasonally reproduce in nearshore areas may also be vulnerable (e.g., herring laying eggs within the intertidal zone). Species differ in their resilience, and northern species with more seasonal periods of reproduction are more likely to demonstrate longer lasting effects than are tropical species that are more frequently reproductive. The higher temperatures and light levels of the tropics should increase oil weathering rates, thus speeding recovery.

In addition to the acute effects of a spill and the cascading effects those create, the persistence of oil in the environment can cause continued mortalities, sublethal effects, and may hamper the ability of species to successfully recruit into contaminated areas. Effects may change as the oil weathers. For example, there may be a lag in the expression of mortality; chronic releases of oil that incur new mortalities; effects from chronic exposure that act in a cumulative manner (e.g., mutagenic effects); low-level, periodic chronic exposures; and persistent sub-lethal responses. Effects will be influenced by the type, concentration, and degree of weathering of the oil to which an organism is exposed. Extensive sublethal effects could ultimately be more important than acute mortalities (Southward, 1982; Jackson et al., 1989).

The timing of a spill also can have a tremendous influence on which species are affected and the magnitude of effect. This is most obvious for seasonally abundant, vulnerable species, such as seabirds aggregating at breeding colonies. The occurrence of the *Exxon Valdez* oil spill in late March, a biologically active time, contributed to its wide-ranging effects. Numerous species were congregating in coastal environments and as a result were highly vulnerable. Other large spills have had fewer effects simply because of timing.

### Effects on Biological Communities

Oil spills cause direct mortality through toxic effects, smothering, and breakage due to increased drag. Quite frequently, massive mortalities of benthic or nearshore species have been reported (e.g., NRC, 1985). The first acute phases of toxicity (there are sometimes lag effects) result in the loss of individuals (and/or species), opening up space or habitat for recolonization. Both rocky and fine sediment environments may become dominated by opportunistic species (e.g., ephemeral algae, certain polychaetes). In areas where the substrata remain contaminated, the success of colonizers may be determined by their tolerance to oil, and even within the suite of opportunistic species there may be a succession of genotypes that reflect a range of tolerance to oil (Southward, 1982). Recovery then can be affected by continuing contamination of the environment, but generally, the processes of recruitment, competition and predation interact in the return of more normal patterns of species abundances.

Oil spills disturb naturally occurring communities. It is pertinent to question whether their effects are similar to other disturbance agents, since selective mortality, sublethal (toxic) effects, and the potential for chronic effects are involved. Results from different spills may not be similar, since the type and volume of oil and timing of the spills varies, and different suites of life history stages and processes may be vulnerable. Peterson (2000) contends that the *Exxon Valdez* oil spill has had the characteristics of both a press and pulse disturbance to the coastal ecosystem, based on the occurrence of both acute and chronic effects. The nature of the responses will be dependent upon the magnitude of the initial disturbance, its selectivity, and linked effects throughout the nearshore environment.

Disproportionately large effects can occur when spills affect species that play key functional roles.

### *Structuring Species (Biogenic Species)*

Included within this category are taxa like *Fucus* and *Mytilus*, that provide habitat for other smaller species, as well as those which create the primary structure (e.g., mangroves, corals, large kelps).

#### *Fucus*

In littoral areas affected by the *Exxon Valdez* oil spill, *Fucus* previously comprised the highest biomass in the intertidal zone. Fucoids provide structural habitat for other algal species and small invertebrates (Notini, 1980). The response of *Fucus* spp. to spills is quite variable. Even when adult plants prove resistant to oil, their associated macrofauna may be decimated, as after the *Tsesis* spill (Notini, 1980).

#### *Mytilus*

Mussels dominate many shores, providing habitat for a species-rich community of invertebrates. Adult mussels may be relatively tolerant to oil exposure (Babcock et al., 1996), although following the *Exxon Valdez* spill, mussel abundance was significantly reduced at a number of sites (Highsmith et al, 1996). Smaller mussels are generally more sensitive than adults. Mussel beds may be areas of highly persistent oil contamination (Babcock et al., 1996, 2000).

#### *Kelp species*

Large subtidal algae rarely have been reported to be significantly affected by oil. However, following a spill of diesel fuel in Northern Mexico, grazing molluscs and echinoderms died, allowing the giant kelp, *Macrocystis*, to become abundant and form an extensive bed. In this case, "structure" and possibly species diversity were increased through reductions in important grazers (North et al., 1965). Dean et al. (1996) found some effects on subtidal kelp species following the *Exxon Valdez* spill, but the specific cause(s) of the effects were not clear, and could have resulted from cleanup activities.

#### *Mangroves*

Oil spills have had significant negative effects on mangrove trees (e.g., Getter et al., 1984; Duke et al., 1997) as well as on the biota that live on the prop roots (e.g., Jackson, et al., 1989; Garrity et al., 1993; Burns et al., 1993). A large spill (75,000–100,000 barrels) of medium-weight crude oil occurred on the Caribbean coast of Panama in 1986 when a storage tank at the Refinería Panama ruptured. This oil affected nearby coral reef, mangrove and seagrass beds, including the Punta Galeta marine lab and biological reserve of the Smithsonian Tropical Research Institute. The combination of prior research on epibiota of mangrove prop roots (and other coastal communities) and an integrated science approach to broader effects of the spill, provided an unusual ability to understand effects. Garrity et al. (1993) demonstrated long-term (at least five-year) effects on the physical structure of the mangrove (*Rhizophora mangle*) forests, different effects and recovery rates in three mangrove habitats (open coast, channel and stream), and differential effects of the spill on predominant epibiota of different habitats (e.g., the edible oyster, *Crassostrea virginica*, in channels, and the false mussel, *Mytilopsis sallei*, in streams).

Since mangroves often grow in anoxic sediments in low-energy environments, the potential for persistent oil spill effects is high. In fact, oil persistence following the refinery tank rupture is estimated to be 20 years or longer (Burns et al., 1993), based on the rates of change of hydrocarbon concentrations in tissues and sediments for the five years following the 1986 spill, and sediment cores that showed the signature of another spill (the *Witwater*) that occurred in the same area 17 years earlier. The chronic release of hydrocarbons from eroding sediments kept hydrocarbon levels high in *Crassostrea* and *Mytilopsis* for five years (Garrity et al., 1993). The initial weathering of the oil was fairly rapid: most volatile hydrocarbons and the marker alkanes were gone by six months. However this biodegradation rate was not maintained in the disappearance rate of aromatic hydrocarbons (Burns et al., 1993). The estimated 20 year span for oil persistence is similar to estimates for an oiled Puerto Rican mangrove system (Corredor et al., 1990) and a temperate salt marsh (Teal et al., 1992). Another consequence of prolonged oil exposure may be increasing mutation rates of exposed organisms. In coastal mangrove populations in Puerto Rico, Klekowski et al. (1994) found a strong correlation between the incidence of recessive lethal traits (chlorophyll-deficient mutations) and polynuclear aromatic hydrocarbon concentrations in the underlying sediments.

#### *Coral reefs*

Emergent reef flats dominated by algae and sessile invertebrates are vulnerable and have been affected by spills (Jackson et al., 1989; Cubit and Connor, 1993). Numerous mobile species have died following spills in nearshore tropical waters (e.g., Mignucci-Giannoni, 1999). Effects on subtidal coral mortality, abundance, reproduction and recruitment have been documented (e.g., Loya and Rinkevich, 1980; Jackson et al., 1989; Guzmán et al., 1991). In the Refinería Panama spill, extensive mortality of subtidal corals was recorded. Numbers of corals, total coral cover, and species diversity based on percent cover estimates, all decreased significantly with increasing amounts of oiling. Coral species were affected differentially. *Acropora palmata*, a branching coral, suffered the highest percent cover losses, while massive corals showed increased frequency and size of recent injuries that increased with level of oiling. Long-term effects are expected due to chronic oiling from eroding mangrove sediments (Burns and Knap, 1989), vulnerability of coral gametes and larvae, and slow growth and repopulation. Several decades might be necessary before populations recover to pre-spill abundances (Guzmán et al., 1991). Hydrocarbon concentrations in coral tissue were correlated with increased mortality rates as measured by decrease in area coverage (Burns and Knap, 1989).

Corals have not always shown marked effects from oil spills. Possibly the differences are related to the availability of hydrocarbons to corals, either through contaminated sediments or seawater. In the Arabian Gulf, corals were apparently little affected by the largest oil spill recorded, the 6–8 million barrels released as a consequence of the 1991 Gulf War (e.g., Vogt, 1995). Analysis of sediment hydrocarbons one year after the spill showed no evidence of large-scale sinking of oil (Michel et al., 1993). Following the Refinería Panama spill, elevated hydrocarbon concentrations were found in both water and subtidal sediments (Guzmán et al., 1991, Burns and Knap, 1989). Relatively small amounts of dispersants (estimated <21,000 l of Corexit 9527) were sprayed in a limited area following the Panama spill, and its use was thought to be inadequate to produce the extensive areal coral mortality that was observed (Burns and Knap, 1989). Analysis of the hydrocarbons in coral tissue compared to that of seawater and sediments, suggested that the corals appeared to take up the hydrocarbons primarily from the water column. This continued contamination is thought to derive from the highly oiled mangrove sediments, which contaminate the seawaters associated with the mangroves (Burns and Knap, 1989).

### *Consumers (predators and herbivores)*

Oil spills have frequently been noted to kill limpets, periwinkles and other herbivores (e.g., North et al, 1965; Southward and Southward, 1978; Teal and Howarth, 1984; Jackson et al., 1989), resulting in a bloom of ephemeral algae or subtidal kelps (North et al., 1965). These opportunists may persist if there is sufficient contamination to keep herbivore populations reduced. However, at least in hard substrate environments, eventually either later successional species colonize or come to dominate space through asexual reproduction, or herbivores start rebounding (e.g., Southward, 1982; Jackson et al., 1989). Following the *Torrey Canyon* spill, a combination of oil plus intensive dispersant use decimated populations of limpets, *Patella* spp., resulting in a bloom of ephemeral algae followed by colonization and dominance by *Fucus* in exposed rocky habitat (Southward and Southward, 1978; Hawkins and Southward, 1992). This pattern of ephemeral algae coming in after grazers have been reduced following a spill has not uniformly been noted, although the season of the spill and the timing of studies could affect the ability to detect such a response.

Some predatory species have also been negatively affected by oil spills, although fewer consistent observations have been made regarding these losses and resultant effects. Sea otter (*Enhydra lutris*) populations in the western part of Prince William Sound were reduced approximately 50% by the *Exxon Valdez* oil spill. The keystone role of these predators on nearshore ecosystems has been well documented. Sea otter populations in the oil-affected area have been depressed for nine years (Monson et al., 2000). Reductions in sea otters may have led to an increase in the size of sea urchins within oiled areas (Dean et al., 2000). Effects on another benthic invertebrate-feeding predator, the harlequin duck (*Histrionicus histrionicus*) were also examined. Populations of these have been depressed in oiled areas of Prince William Sound for 10 years (Esler et al., 2000).

### *Prey Species*

Usually it is difficult to isolate effects due to depression of a particular prey species, as frequently whole suites of species are negatively affected by a spill. An example is the mussel, *Mytilus trossulus*, following the *Exxon Valdez* oil spill. Mussel abundance declined in many oiled sites, and more importantly, mussel beds trapped oil in their underlying sediments and byssal threads. Examination of the hydrocarbon concentrations in mussel tissues and underlying sediments revealed extensive contamination, extending more than 300 km from the spill's origin (Babcock et al., 1996). Studies of nearshore vertebrate predators in Prince William Sound that had not recovered after five years indicate that several species that feed on mussels (sea otters, harlequin ducks, Barrow's goldeneye) have experienced continuing exposure to hydrocarbons, as revealed by elevated P450-1A cytochrome values (Holland-Bartels, 1999; Trust et. al., 2000). Although the route of exposure is not certain, oiled mussel beds are a possible source.

### *Effects on Sensitive Species with Localized Recruitment*

When oil severely affects populations of species that have only limited dispersal of propagules, then the effects of a spill can be prolonged. A notable example is amphipods. Amphipods, generally, are highly sensitive to oil, and their populations have decreased dramatically following spills (e.g., Teal and Howarth, 1984). Subtidal populations affected by the *Amoco Cadiz* spill (Dauvin, 1998; Dauvin and Gentil, 1990) were severely reduced for most species, especially the ampeliscids. Amphipods brood their young, so their dispersal is constrained. This fact, combined with the relative isolation of the embayments with the fine sands preferred by amphipods, means that oil effects on particular bays affected isolated populations. Resampling for approximately two years after the spill, then 10 and 20 years after the spill, has indicated that many of the populations had recovered or were recovering by 10 years (Dauvin and

Gentil, 1990), and after 20 years, only a few had not recovered to pre-spill levels (Dauvin, 1998).

Other sensitive species with localized recruitment include fucoid algae. When *Fucus* (*Ascophyllum or Pelvetia*) spp. are severely affected (as following the *Arrow, Amoco Cadiz*, and *Exxon Valdez* spills), recovery can be slow due to the limited dispersal distances of the young. Experimental studies following the *Exxon Valdez* spill indicated that most propagules are dispersed within 20 cm of the adult plant (van Tamelen and Stekoll, 1995). Part of the severity of the effect on *Fucus* following the *Exxon Valdez* spill appears to be due to the use of high-pressure-hot-water washing as a treatment (e.g., Houghton et al., 1997), which affected broad expanses of *F. gardneri*. Along both North Atlantic and North Pacific Ocean coasts, when nearby *Fucus* or *Ascophyllum* plants are available as sources of propagules, recovery can proceed more quickly (e.g., van Tamelen and Stekoll, 1995; van Tamelen et al., 1997; Teal and Howarth, 1984).

A secondary effect of *Fucus* recovery derives from the creation of even-aged stands of *Fucus*, brought about by successful settlement of a year-class following the oil spill disruption. After the *Exxon Valdez* oil spill, the senescence and decline of *Fucus* after 4–6 years led to decreased abundances of *Fucus*, followed by another wave of recruitment and then increasing abundance as these germlings grew. This same pattern of damped oscillating *Fucus* abundance was observed following the *Torrey Canyon* spill, but occurred as a result of the decline of limpets, *Patella* spp., and was an indirect effect rather than a direct recovery response (Southward and Southward, 1978; Hawkins and Southward, 1992).

#### *Alterations in the Pattern of Succession and Dominance*

When the pattern of species replacements through time is affected by oil, then the potential exists for longer-term effects. This is particularly true when long-lived species pre-empt annuals, or an alternative set of long-lived species, after oil spills. This may occur when oil affects the timing or extent of cleared space available, the success of different colonizers, the relative success of different plant species through the elimination of grazers, etc. When long-lived species replace ephemerals or annuals, the mechanisms of succession may be altered. A good example is that involving severe reductions of limpets, *Patella* spp., from exposed rocky coasts of England following the *Torrey Canyon* spill (Southward and Southward, 1978; Hawkins and Southward, 1992). *Fucus* colonized and came to dominate many sites, even though it is usually absent in these exposed environments. In this case, a long-lived perennial replaced ephemeral algae and other relatively long-lived animals (a limpet grazer and barnacles). *Fucus* dominated the community for 2–5 years, and then had several pulses of abundance; the community still had not fully recovered after 10–15 years (Hawkins and Southward, 1992).

A second example involves species replacements low in the intertidal zone following the *Exxon Valdez* oil spill. In control areas, the lower intertidal is dominated by a mixture of rapidly growing annual kelps (e.g., *Alaria* spp.) and other annual and ephemeral algae. Two years post-spill, in some of the geographical areas, *Fucus* had come to predominate in the lower intertidal zone (Highsmith et al., 1996), even though it is usually most abundant in the mid and upper regions of the intertidal. The mechanisms of this replacement are not known, although one suspects that a reduction in grazers and annual algae, perhaps combined with availability of *Fucus* propagules, led to successful colonization by *Fucus*. The potential for long-term effects is derived in part from the relatively long life-span of *Fucus*. If *Fucus* can successfully replace itself in the low intertidal, then the effects could be quite long-term. The stability of such a replacement is not known. Damped oscillations in abundance, as described above, could occur, and may create opportunities for faster-growing annuals to recolonize and outcompete *Fucus*.

## RECOVERY OF BIOTA AND THE IMPORTANCE OF PERSISTENT OIL

Following the acute phase of an oil spill, populations begin the process of recovery. Recovery may be defined (Hawkins and Southward, 1992) as: "a return to normal levels of spatial and temporal variation".

The challenges of determining recovery are several. First, without pre-spill data it is difficult to quantify effects and recovery. Some of the most compelling assessments of effect have occurred precisely because there were good pre-spill data coupled with some understanding of the processes affecting assemblages of organisms (e.g., Southward and Southward, 1978; Hawkins and Southward, 1992; Dauvin, 1998; Dauvin and Gentil, 1990; Jackson et al, 1989; Garrity et al., 1993). Even with pre-spill data, control sites are important because other, non-spill-related environmental changes may occur simultaneously and complicate determination of oil spill effects. Even with controls, background changes may complicate the analysis of oil effects (e.g, Lewis, 1982; Holland-Bartels et al., 1999; Jackson et al., 1989).

### Persistent Oil Effects and Their Causes

Shoreline types have been rated (via Environmental Sensitivity Indices, ESIs; Hayes et al., 1979) according to their expected retention of oil and, to some extent, biological effects have been thought to be aligned with oil persistence (but see Southward and Southward, 1978). This appears to be true in some low-energy environments (e.g., mangroves, salt marshes), perhaps due to the extreme contamination that can occur and persist in these environments. Oil has been found or estimated to persist for at least 17–20 years in these environments (Corredor et al., 1990; Teal et al., 1992; Baker et al., 1993; Burns et al., 1993), and where there has been further injury due to clean-up activities (Ile Grande Marsh in France), recovery has been estimated to take from 8 to 100 years (Long and Vandermuelen, 1979; Baca et al., 1987). The *Metula* spill in Chile is unusual, in that there was no clean-up activity after the spill. There, oil on low-energy sandy and gravel beaches is expected to persist for 15–30 years and for more than 100 years on sheltered tide flats and marshes (Gundlach et al., 1982).

In low-energy environments, persistent effects on biota appear to be related to continuing oil presence. In marshes and soft-sediment habitats, oil has been noted to cause extensive acute effects, some of which were observable for only a very short time following a spill, as the dead, soft-bodied invertebrates quickly disappeared (e.g., Sanders et al., 1980). Opportunist species (e.g., capitellid and cirratulid polychaetes) often colonize the depopulated areas, and are gradually replaced by species that were abundant previously (e.g., Southward, 1982). Within these low-energy environments, persistent oil effects may include: altered behaviours that increase mortality; increased mortality to immigrating individuals that move into contaminated sediments (Teal and Howarth, 1984); increased hydrocarbon burdens in various organisms (e.g., clams, fish, crabs); and induction of enzyme systems (e.g., cytochrome P450 1A) that indicate exposure to hydrocarbons or other pollutants (e.g., Teal et al., 1992).

Effects on marsh vegetation can be severe (e.g., Hampson and Moul, 1978; Baca et al., 1987; Baker et al., 1993). Plant species are differentially affected, with annual plants apparently more sensitive to oil. Mortality of perennials has also been reported, but the underground, undamaged portions of these plants can start producing new growth relatively quickly. Recovery is impeded by high hydrocarbon concentrations in or on the sediments (e.g., Hampson and Moul, 1978; Baker, et al., 1993). Rates of recovery are also affected by the degree of oil weathering; in the Puerto Espora marshes oiled by the *Metula* spill, there was no vegetative recovery in areas that still had thick deposits of mousse 16 and 17 years later. The interior portions of this oil were still quite fresh chemically (Baker et al., 1993). Another side effect of death of marsh plants is increased erosion (e.g., Hampson and Moul, 1978). Clean-up activities in marshes may accelerate rates of erosion and retard recovery (Long and Vandermuelen, 1979). Rates of recovery for marsh vegetation have been variously reported to occur from years (Baca et al., 1987) to decades (Baker et al., 1993) following a spill.

Persistence of oil on exposed rocky shores has been estimated in ESIs to range from days to a few weeks, but effects to biota can be much longer lasting, and are more likely when long-lived species are negatively affected or when they come to replace other species.

Armouring of beaches by cobbles/boulders or mussels also has increased the persistence of oil on both exposed and moderately exposed beaches following the *Exxon Valdez* oil spill (e.g., Babcock et al., 1996, 2000; Hayes and Michel, 1999; Irvine et al., 1999). Recent observations (1999, G. Irvine) indicate that relatively unweathered oil still persists in exposed, boulder-armoured beaches and in some of the more protected mussel beds. Extrapolations based on the rate of weathering of oil samples taken in 1999, 10 years post-spill, suggest oil will persist in a relatively unweathered state on the high-energy, boulder-armoured beaches for decades, until an unusual storm event disturbs the boulder armouring. Effects on biota are expected to be slight until the oil gets disturbed, when it is more likely to come into contact with biota (Irvine et al., 1999). In oiled mussel beds, hydrocarbon levels are expected to approach background levels in most cases within 30 years following the spill (Babcock et al., 2000).

## EFFECTS OF TREATMENTS

A variety of techniques have been used in attempts to collect oil once it has been spilled, clean it up, or treat it to minimize its effects. If oil were transparent, perhaps humans would not have chosen some of the various treatment techniques used in past spills to get rid of oil before and after it strands on beaches. One is struck sometimes by the push to get the oil "out of sight", without full consideration of how that might transfer the potential effects elsewhere.

### Dispersants

Perhaps one of the best known cases of treatment is the widespread, intensive use of dispersants following the *Torrey Canyon* spill. The treatment probably had worse effects than the oil (Southward and Southward, 1978; Hawkins and Southward, 1992), and intensive dispersant use fell into disfavour following that spill. Newer versions of dispersants were formulated that reduced the toxicity of the carrier solvent. Perhaps the greatest concern is the transference of oil effects into the water column and into benthic subtidal habitats. Since oil (and oil + dispersant) effects are more difficult to detect in these environments and have been less studied, the significance of such choices is not known. Of particular concern are effects on primary producers, and eggs and larvae of fishes and invertebrates. Dispersants have been applied to oil spills that occurred after the *Torrey Canyon* spill (e.g., the *Amoco Cadiz* and Refinería Panama spills), and subtidal effects were more pronounced after these spills than for many other spills. It is unclear whether even low-volume dispersant use can lead to significant subtidal effects.

### Manual Removal of Oil

This technique has been employed after many spills. It is very labour-intensive and can result in trampling effects on sessile biotic communities. In situations where subsurface oiling is significant, manual removal is usually ineffective. The greatest benefits of manual oil removal accrue in those areas where wave energy is low (sheltered environments), and in removal of oiled carcasses to reduce the spread of oil to scavenging species. In sheltered environments where asphalts are likely to form, efforts should be made to remove the asphalts, since their long-term persistence is so great (as illustrated by the *Metula* spill, where asphalt pavements have changed little in 12–17 years; Owens, 1993; Baker et al., 1993).

### Berm Relocation and Other Mechanized Efforts on Beaches

Heavy equipment has been used to remove or move oiled sediments following several spills (e.g., *Amoco Cadiz, Exxon Valdez*). In the case of the Ile Grande Marsh in France, these efforts caused changes in the erosional patterns in the marsh that were expected to last for perhaps a century (Long and Vandermuelen, 1979), although more recent estimates (made after extensive replanting efforts) suggest recovery after 7–8 years (Baca et al., 1987). Following the *Exxon Valdez* spill, berm relocations were used to move heavily oiled sediments from high on the beach to lower in the intertidal where they would be subjected to greater wave energies. This type of technique is more appropriate for beaches with low concentrations of biota, where the disruption and burial of sediments are likely to have fewer biological effects. It can be effective in reducing persistent, subsurface oil, but it is intrusive and its use should be carefully evaluated (Owens, 1998; Hayes and Michel, 1999).

### Sand-blasting and High/Low-Pressure-Water Techniques

Sandblasting has been used in low biota, high human use areas. When the scale of the spill is small, and effects on biota are likely to be ameliorated by settlement from nearby unoiled areas, then this may be a viable technique, although re-oiling can occur if oil buried in nearby sediments becomes remobilized. High-pressure-hot-water washing (plus cold-water and moderate-pressure washing) of shores with concomitant booming of the shore being treated, and removal of floating oil within the booms, was used extensively within Prince William Sound following the *Exxon Valdez* oil spill. Fortunately, studies were set up to investigate treatment effects (e.g., Houghton et al., 1997) through comparisons of biota on oiled, oiled + treatment, and unoiled sites. The greatest effects and slowest recovery were observed at the oil + treatment sites (Houghton et al., 1997). In 1998, nine years post-spill, recovery of some sites still had not occurred, although some scientists were arguing that unequal, but parallel patterns of abundances did represent recovery. In sedimentary habitats, high-pressure, hot-water flushing has apparently reduced the fine sediments necessary for successful settlement of infaunal clams, thus creating a long-term disruption of this community (Peterson, 2000). Use of this technique, based on observed results, seems contraindicated.

Low pressure, ambient-temperature water flushing has been used to remove oil from affected mangroves in the Arabian Sea, reportedly with good results (Watt et al., 1993). The effectiveness of low pressure, cold water was tested with good results on oiled settlement plates placed in rocky habitats with differing exposures (Menot et al., 1999), although some results were clouded by micro-placement effects.

### Bioremediation

Petroleum hydrocarbons are degraded in the environment by naturally occurring bacteria. The effectiveness of this process is dependent, at least in part, upon concentrations of the bacteria, temperature, availability of nutrients, and the extent of oil surface available for microbial action. Several approaches have been taken to enhance the effectiveness or rate of biodegradation. One has been the application of nutrients, another has been the application of nutrients combined with an oleophilic carrier. These techniques increased rates of biodegradation 3–5 times when used following the *Exxon Valdez* oil spill (e.g., Lindstrom et al., 1991; Atlas, 1995).

### Additional Treatment Considerations

Although these treatments (and others) have been used with varying success following spills, they come with their own effects and costs. Some of these effects may be broader in scope—such as the increased human presence on nearshore and onshore environments brought about by spill assessments and clean-up activities. Rarely are these comprehensively examined. Treatment effects seem to be pronounced in low-energy environments, perhaps due to the low delivery rates of sediments.

The choice to use particular treatments should be carefully evaluated, and to further informed decision-making, additional well-designed studies are needed to evaluate the effectiveness of techniques in situations where there are concerns about the effects of persistent oil. Some say that doing nothing is best, but inaction requires as much assessment as active choices (Foster et al., 1990). Results of long-term studies, like those following the *Exxon Valdez* spill, indicate that concern about ongoing, chronic effects of spills is justified, and that action to remove persistent oil may be warranted, especially in some environments.

## SYNTHESIS

Determining the effects of a spill can be difficult, but predicting the effects may be impossible. The difficulties are due to the superposition of three layers of complexity: the qualities of the spilled oil, the physical situation, and the biological panorama. These are neither independent nor static. The oil starts changing in chemical composition immediately following its release. The rate of change, the form it takes (e.g., whether it becomes emulsified), and its consequent toxicity will be influenced by both physical and biological factors. The physical conditions at the time of a spill and thereafter, direct its trajectory, whether and where it strands, and its persistence. The oil affects biota, and is itself affected by the presence and actions of biota. The magnitude and direction of acute and indirect effects are influenced by which species are present, their vulnerabilities, and relative sensitivities.

Acute, sublethal and chronic effects from oil spills may occur over broad geographical and temporal scales that make assessments of the effects of spills very difficult. The design and execution of appropriate studies to investigate the effects can be constrained by the geographical scale of the event, the difficulty of selecting appropriate controls, lack of pre-spill data, and money. The geographic scale of the spill will affect species differentially and may influence how well different species recover, since modes of reproduction and the sources of propagules that fuel recruitment vary among species. Control areas could be difficult to judge *a priori*, as it may be difficult to determine which areas were contaminated. Broad-scale, integrated studies may reveal the tremendous range of direct and indirect effects resulting from a spill (Peterson, 2000), as well as the complexity of separating spill effects from other changes occurring to populations (e.g., assessing spill effects on species whose populations are already in decline (Holland-Bartels, 1999).

### Common Themes

Several themes emerge. First, oil and treatment effects are often intertwined and difficult to separate. Second, effects in one habitat can spill over, or have consequences in another. Following the Refinería Panama spill, hydrocarbon contamination from eroding sediments in mangroves appeared to be the source of contamination affecting subtidal corals (Burns and Knap, 1989). Third, there are many direct and indirect effects which are overlooked unless numerous species are examined and unless studies are designed with oiled and control sites (Holland-Bartels, 1999; Peterson, 2000). Fourth, persistent biological effects can occur without the persistence of oil due to effects on key species or processes. Fifth, effects generally seem longer-term in many habitats than were first expected, with a number of studies now showing continuing contamination, or continuing effects for periods of 10–20 years post-spill, and in a few extreme cases, for 50–100 years. However, many effects of spills are relatively short-lived and recovery of affected species and communities occurs within several years (e.g., Teal and Howarth, 1984). Sixth, biota can influence the persistence and rate of weathering of oil, either passively or through feeding and burrowing activities. Seventh, residual oiling of habitats plays a large role in continued effects on biota, but the effects and sources of contamination are not limited to habitats where oil has traditionally been thought to persist (e.g., low-energy environments). Oil may persist in moderate to high-energy environments when armouring by larger substrates or biota (mussels) occurs. Further understanding of the conditions that encourage persistence of subsurface oiling and the rates of oil weathering is needed.

The knowledge that oil often persists longer than expected in a number of habitats suggests that we should be more concerned about its continued biological effects, how they may be changing through time, and the thresholds for different types of effects. For example, long-term exposure of organisms to oil may lead to mutagenic effects (e.g., to mangroves, Klekowski et al., 1994). Shorter-term exposure (days to months) of eggs and larvae to low levels (ppb) of oil can lead to malformations, genetic damage and increased mortality (herring and salmon, Carls et al., 1999; Heintz et al., 1999). These types of effects may be caused by weathered oil, which is more toxic per unit mass than unweathered oil because the most toxic compounds are also the most refractory (Carls et al., 1999; Heintz et al., 1999). Those situations that foster persistence of oil and reduced rates of oil weathering, extend the potential for long-term biological effects. Thus, longer-term studies are still needed, especially to extend studies with pre-spill data, repeated sampling through time, and lack of recovery of habitats, populations, or communities.

Our accumulated and increasing understanding of the potential effects of persistent oil on biota should lead us to be more concerned about the spilling of oil into the world's oceans. The lengthier persistence of oil, and intricate complex of effects that spilled oil initiates are cause for concern. If oil is spilled, we need to try to minimize effects to sensitive or vulnerable species or habitats, and we need to carefully evaluate treatment and removal of oil from those situations where we expect oil to persist, or where pronounced biological effects are likely. Further well-designed studies to examine treatment options for different types of habitats and communities are needed. Prevention of spills should be of paramount importance. Legislation (such as the U.S. Oil Pollution Act of 1990) is one avenue for effecting societal change with the intent of reducing oil spills in the marine environment.

## ACKNOWLEDGMENTS

I would like to thank Dan Mann for stimulating discussions about oil persistence and his indulgence in reviewing my foray into the physical/geomorphological realm; Eric Knudsen for reviewing the manuscript; Amy DeLorenzo for valuable library assistance; and Charles

Sheppard for presenting the challenge. Support was provided by the Alaska Biological Science Center of the U.S. Geological Survey and the *Exxon Valdez* Oil Spill Trustee Council. However, the views expressed by the author are her own and do not necessarily reflect those of the Trustee Council.

## REFERENCES

Atlas, R.M. (1995) Petroleum biodegradation and oil spill bioremediation. *Marine Pollution Bulletin* **31** (4–12), 178–182.

Babcock, M.M., Carls, M.G., Harris, P.M., Irvine, G.V., Cusick, J.A. and Rice, S.D. (2000) Persistence of oiling in mussel beds after the *Exxon Valdez* oil spill. *Marine Environmental Research*, (in press).

Babcock, M.M., Irvine, G.V., Harris, P.M., Cusick, J.A. and Rice, S.D. (1996) Persistence of oiling in mussel beds three and four years after the *Exxon Valdez* oil spill. In *Proceedings of the Exxon Valdez Oil Spill Symposium: American Fisheries Society Symposium 18*, eds. S.D. Rice, R.B. Spies, D.A. Wolfe and B.A. Wright, pp. 286–297. American Fisheries Society. Bethesda, Maryland.

Baca, B.J., Lankford, T.E. and Gundlach, E.R. (1987) Recovery of Brittany coastal marshes in the eight years following the *Amoco Cadiz* incident. In *Proceedings of the 1987 International Oil Spill Conference,* pp. 459–464. American Petroleum Institute. Washington, D.C.

Baker, J.M., Guzman M., L., Bartlett, P.D., Little, D.I. and Wilson, C.M. (1993) Long-term fate and effects of untreated thick oil deposits on salt marshes. In *Proceedings of the 1993 International Oil Spill Conference,* pp. 395–399. American Petroleum Institute. Washington, D.C.

Bragg, J.R. and Yang, S.H. (1995) Clay–oil flocculation and its role in natural cleansing in Prince William Sound following the *Exxon Valdez* oil spill. *Exxon Valdez Oil Spill: Fate and Effects in Alaskan Waters, ASTM STP 1219*, eds. P.G. Wells, J.N. Butler and J.S. Hughes. American Society for Testing and Materials (ASTM), Philadelphia. pp. 178–214.

Burns, K.A., Garrity, S.D. and Levings, S.C. (1993) How many years until mangrove ecosystems recover from catastrophic oil spills? *Marine Pollution Bulletin* **26** (5), 239–248.

Burns, K.A. and Knap, A.H. (1989) The Bahía las Minas oil spill: hydrocarbon uptake by reef building corals. *Marine Pollution Bulletin* **20** (8), 391–398.

Carls, M.G., Rice, S.D. and Hose, J.E. (1999) Sensitivity of fish embryos to weathered crude oil: Part I. Low level exposure during incubation causes malformations, genetic damage and mortality in larval Pacific herring (*Clupea pallasi*). *Environmental Toxicology and Chemistry* **18**, 293–305.

Corredor, J.E., Morell, J.M. and Del Castillo, C.E. (1990) Persistence of spilled crude oil in a tropical intertidal environment. *Marine Pollution Bulletin* **21** (8), 385–388.

Cubit, J. and Connor, J.L. (1993) Effects of the 1986 Bahía las Minas oil spill on reef flat communities. In *Proceedings of the 1993 International Oil Spill Conference*, pp. 329–334. American Petroleum Institute. Washington, D.C.

Dauvin, J.-C. (1998) The fine sand *Abra alba* community of the Bay of Morlaix twenty years after the *Amoco Cadiz* oil spill. *Marine Pollution Bulletin* **36** (9), 669–676.

Dauvin, J.-C. and Gentil, F. (1990) Conditions of the peracarid populations of subtidal communities in Northern Brittany ten years after the *Amoco Cadiz* oil spill. *Marine Pollution Bulletin* **21** (3), 123–130.

Dean, T.A., Bodkin, J.L., Jewett, S.C., Monson, D.H. and Jung, D. (2000) Changes in sea urchins and kelp following a reduction in sea otter density as a result of the *Exxon Valdez* oil spill. *Marine Ecology Progress Series* (in press)

Dean, T.A., Stekoll, M.S., and Smith, R.O. (1996) Kelps and oil: the effects of the *Exxon Valdez* oil spill on subtidal algae. In *Proceedings of the Exxon Valdez Oil Spill Symposium: American Fisheries Society Symposium 18*, eds. S.D. Rice, R.B. Spies, D.A. Wolfe and B.A. Wright, pp. 412–423. American Fisheries Society. Bethesda, Maryland.

Duke, N.C., Pinzon, Z.S. and Prada, M.C. (1997) Large-scale damage to mangrove forests following two large oil spills in Panama. *Biotropica* **29** (1), 2–14.

Esler, D., Schmutz, J.A., Jarvis, R.L., and Mulcahy, D.M. (2000) Winter survival of adult female harlequin ducks in relation to history of contamination by the *Exxon Valdez* oil spill. *Journal of Wildlife Management* **64** (in press)

Foster, M.S., Tarpley, J.A., and Dearn, S.L. (1990) To clean or not to clean: the rationale, methods, and consequences of removing oil from temperate shores. *Northwest Environmental Journal* **6**, 105–120.

Garrity, S.D., Levings, S.C., and Burns, K.A. (1993) Chronic oiling and long-term effects of the 1986 Galeta spill on fringing mangroves. In *Proceedings of the 1993 International Oil Spill Conference: Prevention, Preparedness, Response*, pp. 319–324. American Petroleum Institute. Washington, D.C.

GESAMP (IMO/FAO/UNESCO/WMO/WHO/IAEA/UN/UNEP Joint Group of Experts on the Scientific Aspects of Marine Pollution). (1993) Impact of Oil and Related Chemicals and Wastes on the Marine Environment. GESAMP Reports and Studies, Vol. 50, 180 pp.

Getter, C.D., Cintron, G., Dicks, B., Lewis, R.R. III, and Seneca, E.D. (1984) The recovery and restoration of salt marshes and mangroves following an oil spill. *Restoration of Habitats Impacted by Oil Spills*, eds. J. Cairns, Jr. and A.L. Buikema, Jr. Butterworth Publishers, Boston. pp. 65–113.

Gordon, D.C. Jr., Dale, J. and Keizer, P.D. (1978) Importance of sediment working by the deposit-feeding polychaete *Arenicola marina* on the weathering rate of sediment-bound oil. *Journal of the Fisheries Research Board of Canada* **35**, 591–603.

Gundlach, E.R., Domeracki, D.D., and Thebeau, L.C. (1982) Persistence of METULA oil in the Strait of Magellan six and one-half years after the incident. *Oil and Petrochemical Pollution* **1** (1), 37–48.

Guzmán, H.M., Jackson, J.B.C., and Weil, E. (1991) Short-term ecological consequences of a major oil spill on Panamanian subtidal reef corals. *Coral Reefs* **10**, 1–12.

Hampson, G.R. and Moul, E.T. (1978) No. 2 fuel oil spill in Bourne, Massachusetts: immediate assessment of the effects on marine invertebrates and a 3-year study of growth and recovery of a salt marsh. *Journal of the Fisheries Research Board of Canada* **35**, 731–744.

Harper, J.R., Miskulin, G.A., Green, D.R., Hope, D. and Vandermeulen, J.H. (1985) Experiments on the fate of oil in low energy marine environments. In *Proceedings of the 8th Arctic and Marine Oilspill Program (AMOP) Technical Seminar*, pp. 383–399. Environment Canada.

Hawkins, S.J. and Southward, A.J. (1992) The *Torrey Canyon* oil spill: recovery of rocky shore communities. *Restoring the Nation's Marine Environment*, ed. G.W. Thayer. Maryland Sea Grant College, College Park, Maryland. pp. 583–631.

Hayes, M.O. (1996) An exposure index for oiled shorelines. *Spill Science & Technology Bulletin* **3** (3), 139–147.

Hayes, M.O., Gundlach, E.R., and D'Ozouville, L. (1979) Role of dynamic coastal processes in the impact and dispersal of the *Amoco Cadiz* oil spill (March 1978) Brittany, France. In *Proceedings of the 1979 International Oil Spill Conference*, pp. 193–198. American Petroleum Institute. Washington, D.C.

Hayes, M.O. and Michel, J. (1999) Factors determining the long-term persistence of *Exxon Valdez* oil in gravel beaches. *Marine Pollution Bulletin* **38** (2), 92–101.

Hayes, M.O., Michel, J., Montello, T.M., Aurand, D.V., Sauer, T.C., Al-Mansi, A. and Al-Momen, A.H. (1995) Distribution and weathering of oil from the Iraq–Kuwait conflict oil spill within intertidal habitats—two years later. In *Proceedings of the 1995 International Oil Spill Conference*, pp. 443–452. American Petroleum Institute. Washington, D.C.

Heintz, R.A., Short, J.W. and Rice, S.D. (1999) Sensitivity of fish embryos to weathered crude oil: Part II. Increased mortality of pink salmon (*Oncorhynchus gorbuscha*) embryos incubating downstream from weathered *Exxon Valdez* crude oil. *Environmental Toxicology and Chemistry* **18** (3), 494–503.

Highsmith, R.C., Rucker, T.L., Stekoll, M.S., Saupe, S.M., Lindeberg, M.R., Jenne, R.N. and Erickson, W.P. (1996) Impact of the *Exxon Valdez* oil spill on intertidal biota. In *Proceedings of the* Exxon Valdez *Oil Spill Symposium: American Fisheries Society Symposium 18*, eds. S.D. Rice, R.B. Spies, D.A. Wolfe and B.A. Wright, pp. 212–237. American Fisheries Society. Bethesda, Maryland.

Holland-Bartels, L.E. (1999) Mechanisms of Impact and Potential Recovery of Nearshore Vertebrate Predators Following the *Exxon Valdez* Oil Spill. U.S. Geological Survey. *Exxon Valdez* Oil Spill Restoration Project Final Report 95025-99025. Anchorage, Alaska.

Houghton, J.P., Gilmour, R.H., Lees, D.C., Driskell, W.B. and Lindstrom, S.C. (1997) Long-term recovery (1989–1996) of Prince William Sound littoral biota following the *Exxon Valdez* oil spill and subsequent shoreline treatment. National Oceanic and Atmospheric Administration Technical Memorandum. NOAA. NOS ORCA 119. Seattle, Washington.

Irvine, G.V., Mann, D.H. and Short, J.W. (1999) Multi-year persistence of oil mousse on high energy beaches distant from the *Exxon Valdez* spill origin. *Marine Pollution Bulletin* **38** (7), 572–584.

Jackson, J.B.C., Cubit, J.D., Keller, B.D., Batista, V., Burns, K., Caffey, H.M., Caldwell, R.L., Garrity, S.D., Getter, C.D., Gonzalez, C., Guzman, H.M., Kaufmann, K.W., Knap, A.H., Levings, S.C., Marshall, M.J., Steger, R., Thompson, R.C. and Weil, E. (1989) Ecological effects of a major oil spill on Panamanian coastal marine communities. *Science* **243**, 37–44.

Klekowski, E.J.Jr., Corredor, J.E., Morell, J.M. and Del Castillo, C.A. (1994) Petroleum pollution and mutation in mangroves. *Marine Pollution Bulletin* **28** (3), 166–169.

Lee, K., Stoffyn-Egli, P., Wood, P.A. and Lunel, T. (1998) Formation and structure of oil-mineral fines aggregates in coastal environments. In *Proceedings of the 21st Arctic and Marine Oilspill Program (AMOP) Technical Seminar*, pp. 911–922. Environment Canada.

Lewis, J.R. (1982) The composition and functioning of benthic ecosystems in relation to the assessment of long-term effects of oil pollution. *Philosophical Transactions of the Royal Society of London*, Series B, **297**, 257–267.

Lindstrom, J.E., Prince, R.C., Clark, J.C., Grossman, M.J., Yeager, T.R., Braddock, J.F. and Brown, E.J. (1991) Microbial populations and hydrocarbon biodegradation potentials in fertilized shoreline sediments affected by the T/V *Exxon Valdez* oil spill. *Applied and Environmental Microbiology* **57** (9), 2514–2522.

Little, D.I. and Scales, D.L. (1987) The persistence of oil stranded on sediment shorelines. In *Proceedings of the 1987 International Oil Spill Conference*, pp. 433–438. American Petroleum Institute. Washington, D.C.

Long, B.F. and Vandermeulen, J.H. (1979) Impact of clean-up efforts on an oiled saltmarsh (Ile Grande) in North Brittany, France. *Spill Technology Newsletter*, pp. 218–229.

Long, B.F.N., Vandermeulen, J.H., and Ahern, T.P. (1981) The evolution of stranded oil within sandy beaches. In *Proceedings of the 1981 International Oil Spill Conference: Prevention, Behaviour, Control, Cleanup*, pp. 519–524. American Petroleum Institute. Washington, D.C.

Loya, Y. and Rinkevich B. (1980) Effects of pollution on coral reef communities. *Marine Ecology Progress Series* **3**, 167–180.

Menot, L., Chassé, C. and Kerambrun, L. (1999) Assessment of natural cleaning and biological colonization on oiled rocky shores: in situ experiments. In *Proceedings of the 1999 International Oil Spill Conference*, American Petroleum Institute. Washington, D.C.

Michel, J., Hayes, M.O., Keenan, R.S., Jensen, J.R. and Narumalani, S. (1993) Oil in nearshore subtidal sediments of Saudi Arabia from the Gulf War spill. In *Proceedings of the 1993 International Oil Spill Conference: Prevention, Preparedness, Response*, pp. 383–388. American Petroleum Institute. Washington, D.C.

Mignucci-Giannoni, A.A. (1999) Assessment and rehabilitation of wildlife affected by an oil spill in Puerto Rico. *Environmental Pollution* **104**, 323–333.

Monson, D.H., Doak, D.F., Ballachey, B.E., Johnson, A. and Bodkin, J.L. (2000) Long-term impacts of the *Exxon Valdez* oil spill on sea otters, assessed through age-dependent mortality patterns. In *Proceedings of the National Academy of Science*, in press.

NRC (National Research Council) (1975) *Petroleum in the Marine Environment*. National Academy of Sciences, Washington, D.C.

NRC (National Research Council) (1985) *Oil in the Sea: Inputs, Fates and Effects*. National Academy Press, Washington, D.C. 601 pp.

North, W.J., Neushul, M. and Clendenning, K.A. (1965) Successive biological changes observed in a marine cove exposed to a large spillage of mineral oil. *Pollutions marines par les microorganisms et les produits petrolies: Symposium de Monaco*, pp. 335–354. Commision Internationale pour l'Exploration Scientifique de la Mer Mediterranee. Monaco.

Notini, M. (1980) Impact of oil on the littoral ecosystem. In The *Tsesis* Oil Spill (Report of the first year scientific study, Oct. 26, 1977 to Dec. 1978), pp. 129–165. National Oceanic and Atmospheric Administration. Boulder, Colorado.

Owens, E.H. (1978) Mechanical dispersal of oil stranded in the littoral zone. *Journal of the Fisheries Research Board of Canada* **35**, 563–572.

Owens, E.H. (1985) Factors affecting the persistence of stranded oil on low energy coasts. In *Proceedings of the 1985 International Oil Spill Conference*, pp. 359–365. American Petroleum Institute. Washington, D.C.

Owens, E.H. (1993) An A-B-C of oiled shorelines. In *Proceedings of the 16th Arctic and Marine Oilspill Program (AMOP) Technical Seminar*, pp. 1095–1110. Environment Canada.

Owens, E.H. (1998) Sediment relocation and tilling—underused and misunderstood techniques for the treatment of oiled beaches. In *Proceedings of the 21st Arctic and Marine Oilspill Program (AMOP) Technical Seminar*, pp. 857–871. Environment Canada.

Owens, E.H., Robson, W. and Humphrey, B. (1986) Data on the character of asphalt pavements. In *Proceedings of the 9th Annual Arctic and Marine Oilspill Program (AMOP) Technical Seminar*, pp. 1–17. Environment Canada.

Owens, E.H., Sergy, G.A., McGuire, B.E. and Humphrey, B. (1993) The 1970 *Arrow* oil spill: what remains on the shoreline 22 years later? In *Proceedings of the 16th Arctic and Marine Oilspill Program (AMOP) Technical Seminar*, pp. 1149–1167. Environment Canada.

Peterson, C.H. (2000) The web of ecosystem interconnections to shoreline habitats as revealed by the *Exxon Valdez* oil spill perturbation: a synthesis of acute direct vs. indirect and chronic effects. *Advances in Marine Biology*, in press.

Sanders, H.L., Grassle, J.F., Hampson, G.R., Morse, L., Garner-Price, S. and Jones, C.C. (1980) Anatomy of an oil spill: long-term effects from the grounding of the barge *Florida* off West Falmouth, Massachusetts. *Journal of Marine Research* **38** (2), 265–380.

Southward, A.J. (1982) An ecologist's view of the implications of the observed physiological and biochemical effects of petroleum compounds on marine organisms and ecosystems. *Philosophical Transactions of the Royal Society of London*, Series B, **297**, 241–255.

Southward, A.J. and Southward, E.C. (1978) Recolonization of rocky shores in Cornwall after use of toxic dispersants to clean up the *Torrey Canyon* spill. *Journal of the Fisheries Research Board of Canada* **35**, 682–706.

Suchanek, T.H. (1993) Oil impacts on marine invertebrate populations and communities. *American Zoologist* **33**, 510–523.

Teal, J.M., Farrington, J.W., Burns, K.A., Stegeman, J.J., Tripp, B.W.,

Woodin, B. and Phinney, C. (1992) The West Falmouth oil spill after 20 years: fate of fuel oil compounds and effects on animals. *Marine Pollution Bulletin* **24** (12), 607–614.

Teal, J.M. and Howarth, R.W. (1984) Oil spill studies: a review of ecological effects. *Environmental Management* **8** (1), 27–44.

Trust, K.A., Esler, D., Woodin, B.R. and Stegeman, J.J. (2000) Cytochrome P450 1A induction in sea ducks inhabiting nearshore areas of Prince William Sound, Alaska. *Marine Pollution Bulletin* **39** (in press.)

van Tamelen, P.G. and Stekoll, M.S. (1995) Recovery mechanisms of the brown alga, *Fucus gardneri*, following catastrophic disturbance: lessons from the *Exxon Valdez* oil spill. In *Proceedings of the Third Glacier Bay Science Symposium*, ed. D.R. Engstrom, pp. 221–227. National Park Service. Anchorage, AK.

van Tamelen, P.G., Stekoll, M.S. and Deysher, L. (1997) Recovery processes of the brown alga *Fucus gardneri* following the *Exxon Valdez* oil spill: settlement and recruitment. *Marine Ecology Progress Series* **160**, 265–277.

Vandermeulen, J.H. (1977) The Chedabucto Bay spill, *Arrow* 1970. *Oceanus* **20**, 32–39.

Vogt, H.P. (1995) Coral reefs in Saudi Arabia: 3.5 years after the Gulf War oil spill. *Coral Reefs* **14**, 271–273.

Watt, I., Woodhouse, T. and Jones, D.A. (1993) Intertidal clean-up activities and natural regeneration on the Gulf coast of Saudi Arabia from 1991 to 1992 after the 1991 Gulf oil spill. *Marine Pollution Bulletin* **27**, 325–331.

## THE AUTHOR

**Gail V. Irvine**
*U.S. Geological Survey,*
*Alaska Biological Science Center,*
*1011 E. Tudor Rd., Anchorage, AK 99503, U.S.A.*

Chapter 127

# REMOTE SENSING OF TROPICAL COASTAL RESOURCES: PROGRESS AND FRESH CHALLENGES FOR THE NEW MILLENNIUM

Peter J. Mumby

Interest in mapping tropical coastal resources using remote sensing has never been so extensive. Tropical coastal resources are vital to many national economies but are threatened by development, over-fishing and large-scale climatic phenomena including the El Niño-Southern Oscillation. Remote sensing is the only practicable means of making large-scale synoptic assessments of coastal resources and we are entering an age of unprecedented access to remotely sensed data. Generally speaking, airborne sensors with high spectral and spatial resolution distinguish finer biological detail than coarser resolution satellite sensors. The current mapping capabilities for the major tropical systems of mangrove, seagrass and reefs are: mangrove boundaries, mangrove canopy height and closure, mangrove leaf area index, seagrass boundaries, seagrass density (three classes), seagrass standing crop, reef geomorphology, reef habitats (assemblages of species and associated substrata), and living/non-living coral colonies in very shallow clear water (<1 m). Not all parameters of interest to scientists and managers can be mapped. Further research, and possibly new sensors, are needed to map coral versus algal cover and seagrass species composition. Ecologists face the important challenge of embracing spatially explicit data on coastal resources. Spatial data need to be incorporated into ecosystem process models to help bridge the gap in scale between most field studies and functional ecosystems. As remote sensing of coastal resources becomes more sophisticated (i.e. better data and analyses), the benefits of obtaining such data will grow through development of more advanced, and potentially useful, applications. I argue that the economics of undertaking remote sensing should be assessed in parallel with such developments in order to prevent over-selling or under-exploitation of the technology.

*Seas at The Millennium: An Environmental Evaluation (Edited by C. Sheppard)*

## INTRODUCTION

Virtually every ecosystem on the planet has been studied using remote sensing, and tropical coastal resources such as mangroves, seagrass beds, and coral reefs have probably received as much attention as any other system. Their popularity among the remote-sensing community stems in part from the characteristic shallow, clear water associated with many tropical coastal areas. Clear water transmits relatively high levels of photosynthetically active radiation (PAR) which is a biological necessity for photosynthetic organisms inhabiting the benthos (e.g. the algal symbionts of hermatypic corals). Clear water also allows remote-sensing instruments to measure the light reflected from the benthos and has stimulated intensive coastal research since the launch of the multispectral scanner (MSS) sensor on the Landsat satellite in 1972.

A great deal of progress has been made since the mid-1970s so why is the remote sensing of tropical coastal resources more popular than ever? There are probably three principal reasons for the subject's popularity. Firstly, tropical coastal resources provide food, income, and coastal protection for hundreds of millions of people and are vital to many national economies. However, coastal areas are threatened globally from anthropogenic development and natural disturbances. A good example of the latter is the 1997/98 El Niño-Southern Oscillation which elevated sea surface temperature (and possibly penetration of PAR), eliciting widespread coral bleaching which resulted in high levels of coral mortality at many sites, particularly in the Indo-Pacific (Wilkinson et al., 1999). Such global phenomena cannot be recorded adequately *in situ* so remote sensing is the only means of providing large-scale synoptic data on large-scale phenomena. Secondly, managers and scientists require a suite of information on coastal resources, but because remote sensing does not presently satisfy all their needs, much research is dedicated to developing methods of extracting more information from remotely sensed data. For example, it is not currently possible to distinguish seagrass species in a mixed-species bed, but this problem may be solved eventually. Thirdly, we are entering an age of unprecedented access to information where commercial high-resolution satellite sensors are being launched regularly and electronic communications are highly sophisticated. In short, remote sensing is becoming increasingly accessible to a wide community.

The aim of this chapter is to describe the current status of remote sensing in the tropical ecosystems of mangrove, seagrass, and coral reefs, and to highlight some future research needs. I begin by commenting briefly on the variety of data sources available but then focus on individual ecosystems. Finally, I make some comments regarding the economics of remote sensing and their implications for the new millennium.

## SENSORS RELEVANT TO MAPPING TROPICAL COASTAL ZONES

By the end of 1999 a large number of sensors were available for mapping tropical coastal resources. Their names and specifications are too numerous to list here. Readers looking for a more detailed treatise of the subject should access the Internet or read reviews by Green et al. (1996; 2000). Most research into tropical coastal remote sensing has utilised one or more of the sensors listed in Table 1, which are mounted on either satellite or aircraft platforms. Although other sensors exist (e.g. the new Landsat ETM with an enhanced panchromatic mode), details of advances in remote sensing have yet to reach the literature.

Most new commercial satellite sensors (e.g. Earlybird and Quickbird) will have high spatial resolution (metres) but limited spectral resolution often with only three fairly broad spectral bands. Alternatively, satellite-based hyperspectral data (many bands) will be available from sensors like ARIES (Australian Resource Information and Environment Satellite) which is scheduled for launch in 2000 and will have approximately 100 spectral bands covering the visible, near-infrared and shortwave infrared spectrum with a spatial resolution of 30 m. An important development in airborne remote sensing is the use of multiple digital video cameras, each with a distinct spectral filter over the lens so that each camera captures a different spectral band. The advantage of such systems is that image data can be supplied to the user quickly, but the selection of appropriate filters remains an area for further research.

Table 1

Principal specifications of sensors most commonly used for tropical coastal resources assessment MSS = Multispectral Scanner, TM = Thematic Mapper, SPOT = Système Probatoire de l'Observation de la Terre, XS = Multispectral Scanner, Pan = Panchromatic, CASI = Compact Airborne Spectrographic Imager.

| Specification | Satellite-borne Sensors | | | | Airborne Sensors | |
|---|---|---|---|---|---|---|
| | Landsat MSS | Landsat TM | SPOT XS | SPOT Pan | CASI | Aerial photography |
| Spatial resolution (m) | 80 | 30 | 20 | 10 | 10–0.5 | variable –>0.2 |
| No. spectral bands available | 4 | 6 | 3 | 1 | up to 21 user defined | 1 broad band |
| Radiometric resolution | 6-bit | 8-bit | 8-bit | 6-bit | 12-bit | analog |
| Area covered (km) | 185×172 | 185×185 | 60×60 | 60×60 | variable | variable |

## TYPES OF DATA ACHIEVABLE FROM REMOTE SENSING

The spatial, spectral and radiometric resolutions (Mather, 1997) of a sensor dictate how data are recorded and the information obtainable from remote sensing will vary from sensor to sensor. Assuming that information can be categorised hierarchically in terms of detail, some sensors will provide more detailed information than others. Since the detail of information is a central theme to resource assessment, it has been defined using the term "descriptive resolution" (Green et al., 1996). Descriptive resolution essentially describes the net effect of a sensor's specifications (spatial, spectral and radiometric resolution) on its ability to provide a particular type of data. For example, a hierarchy of descriptive resolution for a coral reef would include (from coarse to fine resolution): geomorphology, benthic assemblages, percentage coral cover, and the health of individual coral colonies.

## REMOTE SENSING OF MANGROVES

Mangroves often grow in dense stands and have complex aerial root systems, which make extensive field sampling logistically difficult. Remote sensing offers a rapid and non-intrusive means of making large-scale measurements on a variety of mangrove parameters. Maps of mangrove have been used to plan aquaculture (Biña et al., 1980; Populus and Lantieri, 1991), assess deforestation rates (Ibrahim and Yosuh, 1992), examine mangrove zonation (Vibulstreth et al., 1990), and predict the sensitivity of mangal to oil spills (Jensen et al., 1991). My aim here is to provide a brief overview of the remote-sensing sensors which are most widely used for mangrove assessment and comment on the type of information which may be achieved.

The simplest and most easily attainable mapping objective is locating mangrove areas and discriminating them from non-mangrove systems. Both satellite and airborne methods appear to achieve this objective with reasonable accuracy. Segregating mangrove into its constituent assemblages is more problematic. In general, satellite sensors can discriminate mangrove zonation on the basis of tree height, but not species composition (Gray et al., 1990; Green et al., 1998a). To discriminate and map more detailed zonation, aerial photography or multispectral airborne remote sensing are required. For example, Green et al. (1998b) found that the airborne multispectral instrument, CASI, could distinguish the following mangrove habitat classes with high (>70%) thematic accuracy: "*Conocarpus erectus*", "*Avicennia germinans*", "Short, high density, *Rhizophora mangle*", "Tall, low density, *Rhizophora mangle*", "Short mixed mangrove, high density", "Tall mixed mangrove, low density", and "*Laguncularia* dominated mangrove". High resolution instruments are also needed to map fringes of mangrove propagules which may constitute a dynamic zone of encroachment or mangal retraction (Fig. 1).

Fig. 1. Young red mangrove (*Rhizophora mangle*) in the Turks & Caicos Islands. The seaward edge of the mangrove forest is particularly dynamic due to growth and recruitment of new trees.

A slightly more advanced objective of remote sensing is the detection of change in mangrove resources over time. The degree of habitat detail, accuracy of the habitat map, and spatial resolution are critical limiting factors in such analyses. A successful geometric correction (registration of the imagery to map coordinates) usually has a root mean square error of ca 0.5–1 pixel widths. Therefore, if Landsat MSS was used to detect the encroachment rate of a mangal in, say, two images spaced one year apart, a positional error of 40–80 m would be expected in *each* image before encroachment could be estimated. If airborne methods were used, positional errors would be reduced to metres. Secondly, if habitat classes are not mapped accurately (e.g. 50% of pixels are misidentified, which is not uncommon), an analysis of change in habitat abundances would be highly uncertain.

One of the most sophisticated measurements obtainable from multispectral remote sensing is mangrove leaf area index (LAI). LAI is defined as the single-side leaf area per unit ground area, and is a dimensionless number. Many methods are available to measure LAI directly and are variations of either leaf sampling or litterfall collection techniques (English et al., 1997). These methods tend to be difficult to carry out in the field, are labour intensive, require many replicates to account for spatial variability in the canopy, and are thus costly. Consequently, many indirect methods of measuring LAI have been developed (see references in Nel and Wessman, 1993). Gap-fraction analyses are appropriate for use with remote sensing and assume that leaf area can be calculated from the *canopy transmittance* (the fraction of direct solar radiation which penetrates the canopy). This approach to estimating LAI involves a relatively short field period. Fortunately a linear relationship exists between mangrove LAI and the normalised difference vegetation index (Ramsey and Jensen, 1995, 1996; Green et al., 1997, 1998b) which is derived from multispectral image data with red and infrared spectral bands. Field data are used to calibrate the remotely sensed images and LAI is then predicted throughout the imagery. Landsat TM, SPOT XS and CASI sensors have the spectral bands needed to predict mangrove LAI and all are appropriate for mapping LAI.

Cheap sources of satellite imagery (e.g. Landsat MSS, Landsat TM greater than ten years old) are adequate for providing a general inventory of mangrove resources and for detecting large-scale changes over time such as mass deforestation. High-resolution and recent satellite imagery (e.g. SPOT, Landsat TM, Landsat ETM) are more appropriate for detecting medium-scale changes in mangrove habitat and predicting LAI. For accurate and small-scale (usually <100s km) measurements of change, for example if monitoring changes in undisturbed mangrove forests, airborne remote sensing is often required. Multispectral airborne sensors will provide the greatest biological resolution on mangrove canopies but the cost is greatest (ca £35,000 for 150 km$^2$).

### Future Challenges in Mangrove Remote Sensing

Measurements of leaf area index are valuable in modelling the ecological processes occurring within a forest and potentially in predicting ecosystem responses to disturbance. LAI is related to a range of ecological processes such as rates of photosynthesis, transpiration and evapotranspiration (McNaughton and Jarvis, 1983; Pierce and Running, 1988), and net primary production (Monteith, 1972; Gholz, 1982). Measurements of LAI have been used to predict future growth and yield (Kaufmann et al., 1982) and to monitor changes in canopy structure due to pollution and climate change. Since remote-sensing methods can predict mangrove LAI, mangrove ecologists face the challenge of building spatial components to physiological process models that use LAI. Conceivably, remote-sensing outputs will be used to test the predictions of such models and help build the bridge in scale between field studies and the processes affecting entire mangrove ecosystems.

## REMOTE SENSING OF TROPICAL SEAGRASS ECOSYSTEMS

Ecologists and managers of seagrass systems require a suite of data on the status of seagrass habitats (for an overview see Phillips and McRoy, 1990). Seagrass boundaries have been mapped extensively with aerial photography (Kirkman et al., 1988; Ferguson et al., 1993; Sheppard et al., 1995) and by showing the location and extent of seagrass habitat, they may be indicative of the health of coastal systems (Dennison et al., 1993). Similarly, SPOT XS imagery has been used to highlight the seagrass die-off in Florida Bay (Robblee et al., 1991).

Seagrass cover has been mapped semi-quantitatively using an airborne multispectral scanner (Savastano et al., 1984) and the satellite sensors, Landsat TM (Luczkovich et al., 1993; Zainal et al., 1993) and SPOT XS (Cuq, 1993). A more detailed study in the Bahamas obtained a quantitative empirical relationship between Landsat TM and seagrass biomass (Armstrong, 1993). Mumby et al. (1997a) obtained similar quantitative relationships between seagrass standing crop and data from Landsat TM, SPOT XS, and CASI. The radiative transfer processes underlying this empirical relationship were then modelled by Plummer et al. (1997). The most ambitious goal for remote sensing of seagrass has been species composition. However, recent evidence suggests that this goal is unlikely to be realised because spectral sensors lack the sensitivity to distinguish the spectra of different species (Jernakoff and Hick, 1994).

The main limiting circumstances for optical remote sensing of seagrass are cloud cover (particularly for satellite imagery), turbid or deep water, and macroalgae, which may strongly resemble seagrass. Excessive cloud cover in a region will reduce the chance of obtaining suitable imagery, and severe light attenuation in deep and turbid water can make optical remote sensing inappropriate. Under such circumstances, acoustic (Sabol et al., 1997) or videographic (Norris et al., in press) remote-sensing methods are preferable, although both are currently under development. Distinguishing seagrasses from macroalgae requires spectral separation of their principal chlorophyll and accessory pigments and will probably require good multispectral or hyperspectral sensors.

SPOT XS and Landsat TM imagery date back to the mid-1980s and can, in principle, be used to assess seagrass dynamics over long periods. However, the pixel dimensions of these sensors are 20 m and 30 m respectively, so neither sensor is likely to be sensitive to changes in seagrass distribution unless the change constitutes several pixels (ca 100 m). Added to this is the uncertainty of correctly classifying a pixel as seagrass—particularly at the edge of a seagrass bed where spectral confusion with neighbouring habitats is greatest. Under these circumstances, it is arguable whether remote sensing would be required to measure such extensive changes in cover (i.e. local observations may suffice). Airborne methods such as photography or multispectral scanning are needed to measure small scale (~10 m) dynamics of seagrass. Most monitoring of seagrass boundaries has been based on aerial photography, although digitisation times can be lengthy and the relative merits of aerial photography and airborne multispectral scanning are yet to be evaluated in this context. For monitoring the distribution of seagrass biomass with high spatial resolution (i.e. metres), airborne multispectral scanners are appropriate.

### Future Challenges in Seagrass Remote Sensing

Under favourable conditions remote sensing can provide useful ecological data on the location of seagrass boundaries and distribution of seagrass standing crop. The challenge now is to take advantage of such data and extend the ecology of seagrass communities to explicit spatial scales. Spatial statistics such as variograms and fractals (Farina, 1998) can be calculated for maps of seagrass standing crop and will reveal the spatio-temporal heterogeneity of seagrass communities. Spatial analyses may be purely exploratory to begin with, but will inevitably reveal patterns that generate testable hypotheses concerning the processes governing seagrass community structure.

## REMOTE SENSING OF CORAL REEF ECOSYSTEMS

Satellite investigations of coral reef structure date back to 1975, shortly after the launch of Landsat MSS (Smith et al., 1975). Since then, the search for applications of satellite imagery to reef science and management has been almost exhaustive (see reviews by Jupp, 1986; McCracken and Kingwell, 1988; Green et al., 1996). Satellite imagery has been used for cartographic base mapping (Jupp et al., 1985), detecting change in coastal areas (Loubersac et al., 1989; Zainal et al., 1993), environmental sensitivity mapping (Biña, 1982), charting bathymetry (Benny and Dawson, 1983), fisheries management (Populus and Lantieri, 1991) and even stock assessment of commercial gastropods (Bour et al., 1986). The most widespread use of satellite imagery has been the mapping and inventory of coastal resources (e.g. Kuchler et al., 1986; Bastin, 1988; Luczkovich et al., 1993; Mumby et al., 1995).

Remote sensing provides information on several parameters which are of importance to reef science and management (see hierarchy in Table 2). From a remote-sensing point of view, the easiest of these to map are coral reef boundaries. The next level of sophistication is to distinguish principal geomorphological zones of the reef (e.g. reef flat, reef crest, spur and groove zone). For management purposes, such information may be used to provide a background for planning but these maps may also have more sophisticated ecological uses which include the stratification of field sampling regimes. Maps of reef geomorphology have been made with considerable success (Table 2), although attempts to make detailed comparisons between

Table 2

The descriptive resolution achieved for coral reefs using high resolution sensors and photography. A broad hierarchy of descriptive detail is given for each habitat. Numbers denote number of papers reviewed by Green et al. (1996) in each category and broadly reflect the popularity and limitation of each sensor. API = aerial photography.

| Type of class | Landsat MSS | Landsat TM | SPOT XS | SPOT Pan | Air-borne MSS | API |
|---|---|---|---|---|---|---|
| Coral reef (general) | – | 3 | 1 | – | – | 2 |
| Reef geomorphology | 10 | 2 | 6 | 1 | – | 7 |
| Reef habitats | – | 5 | 3 | – | 1 | 1 |
| Coral colony density (shallow reef flats only) | – | – | – | – | – | 3 |
| Live coral cover (shallow reef flats only) | – | – | – | – | – | 3 |
| Species determination | – | – | – | – | – | – |

the ability of different sensors are difficult because of the differences in reef terminology, study sites and the objectives employed in each study. The most general (and obvious) rule to emerge is that sensors with higher resolution will offer greater detail (Bainbridge and Reichelt, 1989; Ahmad and Neil, 1994).

Mapping the ecological components of coral reefs appears to be considerably more difficult for remote sensing (Bainbridge and Reichelt, 1989). Ecological components may be defined in various ways including assemblages of coral species (e.g. Done, 1983), assemblages of coral and non-coral species, or assemblages of species and substrata (e.g. Sheppard et al., 1995; Mumby et al., 1997b; Mumby and Harborne, 1999). The choice depends on the ecological objective and physiognomy of the area. For example, coral species assemblages would be appropriate in places where coral cover was high, but perhaps less appropriate where coral cover rarely exceeded 20%. Irrespective of their foundation, maps with an ecological basis will be referred to as "habitat maps". Maps of reef habitat are a useful planning tool which, among other uses, allow management boundaries to be located (Kenchington and Claasen, 1988) and the identification of representative reef habitats (McNeill, 1994). Colour aerial photography has been used successfully to map tropical marine habitats (Sheppard et al., 1995), whereas satellite-borne sensors seem to be unable to map habitats in detail (Fig. 2) (Mumby et al., 1997b, 1998b). For example, a number of authors have reported difficulties using satellite data to distinguish coral habitats from seagrass beds (e.g. Luczkovich et al., 1993; Zainal et al., 1993). The specifications of most satellite sensors (Table 1) appear to be inadequate for distinguishing marine habitats below hectare scales. For example, Mumby et al. (1999) measured the average patch diameter of reefal habitats in the Turks and Caicos Islands and found it to be ca 20 m with a range

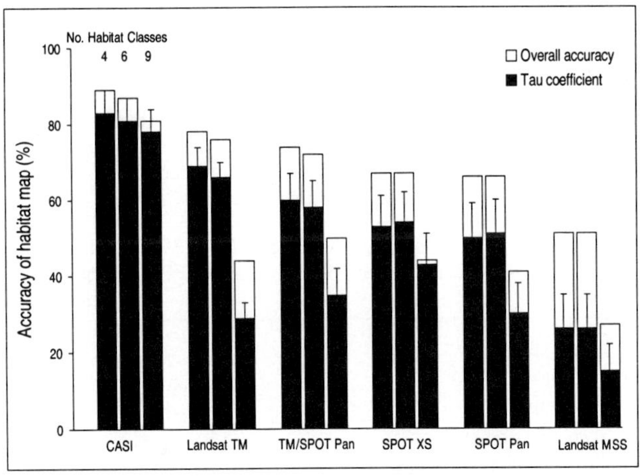


Fig. 2 Thematic accuracies of the airborne multispectral scanner CASI and five types of satellite imagery for mapping reef habitats in the Turks and Caicos Islands, West Indies. Three levels of descriptive resolution are given, the coarsest being simply coral, seagrass, algal and sand classes. The finest level of descriptive resolution includes nine reefal classes. Note that only CASI imagery manages to retain high levels of accuracy at this level of habitat discrimination. Error bars denote standard error of Tau coefficient. See Mumby et al. (1997b; 1998a) for further details.

of <10–50 m. Thus, the spatial resolution of satellite sensors with the greatest number of spectral bands (Landsats MSS and TM) is too coarse (80 m and 30 m) to represent individual habitats in each pixel (i.e. most pixels represent a mixture of habitats). Further, those sensors with higher spatial resolution (SPOT XS and SPOT Pan) have the fewest spectral bands which inhibits their ability to distinguish reef spectra (spectral signatures). However, with appropriate image processing, airborne multispectral data are able to distinguish and map reef habitats to a depth of ca 18 m (Fig. 2) (Mumby et al., 1998a,b). The importance of this result stems from the way in which habitats were categorised. Extensive field data were collected using standard ecological methods (percentage cover of biota and substrata in 1 $m^2$ quadrats) and then categorised objectively using multivariate methods (Mumby and Harborne, 1999). Therefore, the resulting classes were ecological representations of habitat and remote sensing permitted these habitats to be mapped accurately. Such mapping explicitly links the scale of individual habitat patches and the overall seascape, but the ecological applications of such maps are yet to be explored.

Moving beyond mapping reef habitats to the status of individual coral colonies, it seems likely that only high-resolution (low-altitude) airborne methods will be successful. Catt and Hopley (1988) and Thamrongnawasawat and Catt (1994) have used infrared aerial photography to estimate percent live coral cover. Healthy corals strongly reflect near-infrared radiation, so these wavelengths are particularly useful for discriminating coral growth from some algae. However, infrared radiation will only penetrate clear water to approximately 1 m, so the major limitation of the technique is the poor depth of penetration which renders its use largely confined to assessing the coral cover of emergent and extremely shallow reef flats at low tide. Coral or algal species discrimination does not appear to be possible although this is barely surprising considering the spatial resolution of most sensors and the arguments made above.

### Future Challenges in Coral Reef Remote Sensing

Most deleterious processes on coral reefs such as nutrification and overfishing result in reduced coral cover and increased algal cover (Hughes, 1994; Lapointe, 1997). Detecting changes in coral and algal cover is therefore a key goal of remote sensing. However, no remote-sensing method has been proven to have the sensitivity required to detect quantitative changes in coral and algal cover (e.g. losses of coral cover from 10% to 3% cannot be measured using remote sensing). Perhaps the greatest potential for resolving these issues is the use of hyperspectral remote sensing with high spatial resolution. Several studies have undertaken the measurement of coral and algal spectra *in situ* (Holden and Ledrew, 1998, 1999; Myers et al., 1999; Clark et al., 2000) and there appears to be reasonable grounds to expect differentiation of bleached versus non-bleached coral, recently dead versus long-dead coral, and coral versus macroalgae. The key question is whether such spectral differences can be resolved by remote-sensing instruments given that the distinguishing spectra are often confined to narrow portions of the electromagnetic spectrum and often have high attenuation in water. Modelling studies have now begun which take a physical approach to the question of spectral discrimination and will help design future sensors for reef remote sensing.

A surprising additional challenge for the remote-sensing community is the basic mapping of the World's coral reefs. Despite 30 years of reef remote sensing, many reefs have never been mapped although such maps are needed to understand the connectivity between reef systems (Mumby, 1999) and develop transboundary coastal management initiatives. Efforts to map the world's reefs are being spear-headed by *Reefbase*, the global database on coral reefs (McManus and Alban, 1997).

## ECONOMIC CONSIDERATIONS FOR REMOTE SENSING

Two years ago, my colleagues and I questioned 60 coastal managers from tropical areas on the cost-effectiveness of remote sensing and found that 70% believed cost to be the main hindrance to remote sensing. We then assessed the cost of producing habitat maps without using remote sensing (i.e. using boat-based surveys) and found it to be considerably more expensive than using imagery. For example, to map the Caicos Bank using boat-based surveys would cost around £380,000 and would take a team of three surveyors over eight years to complete! Remote-sensing methods would cost less than £4000 and would take weeks.

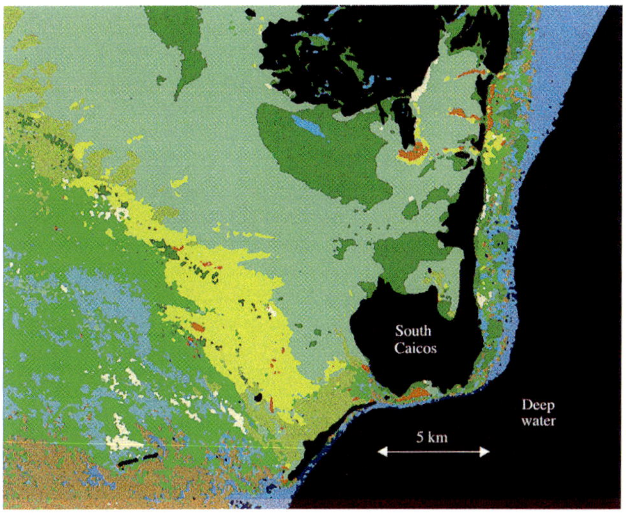


Fig. 3. Coastal habitat map from the Turks & Caicos Islands derived from Landsat TM data. Although the map looks "attractive", it really attempts to show too much detail as the overall accuracy of the map is only 37%. A more appropriate map would only show four classes representing a more general classification of habitats (i.e. coral reef, seagrass bed, algal bed, and sand). In this case, the four-class map had an overall accuracy of 73%.

We concluded that habitat mapping is expensive, but in fact remote sensing is the most cost-effective tool to achieve the objective (Mumby et al., 1999). Having said that, some cheaper remote-sensing products can be seductively detailed but surprisingly misleading. To demonstrate the point, a detailed habitat map is shown of the Caicos Bank created from Landsat TM data (Fig. 3). The map is fairly detailed which intuitively tends to suggest that it must also be accurate. However, the overall accuracy of the map in terms of habitat labels is only 37%. If this map was used as a basis for planning, management decisions might be flawed, possibly having economic and environmental consequences. However, if a more accurate map had been created using alternative (probably more expensive) technology, the decisions might have been different and accurate. This begs the question, "What is the economic value of accurate spatial data on coastal resources?". I see this as a major issue to be resolved in the foreseeable future. At the end of the 20th century, habitat maps do not have particularly sophisticated applications in coastal management, but with the plethora of improved technologies under development, better quality maps will become available permitting more sophisticated management applications, particularly in monitoring contexts. At this point, the economics of remote sensing will warrant renewed scrutiny.

## ACKNOWLEDGEMENTS

I would like to thank my groundbusting co-authors over the last few years who have made remote sensing so enjoyable: Al Edwards, Ed Green, Chris Clark, and Al Harborne. Charles Sheppard is thanked for inviting me to write this review. I am funded by a NERC post-doctoral fellowship under the Earth Observation Science Initiative.

## REFERENCES

Ahmad, W. and Neil, D.T. (1994). An evaluation of Landsat Thematic Mapper (TM) digital data for discriminating coral reef zonation: Heron Reef (GBR). *International Journal of Remote Sensing* **15**, 2583–2597.

Armstrong, R.A. (1993) Remote sensing of submerged vegetation canopies for biomass estimation. *International Journal of Remote Sensing* **14**, 10–16.

Bainbridge, S.J. and Reichelt, R.E. (1989) An assessment of ground truthing methods for coral reef remote sensing data. In *Proceedings of the 6th International Coral Reef Symposium*, Vol. 2, eds. J.H. Choat et al., pp. 439–444. Sixth International Coral Reef Symposium Committee, Townsville.

Bastin, J. (1988) Measuring reef areas using satellite images. *Proceedings of the Symposium on Remote Sensing of the Coastal Zone*, Gold Coast, Queensland. Department of Geographic Information, Brisbane, pp. VII1.1–VII1.9.

Benny, A.H. and Dawson, G.J. (1983) Satellite imagery as an aid to bathymetric charting in the Red Sea. *Cartography Journal* **20**, 5–16.

Biña, R.T. (1982) Application of Landsat data to coral reef management in the Philippines. *Proceedings of the Great Barrier Reef Remote Sensing Workshop*, James Cook University, Townsville, Australia. pp. 1–39.

Biña, R.T., Jara, R.B. and Roque, C.R. (1980) Application of multi-level remote sensing survey to mangrove forest resource management in the Philippines. *Proceedings of the Asian Symposium on Mangrove Development: Research and Management*, University of Malaya, Kuala Lumpar, Malaysia

Bour, W., Loubersac, L. and Rual, P. (1986) Thematic mapping of reefs by processing of simulated SPOT satellite data: application to the *Trochus niloticus* biotope on Tetembia Reef (New Caledonia). *Marine Ecology Progress Series* **34**, 243–249.

Catt, P. and Hopley, D. (1988) Assessment of large scale photographic imagery for management and monitoring of the Great Barrier Reef. *Proceedings of the Symposium on Remote Sensing of the Coastal Zone*, Gold Coast, Queensland. Department of Geographic Information, Brisbane. pp. III.1.1–1.14.

Clark, C.D., Mumby, P.J., Chisholm, J.R.M. and Jaubert, J. (2000) Spectral discrimination of coral mortality states following a severe bleaching event. *International Journal of Remote Sensing*, in press.

Cuq, F. (1993) Remote sensing of sea and surface features in the area of Golfe d'Arguin, Mauritania. *Hydrobiologica* **258**, 33–40.

Dennison, W.C., Orth, R.J., Moore, K.A., Stevenson, J.C., Carter, C., Kollar, S., Bergstrom, P.W. and Batiuk, R.A. (1993) Assessing water quality with submersed aquatic vegetation: habitat requirements as barometers of Chesapeake Bay health. *BioScience* **43**, 86–94.

Done, T.J. (1983) Coral zonation: its nature and significance. In *Perspectives on Coral Reefs*, ed. D.J. Barnes. Brian Clouston Publishing, Canberra. pp. 107–147.

English, S., Wilkinson, C. and Baker, V. (1997) *Survey Manual for Tropical Marine Resources*. 2nd Edition. ASEAN–Australia Marine Science Project: Living Coastal Resources. AIMS, Townsville.

Farina, A. (1998) *Principles and Methods in Landscape Ecology*. Chapman & Hall, London.

Ferguson, R.L., Wood, L.L. and Graham, D.B. (1993) Monitoring spatial change in seagrass habitat with aerial photography. *Photogrammetric Engineering & Remote Sensing* **59**, 1033–1038.

Gholz, H.L. (1982) Environmental limits on above-ground net primary production, leaf area and biomass in vegetation zones of the Pacific Northwest. *Ecology* **63**, 469–481.

Gray, D., Zisman, S. and Corves, C. (1990) Mapping of the mangroves of Belize. Technical Report, University of Edinburgh, Geography Department, UK.

Green, E.P., Clark, C.D., Mumby, P.J., Edwards, A.J. and Ellis, A.C. (1998a) Remote sensing techniques for mangrove mapping. *International Journal of Remote Sensing* **19**, 935–956.

Green, E.P., Mumby, P.J., Edwards, A.J., Clark, C.D. and Ellis, A.C. (1998b) The assessment of mangrove areas using high resolution multispectral airborne imagery (CASI). *Journal of Coastal Research* **14**, 433–443.

Green, E.P., Mumby, P.J., Edwards, A.J., Clark, C.D. (1996) A review of remote sensing for tropical coastal resources assessment and management. *Coastal Management* **24** (1), 1–40.

Green, E.P., Mumby, P.J., Ellis, A.C., Edwards, A.J. and Clark, C.D. (1997) Estimating leaf area index of mangroves from satellite data. *Aquatic Botany* **58**, 11–19.

Green, E.P., Mumby, P.J., Edwards, A.J. and Clark, C.D. (2000) *Remote Sensing Handbook for Tropical Coastal Management*. UNESCO, Paris. In press.

Holden, H. and LeDrew, E. (1998) Spectral discrimination of healthy and non-healthy corals based on cluster analysis, principal components analysis, and derivative spectroscopy. *Remote Sensing of Environment* **65**, 217–224.

Holden, H. and LeDrew, E. (1999) Hyperspectral identification of coral reef features. *International Journal of Remote Sensing* **20**, 2545–2563.

Hughes, T.P. (1994) Catastrophes, phase shifts and large-scale degradation of a Caribbean coral reef. *Science* **265**, 1547–1551.

Ibrahim, M. and Yosuh, M. (1992) Monitoring the development impacts on the coastal resources of Pulau Redang marine park by remote sensing. In *Third ASEAN Science & Technology Week Conference Proceedings*, Vol. 6 Marine Science: Living Coastal Resources, eds. L.M. Chou and C.R. Wilkinson. University of Singapore, Singapore. pp. 407–413.

Jernakoff, P. and Hick, P. (1994) Spectral measurements of marine habitat: simultaneous field measurements and CASI data. *Proceedings of the 7th Australasian Remote Sensing Conference*,pp. 706–713.

Jensen, J.R., Ramsey, E., Davis, B.A. and Thoemke, C.W. (1991) The measurement of mangrove characteristics in south-west Florida using SPOT multispectral data. *Geocartography International* **2**, 13–21.

Jupp, D.L.B. (1986) The application and potential of remote sensing in the Great Barrier Reef region. Great Barrier Reef Marine Park Authority Research Publication.

Jupp, D.L.B., Mayo, K.K., Kuchler, D.A., Claasen, D.V., Kenchington, R.A. and Guerin, P.R. (1985) Remote sensing for planning and managing the Great Barrier Reef, Australia. *Photogrammetria* **40**, 21–42.

Kaufmann, M.R., Edminster, C.B. and Troendle, C., 1982. Leaf area determinations for subalpine tree species in the central Rocky Mountains. U.S. Dep. Agric. Rocky Mt. For. Range. Exp. Stn. Gen. Tech. Rep., RM-238.

Kenchington, R.A. and Claasen, D.R. (1988) Australia's Great Barrier Reef—management technology. *Proceedings of the Symposium on Remote Sensing of the Coastal Zone*, Gold Coast, Queensland, Department of Geographic Information, Brisbane. KA.2.2–2.13.

Kirkman, H., Olive, L. and Digby, B. (1988) Mapping of underwater seagrass meadows. *Proceedings of the Symposium Remote Sensing Coastal Zone*, Gold Coast, Queensland. Department of Geographic Information, Brisbane. VA.2.2–2.9.

Kuchler, D.A., Jupp, D.L.B., Claasen, D.R. and Bour, W. (1986a) Coral reef remote sensing applications. *Geocarto International* **4**, 3–15.

Lapointe, B.E. (1997) Nutrient thresholds for bottom-up control of macroalgal blooms on coral reefs in Jamaica and southeast Florida. *Limnology & Oceanography* **42**, 1119–1131.

Loubersac, L., Dahl, A.L., Collotte, P., LeMaire, O., D'Ozouville, L. and Grotte, A. (1989) Impact assessment of Cyclone Sally on the almost atoll of Aitutaki (Cook Islands) by remote sensing. *Proceedings of the 6th International Coral Reef Symposium*, Vol. 2, eds. J.H. Choat et al., pp. 455–462. Sixth International Coral Reef Symposium Committee, Townsville.

Luczkovich, J.J., Wagner, T.W., Michalek, J.L. and Stoffle, R.W. (1993) Discrimination of coral reefs, seagrass meadows, and sand bottom types from space: a Dominican Republic case study. *Photogrammetric Engineering & Remote Sensing* **59**, 385–389.

Mather, P.M. (1997) *Computer Processing of Remotely-sensed Images: An Introduction*. 2nd Edition. John Wiley & Sons, Chichester.

McCracken, K.G. and Kingwell, J. (1988) *Marine and Coastal Remote Sensing in the Australian Tropics*. CSIRO Office of Space Science and Applications, Canberra.

McManus, J.W. and Ablan, M.C. (1997) Reefbase: A global database on coral reefs and their resources. International Center for Living Aquatic Resources Management, Manila.

McNaughton, K.G. and Jarvis, P.G. (1983) Predicting effects of vegetation changes on transpiration and evaporation. In *Water Deficits and Plant Growth*. Vol. 7, ed. T.T. Kozlowski. Academic Press, London, UK. pp. 1–47.

McNeill, S.E. (1994) The selection and design of marine protected areas: Australia as a case study. *Biodiversity & Conservation* **3**, 586–605.

Monteith, J.L. (1972) Solar radiation and productivity in tropical ecosystems. *Journal of Applied Ecology* **9**, 747–766.

Mumby, P.J. (1999) Can Caribbean coral populations be modelled at metapopulation scales? *Marine Ecology Progress Series* **180**, 275–288.

Mumby, P.J., Clark, C.D., Green, E.P. and Edwards, A.J. (1998a) The practical benefits of water column correction and contextual editing for mapping coral reefs. *International Journal of Remote Sensing* **19**(1), 203–210.

Mumby, P.J., Green, E.P., Clark, C.D. and Edwards, A.J. (1998b) Digital analysis of multispectral airborne imagery of coral reefs. *Coral Reefs* **17** (1), 59–69

Mumby, P.J., Gray, D.A., Gibson, J.P. and Raines, P.S. (1995) Geographic Information Systems: a tool for integrated coastal zone management in Belize. *Coastal Management* **23**, 111–121.

Mumby, P.J., Green, E.P., Edwards, A.J. and Clark, C.D. (1999) The cost-effectiveness of remote sensing for tropical coastal resources assessment and management. *Journal of Environmental Management* **55**, 157–166

Mumby, P.J., Green, E.P., Edwards, A.J. and Clark, C.D. (1997a) Coral reef habitat mapping: how much detail can remote sensing provide? *Marine Biology* **130** (2), 193–202.

Mumby, P.J., Green, E.P., Edwards, A.J. and Clark, C.D. (1997b) Measurement of seagrass standing crop using satellite and digital airborne remote sensing. *Marine Ecology Progress Series* **159**, 51–60.

Mumby, P.J. and Harborne, A.R. (1999) Development of a systematic classification scheme of marine habitats to facilitate regional management of Caribbean coral reefs. *Biological Conservation* **88**(2), 155–163.

Myers, M., Hardy, J.T., Mazel, C. and Dustan, P. (1999) Optical spectra and pigmentation of Caribbean reef corals and macroalgae. *Coral Reefs* **18**, 179–186.

Nel, E.M. and Wessman, C.A. (1993) Canopy transmittance models for estimating forest leaf area index. *Canadian Journal of Forest Research* **23**, 2579–2586.

Norris, J.G., Wyllie-Echeverria, S., Mumford, T., Bailey, A. and Turner, T. (1997) Estimating basal area coverage of subtidal seagrass beds using underwater videography. *Aquatic Botany* **58** (3–4) 269–287.

Phillips, R.C. and McRoy, C.P. (1990) *Seagrass Research Methods*.

Monographs on Oceanographic Methodology 9. UNESCO, Paris.
Pierce, L.L. and Running, S.W. (1988) Rapid estimation of coniferous forest leaf area index using a portable integrating radiometer. *Ecology* **69**, 1762–1767.
Plummer, S.E., Malthus, T.J. and Clark, C.D. (1997) Adaptation of a canopy reflectance model for sub-aqueous vegetation: definition and sensitivity analysis. *Proceedings of the 4th International Conference Remote Sensing for Marine & Coastal Environments*, Orlando, 1, pp. 149–157.
Populus, J. and Lantieri, D. (1991) Use of high resolution satellite data for coastal fisheries. *FAO RSC Series* No. 58, Rome.
Ramsey, E.W. and Jensen, J.R. (1995) Modelling mangrove canopy reflectance by using a light interception model and an optimisation technique. In *Wetland and Environmental Applications of GIS*, Lewis, Chelsea, Michigan, USA.
Ramsey, E.W. and Jensen, J.R. (1996) Remote sensing of mangrove wetlands: relating canopy spectra to site-specific data. *Photogrammetric Engineering and Remote Sensing* **62** (8), 939–948.
Robblee, M.B., Barber, T.R., Carlson, P.R., Durako, M.J., Fourqurean, J.W., Muehlstein, L.K., Porter, D., Yarbro, L.A., Zieman, R.T. and Zieman, J.C. (1991) Mass mortality of the tropical seagrass *Thalassia testudinum* in Florida Bay (USA). *Marine Ecology Progress Series* **71**, 297–299.
Sabol, B., McCarthy, E. and Rocha, K. (1997) Hydroacoustic basis for detection and characterisation of eelgrass (*Zostera marina*). *Proceedings of the 4th International Conference Remote Sensing for Marine & Coastal Environments* 1, pp. 679–693.
Savastano, K.J., Faller, K.H. and Iverson, R.L. (1984) Estimating vegetation coverage in St. Joseph Bay, Florida with an airborne multispectral scanner. *Photogrammetric Engineering & Remote Sensing* **50**, 1159–1170.
Sheppard, C.R.C., Matheson, K., Bythell, J.C., Murphy, P., Blair Myers, C. and Blake, B. (1995) Habitat mapping in the Caribbean for management and conservation: use and assessment of aerial photography. *Aquatic Conservation: Marine & Freshwater Ecosystems* **5**, 277–298.
Smith, V.E., Rogers, R.H. and Reed, L.E. (1975) Thematic mapping of coral reefs using Landsat data. *Proceedings of the 10th International Symposium Remote Sensing of Environment* 1, 585–594.
Thamrongnawasawat, T. and Catt, P. (1994) High resolution remote sensing of reef biology: the application of digitised air photography to coral mapping. *Proceedings of the 7th Australasian Remote Sensing Conference*, Melbourne, pp. 690–697.
Vibulstreth, S., Downreang, D., Ratanasermpong, D. and Silapathong, C. (1990) Mangrove forest zonation by using high resolution satellite data. *Proceedings of the 11th Asian Conference on Remote Sensing* D1-6.
Wilkinson, C., Linden, O., Cesar, H., Hodgson, G., Rubens, J. and Strong, A.E. (1999) Ecological and socioeconomic impacts of 1998 coral mortality in the Indian Ocean: An ENSO impact and a warning of future change? *Ambio* **28**, 188–196.
Zainal, A.J.M., Dalby, D.H. and Robinson, I.S. (1993) Monitoring marine ecological changes on the east coast of Bahrain with Landsat TM. *Photogrammetric Engineering & Remote Sensing* **59**, 415–421.

## THE AUTHOR

**Peter J. Mumby**
*Centre for Tropical Coastal Management Studies,*
*Department of Marine Sciences & Coastal Management,*
*Ridley Building, The University,*
*Newcastle upon Tyne, NE1 7RU, U.K.*

Chapter 128

# SATELLITE REMOTE SENSING OF THE COASTAL OCEAN: WATER QUALITY AND ALGAE BLOOMS

Bertil Håkansson

This review covers ocean colour remote sensing of the coastal ocean. The name 'sea colour' is a better overall concept of the problems related to optical remote sensing of Case 2 waters. This optically complex water type cannot readily be measured from space using current and previous satellite sensors, in contrast with ocean colour remote sensing. Nevertheless, the image data available for the coastal ocean either have very coarse geometric resolution and high temporal coverage (AVHRR, CZCS, SeaWiFS) or the reverse with finer geometric resolution and lower temporal cover (TM and MSS). In both cases the radiometric resolution is very coarse for coastal and Case 2 waters, which often require a more detailed focus than does the large expanse of ocean. However, there are several applications and products at an advanced stage of development, which can be in operation on demand, such as Secchi disk depth, diffusive attenuation coefficient, total suspended matter and algae bloom monitoring. These products are valid on a regional basis, though much effort remains to make them globally valid, and indeed, this may not ever be possible. The next generation of sensors (cf. MODIS, MERIS) are likely to advance the coastal ocean colour problem toward the same level of effectiveness and utility as ocean colour monitoring is today, but this still needs to be demonstrated during the coming decades of the third millennium.

*Seas at The Millennium: An Environmental Evaluation (Edited by C. Sheppard)*

## INTRODUCTION

Remote sensing as a technique to observe the world from a remote and moveable platform stems from the early decades of the twentieth century. The methods and techniques have expanded dramatically since then, not least several oceanographic applications after two *Oceanography from Space* conferences (Ewing, 1965; Gower, 1981) which established possibilities and demands for future applications. The last thirty years have seen many new sensors and applications, especially in global oceanography. For applications in the coastal ocean, evolution has been slower, due to more complicated conditions and an order of magnitude smaller spatial and temporal scales compared to the ones found in the ocean, putting higher technical requirements on the satellite sensors. Some solutions are already at hand. There are several satellite missions to come (IOCCG, 1998) addressing the coastal ocean water quality problem besides pure scientific interests. Research and applications during the first decade of this millennium will show how successful this tool can be for science, management and monitoring in the coastal ocean.

This chapter is limited to applications concerned with environmental problems related to water quality in the coastal ocean and semi-enclosed seas. Satellite remote sensing of electromagnetic radiation in the visual part of the spectrum is most sensitive to inherent and apparent water optical characteristics. In the literature it is often called ocean colour remote sensing, but this has no counterpart in nomenclature regarding the coastal ocean. Here it will be called "sea colour remote sensing", just to stress the difference from the global ocean case. In optical oceanography ocean waters are named Case 1 waters as opposed to the Case 2 waters found in the coastal ocean, due to differences in optically active constituents determining the optical conditions (Jerlov, 1976). Furthermore, this review concentrates on studies which present possibilities for near-future management, and surveillance and monitoring of environmental problems and hazards. The choice was to present a guide in which references and summary reports can be found as well as literature on new sensors and satellites. There are also several other reviews to be found in the literature but the ones to mention here, which are related to the coastal ocean, are Cracknell (1989, 1999). This chapter therefore reviews what has been done during the last 30 years in the field of coastal ocean environment, using examples.

A modern definition of remote sensing is: "The use of electromagnetic radiation sensors to record images of the environment, which can be interpreted to yield useful information". Of extreme importance for the success of remote sensing is its dependence on the evolutionary technical improvement of telecommunications, data storage and computers, both software and hardware, as well as on carrier technology such as satellites, rockets, shuttles and unmanned airborne vehicles (UAVs).

Civil applications of remote sensing have in the past lacked real user-driven needs, with a few exceptions such as in meteorology, whereas in oceanography and in many terrestrial sciences, technology and research has been the driving force. The development of remote-sensing products of geophysical interest has always lagged behind the introduction of new remote-sensing methods. Hence, in many cases, the remote-sensing imagery has been used only qualitatively, i.e. to view the image, and quantitative products have only been available for the last 20 years. Today, there are some cases of operational products for marine biology, but more in physical oceanography related to modelling and monitoring.

During the exploratory period (1970–1995) there was a strong belief that remote sensing would be the technique to bring new knowledge into marine sciences. However, the technique was rather oversold and could not stand up to expectations. It turned out to be cumbersome to extract high quality information from marine remote-sensing data, and this is still by far the most important issue for marine remote-sensing scientists. The development of algorithms, estimating geophysical variables quantitatively, is an iterative process. The algorithms evolve as scientific understanding grows of processes about the interaction of electromagnetic radiation with the surrounding nature. Hence, it is likely that we can expect that re-analysis of image data sets will be on the agenda for a long time to come.

It must also be remembered that satellite data is an expensive affair, including development of scientific and technological ideas, putting the satellite and its sensors into earth orbit, and eventually gathering and distributing reliable data. Imagery costs are generally very high, even for research purposes, and it is only in the last ten years that sufficient data has become available for demonstration and semi-operational applications. Changes in data and cost policy are, however, underway. The European Space Agency recently published a document, detailing a new data distribution and cost policy. In addition, a change for the LANDSAT satellite imagery costs is expected in the near future (Reichardt, 1999). Since the end of the cold war in the early 1990s, ties with civilian needs have been more heavily emphasised, and there is hope for a much improved infrastructure for satellite data in future. There is also a clear trend going from large multi-sensor vehicles with many different purposes towards single-sensor vehicles with dedicated purposes, presumably a quicker and less expensive affair for governmental space agencies and commercial enterprises. It is with great confidence we can look forward to several missions for coastal ocean applications in the near future, from NASA, ESA and NASDA.

Image data sources during the last 30 years have mainly come from the CZCS, AVHRR, TM and MSS sensors. The CZCS was the first ocean colour satellite sensor and there have been several journal special issues devoted to basic results (i.e., *JGR*, Vol. 99, No. C4, 1994 and *International*

*Journal of Remote Sensing*, Vol. 20, No. 7, 1999). Looking into the near-future, the next generation ESA earth-watching satellite, ENVISAT, carries the ocean colour sensor MERIS with a geometrical resolution of 300 m with 11 channels in the visible part of the spectrum (*International Journal of Remote Sensing*, Vol. 20, No. 9). Due to its relatively high geometric and radiometric resolution, it is of great interest for applications in the coastal ocean. The first NASA EOS satellite, Terra, recently put into earth orbit, carries the ocean colour sensor MODIS (Esaias et al., 1998) with a geometrical resolution of 1 km and with 9 spectral channels between 400 and 900 nm. Even more sensors are planned for the near future.

## AREAS OF APPLICATION

Coastal zones and shelf seas are vulnerable areas often exposed to influence from mankind and its activities. Today, about 60% of the world's population lives within a zone not wider than 60 km from the nearest shore, and the population pressure is expected to increase in these areas. A scientific definition of the coastal zone has been given by the Land Ocean Interactions in the Coastal Zone (LOIZE) programme, which is a sub-programme of the International Geosphere Biosphere Programme (IGBP). It is stated that the coastal zone covers the area between the land height contour of 200 m and the bathymetric contour of 200 m depth at sea.

In most cases, demographic pressure in the coastal zone results in nutrient and wastewater pollution from intense land use, tourism and growing cities. The water quality diminishes due to increased primary production, harmful algal blooms and reduced water clarity, resulting in high oxygen consumption in deep waters which may turn into an anoxic state. On shallow bottoms the areal extent of seagrass and bladderwrack diminishes. Corals may become bleached due to loss of the alga living in symbiosis with the coral when exposed to various stress factors (Wilkinson et al., 1999). Over-fishing, destruction of mangrove forests and apparent increase in marine diseases (Harvell et al., 1999) are other examples of coastal water degradation. A recent review of the environmental status of the Baltic Sea (Jansson and Dahlberg, 1999) is one example of how pollution problems in waters with strong positive water balance at high latitudes is changing due to eutrophication. Despite this difficult situation, the coastal zone is also a habitat for tourism with strong demands of water quality. It is also a fact that 90% of the world's fish catch takes place here (IGBP, 1995). Hence, the situation is progressing into a growing conflict between different interests, since restoring water quality is expensive and, in most cases, slow. In this context, remote sensing has the potential to provide information for environmental research, monitoring and surveillance, given that we know the processes governing radiation transfer in the atmosphere and sea.

## SEMI-ANALYTICAL ATMOSPHERIC RADIATION TRANSFER MODELS

Since ocean surface waters act almost as a black body for sun radiation, the radiance leaving water is a minor part of the total signal eventually reaching the satellite sensor, which is instead dominated by atmospheric Rayleigh and aerosol scattering during cloud-free conditions. In typical ocean waters, the atmospheric aerosols are of a single type, greatly simplifying the radiation transfer relations (Gordon, 1978; 1997). The stage of advancement in handling atmospheric corrections on a pixel-by-pixel basis is such that the application of ocean colour is mature, whereas for the sea colour problem (Case 2 waters) the atmospheric conditions are more complex, often with a mixture of marine, terrestrial and urban aerosols. The coastal ocean water particulate matter is also often a mixture of plankton and inorganic material, sometimes coexisting with dissolved coloured organic matter which strongly absorbs radiation, as compared with the open ocean where plankton is the only available particulate matter. A schematic outline of the governing processes is given in Fig. 1. In the scientific community, strong focus is placed on the coastal ocean problem, often using a multi-disciplinary approach, since the analysis of the atmospheric corrections must include optical properties of the ambient water masses (Doerffer and Fischer, 1994; Moore et al., 1999; Schiller and Doerffer, 1999; Sturm et al., 1999). The goal is obvious. The near-future satellite sensors can resolve the coastal ocean temporal and spatial scales with greater accuracy than previously, due to improved sensor radiometric accuracy and with measurements at many, but narrow, wavelength bands (cf. IOCCG, 1998).

## WATER QUALITY PARAMETERS

According to Gordon and McLuney (1975), 90% of the upwelled irradiance ($E_u$), caused by backscattered downwelling irradiance ($E_d$), originates from waters between the surface and the depth where $E_d$ has decreased to 37% of its subsurface value. Using Beers law this corresponds to one diffuse attenuation depth given by the $K^{-1}$ metres, where $K$ is the diffusive attenuation coefficient. Taking into account that the euphotic zone reaches a depth where $E_d$ has decreased to 1% of its subsurface value, it is easily recognised that optical remote sensing provides information of the optical state down to a quarter of the depth of the layer in which the photosynthetic process is active. Going from the clearest ocean waters into the coastal zone, $K$ can vary between 0.025 and 2 $m^{-1}$ and be even larger. If the total water column depth is of the order of $K^{-1}$, it is expected that the bottom reflectance can influence the upwelling irradiance. In the following it is assumed that the bottom depth is greater than this except for the section specifically focused on studies of bottom reflectance.

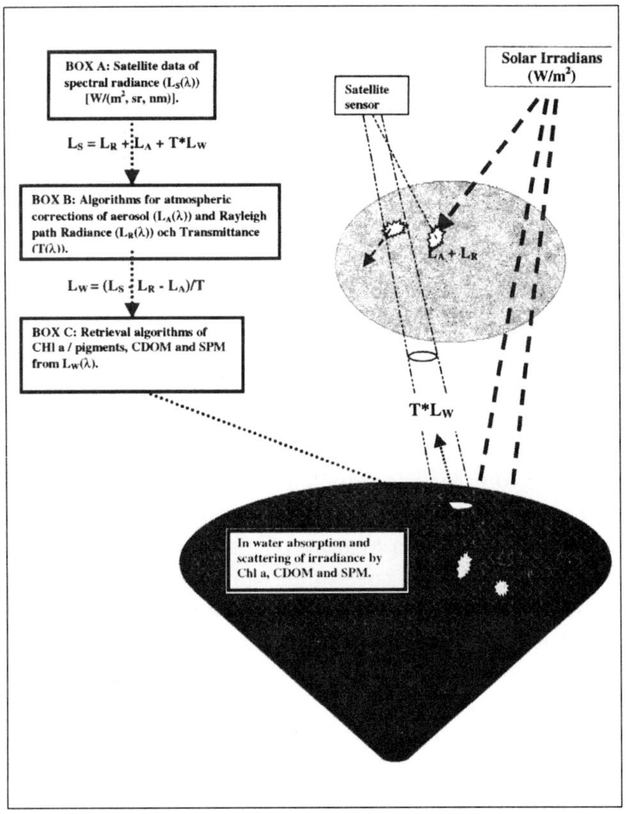


Fig. 1. A schematic view of the governing processes involved in optical remote sensing. The boxes indicate the various steps involved in the analysis.

### Secchi Disk Depth and Diffusive Attenuation Coefficient

The most used and familiar parameter of water quality is the Secchi disk depth. This extremely simple and cost-effective method is almost always used in monitoring and scientific expeditions, and is a crude but robust observation of water quality. According to Preisendorfer (1986) it gives a visual index of water clarity, dependent on both inherent (beam attenuation) and apparent (diffuse attenuation coefficient) characteristics. Nevertheless, it is often possible to find empirical relationships, although weak and of regional relevance, between chlorophyll *a* and Secchi disk depth (i.e. Sandén and Håkansson, 1996). There are many investigations reported in the literature on empirical studies of Secchi depth and satellite remote sensing of water-leaving radiance ($L_w$) at green to red wavelengths. The most often used sensor is LANDSAT TM and MSS, since the instruments have a geometrical resolution (30 and 80 m, respectively) fitting many coastal ocean areas in a geographical context, but not in temporal coverage, taking into account that the repeat cycle of the satellite is 16 days. Many different empirical algorithms exist, but the relationship is more or less the same. The reciprocal of the Secchi depth is linearly dependent on water-leaving radiance values in the green to red part of the spectrum (Lindell et al., 1985; Khorram, 1985; Lathrop and Lillesand, 1986; Pattiaratchi et al., 1994), in which case the water radiance increases with decreasing Secchi depth. This can be explained by the scattering process, which increases with increased content of plankton and inorganic particles, inhibiting the depth of light penetration. The scientific evaluation of retrieving Secchi disk depth from satellite data is empirically mature, though the disk physics is far from straightforward. Most likely, the best way to retrieve this parameter from space is to correlate the water-leaving radiance with *in situ* data, minimising atmospheric influence on the analysis. In this way, an increased knowledge of water clarity is obtained beyond the points of *in situ* measurements. The best achievement is perhaps that the *in situ* monitoring can be made more efficient and cost-effective. A pre-requisite is, of course, that the costs for image data can match the benefits for the user.

On the other hand, satellite data of water-leaving radiance is a direct measurement of the diffuse attenuation coefficient ($[K] = m^{-1}$), a more difficult parameter to measure *in situ* than Secchi disk depth. The diffusive attenuation coefficient is interesting since it directly measures a quantity that is of relevance for physics, biology and modelling in the coastal ocean. On a regional basis it is possible to correlate Secchi depth with the inverse of the diffusive attenuation coefficient (Pilgrim, 1987). In some studies this kind of relationship is used to get information on diffusive attenuation coefficients for evaluation of algorithms for satellite data (cf. Siegel, 1992; Epstep and Arnone, 1993). Some early attempts to find relationships between satellite recorded water-leaving radiance and diffusive attenuation coefficient were carried out by Gordon and McLuney (1975) and Austin and Petzold (1981), using CZCS data. Their results were limited to $K < 0.5\ m^{-1}$ due to limitations in instrument and atmospheric correction accuracy. Ten years later, Stumpf and Pennock (1991) presented a method to retrieve $K$ in the coastal zone where $K > 0.5\ m^{-1}$ and where it was characterised by high content of total suspended matter (up to 60 mg/l). The data source was NOAA-AVHRR, which is much less sensitive to water-leaving radiance in the optical channels compared to CZCS. The method was later applied by Stumpf and Frayer (1997) in Florida Bay, demonstrating the usefulness of AVHRR imagery for long-term trends of water clarity. Further development of the method is given by Woodruff et al. (1999). In a recent study by Bowers et al. (1998) K-maps were derived from AVHRR imagery of the Irish Sea showing values between 0.25 and 0.35 $m^{-1}$ during typical summer conditions. Their approach is slightly different from that of Stumpf and Pennock, but both use semi-analytical modelling. These studies show that it is possible to cover a wide range of K-values from open ocean conditions towards coastal waters, using existing data sets from the CZCS and AVHRR sensors. Merging of analysis methods to achieve global algorithms should be addressed in future investigations, taking into account new sensors, which should then be applied to a wider range of coastal oceans.

### Bottom Depth and Reflectance

Studying coastal zones, specifically the shallow parts, it would be expected that different types of bottom strata may influence the remotely sensed radiance. As mentioned above, the critical depth scale at which an influence is expected is $K^{-1}$ metres. An early attempt to classify sea-bottom types using remote sensing was proposed by Lyzenga (1978). His method was later improved and applied to various areas. An approximate formula for the total subsurface reflectance ($R_t$), yields:

$$R_t = R_w + (R_b - R_w) * e^{-2Kz}$$

Here $R_w$ is the water-leaving reflectance for infinite bottom depth (z), whereas $R_b$ is the sea-bottom reflectance (Siegel and Seifert, 1985; Siegel, 1992; Spitzer and Dirks, 1987; Maritorena et al., 1994; Jaquet et al., 1999). The subsurface reflectance $R_w$ is equal to the ratio of upwelling and down-welling irradiance ($E_u/E_d$) and it can be shown that $R_w$ is governed by the inherent optical properties of the water mass. The satellite sensor, on the other hand, measures upwelled radiance ($L_w$), which for the sake of simplicity is here assumed to be directly proportional to upwelled irradiance ($E_u$). Using the well-known extraterrestrial downwelling irradiance for ($E_d$) closes the relationship between satellite radiance reflectance and the inherent properties.

The influence of bottom depth and bottom reflectance might be determined by the condition that the subsurface total reflectance should at least be larger than twice the water-leaving reflectance for infinite bottom depth (Maritorena et al., 1994). In this case a critical bottom depth is defined by the following formula:

$$z_c = \ln((R_b - R_w)/R_w) / (2K)$$

In the simplest case, when $K$ and thus $R_w$ are horizontally homogenous, the bottom depth is known and $z < z_c$ the bottom-type reflectance can be determined. The technique has been applied at the Oder bank and in the Greifswald Bay in the southern Baltic (Siegel and Siefert, 1985; Siegel, 1992), at the Kerkennah shelf in the southern Mediterranean (Jaquet et al., 1999) and in the lagoons of French Polynesia (Maritorena et al., 1994). Tassan (1996) proposed an extended method to take into account variable water constituents, i.e. that $K$ and $R_w$ vary in the area of interest. Nevertheless, the method is sensitive to $K$ and $R_w$ and, in addition, atmospheric corrections are needed of the colour imagery. Probably the method is most suitable in clear water areas (Case 1 waters), where both $R_w$ and $K$ are small or at bottom depths much shallower than the critical depth ($z_c$). One near-future and emergent application can be to develop the method for mapping bleached or unhealthy coral reefs, which have recently been recognised to decline in biodiversity, and in some cases to show signs of total decline (Sheppard, 1999; Wilkinson et al., 1999). In this very new and difficult subject for optical remote sensing, Holden and Ledrew (1999) demonstrated a novel method of detecting bleached corals based on *in situ* optical measurements. The applicability remains to be tested with remote sensing data.

### Chlorophyll Pigment, Coloured Dissolved Organic Matter and Total Suspended Matter

The optically active water constituents determine the water type (Jerlov, 1976): simply speaking plankton pigments (often measured as Chlorophyll *a*), seston or particle content and type (TSM, total suspended matter) and dead organic material such as detritus and dissolved organic matter (CDOM, coloured dissolved organic matter). From the early days of remote sensing, the goal has been to retrieve these parameters from space. The development took place in ocean waters (Case 1 water) where only plankton dominates the optical conditions, besides the water itself. Hence this is a fairly simple water mass to investigate and at the same time the one which was least known regarding distribution in time and space of it biomass. The first problem to retrieve water-leaving radiance from space data, i.e. to eliminate the atmospheric contributions of path radiance and transmittance (Fig. 1), was first solved by Gordon (1978) and applied to CZCS imagery. This basic approach is still the correction model used for present and planned near-future ocean/sea colour sensors.

The second problem is to retrieve the chlorophyll content from the spectral radiance. The most common method is to use band ratio, which is empirically fitted to *in situ* measurements of pigments sampled during a satellite overpass. The ratio consists of one spectral band covering a part of the spectrum which plankton select for energy absorption, and another band where this absorption is at a minimum. In ocean waters, wave bands around 445 and 550 nm are typically chosen. The band ratio is sensitive to absorption and scattering, which are positively correlated in ocean waters. This is not necessarily so in coastal ocean waters, where inorganic particles blur the correlation between TSM and Chl *a*. In addition, the terrestrial leakage of dissolved organic matter, which can be extensive, discolours water toward yellow and brown (CDOM). This state of affairs complicates the sun absorption, disconnecting relationships between chlorophyll, CDOM and total suspended matter. Nevertheless, in an encouraging study by Sathyendranath et al. (1989) it was shown possible to determine chlorophyll content in Case 2 waters, using a three-component optical model of sea colour. They suggested that in addition to the classical remote-sensing wavelengths, the inclusion of the 400 nm channel can help to distinguish between chlorophyll content and CDOM. On the other hand, airborne spectrometer studies of lake waters reveal that longer wavelengths can be of more importance, using the narrow chlorophyll absorption peak centred at 676 nm in comparison with the low-absorption peak at 706 nm (cf. Dekker, 1993). This might be an advantage since atmospheric disturbance is less strong at the red end of the visible spectrum than it is at the blue end.

There is a growing need to develop methods for sea colour remote sensing, taking into account the availability of imagery from near-future satellites that can resolve the water-leaving radiance at high spectral resolution. In this case, other spectral peaks in plankton absorption can be exploited than are traditionally used for ocean colour. There are few, if any, convincing studies demonstrating the possibility of determining chlorophyll pigment in the coastal ocean from space, other than those which are valid at local or regional scales only. Better results have been gained with TSM, but still these are often regionally valid (i.e. Tassan, 1987; Bowers et al., 1998).

## ALGAL BLOOM DETECTION

In the coastal ocean there is an increasing problem of harmful algal blooms (Hallegraeff, 1993) causing, for example, paralytic shellfish poisoning (Andersson, 1997). These are often named red and brown tides in older literature, whereas modern nomenclature refers to harmful algal blooms (HAB). As expressed by their old names, these blooms are often strongly coloured and differ in many respects from spring (upwelling) diatom-dominated blooms (Smayda, 1997 a,b). Simply speaking, ocean and sea colour remote sensing mainly addresses the diatom-dominated blooms by their characteristic absorption of chlorophyll pigments, whereas the algal blooms addressed here have strong colour and most probably the optical conditions are governed by scattering instead of anomalous absorption. However, scientific evidence is still scarce, although there are some recent studies indicating the importance of backscatter as the principle mechanism. Perhaps algal-bloom detection is the most interesting and advanced application of remote sensing in coastal ocean waters today, since these blooms are easily detectable from satellite imagery in otherwise optically complicated waters.

### Red and White Tides (HAB)

This nomenclature is perhaps not very exact, as the 'tide' reflects early observations of reddish algae groups blooming in some kind of phase with lunar tidal motions. The red and brown blooms are mainly caused by mass growth of dinoflagellates (Friedrich, 1973) and occur frequently in the coastal ocean. The bloom can be toxic and in shallow areas can cause anoxia. It is a growing water quality problem in many coastal areas of the world. The remote sensing community has often tried to find methods to detect and monitor these events. Previously the colour was thought to be due to anomalous cell absorption characteristics. Recently it was found (Karhu and Mitchell, 1998) that only at shorter wavelengths (< 400 nm) could a difference be established for the blooming dinoflagellate *Lingulodinium polyedra* compared to the no-bloom situation. They recommended that the forthcoming Japanese Global Imager (GLI) could be used to exploit the findings, taking into account the UV part of the spectrum to be measured by this sensor. If the strong colours of red and brown tides are not governed by absorption, perhaps backscattering can. Roesler and McLeroy-Etheridge (1998) present results, based on bio-optical modelling and reflectance measurements, indeed showing that anomalous backscattering, relative absorption characteristics can explain the colour. Furthermore, in the Adriatic Sea, a diatom bloom occurs frequently during special atmospheric conditions. The blooms are a nuisance for the tourism industry. The algae produce secretions which eventually surface and become visible to AVHRR and TM sensors (Tassan, 1993). Here the basic reflectance mechanism is also governed by backscattering.

### Coccolithophores

Coccolithophores are a group of marine phytoplankton having external calcite plates, which strongly scatter sunlight and are observed as high reflectance patches in AVHRR and CZCS imagery. This plankton is an important transport link of inorganic carbon between ocean waters and sediments besides foraminifera and pteropods in the ocean (Groom and Holligan, 1987). The possibility of detecting and mapping the coccolithophore bloom with satellite data was presented by Holligan et al. (1983). The mechanism by which the algae are detected is through backscattering of detached coccoliths together with absorption by plankton cells. According to Groom and Holligan (1987), the coccoliths accumulate as long as the bloom continues. Using spectral characteristics of the coccolithophore bloom, it might be possible to retrieve chlorophyll concentrations from satellite data. Another example is from the Gulf of Maine using AVHRR imagery (Ackleson et al., 1989; Ackleson and Holligan 1994).

### Cyanobacteria

Another phytoplankton (or bacteria) is the blue-green or cyanobacteria common in tropical ocean lakes and in brackish water bodies. This plankton group can fix atmospheric nitrogen, and this process is a direct source of new nitrogen to surface waters, whose load is relatively unknown. Sellner (1997) and Kohonen (1992) review marine and brackish water cyanobacteria, respectively. Like the coccolithophores, the blooms generally occur when surface waters are drained of dissolved nitrogen and a stable stratification is established, however the exact conditions favouring a bloom are still not clear. Usually the bloom is of *Nodularia spumigena* during strong influx of irradiance and when water temperature is above 15°C, while other groups may

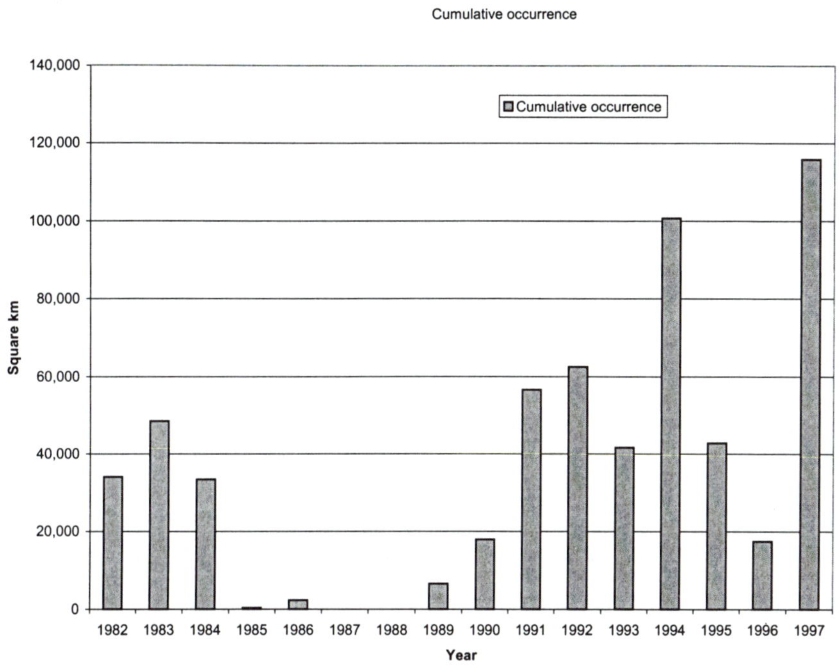


Fig. 2. The time series of annual cyanobacteria bloom coverage from 1982 to 1997 in the Baltic Sea based on time series of NOAA-AVHRR imagery (Karhu et al., 1994). This time series is not corrected for the bias caused by irregular sampling frequency. Copyright Ove Rud, Department of Physical Geography, Stockholm University.

bloom early in spring. Besides fixing nitrogen, the algae can also be toxic and harmful to animals. At many bloom events the cyanobacteria is a nuisance to local authorities and to tourists, and can be a sign of eutrophication. In the Baltic these summer blooms were reported to occur frequently long before the eutrophication was recognised as being a severe problem. Historical notes date back to 1850 (Wallström, pers. comm.), while satellite imagery of surface filamentous cyanobacteria was detected, apparently for the first time, by Landsat-2 imagery in 1975 (Jansson and Nyqvist, 1976). Even earlier, it was found by Öström (1976) that ERTS-1 had made multispectral scanning (MSS) registrations of a bloom event during the summer 1973, demonstrating for the first time that the cyanobacteria bloom covered the main parts of the Baltic Proper. Typically the length-scale of blooms is 10 to 100 km, and they last for about one week to a month.

Since the very early remote-sensing studies in the 1970s, it took another 15 years before it was noted how well AVHRR could detect these blooms (Hakansson and Moberg, 1994). A systematic investigation of cyanobacteria blooms in the Baltic using AVHRR imagery was presented by Kahru et al. (1994) and Kahru and Rud (1995), showing bloom events in the Baltic Proper since 1982. The developed algorithm for bloom detection was later presented in Karhu (1997). He also corrected the time series for the bias due to irregular sampling frequency caused by cloud occurrence. In this case, no particular trend was found during the period 1982 to 1994, but the data indicated large-scale (decadal) temporal variability, and new areas were found to host cyanobacteria blooms. An updated time series of the annual area coverage in the Baltic Sea is shown in Fig. 2.

The surface filaments are easily detected in imagery due to strong scattering of irradiance by gas vesicles, and appear white to yellow in colour. Kirk (1994) presents some laboratory measurements on specific backscatter coefficients indicating that it can reach 0.12 $m^2$ (mg Chl *a*)$^{-1}$ during the cyanobacteria bloom. However, there is hardly any scientific evidence on the details of the scattering process, which one would like to know for modelling purposes to estimate total nitrogen fixation during bloom events. It was noted by Kahru (1997) that his AVHRR algorithm for detection of cyanobacteria works well in the red and near-infrared wavebands, since the filaments almost surface during the bloom. This is in contrast with the coccolithophores, which should be opaque at these wavebands, since the coccoliths are submerged in the water column below the depth where red light and longer wavelengths can penetrate. In tropical waters, *Trichodesmium* is a nitrogen-fixing cyanobacteria, which has unique absorption and scattering properties according to Subramaniam and Carpenter (1994) who formulated an empirical protocol, identifying this group using CZCS imagery. The protocol was not successful in identifying the Baltic *Nodularia spumigena* (Karhu, 1997), demonstrating the difference in optical properties that can be found between different cyanobacteria species.

One can always argue that AVHRR imagery can only detect the surfacing blooms, leaving the early stage of development unobserved. However, the new ocean/sea colour sensors will provide the opportunity to extend the remote-sensing methods to cover this part as well. This is made possible due to dedicated sensors with much higher signal-to-noise ratios, and with more and much less broad wavebands (IOCCG, 1998; 1999). In some cases (MODIS and MERIS) the sensor will provide almost hyperspectral data. It is left to the coming decades of research to exploit this imagery for new marine applications and to improve existing ones.

## CONCLUSIONS

Sea colour remote sensing of coastal waters is not as advanced as that for ocean colour, but there are several methods and satellite data available concerning the variables Secchi disk depth, diffusive attenuation coefficient, total suspended matter and algae detection. The first two variables represent the overall state of water quality, whereas the last two are specific variables mainly determined by sunlight scattering characteristics.

It might be fair to state that the period 1970–1995 can be defined as an exploratory phase of ocean/sea colour remote sensing. In August 1997 the SeaWiFS sensor was launched, representing a very late follow-up of the first ocean colour sensor CZCS (1978–1986). For the coastal ocean, the community is still waiting for almost hyper-spectral sensors such as MERIS and MODIS. If all planned missions are realised, there will be extensive data sets available with global coverage. Probably the coastal ocean sea colour problem will stay in the exploratory phase for another decade, involving these sophisticated sensors. For satellite sensors such as AVHRR and SeaWiFS, imagery with high temporal and long-lasting coverage, there are several coastal ocean products to be exploited by scientists and managers.

## ACKNOWLEDGEMENT

This work was supported by the Swedish Foundation for Strategic Environmental Research (MISTRA).

## ABBREVIATIONS

| | |
|---|---|
| AVHRR | Advanced Very High Resolution Radiometer |
| CDOM | Coloured Dissolved Organic Matter |
| CZCS | Coastal Zone Colour Sensor |
| EOS | Earth Observing System |
| ESA | European Space Agency |
| IGBP | International Geosphere-Biosphere Programme |
| IOCCG | International Ocean Colour Co-ordinating Group (www.ioccg.org) |
| LOIZE | Land-Ocean Interactions in the Coastal Zone |
| MERIS | Medium Resolution Imaging Spectrometer |
| MODIS | Moderate Resolution Imaging Spectroradiometer |
| MSS | Multi-Spectral Scanner |
| NASA | National Aeronautics and Space Administration |
| NASDA | National Space Development Agency of Japan |
| NOAA | National Oceanic and Atmospheric Administration |
| SeaWiFS | Sea-viewing Wide Field-of-view Sensor |
| TM | Thematic Mapper |
| TSM | Total Suspended Matter |

## REFERENCES

Anderson, D. (1997) Bloom dynamics of toxic *Alexandrium* in the northeast U.S. *Limnology and Oceanography* **42** (5, part 2), 1009–1022.

Ackleson, S.G. and Holligan, P.M. (1994) Response of water-leaving radiance to particulate calcite and chlorophyll *a* concentrations: A model for the Gulf of Maine coccolithophore blooms. *Journal of Geophysical Research Oceans* **99**, 7483–7499.

Ackleson, S.G., Balch, W.M. and Holligan, P.M. (1989) AVHRR observations of a Gulf of Maine coccolithophore bloom. *Photogrammetric Engineering Remote Sensing* **55**, 473–474.

Austin, R.W. and Petzold, T.J. (1981) The determination of the diffusive attenuation coefficient of sea water using the coastal zone color scanner. In *Oceanography from Space*, ed. J. Gower. Plenum, New York, pp. 239–256.

Bowers, D.G., Boudjelas, S. and Harker, G.E.L. (1998) The distribution of fine sediments in the surface waters of the Irish Sea and its relation to tidal stirring. *International Journal of Remote Sensing* **19** (14), 2789–2805.

Cracknell, A.P. (1989) Remote sensing in estuaries: an overview. In *Developments in Estuarine and Coastal Study techniques, EBSA 17 Symposium*, eds. J. McManus and M. Elliott. Olsen & Olsen, Frendesborg. pp. 7–13.

Cracknell, A.P. (1999) Remote sensing techniques in estuaries and coastal zones—an update. *International Journal of Remote Sensing* **19** (3), 485–496.

Dekker, A.G. (1993) Detection of optical water quality parameters for eutrophic waters by high resolution remote sensing. PhD thesis, Vrije Universiteit, Amsterdam.

Doerffer, R. and Fischer, J. (1994) Concentrations of chlorophyll, suspended matter and gelbstoff in Case II waters derived from satellite coastal zone colour scanner data with inverse modelling methods. *Journal of Geophysical Research* **99** (C4), 74–83.

Epstep, L. and Arnone, R. (1993) Correlation of CZCS surface ks with ks derived from Secchi disk. *Photogrammetric Engineering Remote Sensing* **59**, 345–350.

Esaias, W.E., Abbott, M.R., Barton, I., Brown, O.B., Campbell, J.W., Carder, K.L., Clark, D.K., Evans, R.L., Hodge, F.E., Gordon, H.R., Balch, W.P., Letelier, R. and Minnet, P.J. (1998) An overview of MODIS capabilities for ocean science observations. *IEEE Trans. Geosci. Remote Sensing* **36**, 1250–1265.

Ewing, G.C. (ed.) (1965) *Oceanography from Space*. Proceedings of the Conference on the Feasibility of Conducting Oceanographic Explorations from Aircraft, Manned Orbital and Lunar Laboratories, held at Woods Hole, Massachusetts, 24–28 August 1964.

Friedrich, H. (1973) *Marine Biology*. Sidgwick and Jackson, London, 474 pp.

Gordon, H.R. (1978) Removal of atmospheric effects from satellite imagery of the oceans. *Applied Optics* **17**, 1631–1636.

Gordon, H.R. (1997) Atmospheric correction of ocean colour imagery in the Earth Observing System era. *Journal of Geophysical Research* **102**, 17081–17106.

Gordon, H.R. and McLuney, W.R. (1975) Estimation of the depth of sunlight penetration in the sea for remote sensing. *Applied Optics* **14**, 513–416.

Gower, J.F.R. (ed.) (1981) *Oceanography from Space.* Proceedings of the COSPAR/SCOR/IUCRM Symposium on Oceanography from Space, May 26–30, 1980, Venice, Italy. Plenum Press, N.Y.

Groom, S.B. and Holligan, P.M. (1987) Remote sensing of coccolithophore blooms. *Advances in Space Research* 7 (2), 273–278.

Hakansson, B.G. and Moberg, M. (1994) The algal bloom in the Baltic during July and August 1991, as observed from the NOAA weather satellites. *International Journal of Remote Sensing* **15**(5), 963–965.

Hallegraeff, G.M. (1993) A review of harmful algal blooms and their apparent global increase. *Phycologia* **32**, 79–99.

Harwell, C.D., Kim, K., Burkholder, J.M., Colwell, R.R., Epstein, P.R., Grimes, D.J., Hofmann, E.E., Lipp, E.K., Osterhaus, A.D.M.E., Overstreet, R.M., Porter, J.W., Smith, G.W. and Vasta, G.R. (1999) Emerging marine diseases—climate links and anthropogenic factors. *Science* **285** (3), 1505–1510.

Holden, H. and Ledrew, E. (1999) Hyperspectral identification of coral reef features. *International Journal of Remote Sensing* **20** (13), 2545–2563.

Holligan, P.M., Viollier, M., Harbour, D.S., Camus, P. and Champagne-Phillipe, M. (1983) Satellite and ship studies of coccolithophore production along a continental shelf edge. *Nature* **304**, 339.

IGBP (1995) Land–ocean interactions in the coastal zone—Implementation Plan. Global Change Report No. 33, eds. J.C. Pernetta and J.D. Milliman.

IOCCG (1998) Minimum Requirements for an operational ocean-colour sensor for the open ocean. Report No. 1, 46 pp.

IOCCG (1999) Status and plans for ocean-colour missions: considerations for complementary missions. Report No. 2, 43 pp.

Jansson, B-O. and Nyqvist, B.G. (1976) Dynamics and energy flow in the Baltic ecosystems—remote sensing. Asko Laboratory, Box 58, S-150 13 Trosa, Sweden. Final Report of Landsat-2 investigation No. 28470 prepared for NASA, USA.

Jansson, B-O. and Dahlberg, K. (1999) The environmental status of the Baltic sea in the 1940s, today and in the future. *Ambio* **28** (4), 312–319.

Jaquet, J.-M., Tassan, S., Barale, V. and Sarbaji, M. (1999) Bathymetric and bottom effects on CZCS chlorophyll-like pigment estimation: data from the Kerkennah Shelf (Tunisia). *International Journal of Remote Sensing* **20** (7), 1343–1362.

Jerlov, N.G. (1976) *Marine Optics.* Elsevier, Amsterdam.

Kahru, M. and Rud, O. (1995) Monitoring the decadal-scale variability of cyanobacteria blooms in the Baltic Sea by satellites. *Proc. Third Conference on Remote Sensing for Marine and Coastal Environments*, **2**, 76–83.

Kahru, M. (1997) Using satellites to monitor large-scale environmental change: A case study of cyanobacteria blooms in the Baltic Sea. In: *Monitoring Algal Blooms. New Techniques for Detecting Large-scale Environmental Change*, eds. M. Kahru and C.W. Brown. Springer-Verlag Berlin, Heidelberg, New York, pp. 43–61.

Kahru, M., Horstmann, U. and Rud, O. (1994) Satellite detection of increased cyanobacterial blooms in the Baltic Sea: Natural fluctuation or ecosystem change? *Ambio* **23** (8), 469–471.

Karhu, M. and Mitchell, B.G. (1998) Spectral reflectance and absorption of a massive red tide off southern California. *Journal of Geophysical Research* **103** (C10), 21601–21609.

Khorram, S. (1985) Development of water quality models applicable throughout the entire San Francisco Bay and delta. *Photogrammetric Engineering Remote Sensing* **51** (1), 53–62.

Kirk, J. (1994) *Light and Photosynthesis in Aquatic Ecosystems.* 2nd edn. Cambridge University Press.

Kohonen, K. (1992) Dynamics of the toxic cyanobacteria blooms in the Baltic Sea. *Finish Marine Research* **261**, 3–36.

Lathrop, R.G. and Lillesand, T.M. (1986) Use of thematic mapper data to assess water quality in Green Bay and central Lake Michigan. *Photogrammetric Engineering and Remote Sensing* **52**, 671–680.

Lindell, L.T., Steinvall, O., Jonsson, M. and Claesson, T. (1985) Mapping of coastal-water turbidity using Landsat imagery. *International Journal of Remote Sensing* **6** (5), 629–642.

Lyzenga, D.R. (1978) Passive remote sensing techniques for mapping water depth and bottom features. *Applied Optics* **17**, 379–383.

*Marine Biology*, ed. H. Friedrich, Sidgwick & Jackson, London.

Maritorena, S., Morel, A. and Gentili, B. (1994) Diffuse reflectance of oceanic shallow waters: Influence of water depth and bottom albedo. *Limnology and Oceanography* **39** (7), 1689–1703.

Moore, G.F., Aiken, J. and Lavender, S.J. (1999) The atmospheric correction of water colour and the quantitative retrieval of suspended particulate matter in Case 2 waters: application to MERIS. *International Journal of Remote Sensing* **20** (9), 1713–1733.

Öström, B. (1976) Fertilization of the Baltic by nitrogen fixation in the blue-green alga *Nodularia spumigena. Remote Sensing of the Environment* **4**, 305–310.

Pattiaratchi, C., Lavery, P., Wyllie, A. and Hick, P. (1994) Estimates of water quality in coastal waters using multi-date Landsat Thematic Mapper data. *International Journal of Remote Sensing* **15** (8), 1571–1584.

Pilgrim, D.A. (1987) Measurement and estimation of the extinction coefficient in turbid estuarine waters. *Continental Shelf Research* **7**, No. 11/12, 1425–1428.

Preisendorfer, R.W. (1986) Secchi disc science: Visual optics of natural waters. *Limnology and Oceanography* **31**, 909–926.

Reichardt, T. (1999) *Nature* **400**, 19 August, 702.

Roesler, C.S. and McLeroy-Etheridge, S.L. (1998) Remote detection of harmful algal blooms. Ocean Optics XIV, Kailua-Kona, Hawaii, USA, 10–13 November 1998.

Sandén, P. and Håkansson, B. (1996) Long-term trends in Secchi depth in the Baltic Sea. *Limnology and Oceanography* **41** (2), 346–351.

Sathyendranath, S., Prieur, L. and Morel, A. (1989) A three-component model of ocean colour and its application to remote sensing of phytoplankton pigments in coastal waters. *International Journal of Remote Sensing* **10** (8), 1373–1394.

Schiller and Doerffer, R. (1999) Neural network for emulation of an inverse model-operational derivation of case 2 water properties from MERIS data. *International Journal of Remote Sensing* **20** (9), 1735–1746.

Sellner, K.G. (1997) Physiology, ecology and toxic properties of marine cyanobacteria blooms. *Limnology and Oceanography* **42** (5, part 2), 1089–1104.

Sheppard, C.R.C. (1999) Coral decline and weather patterns over 20 years in the Chagos archipelago, central Indian Ocean. *Ambio* **28** (6), 472–478.

Siegel, H. (1992) On the influence of sediments and phytobenthos on spectral reflectance at the sea surface. *Beiträge Meereskunde, Berlin* **63**, 91–104.

Siegel, H. and Seifert, T. (1985) Influence of the sea bottom on the spectral reflectance in the Oder Bank region. *Beiträge Meereskunde, Berlin* **52**, 65–71.

Smayda, T.J. (1997a) What is a bloom? A commentary. *Limnology and Oceanography* **42** (5, part 2), 1132–1136.

Smayda, T.J. (1997b) Harmful algal blooms: Their ecophysiology and general relevance to phytoplankton blooms in the sea. *Limnology and Oceanography* **42** (5, part 2), 1137–1153.

Spitzer, D. and Dirks, R.W.J. (1987) Bottom influence on the reflectance of the sea. *International Journal of Remote Sensing* **8** (3), 279–290.

Stump, R.P. and Pennock, J.R. (1991) Remote sensing of the diffusive attenuation coefficient in a moderately turbid region. *Remote Sensing of the Environment* **38**, 183–191.

Stumpf, R.P. and Frayer, M.L. (1997) Use of AVHRR imagery to examine long-term trends in water clarity in coastal estuaries: Example in Florida Bay. In *Monitoring Algal Blooms. New Techniques for Detecting Large-scale Environmental Change*, eds. M. Kahru and C.W. Brown. Springer-Verlag, Berlin, Heidelberg, New York, pp. 1–23.

Sturm, B., Barale, V., Larkin, D., Andersson, J.H. and Turner, M. (1999) OCEANcode: the complete set of algorithms and models for the level-2 processing of European CZCS historical data. *International Journal of Remote Sensing* **20** (7), 1219–1248.

Subramaniam, A. and Carpenter, E.J. (1994) An empirically derived protocol for the detection of blooms of the marine cyanobacterium *Trichodesmium* using CZCS imagery. *International Journal of Remote Sensing* **15** (8), 1559–1569.

Tassan, S. (1987) Evaluation of the potential of the Thematic Mapper for marine applications. *International Journal of Remote Sensing*, 10, 1455–1478.

Tassan, S. (1993) An algorithm for the detection of the white-tide ("mucialge") phenomenon in the Adriatic Sea using AVHRR data. *Remote Sensing of the Environment* **45**, 29–42.

Tassan, S. (1996) Modified Lyzenga's method for macroalgae detection in water with non-uniform composition. *International Journal of Remote Sensing* **17**, 1601–1608.

Wilkinson, C., Lindén, O., Cesar, H., Hodgson, G., Rubens, J. and Strong, A.E. (1999) Ecological and socioeconomic impacts of 1998 coral mortality in the Indian ocean: An ENSO impact and warning of future change? *Ambio* **28** (2), 188–196.

Woodruff, D., Stumpf, R.P., Scope, J.A. and Paerl, H.W. (1999) Remote estimation of water clarity in optically complex estuarine waters. *Remote Sensing of the Environment* **68**, 41–52.

## THE AUTHOR

**Bertil Håkansson**

*Swedish Meteorological and Hydrological Institute, S-60176 Norrköping, Sweden*

Chapter 129

# ENERGY FROM THE OCEANS: WIND, WAVE AND TIDAL

Rebecca J. Barthelmie, Ian Bryden, Jan P. Coelingh and Sara C. Pryor

Virtually endless energy is available from the seas. The question is how to harness it to provide relatively pollution-free energy for use by humankind. This chapter reviews the current state of wind, wave and tidal energy technology and the prospects for the supply of sustainable energy from these sources for the new millennium.

Renewable energies as a whole are an attractive proposition offering clean and sustainable energy supply based on indigenous resources. Offshore wind energy has also been shown to be economically viable and many European countries are already promoting renewable technologies to help meet environmental goals such as reduction of greenhouse gases as agreed under the Kyoto protocol. Wind energy is technologically the most advanced of the three approaches considered, due to its long history of deployment at land sites. Offshore wind energy is a relatively new phenomenon, poised for rapid expansion. The first offshore turbine was installed at Blekinge in Sweden in 1991 with rated power of 220 kW, by 1999 approximately 30 MW of installed capacity of offshore wind turbines were operating, and there are developments planned for the next decade for at least ten individual wind farms of up to 100 turbines each rated between 1 and 2 MW. Most northern European countries with coastlines are considering offshore wind energy as a proven technology for supply of clean energy which can be deployed relatively rapidly but does not attract concern from local residents. The first installations of offshore wind farms are planned in the region 5–15 km from the coast which reduces visual intrusion, optimises the wind resource and keeps transmission cable costs low. However, future developments may include floating turbine structures rather than the fixed foundation types currently in use. Floating turbines are an attractive proposition for deeper waters such as those of the Mediterranean.

The wave energy resource is substantial but has yet to be harnessed in a commercially viable manner due in part to technical and logistical constraints. All wave power devices for deployment in deep water must be large enough to survive ocean wave conditions, but for optimal extraction of the resource the device should be of a similar size to the waves themselves. When it is appreciated that ocean waves can exceed 20 m in height and that, to be economic, a device needs to remain in location for an excess of 15 years, the challenge to the designer can be seen to be considerable. In the last few years a new resurgence of interest has followed economic assessments which suggest that wave energy costs are competitive with conventional fuels in some environments and can supply power in remote areas. Most research is currently directed towards coastal sites for development of projects with rated power outputs up to a few MW.

*Seas at The Millennium: An Environmental Evaluation (Edited by C. Sheppard)*

Tidal energy is more spatially constrained as a viable energy resource, but is already making a contribution to meeting electricity demand. The 240 MW La Rance tidal barrage project in France has been operating successfully for the last 25 years. Other potential tidal barrage sites also have enormous potential with well-developed technology but the high capital costs associated with tidal barrage systems are likely to restrict development of this resource in the near future. Researchers are currently examining the prospects of utilising tidal currents. In the open ocean these currents are typically very small and are measured in cm/s at most. Local geographical effects can, however, result in quite massive local current speeds. In the Pentland Firth to the north of the Scottish mainland, for example, there is evidence of tidal currents exceeding 7 m/s.

The chapter describes in detail the exploitation of these resources in different countries and plans for further development. A technological description is given for each methodology together with a brief overview of the environmental considerations.

---

## RENEWABLE ENERGY FROM THE OCEANS

World energy demand increased by around 60% between 1970 and 1995 (Milborrow, 1996). As the demand for energy increases and readily accessible fossil fuel reserves are depleted, new energy supplies are being sought. Renewable energy sources are an attractive alternative to fossil fuels because they can also help to meet environmental goals such as reduced emissions of greenhouse and other trace gases (e.g. nitrogen and sulphur oxides and volatile organic compounds which are implicated in acid rain, smog and aerosol production in the atmosphere). In addition to providing cleaner sustainable energy sources, renewable energy projects can also provide employment since they typically produce more jobs per energy unit installed than 'conventional' energy sources. A further political motivation for the interest in renewable energy sources is to decrease the dependence on imported energy. In several countries the oil crises in the 1970s and 1980s as well as the Gulf War, provided strong incentives for greater reliance on local energy resources.

According to agreements reached in Kyoto, the European Union (EU) is committed to reducing greenhouse gas emissions by 8%, the US by 7% and Japan (and other industrialised countries) by 6% by the year 2010, relative to 1990 levels. In 1999 a consortium of environment and industry groups warned that the EU will fail to meet these targets and is also losing a potentially valuable opportunity to create thousands of jobs by failing to implement a renewable energy directive. The directive suggested by the consortium has targets of 8% of each country's energy coming from renewables by 2005, increasing to 16% by 2010 and increasing by 2% each subsequent year. The EU White Paper on Renewable Energy (1997) proposes that 12% of energy should be supplied from renewables by 2010. Within this framework the EU target is 40 GW of installed wind capacity.

This review focuses on tapping energy reserves within and above the world's oceans. Ocean Thermal Energy Conversion (OTEC) is based on exploitation of the thermal gradient between warm ocean surface waters and cold water at depths of 600 m or more (Sterrett, 1995). Although OTEC presents a potentially large energy source it is not considered in detail here, in part because the Natural Energy Laboratory of Hawaii has the world's only operating OTEC plant. Instead we focus on three technological solutions which are currently deployed to provide energy from the oceans: wind, wave and tidal. Tidal energy requires a large tidal range and thus the greatest potential is found at a limited number of sites worldwide. Tidal energy projects are usually large-scale ventures such as the 240 MW La Rance project in France which has been operating successfully for the last 25 years. Good wind and wave energy sites are less geographically restricted and this flexibility allows additional factors such as electricity grid connections to be considered, although both offshore wind and wave projects are likely to be sited away from areas with high electricity demands. Wave power offers enormous potential but in 1995 the installed capacity was only 685 kW. According to the European Wind Energy Association, wind energy installation in Europe (mainly onshore) was 6303 MW at the end of 1998 with a small contribution from offshore developments. However, installation both onshore and offshore is expected to be significant over the next few years, 2100 MW were installed during 1998 alone according to the American Wind Energy Association, and the world-wide installed wind power capacity exceeded 10,000 MW during April 1999. Gaudiosi (1999) forecasts up to 20% of wind power (up to 12,000 MW) will be installed offshore by 2010.

The following sections briefly describe the technology used to extract energy from the wind, waves and tides and give an overview of the current utilisation of these technologies and the prospects for the next millennium.

## OFFSHORE WIND ENERGY

### Overview of Current Technology

Power output of a wind turbine (Fig. 1) is proportional to the cube of the wind speed and to the swept area of the rotor. Hence, increasing wind speed and rotor diameter increase power output significantly. However, because

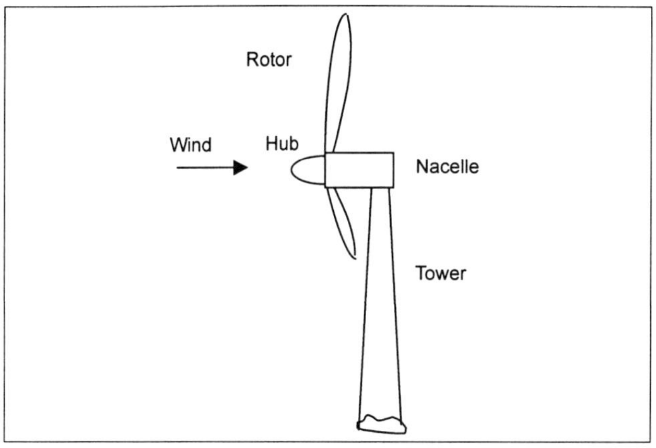


Fig. 1. Schematic of a wind turbine.

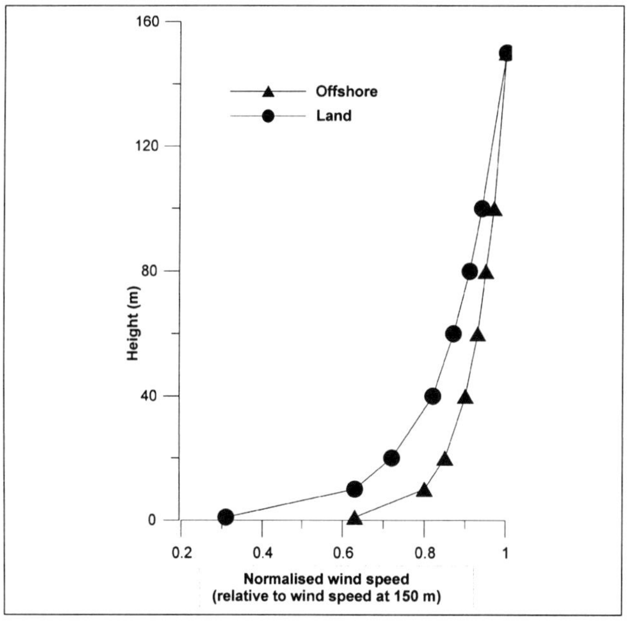


Fig. 2. Variation of wind speed with height over different roughnesses.

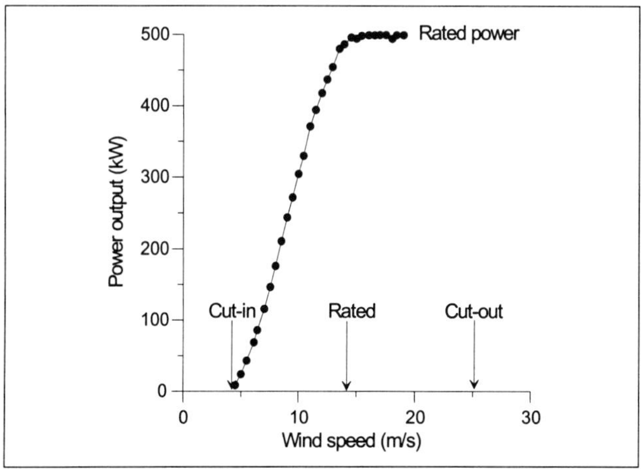


Fig. 3. Measured power curve of a VESTAS V39-500 kW wind turbine from data given in (Energy Centre Denmark, 1994).

momentum is removed at the ground, the wind speed profile with height exhibits a rapid decrease near the surface (Fig. 2) and so the increase in power output from tall towers and large blades is partially offset by increased stress on wind turbine blades, hub and tower which reduces the lifetimes of these components. Thus, the optimal size for wind turbines is a balance between these considerations. The variation of power output with wind speed is commonly known as the power curve (Energy Centre Denmark, 1994) (Fig. 3). Power output is zero at low wind speeds, increases rapidly after the 'cut-in' wind speed is achieved, reaches the rated output of the wind turbine at about 12 m/s and drops to zero again after 'cut-out' to prevent damage to the wind turbine at high wind speeds.

Wind speeds close to the ground, where we experience them, are a result of a balance between momentum transfer from higher wind speeds aloft and dissipation at the ground. The higher the roughness of the surface, the more momentum is lost at the ground. The result is a greater variation of wind speed with height and lower wind speeds near the ground. Surface roughness offshore is usually of the order 0.0002 m whereas for land surfaces roughness varies between 0.03 and 0.3 m for agricultural surfaces and much higher for urban landscapes and forests. Figure 2 illustrates this effect by showing how wind speed below 150 m is reduced over surfaces with roughness of 0.1 m (representing a land surface) relative to a surface with roughness of 0.0002 m (representing offshore). Due to the lower surface roughness, wind speeds offshore are typically higher than on land (except where wind speeds on land are enhanced by topography), and the variation of wind speed with height (wind shear) is lower. Higher wind speeds and lower vertical wind speed variations across the turbine blades increase power output and lead to reduced stress and fatigue on blades (hubs and towers) offshore. Another way of viewing this is that the gain of wind speed with height is lower than over land, reducing the necessity for tall towers (Krohn, 1998). The negative implications of the reduction in turbulence offshore associated with the lower roughness of the sea surface is that wake effects will be higher (i.e. the disturbance of the wind flow experienced by a wind turbine caused by the other wind turbines located 'up stream'). Therefore optimum wind farm design (e.g. turbine spacing) may be quite different from land-based wind farms.

Developing wind farms offshore is a complex affair. Despite many years of experience of developing onshore wind farms and experience gained through offshore gas and oil production, the offshore environment presents a technological challenge. Wind turbines to be deployed in marine environments have to be modified to operate efficiently in a moist environment in which abrasive chemicals such as salt are found, and to withstand stresses due to application of wind and wave or wind and ice loadings in different directions. Engineering for these and other considerations adds to the overall cost of offshore wind energy.

Foundations are one of the largest components of the installation costs. Three design concepts for foundations have been considered:

1. Gravity-based; best suited for low water depths (5–10 m) with relatively low wave loading e.g. sheltered areas of the Baltic Sea, IJsselmeer. Although the offshore wind farms at Vindeby and Tunø Knob in Denmark were installed with concrete foundations, a Danish report concluded that steel can be adequately protected offshore and is significantly more economic (Danish Wind Turbine Manufacturers Association, 1999). Adaptations of this design have also been considered by Svenson and Olsen (1999).
2. Monopile foundation; best suited for deeper waters (>10 m). Monopile foundations were chosen for the Bockstigen wind farm and this gave a considerable cost reduction for this part of the installation (Lange et al., 1999).
3. Floating structures; most expensive and technologically the least developed, these will be necessary in deeper waters, like the Mediterranean Sea.

Offshore wind energy was probably first suggested by Hermann Honnef in the 1930s with his idea for a moored pontoon system (Doerner, 1999). However, current plans for offshore wind farms anticipate use of fixed rather than floating foundations. Fixed foundations need to be able to withstand wave and current loading and in some areas they may also be subject to ice build-up and ice floes affecting foundation design. A second major cost of an offshore wind farm is the installation of undersea cable for electrical grid connections, although development of large wind farms reduces the proportion of this cost per unit output. Wind turbines are currently designed to have an operational lifetime of 20 years, but components such as foundations and cabling may have significantly longer lifetimes and optimisation of designs for the main components (foundations, towers) may increase lifetimes to 50 years (Danish Wind Turbine Manufacturers Association, 1999) leading to a significant cost reduction per kilowatt hours (kWh) generated by offshore wind energy. The final major difference in the economics of offshore wind farms compared to those over land is the issue of access. Modern wind turbines are reliable at about 98%. However, access for both planned and emergency maintenance is more difficult offshore since bad weather may prevent landing by ship (or possibly by helicopter). Careful design of mooring facilities to enhance access times, preventative maintenance and the use of 'smart' turbines that are capable of predicting component failures should reduce outage times. Since offshore wind farms will be developed away from populated areas, and are not staffed apart from short maintenance visits, it has been suggested that these structures could be subject to less strict design criteria than wind farms onshore or oil and gas platforms.

Offshore wind energy remains more expensive than onshore wind energy. Higher wind speeds and lower turbulence offshore are offset by increased installation and maintenance costs. Installation costs are increased by the requirement for seabed foundations, undersea cable connections and current, relatively low, unit production of offshore wind turbines. Maintenance and operation costs are increased by potential losses in grid connections over long distances and restricted times and modes of access for maintenance. However, the wind energy industry has been remarkably innovative in finding solutions to these technical difficulties (e.g. the use of monopile foundations) and more large wind farms offshore will, given economies of scale, continue to improve the economics of offshore wind farms. Development of pilot projects in Denmark, Sweden and the Netherlands (see below) to assess turbine modification and optimisation and good monitoring and analysis programs to effectively assess wind resources have already reduced costs considerably. Offshore wind energy is currently about 15–20% more expensive than onshore wind energy and nearly equivalent at good sites (sites with a good wind resource but where installation costs are expected to be low). Costs have been reduced from 0.086 ECU per kWh at Vindeby, Denmark to an estimated 0.048 ECU per kWh estimated for the large wind farms in the next Danish installations (Svenson and Olsen, 1999; Kuhn and Bierbooms, 1999).

Offshore wind farm development can only be economic when projects are of considerable size (probably at least 100 MW). This means that financing these large projects (several hundreds of millions of ECU/EURO) presents a challenge to developers. The revenues of the wind farms are simply the number of kWh times the average price paid by the buyers. However, in a time when the electricity market is being liberalised and spot markets, even for electricity, are emerging, it is not easy to interest investors. Onshore wind farm development is impeded in several countries due to a complicated legal framework, e.g. in The Netherlands it is not exceptional that the procedures take up to almost 10 years before a final ruling on a proposed wind project is reached. Offshore it may prove somewhat simpler to comply with legal regulations, because there are fewer authorities to address. Moreover, just because of the size of the projects a legal process of several years can be justified in terms of effort and of cost. Another issue connected with the electricity market is the grid-connection. Offshore wind farms will be connected through DC or AC cables with high rated power (tens/hundreds of megawatts). However, as the wind resource is variable no absolute guarantee for the supply of wind-generated electricity can be given. Therefore methods to predict the wind power up to thirty-six hours or three days ahead will be necessary to increase its economic value (see e.g. Hutting and Cleijne, 1999).

A major advantage of installing wind turbines offshore is that very large wind farms can be developed away from major human activity. Although wind turbines themselves take up only about 1% of the land on which they are installed, to maximise power output they must be spaced five to ten rotor diameters apart. Grazing or agriculture can continue around the wind turbines installed on land but in many densely populated countries there are few areas with good wind resources suitable for installation of large wind farms. Offshore wind farms should be complementary to onshore development since many good onshore locations for single turbines or smaller wind farms still remain to be exploited. Offshore areas which are suitable for wind energy are subject to constraints other than the availability of the resource. Many coastal areas are already used extensively, for shipping, sand and gravel production, fishing, for leisure or by the military. Although wind farm development does not prohibit use of these areas for other purposes, these needs have to be carefully balanced. Wind farms on land have frequently been subject to criticism regarding visual intrusion, noise and the possibility of interference with television reception, which can clearly be reduced by moving large wind farms offshore. Visual intrusion can still be a problem if wind farms are installed in areas frequented by tourists, but most large wind farms will be at least 5 km from the coast and, although relatively little is known about propagation of wind turbine noise over water, offshore turbines are unlikely to present a significant visual or noise problem. Thus, developing offshore wind farms reduces many potential impacts on local residents. Additionally, design and operation of onshore wind turbines is frequently constrained by noise considerations which could be relaxed offshore, increasing power output.

## The Status of Offshore Wind Energy at the Millennium

Offshore wind energy development is most advanced in Europe due to the nature of the physical and political environment. The European Union has funded a number of projects aimed at improving offshore wind energy technology and understanding of wind, wave and turbulence climates offshore. The first study (Matthies et al., 1995) examined the potential for offshore wind energy in Europe (based on surface meteorological observations) and surveyed relevant experience in offshore construction. Emphasis was also placed on design criteria and modelling of wind and wave loading. Further projects such as Opti-OWECS (Optimisation of Offshore Wind Energy Converter Systems) (Kuhn et al., 1998; Kuhn and Bierbooms, 1996) extended the state of the art in designing and planning offshore wind farms to reduce power costs. The methodology developed was to consider the wind farm as an integrated system (turbine, support, grid and operation and maintenance) and to provide costs estimates for different designs (Kuhn et al., 1998). Other EU-supported projects include that co-ordinated by ELKRAFT (Danish electricity utility) which combined measurements at prospective Danish wind farm sites with modelling of wakes in large offshore wind farms. Current projects include predicting offshore wind energy resources (POWER), efficient development of offshore wind farms (ENDOW) and wind energy mapping using synthetic aperture radar (WEMSAR) (Svenson and Olsen, 1999).

In the sections below we describe the current status of offshore wind energy developments in two countries in detail.

### *Offshore Wind Energy Production in Denmark*

Denmark is committed to a 20% cut in carbon dioxide emissions by 2005 relative to 1988 levels and has a long-term goal of halving carbon dioxide emissions by 2030. The country is in the process of implementing a governmental action plan called 'Energy 21' which anticipates the installation of 4000 MW of offshore wind capacity before the year 2030 (Krohn, 1998). An 'Action Plan for Offshore Wind Farms in Danish Waters' has been prepared by the Offshore Wind Energy Working Group and the Danish Energy Agency to facilitate this process. With an additional investment in 1500 MW of onshore wind energy, Denmark is planning to supply more than 50% of annual domestic electricity consumption and there will periodically be the potential to export electricity from wind energy sources.

Denmark is well-suited to development of offshore wind farms, possessing both a good wind resource and large areas of relatively shallow water. Two offshore wind farms have already been developed; eleven 450 kW wind turbines were installed off the northwestern coast of Lolland at Vindeby in 1991 (Fig. 4). The separation between the two rows (oriented northwest to southeast) of turbines is 300 m, as is the separation between each turbine in the row (Fig. 5). The wind farm varies from 1.5 km from land for

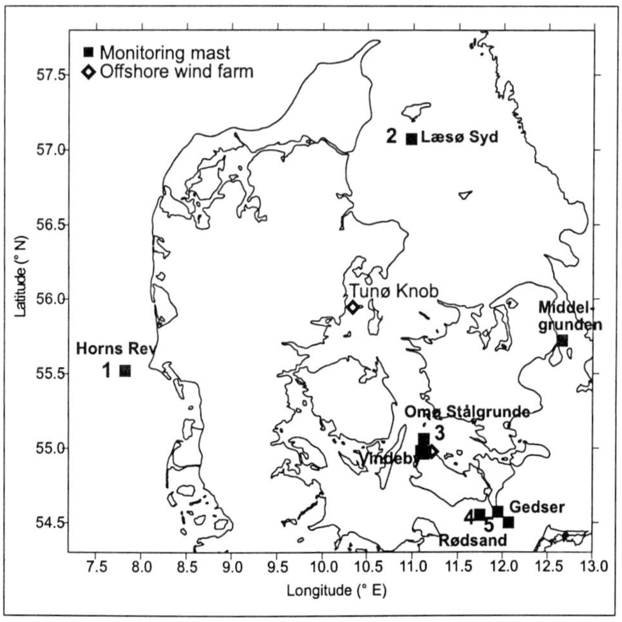


Fig. 4. Danish offshore wind monitoring sites. Development of large wind farms is planned in the areas marked 1–5.

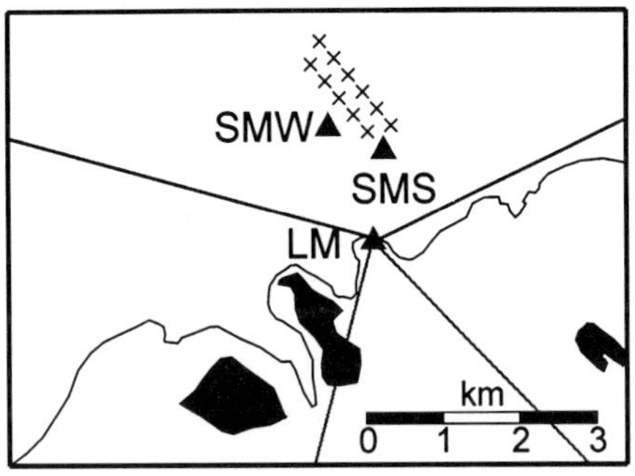


Fig. 5. Map of the Vindeby wind farm. Each wind turbine is represented by × while the monitoring masts (Sea Mast West, SMW; Sea Mast South, SMS; and Land Mast, LM) are marked ▲.

the closest turbine to about 3 km for the farthest turbine. Three meteorological masts (one at the coast, two offshore) were also installed and monitoring of the meteorology and the performance and loads of two turbines in the array have been conducted (Barthelmie et al., 1996; Frandsen et al., 1996). In 1995 a second wind farm was installed at Tunø Knob (Fig. 4). The Tunø Knob wind farm comprises ten Vestas V39 500 kW pitch controlled wind turbines (power curve shown in Fig. 3) and was developed by the Midkraft Energy Group under the electricity utility ELSAM (Madsen, 1996; Anon., 1995). Both wind farms have operated successfully and are exceeding expectations in terms of power supply to the grid. Plans are underway to install further major wind farms of 750 MW installed capacity in the areas shown in Fig. 4. These areas were selected based on their power production potential (combined, it is approximately 8000 MW) and because the sites are relatively remote, thus reducing visual impact, yet have water depths which are less than 15 m. The wind farms will most likely be built in blocks or units of around one hundred wind turbines of greater than 1 MW each, connected to the main electricity grids using 30–33 kV undersea (buried) cables (Danish Wind Turbine Manufacturers Association, 1999). Detailed surveys have been conducted in the ELKRAFT utility areas (eastern Denmark) including meteorological and geophysical surveys (Svenson and Olsen, 1999). Over the last few years, environmental investigations and public acceptability surveys have also been conducted for the development of a smaller (40 MW) wind farm off the coast of Copenhagen at Middelgrunden (Fig. 4). During 1999 the utility ELSAM commenced meteorological monitoring at the two prospective sites in their area (Horns Rev and Læsø Syd).

Denmark has also been actively involved in monitoring and modelling of all aspects of offshore wind energy technology. From detailed monitoring of the Vindeby wind farm (Barthelmie et al., 1996), a number of projects have been undertaken to monitor wind and turbulence characteristics in areas likely to be developed for offshore wind farms (see e.g. Barthelmie et al., 1999) and models have been developed for more accurate assessment of the offshore wind resource. The Wind Atlas Analysis and Application Program (WA$^{S}$P) (Mortensen et al., 1993) accurately predicts offshore wind speeds (Troen and Petersen, 1989) but there are still uncertainties in resource assessment, particularly close to the coast.

### *Offshore Wind Energy Production in The Netherlands*

Studies into the feasibility of offshore wind energy applications commenced in the later 1970s but wind farm development was slow. Currently two (semi-) offshore wind farms are up and running in The Netherlands: the first in the IJsselmeer, wind farm Lely (4 × 500 kW) of utility ENW which has operated since 1994, and the second, Irene Vorrink (19 × 600 kW) which has been operated by the utility NUON since 1996. The term semi-offshore is used here to indicate the fact that the IJsselmeer is a lake without tides, smaller waves and no salt water, in contrast to the North Sea.

In 1997 an initiative was launched for the construction of a 100 MW nearshore demonstration wind farm in the North Sea within 15 miles of the coast with water depths up to 15–20 m. It is hoped that experience gained here can then be used for future offshore wind farms further away from the coast. An Environmental Impact Assessment is currently (mid-1999) being conducted for the proposed wind farm and several locations are under consideration. The turbines should be installed and operating in the year 2000 2001. A 300 MW wind farm is also being proposed which will probably be located in the IJsselmeer or near the Afsluitdijk (i.e. the 30 km dyke which forms the boundary between the IJsselmeer (lake) and the Waddenzee (part of the North Sea). This initiative is also in the process of an Environmental Impact Assessment. It is proposed to deploy 3 MW wind turbines in this development which have yet to be developed.

The government target is to have 1500 MW wind energy capacity installed offshore in the year 2020 (see e.g. official publications of Ministry of Economic Affairs and NOVEM). A study commissioned by NOVEM was conducted by Grontmij to look at a longer term strategy to achieve 8000 MW of offshore wind energy, and a further study commissioned by Greenpeace and carried out by E-Connection evaluated the possibilities to implement 10,000 MW of offshore wind energy.

## Prospects for the Future

### *Plans for Offshore Wind Energy in the Short to Medium Term*

Figure 6 shows offshore wind resources for Europe predicted using the WA$^{S}$P model (Petersen, 1992). The simulation indicates that the best offshore wind resources are concentrated in the northwest of Europe, but significant resources also exist in the Baltic and Mediterranean seas. Figure 7 shows European countries actively engaged in the

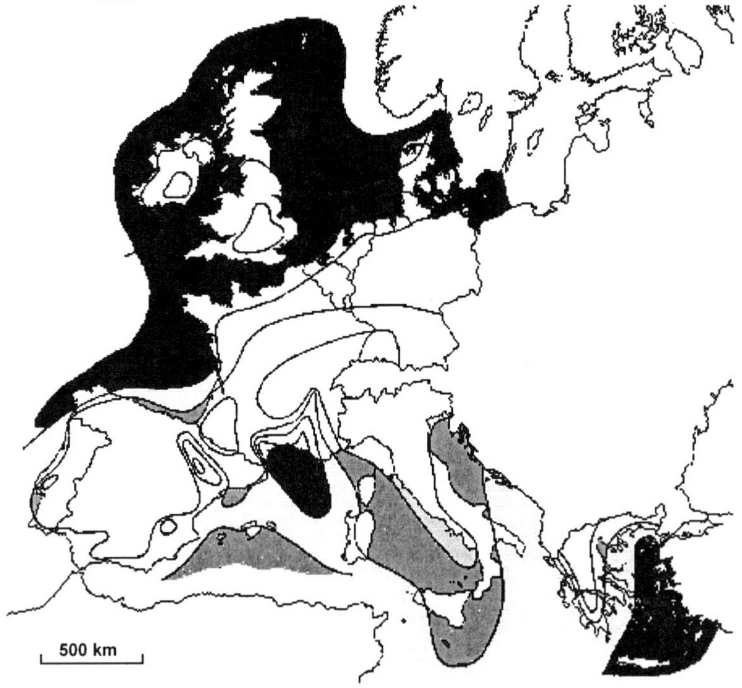


© 1997 Risø National Laboratory, Denmark

| | 10m | | 25m | | 50m | | 100m | | 200m | |
|---|---|---|---|---|---|---|---|---|---|---|
| | $ms^{-1}$ | $Wm^{-2}$ | $ms^{-1}$ | $Wm^{-2}$ | $ms^{-1}$ | $Wm^{-2}$ | $ms^{-1}$ | $Wm^{-2}$ | $ms^{-1}$ | $Wm^{-2}$ |
| | > 8.0 | > 600 | >8.5 | >700 | >9.0 | >800 | >10.0 | >1100 | >11.0 | >1500 |
| | 7.0-8.0 | 350-600 | 7.5-8.5 | 450-700 | 8.0-9.0 | 600-800 | 8.5-10.0 | 650-1100 | 9.5-11.0 | 900-1500 |
| | 6.0-7.0 | 250-300 | 6.5-7.5 | 300-450 | 7.0-8.0 | 400-600 | 7.5-8.5 | 450-650 | 8.0-9.5 | 600-900 |
| | 4.5-6.0 | 100-250 | 5.0-6.5 | 150-300 | 5.5-7.0 | 200-400 | 6.0-7.5 | 250-450 | 6.5-8.0 | 300-600 |
| | < 4.5 | < 100 | < 5.0 | < 150 | < 5.5 | < 200 | < 6.0 | < 250 | < 6.5 | < 300 |

Fig. 6. Offshore wind resources of Europe (from Petersen, 1992).

planning or development of offshore wind farms. A country-by-country review of the position of each country with regard to their offshore wind farm development is given below.

*Belgium*: Governmental support for the development of a 100 MW offshore wind energy demonstration project has been announced. In the middle of 1999 the legal framework for obtaining a concession to develop this wind farm is expected to be completed.

*France*: Interest in offshore wind energy in the Atlantic Ocean (Channel) has been expressed by the government and several industrial parties. Plans are underway for a wind farm at Le Havre (10 × 750 kW of Jeumont wind turbines).

*Germany*: Several scoping studies of the potential for offshore wind farms have been undertaken (Knight, 1995). Two sites were named for further exploration, one at an industrial site in Wilhelmshaven with good grid connections and one which, when developed, could assist efforts to reduce coastal erosion on the island of Sylt. Despite apparently favourable wind resources, environmental protection of much of the north and west coasts (the Wattenmeer nature reserve) and the extensive use of the Baltic coasts by shipping and for tourism has reduced the number of potential development sites. To date no offshore wind farms have been developed, although the project development company WINKRA have announced that they will go ahead with an offshore wind farm consisting of several 1.5 MW wind turbines near Wilhelmshaven. It is proposed that this will be the start of development of 2000 MW offshore wind power capacity in the North Sea and the Baltic Sea over the next seven years.

*Ireland*: A joint study of wind energy resources is currently underway between the Republic of Ireland and Northern Ireland and a consortium is investigating the development of an offshore wind farm off the coast, close to Dublin.

*Sweden*: The Wind World company installed a 220 kW wind turbine off the coast of Blekinge in 1991 and an offshore wind farm comprising five turbines rated at 550 kW each began operating at Bockstigen (off the south coast of Gotland) in 1998 (Lange et al., 1999). Costs of installation at this site were substantially reduced in comparison with those at Vindeby and Tunø Knob in Denmark by the first use of drilled monopile foundations for offshore wind energy. Lange et al. (1999) estimate that costs of wind energy on land are only 15–20% less expensive than at a wind farm such at Bockstigen. Enron Wind are involved in

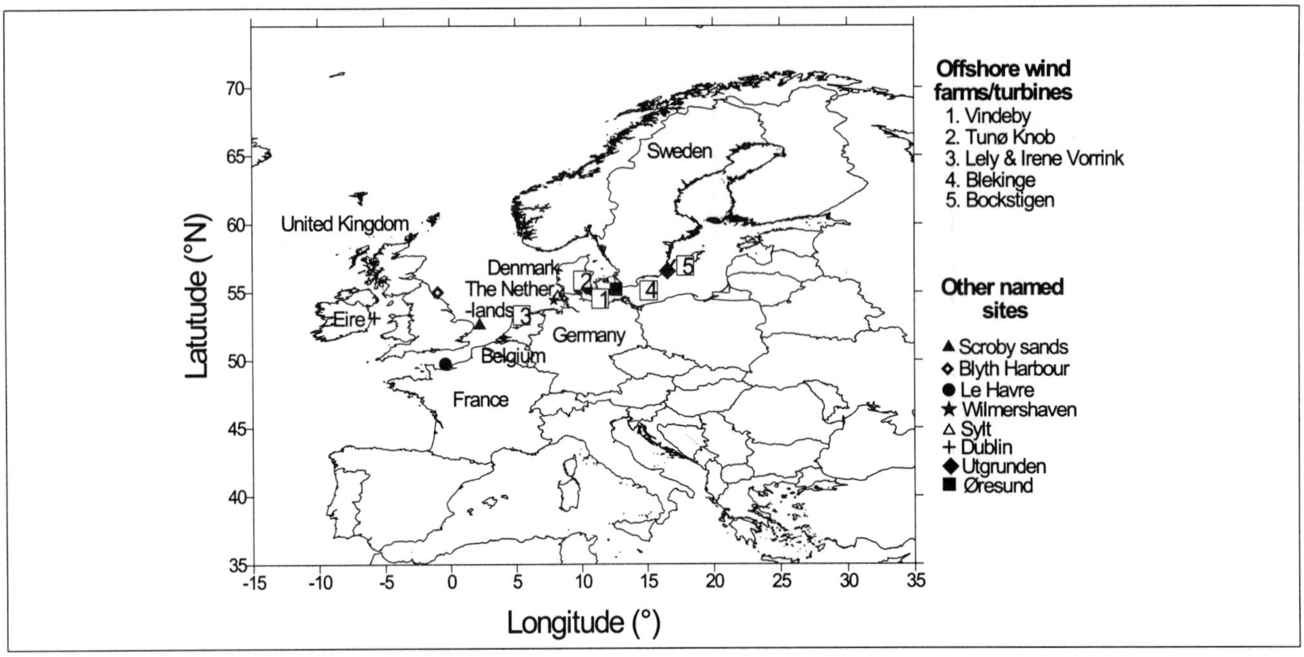


Fig. 7. Locations of current and planned offshore wind farms in Europe.

a 10 MW offshore project in Utgrunden which is due to be constructed in 2000. Other planned developments include a large wind farm in the Øresund between Denmark and Sweden (Caudiosi, 1999). There are also many potential sites in the Baltic and meteorological examination of the special conditions in the Baltic has suggested that offshore wind energy in this environment could be enhanced by the frequent occurrence of low-level jets, although the associated large wind shear could increase loads on the blades (Smedman et al., 1996). Given the Swedish commitment to replace all of their nuclear power plants, offshore wind energy remains a viable and likely alternative.

*United Kingdom*: The UK has been seriously considering offshore wind energy for some time. Offshore wind energy is considered by the Department of Trade and Industry to represent a medium-term technology which will contribute significantly to the commitment to develop the required 10% of energy from renewables by the year 2010 (Department of Trade and Industry, 1999). The UK has set targets of reducing greenhouse gas emissions by 12.5% by 2008–2012 and a goal to reduce carbon dioxide emissions by 20% by the year 2010. The situation regarding offshore wind farms in Britain is perhaps unique, since the Crown Estates (owned by the sovereign but not part of the sovereign's private estates) own about half of the foreshore around the UK and almost all of the seabed between mean low water and the territorial extent (Jacobson, 1998). Since offshore activities are licensed by the Crown Estates, they play a key role in advising developers of potential conflicts with other users including marine dredging. As landowners, the Crown Estates are in the process of developing a procedure to enable fair competition for development of offshore wind farms in consultation with other interested parties, such as the Department of Trade and Industry, Department of Environment, Transport and the Regions and the British Wind Energy Association. The operational rent has been initially suggested as 2% of gross turnover (reduced to 1.5% of estimated gross turnover during construction). The Crown Estates surrenders surplus net rents and profits to the Exchequer (Jacobson, 1998), thus offshore wind energy can be expected to be of direct benefit to the UK economy. The first offshore wind farm was planned for the North Norfolk coast in 1989 (Massy, 1995), but there were no wind farms in UK offshore waters prior to mid-1999. Border Wind has operated a nine turbine wind farm on Blyth Harbour wall in Northumberland since 1992 and has plans to extend this facility to include two offshore turbines (of 1.8 or 2 MW design) on monopile foundations during 2000. Planning and permits for this development are well underway and more detailed investigations of wave characteristics and foundation design are currently in progress (Grainger, 1999, pers. comm.). Offshore monitoring commenced at Scroby Sands (off the east coast of Norfolk) in the mid-1990s and plans to develop a large wind farm containing up to fifty 1.8 MW turbines are in place. In addition, five sites have been granted licences by the Crown Estates to commence meteorological monitoring in early 1999. In mid-1999 the UK Department of Trade and Industry review was in consultation and it was expected that, as part of the Non-Fossil Fuel Obligation, a special Offshore call for proposals would be issued during 2001 or 2002.

*Others*: Wind energy has also been adopted by a number of Mediterranean countries. Spain has the third largest installed capacity in Europe. However, offshore development has not proceeded at the same pace as in Northern Europe due to less pressure for land, more uncertainty in wind resource estimates and the technological challenge posed by the deeper waters of the Mediterranean. Gaudiosi (1999) reports that Italy proposes 1000 MW of offshore wind development by 2030 in order to cut greenhouse gas emissions. In addition a number of feasibility studies have been undertaken including several on breakwaters. In 1999 a report was released (Fioravanti, 1999) which concluded that development of offshore wind energy would be faster, about 40% cheaper and provide more energy security than using plutonium reactors. The study was based on Japan but the conclusions were applied to other countries with a large potential for the development of offshore wind energy, including Russia and the United States.

### *Prospects for New Wind Energy Technology and Research Needs*

Given the constraints on offshore wind farms described above, development of offshore wind energy is focused on large wind farms with large wind turbines. Since the installation of 450 kW wind turbines at Vindeby in 1991, 500 kW wind turbines at Tunø Knob in 1995, and 550 kW wind turbines at Bockstigen in 1998, most planned offshore wind farms over the next five to ten years will use turbines with rated power in the range of 1–2 MW. In early 1999 Vestas reported development of a new 1.8 MW turbine with modification for offshore operation and in The Netherlands a large multi-year research project will begin in October 1999 to develop a 5 MW offshore wind turbine, under financing by the Economy, Ecology and Technology (EET) programme of the Ministry of Economic Affairs.

To install wind turbines in deeper water (more than 25 m) new technology will be required, probably using floating platforms. These ideas have been studied (Harmer, 1994) but remain longer-term options. So far there has been relatively little consideration given to the possibility of developing combined technologies offshore, such as the use of wind–wave or wind–tidal systems, although as discussed by Lakkoju (1996), wind–wave energy technologies provide the potential for a more continuous power supply. Combined wind and wave could also proportionally reduce transmission costs per unit of installed energy but there is also the problem that areas with high wave range or a large tidal range will present even greater challenges for foundation or mooring design for wind turbines. Innovative design solutions including use of lighter materials for turbine components such as carbon or glass fibre may provide less expensive but more productive and durable wind turbines. In addition, energy storage solutions to weak grid or loss in transmission problems may also assist in the exploitation of remote energy sources. Further technological developments such as self-monitoring components or turbine access via hydraulic ramps will decrease turbine maintenance requirements while increasing access times.

More research using existing and new measurement techniques and model development is also required for prediction of wind speeds, turbulence and wake effects offshore, particularly in complex coastal areas for long time-scales (resource estimation, extreme events) and for short-term prediction (forecasting, maintenance). Since the relationship between wind and waves (in terms of separate and combined loads on foundations and towers) is not yet well understood, this area also requires major research initiatives. Continued monitoring and detailed investigation of these effects at the first large offshore wind farms will provide invaluable data for use in better evaluating and harnessing the offshore wind resource.

Although this review has not addressed environmental aspects of wind energy developments in great detail, it is worth noting that, although offshore areas are frequently utilised by both migrating and resident bird life and also by marine mammals and fish, studies to date have not suggested that installation of offshore wind farms cause negative impacts on marine wildlife. A detailed study of bird life undertaken at Tunø Knob before and after construction of the wind farm concluded that eider ducks were able to avoid the wind farm but were not greatly disturbed by it (Danish Wind Turbine Manufacturers Association, 1999). However, the next generation of offshore wind farms will be developed on a much larger scale and environmental aspects will need to be carefully considered.

## WAVE ENERGY

### Extraction of Energy from Waves

It is impossible to consider the technology for extracting energy from ocean waves without considering the nature of the resource itself. Waves at sea are generated by "frictional" forces between the wind and the water surface. This results in a transfer of energy from the wind into the water (Fig. 8). If the water is perfectly calm, this will result in motion in the upper layers, although if the wind blows for an extended period the Coriolis effect will result in a non-zero angle between the direction of the wind and the direction of the water flow. Once kinetic energy has been transferred to the seawater in this manner, the shape of the sea surface itself becomes unstable. If anything distorts the surface, such as a seismic disturbance or a thermal fluctuation, then the flow of the wind will result in a differential pressure distribution as shown in Fig. 9. It is this mechanism which results in the growth in ocean waves. Waves do not grow in height indefinitely, however; eventually, energy will be lost from the waves as a result of white-capping and breaking, as well as other mechanisms. Generally wave height will rise rapidly in the early period of a storm but will, in effect, reach a maximum height, which is largely dependent upon the wind speed irrespective of the duration of the storm (Fig. 10). The term $H_{sig}$, or Significant Height, is frequently used to describe the 'size' of storm waves. The height of an individual wave is defined as the vertical distance between the crest and the trough. Use of

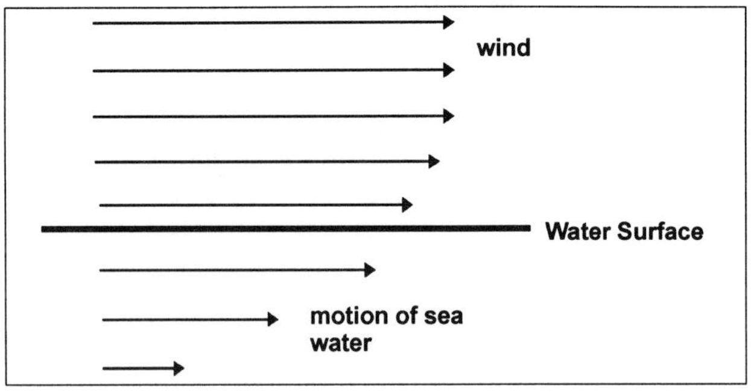


Fig. 8. Transfer of kinetic energy from wind to sea-water.

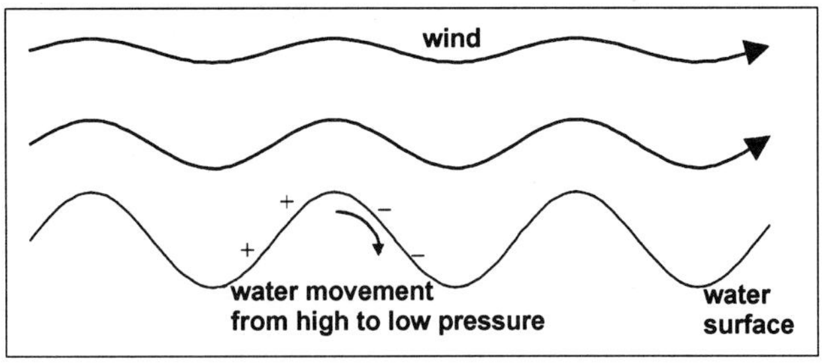


Fig. 9. Differential pressure distribution which causes wave growth.

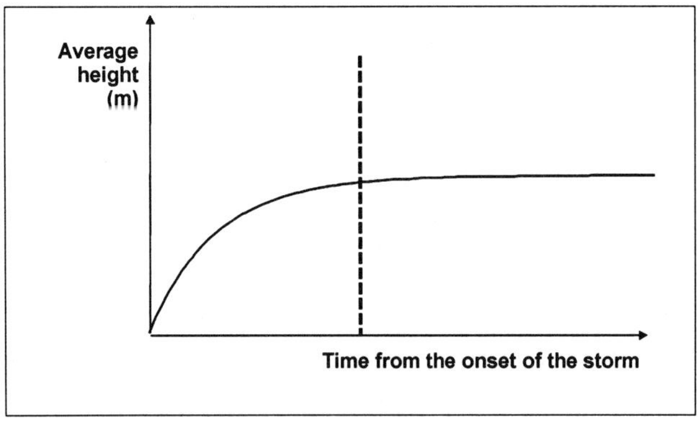


Fig. 10. Growth of wave height with time of onset of storm.

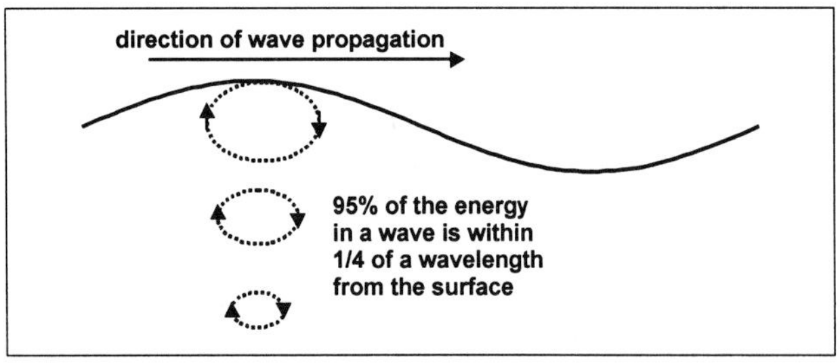


Fig. 11. Change of orbital size with water depth.

the significant height recognises that waves at sea are not regular and that they are a totally random process. In effect the significant height is the average value of the biggest third of the observed waves. Why this parameter is used instead of the simpler "average" wave height is open to some historical debate outside the scope of this text.

Energy in the wind is essentially kinetic. This is not the case in a wave. Within the water itself, energy is being continually transformed from kinetic to potential and back again. A simple theory of ocean waves devised by Airy (1845) implies that the water under a regular ocean wave can be described as moving in circular paths. These paths are known as orbitals. Similarly Airy (1845) suggested that the diameter of the orbitals decreases rapidly as one descends below the sea surface (Fig. 11). It is clear, therefore, that any device designed to harness energy in waves should lie at, or close to, the sea surface.

Airy's theory can also be used to predict the energy in a wave. The energy per square meter of sea surface, due to a wave of height $H$, can be shown to be

$$e = \rho g \frac{H^2}{8},$$

where $E$ is measured in Joules, $\rho$ is the seawater density, which is approximately 1025 kg/m$^3$, and $g$ is the acceleration due to gravity, which is approximately 9.81 m/s$^2$.

The influence of height upon energy density can be seen in Fig. 12. Waves do not, of course, stand still. Their motion carries their energy across the sea surface. In interpreting the influence of such motion it is necessary to consider the interception of waves by an energy converter as shown in Fig. 13. The energy intercepted per second by the device, if it has a width of 1 m, can be approximated as shown in Fig. 14. This curve takes into account the anticipated variation in wavelength with wave-height, which is not predicted by Airy's theory. It is obvious that there is considerable energy available in waves. The extraction of this energy is not, however, a trivial engineering exercise.

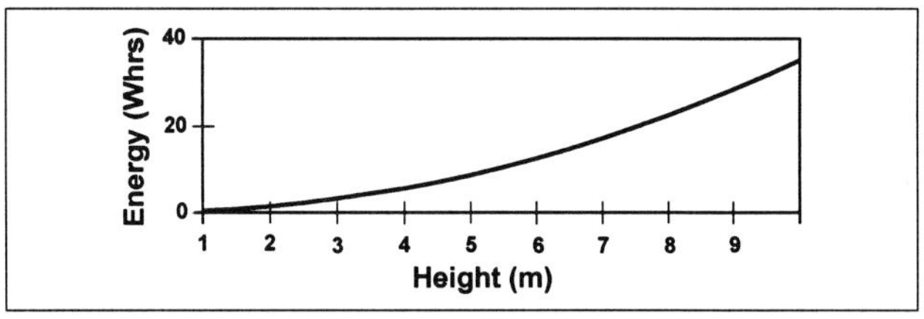


Fig. 12. Energy density of waves compared with wave height.

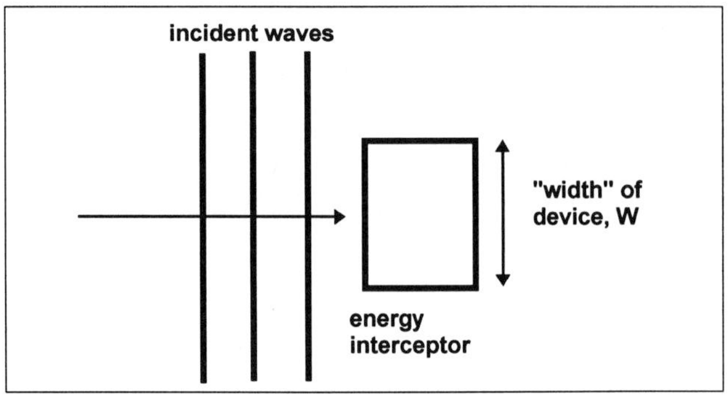


Fig. 13. Interception of waves by a wave energy device.

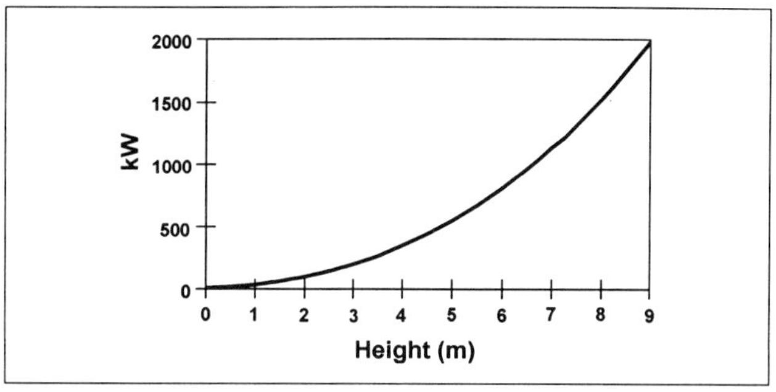


Fig. 14. Energy interception according to wave height for a device with a width of 1 m.

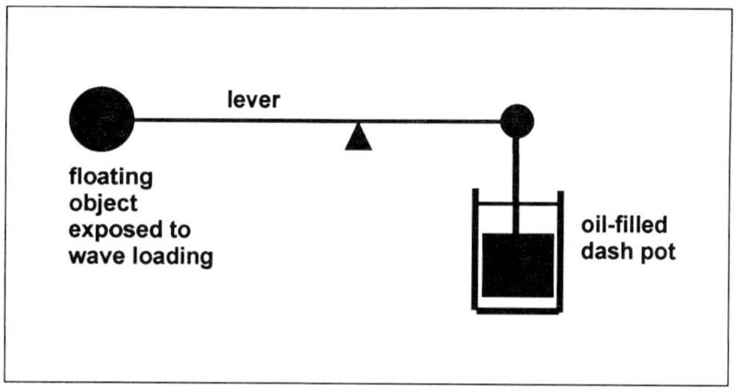


Fig. 16. The concept of converting mechanical energy to heat.

## Wave Energy: The Resource

The wave energy resource is substantial. Figure 15 shows the wave power resource in deep water for a selection of locations around Scotland. Although economic exploitation of energy in such locations remains beyond present technology, the principal design criteria for wave-power devices may be summarised as follows. All wave-power devices for deployment in deep water must be large enough to survive ocean wave conditions, but for optimal extraction of the resource the device should be of a similar size to the waves themselves. When it is appreciated that ocean waves can exceed 20 m in height and that, to be economic, a device needs to remain in location for an excess of 15 years, the challenge to the designer can be seen to be considerable. Before a device can convert wave energy into a useful form, it is necessary that it is constrained in some way, so that the forces created by the waves can be usefully resisted by the energy converter. There would, for example, be no way of extracting energy from a device freely bobbing on the wave surface, but the attachment of the device to a form of lever would, in principle, allow the conversion of wave energy into another form of energy. In the concept shown in Fig. 16, the lever is attached to an oil-filled dash pot so that the mechanical energy is converted to heat. Obviously this would be a very impractical design and many alternative forms have been suggested. Most, however, follow a small range of basic concepts which are described below:

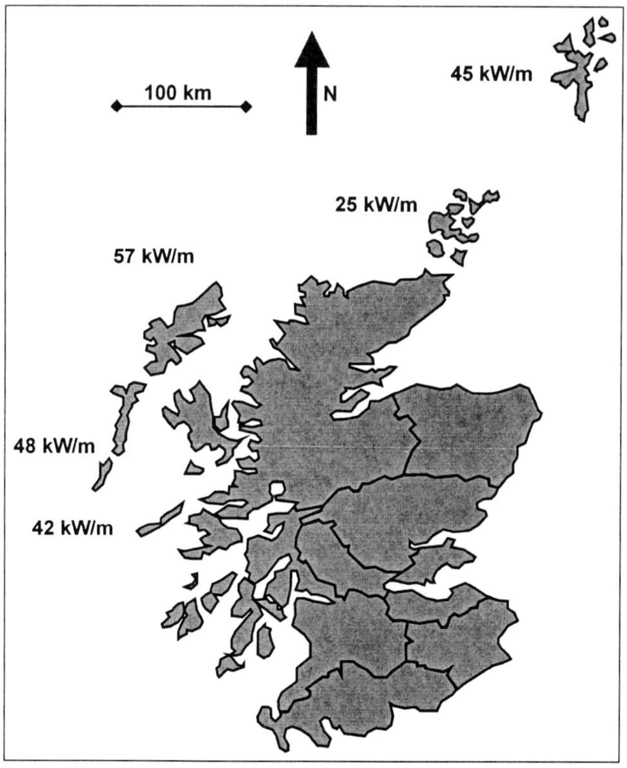


Fig. 15. Map of deep water wave power availability for Scotland.

*Tethered Buoyant Structures*: These are essentially free-floating objects whose motion is restricted by a mooring system, which incorporates some form of power extraction system (Fig. 17a).

*Relative Motion Devices*: These are devices in which energy is extracted via the relative motion of different parts of an extended structure. An example of this kind might be an articulated raft of hinged pontoons (Fig. 17b). A more exotic version of this principle might involve the use of motion relative to internal gyroscopes, as proposed for the Salter Duck (Peatfield, 1987).

*Flexible Membrane Devices*: In this family of devices, characterised by the SEA Clam (Edinburgh Wave Energy Group, 1979), the pressure under a wave is used to compress air, which is then driven through low-pressure turbines (Fig. 17c).

*Enclosed Water Column Devices*: These devices are philosophically related to the flexible membrane types, in that they use water pressure to drive air through a turbine, but they rely on the air–water interface itself to act as the pressure mechanism (Fig. 17d). This type of device is already in commercial use, most notably in Japan, to power navigation buoys.

In addition to the conversion principles, a deep-water device, or combination of devices, can also be characterised by its orientation to the waves. If individual devices are positioned in arrays parallel to wave crests, they are described as being in "absorber" mode (Fig. 18a). The

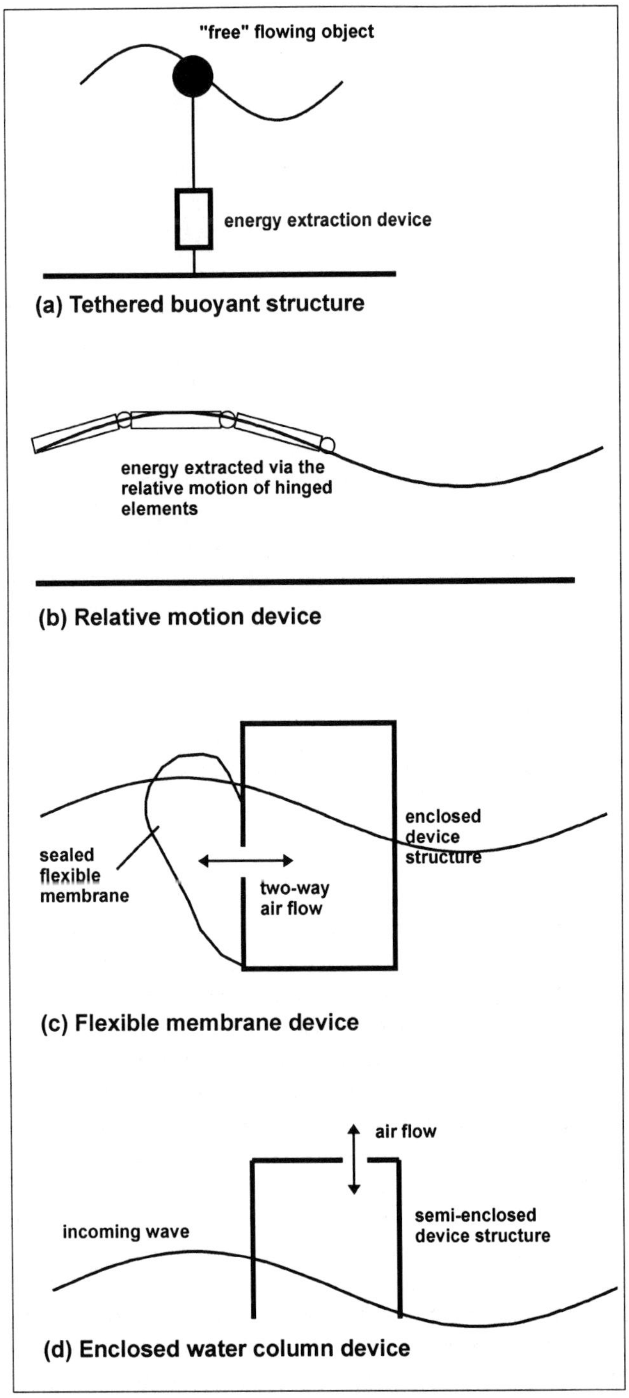


Fig. 17. Schematic of devices for extracting wave energy.

alternative is to position long devices which gradually absorb waves over a distance approximating to a wavelength. This is attenuator mode (Fig. 18b). To maximise the effectiveness of this configuration, the devices would need to be over 100 m in length!

Research into the design of offshore wave-energy conversion systems is now very limited and attention has been transferred to coastal exploitation. This is because of the engineering difficulties associated with mooring devices in deep water, while exposed to extreme ocean conditions,

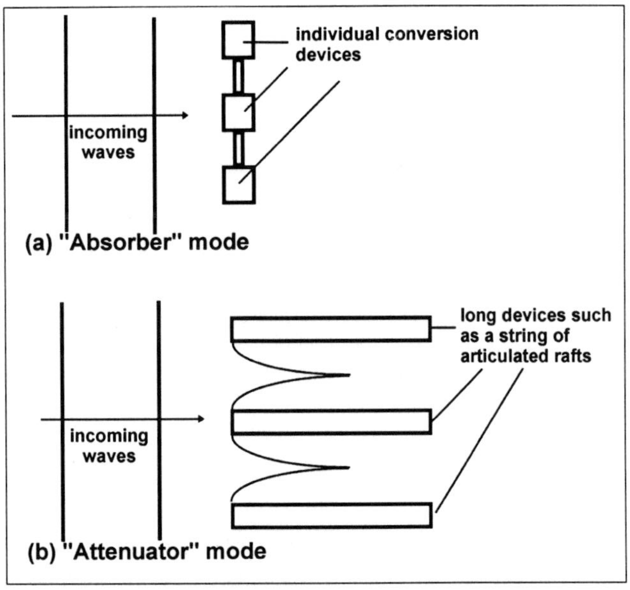


Fig. 18. Wave extractor device positioning.

and the not inconsiderable problems associated with bringing offshore electricity back onto land for distribution. Coastal locations have some very distinct advantages over deep water:

- devices can be more readily installed, thus reducing commissioning costs;
- devices might be rigidly fixed to the seabed or to coastal cliffs, thus providing more robust opportunities for energy extraction;
- electricity distribution costs will be lower because of the reduced need for expensive seabed power cables;
- maintenance costs will be lower because of greater accessibility.

There are, of course, also possible disadvantages. These include:

- wave energy at the coast is generally considerably less than in the open sea, although natural focusing could be used to counter this (Fig. 19);
- major engineering structures could spoil otherwise attractive coastlines with serious effects on tourism and wildlife.

There are two principal technologies which appear particularly suited to coastal wave energy developments. These are the enclosed water column devices (Whittaker et al., 1991) and tapered channel systems. In shallow water, *enclosed water column devices* might be firmly fixed to the seabed or onto a cliff wall. This would give considerable stability advantages when compared to a system moored in deep water. In addition, the device could be located in a gully to fully maximise the effect of wave focusing (Fig. 20). The enclosed water column principle is currently in use in the wave-power system currently operational in Islay, Scotland (Whittaker et al., 1991). The *tapered channel concept* is uniquely applicable to a coastal development. The energy of a wave is used to lift water up an artificial channel into an

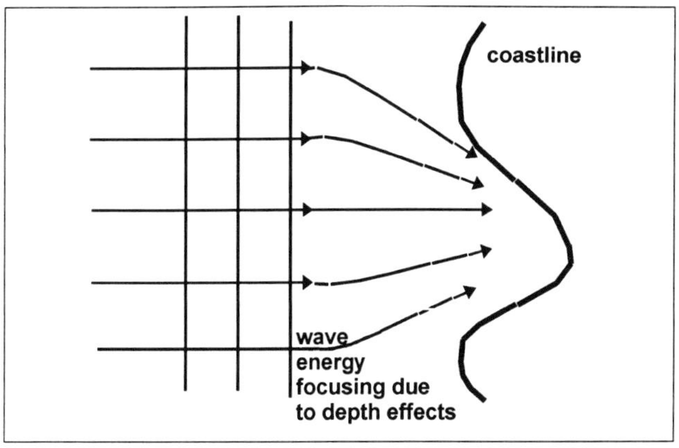


Fig. 19. Wave energy focusing due to coastal topography.

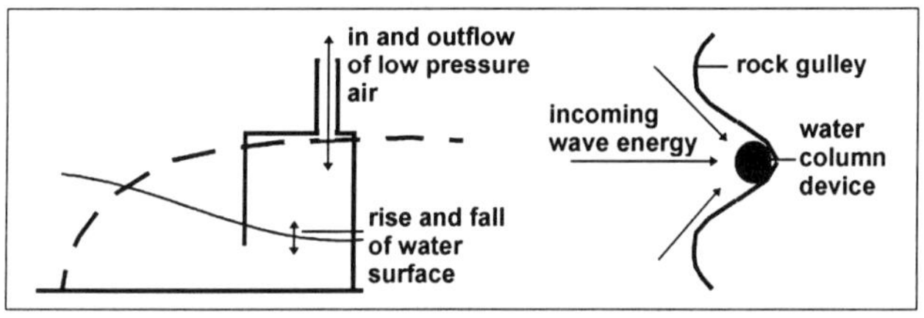


Fig. 20. Enclosed water column devices in coastal waters.

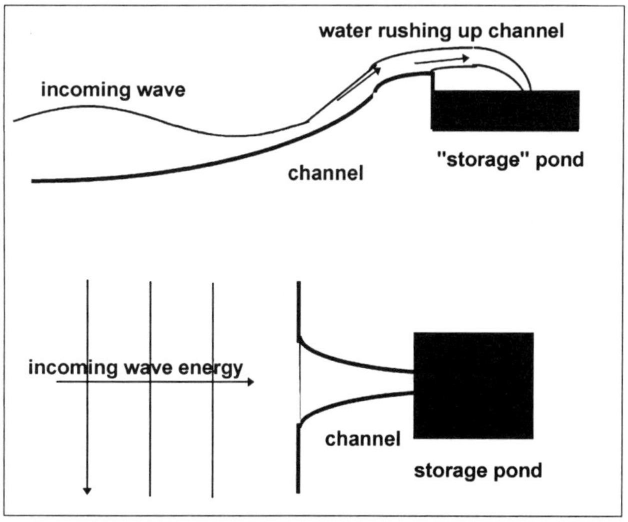


Fig. 21. Tapered channel concept for use in coastal waters.

artificial pond, from which it is allowed to drain back down to sea level through a low-head water turbine (Fig. 21). This approach has the advantage that the storage pond does allow some energy storage against the time when wave energy is not available. Disadvantages include its extreme sensitivity to tidal sea-level changes and the requirement for a very specific type of coastline, suitable for the kind of civil engineering required for the creation of the channel and storage pond. A 350 kW tapered channel device has been in use in Norway, 40 km northwest of Bergen since 1985.

### Current and Future Prospects

It is likely that small-scale operations, using enclosed water columns, will continue to be used to power specialised applications such as marine buoys. Similarly, coastal stations, with rated power outputs up to a few MW, will be commissioned for several locations around the world, most notably the Azores and Scotland. Initially these will be considered as extended research projects aimed at identifying the operational limitations and opportunities. Eventually, however, as has already happened with wind power, fully commercial coastal wave-power facilities will make a valuable contribution to the generation capacity in remote areas and islands, where fuel and electricity imports increase the costs associated with more conventional sources. It is likely, however, that due to the likely competition for suitable coastal sites between wave power and other uses and priorities (tourism, visual amenity etc.), coastal wave power is likely to remain a local source only and is unlikely to become a major strategic source. In Europe, of particular note in the near future are the Osprey 2000, which will be situated in 15 m of water 1 km from the shoreline and have a rated power of 2 MW, and an extension to the Islay project which will be rated at 0.5 MW. Similarly it is anticipated that there will be megawatt scale projects in the Azores, California, Greece, Indonesia and elsewhere. Several prototype wave-energy converters are in operation, and in Norway a 600 kW oscillating water device and a 350 Tapchan device have been demonstrated (Duckers, 1998). In the last few years a new resurgence of interest has followed economic assessment which suggests that wave-energy costs are competitive with conventional fuels in some environments and can supply power in remote areas. A new test Basin has been developed in Denmark at Nisum Bredning. Twenty million Danish kroner have been committed by the Danish Energy Agency to development and evaluation of wave-energy technologies at the new facility (*Renewable Energy World*, November 1998).

The situation with respect to deep-water wave-power generation is less clear. In the long term, it is likely that, if wave power is to be a major factor in future world energy plans, the offshore resource must be harnessed. There are still engineering problems to be solved before the resource can be safely harnessed. If these problems can be solved, then it is likely that offshore wave power could, like wind power, become a true competitor in the global energy market.

## TIDAL ENERGY

### Extracting Energy from the Tides

Tides are cyclic variations in the level of the seas and oceans which are largely (but not solely) due to the gravitational force exerted by the moon. Over the lunar month (the

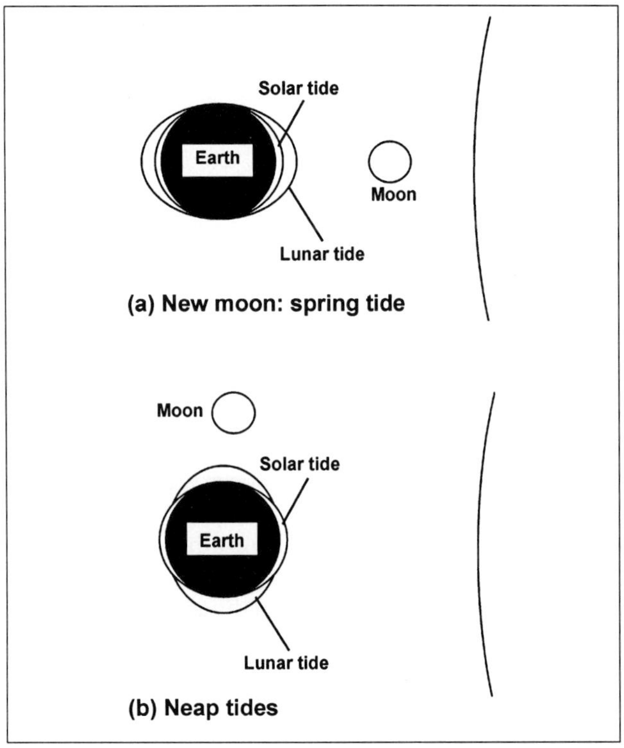


Fig. 22. Position of the sun–moon–earth system producing 'spring' and 'neap' tides.

rotational period of the moon around the earth) there are variations in the lunar tide influence. The lunar orbit is elliptical in form and the tide-producing forces vary by approximately 40% over the month. Similarly, the moon does not orbit around the earth's equator. Instead there is 28° between the equator and the plane of the lunar orbit. This also results in monthly variations. The earth–sun system is also elliptical but with only a 4% difference between the maximum and minimum distance from the earth to the sun. The relative positions of the earth, moon and sun produce the most noticeable variations in the size of the tides (Fig. 22). Figure 22a shows the earth–moon–sun where the influence of the moon and sun reinforce each other to produce the large tides known as Spring Tides, or Long Tides. A similar superposition also exists at the time of full moon. When the sun and moon are at 90° to each other, the effect is of cancellation as shown in Fig. 22b. This configuration results in Neap Tides also known as Short Tides.

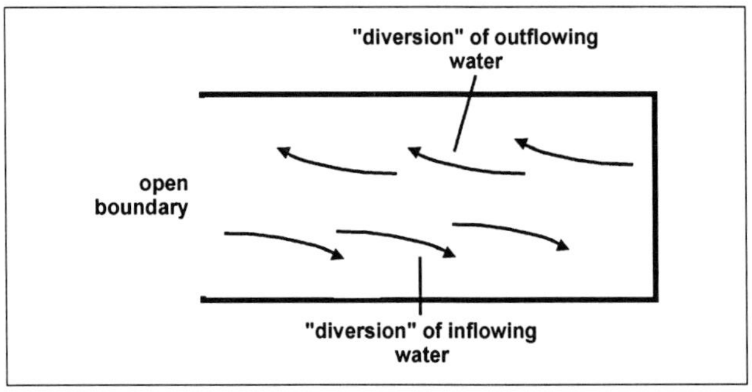


Fig. 23. Coastal influences on tidal range.

If the earth were covered entirely by water of a constant depth, the equilibrium theory of tides outlined above would give a perfectly reasonable description of water behaviour. Unfortunately, the oceans are not all of a constant depth and the presence of continents and islands severely influences the behaviour of the oceans under tidal influences. The influence of the Coriolis force, in the presence of land, can be considered, as shown in Fig. 23, by visualising water flowing into and out of a semi-enclosed basin in the Northern hemisphere under the influence of tidal effects. On the way into the channel, the water is diverted to the right towards the lower boundary. When the tidal forcing is reversed, the water is diverted towards the upper boundary. This simple example shows how coastal influence can significantly affect the size of the tidal ranges. At both the upper and lower boundaries the difference in level between high and low tides will be considerable greater than at the centre of the channel.

## Harnessing the Energy in the Tides

There are two fundamentally different approaches to exploiting tidal energy. The first is to exploit the cyclic rise and fall of the sea level using barrages and the second is to harness local tidal currents in a manner somewhat analogous to wind power.

### *Tidal Barrage Methods*

There are many places in the world in which local geography results in particularly large tidal ranges. Two sites which have been the subject of previous proposals for development are the Bay of Fundy in Canada, which has a mean tidal range of 10 m, and the Severn Estuary between England and Wales, with a mean tidal range of 8 m. The largest operational plant is located in La Rance (Brittany) which has a mean range of 7 m and has been in service since 1966 (Banal and Bichon, 1981). This plant, which is capable of generating 240 MW, incorporates a road crossing of the estuary. Other operational barrage sites are at Annapolis Royal in Nova Scotia (18 MW), The Bay of Kislaya, near Murmansk (400 kW) and at Jangxia Creek in the East China Sea (500 kW).

Essentially the principles of operation are always the same. An estuary or bay with a large natural tidal range is identified and then artificially enclosed with a barrier. This would typically also provide a road or rail crossing of the gap in order to maximise the economic benefit. There are a variety of suggested modes of operation. These can be broken down initially into Single Basin Schemes and Multiple Basin Schemes. The simplest of these are the *Single Basin Schemes* which, as the name implies, require a single barrage across the estuary (Fig. 24). There are, however, three different methods of generating electricity with a single basin.

*Ebb Generation*: As illustrated in Fig. 25a, during the flood tide, incoming water is allowed to flow freely through

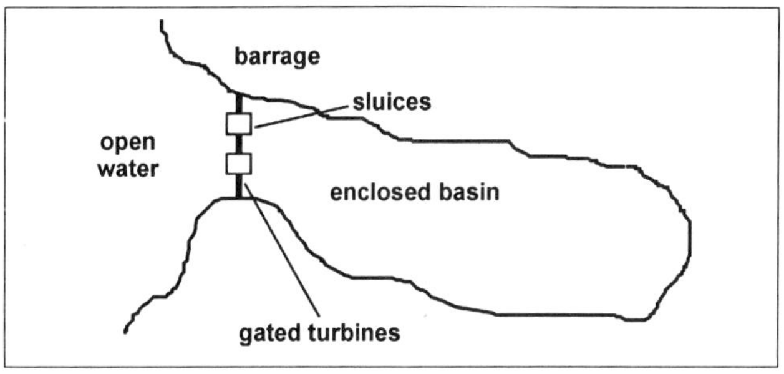


Fig. 24. Single barrage approach to tidal energy.

sluices in the barrage. At high tide, the sluices are closed and water retained behind the barrage. When the water outside the barrage has fallen sufficiently to establish a substantial head between the basin and the open water, the basin water is allowed to flow out through low-head turbines and to generate electricity. This method will generate electricity for, at most, 40% of the tidal cycle.

*Flood Generation*: The sluices and turbine gates are kept closed during the flood tide to allow the water level to build up outside of the barrage (Fig. 25b). As with ebb generation, once a sufficient head has been established the turbine gates are opened and water can, in this case, flow into the basin generating electricity. This approach is generally viewed as less favourable than the ebb method as keeping a tidal basin at low tide for extended periods could have detrimental effects on the environment and shipping. In addition, the energy produced would be less as the surface area of a basin would be larger at high tide than at low tide, which would result in a rapid reduction in the head during the early stages in the generating cycle.

*Two-Way Generation*: In this mode of operation, use is made of both the flood and ebb phases of the tide (Fig. 25c). Near the end of the flood generation period, the sluices would be opened to allow water to get behind the barrage and would then be closed. When the level on the open-water side of the barrage had dropped sufficiently, water would be released through the turbines in ebb generation mode. Unfortunately, computer models do not indicate that there would be a major increase in the energy production. In addition, there would be additional expenses associated in having a requirement for either two-way turbines or a double set to handle the two-way flow. However, advantages include a reduced period with no generation and the peak power would be lower, allowing a reduction in the cost of the generators.

All single basin systems suffer from the disadvantage that they only deliver energy during part of the tidal cycle and cannot adjust their delivery period to match the requirements of consumers. *Double basin systems* (Fig. 26) have been proposed to allow an element of storage and to give time control over power output levels as illustrated in Fig. 27. The main basin would behave essentially like an ebb generation single basin system. A proportion of the electricity generated during the ebb phase would be used to pump

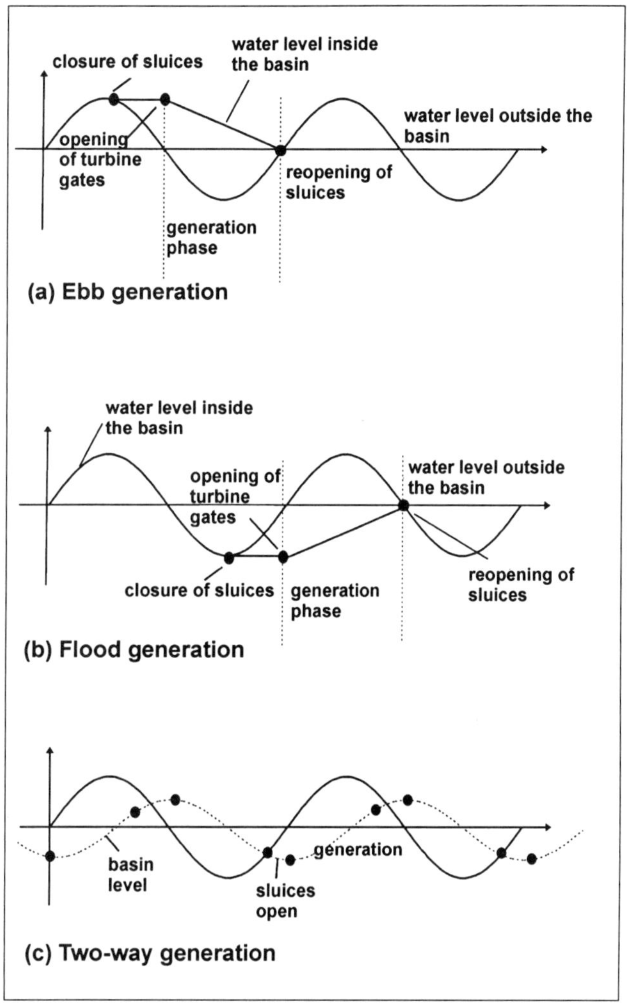


Fig. 25. Operation of different types of single basin tidal generation.

water to and from the second basin to ensure that there would always be a generation capability. It is not a simple procedure to design double basin systems as the construction of the barrage will inevitably change the nature of the tidal environment. It is theoretically possible, for example, for the tidal range to be very much less once a barrage has been constructed, than it was prior to construction. If an expensive mistake is to be avoided, it is necessary that reliable computer models are used to predict future behaviour as much as possible. Such models are available and can be used with a great degree of confidence.

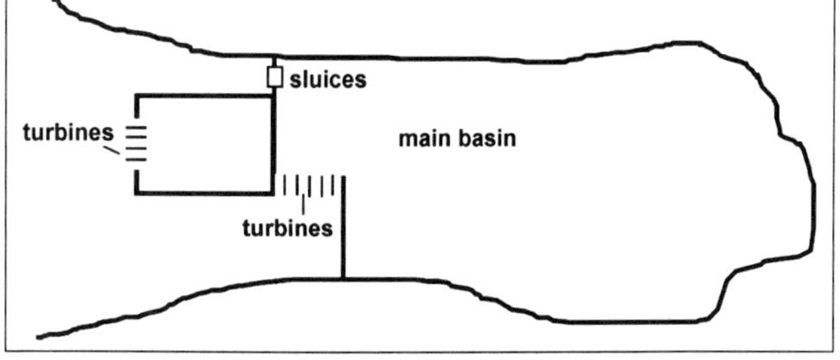


Fig. 26. Double basin systems.

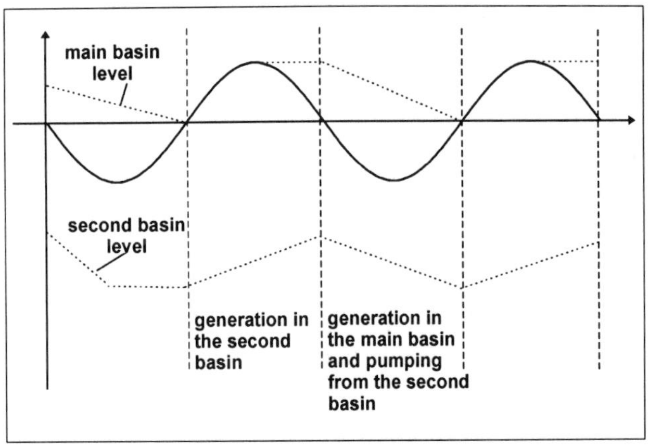


Fig. 27. Energy generation from double basin systems.

The construction of a barrage across a tidal basin will change the environment considerably. An ebb system will reduce the time during which tidal sands are uncovered. This would have considerable influences on the lives of wading birds and other creatures. The presence of a barrage will also influence maritime traffic and it will always be necessary to include locks to allow vessels to pass through the barrage. This will be much easier for an ebb system, where the basin is potentially kept at a higher level, than it would be with a flood generation system, in which the basin would be kept at a lower than natural level.

The public perception of tidal power is most definitely that of large barrage schemes such as that at La Rance or proposed for the Severn Estuary. These schemes can produce a vast amount of energy. It has been suggested that the Severn Estuary could provide in excess of 8% of the UK's requirement for electrical energy (Department of Energy, 1989). However, all barrage schemes are constrained by the massive engineering operations, and associated financial investment, that would be required. Construction would take several years; La Rance, for example, took six years, and no electricity would be generated before the total project was completed. This is a major disincentive for commercial investment.

### *Tidal Current Generation*

In addition to changes in sea surface level tides also generate water currents. In the open ocean these currents are typically very small and are measured in cm/s at most. Local geographical effects can, however, result in quite massive local current speeds. In the Pentland Firth to the North of the Scottish mainland, for example, there is evidence of tidal currents exceeding 7 m/s. The kinetic energy in such a flow is considerable. Other sites, in Europe alone, with large currents, include the Channel Islands and The Straits of Messina. Indeed it has been estimated that the European resource could substantially exceed 12,500 MW (CENEX, 1993). In addition to major sites such as the Pentland Firth, there are numerous local sites which experience very rapid currents capable of generating electricity with suitable technology.

There are two fundamental concepts for a current converter. These are vertical axis and horizontal axis turbines and mirror the development of wind turbines. In the case of the vertical axis turbine, the rotational axis of the system is perpendicular to the direction of water flow (Fig. 28a). A horizontal axis turbine has the traditional form of "fan" type system familiar in the form of windmills and wind energy systems (Fig. 28b). In this case, the rotational axis is parallel to the direction of the water flow.

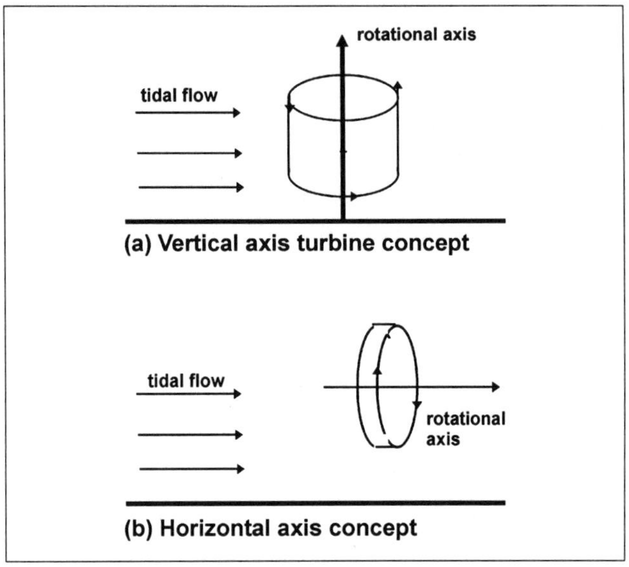


Fig. 28. Different concepts of tidal current generation.

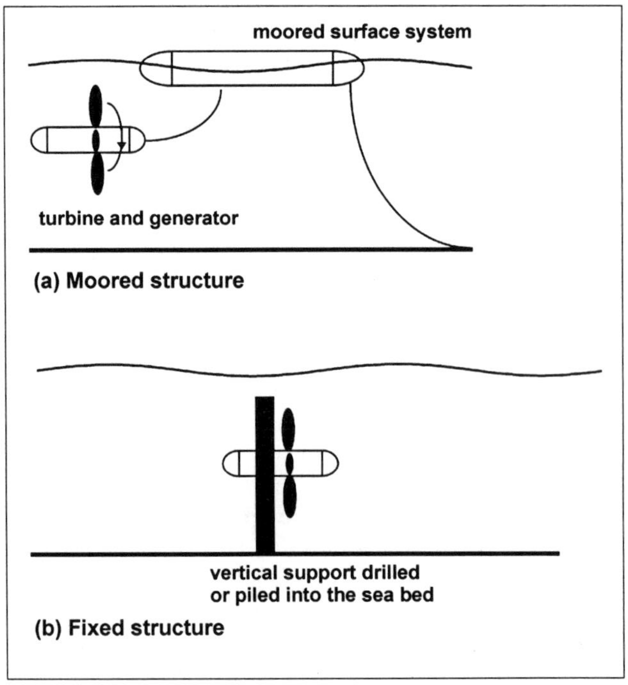


Fig. 29. Schematic of moored or fixed tidal current generators.

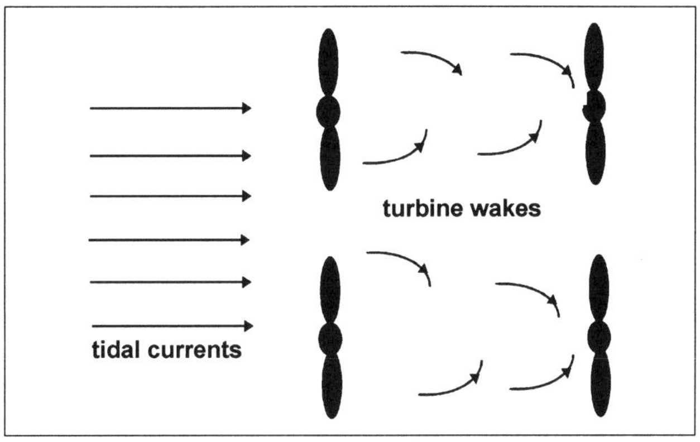


Fig. 30. Tidal farm concept.

In addition to the turbine type, there is also a question as to how the device is to be fixed in position. Is it to be suspended from a floating structure or fixed to the seabed? Moored surface systems (Fig. 29a) have advantages of mobility and accessibility. There are, however, possible problems concerning the stability of the surface pontoon and the generator/turbine. The alternative concept of fixing the turbine to the seabed (Fig. 29b) could provide a stable platform but the construction and installation costs could be very much larger. It is likely that, if tidal currents are to be commercially exploited, the generators will have to be mounted in clusters. If this is done, then, as with wind turbines, the devices will have to be sufficiently dispersed to ensure that the turbulence from individual devices does not interfere with others in the cluster (Fig. 30).

At the time of writing, there is no commercial generation of electricity from tidal currents anywhere in the world. It is anticipated, however, that the European Union will fund the first such turbine in 2000/2001 under the SeaFlow project. The device is likely to be located in South West England.

## Current and Future Prospects

At present, there is little worldwide experience in harnessing tidal current energy. Experimental turbines have been used in Japan and Scotland but, as yet, they have never been used to generate electricity for consumption. The Scottish trials in Loch Linnhe, using a floating horizontal axis system, demonstrated that the technology is feasible, while the Japanese experience has been largely with vertical axis systems. Until a large-scale device is commissioned there will be uncertainties about the long-term reliability and maintenance. Access to machines located in high energy tidal streams will always be problematic and could prove to require expensive maintenance procedures. Similarly the rapid tidal currents themselves could provide difficulties in transferring energy via seabed cables to the coastline. Indeed some studies have suggested that the cable and its installation could be responsible for more than 20% of the total cost of a tidal current system (Bryden, 1995).

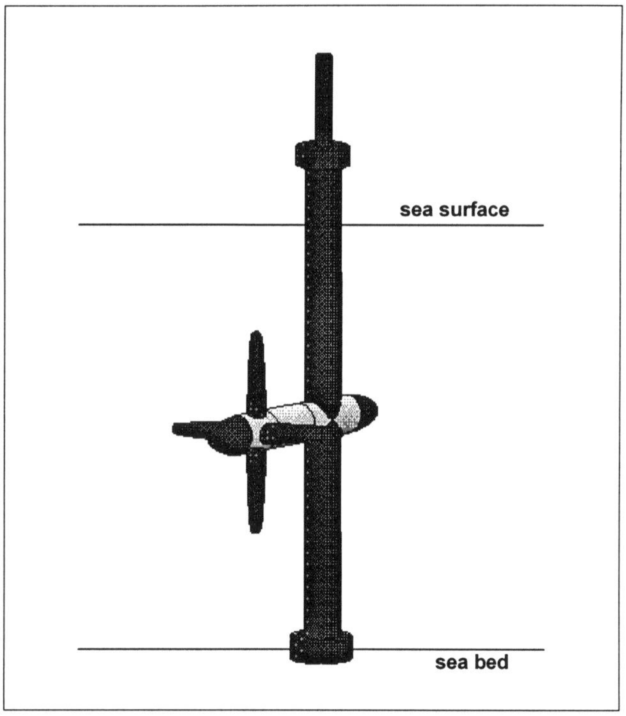


Fig. 31. Horizontal tidal current system proposed for large scale development.

In the near future it is likely that tidal current systems will appear in experimental form in many places around the world. Initially these will be rated at around 300 kW, such as that planned under the Sea Flow Project for the south of England. If this scheme proves a success, then the first truly commercial development should appear in the first decade of the 21st century and will probably take the form of a multiple array of fixed horizontal axis devices (Fig. 31). Current systems may not have the strategic potential of barrage systems but in the short term at least, they do offer opportunities for supplying energy in rural coastal and island communities.

Environmental impacts from tidal current systems should be minimal. Very energetic tidal channels do not tend to be home to many aquatic species and, although some species of marine mammals use tidal channels in their migration, the slow motion of the turbines and careful design should ensure minimal environmental impact.

The high capital costs associated with tidal barrage systems are likely to restrict development of this resource in the near future. What developments do proceed in the early 21st century will most likely be associated with road and rail crossings to maximise the economic benefit. In a future in which energy costs are likely to rise, assuming that low-cost nuclear fusion does not make an unexpectedly early arrival, then tidal barrage schemes could prove to be a major provider of strategic energy in the late 21st century and beyond. The technology for barrage systems is already available and there is no doubt, given the experience at La Rance, that the resource is substantial and available.

## SUMMARY

Three technologies have been described which are designed to extract energy from oceanic regions. In mid-1999 in terms of installed capacity these are: offshore wind energy close to 30 MW, wave energy just under 1 MW and tidal energy 260 MW. Wind and wave energy devices are typically smaller units (of up to 2 MW) and can be rapidly installed to meet energy needs. As the new technology of exploiting tidal currents is developed this will be within a similar small unit framework while barrage-based tidal projects are likely to continue to be large-scale engineering projects. In the first ten years of the next decade the status of these systems is expected to change rapidly. Advances in wind energy in recent years, coupled with ongoing research projects regarding offshore wind climatology and design of large wind farms mean that offshore wind is reaching the end of its development phase and can advance rapidly to large-scale installation. Installed offshore wind capacity will likely exceed 500 MW by 2010. Prospects for wave and tidal energy are less certain since the technology remains in its developmental stage. However, all three technologies are poised to make a substantial contribution to production of clean and sustainable energy for the new millennium.

## ACKNOWLEDGEMENTS

The authors extend grateful thanks to the following people who have contributed information for inclusion in this chapter: Paul Hannah, National Wind Power (UK); Guy Wilson, Energiekontor (Germany), Neil Jacobson, Crown Estates (UK); Heiner Doerner, University of Stuttgart (Germany); Bill Grainger, Border Wind (UK); Herman Busschots; Gunner Larsen, Risø National Laboratory (Denmark); Roger Wilson, Smit International (UK) Ltd; Lars Landberg, Risø National Laboratory (Denmark).

## REFERENCES

Airy, G.B. (1845) *Tides and Waves*. Encyclopaedia. Metropolitan, 192, pp. 241–396.

Anon. (1995) Heading offshore. *Windpower Monthly* (June), 23.

Banal, M. and Bichon, A., (1981) Tidal energy in France. The Rance tidal power station—some results after 15 years in operation. Proceedings of the Second International Symposium on Wave and Tidal Energy, Cambridge.

Barthelmie, R.J., Courtney, M., Lange, B., Nielsen, M., Sempreviva, A., M., Svenson, J., Olsen, F. and Christensen, T. (1999) Offshore wind resources at Danish measurement sites, 1999 European Wind Energy Conference and Exhibition, Nice.

Barthelmie, R.J., Courtney, M.S., Højstrup, J. and Larsen, S.E. (1996) Meteorological aspects of offshore wind energy—observations from the Vindeby wind farm. *Journal of Wind Engineering and Industrial Aerodynamics*, **62** (2–3), 191–211.

Bryden, I.G., Bullen, C.R., Bain, M. and Paish, O. (1995) Generating Electricity from Tidal Currents in Orkney and Shetland. *Underwater Technology* **21** (2).

CENEX (1993) Tidal and Marine Currents Energy Exploitation, European Commission, DGXII, JOU2-CT-93-0355.

Danish Wind Turbine Manufacturers Association (1999) Web site www.windpower.dk.

Department of Energy (1989) The Severn Barrage Project Summary: General Report. Energy Paper 57, HMSO.

Department of Trade and Industry (1999) *New & Renewable Energy: Prospects for the 21st Century*. URN 99/744, DTI Publications, London.

Doerner, H. (1999) Milestones of wind energy utilization. http://www.ifb.uni-stuttgart.de/~doerner/ewindenergie.html

Duckers, L. (1998) Power from the waves. *Renewable Energy World* (November), 52–55.

Edinburgh Wave Energy Group (1979) Fifth Year Report. ETSU WV 1512 Part 4A, ETSU Report.

Energy Centre Denmark (1994) European wind turbine catalogue. European Commission Directorate-General for Energy, Copenhagen.

Fioravanti, M. (1999) Wind power versus plutonium: an examination of wind energy potential and a comparison of offshore wind energy to plutonium use in Japan. Institute for Energy and Environmental Research, Takoma Park, Maryland.

Frandsen S. (ed.), Chacon, L., Crespo, A., Enevoldsen, P., Gomez-Elvira, R., Hernandez, J., Højstrup, J., Manuel, F., Thomsen, K. and Sørensen, P. (1996) *Measurements on and Modelling of Offshore Wind Farms*. Risø-R-903(EN), Risø National Laboratory, Denmark.

Gaudiosi, G. (1999) Offshore wind energy prospects. *Renewable Energy* **16**, 828–834.

Harmer, M. (1994) Floating wind farm awaits fans. *New Scientist* **142**, 21.

Hutting, H.K. and Cleijne, J.W. (1999) The price of large scale offshore wind energy in a free electricity market. 1999 European Wind Energy Conference and Exhibition, Nice.

Jacobson, N. (1998) The role of the Crown Estates in the development of offshore wind farms. Proceedings of the British Wind Energy Association Conference.

Knight, S. (1995) Heading offshore. *Windpower Monthly* (January), 46–47.

Krohn, S. (1998) Offshore wind energy: full speed ahead. *CADET* (December), 16–18.

Kuhn, M. and Bierbooms, W. (1999) Offshore wind energy—a future market under rapid development. http://www.nemesis.at/publication/ gpi_98_2/articles/56.html.

Kuhn, M. and Bierbooms, W.A.A.M. (1996) Structural and economic optimisation of bottom mounted offshore wind energy converters (OPTI-OWECS). *European Wind Energy Conference and Exhibition*. H.S. Stephens and Associates, Goteborg.

Kuhn, M., Bierbooms, W.A.A.M., van Bussel, G.J.W., Ferguson, M.C., Göransson, B., Cockerill, T.T., Harrison, R., Harland, L.A., Vugts, J.H. and Wiecherink, R. (1998) *Structural and Economic Optimization of Bottom-mounted Offshore Wind Energy Converters*. JOR3-CT95-0087 (five vols. plus summary), Institute for Wind Energy, Delft University of Technology, Delft.

Lakkoju, V.N.M.R. (1996) Combined power generation with wind and ocean waves. *Renewable Energy* **9**, 870–874.

Lange, B., Aagard, E., Andersen, P.E., Møller, A., Niklassen, S., Wickman, A. (1999) Offshore wind farm Bockstigen—installation and operation experience, European Wind Energy Conference and Exhibition, Nice, May 1999.

Madsen, P.S. (1996) *Tunoe Knob offshore wind farm, 1996 European Wind Energy Conference*. H.S. Stephens and Associates, Goteborg, pp. 4–7.

Massy, J. (1995) Race to be the first offshore. *WindPower Monthly* (December), 29–30.

Matthies, H.G., Garrad, A., Scherweit, N., Nath, C., Wastling, M.A., Siebers, T., Schellin, T. and Quarton, D.C. (1995) *Study of Offshore*

*Wind Energy in the EC.* Verlag Natürliche Energie, Brekendorf,.

Milborrow, D. (1996) Feeding a power hungry world. *Windpower Monthly* (January), 26–30.

Mortensen, N.G., Landberg, L., Troen, I. and Petersen, E.L. (1993) Wind Analysis and Application Program (WASP). Risø-I-666 (EN)., Risø National Laboratory, Roskilde, Denmark.

Peatfield, A.M. (1987) Wave energy—a British way forward with the circular SEA Clam, Proceedings of the Fifth International Conference on Energy Options, The University of Reading, UK, 7–9 April 1987, pp. 194–197.

Petersen, E.L. (1992) Wind resources of Europe (the offshore and coastal resources), EWEA Special Topic Conference '92, September 8–11 1992, Herning, Denmark.

Smedman, A.S., Hogstrom, U. and Bergstrom, H. (1996) Low level jets—a decisive factor for off-shore wind energy siting in the Baltic Sea. *Wind Engineering* **20** (3), 137–147.

Sterrett, F.S. (1995) *Alternative Fuels and the Environment.* Lewis Publishers, Boca Raton, FL, 276 pp.

Svenson, J. and Olsen, F. (1999) Cost optimising of large scale offshore wind farms in the Danish waters, European Wind Energy Conference and Exhibition, Nice.

Troen, I. and Petersen, E.L. (1989) *European Wind Atlas.* Risø National Laboratory, Roskilde, Denmark, 656 pp.

Whittaker T.J.T., Long, A.E., Thompson, A.E. and McIlwaine, S.J. (1991) Islay Gulley Shoreline Wave Energy Device Phase 2: Device Construction and Monitoring. ETSU WV1680, ETSU.

## APPENDIX 1: DEFINITIONS

Installed energy capacity is given in watts where:

1 kilowatt (kW) = $10^3$W = 1000 W

1 megawatt (MW) = $10^6$W = 1000000 W

1 gigawatt (GW) = $10^9$W = 1000000000 W

*Rated* wind speed of a wind turbine is the wind speed at which the turbine operates at its nominal power output

*Cut-in* wind speed of a wind turbine is the minimum wind speed at which the turbine operates and begins to produce power

*Cut-out* wind speed of a wind turbine is the maximum wind speed at which the turbine operates and produces power. At higher wind speeds the wind turbine is braked and/or decoupled from electricity generation to prevent damage to the components.

A *watt hour* is the energy used in one hour by a device consuming one joule per second. A typical electric fire uses 1000 watts, i.e. 1000 joule per second.

## THE AUTHORS

**Rebecca J. Barthelmie**
*Department of Wind Energy and Atmospheric Physics,*
*Risø National Laboratory,*
*4000 Roskilde, Denmark*

**Ian Bryden**
*The Robert Gordon University,*
*School of Mechanical and Offshore Engineering,*
*Aberdeen, AB10 1FR, Scotland*

**Jan P. Coelingh**
*ECOFYS energy and environment,*
*P.O. Box 8408, NL-3503 RK Utrecht,*
*The Netherlands.*

**Sara C. Pryor**
*Atmospheric Science Program, Department of Geography,*
*Indiana University,*
*Bloomington, IN 47405, USA*

Chapter 130

# MULTINATIONAL TRAINING PROGRAMMES IN MARINE ENVIRONMENTAL SCIENCE

G. Robin South

The global training programmes of the United Nations' system and of the International Ocean Institute Network are described. The programmes have been developed in the post-UNCLOS and UNCED eras, and in response to the needs of developing countries in human resource development in a number of areas, including coastal zone management (the TRAIN-SEA-COAST Programme), and climate change (CC:TRAIN). The UN programmes use a common methodology, the TRAIN-X methodology, and common principles of quality control, production of high quality training materials and networking of course modules. The IOI programmes are less rigid in format to those of the UN, but address the same needs. These global programmes represent a new era of cooperation among donors, training institutions and developing countries and are intended to reduce duplication of effort and to take advantage of the modern era of electronic communications in the area of training. While not the only examples of global training initiatives, they represent a concerted effort to address urgently needed human resources needs of developing countries in the marine sector.

*Seas at The Millennium: An Environmental Evaluation (Edited by C. Sheppard)*

## INTRODUCTION

The convergence of the political will of governments, as demonstrated by the United Nations Conference on Environment and Development (UNCED) with the emphasis on education and training permeating Agenda 21, set the stage for the development of global strategies in training and education for the 1990s. Faced with the demands and responsibilities inherent in the various Conventions and Agreements that arose from UNCED, countries found themselves at a crossroads concerning human resources development (HRD) in the fields of coastal zone and ocean management, climate change, and fisheries conservation. As a consequence, they appealed for ways and means to develop the human and financial resources needed to enhance their capabilities to effectively deal with these issues in the years to come.

The United Nations system has responded to the pressing need for the development of HRD in these and other areas through the establishment of global training programmes, which share a common methodology in course design, delivery and networking. In addition, the International Ocean Institute (IOI), through its global network of ten operational centres, has developed training programmes of considerable scope, and of global significance, including courses embracing issues of importance to environmental marine science. It is the purpose of this chapter to discuss the background leading to the establishment and implementation of the UN and IOI programmes, and to examine their advantages and limitations.

## THE UNITED NATIONS TRAINING PROGRAMMES – THE TRAIN-X STRATEGY

The fundamental strategy underlying UN training programmes was established in the 1960s with the development of the first training programme in the field of telecommunications. Subsequently, the standardised methodology for course design (referred to here as the TRAIN-X strategy) has expanded into a family of seven training networks (Fig. 1) in the following fields of specialisation: telecommunications (CODEVTEL/ITU); maritime transport (TRAINMAR/UNCTAD); civil aviation (TRAINAIR/ICAO); trade (TRAINforTRADE/UNCTAD); coastal and ocean management (TRAIN-SEA-COAST/UN/DOALOS); climate change (CC:TRAIN/UNITAR); and postal services (TRAINPOST/UPU). The strategy is based on the creation of cooperative training and HRD networks for the development and sharing of high quality course materials. The rigorous course development methodology leads to broad improvements in training centres and their associated policies and programmes.

The main elements of the TRAIN-X Strategy are:

1. Training centres with similar training needs agree to join a network and cooperate in sharing the training development task and related costs. The majority of centres are located in developing countries, but the networks are also composed of centres in developed countries and in countries in transition.
2. In order that courses are readily exchangeable, a common set of standards, both in the methodology of course design and in the form of presentation of training materials is applied throughout the network of Course Development Units (CDUs). Trainee and instructor materials are fully documented.
3. The courses are freely exchanged between the members of the network, with local adaptation only when justified. This for maximum utilisation, exchange and distribution of training courses and materials worldwide.
4. The course development activities are centrally coordinated to ensure quality control, to meet common needs, and to avoid duplication.
5. Exchange of instructors and materials between centres is organised whenever this is the most cost-effective way.
6. To ensure proper implementation at the national level, HRD policies are defined and implemented through the training of HRD managers.

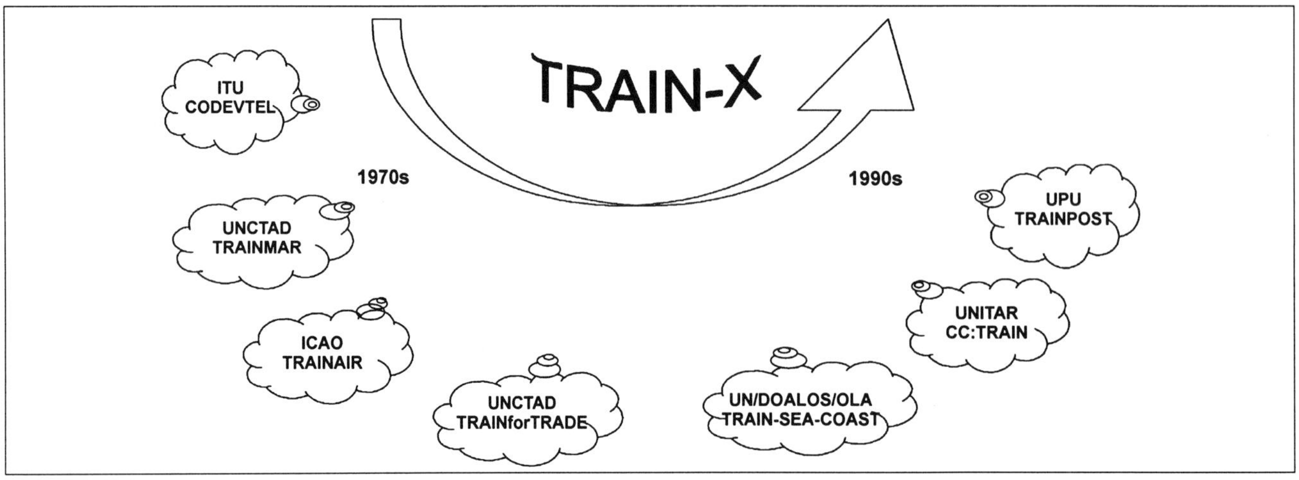


Fig. 1. The TRAIN-X family of training networks.

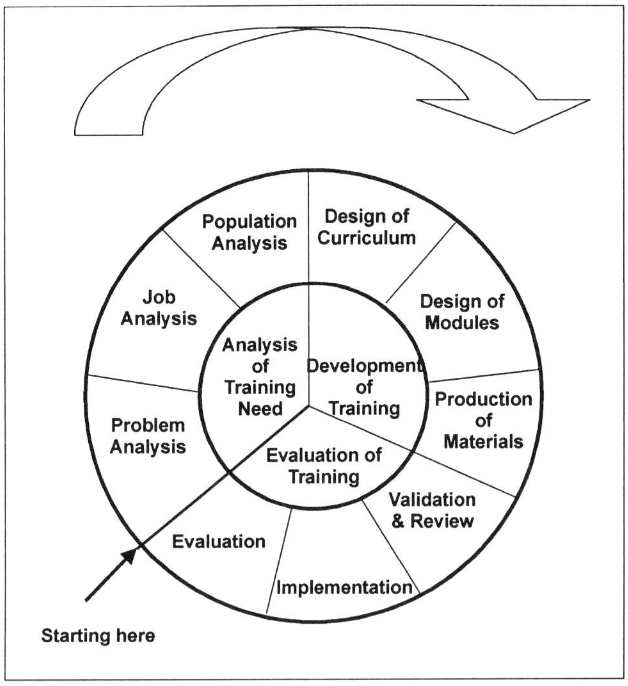


Fig. 2. The steps in training development in the TRAIN-X Strategy.

The TRAIN-X methodology involves three stages and nine phases (Fig. 2). The first stage consists of an analysis of the training need for a specific target training group, the second is concerned with development and delivery of the training, and the third focuses on evaluation of the training. The curriculum is developed only after there has been a thorough analysis of the jobs carried out by the target group and by an analysis of the target group themselves. In the development phase the curriculum is designed, is then organised into modules, and then the training materials are produced. Finally, the training is validated (i.e. is conducted for the first time), following which it can be implemented. If a standard training package (STP) is to be networked, then it is up to the recipient institution to adapt the materials to suit local needs and conditions.

These training networks currently comprise several hundred TRAIN-X partner institutions in over 100 countries worldwide. Two of them are directly relevant to marine science: TRAIN-SEA-COAST, and CC:TRAIN.

## TRAIN-SEA-COAST

The rapid development of national initiatives and international projects in the field of integrated coastal and ocean management worldwide presents many challenges to institutions and individuals involved in HRD. In 1992 UNCED made an urgent call to States to promote and facilitate education and training efforts in integrated coastal and ocean management, challenging them to enhance their capabilities of effectively dealing with coastal and marine issues in the years to come. This includes the development of technical knowledge and management skills over a wide range of complementary areas, from policy making to socio-economic and environmental aspects and the methodologies to develop and implement coastal and ocean management plans. Furthermore, the concept of integrated management of coastal and marine areas has become a central organising concept in a number of the post-UNCED related conferences and international agreements, with important implications in terms of emerging training needs. The scope of training is thus expanded to a much wider range of issues, e.g. from dealing with the impacts of climate change on low-lying coastal areas to addressing the human impacts on marine and coastal biological diversity, to the control of land-based sources of marine pollution.

Institutions delivering training all over the world are facing enormous challenges. In response to these, the UN/DOALOS embarked on a programme of action in 1993 to help developing countries tackle their training problems in coastal and ocean management. The underlying philosophy is that: (a) the problem needs to be sized by developing countries themselves and training provided through specialised training centres at the national level; (b) the scale of the problem is simply not compatible with the resources of any single organisation; and (c) the time has come for the design of global strategies for HRD that are rooted at the subregional, national and local levels.

A major development leading to global strategies was the 'Consultative meeting on Training in Integrated Management of Coastal and Marine Areas for Sustainable Development'. This meeting was convened by the UN Division for Ocean Affairs and the Law of the Sea, Office of Legal Affairs (UN/DOALOS) in Sasari, Sardinia, Italy from 21–23 June 1993. In attendance were UN officials as well as representatives of non-UN organisations, all of which have interest, experience and competence in the field of training. An important result of the meeting was the 'Action Plan for Human Resources Development and Capacity Building for the Planning and Management of Coastal and Marine Areas'. This was a milestone in a concerted approach to training. It contains a training strategy, specific action areas and preliminary course proposals in support of a new cost-effective strategy that would be widely beneficial for the countries concerned. Two activities resulted from the Action Plan: first, the development of a database on training programmes, and second, the launching in 1993 of the TRAIN-SEA-COAST Programme. The latter was an initiative of UN/DOALOS, with the support of the United Nations Development Programme, Science, Technology and Private Sector division (UNDP/STAPS), and in collaboration with UN and non-UN organisations involved in course development.

The TRAIN-SEA-COAST Programme is the primary instrument through which the UN/DOALOS aims at strengthening the existing capabilities of qualified training and educational institutions and individuals who have responsibilities in the field of coastal and ocean management.

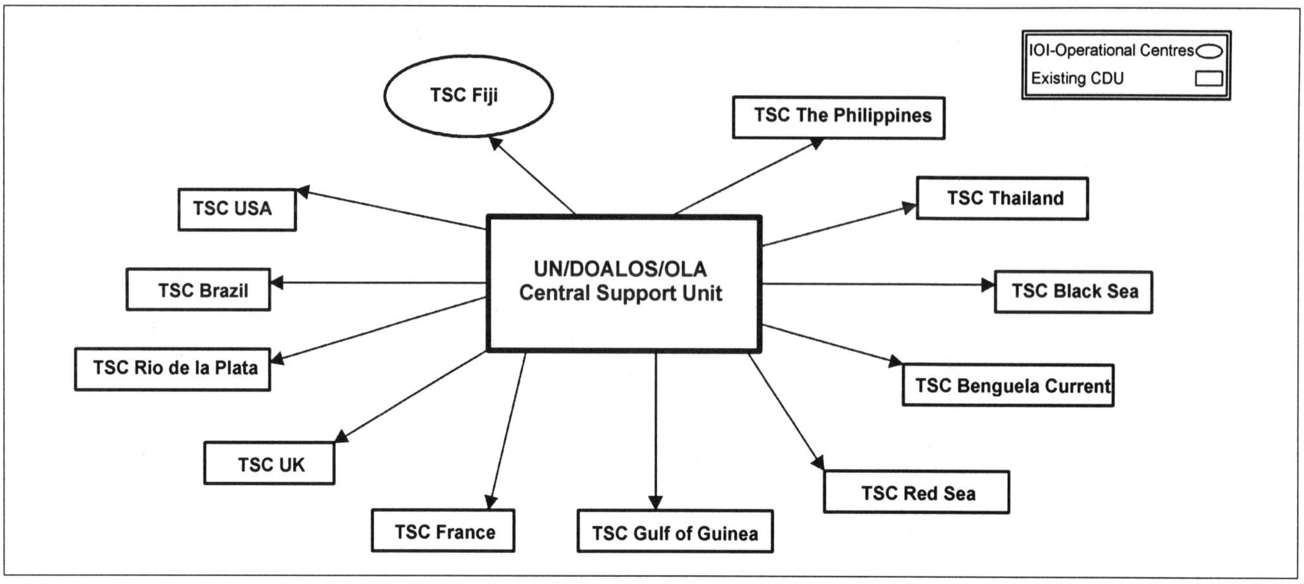


Fig. 3. Train–Sea–Coast Network.

The instruments for capacity building are the same as those identified above under the TRAIN-X strategy.

The TRAIN-SEA-COAST Network (Fig. 3) currently consists of 13 CDUs. Between 1995 and 1997, 11 CDUs were established. In 1998, five additional CDUs were created in association with GEF International Waters Projects. In 1999 the membership of three CDUs was revoked because they were unable to become operational. The current CDUs are: TSC/France (Université de Nice); TSC/Brazil (Fundacao Universidade do Rio Grande); TSC Philippines (International Center for Living Aquatic Resources Management, ICLARM and Philippines Council for Aquatic and Marine Research Development, PCAMRD); TSC/Thailand (Coastal Resources Institute of Prince Songkhla University); United States of America (University of Delaware); United Kingdom (University of Wales) and the following IOI Centres: TSC/Fiji (The University of the South Pacific). Currently, a further five units, linked with GEF International Waters Projects, are under development: TSC/Gulf of Guinea (Center for Environment and Development in Africa), Red Sea, TSC/Benguela Current (University of the Western Cape, South Africa), TSC/Black Sea (Ovidius University Constanza, Romania), and TSC/Rio de la Plata (Programa de Conservacion de la Biodiversidad y Desarrollo Sustantable en los Humedales del Este; PROBIDES).

The Central Unit is based at UN/DOALOS/OLA New York. It provides the link between training centres through programme management, monitoring and coordination. The Central Unit is responsible for:

1. quality control to member institutions at all stages of course preparation;
2. HRD in the form of a series of courses for training of course developers, instructors and training managers;
3. backup support facilities to provide, if necessary, technical advise to centres participating in the network;
4. periodic meetings to monitoring the activities of the network.

The rules of the network were agreed at the first TSC Coordination Conference. Participating host institutions are required to sign a letter of agreement as to the terms of their participation.

The TRAIN-SEA-COAST Network has the capability of responding to emerging needs and institutional changes that have taken and will continue to take place in the coastal and ocean sectors. Senior policy makers and planners at the national and subregional levels need to be sensitised to the underlying principles of integrated management of coastal and marine areas. Programme and project managers need the knowledge, skills and attitudes required in the practice of coastal and ocean management, whether plans or programmes are set up on an integrated basis or are tied to sectoral components. Operational agents, including coastal and ocean management practitioners, require specialised training in the specific technical, legal and managerial requirements of users. Implementors, operators and resource owners in local communities, fisherfolk, etc. require technical skills and enhancement of awareness of issues of conservation and sustainable development.

The provision of training in the field of coastal and ocean management must adapt to the needs of the users. Training is rarely designed and delivered in sufficient alignment with management priorities to ensure that it is meeting the real needs, that it is followed by the people most needing it, and that trainees can apply their knowledge and skills when they return to the workplace. TRAIN-SEA-COAST course development units are constantly faced with new demands that must be tailored to the specific training priorities of, for example, national programmes, local administrations, project personnel, and the private sector.

Short courses (one week or less) based on high-quality materials geared to the client's needs, and flexible schedules outside working hours, are in high demand. Time availability and increasing budget reductions for travelling away from home to take courses are also creating a demand for new modes of learning that can be conducted outside the classroom, such as computer-assisted and distance learning.

The following gives some idea of the range of training courses under development in the TSC network:

- management of marine protected areas
- the role of women fishers within coastal communities
- marine pollution control
- Law of the Sea and ocean policies
- Integrated Coastal Zone Management
- exchange and inter-relationships between the river basins, coastal lagoons and adjacent marine areas
- identification of coastal and marine sensitive areas
- nutrient pollution from agriculture
- approaches to conflict management in ocean and coastal management.

It is noteworthy that some of these topics overlap with those that are covered by the IOI Network. Since some of the same institutions are involved in two or three of these networks, it is evident that cross-fertilisation of networks is a real possibility.

## CC:TRAIN

The UN Framework Convention on Climate Change (UNFCCC) was signed in Rio de Janeiro in 1992, and ratified by three-quarters of the world's countries. The first Conference of the Parties was convened in April 1995. The Convention calls for a global partnership to limit emissions of the greenhouse gases believed to cause climate change. It includes a financial mechanism to support the efforts of developing countries and commitments to facilitate scientific and technological cooperation among nations. In addition, all Parties to the Convention have a committee to provide information on, amongst other issues, their vulnerability and adaptation to climate change. This information is reviewed by Convention bodies.

Implementing the Climate Change Convention poses many challenges. It requires, at the national level, the reconciliation of priorities such as economic growth, employment and poverty alleviation in the search for sustainable development for all countries, rich and poor. It also implies a collective effort on the part of governments, firms, industries, NGOs and international organisations to find new and innovative ways of addressing climate change while promoting economic and social development.

In many countries, however, neither climate change nor the Climate Change Convention is clearly understood. Against this background, the Secretariat of the Convention and the United Nations Institute for Training and Research (UNITAR) has developed CC:TRAIN, a training programme to support the implementation of the Convention. Designed and implemented with funding from the Global Environment Facility (GEF) and the support of the United Nations Development Programme (UNDP), the programme seeks to mobilise countries in implementing the global effort to curb climate change at the national level, and to provide the training required to enable countries to meet their reporting obligations under the Convention. In developing training materials, CC:TRAIN collaborates with regional partners from a global network. These regional partners are: Environnement et Développement du Tiers-Monde (ENDA-TM), Dakar Sénégal (serving Bénin, Chad, Sénégal and Zimbabwe); Fundación Latinoamericano (FFLA), Quito, Ecuador (serving Bolivia, Cuba, Ecuador, Paraguay and Peru); and the South Pacific Regional Environment Programme (SPREP) in Apia, Samoa (serving Cook Islands, Fiji, Kiribati, Marshall Islands, Nauru, Solomon Islands, Tuvalu, Vanuatu and Samoa).

CC:TRAIN is coordinated at the United Nations Institute for Training and Research (UNITAR, Switzerland), and operates under the same TRAIN-X principles as the other members of the TRAIN-X family. CC:TRAIN is a part of UNITAR's overall approach to assist countries in the implementation of the UNFCCC. An example of the levels of cooperation involved in the development of CC:TRAIN training materials is that operated in the Pacific.

UNITAR employs an approach characterised by:

1. Country Team Approach
2. Network of Regional Partners
3. TRAIN-X Training Development Method
4. Partnerships

The Country Team Approach involves

- inviting and supporting governments to establish multi-disciplinary, multi-sectoral teams which can be made responsible for developing plans and strategies to implement the UNFCCC;
- focusing on the provision of assistance and clarifies what institutions and processes are being supported;
- inspiring local ownership and adaptation to local conditions.

The concept of the country team works through national governments, who identify a host ministry (e.g. Environment, Economic Planning) and who select a Country Team Coordinator from that ministry. The Country Team Coordinator develops a Country Team whose membership may include other ministries, environment and development NGOs, private sector and members of the academic community. The Country Team is responsible for vulnerability and adaptation assessments, a greenhouse gas inventory, and mitigation studies.

The role of the CC:TRAIN network is to assist countries in the development of HRD through training materials and courses. An example of a training programme is the Vulnerability and Adaptation Assessment (V&A) certificate course currently run by the Marine Studies Programme at the

University of the South Pacific in Suva, Fiji. The four-month long training course is offered to participants from Pacific Island nations who are signatories to the UNFCCC. It is coordinated by the Pacific Islands Climate Change Assistance Programme (PICCAP) of the South Pacific Regional Environment Programme (SPREP), based in Apia, Samoa. The original course materials were developed through cooperation between UNITAR and the University of Waikato (Hamilton, New Zealand) International Global Change Institute (IGCI). The entire programme is designed to train participants in how to develop a national Vulnerability and Adaptation Assessment Plan for their respective countries. The course is offered in modules, and is focused on an imaginary country called Vanda. The largely computer-based course is designed to facilitate "learning by doing" through simulation. By the end of the course, the cumulative exercises will have enabled the participants to complete a "mini assessment" of vulnerability and adaptation for the country of Vanda. They can then apply this to the real situation when they return to their home country.

The V&A training course is designed for participants with diverse backgrounds and different levels of understanding of the technical aspects. It may include individuals from government, universities, and firms or business groups. The course is also designed for use in different countries.

Other training packages being developed under the UNITAR Climate Change Programme include:

- Workshop Package on climate change and the UNFCCC: Challenges and Opportunities (1–2 days, intended for national awareness raising)
- Training Package on Vulnerability and Adaptation Assessment (10–14 days)
- Training Package on National GHG Inventory (4–5 days)
- Training Package on Mitigation Analysis (5–8 days)
- Workshop Package on National Communications and Non-Annex I Parties (1 day)
- Training Package on National Implementation Strategy (5 days)
- Training Package on Using Information systems for Assessing Impacts (5 days)
- Certificate on Vulnerability and Adaptation Assessment (4-6 months).

The CC:TRAIN programme has initiated a library system to compile and disseminate climate change-related training and information resources using CD-ROM and the Internet, in cooperation with the Global Environmental Information Center of the United Nations University, the Japanese Environmental Agency, and the Climate Change Secretariat.

## THE INTERNATIONAL OCEAN INSTITUTE TRAINING PROGRAMME

Established in 1972 by its Founder and Honorary Chair, Professor Elisabeth Mann Borgese, the IOI is the only international NGO entirely devoted to the oceans. IOI is headquartered in Malta and has a total of ten operational centres worldwide (Fig. 4): (IOI-Canada, IOI-Malta, IOI-India, IOI-Pacific Islands, IOI-Southern Africa, IOI-Senegal, IOI-Costa Rica, IOI-Black Sea, IOI-China, and IOI-Japan). Training programmes are integral to the IOI, and more than 10,000 alumni have participated in IOI courses. The longest-running course is the 10-week course on Management of the EEZ offered by IOI-Canada. The expansion of IOI's network to the current number of operational centres began in 1993, with funding from the UNDP under the Global Environment Facility. Four centres were added at that time: IOI-Pacific Islands, IOI-Senegal, IOI-India and IOI-Costa Rica. Important in this expansion was the

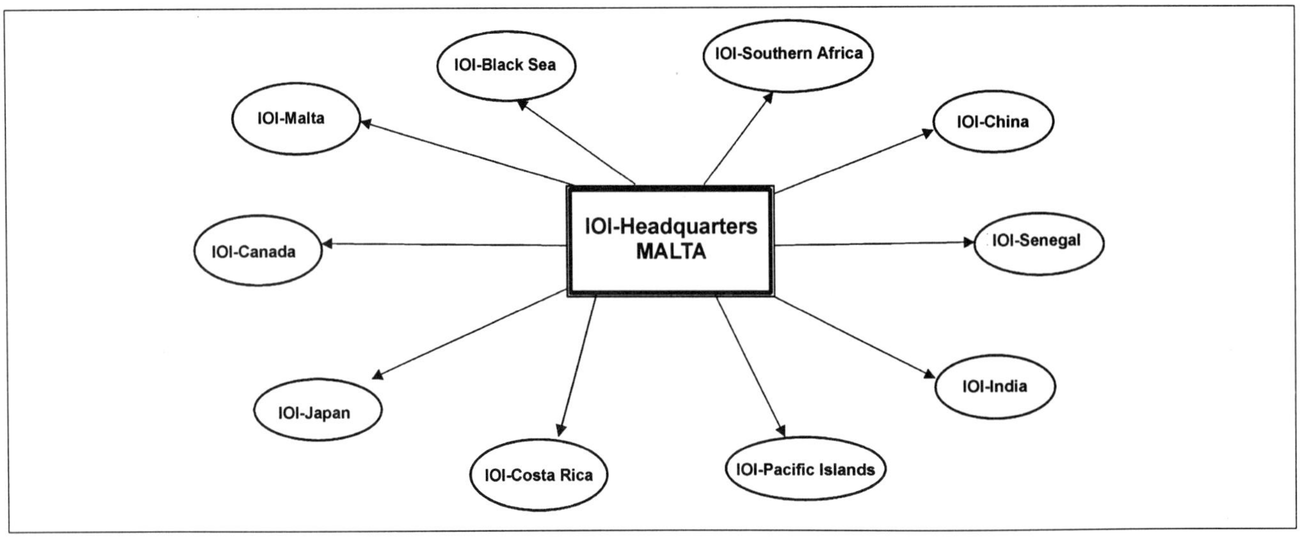


Fig. 4. The International Ocean Institute Network.

recognition of the need for the development of regionally relevant training courses, and new courses include:
- Management and Development of Coastal Fisheries
- Environmental and Resource Economics
- Sustainable Use of Fisheries
- Interactions between Tourism and Fisheries
- Deep Seabed Mining
- Small Islands
- Integrated Coastal Zone Management
- Quantitative Resource Biology
- GIS and Fisheries Assessment

While networking of these courses is the ultimate goal, this has not as yet occurred to any extent. One unique strategy of the IOI's training programmes, however, has been the delivery of Leaders Seminars. These are usually one-day events that target senior decision-makers in government, at the minister level if possible, but at the least the most senior bureaucrat level. The objective is to sensitize these leaders on significant ocean issues, such as the importance of cross-disciplinary decision-making on coastal and ocean issues, and the setting up of appropriate consultative processes that cut across traditional vertically organised departments. While the themes are nationally or regionally driven, the concept is globalised through the IOI Network.

Regardless of the training programme concerned, the driving force behind the IOI's global training programmes is the need to directly impact decision-makers from government, the private sector and NGOs with ocean or coastal mandates, on their nation's obligations under UNCLOS, and under the various conventions and agreements emanating from UNCED. Thus, there are common iterative themes in the training programmes that place the IOI stamp upon them.

## SYNTHESIS

The training programmes outlined here are by no means the only ones targeting global needs in HRD in the marine sector. A number of institutions have individually taken up this challenge and are offering internationally available courses in marine-related HRD training and education. Nor, it should be emphasised, are the nationally or regionally generated responses to training needs the only mechanisms whereby training needs are identified. Most importantly, it is clear that many governments and other organisations mandated to develop the marine sector have yet to create the proper cross-sectoral and interdisciplinary decision-making procedures inherent in their compliance requirements in the post-UNCLOS and post-UNCED era. Finally, it should be evident that the boundaries between the three global training programmes described here are blurred and that there is ample room for sharing and collaboration. There are, for example, many commonalties between programmes, and mechanisms should be developed, where they do not yet exist, for the sharing of modules among them.

The UN and IOI programmes stress the importance of sharing and networking courses and modules; this is an excellent way of avoiding duplication of effort. In the UN system mechanisms exist for the adaptation of materials for a specific location, and for their translation to other languages. The Coordination Units of all the UN programmes offer assurance of quality control, facilitate networking, and assist in the evaluation of materials. CC:TRAIN has an additional strategy of establishing national teams within their member countries, so that there is continuous and meaningful feedback, as well as the involvement of as wide as possible a cross-section of the stakeholder community. All of the training programmes described here are hosted by academic or training institutions, who benefit from the learned methodology and from the opportunity to adapt the training courses to their own specific needs. An example would be the Marine Studies Programme of the University of the South Pacific, where two courses developed by IOI-Pacific Islands on Integrated Coastal Management, and Environmental and Resource Economics, have subsequently been adapted as undergraduate courses. Furthermore, the TRAIN-X methodology, while radically different from the usual teaching methods at a university, has proved to be a very valuable tool in the improvement of teaching by those staff who have undertaken TRAIN-X training.

Over time, it is anticipated that the programmes described here, plus any others that may develop in response to UN Conventions and Agreements, will provide an invaluable service to the HRD of governments of developing countries worldwide. There are some obvious shortcomings, however, the most important of which is the potentially high cost of producing course materials. Other difficulties occur when the target group of trainees is heterogeneous, or where the required training is poorly defined. The great success of the earlier UN training programmes was because they address very specific jobs or tasks that are universally applicable (such as in TRAINAIR). To an extent, this is also true of CC:TRAIN, but it is not the case with TRAIN-SEA-COAST, or with some of the IOI Training Courses, where the topics to be covered are highly interdisciplinary and cross-sectoral, and where the clientele are more difficult to define.

Several important international fora have looked at the training needs in the marine sector, and two are worth mentioning here. One was a result of an international workshop held at the University of Rhode Island in 1995, from which a 'Call to Action' was developed on educating coastal managers. The workshop recommended two strategies to meet the growing demand for coastal management professionals:
- short-term training programmes to enhance the skills of today's coastal resource management professionals; and
- a longer-term programme to strengthen university curricula to provide greater depths in theory and practice for future generations of coastal management practitioners.

The first of these strategies is now well underway, as described in this chapter, but the second requires much greater attention.

A second example was the result of an international workshop on 'Planning for Climate Change through Integrated Coastal Management' held in Taipei in February 1997. A result of this workshop was the development of guidelines for integrating coastal management programmes and national climate change action plans. This is an excellent example of pointing the way on how collaboration should develop between training programmes, in this case between TRAIN-SEA-COAST and CC:TRAIN.

There are some inherent shortcomings in all of these programmes. They all depend on collaboration between education and training institutions mostly located in developing countries. In addition, they require a substantial commitment in time, space and expertise by those institutions, with minimal funding to meet the costs of such in-kind inputs. Some essential equipment such as computers is supplied, at least at the beginning, and funding is usually made available to support the validation (first offering) of training packages. Subsequent course offerings may, however, depend on aid funds to assist participants travel to training centres, or for the trainers to travel to them, and these funds are becoming difficult to obtain. Unlike professional courses such as those offered through TRAINAIR, courses offered by the programmes described here do not have well-endowed private sector support for training. The building of training costs into major programmes funded through GEF is now recognised as an important way of supporting identified associated training needs, and should be encouraged. Ultimately, the success of most global training programmes in the marine sector will depend on donor support.

These global programmes represent a new era of cooperation among donors, training institutions and developing countries and are intended to reduce duplication of effort and to take advantage of the modern era of electronic communications in the area of training. While not the only examples of global training initiatives, they represent a commendable and concerted effort to address urgently needed human resources needs of developing countries in the marine sector.

## ACKNOWLEDGEMENTS

The opinions expressed here are those of the author and do not necessarily represent those of the United Nations or of the International Ocean Institute. I would like to thank Stella Maris Vallejo (TRAIN-SEA-COAST Coordinator) and Gao Pronove (CC:TRAIN Coordinator) for their critical reviews of the manuscript and for providing updates on their respective training programmes, and Posa Skelton of the Marine Studies Programme, the University of the South Pacific, for assistance in the preparation of the figures and for proof reading. Support was provided by the Canada-South Pacific Ocean Development Program, Phase II.

## REFERENCES

Cicin-Sain, B., Ehler, C.N., Knecht, R., South, R. and Weiher, R. (1997) Guidelines for Integrating Coastal Management Programs and National Climate Change Action Plans. Chinese Taipei, February 24–28, 1997. National Oceanic and Atmospheric Administration. 37 pp.

Coastal Resources Center, University of Rhode Island (1995) Call to Action from the Rhode Island Workshop on Educating Coastal Managers. 11 pp.

International Ocean Institute (1999) The International Ocean Institute. The Future of the Oceans. Brochure.

International Ocean Institute (undated). Various materials describing the IOI training programme and courses, drawn from records of IOI meetings.

South, G.R. (1998) The UN/DOALOS TRAIN-SEA-COAST Programme: Training coastal managers for the next millennium. In *Fisheries and Marine Resources. Proceedings of Symposium 8, the VIIIth Pacific Science Association Inter-Congress, Suva, Fiji*, eds. J. Seeto and N. Bulai. The University of the South Pacific Marine Studies Programme Technical Reports, 1998 (3), pp. 16–25.

South Pacific Regional Environment Programme (undated) Various materials describing the Pacific Islands Climate Assistance Programme, and CC:TRAIN. Apia, Samoa.

United Nations (1994) Action Plan for Human Resources Development and Capacity Building for the Planning and Management of Coastal and Marine Areas 1993–1997. First Edition. i + 58 pp.

United Nations Institute for Training and Research (1997) CC:TRAIN Training Package Guide. Climate Change Vulnerability and Adaptation Assessment. 29 pp.

United Nations Institute for Training and Research (1998) Programme 1998–1999. 20 pp.

University of the South Pacific (1998) Brochure: Certificate in Climate Change Vulnerability and Adaptation Assessment.

Vallejo, S.M. (1996) TRAIN-SEA-COAST Programme. Report of the Regional Consultation Meeting. Nuku'alofa, Tonga. 10 May 1996 (verbal presentation, unpublished).

Vallejo, S.M. (1996) Human Resources Development: Challenges for the year 2000. Third Regional Meeting of the Brazilian Society for Advancement of Science (SBPC) "Coastal Ecosystem: from Knowledge to Management: Florianopolis, Santa Catarina, Brazil. 1–4 May, 1996. 13 pp (mimeographed).

## THE AUTHOR

**G. Robin South**
*International Ocean Institute – Pacific Islands*
*The University of the South Pacific*
*P.O. Box 1168, Suva,*
*Republic of the Fiji Islands*

Chapter 131

# GLOBAL LEGAL INSTRUMENTS ON THE MARINE ENVIRONMENT AT THE YEAR 2000

Milen F. Dyoulgerov

As part of its rich legacy, the 20th century will leave a well-defined body of international environmental law, including a body of global legal instruments dealing with the marine environment. Some of these instruments came about initially as individual responses to sectoral issues. However, in the aftermath of the Earth Summit and within the framework provided by the 1982 United Nations Convention on the Law of the Sea, all of them can be viewed as an increasingly interrelated legal structure building a broad foundation on which individual states can effectively pursue protection of the marine environment and, ultimately, the goal of sustainable development. In a concurrent development, we have also witnessed a move away from the traditional treaty model in favour of framework agreements and "soft-law" instruments, with a corresponding shift from normative to programmatic approaches to the protection and sustainable use of the marine environment.

While much has been achieved, some "grey areas" of partially addressed environmental concerns remain. The imperatives of comprehensiveness, integration, and effectiveness—comprehensivess in coverage, legal and functional integration of the instruments within and among related marine environmental sectors, and effectiveness in individual instrument implementation—are central in addressing these concerns. With a broad legal foundation more or less in place, coming to grips with how to meet the needs of global institutional and technical capacity building, including training, technical, and financial assistance for the countries in need of such, might be the single most important step toward increasing the effectiveness of international environmental law. At the same time, we still face the need to formalize a learning-based approach to global codification by coming up with a comprehensive evaluation mechanism on the effectiveness of the already adopted global legal instruments.

*Seas at The Millennium: An Environmental Evaluation (Edited by C. Sheppard)*
*Published by Elsevier Science Ltd.*

## MARINE ENVIRONMENTAL LAW AT THE END OF THE CENTURY

The 20th century leaves a complex and evolving body of public international environmental law, including a well-established branch that focuses on the marine environment. The unprecedented wave of global environmental treaty making and institutional capacity building that took place in the second half of the century has largely codified the fundamental principles of this body of law, shaped its foundation of sources and evidence, and established mechanisms to determine and allocate liability and compensation. Modern marine environmental law rests on some thirty global, hundreds of regional, and thousands of bilateral agreements now acting within the overall framework established by the 1982 United Nations Convention on the Law of the Sea (LOS). Within this framework, the global legal instruments on the marine environment function as part of a dynamic, interdependent, and complementary system of soft and hard law—a system that encompasses countries and international governmental organizations as well as sub-national, national, and international non-governmental entities. Perhaps more than any other area of environmental law, marine environmental law traces its origins to the growing realization of a global interdependence with respect to the detrimental environmental degradation that could result from increasingly invasive technologies and human practices associated with the rapid economic and population expansion of the 20th century.

The development and adoption of marine environmental law instruments parallel the fundamental changes in our environmental and social awareness. From a historical perspective, two major waves of treaty making have shaped modern international environmental law: the first emanating from the changes in environmental awareness that led to the United Nations Conference on the Human Environment, Stockholm, 1972, and the second associated with the United Nations Conference on Environment and Development, Rio de Janeiro, 1992. Growing recognition of scientific uncertainty has displaced established deterministic perceptions in favour of the precautionary principle, and the single sector/single issue approach of the early instruments has yielded to the recognized need for ecosystem-based, integrated management. From the 1946 International Convention for the Regulation of Whaling to the 1992 problem-oriented Framework Conventions on Climate Change and Biological Diversity, the evolution of the existing global instruments dedicated to marine environmental protection has been driven in turn by concerns over individual commercially valuable stocks; awareness of the Earth's biosphere integrity and the need to target pollution sources and categories for monitoring and regulation; efforts to mitigate and, eventually, prevent pollution effects; and the realization that development and environmental protection are intrinsically connected in addressing emerging global environmental threats.

As an integral part of the body of general international law, the development of international marine environmental law has also been affected by the changing nature of that body. Originating within a legal system based on the concept of absolute state sovereignty, in which states were the only actors, international marine environmental law is now a part of a body of law based on the concept of sovereignty as "an international social function" (Corfu Channel Case 1949) exercised in accordance with the broader interests of mankind and within a radically evolved decision-making and information-dissemination system that involves and is affected by a growing number of actors other than sovereign states. Within these developments, we have witnessed the growth of an international dispute-resolution mechanism as related to the marine environment, a process that has culminated in the LOS with its "unequivocal" commitment to compulsory settlement of environmental disputes: "a novelty among even the most ambitious of environmental treaties" (Boyle, 1999). We are also increasingly seeing a shift in global marine environmental law from re-active codification to pro-active, progressive policy setting.

## TAKING STOCK

The existing body of global instruments on the marine environment has traditionally developed to cover two major sectors: (1) protection of the marine environment (ecosystems as well as organisms) from pollution and (2) conservation and protection of marine organisms. The former sector comprises mainly instruments dealing with marine pollution according to its source. Until the adoption of the 1992 Convention on Biological Diversity, the latter was dominated by development-oriented instruments (mostly quasi-global, regional, and bilateral), aimed at conservation of commercially valuable stocks. A third sector addresses the protection of the marine environment from damage due to activities of a military nature (military activities). Some of these instruments are exclusively dedicated to marine environmental issues, others address such issues explicitly within a broader scope of objectives. The objectives of yet another group bear a direct relevance to recognized marine environmental problems. With the LOS coming into force, all these instruments have to be considered within the jurisdictional framework and the general principles codified by that Convention. Table 1 presents a summary of the presently active global legal instruments on the protection of the marine environment—the instruments that are to serve as a worldwide legal foundation for our efforts to protect the marine environment in the new millennium.

## THE LAW OF THE SEA

A culmination of more than seven decades of global codification efforts, the 1982 United Nations Convention on the Law of the Sea (LOS) now provides the global legal framework for all existing and foreseeable future codification and

Table 1

| Conventions | Problem address (as related to the marine environment) | Time & place of adoption | Entry into force | Ratification & accessions* | Secretariat | Internet* (http://) |
|---|---|---|---|---|---|---|
| United Nations Convention on the Law of the Sea (UNCLOS) | States' legislative and enforcement jurisdiction, marine pollution; unsustainable use of marine living resources; compulsory dispute settlement. | Montego Bay, Dec. 10, 1982 | Nov. 16, 1994 | 132 | UN Division for Ocean Affairs and the Law of the Sea (UN DOLAS) | www.un.org |
| | | **MARINE POLLUTION** | | | | |
| | | **Vessel-source Pollution** | | | | |
| Convention on the International Maritime Organization (IMO), as amended** | safety of ships and marine pollution | Geneva, March 6, 1948 | March 17, 1958 | 157 | IMO | www.imo.org |
| International Convention for the Prevention of Pollution from Ships, 1973, as modified by the Protocol of 1978 relating thereto (MARPOL 73/78) , as amended | marine pollution caused by ship operatoins and accidents | London, Nov. 2, 1973 & London, Feb. 17, 1978 | Annex I/II: Oct. 2,1983<br>Annex III: July 1, 1992<br>Annex IV: **not yet in force**<br>Annex V: Dec. 31,1988<br>Annex VI: **not yet in force** | 108<br>91<br>75<br>94<br>2 | MARPOL c/o IMO | www.imo.org |
| International Convention for the Safety of Life at Sea, 1974, (SOLAS 74), & Protocol, as amended | safety of navigation & marine environmental protection | Nov. 1, 1974 | May 25, 1980 | 139 | IMO | www.imo.org |
| International Convention Relating to Intervention on the High Seas in Cases of Oil Pollution Casualties, (Intervention Convention) & Protocol, as amended | marine pollution by oil resulting from ship accidents | Brussels, Nov. 29, 1969 | May 6, 1975 | 73 | IMO | www.imo.org |
| International Convention on Oil Pollution, Preparedness, Response, and Co-operation, (OPRC) | oil pollution caused by off shore operations and accidents | London, Nov. 30, 1990 | May 13, 1995 | 45 | IMO | www.imo.org |
| | | **Liability and Compensation** | | | | |
| International Convention Civil Liability for Oil Pollution Damage (CLC) & Protocols, as amended | oil pollution caused by ship operations and accidents | Brussels, Nov. 29, 1969 | June 19, 1975 | 75 | None (depository functions at the IMO) | www.imo.org |
| International Convention on the Establishment of and International Fund for Compensation for Oil Pollution Damage, (Fund Convention) & Protocols, as amended | oil pollution caused by ship operations and accidents | Brussels, Dec. 18, 1971 | Oct. 16, 1978 | 50 | IMO-IOPC Fund | www.imo.org |
| International Convention on Liability and Compensation for Damage in connection with the Carriage of Hazardous and Noxious Substances by Sea (HNS) | pollution with hazardous and noxious substances caused by ship operations and accidents | London, May 2, 1996 | **not yet in force** | | IMO | www.imo.org |
| Convention Relating to Civil Liability in the Field of Maritime Carriage of Nuclear Material (Nuclear) | radioactive contamination caused by ship operations and accidents | Brussels, Dec. 17, 1971 | July 15, 1975 | 14 | IMO | www.imo.org |

continued....

Table 1 (continued)

| Conventions | Problem address (as related to the marine environment) | Time & place of adoption | Entry into force | Ratification & accessions* | Secretariat | Internet* (http://) |
|---|---|---|---|---|---|---|
| | | **Related Instruments** | | | | |
| International Convention on Load Lines (LL), as amended | safety of navigation | London, April 5, 1966 | July 21, 1968 | | IMO | www.imo.org |
| International Convention on Standards of Training, Certification and Watchkeeping for Vessel Personnel (STCW), as amended | safety of navigation | London, Dec. 1, 1978 | April 28, 1984 | 133 | IMO | www.imo.org |
| International Regulations for Preventing Collisions at Sea (COLREG), as amended | safety of navigation | London, Oct. 20, 1972 | July 15, 1977 | 133 | IMO | www.imo.org |
| International Convention on Salvage (SALVAGE) | safety of navigation | London, April 28, 1989 | July 14, 1996 | 30 | IMO | www.imo.org |
| | | **DUMPING** | | | | |
| Convention on the Prevention of Marine Pollution by Dumping of Waste and Other Matter (London Convention, 1972), as amended | marine pollution caused by dumping of wastes generated on land | London, Mexico City, Moscow, & Washington DC, Dec. 29, 1972 | August 30, 1975 | 77 | c/o IMO | www.imo.org |
| | | **Related Instruments** | | | | |
| Convention on the Control of Transboundary Movements of Hazardous Wastes and their Disposal, (Basel Convention) | hazardous wastes pollution | Basel, March 22, 1989 | May 5, 1992 | 133 | UNEP / Secretariat for the Basel Convention | www.unep.ch /basel/index/ html |
| | | **LAND-BASED SOURCES** | | | | |
| | | **Related Instruments** | | | | |
| Convention on Wetlands of International Importance, Especially as Waterfowl Habitat, (Ramsar), as amended | coastal and marine pollution caused by point and non-point disposal of wastes generated on land | Ramsar, Feb. 2, 1971 | Dec. 21, 1975 | 146 | IUCN / Ramsar Bureau | www.cites.org |
| | | **MARINE ORGANISMS** | | | | |
| International Convention for Regulation of Whaling, (ICRW) & Protocol, as amended | unsustainable use of marine living resources; loss of species | Washington DC, Dec. 2, 1946 | Nov. 10, 1948 | 42 | International Whaling Commission (ICW) | |
| Agreement for the Implementation of the provisions of the United Nations Convention on the Law of the Sea of 10 December 1982 relating to the Conservation and Management of Straddling Fish Stocks and Highly Migratory Fish Stocks (Fish Stocks Agreement) | unsustainable use of marine living resources; loss of species | New York, Aug. 4, 1995 | **not yet in force** | 24 | – | www.un.org/ Depts/los/ index.htm |
| Conventional on Biological Diversity, (CBD) | loss of species and habitats | Nairobi, May 22, 1992 | Dec. 20, 1993 | 176 | CBD Secretariat | www.biodiv. org |
| Convention on International Trade in Endangered Species of Wild Fauna and Flora, (CITIES), as amended | loss of species | Washington DC, March 3, 1973 | July 1, 1975 | 146 | UNEP / CITES | www.cites.org |

Table 1 (continued)

| Conventions | Problem address (as related to the marine environment) | Time & place of adoption | Entry into force | Ratification & accessions* | Secretariat | Internet* (http://) |
|---|---|---|---|---|---|---|
| Convention on the Conservation of Migratory Species of Wild Animals (CMS or Bonn Convention) | loss of species | Bonn, 23 June, 1979 | Nov. 1, 1983 | 65 | UNEP | www.wcmc.org.uk/cms/ |
| | | **Related Instruments** | | | | |
| Framework Convention on Climate Change, New York, 1992 (FCCC) | global warming: loss of species | New York, May 9, 1992 | March 21, 1994 | 180 (16 to the Kyoto Prot.) | Climate Change Secretariat (UNFCCC) | www.unfccc.de/ |
| Vienna Convention for the Protection of the Ozone Layer | ozone depletion: loss of species due to increase in UV radiation | Vienna, Mar. 22, 1985 | Sept. 22, 1988 | 172 | UNEP / Ozone Secretariat | www.unep.ch/ozone/ |
| Montreal Protocol on Substances that Deplete the Ozone Layer, as amended | | Montreal, Sept. 16, 1987 | Jan. 1, 1989 | 171 | | |
| | | **PROTECTION FROM MILITARY ACTIVITIES** | | | | |
| Treaty Banning Nuclear Weapons Tests in the Atmosphere, in Outer Space, and Under Water | radioactive contamination and other ecosystem damage resulting from nuclear tests | Moscow, Aug. 5, 1963 | Oct. 10, 1963 | 123 (1998) | None (depository functions at the Russian Federation) | – |
| Treaty on the Prohibition of the Emplacement of Nuclear Weapons and other Weapons of Mass Destruction on the Seabed and Ocean Floor and in the Subsoil Thereof Under Water | radioactive contamination and other ecosystem damage resulting from military activities | London, Feb. 11, 1971 | May 18, 1972 | 98 (1998) | None (depository functions at the USA) | – |
| Convention on the Prohibition of Military or Any Other Hostile Use of Environmental Modification Techniques | ecosystem damage resulting from military activities | Geneva, May 18,1977 | Oct. 5, 1978 | 67 (1998) | UN Department of Disarmament Affairs | – |
| | | **ANTARCTIC TREATY SYSTEM** | | | | |
| Antarctic Treaty | marine pollution from military and land-based activities | Washington DC, Dec. 1, 1959 | June 23, 1961 | 42 | none | |
| Protocol on Environmental Protection to the Antarctic Treaty (Madrid Protocol) | loss of species and marine pollution from ships and land-based activities | Madrid, Oct. 4, 1991 | Jan. 1998 | 27 | – | – |
| Convention on the Conservation of Antarctic Marine Living Resources (CCAMLR) | loss of species | Canberra, May 20, 1980 | April 7, 1982 | 29 | Commission for Conservation of Antarctic Marine Living Resources | www.ccamlr.org |
| Convention for the Conservation of Antarctic Seals | loss of species | London, June 1, 1972 | March 11, 1978 | 16 | – | – |

* Information available as of November 3, 1999.

** With the entry into force of the amendments adopted by the IMCO Assembly Resolutions A.358 (IX) of Nov. 14, 1975 and A.371 (X) of Nov. 9, 1977, the name of the Intergovernmental Maritime Consultative Organization (IMCO) has been changed to "International Maritime Organization" (IMO), with the title of the convention being modified accordingly.

implementation efforts to protect and preserve the world oceans and their living resources. Within a broader "Constitution for the Oceans," the Convention codifies for the first time the general obligation of all States to protect and preserve the marine environment from all recognized pollution sources and defines the jurisdictional rights and obligations of the Coastal, Flag, and Port States to legislate and enforce anti-pollution rules, standards and measures (UN, 1983). It further stipulates the States' general obligations for global and regional cooperation, technical assistance, monitoring and environmental assessment, and responsibility and liability in the area of marine environmental protection and preservation. Codifying the rights of States to explore and exploit marine resources under their jurisdiction and on the high seas, the Convention goes on to counterbalance States' rights by obligations for management and conservation of those same resources.

The importance of the LOS codification of the Coastal, Flag, and Port States' competence and duties in prescribing and enforcing applicable standards cannot be overestimated for its role in strengthening global implementation and enforcement efforts, particularly in the areas of protection of the marine environment from vessel-source pollution and protection of marine living resources. At the same time, with the coastal states being granted custody over the most productive areas of the ocean, which account for approximately 23% of primary productivity and 40% of fish production (Houde and Rutherford, 1993), we have to bear in mind that, while global regulations and standards can provide guidance, it is the efforts at national and subnational levels that make all the difference. Accordingly, increasing the awareness, mobilizing and providing resources, and building and strengthening the capacity at those levels, especially in the newly industrialized and developing countries, will continue to be a major task well into the future.

The Convention has now emerged from years of uncertainty and arduous negotiations as a widely accepted and universally recognized instrument, with even some of its innovative provisions already established as customary international law. Recent decisions of the International Tribunal for the Law of the Sea (ITLOS) on the Southern Bluefin Tuna Cases (Southern Blufin Tuna Case, 1999) have further demonstrated the LOS function in strengthening horizontal enforcement under international law and the potential of its comprehensive, yet versatile, dispute-resolution mechanism to advance progressive principles and objectives of marine environmental protection.

## MARINE POLLUTION

### Shipping

According to the International Maritime Organization (IMO), oil is the most serious pollutant in terms of volume to enter the marine environment as a result of normal vessel operations or vessel accidents (IMO, 1998). US National Academy of Science studies have estimated that shipping activities contributed approximately 35% (2,133,000 t) and 45% (1,470,000 t) for 1973 and 1981, respectively, of the oil entering the marine environment (USNAS, 1975; 1985). The updated estimates for 1989 continue to scale back in absolute terms this contribution to approximately 568,800 tons (IMO, 1990). In addition, accidents and normal vessel operations can result in contamination of the marine environment by chemicals in liquid or bulk form, garbage and sewage, radioactive materials, or organotin compounds (used in ship anti-fouling systems). Introduction of harmful aquatic organisms and pathogens ranks also among the potential adverse effects of shipping.

As seen in Table 1, all but the last two of the above listed aspects of vessel-related marine environmental pollution are addressed in a comprehensive manner by an established body of global conventions adopted under the auspices of the IMO. Among themselves, these instruments codify global standards and measures to reduce or eliminate pollution from ships and, in the context of the LOS legal framework, outline state responsibilities for the prescription and effective enforcement of these standards. They provide for detailed measures to avoid pollution resulting from normal operation and accidents and establish a legal and technical basis for co-operation in dealing with emergencies and reduction of pollution casualties. A number of conventions codify a liability and compensation scheme for environmental casualties of vessel-source oil pollution or radioactive contamination—still the only existing scheme for civil liability for damage of the marine environment. This scheme is to be further enhanced by a new convention for liability and compensation for damage caused by oil from ships' bunkers, currently under preparation at the IMO. The function of the existing conventions is complemented by a number of IMO mandatory and voluntary resolutions, technical codes, and recommendations (IMO, 1998). Draft conventions addressing the two remaining gaps—controlling the use of anti-fouling systems and preventing the transfer of harmful aquatic organisms in ballast waters—are also presently under negotiation and are expected to be adopted during the 2002–2003 biennium. The global regulatory scheme for protection of the marine environment from vessel-source pollution is supplemented by a number of regional conventions and conventional instruments. It is further strengthened by the IMO regulatory and programmatic mechanism for training, technical assistance, and capacity building.

### Seabed Activities: Peaceful Exploration and Exploitation

At present, pollution threats to the marine environment resulting from seabed activities are caused primarily by

# Oil Pollution

Oleg Khalimonov

*International Maritime Organization, London, UK*

Oil pollution of the sea, especially in ports and harbours, was first recognized as a problem before the First World War and, during the 1920s and 1930s, various countries introduced measures to control discharges of oil within their territorial waters. No international agreement had been reached before the outbreak of the Second World War.

In 1954, the United Kingdom organized a conference which resulted in the adoption of the International Convention for the Prevention of Pollution of the Sea by Oil, 1954 (OILPOL 1954). It recognized that most oil pollution resulted from routine shipboard operations such as the cleaning of cargo tanks, a normal practice being to wash tanks out with water and pump the resulting mixture into the sea. OILPOL 1954 prohibited this within a certain distance from land and in 11 "special areas" where the danger to the environment was especially acute. However, growth in maritime transport of oil and tanker size, the increasing amount of chemicals being carried at sea, and a growing concern for the world's environment made many countries feel that OILPOL was no longer adequate, despite various amendments.

In 1969, the IMO Assembly decided to convene to adopt a new comprehensive instrument (MARPOL 1973). This Convention incorporated much of OILPOL 1954 and its amendments into Annex 1, covering oil, while other annexes covered chemicals, harmful substances carried in packaged form, sewage and garbage. Although it was hoped that MARPOL 1973 would enter into force quickly, progress was very slow. This was due largely to a number of technical difficulties, in particular those associated with Annexes I and II.

To rectify those difficulties and in response to a spate of tanker accidents in 1976–1977, in 1978 the Conference adopted a Protocol to the 1973 MARPOL Convention. The changes envisaged involved mainly Annex I and it was, therefore, decided to adopt the agreed changes and, at the same time, allow Contracting States to defer implementation of Annex II for three years after the date of entry into force of the Protocol. By then, it was expected that the technical problems would have been solved. The 1973 Convention and the 1978 MARPOL Protocol should be read as one instrument, usually referred to as MARPOL 73/78. It finally entered into force on 2 October 1983 (for Annexes I and II). Annex V, covering garbage, achieved sufficient ratifications to enter into force on 31 December 1988, while Annex III, covering harmful substances carried in packaged form, entered into force on I July 1992. Annex IV, covering sewage, has received 75 ratifications (by November 1999), representing 43.11% of world shipping tonnage, out of 50% needed for entry into force. In 1997, a new Annex VI on prevention of air pollution from ships was added.

After the 1978 conference on Tanker Safety and Pollution Prevention, which both strengthened provisions for tanker safety and removed the obstacles that were preventing the entry into force of the Convention, the twin aims of "Safer Shipping and Cleaner Oceans" became the dual objective of IMO's work. When MARPOL 73/78 entered into force, it proved that countries were prepared to implement measures to protect the marine environment. Today, MARPOL is recognized as the most important set of international regulations for the prevention of marine pollution by ships and has been credited for its decline over the years.

In 1990, the National Research Council Marine Board of the United States credited MARPOL 73/78 with making "a substantial positive impact in decreasing the amount of oil that enters the sea." A study carried out by the Board showed that in 1981, some 1,470,000 tons of oil entered the world's oceans as a result of shipping operations. Most came from routine operations, such as discharges of machinery wastes and tank washings from oil tankers (the latter alone contributed 700,000 tons). Accidental pollution contributed less than 30% of the total. By 1989, it was estimated that oil pollution from ships had been reduced to 568,800 tons. Tanker operations contributed only 158,000 tons of this. Moreover, although the 1978 Protocol did not enter into force until 1983, many of its requirements were already being implemented—requirements that were also recognized by the industry as improving its own efficiency.

IMO measures have been successful in reducing pollution from ships from over 2 million tons in 1973 to just over 500,000 tons in 1990 (Fig. 1). Perhaps the biggest improvement has been made in reducing operational pollution, through measures such as load on top and crude oil washing. Improved safety standards have also led to a significant reduction of major pollution incidents. According to the International Tanker Owners Pollution Federation, the number of spills of over 700 tons at the end of the 1980s was only a third of the total a decade earlier.

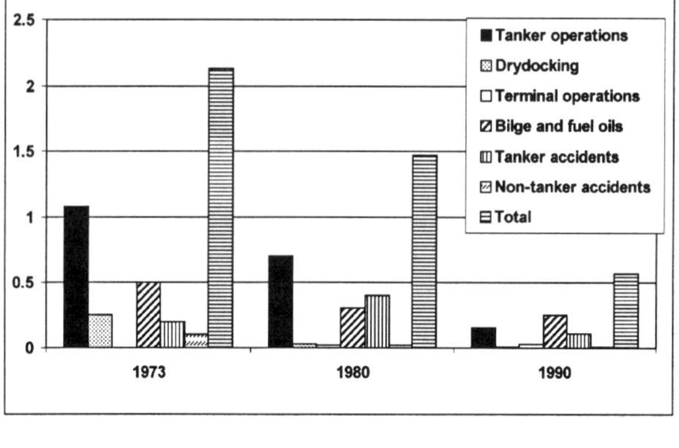


Fig. 1. Oil pollution from ships from various categories. (Based on information from the United States National Academy of Science.)

operations for exploration and exploitation of the seabed mineral resources (solid, liquid, and gaseous). The above-listed USNAS studies have estimated the contribution to marine environmental pollution by oil from offshore production activities to 80 tons and 50 tons for 1973 and 1981, respectively. Continuous technological improvements have reduced the pollution from normal operation to insignificant levels and now accidents constitute the major risk related to seabed operations. The Law of the Sea Convention is the only binding instrument to address the issue at the global level and it does that in most general terms. Accidental pollution from seabed activities is also addressed under some regional agreements, whether explicitly or in general provisions, as well as by non-mandatory IMO technical code (IMO, 1979) and UNEP set of guidelines (UNEP, 1982).

In respect to operations conducted in the territorial sea and on the continental shelf, LOS provides general obligations for the Coastal States to adopt and enforce regulations to prevent, reduce, and control pollution resulting from seabed activities. Such regulations should be "no less effective than international rules, standards and recommended practices and procedures" (Art. 208(3)). States are further asked to harmonize their policies at the appropriate regional level and to establish global and regional rules, standards, and recommendations. However, so far the effort invested to this end, particularly at the global level, is less than optimal. To a large extent this is due to the growing sensitivity of the oil industry to environmental problems; the effectiveness of the industry could be judged by the fact that the industry liability and compensation scheme for the northwest European Waters, the Offshore Pollution Liability Agreement, has resulted in the 1977 London Convention on Civil Liability for Oil Pollution Damages Resulting from Exploration and Exploitation of Seabed Mineral Resources not being ratified.

The ensuing problem, however, is that all too often different standards are applied to different regions. As UN (1999) indicates, the debate for and against global regulations is to continue into the new millennium. Regarding operations within the International Seabed Area, LOS Article 145 provides that the International Sea Bed Authority is to adopt rules to prevent pollution from deep-sea mining. Such rules are presently in the process of formulation as part of the on-going readings of the draft mining code being developed by the Authority.

Ironically, the most extensive global legal coverage has been allotted to pollution presenting the least problem at present—that from seabed activities other than offshore explorations, exploitation, and processing of minerals. In addition to being subject to general provisions of LOS, these activities are also regulated under MARPOL 73/78, Annex I of which regulates discharge of oil in such operations, Annex IV (if and when it enters into force) will regulate sewage, and Annex V regulates garbage discharges.

### Dumping

International law has come to treat dumping as a source of pollution separate from shipping. The 1972 London Convention defines "dumping" as the deliberate disposal at sea of wastes or other matter from vessels, aircraft, platforms or other man-made structures, as well as the deliberate disposal of vessels, aircraft, platforms or other man-made structures themselves (Art. 3). Unlike other pollution from ships, dumping is considered always deliberate and is viewed as pollution originating exclusively on land, though disposed in areas jurisdictionally different from land (Churchill and Lowe, 1988).

At present, the 1972 London Convention, as applied within the LOS legal framework, provides the foundation of the global regime for protection of the marine environment from dumping. It is further expanded by a number of related binding regional instruments and is complemented by the 1989 Basel Convention in the areas of controlling the generation and transboundary movements of hazardous waste, its environmentally sound management and disposal, as well as in providing technical assistance and training. The 1972 London Convention has a preventive purpose: dividing waste into three categories, it prohibits dumping of listed substances (Annex I); subject to the provision of special prior permits, it allows dumping of the substances listed in Annex II; and allows dumping of all the remaining substances upon provision of general permits on the basis of criteria set out in Annex III. Reflecting changes in attitudes towards the disposal of waste at sea and changes in corresponding practices, the 1993 amendments to the Conventions codified permanent bans on dumping of nuclear waste, the practice of incineration at sea, and the dumping of industrial waste.

Perhaps more than any other global instrument, the evolution of the London Convention reflects the journey that global environmental law has undertaken to date. Emanating from the recommendations of the 1972 Stockholm Conference, the Convention has evolved from a 'dilute and disperse' philosophy of monitoring and control to codifying prevention and prohibition of waste release in the marine environment. NGO scientific and policy contributions have substantially assisted this transition (see Stairs and Taylor, 1992). As part of the UNCED process, in the early 1990s the Parties to the Convention launched a review process aimed at improving the strategies and implementation of the Convention, bringing it up to date with modern waste management approaches, and outlining future directions. This process led to the development of the Waste Assessment Framework, introducing a regulatory mechanism for step-by-step waste evaluation that extends the Convention's scope well beyond the particular activity of dumping. The Global Waste Survey (IMO, 1995), the review's final report, demonstrated the connection between uncontrolled waste generation, treatment, disposal, and stockpiling on land and the disposal of waste at sea. The

Survey concluded that stockpiling of industrial waste is equivalent to open dumping in many developing and newly industrialized states. It further pointed out that if many such states have not widely relied on ocean dumping so far, it is because they have been able to directly dispose such waste on land or in the coastal environment at low or no cost. Recognizing that the elimination of ocean dumping of industrial waste cannot and will not happen simply as a result of a global ban, the review underlined the necessity for the London Convention to adopt a broader and more comprehensive perspective on waste management, a task achieved with the adoption of the 1996 Protocol to the London Convention.

When the 1996 Protocol enters into force, it will not only consolidate the number of amendments made to the Convention over the years and establish a dispute-resolution mechanism, but it will also introduce a major change in the Convention's fundamental premises under the so-called "reverse list" approach—from regulation of controlled dumping to a general prohibition of all dumping with six codified exceptions. Expanding the scope of the Convention beyond control to include mitigation, precaution, and prevention, the 1996 Protocol makes the precautionary approach mandatory and promotes the "polluter pays" principle. It reaffirms the LOS prohibition to transform one type of pollution into another or to transfer pollution damage from one part of the environment to another and calls for "waste prevention audits" to provide a basis for integrated waste management solutions. Thus, with its entry into force, the 1996 Protocol will provide the legal basis for moving the scope of the original London Convention landward, relating it to waste and pollution policy and management issues on land in addition to the waste disposal at sea. The task now is to bring the 1996 Protocol into force and to face the policy, management, capacity-building and technical assistance challenges in order to put it into operation. The logical, though quite difficult, step then will be to fully integrate sea-disposal and land-based pollution within a comprehensive global legal regime.

### Dismantling of Ships

A textbook example of exporting polluting industries, the present ship dismantling practices "continue to give rise to grave concern" (UN, 1999) for pollution of the marine environment from contamination with hydrocarbons, chemicals, asbestos, polychlorinated biphenyl (PCB), heavy metals, ozone-depleting substances, and others. While 90% of the operations are concentrated in four developing economies—India and Bangladesh account for 68% and Pakistan and China for another 22% (UN, 1999)—the demand-and-supply factors necessitate a global approach, as is the case with the rest of the shipping industry. At present, however, ship breaking falls between the cracks in the body of global legal instruments on prevention of pollution of the marine environment. It will continue to do so until a decision is reached on the question of whether a ship, once removed from service and taken off registers in preparation for scrap, continues to be regulated as a "ship" or whether her dismantling is a shore industry activity, outside of the IMO purview. Meanwhile, collaborative efforts to address the problem continue, involving the IMO, the 1989 Basel Convention, and the 1972 London Convention.

### Land-Based Pollution

By far the single greatest and most important contributor to marine pollution, is land-based pollution, which is also the source least regulated at the global level. With the exception of the general obligations imposed on states by various LOS Articles, and the provisions for wetland protection under the 1971 Ramsar Convention, no global binding instrument has ever been adopted. Instead, international efforts for regulation and management of marine pollution from land-based sources have relied exclusively on regional conventions and conventional instruments, whether specifically addressing land-based sources or regulating transboundary air or river pollution. Thus, in addition to the 1974 Paris Convention on the Protection of Marine Pollution from Land-based Sources and the 1974 and 1992 Helsinki Conventions on the Protection of the Marine Environment of the Baltic Sea Area, the conventions under the UNEP Regional Seas Program have provisions or dedicated Protocols on protection against pollution from land-based sources. In drawing the global framework for those efforts, the LOS provisions have been, in effect, made operational by non-mandatory, "soft law" instruments—initially, by the 1985 UNEP Montreal Guidelines for the Protection of the Marine Environment Against Pollution from Land-based Sources and currently, by the 1995 Declaration and Global Program of Action (GPA) for the Protection of the Marine Environment from Land-based Activities. Unlike shipping or seabed activities, no global or regional instrument has been negotiated to establish a liability and compensation mechanism with respect to land-based pollution.

At the same time, the Global Waste Survey under the 1972 London Convention has put waste disposal at sea in the broader waste assessment context, and with the 1996 Protocol, we now have at least one of the elements to place land-generated marine pollution under an integrated global regulatory framework. Another element might be added early in the new millennium, if the negotiations on an international legally binding Instrument for Implementing International Action on Certain Persistent Organic Pollutants, initiated by UNEP in 1997, reach successful conclusion.

As Churchill and Lowe (1988) noted, in light of the fact that land-based pollution is the most "national" of all sources of marine pollution, it comes as no surprise that the problem has been addressed less extensively and more slowly than the other sources of marine pollution. The ensuing 12 years have hardly challenged the shield of sovereignty that covers land-based pollution, despite the increasing "democratization" of international law. If anything, the achievements in strengthening national and regional action under the GPA and some of the regional instruments have come as a result of changes in developmental and technical assistance practices as well as NGO initiatives and public pressure. The very nature of the problem—addressing a wide range of pollutants that are often less prominent but with more severe cumulative impact, are produced by a wide variety of activities, and enter the ocean from a myriad of point and non-point sources—further adds to the exceptional difficulty of establishing a legal and institutional framework at a supranational level. At the same time, the damage of these pollutants can transcend national boundaries to gain region-wide importance, with the problem particularly acute in enclosed and semi-enclosed seas.

Reflecting on these factors and building on the accumulated experience under the UNEP Regional Seas Program, a consensus has emerged among scholars on the merits of the regional approach in dealing with land-based source pollution (see Vallega, 1994). Bearing in mind that the 1995 choice of a non-treaty, programmatic instrument came out of UNEP-led deliberations over a number of options (to include a global convention or a global convention supplemented by an action plan) for strengthening the then existing regime, it is unlikely that the currently existing non-binding legal framework will be supplanted by a binding global instrument in the foreseeable future.

### Atmospheric Pollution

The LOS treats atmospheric pollution as a separate source of marine pollution, while imposing on States the same broad obligations to adopt and enforce measures and regulations to prevent, reduce, and control pollution of the marine environment from or through the atmosphere, as provided for the other pollution sources covered by the Convention. On a global scale, operationalization of these obligations has been achieved only with respect to pollution emanating from activities aboard ships. Regulating atmospheric pollution resulting from incineration of industrial waste at sea, the 1993 amendments to the London Dumping Convention, 1972, have imposed a permanent ban on that practice. Measures for the reduction of ship engine atmospheric emissions of sulphur and nitrogen oxides are being introduced by the new Annex VI to the MARPOL Convention adopted in 1997.

By far the greatest share of pollution of the marine environment (coastal waters in particular) from or through the atmosphere comes from land-based sources. Respectively, the existing regulatory framework for such pollution coincides or closely matches the one for land-based sources. Internationally binding commitments exist only at the regional level, whether in the form of a regional agreement dealing with the issue of atmospheric pollution in general, as is the case with the 1979 Geneva Convention on Long-Range Transboundary Air Pollution, or as a part of regional conventions or conventional instruments on the prevention of marine pollution from land-based sources. A promise that at least some of the marine environmental implications of regional and global air pollution transport will be comprehensively addressed at the global level comes from the on-going negotiations on Certain Persistent Organic Pollutants.

## MARINE ORGANISMS

Looking at the body of global instruments regulating the protection of marine organisms, one can hardly see the systematic, source-by-source coverage that characterizes the marine pollution area. The same holds true in terms of comprehensiveness and timing. In fact, apart from the general LOS provisions on conservation and management of living resources in the EEZ, the Continental Shelf, and on the High Seas, and the Antarctic Treaty System instruments, at present only one global binding instrument dedicated exclusively to marine organisms is in force: the 1946 International Convention for Regulation of Whaling. It might take several more years for the 1995 Fish Stocks Agreement to enter into force, and a mandatory instrument to address the transfer of harmful aquatic organisms (to the extent that this also contributes to the protection of marine living resources) is yet to be negotiated. Until such time, the objective of conservation and protection of marine species will remain codified at the global level predominantly within related agreements with broader nature-protection and preservation scope: the 1971 Ramsar Convention, the 1979 Bonn Convention, CITES, and the 1992 Biodiversity Convention. In addition, the 1992 Framework Convention on Climate Change (FCCC) and the 1985 Vienna Convention, including the 1987 Montreal Protocol, address human-induced changes in the Earth's atmosphere that also have detrimental effects on marine organisms and ecosystems.

Besides the hundreds of bilateral agreements, the overwhelming body of international law in this area is concentrated in regional or quasi-global fishery agreements. The latter consist of a number of fishery agreements with regional geographic scope that permit conditional or unconditional participation of Parties from outside the region. This state of development is hardly surprising, considering the fact that up until UNCED all international instruments dealing with marine organisms were development-oriented. Few areas of international legislation have proved as politically sensitive and as unyielding to international regulations as fisheries and even fewer have been characterized by the ideological clashes accompanying the efforts to protect marine mammals.

### Regulation of Whaling

Still the only binding instrument to specifically address protection of marine species at the global level, the 1946 Whaling Convention is emblematic of the changing dynamics of international law over the second part of the 20th century. The convention was adopted by 15 whaling nations as an industry conservation scheme aimed at sustaining the whaling industry well before the waves of environmental awareness in the 1960s and the 1970s. In a classic example of the growing importance of non-state actors in international law, however, the combined efforts of a variety of NGOs and other constituent groups have transformed the convention into a powerful preservationist tool—a process culminating in the 1985–1986 moratorium on all commercially exploited stock that has remained in force ever since. Meanwhile, the International Whaling Commission (IWC), the principal organization under the convention, has expanded the scope of its work to address the effects that other human activities (such as whale watching) and human-induced habitat changes have on cetaceans.

So far the whaling regime has demonstrated a considerable capacity to adapt in response to the changing political, economic, and environmental circumstances. However, this regime has also reached a point at which its future depends on the willingness of its Parties to bridge what so far has proved an unbridgeable fundamental difference of values between permissible as opposed to impermissible whaling (Young, 1994). Today it is still unclear whether the global whaling regime will disintegrate under the pressure of ideology or whether it will be possible for the principle of sustainable development to be worked out in a meaningful way within its framework.

### Fisheries

Most people now agree that the origin of fishery failures is institutional. It also appears now that, among themselves, the LOS, the 1995 Fish Stocks Agreement, and the non-mandatory FAO Code of Conduct for Responsible Fisheries (FAO Code) have the potential for providing the global legal foundation for this problem to be addressed. With the codification of the "enclosure" of approximately 40% of the world oceans, LOS has effectively addressed the problem of "open access" at an international level by placing the overwhelming majority of the world's fish stocks (and other marine living resources) under the sovereign jurisdiction of the coastal states. The LOS dispute-resolution mechanism further broadens the discretionary powers of the coastal state with respect to both conservation and management of these resources, to the exclusion of all other states. As Hardin (1998) noted, now "the devil is in the details." It is now up to the coastal states to fill in these details in order to translate the LOS legal framework into sustainable management of the fish stocks under their jurisdiction, for it seems highly unlikely that any new binding global legal instruments will be adopted to this end.

Compensating for the preservation of open access on the High Seas, the 1995 Fish Stocks Agreement codifies much of the FAO Code and strengthens and further develops the LOS rules as related to straddling and highly migratory stocks (see Hayashi, 1995; Speer and Chasis, 1995; Fontaubert, 1995). If and when it enters into force, the 1995 Agreement will offer the much needed legal foundation for dealing with the "fishing" flags of convenience, a problem exacerbated by the recent ITLOS decision on the M/V "Saiga" (No. 2) Case (M/V Saiga Case No. 2, 1999). Most importantly, the 1995 Fish Stocks Agreement holds the potential to link the LOS and the existing regional fishery schemes into an integrated global legal mechanism for conservation and sustainable management of migratory and straddling stocks.

### The Convention on Biological Diversity Framework

Exemplifying the change brought by UNCED toward a holistic approach in international environmental law codification, the 1992 Convention on Biological Diversity (CBD) marks a move away from specific, single-sector instruments and toward framework agreements with increasing reliance on "soft-law" instruments for their operationalization. The Convention's global program of action in the marine environment area, the Jakarta Mandate on Marine and Coastal Biodiversity, identifies five priority areas: integrated marine and coastal area management (ICAM), marine and coastal protected areas, sustainable use of marine and coastal living resources, mariculture, and alien species. It further emphasizes the need for an integrated approach for achieving biodiversity protection (de Fontaubert et al., 1996).

As evident from its programmatic scope, the CBD coastal and marine components have the potential to serve as a comprehensive bond, bringing together all global instruments on protection of marine living resources. Furthermore, within its ICAM component, it can provide for the integration of the implementation efforts under these instruments at national, sub-regional, and regional levels. It is difficult to envision, however, how this will be achieved since the Jakarta Mandate is but one of the Convention's programmatic areas and considering the overall modest institutional capacity dedicated to it. After a sluggish start, the CBD marine component is now gaining momentum under the program on marine and coastal biodiversity, adopted by the fourth Conference of the Parties in 1998, both in the implementation of the program's operational objectives and in addressing substantive issues of marine biodiversity protection. Whether the full potential of the Convention in the area of coastal and marine environment will be achieved is up to our future efforts.

## PROTECTION OF THE MARINE ENVIRONMENT FROM MILITARY ACTIVITIES

An inheritance from the Cold War years, the three conventions listed in Table 1 have established a comprehensive (and so far functioning) regime for the protection of the marine environment from radioactive contamination resulting from development or stationary deployment of nuclear weapons. The 1963 Nuclear Test Ban Treaty protects the ocean both from direct contamination and from radioactive atmospheric pollution. Bilateral agreements, regional treaties such as the Treaty for the Prohibition of Nuclear Weapons in Latin America, and the South Pacific Nuclear Free Zone Treaty, together with a series of UN General Assembly resolutions, complement this regime. Currently, the only substantiated threat of nuclear contamination related to marine activities stems from potential accidents involving nuclear-powered ships and submarines with deployed nuclear weapons, none of which are covered by any global regulations. While a number of accidents (emergencies, collisions, and casualties) involving Russian, US, and British vessels have demonstrated that such potential is not insignificant, it is unlikely that those activities will be regulated in the foreseeable future. Finally, the 1977 Geneva Convention addresses both nuclear and conventional threats to the marine environment arising from hostile use of environmental modification techniques—threats all too real as demonstrated by the events of the 1990 Gulf War. These events have underlined the need for further efforts to broaden the participatory base of the above-mentioned treaties. They also remind us of the limits of the international legal system as it exists today.

## ANTARCTIC REGIME

A quasi-global legal regime, the Antarctic Treaty System arguably constitutes one of the most comprehensive and successful regimes in modern international environmental law. The same holds true with respect to its conventions and conventional instruments dedicated to the protection of the marine environment, which have also been at the forefront of progressive codification of marine environmental law. Thus, the Convention on the Conservation of Antarctic Marine Living Resources remains not only the most comprehensive legal instrument dedicated to the conservation of marine living resources, but presents the only formal legal adoption of the ecosystem approach to date.

The 1991 Madrid Protocol with its five annexes has further consolidated and fortified the legal foundation for protection of the Antarctic environment and its ecosystems. As Joyner (1996) points out, "the Protocol unmistakably mandates that marine pollution must be prevented from befouling the Antarctic marine ecosystem". The Protocol has also brought a resolution to the contentious issue of commercial Antarctic mineral exploration and exploitation until at least 2041 by placing a flat prohibition on all such activities. Marine pollution from vessels under sovereign immunity still needs to be fully addressed, however, as does the uncertainty carried out into the new millennium by the temporary ban on mineral activities. Both issues are intrinsically linked to the persisting fundamental questions of broadening the participatory base of the Antarctic regime and, ultimately, achieving a permanent resolution on the status of the Antarctic continent.

## FROM THE PRESENT TO THE FUTURE

As this brief review demonstrates, a substantial body of international agreements dedicated to the protection of the marine environment has been adopted or is in the process of becoming adopted at the global level, within the overall legal framework created by the LOS. However, this body is still far from being complete, with a number of remaining "gray areas" of partially addressed environmental concerns in addition to the "white areas" of new challenges. As the global codification efforts in the past decade have demonstrated, we continue to improve our promptness and creativity in responding to newly emerging problems. Filling in the deficiencies and consolidating the already "charted" areas of environmental concerns, however, appear to be a major challenge at the beginning of the new millennium. Paramount to addressing this challenge are the imperatives of comprehensiveness, integration, and effectiveness: comprehensiveness in coverage; legal and functional integration of the instruments within and among related marine environmental sectors; and effectiveness in individual instrument implementation. This is particularly relevant for the global body of law on marine pollution and marine organisms where the problems are not self-contained, as is the case of the protection of the marine environment from military-related activities.

As UNCED has taught us, the importance of the consolidation and integration of new and existing global legal instruments within a common approach to marine environmental protection cannot be overestimated. This is imperative if we are to meet the challenge currently facing all the marine environmental sectors alike. The way marine environmental law has evolved, however, makes it difficult, if not impossible, for this to be achieved in a unified fashion.

Thus, within the marine pollution sector, only the global instruments regulating pollution from shipping could be regarded as providing comprehensive coverage and as being fully integrated into a functional mechanism. A number of initiatives currently under way at the IMO, including the three already mentioned conventions under negotiation, give ground to expectations that this regime will be "complete" within the next half a decade. With the 1996 Protocol entering into force, the same will be achieved for the global ocean dumping regime. It is still unclear, however, whether a non-binding instrument will be successful in effectively consolidating the existing international

## Reducing Vessel-source Pollution

Oleg Khalimonov

*International Maritime Organization, London, UK*

The MARPOL 73/78 regulations for the prevention of pollution by oil are generally considered as 'complete.' Nevertheless, IMO is continuing consideration of some outstanding issues, such as a phasing-in period for double hulls on existing tankers and the problem of the inadequacy of reception facilities for the ship-generated wastes in ports. At present, MARPOL 73/78 Annexes I & II are undergoing a thorough review aimed at making them in line with the development of the precautionary approach and facilitating their more effective implementation. This work should be completed by 2002. The final draft of the reformatted International Maritime Dangerous Goods Code (IMDG Code) is expected to enter into force in January of 2001, thus further strengthening Annex III of the Convention. Work on updating and revising Annex IV continues with the goal to encourage further ratifications, including consideration on mandatory requirements for all or most ships to carry sewage treatment plants as an alternative to port reception facilities. Finding financial and technical means to assist states in providing port reception facilities for garbage is central to efforts to extend the overall participation in MARPOL 73/78 while facilitating the implementation and enforcement of Annex V. The IMO has also been dealing with the problem of prevention of marine pollution by less traditional pollutant categories such as unwanted aquatic organisms in ballast waters and toxic anti-fouling paints. Draft legal instruments to regulate these issues are being developed and will be available for adoption within 2–3 years.

There are still concerns over pollution entering the world's oceans from marine transportation activities. The key to pollution prevention is implementing and updating the IMO Conventions. IMO is focusing on this through its Committees and Sub-Committees. Through its Technical Cooperation Programs, IMO aims to assist developing countries in acquiring the infrastructure and trained personnel necessary to achieve ratification and implementation of the international regulations. A major development in technical co-operation, the World Maritime University has graduated 1400 qualified professionals from 132 nations since its inception in 1983.

Other important contributions to preventing marine pollution include port State control, the introduction of the International Safety Management (ISM) Code, and the 1995 amendments to the 1978 STCW Convention. Many of IMO's most important technical conventions contain provisions for ships to be inspected when they visit foreign ports to ensure that they meet IMO requirements.

The ISM Code became mandatory for passenger ships, oil and chemical tankers, bulk carriers, gas carriers and cargo high speed craft of 500 gross tonnage and above on July 1, 1998 and will be extended to other ships in 2002. This is aimed at ensuring that ships are properly managed and operated; the objectives, stated clearly in the Code, are to "ensure safety at sea, prevention of human injury or loss of life, and avoidance of damage to the environment, in particular to the marine environment and to property."

The human factor is also being addressed by the 1995 amendments to the International Convention on Standards of Training, Certification and Watchkeeping for Seafarers (STCW Convention). These amendments entered into force in February 1997. By August 1, 1998, all Parties to the Convention had to submit documentation to IMO showing that their training institutions complied with the requirements of the revised Convention.

legislation on land-based pollution in a similarly comprehensive global regime. This is also the case with respect to the global instruments dealing with marine living resources, the high seas fisheries regime as envisioned by the 1995 Fish Stocks Agreement being an exception. In both cases there is still uncertainty as to whether we have come up with a reliable global functional mechanism, in the form of an overarching regulatory instrument or as an adequate institutional arrangement at the global level, which could effectively bind all existing pieces together within the jurisdictional rights and obligations defined by the LOS.

The discrepancy stems from changing codification approaches as much as from the different nature of the problems and the factor of time. As already noted, the 1972 United Nations Conference on the Human Environment and the 1992 United Nations Conference on Environment and Development mark two major, yet distinctive, cycles of treaty making. Within the first cycle, the global legal regimes on vessel-source pollution and dumping have evolved while relying exclusively on traditional treaty making: to include rigid universal codification and reliance on public (i.e. "peer") pressure for implementation. Characteristic for these regimes is a comprehensive codification of marine environmental law at the global level, with "soft law" and binding regional instruments having supplemental functions. Here, the integration and consolidation focus is on the global mandatory instruments. To consider prevention of pollution from ships as a model (see Table 1): MARPOL 73/78 and SOLAS '74 codify global standards and measures to reduce or eliminate pollution from ships and, within the LOS legal framework, outline state responsibilities for the prescription and enforcement of these standards. The two conventions, along with COLREG and STCW and supplemented by a number of technical codes as well as the continuous regulatory work coming out of the IMO committees and sub-committees, collectively contribute to reducing the number of ship accidents and the severity of associated environmental damage. The Salvage and the Intervention Conventions provide the legal basis for co-operation in dealing with emergencies and for reducing the pollution casualties, while OPRC facilitates international cooperation and assistance in preparing for and responding to major off-shore oil spills and encourages states to establish adequate institutional and technical capacity to deal with such spills. Yet another growing set of instruments, which now includes the CLC, Fund, and Nuclear Conventions, provides for determining liability and compensation in case of environmental casualties from vessel operations. Taken alone, any of these instruments covers but a fraction of the complex problem of regulating globally the pollution impact from marine transportation activities. Integrated within the functional mechanism established by the IMO Convention, these instruments form a comprehensive and increasingly effective global legal regime.

In an attempt to more effectively integrate science, technology, and economics within a broader participatory and more transparent decision-making process, the UNCED cycle marked a move away from the traditional treaty model in favour of framework agreements and "soft-law" instruments. It also introduced substantial diversity of regulatory approaches in the area of marine environmental protection. The net result of this second cycle for the marine area has been a pronounced shift from normative to programmatic approaches at the global level, with a corresponding shift of codification of command and control policies based on direct regulations from global to regional, sub-regional, and national levels. Thus, global plans and programs of actions as exemplified by the GPA, the Global Plan of Action for the Conservation, Management, and Utilization of Marine Mammals (MMAP), the FAO Code of Conduct for Responsible Fisheries and its supporting International Plans of Action for the Management of Fishing Capacity, Reducing Incidental Catch of Seabirds in Longline Fisheries, and Conservation and Management of Sharks, are substituting for traditional treaty making in the marine environmental area. Regional instruments do not have supplementary but complementary function with respect to both land-based pollution and fishery management; in fact they carry the thrust of regulatory action. Respectively, regional and sub-regional levels are also the ones at which consolidation and integration of the international law regulating marine living resources and pollution from land-based sources will have to take place in the absence of renewed attention to binding global codification efforts. This development holds the promise of better on-the-ground outcomes as a result of an increased emphasis on education and more flexible use of economic instruments that, ultimately, bring pressing marine environmental issues closer to those on whom implementation and compliance depend. However, overreliance on programmatic approaches is not without deficiencies: one only needs to recall that, in the absence of an overarching legal mandate, it took two and a half decades for the Regional Seas Programs just to come together to discuss and coordinate their efforts on a global scale.

On the one hand, the move from rigid global codification to non-binding, program-oriented efforts signalled a move from targeting the sources of environmental threats to dealing with the problems associated with those threats. On the other hand, this move came to introduce an order of flexibility aimed at improving participation, consensus-building, and compliance levels as well as at enhancing adaptation to changes in science and technology (see Sand, 1990; Brown Weiss, 1992; Susskind, 1994). Recognizing that it is ultimately at the sub-national level that the effectiveness of international law is determined, the programmatic approach allows for increased public engagement that will eventually lead to growing public pressure for actions. Most importantly, the move away from rigid global codification allows for further increase in NGO participation in all aspects of decision-making and implementation. At the same time, it avoids the compliance "Catch 22" of existing international law: the more stringent the compliance mechanism, the less willing the nations are to participate in a binding agreement. Expanded participation also facilitates building sub- and supra-national public and public–private partnerships to strengthen implementation efforts in marine environmental sectors, for which it is not as easy as it is in shipping to make noncompliance economically unacceptable.

A related development accelerated by UNCED, namely the progress achieved in the implementation of the concept of integrated marine and coastal area management (ICAM), might hold significant importance for the future of marine environmental law. The ICAM concept has already been adopted as a central organizing concept of the marine component of a number of international agreements and post UNCED conferences that include CBD, FCCC, GPA, the International Coral Reef Initiative, and the Global Conference on Sustainable Development of Small Island Developing States (Circin-Sain and Knecht, 1998). Furthermore, it envisions an integrated approach to multilateral marine environmental instruments to be realized within a functioning institutional mechanism at the national level—an alternative to integration at the global level that may or may not be fully attained in all marine environmental sectors. Most significantly, by rationalizing national development and environmental protection needs within a comprehensive management process, ICAM offers an opportunity for streamlining institutional and technical capacity-building efforts, including training, technical, and financial assistance for the countries.

There are some inherent problems, however, in moving away from the more structured mechanisms of global binding instruments to more flexible and dynamic arrangements. As has been demonstrated time and again by the existing global marine environmental agreements, the originally adopted objectives and measures have proved to be less significant than the institutional arrangements within which an instrument has been set for its operation, review, and adjustment in response to changes in distribution of power, economic interests, and environmental concerns across nations. And, if the IMO experience has taught us anything, they are not nearly as significant as the codified institutional mechanism for the integration of an instrument with its other related instruments. Thus, an instrument falling under the auspices of a specialized international organization has better chances for effective implementation, review, and adjustment than a stand-alone agreement with an autonomous or semi-autonomous secretariat. Similarly, a dedicated multiforum mechanism, which operates throughout the year and allows for functional integration of its respective instruments, holds significant advantages to periodic meetings conducted by conferences of the parties and their subsidiary bodies every two years or so, as stand-alone events or under an institutional mechanism which, by its charter, focuses on co-ordination at best, and which is dedicated to a much broader and more fluid array of issues.

A look at the institutional mechanisms established under the IMO and the UNEP, the two UN agencies under the auspices of which the predominant body of international instruments on marine environmental protection is concentrated, provides ample illustration of this point. The former offers a forum for interested countries to exercise strong pressure on a continuous basis and allows for environmental issues to be addressed across a number of related instruments that are institutionally and functionally integrated within the IMO committee structure. This allows for scientific, technical and enforcement provisions across individual instruments to be brought functionally together to address specific environmental objectives (as was the case with recent efforts for protection of the North Atlantic right whales and for adopting precautionary measures against pollution by oil or other hazardous substances, both of which were dealt with under SOLAS). In contrast, while the marine environment instruments under the UNEP auspices also allow for similar integration along their respective functional sectors, this cannot be achieved at present due to the limitations associated with the mandate and the organization of that body. Established to promote co-operation and provide for co-ordination across the environmental activities of all UN bodies, the UNEP mechanism is too scattered in order to make up for what should properly be the function of one or more dedicated specialized agencies.

The experience since UNCED has also demonstrated that the move away from traditional global binding commitments with the ensuing diversification of decision-making and implementation responsibilities to a variety of non-state/sub-state players introduces a measured lack of continuity and stability. Whether and how individual substantive issues will be addressed has become increasingly dependent on key individuals holding positions at various supranational organizational levels. The overall result is increasing vulnerability to changing priorities, changing institutional agendas, changing program officers, and changing government officials—a trend especially prominent with respect to implementation efforts in developing countries and to technical assistance to enhance such efforts.

At the same time, coming to grips with how to meet the needs of building institutional and technical capacity, including training, technical, and financial assistance for the countries in need of such, might be the single most important achievement to further future development and effectiveness of international environmental law, including global marine environmental rule-making. It has been demonstrated again and again in global codification efforts that, even when all parties come to support given measures, we cannot expect to have the benefits of global participation, implementation, and compliance without providing for the benefits of adequate financing, capacity building, and technology—something that will continue to translate into technical and financial assistance for an overwhelming majority of nations well into the new millennium.

Thus, port and coastal state control and, hence, enforcement in the shipping and fisheries sectors, present a meaningless notion, if a state does not have or cannot afford to maintain a sufficient contingent of qualified personnel. According to both the IMO Marine Environmental Protection Committee and the shipping industry, the lack of reception facilities in some areas of the world, including MARPOL 73/78 specially designated areas, is the main reason for the continuous pollution of the marine environment by oil, chemicals, and garbage as well as for MARPOL Annex IV not yet coming into force. To take the Black Sea region as an example, it was only after external assistance became available in the late 1990s that the states there were able to move to address the issues of reception facilities and port state control—more than two decades after the Black Sea was designated a special area. No different is the situation in the Wider Caribbean Region, where the lack of port reception facilities for garbage has proved to be a major impediment for the wider ratification of this convention. In the early 1990s, 75% of the countries outside OECD that operated ports and harbours did not have the necessary infrastructure in place to properly handle wastes from ships and the situation has hardly improved since then. It will not come as a surprise if the global conference on sewage, planned under the GPA by UNEP for the year 2000, reaches a similarly disturbing conclusion with regard to sewage treatment plants. As the Global Waste Survey has pointed out, implementation efforts under the 1972 London Convention and GPA are subjected to the same capacity-building, technology, and financing constraints.

The UNCED process has brought some measured improvements in the redirection of resources to environment and development resulting from changes in UN practices, restructuring of the Global Environmental Facility (GEF) and the World Bank, and the GEF replenishment (Cicin-Sain, 1996). However, the overall financial flow dedicated to the environment has not increased substantially. Moreover, at the global level, funding has followed the industrialized countries' priorities and not necessarily the needs of the developing countries to meet global standards and obligations. Environmental and other assistance has been further hampered by the strain of conflicts in the last decade of the 20th century. Yet environmental protection decision-making, implementation, and enforcement are going to become even more dependent on global financial and technology flows, to the extent that future environmental policy could just as well be negotiated in the World Trade Organization, if recent CBD and FCCC negotiations are of any indication.

It is also important to keep in mind that the developments that are presently shaping the future of the body of global legal instruments dedicated to the marine environment trace their roots to the post-Cold War vision of a global economic and political "village" and of the further erosion of state sovereignty in international law. Considering the post-UNCED world, however, this vision has hardly materialized and it seems that Mgonigle's (1980) observation from two decades ago that "where economic interests are strong, only clear legal obligations can provide any measure of future security" still remains relevant at the threshold of the new millennium.

### A Learning-based Approach to Global Codification

We need to also keep in mind that even the most comprehensive and fully integrated system is only as good and effective as each of its individual building blocks—the existing global legal instruments on the marine environment in this case. What has been the effectiveness of these instruments? Do they lay down a legal foundation that is adequate for meeting the needs of sustainable development? Have they managed to trigger the substantive changes in behaviour required to address the pressing problems of environmental degradation? The ultimate answer to the first question will always depend on the individual's ideas of the purpose and nature of international law. As Susskind (1994) succinctly illustrates, an "idealist" will see continued environmental degradation where a "pragmatist" will find substantial progress in terms of commitments, learning, and capacity building. However, a growing body of knowledge is emerging to provide more definite conceptual directions for objective evaluation of the effectiveness of the global environmental legislation codification process and its products, including global marine and coastal legislative efforts (see Peet, 1992; Mitchell, 1995; Brown Weiss and Jacobson, 1999). Building that knowledge into comprehensive, in-depth, and on-going assessment of the effectiveness of the existing global legal instruments to serve as a learning vehicle for better protection of the marine environment is a task that still needs to be accomplished.

Agreeing on criteria for defining "effectiveness" is fundamental to such assessment. One of the Rio process' major contributions to the evolution of international environmental law was coming up with such an agreement. In the course of a comprehensive survey aimed at determining the adequacy and effectiveness of the then-existing international legal mechanisms in the field of environment, the UNCED Preparatory Committee formulated a set of some 32 process-oriented criteria grouped under six functional areas: "Objectives and Achievement", "Participation", "Implementation", "Information", "Operation, review, and adjustment", and "Codification programming" (Sand, 1992). Using these criteria as a framework, a survey of 124 multilateral environmental instruments was conducted drawing on a number of evaluations of individual instruments and comparative analyses that were conducted by Parties, secretariats, UN bodies, or independent observers in preparation for the Rio Summit. Its results were presented at the Earth Summit and were ultimately reflected in Agenda 21. "A benchmark study that will need to be carried forward and refined by future, more in-depth re-assessment" (Sand, 1992), the 1992 Survey still remains the single effort to look at the effectiveness of international environmental agreements in a comprehensive, if not sufficiently in-depth, way.

The necessity of future refinement was most recently re-iterated by (UNEP, 1999) pointing to the difficulty in assessing accurately the effectiveness of multilateral environmental agreements and non-binding instruments due to the lack of accepted indicators. Most of the institutional and scholarly attention to the effectiveness of international legal instruments has so far been limited to process-oriented evaluation. This reflects partly the fact that the objectives of most agreements are formulated in highly general and abstract terms, as the 1992 survey pointed out (Sand, 1992), and partly the lack of systemized baseline data for the majority of the agreements. At the same time, all too often, diligent reporting and full compliance does not guarantee (or demonstrate) the desired impact on the actual state of the environment. A direction for overcoming these limitations as related to the area of marine environmental protection has been suggested by some recent national and international efforts to study effectiveness at a program level in the area of coastal and ocean management. Recognizing the need of outcome–performance evaluation, 15 bilateral and multilateral donor-agencies have initiated the development of a common evaluative framework as a vehicle for learning from ICAM initiatives (see Lawry et al., 1999; Olsen et al., 1999). This work has added to the growing recognition of "learning" as both a legitimate objective and an outcome defining effectiveness.

Lessons could also be learned from the most comprehensive attempt to date for an independent evaluation of the program effectiveness under the United States Coastal Zone Management Act (Hershman et al., 1999). The methodology developed for this evaluation measures effectiveness in terms of on-the-ground outcomes, processes used to achieve these outcomes, and the relative importance given to the issue. While introduced in program evaluation, this approach could be adapted for further refinement of the 1992 Survey evaluation framework, so that a comprehensive common evaluation methodology could be applied both across the individual instruments and the functional sub-areas of an increasingly multifarious, yet inter-related, body of international environmental law.

Indeed, there are telling linkages between the US CZM evaluation model and the 1992 UNCED survey. In both cases insufficient outcome data and lack of adequate monitoring present major evaluation difficulty in addition to the lack of precisely stated objectives. The US study concluded that there is a pressing need for institutionalizing outcome monitoring and performance evaluation system. If the present trend of "opening" international environmental law to multiple actors continues, institutionalizing outcome monitoring and a compliance evaluation system appears practical, even warranted, in future instruments. Also, the 1992 Survey looked only at the individual instruments within the various major areas of environmental protection, without considering their interrelation and combined effect. However, as a result of the implicit decision to approach different environmental problems (or, sometimes, even one and the same problem) through separate instruments, we are now faced by environmental sub-areas being regulated by clusters of such instruments—the sub-areas covering pollution from shipping or from dumping are an excellent illustration. In this situation an instrument-by-instrument analysis will inevitably face an attribution problem and will have difficulty capturing the factors for improving the overall effectiveness within a given sub-area. The US CZM study presents a ready operational model for the evaluation of such a multi-element system.

The more complex and dynamic the body of international environmental law becomes in the coming years, the stronger the need for a formal evaluation process to provide parties and appropriate institutions with the opportunity to follow the implementation and the relative impact of the existing instruments and to adopt lessons on an on-going basis. The time factor is also of significance. Ten years between studies assessing the on-the-ground impact of individual instruments (as is the case with MARPOL 73/78) and twenty years between comprehensive instrument-by-instrument and area-by-area reviews (assuming that the 1992 Survey will be repeated for 2012) are not acceptable time frames if we would like to shorten the present average periods of two decades for conventions to mature to effectiveness and a decade for lessons to be adopted.

## CONCLUSION

In our efforts to protect and preserve the World Ocean and its living resources into the new millennium we can rely on a substantial, well-established, but also increasingly amorphous, complex, and dynamic body of marine environmental law. While we are close to achieving comprehensive and integrated legal regimes in the areas of vessel-source pollution and ocean dumping, we still face a formidable challenge in achieving the same for land-based sources and marine living organisms. Universal participation continues to be a major objective under all related instruments. Achieving universal participation, implementation, and compliance for the marine pollution and marine living resources sectors translates first and foremost into the challenge to meet the financial, institutional, and technical-capacity needs of the newly industrialized and developing nations. With accumulating empirical data on, and growing scholarly attention to, treaty effectiveness, we now have the opportunity to put in place a formal evaluation process that provides information on already adopted environmental instruments, thus empowering us with a common, learning-based approach to global codification.

UNCED has given us the promise that we will leave the twentieth century with an increased environmental awareness: recognizing the global nature of the most pressing environmental problems, their mutual interrelation and their interdependence with social and economic development, as well as the need for a holistic approach within adequate global legal, institutional, and financial frameworks. The challenge of UNCED is to make the existing system work by consolidating and integrating it: legal instruments, institutions, programs, financial and technical flows alike. It will be up to us to face this challenge in the years to come.

## ACKNOWLEDGEMENTS

I would like to thank Oleg Khalimonov for his contributions and Robert Knecht, Biliana Cicin-Sain, Charles Ehler, Lindy Johnson, Nedalina Dineva, and Capt. Robert Hunt for their comments and support.

*All views expressed are strictly those of the author and should not be associated with the NOS, NOAA or any other Federal agency.*

## REFERENCES

Boyle, A.E. (1999) Problems of compulsory jurisdiction and the settlement of disputes relating to straddling fish stocks, *International Journal of Marine and Coastal Law* **14** (1), 17.

Brown Weiss, E. (ed.) (1992) *Environmental Challenge and International Law: New Challenges and Dimensions*, UN University Press, Tokyo.

Brown Weiss, E. and Jacobson, H.K. (1999) Getting countries to comply with international agreements *Environment* **41** (6), 16.

Churchill, R.R. and Lowe, A.V. (1988) *The Law of the Sea*. Manchester University Press.

Cicin-Sain, B. (1996) Earth Summit implementation: progress since Rio *Marine Policy* **20** (2), 123–143.

Cicin-Sain, B. and Knecht, R.W. (1998) *Integrated Coastal & Ocean Management: Concepts and Practices*. Island Press, pp. 95–117.

Corfu Channel Case (1949) The Corfu Channel Case (1949), ICJ Rep. 4.

Fontaubert de, A.C. (1995) The politics of negotiation at the United Nations conference on straddling fish stocks and highly migratory fish stocks. *Ocean and Coastal Management* **29**, 79–93.

Fontaubert, de, A.C., Downes, D.R. and Agady, T.S. (1996) *Biodiversity in the Seas: Implementing the Convention on Biological Diversity in Marine and Coastal Habitats*, IUCN Environmental Policy and Law Paper No. 32. IUCN, Switzerland

Hardin, G. (1998) Extensions of The Tragedy of the Commons. *Science* **280** (5364) 682.

Hayashi, M. (1995) Agreement on the conservation and management of straddling and highly migratory fish stocks: significance for the Law of the Sea Convention. *Ocean and Coastal Management* **29**, 51–69.

Hershman, M.J. et al. (1999) The effectiveness of coastal zone management in the United States. *Coastal Management* **27**, 113–138.

Houde, E.D. and Rutherford, E.S. (1993) Recent trends in estuarine fisheries: prediction of fish production and yield. *Estuaries* **16** (2), 161–176.

IMO (1995) *Global Waste Survey Final Report* (Ross, S.A). IMO Publication.

IMO (1979) Code for the Constructions and Equipment of Mobile Offshore Drilling Units, IMO Assembly Resolution A. 414 (XI), 1979.

IMO (1998) Focus on IMO: Basic facts about IMO, IMO Secretariat, London, 13. Available at http://www.imo.org.

IMO (1998) Focus on IMO: Preventing Marine Pollution, IMO Secretariat, London. Available at http://www.imo.org.

IMO (1990). Document MEPC 30/INF

Joyner, C.C. (1996) The 1991 Madrid Environmental Protection Protocol. *Marine Policy* **20** (3), 187.

Lawry, K. et al. (1999) Donor evaluation of ICM initiatives: what can be learned from them. *Ocean and Coastal Management* **42** (9).

M/V Saiga Case No. 2 (1999) The M/V "Saiga" Case No. 2. Saint Vincent and the Grenadines v. Guinea. July 1,1999. para 82. Available at http://www.un.org/Depts/los/Judg_E.htm.

Mgonigle, R.M. (1980) The Economizing of Ecology: Why Big, Rare Whales Still Die. *Ecology Law Quarterly* **9**, 221.

Mitchell, R.B. (1995) Compliance with International Treaties: Lessons from International Oil Pollution. *Environment* **37** (4).

Olsen, S. et al. (1999) *The Common Methodology for Learning: A Manual for Assessing Progress in Coastal Management*, Coastal Management Report # 2211, University of Rhode Island.

Peet, G. (1992) The MARPOL Convention: Implementation and Effectiveness, *International Journal of Estuarine and Coastal Law* **7** (4), 277–295.

Sand, P.H. (1990) *Lessons Learned in Global Environmental Governance*. World Resource Institute, Washington.

Sand, P.H. (ed.) (1992) *The Effectiveness of International Environmental Agreements: A Survey of Existing Legal Instruments*, Grotius Publications Ltd., Cambridge, UK.

Southern Blufin Tuna Case (1999) The Southern Bluefin Tuna Cases (ITLOS cases No. 3 & 4: New Zealand v. Japan & Australia v. Japan), ITLOS Order from August 27, 1999. Also available at http://www.un.org/Depts/los/ITLOS/Tuna_cases.htm.

Speer, L. and Chasis, S. (1995) The Agreement on the conservation and management of straddling and highly migratory fish stocks: an NGO perspective, *Ocean and Coastal Management* **29**, 71–77.

Stairs, K. and Taylor, P. (1992) Non-Governmental Organizations and the Legal Protection of the Oceans: A Case Study. In *The International Politics of the Environment*, eds. A. Hurrell and B. Kingsbury. Clarendon Press, Oxford, pp. 110–142.

Susskind, L.E. (1994) *Environmental Diplomacy: Negotiating More Effective Global Agreements*, Oxford University Press.

UN (1999) Oceans and the Law of the Sea, Report of the UN Secretary General, Adv. A/54/429, para 345-360. Available at http://www.un.org/Depts/los/.

UN (1983) *The Law of the Sea*. UN Publication, New York.

UNEP (1982) UNEP Guidelines Concerning the Environment Related to Offshore Mining and Drilling Within the Limits of National Jurisdiction.

UNEP (1999) *Global Environmental Outlook 2000*. UNEP Publication, Earthscan Publications Ltd. Also available at http://www.unep.org/unep/eia/geo2000/.

USNAS (1985) *Oil in the Sea: Inputs, Fates and Effects*. National Academy Press, Washington, DC.

USNAS (1975) *Petroleum in the Marine Environment*. National Academy Press, Washington, DC.

Vallega, A. (ed.) (1994) Special Issue to the Regional Approach to the Oceans. Concept and Policy, *Ocean and Coastal Management* **24**.

Young, O.R. (1994) Subsistence, sustainability, and Sea Mammals: Reconstructing the International Whaling Regime.


## THE AUTHOR


**Milen F. Dyoulgerov**

*International Program Office, National Ocean Service,*
*National Oceanic and Atmospheric Administration,*
*1305 East West Highway,*
*Silver Spring, MD 20910, U.S.A.*

Chapter 132

# COASTAL MANAGEMENT IN THE FUTURE

Derek J. McGlashan

Management of coastal resources will not improve unless tough questions are asked, and areas of conflict addressed. In the discipline of coastal management, the question needs to be asked where coastal management is going and whether it is achieving what it aims to achieve. The need for the motivation of political will is addressed, as is the inclusion of economic issues in management programmes. Using examples from numerous countries, the need for natural processes to be the driving force for coastal management units is stressed. A number of legal issues are discussed which may aid the quest for sustainability in a voluntary system along with data collection and storage. Stakeholder involvement and inclusivity are considered in the context of achieveability, and a number of transboundary problems are highlighted. The final area considered is the financing of coastal initiatives.

*Seas at The Millennium: An Environmental Evaluation (Edited by C. Sheppard)*

## COASTAL MANAGEMENT AND POLICY EVOLUTION

Coastal management has evolved over the years from a basic concept of protecting the coast at all costs to one in which the coast should be managed in accordance with natural processes rather than against them. O'Riordan and Vellinga (1993) have categorised this evolution in the United Kingdom into four phases starting in the 1950s: phase one (1950–1970) was sectoral, had limited participation, rarely included a conservation perspective and was reactive; phase two (1970–1990) was a period in which environmental assessment increased, there was an increase in coordination and participation, but the engineering dominance remained; phase three (1990–present) involves focus on sustainable development and comprehensive environmental management and restoration with an emphasis on public participation; and phase four (future) should see coastal area management which focuses on shared governance, the precautionary principle and ecological empathy. The extent to which most countries are at phase three is debateable, and whether phase four will emerge in the future remains to be seen.

The current paradigm in coastal management is one of 'sustainability' (Kay and Alder, 1999). Definitions of sustainability vary, but tend to follow the Brundtland Report definition (World Commission on Environment and Development, 1987, p. 40): "sustainable development seeks to meet the needs and aspirations of the present without compromising the ability of those of the future". This effectively follows the principles of social justice, environmental justice and equity (Younge, 1992; Buckingham-Hatfield and Evans, 1996; Common, 1995). Integrated Coastal Zone Management (ICZM) is the latest incarnation in the evolution of coastal management programmes and policies and has really only emerged post-Brundtland. As such, ICZM should follow some of the guiding principles in the UNCED (1992) Agenda 21 document, including the precautionary principle, polluter pays, resource accounting, transboundary responsibility, and intergenerational equity (Kay and Alder, 1999).

ICZM is seen as a saviour for coastal regions by many, but few programmes which claim to be integrated truly are. Kenchington and Crawford (1993) note that integrated management is 'complete or unified', though it may have individual elements. Cicin-Sain (1993) suggests a five stage continuum of integrated coastal management: fragmented, communication, coordination, harmonisation, and integration. Most current projects have still to reach 'integration' (even in a broad sense), but as Cicin-Sain suggests, integration can require 'institutional reorganisation'. It is interesting to compare the evolutionary model of O'Riordan and Vellinga (1993) with Cicin-Sain (1993), the former appears to show that ICZM has evolved further than the latter; Cicin-Sain (1993) is probably a little closer to the current situation, whereas O'Riordan and Vellinga (1993) are possibly a little optimistic. This is where political will is especially important. Kay and Alder (1999) believe that "successful coastal management programmes use a systematic process to develop guiding statements" which then "integrate the views of both the top and bottom decision-making levels with those of the stakeholders in the management of the coast". This emphasises the point that ICZM is a process and not a solution. It is not the answer to conservationists' or industrialists' dreams, but should rather be used better to balance the pull between development and conservation, and hopefully foster the sustainable management of the world's coasts, with sensitive development, resource use and conservation.

If coastal management is to be integrated it must be integrated spatially, temporally, vertically and horizontally. Spatial integration includes marine, terrestrial and boundary issues considered simultaneously and given equal weighting. Temporal integration includes issues of lag times etc. so that when making decisions, the impact over time is considered. Vertical integration means that all levels must cooperate and integrate (from individual site plans through local and regional management to national and international policies, local councils to national governments and international bodies). Horizontal integration includes the holistic nature that many management programmes are currently aiming at, for example conservation, coastal defence, economic development, and departments within organisations working with each other. A further issue in the current ICZM sustainable development paradigm is inclusivity. ICZM must be participatory and wholly inclusive. Is this achievable?

Resource management in the United States (as with many countries) is currently in "two parallel and potentially conflicting paradigms" (Crozier, 1998): top-down (the traditional approach) and bottom-up (the stakeholder approach). The 'bottom-up' is the new participatory system which includes stakeholder involvement. However, although it allows local populations to learn about the environment and engender 'ownership', it can often be difficult to get individuals to think strategically (which is the advantage of top-down); each has their own interests and agendas. In such a situation, can inclusivity ever generate strategic and sustainable coastal management? Are the principles of inclusivity compatible with practical delivery of ICZM rather than just the theoretical compatibility?

Kay and Alder (1999) note that "a critical issue when considering scales of integrated coastal planning is how [coastal zone management] plans fit together", and further note that "there is a danger in thinking that a plethora of local area coastal plans will achieve the same objectives as higher order plans; they cannot". This is a key point: one national level plan cannot achieve the same as a number of local and regional plans, however, policy is usually required at the national or international level to guide local actions.

The issue with sustainable development (and with ICZM) is how it is applied to society. El-Sabh et al. (1998)

believe that there are three cornerstones of sustainable development: environmental sustainability, social equity, and economic viability. Sustainable development in coastal zones should be based on avoiding future problems, an eco-efficiency concept (DeAndreaca and McCready, 1994), and the precautionary principle. This is therefore where ICZM should be heading, to inform and to aid precautionary decision-making, and move away from the *ad hoc* nature of decision-making found in many nations, which is the crux of the UNCED (1992) Agenda 21 concept.

Coastal management cannot succeed without 'political will'. Agenda 21 may have raised awareness, but unfortunately political will is lacking in many areas, with politicians keen to treat the symptoms rather than cure the cause of local problems, which gives rise to short-term *ad hoc* solutions that can been seen within a re-election term. This short-sighted behaviour must be addressed and the political will found to push sustainable coastal management to the fore, whether it is in the form of ICZM or some other name. Nichols (1999) states that "Integrated Coastal Management follows the... neoliberal view ... that economic development and conservation goals are mutually supportive under the right circumstances of regulation and resource privatisation". Whether this can be achieved or not remains to be seen. But few coastal management/planning projects have even attempted to look at economic development. There have been examples from small island developing states (Maul, 1996) which mostly focus on eco-tourism, fisheries (particularly with dynamite and poisons) and coral fishing. There are few from the developed world. This may be because the issues are too complex, but probably is also due partly to a lack of political will.

Increasingly, companies and organisations are accepting that they have to participate in sound environmental management. In the United States, the Bank of America (1993) proposed three ways in which the environment contributes to the economy: quality of life improvement, better resource use and management, and sustained long run growth. Hale (1993) states "the necessity of economic development is apparent, and the link between productive, healthy ecosystems and both the economy and people's overall well-being is also clear" and then goes on to say "the proverbial pie is getting smaller and must be divided among more people". These theories still need to be applied to the coast, to bring true sustainable economic development into coastal management.

The race is on, as there is no doubt that the 'pie' is getting smaller. The problem that remains is that although economic development provides jobs, it tends to be dependant on a healthy environment (e.g. fisheries and tourism) yet has the potential to further degrade the environment if insensitively developed (e.g. energy production, heavy industry etc.). "Hence the challenges of coastal ecosystem management and of economic development are inextricably intertwined" (Hale, 1993). This creates a dilemma which can be overcome with dialogue and careful planning and management, which is why economic development is a key component of ICZM.

In Scotland, the Forth Estuary Forum (a voluntary partnership whose aim is to 'promote the wise and sustainable use of the Firth of Forth') commissioned an economic appraisal of the Forth Estuary and surrounding environs (Firn Crichton Roberts Ltd., 1998). This made a number of recommendations and highlighted areas for future thought. Economic development has now become one of the Forth Estuary Forum 'Flagship Projects' and they are currently examining how to include economic development in ICZM for the Forth Estuary. Part of the problem is encouraging industry and the economic development groups to sit down with the forum and participate in 'the wise and sustainable use of the Firth of Forth'.

In the United States, "promoting new economic activities" (NOAA, 1998) is seen as a key area. However, this report only considers fishery stocks and aquaculture, hardly holistic inclusion of economic development issues. If ICZM is going to be successful it must include economic development, as how else can it claim to have 'sustainable development' as the running paradigm, and more importantly how can the management programmes be 'integrated' and 'fully inclusive'? This is a point for developing and developed countries alike, coastal managers and academics are going to have to forge links with industry, commerce and development bodies if 'sustainable development' is to be achieved. This must then become integrated into the coastal decision-making process and not just brushed over when projects are being planned.

Currently, many coastal management schemes are still at the planning stage, like the Forth Estuary Forum. The latter was formed as part of the Scottish Natural Heritage 'Focus on Firths' project in the early 1990s. The strategy document was only published in the summer of 1999 following seven years of working towards that point, and as it is not statutory it does not have to be followed and can only act as a guide. In England and Wales, Shoreline Management Plans (SMPs) have been produced based upon natural processes (the coastal cells as designated by Motyka and Brampton, 1993). These are plans which the UK government 'encourage' Local Authorities to commission. However, if a coastal authority wished to build a coastal defence, the project would not attract central government grant aid unless it complied with the SMP. As these SMPs are not statutory, new developments do not have to comply with the guidance in the SMP (unless seeking central government finance), so it comprises a 'carrot and stick' approach, which can be bypassed with enough private money. Jewell et al. (2000) state that "the non-statutory SMP process was invaluable in bringing together planners, engineers, ecologists and archaeologists in the context of coastal defence... there is a greater understanding of problems and in addition the SMP process has encouraged all parties to think more strategically about coastal defence issues". Given the considerable volume of data and information in

SMPs it is disappointing that some decisions are still made at the coast which ignore these (relatively) holistic documents. Again, this shows the vulnerable and *ad hoc* nature of coastal management which still exists despite many decades of coastal management initiatives in varying forms.

In many ways, the above paragraph gives a classic example of rhetoric. The UK government has signed up to Agenda 21 (UNCED, 1992), as many governments have. Agenda 21 states that coastal states must promote ICZM, but in effect the UK government has only enhanced the potential for improved coastal decision-making utilising SMPs, a move in the right direction, but not ICZM. Following the 'Rio Earth Summit' and the adoption of Agenda 21 and the phrase "think global act local" little has changed in many countries. National policy evolves slowly unless there is political will, which is most readily available from 'knee jerk reactions'. An example of this is the recent production of a 'flood warning system' for the Firth of Clyde in Scotland. The political will was only found for this following a serious flood in 1991 which cost an estimated £7,000,000 (SEPA, 1999), causing a major public outcry. However, it could be argued that the buildings (many of which were residential) that suffered the flood damage should never have been built in an area prone to flooding, and may not have been had the Firth of Clyde had ICZM. The recent UNEP (1999) GEO 2000 report stated that "the global system of environmental policy and management is moving in the right direction but much too slowly. Inspired political leadership and intense co-operation across all regions and sectors will be needed to put both the existing and new policy instruments to work."

## PHYSICAL SYSTEMS AND MANAGEMENT

A key point at the coast is the need to understand physical systems, Charlier (1989) stated "any study of a 'coastal zone' requires knowledge of some geomorphological concepts". The majority of policy-makers in civil services around the world are not coastal geomorphologists, oceanographers, marine ecologists or even scientists. Those who develop policies must fully understand the physical systems for which they are deciding the future. At the coast, sediment availability and movement is the key to understanding: the geomorphology is the reason for the shape of the coast and how it evolves. If there is no supply of fine sediment there will be no saltmarsh or mangrove habitats. If there is no supply of sand on a beach of a suitable width to allow the beach sand to dry at low tide there would be no sand dunes, no sand dune habitats, no grasses to colonise and stabilise the dunes, therefore no dune ecosystems, and so on. Consequently, sediment availability is the key to understanding coastal systems. The erosion, transport and deposition of sediment controls the shape of the coast and the habitats within it. However, biological communities can also have a major impact on the stability of coastal and marine sediments (Pender et al., 1994). The knowledge of these sediment transport routes (and interactions with ecology) are of vital importance in understanding the evolution of coastal environments in the past and future. It is therefore vital that coastal management looks at sediment circulation and uses sediment cells (Motyka and Brampton, 1993) as a technique for deciding upon management units, much as the Shoreline Management Planning system in England and Wales. The SMP process is a good start, a foundation upon which coastal management can be built.

Sediment depletion in the coastal system is an increasing problem, particularly in developed nations, and understanding of sediment budgets within cells is essential. Many sources of sediment to the coastal system are being reduced, rivers are dammed, reducing fluvial input. An example of this is in the Turkish northern Aegean where construction of an irrigation dam has decreased sediment supply to the coast over seven years, the result of which has been a recession of the shoreline by 200–250 m and the loss of 50,000 m$^2$ of beach (Geerders et al., 1998). Erosional coasts are being protected and claimed from the sea, further reducing sediment availability. Sediments are being extracted from the sea bed for aggregate and beach recharge schemes. This is compounded by rising relative sea level in many coastal locations (causing the wave base to become higher relative to the seabed) which reduces the amount of sediment available to the coastal system from the seabed. Increasingly, beach recharge is seen as a popular management option to combat erosion. However, where does the recharge sediment come from? Beach recharge material is in great demand; should it come from marine extraction? If this takes place within an estuary it affects the tidal prism (Pethick, unpublished). There is also the question of how sustainable beach recharge is in the longer term, and how long can it continue. Knowing the movements of sediment is not only important for planning at the coast relating to erosion and accretion, but also for identifying pollutant transport routes as many pollutants attach themselves to sediment particles. However, the pollutants in the sediment can also be used as tracers (French, 1997). This therefore is forwarding the cause of two types of knowledge, pollutant and sediment life cycles. Once the sediment resting place is known, further work can look at biological uptake, etc.—a positive spin on a negative human intervention.

Another human intervention evident on the coast are coastal structures, the impacts of which have been documented as far back as 1934 (by Mathews). In recent years there has been a wealth of literature on the topic, including journal special issues (notably Kraus and Pilkey, 1988) and a number of important literature reviews (Kraus, 1988; Kraus and MacDougal, 1996; MacDougal et al., 1996; Tait and Griggs, 1990) throughout the coastal management, geomorphological and engineering literature. Over time, the need for more sensitive management has been slowly recognised and, as such, beach feeding and managed realignment has developed into coastal management options which work with nature rather than against it. There is no point is reviewing the literature relating to the issues mentioned; the reader is referred to French (1997) for further information. However, the relocation of Belle Tout Lighthouse at Beachy Head on the south coast of the UK opened up another option. Belle Tout lighthouse is an historical monument, but no longer functioning as a lighthouse. Until November 1998, it lay approximately 14 m from the edge of a 100 m vertical chalk cliff; there was then a large collapse of the cliff, which took the building to within 3 m of the edge. Over the following months a substantial engineering project was undertaken (involving 22 men working 12 hours a day, 7 days a week), which saw the lighthouse being jacked up intact, slid along rails and placed down on new foundations 12 m further inland on 17–18 March 1999. Belle Tout had been 'relocated'. This is an option which merits further examination in a number of settings, particularly where there is an individual building of historic interest on an otherwise natural coast, and especially where there are high cliffs involved, where stabilising the cliffs would not only be undesirable but impractical. However, it does require a suitable amount of available space inland and a substantial financial outlay. The relocation of Belle Tout is thought to have cost approximately £250,000, a fraction of the cost of losing the historic property or trying to stabilise the cliff. However, there are no ongoing regular maintenance charges (as long as it is moved far enough inland), no interruptions to the natural sediment dynamics and no geomorphologically and ecologically damaging protection structures.

In many coastal areas, particularly estuaries where there are important habitats and urbanised/industrialised areas in close proximity, coastal defences have caused 'coastal squeeze' where habitats are being reduced as they cannot migrate inland naturally with a rising relative sea level. In these cases, managed realignment (or managed retreat) has been advocated by many. Managed realignment usually consists of beaching (or removing) coastal defences, allowing the land behind to flood (often building a secondary, inland defence); it allows habitats to begin to move inland to areas predominantly of agricultural use (though in many cases this land was 'claimed' from the sea originally) and lets the natural system begin to re-evolve. Also, careful site selection can reduce the maximum tidal heights in an estuary (Pethick, unpublished).

Development decisions at the coast should be undertaken with the full awareness of the sensitivity of that coast. Gemmell et al. (1996) and Hansom and McGlashan (1998) propose the use of a Conservation Sensitive Management Strategy (CSMS) for assessing the appropriate developments on open coasts and lake shores. The CSMS analyses the geomorphological sensitivity of a designated coast and compares this to a set of predetermined management options, which are ranked by order of sensitivity. The more sensitive the coast, the less options are compatible with CSMS. This concept should be expanded to consider ecological and social impacts as well as geomorphological effects.

One management option is the use of set-back lines, which have been used particularly in a number of European countries (Denmark, France, Greece, Italy, Norway and Sweden). They are an effective and sensible planning tool when used properly and adhered to. A calculation is made which relates the average erosion rate of a given section of coast to an acceptable planning timescale (usually 50–100 years) and then the two are multiplied and a safety factor is added. However, it is not unusual that a blanket distance is used for a whole country (for example 300 m in Italy). Usually it is a planning tool, whereby no new construction is allowed within the 'set back distance' of the mean high water mark; in some cases existing buildings can be served with demolition orders if the erosion rate brings them well within this zone.

## LEGAL ISSUES

Coastal management and planning programmes of many countries vary from voluntary agreements and government guidance—which encourage moves towards sustainable management, for example the UK—to the opposite approach on the scale where Sri Lanka has taken the entirely regulatory route. There are advantages and disadvantages to both. However, one issue that does need to be considered is the definition of the coast. Sri Lanka has a very rigid definition of the coast from a legal perspective through the Coast Conservation Act 1981. The seaward limit of the coastal zone extends to 2 km seaward of the 'Mean Low Water Line', landward it extends 300 m inland from the 'Mean High Water Line', except in "the case of rivers, streams, lagoons or any other body of water connected to the sea, either permanently or periodically, where the landward boundary extends to a limit of 2 km, measured perpendicular to the straight base line drawn between the natural entrance points" to the sea (Hettiarchchi, 2000). This is a very fixed legal definition, if fixed on dynamic boundaries. However, it is interesting to note that it only extends 2 km inland, and not 300 m round estuarine coasts. Although the definition in estuaries and lagoons may look like a flaw, many countries do not have any legal definition of the coast, for example the United Kingdom. The UK has a plethora of laws which have some form of relevance to the coast, but no one piece of legislation. A recent review of legislation relating to the coastal and marine environment in Scotland found 79 relevant Acts of Parliament between 1861 and 1992 (Cleator and Irvine, 1995). This does not even begin to cover the role of the common law governing the traditional use of property rights. At least in recent years the government has simplified the number of regulatory bodies that deal with pollution and water quality issues. However, this does not change the fact that there is no legal definition of the coastal zone, or one act which regulates activities, unlike Sri Lanka.

In Sri Lanka any development activity, which is "any activity which is likely to alter the physical nature of the coastal zone in any way" (Hettiarchchi, 2000) requires a permit.

Although the UK and Sri Lankan approaches to coastal management vary, recent (non-coastal) legal cases have highlighted a potential route by which coastal actors could encourage others to develop the coast in a more sustainable way through the common law system in the UK (McGlashan and Fisher, 2000), and potentially other similar legal systems (e.g., Australia, Canada and New Zealand). The concepts used are drawn from common law principles, but have yet to be tried in the courts to any great extent. The assumption here is that coastal structures can have a measurable impact on sediment movement within the coastal cell. In this case, there are two adjacent coastal proprietors 'A' and 'B', 'B's land is down-drift of 'A's (Fig. 1). If coastal proprietor 'A' builds an insensitive structure which impacts upon sediment movement creating (or enhancing) erosion land belonging to coastal proprietor 'B', 'B' may be able to sue 'A' on the specific ground of nuisance to stop the 'nuisance' and potentially also seek damages. However, 'B' must be able to show a loss or injury to his land, and must act within 20 years of the construction of the insensitive structure. The law must analyse whether the structure caused the loss, which poses two questions: (1) would the loss have occurred without the construction of the structure; and (2) whether the structure was the legally significant cause. The first question is a matter of fact and relies on an expert testimony. The second is more complex and effectively depends upon whether the defendant could have reasonably foreseen the consequences of the action. However, given that the coastal literature has many papers detailing the impacts of coastal structures stretching back to the early twentieth century, this should not be a bar to liability. McGlashan and Fisher (2000) provide a more detailed treatment of this brief discussion.

In general, legal systems tend to be terrestrially based, often biased towards property protection, which is often not ideal when considering natural environments. The coastal environment is a particularly dynamic one, constantly changing on many timescales. This does not merge well with most legal systems which are static and tend to view boundaries as static lines on maps, rather than as dynamic and time-transgressive. It is for this reason that many legal systems may need to be revised, or at the very least have holistic, all inclusive, definitions of the coastal zone extending far enough inland and seawards to be effective. The common law approach noted above illustrates that there can be alternative methods which achieve, or work towards more sustainable coastal management, even in complex legal systems like those in the UK.

The importance of property rights in the regulation of the coast even in the modern state can be seen from the example of the US Supreme Court's 1992 decision in the case of *Lucas v. South Carolina Coastal Council* (505 US 1003). The case involved a landowner's challenge of the state

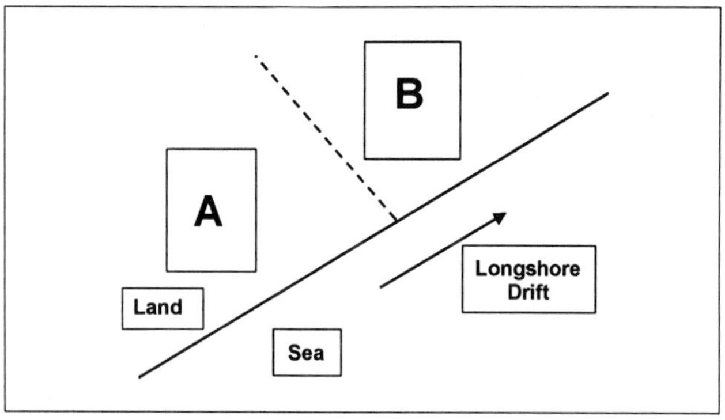


Fig. 1. Coastal proprietors 'A' and 'B' and their interaction.

regulations for control of the coastal zone which provided for a setback line. The Fifth Amendment to the US Constitution prohibits (as does Article 1 Protocol 1 of the European Convention on Human Rights, for example) the appropriation of private property by the authorities without the payment of compensation. The challenge was on the basis that the regulation effectively destroyed the value of the land, highlighting the conflict between property rights and state regulation in the public interest or in the interest of the environment. The state had been able to justify the regulations in its own courts on the basis that "harmful ... use" of the land was thereby prohibited, but the Supreme Court disagreed: it explicitly stated that the justification must arise from traditional notions of the common law interaction of property rights through the law of nuisance. Thus, in the US also, traditional legal rights contained in the common law retain their importance, and top-down rational environmental regulation must compete with other value systems inherent in the legal system. There may be increasing rational legislative intervention today and in the future. However, there is also greater global attention given to what are ultimately human rights issues, and the opportunities for conflict abound. This case also illustrates the conflict between the national and the local at the coast: the Supreme Court effectively rules in favour of the local, and against any prospect of truly integrated management.

As part of 'integration', all departments within an organisation (whether governmental or not) must work together and not contradict each other (the difficulties in this are highlighted in the paragraph above); how practical this is remains to be seen. In the UK, Scottish Natural Heritage and English Nature agree that the coastline is dynamic in nature, and must (where possible) be free to adapt and change in its own way. As such, the EU Habitats Directive was seen as a suitable tool in achieving this. However, the way the directives have been implemented in the UK (and other countries) has often been in direct conflict with these aims. In fact, the UK regulations attempt "to *preserve* conservation sites in their current form rather than *conserve* the interesting features as part of a dynamic system" (John and Leafe, 2000). What this legislation fails to take into account is the change which naturally occurs over time as part of the dynamic nature of the coast. This is not to say the Habitats Directive has had a negative effect on the coast. It has brought many benefits and created more informed decision-making. The implementations of these directives in member countries must be flexible enough to allow natural systems to evolve in the naturally dynamic way. From another perspective, the drive to have aesthetically pristine beaches has resulted in many coastal councils cleaning beaches regularly, often using mechanical devices. This may result in a clean beach, but it also removes natural beach litter (for example seaweed) which often becomes the foci for new sand dune formation (Hansom, 1988), and so, although there are clean beaches, the chances of new dunes forming (and hence allowing the natural sand dune evolution to continue) are reduced. Furthermore, the sediment is compacted by the machinery, reducing the dynamism of the beach and its ability to adapt to changing conditions.

## DATA AND INCLUSIVITY

Many countries require a body to collect data relating to the coast and to make it available to allow informed decision-making. GIS would appear to be the perfect instrument if one body in each country can pick up the role of coastal data coordinator. In such a case, a form of international data collection standardisation would be required to allow for accurate comparisons, or records being kept detailing collection methods. Lee (1993) believes that better procedures are required to integrate research results into decision-making; such a database could help foster this. In the current coastal decision-making framework many coastal projects require an environmental assessment. Particularly for larger projects, the results of these must go out to consultation and require 'stakeholder' involvement. This is good in theory, but by the time the 'stakeholders' have become involved, the decision has often already been made. The environmental assessment which is presented often just highlights the mitigation measures. In many cases, this is the point where the geomorphologist becomes involved, post-decision, to mitigate impacts. A more sensible option would be for geomorphologists (and others) to be included in the decision-making process; instead of undertaking a full environmental assessment, the geomorphologist could use 'knowledge-based tools' to aid decision-making. These knowledge-based tools involve the use of expertise, rather than numerical modelling to predict potential problems. This allows the geomorphologist to suggest what the likely impacts would be of different options. This would be presented in a non-technical way and would allow for earlier and more informed consultation with stakeholders. The time required to produce the information would be minimal and so fairly inexpensive, would make the consultees feel more included in the decision process, and would allow them to make more meaningful observations. The process would also be more inclusive, as some people who would feel oppressed by the usual round of consultation would be able to participate, and not be required to be highly literate, or wade through piles of documents.

## TRANSBOUNDARY ISSUES

Many coastal and marine problems are transboundary. This covers sediment (hence the advocation here of a management policy following coastal cells), pollutants, fish, marine mammals and birds. One example is Laguanas Bay on the Greek is island of Zakynthos (in the Ionian Sea), which is an important breeding ground for the loggerhead turtle (*Caretta caretta*). However, this important nesting site is suffering from continued development pressure which is impacting upon the breeding site of this endangered species (Smith, 1999). Greece has ratified four international conventions and has had two presidential decrees relating to the protection of the turtles. One of the presidential decrees includes measures for Nature Reserves, which prohibit tourist development, restrict construction and use of lights, etc. However, these and other regulations are being openly flaunted, with illegal development on the beach (which is not being prosecuted). Zakynthos National Marine Park is expected to be created soon. However, "given the lack of progress to date", Smith (1999) expects it to become another 'paper park' "capable of producing only theoretical solutions and failing to meet the practical expectations of a marine protected area." There are similar problems relating to many marine parks and conservation areas throughout the world.

A major coastal transboundary issue is one of alien species and diseases carried around the world in the ballast waters of major ships. Gollasch and Leppakoski (1999) note that where aquaculture is sited near shipping routes there is a "high risk of disease transfer via ballast water". They go on to note that the problems are not only in the ballast water and the associated sediments, but with organisms fouling the hulls of ships. They also note that between 3000 and 4000 species are transported by ships every day. At the time of writing, Rio de Janeiro was suffering a cholera outbreak thought to have been introduced by ballast water (Burbridge, unpublished). With coastal shipping being increasingly seen as a more 'environmentally friendly' form of transport, being approximately two and a half times more fuel efficient than trains and five times more fuel efficient than road haulage (Kolle et al., 1991), this issue requires urgent attention.

## SOLITARY WAVES

Development at the coast has taken place on the marine side as well as the terrestrial side. An issue which has recently been highlighted relates to the so-called 'fast ferries'. Hamer (1999) gives examples of 5 m waves which "appear out of nowhere". These waves are called solitary waves and have a long wavelength. However, when they reach the shore they steepen up and can create giant waves. Part of the problem is that when they form, they develop a crest, but no trough, which means the crest cannot collapse on breaking as a normal wind-generated wave would. These waves are generated when 'fast ferries' reach a critical speed (which is dependant on the depth of the water). The solitary waves have been blamed for a number of incidents involving damage to pleasure craft and loss of life, and at Belfast sea walls have been heightened after a pram was almost washed away (Hamer, 1999). In the Hamer (1999) article, Ian Dand (of British Maritime Technology) was quoted as saying "a solitary wave is first cousin to a tsunami". There are obviously coastal erosion and habitat destruction issues, but problems are not just onshore; offshore there are potential problems relating to erosion and settlement, where the solitary waves can generate sediment suspension where wind waves would not (Danish Maritime Authority, 1998). Therefore, fast-ferries can potentially affect the natural sediment dynamics of a section of coast, as well as create erosion and other safety problems on land.

## FINANCING

For ICZM (or any other form of coastal management) to succeed it must be suitably financed. The European Union have recently funded 35 demonstration programmes on ICZM (European Commission, 1999). The funding for these programmes has now run out. In Scotland, the Forth Estuary Forum had three project officers until recently. Due to its non-statutory status, the forum will have to raise all of its funds from member organisations, which has meant radically rethinking its role. This is hardly the way to run ICZM, and this is only a few months after the production of the strategy document. Yet all parties acknowledge that the strategy document is only the start; it lays out a framework which estuary actors should aim to follow. Similar problems have been noted elsewhere (including many of the EU demonstration projects), whereby funds are provided to set up a coastal management programme, but not continued to allow the programme to evolve into the implementation phase. If any coastal management techniques are to succeed (whether ICZM or some other form of coastal management), they must be funded in a sustainable long-term way. The current system (particularly in the UK) does not provide an ideal platform.

## CONCLUSIONS

The 1990s have shown that integrated management and sustainable development are of key importance to the world's coasts following the Brundtland report (World Commission on Environment and Development, 1987). During this period, the balance between economic development and environmental needs have become a 'paper goal', but remain a fundamental need. Hanson (1998) notes that "despite the significance of sustainable development in the treatment of our relationship to the oceans [and coasts] few—if any—nations currently base their ocean and coastal zone management laws and policies explicitly and comprehensively around principles of sustainable development". This is a point which needs to be stressed. Coastal experts have a duty to push these issues into the political arena and generate the political will that will be required to accelerate these programmes from paper plans and policies to implementation reality.

If coastal planning and management is to be effective, decision-makers must have access to the relevant information at an appropriate level to be of use. Kay and Alder (1999) suggest that, once down to a site scale, "detailed site maps indicating individual plants, species lists with possible densities, and detailed geology and geomorphological characteristics" would be an appropriate level of information. Again this takes us back to the SMP concept, although there is no point in looking at the site in extreme detail if the decision-makers cannot fit the details into the bigger picture at a coastal cell level. Malafant and Radke (1995) state that reliable comprehensive environmental databases (preferably electronic) should be set up and distributed. An SMP is almost sufficient; it would merely require to be produced in an electronic form. However, although the information contained in SMPs is often guarded, for example with access given to view documents in a set location, the executive summary may occasionally be purchased.

Until coastal decision-makers have access to high quality, relevant data, and operate in an integrated process, decisions will continue to be taken in the current *ad hoc* manner, and research may continue to be undertaken in a piecemeal way which is inappropriate to decision-making needs (Hansom and Kirk, 1991). Certainly in the UK the problem is one of the planning system being a control mechanism, which regulates and legislates for "defined *activities* rather than for defined *environments*" following a pro-active process (Hansom, 1995). To be part of ICZM, a pro-active planning process needs to be part of the coastal management set-up, whether voluntarily or statutorily. However, as noted earlier, the research conducted so far suggests that ICZM requires government legislation and must clearly specify the limits of the coastal zone (Hansom, 1995; Carter, 1988).

Is ICZM achievable? In the current political climate the jury is out. It will require considerably more political will than is presently available in most countries around the world. Individuals, councils, governments and intergovernmental panels and organisations have to act, instead of simply acknowledging. This acting will have to be in strategic, unselfish ways. This will require institutional change at all levels of government and in many organisations. Economic development issues will have to be tackled. Education regarding the natural environment needs to be promoted both to the public at large, and between academics and coastal management practitioners; this has already started, but these links need to improve. Long-term strategic decisions must be made to find the cure to problems, rather than just tinkering with the symptoms. Funding of coastal management projects must be long term—a system of hand-to-mouth subsistence funding is not acceptable. More countries should consider adopting an approach which follows coastal processes: the coastal cells concept. A definition of the coast is required in many countries, but one which is flexible enough to allow for the dynamic nature of the coast. Legal regimes in most countries need to develop an awareness of the dynamism at the coast and reflect this in its laws and judicial decisions. The coastal 'public good', or 'collective good' must come before private property rights, this may require shifts in legal paradigms in a number of countries. Legal systems must not only be able to identify the coast, but understand it as the interface between land and sea: a dynamic time-transgressive interface. Politicians and the public must be educated, political will must be generated, and where necessary alternative and novel methods must be found to try and encourage sustainable management.

Developing nations should learn from the mistakes made by developed nations. Many developed nations have taken up coastal management retrospectively. Developing nations have the chance to allow their coastal zones to develop without suffering the same problems as many developed nation coastal zones which suffer from coastal squeeze, insensitive development, planning which did not consider the natural system and contradictory legal systems. If developing nations can learn from this and resist the pressure to build on potentially vulnerable coastal land, they will be able to avoid having to make the tough and costly choices already facing many developed coastal nations. Although developed countries are beginning to tackle some tough questions, they are only in the lifetime of one or two generations, and this timescale needs to become much longer. Tackling the issues which lie ahead sustainably and in an integrated fashion is not going to be easy, or cheap, but all these issues need to be considered soon.

## ACKNOWLEDGEMENTS

The author wishes to thank Milen Dyoulgerov for assistance and Mr. Graham Fisher for comments on the earlier drafts.

## REFERENCES

Bank of America (1993) Economics and business. *Outlook* June/July. As cited in P.H. Templet (1996) Coastal management, oceanography and sustainability of Small Island Developing States. In *Small Islands: Marine Science and Sustainable Development*, ed. G.A. Maul. Coastal and Estuarine Studies 51, American Geophysical Union, Washington DC, pp. 366–384.

Buckingham-Hatfield, S. and Evans, B. (1996) Achieving sustainability through environmental planning. In *Environmental Planning and Sustainability*, eds. Buckingham-Hatfield and Evans. Wiley, Chichester, pp. 1–18.

Burbridge, P.R. (unpublished) Verbal presentation to the Scottish Coastal forum and others on the EU ICZM demonstration programme workshop at Battleby near Perth, UK, 6 September 1999.

Carter, R.W.G. (1988) *Coastal Environments*. Academic Press, London.

Charlier, R.H. (1989) Coastal zone: occupance, management and economic competitiveness. *Ocean and Shoreline Management* **12**, 383–402.

Cicin-Sain, B. (1993) Sustainable development and integrated coastal zone management. *Ocean and Coastal Management* **21**, 11–44.

Cleator, B. and Irvine, M. (1995) A review of legislation relating to the coastal and marine environment in Scotland. *Scottish Natural Heritage Review*, No 30.

Common, M. (1995) *Sustainability and Policy*. Cambridge University Press, Cambridge.

Crozier, G.F. (1998) Fallacies associated with bottom-up management. Poster presentation presented at Soil and Water Conservation Society meeting, San Diego, Summer, 1998.

Danish Maritime Authority (1998) The environmental impact of high-speed ferries: How to assess maximum safe speeds for surface-effect vessels in close proximity to land. *Seaways* (October), 9–13.

DeAndreaca, R. and McCready, K.F. (1994) Internationalizing environmental costs to promote eco-efficiency. Business Council for Sustainable Development, Gehera.

El-Sabh, M., Demers, S. and Lafontaine, D. (1998) Coastal management to sustainable development: from Stockholm to Rimouski. *Ocean and Coastal Management* **39**, 1–24.

European Commission (1999) *Lessons from the European Commission's demonstration programme on Integrated Coastal Zone Management* (ICZM). European Commission, Luxembourg.

Firn Crichton Roberts Ltd. (1998) An economic appraisal of the Forth Estuary and Firth of Forth. Report to the Forth Estuary Forum.

French, P.W. (1997) *Coastal and Estuarine Management*. Routledge, London.

Geerders, P., Klöditz, C. and Uslu, O. (1998) Mapping Tourism in the Coastal Zone with Digital Aerial Photography. *Intercoast Network: International Newsletter of Coastal Management* **31** (Spring), 10–11.

Gemmell, S.L.G., Hansom, J.D. and Hoey, T.B. 1996. The geomorphology, conservation and management of the River Spey and Spey Bay SSSIs, Moray. Scottish Natural Heritage Research, Survey and Monitoring Report.

Gollasch, S. and Leppakoski, E. (1999) *Initial Risk Assessment of Alien Species in Nordic Coastal Waters*. Stationary Office.

Hale, L.Z. (1993) Why coastal zone management and economic development are complimentary and mutually beneficial. In *Are Coastal Zone Management and Economic Development Sustainable in Sri Lanka*, eds. A.T. White and M. Wijeratne. Proceedings of a seminar, Colombo, Sri Lanka.

Hamer, M. (1999) Solitary Killers. *New Scientist* (28 August), 18–19.

Hanson, A.J. (1998) Sustainable development of the oceans. *Ocean and Coastal Management* **39**, 167–177.

Hansom, J.D. and Kirk, R.L. (1991) Change on the coast: the need for management. In *Aspects of Environmental Change*, eds. T.T.R. Johston and J.R. Flenley. Massey University Press, Palmerstone, pp. 49–62.

Hansom, J.D. (1988) *Coasts*. Cambridge University Press, Cambridge.

Hansom, J.D. (1995) Managing the Scottish Coast. *Scottish Geographical Magazine* **111** (3), 190–192.

Hansom, J.D. and McGlashan, D.J. (1998) Impacts of Bank Protection on Loch Lomond. Scottish Natural Heritage Research, Survey and Monitoring Report.

Hettiarchchi, S.S.L. (2000) Planning and Implementation of Coastal Zone Management in Sri Lanka. In *Coastal Management: Integrating Science, Engineering and Management*, ed. C. Fleming. Thomas Telford, London.

Jewell, S., Roberts, H. and McInnes, R. (2000) Does coastal management require a European Directive? The advantages and disadvantages of a non-statutory approach. In *Coastal Management: Integrating Science, Engineering and Management*, ed. C. Fleming. Thomas Telford, London.

John, S.A. and Leafe, R.N. (2000) Coping with dynamic change on the coast - do we have the right regulatory system? In *Coastal Management: Integrating Science, Engineering and Management*, ed. C. Fleming. Thomas Telford, London.

Kay, R. and Alder, J. (1999) *Coastal Planning and Management*. E. and F.N. Spon, London.

Kenchington, R.A. and Crawford, D. (1993) On the meaning of integration of coastal zone management. *Ocean and Coastal Management* **21**, 109–127.

Kolle, L., Melhus, Ø., Bremnes, K. and Fiskaa, G. (1991) Emissions from inland coastal shipping and potential for improvement. In *Freight Transport and the Environment*, eds. M. Kroon, R. Dmit and J. van Han. Elsevier Science, Amsterdam, pp. 163–173.

Kraus, N.C. (1988) The effects of seawalls on the beach: An extended literature review. In Kraus and Pilkey (eds.). *Journal of Coastal Research Special Issue* **4**, 1–28.

Kraus, N.C. and MacDougal, W.G. (1996) The effects of sea walls on the beach: Part I An updated literature review. *Journal of Coastal Research*, **12**, 691–701.

Kraus, N.C. and Pilkey, O.H. (eds.) (1988) The effects of sea walls on the beach, *Journal of Coastal Research Special Issue* **4**.

Lee, E.M. (1993) The political ecology of coastal planning and management in England and Wales: policy responses to the implications of sea level rise. *Geographical Journal* **159**, 168–178.

MacDougal, W.G., Kraus, N.C. and Ajiwibwo, H. (1996) The effects of sea walls on the beach: Part II, Numerical modelling of SUPERTANK sea wall tests. *Journal of Coastal Research* **12**, 702–713.

Malafant, K. and Radke, S. (1995) The terabyte problem in environmental databases. In *Proceedings of the PACON conference, Townsville, Australia*, eds. O. Ballwood, H. Choat and N. Saxena. As cited in: Kay and Alder (1999).

Mathews (1934) *Coastal erosion and protection*. Charles Griffin and Co. Ltd., London.

Maul, G.A. (ed.) 1996. *Small Islands: Marine Science and Sustainable Development*. Coastal and Estuarine Studies 51, American Geophysical Union, Washington DC.

McGlashan, D.J. and Fisher, G.R. (2000) The legal and geomorphic impacts of engineering decisions on integrated coastal management. In *Coastal Management: Integrating Science, Engineering and Management*, ed. C. Fleming. Thomas Telford, London.

Motyka, J.M. and Brampton, A.H. (1993) Coastal Management: Mapping of littoral cells. HR Wallingford, Report SR 328.

Nichols, K. (1999) Coming to terms with "Integrated Coastal Management": Problems of meaning and method in a new arena of resource regulation. *Professional Geographer* **51**(3), 388–399.

NOAA (1998) Coastal Stewardship: Towards The New Millennium, 1996–1997. The Biennial Report to Congress on Administration of The Coastal Zone Management Act. National Oceanic and Atmospheric Administration and the US Department of Commerce, April, 1999.

O'Riordan, T. and Vellinga, P. (1993) Integrated Coastal Zone Management: the next steps. In *World Coast '93*, eds. P. Beukenkamp, P. Gunther and R. Klien. National Institute for coastal and Estuarine Management, Coastal Zone Management Centre, Noordwijk, The Netherlands, pp. 406–413.

Pender, G., Meadows, P.S. and Tait, J. (1994) Biological impact on sediment processes in the coastal zone. *Proceedings of the Institution of Civil Engineers: Water, Maritime and Energy* **106**, March, pp. 53–60.

Pethick, J. (unpublished) Managed realignment: Coastal wetland restoration policy and practice in the UK. Presentation to the seminar Flood risk and managed retreat in the Forth Estuary, Falkirk Lesser Town Hall, 20th October, 1999.

SEPA (1999) Firth of Clyde Flood Warning System Goes Live. Scottish Environment Protection Agency Press Release, 15 October, 1999 (ref 29/99).

Smith, K.S. (1999) Marine protected areas for marine turtle conservation: a case study of Zakynthos. Unpublished Masters Thesis, Graduate School of Environmental Studies, University of Strathclyde.

Tait, J.F. and Griggs, G.B. (1990) Beach response to the presence of a sea wall: A comparison of field observations. *Shore and Beach* **59**(4), 11–28.

UNCED (1992) Agenda 21—United Nations Conference on Environmental and Development: Outcomes of the conference, Rio de Janeiro, Brazil, 3–14 June.

UNEP (1999) Global Environment Outlook 2000. United Nations Environment Programme, Nairobi, Kenya.

World Commission on Environment and Development (1987) *Our Common Future*. Oxford University Press, Oxford.

Young, M.D. (1992) *Sustainable investment and resource use: equity, environmental integrity and economic efficiency*, UNESCO, Paris.


## THE AUTHOR


**Derek J. McGlashan**

*Graduate School of Environmental Studies,*
*Wolfson Centre, 106 Rottenrow East,*
*University of Strathclyde,*
*Glasgow, G4 0NW, U.K.*

Chapter 133

# SUSTAINABILITY OF HUMAN ACTIVITIES ON MARINE ECOSYSTEMS

Paul Johnston, David Santillo, Julie Ashton and Ruth Stringer

Human activity currently exerts multiple impacts on oceans and coastal seas which may be severe on a local or regional basis and which are often also detectable at a global level. There is little doubt that such impacts will increase in the future. This chapter provides an overview of human interactions with marine ecosystems from the perspective of sustainability, the latter requiring that:

1. substances from the earth's crust must not systematically increase in the ecosphere;
2. substances produced by society must not systematically increase in the ecosphere;
3. the physical basis for productivity and diversity of nature must not be systematically diminished;
4. there should be a fair and efficient use of resources with respect to meeting human need.

The case is presented that man's influence on the seas at the turn of the millennium is currently unsustainable. For example, the release of metals and persistent organic pollutants has resulted in a systematic increase of contaminants in the marine environment. Bioaccumulation is of particular concern. In addition to violating the first two criteria of sustainability, the toxicity of many of these substances also threatens the third. Further, the exploitation of theoretically "renewable" resources, particularly fisheries, impacts biological productivity and diversity. Overfishing is a worldwide problem which threatens the sustainability of both target and non-target species. The third criterion (and from many perspectives, the fourth) is thus violated.

To address currently unsustainable practices, a fundamental shift is required in the way in which we view and interact with the marine environment. A precautionary approach, in which systematic environmental degradation is avoided in advance wherever possible, will be essential if sustainability is to be achieved. To continue with "business as usual" will undoubtedly become increasingly unsustainable, and will run contrary to our responsibilities to protect future generations.

*Seas at The Millennium: An Environmental Evaluation (Edited by C. Sheppard)*

## INTRODUCTION: HUMAN DEMANDS ON MARINE ECOSYSTEMS

Human activities exert a diverse range of impacts, both direct and indirect, on marine ecosystems. Given projected increases in population, particularly through expansion of coastal urbanisation, these impacts may be expected to increase further in the future. The sea represents a vital resource for a large proportion of the global population, and has commonly been viewed as providing an array of "ecosystem services". This has led to a systematic contamination of the marine environment and the over-exploitation of its ecological "capital". In turn, these trends threaten not only the continued derivation of human benefit from the oceans, but also the very basis of biological productivity and diversity itself.

Most human exploitation of the oceans concentrates on coastal and continental shelf resources. Nevertheless, the footprint of human activities, especially the distribution of many anthropogenic contaminants and influences on global carbon and energy cycles, extends well beyond these regions.

## THE GLOBAL COMMONS

Costanza et al. (1997) identified five major concerns in relation to human impacts on the oceans: land-based sources of pollution, ocean disposal of wastes (and spills), overfishing, destruction of coastal ecosystems and climate change. Despite long-standing recognition of the existence of these problems, and the development of numerous progressive international initiatives aimed at addressing them, they remain substantial threats on a global scale at the turn of the millennium. In addition to the potential loss of biodiversity, and the moral and ethical issues which such losses raise, continued perturbation of marine systems threatens also to violate the fundamental principle of transgenerational responsibility enshrined within the Rio Declaration (UNCED, 1992):

> "The right to development must be fulfilled so as to equitably meet developmental and environmental needs of present and future generations".

How then to assess the sustainability of current and projected future practices, such that appropriate and effective regulatory mechanisms can be implemented? Attempts have been made to approach this problem by assigning monetary values to marine ecosystems. Such approaches assume that the economic benefits derived from resource exploitation may be balanced against the costs of the environmental damage that may result, such that the potential consequences of different management options may in some way be compared quantitatively. They do not give an estimate of the costs associated with replacing an "ecosystem service" by another, technological means, particularly as many of the "services" are irreplaceable (Cairns and Dickson, 1995).

Beyond demonstrating the nominally enormous value of the oceans within the global economy, valuation approaches suffer from a number of fundamental limitations, not least of which are the high (and, in many cases, irreducible) uncertainties surrounding a valuation of such complex entities. Moreover, such attempts at economic balancing are unlikely ever to lead to the development of measures which will ensure their sustainability.

As an alternative, Cairns (1997) presents four fundamental criteria against which sustainability may be gauged:

1. Substances from the earth's crust must not systematically increase in the ecosphere.
2. Substances produced by society must not systematically increase in the ecosphere.
3. The physical basis for productivity and diversity of nature must not be systematically diminished.
4. Fair and efficient use of resources with respect to meeting human needs.

In the following, the nature of various human interactions with marine ecosystems is examined to provide an evaluation of the extent to which the above criteria are currently being met or violated. The intention is to provide a very brief overview of threats to the oceans from human activities on a global basis, with a focus on intentional and unintentional introduction of pollutants, exploitation of living resources (including coastal habitats) and the consequences of global climatic change (Johnston et al., 1998).

## INTRODUCTION OF HAZARDOUS SUBSTANCES AND RADIOISOTOPES TO THE MARINE ENVIRONMENT

Many human activities have the potential to redistribute naturally occurring substances in the biosphere and to release synthetic substances. In relation to the criteria for sustainability outlined above, the greatest concerns relate to substances which, when released, tend to persist long enough to build up in the environment and which may have significant deleterious effects. These chemicals, including several metals, radioisotopes and a diversity of persistent organic pollutants (POPs), also have the greatest potential for long-range dispersal.

By far the greatest source of contaminants entering the seas from human activities arise from land-based sources (approx. 77%), either through direct discharges and releases to surface waters (44%) or indirect inputs, e.g. via the atmosphere (33%). Inputs of pollutants from dumping of wastes at sea account for a further 10%, with spillages and discharges from shipping and the offshore oil and gas sector accounting for approximately 12% and 1% respectively (ICS, 1997) (Fig. 1). Some minor revision of these data may be necessary in the light of the prohibition on the dumping of industrial wastes at sea (since January 1996) under the London Convention, and trends in releases from other sources.

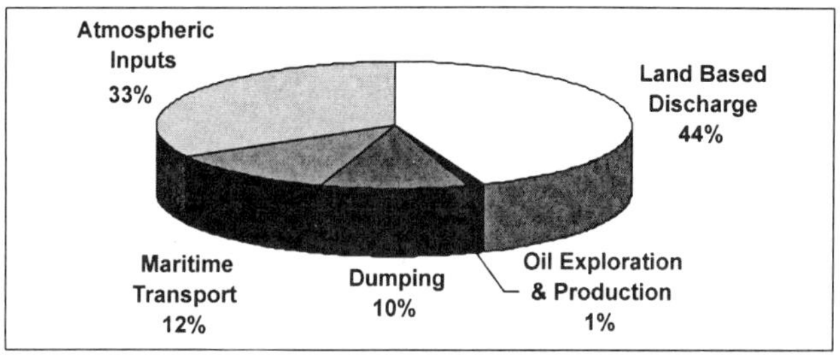


Fig. 1. Proportions of pollutants introduced to the marine environment from various sources (ICS, 1997).

## Land-based Sources

### *Metals*

Metal fluxes from human activities now equal or exceed inputs from natural sources in many cases (Nriagu, 1992). Estimates of metals reaching the ocean from the atmosphere, the primary transport and distribution medium, show a strong regional variation. Ocean waters closest to intense industrial activity, unsurprisingly, receive the greatest inputs. Currently, for example, the average lead concentration in the mixed surface waters of the North Atlantic is approximately 26 times higher than in South Pacific surface waters.

Emissions of some metals are estimated to have trebled or quadrupled between 1900 and 1990, and this increase is detectable remote areas. As an example, recently deposited coral skeletons in the Pacific contain 15 times more lead than those laid down one hundred years ago. These general trends seem set to continue. Metals are central to current global economies and controls upon emissions in industrialised countries are likely to be offset by those made as industry develops in regions such as the African continent.

### *Synthetic Organic Chemicals*

Synthetic chemicals derived from combustion or chemical processing of petroleum began to increase in the environment with the advent of widespread industrialisation. A key development was the emergence of a chemicals market based upon chlorine chemistry, which led to a range of synthetic chemicals being introduced from the 1920s onwards. Woodhead et al. (1999) report a correlation between levels of polycyclic aromatic hydrocarbons (PAHs) in sediments and degree of catchment industrialisation for estuaries around the UK. As with emitted metals, however, the footprint of the organic chemical industry can be traced globally. Residues of chlorinated pesticides and other POPs can be found in the tissues of marine organisms from all parts of the world. Appreciable quantities are even found in cod-liver oil and other fish oils sold as dietary supplements (Jacobs et al., 1998), despite the relatively remote sources of many of the fish stocks from which the oils are derived.

Many persistent chemicals emitted in temperate and tropical regions are transported in the atmosphere towards colder regions of the globe (Wania and Mackay, 1993). Here, they "condense" out (Fig. 2). In summarising their extensive review prepared under the Arctic Monitoring and Assessment Programme (AMAP), de March et al. (1998) note that "All POPs considered under the AMAP monitoring programme have been found in air, snow, water, sediments and/or biota in the Arctic. In some cases, a number of Arctic species have POPs levels high enough to cause effects". Understanding of the transport processes involved, however, remains very limited (Fig. 2).

Marine mammals are particularly susceptible to accumulation of these chemicals as a result of their large blubber reserves and their apparent limited capacity to metabolise such chemicals. Burdens of PCBs appear to be especially high in polar bears and other Arctic top predators (Fig. 3), with measured tissue concentrations greater than those demonstrated to cause adverse effects in other species (de March et al., 1998). Recent observations have even been made of hermaphroditism in Svalbard polar bears. Indeed, elevated tissue levels of organochlorine and other persistent contaminants may play a role in a range of reproductive and developmental effects, either at individual or at population level (Allsopp et al., 1999). High organochlorine burdens are thought to be slowing the recovery of Baltic populations of grey and common seals, depleted through hunting and overfishing. There is also increasing evidence that immune function in marine mammals can be adversely effected by exposure to such contaminants, reducing disease resistance and probably contributing to disease-related mass mortalities.

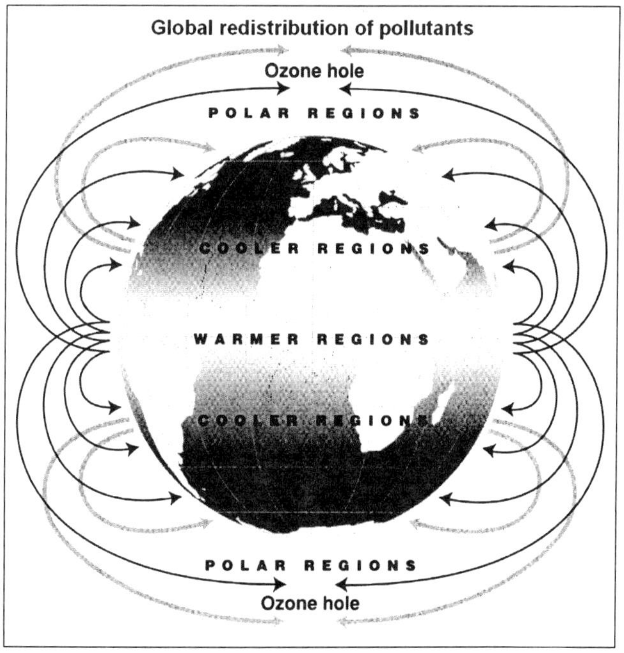


Fig. 2. Representation of global atmospheric circulation patterns and their role in poleward transport of persistent organic pollutants. Chemicals which volatilise in regions of relatively high ambient temperatures are transported to cooler regions where they condense (Wania and McKay, 1993).

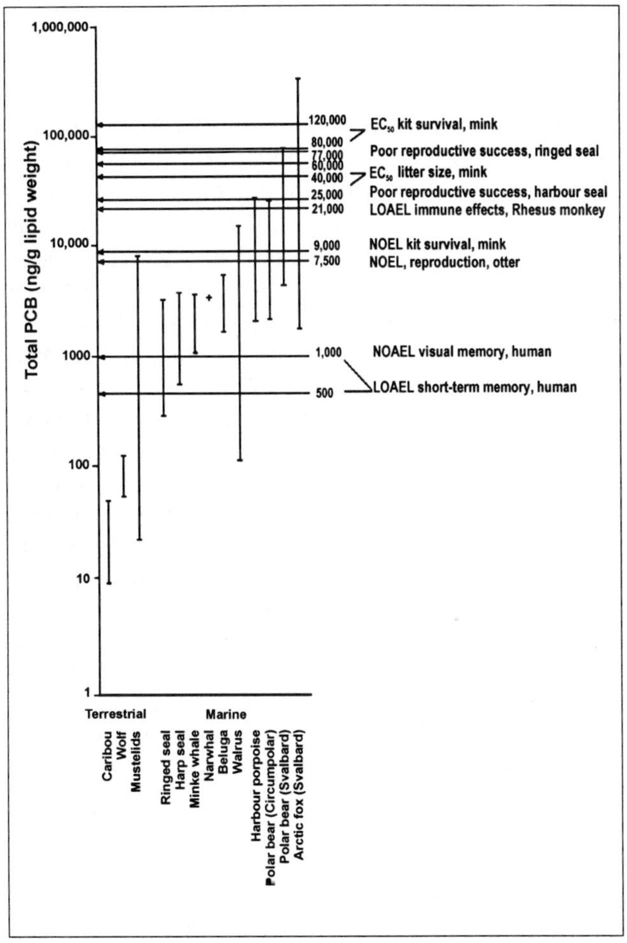


Fig. 3. Concentration ranges reported for PCBs in a range of Arctic species, compared to relevant effect or no-effect levels from laboratory studies. $EC_{50}$, concentration yielding an effect in 50% of test individuals; LOAEL, lowest observed adverse effect level; NOEL, no observed effect level; NOAEL, no observed adverse effect level (de March et al., 1998)

Environmental contamination with such chemicals represents a long-term problem. There are strong indications that, despite phase-out of production, environmental levels of certain POPs are declining only very slowly. Controls on the production and release to the environment of PCBs, banned in many countries by the late 1970s, have done little more than halt their upward trend in biota in Arctic and other marine systems (e.g. de March et al., 1998). This persistence may be enhanced by a combination of continued release (e.g. from waste dumps, obsolete equipment, continued production as by-products), and ongoing redistribution and equilibration of PCBs already in the biosphere. For many POPs, production and use are a continuing problem in lower latitudes and newly industrialising countries.

While the contamination of remote regions with organochlorines has been documented for some years, evidence for widespread contamination with similar brominated chemicals—many still in widespread production and use—is now beginning to emerge. Early reports of residues of polybrominated diphenyl ethers (PBDEs, used as fire retardants in a wide range of consumer goods) in Arctic seals have been followed more recently by confirmation of their presence in deep-sea-feeding sperm whales and pilot whales taken off the Faroe Islands (e.g. Lindstrom et al., 1999). Like many of the organochlorines, these brominated compounds are passed from mother to offspring. Time trends are not yet available for these compounds in the marine environment, but systematic increases in concentrations have been reported in human tissues (Allsopp et al., 1999).

For many organic chemicals discharged as by-products of industrial processes, toxicological data are sparse or lacking. Many others, often discharged as components of complex mixtures, simply cannot be reliably identified, rendering more complete chemical and toxicological evaluation impossible. For example, a large proportion of the organochlorine chemicals isolated from the tissues of Arctic beluga have yet to be identified (Kiceniuk et al., 1997).

### *Radioactive Elements*

Radioactive elements are also systematically increasing in the ecosphere, although the radioactive inventory in the oceans is dominated by natural radionuclides. As with synthetic chemicals, however, many artificial radionuclides have no natural counterparts and extremely long lifetimes.

Facilities in the UK and France are currently the most important point sources of radioactivity into the oceans. Radioactivity is transported in ocean currents and can be traced into remote Arctic regions (Strand, 1998). Other radionuclides have become localised in the sediments close to the plants, acting as an unpredictable source of particularly long-lived isotopes into adjacent ecosystems. The nuclear industry generally continues to emit artificial radionuclides, while, in addition, certain mineral-processing industries emit substantial quantities of natural radioactive elements.

### *Regulation of Land-based Sources*

The significance of land-based sources of marine pollution has long been recognised. Numerous initiatives exist to encourage or enforce the reduction of inputs of hazardous substances and radioactive elements to the marine environment, although types and levels of regulation and enforcement vary greatly from region to region.

Recognition of the scale of the problems presented by certain metals and POPs, the impossibility of managing such chemicals once they are released, and the need, therefore, for a more precautionary approach, led to the adoption by ministers of North Sea States in 1995 of the target of cessation of all discharges, emissions and losses of hazardous substances within one generation (i.e. by 2020). More recently, the same target was adopted by all 15 contracting parties to the OSPAR Convention (OSPAR 1998).

The setting of these progressive targets is clearly an enormous step forward. OSPAR has already identified a list of 15 hazardous substances or groups of substances which are recognised as priorities (including *inter alia* PCBs, dioxins and brominated flame retardants), and is currently developing a mechanism to update this list further in 2000, with the intention of addressing all hazardous substances in some way by 2020. Work towards the effective implementation of the strategy is only now beginning. While it is clear that it will be no easy task, in some cases requiring a fundamental re-evaluation of the way in which we produce, use and dispose of chemicals, it is, nevertheless, an essential one if the protection of the North East Atlantic region is to be assured.

Other regional seas programmes have made perhaps less progress to date. The Barcelona Convention 1976, aiming to protect the Mediterranean Sea, has a Protocol to address land-based sources (yet to enter into force) and an interim Strategic Action Plan, which sets targets for the progressive reduction or elimination of releases of a number of key heavy metal and organic pollutants. Nevertheless, the economic and developmental situation in most Mediterranean states is such that acute, point-source related pollution problems (often attracting "end-of-pipe" solutions) remain the priority.

Regional programmes for the protection of the marine environment can, if effectively implemented, have substantial benefits. It is to be hoped that other regional seas programmes will adopt a similar progressive approach to that of OSPAR. Nevertheless, the potential presented by many key contaminants for transboundary pollution ultimately demands global action to address sources of hazardous chemicals to the environment. The developing UNEP global "POPs" Convention, designed to address 12 priority persistent organic pollutants (all chlorinated), is an important start to this process. Progress within the International Negotiating Committees is scheduled to lead to a signing of the Convention in 2001, with entry into force perhaps in 2004. The effectiveness of the Convention in reducing overall inputs of such contaminants to the environment remains, of course, to be seen.

Notwithstanding the undoubted progress being made, it is clear that current inputs of hazardous substances and radioisotopes from land-based sources violate the first two principles of sustainability outlined by Cairns (1997), in that they result in systematic increases in contaminants in the environment. Furthermore, the potential or observation of adverse impacts indicates that, in turn, the third principle is also violated.

## Operational Discharges from the Offshore Oil and Gas Sector

To date, the use and release of chemicals by the offshore oil and gas industry has been relatively poorly controlled and documented, other than for a very limited range of parameters. Although some of the agents used in the drilling or workover of wells and the production of oil and gas are chemically inert, many are hazardous.

There is increasing evidence, for example, that the discharge of contaminated drill cuttings has resulted in significant impacts on the chemistry and biology of the benthos, even at distances of several km from the well site. While major impacts may generally be restricted to a relatively limited area surrounding individual platforms, subtle but significant alterations to benthic community structure have been detected within a radius of 2–6 km from disused oil platforms in the North Sea (Olsgard and Gray, 1995).

Concerns over the toxicity of oil-based drilling fluids led to action within the former Paris Convention (now subsumed under OSPAR) to restrict the use and disposal of these muds. Nevertheless, evidence is now increasing that replacement synthetic muds, including those with olefin or ester bases, also exert substantial impacts on marine benthic communities. Some components resist breakdown and accumulate in tissues. Others which do degrade more easily can cause de-oxygenation of the sediments, and loss of biodiversity, as a result. Even water-based muds, though relatively chemically inert, have been demonstrated to have significant biological impacts in some systems (Raimondi et al., 1997).

Similarly, the effects of oil and other chemicals in produced water (derived from water in the oil or gas formation) may also be significant. Input volumes are generally high (estimated at 340 million $m^3$ to the North Sea alone in 1998) and discharges represent a direct introduction of contaminants to the water column. Approximately 87% of discharges in this region occur in the UK sector. Despite gradual improvements in oil/water separation, oil discharges in produced water have increased substantially over the last 10–15 years as a result of the increases in volumes of water recovered as fields are exploited (Fig. 4).

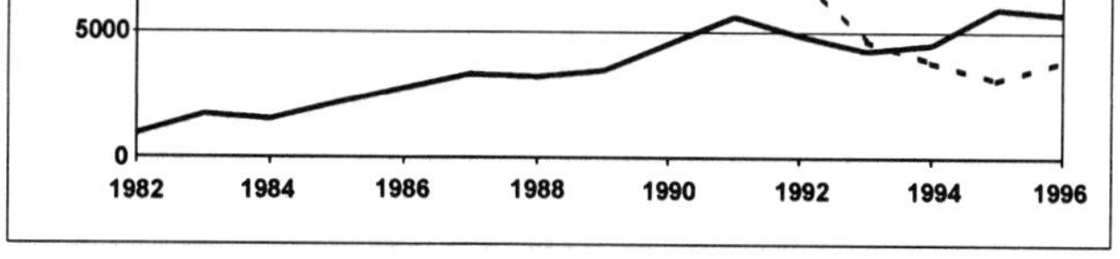


Fig. 4. Discharges of oil from offshore oil and gas platforms in the UK sector. The decline in oil discharged on drill cuttings results largely from restrictions on the use of oil-based drilling fluids under OSPAR. The increase in produced water-related discharges reflects the overall increase in volume of produced water discharged in this sector since the early 1980s (EA, 1998).

While the greatest impacts of contaminants in produced water might be expected in the immediate vicinity of the platform, far-field effects cannot be ruled out. Recent evidence after the Exxon Valdez spill has indicated that some aromatic compounds, also present in produced water, can exert adverse biological effects at levels considerably lower (low ppb) than previously considered harmful (Carls et al., 1999).

Produced water discharges can also introduce substantial quantities of ammonia, hydrogen sulphide, organic acids and alkalis and particulates (Sauer et al., 1997). The impacts of such discharges, particularly to shallow or enclosed waters, could be severe. Discharges of radioisotopes, occurring at naturally enhanced levels in production wells, may also be substantial in some areas. Nevertheless, the build up of radioactive elements as "low-specific activity scale" on the inside of pipe-work, presenting significant waste problems during installation decommissioning, may be a greater concern.

Many other chemical preparations are used offshore, including emulsion breakers, corrosion and scale inhibitors, flocculants and bactericides (Sauer et al., 1997). Information on the quantities of such chemicals used and discharged is extremely limited for all regions. Indeed, the monitoring and control of operational discharges from the offshore sector as a whole remains very limited. To some degree, controls are stronger within the OSPAR region than in other maritime areas (e.g. the "one generation" cessation target applies equally to offshore operations). Existing OSPAR measures should become further strengthened through the implementation of an integrated strategy for the offshore sector, adopted by OSPAR in 1999. Nevertheless, it is likely that measures will continue to focus on a very limited range of contaminants for some time to come. At the same time, operational discharges in other regions, particularly in the developing oil and gas fields of Asia, may be expected to remain relatively poorly controlled.

Operations of the offshore oil and gas sector therefore violate (or threaten to violate) the first two criteria for sustainability and possibly also the third. The continued exploration and exploitation of oil and gas raises further concerns regarding sustainability in general.

## Inputs of Contaminants from Shipping

Currently, there are an estimated 40,000 merchant ships with a gross registered tonnage of about 520 million tonnes (ICS, 1997). Over 50% of the diverse materials carried may be hazardous or harmful from the point of view of safety or environmental protection. Apart from the losses and spills of cargoes, accidents, routine use and discharge of chemicals can also result in significant inputs of hazardous materials. The precise contributions of shipping to pollution is difficult to estimate, with very few data in particular relating to atmospheric emissions. The best data set relates to inputs of oil, for which operational discharges, not spills, account for 80% of inputs from shipping, and approximately 20% of all inputs of oil to the marine environment (ICS, 1997).

Operational discharges of fuel-oil sludge and oil from machinery spaces, are regulated under MARPOL 73/78 (under the auspices of IMO). The effectiveness of MARPOL is illustrated by the reported fall in global discharges from 1.47 to 0.57 million tonnes between 1981 and 1989. Nevertheless, substantial problems remain, including illegal discharges. Large slicks (100 tonnes oil and greater) of unknown origin, and extensive beach contamination are still commonplace in many regions subject to heavy shipping traffic, including the North Sea, Red Sea, Arabian Gulf, Caribbean, and the coastal waters of Indonesia and the Philippines. There is evidence of a reduction in the frequency of seabird oiling in certain regions—e.g. Wadden Sea (Camphuysen, 1998)—but mortalities are undoubtedly still significant.

Loss of non-oil chemical cargoes and packaged goods also remains significant. Carriage and loss of pesticides has perhaps received more attention than most other cargoes (see Johnston et al., 1997). Marine litter and debris represent a further widespread and persistent shipping-related problem. Concerns have also increased over introductions of alien species and pathogens through the discharge of ballast water.

Other highly significant inputs from shipping include emissions to air and the leaching of antifouling agents during normal operations. Despite generating among the lowest atmospheric emissions per unit mass of goods carried by any form of transport, emissions of nitrogen and sulphur oxides from shipping are among the highest per tonne of fuel used. This results principally from the use of low-quality fuels (often heavy oils with high sulphur and heavy metal contents) (Corbett and Fishbeck, 1997). As a result, shipping contributes an estimated 14% of total anthropogenic nitrogen emissions from fossil fuel combustion, 16% of sulphur and 1.2% of $CO_2$.

Inputs of antifouling agents from shipping traffic are diffuse and difficult to quantify. TBT (tributyl tin), the most widely used agent, has long been known to be acutely toxic to a wide range of marine organisms. Development of imposex in prosobranch molluscs, thought to be mediated through interference with testosterone production (Mathiessen and Gibbs, 1998), is perhaps the best documented, but by no means only, effect.

Current inputs of TBT arise principally from use on larger vessels, following earlier national bans on its use on craft smaller than 25 m in length. While such restrictions have led to a decline in inputs of TBT, decline in coastal sediment concentrations and recovery of mollusc populations can be very slow (e.g. Minchin et al., 1997 ). At the same time, inputs continue from use on larger vessels. Organotin residues are detectable levels in fish, birds and marine mammals, including in specimens feeding in remote offshore regions (Iwata et al., 1997). Whereas there is some evidence for atmospheric transport, the global nature of the shipping industry may be the key factor behind widespread organotin distribution.

Increasing concerns over TBT contamination and effects have culminated in an IMO resolution to ban all applications by 2003 and all uses by 2008, under the auspices of its Marine Environmental Protection Committee (MEPC). While this is clearly a significant step forward, several problems remain to be addressed. Firstly, it is likely that some flag states and sectors of industry will strongly oppose such measures. Secondly, even if the ban can be agreed and implemented, the persistence of TBT will ensure it remains a problem for many years to come. Moreover, some of the proposed alternatives to TBT also rely on the addition of highly toxic biocides, some of which (e.g. Irgarol) are increasingly being reported to be detectable in coastal marine systems and which may be toxic to non-target organisms.

## Dumping of Wastes at Sea

In addition to inputs from direct or indirect discharges and emissions, the deliberate dumping of wastes at sea has accounted for substantial inputs of organic matter and contaminants for many decades. Regulation and practice have changed markedly over the past 20 years, however, largely as a result of the developments within the London Convention (1972), a global instrument which regulates sea dumping. During this time, the LC has evolved from a body largely responsible for permitting dumping operations (originally the London Dumping Convention) to one within which the dumping of wastes is severely restricted, and must be justified on the basis that no other viable options exist.

In early 1990s, the dumping of all radioactive material, and the dumping and incineration of industrial wastes at sea, were finally prohibited. These and other precautionary developments in the Convention are enshrined within its 1996 Protocol, which specifies a very limited range of materials which may be considered for dumping, providing they are well characterised and contain contaminant levels below national trigger levels. Moreover, detailed consideration must be given to the potential for waste minimisation and prevention. The Protocol has yet to enter into force but will undoubtedly represent a further step forward in the protection of the marine environment.

Despite progress on the regulation of sea dumping, and the substantial degree of protection which the Convention affords, a number of wastes may still be permitted for dumping under the Convention, including obsolete vessels, offshore platforms, sewage sludge and dredge materials (although some regions, e.g. OSPAR, EU, operate more restrictive regimes). In relation to overall quantities of contaminants introduced, two key concerns are the dumping of harbour dredge spoils and dumping operations (both reported and unreported) in breach of the Convention.

Table 1

Total quantities of dredged material permitted for dumping under the London Convention in 1996 (source: IMO, 1999)

| Country | Quantity licensed (mt) | Location of dump sites |
|---|---|---|
| Australia | 5,596,000 | SW Pacific, Indian Ocean and Tasman Sea |
| Belgium | 5,520,000 | North Sea |
| Canada | 5,484,765 | NW Atlantic, NE Pacific and Quebec area |
| Chile | 136,135 | SE Pacific Ocean |
| Denmark | 616,000* | North Sea and Baltic Sea |
| Germany | 19,123,000 | North Sea |
| Hong Kong (form. UK) | 47,998,957 | East Asian Seas |
| Iceland | 220,698 | Atlantic Ocean |
| Ireland | 1,372,734 | Irish Sea |
| Japan | 11,872,776 | East Asian Seas |
| Netherlands | 8,016,381 | North Sea |
| New Zealand | 2,488,379 | SW Pacific Ocean |
| Norway | 399,716 | Atlantic Ocean and North Sea |
| Poland | 331,000 | Baltic Sea |
| Portugal | 7,991,650 | Atlantic Ocean |
| Russian Federation | 444,600 | Baltic Sea |
| South Africa | 4,296,784 | Indian Ocean |
| Spain | 2,055,148 | Atlantic Ocean |
| Sweden | 3,308,608 | North Sea and Baltic Sea |
| UK | 51,251,367 | Atlantic, North Sea, English Channel and Irish Sea |
| USA | 46,303,955 | NW Atlantic, NE Pacific and Gulf of Mexico |

Dredge materials comprise the vast bulk of waste dumped by Contracting Parties to the Convention (Table 1); for example, more than 51 million tonnes were permitted for dumping by the UK alone in 1996 (IMO, 1999). Concerns relate, however, less directly to the quantity of material than to the quantities of contaminants which dredge spoils contain. Maintenance dredging is unavoidably concentrated in the busiest and potentially most contaminated ports and harbours, often located in estuaries of highly industrialised and/or urbanised catchments. Much of the contaminant load may arise from upstream sources, largely beyond the direct control of the Convention. Measures to achieve tighter controls on such sources, and even cessation of discharges, essential if the objective of the London Convention is to be further pursued, will necessitate concerted action between international fora addressing different aspects of environmental protection and waste prevention.

Secondly, reporting of dumping activities remains very incomplete, such that compliance even with the 1972 Convention is extremely difficult to monitor. In other cases, limited reporting has highlighted some important differences in interpretation of the prohibition on the dumping of industrial waste. These difficult and sometimes sensitive issues remain to be resolved.

A third and emerging issue is the recently renewed interest in the disposal of $CO_2$ at sea as a climate change mitigation strategy. Besides the enormous uncertainties surrounding such proposals, regarding their likely effectiveness and the potential for adverse impacts at local, regional and global level, the dumping of fossil fuel-derived $CO_2$ is prohibited under the terms of the London Convention (Johnston et al., 1999). Notwithstanding these seemingly insurmountable limitations, sea dumping (or enhanced sequestration) of $CO_2$ is increasingly being portrayed not simply as a mitigation strategy, but as a way to permit continued exploitation of fossil fuel reserves. Such approaches may compound already substantial threats of adverse changes to global climate.

## INTRODUCTION OF ELEVATED NUTRIENT LOADS TO THE MARINE ENVIRONMENT

While many synthetic chemicals have no natural analogues, nutrients (e.g. nitrogen, phosphorus) are essential for marine ecosystem function. Nevertheless, dysfunction can result when inputs become excessive or lead to a shift in the ratio of nutrient supply. As in the case of hazardous substances, estuarine and coastal waters are generally the most severely affected by anthropogenic nutrient loading, although the impacts of eutrophication are by no means restricted to these waters.

Global estimates of anthropogenic contributions to river-borne nutrient loads vary widely, although there is little disagreement that it is substantial. GESAMP concluded in 1990 that "globally, present inputs of nutrients from rivers due to man's activities are at least as great as those from natural processes" and have led to "clearly detectable and sustained increases in nutrient concentrations in the [receiving] waters". Sources to surface freshwaters include discharges of domestic sewage, run-off of inorganic fertilisers and inputs from certain industrial processes. As part of a review of inputs from 14 catchments to the North Atlantic, Howarth et al. (1996) estimated that riverine nitrogen flux had increased by between 2 and 20 fold above background in urbanised, agricultural and industrialised catchments. Other reports note increases of 2–4 fold in nitrogen and phosphorus concentrations, respectively, between the late 1960s and early 1980s in both the Northern Adriatic Sea and the Gulf of Mexico.

Estimating influences on offshore nutrient budgets is more difficult and is fraught with uncertainties, particularly in relation to atmospheric deposition, a primary pathway for the introduction of elevated nitrogen inputs (arising from both stationary and transport-related sources) to more remote regions.

Although, by definition, eutrophication involves an increase in primary productivity, it does not necessarily follow that increases will also be realised at higher trophic levels. Impacts of eutrophication may include quantitative and qualitative shifts in ecosystem composition, and possibly development of toxic blooms. Widespread anoxia in the water column and benthos (with potentially serious consequences for benthic communities), has been documented in several seas, such as the Baltic, Black Sea and Gulf of Mexico (Lohrenz et al., 1997).

Ecosystem response to increase nutrient inputs in any one location will likely remain extremely difficult to predict. Changes could, however, be long-lasting, involve fundamental alterations to pathways of carbon and energy transfer over extensive areas, and in terms of the simple criteria for sustainability, anthropogenic inputs of nutrients may be seen directly to violate Criterion 2 (particularly, though not exclusively, in coastal waters). Where inputs lead to eutrophication, they threaten to violate the third criterion also. Understanding of the consequences of eutrophication for global gas exchange and climatic trends remains poor.

## EXPLOITATION OF ECOLOGICAL RESOURCES

Impacts on biodiversity from the exploitation of fish populations and other ecological resources or habitat destruction/modification are more direct. The extent to which resource exploitation and impacts of pollutants exert synergistic effects is poorly understood, although the potential clearly exists. In an increasing number of regions, it is clear that over-exploitation and habitat destruction have already had devastating effects on both target and non-target populations of marine organisms, and on ecosystem structure and function as a whole.

### Marine Capture Fisheries

Marine fishing generates around 1% of the visible global economy and supports the livelihoods of some 200 million people. Between 1945 and 1992 catches rose from 17.7 to 85 million tonnes year$^{-1}$, with discarded by-catch adding a further 27 million tonnes (Table 2).

Fisheries rely on relatively few species. Five species groups make up 50% of the global total caught, while 200 major stocks account for 77% of landings. Of these, 35% are regarded as overfished, 25% are fully exploited, and 40% are under development. It is estimated that over 60% of world stocks require urgent, appropriate management, a

Table 2

Total discards from fisheries in different ocean regions and regional seas (source: Alverson et al., 1994)

| Area | Discard weight (mt) |
|---|---|
| Northeast Atlantic | 3,671,346 |
| Northwest Atlantic | 685,949 |
| Southeast Atlantic | 277,730 |
| Southwest Atlantic | 802,884 |
| East Central Atlantic | 594,232 |
| West Central Atlantic | 1,600,897 |
| Mediterranean and Black Sea | 564,613 |
| Northeast Pacific | 924,783 |
| Northwest Pacific | 9,131,752 |
| Southeast Pacific | 2,601,640 |
| Southwest Pacific | 293,394 |
| East Central Pacific | 767,444 |
| West Central Pacific | 2,776,726 |
| East Indian Ocean | 802,189 |
| West Indian Ocean | 1,471,274 |
| Antarctic Atlantic | 35,119 |
| Antarctic Pacific | 109 |
| Antarctic Indian Ocean | 10,018 |
| Total | 27,012,099 |

Table 3

Temporal changes in catch statistics in the Canadian Atlantic cod (*Gadus morhua*) fishery, 1805–1992. Stock collapse (Fig. 5) led to closure of the fishery in 1992 (source Hutchings and Myers, 1994)

| Year or period | Catch (mt) | Comments |
|---|---|---|
| 1805 | 100,000 | |
| 1850 | 150,000 | |
| Late 1800s | >200,000 | |
| 1900–1960 | freq. >250,000 | |
| 1959 | 360,000 | |
| 1968 | 810,000 | Influx of European factory trawlers |
| 1978 | 140,000 | Maximum recorded yield |
| 1988 | 270,000 | Canadian management begins |
| 1991 | 97,000 | Fishery closed |
| 1992 | 0 | |

point underscored by the fact that 8 million tonnes of fish were landed in 1994 from resources defined as depleted or over-exploited (FAO 1997). A significant contributor to overfishing is the overcapacity in the world fleet, estimated at at least 50%. In addition, misdirected government subsidies to the industry encourage continued exploitation at net economic loss.

Intensive fishing activity, leading to overfishing has caused some well publicised collapses of fish stocks, e.g. North Sea herring and mackerel stocks. Currently, all the major roundfish stocks in the North Sea are intensively exploited or over-exploited and the size of stocks and catches is highly dependent on young fish (Serchuk et al., 1996). The Canadian cod stock, a mainstay of the regional economy, collapsed and the fishery was closed in 1992 (Table 3 and Fig. 5). Cod stocks off the eastern United States are at unprecedented low levels and Icelandic stocks are considered to be severely depleted.

These collapses took place against a background of intensive fisheries management supported by sophisticated fish population models. A variety of factors have been involved. Overly simplistic models using poor-quality data, poor understanding of the fish population dynamics, coupled with a failure to heed scientific advice in setting catch limits have all contributed. The significance of by-catch is not incorporated into most fisheries management models. In open ocean fisheries, the picture is similar. For example, the spawning stock biomass of the Southern Bluefin Tuna is currently between 6 and 11% of its level in 1960 (effectively collapsed) and projections suggest that there is a high probability (60%) that the spawning stock biomass will be reduced to zero by the year 2020 if current catch rates are maintained (Klaer et al., 1996).

Insofar as these impacts diminish the production of fish as food and deplete the biological productivity of ecosystems, they clearly violate the third principle of sustainability. But, in focusing upon the single species target populations themselves, relatively little attention has been directed at impacts upon their wider ecosystems. By the early 1980s it had become clear that replacement of long-lived fish species by smaller shorter-lived species had taken place in the Gulf of Thailand, the North Sea and off the west coast of Africa. Ecosystems of the intensively fished Georges Banks, where cod populations collapsed due to over-exploitation, have shown a trend towards an increase in fish of low commercial value, specifically shark and ray species (Fogarty and Murawaski, 1998).

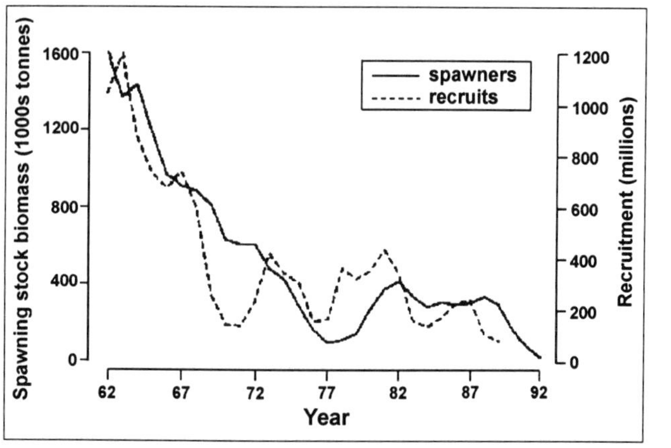


Fig. 5. Changes in recruitment (numbers of 3-year-old cod) and spawning stock biomass (individuals age 7 or over) of northern cod in the Canadian Atlantic fishery (Hutchings and Myers, 1994).

A recent re-analysis of FAO catch statistics (Pauly et al., 1998) documents progressive "fishing down" of food chains as fishing effort responds to depletion of original target stocks. A consistent downward trend in the trophic level occupied by commercially fished species has been identified in global fisheries as predatory demersal fish are superseded by pelagic fish which feed lower in the food chain. The significance of these actual and potential shifts is a subject of debate. The view that these changes result in an ecologically acceptable, though economically less valuable fish community, ignores transgenerational responsibilities and also contrasts with the view that regime shifts may result in the establishment of a different, undesirable but stable ecosystem state.

Added to pressures of overfishing and the by-catch of non-target organisms (including seabirds, turtles and cetaceans), are impacts of physical disturbance by fishing gear. Some areas of the southern North Sea, for example, are swept by beam trawls up to seven times a year causing extensive damage to benthic ecosystems.

Uncertainties in the models used, poor knowledge of multi-species impacts, the use of inappropriate management targets (exemplified by the use of the concept of minimum biologically acceptable level (MBAL) in the North Sea), fishing fleet over-capacity, and misguided government subsidisation of the industry have jointly contributed to the current crisis in marine fisheries. Increasingly, scientists are calling for fundamental changes to the current fisheries management paradigm which is seen as fatally flawed. While it is thought that some populations will be unable to rebuild from low stock levels, many effects of overfishing should be reversible at the present point. The key to achieving this and meeting transgenerational responsibilities is rigorous and protective management of marine resources.

Parallels with whaling are striking, discussed fully in other chapters.

### Aquaculture

The rapid growth in demand for aquaculture products, particularly high-value marine products such as salmon and shrimp, has led to widespread development of intensive production facilities throughout the world. Most recent estimates (FAO, 1997) indicate total aquaculture production in 1994 reached a record 25.5 million tonnes, an increase in production of 11.8% over 1993. Asia dominates the global aquaculture market, supplying approximately 80% of total production by mass.

While culture of marine finfish and crustaceans account for relatively small proportions of total biomass production (1.8% and 4.2% respectively), their commercial contribution is high. Moreover, as the vast majority of marine aquaculture is located on the coastal fringe and in estuaries, potential and actual impacts can be substantial.

Table 4

The development of intensive shrimp aquaculture in Thailand and in Asia as a whole over the last 30 years (reproduced from Flaherty and Karnjanakesorn, 1995)

| Region/Year | No. facilities | Total area (ha) | Production (mt/y) |
|---|---|---|---|
| Thailand | | | |
| 1972 | 1,154 | 9,000 | 1,000 |
| 1990 | 15 278 | 64,081 | 118,227 |
| 1993 | nd | nd | 155,000 |
| 1994 | nd | nd | 225,000 |
| Asia (total) | | | |
| 1984 | nd | nd | 100,000 |
| 1994 | nd | nd | 650,000 |

Much of the extremely rapid development of marine aquaculture in Asia and Latin America over the last 10 to 20 years has been in intensive and semi-intensive production of shrimps (*Penaeus* spp.) in artificially constructed coastal ponds (Table 4). In general, little consideration has been given to protection of coastal habitat or water quality. The common goal has been to maximise profits in short-term intensive operations, often involving the abandonment of ponds as soon as yields begin to fall, frequently as a result of inadequate waste management, poor husbandry and the inevitable spread of disease. In the north of Thailand, 10,120 ha of intensive shrimp ponds developed from mangrove were abandoned between 1989 and 1991 as a result of declining water quality or disease outbreaks (Flaherty and Karnjanakesorn, 1995). Such unsustainable "shifting aquaculture" supplies a large proportion of global shrimp production.

Aquaculture may cause a range of direct and indirect impacts on habitat and biodiversity, including habitat destruction or modification and the depletion of natural populations through collection of brood and feedstock. Most notable is the large-scale clearance of mangroves for the construction of intensive aquaculture facilities. For example, in the Philippines, aquaculture has contributed significantly to the removal of 75% of the original mangrove habitat (Iwama, 1991).

While the areas of the ponds themselves are substantial, the total area impacted through development of shrimp culture (i.e. its "ecological footprint") may be much larger (almost 200 times greater than the facility itself). Flaherty and Karnjanakesorn (1995) highlight the potential for negative impacts on inshore fisheries through removal or modification of nursery grounds, estimating that the production of 120,000 tonnes of shrimp in Thailand may have resulted in a loss of potential yield from inshore fisheries of up to 800,000 tonnes.

Intensive culture of finfish and shellfish is expensive in terms of energy use. Coastal pond culture of marine and brackish water species also presents high demands for water. The production of 1 tonne of shrimp can utilise and contaminate 50–60 million litres of seawater, or a mixture of sea and freshwater. Moreover, inputs of organic matter and nutrients can be locally and regionally substantial. Increased nutrient inputs from ponds in Thailand have been estimated as equivalent to that expected from a 50–100% increase in coastal population, with no provisions for treatment (Dierberg and Kattisimkul, 1996).

Diseases which, in natural populations, may account for significant mortality can devastate stocks of intensively farmed finfish or shrimp over wide areas. An array of chemical agents is applied in order to control viral, bacterial, fungal and other pathogens. These chemicals, including antibiotics, pesticides and detergents, are used in large quantities by the aquaculture industry globally. Although there are some controls on use, information on the types and quantities of chemicals employed in many countries is very scarce, and their environmental fate is even less well understood. Undoubtedly, substantial proportions are lost to the environment during water exchange, harvesting or disposal of waste sludges.

It is estimated that 37 tonnes of antibiotics were used in marine fish cages in Norway in 1990 alone. The figure for Denmark in 1995, nearer 3.5 tonnes, nevertheless represented approximately 10% of the total quantity of antibiotics administered for therapeutic purposes to the whole of Denmark's 5.2 million inhabitants (Halling-Sorensen et al., 1998). Contamination of the surrounding ecosystem can lead to toxic effects in other organisms and the development of bacterial strains resistant to antibiotics.

A very high proportion of wild fish caught in the vicinity of Norwegian fish farms has been found to contain significant residues of antibiotics (Ervik et al. 1994). The commonly used antibiotic oxytetracycline breaks down extremely slowly in organic-rich, anaerobic sediments, such as those in intensive shrimp ponds, and is known to bioaccumulate in some crustaceans. The accumulation of antibiotic residues in aquaculture products themselves is increasingly recognised as a public health concern.

A wide range of pesticides and other chemicals is reportedly used to control disease and pest species in commercial aquaculture, although data on use patterns are scarce. Many of the chemicals employed (e.g. trifluralin, malathion) are toxic to marine life and present substantial hazards to humans during application.

Aquaculture has also led to widespread introductions of cultured organisms, and frequently non-indigenous species, to the wild. Such individuals may be capable of interbreeding with wild populations, introducing foreign genetic material to local gene pools, or may simply compete with wild stocks for limited resources. Thompson et al. (1995) estimated that up to 20% of salmon caught by Norwegian fisheries may be escapees; this figure may be as high as 30–40% in some years. The potential for introduction of novel diseases is also of serious concern, as is the release of newly developed genetically modified aquaculture stock.

## GLOBAL TRENDS

### Climate Change

Due to the greenhouse effect, global average surface air temperatures are projected to increase by between 1 and 3.5°C (relative to 1900) by the year 2010. Estimates vary, but seem set to be greater than any historical changes since the ending of the last ice age, 10,000 years ago. Already, there is considerable evidence that greenhouse warming is a real phenomenon (IPCC, 1996) with increasing numbers of global temperature observations supporting predictions of computer models, clear trends emerging from the "noise" imposed by natural variation, and observations of phenomena such as glacier retreat and smaller sea-ice extent in the Arctic and Antarctic.

Some of the potential impacts of this temperature rise are obvious, others less so, and include sea-level rise, significant land losses in some coastal nations, implications for human food supply and coastal developments. Recent modelling studies have suggested higher sea-surface temperatures will lead to increasing numbers and intensities of tropical cyclones which could amplify the impacts of sea-level rise.

Many marine ecosystems are likely to be impacted through changes in salinity and sedimentation patterns. Widespread changes in ocean productivity and fisheries may also be expected, and these are likely to exacerbate problems caused by overfishing. It is possible that there will be a general decrease in the biological productivity of the oceans as climate changes (IPCC, 1996). Temperature regimes are an important determinant of organismal and whole ecosystem function and changes in temperature are likely to alter species composition and geographical extent of habitats.

The oceans, through their fluid motion, high heat capacity and biotic processes play a pivotal role in shaping the climate. As a result of increasing temperatures at the poles and a decrease in equator–pole temperature gradients, it has been suggested that trade winds and upper ocean currents will weaken and the intensity of upwelling will be reduced (IPCC, 1996). Some studies have pointed to the possibility that thermohaline circulation could be reduced or stopped (Rahmstorf et al., 1996) (Fig. 6). Increased precipitation and melting ice could make Arctic ocean waters less saline, affecting the formation of the deep water which drives global ocean circulation patterns (the Ocean Conveyor). Alternatively, the site of deep-water formation could shift southwards, ending the influence upon European and Scandinavian climate of the warm waters of the Gulf Stream and North Atlantic drift. In short, the response of the thermohaline circulation, which seems to have been stable over the last 10,000 years, is highly unpredictable.

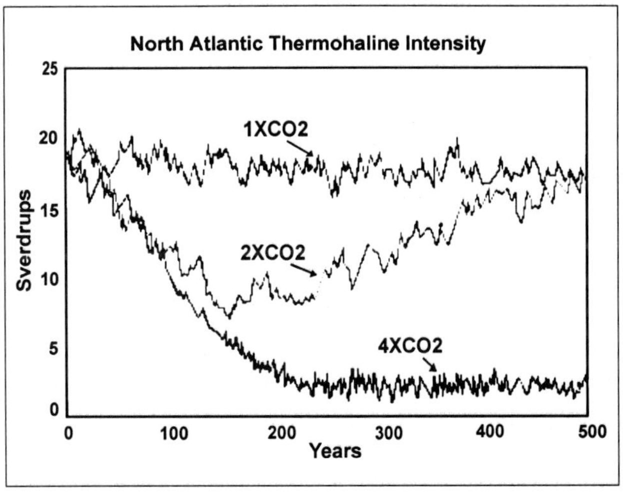


Fig. 6. Possible behaviour of the thermohaline circulation under different atmospheric $CO_2$ emission scenarios. With a doubling of $CO_2$ concentration, the circulation is predicted to decline sharply over time and recover gradually over 500 years. At quadrupled $CO_2$ concentrations, the thermohaline circulation could shut down completely. Recent work has suggested that not only is the final atmospheric concentration important, but also the rate at which it increases (Johnston et al., 1998).

The potential effects are illustrated by the profound climatic changes resulting from the El Niño-Southern Oscillation (ENSO). There is some evidence to suggest that natural ENSO events may become more intense and more frequent under conditions of global climate change (Trenberth and Hoar, 1997). Irrespective of this, the ENSO phenomenon provides an indication of the far-reaching changes which could result from human-induced climate change.

In the sense that the globe is committed to a temperature rise in the future as a result of human activity—especially fossil-fuel use—intergenerational responsibilities have been abdicated to a significant degree. In allowing the build up of carbon dioxide in the atmosphere, arguably both a societal product and a component of the earth's crust, the first and second principles of sustainability have been breached. The subsequent temperature rise threatens to cause numerous violations of principles three and four. The question at present is not one of how climate change can be prevented but rather one of keeping the impacts within certain limits in order to allow marine systems to accommodate the predicted changes and to preserve their functioning to the maximum extent possible. This will require considerable discipline to be developed in the exploitation of marine resources to ensure that pressures of climate change added to other existing pressures do not fatally compromise them.

### Ozone Depletion

Every spring for over 20 years a sizeable portion of the stratospheric ozone over Antarctica has disappeared. Substantial depletion of the ozone layer is now being recorded over the Arctic and at temperate and tropical latitudes. Even models based on fairly optimistic assumptions predict continued declines in stratospheric ozone for at least the next few years and recovery will be slow over much of the next century (Madronich et al., 1995).

There is overwhelming evidence that production and use of various chlorinated and brominated chemicals have contributed greatly to the depletion of ozone in the lower stratosphere, and that this depletion has resulted in increased levels of UV radiation incident at the earth's surface. In addition, there is clear evidence that UV-B radiation is detrimental to various marine species in the upper layers of the ocean and reduces primary production. However, there is uncertainty of the significance of this at the scale of communities and ecosystems (Hader et al., 1995).

Even without the loss of stratospheric ozone, many marine organisms are exposed to UV-B on a daily basis. Ambient levels at any latitude are potentially detrimental and evidence suggests that normal levels of solar UV are an important ecological factor in oceanic processes.

Harmful effects can be countered to some degree by mechanisms such as UV-absorbing compounds, DNA repair pathways and the ability to resynthesize damaged proteins and/or pigments. However, such mechanisms entail considerable metabolic costs.

While studies of effects of enhanced UV exposure in marine organisms have focused principally on phytoplankton, the viability of other organisms may also be directly affected, including non-photosynthetic bacterioplankton and higher organisms. Increased UV-B can cause death, decreased reproductive capacity, reduced survival and impaired larval development in various zooplankton and fish species (Hader et al., 1995). The larvae of many fish and even benthic invertebrates, have planktonic phases during which they remain close to the surface. Exposure to ambient UV-B increases the frequency of underdeveloped and malformed Antarctic sea urchin embryos and mortality in early life-cycle stages (Karentz, 1994). Earlier studies had demonstrated that as little as a 16% reduction in stratospheric ozone would give rise to an increase in incident UV-B sufficient to cause 50–100% mortality in anchovy larvae at 0.5 m depth. The impacts of enhanced UV-B exposure on other organisms appear to have been poorly investigated.

As the phytoplankton are the basis of marine food webs, any loss in overall biomass or changes in species composition could cause the reduction in biomass at higher trophic levels. Direct effects of UV-B on zooplankton and on fish eggs and larvae could also cause a significant impact at higher trophic levels.

Given the wide range in UV-B sensitivity among phytoplanktonic species and taxonomic groups, UV-B could play an important role in species selection and in lower food web processes. Such selection, conceptually similar to that relating to shifts in nutrient availability, could ultimately result in fundamental changes in community structure at higher trophic levels with far-reaching implications. The precise ramifications of what might at first appear quite subtle changes in rates of, or organisms responsible for, primary production are practically impossible to predict. Moreover, as marine phytoplankton are a major biological sink for atmospheric $CO_2$, any decrease in their overall productivity could lead to further enhancement of the greenhouse effect.

## TOWARDS SUSTAINABILITY: THE CHALLENGE FOR THE NEW MILLENNIUM

The preceding discussion has highlighted numerous and diverse ways in which current human activities are unsustainable. While the picture which emerges clearly raises grave concerns, we are also capable of changing our behaviour such that we might increasingly avoid environmental degradation in the future. Restoration of existing systems might also be feasible, although in some cases long-term and even irreversible changes have undoubtedly already occurred.

If we are to do so, however, it will be necessary to step outside the narrow management paradigms which drive much of human "development" and embrace some more fundamental concepts. For example, much has been written about the principle of precautionary action (see e.g. Santillo et al., 1998) and there is considerable debate as to its practical application. In simple terms, however, it is an expression of the ethic that "prevention is better than cure". As such, it can be seen to have application in all fields of human interactions with the seas and, indeed, has been incorporated as a guiding principle in the majority of the conventions concerned with marine environmental protection. Its application involves *inter alia* the use of scientific knowledge in order to assist in the identification of the potential for serious or irreversible adverse impacts before they occur, such that preventative rather than restorative measures can be taken. In turn, it requires the explicit recognition of, and allowance for, limitations to our understanding which stem from the complexity of natural systems and our interactions with them. Further research will undoubtedly enable us to reduce some of the uncertainties which prevail, but much will remain beyond analytical reduction by its very nature. Application of the precautionary principle implies what Stirling (1999) terms a greater "humility" about what we know and, equally, what we may not know.

The criteria for sustainability outlined at the beginning of this document similarly provide an alternative framework to the essentially economic indicators which are currently used to measure "development", for the evaluation of current practice and decision-making. The sustainability of marine ecosystems is not an ideal which must be balanced against economic gain, but an essential foundation on which the functioning of global systems rests. Ultimately, current trends of systematic degradation of the marine environment will impact on the availability of the very "services" on which we rely, and on the ability of future generations to meet their own needs. Even in the most anthropocentric evaluations, this must clearly be seen to violate our transgenerational responsibilities.

## ACKNOWLEDGEMENTS

We acknowledge all those who contributed to the production of the Greenpeace Report on the World's Oceans (Johnston et al., 1998) on which this chapter is based.

## REFERENCES

Allsopp, M., Santillo, D., Johnston, P. and Stringer, R. (1999) *The Tip of the Iceberg? State of knowledge on persistent organic pollutants in Europe and the Arctic.* Greenpeace International, August 1999, ISBN 90-73361-53-2, 76 pp.

Alverson, D.L., Freeberg, M.H., Murawski, S.A. and Pope, J.G. (1994) A global assessment of fisheries bycatch and discards. FAO Fisheries Technical Paper 339, Food and Agriculture Organisation of the United Nations, Rome, ISBN 92-5-103555-5, 233 pp.

Cairns, J. (1997) Defining goals and conditions for a sustainable world. *Environmental Health Perspectives* **105** (11), 1164–1170

Cairns, J. and Dickson, K.L. (1995) Individual rights versus the rights of future generations: ecological resource distribution over large temporal and spatial scales. In *An Aging Population, an Aging Planet, and a Sustainable Future*, eds. S.R. Ingman, X. Pei, C.D. Ekstrom, H.J. Friedsam and K.R. Bartlett, Chap. 11. Texas Institute for Research and Education on Aging, Denton, TX.

Camphuysen, K. (1998) Beached bird surveys indicate decline in chronic oil pollution in the North Sea. *Marine Pollution Bulletin* **36** (7), 519–526.

Carls, M.G., Rice, S.D. and Hose, J.E. (1999) Sensitivity of fish embryos to weathered crude oil: Part I. Low level exposure during incubation causes malformations, genetic damage and mortality in larval Pacific herring (*Clupea pallasi*). *Environmental Toxicology and Chemistry* **18** (3), 481–493.

Corbett, J.J. and Fischbeck, P. (1997) Emissions from Ships. *Science* **278**, 823–824.

Costanza, R., d'Arge, R., de Groot, R., Farber, S., Grasso, M., Hannon, B., Limburg, K., Naeem, S., O'Neill, R.V., Paruelo, J., Raskin, R.G., Sutton, P. and van den Belt, M. (1997) The value of the world's ecosystem services and natural capital. *Nature* **387**, 253–260.

de March, B.G.E., de Wit, C.A. and Muir, D.A. (1998) Persistent organic pollutants. In *AMAP Assessment Report: Arctic Pollution Issues*, eds. S.J. Wilson, J.L. Murray and H.P. Huntington. Arctic Monitoring and Assessment Programme (AMAP), Oslo, ISBN 82-7655-061-4, pp. 183–371.

Dierburg, F.E. and Kiattisimkul, W. (1996) Issues, impacts and implications of shrimp aquaculture in Thailand. *Environmental Management* **20** (5), 649–666.

EA (1998) *Oil and Gas in the Environment.* Environment Agency for England and Wales, ISBN 0-11-310152-X, 104 pp.

Ervik, A., Thorsen, B., Eriksen, V., Lunestad, B.T. and Samuelsen, O.B. (1994). Impact of administering antibacterial agents on wild fish and blue mussels *Mytilus edulis* in the vicinity of fish farms. *Diseases of Aquatic Organisms* **18**, 45–51.

FAO (1997) *The State of World Fisheries and Aquaculture 1996.* Food and Agriculture Organisation of the United Nations, Rome.

Flaherty, M. and Karnjanakesorn, C. (1995) Marine shrimp aquaculture and natural resource degradation in Thailand. *Environmental Management* **19** (1), 27–37.

Fogarty, M.J. and Murawski, S.A. (1998) Large scale disturbance and the structure of marine systems: Fishery impacts on Georges Bank. *Ecological Applications* **8** (1), Supplement 6–22

Hader, D.-P., Worrest, R.C., Kumar, H.D. and Smith, R.C. (1995) Effects of increased solar ultraviolet radiation on aquatic ecosystems. *Ambio* **24** (3), 174–180.

Halling-Sorensen, B., Nors Nielsen, S., Lanzky, P.F., Ingerslev, F., Holten Lutzhoft, H.C. and Jorgensen, S.E. (1998). Occurrence, fate and effects of pharmaceutical substances in the environment—a review. *Chemosphere* **36** (2), 357–393.

Howarth, R.W., Billen, G., Swaney, D, Townsend, A., Jaworski, N., Lajtha, K., Downing, J.A., Elmgren, R., Caraco, N., Jordan, T., Berendse, F., Freney, J., Kudeyrov, V., Murdoch, P. and Zhu, Z.L. (1996) Regional nitrogen budgets and riverine N and P fluxes for the drainages of the North Atlantic Ocean—natural and human influences. *Biogeochemistry* **35** (1), 75–139.

Hutchings, J.A. and Myers, R.A. (1994) What can be learned from the collapse of a renewable resource? Atlantic cod, *Gadus morhua*, of Newfoundland and Labrador. *Canadian Journal of Fisheries and Aquatic Sciences* **51** (9), 2126–2146.

IPCC (1996) Climate Change 1995 Impacts, Adaptations and Mitigation of Climate Change: Scientific-Technical Analyses. Contribution of Working Group II to the Second Assessment report of the Intergovernmental Panel on Climate Change. Cambridge University Press.

ICS (1997) Shipping and the Environment: A Code of Practice. International Chamber of Shipping, London, 24 pp.

IMO (1999) Final report on permits issued in 1996. Document LC 21/4/1, Twenty-first Consultative Meeting of Contracting Parties to the Convention on the Prevention of Marine Pollution by Dumping of Wastes and Other Matter (London Convention), 4–8 October 1999, International Maritime Organisation, 40 pp.

Iwama, G.K. (1991) Interactions between aquaculture and the environment. *Critical Reviews in Environmental Control* **21** (2), 177–216

Iwata, H., Tanabe, S., Mizuno, T. and Tatsukawa, R. (1997) Bioaccumulation of butyltin compounds in marine mammals: the specific tissue distribution and composition. *Applied Organometallic Chemistry* **11**, 257–264.

Jacobs, M.N., Santillo, D., Johnston, P.A., Wyatt, C.L. and French, M.C. (1998) Organochlorine residues in fish oil dietary supplements: comparison with industrial grade oils. *Chemosphere* **37** (9–12), 1709–1721.

Johnston, P.A., Stringer, R.L. and Santillo, D. (1996) Cetaceans and Environmental Pollution: The Global Concerns. In *The Conservation of Whales and Dolphins: Science and Practice*, eds. M.P. Simmonds and J.D. Huthinson. Wiley, New York, pp. 220–261

Johnston, P.A., Marquardt, S., Keys, J. and Jewell, T. (1997) Shipping and Handling of Pesticide Cargoes: The Need for Change. *Journal of the Chartered Institution of Water and Environmental Management* **11** (3), 157–163.

Johnston, P., Santillo, D., Stringer, R., Ashton, J., Mckay, B., Verbeek, M., Jackson, E., Landman, J., van den Broek, J., Samson, D. and Simmonds, M. (1998) Report on the World's Oceans. Greenpeace Research Laboratories Report, May 1998. ISBN 90-73361-45-1, 154 pp. (available at http://www.greenpeace.org/~oceans/reports/sotofull.pdf)

Johnston, P., Santillo, D., Stringer, R., Parmentier, R., Hare, B. and Krueger, M. (1999) Ocean disposal/sequestration of carbon dioxide from fossil fuel production and use: an overview of rationale, techniques and implications—executive summary. *Environmental Science and Pollution Research* **6** (4), 245–246.

Karentz, D. (1994) Considerations for evaluating ultraviolet radiation-induced genetic damage relative to Antarctic ozone depletion. *Environmental Health Perspectives* **102** (Suppl. 12), 61–63.

Kiceniuk, J.W., Holzbecher, J. and Chatt, A. (1997) Extractable organohalogens in tissues of Beluga Whales from the Canadian Arctic and the St. Lawrence estuary. *Environmental Pollution* **97** (3), 205–211.

Klaer, N., Polachek, T., Sainsbury, K. and Preece (1996) Southern Bluefin Tuna Stock and Recruitment Projections. Document SC/96/17 CCSBY Scientific Committee.

Lean,, G. and Hinrichsen, D. (eds.) (1992) *The Atlas of the Environment*. Helicon Publishers, London, 192 pp.

Lindstrom, G., Wingfors, H., Dam, M. and Bavel, B.V. (1999). Identification of 19 polybrominated diphenyl ethers (PBDEs) in Long-Finned Pilot whale (*Globicephala melas*) from the Atlantic. *Arch. Environmental Contaminant Toxicology* **36**, 355–363.

Lohrenz, S.E., Fahnstiel, G.L., Redalje, D.G., Lang, G.A., Chen, X.G. and Dagg, M.J. (1997) Variations in primary production of the northern Gulf of Mexico continental shelf waters linked to nutrient inputs from the Mississippi River. *Marine Ecology Progress Series* **155**, 45–54.

Madronich, S., McKenzie, R.L., Caldwell, M.M. and Bjorn, L.O. (1995) Changes in ultraviolet radiation reaching the Earth's surface. *Ambio* **24** (3), 143–152.

Matthiessen, P. and Gibbs, P.E. (1998) Critical appraisal of the evidence for tributyltin-mediated endocrine disruption in mollusks. *Environmental Toxicology and Chemistry* **17** (1), 37–43.

Minchin, D., Bauer, B., Oehlmann, J., Schulte-Oehlmann, U. and Duggan, C.B. (1997) Biological indicators used to map organotin contamination from a fishing port, Killybegs, Ireland. *Marine Pollution Bulletin* **34** (4), 235–243.

Nriagu, J.O (1992) Worldwide contamination of the atmosphere with toxic metals. Proceedings of a Symposium: *The Deposition and Fate of Trace Metals in our Environment*, Philadelphia, October 8 1991. General Technical Report NC-150, United States Department of Agriculture, Forest Service. pp. 9–21

Olsgard, F. and Gray, J.S. (1995). A comprehensive analysis of the effects of offshore oil and gas exploration and production on the benthic communities of the Norwegian continental shelf. *Marine Ecology Progress Series* **122**, 277–306.

OSPAR (1998) The Sintra Statement (Final Declaration of the Ministerial Meeting of the OSPAR Commission, Sintra, 20–24th July 1998). OSPAR 98/14/1 Annex 45. OSPAR Convention for the Protection of the Marine Environment of the North-East Atlantic.

Pauly, D., Christensen, V., Dalsgaard, J., Froese, R. and Torres, F. (1998) Fishing down marine food webs. *Science* **279**, 860–863.

Rahmstorf, S., Marotzke, J. and Willebrand, J. (1996) Stability of the thermohaline circulation. In *The Warm Water Sphere of the North Atlantic Ocean*, ed. W. Krauss, Chap. 5. Borntraeger, Stuttgart .

Raimondi, P.T., Barnett, A.M. and Krause, P.R. (1997) The effects of drilling muds on marine invertebrate larvae and adults. *Environmental Toxicology and Chemistry* **16** (6), 1218–1228.

Santillo, D., Stringer, R., Johnston, P. and Tickner, J. (1998) The Precautionary Principle: Protecting against failures of scientific method and risk assessment. *Marine Pollution Bulletin* **36** (12), 939–950.

Sauer, T.C., Costa, H.J., Brown, J.S. and Wards, T.J. (1997) Toxicity identification evaluations of produced water effluents. *Environmental Toxicology and Chemistry* **16** (10), 2020–2028.

Serchuk, F.M., Kirkegaard, E. and Daan, N. (1996) Status and trends of the major roundfish, flatfish and pelagic fish stocks in the North Sea: A thirty year overview. *ICES Journal of Marine Science* **53**, 1130–1145.

Stirling, A. (1999) On science and precaution in the management of technological risk. Synthesis report of studies conducted by O. Renn, A. Klinke, A. Rip, A. Salo and A. Stirling, EC Forward Studies Unit/ESTO Network, European Commission, May 1999.

Strand, P. (1998) Radioactivity. In *AMAP Assessment Report: Arctic Pollution Issues*, eds. S.J. Wilson, J.L. Murray and H.P. Huntington. Arctic Monitoring and Assessment Programme (AMAP), Oslo,

ISBN 82-7655-061-4, pp. 525–619.
Stroud, C. and Simmonds, M. (1998) The future of commercial whaling and the IWC: A strategy for the UK. Paper presented at the Meeting on Marine Environmental Management Review of 1997 and Future Trends, London, January 1998, 6 pp.
Thompson, S., Treweek, J.R. and Thurling, D.J. (1995) The potential application of strategic environmental assessment (SEA) to the farming of Atlantic salmon (*Salmo salar* L.) in mainland Scotland. *Journal of Environmental Management* **45**, 219–229.
Trenberth, K.E. and Hoar, T.J. (1997). El Niño and climate change. *Geophysical Research Letters* **24**, 3057–3060.
UNCED (1992) *Rio Declaration on Environment and Development*. United Nations Commission on Environment and Development, ISBN 9-21-100509-4.
Wania, F. and Mackay, D. (1993) Global fractionation and cold condensation of low volatility organochlorine compounds in polar regions. *Ambio* **22** (1), 10–18.
Woodhead, R.J., Law, R.J. and Matthiessen, P. (1999) Polycyclic aromatic hydrocarbons in surface sediments around England and Wales, and their possible biological significance. *Marine Pollution Bulletin* **38** (9), 773–790.

## THE AUTHORS

**Paul Johnston**
*Greenpeace Research Laboratories,*
*University of Exeter,*
*Exeter EX4 4PS, U.K.*

**David Santillo**
*Greenpeace Research Laboratories,*
*University of Exeter,*
*Exeter EX4 4PS, U.K.*

**Julie Ashton**
*Greenpeace Research Laboratories,*
*University of Exeter,*
*Exeter EX4 4PS, U.K.*

**Ruth Stringer**
*Greenpeace Research Laboratories,*
*University of Exeter,*
*Exeter EX4 4PS, U.K.*

Chapter 134

# Marine Reserves and Resource Management

Michael J. Fogarty, James A. Bohnsack and Paul K. Dayton

The application of marine reserves has generated considerable recent interest as a tool for conservation and resource management. Particular attention has been placed on the potential utility of areas within which all extractive activities are prohibited (No-Take Zones, NTZs) for management. The no-take reserve concept builds on and extends traditional closure strategies in fishery management by providing more inclusive protection for populations, communities, and habitats. Marine reserves in general have now been implemented on a global basis but relatively few provide the full protection of no-take zones and many currently lack effective management plans and enforcement. Established no-take reserves often demonstrate increases in biomass or abundance and average size within reserve boundaries. In some instances, benefits in terms of yield and resilience to exploitation have been documented, although much greater attention to this issue will be required as the experience with reserves increases. Predicted performance of reserves is strongly dependent on underlying model structures and assumptions. Whole-life-cycle models with sedentary adults and a dispersive pre-recruit stage often indicate important benefits of reserves in terms of resilience to exploitation. Partial life-cycle models of the harvestable component of the population show a relative loss of yield per recruit with reserves when dispersal rates are high, except when fishing mortality is also high. In all cases, the uncertainty associated with actual dispersal processes at different life stages is the most challenging aspect of predicting reserve performance. We conclude that no-take zones do hold considerable promise as part of an integrated management strategy for ecosystem-based management. The ability of NTZs to meet specific fishery management objectives strongly depends on the life history and dispersal characteristics of the species.

*Seas at The Millennium: An Environmental Evaluation (Edited by C. Sheppard)*
*Published by Elsevier Science Ltd.*

## INTRODUCTION

The scope of marine conservation and resource management encompasses multiple objectives including preservation of biodiversity and ecosystem integrity, habitat protection, and sustainable resource use. The development of management techniques to meet these requirements presents unique challenges. Considerable attention has been devoted to the potential for marine reserves to meet diverse management objectives (Agardy, 1997). Particular emphasis has recently been placed on the utility 'no-take' zones and networks of reserves in which all extractive activities are prohibited with the objectives of preserving ecosystem integrity and serving as a source for population replenishment in adjacent areas (Murray et al., 1999). In addition to the objective of helping to ensure sustainability and the maintenance of intact marine ecosystems, protected areas can serve as relatively pristine research sites and baseline systems against which environmental change can be evaluated. These areas can also serve as important locations for eco-tourism and the aesthetic appreciation of natural systems.

Concerns about the direct and indirect effects of fishing practices on marine ecosystems can be traced to the 14th century (Anon., 1921). In 1366, a petition was placed before the Commons in Britain to ban the Wondyrchoun, a type of dredge said to press *"...so hard on the ground when fishing that it destroys the ... plants growing on the bottom under the water, and also the spat of oysters, mussels, and other fish, by which the large fish are accustomed to live and be nourished"*. In 1499, the use of trawls was prohibited in Flanders because *"... the trawl scraped and ripped up everything it passed over in such a way that it rooted up and swept away the seaweeds which served to shelter the fish; it robbed the beds of their spawn or fry..."*. The Dutch banned the use of trawls in 1583 in their estuarine waters and trawling was made a capital offense in France in 1584. In Britain in 1631, the use of *"...traules* [was] *forbidden as well as of other nets which shall not have the meshes of the size fixed by law and orders..."*. The impact of different types of fishing gears on marine ecosystems remains an important source of concern that has intensified with the development of new harvesting technologies and the advent of destructive fishing practices, such as the use of explosives and toxins. A principal advantage of the use of marine reserves is that with proper enforcement, negative impacts of fishing gears and practices can be eliminated in defined areas.

Attempts to determine effects of fishing on marine fish populations experimentally were initiated as early as 1886 in Scotland in comparisons of areas closed and open to fishing (Garstang, 1900). These studies involved monitoring the open and closed areas within each site by trawling at planned monthly intervals over ten years. The principal target fish species for the experiment, plaice (*Pleuronectes platessa*), increased significantly in the closed area following the cessation of fishing. This study remains an important precursor to contemporary efforts to establish the potential utility of marine reserves. In the late 19th century, several proposals were made to set aside reserve areas in the North Sea to permit recovery of exploited stocks, but were not adopted because of lack of international agreement. That exploited marine populations respond to reductions in fishing pressure and to the spatial distribution of fishing effort was subsequently demonstrated on a large scale during the two world wars when curtailment of fishing activities was followed by marked increases in fish populations in the North Sea (Beverton and Holt, 1957). Not only did catch rates increase, but the average size in the catch was greater, foreshadowing results observed in later studies of the potential effectiveness of marine reserves.

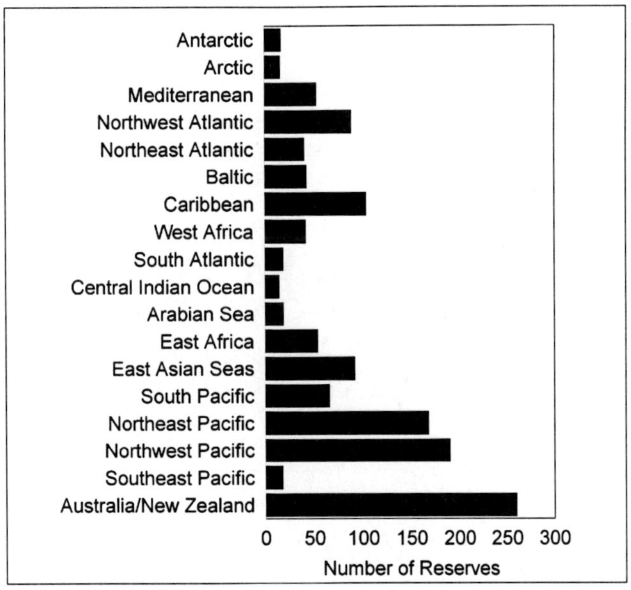


Fig. 1. Global distribution of marine reserves by geographical regions (based on Kelleher et al., 1995).

Marine reserves have now been implemented on a global basis. Kelleher et al. (1995) list 1306 subtidal marine reserves distributed from tropical to polar regions (Fig. 1). Despite this, their total area is relatively small (the median reserve size listed by Kelleher et al. (1995) is just 16 km$^2$) and much less than 1% of the coastal zone is now under their protection. Further, in many, active management and enforcement is not practised or is weak and ineffective.

A number of types of marine reserves or protected areas can be identified. Many marine reserves now in place have been established on an opportunistic basis and the theoretical framework for the design of no-take marine reserves and networks is still evolving. Accordingly, we will examine both empirical evidence for their effectiveness and the issues involved in their design and implementation. In this review, we focus on the role of marine reserves as fishery management tools within the broader context of ecosystem-based management. We seek to complement recent reviews of marine reserves in a fishery management context (e.g. Bohnsack, 1996; Murray et al., 1999; Guènette et al., 1998) and on the direct and indirect effects of fishing on marine systems (Dayton et al., 1995, in press; Hall, 1998).

## Classification of Marine Protected Areas

Agardy (1997) defines Marine Protected Areas (MPAs) as: "... any area of the coastal zone or open ocean conferred some level of protection for the purpose of managing use of resources and ocean space or protecting vulnerable or threatened habitats and species." In this article, we shall use the terms marine protected areas and marine reserves interchangeably.

Types of MPAs recognized by Agardy include:

- **Closed Areas**: regions designated as harvest refugia or other zones in which human activities are restricted with the objective of ensuring sustainability of resources on a seasonal, temporary or longer term basis.
- **Research and Monitoring Areas**: regions designated as scientific control sites or monitoring locations in support of basic research in marine science.
- **Sensitive Sea Areas**: regions requiring special protection because of ecological or socioeconomic significance and vulnerability to damage by maritime activities (designated by the International Maritime Organization).
- **Marine Sanctuaries and Parks**: regions established to permit specified activities while preserving and protecting the ecological integrity of the area.
- **Regional Seas and Large Marine Ecosystem Areas**: designated enclosed or semi-enclosed seas jointly managed by more than one nation (Regional Seas) and large-scale regions representing coherent ecological units or biogeographical zones (Large Marine Ecosystems) which may form the basis for Regional Seas agreements and jurisdiction.
- **Integrated Management Areas**: governmental coastal zone planning areas or exclusive economic zones.
- **High Seas (U.N. Law of the Sea Treaty)**: regions outside coastal state exclusive economic zones subject to international treaties and cooperative management regimes for signatory nations

We trace the antecedents of no-take marine reserves to closed-area management strategies employed in many fisheries and highlight the impacts of harvesting on marine ecosystems and the potential role of marine reserves in mitigating these effects in specified areas. Critical issues in the effective design of marine reserves are reviewed, as is the empirical evidence for reserve performance and attempts to develop models of marine reserves and networks. Finally, we consider the human dimension in the success of reserves.

## ECOSYSTEM MANAGEMENT AND THE RESERVE CONCEPT

Consideration of the production potential of marine fisheries and recent trends in global catch levels suggest that we are near or have exceeded the limits to overall yield from coastal and continental shelf regions of the world ocean. Estimates of the production capacity of the oceans suggest limits to catch of 100–200 million t depending on whether the deep ocean basins are included. Reported landings have levelled off at approximately 85 million t since 1989, following steady increases since 1950. Recent estimates of fish caught but discarded at sea are as high as one third of the landed catch (Alverson et al., 1994). An evaluation of the global status of fishery resource species in 1994 listed 31% as lightly to moderately exploited, 44% as fully to intensively exploited, 16% as overexploited, 6% as depleted and 3% as recovering (Garcia and Newton, 1997). General patterns are of declining yields for 35% of these species, with 25% showing relatively stable yields at high exploitation levels, and 40% with some development potential (FAO, 1997). Overcapacity of world fishing fleets has placed excessive stress on the productivity of marine resources while reducing the profitability of global fishing enterprises. Expansion of the species under exploitation and fishing at the limits to sustainable production clearly increases the risk of interference with ecosystem processes and functions.

Escalation of fishing pressure on fishery resources and increased recognition of problems associated with incidental catch, disruption of food web structure, and habitat disturbance by fishing gear has underscored the need for a more holistic approach to management based on ecosystem principles. Christensen et al. (1996) define ecosystem management as:

> "... management driven by explicit goals, executed by policies, protocols, and practices, and made adaptable by monitoring and research based on our best understanding of the ecological interactions and processes necessary to sustain ecosystem composition, structure, and function."

Marine reserves, used with other management measures, hold the potential of addressing fundamental issues in the development of ecosystem-based management strategies. In particular, no-take zones (NTZs) uniquely afford the possibility of simultaneously reducing fishing mortality on target species, preventing habitat disturbance due to fishing practices, protecting non-target organisms, and preserving biodiversity. The effectiveness of marine reserves in meeting these different objectives in specific situations will depend critically on life history characteristics of the species involved, the oceanographic setting, and factors such as the level of compliance and enforcement.

With a significant fraction of marine fishery resources classified as overexploited, depleted, or recovering, attention has turned to the question of whether marine reserves can provide a more robust approach to management in practice. Failures in management can be linked to uncertainties due to incomplete knowledge of marine populations, communities and ecosystems. The broader application of no-take marine reserves may serve as a hedge against uncertainty in our understanding (Bohnsack, 1996; Murray et al., 1999) and in implementing controls on fishing pressure (Lauck et al., 1998) although uncertainty associated with critical factors in reserve performance itself should not be under-estimated.

Marine reserves provide the opportunity to devise management approaches directed at understanding and protecting ecosystem function. However, many failures in resource management are directly linked to structural problems such as conflicting conservation and social/economic goals, the prevalence of open-access fisheries, and a tradition in which the burden of proof of damage to natural resources has rested with scientists and managers. Conflicting goals such as resource conservation and the preservation of broad employment opportunities in the fishing industry has resulted in compromises that have consistently undermined conservation efforts. Open-access fisheries contribute substantially to this problem, creating inexorable pressure to permit continued harvests as resources decline. Finally, scientific uncertainties have been used as an excuse to permit continued high harvesting levels. These common characteristics have led to strong political pressures that have resulted in resource depletion. It must be recognized that unless these underlying problems are addressed, attempts to establish marine reserve systems will be also undermined by compromises in the size and placement of reserves which degrade their potential effectiveness.

## CLOSED AREAS AND FISHERY MANAGEMENT

Regulations specifying temporal and/or spatial closures of fishing grounds with the objective of protecting juvenile fish, nursery grounds, and spawning aggregations have an extensive history and remain widely employed in marine fishery management, often in concert with other management approaches such as quotas, effort limitations and gear regulations. Virtually all management systems for artisanal fisheries employed some form of stricture on the time and place of harvest and the recorded use of fishery closures in western societies extends to the middle ages (Anon., 1921). The migration circuit concept (Fairweather, 1991) (Fig. 2) provides a useful illustration of key considerations underlying traditional spatial and temporal closure strategies in fishery management. Many marine fish and invertebrate species have well defined spawning locations. During spawning, adults are often aggregated and vulnerable to capture. Closures of these grounds during the spawning season is often one of a suite of management measures employed to protect stocks and/ or reduce exploitation rates. Transport of planktonic eggs and larvae in ocean currents to nursery areas with favourable habitat and/or food resources is common in many species. Restrictions on fishing in nursery areas in many systems has been undertaken to protect small fish and to preserve vulnerable habitats. Closures of these types have typically been on a seasonal rather than a long-term basis. It is clear that unless harvest is controlled at points other than the spawning and nursery areas, the stock can be depleted. Management strategies that attempt to limit catch, effort, and/or the size of individuals harvested are aimed at controlling the impact of harvesting on other components of the life cycle. For seasonal closure systems, habitat destruction by fishing practices in sensitive areas at other times of the year can adversely affect recruitment processes. The development of no-take marine reserves involves the establishment of long-term closures at key locations to prevent this disruption and to protect other components of the system, including ecologically important species.

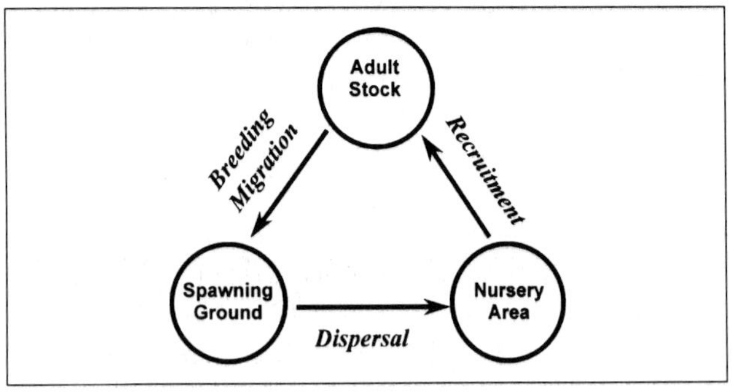


Fig. 2. Migration circuit concept for marine organisms (adapted from Fairweather, 1991).

The current emphasis on no-take marine reserves therefore extends the more traditional forms of fishery closures to encompass broader objectives, more inclusive levels of protection, and broader time horizons. It places the strategy of fishery closures within the context of ecosystem-based management and the need to understand system-wide effects. The importance of maintaining representative habitat types and loci of high abundance and species diversity assumes high priority. The tradition of closed area management may serve as a guide and stepping-stone to the broader implementation of no-take zones. By building on the experience gained in the establishment of closed areas for more limited objectives and on the acceptance of traditional closure strategies among fishers, it may be possible to promote the broader use of NTZs as effective management tools.

## ECOSYSTEM EFFECTS OF FISHING

The high variability of marine ecosystems at most scales means that it is difficult (but all the more important) to tease apart the physical and biotic driving functions and to separate natural from human impacts; any measure of change in a system must be understood or grounded in a well-defined natural standard or benchmark against which the measured changes can be evaluated. It is clear that ecosystem effects of fishing can be profound; for recent comprehensive reviews see Dayton et al. (1995; in press) and Hall (1998).

The effects of fishing on biodiversity range from biological extinction to shifts in species composition. Some marine organisms have in fact been driven extinct by human activities (Atlantic gray whale, great auk, Caribbean monk seal, etc.). Others are probably close to extinction; for example, northern right whales, Mediterranean and Hawaii monk seals, Irish ray, and white abalone. Many others, although not biologically extinct, are functionally so.

### By-catch and Incidental Take

By-catch is perhaps the most serious general environmental consequence of fishing (Alverson et al., 1994). Most fisheries have some by-catch, but net, trawl and long-line fisheries have particularly important by-catch, especially of vulnerable species. While technical improvements have vastly reduced the incidental take, the best-known example was the tuna purse seine fishery in the Pacific that may have killed over 6 million porpoise by 1987. Approximately 5 and 20% of the Bering and Western North Pacific Dall porpoise stocks were taken in 1987 in the Japanese salmon drift-net fishery. Sea turtles are vulnerable to being killed in various fisheries, and these species are considered highly endangered. Many seabirds also are vulnerable in many types of gill nets and even long-lines such as the tuna long-line fishery that may have taken tens of thousands of albatrosses. Because of low growth and reproductive rates, populations of sharks and rays are particularly vulnerable. While data on the massive takes of sharks and rays are poor, evidence of severe depletion, local extinctions and destruction of pupping habitats is increasing.

Dredging and bottom trawling can be very destructive, but because so few studies documented marine habitats before they were trawled, we may never know how many species may have been extirpated. We know, for example, that there used to be dense populations of deep-sea corals in many parts of the world; few studies describe these habitats but we know they are diverse. Koslow and Gowlett-Holmes (1998) report at least 299 species from a single short seamount cruise near Tasmania; 24–43% of these were new to science, and the benthic biomass from fished seamounts was 83% less than from lightly fished or unfished habitats. These authors point out that, after 100 years of exploration, only 598 seamount species had been described worldwide, but over 1300 species have been found in the Coral Sea alone over the past decade; over 60% of these were new to science, and 5–10% were endemic to a small area. The life history of most of these species is unknown, but they appear to have exceptionally slow growth and reproductive rates. *Lophelia* banks off Norway and the Faroe Islands have several hundred species in association, and with the exception of small areas of Norway, most have been heavily damaged. A small part of the *Oculina* reefs off Florida have been saved, but most have been reduced to rubble by fishers. Reserves cannot, of course, bring back biological extinct species, but they may help the recovery of functionally extinct ones and prevent the loss of others.

One poorly appreciated impact of by-catch is that there are numerous indirect effects. For example, trophic relationships are affected by disturbing the sediment such that infaunal species, normally buried, are brought to the surface where other species consume them. Another indirect effect is the sometimes large amount of sedimentation resulting from the trawling activities. Finally, perhaps most importantly, relatively large amounts of by-catch is simply discarded back to the sea (for example, Australian prawn trawlers discarded some 3000 tons of material for each 500 tons of prawns) and for many years the by-catch ratio often was 15:1. These affect marine ecosystems by creating large patches of food that alter the foraging behaviour of other species. In addition, this massive input of organic pollution can deplete the oxygen over remarkably large areas (Dayton et al., 1995).

### Habitat Destruction

Trawling and dredging destroy vulnerable benthic communities by killing juveniles of the target species, killing epibenthic species (many of which are likely to be important nurseries or food sources for other species), and altering the substratum in ways that open the community to colonization by opportunistic species (Hall, 1998) (see Fig. 3). These destructive impacts are so widespread and

Fig. 3. Example of benthic fauna in undisturbed (upper) and heavily impacted (lower) sites in scallop fishing grounds in the Gulf of Maine, USA (courtesy Page Valentine USGS, Woods Hole, MA).

complete that little, if any, unfished habitats with economically exploitable stocks remain for use as controls. Despite the well-known destructive nature of trawling and dredging, it is rare for management to exclude them from any areas except as a response to fishing gear conflicts or over by-catch of other commercial species.

### Debris and Ghost-Fishing

Many forms of fishing generate large amounts of debris that are lost or discarded at sea. For example, while the north Pacific drift gill net fishery was active, it lost perhaps 20% of the some 30–40,000 km of nets a day, much of which continues to fish almost indefinitely ('ghost-fishing'). Crab pots in the northern ocean are also commonly lost and continue to kill animals. This ghost fishing is poorly understood but probably kills substantial numbers of mobile benthic animals. Diving sea birds are also vulnerable to entanglement in fishing gear.

## DESIGN OF MARINE RESERVES

Design considerations for marine reserves and networks includes the individual size, configuration, placement and number of reserves. The most appropriate design for reserves and/or network of reserves depends on the specific objectives for their establishment. The development of designs for terrestrial reserves was initially guided by consideration of species-area relationships and the equilibrium theory of biogeography. Subsequent theoretical and pragmatic considerations in conservation biology centred on the determination of minimum viable populations sizes, the impacts of habitat fragmentation, and issues such as the establishment of migratory corridors linking reserves. Critical considerations for the development of marine reserves, however, differ substantially and include the recognition of the wide dispersal characteristics of many marine organisms and the importance of environmental variability on a broad range of time scales. Different theoretical constructs are necessary to deal with the distinctive characteristics of marine systems.

Schwartz (1999) distinguished between 'fine filter' conservation reserves targeted at particular species or populations in danger of decline or extinction and 'coarse filter" reserves aimed at protecting target species, habitats, and ecosystem processes and linkages. Much of the effort devoted to defining the required size of marine reserves based on minimum biomass constraints for fishery management has been implicitly based on fine-filter, single-species analyses. In contrast, the broader dialogue concerning the potential usefulness of marine reserves has centred around coarse filter considerations.

> **Guidelines for the Development of Marine Reserves**
>
> - Clearly specify the goals, objectives, and expectations for the reserve including the species, communities, and habitats to be protected and the role of the reserve within a network if appropriate.
> - Choose reserves to represent a broad spectrum of (a) environmental conditions, (b) habitat types and quality (c) biotic communities, (d) oceanographic features including current systems, upwelling zones, and retentive features (e) depth and latitude within biogeographic regions, and (f) levels of human impact.
> - Design reserves to match the scale of ecological and oceanographic processes. Reserves should be sufficiently large to be self sustaining and to minimize edge effects. Placement of reserves within networks should consider dispersal and transport among sites in oceanographic features
> - Replicate reserves of similar habitat types and biotic communities as a hedge against local catastrophic events at any one individual site and to allow effective and statistically valid evaluation of reserve performance.
> - Apply principles of adaptive management in reserve design and evaluation and retain flexibility to accommodate changing conditions.
>
> (adapted from Murray et al., 1999)

### Dispersal

Understanding and quantifying the dispersal characteristics of marine populations at different life history stages remains the most difficult and challenging issue in establishing effective marine reserves, but is the least developed and the greatest source of uncertainty.

If a reserve is to be self-sustaining, then careful attention to dispersal patterns is essential. Consideration of the reproductive strategies of the target organisms and the oceanographic features in the vicinity of the reserve is paramount. The probability of retention of eggs and larvae within the reserve is a function of the mode of reproduction, duration of early life stages, stage-specific characteristics and behaviour (demersal or pelagic eggs, depth preferences and locomotion of larvae, etc.), and hydrodynamic factors such as current speed and direction.

Species characterized by egg brooding behaviour, direct development, viviparous reproductive patterns or limited egg/larval stage duration and limited dispersal are good candidates for small scale, self-sustaining reserves. Examples of species exhibiting this patterns include abalone populations and some coral species. In contrast, eggs and larvae of other species can be expected to be transported some distance from the reserve with varying degrees of dispersion around an average settlement distance.

The oceanographic setting is crucial, especially for the egg and larval stages. Careful attention to circulation patterns and larval behaviour is essential, especially with respect to issues such as retention and loss from the vicinity of the reserve (Fig. 4). Various types of retentive physical systems often are utilized by marine organisms in their overall reproductive strategy. The placement of reserves in persistent current/counter-current and gyre systems can enhance the probability of retention and return and increase the likelihood of successful self-seeding of the reserve. This can be important for species with relatively long egg and larval stages. In advective systems, a series of reserves in the flow field may be required depending on the variability in the larval dispersal distance. Such a system may effectively operate as a single-source system in a region and can only persist if some propagules are retained within the source site. In more diffusive systems, there may be multiple exchanges among adjacent reserves (Fig. 4).

Direct empirical information on dispersal distances during the early life stages is available for very few marine populations. The technical challenges in tracking these stages are formidable, although advances in mark-recapture methodology (e.g. coded microwire tags, transmitting tags) and differentiation possible with new genetic techniques and biogeochemical markers offer considerable promise. Some insights into the *potential* dispersal ranges are possible based on consideration of the stage durations. Species characterized by short stage durations have a higher probability of retention near the reserve sites while longer stage durations may result in greater dispersal. Houde and Zastrow (1993) provide information on larval stage durations for representative fish species in different system types (Fig. 5). The median larval stage durations are lowest among coral reef fishes and large pelagic species (e.g. tunas) in open oceanic environments. In contrast,

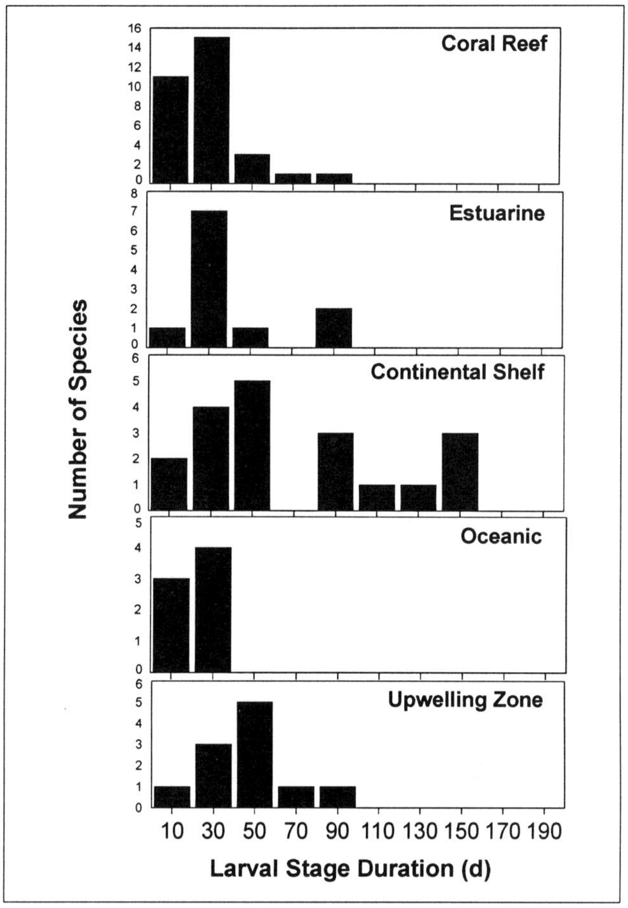


Fig. 5. Estimates of larval stage duration for fish species occupying five general habitat/system types (based on Houde and Zastrowe, 1993).

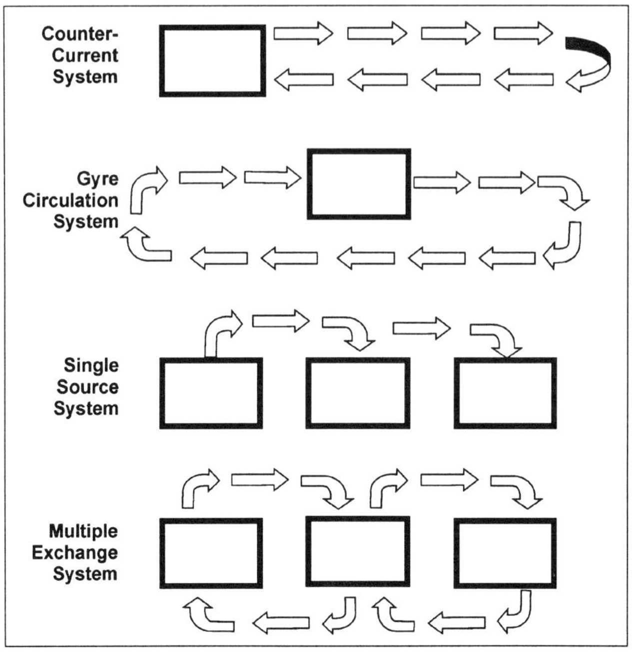


Fig. 4. Representation of four system types relevant to the placement of marine reserves including two retentive systems (counter-current and gyre type), an advective system, and a diffusive system.

some of the longest durations are found among fishes inhabiting continental shelf environments, implying the greatest *potential* for dispersal. Attwood and Bennett (1995) employed estimated residual transport of surface waters and larval duration times to estimate potential larval diffusivity.

The application of hydrodynamic models has assumed increasing importance in the study of recruitment processes of marine organisms. Coupled physical–biological models will undoubtedly assume a central role in modelling the potential efficacy of reserves. Lewis (pers. comm.) employed this approach in assessing the role of closed areas on Georges Bank off the northeastern United States as larval source sites for Atlantic Sea scallop. Examination of the predicted settlement locations of post-larvae originating from two closed areas indicated that in this gyre-type system, the closed areas were self-seeding. The predicted trajectories also indicated cross-seeding and sources of new recruits in open areas (Fig. 6).

## Reserve Size and Number

Consideration of required sizes and optimal configurations of marine reserves include the desirability of establishing

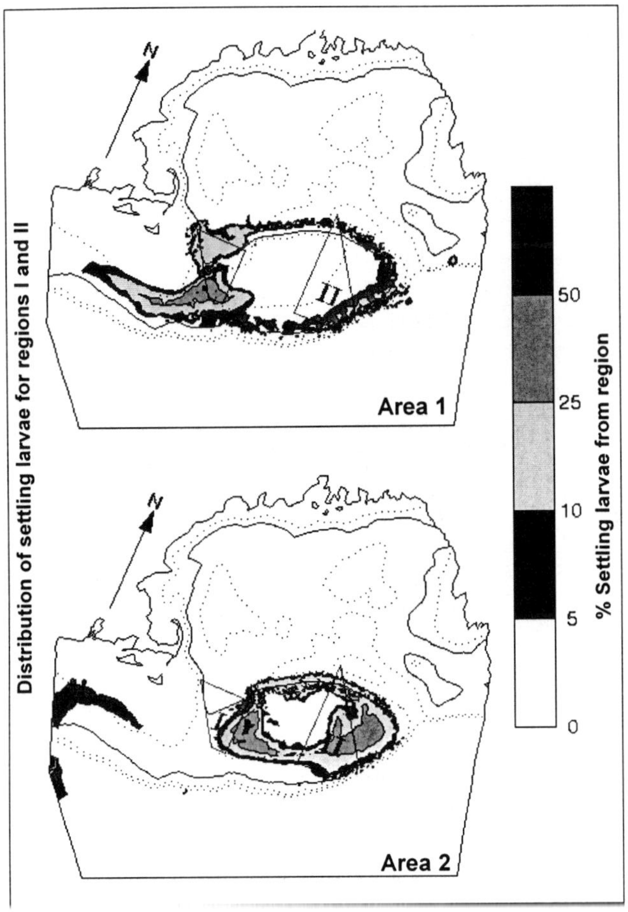


Fig. 6. Results of hydrodynamic model coupled with a larval behaviour component for Atlantic sea scallop depicting the predicted distribution of settlement-stage post-larvae origination from two closed areas on Georges Bank off the northeastern United States (polygons designated by roman numerals) from (Craig Lewis, Dartmouth College, Hanover, NH, USA, pers. comm.).

protected areas that are sufficiently large to contain self-sustaining populations and encompass diverse biological communities. Consideration of 'edge' effects range from concerns about impacts from adjacent areas across the reserve boundaries to enforcement issues and concerns over illegal encroachment. Large reserves hold the advantage that species with broad-scale dispersal patterns at one or more life stages have a greater potential for protection, more species can be expected to fall within the reserve boundaries, and larger reserves have a smaller perimeter to area ratio. Circumstances favouring small reserves include situations where a species under protection occurs in relatively small discrete areas and exhibits limited dispersal or where specific limited habitat types require protection. Smaller reserves also generally encounter less opposition in development and implementation

General guidelines for the requisite size of marine reserves depend on the specific objectives for their establishment. For certain fishery management objectives, the issue is more properly framed as protecting a certain proportion of the adult biomass of a stock than protecting a specified proportion of the area within a region. For other objectives, protecting designated habitat types and or representative communities may be most important. Many modelling studies of the necessary size of marine reserves for single-species populations have been based on the implicit assumption of relatively homogeneous distribution and/or broad dispersal and mixing patterns that permit a direct translation from the area to the proportion of the population protected. This assumption requires careful scrutiny in application to actual situations where heterogeneous distribution patterns occur.

### Minimum Viable Biomass

Important reference points widely used in fishery management can be classified into (a) limits to exploitation and (b) target exploitation rates. In the former, an attempt is made to specify levels of exploitation and/or minimum biomass at which the probability of recruitment overfishing and stock collapse is high, while in the latter, exploitation rates that optimize yield in some specified fashion are determined for a particular age or size-specific exploitation pattern. Limit reference points therefore define danger points to be avoided, while target reference points define specific goals to be met. These reference points can be readily recast in terms of population size or biomass. Both limit and target reference points have been addressed in the context of the potential effectiveness of marine reserves. We will focus on estimates of the minimum viable population biomass to provide general advice on the required size of marine reserves.

If the objective is to preserve at least a minimum adult biomass, limit reference points for conventional management measures can be employed. These estimates can serve as a counterpart to the development of minimum viable population sizes applied more broadly in terrestrial systems. The rate of recruitment at low spawning biomass is the primary determinant in considering a maximum exploitation rate or a minimum adult biomass level. Empirical evidence based on analyses of temperate and boreal fish populations has led to a rule of thumb that a minimum of 20–35% of the spawning biomass should be preserved (Mace and Sissenwine, 1993). A suggestion that at least 20% of the adult population in reef systems should be protected by marine reserves (reviewed in Bohnsack, 1996) is based on this general approach.

There are, however, substantial differences among taxa in the required level of protection based on life history characteristics and other considerations. Further, within a species, different populations in different environmental regimes can exhibit wide variation in the minimum required spawning biomass level. In particular, populations located at the extremes of the range of a species may exhibit lower levels of resilience and may require higher minimum biomass levels (expressed as a proportion of the virgin level). Estimates of the minimum adult biomass (proportion

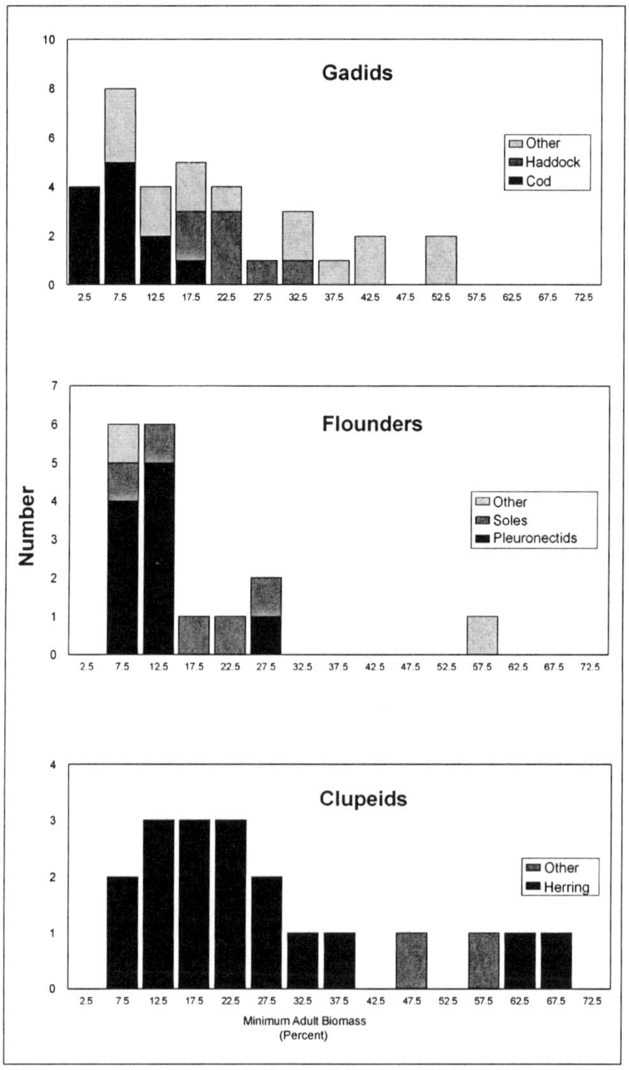


Fig. 7. Estimates of minimum viable adult biomass required for stock replacement for three groups of temperate-boreal water fishes (after Mace and Sissenwine, 1994).

of virgin stock) for three major temporate and boreal fish groups (Mace and Sissenwine, 1994) show substantial variation within and between groups (Fig. 7). For example, among the cod-like fishes (gadids), Atlantic cod (*Gadus morhua*) generally shows high levels of resilience relative to Atlantic haddock (*Melanogrammus aeglefinus*) while other gadids exhibit much lower levels of resilience still (Fig. 7). Estimates for individual populations of cod and haddock show considerable within-species variation. Comparison among the gadids, flounders, and herring-like fishes suggest generally higher levels of resilience for flounders than gadids and the modal level of resilience is lowest for the clupeids. These observations indicate that choices concerning target levels of minimum adult biomass must be tailored to the individual populations and locations. The reserve design must also take into account the dispersal characteristics of the species to provide even minimum protection to the stock.

## Optimum Reserve Size

Estimates of the reserve size needed for more optimal exploitation indicate that nearly 50% of the area must be protected by the reserve(s) based on metapopulation theory (Man et al., 1995). Other modelling studies have indicated that the optimal level of exploitation is linked to the transfer rates between open and closed areas with higher transfer rates requiring larger reserve (protected population) sizes; again reserve sizes ranging from approximately 35% (Attwood and Bennet, 1995) to 75% (Sladek Nowlis and Roberts, 1998) of the area (population) were required to optimize yield (see Guènette and Pitcher, 1999). Lauck et al. (1998) reported that reserves of at least 50% of the area would be necessary as a hedge against uncertainty in a system in which exploitation rates could not be carefully controlled.

## Edge Effects

It will often be desirable to minimize the perimeter to area ratio of a reserve to reduce edge effects. Reserves with large perimeters relative to area may be adversely affected by events outside the reserve boundaries. While it would be desirable for the boundaries of a reserve to be defined by natural features (Agardy, 1997), the need to demarcate reserve boundaries simply and reliably in the field, particularly when set away from the shore, has resulted often in the use of simple reserve configurations. For a simple rectangular design, the perimeter/area ratio decreases hyperbolically with increasing area (Fig. 8). The shape of the reserve also affects the perimeter to area ratio. Long narrow reserves have a higher perimeter to area ratio than more symmetrical reserves; a circular configuration would of course have the minimum possible perimeter/area ratio for any possible planar design. The issue of edge effects is also connected to the question of the appropriate number

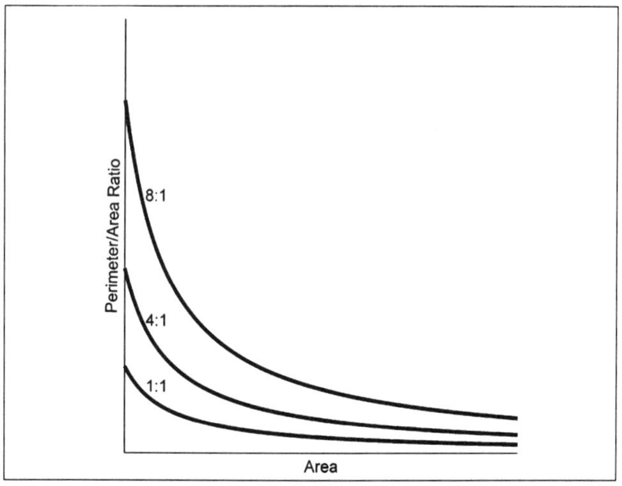


Fig. 8. Illustration of the perimeter to area ratio as a function of the reserve area for three rectangular shapes corresponding to 1:1, 4:1 and 8:1 length to width ratios.

of reserves to be placed within a region. Multiple reserves necessarily have a higher overall perimeter to area ratio than a single reserve of the same area as the sum of the individual smaller reserves.

In species with limited dispersal characteristics, it may in fact be desirable to have a somewhat higher perimeter to area ratio from a fisheries management perspective to permit some benefits accruing from spill-over effects or larval replenishment. In supply-limited recruitment situations, increasing the reserve perimeter, coupled with effective placement with respect to currents and oceanographic features can be important in 'capturing' propagules.

### Number of Reserves

The number of reserves to be placed in a specified geographical region involves tradeoffs between the area of individual reserves as specified above and factors such as spreading the risk that catastrophic impacts will adversely affect an individual reserve. Other considerations include the requirement to protect diverse habitat types or species assemblages that are not contiguous, possible small concentrations of high biological diversity, and desirability of replication for evaluation of reserve performance (Murray et al., 1999). The question of whether to have a single large, or several small, reserves is still under debate (Schwartz, 1999), but many commentators now generally favour larger reserves where feasible (although not necessarily a single large reserve). The marginal decreases in the perimeter to area ratio with increasing reserve area declines at larger sizes (Fig. 8) suggesting diminishing returns at large reserve sizes in circumstances where it is desirable to limit the exposure at the boundaries. We suggest a minimum of two reserves in each representative system type.

Daan (1993) examined the question of whether a single large reserve would be more effective than a larger number of small reserves and reported that a larger protected area resulted in higher survival rates in a model of juvenile-adult movement patterns and effort allocation. DeMartini (1993) provided modelling results for three species-types with varying life history dispersal characteristics and reported that a single large reserve was more effective for species with higher dispersal rates. However, for species with fine scale dispersion patterns, numerous small reserves are more effective (DeMartini, 1993; Quinn et al., 1993). Quinn et al. (1993) noted that the optimal reserve spacing in such a situation is less than the dispersal distance.

## ASSESSING THE EFFECTIVENESS OF MARINE RESERVES

The experience with established marine reserves provides an important basis for evaluating the general utility of the reserve concept. Because many of the existing reserves have been established on an *ad hoc* basis, however, clear interpretation of their effects can be difficult. While examination of changes within the reserve boundaries is relatively straightforward, determination of the effects of the reserve through spill-over of juveniles and adults or export of eggs and larvae is far more challenging. Consideration of the potential effectiveness of marine reserves for fishery management purposes is necessarily centred on these issues. A reserve that does not contribute to the productivity of areas open to harvest through some level of dispersal at one or more life stages, cannot enhance yield or resilience in these areas. In some instances, it may be feasible to use rotating harvest schemes in which areas are sequentially opened and closed if habitat damage and other factors do not accompany the harvest. Models of potential reserve performance have been employed to evaluate key issues in reserve design and to complement the information based on empirical performance of reserves. The many potential benefits of reserves (see box) has contributed to tremendous interest in their observed and predicted performance.

> **Potential Benefits of Marine Reserves**
>
> Fishery-related
> - Increase abundance, average size of target organisms, reproductive output, and genetic diversity
> - Enhance fishery yield in adjacent grounds
> - Provide simple and effective management tool which is readily understood and enforced
> - Guard against uncertainty and reduce probability of overfishing and fishery collapse
> - Protect rare and valuable species
> - Provide opportunities for increased understanding of exploited marine systems
> - Provide basis for ecosystem management
>
> General
> - Increase species diversity and community stability
> - Increase habitat quality
> - Provide scientific control sites and undisturbed monitoring sites for assessing human impacts
> - Create or enhance non-extractive uses, including tourism
> - Reduce user conflicts
> - Improve public awareness, education, and understanding
> - Create areas with intrinsic value
>
> (after Bohnsack, 1996)

### Within-Reserve Effects

Dugan and Davis (1993) reviewed 31 studies and reported that 24 showed greater population abundance for exploited species in reserves, while four reported reduced abundance and three did not examine abundance. These general conclusions hold with the inclusion of 37 newer studies (Table

Table 1

Summary of additional studies since Dugan and Davis (1993) showing evidence of effects of marine reserves on populations of exploited species in terms of increased abundance, individual size, reproductive output, recruitment and fishery yield. — = not applicable.

| Taxa | Area | Abundance | Size | Reprod-uction | Recruit-ment | Yield | Species richness | Reference |
|---|---|---|---|---|---|---|---|---|
| Estuarine Fishes | Florida, USA | Yes | Yes | Yes | — | — | Yes | Johnson et al. 1999 |
| Temperate Fishes | South Africa | — | Yes | Yes | — | — | — | Buxton 1993 |
| Temperate Fishes | South Africa | — | — | Yes | Yes | Yes | — | Tilney et al. 1996 |
| Temperate Fishes | Tasmania, Australia | No | Yes | — | — | — | Yes | Edgar & Barrett 1999 |
| Temperate Fishes | France | Yes 6, No 4 | Yes | — | — | — | — | Dufour et al. 1995 |
| Temperate Fishes | France | Yes | Yes | Yes | Yes | — | Yes | Harmelin, et al. 1995 |
| Temperate Fishes | Gulf of Maine | No | No | — | — | — | — | Cadrin et al. 1995 |
| Temperate Fishes | North Sea | Yes | Yes | | | | | Horwood et al. 1998 |
| Temperate Fishes | Georges Bank | Yes | Yes | — | — | — | — | Murawski et al. in press |
| Coral Reef Fishes | Caribbean | Yes | Yes | Yes | — | — | — | Polunin & Roberts 1993 |
| Coral Reef Fishes | Barbados | Yes | Yes (18 of 24 sp.) | — | — | — | — | Rakitin & Kramer 1996 |
| Coral Reef Fishes | Barbados | Yes | Yes | Yes | No | — | — | Tupper & Jaunes 1999 |
| Coral Reef Fishes | Barbados | Yes | Yes | — | — | — | — | Chapman & Kramer 1999 |
| Coral Reef Fishes | Saba, Netherland Antilles | Yes | Yes | — | — | — | — | Roberts 1995 |
| Coral Reef Fishes | Caribbean | Yes | Yes | — | — | No | — | Roberts & Polunin 1993 |
| Coral Reef Fish | Bahamas | Yes | Yes | Yes | — | — | — | Sluka et al., 1997 |
| Coral Reef Fishes | Belize | Yes | Yes | Yes | — | — | Yes | Roberts & Polunin 1994 |
| Coral Reef Fishes | Belize | Yes | — | — | — | — | Yes | Sedberry et al. (in press) |
| Coral Reef Fishes | Kenya | Yes | — | — | — | — | Yes | McClanahan 1994 |
| Coral Reef Fishes | Kenya | Yes | — | — | — | Yes | — | Watson et al., 1997 |
| Coral Reef Fishes | Kenya | Yes | — | — | — | Yes | — | McClanahan et al. 1999 |
| Coral Reef Fishes | Philippines | Yes | Yes | — | — | Yes | — | Russ & Alcala 1996a |
| | | Yes | Yes | — | — | Yes | — | Russ & Alcala 1996b |
| | | Yes | Yes | — | — | Yes | — | Russ & Alcala 1998 |
| Coral Reef Fish | Great Barrier Reef, Australia | Yes | No | — | — | — | — | Ferreira and Russ 1995 |
| Coral Reef Fishes | New Caldonia | Yes | No (6), Yes (2) | — | — | — | Yes | Wantiez et al. 1997 |
| Coral Reef Fishes | Seychelles | Yes | — | — | — | — | Yes | Jennings et al., 1996 |
| Lobster, *J. Edwardsii*, | New Zealand | Yes | Yes | Yes | — | — | — | MacDiarmid & Breen 1992 |
| *J. verreauxi* | | No | — | | | | | |
| Lobster | Mexico | Yes | Yes | Yes | Yes | — | — | Lozano-Alvarez et al. 1993 |
| Lobster | Tasmania, Australia | Yes | Yes | Yes | — | — | — | Edgar and Barrett 1999 |
| Queen conch | Bahamas | Yes | Yes | Yes | Yes | Yes | — | Stoner and Ray 1996 |
| Abalone | Tasmania, Australia | Small -No Large -Yes | Yes | Yes | No | — | — | Edgar and Barrett 1999 |
| Abalone | California, USA | No | Yes | Yes | Yes | Yes | — | Karpov et al. 1998 |
| Abalone | British Columbia, Canada | Yes | Yes | Yes | — | — | — | Wallace 1999 |
| Scallops | Georges Bank | Yes | Yes | — | — | — | — | Murawski et al. in press |
| Urchins | Tasmania, Australia | No | — | — | — | — | — | Edgar and Barrett 1999 |
| Algae | Tasmania, Australia | Yes (cover) | — | Yes | Yes | — | Yes | Edgar and Barrett 1999 |

1). Several studies also reported larger individual sizes, total biomass, or greater species biodiversity in no-take reserves.

While changes inside protected areas are increasingly well documented, few studies have shown impacts in surrounding areas, partly because few reserves are large or old enough to show any measurable effects. Tagging studies have shown fish moving from reserves to surrounding fishing grounds in South Africa, Australia, Philippines, and the U.S.

Efforts to determine the effectiveness of marine reserves would benefit strongly from planned experiments in which prior baseline measurements are made, valid control sites are identified, rigorous statistical treatment is applied, and adaptive management is practised.

## Regional Summaries of Reserve Implementation

Although commonly used on land, no-take protection has only recently been applied to the ocean. Marine reserves in general have now been established on a global basis; we will focus on some of the better studied systems and areas. Many tropical island nations had no-fishing zones in the past but lost those traditions. Hawai'i, for example, had an extensive network of no-fishing 'kapu' zones in which violation was a serious offense punished by death. In modern times, Hawai'i has led in re-establishing reserves when it began prohibiting fishing in conservation zones such as Hanauma Bay, Oahu in the 1960s. In the 1990s fishing was prohibited in much of the northwest Hawaiian Islands. In 1998, Hawai'i established 19 reserves around the main islands that protect 20–30% of the deep reef habitat.

Although use of no-take reserves varies considerably by region and country, certain trends are common. Initially MPAs are created to protect unique or special areas with particular importance for recreation, tourism, science or conservation. In the U.S., for example, the first MPAs protected coral reefs in John Pennekamp Coral Reef State Park, Florida in 1960 and Buck Island National Monument, U.S. Virgin Islands in 1961. Over time, new, larger, and more representative areas are added and the level of protection tends to increase until eventually no-take protection is applied. In the Bahamas, for example, the 456 km$^2$ Exuma Land and Sea Park was first established in 1958 and went through several stages before becoming a no-take zone in 1986. In New Zealand the first no-take reserve was established at Leigh after ten years of effort. Since then, 16 no-take reserves have been established and more are being planned. A similar pattern has occurred in the U.S., Spain, Bahamas, and elsewhere.

Although most existing no-take reserves were started in tropical coral reef areas, their use is expanding rapidly into temperate and boreal regions. Their use also has expanded to include reducing conflicts between incompatible uses, protecting marine biodiversity, and enhancing scientific research. One reserve in Alaska was intended to protect food resources for dwindling pinniped populations from conflicts with fishing. Interest in using reserves for fishery management purposes in coral reef systems developed partly because other methods used to regulate catch and fishing effort have been ineffective and because large population increases among exploited species were observed in some reserves. Large areas of Georges Bank have been effectively closed to fishing to protect depleted groundfish stocks. Reserves also have been established that prohibit bottom fishing to protect critical fish habitat at high relief pinnacles in Alaska and approximately 293 km$^2$ of *Oculina* coral habitat off the east coast of Florida.

Little correlation exists between establishment of no-take reserves and a country's wealth, level of scientific knowledge, or public literacy. Many reserves exist in developing countries in Central America, the Caribbean, Africa, and the Indo-Pacific island nations. such as the Philippines and Indonesia. Ho Chan Reserve, Belize and Apo and Sumilon Island reserves, Philippines are perhaps the most famous examples. Among more developed countries reserves exist in the U.S., Australia, Spain, New Zealand, and South Africa. South Africa has one of the largest networks with 112 MPAs covering 17% of the coastline, although only six, covering 4.9% of the coastline, provide no-take protection. In contrast, other "developed" countries including the United Kingdom, most of northern Europe, Japan, and mainland Asia have essentially no reserves. A few of the 22 countries in the Mediterranean have mostly small reserves. Italy, for example, currently has only about 10 km out of its 7550 km coast protected by MPAs, while Spain leads with a network of 11 reserves developed after establishing its first at Tabarca in 1986. Canada established its first reserve at Whytecliff Park, British Columbia in 1993. It currently has only two no-take reserves but this may change since it is in the process of developing a national policy.

The U.S. currently has mostly small reserves developed independently by various agencies at state and regional levels. One of the first was not intended to protect marine resources. It started in 1962 when approximately 40 km$^2$ (22% of aquatic areas) of estuarine habitat at the Merritt Island National Wildlife Refuge were closed to all public access for security of the Kennedy Space Center at Cape Canaveral. This no-take area was later shown to have greater fish biodiversity, and abundance, size, and age structure among exploited fishes than similar habitats in surrounding fished areas (Johnson et al., 1999). The first planned network was established in 1997 as part of the Florida Keys National Marine Sanctuary. It included 18 small (mean 0.87 km$^2$) no-take zones that protected coral reefs and one larger (31 km$^2$) no-take ecological reserve that includes a variety of habitats. Although only 46 km$^2$ (0.5%) of the Sanctuary are no-take, plans are in progress to significantly expand the network to as much as 5% of the Sanctuary in 2000. Other reserves were established in 1999 in the U.S. Virgin Islands and Puerto Rico and several are proposed off Florida. On the west coast, California has 104 MPAs but only 0.2% of coastal waters are protected by no-take provisions. Additional small reserves exist in Washington and Alaska.

## Predicting the Effect of Marine Reserves

Models developed to evaluate the potential performance of marine reserves can be classified along several principal dimensions including whether (a) part or all of the life cycle is modeled (b) adults are sedentary or mobile, (c) dispersal at one or more life stages is explicitly modeled and (d) the age or size composition of the population is considered. The structure of the models and the assumptions employed strongly influence the conclusions drawn. Different modelling efforts have also made varying assumptions concerning the redistribution of fishing pressure from closed to open areas. In the absence of other controls to reduce fishing pressure, it is probable that the fishing intensity per unit area outside the reserve will increase and this must be considered in the overall evaluation of reserve impacts.

A conceptual model structure is provided in Fig. 9. This model has a simplified age structure (recruits and adults) for a species which becomes vulnerable to the fishery as juveniles (here designated as recruits) and is distributed inside and outside of a hypothetical NTZ with interchange from the closed to the open and areas. The transition between the recruit and adult stages is given by the survival rate in the reserve ($s_r$) and in the open area ($s_0$); the survivorship in the open area is affected by the fishery exploitation rate. The recruitment is a function of reproductive output by the adults in both areas and the survivorship during the pre-recruit phase of the life cycle ($f_r$ and $f_0$ for the reserve and open populations respectively). Transfer between the open and closed areas can be effected by dispersal of eggs, larvae, and juveniles and by movement and migration of the adults. In the simple example depicted, the reserve supplies adults to the open area through movement at rate $T_1$ and eggs and larvae are transferred from the reserve to the open area at rate $T_2$ and then subjected to mortality. More complex patterns of interchange involving two-way transfer would of course be possible.

Three general classes of models have been widely used to examine the potential effectiveness of reserves. Following the terminology of Guènette et al. (1998), the types and general characteristics of each are:

1. *Logistic* (and related) population models do not explicitly consider the age or stage structure of the population. Recruitment, individual growth and mortality are represented in an aggregate production function. Applications of these models for evaluation of reserve performance have typically dealt implicitly with dispersal processes (but see Quinn et al., 1993).
2. *Yield per recruit* models specify the expected yield over the lifetime of a cohort under different exploitation patterns and consider only growth, survival and dispersal from the recruitment stage through the remainder of the life cycle. Extensions include determination of the expected lifetime reproductive output (spawning biomass or egg production per recruit) and economic value per recruit.
3. *Dynamic pool* models incorporate yield and spawning biomass recruit models with a stock-recruitment component to model the complete life cycle and incorporate dispersal processes for both pre- and post-recruits. Age or stage-specific models with an equivalent structure have also been specified in matrix form.

Other model types have also been employed, including a metapopulation model in which the fraction of patches or cells occupied, rather than population size or biomass *per se*, is considered (Man et al., 1995). Metapopulation models typically assume an equal interchange among patches through dispersal processes.

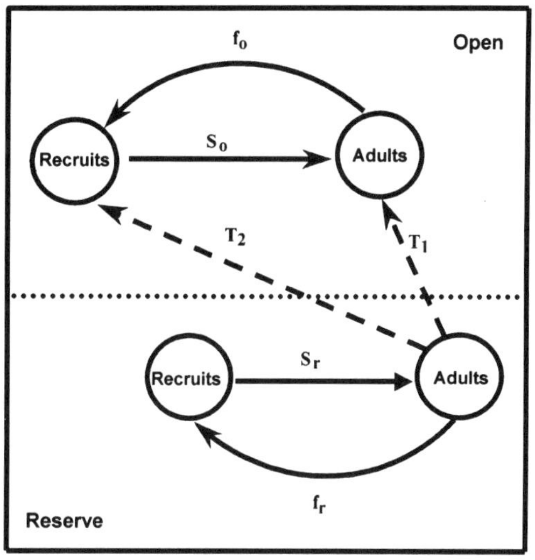


Fig. 9. Conceptual model for a simple age structured system for a population distributed within and outside a reserve.

Mangel (1998) considered a simple logistic model structure in which a fraction of the area occupied by the population is specified and all harvesting activity is prohibited. In this model, the reserve protects a specified fraction of the population and reduces the overall exploitation rate. In this deterministic model, various combinations reserve area and exploitation rates outside the reserve can be used to achieve the same overall target exploitation rate (or population size) for a specified intrinsic rate of population increase; see Fig. 10). The same model structure was used by Lauck et al. (1998) to examine the implications of uncertainty in the harvest rate achieved outside the reserve.

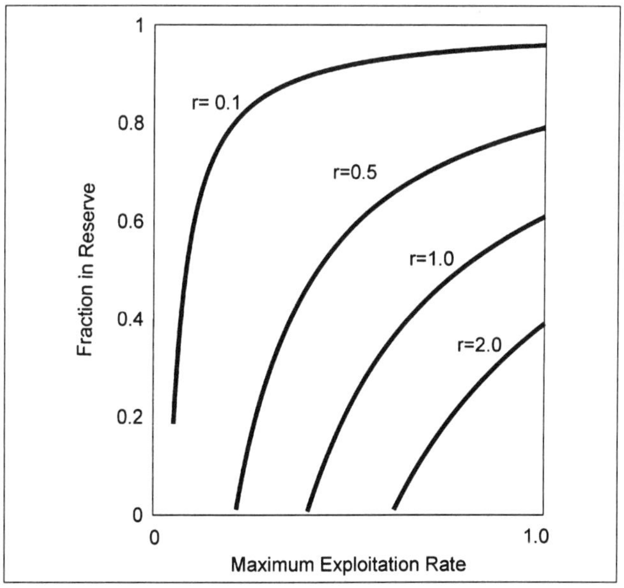


Fig. 10. Isolines of combinations of maximum exploitation rate and fraction of the population protected by marine reserves for four levels of the intrinsic rate of increase (r) of the population (based on Mangel, 1998). The lines represent combinations of the maximum exploitation rate and the area held in reserves that give the same result relative to a target overall exploitation rate.

Hastings and Botsford (1999) demonstrated that for a model with simple age structure, sedentary adults, and implicit dispersal of progeny, marine reserves can provide protection equivalent to reductions in exploitation rates and do not necessarily entail loss in yield.

Beverton and Holt (1957) pioneered the development of spatial harvesting models. They modeled the effects of closed (or otherwise inaccessible areas) on yield per recruit. This general strategy has been followed in several subsequent modelling efforts (Table 2) and extended to include consideration of reproductive output and economic value. General results from these analyses indicate that the yield per unit recruitment is generally lower than for the case of no reserves when dispersal rates are high except when fishing mortality rates are also high. However, positive benefits of reserves in terms of spawning biomass per recruit and value are generally predicted and these increases can be substantial when transfer rates between open and closed areas are relatively low or moderate. At high exchange rates, these benefits are sharply diminished. In general, at higher exchange rates, larger reserve areas are necessary to achieve benefits in spawning biomass per recruit. Because these models represent only the recruited portion of the life cycle, questions related to the potential benefits of dispersal of spawning products cannot be addressed. These models also cannot represent the effects of fishing on recruitment and therefore do not fully represent the impact of harvesting on the stock(s).

Table 2

Model types and characteristics employed in evaluating the potential effects of marine reserves in a fishery management context (Adapted from Guenette et al. 1998). For description of model types see text. Y/R = yield per recruit; SSB/R = spawning stock biomass per recruit; E/R = eggs per recruit; V/R = value per recruit.

| Species | Region | Model type | Life cycle representation | Output variables | Dispersal characteristics | Original source |
|---|---|---|---|---|---|---|
| Plaice | North Sea | Yield per recruit | Post-recruits | Y/R | Post-recruit diffusion | Beverton and Holt (1957) |
| Atlantic Cod and Haddock | Georges Bank | Yield per recruit | Post-recruits | Y/R SSB/R | Post-recruit diffusion | *Polacheck (1990) |
| Atlantic Cod | N/A | Dynamic pool | Full | Total yield | Pre-recruit dispersal/ adult movement | *Mohn (1996) |
| Atlantic Cod | Barents Sea | Dynamic pool | Full | Total yield | Pre-recruit dispersal/ adult movement | *Sumaila (1997) |
| Peneid Shrimp | Australia | Yield per recruit | Post-recruits | Y/R E/R V/R | Post-recruit diffusion | *Die and Watson *(1992) |
| Peneid Shrimp | Australia | Yield per recruit | Post-recruits | Y/R E/R V/R | Post-recruit diffusion | *Watson et al. (1993) |
| Jack+, DamselFish+, Surgeonfish + | Tropical Pacific | Yield per recruit | Post-recruits | Y/R E/R | Post-recruit diffusion | *DeMartini (1993) |
| Reef Fish | Philippines | Yield per recruit | Post-recruits | Y/R E/R | Post-recruit diffusion | *Russ et al. (1992) |
| Reef Fish | N/A | Metapopulation | Full | No. patches occupied | Nondirectional transport | *Man et al. (1995) |
| Reef Fish | Carribean | Dynamic pool | Full | Total yield/ value | Pre-recruit dispersal/ sedentary adult | Sladek Nowlis and Roberts (1998) |
| Red Sea Urchin | Pacific Coast, USA | Logistic w/ Allee Effect | Full | Total yield | Pre-recruit dispersal/ sedentary adult | *Quinn et al. (1993) |
| Surf Zone Fishes | South Africa | Dynamic pool | Full | Total yield | Pre-recruit dispersal/ adult movement | *Attwood and Bennett (1995) |
| Cod+ | North Sea | Movement/mortality | Post-recruits | Survival | Post-recruit diffusion | *Daan (1993) |
| Red Snapper | Gulf of Mexico | Dynamic pool | Full | Total yield/ value | Pre-recruit dispersal/ adult movement | *Holland and Brazee (1996) |
| N/A | N/A | Discrete logistic | Full | Total yield | N/A | Lauck et al. (1998) |
| N/A | N/A | Delay-difference | Full | Total yield | Pre-recruit dispersal/ sedentary adult | Hasting and Botsford (1999) |
| N/A | N/A | Discrete logistic | Full | Total yield | N/A | Mangel (1998) |
| Cod | Georges Bank | Mortality | Pre-recruits | Cohort survival | Pre-recruit dispersal | Lindholm (1999) |
| Cod | Georges Bank | Dynamic pool | Full | Total yield | Pre-recruit dispersal/ adult movement | Guènette and Pitcher (1999) |
| Sole | North Sea | Yield per recruit | Post-recruits | Yield-per-recruit | Post-recruit dispersal | Horwood et al. (1998) |

+Parameters chosen to be representative of the fish types.
*Space limitations preclude providing full citations in reference section. See Guènette et al. (1998) for journal citations.

Dynamic pool models combine yield and spawning biomass per recruit models with stock-recruitment models to define the entire life cycle or employ alternative formulations such as matrix population models for all life history stanzas. These models typically include consideration of dispersal at the early life stages (particularly eggs and larvae) as well as movement and migration of later life stages as appropriate. In contrast to the yield per recruit models, these models can account for the full effects of harvesting on the population. These models have indicated potential benefits in terms of total yield, principally resulting from increased recruitment as a result of enhanced spawning biomass (or egg production) with the use of reserves. These models also demonstrate that the overall resilience of the population to harvesting can be substantially enhanced when a fraction of the population is protected and serves as a source of progeny to other areas. To date, these models have generally employed simplified assumptions about dispersal of progeny and have not been tailored to specific situations using realistic flow fields (but see Attwood and Bennett, 1995).

## DEALING WITH UNCERTAINTY

The potential utility of marine reserves as a hedge against uncertainty in the assessment and management of fishery resources has received considerable attention. Sources of uncertainty in the analysis of fishery systems include (a) measurement error, (b) incomplete information concerning the population, community, and ecosystem dynamics (c) errors in model specification used to assess living marine resources and (d) implementation error in the execution of fishery management regulations due to an inability to control harvest rates effectively and circumvention of regulations.

Lauck et al. (1998) illustrated the impacts of implementation error in controlling the exploitation rate in a system in which both marine reserves and (imperfect) controls on harvest are employed to achieve a target population level. The exploitation rate was assumed to be a random variable following a beta probability distribution. As noted above, Lauck et al. demonstrated that reserve areas protecting at least 50% of the population were generally necessary to counter the effects of uncertainty in achieving the target population levels in this analysis. In this model, the effect of the reserve was assumed to be known with certainty (i.e., it protected a known specified fraction of the population). This situation is applicable for systems in which the harvested stock is sedentary and its distribution can be specified with certainty.

The potential impact of uncertainty in the protection afforded by reserve(s) has received far less attention. It is clear, however, that for more mobile species, the issue of uncertainty in dispersal and distribution and their implications for the protection afforded by a reserve must also be considered and will be central to the predicted performance of the reserve. If variability in the distribution of the stock with respect to reserve boundaries is important for mobile species, the overall uncertainty in the management system is compounded. An illustration of the variability induced by uncertainty in the exploitation rate alone and with added uncertainty in the proportion of stock protected by the reserve is provided in Fig. 11). It is, of course, expected that substantially larger population variability will occur when both types of uncertainty are considered. In these simulations, however, the mean population level is lower under the combined sources of variability and the risk to the population is greater.

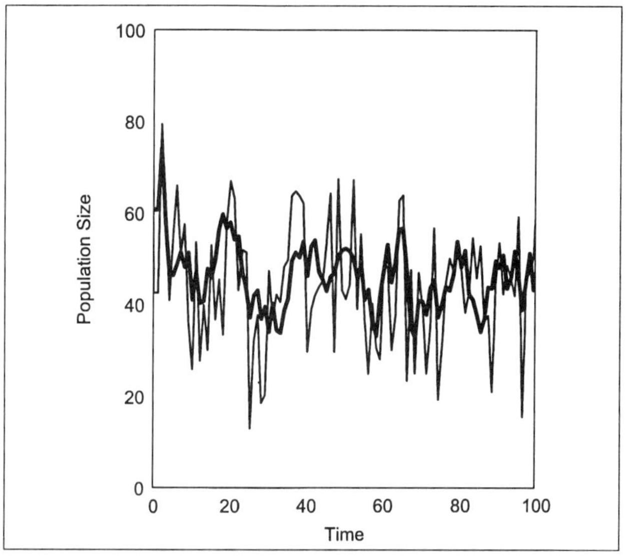


Fig. 11. Simulated population trajectories over 100 time steps for the case of a logistic harvesting model in which the exploitation rate cannot be precisely controlled (heavy line) and the case where implementation error is compounded by variability in the effectiveness of a designated reserve to protect a segment of the population. Parameters are as in Lauck et al. (1998) The exploitation rate and the effective reserve fraction (light line only) are random variables drawn from identical beta distributions with a mean of 0.5.

## SOCIAL AND ECONOMIC CONSIDERATIONS

The ultimate success in the development and implementation of marine reserves will depend greatly on social considerations such as the perceived impacts on user groups of the reserve and on levels of compliance and enforcement. Effective communication and consultation with those potentially affected by the placement of the reserve will be essential in reducing conflict and resistance.

### Zoning

MPAs and marine reserves are a form of marine zoning. Zoning has been widely used in the ocean to protect resources from various forms of human disturbance and to reduce user conflicts by segregating incompatible activities. Zones are used in Florida, for example, to separate stone crab trapping from shrimp trawling to allow both fisheries to coexist. No-spearfishing zones have been used for decades in the Florida Keys to reduce conflicts with line fishing and non-extractive recreational diving. Also, no-take zones reduce conflicts between divers and fishers and support scientific research.

## Enforcement and Compliance

The effectiveness of marine reserves depends on public acceptance, understanding, and compliance which vary greatly around the world. Patterns in New Zealand, Spain, and the U.S. suggest that public acceptance is a sociological process that develops with direct local experience and evidence of effectiveness. Successful reserves often have considerable effort devoted to increasing public awareness and education. Public support leads to stronger regulations, better enforcement, and improved compliance. Reserves established in 1997 in the Florida Keys, for example, are somewhat self-enforcing with good compliance in part because of extensive education efforts, while an experimental reserve at the Oculina Banks off the east coast of Florida was initially unsuccessful with poor compliance because much of the public was unaware of the regulations or confused about their purpose. Allowing surface fishing at the Oculina Bank reserve off Florida also made poaching easier and enforcement more difficult.

A review of Florida marine reserves concluded that successful reserves tended to follow key design principles that included permanent no-take protection and allowing public access. The no-take provision is easier to understand when it applies to all extractive users. A proposed "ecological reserve" off Key Largo, Florida was rejected by the public, partly because it violated the no-take rule in allowing certain kinds of fishing which conflicted with the stated goals of an ecological reserve. Allowing public access can facilitate public understanding and appreciation of management and can improve surveillance and compliance by discouraging deliberate violations. In contrast, a biologically successful Cape Canaveral reserve has largely gone unrecognized because public access has been prohibited.

The more successful a reserve is in building stock biomass and abundance, the more desirable it becomes as a target for poachers. Poaching can dissipate reserve benefits, undermine public support, and requires enforcement. In many cases the rewards from poaching greatly outweigh the chances and cost of being detected. The ability to obtain sufficient resources to enforce marine reserves partly depends on cultural attitudes about the seriousness of resource violations. Some juries in the U.S. have failed to convict violators because they did not consider poaching a serious offense. Ideally, a threat assessment should be conducted when creating marine reserves to assess potential problems and to develop an appropriate response plan.

Marking reserve boundaries can facilitate compliance and enforcement but can be costly or not feasible in many cases. Boundaries in the Florida Keys are marked by buoys at one nautical mile intervals, partly to help a large number of users who are not familiar with local waters. In New Zealand boundaries are not marked under the presumption that boat operators have a responsibility to know where they are. Reserve boundaries in Kenya could be distinguished by the large numbers of fishing buoys distributed along the edge. Improved technology now allows fishers to know their precise location, even far offshore. Monitoring and enforcement on the high seas is becoming easier because transponders, satellite tracking devices, and other vessel monitoring systems are becoming increasingly affordable and reliable. An example is provided in Fig. 12. Satellite tracking was successfully used to monitor long-line vessels' compliance with no-fishing zones in the remote northwestern Hawaiian Islands.

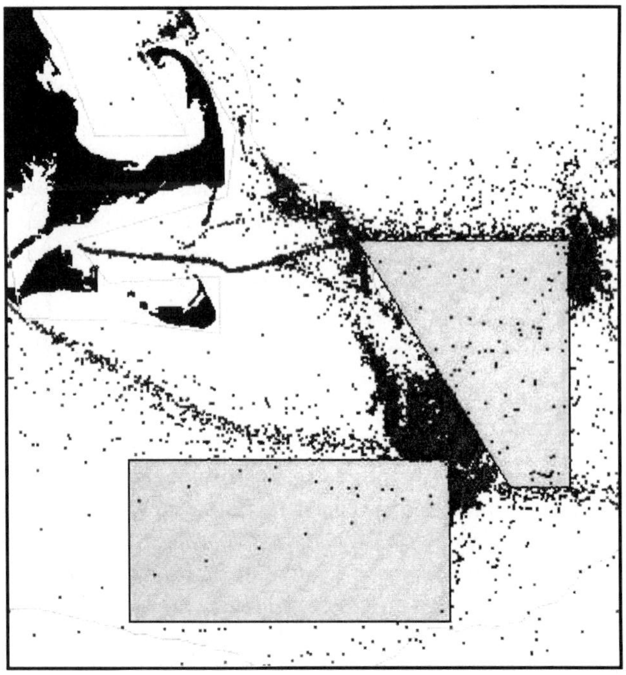

Fig. 12. Example of concentration of fishing effort near boundary of closed areas in the Georges Bank region off the northeastern United States. Each point represents a satellite-derived position for scallop fishing vessels on an hourly basis for the period April 14–June 14, 1999. Points within the interior of the closed areas represent legal traversal of the closed area with stowed fishing gear. Concentration of points at the perimeter of the closed areas represent potential encroachments. (Courtesy of Michael McSherry, NOAA, NMFS, Woods Hole, MA.)

## User Participation in Management Process

Reserves established solely by government decree tend to have little success. Broad agreement exists that user participation in developing and managing marine reserves is critically important for success. Users have unique local knowledge that can be used to pick the good sites and avoid areas that would have unacceptable impacts on users. Users are also more likely to comply with regulations when they have a chance to participate in the management process. Unfortunately there is no one model for getting effective public participation. New Zealand and the Philippines have successfully used local initiatives. In the Florida Keys, a professionally facilitated consensus building approach was used.

## CONCLUSIONS

We conclude that marine reserves do hold considerable promise, particularly within an ecosystem context where habitat protection and preservation of biodiversity and ecosystem integrity are cornerstones of the management objectives. Substantial evidence exists that abundance or biomass and average size does increase within reserve boundaries. Ultimately, the more important question from a fishery perspective is whether overall levels of yield and resilience will increase. This will strongly depend on the life history features of the target species and their dispersal characteristics at different life stages. Reserves can play an essential role in conservation and resource management as an integral component of an overall strategy in which exploitation rates are actively constrained.

## REFERENCES


Agardy, M.T. (1997) *Marine Protected Areas and Ocean Conservation.* Academic Press, San Diego, CA.

Alverson, D.L., Freeberg, M.H., Murawski, S.A. and Pope, J.G. (1994) A global assessment of fisheries bycatch and discards. *Fish. Tech. Paper* **339**, FAO, Rome.

Anon. (1921) The history of trawling: Its rise and development from the earliest times to the present day. *Fish. Trades Gaz.* **21**.

Attwood, C.G. and Bennett, B.A. (1995) Modelling the effect of marine reserves on the recreational shore-fishery of the southwestern cape, South AFRICA. *South African Journal of Marine Science* **16**, 227–240.

Beverton, R.J.H. and Holt, S.J. (1957) On the dynamics of exploited fish populations. Fish. Invest. (London) Ser. 2.

Bohnsack, J.A. (1996) Maintenance and recovery of reef fishery productivity. In *Reef Fisheries*, eds. N.V.C. Polounin and C.M. Roberts. Chapman and Hall, London, pp. 283–313.

Brothers, N. (1991) Albatross mortality and associated bait loss in the Japanese longline fishery in the Southern Ocean. *Biological Conservation* **55**, 255–268.

Buxton, C.D. (1993) Life-history changes in exploited reef fishes on the east coast of South Africa. *Environment Biology Fish.* **36**, 47–63.

Cadrin, S.X., Howe, A.B., Correia, S.J. and Currier, T.P. (1995) Evaluating the effects of two coastal mobile gear fishing closures on finfish abundance off Cape Cod. *North American Journal of Fisheries Management* **15**, 300–315.

Chapman, M.R. and Kramer, D.L. (1999) Gradients in coral reef density and size across the Barbados Marine Reserve boundary: effects of reserve protection and habitat characteristics. *Marine Ecology Progress Series* **181**, 81–96.

Christensen, N.L. et al. (1996) The report of the Ecological Society of America on the Scientific Basis for Ecosystem Management. *Ecological Appl.* **6**, 665–691.

Daan, N. (1993) Simulation study of effects of closed areas to all fishing with particular reference to the North Sea ecosystem. In *Large Marine Ecosystems: Stress, Mitigation, and Sustainability*, eds. K. Sherman, L.M. Alexander and B.D. Gold. AAAS Press.

Dayton, P., Sala, K., Tegner, M.J. and Thrush, S. Marine protected areas: parks, baselines, and fishery enhancement. *Bulletin of Marine Science.* In press.

Dayton, P.K., Thrush, S.F., Agardy, M.T. and Hofman, R.J. (1995) Environmental effects of marine fishing. *Aquatic Conservation Marine Freshwater Ecosystems* **5**, 205–232.

DeMartini (1993) Modelling the potential of fishery for managing Pacific coral reef fishes. *Fishery Bulletin* **91**, 414–427.

Dufour, V., Jouvenel, J-Y. and Galzin, R. (1995) Study of a Mediterranean reef fish assemblage. Comparisons of population distributions between depths in protected and unprotected areas over one decade. *Aquatic Living Res.* **8**, 17–25.

Dugan, J.E. and Davis, G.E. (1993) Introduction to the international symposium on marine harvest refugia. *Canadian Journal of Fisheries & Aquatic Science* **50**, 1991–1992.

Edgar, G.J. and Barrett, N.S. (1999) Effects of the declaration of marine reserves on Tasmanian reef fishes, invertebrates and plants. *Journal of Experimental Marine Biology and Ecology* **242**, 107–144.

FAO (1997) *The State of the World Fisheries and Aquaculture.* United Nations, FAO, Rome.

Fairweather, P.G. (1991) Implications of 'supply-side' ecology for environmental assessment and management. *TREE* **6**, 60–63.

Ferreira, B.P. and Russ, G.R. (1995) Population structure of the leopard coralgrouper, *Plectropomus leopardus*, on fished and unfished reefs off Townsville, central Great Barrier Reef, Australia. *Fishery Bulletin, U.S.* **93**, 629–624.

Garcia, S.M. and Newton, C. (1997) Current situation, trends, and prospects in world capture fisheries. In *Global Trends: Fisheries Management*, eds. E.K. Pikitch, D.D. Huppert, M.P. Sissenwine and M. Duke (Eds.) American Fisheries Society Symposium, 20, pp. 3–27.

Garstang, W. (1900) The impoverishment of the sea. *Journal of the Marine Biological Association of the U.K.* **VI**, 1–69.

Guènette, S. and Pitcher, T.J. (1999) An age-structured model showing the benefits of marine reserves in controlling overexploitation. Fisheries Research **39**, 295–303.

Guènette, S., Lauck, T. and Clark, C. (1998) Marine reserves: from Beverton and Holt to the present. *Rev. Fish. Biol. Fish.* **8**, 251–272.

Hall, S.J. (1998) *The Effects of Fisheries on Marine Ecosystems and Communities.* Blackwell Science, London.

Harmelin, J.-G., Bachet, F. and Garcia, F. (1995) Mediterranean marine reserves: fish indices as tests of protection efficiency. *Marine Ecology Progress Series* **16**, 233–250.

Hastings, A. and Botsford, L.W. (1999) Equivalence in yield from marine reserves and traditional fisheries management. *Science* **284**, 1537–1538.

Horwood, J.W., Nichols, J.H. and Milligan, S. (1998) Evaluation of closed areas for fish stock conservation. *Journal of Applied Ecology* **35**, 893–903.

Houde, E.D. and Zastrow, C.E. (1993) Ecosystem- and taxon-specific dynamic and energetics properties of larval fish assemblages. *Bulletin of Marine Science* **53**, 290–335.

Jennings, S., Marshall, S.S. and Polunin, N.V.C. (1996) Seychelles marine protected areas: Comparative structure and status of reef fish communities. *Biological Conservation* **75**, 201–209.

Johnson, D.R., Funicelli, N.A. and Bohnsack, J.A. (1999) Effectiveness of an existing estuarine no-take fish sanctuary within the Kennedy Space Center. *North American Journal of Fisheries Management* **19**, 436–453.

Karpov, K.A., Haaker, P.L., Albin, D., Taniguchi, I.K. and Kushner, D. (1998) The red abalone, *Haliotis rufescens*, in California: Importance of depth refuge to abalone management. *Journal of Shellfish Research* **17** (3), 863–870.

Kelleher, G., Bleakley, C. and Wells, S. (eds.) (1995) *A Global Representative System of Marine Protected Areas.* Vols. 1–4. World Cons. Union (IUCN), Washington, D.C.

Koslow, J.A. and Gowlett-Holmes, K. (1998) The seamount fauna off southern Tasmania: benthic communities, their conservation and impacts of trawling. Final Report to Environ. Aust. and the Fish. Res. Dev. Corp. FRDC Proj. 95/058. 104 pp.

Lauck, T., Clark, C.W., Magel, M. and Munro, G.R. (1998) Implementing the precautionary principle in fisheries management through marine reserves. *Ecological Application* **8** (Suppl.) S72–S78.

Lindholm, J.B.X. (1999) Habitat-mediated survivorship of juvenile Atlantic cod (*Gadus morhua*): fish population responses to fishing induced alteration of the se floor in the Northwest Atlantic and implications for the design of marine reserves. Ph.D. Dissertation, Boston University, Boston, MA.

Lozano-Alvarez, E., Briones-Fourzan, P. and Negrete-Soto, F. (1993) Occurrence and seasonal variations of spiny lobsters, *Panulirus argus* (Latreille), on the shelf outside Bahia de la Ascension, Mexico. *Fishery Bulletin, U.S.* **91**, 808–815.

MacDiarmid, A.B. and Breen, P.A. (1992) Spiny lobster population change in a marine reserve. In *Proceedings of the 2nd International Temperate Reef Symposium, 7–10 Jan 1992, Auckland, New Zealand*, ed. C.N. Battershill, pp. 47–56. NIWA Marine, Wellington, New Zealand.

Mace, P.M. and Sissenwine, M.P. (1993) How much spawning per recruit is enough? *Canadian Special Publication Fish. Aquatic Science* **120**, 101–118.

Man, A., Law, R. and Polunin, N.V.C. (1995) Role of marine reserves in recruitment to reef fisheries: a metapopulation model. *Biological Conservation* **71**, 197–204.

Mangel, M. (1998) No-take areas for sustainability of harvested species and a conservation invariant for marine reserves. *Ecological Letters* **1**, 87–90.

McClanahan. T.R. (1994) Kenyan coral reef lagoon fish: effects of fishing, substrate complexity, and sea urchins. *Coral Reefs* **13**, 231–241

McClanahan, T.R., Muthiga, N.A., Kumukuru, A.T., Machano, H. and Kiambo, R.W. (1999) The effects of marine parks and fishing on coral reefs of northern Tanzania. *Biological Conservation* **89**, 161–182.

Murawski, S.A., Brown, R., Lai, H.-Lin, Rago, P.J. and Hendrickson, L.J. Large-scale closed areas as a fishery management tool in temperate marine systems: the Georges Bank experience. *Bulletin of Marine Science*. In press.

Murray, S.N. and 18 co-authors (1999) No-take reserves networks: sustaining fishery populations and marine ecosystems. *Fisheries* **11**, 25.

Polunin, N.V.C. and Roberts, C.M. (1993) Greater biomass and value of target coral reef fishes in two small Caribbean marine reserves. *Marine Ecology Progress Series* **100**, 167–176.

Quinn, J.F., Wing, S.R. and Botsford, L.W. (1993) Harvest refugia in marine invertebrate fisheries: models and applications to the red sea urchin, *Strongylocentotus franciscanis. American Zoologist* **33**, 537–550.

Rakitin, A. and Kramer, D.L. (1996) Effect of a marine reserve on the distribution of coral reef fishes in Barbados. *Marine Ecology Progress Series* **131**, 97–113.

Roberts, C.M. (1995) Rapid build-up of fish biomass in a Caribbean marine reserve. *Conservation Biology* **9**, 816–826.

Roberts, C.M. and Polunin, N.V.C. (1993) Marine reserves: Simple solutions to managing complex fisheries? *Ambio* **22** (6), 363–368.

Roberts, C.M. and Polunin, N.V.C. (1994) Hol Chan: demonstrating that marine reserves can be remarkably effective. *Coral Reefs* **13**, 90.

Russ, G.R. and Alcala, A.C. (1996a) Do marine reserves export adult fish biomass? Evidence from Apo Island, Central Philippines. *Marine Ecology Progress Series* **132**, 1–9.

Russ, G.R. and Alcala, A.C. (1996b) Marine reserves: Rates and patterns of recovery and decline of large predatory fish. *Ecological Applications* **6** (3), 947–961.

Russ, G.R. and Alcala, A.C. (1998) Natural fishing experiments in marine reserves 1983–1993: community and trophic responses. *Coral Reefs* **17**, 383–397.

Schwartz, M.W. (1999) Choosing the appropriate scale of reserves for conservation. *Annual Reviews of Ecological Systems* **30**, 83–108.

Sladek Nowliss, J. and Roberts, C.M. (1998) Fisheries benefits and optimal design of marine reserves. *Fishery Bulletin* **97**, 604–616.

Sedberry, G.R., Carter, J., and Barrick, P.A. A comparison of fish communities between protected and unprotected areas of Belize reef ecosystem: Implications for conservation and management. *Proc. Gulf. Carib. Fish. Inst.* **45** (in press).

Sluka, R., Chiappone, M., Sullivan, K.M. and Wright, R. (1997) The benefits of marine fishery reserve status for Nassau grouper *Epinephelus striatus* in the central Bahamas. *Proc. 8th Int. Coral Reef Symposium* **2**, 1961–1964.

Stoner, A.W. and Ray, M. (1996) Queen conch, *Strombus gigas*, in fished and unfished locations of the Bahamas: effects of a marine fishery reserve on adults, juveniles, and larval production. *Fishery Bulletin, U.S.* **94**, 551–565.

Tilney, R.L., Nelson, G., Radloff, S.E. and Buxton, C.D. (1996) Ichthyoplankton distribution and dispersal in the Tsitsikamma National Park Marine Reserve, South Africa. *South African Journal of Marine Science* **17**, 1–14.

Tupper, M. and Juanes, F. (1999) Effects of a marine reserve on recruitment of grunts (Pisces: Haemulidae) at Barbados, West Indies. *Environmental Biology Fish.* **55**, 53–63.

Wallace, S.S. (1999) Evaluating the effects of three forms of marine reserve on northern abalone populations in British Columbia, Canada. *Conservation Biology* **13**, 882–887.

Watson, M., Ormond, R.F.G. and Holliday, L. (1997) The role of Kenya's protected areas in artisanal fisheries management. *Proc. 8th Int. Coral Reef Symposium* **2**, 1955–1960.

Wantiez, L., Thollot, P. and Kulbicki, M. (1997) Effects of marine reserves on coral reef fish communities from five islands in New Caledonia. *Coral Reefs* **16**, 215–224.

## THE AUTHORS

**Michael J. Fogarty**
*University of Maryland Center for Environmental Science, Chesapeake Biological Laboratory, Solomons, MD 20688, U.S.A.*
*Present Address:*
*National Marine Fisheries Service, Northeast Fisheries Science Center, 166 Water St., Woods Hole, MA 02543, U.S.A.*

**James A. Bohnsack**
*National Marine Fisheries Service, Southeast Fisheries Science Center, 75 Virginia Beach Drive,, Miami, FL 33149, U.S.A.*

**Paul K. Dayton**
*Scripps Institution of Oceanography, 9500 Gilman Dr., La Jolla, CA 92093, U.S.A.*

Chapter 135

# THE ECOLOGICAL, ECONOMIC, AND SOCIAL IMPORTANCE OF THE OCEANS

Robert Costanza

The oceans have long been recognized as one of humanity's most important natural resources. Their vastness has made them appear to be limitless sources of food, transportation, recreation, and awe. The difficulty of fencing and policing them has left them largely as open-access resources to be exploited by anyone with the means. But in recent times we have begun to reach the limits of the oceans and must now begin to utilize and govern them in a more sustainable way. This chapter summarizes emerging information on the interrelated ecological, economic, and social importance of the oceans, and on developing institutions for their sustainable governance.

In addition to their traditional importance as sources of primary and secondary production, and biodiversity, the importance of the oceans in global material and energy cycles is now beginning to be better appreciated. Integrated models of the global ocean–atmosphere–terrestrial biosphere system reveal the critical role of the oceans in atmospheric gas and climate regulation, and for water, nutrient, and waste cycling.

Recent estimates of the economic value of the marketed and non-marketed ecosystem services of the oceans indicate a huge contribution to human welfare from the functions mentioned above plus raw materials, recreational and cultural services. The oceans have been estimated to contribute a total of about US$ 21 trillion/year to human welfare (compared to a global GNP of about $25 trillion), with about 60% of this from coastal and shelf systems and the other 40% from the open ocean, and with the oceans contributing about 60% of the total economic value of the biosphere (Costanza et al., 1997).

The social importance of the oceans for global transportation and as a unifying element in the cultures of many coastal countries cannot be overestimated. But the cultural traditions of open access must be replaced with more appropriate property rights regimes and governance structures. Some alternative sustainable governance ideas are briefly discussed, emphasizing the need for an expanded deliberative process to develop a shared vision of a sustainable use of the oceans.

*Seas at The Millennium: An Environmental Evaluation (Edited by C. Sheppard)*

## ECOLOGICAL IMPORTANCE OF THE OCEANS

The fact that 71% of the earth's surface is ocean determines a significant part of its climate and ecology. The hydrologic cycle is dependent on the vast amounts of water evaporated by solar energy from the oceans and deposited as rain on the land. Without this vast reservoir of open water, the earth would quickly become a desert. The oceans also provide a sink for nutrients eroded from the land, and the seas regulate the global climate by serving as an enormous thermal mass for heat storage and as a reservoir for $CO_2$. From a purely physical point of view, the presence of the oceans can be seen as essential for a climate on earth suitable for human life.

The seas were not always a part of the earth's surface. When the earth first formed, its surface had little water or atmosphere. Early models of the formation of the earth's atmosphere were based on the idea of volcanic outgassing slowly building up a primitive atmosphere containing mainly methane, ammonia, water vapour and hydrogen (but no oxygen). As the earth cooled, the water vapour condensed and formed the early seas. More recent models hypothesize that accretion of material from impacts continued until around 4.5 billion years ago, after which the steam atmosphere that had built up rained out to form the oceans (Kasting, 1993).

Life on earth probably began in these primitive seas. Although there is some controversy, the most likely scenario for the formation of life on earth has it evolving from organic molecules to primitive single-celled organisms about 3.5 billion years ago (Kasting, 1993). A key next step was the development of photosynthesis, enabling both the utilization of light energy and the production of oxygen as a by-product. Life on dry land was only possible much later after many eons of prolific photosynthetic production in the sea had increased the oxygen level in the atmosphere to something close to current levels and produced a high ozone layer to screen out damaging ultraviolet light.

After preparing the atmosphere and the land physically, the seas also contributed the genetic stock that would ultimately evolve into the enormous variety of life we now see in the terrestrial biosphere. Even now, almost all life on earth, both on land and in the seas, takes place in an internal aqueous medium, not much different from the chemical composition of the oceans. In several very real senses, the oceans are the source of all life on earth.

The oceans have been estimated to produce more than 35% of the primary production of the planet (Lalli and Parsons, 1993). But they also provide almost 99% of the "living space" on the planet. This is true because on land, plants and animals live in a shallow vertical zone from a few meters below the surface to perhaps a hundred meters above the surface, while life in the seas extends from the surface to depths of 13,000 meters. So, while the surface area of the oceans is 71% of the planet, their volume represents almost 99% of the available living space. Most of this space is only now being explored.

Some of the terrestrial life made possible by the oceans eventually made its way back to the seas as well, including a species of primate just recently arrived on the evolutionary scene. This clever primate could not only take fish and other food resources from the seas, but could also use the seas for migration and transport routes. Its success at exploiting both the oceans and the land caused the population of this species to explode, ultimately threatening its own resource base. But more on that later.

### Science and the Seas

Because of the relative vastness and inaccessibility of the oceans, their scientific exploration had, in many senses, lagged behind the study of terrestrial systems. But in recent times, new monitoring and remote sensing technologies have led to an explosion of new scientific information about the oceans. Deep-sea diving submersibles have allowed exploration of the deepest regions of the oceans, with the dramatic discovery of completely new life forms inhabiting deep-sea thermal vents (Grassle, 1985). Satellite remote sensing has allowed the observation and mapping of the entire ocean surface. This has led to the discovery and mapping of important new features of ocean structure and circulation, such as mesoscale eddies (Brown et al., 1989) and the complex spatial patterns of marine photosynthesis (Perry, 1986).

This explosion of new information has allowed the development of sophisticated models of various aspects of the ocean system and its links to the atmosphere and terrestrial biosphere. This activity has been stimulated by growing interest in the problem of global climate change and the important role of the oceans in moderating the climate and exchanging greenhouse gases with the atmosphere.

Several large, long-term, international, interdisciplinary research projects are now underway on the oceans as part of the IGBP (International Geosphere–Biosphere Program). One example is GLOBEC (Global Ocean Ecosystem Dynamics), a program that was developed and sponsored by the Scientific Committee on Oceanic Research (SCOR), the Intergovernmental Oceanographic Commission (IOC), the International Council for the Exploration of the Sea (ICES), and the North Pacific Marine Science Organization (PICES). GLOBEC's goal is to advance our understanding of the structure and functioning of the global ocean ecosystem, its major subsystems, and its response to physical forcing, to the point where we can develop the capability to forecast the marine upper trophic system response to scenarios of global change. In pursuing this goal, GLOBEC concentrates on zooplankton population dynamics and its response to physical forcing. It bridges the gap between phytoplankton studies, such as are being pursued by projects such as JGOFS (Joint Global Ocean Flux Study), and predator-related research that more closely pertains to fish recruitments and exploration of living marine resources.

### Economic Importance of the Oceans

The services of ecological systems and the natural capital stocks that produce them are critical to the functioning of the earth's life-support system, as described above. They contribute significantly to human welfare, both directly and indirectly, and therefore represent a significant portion of the total economic value of the planet. Costanza et al. (1997) estimated the current economic value of 17 ecosystem services for 16 biomes, based on a synthesis of published studies and a few original calculations. For the entire biosphere, the value (most of which is outside the market) was estimated to be in the range of \$16–54 trillion/year, with an average of \$33 trillion/year. Because of the nature of the uncertainties, this must be considered a minimum estimate. Coastal environments, including estuaries, coastal wetlands, beds of seagrass and algae, coral reefs, and continental shelves are of disproportionately high value. They cover only 6.3% of the world's surface, but are responsible for 43% of the estimated value of the world's ecosystem services. These environments are particularly valuable in regulating the cycling of nutrients which control the productivity of plants on land and in the sea.

### GNP versus Sustainable Welfare

If ecosystem services were actually paid for, in terms of their value contribution to the global economy, the global price system would be very different than it is today. The price of commodities utilizing ecosystem services directly or indirectly would be much greater. The structure of factor payments, including wages, interest rates, and profits would change dramatically. World GNP would be very different in both magnitude and composition if it adequately incorporated the value of ecosystem services. One practical use of the estimates mentioned above is to help modify systems of national accounting to better reflect the value of ecosystem services and natural capital. Initial attempts to do this paint a very different picture of our current level of economic welfare than conventional GNP. For example, Daly and Cobb (1989) calculated an Index of Sustainable Economic Welfare (ISEW) for the US economy from 1950 to 1986 which incorporates income distribution effects, congestion effects, and the loss and damage to natural capital. Since then, ISEW has been updated for the US and calculated for several other countries. These results are shown in Fig. 1. While GNP/capita continued to rise over the entire interval for the countries shown, ISEW/capita paralleled GNP/capita during the initial period, but then levelled off and in some cases began to decline. When exactly this levelling occurred varies by country, but it has occurred in all the countries studied so far. Max-Neef (1995) has postulated that this is evidence for the "threshold hypothesis", that economic growth increases welfare only until a threshold is reached where the costs of additional growth begin to outweigh the benefits. ISEW, by doing a better job of including

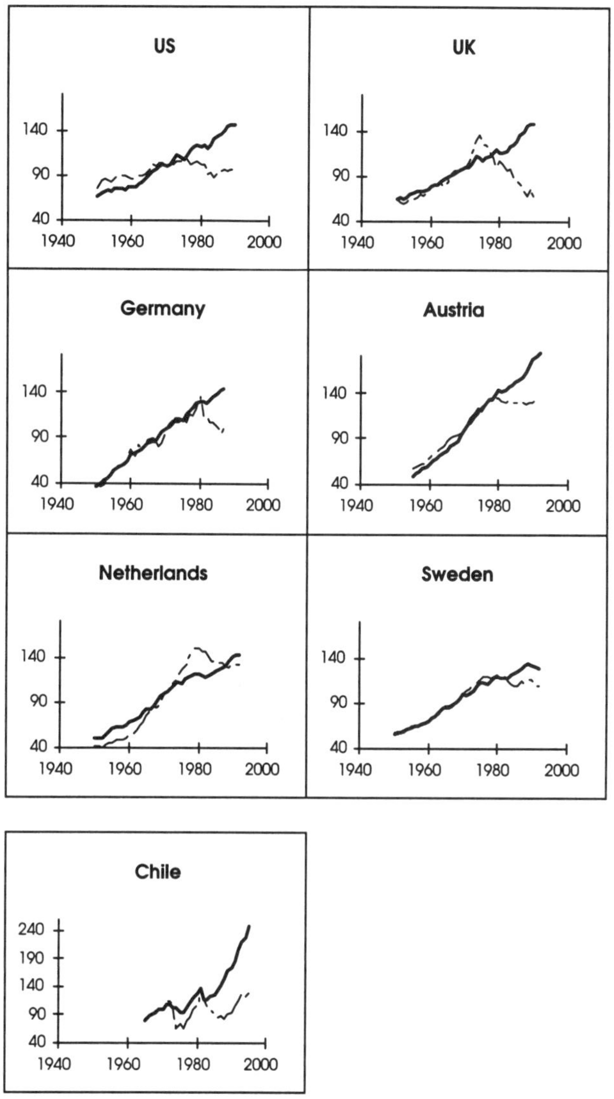


Fig. 1. Indices of GNP and ISEW (1970=100) for seven countries (from Max-Neef, 1995).

both the costs and benefits of growth, can clearly show when this threshold has been passed. In the US it was around 1970. In the UK it was around 1975, and in the other cases (Germany, Netherlands, Austria) around 1980.

The ISEW and similar measures are based on national accounts and do not adequately incorporate international resources such as the oceans. If we have passed the sustainable threshold without even taking the oceans into account, our real position is almost certainly much worse.

## SOCIAL IMPORTANCE OF THE OCEANS

The social importance of the oceans for global transportation and as a unifying element in the cultures of many coastal countries cannot be overstated. The oceans are so large that during the development of most of the world's cultures they could be considered to be almost infinite, with little risk of their over-exploitation. But the cultural traditions of open access that developed during this period are no longer adequate in the "full world" in which we now find ourselves, where humans and their artifacts are beginning to stress the very life-support functions of the biosphere. Open access to the oceans must be replaced with more appropriate property rights regimes. This has already begun to happen with the extension of territorial waters to 200 miles offshore, the Law of the Sea treaty, and the development of coastal zone management agencies and plans in most countries. But there is much more to be done, as outlined in the following section.

## Population and Consumption

The global human population has been growing exponentially. For the first 99.9% of human history, the life-support functions of the biosphere had to be shared among 10 million people at most. The human population did not reach 100 million until around 500 BC (Weber and Gradwohl, 1995). But, advances in health and agriculture have removed many of the natural checks on human population growth. In 1998 the life-support functions of the biosphere had to be shared among 6 billion humans. If current fertility and mortality rates were to remain unchanged, the world will have to be shared by more than 40 billion humans by the year 2100, when a few of the children born today will still be alive (Fig. 2).

Many believe that a human population of 40 billion would be either unsupportable, or at least highly undesirable, given the potential implications for the average standard of living and quality of life (Barney, 1993). Of course, fertility and mortality rates are not expected to

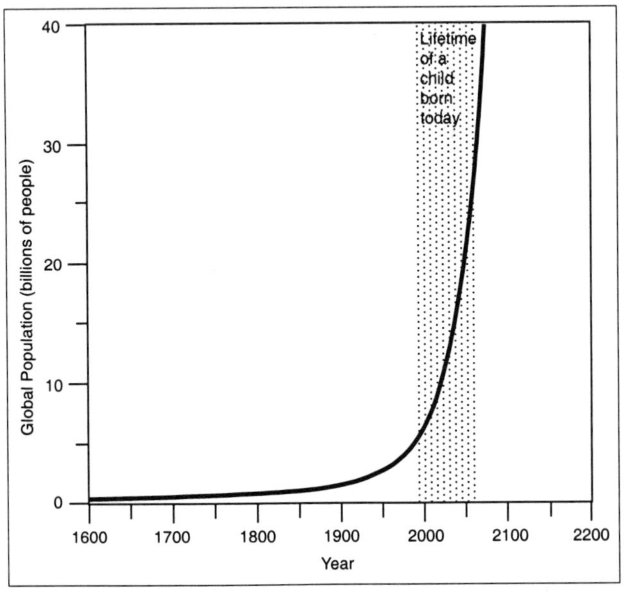


Fig. 2. Human population from 1600 to present and projected to 2100 if current fertility and mortality rates remain unchanged (after Barney, 1993).

remain unchanged, but we face hard choices about what level of human population is possible and what level is desirable.

In addition, we face hard choices about how resources are to be shared. Currently, the distribution of economic income is highly skewed in the shape of a "champagne glass", with the richest fifth of the world's population receiving 82.7% of the world's income while the poorest fifth received only 1.4% (UNDP, 1992).

In terms of impact on the environment and the carrying capacity of the environment for humans, the level of consumption per capita is at least as important as the total number of people. Cultural evolution has an interesting effect on human impacts on the environment. By changing the learned behaviour of humans and incorporating tools and artifacts, it allows individual human resource requirements and their impacts on their resident ecosystems to vary over several orders of magnitude. Thus it does not make sense to talk about the "carrying capacity" of humans in the same way as the "carrying capacity" of other species since, in terms of their carrying capacity, humans are many subspecies. Each subspecies would have to be culturally and temporally defined to determine levels of resource use and carrying capacity. For example, the global carrying capacity for *homo americanus* would be much lower than the carrying capacity for *homo indus,* because each American consumes much more than each Indian does. And the speed of cultural adaptation makes thinking of species (which are inherently slow changing) misleading anyway. *Homo americanus* could change its resource consumption patterns drastically in only a few years, while *Homo sapiens* remains relatively unchanged. I think it best to follow the lead of Herman Daly (1977) in this and speak of the product of population and per capita resource use as the *total impact* of the human population. It is this total impact that the earth has a capacity to carry, and it is up to us to decide how to divide it between numbers of people and per capita resource use. This complicates population policy enormously, since one cannot simply state an optimal *population,* but rather must state an optimal number of *impact units.* How many impact units the earth can sustain and how to distribute these impact units over the population are very dicey problems indeed.

There is also one other important complicating factor of particular relevance to the oceans. The geographic distribution of humans over the face of the earth is nowhere near homogenous. Most of the human population lives near the coast, where the impacts on the ocean environment are greatest, and this percentage is increasing.

## Property Rights Regimes and the Oceans

It is fairly easy to assign and enforce property rights to some resources and ecosystems such as agricultural fields, trees or a lake, because excluding non-owners from using the resource is fairly straightforward. However, it is much more difficult to assign and enforce property rights to resources such as migrating fish populations, biological diversity, nutrient cycles, water cycles, and many other ecological services. The reason is that it is either too expensive or literally impossible to exclude non-owners from using these resources and services, partly because they are highly interconnected with other ecosystems, thereby transcending several property rights regimes.

The oceans are the classic case of an open-access (i.e. *no* property rights) resource because of their fluid interconnectedness, their vast size, and the resulting difficulty of enforcing property rights to any particular area or resource. But even the vast oceans are gradually coming under various property rights regimes as the technology to monitor and enforce these regimes advances. The extension of territorial waters to 200 miles, the Law of the Sea, international fishing commissions, and various other institutions are beginning to establish property rights regimes on various parts of the ocean.

There is also a growing recognition that property rights regimes are complex social institutions, encompassing much more than simply establishing and enforcing boundaries (Hanna et al., 1996). These institutions must be matched to the complexity of the behaviour of the ecological systems they are attempting to manage. In this regard, various forms of community ownership (such as share-based ownership of fisheries as discussed further below) are proving to be better adapted to complex systems like the oceans. The real challenge in the sustainable governance of the oceans is in designing an appropriate set of institutions, including property rights regimes and other management institutions, that can adequately deal with the complexities of both the ocean system itself and the humans involved.

## SUSTAINABLE GOVERNANCE

The special characteristics of the oceans mentioned above lead to several unique problems that need to be addressed if the oceans are to be governed sustainably. These include:

1. The open access and common property characteristics of the oceans requires that special measures need to be taken to regulate access. Some possibilities are discussed below.
2. The role of the oceans in the global ecological system, as discussed above, favours a tendency to free ride on conservation issues. This means that some countries or actors can benefit from the system without having to pay the cost of using it.
3. The intergenerational and interspatial effects of the use of ocean resources result in a tendency to ignore effects that might be distant in time and space. There is a need to change the way such effects are handled.
4. The impact of human activity on the oceans is subject to fundamental uncertainty about the behaviour of the system, partly because of its complexity. This calls for new models of decision-making and different management rules based on maintaining the system within sustainable bounds and on exercising the precautionary principle in order to keep uncertainty within acceptable limits.
5. All of the above lead to "market failure." Hence, market prices are inadequate measures of the social value of ocean assets and require corrective incentives to guide behaviour.
6. Relative poverty is exacerbated by forms of globalization that ignore environmental externalities. Ocean use is particularly susceptible to this problem.

Several principles of sustainable ocean governance have been developed (Costanza et al., 1998). Below are some ideas about how to implement these principles.

### The Deliberative Process in Governance

What we are learning about the change process in various kinds of organizations and communities is that the most effective ingredient to move change in a particular direction is having a clear vision of the desired goal which is also truly shared by the members of the organization or community (Senge, 1990; Weisbord, 1992; Weisbord and Janoff, 1995).

In another context, Yankelovich (1991) has described the crisis in governance facing modern societies as one of moving from public *opinion* to public *judgment*. Public opinion is notoriously fickle and inconsistent on those issues for which the public has not confronted the system-level implications of their opinions. Coming to judgment requires the three steps of: (1) consciousness raising; (2) working through; and (3) resolution. A prerequisite for all three of these steps is the breaking down of the gap between expert knowledge and opinion and the public—a breaking down of what Yankelovich calls the "culture of technical control". Information in the modern world is compartmentalized and controlled by various elites who do not communicate with each other. This allows experts from various fields to hold contradictory opinions and the public to hold inconsistent and volatile opinions. Coming to judgment is the process of confronting and resolving these inconsistencies by breaking down the barriers between the mutually exclusive compartments into which knowledge and information have been put. For example, many people in opinion polls are highly in favour of more effort to protect the environment, but at the same time are opposed to any diversion of tax revenues to do so. Coming to judgment is the process of resolving these conflicts.

According to Yankelovich, one of the most effective ways to start the dialogue and move quickly to public judgment is to present issues in the form of a relatively small number of "visions" which lay bare the conflicts and inconsistencies buried in the technical information. The decisions we face today about the future of the oceans (and the planet

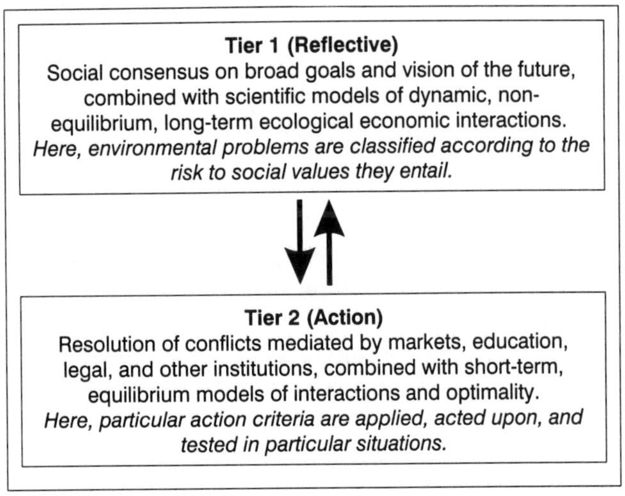

Tier 1 (Reflective)
Social consensus on broad goals and vision of the future, combined with scientific models of dynamic, non-equilibrium, long-term ecological economic interactions.
Here, environmental problems are classified according to the risk to social values they entail.
Tier 2 (Action)
Resolution of conflicts mediated by markets, education, legal, and other institutions, combined with short-term, equilibrium models of interactions and optimality.
Here, particular action criteria are applied, acted upon, and tested in particular situations.


Fig. 3. Two-tiered social decision structure (from Norton et al., 1998).

as a whole) are by far the most complex we have ever faced, the technical information is daunting (even to the experts), and we have very little time to come to public judgment. Integrated, participatory modelling and analysis of the problems is one way to pull the disparate bits of the problem together into a coherent picture that can help move to judgment (van den Belt, 1997).

How does one integrate these goals and visions and their related forms of value into a social choice structure that preserves democracy? A two-tiered conceptual model (Costanza et al., 1997) makes value formation and reformation an endogenous element in the search for a rational policy for managing human activities. Figure 3 outlines this process. Tier 1 is the "reflective" level, where social discourse and consensus is built about the broad goals and visions of the future and the nature of the world in which we live. This consensus then motivates and mediates the second, or "action" tier, where various institutions and analytical methods are put in place to help achieve the vision. There is feedback between the two tiers and the process of envisioning, goal setting, and value formation is an ongoing and critical one. There is a critical connection between value formation and decision-making, but the very existence and necessity of tier 1 is often ignored. The "culture of technical control" (Yankelovich, 1991) which dominates our current social decision making, views the problem as merely a tier 2 implementation of fixed values.

Conventional social choice theory has, in general, also tended to avoid this issue of the connection between value formation and the decision-making process. As Arrow (1951, p. 7) put it: "we will also assume in the present study that individual values are taken as data and are not capable of being altered by the nature of the decision process itself." But this process of *value formation through public discussion*, as Sen (1995) suggests, is the essence of democracy. Or, as Buchanan (1954, p. 120) puts it: "The definition of democracy as 'government by discussion' implies that individual values can and do change in the process of decision-making." Limiting our valuations and social decision-making to a fixed set of goals based on fixed preferences prevents the needed democratic discussion of values and future options and leaves us with only the "illusion of choice" (Schmookler, 1993).

## Integrated Ecological–Economic Modelling and Assessment

Once the goal of ecological sustainability has been established in tier 1, addressing it in tier 2 requires a large measure of scientific assessment and modelling (Faucheux et al., 1996). The process of integrated ecological economic modelling can help to build mutual understanding, solicit input from a broad range of stakeholder groups, and maintain a substantive dialogue between members of these groups. In the process of adaptive management, integrated modelling and consensus building are essential components (Gunderson et al., 1995).

A recent SCOPE project on Integrated Ecological Economic Modelling and Assessment (IA for short) developed the following basic framework (Costanza and Tognetti, 1996).

The general principles and defining characteristics are that:

- *Predictability is limited.* We should not confuse our models of the system with reality. There is no one correct way to model the world, nor are there any absolute answers.
- There will be *multiple assumptions held by the different stakeholders* within different cultures, which will affect the articulation of visions, goals, problems, and perceptions of reality.
- *Different paradigms* will be involved in shaping and interpreting problems and solutions.
- *Transdisciplinary approaches* (i.e. transcending disciplines rather than monodisciplinary or interdisciplinary) will be necessary to achieve the horizontal linkages necessary for IA.
- The *process is as important as the product,* because it achieves participation and builds consensus and can overcome power differentials.
- *Legitimacy is derived from the process of involvement* of all stakeholders and the development of an overlapping consensus that can be used to resolve or reduce conflicts. It can also help to harmonize bottom-up and top-down approaches and to educate and build capacity to handle future problems.
- *Researchers are included as stakeholders.* They cannot be neutral or unbiased and must consider their own role in the process.
- It is important to acknowledge and deal with the many forms of *uncertainty* inherent in complex systems. This includes parameter uncertainty, process uncertainty, and data quality.

The framework is seen as a creative and learning process rather than as a purely technical tool, within which a well-rounded decision can be achieved through the consensus of stakeholders. Within this framework, a process consisting of 12 steps was developed to implement integrated assessment. The process assumes feedback loops from later steps to earlier steps in an adaptive management context, where policy recommendations are viewed as experiments rather than as answers (Holling, 1978; Walters, 1986):

1. *Define the focus of attention.* This would likely result from a proposed development opportunity and/or an ecological concern.
2. *Identify stakeholders.* These typically would include government, business, land-owners, non-governmental organizations, funding agencies, community-based organizations, researchers, etc.
3. *Establish techniques to bring stakeholders together (e.g. roundtable).* This step presupposes that one or more of the stakeholders has sufficient interest to draw the remaining stakeholders to a meeting. It may be that specific stakeholders need to be persuaded that it is in their best interest to convene such a roundtable. Other stakeholders may need to convince them of the value of developing a participatory approach.
4. *Seek agreement on an acceptable facilitator.* Ideally such a person should be as neutral and unbiased as possible and without a stake in the outcome of the process. The facilitator should nevertheless be committed to the process and be able to balance the differing powers of the stakeholders.
5. *Define stakeholder interests.* Prior to the roundtable meeting, stakeholder groupings should be encouraged to meet and discuss their own interests.
6. *Hold roundtable.* The roundtable should ideally be convened jointly by several stakeholders. The agenda should include opportunities for:
   - sharing of individual visions;
   - identifying complementarity and conflicts;
   - agreeing that a process is necessary to address conflicts;
   - seeing that integrated assessment is a way forward with the potential to develop consensus and arrive at a "win-win" situation;
   - establishing a structure for ongoing dialogue including a stakeholder committee to oversee the process and feedback opportunities to the stakeholder groups and to all stakeholders collectively.
7. *Undertake a scoping exercise.* This process is necessary to identify the key issues, questions, data/information availability, land-use patterns, proposed developments, existing institutional frameworks, timing and spatial consideration, etc. It provides a means to determine whether a specific action will have significant effects on expressed values and to link the model with those values. This scoping exercise is also seen as building trust among the stakeholders and an acceptance of the process. The stakeholders build upon knowledge and capacity.
8. *Build and run a scoping model.* A scoping model provides a relatively quick process of identifying and building in the key components to:
   - generate alternative scenarios;
   - identify critical information gaps;
   - understand the sensitivity of the scenarios to uncertainty;
   - identify and agree on additional work to be undertaken by one or more methods of detailed modelling.

   Stakeholders participate in the development of the scoping model.
9. *Commission detailed modelling.* Additional information is gathered and the chosen model(s) are modified, extended, and run.
10. *Present models* and results of model scenarios and discuss findings among stakeholders.
11. *Build consensus recommendations.*
12. *Proceed with, and monitor the development of, the preferred scenario.* Learn from the results and iterate the IA process as necessary. Perceptions change as things actually happen, thus the process must reflect changing values in the modelling process that can have a feedback to decisions at each step in the process. As iterations occur, the scenario conception changes, leading to new issues for resolution among groups.

## New Property Rights Regimes

There is a major challenge in designing institutions and property rights regimes that are in tune with the functions of ecosystems and the goods and services that they generate. How do we design institutions and property rights regimes that account for the complex flows and feedback between systems, and that maintain the buffer capacity to ensure a continuation of these flows? Luckily, there are design principles derived from studies of long-enduring institutions that have, at least to some extent, been successful in managing ecological resources in a sustainable fashion (see e.g. Ostrom and Schlager, 1996). The design principles include: clearly defined boundaries for the use of the resources, as well as clearly defined individuals or households with rights to harvest the resources; rules specifying the amount of harvest by users related to local conditions and to rules requiring labour, materials, and/or money inputs; collective-choice arrangements; monitoring of resource conditions and user behaviour; graduated sanctions when rules are violated; conflict-resolution mechanisms; long-term tenure rights to the resource and rights of users to devise their own institutions without being undermined by governmental authorities; and for resources that are parts of larger systems, appropriation, provision, monitoring, enforcement, conflict resolution, and governance activities need to be organized in multiple layers of nested enterprises (Becker and Ostrom, 1995).

Some of the most sophisticated property rights institutions are found in areas in which these systems have developed over a long period of time, on the order of hundreds of years. Examples include Spanish *huertas* for irrigation, Swiss grazing commons, and marine resource tenure systems in Oceania (see Berkes, 1996). Yet other systems have collapsed and recovered over a period of time, sometimes more than once. In contrast, many traditional local communities have recognized the necessity of the coexistence of gradual and rapid change. In their institutions they have accumulated a knowledge base for how to respond to feedbacks from the ecosystem. Holling et al. (1995) argue that these societies are successful in managing their resources sustainably because they have developed social mechanisms that interpret the signals of creative destruction and renewal of ecosystems and cope with them before they accumulate and challenge the existence of the whole local community. Disturbance has been allowed to enter at smaller scales, instead of being blocked out as is often the case in contemporary society. There is a culturally evolved monitoring system that reads the signals, the disturbances, and thereby is more successful in avoiding the build up of an internal structure that will become brittle and invite large-scale collapse. The local institutions have evolved so that renewal occurs internally while overall structure is maintained. The accumulation and transfer of this knowledge between generations has made it possible to be alert to changes and continuously adapt to them in an active way. It has been a means of survival (Holling et al., 1995).

We can learn from those local institutions that do not undermine their existence by degrading their ecological life-support system, thereby losing ecological and institutional resilience. A major task for modern society is to find similar ways of responding to changes in ecosystems. At present there is a pervasive lack of social mechanisms for dealing with changing environmental conditions.

Sustainability requires that human social systems and property rights regimes are adequately related to the larger ecosystems in which they are embedded. Understanding the complex evolutionary dynamics of these ecosystems is essential, but we must acknowledge and deal with the inherently limited predictability of complex systems. Because our knowledge of the structure and function of ecological systems is limited, and because we do know that sustainability depends on these systems, we must take a precautionary approach to their management (O'Riordan and Jordan, 1995). Complex adaptive systems require "adaptive management" (Holling, 1978). This means that we need to view the implementation of policy prescriptions in a different, more adaptive way that acknowledges the ever-present uncertainty and allows participation by all the various stakeholder groups. Adaptive management views regional development policy and management as "experiments", where interventions at several scales are made to achieve understanding and to identify and test policy options (Holling, 1978; Walters, 1986; Lee, 1993; Gunderson et al., 1995), rather than as "solutions."

There are several recent examples of new property rights systems employing adaptive management. One particularly relevant example is the recently passed legislation in the Australian state of New South Wales (NSW) to introduce a fishery share system (New South Wales, 1994). It is similar in general form and purpose to the "ITQ" (or individual transferable quota) fishery management systems found in New Zealand, Iceland, Australia, Canada, and other countries, but has special design features, including the allotment of shares for "fisheries" that include many different species (Young, 1992). The system is designed to give fishers security within the context of an adaptive resource management system designed to ensure that fishery use is sustainable and consistent with social objectives as they change through time.

Young (1999) describes the NSW share system and its relationship to quota systems as follows:

> "Most individual transferable quota regimes are for single species; accordingly, they do not encourage fishers to recognize the interdependence of species. Moreover, it is arguable that they neither create a strong sense of industry responsibility for the state of a fishery nor encourage participation in the management process. Weighing these and other considerations, the NSW system grants each fisher a guaranteed opportunity and compensable right to a proportional share of all the commercial opportunities in the fisheries they use. The term "share" is used intentionally to stress the idea that each shareholder owns a legally enforceable share of each fishery's commercial opportunities. The legislation establishes a "core property right" as a legally transferable entitlement to a proportional share of all the commercial fishing opportunities associated with the fishery.
>
> Wherever possible, corporate-like administrative structures are used as these are well understood by fishers. Effectively, each person is given a guaranteed share of the opportunity set out in a periodically revised management plan for the species that comprise the fishery. Formally, each fisher is entitled to a share of any allocation of quota and gear or input restriction in proportion to the number of shares they hold. If they want to use a larger boat or bigger net, then they must buy shares from people already in the fishery. Similarly, allocation of any quota is in proportion to the number of shares held. The corporatised structure enables reference to corporate management experiences and enables both input and output controls to be varied equitably without affecting resource security."

The NSW share system's conceptual framework is of relevance to other fisheries and also many other natural resources. It represents an approach to adaptive management and common property regimes that appears to be efficient, fair, and sustainable. Because it conforms to the principles discussed above and in Costanza et al. (1998), it may be a viable model for broader ocean governance issues.

### Taxes and Other Economic Incentives

Using economic incentives to achieve environmental goals can be much more efficient than traditional command and control regulation, if the incentives can be put in place and enforced at relatively low cost. This is a key point for the oceans since one of their main characteristics has been the difficulty of monitoring and policing almost any kind of intervention. This situation is changing, however, and the time may be ripe for various kinds of economic incentives, especially if they can be incorporated into community-based management and co-management approaches as discussed above. Using self-policing, share-based common property systems, and trust-building mechanisms can significantly lower the costs of implementing any incentive scheme, or any direct regulation scheme for that matter.

The key point is that taxes or other economic incentives must be seen as one instrument from which the community can choose in designing its ecosystem management systems. While it is not a panacea, this instrument can be quite effective in the appropriate situations. For example, one such situation being discussed recently is the idea of "ecological tax reform".

There is a growing consensus among a broad range of stakeholder groups in the US, and even more so in Europe, concerning the need to reform tax systems to tax "bads" rather than "goods." Taxes have significant incentive effects which need to be considered and utilized more effectively. The most comprehensive proposed implementation of this idea is coming to be known under the general heading of "ecological tax reform" (von Weizsäcker and Jesinghaus, 1992; Passell, 1992; Repetto et al., 1992; Hawken, 1993; Costanza, 1994). Earlier discussions of similar schemes were given by Page (1977), who considered a national severance tax, and Daly (1977), who discussed a depletion quota auction.

The basic idea is to limit the throughput flow of resources to an ecologically sustainable level and composition, thus serving the goal of a sustainable scale of the economy relative to the ecosystem, a goal until recently neglected. The more traditional goal of efficient allocation of resources is also served by this instrument because it raises the tax on "bads" and lowers the tax on "goods"—it internalizes externalities. The third goal of distributive equity is both helped and hindered. Since the throughput tax is basically a capturing for public purposes of the scarcity rent to natural capital as economic and demographic growth increases its value, it has some of the equity appeal of Henry George's rent tax. However, like all consumption taxes, it is regressive. This could be counteracted by retaining the income tax at the extremes—a positive income tax for high incomes, a negative income tax for very low incomes, and a negligible income tax between the extremes. Of the three major goals of economic policy (sustainable scale, efficient allocation, and just distribution), the ecological tax reform serves the first two quite well and the third partially, requiring some supplement from an attenuated income tax structure.

The idea is to gradually shift much of the tax burden away from "goods" like income and labour, and toward "bads" like ecological damages and consumption of nonrenewable resources.

Such a system would need three components:

1. A *natural capital depletion tax* aimed at reducing or eliminating the destruction of natural capital. Use of nonrenewable natural capital would have to be balanced by investment in renewable natural capital in order to avoid the tax. The tax would be passed on to consumers in the price of products and would send the proper signals about the relative sustainability cost of each product, moving consumption toward a more sustainable product mix.
2. The *precautionary polluter pays principle (4P)* (Costanza and Cornwell, 1992) would be applied to potentially damaging products to incorporate the cost of the uncertainty about ecological damages as well as the cost of known damages. This would give producers a strong and immediate incentive to improve their environmental performance in order to reduce the size of the environmental bond and tax they would have to pay.
3. A system of *ecological tariffs* aimed at allowing individual countries or trading blocks to apply 1 and 2 above without forcing producers to move overseas in order to remain competitive. Countervailing duties would be assessed to impose the ecological costs associated with production fairly on both internally produced and imported products. Some of the revenues from the tariffs could be reinvested in the global environment, rather than added to general revenues of the host country.

Such a system would have far-reaching implications, and would simultaneously encourage employment and income, reduce the need for government regulation, and promote the sustainable use of natural resources and ecosystems. Reducing taxes on income and labour encourages employment because it reduces the cost of labour to employers. It also encourages work because it increases net pay for workers. Both are good for the economy. Taxing depletion of natural resources and pollution effectively works them into the market system so polluters and depletors pay for their actions, and have a reason to lighten their impact on the environment. Because of the revenue neutral aspect of the tax shift, it does not raise costs for business, but rather gives businesses appropriate incentives to develop new technology, improve production efficiency, and improve their environmental performance.

Such a tax shift could work well for national economies, and by extension their exclusive economic zones in the oceans. But the problem remains of what to do about the fact that the open oceans are under no country's exclusive jurisdiction. This will require new international agreements and institutions. But, as indicated above, the possibilities for using taxes and other economic incentives should be among the tools available to the international community as it begins to design a sustainable governance system for the oceans.

## CONCLUSIONS

The oceans are ultimately the heritage of all of humanity. Their role and value in supporting human life are only now becoming fully recognized. Ecological sustainability, economic efficiency, and social fairness need to become joint objectives in order to adequately maintain the oceans as humanity's common heritage. The oceans are too important to humanity's survival to allow their continued exploitation as if they were infinite.

Governance systems, property rights regimes, economic incentives, and other institutions that can adequately deal with the inherently common property nature of the oceans are sorely needed. Creative deliberation and consensus building among the various stakeholder groups is an essential, and still largely missing, element. Innovative common property regimes, like the "share-based" fishery system in New South Wales, may provide models that take this approach and simultaneously meet the joint goals of sustainability, fairness and efficiency.

Movement toward these goals at larger scales is being impeded, not so much by lack of knowledge, but by a lack of a coherent, relatively detailed, shared vision of what a sustainable society would actually look like. Developing this shared vision is an essential prerequisite to generating any movement toward it. The default vision of continued, unlimited growth in material consumption is inherently unsustainable, but we cannot break away from this vision until a credible and desirable alternative is available. The process of collaboratively developing this shared vision can also help to mediate many short-term conflicts that will otherwise remain irresolvable. Envisioning and "future searches" have been quite successful in organizations and communities around the world (Weisbord, 1992; Weisbord and Janoff, 1995). This experience has shown that it is quite possible to have disparate (even adversarial) groups collaborate on envisioning a desirable future, given the right forum. The process has been successful in hundreds of cases at the level of individual firms and communities up to the size of large cities. The challenge is to scale up to whole states, nations, and the world, and to make it a routine part of the democratic process, as described above in the "two-tier" decision process. This new paradigm of governance holds some promise for actually achieving the goal of sustainable governance of the oceans, and the rest of the planet as well.

## REFERENCES


Arrow, K.J. (1951) *Social Choice and Individual Values.* Wiley, New York.

Barney, G.O. (1993) *Global 2000 Revisited: What Shall We Do?* Millennium Institute, Arlington, VA.

Becker, C.D. and Ostrom, E. (1995) Human ecology and resource sustainability: the importance of institutional diversity. *Annual Review of Ecology and Systematics* **26**, 113–133.

Berkes, F. (1996) Social systems, ecological systems, and property rights. In *Rights to Nature: Ecological, Economic, Cultural, and Political Principles of Institutions for the Environment*, eds. S. Hanna, C. Folke, and K-G. Mäler, pp. 87–110. Island Press, Washington, DC. 298 pp.

Brown, J., Colling, A., Park, S., Phillips, J., Rothery, D. and Wright, J. (1989) *Ocean Circulation.* The Open University and Pergamon, Oxford.

Buchanan, J.M. (1954) Social choice, democracy, and free markets. *Journal of Political Economy* **62**, 114–123.

Costanza, R. (1994) Three general policies to achieve sustainability. In *Investing in Natural Capital: The Ecological Economics Approach to Sustainability*, eds. A.M. Jansson, M. Hammer, C. Folke and R. Costanza, pp. 392–407. Island Press, Washington DC, 504 pp.

Constanza, R. and Cornwell, L. (1992) The 4P approach to dealing with scientific uncertainty. *Environment* **34**, 12–20, 42.

Costanza, R., Andrade, F., Antunes, P., van den Belt, M., Boersma, D., Boesch, D.F., Catarino, F., Hanna, S., Limburg, K., Low, B., Molitor, M., Pereira, G., Rayner, S., Santos, R., Wilson, J. and Young, M. (1998) Principles for sustainable governance of the oceans. *Science* **281**, 198–199.

Costanza, R., d'Arge, R., de Groot, R., Farber, S., Grasso, M., Hannon, B., Naeem, S., Limburg, K., Paruelo, J., O'Neill, R.V., Raskin, R., Sutton, P., and van den Belt, M. (1997) The value of the world's ecosystem services and natural capital. *Nature* **387**, 253–260.

Costanza, R. and Folke, C. (1996) The structure and function of ecological systems in relationship to property rights regimes. In *Rights to Nature: Ecological, Economic, Cultural, and Political Principles of Institutions for the Environment*, eds. S. Hanna, C. Folke, and K-G. Mäler, pp. 133–134. Island Press, Washington, DC. 298 pp.

Costanza, R. and Tognetti, S. (eds.) (1996) Integrated adaptive ecological and economic modelling and assessment—a basis for the design and evaluation of sustainable development programs. DRAFT Synthesis paper. Scientific Committee on Problems of the Environment (SCOPE), 51 Bld de Montmorency, 75016 Paris, France.

Daly, H.E. and Cobb, J. (1989) *For the Common Good: Redirecting the Economy Towards Community, the Environment, and a Sustainable Future.* Beacon Press, Boston. 482 pp.

Daly, H.E. (1977) *Steady State Economics.* W.H. Freeman, San Francisco, CA. 185 pp.

Faucheux, S., Pearce, D. and Proops, J. (1996) *Models of Sustainable Development.* Edward Elgar Press, Cheltenham, UK. 365 pp.

Grassle, J.F. (1985) Hydrothermal vent animals: distribution and biology. *Science* **229**, 713–717.

Gunderson, L., Holling, C.S. and Light, S. (eds.) (1995) *Barriers and Bridges to the Renewal of Ecosystems and Institutions.* Columbia University Press, New York.

Hanna, S., Folke, C. and Mäler, K-G. (eds.) (1996) *Rights to Nature: Ecological, Economic, Cultural, and Political Principles of Institutions for the Environment.* Island Press, Washington, DC. 298 pp.

Hawken, P. (1993) *The Ecology of Commerce: A Declaration of Sustainability.* Harper Business, New York. 250 pp.

Holling, C.S. (ed.) (1978) *Adaptive Environmental Assessment and Management.* Wiley, London.

Holling, C.S. (1994) New science and new investments for a sustainable biosphere. In *Investing in Natural Capital: The Ecological Economics Approach to Sustainability*, eds. A.M. Jansson, M. Hammer,

C. Folke and R. Costanza, pp. 57–73. Island Press, Washington DC. 504 pp.

Holling, C.S., Berkes, F. and Folke, C. (1995) Science, sustainability, and resource management. Beijer Discussion Paper 68. Beijer International Institute for Ecological Economics, Stockholm, Sweden.

Kasting, J.F. (1993) Earth's early atmosphere. *Science* **259**, 920–926.

Lalli, C.M. and Parsons, T.R. (1993) *Biological Oceanography: An Introduction*. Butterworth-Heinemann, Oxford.

Lee, K. (1993) *Compass and the Gyroscope*. Island Press, Washington DC.

Max-Neef, M. (1995) Economic growth and quality of life: a threshold hypothesis. *Ecological Economics* **15**, 115–118.

New South Wales (1994) Fisheries Management Act, 1994. NSW Government Printer, Sydney.

Norton, B., Costanza, R. and Bishop, R. (1998) The evolution of preferences: why "sovereign" preferences may not lead to sustainable policies and what to do about it. *Ecological Economics* **24**, 193–211.

O'Riordan, T. and Jordan, A. (1995) The precautionary principle in contemporary environmental politics. *Environmental Values* **4**, 191–212.

Ostrom, E. and Schlager, E. (1996) The formation of property rights. In *Rights to Nature: Ecological, Economic, Cultural, and Political Principles of Institutions for the Environment*, eds. S. Hanna, C. Folke and K-G. Mäler, pp. 127–156. Island Press, Washington, DC. 298 pp.

Page, T. (1977) *Conservation and Economic Efficiency*. Johns Hopkins University Press, Baltimore, MD.

Passell, P. (1992) Cheapest protection of nature may lie in taxes, not laws. *New York Times*, Nov. 24, 1992.

Perry, M.J. (1986) Assessing marine primary production from space. *BioScience* **36**, 461–467.

Repetto, R., Dower, R.C., Jenkins, R. and Geoghegan, J. (1992) *Green Fees: How a Tax Shift Can Work for the Environment and Economy*. World Resources Institute, Washington, DC.

Schmookler, A.B. (1993) *The Illusion of Choice: How the Market Economy Shapes Our Destiny*. State University of New York Press, Albany, NY, 349 pp.

Sen, A. (1995) Rationality and social choice. *American Economic Review* **85**, 1–24.

Senge, P.M. (1990) *The Fifth Discipline: The Art and Practice of the Learning Organization*. Currency-Doubleday, New York.

UNDP (1992) *Human Development Report*. Oxford University Press, New York.

van den Belt, M.J., Janson, A. and Deutsch, L. (1997) A consensus-based simulation model for management of the Patagonian coastal zone. *Ecological Modelling*.

von Weizsäcker, E.U. and Jesinghaus, J. (1992) *Ecological Tax Reform: A Policy Proposal for Sustainable Development*. Zed Books, London.

Walters, C.J. (1986) *Adaptive Management of Renewable Resources*. McGraw Hill, New York.

Weber, M.L. and Gradwohl, J.A. (1995) *The Wealth of Oceans*. Norton, New York.

Weisbord, M. (ed.) (1992) *Discovering Common Ground*. Berrett-Koehler, San Francisco. 442 pp.

Weisbord, M. and Janoff, S. (1995) Future search: an action guide to finding common ground in organizations and communities. Berrett-Koehler, San Francisco.

WCED (1987) *Our Common Future: Report of the World Commission on Environment and Development*. Oxford University Press, Oxford, UK.

Yankelovich, D. (1991) *Coming to Public Judgement: Making Democracy Work in a Complex World*. Syracuse University Press. 290 pp.

Young, M.D. (1992) The design of resource right systems: Allocating opportunities for sustainable investment and resource use. Chapter 4 in: Sustainable Investment and Resource Use: Equity, Environmental Integrity and Economic Efficiency. Parthenon Press, Carnforth and UNESCO-MAB, Paris.

Young, M.D. (1999) The design of fishing-right systems—the NSW experience. *Ecological Economics* **31**, 305–316.

## THE AUTHOR

**Robert Costanza**
*Center for Environmental Science and Biology Department, and*
*Institute for Ecological Economics,*
*University of Maryland, Box 38,*
*Solomons, MD 20688-0038, U.S.A.*

# INDEX

*Page numbers in bold refer to tables, in italics to illustrations. Roman numerals indicate volume number.*


abalone, recreational harvesting, South Africa II-139
abalone fishery II-27, II-137
— Australia II-586
— dive fishery, Victoria Province II-668
— Tasmania II-655–6
Aboriginal people, Australia II-583–4, II-619–20
— dugongs a festive food for II-619
— southern and northern Nullabor Plain II-682
aboriginal subsistence whaling III-74
Abrolhos Archipelago and the Abrolhos Bank, Brazil *I-732*
— coral reefs *I-720*, I-724, I-726
Abrolhos Bank–Cabo Frio, Southern Brazil I-734
— Brazil Current I-734
— environmental degradation I-740
— eucalyptus plantations for pulp production I-738
— industrial pollution from pulp production I-740
— major activities I-737–8
— shallow water marine and coastal habitats I-735–6
— Vitória Bay
— environmental degradation I-740
— oil spill risk for surrounding fragile areas I-739
Abrolhos National Marine Park, Brazil *I-720*, I-726–7, I-728
abundance, change in may be due to new competitor I-257
abundances, damped oscillations in III-275
accidental introductions *see* alien/accidental/exotic/introduced organisms/species
accretion, Gulf of Guinea coast I-778
acid mine drainage, Tasmania II-654
acid sulphate soil conditions, associated with mangrove clearance II-369, II-395
acid sulphate soil run-off
— a critical issue in eastern Australia II-638
— Great Barrier Reef region II-620–1
adaptive management III-400
Aden, Gulf of *II-36*, II-38, II-47–61, *II-48*
— coastal erosion and landfill II-58
— defined II-49
— effects from urban and industrial activities II-58–9
— major shallow water marine and coastal habitats II-51–4
— offshore systems II-54–5
— populations affecting the area II-55–6
— ports **II-55**
— production higher II-42
— protective measures II-59–60
— conventions signed relating to the Marine Environment **II-59**
— rural factors II-56–8
— seasonality, currents, natural environmental variables II-49–51
Aden port II-55
— sand and mud flats II-53
Adriatic Sea *I-268*
— agricultural pollution load I-275
— circulation and water masses I-270
— described I-269–70
— differences between east and west coasts I-269, I-272
— effects of urban and industrial activities I-277–8
— lagoon ports and canal harbours I-277
— nuisance diatom bloom III-298
— salinity and tidal amplitude I-269
— seasonality, currents and environment variables I-270–1
adult biomass, minimum viable, required for stock replacement III-382–3, *III-383*
adverse health effects, of toxic chemicals II-456–8
— coplanar PCBs most suspect contaminants II-457
Aegean Sea I-233–52
— Aegean basins I-235
— partitioned by islands I-237
— behaviour of water masses I-236–7
— coastal erosion and landfill I-245
— defined I-235–6
— Miocene—present evolution I-235
— effects from urban and industrial activities I-245–9
— human populations I-241–2
— northern, Turkish, sediment decreased through dam construction III-352
— offshore systems I-240–1
— protective measures I-245–9
— rural factors I-242–5
— seasonality, currents and natural environmental variables I-236–7
— shallow water ecosystems and biotic communities I-237–40
aerial surveys, for bird distributions III-109
Africa, Eastern, coral bleaching and mortality events III-48
Africa, southwestern I-821–40
— Benguela Current Large Marine Ecosystem I-823
— commercial fisheries I-833
— effects of urban and industrial activities I-835
— major shallow water marine and coastal habitats I-828–30
— the next millennium I-835–7
— fisheries management, past, present and future I-836–7
— oil spill contingency plans I-837
— other development and commercial activities I-837
— offshore systems I-830–5
— physical environment I-823–8
— status of selected Convention, Agreement and Codes of Conduct **I-836**
African dustfall
— an ecological stressor in the Florida Keys I-410
— potential cause of other Caribbean coral diseases I-410
Agalega (Mascarenes) II-255
— coconut plantations II-257
— population II-260
— upwelling increases nutrients II-259–60
Agenda 21 III-351
— and promotion of ICZM III-352
— evolution of III-352
— on Protection of the Oceans I-58
— on rights and duties of fishing III-158
— using two UNCED principles I-459
aggregate extraction
— offshore sites, Irish Sea I-95–6
— sand for beach nourishment I-53–4, I-96
— sand and gravel, off UK coasts I-53, I-74
aggregate mining, wadi beds II-27
agrarian reform, Nicaragua, created environmental problems I-538–9
agricultural pollution II-376
— Malacca Strait II-338
— Mauritius II-261
— problem of, Adriatic and Tyrrhenian basins I-275–6

— southern Spain I-174
— Sri Lanka II-184
agriculture
— American Samoa, Tutuila Island II-770
— Argentina I-758
— Buenos Aires Province, fertilisers and pesticides I-762
— Australia II-584
— poor agricultural practices II-584
— in the Bahamas I-423–4
— Bangladesh II-292
— Belize I-508
— commercial, effects of I-508
— potential impacts on the marine system I-508
— Borneo II-374–5
— Cambodia, traditional II-576
— Chesapeake Bay
— best management practices I-347
— contributes N and P to pollution I-345
— China's Yellow Sea coast, increased output II-492
— coastal
— Mozambique II-107
— subsistence, Gulf of Aden II-57–8
— coconut plantation, Marshall Islands II-781
— Colombian Pacific Coast I-682
— the Comoros II-247
— Coral, Solomon and Bismark Seas region II-435
— Côte d'Ivoire, basis of the economy I-816
— creating problems, Andaman, Nicobar and Lakshadweep Islands II-193
— eastern Australian region II-636–8
— central and southern New South Wales II-638
— northern New South Wales II-637–8
— south-east Queensland II-637
— effects of in Tanzania II-88–9
— El Salvador I-551
— poor land-use and agricultural practices I-551–2
— extensive cattle industry I-479
— Fiji Islands, and its impact II-758
— Great Barrier Reef region II-620–1
— land clearing II-620–1
— sedimentation and nutrients II-621
— Gulf of Guinea, use of agrochemicals I-784–6
— Gulf of Papua II-600–1, II-604
— Hawaiian Islands II-801
— Hong Kong, declining II-542
— influenced by water availability, Arabian Peninsula southern coast II-25–6
— intensive
— Greece and Turkey I-242
— Guadalquivir valley I-174
— Jamaica I-568
— Lesser Antilles I-635
— export crops I-635
— Madagascar II-123–4
— hill rice planting leads to soil loss II-123
— the Maldives **II-208**
— environmental effect of II-207
— Marshall Islands II-778
— Mascarene Region
— Mauritius II-260–1
— Reunion II-261
— Rodrigues II-261
— Mexican Pacific coast I-489
— and the green revolution I-492
— Mozambique coast, inappropriate techniques II-109
— N and P to Baltic Sea I-104, *I-104*
— New Caledonia II-729, **II-729**
— Nicaragua
— Caribbean coast I-523
— Pacific coast I-538–9, I-540
— Peru, use of organochlorine pesticides I-696
— the Philippines II-412–13
— plant nutrients from I-93
— poor farming practice leads to estuarine sedimentation and turbidity II-136
— Samoa II-716
— Sea of Okhotsk, poorly developed II-469
— the Seychelles
— deforestation for coconut plantations II-238
— poor agricultural practice and unwise development activity II-238–9
— Somalia
— few crops 179, II-77
— nomad/semi-nomad II-77
— southern Brazil
— subsistence I-738, I-739
— sugar cane cultivation I-737–8
— Southern Gulf of Mexico, plantations I-476
— in the Sundarbans II-153
— Taiwan west coast II-503
— Tonga II-716
— Turks and Caicos Islands, small amount only I-591–2
— Vanuatu
— and its impacts II-743
— Pilot Plantation Project II-743
— Vietnam, rice growing II-563
— western Indonesia II-396
agrochemical pollution
— Malacca Strait II-318–19
— south coast, southern Brazil I-739
agrochemicals
— use of
— China II-492
— Vanuatu II-743
— western Indonesia II 396
Agulhas Bank II-135, II-137
Agulhas current *I-822*, I-824, *II-101*, *II-134*, II-135
Agulhas Gyre *II-101*
Agulhas Province *II-134*
air masses
— main winter transport routes over the Arctic *I-13*
— transport of from industrialised areas to the Arctic I-5
air pollution
— by ships, prevention of III-337
— Chesapeake Bay I-341
— locally high, Jamaica I-570
— Rudnaya River Valley, eastern Russia II-485
— Sea of Japan II-476–7
— industrial contributions II-476–7
— sources of II-476
— southern Brazil
— Rio de Janeiro metropolitan area I-741
— Santos Bay I-743
— through forest burning, Borneo II-375
air temperatures
— Côte d'Ivoire I-809
— Gulf of Guinea I-776
Air-Ocean Chemistry Experiment (AEROCE) I-224, I-229
airborne remote sensing, use of multiple digital video cameras III-284
airmasses, contribution to pollution over the Sea of Japan *II-476*
airport construction, Bermuda, effects of I-228
airports, environmental problems of II-214
Airy's theory, to predict energy in a wave III-312
Al Batinah *II-18*, II-19, II-26
— coastal erosion acute II-28
— dams across wadis keep sediment from coast II-27
— a wadi plain II-27
Alaska, protection for high relief pinnacles III-386
Alaska Coastal Current *I-374*, I-375
— and Ekman drift I-378
Alaska Current *I-374*, I-375, *III-180*
Alaska, Gulf of I-373–84, III-128

— anomalous along-shore flow III-184
— baseline studies, pollutants I-38I-I-383
— climatic changes III-179–86
  — changes in physical properties III-181–3
  — impact of El Niño events III-183–4
  — implications of observed changes III-184–5
  — physical properties III-181
  — possible biological effects: salmon as an example III-185–6
— effects from urban and industrial activities I-383
— the *Exxon Valdez* oil spill I-382
— Forest Practices Act I-383
— geographic setting I-375
— human populations I-381
— major shallow water marine and coastal habitats I-375
— overfishing of virgin groundfish stocks, change to fishing other species III-120–1
— oxygen depletion impacts fisheries III-219
— physical oceanography I-375, I-378–9
  — shelf hydrography and circulation vary seasonally I-378
— primary productivity and nutrient cycles I-379–80
  — biomass doubling round perimeter, hypotheses I-380
— rural factors I-383
— 'strip-mining' of Pacific Ocean Perch III-120, III-121
— tidal propagation III-188–90
— trophic shift in marine communities I-376–8
Alaska Gyre *III-180*
Alaska Peninsula *I-374*
Alaskan Stream *I-374*, I-375, *III-180*
albacore I-141
albatrosses, wandering I-757
Albermarle Sound *I-352*, I-360
— environmental problems for bottom communities I-360
Albermarle—Pamlico Estuarine System *I-352*, I-353, I-362
— low benthic diversity and high seasonality in abundances I-360
— supports extensive shellfisheries I-360
Albufeira Lagoon *I-152*, I-154
— phytoplankton community I-156
— purification of mussels I-158
— RAMSAR site I-162, *I-163*
Aleutian Islands *I-374*
Aleutian Low Pressure region/system I-376, II-465
— linked to Alaskan Basin and shelf I-375, I-378
Alexander Archipelago *I-374*, I-375
algae I-590, I-649, I-754
— Andaman and Nicobar Islands II-192
— benthic, degraded, Black Sea I-291
— blue-green I-106
— brown I-47, I-70, I-109, I-125, I-159, I-290, I-669, II-23, II-42, II-664
— calcareous, maërl I-69–70, I-140
— calcareous, exploitation of, Abrolhos Bank–Cabo Frio, Southern Brazil I-737
— coralline II-370, II-711, II-741
  — crustose II-72, II-796
— drift, long-range dispersal, Western Australia II-695
— encrusting I-21
— epigrowth I-21
— epiphytic II-665, III-2
  — reduce diffusion of gases and nutrients to seagrass leaves II-701
— fucoid I-70
  — sensitive to oil III-275
— Galician I-137
— Great Barrier Reef region II-615
  — cross-shelf difference in communities II-615
— green I-70, I-71, I-194
  — *Halimeda* II-615
— Gulf of Maine, comparable to northern Europe I-311
— Gulf of Mannar II-164
— marine
  — in the Azores I-205–6
  — Fiji Islands II-754
  — Marshall Islands II-779
— Mexican Pacific coast, diversity of I-490
— red I-47, I-70, I-109, I-125, I-159, I-290, I-754
  — endemism, Australian Bight II-678
— reefal I-599–600
— the Seychelles II-236
— The Maldives II-204
— toxic
  — in English Channel I-69
  — Irish Sea I-90
— Vietnam II-564
— Wadden Sea, changes in I-49
algal assemblages, Western Australia II-695
algal beds
— green, New Caledonia II-726
— Hawaiian Islands II-796
algal blooms I-743, I-821, III-198
— Baltic Proper I-126, I-128
— Baltic Sea I-125
— Bay of Bengal II-275
— Black Sea, intensifying I-290
— detection of III-298–300
  — coccolothophores III-298
  — cyanobacteria III-298–300
  — red and white tides (HABs) III-298
— English Channel I-69
— entrainment blooms II-20
— harmful (HAB) I-827
  — and fish kills II-320
  — Hong Kong coastal waters II-541
  — increase in I-20
  — may be associated with ENSO events II-408
  — Yellow Sea II-495
  — *see also* harmful algal blooms (HABS)
— lakes, central New South Wales II-642
— novel III-259
— nuisance, creating anocic/hyppoxic disturbances III-219
— potentially harmful II-10–11, II-13
— red, Zanzibar II-92
— toxic
  — Australia's inland waters II-585
  — Carolinas coast I-356
  — increase may be due to ballast water I-26
  — Portuguese coastal waters I-160
— toxic and nuisance
  — Irish Sea I-90
  — North Sea I-53
algal mats I-131
— cyanophyte mat formers I-357
— may cause anaerobic condition I-53
algal reefs, Red Sea II-41
algal ridges, Fangataufa atoll, French Polynesia *II-818*
algal turf I-205, I-583
— Red Sea II-42
algal-vermetid reefs (boilers) I-227
alien/accidental/exotic/introduced organisms/species I-26, II-587, II-588–9, II-623, II-641
— accidental introduction of III-86
— Adriatic and Tyrrhenian seas I-282
— Argentine coastal waters I-765
— the Azores I-209
— Baltic Proper I-110
— Baltic Sea I-129
— Bay of Bengal II-279–80
— Black Sea I-290, I-291
— carried in ballast water II-319, III-227
— a coastal transboundary issue III-355
— a concern in Hong Kong II-542
— effects on seabirds III-114
— English Channel I-70
— Gulf of Maine I-312
  — and Georges Bank I-310

— Hawaiian Islands II-802
— Irish Sea I-89
— a major impact on the Tasmanian marine environment II-657–8
— Mediterranean mussel, southwestern Africa I-830
— the Mediterranean via the Suez Canal I-257–8
— North Sea I-47–8
— Victoria Province, Australia II-661, II-669
— Western Australia II-701
— *see also* new, novel occurrences and invasive disturbances
Alisios Winds *see* Trade Winds
alkyl lead I-95
alluvial fans II-37
alluvial plains, Borneo II-366
Altata-Ensenda del Pabellón, Mexican Pacific coast I-493
— pesticides causing environmental stress I-493
aluminium contamination, from bauxite mining II-395–6
Amazon River water, entering the Caribbean I-579
American flamingo, at Laguna de Tacarigua, Venezuela I-652
American Samoa II-765–72
— coastal habitats II-767–9
— coral reefs II-712, II-767–8
  — degradation of II-712
— effects from urban and industrial activities II-770–1
— environment II-767
— offshore systems II-769
— population II-714, II-769
  — must import to support population II-769
— protective measures II-771
  — marine protected areas II-771, **III-771**
— rural factors II-770
Americas, Pacific Coast, coral bleaching and mortality events III-53–4
americium I-94
Amirantes Bank *II-234*, II-235
ammonia, exports from Trinidad and Tobago I-638
amnesic shellfish poisoning (ASP) III-221
Amnesic Shellfish Toxin contamination, Scottish scallop fishery I-90
*Amoco Cadiz* oil spill
— change to *Fucus* III-272
— migration of oil layers downward within beach sediments III-270–1
amphidromic points *I-44*, I-45
amphipods, highly sensitive to oil III-274–5
Amursky Bay, eastern Russia
— chemical pollution **II-484**
— decline in annual biomass II-484
— ecological problems II-483–5
— the ecosystem is decaying II-484–5
— heavy metals in bottom sediments II-484
— metal concentrations, high in suspended matter II-484
— organic substances in wastewater II-483
anchialine pools, Hawaii II-796
anchor damage
— fishing and diving II-806
— ship groundings, Florida Keys I-409
— to corals I-605–6, I-637
anchovy
— beach seining, southern Brazil I-739
— Peru, replaced by sardines during an El Niño event I-693
anchovy fishery I-141
— Black Sea
  — collapse of I-290
  — Turkey I-292
— larval fishery, Taiwan Strait II-504–5
— Mexican Pacific waters I-490
— Peru I-691, I-693
— southwestern Africa I-832, I-833
Andaman Islands *II-270*
Andaman and Nicobar Center for Ocean Development (ANCOD) II-280
Andaman, Nicobar and Lakshadweep Islands II-189–97
— biodiversity II-192
— climate and coastal hydrography II-191
— coastal ecosystems II-191–2
— conservation measures II-195–6
— fish and fisheries of the Andamans II-192
— impacts of human activities on the ecosystem II-193, II-195
— islands of volcanic origin II-191
— Lakshadweep Islands II-194–5
— national parks and wildlife sanctuaries II-195
— population II-193
Andaman Sea *II-270, II-298, II-310*
— coral reefs II-302
— poor reef status II-303
— surfaces water influenced by freshwater continental runoff II-191
Andros Island *I-416*, I-431
Angola, development limited by civil war I-837
Angola Current *I-822*, I-824
Angola–Benguela front *I-822*, I-825
— southward displacement I-827
anguilla I-226, *I-616*, I-617
— beaches I-620
— climate I-617
— coastal protection I-624
— coral reefs I-618
— Exclusive Fisheries Zone I-621
— hurricane damage I-622
— impact of urban development I-622–3
— increased tourism I-620
— mangroves I-619
— marine parks I-625
— rocket launching site, environmental impact assessment criticised I-623
— seagrass beds I-618, *I-619*
— sustainable yields for fisheries I-621
— tourism I-620, I-622
Anguilla Bank I-618
Anjouan (Ndzouani/Johanna)
— coconut and ylang-ylang II-246
  — fringing reefs II-246
— reef front changed by quarrying II-248
— young island II-246
Annapolis Royal, Nova Scotia, tidal energy project III-316
Anole, Lac and Lac Badana *II-67*, II-68
— ecological profiles *II-76*
anoxia I-337, III-366
— due to decay of large plant biomass III-259
— from HABs III-298
— summer, Thessaloniki I-241
anoxic basins, confined, for dumping of contaminated sediments I-27
anoxic water
— Oslofjord *I-27*
— shallow-silled fjords I-17
anoxic/hypoxic disturbances III-219–21, *III-220*
— fish kills due to hypoxia III-220
Antarctic Circumpolar Current *II-581*, II-649, II-650, *II-674*
Antarctic Convergence II-583
Antarctic ecosystem, diversity and abundance in I-706–7
Antarctic Intermediate Water I-579, I-632
— Mascarene Region II-255
Antarctic polar front I-721
Antarctic Treaty system **III-335**, III-342
— Madrid Protocol III-342
Antarctica
— birds of III-108
— the grave of whaling III-75
anthropogenic disturbance, imposed on seagrasses III-10–11
anthropogenic influences, Norwegian coast I-24
anti-cancer/anti-infective agents, natural or modelled on natural products III-38–9
antibiotics, used in marine fish cages III-369
anticyclones
— Azores I-188, I-203

— Azores—Bermuda High I-439
— Bermuda High I-223
— Great Australian Bight II-675
— Mascarene II-116
— North Pacific High I-378
— Siberian High I-287, I-376, II-465
— south Atlantic I-823
antifouling paints I-788
— *see also* marine antifoulants; organotins; tributyltin (TBT)
Antilles Current I-419, *I-628*
Apo Reef *II-406*
Aqaba, Gulf of *II-36*, II-39
— minerals as pollutants *II-36*, II-39
aquaculture I-52, I-162, I-366, I-835, II-558–9, III-368–9
— Adriatic coasts I-275
  — lagoons and small lakes I-277
— Asian, significance of III-166
— Australia II-587
— the Bahamas, unsuccessful I-427
— Cambodia II-576–7
— can be a perfect environment for epidemics III-221
— Chilean fjords I-712
— the Comoros II-249
— Côte d'Ivoire I-818
— diseases, from and among facilities III-226
— east coast, Peninsular Malaysia II-352–3
  — changing water quality II-353
— effective planning and management to control pollution and disease III-170
— French Polynesia II-820
  — raising barramundi II-820
— Galicia I-144
— global, overview III-166
— Godavari-Krishna delta II-171
— limited in the Mascarenes II-262
— Malacca Straits, vulnerable to oil spill damage II-319–20
— marine, Tasmania II-656
— marine, Argentina I-764–5
  — biological species introduction I-764
  — environments: characteristics and adaptability I-764
  — harmful algae I-764–5
— marine, overview III-166–71
  — culture facilities III-170
  — fish aquaculture, development of III-168–9
— Marshall Islands II-781–2
— New Caledonia II-729
— oyster farming, eastern Australia II-641
— the Philippines, issues associated with II-413
— ponds in marine and brackish water, high demands on water III-368–9
— scenarios for opportunities to secure and increase production III-171
— seaweed, Fiji Islands II-754
— small-scale II-438
— sustainable II-375
— sustainable, Chinese and Thai experience III-171–7
  — national perspectives on issues and challenges III-171–2
  — national support for III-172
— The Maldives II-207
— Tyrrhenian coast I-279
— Venezuela I-653–4
— Victoria Province, Australia II-668–9
— west coast of Malaysia II-339
— Western Australia II-700
— western Indonesia II-394–5
  — high fluxes of nutrients and sediments into nearshore waters II-394–5
— Yellow Sea coastal waters II-494
— *see also* mariculture; shrimp aquaculture; shrimp farming
aquarium fish
— from Mauritius II-263
— trade in I-510
aquarium fish collection
— Cambodia II-576
— Great Barrier Reef II-617
— Hawaiian Islands II-803
— Marshall Islands II-784
— western Indonesia II-400
aquarium fish trade, New Caledonia II-734
aquasports, non-consumptive resource use II-141
aquatic ecosystems, atmospheric pathway a significant pathway of pollutant and nutrient fluxes III-201, III-207
aquatic organisms, farming of *see* aquaculture
aquatic plant farming III-167–8
aquatic vegetation
— decline of, Tangier Sound I-345, I-347
— improving, Chesapeake Bay I-347
Arabian Gulf II-1–16, *II-18*
— autumn seabird breeding season III-110
— coral bleaching and mortality events III-47
— corals little affected by Gulf War oil spills III-274
— defined II-3–4
— development issues II-9–11
— evaporation in excess of precipitation II-3
— future marine studies II-13
— major coastal habitats and biodiversity II-6–9
— marine studies II-3–4
— natural environmental variables II-4–6
— oil II-3
  — oil spills and the Gulf War II-11–13
— protective measures II-13
Arabian Gulf Co-operative Council (AGCC), Marine Emergency Mutual Aid Centre (MEMAC) II-29
Arabian Oryx Sanctuary II-31
Arabian Plate II-37
Arabian Sea
— abundance of meso-pelagic fish II-24–5
— coral bleaching and mortality events III-47–8
— crossed by major trading routes II-25
— a cyclone-generating region II-20
— erosion–deposition cycle II-27
— extreme marine climate II-20–1
— floor of II-19
— low pressure, ambient-temperature water flushing removes oil from mangroves III-277
— mangrove stands II-23
— northern, no freshwater inflow from Arabian Peninsula II-23
— suboxic conditions II-21
Arcachon Bay, French coast, serious environmental impacts of TBT III-250
arctic haze I-13, I-14
Arctic Mediterranean I-33
Arctic Monitoring and Assessment Programme (AMAP) I-9, I-26
— POPs ubiquitous III-361
areal surveillance, in fisheries management III-160
Argentina
— co-operation with Uruguay I-463
— laws and decrees undertaken by **I-766**, I-767
— main coastal management concerns I-751
— provincial and municipal ordinances **I-767**
Argentine coast
— estuaries and salt marshes I-755–6
  — Mar Chiquita coastal lagoon I-755
Argentine Sea *see* Southeast South American Shelf Marine Ecosystem
Argentinian continental shelf
— circulation I-754
— coastal system I-753
— Malvinas/Falklands system I-753
— subantarctic shelf waters system I-753
arid soil extraction, for beach regeneration, southern Spain I-177
Arkona Basin I-101, I-106
— deterioration of oxygen conditions I-105

armouring
— of beaches
— and oil penetration III-269–70
— and oil persistence III-276
*Arrow* oil spill, effect on *Fucus* spp. III-272
arsenic II-12
arsenic contamination, Bangladeshi groundwater II-293
artificial islands, for airport construction I-54
artificial reef structures I-347–8
artisanal fishing, impact of, New Caledonia II-730–1
Aruba *I-596*, I-597
— coastal and marine habitats *I-599*
— coastal urbanization I-606
— fishery I-603
— landfill I-602
— limited reef development I-599
— mass tourism I-598
— sewage discharge I-606
ASEAN Council on Petroleum (ASCOPE) II-325
Ashmore Reef National Nature Reserve, Western Australia II-696
Asia Minor Current I-236
Asian developing regions: persistent organic pollutants in the seas II-447–62
— coastal waters II-449–54
— open seas II-454–60
asphalts, formation of III-270
astronomical tidal forcing III-190, III-193
Aswan Dam, effects of I-256
Atacama Desert I-704
Atchafalaya River *I-436*, I-443
Athens *I-234*, I-241
— primary sewage treatment I-245–6
Atlantic basin
— and ENSO events I-223–4
— and tropical cyclones I-224
Atlantic Conveyor, postulated reversal of III-86–7
Atlantic Niños, related to Pacific El Niños, effects of I-781–2
Atlantic Rainforest
— Abrolhos Bank–Cabo Frio, Southern Brazil I-735
— Brazilian tropical coast I-725–7
Atlantic Water I-255
— entering the Tyrrhenian circulation I-272, I-274
— flowing into the North Sea I-45, *I-45*
atmosphere—surface fluxes, determined using surface analysis methods III-204
atmospheric nuclear experiments, fallout from I-11
'atmospheric particles'
— described III-199
— origins of III-199
— primary and secondary particles III-199
— size distribution in space and time III-199–200
— can change as a result of physical and chemical processes III-200
Australia
— an island continent II-581
— biogeography II-581
— coastal erosion and landfill II-585
— continental shelf II-582
— coral bleaching and mortality events III-52–3
— Western Australia (1998), variable III-3
— degradation of the Great Barrier Reef III-37
— EEZ II-581
— effects of urban, industrial and other activities II-586–9
— human populations affecting the area II-583–4
— laws protect flatback turtle III-64
— major shallow water marine and coastal habitats II-582
— mangroves III-20
— marine biogeographical provinces *II-675*
— offshore systems II-583
— protective measures II-589–91
— general marine environmental management strategies II-590
— International arrangements and responsibilities II-590
— marine protected areas II-590
— oceans policy II-590–1
— a regional overview II-579–92
— status of scientific knowledge of the marine environment II-584
— rural factors II-584–5
— seasonality, currents, natural environmental variables II-581–2
— status of the marine environment and major issues II-591
Australia, eastern: a dynamic tropical/temperate biome II-629–45
— biogeography II-631
— coastal erosion and landfill II-639
— effects of urban and industrial activities II-639–42
— human populations affecting the area II-635–6
— major shallow water and coastal habitats II-632–5
— protective measures II-642–3
— evaluation of protected areas II-643
— legislation and responsibilities II-642
— marine protected areas II-643
— protected species II-643
— rural factors II-636–8
— seasonality, currents, natural environmental variables II-631–2
— status of the marine environment II-643
Australia, northeastern, Great Barrier Reef region II-611–28
— biogeography II-613
— coastal erosion and landfill II-622
— effects of urban and industrial activities II-622–4
— protective measures II-624–7
— major shallow water marine and coastal habitats and biota II-614–19
— offshore systems II-619
— populations affecting the area II-619–20
— rural factors II-620–1
— seasonality, currents, natural environmental variable II-613–14
Australian Sea Lion II 687
— on IUCN Red List II-681
— threats to recovery of II-685–6
AVHRR imagery
— detection of cyanobacteria blooms in the Baltic III-299–300
— algorithm for bloom detection III-299
— sea surface temperature (SST), Venezuela *I-647*
— for trends in water clarity III-296
The Azores I-201–19
— climate I-203–4
— coastal erosion and landfill I-211
— effects from urban and industrial activities I-212–13
— major shallow water marine and coastal habitats I-205–9
— importance of harbours I-209
— offshore systems I-209–10
— populations I-210–11
— protective measures I-213–17
— in word but not deed I-214–15
— the region defined I-203
— rural factors I-211
— seasonality, currents, natural environmental variables I-203–5
Azores anticyclone I-188, I-203
Azores Current I-204–5, *I-204*
Azores Microplate I-203
Azores—Bermuda High I-439
Azov, Sea of *I-286*, I-287
— suffering from hypoxia I-291

Bab el Mandeb *II-36*, II-38
— flow through not clear II-39–40
— traversed by oil tankers II-59
back water effect, Bangladesh II-273
bacteria
— on the Southern Californian shoreline I-399
— sulphur-oxidizing III-263, *III-263*
— zooplankton and phytoplankton as reservoirs for III-221
bacterial decomposition, Arabian Sea II-24

Baffin Bay *I-6*, I-7, I-8
— increased mercury in upper sediments I-9
Bahama Bank Platforms, calcium carbonate I-419
Bahama Banks
— biota of Caribbean origin I-418
— fossil reefs, used to date sea-level change I-417
— islands formed during sea-level lowstands I-417
— origins of debated I-589
The Bahamas I-415–33
— agriculture I-423–4
— aquaculture I-437
— artisanal and commercial fisheries I-424–5
— coral bleaching and mortality events III-54
— effects from urban and industrial activities I-427–30
— Landsat TM–seagrass biomass relationship III-286
— major shallow water marine and coastal habitats I-419–22
— Marine and Coastal National Parks of the Bahamas **I-431**
— offshore systems I-422–3
— origin of I-417–18
— other fisheries resources I-426–7
— population I-423
— protective measures I-430–2
— seasonality, currents, natural environmental variables I-418–19
— tourism and its effect on the population I-423
Bahamas National Trust I-432
Bahamas National Trust Park *I-416*
Bahamas Reef Environment Education Foundation I-432
Bahamian Banks *see* Bahama Banks
Bahamian Exclusive Economic Zone (EEZ) I-422
Bahrain *II-2*
— coral bleaching and mortality events III-47
Baie de Seine *I-66*, I-70
— dumping of metals I-75
— PCBs and organochlorines in meio- and macrofauna I-75
Baie de Somme *I-66*
— marsh with ponds I-69
baiji (Yangtze river) III-90
— will be affected by Three Gorges Dam III-97
Baird's beaked whale III-76, III-92, III-93
Baja California I-486
— marine habitat I-487
Baja California Sur I-486, I-487
Bajuni Islands (Archipelago) *II-67*, II-68, *II-75*
— coral carpet development II-72
— development of shelf since isotope stage 5e *II-73*
— fringing reefs II-72, II-74
— shows features of barrier island complex II-69, *II-73*, *II-75*
  — build-up of sand bodies II-69, *II-71*
Bajuni Sound
— coral knobs, patch reefs and seagrass beds II-74
— intertidal abraded flats facing channels II-77
— mixture of habitats *II-71*, *II-75*, II-76, *II-78*
"balance of nature" III-86
baleen (rorqual) whales III-74, III-76
— estimates of numbers III-82–3
— "safe catch" limit calculations III-79
ballast water
— dumping of II-295
— environmental risks from II-319, III-226–7
Baltic ecosystem, pollution sensitivity of I-129–30
Baltic Proper *I-122*
— biota
  — alien species I-110
  — birds and mammals I-109
  — main coastal and marine biotopes I-108–9
  — pelagic and benthic organisms I-109
— central parts permanently stratified I-123
— chemical munitions dumped in I-117
— eutrophication I-103–8
  — biological effects of I-109–10
  — inputs from land and atmosphere I-103–4
  — oxygen I-104–6
  — temporal and spatial variability in nutrients I-106–8
— major oceanic inflows I-101–2, *I-102*
  — environmental conditions dependent on I-124
— natural immigrants I-108
— persistence of anoxic zones I-105–6
— reasons for problems I-101
— shallow banks providing spawning and nursery areas I-124
— standing stock and carbon flow I-126, *I-127*
Baltic Sea *I-100*, *I-122*, III-205
— anoxic areas III-258
— basins *I-12*Л-I-123
— cyanobacteria blooms III-299
  — investigation using AVHRR imagery III-299
— described I-123
— freshwater immigrants I-125
— including Bothnian Sea and Bothnian Bay I-121–33
  — coastal and marine habitats I-124–8
  — effects of pollution I-130–1
  — environmental factors affecting the biota I-123–4
  — fish and fisheries I-128–9
  — pollution sensitivity of the Baltic ecosystem I-129–30
  — protective measures I-132
— southern, eutrophication of coastal inlets III-260–3
— southern and eastern regions I-99–120
  — biota in the Baltic Proper I-108–10
  — condition of the Baltic Proper I-103–8
  — environmental pollutants I-112–17
  — fish stocks and fisheries I-110–12
  — regional setting I-101–3
— a young sea I-101, I-108
Baltic Water *I-18*, I-23
band ratio, for retrieval of chlorophyll content from spectral radiance III-297
Bangladesh II-285–96
— coastal environment and habitats II-289–91
— cyclones and storm surges II-288
— fisheries resources II-294
— legal regime II-295
— Naaf estuary Ransar site II-289
— need for integrated coastal management II-295–6
— offshore system and fisheries resources II-292–5
— physical setting of the Bay of Bengal II-287
— population and agriculture factors in the coastal areas II-291–2
— seasonality, currents, and natural environmental variables II-287–8
bank reefs, Brazil I-723, I-723–4
banks and shoals
— Madagascar II-118
— Mascarene Plateau, habitats poorly known II-256
Bar al Hickman *II-18*
Barbados Marine Reserve I-639
Barcelona Convention I-249, I-262, III-363
Barents Sea, effects of collapse of capelin stock III-129
barnacles I-71, I-194
— barnacle belt, Spanish north coast I-140
— bioindicators of Cu, Zn, and Cd concentrations I-213
— Deltaic Sundarbans II-156
— Mauritius II-258
Barnegat Inlet *I-322*
barramundi fishery, Gulf of Papua II-602–3, II-607
barrier island lagoons I-437
barrier island-sound systems, diverse communities I-361
barrier islands I-353
— provide shelter for seagrass communities I-358
barrier reefs II-370
— Andros Island I-421
— Belize *I-502*, I-503, I-504
  — affected by hurricanes I-504
  — Barrier Reef Committee I-512–13
  — coral mortality and macroalgae increase I-505

— system may be damaged by sediment, nutrients and contaminants I-508
— Fiji Islands II-756, **II-756**
— French Polynesia II-817
— Madagascar II-118
— Mauritius II-258
— New Caledonia II-725
— off northeast Kalimantan II-388
— San Andrés and Providencia Archipelago I-668
— *see also* Great Barrier Reef
basin water exchange processes I-21
basking sharks, Irish Sea I-91
Basque Country
— heavy industry decline, leaving environmental damage I-142
— recovery of coastal water quality I-142
Bass Strait *II-648*, *II-662*, II-663
— local sea floor pollution from offshore rigs II-669
— oscillatory tidal currents II-650
— species richness II-651
— tidal currents II-664
— topography II-663
Bassian Province II-649
bathing water quality I-51, I-76
— Aegean Sea I-248
— Israeli coast I-261
— Turkish Black Sea coast I-302–3
Bay of Biscay Central Water, off Ortegal Cape I-139
Bay of Fundy, Canada, tidal barrage III-316
bays, mudflats and sand spits
— southwestern Africa I-828–9
  — saltpans I-828
  — Walvis Bay and Sandwich Bay, Ramsar wetlands I-829
Bazaruto Islands, Mozambique II-104–5
— fishing a key activity II-109
— shallow and shelf waters II-104
beach angling I-835
beach armour, or beach nourishment III-66
beach erosion
— Australia II-585
— the Bahamas I-428
— Belize I-510
— Fiji Islands II-760
— human-induced I-211
— and sand drift, NSW, case study II-640
— South Florida I-428
— south-east Queensland and New South Wales II-639
— Tanzania, possible reasons for II-93
  — buffer zone principle II-93
— through sand mining II-782
beach forest vegetation II-777
beach formation, and longshore drift, east coast of Madagascar II-117
beach loss, in the Comoros II-248
beach mining *see* sand mining
beach nourishment/replenishment I-604, II-356, III-352
— Gulf of Guinea I-787
— sand for I-53–4, I-96, I-176, I-366, I-428
beach sands, quarrying of II-248
beach seining, Great Barrier Reef region II-623
beach vegetation, east coast, Peninsular Malaysia II-349
beach-rocks habitat, possible origin of I-273
beaches
— Adriatic, more severe storm patterns I-276
— Anguilla I-620
— artificial, for tourists I-604
— Curaçao, littered I-607
— impermeable layers affect oil penetration and persistence III-270
— loss of sand from, Jamaica I-569
— Malacca Straits II-312
— Maldivian II-203
— not meeting EU standard, Irish Sea coast I-95
— oil contamination of III-364
— and soft substrates, Victoria Province, Australia II-665–6
— Southern California, recreational shoreline monitoring I-399
— Sri Lanka, squatters on II-183
— Turkish Black Sea coast I-302, **I-302, I-303**
— *see also* sandy beaches
beaked whales III-92
Beaufort Sea, summer ice thickness and residual circulation III-190, *III-190*
Beaufort's Dyke *I-84*, I-85
bêche-de-mer fishery, Coral, Solomon and Bismark Seas region II-437
bêche-de-mer production, New Caledonia II-730
Belfast Lough *I-84*
Belgium, offshore wind farm planned III-309
Belize I-501–16
— after the 1998 bleaching and hurricane Mitch III-55
— coastal erosion and landfill I-509–10
— Coastal Zone Management Authority I-512, I-514
— Coastal Zone Management Unit I-512
— effects from urban and industrial activities I-510–12
— geology of I-504
— major shallow water marine and coastal habitats I-504–6
— Marine Protected Areas Committee I-513
— National Coral Reef Monitoring Working Group I-513
— offshore systems I-506
— populations affecting the area I-506–7
  — population and demography I-506–7
  — use of the coastal zone I-507
— protective measures I-512–14
  — challenges I-514
  — policy development and integration I-512–13
  — regulation of development I-513
— rural factors I-508–9
— seasonality, currents and natural environmental variables I-503–4
— source area for fish, coral and other larvae I-503
Belize Barrier Reef *I-502*, I-503, I-504
Belize City, habitat destruction during growth of I-510
Belle Tout Lighthouse, Beachy Head, relocation of III-353
beluga whales III-76, III-90
— contaminant-induced immunosuppression III-95
— hunted in Arctic and sub-Arctic III-92
Bengal, Bay of (northwest coast) and the deltaic Sundarbans II-145–60
— effects from urban and industrial activities II-154–6
— major rural activities and their impact II-153–4
— major shallow water marine and coastal habitats II-149–51
— protective measures II-156–9
  — captive breeding programmes II-158–9
  — conservation of biological resources II-156–7
  — conservation policies II-157–8
— seasonality, currents and natural environmental variables II-148–9
— social history and population profile II-151–3
Bengal, Bay of II-269–84
— chemical features of the water II-274
— coastal habitats and biodiversity II-274–7
— defined II-271
— effects from urban and industrial activities II-279–80
— marine fisheries II-277–8
— mining, erosion and landfill II-279
— natural environmental variables II-271–4
— new millennium: need for east coast zone management authority II-280–1
— physical setting II-287
— populations affecting the area II-278
— protective measures II-280
  — Coastal Ocean Monitoring and Prediction System, Indian Coast II-280
  — public awareness II-280
— rural factors II-278–9
Bengal Deep Sea Fan II-287
Bengal tiger II-150, II-290
— Project Tiger II-156, II-157

Bengkali Strait II-311
Benguela Current *I-822*, I-825
— eastern boundary current I-823
— fish kills, from upwelling, blooms and oxygen depletion III-129
— red tides III-218
Benguela ecosystem *II-134*
— upwelling II-135
Benguela Environment Fisheries Interaction and Training Programme I-837
Benguela Niño years I-825, I-826, I-827
benthic assemblages, offshore, Côte d'Ivoire I-814
benthic biomass, South China Sea coast **II-554**
benthic communities
— affected by turbidity and oxygen depletion III-259
— Gulf of Guinea I-782, **I-782**
— Irish Sea
— linked to sediment type *I-86*, I-89
— threats to I-89, *I-94*
— macrofaunal, English Channel bed I-71
— offshore, Carolinas coast I-362
— Portuguese coastal waters, temporal changes due to natural causes I-159–60
benthic fauna
— Aegean Sea I-240
— Baltic Proper, adverse changes below the halocline I-110
benthic microbial communities, effects of eutrophication on, the Bodden III-261–2
benthic monitoring III-243
benthic organisms
— as biomonitors of radionuclides I-116
— West Guangdong coast II-554
benthic species, high mortality rate from trawling III-123
Benthic Surveillance Project (USA) I-452
benthic vegetation, Portuguese coastal waters I-159–60
benthos
— Campeche Sound I-474
— Côte d'Ivoire I-813
— infaunal, filtering ability, Chesapeake Bay I-342
— Marshall Islands II-780
— Southern California Bight, effects of anthropogenic inputs I-395–6
Bergen *I-18*, I-27
Bering Sea *I-374*
— by-catch as an issue III-146
— decline in some marine mammal populations through the pollock fishery III-129
— Eastern, change in groundfish species composition III-128
— residual currents III-190
Bering Sea ecosystem, example of cumulative and cascading impacts III-98–9
Bermuda
— acid rain I-229
— Bermuda Atlantic Time-series (BATS) program I-224–6
— conservation laws I-230
— Hydrostation S program I-224–5
— threats to reef environment I-228
Bermuda High I-223
Bermuda Platform I-227
*Biddulphia sinensis*, an Asian introduction I-26
Bien Dong Sea II-563
— great natural resource potential II-567
— *see also* South China Sea
Bijagos Archipelago, Guinea-Bissau I-781
— breeding and nursery ground, fish and crustaceans I-781
— diverse tidal habitats I-781
Bikini Atoll, nuclear tests at, costs of clean up II-783
Bimini *I-416*
— deep water sport fishing I-422
— mangroves stunted I-422
— sand mining/dredging I-428
bioaccumulation
— of heavy metals II-156
— in seabirds and fish I-212
— of pesticides, northern Gulf of Mexico I-444
— in the Southern California Bight I-394
— of DDTs and PCBs in seabirds I-398
— effects of in fish I-397
— in marine mammals I-399
— of TBT III-250
bioavailability, of contaminants in sediments I-25, I-26
biocides
— antifouling, the ideal III-249
— in marine antifoulant paints III-248
biodiversity II-336
— Andaman, Nicobar and Lakshadweep Islands II-192, **II-193**
— Anguilla I-620
— Arabian Gulf II-6–7
— Australia II-583
— of the Baltic Sea I-124–5
— marine species decrease to the north I-124–5
— Bay of Bengal **II-274**
— benefits to tourism and recreation II-326
— British Virgin Islands I-620
— coral reefs III-34
— endangered III-36–7
— Coral, Solomon and Bismark Seas region II-431
— Deltaic Sundarbans II-150, **II-150**
— English Channel
— fauna I-70–1
— flora I-69–70
— Fiji Islands II-757
— marine **II-754**, II-762
— Godavari-Krishna delta II-170
— Great Barrier Reef II-613
— Gulf of Aden II-49
— Gulf of Guinea LME I-794
— Gulf of Mannar II-164, **II-164**
— high
— macroalgae and invertebrates, the Quirimbas II-102, II-104
— New Caledonia II-726, **II-726**
— hunted in Arctic and sub-Arctic III-92
— increase from fjord head to coast I-22
— Lakshadweep Islands II-194
— Malacca Straits II-312
— marine
— Australian Bight II-678–9
— Hawaiian Islands II-794
— southern Australia II-582
— Vanuatu **II-741**
— Vietnam II-564
— western Indonesia, threats to II-389–90
— Palk Bay **II-166**
— Palk Bay–Madras coast II-169
— Patagonian shores I-756
— Peru I-691–3
— biological effects of El Niño I-692–3
— wetlands and protected areas I-691–2
— the Philippines II-408–10
— potential loss of in the Comoros II-251
— range of effects of fishing III-378–9
— rocky shores, eastern Australian region II-633
— South African shores II-135
— in the Sundarbans II-291
— super-K species III-39
— and system integrity, mangroves III-28
— in terms of higher taxa, mine tailings III-241–2
— threatened by by-catch III-136
— western Indonesia II-389–90
— Wider Caribbean I-589
— within the Belize coastal zone I-506
— Yellow Sea, loss of II-497
biodiversity management, Great Barrier Reef Marine Park II-625–6
biodiversity recovery, mine tailings III-241

bioerosion II-69
— contributes to destruction of the reef matrix II-397
— the Maldives, after coral reef bleaching event II-214
biofilms
— as indicators for eutrophication III-263
— and microorganism habitats III-261
— oxygen supply to III-263
— photoheterotrophic III-263
— toxic III-219
biogenic species, effects of oil spills III-273–4
biogeochemistry, around Bermuda I-225
biogeography
— Australia II-581
— eastern Australian region II-631
— equilibrium theory of III-380
— Fiji Islands II-753
— Vanuatu II-739
— Western Australian region II-695
biological communities
— and coastal stability III-352
— effects of oil spills III-272–5
  — alterations in pattern of succession and dominance III-275
  — on consumers (predators and herbivores) III-274
  — on prey species III-274
  — on sensitive species with localized recruitment III-274–5
  — on structuring communities III-2734
biological factors, the key to the Sundarban coast II-148
biological invasions *see* alien/accidental/exotic/introduced organisms/species
biological oxygen demand (BOD)
— direct increases in through nutrient loading I-367
— high, leads to hypoxia/anoxia I-364
biological production
— increased, Baltic Sea I-130
— Portuguese coastal waters I-153–4
biology, evolutionary, and "punctuated equilibrium" III-86
biomagnification I-5
— of Cd and Hg, Greenland I-9, *I-10*
— of mercury in the biota, Jakarta Bay II-396
"biophile" elements I-116
biophysical features, gradients in
— Hawaii II-795
— the Maldives II-202, **II-202**
biopollution *see* alien/accidental/exotic/introduced organisms/species
bioregions, Tasmania II-651
biosphere, value of ecosystem services III-395
Biosphere Reserves
— Gulf of Mannar II-163
— Mananara Nord, Madagascar II-127
— Nancowrie Biosphere Reserve II-196
— Odiel saltmarshes I-170
— Rocas Atoll, Brazil I-724
— Sikhote-Alin Biosphere Reserve, eastern Russia II-481–2
— Sundarban Biosphere Reserve II-147, II-156–7
biota
— influence on oil persistence III-271
— Madagascan, effects of climate on II-116–17
  — migratory patterns of some species II-116
  — shallow-water assemblages differ from North to South II-116
biotoxin and exposure disturbances III-218–19, *III-220*, **III-220**
biotoxins
— causing mortalities III-218
— cyanobacterial, implicated in chronic diseases III-218
— effects of direct exposure to III-219
birds
— Asian, organochlorine pollution in II-452, *II-454*
— commercially reared III-225
— Doñana National Park as breeding site and migratory stop I-171–2
— El Salvador I-549
— Gulf of Mexico Coast *I-473*, **I-473**, I-474
— land and sea, Marshall Islands II-778
— Mai Po marshes, Hong Kong, high species diversity II-539
— migratory
  — disease among III-225
  — exposed to HCHs and PCBs in India II-452
  — Sarawak II-368
— *see also* marine birds; seabirds; shorebirds
Biscay, Bay of, weak circulation I-138
Biscayne Bay *I-406*, I-407, III-2, III-9
Biscayne National Park I-412
Bismark Sea *II-426*
Bitter Lakes, kept Suez Canal salinity high II-38
bivalve molluscs, particle-feeding III-240
bivalve mortalities III-226
bivalves
— Aegean I-239
— Bothnian Bay I-125
— filter feeding activity reduced by brown tides III-219–20
— Portuguese coastal waters, some bacterial problems I-158
— southern Baltic I-109
BKD (kidney disease) I-35
Black River Morass, Jamaica I-565
— mangroves I-565
Black Sea I-285–305
— anoxic interface I-287
— climate I-287
— coastal development expected to continue I-303
— effects of recurring hypoxic conditions III-219
— environmental policy difficult to implement I-304
— land-based pollution I-296–303
— major shallow water marine and coastal habitats and offshore systems I-290–4
— residence time of waters I-303
— seasonality, currents, natural environmental variables I-287–9
— southern Black Sea, Turkey I-294–6
— surface circulation I-287–8, *I-288*
  — upper layer general circulation *I-289*
— world's largest anoxic water mass I-287
Black Sea Environmental Programme I-290, I-294–5
— Black Sea Action Plan I-295, I-303
— financing of I-295
Black Sea Water I-236, I-243
blacklisting, of fishing vessels III-160–1
blast fishing *see* destructive fishing
blast fishing, Indonesian seas II-382
blue crab pot fishery I-362
— Pamlico and Neuse estuaries I-360
Blue Mountains, Jamaica, effects of I-561
blue mussels I-52, I-131
— Baltic Proper I-126
— Greenland, high lead concentrations I-12
blue whales III-82, III-82–3
— secretly killed III-77, III-86
BOD *see* biological oxygen demand (BOD)
Bodden, southern Baltic Sea
— changes of ecosystem structure and function following eutrophication III-264, **III-265**
— characteristics of III-260
  — high filter and buffer capacity III-260
— eutrophication of III-260–1
  — buffer capacity exhausted III-261
  — causes of III-260
  — effects on benthic microbial communities III-261–2
  — investigations of the impact on the nitrogen cycle III-262–3
— Nordrügensche Bodden, effects of increasing eutrophication III-261–2
— remediation possibilities III-263–4
bolide impact, effects of Chesapeake Bay I-337
"Bolivian winter" I-705
Bonaire *I-596*, I-597
— coastal development I-606
— coastal and marine habitats *I-599*

— dive tourism I-597
— excavation for construction, ruined groundwater quality I-602
— fishery I-603
— Flamingo Sanctuary and Washington Park I-609
— sewage discharge I-606
— turtle grass I-601
Bonifacio Strait I-273
Bonn Convention III-340
Borneo II-361–79
— erosion and landfill II-375
— habitat types II-364–5
— human populations II-373–4
— major coastal habitats II-366–72
— management objectives for marine protected areas II-377
— marine conservation areas II-376–8
— natural environmental variables II-365–6
— offshore systems II-372–3
— regional extent II-363–5
— rural factors II-374–5
— urban and industrial pollution II-375–6
Bornholm Basin I-101
— deterioration of oxygen conditions I-105
Bornholm Deep I-102, I-106
Bosphorus I-198, *I-286*
— connects Black Sea to the Mediterranean I-287
Boston *I-308*, I-312, I-313
Boston Harbor I-314, I-317
— a history of lead I-315
Bothnian Bay *I-100*, *I-122*, I-123, I-126
— heavy metals I-131
— standing stock and carbon flow I-126, *I-127*
Bothnian Sea *I-100*, *I-122*
— fresh water species I-125
— heavy metals I-131
— standing stock and carbon flow I-126, *I-127*
bottlenose whales III-76, III-83, III-92
bottomland forest wetlands, Gulf Coast, USA I-440
boulder/cobble shores, the Azores I-207
Boundary Current, North and East of Madagascar *II-114*
boundary-layer flow III-206
— *see also* internal boundary layers
bowhead whales III-76
— catch limits III-83
— distinct stocks III-81
— occasional Canadian aboriginal kills III-81
brackish-water conditions, adaptation to in Baltic Sea I-125
Brahmaputra River *II-146*
braided channels, Somalia
— biota in II-77
— channel levees II-77, *II-78*
— encrusted hard bottoms *II-76*, II-77
Brazil
— degradation of coastal areas I-463
— education to solve some problems I-463
— mangroves III-20
— Special Management Zone, Bahia de Caraquez I-463
Brazil Current I-722, I-751, I-753
— Abrolhos Bank–Cabo Frio region I-734
Brazil Current–Falklands/Malvinas Current confluence zone I-751
Brazil, southern I-731–47, I-744–5
— activities I-737–40
— degradation I-740–3
— east coast I-733
  — Abrolhos Bank–Cabo Frio I-734
  — habitats I-735–6
— environmental laws I-744–5
  — territorial sea defined I-744
— historical setting I-733–4
— major shallow water marine and coastal habitats I-735–7
— physical description I-734–5
— south coast I-733
  — habitats I-737
  — large coastal plain I-737
  — Southern Brazil Shelf I-735
— southeast coast I-733
  — habitats I-736–7
  — South Brazil Bight I-734–5
Brazil, tropical coast of I-719–29
— major coastal habitats I-725–7
— major environmental concerns and preservation I-727–8
— major marine habitats I-722–5
— oceanographic parameters I-721–2
— the region I-721
— Rocas Atoll I-723
  — Biosphere Reserve I-724
— Southern Bahia, coral reefs I-724
breaking wave forces, increased with rising sea level III-190
bridges, may obstruct water flow in estuaries II-135
brine pools
— contain hydrogen sulphide I-442
— hot, Red Sea II-39
brine rejection, due to freezing I-33
British Indian Ocean Territories (BIOT)
— included in UK's ratification of conservation and pollution Conventions II-231
— *see also* Chagos Archipelago, Central Indian Ocean
British Virgin Islands *I-616*, I-617
— Coast Conservation Regulations I-624
— coral communities I-618
— effects from urban and industrial activities I-623
— Exclusive Fishing Zone I-621
— expansion of tourism I-620
— hurricane damage I-622
— mangroves I-619
— marine protected areas I-625
— mooring system I-624
— National Integrated Development Plan I-624
— tourism I-620
Brittany coast *I-66*, I-69, I-74
Broad River Estuary *I-352*, I-354
brominated chemical, widespread contamination by III-362
brown tides I-439, III-1298
— anoxic impact III-219–20
— cause persistent economic fisheries losses III-220
— high bivalve mortality III-220
Browns Bank *I-311*
Brunei *II-362*, II-365
— population II-374
Bryde's Whale II-676, III-78
bubble burst activity 202
— change in deposition velocity due to III-202–3, *III-203*
Buckingham Canal, southeast India, a health hazard II-170
Buenaventura Bay *I-678*
— high hydrocarbon levels in bivalves I-683
Buenos Aires Province I-751
— coast *I-752*
— coastal erosion I-762
— effects of harbour construction I-762
Burmeister's porpoise III-90
burning
— of grassland, Madagascar, leads to topsoil loss II-123–4
— of secondary forest, produced air polluting haze, Borneo II-375
— *see also* fire
burrowing animals, effects on oil spills III-271
Busc Busc Game Reserve II-80
Busc Busc, Lac *II-67*, II-68
by-catch I-445, II-57, II-59
— black-browed albatross as I-758
— by-catch reduction programs III-140–1
  — reductions in BPUE III-141
  — reductions in effort III-140–1
— as a component of fishing mortality III-140

— creating conservation problems III-136
— defined III-137
— and discard mortality III-123–5
  — reasons for discard III-123
— Great Barrier Reef region II-623
— history of the issue: some early examples III-141–7
  — coastal gillnets and seabirds III-146
  — discards in shrimp and prawn trawls III-143
  — gillnets and cetaceans III-144–5
  — high seas drift nets III-145
  — longlines and sea turtles III-146
  — longlines and seabirds III-145–6
  — Northeast Pacific groundfisheries III-146–7
  — shrimp–turtle problem III-143
  — trawls and cetaceans III-145
  — tuna–dolphin problem III-141–3
— and incidental take III-379
— includes many dolphins III-91
— indirect results of III-379
— a main fisheries issue III-161
— of non-target organisms III-368
— not incorporated in most fisheries management models III-367
— originally ignored, now important III-136–7
— prawn trawling Torres Strait and Gulf of Papua II-604
— problems and solutions III-135–51
  — by-catch classification: why is it useful? III-139–40
  — definitions III-137
  — into the 21st century III-147–8
  — reasons for discarding III-137–8
  — regulations and guidelines III-138–9
— of seabirds III-113
— shrimp fishing/trawlers I-490, I-536, I-539, I-554, II-121
— solutions to III-147
— squids, southern Brazil I-739
— and technology III-148
— wasteful I-3
by-catch-per-unit effort (BPUE), reductions in III-141
— deployment and retrieval changes III-141
— management action III-141
— training III-141
by-catch-reduction devices (BRDs) III-144, III-145, III-146
bycatch quotas III-156
Bylot Sound, Thule, nuclear weapon accident, plutonium contamination I-11

Cabo de Santa Marta Grande-Chui, southern Brazil
— environmental problems I-743
  — primary problems I-743
— major activities I-739–40
  — tourism I-739
— marine and coastal habitats I-737
— Port of Rio Grande I-740
Cabo Delgado, dividing point for South Equatorial Current II-101
Cabo Frio–Cabo de Santa Marta Grande, southern Brazil I-736–7
— bays
  — Rio de Janeiro coast I-736
  — Sao Paulo State I-736
— environmental degradation I-738
  — Cubatao Pollution Control Project I-743
  — from unregulated urbanisation I-740
  — in Guanabara Bay I-741
  — Santos estuary I-742–3
  — in Sepetiba Bay I-741
— estuarine–lagoon complex, Iguape–Cananéia–Paranaguá
  — important littoral ecological system I-736–7
  — subsistence agriculture and fisheries I-739
— major activities I-738–9
— mangroves degraded, south of Caraguatatuba Bay I-736
cachelot *see* sperm whales/whaling
Cadiz Bay *I-168*
— commercial port I-176
— saltmarshes in decline I-177
— urban–industrial development environmentally hazardous I-174
Cadiz, Gulf of *I-168*
— connects Atlantic and Mediterranean trading ports I-181
— contamination related to heavy industrial activity I-179, **I-180**, *I-181*
— meeting of water masses I-169
— monitoring quality of coastal waters I-179
— rich fishing grounds I-173
— use of traditional fishing techniques I-175–6
Cadiz–Tarifa arc
— rocky I-172
— rugged bottoms, Tarifa I-172
cadmium (Cd) I-5, I-75, I-260–1, I-301
— Baja California I-495, **I-496**
— concentrations in the Azores I-213
— in Greenland seabirds and mammals I-9, *I-10*, I-14
— levels unsafe in Norwegian mussels and fish liver I-25
— off Cumbrian coast I-94
— in whales I-230
caesium (Cs), conservative behaviour I-13
caesium-137($^{137}Cs$)
— from Chernobyl I-11
— from Sellafield
  — by long-distance marine transport I-11
  — in Irish Sea water I-94, **I-95**
— outflow from the Black Sea I-249
— Sea of Japan II-479
Caicos Bank *I-588*
— reef areas I-590
Caicos Passage *I-416*, I-417
calamari fishery, Victoria Province, Australia II-668
calcification, corals
— enhanced by algal turf communities III-38
— and nutrient uptake III-35–6, *III-36*, III-38
— physiology of III-35, *III-36*
California Current *I-374*, *I-386*, I-486, I-487, I-488, I-489, I-543, III-113, *III-180*
— affected by El Niño and La Niña events I-387
California sea lion I-399
Californian Coastal Province, Mexico I-486–7
— environmental importance of I-487
— features of I-486
— habitats I-486–7
— monitoring work I-487
Calvados Coast *I-66*, I-69
Calvert Cliffs *I-336*, I-346
Cambodia II-299
Cambodian Sea II-569–78
— coastal deterioration due to erosion and landfill II-576
— coastal habitats II-573–4
— coastal population II-575
— defined local marine environment II-571
— effects of the rural sector II-575–6
— effects of urban and industrial development II-576–7
— offshore habitat II-574–5
— physical and chemical conditions in surface waters II-572–3, **II-573**
— protective measures II-577–8
  — limited perception of environmental impact II-577
  — status of marine environment and habitats protection measures **II-577**
— seasonal variability of the natural environment II-571–3
Cameroon
— conservation concerns I-790, *I-791*
  — coastal environment assessment I-790
— increasing waste contamination I-789
— rivers I-777
Campeche Sound *I-468*
— oil industry I-476
— primary production I-474
Canadian Archipelago I-5, *I-6*

Canary Current I-204, *I-204*, I-223
Canary Islands I-185–99
— climate I-188
— development pressures and protective measures I-196–8
— ecosystem characteristics and impacts on coastal communities **I-197**
— fishing resources I-195–6
— marine ecosystems I-193–5
— physical and geological background I-187–9
— geological evolution of I-187
— seasonality, currents and natural environmental variables I-189–92
Canary Islands Counter Current I-189
Canary Islands Stream I-189
canneries, Fiji and American Samoa II-718, II-770
canning industry, Venezuela I-653
Cantabrian Sea *I-136*
— summer subsurface chlorophyll maximum I-141
Cantabrian Shelf, sand and silt *I-136*, I-137
Cap-Breton Canyon *I-136*
Cape Cod *I-308*, *I-311*, I-314
Cape Fear Estuary I-353, I-362
— source of chronic BOD load I-364
— well flushed I-363
Cape Fear River *I-352*, I-354, I-361, I-364
— turbidity, faecal coliforms and BOD, correlation I-364, **I-365**
Cape Hatteras *I-352*
— associated with geographic division of plankton I-356
Cape Horn Current I-701
Cape Johnson Trench *II-426*
Cape Sable *I-308*
capelin I-376
— industrial fishery based on I-18
capelin fishery, collapse of affected small cetaceans III-97
Capelinhos Mountain, Faial I-203, I-207
captive breeding programmes, Sundarbans
— estuarine crocodile II-158
— horseshoe crabs II-158–9
— Olive Ridley turtle II-158
carbon dioxide ($CO_2$)
— anthropogenic, elevated III-38
— atmospheric
— increase in III-188
— reduces oceanic $CaCO_3$ supersaturation III-36
— disposal at sea to mitigate climate change III-366
— from conversion of bicarbonate III-35, III-35–6
carbon dioxide pollution, transferred to deep sea I-7
carbonate platforms, the Bahamas I-417
carbonate sediments
— dominate tropical Brazilian middle and outer shelves I-722–3
— open shelf, Australian Bight II-681
carbonates, biogenic and chemically precipitated, Borneo II-366
Cardigan Bay *I-84*
— shore communities I-88
— Special Area of Conservation I-90–1
Cariaco Gulf *I-644*
— *Thalassia* beds I-650
Caribbean, collapse of coral reefs III-37, *III-37*
Caribbean and Atlantic Ocean, coral bleaching and mortality events III-54–6
Caribbean Basin, eddies within I-580
Caribbean Coastal Marine Productivity (CARICOMP), research and monitoring network I-634
Caribbean Current I-580, I-617–18, *I-628*, I-665
Caribbean Lowlands, Nicaragua I-524
— ethnic hierarchy I-524
— home to Miskito, Creoles and mestizos I-524
Caribbean Oceanographic Resources Exploration (CORE) I-634
Caribbean Sea I-579
Caribbean Small Island Developing States I-617
Caribbean Surface Water, density, temperature and salinity I-579
Caribbean–North Atlantic convergence I-497, **I-497**
Carlsberg Ridge II-49
Carolina Coasts, north and south I-351–71
— better management on non-point source runoff essential I-367
— coastal rivers, problems of I-361
— environmental concerns in estuarine and coastal systems I-363–6
— flora and fauna of the coastal waters I-354–63
— limits needed on nutrient inputs to rivers and estuaries I-367
— physical setting I-353–4
— sources of pollutants I-354
— prognoses for the future I-366–7
carrying capacity III-86
Cartagena Convention
— protocol on Specially Protected Areas and Wildlife (SPAW) I-608, I-640
— SPAW protocol III-68
— and whale protection III-85
Carysfort Reef
— continuing decline in deep and shallow waters I-410, I-412
— coral vitality: long-term study I-410–12
— reaching stage of ecological collapse I-412
— sediments of I-409
"cascade hypothesis" III-99
catchment impacts, Tasmanian coastline II-654
categorical correlation matrix, from HEED survey *III-217*
catfish, mysterious mortalities III-223
Cay Sal Bank I-417, I-421
Cayman Islands, coral bleaching and mortality events III-54
cays
— Belize I-505
— shifting populations I-507
CC:TRAIN III-327–8
— operates under TRAIN-X principles III-327
— role of III-327
— Vulnerability and Adaptation Assessment (V&A) COURSE III-327–8
CDOM (coloured dissolved organic matter) III-297
cellulose industry, Chile, disposal of liquid waste I-712–13
Central Adriatic I-269
— surface circulation I-271
— water column divisions I-270–1
Central America
— analysis of coastal zones I-463
— described I-460
— institutional issues in coastal resource management I-460
— major coastal resource management issues **I-460**
Central Bass Strait Waters II-664
central Bight water mass, salinity of II-676
Central Equatorial Water I-679
central south Pacific gyre ecosystem I-706
Central South Pacific Ocean *see* American Samoa
cephalopods, important in Gulf of Thailand II-306
'Certain Persistent Organic Pollutants', negotiations on III-340
cetaceans
— caught by gillnets III-144–5
— caught in midwater trawls III-145
— in the English Channel I-70
— Great Australian Bight II-680
— Gulf of Aden II-54
— increased strandings in the North Sea I-50
— organochlorine residue levels, western Pacific II-455, *II-459*
— Patagonian coast I-756–7
— PCB concentrations II-455, *II-459*
— Southern California I-399
— vulnerable to undersea noise II-685
— *see also* dolphins; whales
Chagos Archipelago, Central Indian Ocean II-221–31
— available for defence purposes, with conservation provisions II-230
— biogeographic position in the Indian Ocean II-224–5
— a British Indian Ocean Territory II-223
— geographical and historical setting II-223–4

— importance of Chagos II-231
— major shallow water marine and coastal habitats II-226–9
— offshore systems II-229–30
— population, urban and industrial activities II-230
— a pristine environment II-230
— protective measures II-230–1
— reef studies II-223
— seasonality, currents, natural environmental variables II-225–6
Chagos Bank *see* Great Chagos Bank
Chain Ridge *II-64*, II-67
chalk cliffs, erosion of I-68
Challenger Bank I-227
Chang Yun Ridge, Taiwan Strait, effects on pollutants II-501–2
chank fishery, Sri Lanka II-183
Channel Islands I-71
Char Bahar *II-18*, II-24
Charleston Harbor *I-352*, I-353
chemical cargoes, loss of from shipping III-364
chemical contamination/pollution
— Amursky Bay, eastern Russia **II-484**
— Bay of Bengal II-295
— English Channel, and its impacts I-74–6
— Faroes I-31, I-36
— northern New South Wales catchments II-637
— Norway I-24–5
— southern Brazil, Paranaguá Port I-743
chemical energy, from natural oil seeps I-442
chemical industry, southern Brazil I-738
chemical spills, toxic I-72–3
chemical warfare agents, dumped in the Baltic Proper I-117
chemical wastes, Bangladesh, disposal of II-295
Chernobyl
— contamination from I-11, I-55
— in Baltic seawater I-116–17
Chesapeake Bay I-335–49, III-207
— C and D canal *I-336*, I-339
— coastal erosion and landfill I-346
— the defined region I-337–9
— eutrophication and remediation III-205, III-206
— land use *I-338*
— largest estuarine system in the USA I-337
— major shallow water marine and coastal habitats I-341–3
— offshore systems I-343–4
— pesticides affected eelgrass III-9
— populations affecting the area I-344–5
— principal rivers entering I-337, I-339, **I-339**
— protective measures I-346–8
— nutrient management plans for manures I-347
— pollution clean up I-346
— removal of eelgrass by rays III-8
— rural factors I-345
— seasonality, currents, natural environmental variables I-339–41
Chesapeake Bay ebb tidal plume I-344
— USEPA Chesapeake Bay Program I-346
Chesapeake Bay ecosystem, loss of keystone species leads to ecosystem shift III-227
Chesil Beach *I-66*, I-68
Chichester Harbour *I-66*, I-74
Chile
— industrial fishery development I-463
— possibilities for coastal management plan I-463
— small cetaceans as fish bait III-92–3
Chile–Peru Current I-705
Chilean Coast I-699–717
— advances in control and pollution abatement I-715
— coastal marine ecosystems I-705–9
— major determinants of distribution/abundance of marine species I-708–9
— human coastal activity I-709–15
— large and mesoscale natural variability I-702–5
— long-term variations: El Niño Southern Oscillation I-702–4
— seasonal variations I-704–5
— physical setting I-701
— oceanic islands I-701
— water masses I-701–2
Chilean Coast oceanic islets I-701
Chilean Trench I-701
Chilka Lake, India, and environmental problem II-279
China
— aquaculture
— allocation and utilisation of natural resources III-174
— changes in living standards and consumer preferences III-172–3
— changing scenarios in development III-172–5
— freshwater aquaculture expansion III-173
— production diversification III-174–5
— provincial expansion III-173–4
— short comings in the sector III-175
— coastal population, Yellow Sea II-491
— increased agricultural output, Yellow Sea coast II-492
— legislation and regulations concerning water quality III-175, **III-175**
— mortalities in mariculture attributed to red tides III-219
— oil spills II-495
— species used in aquaculture III-171, III-174–5
China Sea, effects of sea level rise, cases considered III-190, *III-192*, III-193
chlorinated compounds, Great Barrier Reef region II-623
chlorinated polycyclic aromatic hydrocarbons (Cl-PAHs), in the Baltic I-114
chlorobiphenyl congeners, decreased concentrations, Baltic Sea I-114
chlorofluorocarbons (CFCs) III-188
chlorophyll
— Argentine Sea I-757, *I-757*
— concentrations, Côte d'Ivoire I-814
— deep chlorophyll maximum, Levantine Basin I-256
— southern Aegean Sea I-240
chlorophyll *a* III-297
chlorophyll content
— Case 2 waters, determination of III-297
— distinguished from CDOM III-297
chlorophytes I-821
Chocó Current I-679
Chokoria Sundarbans II-289, II-290
Christiaensen Basin, New York Bight, highly contaminated I-325
Christmas Island II-583, II-696
*Chrysochromulina leadbeteri* bloom I-26
*Chrysochromulina polylepis* bloom (spring 1988) I-20
Chumbe Island, Tanzania
— Marine Park II-94
— environmental education programme II-94
Chwaka Bay, Tanzania, effects of herbicide use II-89
ciguatera poisoning, Mascarene Region II-261–2, II-264
Ciguatoxic Fish Poisoning (CFP) III-221
circulation
— Aegean I-236
— Argentinian continental shelf I-754
— Bay of Bengal, monsoonal II-273, *II-273*
— English Channel I-67
— fresh water-induced, off estuaries I-46
— Gulf of Mexico, Loop Current system I-469
— largest Baltic Sea basins I-123
— northern Gulf of Mexico I-439
— Yellow Sea and East China Sea II-489
circulation patterns
— Coral, Solomon and Bismark Seas Region II-428, *II-428*
— the Maldives II-201
— Polynesian South Pacific II-816, *II-816*
CITES III-340
— Convention on Migratory Species III-68
cities
— Guangdong II-559

— Indonesia **II-384**
— the Philippines II-416–17
clams, Maputo Bay, *Vibrio* contamination II-109
clay–oil flocculation, reduces oil retention in fine sediments III-271
clear water, assists remote-sensing III-284
cliffs, high, central Peruvian coast I-689
climate
— Arabian Gulf II-4
— winter and summer monsoons II-4
— Arabian Sea coastal areas II-20
— Bahamas archipelago I-418–19
— Chesapeake Bay I-339–41
— tornadoes and hurricanes I-340–1
— and coastal hydrography, Andaman and Nicobar Islands II-191
— coastal, tropical Brazil I-721
— Coral, Solomon and Bismark Seas Region II-428
— Côte d'Ivoire I-808–9
— east coast, Peninsular Malaysia II-347
— eastern Australian region II-631
— effect on Madagascar's biota II-116–17
— El Salvador I-547
— Fiji II-753
— Florida Keys, hurricanes I-408
— Great Australian Bight II-675
— Great Barrier Reef region II-613
— Gulf of Guinea I-775–6
— Hawaiian Islands II-795
— leeward and windward Dutch Antilles *I-596*, I-598
— Mexican Pacific Coast I-486, I-487, I-488, I-489
— Nicaragua I-519, I-534
— northern Gulf of Mexico I-439
— the Philippines II-407, *II-407*
— Red Sea, driven by migration of the Inter-Tropical Convergence Zone II-39
— Sea of Okhotsk, similar to arctic seas II-465
— and seasonal rainfall, Chagos Islands II-225
— seasonal variation, Chilean coast I-704–5
— the Seychelles
— controlling factors II-235
— humid tropical II-235
— Somalian Indian Sea coast
— bi-modal rainfall II-65–6
— temperature II-66
— Vanuatu II-739
— (weather), Borneo II-365
— (weather), French Polynesia II-815
— western Sumatra II-386
— Xiamen region, China II-515
climate change III-369–70, III-394
— and changes in marine mammals and seabirds I-9
— and changes in thermohaline circulation I-33
— combined with stressors, possible effects of III-230
— and coral reef degradation III-44
— *see also* coral reef bleaching
— Global Seagrass Declines and Effects of Climate Change III-10–11
— and greenhouse gas emissions III-47
— Gulf of Alaska III-179–86
— monitoring for Caribbean Planning for Adaptation to Climate Change project, Jamaica I-571–2
— and the North Sea I-47
— past, west coast, Sea of Japan II-483
— and sea level change, effects on coastal ecosystems III-187–96
— and warming, Chagos Islands II-225
Climate Change Convention III-327
— implementing the challenges III-327
Climate Change Project (GEF), the Maldives II-215, II-216
climate disturbances, significant III-224
climate influences, act alongside global-scale environmental change III-225
climate shift, Gulf of Alaska I-377
cloud cover, limiting factor for remote sensing III-286
co-management
— in fisheries III-159
— Fisheries Master Plan, Mozambique II-111
— from management to co-management: the *Pomatomus saltatrix* fishery II-138
co-occurring biological anomalies III-214, III-217
coagulation, of particles III-200
coal mining
— Great Barrier Reef hinterland II-622
— southern Brazil I-739, I-743
coastal accretion
— Borneo II-366, II-375
— western Indonesia II-384
coastal area management, integrated, need for III-170–1
coastal area management systems, traditional, Coral, Solomon and Bismark Seas region II-442–3
coastal areas
— changes, tides and long waves through sea level rise III-188
— propagation of tides in III-188
— sea breezes III-207
— showing signs of eutrophication III-199
— Venezuela
— areas under Special Regulation **I-658**
— relevant planning regulations **I-657**
Coastal Biodiversity Action Plans I-77
coastal cold water *I-688*
coastal construction
— Hawaiian Islands II-805
— may provide additional solid substratum II-699
— Xiamen region II-519
coastal currents, Puerto Rico and US Virgin Islands I-580
coastal data coordination III-355
coastal defences
— English Channel coast I-78–9
— soft and hard, loss of, Mozambique II-108
coastal development
— in the Comoros II-247–8
— impacts of, the Bahamas I-427–9
— Western Australia, leading to habitat loss and alienation II-699
coastal dunes, heavy metal mining II-639
coastal ecosystems
— Andaman and Nicobar Islands II-191–2
— changes in, Xiamen region II-524–5
— Maluan Bay II-524
— Tong'an Bay II-524–5
— complex, Grande Island Bay with Septiba Bay, southern Brazil I-738
— Coral, Solomon and Bismark Seas region II-430
— effects of climate change and sea level on III-187–96
— coastal effects III-188
— effects on coastal storm surges and estuarine flood risk III-190, III-193–4
— effects on tides and tidal currents III-188–90
— other effects III-194–5
— Marshall Islands, threatened II-784
— Palk Bay II-166–7
— Palk Bay–Madras coast II-168–9
— lagoon ecosystem II-169
— southeast India, degradation of II-172
— Taiwan, problems of ignorance and lack of public awareness II-511
coastal environment, and habitats, Bangladesh II-289–91
— beaches II-289
— mangroves II-290–1
— Matamuhuri delta and coastal islands II-289
— St Martin's Island II-289–90, **II-290**
— seagrasses II-290
Coastal Environment Program, the Philippines II-419
coastal erosion
— American Samoa II-770
— Ancash, Peru I-689

— Australia
  — and landfill II-585
  — and sea-level change II-585
— the Azores I-211
— the Bahamas I-428–9
— Bay of Bengal II-272
— Brunei II-375
— Cambodia II-576
— Chesapeake Bay I-337
— China's Yellow Sea coast, causes of II-493
— Colombian Caribbean Coast I-672–3, **I-673**
— Colombian Pacific Coast I-682
— Côte d'Ivoire I-816
— and deposition, El Salvador I-552, *I-552*
— east coast, Peninsular Malaysia II-353, II-356–7
— English Channel
  — English coast I-68–9
  — French coast I-69
— Great Australian Bight II-684
  — problem of uncontrolled vehicle access II-684
— Gulf of Guinea
  — and sediment supply I-777
  — through anthropogenic activities I-786
— Gulf of Mexico I-476
— Gulf of Thailand, west coast II-307
— Hawaiian Islands
  — now a socio-economic problem II-805
  — through subsidence II-804–5
— and landfill
  — Adriatic Sea I-276
  — Argentine coastlines I-762–3
  — around the North Sea I-53–4
  — Belize I-509–10
  — Bermuda I-228–9
  — Chesapeake Bay I-346
  — eastern Australian region II-639
  — Great Barrier Reef region II-622
  — Gulf of Aden II-58
  — Hong Kong II-542–3
  — Irish Sea Coast I-93
  — Jamaica I-569
  — Lesser Antilles I-636
  — Madagascar II-124–5
  — Marshall Islands II-782–3
  — Mozambique II-108
  — northern Gulf of Mexico I-445–8
  — northern Spanish coast I-144–5
  — Oman II-27–8
  — The Seychelles II-239
  — South Western Pacific Islands II-716–18
  — southern Spanish coast I-176–7
  — Sri Lanka II-184
  — Torres Strait II-603
  — Turks and Caicos Islands I-592
  — Tyrrhenian Sea I-276–7
— landfill, and effects from urban and industrial activities, West Australia II-700–2
— and landfill, Fijian Islands II-760
— Malacca Straits II-317
— The Maldives II-208–10
  — from pleasure boats II-209–10
— Mediterranean coast of Israel I-258–9
— New Caledonia II-731
— northern end of Sumatra II-339
— Peru I-696
— Portuguese coast I-161
— potentially a serious problem in the Mascarenes II-262
— Puerto Rico I-584
— Tanzania II-93
— Tasmania II-655
— Turks and Caicos Islands I-592
— Vanuatu II-744
— Victoria Province, Australia II-667
— western Indonesia, effects of shrimp ponds on II-395
— western Taiwan II-506
— Xiamen region II-517, **II-517**
coastal flooding, and sea level rise III-190
coastal forests
— Madagascar east coast, alleviate mangrove problem II-125
— Malacca Straits II-312
— Tanzania II-88, II-89–90
coastal habitats
— affected by livestock II-26
— American Samoa II-767–9
  — changes to II-770
— Australian Bight II-678
— Bay of Bengal II-274–7
  — loss of on east coast of India II-279
— and biodiversity, Arabian Gulf II-6–9
— Borneo II-366–72
  — coral reefs II-370–2
  — mangroves II-366–9
  — rocky shores II-369
  — sandy shores II-369–70
  — seagrass and algae II-370
— Brazilian tropical coast I-725–7
  — Abrolhos National Marine Park I-726–7
  — Atlantic Rainforest (maritime forest) I-725–7
  — restinga I-727
  — wetlands I-727
— Cambodia II-573–4
  — Botum Sakor National Park II-573–4
  — Kampot Bay habitat II-574
  — Koh Kong Bay II-573
  — Kompong Som semi-enclosed bay habitat II-574
— Côte d'Ivoire, depletion and degradation of I-817–18
— Dutch Antilles **I-597**
  — under environmental pressure I-612
— Gulf of Thailand II-301–4
— Oman and Yemen, effects of cold nutrient rich upwellings II-50
coastal lagoons I-645
— Adriatic Sea I-272, I-276
  — intrusion of allochthonous species I-282
— Baltic I-124
— and basin estuaries, Sri Lanka II-178–9
— and coastal lakes, Baltic Proper I-109
— Côte d'Ivoire I-811, *I-811*
  — Ebrié lagoon I-811–12
  — and estuaries, depletion and degradation of habitats I-817–18
— Curaçao, habitat destruction I-602
— and estuaries, Tasmania, depauperate II-653
— forming New River Estuary I-353
— Gulf of Guinea coast, suffering from eutrophication I-788–9
— Mar Chiquita coastal lagoon, Argentina I-755
  — changes made by alien reef-builder I-762
  — experimental aquaculture hatcheries I-754
  — heavy metals I-762
— Mexican Pacific Coast I-482, I-486, I-490
  — Huizache y Caimanero lagoon system I-493–4
  — Teacapán—Agua Brava lagoon system I-494–5, I-497
— Muthupet Lagoon, Palk Bay II-167
— Nicaragua, Caribbean coast I-522
  — depth preferences, fish and shrimp I-522
— Pulicat Lake, Palk Bay–Madras coast II-169
— saltwater, Coral, Solomon and Bismark Seas region II-430
— Venezuela I-651–3
  — affected by natural and anthropogenic factors I-651
  — Laguna de Tacarigua I-652
  — legal protection for some I-651
— west Taiwan II-502
  — industrial parks II-509
coastal lowlands, Nicaraguan Pacific coast I-533

coastal management
— Argentina, main concerns I-751
— Latin America, regional examples I-463–4
— rational, lack of, Adriatic and Tyrrhenian seas I-281
— southern Brazil I-744
— active environmental groups I-745
— Coastal Management Law not entirely successful I-745
— National Programme of Coastal Management, proposed de-centralization I-744
— preservation and conservation areas I-745
— Sao Paulo I-744–5
— southern Brazil Rio de Janeiro State Environmental Agency I-745
— techniques, classification of I-464–5
— working with nature III-352–3
— Xiamen region, an integrated approach to II-525–32
coastal management, in the future III-349–58
— coastal management and policy evolution III-350–2
— data and inclusivity III-355
— financing III-356
— integration in III-350
— legal issues III-353–5
— physical systems and management III-352–3
— relocation of historic buildings III-353
— solitary waves III-355–6
— success needs political will III-351
— transboundary issues III-355
coastal management and planning
— effectiveness depends on access to relevant information III-356
— legal issues III-353–5
coastal management professionals, training programmes for III-329–30
coastal management programmes, integration with national climate change action plans III-330
coastal management schemes, at the planning stage III-351–2
coastal marine areas, self-management of II-442
coastal and marine ecosystems
— Malacca Straits II-312–14
— nearshore, El Salvador, ecological important I-547
coastal marine ecosystems, Chile I-705–9
— biogeography of the pelagic system I-707–8
— a biological perspective I-707–9
— south Pacific eastern margin I-705–7
coastal and marine problems, frequently transboundary III-355
coastal marine waters
— eastern Korea, wastewater pollution II-478
— western coast, Sea of Japan II-477–8
— Amur lagoons II-477
— Northern Sakhalin II-477
— pollution
— in Amursky, Nakhodka and Ussuriysky Bays II-478
— from ore mining and chemical production, Zolotoy Cape–Povorotny Cape II-477–8
— southern region, sporadic water pollution II-478
coastal models, for areas associated with sea-level rise III-195
coastal morphology, Colombian Caribbean coast I-666–7
coastal plains, fertile, Al Batinah and Salalah II-26
coastal platform, raised I-170
coastal pollution, Chile, related to geography I-709
Coastal Protected Areas (Proposed), Mozambique II-110, **II-110**
coastal protection, Sri Lanka II-184
coastal reclamation I-93
coastal reef terrace, Somalia, diversified ahallow marine environments II-69, *II-70*
coastal reefs, Western Australia, macroalgal and invertebrate communities II-693
coastal region, Southern Gulf of Mexico, recognition of importance of swamps I-472
Coastal Regulation Zones (CRZ), West Bengal II-157–8, **II-158**
Coastal Resource Management Program, the Philippines II-419
coastal resources
— Australian Bight II-682
— east coast, Peninsular Malaysia, conservation legislation II-355–6
— southern Brazil
— destructive use of I-740–3
— economic, social and environmental differences causing problems I-744
— Tasmania, commercial usage of II-655–6
— use in Coral, Solomon and Bismark Seas region II-434
coastal seas
— Coral, Solomon and Bismark Seas region
— fisheries II-440
— impacts of large urban areas II-439–40
— industrial-scale impacts II-439
— shipping and offshore accidents and impacts II-440–1
— village-level impacts II-438
coastal and shallow water habitats, Puerto Rico I-582–4
coastal squeeze, causing habitat reduction III-353
coastal states, responsibility for pollution by seabed activities III-338
coastal terrestrial vegetation, Madagascar, varies II-119
coastal uses, Hawaiian Islands II-802–4
Coastal Warm Drift I-754
coastal waters
— Carolinas, flora and fauna of
— benthic microalgae I-357–8
— coastal benthic invertebrate communities I-360–2
— estuarine and coastal finfish communities I-362–3
— estuarine phytoplankton I-354–6
— estuarine zooplankton I-359
— macroalgae I-356–7
— marine phytoplankton I-356
— marine zooplankton I-360
— offshore benthic communities I-362
— seagrasses and other rooted submersed aquatic vegetation I-358–9
— Coral, Solomon and Bismark Seas region
— impacts of land use II-435–6
— subsistence and artisanal fisheries II-436–7
— threats to sustainability II-437–8
— Norwegian, constituents of I-20
— West Guangdong coast, water quality II-555
— Xiamen region
— bacterial pollution II-522–3
— mainly in good condition II-521–2
— Maluan Bay sediments a secondary pollution source II-524
— oil pollution II-523
— organic pollution, eutrophication and red tides II-522
— sea dumping and disposal of solid waste II-523–4
— threatened by rapid economic development II-522
coastal wave energy development
— current and future prospects III-315
— Osprey 2000 III-315
— enclosed water column devices III-314, *III-315*
— location advantages and disadvantages III-314
— tapered channel concept III-314–15
coastal wetlands
— Argentina I-755–6
— Black Sea, changes in ecology of I-293
— anthropogenic influences resulting in loss/degradation of I-293
— Côte d'Ivoire I-810–11
— Hueque, Venezuela, drainage and deforestation I-655
— loss of
— northern New South Wales II-638
— Yellow Sea II-493–4
— loss of, Southern California I-388
— affecting seabirds I-398
— lost to landfill, Hawaiian Islands II-805
— mangrove destruction, the Bahamas I-428
— northern Gulf of Mexico, potential loss of I-446
— west Taiwan II-502–3
— *see also* khawrs; swamps; wetlands
coastal wilderness, Dutch Antilles I-601
Coastal Zone Management Act (USA) I-451–2

— National Estuarine Research Reserve System I-452
Coastal Zone Management Plan (CZMP), Sri Lanka II-185–6
— Special Area Management Plans II-186
coastal zone, Somali coast
— alterations of II-80
— geomorphic features II-68–72
  — Bajuni barrier islands II-69, *II-73*, *II-75*
  — braided channelized coast II-69–72
  — coastal reef terrace II-69, *II-70*
  — Merka Red Dune Complex II-68–9, *II-70*
  — rivers and alluvial plains II-68
— suffering degradation through lack of protective measures II-80
coastal zones I-4
— Argentina, main divisions I-751–2
— artisanal and non-industrial uses
  — El Salvador I-553
  — Madagascar II-125
  — Mozambique II-108–9
  — Taiwan II-509
— atmospheric deposition to III-205–7
  — atmospheric flow in, and atmosphere—surface exchange III-206–7
  — impacts in III-205–6
— Bay of Bengal
  — Calcutta and Howrah a significant detriment to II-148
  — mangroves II-149
— Belize I-503
  — aquaculture and fishing I-511
  — artisanal use of I-507
— and non-industrial use I-510–11
  — assessment of anthropogenic impacts I-507
  — effects of coastal erosion and landfill I-510
  — management of I-512
  — protected areas **I-513**
  — tourism I-511
— Brazilian tropical coast, for recreation and tourism I-728
— Cambodia traditional agriculture II-576
— Chile
  — flora and fauna I-707
  — industrial activity I-710–15
  — passage of rivers through I-704–5
  — southern marine habitats I-708, **I-708**
  — upwelling ecosystems and embayment ecosystems I-706
  — upwellings *I-705*
— Colombian Caribbean Coast, influence of Sierra Nevada de Santa Marta I-672
— community-based management, Hawaiian Islands II-809
— the Comoros, urban and industrial impacts II-251
— Côte d'Ivoire
  — cultural and historic sites I-818
  — mining of construction materials from I-816
— definitions of I-458
— east coast, Peninsular Malaysia
  — assessment of liquid and airborne pollution II-351
  — changes in catchment land use II-351
  — development of tourism II-351
  — relevant legislation and guidelines **II-356**
— eastern Australian region agriculture in II-636–8
— eastern Taiwan Strait II-501
— El Salvador I-547
  — coastline meso-tidal I-547
  — industrial uses I-554–5
— Guinea, pressure on natural resources I-801
— industrial effects I-511
— Integrated Management Plan for the Coastal zones of Brunei II-379
— legal definition in need of revision III-354
— Lesser Antilles, major features **I-632**
— Madagascar
  — cities II-125
  — industrial uses II-125
  — shipping and offshore accidents II-125
— Mozambique
  — industrial uses II-109
  — Mecúfi Coastal zones Management Projects II-111
  — a priority area II-110
  — Xai-Xai Sustainable Development Centre for the Coastal zones II-111
— need to understand physical systems III-352
— New Caledonia, potential effects of mining discharges II-731
— Nicaragua, Caribbean coast I-520
  — recreational and tourist destinations I-523
— Nicaragua, Pacific coast I-543
— Peru
  — arid semi-desert I-689
  — characteristics of I-689–90
  — climate influenced by Peruvian current I-690
  — 'wet desert' climate I-690
— pressure on and degradation of III-295
— shallow, human impact on III-7
— southeast India, domestic sewage a problem II-172
— southern Brazil
  — urbanization, tourism and industrialization I-734
  — very little is protected I-745
— Southern Gulf of Mexico, land classification I-475
— Sri Lanka
  — Coastal zones Management Plan (CZMP) II-185–6
  — pressures on II-181
— Taiwan
  — abused by public and private sector II-511
  — coastal land subsidence II-507
  — longshore currents II-506
— Tanzania II-85
— Tasmania
  — recreational use of II-655
  — urban development II-655
— tropical, sensors relevant for mapping III-284
— tropical Brazil, transgressive episodes I-722
— Venezuela, development pressure I-645
— Victoria, Australia
  — indigenous peoples II-666
  — white settlers II-666
coastal/marine structures
— effects on beach morphology I-258–9, I-276
— *see also* dikes and breakwaters
coastguards, for the Dutch Antilles I-609
coastline
— Gulf of Guinea, low-lying and swampy I-775
— Jamaica I-561
— Puerto Rico, shelf morphology I-581
— Venezuela
  — economic activities I-654–5
  — influential because accessible I-654
  — regions of I-645–6
— West Africa I-775
coastline modification
— Coral, Solomon and Bismark Seas region II-438
— French Polynesia II-821
coasts
— braided and channelized, southern Somalia II-69–72
— erosion, protection reduces sediment availability III-352
— North Sea, urban and artisanal use of I-54
Cobscook Bay, marine ecosystem bibliography I-316
coccolithophores, and algal bloom detection III-298
cockle harvesting/fisheries I-51–2, I-73
— disrupts the environment I-92
coconut tree replanting, Marshall Islands, benefits vs. impacts II-785
Cocos (Keeling) Atoll II-583
cod I-35, I-49, I-313
— Baltic Sea I-110
  — growth conditions less favourable I-128
— concentrations of PCBs and DDT in **I-37**

"cold pool", connection to Georges Bank I-324
cold water dome, persistent, East China Sea II-508
cold-water plumes, Venezuelan coast I-646
collisional tectonics, Aegean area I-235
Colombia
— coral bleaching and mortality events III-54
— rehabilitation of key mangrove system I-463
Colombia, Caribbean Coast I-663–75
— coastal erosion I-672–3
— effects from urban and industrial activities I-673
— offshore systems I-669–71
— populations affecting the area I-671
— protective measures I-673
— rural factors I-671–2
— seasonality, currents, natural environmental variables I-665–7
— bimodal wet–dry seasonality I-665
— shallow water marine ecosystems and coastal habitats I-666–9
Colombia, Pacific Coast I-677–86
— coast is tectonically active I-682
— coastal erosion and landfill I-681–2
— effects from urban and industrial activities I-682–4
— major shallow water marine and coastal habitats I-679–80
— offshore systems I-680
— populations affecting the area I-680–1
— protective measures I-684–5
— Gene Bank of Fishing and Aquarian Resources I-684
— National Contingency Plan against Hydrocarbon Spills in Marine Waters I-685
— rural factors I-680–1
— seasonality, currents, natural environmental variables I-679
Colombian Current *I-678*
colonial powers, and the Malacca Strait II-337
colonisation, of North Sea still going on I-45
Commission de l'Océan Indien (COI) *see* Indian Ocean Commission
common dolphins, northern stock III-91
Common Fisheries Policy I-57
Common Market for Eastern and Southern Africa II-80
community-based coastal resources management, the Philippines II-418
Comores *see* Comoros Archipelago
Comoros Archipelago *II-114*, II-243–52, *II-244*
— effects from urban and industrial activities II-251
— major shallow water marine and coastal habitats II-245–6
— Anjouan II-246
— Grande Comore II-245
— Mayotte II-246
— Moheli II-246
— offshore systems II-246–7
— populations affecting the area II-247
— protective measures II-251–2
— regulations for environmental protection and management II-251
— watershed improvements II-252
— seasonality, currents, natural environmental variables II-245
— threats to the environment II-247–51
conch fishery I-591, I-635, I-637
— Belize I-509
— Jamaica I-572
conservation
— efforts in Belize I-512, I-514
— of whale stocks III-75–7
— Yellow Sea area, inhibiting factors II-495–6
conservation measures/policies
— Andaman, Nicobar and Lakshadweep Islands II-195–6
— for the Sundarbans II-157–8
Conservation Sensitive Management System (CSMS), analysis of coastal geomorphological sensitivity III-353
construction, and infilling of coastal mangroves I-637
consumption, and population III-396
contaminants
— chemical, safety limits in Norwegian fish and shellfish I-25
— chemical, secondary sources I-24
— entering the Southern California Bight
— in biota I-394
— DDT contamination I-390
— processes undergone I-390–1
— in sediments I-392–4
— in the water column I-391–2
— Gulf of Maine I-314–16
— input, transport and biological responses of, North Sea I-56
— levels in Greenland marine ecosystem I-5
— future trends I-14–15
— indirect evidence for I-9
— in the marine environment I-450
— Norway, within reach of tidal activity I-28
contamination I-158
— bacterial, Tagus estuary I-158
— biocide I-538
— chemical, from sewage, Arabian Gulf II-10
— Colombian Pacific coast, mainly transitory I-683
— marine, El Salvador I-555
— toxic, North Carolina estuaries I-360–1
— *see also* mercury contamination; microbiological contamination
continental collision II-19
continental seas, western Indonesia, uniqueness of II-383
continental shelf
— Norwegian I-23
— bottom fauna communities I-23
— bottom substrate I-23
— *see also* Argentinian continental shelf; South Brazil Bight
Continental Shelf Alternative (CSA) sites I-327
Convention on Biological Diversity I-608, III-341
Convention on the Conservation of Antarctic Marine Living Resources III-342
Convention for the Protection of the Black Sea against Pollution I-294
Convention on the Protection of the Marine Environment of the Baltic Sea Area (Helsinki Convention) I-101
Cook Inlet *I-374*, I-375
copepods I-814, I-831, II-54
— diapausing populations II-24
— Gulf of Alaska I-379–80
— North Carolina I-359
— Southern Bight I-48
copper (Cu) I-131
— Azorean amphipods I-213
— surficial sediments, Penobscot Bay *I-316*
copper levels, Island Copper Mine III-239
copper mining, Chile I-710
— lessons for the future, a case study I-713
— lessons for the future, case study, complaints about pollution and construction of new tailings lagoon I-713
— *see also* Island Copper Mine, Canada
copra, from Chagos Islands II-223, II-229
coprostanol, west Taiwan coast II-506, II-509
coquinas II-77
coral assemblages, Arabian Sea II-22
coral atolls
— American Samoa II-769
— Belize I-504–5
— Borneo II-370
— Chagos Archipelago II-223
— Lakshadweep Islands II-194
— Marshall Islands II-775
— the Seychelles II-236
— peripheral reefs II-236
— Western Australia II-693
— shelf edge
coral barrier, reef beaches, Cambodia II-574
coral bleaching I-419
— August 1998 I-421
— Belize reefs I-504
— related to water temperature I-409–10

coral carpets, Somali coast II-72
coral cays
— Borneo, Pulau Sangalaki II-371
— western Indonesia II-388–9
coral collection, Hawaiian Islands II-803
coral colonies, use of high-resolution airborne methods for status of III-288
coral communities
— Aruba, affected by oil pollution *I-596*, I-597
— British Virgin Islands I-618
— effects of hurricanes I-618
— coastal reef terrace, Somalia II-69, *II-70*
— eastern Australian region II-633
— Gulf of Aden II-52
— natural stressors II-52
— Gulf of Thailand II-304
— offshore islands, Taiwan Strait II-504
— southern Somali coast II-72
— West African coast I-780
— on Yemeni black basalt effusions II-51
coral degradation, tourist areas, Gulf of Thailand II-304
coral diseases
— global epidemic of III-226
— Hawaiian Islands II-807
coral harvesting, prohibited, the Bahamas I-430
coral mining II-315
— affecting reefs of Pulau Seribu II-397
— the Comoros II-248
— uses of corals II-249
— contributes to reef and forest degradation, Tanzania II-91
— El Salvador I-554
— illegal, western Indonesia II-391
— Madagascar II-125
— the Maldives II-208
— reefs show little sign of recovery II-208, *II-209*
— sea-level rise effects exacerbated by II-215
— Mozambique II-108
— *see also* coral quarrying
coral mortality
— Colombian Caribbean Coast I-668
— due to Crown-of-Thorns starfish outbreaks II-616
coral pinnacles
— Bajuni Islands II-74
— Brazil I-723
coral quarrying, Gulf of Mannar II-165
coral reef areas
— no-take reserves III-386
— US Marine Protected Areas III-386
coral reef bleaching III-37
— 1998 event
— Madagascar II-116
— Sri Lanka II-180
— American Samoa II-712
— Chagos Archipelago
— evidence of cover decline pre-1998 II-228, *II-228*
— massive coral mortality after 1998 event II-228, *II-228*
— the Comoros
— 1983 in Mayotte II-249
— 1998 event, mass mortality II-249
— and coral death III-38
— Coral, Solomon and Bismark Seas region II-429
— Fiji II-711
— followed by coral disease III-224
— French Polynesia
— 1991, 1994 and 1998 bleaching events II-822
— problem of remoteness from coral recruitment II-822
— Great Barrier Reef II-617
— Gulf of Thailand II-304
— Hawaiian Islands II-807
— the Maldives II-202–3
— 1998 event very extensive II-214
— consequences of on the socio-economic welfare of communities II-214
— Marshall Islands II-777
— and mortality, 1998 event III-43–57
— Arabian Region III-47–8
— Caribbean and Atlantic Ocean III-54–6
— Central and Eastern Pacific Ocean III-53
— Chagos Archipelago III-49
— East Asia III-52
— Indian Ocean III-48–9
— interpretations and conclusions III-46–7
— mechanisms of III-44
— Pacific coast of the Americas III-53–4
— Pacific Ocean, Northwest and Southwest III-52–3
— Singapore, Thailand, Vietnam III-52
— Southeast Asia III-49–52
— and mortality, Socotra II-52
— the Philippines II-408
— recent El Niño years, time too long for reef tolerances III-224
— several periods, Andaman Sea II-303
— the Seychelles
— 1998 event II-236
— massive mortalities II-237
— stress thought to be main reason with other factors II-137
— Vanuatu II-742
— western Indonesia II-389
coral reef communities
— east coast, Peninsular Malaysia II-348
— Moorea Island, French Polynesia II-819
coral reef ecosystems
— Andaman and Nicobar Islands II-191–2
— remote sensing of III-287–8
— degrees of sophistications III-287
— future challenges III-288
— reef habitat maps III-287–8
— representation of individual habitats III-287–8
— use of colour aerial photography III-287
coral reef fish II-117, II-180
— The Maldives II-202
coral reef habitats, Sri Lanka II-180
coral reef microcosms, abundance in III-40
Coral Reef Monitoring Project, USEPA I-412
Coral Reef Rehabilitation and Management Project (COREMAP), western Indonesia II-393, II-399–400
coral reef species, Tanzania II-86–7
Coral Reef Symposia III-44
— late awareness of degradation III-37
coral reef zones
— Bermuda Islands I-227
— rim and terrace reefs I-227–8
coral reefs II-693, III-33–42
— affected by hurricanes I-631
— American Samoa II-712, II-767–8
— recovering from series of natural disturbances II-768
— reef fish assemblage II-768
— Anguillan shelf I-618
— Australia II-582
— vulnerable to eutrophication and sedimentation II-585
— the Bahamas I-421
— near-shore health of I-429
— Bay of Bengal II-277
— biodiversity of III-34
— based in calcium framework building III-35
— Borneo II-370–2
— Berau Barrier reef system II-370
— erosion of, Kota Kinabalu Bay and Tunku Abdul Rahman Park II-376
— reef flats II-370–1
— Brazilian I-722, I-723–4
— differ from well known coral reef models I-723
— formed by coalescence of 'chapeiroes' I-723

— reef types I-723
— Cambodia II-574
— cf. rainforests **III-34**
— Chagos Islands II-223
— biological patterns on the reef slopes II-227–8
— changes over twenty years II-228–9
— ecology of II-226–7
— island ecology II-229
— reef flats, algal ridges, spur and groove systems II-227
— submerged banks and drowned atolls II-223, **II-224**
— Colombian Caribbean Coast I-668
— best development in southwest I-668
— general degradation I-668
— Colombian Pacific Coast I-680
— conservation, an international priority III-40–1
— damaged, Morrocoy National Park, Venezuela I-655
— decline in coral cover III-37
— degradation
— early III-37
— primary bases for III-40
— 'design' makes for vulnerability to changing environmental conditions I-413
— destructive fishing practices III-123
— Dutch Antilles
— natural disasters I-607
— and reefal algal beds I-599–600
— east coast, Peninsular Malaysia II-355
— effects of creating Marine Parks II-355
— effects of sewage not well documented II-804
— endangered III-36–7, III-41
— loss through destructive fishing methods III-37
— lost through siltation and eutrophication III-37
— Fiji Islands II-755–7
— reef provinces **II-756**
— reef types II-756, **II-756**
— studies on Suva reef II-756
— Florida Keys I-407–8
— Carysfort Reef I-410–12
— coral diseases I-410, *I-411*
— degradation from agricultural and urban factors I-409
— Florida Bay Hypothesis I-407
— recruitment low I-410
— stressed by environmental change I-408–9
— fossil, Red Sea coastline II-38
— French Polynesia II-817–19
— Fangataufa atoll gastropod assemblage studies II-817
— reef monitoring networks II-824
— reef restoration schemes II-824
— status of II-823
— Great Barrier Reef region II-616–17
— Crown-of-Thorns Starfish outbreaks II-616–17
— little evidence of long-term decline II-616
— pressures and status II-616
— Gulf of Aden II-52
— Gulf of Mannar II-163–4
— and the Gulf islands II-164
— Gulf of Mexico
— impacts on some reefs **I-471**
— principal characteristics **I-471**
— and hard bottoms, northern Gulf of Mexico I-442
— decline of I-445
— Hawaiian Islands II-795, II-796–7
— benthic reef life II-799–800
— better management needed in northwestern islands II-810
— Kane'ohe Bay, reef restoration II-796, II-809
— monitoring of II-796
— indicators of oceanic health and global climate change I-413–14
— Jakarta Bay, once beautiful now almost destroyed II-397
— Jamaica I-562–4, I-566, *I-567*
— changes in I-562
— *Diadema* mass mortality I-563–4
— differences in I-562
— hurricane damage I-562, I-563
— impacts of fishing I-568
— impediments to recovery I-572
— reef deterioration I-563
— reef zonation I-562, I-562–3
— some recovery I-563, I-564
— Kenya, result of predator overfishing III-128
— Lesser Antilles I-631
— Madagascar II-118
— ancient II-118
— Malacca Strait II-336
— the Maldives II-202–3
— need to keep reefs healthy II-218
— protected from mining by tourism II-217
— and marine environments, pharmaceuticals from III-39
— Marshall Islands II-778–9
— Mauritius II-258
— natural perturbations III-37
— natural products, identification and extraction from III-40
— New Caledonia II-725–6
— Nicaragua, Caribbean coast I-520–2
— *Diadema* mass mortality I-520
— Miskito Coast Marine Reserve I-520, *I-521*
— reef fish I-524, I-525
— Norwegian coast I-21–2
— Palk Bay II-166–7
— Papua New Guinea coastline II-429
— the Philippines II-408–9
— effects of overfishing, sedimentation and destructive fishing II-415
— primary productivity II-409
— reef health II-408–9
— primary productivity III-35
— Puerto Rico I-582–4
— fringing reefs I-582
— patch reefs I-582
— shelf reefs I-582, I-583–4
— Queensland Shelf II-619
— Quirimba Archipelago II-102
— Red Sea II-37, II-40–1
— alignment of II-39
— coral distributions II-41
— effects of oil pollution II-43
— fringing reefs II-40
— Reefs at Risk analysis III-44
— risk criteria classification evaluates potential risk from ports and harbours II-417–18
— role of nutrients in degradation III-38
— sensitivity of to N, P and $CO_2$ III-38
— the Seychelles II-236
— threatened II-236
— Singapore, smothered by siltation II-339
— social and economic value III-38, III-40–1
— Solomon Islands II-429–30
— Somali coast II-65
— fringing reefs II-65
— shelf and fringing II-72–4, II-80
— South Western Pacific Islands II-711–12
— Southern Gulf of Mexico I-469
— and protected zones I-471–2, *I-472*
— Vera Cruz reef system I-471
— Sri Lanka II-180
— Straits of Malacca II-314
— and submerged banks, Lakshadweep Islands II-194
— degradation fromsiltation and sponge infestation II-194
— Taiwan II-503, II-504
— Taiwan Strait, increasingly threatened II-504
— Tanzania II-85–7
— degraded sites II-86
— restoration project, Dar es Salaam II-89

— Thailand
  — Andaman Sea II-302
  — Gulf of Thailand II-302–3
— threats pre-1998 III-44
— Torres Strait II-597
— true, Bar Al Hackman II-22
— Turks and Caicos Islands I-590
— Vanuatu II-741–2
  — condition/special features of coral communities **II-742**
  — status of II-742
— Venezuela I-650–1
  — coastline reefs, less diverse and under pressure I-650
  — Mochima Bay I-651
  — on offshore islands, pristine condition I-650
  — Turiamo Bay, diverse reef fauna I-650–1
— Vietnam II-565
— vulnerable to oil spills and their effects III-273–4
— Western Australia II-696
— western Indonesia II-388–9
— western Sumatra II-387
— *see also* coral atolls; fringing reefs; patch reefs; reefs
Coral Sea *II-426*, II-427, *II-612*
— Chesterfield Islands and Bellona reef II-711
— nutrient and sediment loads II-614
— western, circulation in II-614
Coral Sea Basin II-427
Coral Sea Coastal Current II-428, *II-594*
Coral Sea Island Territories, inclusion in Great Barrier Reef Marine Park? II-627
Coral Sea water II-581
Coral, Solomon and Bismark Seas Region II-425–46
— coastline change II-438
— human impacts on coastal seas II-438–41
— land and sea use factors impacting on coastal waters II-435–8
  — inability to manage stocks for sustainability II-437
— offshore systems II-431–3
— people, development and change II-433–5
  — inadequate information for establishing any form of baseline II-435
— provisions for the management and protection of coastal seas II-441–3
  — community-based management II-442–3
  — national administrative and legal arrangements II-441
  — protected species, habitats and areas II-441–2
  — regional cooperation II-442
— seasons, currents, seismicity, volcanicity and cyclonic storms II-428–9
— shallow water marine and coastal habitats II-429–31
coral-rubble mining I-604
coralligenous formations, Tyrrhenian Sea I-273–4
coralline coast, Mozambique *II-100*, II-101, II-104
corals
— Andaman and Nicobar Islands II-191–2
— Arabian Gulf II-7–8
— Brazilian, some endemics I-723
— Chagos Islands
  — most diverse site in the Indian Ocean II-225
  — soft corals II-227
  — a stepping stone for corals II-224–5
— collection for aquarium and shell trade II-438
— the Comoros, used in building II-249
— deep-sea, little studied III-379
— diversity of, Spratly Islands II-364
— French Polynesia II-819
— Great Barrier Reef, diversity and species assemblages II-616
— Gulf of Oman II-22
— hermatypic I-566, I-590, I-723, II-633
  — the Bahamas I-421
— Maldivian, zooxanthellate and azooxanthellate II-202
— Marshall Islands, great biodiversity II-778
— Papua New Guinea reefs II-431
— physiology of calcification III-35
  — calcification and nutrient uptake III-35–6
  — scleractinian corals III-35
— scleractinian
  — Andaman Sea II-302
  — Fiji II-711
  — Gulf of Mannar II-164
  — Lakshadweep Islands II-194
  — Madagascan, affinities of II-117
  — Palk Bay II-167
  — Vanuatu II-741
  — Venezuela I-650
  — western Indonesia II-389
— the Seychelles, growth affected by southeast trade winds II-236
— species diversity, Grand Récif of Toliara II-118
— Sri Lanka II-180
— stony II-796
  — east coast of Taiwan II-508
  — Fiji Islands II-757
— Taiwan II-504
— Turks and Caicos Islands I-589, I-590
— Vanuatu, similarities with Great Barrier Reef II-741
— varying growth conditions, Puerto Rico I-583
— Western Australia II-696
Coriolis effect III-311
Coriolis Force I-631, II-274, II-288, III-316
Coro, Gulf of *I-644*, I-646
corrales (stockyards), Spanish fishing technique I-176
'Corriente de Navidad' I-139
Corsica
— Lavezzi nature reserve I-280–1
— transboundary park in the Bocche di Bonifacio I-281
Costa Rica Dome Structure I-535
Costa Rican Current I-489, *I-678*, I-679
Côte d'Ivoire I-805–20
— coastal erosion and landfill I-816
— construction of Abidjan harbour, importance of I-815–16
— effects of urban and industrial activities I-817–18
— geological context and geographical limits I-807
— major shallow water marine and coastal habitats I-810–13
— offshore systems I-813–15
— population I-815–16
— protective measures I-818–19
  — International conventions, coastal and marine environment **I-818**
  — National Environmental Action Plan I-818
— rural factors I-816
— seasonality, currents, natural environmental variables I-807–10
— water inputs, from continent and ocean I-807
Cotentin Peninsula *I-66*
Cox's Bazaar sand beach II-289
crab fishery, and yields, Island Copper Mine III-244
crabs, pelagic II-54–5
*Crepidula fornicata*, altered benthic habitats I-47–8
Cretan Sea *I-234*, I-235
crocodiles I-505
Cromwell Current I-705
cross-correlation analyses III-217
Crown-of-Thorns starfish II-303
— a potential environmental problem II-214
Crown-of-Thorns starfish outbreaks II-711, II-712
— causes indeterminate II-617
— east coast, Peninsular Malaysia II-355
— French Polynesia II-823
— Great Barrier Reef II-616–17
— Hawaiian Islands II-807
— Marshall Islands II-777
cruise ship discharge, the Bahamas I-430
crustacean farming III-169, III-170, *III-170*
crustacean fishery
— Alaska, collapse of I-380

— KwaZulu-Natal coast II-137
crustaceans I-34, I-172, II-586
— Côte d'Ivoire I-815
— El Salvador I-550–1
— Faroes I-34, I-35
— fishing for I-73
— glacial relict I-125
— Gulf of Maine I-311–12
— PAHs in I-497
— Peru I-691
— Spanish north coast I-139, I-140
— Tagus estuary I-157
— Vietnam II-564
cryptomonads I-355
crystalline rock habitats, Sri Lanka II-180
cultivated land, decline in, Guangdong II-558
cultural convergence, Sundarbans II-152
cultural evolution, and cultural adaptation III-396
Curaçao *I-596*, I-597
— beach replenishment I-604
— coastal and marine habitats *I-599*
— coastal urbanization I-606
— fringing reefs I-599
— land-use planning I-608, I-609
— livestock decline I-601
— national parks I-609
— reef considered to be overfished I-602–3
— reefs damaged by ship groundings I-606
— sewage discharge I-606
— turtle grass I-601
Curaçao Dry Dock, contamination from I-605
curio trade, marine life collection for 251, II-91, II-108, II-125, II-212
— marine ornamentals trade II-393
— Marshall Islands II-782
Curonian Lagoon I-109
— fishery I-111
currents
— affecting the Bahamas I-418, *I-418*
— Arabian Sea, mirror seasonal wind direction II-20
— Belize, affected by prevailing winds I-503
— Cambodian Sea II-571, *II-572*
— Côte d'Ivoire coast I-807–8
— eastern Australian region II-631
— Great Australian Bight II-676
— Gulf of Aden II-50
  — development of gyres II-50
— Gulf of Mexico I-437
— influencing Marsha, Islands II-775–6
— in the Malacca Straits II-311
— Mexican Pacific coast I-487
— offshore, Taiwan Strait II-501–2
— Sea of Okhotsk II-465–6
— South China Sea II-552
  — surface water patterns II-366, *II-366*
— surface, Java Sea II-383–4
— through the Fijian Groups II-754
— Torres Strait II-596
— Xiamen coastal waters II-516
  — residual currents in the Outer Harbour II-516
Currituck Sound *I-352*, I-360
customary fishing rights
— Fiji II-758, II-761
  — more interest in marine protected areas now II-762
customary marine practices/law
— Coral, Solomon and Bismark Seas region II-434
— Marshall Islands II-779, II-786
Cuulong Project, Mekong Delta II-299
Cuvier's beaked whale III-92
Cuvumbi Island, Bajuni Islands, reef front and reef flat II-74, *II-78*
cyanide, and the live reef fish food trade II-373, II-392–3
cyanide fishing
— Indonesian Seas II-392–3
— the Philippines II-415
cyanobacteria I-743, I-821, III-3, III-221
— blooms III-298–9
— and marine mass mortalities III-218
— The Maldives II-204
cyanophytes II-10–11
— picoplanktonic, Neuse Estuary I-355
Cyclades *I-234*, I-235
cyclone shelters, Bangladesh II-288
cyclones I-439, II-596
— affecting Madagascar II-116, II-117
— Bay of Bengal II-148
  — and storm surges II-288
— and coral cover II-616
— formation of, Coral, Solomon and Bismark Seas region II-429
— French Polynesia, related to abnormal El Niños II-815–16
— Gulf of Thailand II-301
— and high rainfall, Fiji II-753
— influence on Great Barrier Reef II-613
— the Maldives II-201
— Mascarene Plateau II-256
— Mexican Pacific coast I-488, I-493
— occasional
  — Mozambique II-102
  — Sri Lanka II-177
— the Philippines II-407
— South West Pacific Islands
  — damage by associated waves II-710
  — effects of II-709, II-710
— Vanuatu II-739–40
  — damage by II-740
— west Seychelles, infrequent II-235–6
Cyprus *I-254*
cytochrome P450, from *Exxon Valdez* oil spill I-382
cytochrome P450-1A values, elevated, Prince William Sound III-274
cytochrome P450-aromatase systems, inhibited by TBT III-249

dabs I-49, I-70
Dahlak Islands *II-36*
Dall's porpoise III-90
— caught in Japanese salmon drift-net fishery III-379
— hunted in Japanese waters III-94
— possible endocrine disruption II-456–7, *II-460*
Dalmation coast, tourism I-277
dam building
— on major rivers, contributing to coastal erosion II-108
— protests against II-306
Damperien Province, Western Australia II-694
dams
— adverse effect on mangroves III-19
— and canals, effects of construction and maintenance of I-446–7
— effects of, Côte d'Ivoire I-810, I-816
— effects of, Gulf of Guinea main rivers I-789, I-792
— in estuaries, South Korea II-493
— hydroelectric power, reduce salmon populations I-128
— responsible for hydrological change, east coast, Peninsular Malaysia II-353
— Taiwan rivers, increasing coastal erosion II-506
— Tasmania, regulation of freshwater flow II-654
— a threat to freshwater fish II-697
Danish water, total nitrogen input to III-201
Danube River, increased loads of organic and inorganic pollutants I-303
Dardanelles *I-234*, I-243
data mining III-212
— and data models III-214–15, **III-215**, **III-216**
— for disturbance indicator types and pathogen toxin and disease combinations III-214
— and other research, to retrospectively derive new time series III-216

Davis Strait *I-6*, I-7
— cod abundance and temperature I-8
— increased Hg in upper sediments I-9
— trawler fishery for Greenland halibut I-8
Daymaniyat Islands *II-18*, II-22
— Daymaniyat Islands National Nature Reserve II-30
DDT II-545
— air and surface seawater, worldwide II-454, *II-457*
— in animals from the Greenland seas I-9, I-11
— in Asian developing region waters II-449
— contamination in Southern California I-390, I-392
  — total DDT of most concern I-393, **I-393**
  — well preserved in anoxic deep basin sediments I-393
— Faroes I-37–8, **I-37**
— Norwegian west coast, still an environmental problem I-25
— recent use, Vietnam II-565
— in river waters, El Salvador I-552
— as seed dressing I-635
— still reaching Baltic Sea via precipitation I-131
de la Mare, Dr William, on the New Management Procedure (IWC) III-78
debris, non-biodegradable, a danger to sea turtles III-67
decision making process, for coastal management, greater inclusiveness more sensible III-355
Declaration on the Protection of the Black Sea Environment I-294
deep water formation
— Greenland Sea I-7
— Tyrrhenian Sea I-272
deep water renewal, Norwegian fjords I-21
deep-sea smelt, Okhotsk Sea II-466, II-468
deepwater dumpsite 106 *I-322*, I-327
deforestation I-241, I-427–8
— Australian Rainforest II-585
— Belize I-508, I-510
— Borneo II-374–5
— catchment areas of Malacca Strait II-338
— causing soil erosion I-174–5
— in the Comoros II-247
— effects of, Andaman and Nicobar Islands II-193
— from mining, New Caledonia II-731
— Great Barrier Reef region catchments II-620–1
— Guangdong II-558
— Guinea I-801
— Gulf of Guinea I-786
— Jamaica I-568
  — reforestation I-572–3
— Kamchatka peninsula II-469
— and land conversion, cause of high sedimentation rates II-384
— Lesser Antilles I-635, **I-636**
— logging in Gulf of Papua watersheds II-604
— Mozambique II-109
— Nicaragua I-523, I-538
— the Philippines II-412
  — reduction in forest cover II-413
— round Sea of Japan II-475
— and soil erosion, Hawaiian Islands II-801–2
— in southeast India, affects coastal zone II-165
— southern Brazil
  — of Atlantic forest I-741
  — of mangroves I-740, I-741
  — for Rio-Santos road I-742
— Southern Gulf of Mexico
  — and erosion I-476
  — impact of I-479
— Sumatra and Kalimantan II-395
— Tanzania II-89–90
— through livestock grazing, Dutch Antilles I-601, *I-602*
— west coast Taiwan II-503
degradation
— southern Brazil I-740–3
  — main problems I-740
Delaware Bay *I-322*, I-329, *I-336*
Deltaic Sundarbans *II-146*, II-147
— brackish waters support phytoplankton, macrobenthic algae and zooplankton II-150
— effects of seasonal changes II-149
— Hugli estuary, carries industrial discharges from Haldia region II-155
— important morphotypes II-147
— major rural activities and environmental impact II-153–4
— monsoon period II-149
— physico-chemical characteristics II-149, **II-149**
— population extremely poor II-152
— post-monsoon periods II-149
— pre-monsoon period II-148–9
— rich in natural resources II-152
— a unique ecosystem II-156
denitrification I-106, III-262
— water-column, Arabian Sea II-24
Denmark
— Action Plan for Offshore Wind Farms in Danish Waters III-307
— acts on behalf of Faroes I-36
— monitoring and modelling of offshore wind energy technology III-308
  — Wind Atlas Analysis and Application Program (WA$^S$P)
— offshore wind energy production III-307–8
  — Tunø Knob installation III-308
  — Vindeby installation III-307–8
— pilot offshore wind energy projects III-306
Denmark Strait *I-6*, I-7, I-9
D'Entrecasteau Basin *II-426*
D'Entrecasteaux Channel *II-648*
— localised cooling II-651
deposition processes, from the atmosphere III-200–1
Derjugin's Basin *II-464*, II-465
desalination
— Arabian Gulf
  — by-products II-10
  — multistage flash evaporation plants II-10
  — and power plants II-10
— Oman II-28–9
— Red Sea, saline discharges from desalination plants II-43
desalination plants, Bay of Bengal, effects of saline discharge II-279
desert
— Namibia I-823
— Peru I-689–90
developed countries, marine reserves in III-386
developing countries
— a chance for coastal zone development with fewer problems III-357
— marine reserves in III-386
— prawn/shrimp fisheries, high utilisation of catch, poor utilisation of species III-143
development pressure
— English Channel coasts I-71
— Venezuelan coastal zone I-645
Dhofar *II-18*
— excessive cutting of firewood II-26
— limestone cliffs II-19, II-21
*Diadema* mass mortality I-410, I-421, I-503, I-520, I-607, I-632
— co-incidental with El Niño conditions III-223
— Jamaica I-563–4
diamonds, coastal deposits, South Africa II-142
diarrhetic shellfish poisoning (DSP) I-69, I-160
diatom blooms I-155
— Baltic Sea I-125
— Irish Sea I-89
— spring and autumn, English Channel I-67
— toxic, causing debilitating illness III-218
diatom sediment records, Portuguese shelf I-160
diatoms I-355, I-692, I-743, I-783, I-814, I-821, I-830–1, II-24, III-221
— Australia II-583

— Bay of Bengal II-275
— Campeche Sound I-474
— discriminated against III-259
— in northern Gulf of Mexico I-449
— West Guangdong coast II-553
dibutyltin (DBT) II-449
Diego Garcia, Chagos Archipelago *II-221*
— military development imported alien flora II-229
— occasional storms from cyclone fringes II-225
— only inhabited island II-223
— recreational fishing II-230
diffusional transfer 202
diffusiophoresis III-201
diffusive attenuation coefficient
— and water quality III-296–7
  — K-maps III-296
dikes and breakwaters, construction interrupts sand drift I-176
dinoflagellate blooms I-827
— toxic, French Polynesia II-821
dinoflagellates I-692, I-783, I-814, I-831, II-320
— Australia II-583
— ichthyotoxic I-363–4
— increased number in the Arabian Gulf *II-12*, II-13
— toxic I-355, II-304, II-587, III-218, III-221
  — benthic, produce tumour-promoting agents III-219
  — *Pfiesteria piscicida* III-223
— winter-blooming I-355
dioxin
— discharge from old Norwegian magnesium plant I-25
— remobilization of I-25
discards I-146–7, **I-148**
— at sea, percentage of total catch III-377
— and by-catch III-123–5
— fate of III-127
— reasons for discarding III-137–8
disease
— an ecological opportunist III-225
— can devastate aquaculture ponds III-369
— chronic conditions within endangered populations, monitoring of III-227
— increase in extent and impact of III-225
— novel, possible introduction through aquaculture III-369
— reflect perturbations with ecosystems III-225
— transfer via ballast water III-355
— trophodynamically acquired, a global issue III-221
— water-related, Vanuatu II-745
disease disturbances III-225–6, **III-226**
diseases, water-borne, Marshall Islands II-781
Disko Bay, Greenland, shrimp fishing I-8
dispersants *see* oil dispersants
dissolved inorganic carbon (DIC), non-Redfield ratio depletion I-225–6
dissolved organic N (DON), varied origin III-199
dissolved oxygen I-240
— Campeche Sound I-469
— depletion of III-199
— and fish kills I-364
— Gulf of Aden II-51
— levels in the Cambodian Sea II-572–3
— variable, Canary Islands I-191, *I-192*
disturbance
— keystone-endangered and chronic cyclical III-227–8
— as a regular feature of an ecosystem III-227
disturbance categories/types III-217–28
— derivation III-217
— grouping of III-217
disturbance regimes III-227
— better understanding of the natural history of III-230
dive tourism
— artificial dive objects I-604
— Bonaire I-597
— Dutch Antilles I-604
— effects of on Borneo islands II-377–8
— Hawaiian Islands II-806
— SCUBA diving
  — East Kalimantan II-377
  — Sabah II-377
— Sri Lanka II-180
— Vanuatu II-742
diving and snorkelling, impacts from II-213
Djibouti *II-36*, II-49, II-55, *II-64*
— use of mangroves II-58
Dnestr River, pollutants carried I-303
Dogger Bank I-46, I-47
— plankton I-48
dogwhelks
— imposex I-25, I-36–7, I-55, I-75
— severe impact of TBT III-250–1
— southeast England, the Dumpton Syndrome III-252
Doldrums *see* Inter-Tropical Convergence Zone (ITCZ)
dolphin mortality
— by-catch is to some degree controllable III-139
— reduction in through changed purse-seining procedure III-141
— through tuna fishing III-142–3
dolphins I-90, I-834, II-24, II-205, II-368, III-90–1
— caught for shark bait II-57
— declining, Black Sea I-291
— Great Barrier Reef II-618–19
— Gulf of Guninea I-780
— illegal hunting in Black Sea III-94
— Java Sea II-363
— Marshall Islands II-780
— and mercury contamination III-95
— Patagonian coast I-756
— Sri Lanka II-180
— *see also* named varieties
dominance, pattern of altered by oil spills III-275
Doñana National Park, southern Spain *I-168*
— controversy over infrastructure improvements around I-173
— sand and marsh ecosystems I-171–2
*Donax denticulatus*, wash zone, dissipative beaches I-648
Dover *I-66*
Dover Straits *I-66*
— shipping through I-72
downwelling
— Ekman downwelling I-223
— nearshore, Alaskan Gulf I-376
dragnet fishery, southern Brazil, east coast, causing degradation I-740
drainage, a development issue, Arabian Gulf II-9
Drake Passage I-753
dredge and fill
— Gulf coastal plain, USA I-445
  — navigation channels I-447
— Jamaica, damaging seagrasses I-569
— of mangrove swamps III-19
dredge fishery, Victoria Province, Australia II-668
dredge spoil, dumped I-77
dredging I-623, II-124, II-546, II-642, II-655
— Belize I-510
— in Chesapeake Bay I-348
— Curaçao I-602
— damaging to seagrass beds III-7, III-8–9
— destructive III-379
— detrimental effects of I-145
— and dumping, in the Comoros II-247–8
— and earth-moving activities, Marshall Islands II-782–3, *II-782*
— effects of, Tagus estuary I-161
— environmental impacts of, southern Spanish coast I-176–7
— and filling
  — for construction purposes I-601–2
  — Jamaica, damaging seagrasses I-569

— followed by reclamation, French Polynesia II-821
— and habitat destruction III-379–80
— harbour entrances and navigation channels, Buenos Aires I-762–3
— instream, degradation and erosion caused by I-444
— Kuwait II-9–10
— and land fill, Hawaiian Islands II-805
— and landfill, Red Sea II-43
— Malacca Strait II-339
— Maputo and Beira II-108
— northern Gulf of Mexico I-447–8
— now a necessity, Malacca Straits II-317
— to improve Gulf of Aden ports 58
drift netting, most popular, east coast, Peninsular Malaysia II-352
Driftnet Ban, United Nations III-140, III-145
drill cuttings, contaminated, discharge of III-363
drilling fluids, oil-based, toxicity of III-363
drilling waste management I-23–4
"drowned river system" *see* Chesapeake Bay
drowned river valleys, Tasmania II-653
*Drupella* snails, coral-eating II-696
dry deposition I-13
— measurement techniques III-203–5
  — field measurement of, current status III-205
  — use of surrogate surfaces III-204–5
  — wind tunnel experiments III-204
— modelling frameworks and algorithms III-202–3
  — current modelling uncertainties III-202–3
  — mathematical treatment of physical processes III-202
— of particles to water surfaces, processes and consequences III-197–209
  — atmospheric deposition to the coastal zone III-205–7
  — atmospheric particles III-199–200
  — deposition processes III-200–1
  — gas deposition and role of particles III-205
  — importance of in nutrient fluxes III-201
  — nutrient fluxes and aquatic cosystem responses III-198–9
dry deposition velocities III-200–1, III-202, III-204–5
— based on eddy correlation III-205
Dry Tortugas *I-406*
— corals little diseased I-410
— decline in pink shrimp landings I-447
— extreme low temperature, and death of corals I-410
— showing only small decline I-412
Dry Tortugas National Monument I-412
— redesignated as Dry Tortugas National Park I-413
Dry Tortugas National Park, coral bleaching (1998) III-56
duck plague virus III-225, III-227
dugongs II-54, II-90, II-123, II-336, II-431, II-574, II-587, II-601, II-615
— Bazaruto Islands II-104
— eastern Australian region II-637
— endangered in Gulf of Mannar II-165, II-166
— Great Barrier Reef
  — entanglement problems II-619
  — pressures and status II-619
— in the Mayotte lagoon II-250
— Red Sea II-42–3
— Sri Lanka II-180
— Vanuatu II-741
— western Indonesian seas II-401
dumping
— at sea III-338–9, III-342, III-365–6
  — of dredge spoils III-365
  — some prohibitions achieved III-365
— Baltic Proper, of chemical warfare agents I-117
— Irish Sea
  — blast furnace spoil I-94
  — munitions I-96
— New York Bight I-323, I-324
— North Sea
  — of industrial waste and sewage sludge finished I-53, I-55
  — of munitions I-52
— *see also* waste dumping
Dumping Convention **III-334**
Dumpton Syndrome III-252
Dungeness *I-66*, I-68
*Durvillia* habitat, exposed coasts, Tasmania II-652
Dutch Antilles I-595–614
— Aruba *I-596*, I-597
— coral bleaching and mortality events III-54
— definition and description I-597–8
— effects from urban and industrial activity I-502–7
  — artisanal and non-industrial uses I-602–4
  — cities I-606
  — cumulative impacts and ecological trends I-607
  — industrial use I-604–6
— environmental hindrances ordinances I-608
— environmental institutional capacity I-609
— Fishery Protection Law I-608
— government environmental protection and management funding, sparse I-612
— law enforcement is critical I-609
— Leeward Group *I-596*, I-597–8
— major shallow marine and coastal habitats I-598–601
— maritime legislation I-608
— National Parks Foundation of the Netherlands Antilles I-607–8
— overview of implementation of marine environmental policy and legislation I-609, **I-610–11**
— protective measures I-607–12
  — implementation and an evaluation I-609
  — policy development and legislation I-607–9
  — practical management measures I-609
  — prognoses and prospects I-609, I-612
— rural factors, coastal erosion, landfill and excavation I-60–2
— Windward Group *I-596*, I-598
dynamite fishing
— cessation of II-95
  — case study Mtwara, Tanzania II 87, II 95
— Gulf of Thailand II-304

earthquake activity
— the Philippines II-408
— Vanuatu II-744
— Xiamen region II-516–17
East African Coastal Current *II-64*
— important for larval dispersal and downwelling II-85
— influenced by the monsoons II-85
East Arabian Current II-50
— behaviour of II-20
East Australian Current II-581, *II-581*, II-614, II-619, II-631, *II-632*, II-649–50, II-664
— giving distinctive ecosystems around the Kent Group II-651
— varies with El Niño/Southern Oscillation cycle II-650
East China Sea *II-474*, *II-500*, II-508
East Greenland Current I-5, I-7
East India Coastal Current (EICC) II-274
East Madagascar Current *II-101*, II-245
East Pacific Plate I-487
East Sakhalin Current II-466
Easter Island I-708
eastern Atlantic flyway, migratory feeding, North Sea coasts I-49
Eastern North Atlantic Central waters I-153
Ebrié lagoon, Côte d'Ivoire I-811–12
— artificial Vridi canal I-812
— fish I-813
— freshwater inputs I-812
— hydrological zonation I-812
— physical framework I-811–12
— pollution in I-817
— sewage discharged to I-817
echinoderms I-35, I-239, III-242
*Ecklonia radiata* habitat II-652
ecological catastrophe, Aznalcóllar mining spillage disaster I-178

ecological problems, and their causes, west coast, Sea of Japan, Amursky Bay II-483–5
ecological resources
— exploitation of III-366–9
— aquaculture III-368–9
— marine capture fisheries III-366–8
ecological tariffs III-401
ecological tax reform III-401
economic development, and coastal management III-351
economic incentives, to achieve economic goals III-401
economic income, skewed III-396
ecosystem, defined I-705
ecosystem health information III-212
ecosystem instability and collapse, processes typically leading to III-227
ecosystem management, and the reserve concept III-377–8
ecosystem modelling I-50
ecosystems
— changes in structure and function following eutrophication III-263, **III-264**
— connectedness of I-478
— dynamics of III-259
— effects of fisheries on III-117–33
— measuring impact of by-catch on III-144
— refers to processes and functions III-259
— response to increased nutrient loads unpredictable III-366
— retrogression III-259–60
— vulnerable species targetted by opportunistic microorganisms III-227
ecosystems services, value of III-395
ecotourism I-4, II-89, II-95, III-376
— and Brazilian coral reefs I-723
— Fiji II-762
— Great Barrier Reef, Australia II-586
— Marshall Islands II-780
— Midway Atoll II-801
— a non-consumptive resource use II-141
— possible in the mangrove ecosystems of Pacific Colombia I-682–3
— potential for, Mayotte II-251
eddies I-46, I-630
— in Caribbean Basin I-580
— mesoscale, in cold water at Paria Peninsula I-646
— surface, Angola-Benguela front I-825
— SWODDIES I-138
— *see also* Sitka Eddy
eddy accumulation techniques, dry deposition measurement III-203–4
eddy correlation/covariance techniques, dry deposition measurement III-203
eddy currents, Canary Islands I-189
eddy diffusion, between deep and upper waters I-391
eddy systems I-85, I-223
eelgrass, Carolinas I-358
effluent disposal pipes
— KwaZulu-Natal II-142
— using Algulhas Current for dispersal, South Africa II-142
effluents
— from shallow-well injection, migration of I-409
— polluting Norwegian coastal waters I-24
Egypt *II-36*
— dive tourism II-44
— Red Sea coast, coastal pollution by oil II-43
Ekaterina Strait *II-464*, II-465
Ekman downwelling I-223
Ekman transport II-274
— drives coastal upwelling off Oman–Yemen coast II-19
El Niño
— affects Mexican Pacific coast I-486, I-488, I-489
— affects Nicaraguan Pacific coast I-534
— biological effects of, Peru I-692–3
— change to subtropical and tropical species I-693
— causes of I-702
— a complex, multi-level phenomenon I-703
— described I-679
— dramatic effects of I-679
— in El Salvador I-547
— *see also* Atlantic Niño; Benguela Niño years
El Niño events II-776
— affecting the California Current I-387
— affecting Gulf of Alaska mixed layer depth III-182
— biological effects of I-703–4
— bring drought, Coral, Solomon and Bismark Seas Region II-428–9
— and coral bleaching III-38
— extreme
— and the 1997–98 coral bleaching event III-44–6
— areas of bleaching III-46
— impact on Indian and Southeast Asian winds III-44
— interpretation and conclusions III-46
— impacts on Northeast Pacific Ocean III-183–4
— effect of 1997–8 event III-183–4
— importance of I-702
— mechanism governing strength of I-702–3
— northern Chilean coast invaded by exotics I-707
— physical alterations due to I-703
— position of Subtropical Water Mass I-701
El Niño Southern Oscillation (ENSO)
— effects of I-223–4
— effects on climate and oceanography of Fiji Islands II-753–4
— effects on the Great Barrier Reef region II-613–14
— long-term variations I-702–4
— periodic influence in Australia II-582, II-632
El Salvador I-545–58
— coastal erosion and landfill I-552
— effects from urban and industrial activities I-553–6
— geography I-547
— major shallow water marine and coastal resources I-548–51
— offshore systems I-551
— populations affecting the area I-551
— protective measures I-556–7
— Ley del Medio Ambiente (Environmental Law) I-556
— protected areas I-556–7
— rural factors I-551–2
— seasonality, currents, natural environmental variables I-547–8
elasmobranchs, sensitive to ecological change, North Sea I-49
Electronic Chart Display and Information Service (ECDIS), Malacca Straits II-325
Elefsis Bay *I-234*, I-237
— decrease in invertebrate abundances I-239
— industrial pollution I-245
— metal pollution I-246
elephant seals I-756, I-757
Eleuthera *I-416*
EMECS (Environmental Management of Enclosed Coastal Seas) I-56
Emperor Seamounts
— fossil coral reefs II-800
— relationship to Hawaiian Islands II-794
enclosed shores, Hong Kong II-538
endangered species
— Abrolhos Bank–Cabo Frio, Southern Brazil I-736
— consequences of climatic irregularities III-224
— leatherback, green turtles and loggerheads III-62
— loss or imminent loss, Black Sea I-291, **I-292**
— Olive Ridley and Hemp's Ridley critically endangered III-63
— Philippines II-410
— sea turtles III-143
— vaquita III-90
endangered species protection, western Indonesia II-400–1
endemic species
— Coral, Solomon and Bismark Seas region II-431
— Hawaiian Islands II-797, II-800
— Marshall Islands II-779

endemism
— Australian Bight II-678–9
— islands of the southwest Indian Ocean II-251
— low, reefal communities, French Polynesia II-819
— southern African marine biota II-135
— southern Australia II-581
— Tanzanian coastal forests II-88
endocrine disrupters
— potential harm from III-86
— TBT III-249
energy
— from the oceans III-303–21
  — possibility of combined technologies III-311
— from seagrasses, detrital and direct grazing pathways III-3
engineering structures, causing beach erosion II-639
English Channel I-65–82
— anthropogenic impacts on I-74–7
— coastal and marine habitats I-68–71
— defined I-67
— formation of I-67
— lies at boundary of Boreal and Lusitanian biogeograpical regions I-68
— physical and biological environment I-67–8
— protective and remediation measures I-77–9
  — International Conventions and Agreements, and EC Directives **I-78**
— urban and rural populations I-71–4
Eniwetak Atoll, nuclear tests II-783
ENSO events
— 1997/1998, impacted mammals and seabirds in several regions III-225
— 1998 event II-236
— affected by climate change III-370
— effects of, Vanuatu II-740
— effects on seabirds III 110
— influence on New Caledonia II-725
— influencing sea surface temperatures II-384
— Marshall Islands II-776–7
  — coral bleaching II-777
  — Crown-of-Thorns starfish II-777
  — sea-level rise II-777
— the Philippines, manifestations of II-408
— South Western Pacific Islands, affect oceanography and climate II-710
— Teacapán—Agua Brava lagoon system I-494
entanglement
— of mammals and seabirds in floating synthetic debris III-125–6
— usually fatal for small cetaceans III-145
entanglement nets (jarife), Tanzania II-90
environmental assessment, for newer Paupuan mines II-439
environmental coastal protection, Chile I-715
environmental concerns
— Brazilian tropical coast I-727–8
— coral reefs, Hawaiian Islands II-807
— in the Maldives II-206–7, **II-206**
— mariculture, Hawaiian Islands II-803
— Marshall Islands II-780, **II-781**
  — related to cross-sectoral issues II-784–5
environmental conservation, restoration and improvement, Gulf of Mexico I-479
environmental criteria, concepts in II-202
environmental degradation
— Andaman and Nicobar Islands II-193
— Cabo Frio–Cabo de Santa Marta Grande, southern Brazil I-738
— China II-492
— coastal forested watershed, Cambodia II-575
— early, the Seychelles II-239
— east coast, Peninsular Malaysia
  — of coral reefs II-355
  — from tourism II-353–4
— Godavari-Krishna delta II-171–2
— Hong Kong II-543–6
— Lakshadweep Islands II-194
— Palk Bay–Madras coast II-169
— parts of North Spanish coast I-142
— the Philippines II-418
— potential, Tasmanian marine fish farms II-641
— Sabah coral reefs II-371, *II-372*
— southern Brazil I-740–3
— Taiwan's marine and coastal environment II-510
— Tasmanian estuaries II-654–5
— through land reclamation, West Guangdong II-557
— Western Australia, metropolitan areas II-697
environmental effects, of Australian fishing II-587
environmental impact assessment, Hawaiin Islands II-808
environmental impacts
— of alien species in Irish Sea I-89
— Chilean fishmeal industry I-713–14
— of coastal settlements, eastern Australian region II-640
— of coastal zone industries, Chile **I-711**
— and ecological trends, Dutch Antilles I-607
— from repetitive trawling II-624
— industrial fishing, Gulf of Guinea I-787–8
— of intensive fishing
  — on the Irish Sea environment I-92
  — on Irish Sea fisheries I-92
— of mining activities, New Caledonia II-731
— nickel mining, New Caledonia II-713
— of tourism II-212–13, **II-213**
environmental issues
— Australia II-586
  — marine environment **II-589**
— Mascarene Region II-265
— south-east Queensland II-641–2
environmental management
— poor, Vietnam II-566
— sound, expected of companies and organisations III-351
— Torres Strait and the Gulf of Papua II-607–8
environmental management strategy, Marshall Islands II-787
environmental matrices
— Gulf of Mexico
  — extensive cattle pasture I-479
  — offshore oil industry I-478–9
environmental opportunities and prospects, Marshall Islands II-787–8
environmental pollution, in the Bay of Bengal II-154–5
environmental problems
— Cambodia, no clear perception of II-576, II-577
— critical, Xiamen region II-521–4
  — bacterial pollution II-522–3
  — oil pollution II-523
  — organic pollution, eutrophication and red tides II-522
  — pollution from pesticides II-524
  — sea dumping and disposal of solid waste II-523–4
  — sediment deterioration and secondary pollution II-524
— Malacca Straits II-317–21
— Marshall Islands, related to population pressure II-780
— some east coast Indian cities II-278
environmental projects, French Polynesia II-824
environmental quality criteria, use in Norwegian fjords and coastal waters I-27
Environmental Risk Assessment, Malacca Strait II-320–1, *II-321*
— analysis of likelihood of adverse effects II-320–1
— retrospective analysis of decline in key habitats II-320, **II-321**
Environmental Sensitivity Index (ESI) III-269, **III-270**
environmental stress, and altered benthos, Southern California I-396
environmental variability, in fish stocks, southwestern Africa I-832
environments, deep, Red Sea II-39
EOS satellite, with colour sensor MODIS III-295
epi-continental seas I-45
epiphytes, seagrass, productivity of III-2
Equatorial Counter Current I-486, I-489, II-201

Equatorial Countercurrent *I-678*, I-679, I-701, *II-581*, II-614, II-619, *II-724*, II-767, II-775, II-780
Equatorial Current, and El Niño I-703
Equatorial Subsurface Water Mass, low dissolved oxygen content I-702
Equatorial Under Current *I-774*
equity and fairness, in fisheries management 157–8, III-155
— inter-generational equity III-157
— issue of relative deprivation and historical participation III-157
equity theory in fisheries management III-157–8
Eritrea *II-36*
EROD induction I-56
erosion
— of beaches through sand mining I-636
— due to hurricanes I-622
— Dutch Antilles I-601–2
— Nicaraguan Pacific coast I-538
erosion–accretion cycles, Hawaiian Islands II-804–5
erosion–deposition cycles
— Arabian Sea II-27
— Gulf of Aden II-58
— Inhaca Island II-105
ERTS-1 satellite, early recognition of cyanobacteria blooms III-299
ESA satellite ENVISAT, with ocean colour sensor MERIS III-295
*Escheria coli*
— Hong Kong bathing waters II-541
— Penang and Selangor II-319
Espichel, Cape *I-152*, I-153
Essential Fish Habitat (EFH) I-451
estuaries
— east coast, Peninsular Malaysia II-349
— eastern Australia II-634
— Gulf of Maine I-310, *I-311*
— New York Bight, altered condition I-327
— North Sea, importance of I-49
— northeastern Australia II-614
— northern Gulf of Mexico
  — freshwater flushing rate a critical parameter I-447
  — tidal and subtidal oyster reefs I-441–2
— South Africa II-135–6
  — greatest threat will be lack of water II-136
  — resource use II-136
— southwestern Africa I-829
— Taiwan Strait II-502, II-503
— Tasmania II-653
  — degree of sedimentation II-654
  — southern, tannin-stained II-652
estuarine crocodiles
— Great Barrier Reef II-618
— Solomon Islands, depleted II-431
estuarine ecosystem, Palk Bay–Madras coast II-169
estuarine habitats
— Channel coast of England I-68
— Xiamen region, Jiulongjiang estuary II-518
estuary wetland ecosystems, North China, affected by oil and agriculture II-494
EU
— Habitats Directive, omits to take natural change into account III-354–5
— MAST programme I-56
— nature conservation, a shared competency I-57
EU/EC
— Convention on International Trade in Endangered Species (CITES) I-608
— Migratory Species Treaty (Bonn Convention) I-608
EU/EC Directives **I-78**
— on Bathing Water Quality I-76, I-95
— concerning waste-water treatment I-145
— Conservation of Natural Habitats and of Wild Fauna and Flora I-52, I-57, I-77, I-96–7, I-162
— Conservation of Wild Birds I-52, I-57, I-77, I-162, I-214
— on the control of nitrates I-53
— for protection of European Seas **I-57**
— Shellfish Hygiene I-76
— on Shellfish Waters I-76
— to increase urban wastewater treatment, too many derogations I-55
Eucla Bioregion, Australian Bight II-678, II-679
euphotic layer, Côte d'Ivoire I-810
Euphrates, River II-3
Europe
— current and planned offshore wind farms III-309–11, *III-310*
— wind resources predicted III-308, *III-309*
European Environment Agency, arrangements for protecting North East Atlantic in the 1990s **I-57**
European Union, funding for off shore wind energy technology III-307
eustatic movements, tilting European landmass I-47
eutrophication I-137–8, I-174, I-278, II-542, II-770, II-771, III-366
— and algal blooms I-20
— areas of French channel coast contributing to I-74
— in Australian coastal waters II-585
— Baltic Proper I-103–8
  — biological effects of I-109–10
  — rivers bring pollutants I-103
  — and supersaturation of oxygen I-104–5
— Baltic Sea I-130–1
— Bermuda I-228
— Black Sea I-290
  — changing phytoplankton community composition I-290
— of coastal areas, western Indonesia II-396
— concepts for remediation III-263
— cultural, Western Australia II-700
— in fjords with restricted circulation I-24
— from aquaculture I-144
— from fish farming I-26
— Guanabara Bay, southern Brazil I-741
— Gulf of Guinea, coastal waters I-784, I-788–9
— Gulf of Thailand II-307
— Hawaiian Islands II-804
— Irish Sea, possible near English coast I-87
— in its broadest sense III-198–9
— lagoon, Vanuatu II-745
— leads to increase in N and P III-259
— local, Red Sea II-43
— of marine waters: effects on benthic microbial communities III-257–65
  — coastal inlets of the southern Baltic Sea III-260–3
  — definition and sources of pollution III-258
  — effects of on marine communities III-259–60
— in most Carolinas estuaries I-363
  — estuarine fish kills I-363–4
  — measures for reduction I-366
  — more stringent N and P reductions needed I-366
— North Korean artificial lakes II-493
— North Sea I-53
— northern Gulf of Mexico, growing problem I-443, I-450
— in Norwegian coastal waters, dependent on transboundary load I-24
— and nutrients, English Channel I-74
— Patagonia I-765
— the Philippines II-417
— potential for, Jamaica I-564
— in a range of aquatic environments III-205–6
— secondary effects I-108
— some North Carolina estuaries I-356, I-363
Euvoikos Gulf *I-234*, I-237, I-244
— affected by wastes I-246
evaporation
— in excess of precipitation
  — Arabian Gulf II-3
  — Red Sea II-39

— Gulf of California Coastal Province, Mexico I-487–8
evolutionary mechanisms, allowing toxic species to spread III-219
Exclusive Economic Zones (EEZs) I-2
— Andaman and Nicobar Islands II-191
— Azores Archipelago I-203
— the Bahamas I-422
— Cambodia II-571
— and fisheries management III-158
— Lakshadweep Islands II-194
— Mozambique II-105
— North Sea I-51
— the Philippines II-408
— South Africa II-137
— USA I-451
  — utilization of economic resources I-365–6
exotic species/introductions *see* alien/accidental/exotic/introduced organisms/species
exports, from Faroese fishery I-35–6
extinction—recolonization dynamics, Chilean coast I-707
extreme events, The Maldives II-203
Exuma Cays *I-416*, I-419
— patch reefs I-429–30
— shallow marine habitat I-421
Exuma Cays Land and Sea Park
— evidence for positive effect of protective measures I-431–2
— regulations I-431
Exuma Sound I-417, I-428
*Exxon Valdez* oil spill I-382, III-275
— biological effects at low levels of aromatic compounds III-364
— characteristics of III-272–3
— decrease in *F. gardneri* III-272
— oiled mussel beds III-271, III-274
— part of severity of effects on *Fucus* due to high-pressure-hot-washing III-275
Eyre Bioregion, Australian Bight II-678

faecal coliforms I-743
— areas of Gulf of Guinea coast I-789
— contaminates freshwater sources, Marshall Islands II-781
— Dar es Salaam and Zanzibar II-92, II-93
— Hong Kong bathing beaches II-541
— Johore Strait II-335
— Maputo Bay II-109
— Morrocoy National Park I-658
— Peninsular Malaysia west coast II-319, II-340
— Sepetiba Bay, southern Brazil I-741
— Suva region, Fiji II-760–1
  — Suva harbour II-719
— Sydney beaches II-642
— Xiamen coastal waters II-522–3
faecal contamination, Durban beaches II-142
faecal pollution, Guanabara Bay, southern Brazil I-741
Fal estuary *I-66*, I-74, I-75
— closure of shellfishery I-69
Falklands/Malvinas Current I-751
FAO
— catch statistics reveal "fishing down" of food chains III-368
— Code of Conduct for Responsible Fisheries III-138, III-341–2
— Code of Conduct for Responsible Fishing III-159
— data base to maintain records of high seas fishing vessels III-160
— international plan to reduce incidental catch of seabirds in longline fisheries III-146
Farasan Islands *II-36*, II-38, II-41
farming, 'modernization' of III-7
Faroe Islands I-31–41
— coasts and shallow waters I-33–5
— mariculture I-35
— oceanic climate I-33
— offshore resources I-35–8
— pilot whales hunted III-92
— protective measures and the future I-38–40
— typical upper and deeper layer water flows I-33, *I-34*
faroes I-505
'faros', Maldivian mini-atolls II-201
Federal and Islamic Republic of Comoros (FIRC) II-247
Federal Water Pollution Control Act Amendments (USA), Section 404 Program I-452–3
feral animals, impact of Australian Bight coast II-684
fertiliser pollution, from Mozambique's upstream neighbours II-109
fertiliser run-off, Madagascar, impacting coral reefs II-123, II-124
fertilizer consumption, Lesser Antilles I-635, **I-635**
*Ficopomatus enigmatus*, positive effects of I-48
Fiji Islands *II-706*, II-707, *II-708*, II-751–64
— agriculture and fisheries II-715–16
— biogeography II-753
— coastal erosion and landfill II-760
— coastal modifications II-717
— coral reefs II-711
  — degraded by pollution II-711
— effects of urban and industrial activities II-760–1
— island groups II-753
— major shallow water marine and coastal habitats II-754–8
— marine environment critical II-762
— offshore systems II-758
— overfishing of marine invertebrate species II-759
— population II-714, II-758
— protective measures II-761–2
  — private sanctuaries II-761
— rural factors II-758–60
— seasonality, currents, natural environmental variables II-753–4
— sensitive and endangered marine species II-757
Fiji plateau II-753
fin whales, depleted III-78
financial and technology flows, global, importance for environmental decision-making III-346
Findlater Jet *II-18*, II-19
finfish communities, estuarine and coastal, Carolinas coast I-362–3
finfish farming III-170
finfish fishery, Belize I-509
Finisterre Cape *I-136*, I-139
Finland, Gulf of *I-100*, *I-122*
fire, effects of, northern Gulf of Mexico I-449–50
fire cycle
— natural, important in maintaining coastal ecosystems I-449
  — changes in detrimental I-449
fish
— the Azores I-207
— Baltic Sea
  — butyltins in **I-115**
  — larval stages in *Fucus* belt I-125
  — mixture of freshwater and marine species I-128
— Belize lagoonal shelf I-506
— Cambodia II-575
— coastal, eastern Australian region II-634
— Côte d'Ivoire, Ebrié Lagoon I-813
— deep water, "boom and bust" cycles III-120
— El Salvador I-550
  — estuarine nursery habitats I-550
— in English Channel I-70
— fauna, Australian Bight II-679
— Great Barrier Reef II-617
— Gulf of Guinea
  — commercial species **I-784**
  — Guinean Trawling Survey I-783
  — populations I-783
— Gulf of Mannar II-164–5
— Lesser Antilles I-633
— life history traits (*r*- and *k*-selection) III-119
— Mauritius II-259
— nearshore, Hawaiian Islands, depleted II-803
— North Aegean Sea I-238
— North Sea I-49

— North and South Carolina coasts I-362–3
— offshore, Fiji Islands II-758
— organochlorine pollution in, Asian waters II-449–51
— possible extinction of groupers and Humphead wrasse II-373
— problem of size-selective fishing III-121–2
  — changes in size structure, community level III-121–2
— South Aegean Sea I-238
— Southern California, pollution effects on I-397
— southwestern Africa
  — mesopelagic I-832–3
  — surf zone I-830
— species diversity, Gulf of Mexico I-473
— species and groups, Vietnam II-564
— Tagus estuary, nursery grounds I-157
— target species, ecologist-market collision III-158
— vulnerable species, fishing of III-120
Fish Aggregating Devices (FADs) II-372
— New Caledonia coastal fishery II-730
— used in the Comoros II-249
fish barrages, intertidal, Madagascar II-126
fish communities
— natural factors for change I-398
— Southern California, impacts of anthropogenic pollution I-397–8
fish diseases, from pollution I-397
fish diversity
— Baltic, marine, decreases northwards I-109
— demersal, soft grounds, Spanish north coast I-141
fish farming
— Faroes I-35
— Norway I-26
  — and expanding industry I-28
  — mass mortalities in fish cages I-20
— potential for, east coast of India II-278
— round North Sea I-52
— southern Spain I-177
fish farms, marine, Yellow Sea II-494
fish mortality, Tanzania II-89
fish plant, Marshall Islands, benefits vs. impacts II-785
fish processing at sea, dumping of organic material III-127
fish recruitment I-92
fish stock assessments III-156
— and Fishers' insights III-157
— relationship between present and future stocks III-156
— validity of single-species models in multi-species ecosystems III-156–7
fish stocks
— Adriatic Sea I-277
— affected by commercial fisheries III-97
— Arabian Gulf, dwindling II-11
— assessments in the Maldives II-210
— Baltic Sea I-110–12
  — freshwater species I-111
— Bangladesh **II-292**
— Chesapeake Bay I-342–3
  — coastal ocean spawners I-343
— collapsing I-3, III-119–21, III-367
— commercial fisheries, Western Australia **II-698**
  — Fish Habitat Protection Areas II-699
— Coral, Solomon and Bismark Seas region
  — coastal, information on II-434–5
  — fisheries management measures II-442
— declining
  — English Channel I-76
  — Gulf of Maine and Georges Bank I-310
— demersal fish, Gulf of Aden II-54
— dolphin stocks III-42
— evaluation of global status of III-377
— Gulf of Guinea, significant changes in I-783
— Gulf of Maine and Georges Bank I-312
— Gulf of Thailand III-128
— increased off Alaskan coast I-376–7
— Jamaica I-564
  — reefs overfished I-568–9
— Lesser Antilles, shared assessment of I-633
— Malacca Strait II-337
— Malacca Straits II-314
  — declining II-320
— New Caledonia II-730
— New York Bight, overfished I-330
— North Sea, management by EU I-57
— northern Spanish coast *I-147*
  — over-exploited and depleted I-146
  — sardine stock critical I-146
— off coasts of Oman and Yemen II-27
— over-exploited III-367
  — Irish Sea I-91
— Queen Conch I-425
— Sea of Okhotsk, some decrease in II-470
— some declines, Victoria Province II-668
— southern Spain, over-exploited I-173
— Tanzania II-88
Fish Stocks Agreement III-334, III-340, III-341, III-343
— *see also* Straddling Stocks Agreement
fisheries III-366–8
— Adriatic Sea I-277, **I-278**
— Aegean Sea, overfished I-244
— American Samoa, changing local fishing patterns II-770
— Andaman Islands, poor development of II-192, **II-193**
— Andaman and Nicobar Islands, deep-sea, development of recommended II-196
— Anguilla
  — artisanal I-621
  — Fisheries Protection Ordinance and Turtle Ordinance I-624
— Arabian Gulf II-8
— Arabian Sea II-25
  — artisanal II-26–7
  — industrial II-29
— Argentina
  — commercial I-763
  — industrial I-763–4
  — offshore, impact on fauna I-757–8
  — semi-industrial and artisanal I-763–4
— Australia II-586–7
  — Australian Fishing Zone II-586
  — commercial II-584
  — decline through overfishing II-587
— the Azores I-209–10
  — demersal and pelagic *I-209*, I-210
  — little regulation I-214
— the Bahamas
  — commercial and artisanal I-422, I-424–5
  — other resources I-426–7
  — protective regulations I-430–1
  — use of poisonous substances prohibited I-430
— Baltic I-111–12
— Bangladesh II-292, **II-293**
  — resources II-294
— Bay of Bengal
  — artisanal and industrial II-293
  — brackish water II-278
  — marine II-277–8
— Belize
  — artisanal I-508–9, I-511
  — regulation of I-513
— British Virgin Islands
  — commercial trap fishing I-621
  — moratorium on large-scale foreign fishing I-624
— Cambodia II-574
  — foreign poachers II-576
  — freshwater II-575
  — inland, Mekong River II-575
  — inshore, depleted II-576

— multi-fishing practices II-576
— Canary Islands I-195–6
— cetacean fishery I-195
— tuna I-196
— Carolinas I-362–3
— fisheries monitoring programs I-363
— viable, stressed and overfished I-363
— Chesapeake Bay, decline in I-343, I-346
— Chile I-710–12
— cash values I-712
— new management tools I-463
— co-management in III-159
— Colombian Caribbean Coast I-669–71
— artisanal I-671–2
— industrial I-670–1
— related to upwellings in northeast I-670
— *Tarpon atlanticus*, depleted population I-669–70
— Colombian Pacific Coast
— artisanal rights protected I-681
— commercial landings I-684
— industrial I-683
— industrial fishing denounced I-681
— in the mangrove forests I-682
— protective measures I-684
— the Comoros, artisanal II-249
— Coral, Solomon and Bismark Seas region II-440
— problems best solved by community-based initiatives II-442
— subsistence and artisanal II-434, II-436–7
— tuna fishery II-432, *II-433*
— Côte d'Ivoire I-816
— changes in I-815–16
— demersal I-815
— pelagic I-815
— declining yields III-377
— destructive II-374, II-387, II-438, II-566
— the Comoros II-249
— Indonesian Seas II-392–3
— destructive fishing techniques II-377, II-503, III-37, III-122–3
— the Philippines II-415
— *see also* cyanide; dynamite fishing
— discards by ocean region **III-367**
— Dutch Antilles I-602–4
— east coast, Peninsular Malaysia II-357
— artisanal II-352–3
— main pelagic species II-352
— eastern Australian region
— commercial II-640–1
— commercial and recreational II-637
— Ebrié lagoon, Côte d'Ivoire I-813
— effects on ecosystems III-117–33
— El Salvador I-550, I-552–3
— affected by oil spills I-556
— artisanal I-553
— English Channel I-73–4
— impacts arising from fishing I-76–7
— Faroes I-35–6
— collapse then increase I-35
— Fiji
— commercial catch data II-760
— commercial/artisanal II-715–16
— overfishing II-760
— overfishing of marine invertebrates II-759
— subsistence, impacts of II-758
— and the Fish Stocks Agreement III-341
— French Polynesia
— lagoonal species II-819
— subsistence and commercial II-819–20
— Godavari-Krishna delta II-170–1
— Great Australian Bight
— commercial II-682–3, II-684
— inshore commercial fishing II-682–3
— Great Barrier Reef region
— commercial II-623
— live food-fish export II-623
— recreational II-621
— Greenland I-8
— Guangdong
— closed fishing seasons II-560
— outstripping sustainable capacity II-557–8
— Guinea
— artisanal I-800, I-801
— Fish Smoking Center I-800
— industrial I-800–1
— National Fishing Supervisions Center I-800
— pelagic I-800
— traditional I-800
— Gulf of Aden II-56–7
— artisanal II-56
— industrial, illegal and uncontrolled II-59
— Gulf of Guinea
— artisanal I-787
— illegal fishing I-787
— industrial I-787–8
— Gulf of Maine and Georges Bank I-312–13
— Gulf of Mannar II-164–5
— indiscriminate use of small nets II-165
— Gulf of Thailand II-305–6
— conflict between subsistence and light-luring fishermen II-306
— decline in the resource II-305
— dominant demersal fish groups II-305
— dominant pelagic fish groups II-306
— mortalities of shellfish and finfish III-218–19
— multi-gear, multi-species II-305
— Hawaiian Islands
— commercial II-802–3
— derelict fishing gear II-804
— resistance to licensing/monitoring II-809–10
— subsistence, artisanal and recreational II-802
— Hong Kong II-538
— seriously affected by urban development II-543
— impact on seabirds III-111
— impact on small cetaceans III-97–8
— Indonesian II-391–4
— artisanal II-391, II-394
— commercial exploitation of offshore fisheries II-391
— high fishing pressure on major stocks II-391
— reef fisheries II-393
— industrial
— effects of III-112
— sequential overfishing III-120–1
— intensive fisheries management III-367
— Irish Sea I-91–2
— artisanal I-91, I-93
— comparison with North Sea I-92, **I-92**
— issues along the Black Sea Coast I-292, **I-293**
— need for cooperative action I-292
— and its problems I-3
— Jamaica
— management plans I-572
— reef I-566, I-568
— Japan, drive fisheries, small cetaceans III-93
— Java Sea, pelagic catch II-391
— Lakshadweep Islands II-194
— large incidental capture of sea turtles III-66
— late development of, Somalia II-80
— Lesser Antilles **I-633**, I-636–7
— artisanal methods can damage marine habitat I-635
— ongoing assessment I-634
— overfishing of nearshore waters I-633
— long-lived species replaced by shorter-lived species III-367
— Madagascar
— artisanal II-121, II-122–3, II-125

— commercial II-121–2
— offshore, estimated potential 120, **II-120**
— Malacca Strait, mainly artisanal II-338
— Malacca Straits, capture fisheries and coastal mariculture II-314
— the Maldives II-210–12
— coral reef fisheries II-211–12
— lagoonal bait fisheries II-210–11
— management for multispecies, marine harvest refugia I-244–5
— Marshall Islands II-783–4
— artisanal II-784
— maximum sustainable yield not yet reached II-783–4
— pelagic fisheries II-784
— Mascarene Plateau
— artisanal II-261–2
— banks fishery II-256
— effects of artisanal fisheries II-262
— potential for longline and deep-water fishery II-260
— some foreign tuna fishery II-260
— may negatively affect survival of marine mammals and birds III-128–9
— Mediterranean, Israeli I-257
— Mexican Pacific coast I-489–91
— artisanal I-490–1, I-491
— benthic invertebrates I-489–90
— foreign fleets I-490
— Montserrat I-621–2
— Mozambique II-105, II-107
— artisanal II-107
— New Caledonia II-728
— artisanal II-729–31
— Beche-de-Mer and Trochus exploitation II-730
— professional II-729–30
— subsistence reef fisheries II-715, II-729–30
— New York Bight
— "foreign fishing" I-330
— Magnuson Act I-330
— recreational I-323, I-329
— Nicaraguan Caribbean coast I-522, I-523–5
— automatic jigging machines I-523
— Nicaraguan Pacific coast I-534
— artisanal fishing I-537
— big pelagic fish I-536
— demersal fish I-536
— economically important marine resources **I-536**
— estimated biomass **I-536**
— industrial I-539
— North Sea I-51–2
— cockle fishery I-51–2
— detritus from catch-processor ships III-127
— shellfish I-51
— North Spanish coast
— artisanal I-144
— demersal I-138
— Galicia I-142
— negative impacts I-146
— pelagic I-141
— northern Gulf of Mexico I-445
— threatened by effects of navigation channels I-447–8
— Norwegian coast I-25–6
— Oman, artisanal II-21, II-26–7
— Palk Bay II-167
— Peru
— artisanal I-693, *I-694*
— demersal I-693
— pelagic I-693
— the Philippines
— coastal, affected by loss of mangroves II-414
— intrusion of commercial fishers into municipal waters II-416
— offshore II-411
— processing waste III-137
— purse-seine
— by-catch from different methods III-124, *III-124*
— dolphin mortality III-124
— reduction in dolphin mortality III-141
— tuna mortality, Eastern Tropical Pacific III-94
— tuna, Pacific, incidental take III-379
— recreational I-323, I-329, I-362, I-397, I-430–1
— Red Sea II-42
— artisanal II-43
— reef fisheries II-43
— rejects and marketable catch III-137
— Sabah, reef, overfished II-372
— Samoa II-714
— inshore II-716
— Sarawak II-372
— Sea of Okhotsk
— commercial, ecosystem effects II-469–70
— intensive II-470
— selective fishing III-136
— may result in ecosystem imbalance III-139
— Norway III-138
— the Seychelles
— artisanal II-238
— foreign vessels charged for fishing II-238
— reef fisheries II-236
— small-scale, more important than large-scale commercial II-140
— South Africa
— artisanal and subsistence II-139–41
— commercial II-136–8
— optimal utilisation of catch II-138
— recreational angling II-138–9
— target switching common II-137
— South Western Pacific Islands, Industrial II-718–19
— southern Brazil I-737
— artisanal I-737, I-739
— industrial I-739
— Southern California Bight
— commercial I-396–7
— recreational I-397
— Southern Gulf of Mexico I-475
— artisanal I-475–6
— species and production **I-476**
— southern Portugal
— artisanal, Tagus and Sado estuaries I-161
— coastal waters I-159
— southern Spain I-172
— major economic activity, Gulf of Cadiz I-172
— overexploited I-175
— southwestern Africa I-830, I-833
— commercial trawling I-830, **I-830**
— hake I-833–4
— purse seiner catches *I-831*
— secondary effects of fishing I-833
— Spratly Islands, pressure on recently increased II-364
— Sri Lanka
— chank and sea cucumber II-183
— coastal II-182
— marine II-181–2
— marine ornamental II-182–3
— Tagus and Sado estuaries I-159
— Taiwan Strait II-503
— artisanal fishing II-506
— "bull-ard" fishery II-504–5
— coral reef II-504
— decline in fishermen II-505
— seasonal grey mullet fishery II-503–4
— Tanzania
— artisanal II-90–1
— commercial II-90
— target and non-target catches III-137
— Tasmania II-654
— commercial II-655–6

— individual transferable quotas (ITQs) II-656
— offshore II-653
— Tonga, subsistence reef fisheries II-716
— Torres Strait and Gulf of Papua
— commercial II-604
— subsistence and artisanal II-601–3
— tropical seas, many low-value species in catches II-372
— Tyrrhenian Sea I-279
— Vanuatu
— artisanal II-715
— commercial II-744
— community-based management schemes II-745–6, **II-746**
— impacts of II-743–4
— major area for development II-744
— undeveloped II-744
— Venezuela I-653–4
— worker conflicts, artisanal and trawl fisheries I-653
— Victoria Province
— commercial II-667–8
— specialised bay and inlet fishery II-667–8
— Western Australia, commercial II-697–9
— Xiamen region, capture fisheries II-519
— Yemen, artisanal II-56
— Yuzhnoprimorsky region II-480
— *see also* by-catch; discards; recreational fishing
fisheries agreements, regional III-340
fisheries law, global III-158–9
— international law, based on UNCLOS III-158–9
— national approaches to III-158
— regional agreements III-158
fisheries management
— for artisanal and subsistence fisheries, South Africa II-140–1
— based on ecosystem principles III-377
— binational I-463
— and closed areas III-378
— designed for fish conservation III-157
— emergence of "epistemic community" in III-163
— emerging global fisheries laws III-158–9
— exploitation rate, implementation error in III-389
— Great Barrier Reef region, evaluation of II-624
— Hawaiian Islands II-803–4
— need for restrictions II-810
— incentive structures III-155–6
— issues related to enforcement III-160–1
— blacklisting III-160–1
— flag state responsibility III-161
— issues related to legitimacy of III-155, III-157
— substantive and process legitimacy III-155
— issues related to surveillance III-159–60
— observer programs III-160
— limits to exploitation and target exploitation rates III-382
— Malacca Strait, little information to base sustainable use on II-341
— the Maldives, for single species fisheries II-211–12
— ocean dynamics and marine habitat changes becoming important III-185–6
— participation in III-159
— past, present and future, southwestern Africa I-836–7
— precautionary approach III-79
— social aspects of management issues III-161–2
— restrictions on effort III-161
— restrictions on technique III-161
— as a social science problem III-153–64
— equity and fairness III-157–8
— "governance" and "institutional embeddedness" III-155
— scientific realism III-156–7
— Sri Lanka **II-186**
— Torres Strait and the Gulf of Papua II-605–7
Fisheries and Oceans Canada, view of passage of El Niño signal, Northeast Pacific III-184–5
fisheries regulations, Madagascar
— conservation of valuable species II-126
— management and policies II-126
— new local control law (GELOSE) II-126
fisheries resources, New Caledonia, not safe from certain threats II-731
fisheries science
— and Fishers' insights III-157
— a mandated science III-156
fisheries scientists I-2
Fisheries Sector Program (FSP), the Philippines II-418–19
fisheries—ecosystem interaction model III-118–19, *III-118*
Fishery Conservation and Management Act (USA) I-451
fishery systems, sources of uncertainty in analysis of III-389
fishing
— as a business III-155
— ecosystems effects of III-378–80
— by-catch and incidental take III-379
— debris and ghost fishing III-380
— habitat destruction III-379–80
— illegal, "patterned deviant behaviour" III-155
— regulatory discards seen as perverse III-156
fishing discards *see* discards
fishing fleet, global, overlarge III-155
fishing fleets, influencing government over international laws III-158
fishing gear
— cause of seabird and marine mammal mortalities II-656
— derelict, dangerous nuisance II-804
— destructive II-27
— subsistence fishing II-436
fishing mortality
— directed, of target organisms III-118–22
— density-dependent responses III-121
— loss of genetic diversity III-122
— overfishing III-119
— population collapses III-119–21
— population size III-118–19
— size-selective fishing III-121–2
— indiscriminate III-122–6
— by-catch and discard mortality III-123–5
— caused by lost gear, ghost fishing III-125–6
— from physical impacts and destructive practices III-122–3
— natural, indiscriminate changes in III-126–9
— changes mediated by biological interactions: competition and predation III-127–9
— changes mediated by dumped 'food subsidies' III-127
— changes mediated by habitat degradation III-126
— underestimation of, and population depletion III-126
fishing practices
— concerns about direct and indirect effects of on marine ecosystems III-352
— destructive I-365, I-366, I-554
fishing pressure, above optimum, Irish Sea I-92
Fishing Reserves, Mascarene region II-265, **II-265**
fishing techniques
— controls on economically inefficient III-161
— potential for environmental impacts II-29
fishmeal, Chile
— effect on coastal zone I-713–14
— environmental impacts of I-713–14
— product of fishing industry I-710, I-712
fjord circulation
— creates environmentally significant gradients I-19
— two-layer flows I-19
fjord complexes, Greenland I-8
fjords
— as contaminant traps I-19
— described I-19
— Norwegian coast I-17, I-19
— changes in shallow water communities I-22
— deep, contain arctic bottom fauna I-20
— deep water circulation I-19
— main fluvial input at head I-19

— southern, stagnant bottom water I-19
— west coast, large and deep I-19
flash floods II-23
flatback turtles III-63–4
— Great Barrier Reef II-618
— restricted range III-64
— a vulnerable species III-64
Fleet lagoon *I-66*, I-68
Flinders Current *II-581*, *II-674*, II-676
Flinders Region, Victoria Province II-663
Flindersian Province
— southern Australia, marine fish fauna II-679
— Western Australia II-694
flood and coastal hazard maps, Hawaiian Islands II-805
flood risk, and rising sea level III-190–4
flood warning system, Firth of Clyde III-352
Flores Sea *II-382*
— atolls, reefs damaged by destructive fishing practices II-389
Florida
— marine reserves, key design principles III-390
— use of zoning III-389–90
Florida arm, Gulf Stream I-589
Florida Bay *I-406*, I-407, *I-436*
— effects of cyanobacteria III-218
— influx of freshwater, nutrients and sediments I-409
— seagrass die-off
— shown by SPOT XS imagery III-286
— a trend for coastal waters in the new Millennium? III-14–16
— western, importance of to Florida Keys reef tract III-16
Florida Current I-407
Florida Escarpment *I-436*
Florida Keys I-405–14, *I-436*
— agriculture and urban factors I-409
— area of sediment influx I-409
— coral bleaching and mortality events III-56
— coral vitality: long-term study of Carysfort reef I-410–12
— Florida Keys National Marine Sanctuary and Protection Act I-412
— global stresses I-409–10
— grim picture for the future of the reefs I-413
— industrial stresses I-409
— islands composed of calcium carbonate rock I-408
— localized ecological reef stress I-410
— major shallow water, marine and coastal habitats I-407–8
— populations affecting the area I-408
— protective measures I-412–13
— multi-use zoning concept I-413
— stress to the environment and reefs I-408–9
— USEPA coral reef monitoring project I-412
Florida Keys Coral Reef Disease Study I-412
Florida Keys National Marine Sanctuary I-408, I-409, I-452
Florida Keys Reef Ecosystem: Timeline **I-413**
Florida Loop Current I-437, I-439
— carries effects of Mississippi floods I-409, *I-410*
Florida Straits *I-416*, *I-436*, I-437, I-439
Flower Garden Banks *I-436*
— marine sanctuary I-442, I-452
— no real coral bleaching III-56
flushing time, North Sea I-46
fly ash disposal, China II-495
fog and cloud, Peruvian coast I-690
Fonseca, Gulf of *I-532*, I-537–8, *I-546*
— coastal diving birds I-549
— industrial shrimp farming I-538
— over-exploitation of resources I-537–8
Food and Agricultural Organisation *see* FAO
food chains
— energy transfer by zooplankton I-359
— marine, Greenland, heavy metals in I-9
food pollution, Asian developing countries and OCs in fish II-450
food webs, detrital III-3
forest clearance *see* deforestation
forestry
— Australia, environmental considerations II-584
— northern Spanish coastal area, causing soil loss I-142–3
forests
— Chesapeake Bay catchments I-345
— El Salvador I-547
— Gulf coastal plain, USA, conversion to farmland I-444
— Nicaragua I-519, I-520
— degradation of I-539
Forth Estuary Forum
— economic appraisal of the Forth Estuary III-351
— economic development a 'Flagship Project' III-351
Forth Kuril Strait *II-464*, II-465
Frailes Canyon I-487
Fram Strait *I-6*
Framework Convention on Climate Change III-340
France
— influence in French Polynesia II-823–4
— La Rance tidal energy project III-304, III-316
— nitrogen inputs to the English Channel and North Sea I-74
— offshore wind farm planned III-309
— prohibition of organotin-based paints on small boats III-251
franciscana (La Plata river dolphin) III-90
Fraser Island II-634
Freez Strait *II-464*, II-465
freezing, northern Gulf of Mexico, causes mass mortalities I-439
French Polynesia II-813–26
— effects of human activities II-820–3
— exploitation of resources II-819–20
— islands described II-815
— major shallow water marine and coastal habitats II-817–19
— offshore systems II-817
— population and politics II-815
— protective measures II-823–4
— Marine Area Management Plans II-824
— reef restoration schemes II-824
— seasonality, currents, natural environmental variables II-815–17
fresh water, lacking, Namibia I-835
freshet (spring flow), Chesapeake Bay I-340
freshwater
— a critical resource
— the Comoros II-251
— the Mascarene Region II-264–5
— dilution of, Hong Kong coastal waters II-537
— a future problem
— Marshall Islands II-780, II-781
— New Caledonia II-729
— reduced in estuaries, Victoria Province, Australia II-666–7
— resources scarce, Xiamen region II-518
freshwater coral kills, Hawaiian Islands II-807
freshwater flushing
— important for water bodies with poor circulation I-449–50
— rate a critical parameter in northern Gulf of Mexico estuaries I-447
freshwater influx
— affects seasonal patterns, Yellow Sea II-489
— Florida Bay I-409
— Gulf of Alaska I-378
— northern Gulf of Mexico I-437, I-443
— deprived of most of spring runoff I-448
— to Cambodian Sea II-571
freshwater lens, Belize continental shelf I-503
freshwater supplies, the Bahamas I-429
freshwater systems, contamination by nutrients and pesticides I-539
freshwater turtles, east coast, Peninsular Malaysia II-350
fringing reefs II-277
— Andaman and Nicobar Islands II-191, *II-192*
— Borneo II-370
— Brazil I-723
— Colombian Pacific Coast, Gorgona Island I-680
— Comoros Islands II-246
— many threats to II-248

— east coast, Peninsular Malaysia II-350
— Fiji Islands II-756
— French Polynesia II-817, *II-818*
— inner and outer, Tanzania II-85–6
— islands in Torres Strait II-597
— Madagascar II-118
— mainland Belize I-505–6
— Mauritius II-258
— the Philippines II-408
— Puerto Rico I-582
— San Andrés and Providencia Archipelago I-668
— the Seychelles II-236
— western Indonesia II-389
frontogenesis, subtropical I-223
*Fucus* I-70
— Baltic Sea, decreased through eutrophication I-130–1
— lead and zinc in, near major Greenland mines *I-12*
— response to oil spills III-273
*Fucus serratus*, scarcity of in Faroes I-33
*Fucus vesiculosus*
— community changes
— disappearances associated with fish/fisheries decline I-131
— southern Baltic I-110
— hard bottoms, Baltic Sea I-125
Fujeirah *II-18*, II-19
— construction of oily waste reception facilities II-29
Fuma Island, Bajuni Islands, sandy tail *II-75*, II-76
Fundy, Bay of *I-308*, I-311, *I-311*
— cetacean–gillnet interaction III-144–5
fur seals, Patagonian coast I-756

Galapagos Islands, coral bleaching and mortality events III-53
Galicia *I-136*, I-137
— dense coastal population I-142
Ganges River *II-146*
— mouths of *II-286*
gap-fraction analyses, used with remote sensing III-286
garbage disposal, poor, Sri Lanka II-183
garfish, Baltic Sea I-111
gases, solubility of in water, and removal by wet deposition III-200
Gdansk Basin I-101, I-106
Gdansk Deep I-101
— cyclic behaviour, phosphate and nitrate concentration I-108
Gdansk, Gulf of, anthropogenic heavy metals in sediments I-115
*Gelidium sesquipedale* beds
— Cantabrian coast I-140
— Portuguese coast I-159
Gene Bank of Fishing and Aquarian Resources, Colombia I-685
genetic diversity, loss of in exploited fish populations III-121, III-122
Geneva Convention, addresses military threats to the marine environment III-342
geological features, South Western Pacific Islands region II-712
geology, Gulf of Thailand II-299
Georges Bank
— closed areas III-386
— as larval source sites III-381, *III-382*
— and "cold pool" I-324
— nontidal surface circulation *I-324*
— primary production over I-310
*Georges Bank* (R. Backus and D. Bourne) I-309
Georges Basin I-309, *I-311*
Geosphere Biosphere Programme (IGBP), scientific definition of coastal zone III-295
German Bight
— cadmium budget I-54
— effects of oxygen depletion I-53
— organic pollutants in I-54
Ghana, mangroves I-779
ghost fishing III-380
— and fishing mortality III-125
giant clam culture, Marshall Islands II-781, II-782
giant clam fishery, Fiji, overexploited II-759
giant clam hatchery, South Sulawesi II-401
giant clams, endangered, the Philippines II-410
Gibraltar Strait *I-168*, I-169, I-172
gillnets
— and cetaceans III-144–5
— coastal
— give high small cetacean by-catch III-94–5
— and seabirds III-146
— drifting/driftnets, high incidental catches III-124–5
— and ghost fishing III-125
gillnetting
— by-catch from II-57
— a danger to dugongs and dolphins II-180
— and decline in Australian dugongs II-587
— prohibited in Gulf of Alaska I-383
— prohibited in Southern California I-397
— Western Australia II-697
Gippsland Lakes
— Victoria
— adjusting to changed environment II-667
— affected by high mercury levels II-669
glaciation, Antarctic I-706
glass-eels I-157
Global Climate Change (GCC) models, predict increase in extreme events III-47
global codification, learning-based approach to III-346–7
global commons, and sustainability III-360
Global Coral Reef Monitoring Network II-435
Global Ocean Observation System I-56
Global Plan of Action (GPA) III-344, III-345
global stresses, Florida Keys I-409–10
global warming
— an abdication of intergenerational responsibilities III-370
— potential effects on oceans III-98
— and range extension I-88
— response to climate change III-181–2
— and whale distribution III-86–7
Global Waste Survey III-339, III-345
— waste connectivity III-338–9
Glovers Atoll (Reef) *I-502*, I-504–5
— deep reef habitats I-506
GNP versus sustainable welfare III-395
GOA *see* Alaska, Gulf of
Godavari and Krishna deltaic coast II-170–3
— biodiversity II-170
— climate and coastal hydrography II-170
— environmental degradation II-171–2
— fish and fisheries II-170–1
— geological features II-170
— impacts of human activities on the ecosystem II-172
— mangrove ecosystem II-170
— suggestions and recommendations II-172–3
Golfe Normano-Breton, gradient in zooplankton type and biomass I-70
Gorgona Island National Park, Colombia I-681, I-684
Gotland Basin I-101, I-106
Gotland Deep *I-101*, I-102
gradient techniques, dry deposition measurement III-203
Grand Comore (Ngazidja/Ngazidia)
— establishment of a coelacanth park II-251
— small reefs and restricted seagrass beds II-245
— a volcanic island II-245
Grand Turk *I-588*, I-590
— tourists I-591
gravel habitats, sensitive to disturbance III-123
Great Australian Bight II-673–90
— coast of geological significance II-678
— coastal erosion and landfill II-684
— effects from urban and industrial activities II-684–6
— major shallow water marine and coastal habitats II-678–81

— need for greater research and conservation management II-686
— offshore systems II-681
— populations affecting the area II-682
— protective measures II-686–8
  — Great Australian Bight Marine Park II-686–8
  — research II-688
— rural factors II-682–4
— seasonality, currents, natural environmental variables II-675–7
— a single demersal biotone II-675
— Yalata Aboriginal Land Lease II-682
Great Australian Bight Marine Park II-686–8
— Benthic Protection Area II-687
— Commonwealth Marine Park II-687
— Marine Mammal Protection Area II-687
— potential pressures on marine conservation **II-688**
— Whale Sanctuary, Head of Bight II-686
Great Australian Bight Trawl Fishery II-682, II-683
Great Bahama Bank *I-416*, I-421
Great Barrier Reef II-595, *II-612*, III-41
— degradation of III-37
Great Barrier Reef Marine Park II-595
— aboriginal interests in II-620
— does not fisheries management II-624
— a model in large marine ecosystem conservation II-626
— protective measures II-624–7
  — biodiversity conservation II-625–6
  — evaluation of the management II-626–7
  — fisheries management II-626
  — human and financial resources II-624–5
  — impact of tourism II-626
  — management framework II-624
  — management mechanisms II-625
  — shipping and oil spills II-626
  — water quality II-626
— tourism II-622
Great Barrier Reef Marine Park Act II-620
Great Chagos Bank *II-221*
— peat on Eagle Island II-229
— seismic tremors II-225–6
— world's largest atoll II-223
Great Corn Island, volcanic with reefs I-522
Great Lakes Fisheries Commission III-154
Great South Channel I-309, *I-311*
Great Whirl *II-18*, *II-48*, II-49, II-50, II-66
— high speed of currents II-54
Greater Antilles, coral bleaching and mortality events III-54, III-56
Greece
— protected areas I-250
— turtle protection not carried through III-355
green turtle nesting sites, Hawaiian Islands II-799
green turtles I-780, II-54, II-180, II-250, II-289, II-336, II-350, II-598, II-779, II-784
— an endangered species III-62
— Chagos Islands II-229
— death through exposure to okadaic acid III-219
— "edible turtle" III-62
— fibropapilloma incidence on, and turtle mortality III-227
— Fiji, endangered II-757
— Great Barrier Reef II-618
— growth slow III-62
— Hawaiian Islands, with fibropapillomatosis II-799
— The Maldives II-205
green urchins I-312
greenhouse gases
— atmospheric III-188
— commitments to reductions in III-304
greenhouse scenarios, for Australia II-585
greenhouse warming III-369
Greenland icecap, increase in Pb, Cd, Zn and Cu in snow and ice cores I-9
Greenland Sea *I-6*, I-7
Greenland seas I-5–16
— future threats for the marine environment I-14–15
— oceanography I-7–8
  — bedrock shorelines I-8
  — ocean surface circulation *I-7*
— pollutant sources I-11–14
— population and marine resources I-8–9
— present contaminant levels I-9–11
Greenland—Scotland Ridge, acts as partial barrier I-33
Grenada *I-644*
Grenadines, whaling I-636
grey seals I-49–50, III-95
— Cornwall and Scilly Isles I-70
— Irish Sea I-91
grey whales III-74, III-76
— census of III-79
— numbers of III-81
— for subsistence III-83
Grotius, freedom of the oceans concept (1609) I-2
groundings, shipping, South Western Pacific Islands II-718
groundwater
— the Azores I-207
— the Bahamas, vulnerable to contamination I-429
— pollution of, Borneo II-376
groundwater contamination, Nicaraguan Pacific coast I-539
groundwater extraction, Taiwan, causing subsidence and saline intrusion II-507, II-509
groundwater pollution
— Gulf of Cadiz I-174
— round Sea of Japan II-477
grouper fishery, the Maldives II-211–12
growth rates, seagrasses III-2
Guadalcanal Island II-427
— effects of earthquakes II-429
— small-scale aquaculture trial II-438
Guadalquivir River *I-168*
— estuary fishing I-175
— regulation of fishing I-183
Guadimar River *I-168*
— spillage from the Aznalcóllar mining disaster I-178
guano mining I-829
Guatemala, focus on an ecosystem framework for planning and management I-463–4
Guban coastal plain *II-64*
Guinea I-797–803
— continental shelf I-799
— fishing benefits from its EEZ I-801
— major shallow water marine and coastal habitats I-799–800
  — main zones I-799
— offshore systems I-800–1
— populations affecting the area I-801–2
— protective measures I-802
  — National Center of Supervision and Fishing Protection I-802
  — party to London Convention I-802
— seasonality, currents, natural environmental variables I-799
Guinea Current *I-774*, I-778, I-781, I-807–8
Guinea, Gulf of I-773–96
— coastal erosion and landfill I-786–7
— defined region I-775
  — geomorphic features of continental shelf I-775
  — LME divided into subsystems I-775
— effects from urban and industrial activities I-787–93
— islands I-781
— major shallow water marine and coastal habitats I-779–81
— offshore systems I-781–4
— populations affecting the area I-784
— protective measures I-793–4
  — Gulf of Guinea LME projects I-793–4
  — marine and environmental legislation I-793
  — protected areas I-793
— rural factors I-784–6

— seasonality, currents, natural environmental variables I-775–9
Guinea Under Current *I-774*, I-778, I-781
Guinea waters I-777
Gulf of Aqaba
— continuation of the Red Sea rift II-37
— sea-floor spreading II-37
Gulf Area Oil Companies Mutual Aid Organisation II-29
Gulf of California Coastal Province, Mexico I-487–8
— habitats I-487–8
Gulf Coast Fisheries Management Council I-525
Gulf Coast, USA, habitats of I-440
Gulf of Guinea LME projects I-793–4
Gulf of Maine Point Source Inventory I-316
Gulf of Mexico Aquatic Mortality Network III-228
— multi-jurisdictional monitoring III-229, III-231
Gulf of Mexico Fishery Management Council I-451
Gulf of Mexico states (USA), regulations addressing habitat degradation I-452
Gulf Stream I-189, I-204, *I-222*, I-223, I-227, I-356, I-362, I-419
— meanders, off the Carolinas coast I-353
— transient effects I-324
Gulf War
— oil spills II-11–13
  — oil penetration aided by burrowing animals III-271
Gush Dan outfall, sewage sludge I-259, I-261
Guyana Current I-630, I-632
*Gyrodinium aureolum* blooms I-69, I-90

habitat changes, small, dramatic effects on whales III-87
habitat degradation III-130
— by fishing methods, effects of III-126
— and change, greatest effects on inshore small cetaceans III-96–7
— Coral, Solomon and Bismark Seas region, localised concerns II-437–8
— leads to decline in fish populations III-126
— Peruvian coast I-694–5
habitat destruction III-379–80
habitat diversity, northern Gulf of Mexico I-438
habitat infilling, North Sea coasts I-54
habitat loss
— American Samoa II-770
— coastal, through new building I-173
— east coast, Peninsular Malaysia II-353
habitat modification, can damage sea turtle populations III-66
habitats I-3
— restoration, Chesapeake Bay I-347
HABS *see* harmful algal blooms (HABs)
haddock I-35, I-49, I-313
Hadramout *II-18*
Haifa *I-254*
Haifa Bay I-255
— coastal zone outside I-261
— heavy metal monitoring I-262
— land-based pollution sources I-259–61
  — nutrients I-259–60
  — toxic metal and organic pollutants I-260–1
Hainan Current II-537, II-538
Hainan Island *II-550*, II-551
Hajar Mountains, Oman *II-18*, II-21
hake I-313
— Argentine Sea I-757
— in Peruvian fisheries I-693
— silver, contaminants in I-315
— southwestern Africa I-833, I-833–4
hake fishery, Argentina I-763–4
Halaniyat Islands *II-18*
— some coral cover II-22
halibut fisheries, individual quota system, US Pacific Northwest III-155
haline stratification
— Baie de Seine I-67
— off Lancashire–Cumbrian coasts I-86
haling, catch per unit whaling effort (CPUE) indices III-79
halocline, Black Sea, generation of I-288–9
halogenated hydrocarbons (HHCs) III-95
harbo(u)r porpoises I-50, III-90
— by-catch in the North Atlantic III-94
— English Channel I-70
— Irish Sea I-90
harbour seals I-49–50
harbours
— and marinas, accumulation of TBT in sediments III-250
— TBT contamination not reduced III-254
hard bottom habitats
— Norwegian coast I-21
  — gradients of change I-22
— Southern California Bight I-388
— Tyrrhenian Sea I-273
hard bottoms I-442
hard substrate communities, Bahamas Archipelago **I-420**
hard-bottom outcrops, Carolinas coast I-353, I-362
hard-bottom/coral reef communities, Bahamas Archipelago **I-420**
hardgrounds, Puerto Rican shelf I-581–2
hardshores, the Azores I-205, *I-205*
— intertidal zonation pattern I-205, *I-206*
harmful algal blooms (HABs) III-98, III-213, III-298
— cause of mass lethal mortality disturbances III-223
— as indirect cause of morbidity and mortality III-218
— profit from wetland and mangrove destruction III-221–2
harp seals
— effect of commercial fisheries on III-97
— hunted in Greenland I-8
Hawai'i
— marine reserves III-386
— no-fishing 'kapu' zones III-386
Hawai'i Coastal Zone Management Plan II-808
Hawaiian Islands National Wildlife Refuge II-798
Hawaiian Islands (USA) II-791–812
— challenges for the new millennium II-809–10
  — better management in the northwestern islands II-810
  — community-based coastal area management II-809
  — controls over fishing II-810
  — preservation of natural tourism amenities II-810
  — public educations II-809–10
— coastal uses II-802–4
— geological origin of II-793–4
— islands listed **II-794**
— major shallow-water marine and coastal habitats II-795–800
— offshore systems II-800–1
— populations affecting the area II-801
— protective measures II-807–9
  — development plans II-808
  — environmental impact assessment II-808
  — environmental legislation II-807–8
  — marine environmental restoration II-809
  — marine protected areas II-808–9
— region defined II-793–4
— rural factors II-801–2
— seasonality, currents, natural environmental variables II-794
— urbanization factors II-804–7
Hawaiian Ocean Resources Management Plan II-808
hawksbill turtle I-780, II-205, II-229, II-250, II-350, II-779, II-784, II-799, III-62–3
— critically endangered III-62–3
— endangered, Fiji II-757
— Great Barrier Reef II-618
— harvested for tortoise shell III-62
hazardous substances, targets planned by North Sea states III-362–3
hazardous waste
— Marshall Islands, nuclear testing II-783
— shipments of II-320
HCHs II-545

— air and surface seawater, worldwide II-454, *II-456*
— in the English Channel I-75
heat flux, across Line-P, Northeast Pacific III-184
heavy metal contamination
— Dutch Antilles I-605
— riverine and coastal environment, western Indonesia II-396
— West Guangdong and Pearl River delta II-556, **II-556**
heavy metal mining I-738
— Australia II-639, II-641
heavy metal pollution I-279
— Amursky Bay, eastern Russia II-484
— Anzoategui State and Cariaco Gulf, Venezuela I-656
— from mining, Greenland I-11–12
— Johore Strait II-335
— Liverpool Bay I-95
— Malacca Straits II-318
— Morrocoy National Park I-657–8, **I-657**
— northern Spanish coast I-145, **I-146**
— Odiel saltmarshes I-170–1
— Pago Pago harbour, American Samoa II-712
— southern Brazil
  — Guanabara Bay I-741
  — and mangroves I-742
  — Sepetiba Bay I-741, I-741–2
— Tanzania II-92
— west Taiwan, and "green oysters" II-509–10
heavy metals
— affecting small cetaceans III-95
— Arabian Gulf
  — in sediments II-12
  — in sewage II-10
— Australia II-588
— Bay of Bengal
  — in cultured prawn tissue II-155
  — and estuaries II-155
— Bothnian Bay and Bothnian Sea I-131
— coastal and river waters, Gulf of Cadiz **I-180**
— dune mining of, KwaZulu-Natal II-142
— found in coral skeletons II-390
— from pesticides and fungicides, in shellfish I-493
— Great Barrier Reef region II-623
— in harbour muds, Belize City I-511
— in the Hoogly (Hugli) river II-278
— incorporated into estuarine trophic web I-179
— long-range airborne transport of I-115
— Norway I-24–5
— off west coast, Peninsular Malaysia II-341
— offshore, Guinea I-802, **I-802**
— one-hop contaminants I-13
— Patagonian coast I-765
  — localised I-761
— a potential problem, Gulf of Guinea I-792–3
— in sediments, Teacapán—Agua Brava lagoon system I-495, *I-495*
— in shrimps I-493, I-494
— in surficial sediments, Gulf of Thailand II-305, **II-305**
— Suva Harbour, Fiji II-719
Hector's dolphin III-91
— incidental mortality III-94
HEED (Health Ecological and Economic Dimensions) approach III-214
HELCOM Convention, on Baltic Marine Environmental Protection strategy I-132
Helsinki Conventions on the Protection of the Marine Environment of the Baltic Sea Area III-339
herpesvirus, ducks, geese and swans III-225
herring I-49, I-90, I-313
— Baltic Sea I-110, I-110–11, I-128
— Vistula Lagoon I-112
herring fishery, decline in Irish Sea I-91
hexachlorocyclohexane isomers *see* HCHs
Himalayas, and the Bay of Bengal II-287
Hiri Current II-619
Honduras, coral bleaching and mortality events III-54
Honduras, Gulf of *I-502*, I-503
Hong Kong II-535–47
— coastal erosion and landfill II-542–3
— coastline and coastal waters II-537
— effects from urban and industrial activities II-543–6
— Hong Kong Special Administrative Region II-537
— hydrography II-537
— Mai Po marshes II-539–41
— major shallow water marine and coastal habitats II-537–8
  — mixed subtropical fauna and flora II-537–8
— offshore systems II-538
— populations affecting the area II-538–42
— problems in Victoria Harbour II-542
— protective measures II-546
  — oil pollution control ordinances II-546
  — Water Control Zones II-546
— rural factors II-542
— Sewerage Master Plans II-546
Hoogly (Hugli) river, heavily polluted II-155, II-278–9
Hormuz, Straits of II-3, *II-18*, II-19
— corals in Musandam II-22
— important shipping lanes II-28
— inflowing water enriches the Gulf II-6
— water flow through II-4, *II-5*
horse mackerel I-141
— southwestern Africa I-832–3, I-833
horseshoe crabs
— captive breeding programmes, Sundarbans II-158–9
— potential source of bioactive substance II-159
hotspot sediments
— environmental implications of I-25
— Southern California Bight I-392
Houtman Abrolhos reefs, Western Australia II-696
Hudson Canyon *I-322*
Hudson Shelf Valley *I-322*, I-326, *I-326*
— up- and down-welling aids sediment transport I-323–4
Huelva *I-168*
— commercial port I-176
— urban-industrial development environmentally hazardous I-174
— waste dumped into the estuary I-179, **I-179**
Huizache y Caimanero lagoon system I-493–4
— fertilizers and agrochemicals from runoff found in shrimps I-494
— lagoon environmentally managed I-494
— marsh I-493–4
— shrimp culture I-494
human activities
— accelerating eutrophication, and its results III-198–9
— direct and indirect effects of seabirds III-111, III-114
— effects of, French Polynesia II-820–3
— increased spread rate of organisms and changed meaning of distance III-336–7
human breast milk
— contaminated with organochlorines
  — Hong Kong II-545
  — India and China II-452
Human Development Index, low, Papua New Guinea and the Solomon Islands II-434
human impacts, on coastal sediments III-352–3
human-assisted redistributions *see* alien/accidental/exotic/introduced organisms/species
Humber estuary I-54
— seasonal change between nutrient source and sink I-53
Humboldt Current *I-700*, I-701
— *see also* Peruvian Current
Humboldt Current ecosystem *see* south Pacific eastern margin ecosystem
humpback whales III-76
— Bermuda Islands I-226–7
— Hawaiian Islands II-800

— recovery of III-82
hurricane damage, and coral bleaching, Belize III-55
hurricanes I-340–1, I-408
— affecting Belize I-503–4
— Anguilla, British Virgin Islands and Montserrat, damage from I-622
— the Bahamas I-418–19
— cause ecological effects to Gulf of Mexico habitats I-439
— destructiveness of III-224
— dispersing anthropogenic rubbish in the marine habitat I-606
— Dutch Antilles, and oil slick movement I-605
— effects on Carolinas coast I-364
— El Salvador I-547
— Gulf of Mexico I-471, *II-470*
— intensity of increased I-631
— Jamaica I-561
  — damaging the coral reefs I-562, I-563, I-564
— most common path, Lesser Antilles *I-628*
— Nicaragua, Joan and Mitch **I-519**
— northeast Caribbean I-617
— Pacific Mexican coast I-487, I-488, I-489, I-493
— Puerto Rico I-578, I-579, I-580
— Turks and Caicos Islands I-589
— windward Dutch Antilles I-598
hydrocarbon pollution
— Australia II-588
— from tanker off Mozambique II-109
hydrocarbons
— enhanced values near refineries I-75
— surface marine sediments, Taiwan Strait II-510
hydroelectric power
— for heavy industry I-24
— regulation of river flow for I-19
hydrogen sulphide ($H_2S$) I-105–6, *I-105*, I-287
— in prawn ponds II-154
hydrography, Hong Kong II-537
hydrography and circulation, Gulf of Thailand
— forcing mechanisms in coastal seas II-300
— understanding of limited II-299–300
hydrologic cycle III-394
hydrology, Xiamen region, China II-515–16
hydrothermal vents, Guayamas basin I-487
hygroscopicity, of particles III-200
hyper-nutrification II-320
hypersaline conditions, south Texas and South Florida I-438
hypersalinisation, Magdalena Delta soils I-671
hyposalinity, Jiulongjiang estuary surface water II-518
hypoxia
— Black Sea I-201, I-303
— Florida Keys, near rivers and deltas I-408
— New York Bight I-325

ice cover
— Baltic Proper I-102
— in Chesapeake Bay I-340
— Greenland seas I-5
— winter, Sea of Okhotsk II-466
ice scouring, winter I-311
icefjords, Greenland I-8
— halibut fishing I-8
Iceland, ITQ system III-162
ICES *see* International Council for the Exploration of the Sea (ICES)
ICES
— Seabird Ecology working group III-112
— studies on ghost fishing III-125
— study group on bottom trawling III-126
ICZM *see* Integrated Coastal Zone Management (ICZM)
image data sources III-294–5
immigrants, behaviour of, Levantine basin I-257–8
IMO *see* International Maritime Organization (IMO)
imposex
— as an indicator of environmental contamination III-253
— causal link to TBT III-249
— dogwhelks I-25, I-36–7, I-55, I-56, I-158, III-250–1, *III-251*
— from tributyltin
  — Malacca Strait II-339
  — Singapore and Port Dickson II-319
— genetic aspects: the Dumpton Syndrome III-252
— occurrence in southwest England *III-251*
— whelks I-210–11
impoundments, lead to loss of wetland I-445–6, I-447
incentive structures, in fisheries management III-155–6
increased prevalence theory III-226
Index of Sustainable Economic Welfare (ISEW) III-395
India
— conservation policies for the Sundarbans II-157
— production of DDT and HCHs II-449
— Southeast II-161–73
  — Godavari and Krishna deltaic coast II-170–3
  — Gulf of Mannar II-163–6
  — Palk Bay II-166–8
  — Palk Bay-Madras Coast II-168–70
Indian Monsoon Current II-274
Indian NE Monsoon current *II-64*
Indian Ocean
— anticyclonic gyre in Mascarene region II-255
— biogeographic position of the Chagos Islands II-224–5
— Central, coral bleaching and mortality events III-48–9
— sever coral bleaching III-224
— Southern, coral bleaching and mortality events III-48
— western, oceanic platform II-235
Indian Ocean Commission (COI)
— joined by the Comoros II-247
— Madagascar a member II-127
— pilot ICZM operation, Mauritius and Reunion II-266
Indian Ocean Whale Sanctuary II-29, II-180
Indian—Australian plate margin, upthrust II-712
individual quota systems III-155, III-161
— best management tool for many fisheries III-162
individual transferable quotas (ITQs)
— ideal but complex and can cause problems III-162
— will lead to concentration of access rights III-162
Indo-Pacific beaked whale II-54
Indo-West Pacific Marine Province, Pan-Tethyan origin II-707
Indonesia
— coral bleaching and mortality events III-52
— Lembata Island, only subsistence traditional whaling operation III-84
— mangroves III-20
— marine conservation target II-401
— national management systems and legislations and the Malacca Straits II-321–3
— National Oil Spill Contingency Plan II-323
— national parks programme II-342
— northeastern, great species diversity II-389
Indonesia, Western, Continental Seas II-381–404
— land-based processes affecting Western Indonesian seas II-395–9
— major marine and coastal habitats II-385–90
— marine resource extraction II-390–5
— a microtidal region II-384
— oceanography II-383–5
— prospects for the future II-403
— protective measures II-399–403
  — endangered species protection II-400–1
  — Integrated Coastal Zone Management II-399–400
  — marine protected areas II-401, **II-402**
  — PROKASIH Program II-400
Indonesian Throughflow, Coral, Solomon and Bismark Seas Region II-428
Indus river dolphin III-90
industrial activities, Malacca Strait II-339

industrial centres, Russian sector, Sea of Japan basin **II-475**
industrial crowding, Calcutta II-155
industrial development
— coastal zone, Cambodia, no serious environmental pollution II-576–7
— Great Barrier Reef hinterland, loss of coastal habitat and water quality II-622
— limited, Coral, Solomon and Bismark Seas region II-439
— poor in Sundarbans II-153
— Tasmania II-654
— Victoria Province coast II-669
— Xiamen region II-518–19
industrial discharges/effluent
— Cabo Frio–Cabo de Santa Marta Grande, southern Brazil I-741
— Hawaiian Islands II-805–6
— Lesser Antilles I-638–9
— Madras, discharged into the estuaries II-169–70
— and marine pollution, Dutch Antilles I-605
— Tanzania II-92
industrial diversification II-263
industrial pollution
— east coast, Peninsular Malaysia II-351
— Fiji II-745
— Gulf of Guinea coast I-788–9
— Hong Kong II-542
— Malacca Straits II-317, **II-318**
— New Caledonia II-733
— Pago Pago harbour, American Samoa II-771
— Peru I-695
— Sarawak II-376
— southern Brazil, east coast I-740
— Sri Lanka II-184
industrial waste III-339
— dumping of, Tasmania II-654
— poorly handled, the Philippines II-417
— Tasmania
— disposal in coastal waters II-656
— dumping of II-654, II-656
industrialisation, at an early stage, Vietnam II-565
industry
— Argentine coast, Bahía Blanca, petrochemicals I-758
— around the North Sea I-52
— Brazilian tropical coast, effects of I-727–8
— causing pollution, Peruvian coast I-693, I-694–5, *I-694*
— development of round the Aegean I-241
— eastern Australian region, central New South Wales II-636
— El Salvador I-554–5
— English Channel coasts I-72
— expansion and diversification, Venezuelan coastal zone I-645
— growth of, Côte d'Ivoire I-815–16
— Guinea, uses old procedures I-802
— Gulf of Maine I-313
— development and change I-314
— heavy, Noumea and Suva II-718
— marine, Guangdong II-559, **II-559**
— Mexican Pacific Coast I-489
— causing environmental stress I-491
— Nicaraguan Caribbean coast I-525
— northern Spanish shoreline, effects of I-145–6
— Oman
— Muscat and Salalah II-25
— new developments, Sur and Sohar II-25, II-27
— and sources of employment, Gulf of Guinea countries **I-785**
— southern Brazil
— impinging on shallow water coastal and marine habitats I-736, I-737
— industrial complexes I-737–9
— Trinidad I-638
— Vanuatu II-745
infaunal biodiversity assessment, Island Copper Mine III-237
— large infaunal species III-242–3
— juveniles, seen infrequently III-242–3
— sampling design and procedures III-237–8, III-239
— similarity analyses of species and abundance data III-239
— tailings rate tolerable to infauna III-239
infections, in humans, from casual exposure to water III-219
informal settlements, South African coast, a pollution problem II-142–3
infrastructure
— causing coastal modification, Samoa II-717
— interrupting coastal sediments, Gulf of Guinea I-786–7
— leading coastal occupation, Brazilian tropical coast I-727
— the Maldives
— interference with natural erosion-deposition cycles II-208
— interfering with sand movement II-203
— Mascarene Islands, interferes with sand movement II-262
Inhaca and Portuguese Islands Reserves, Mozambique II-105
— slow soil recovery from slash and burn agriculture II-107
— turtle nesting beaches II-105
inorganic nutrients
— Neuse Estuary I-366
— Pamlico, Neuse and New River I-363–4, **I-363**
inshore habitats, Australian Bight II-678
— biogeographical regions II-678
Institute of Marine Affairs (IMA), Trinidad I-634
institutional mechanisms
— established under IMO and UNEP III-345
— importance of III-344–5
integrated coastal area management *see* Integrated Coastal Zone Management (ICZM)
integrated coastal management
— need for in Bangladesh II-295–6
— Xiamen region
— land—sea integration II-531–2
— movement towards II-528
Integrated Coastal Zone Management (ICZM) I-79
— and Agenda 21 III-350
— definitions I-458–9
— finance for III-356
— is it achievable? III-356–7
— must include economic development III-351
— needs to develop in Madagascar II-129
— Nicaragua I-541–2
— critical strategies for I-542
— MAIZCo I-526, I-528, I-534
— Nicaraguan Caribbean coast I-526–8
— the Philippines II-418
— a process not a solution III-350
— western Indonesia II-399–400
— CEPI projects II-399
integrated coastal zone management projects, Tanzania II-94
— constraints on development and implementation of II-96
— programmes attempting to put ICZM into practice II-94–5, **II-94**
— Tanga Coastal Zone Conservation and Development Programme II-94
Integrated Ecological Economic Modelling and Assessment (SCOPE project)
— basic framework III-398
— steps in III-399
integrated marine and coastal management (ICAM) III-341, III-344
integration, different kinds I-459
Inter-American Tropical Tuna Commission (IATTC) I-680
— Tuna-Dolphin Program, training skippers III-142
Inter-Tropical Convergence Zone (ITCZ) I-629, I-679, II-39, II-49, II-66, II-347, II-725, II-753, II-780
— Côte d'Ivoire I-808
— Gulf of Guinea I-776
— over Chagos Islands II-225
— and seasonal changes in the Caribbean sea surface waters I-646
Intermediate Antarctic Water Mass I-702
Intermediate Arctic Water I-679
internal boundary layers III-206–7
— concept III-206

ISM Code *see* International Safety Management (ISM) Code
isostatic rebound I-375
Israel, coast of, and the Southeast Mediterranean I-253–65
— coastal erosion I-258–9
— effects of land-based pollution I-259–61
— natural characteristics I-255–6
— Nature Reserves and National Parks Authority I-262
— population I-258
— protective measures I-262–3
— National Masterplan for the Mediterranean coast I-261
— shallow marine habitats, the Red Sea invaders I-256–8
ITCZ *see* Inter-Tropical Convergence Zone (ITCZ)
IUCN, and Project Tiger II-157
Ivittuut, Greenland *I-6*, I-12
Ivoirian Undercurrent I-808
Ivory Coast *see* Côte d'Ivoire
IWC *see* International Whaling Commission (IWC)
Izmir I-241
Izmir Bay *I-234*, I-236, I-237, I-244
— dinoflagellate red tides I-238
— eutrophication spreading I-244
— industrial development I-246–7
— wastes dumped with no treatment I-247

Jakarta Bay
— heavy metal loading II-396, **II-398**
— 'Where Have All the Reefs Gone'? the demise of Jakarta Bay and the final call for Pulau Seribu II-397–8
Jakarta Mandate on Marine and Coastal Biodiversity III-341
— integrated marine and coastal management (ICAM) III-341
Jamaica I-559–74
— coastal erosion and landfill I-569
— coral bleaching and mortality events III-55
— effects of urban and industrial activities I-569–70
— Environmental Protection Areas, Negril and Green Island watersheds I-571
— geography of I-561
— macroalgal mats smother old reefs III-223
— major shallow water marine and coastal habitats I-561–5
— Montego Bay and Negril Marine Parks I-571
— offshore systems I-565–6
— populations affecting the area I-567
— Portland Bight Fisheries Management Council I-572
— Portland Bight Sustainable Development Area I-572
— protective measures I-570–3
— Council on Ocean and Coastal Zone Management I-570
— Environmental Permit and Licensing system I-570
— International Conventions participation I-570, **I-571**
— National System of Protected Areas *I-571*
— Natural Resources Conservation Authority I-570
— rural factors I-568–9
— seasonality, currents, natural environmental variables I-561
— stresses causing collapse of many reef species III-223
Jamaica coral reef action plan I-571
Jangxia Creek, China, tidal energy project III-316
Japan
— acquisition of whaling technology III-74
— against the Southern Ocean Whale Sanctuary III-85–6
— banned organotin-based paints completely III-251, III-253
— coral bleaching and mortality events III-52
— directed hunts of small cetaceans III-93–4
— mistaken killing of dolphins III-97
— resumed Antarctic whaling III-77
— still killing minke whales III-78
— whale meat
— continued search for III-78
— insatiable demand for III-78
Japan, Sea of II-473–86
— climate change II-483
— ecological problems and their causes II-483–5
— Amursky Bay II-483–5
— Rudnaya River Valley II-485
— landscapes and tourism II-482–3
— natural resources, species and protected areas II-480–1
— protected areas II-481–2
— state of marine, coastal and freshwater environment II-476–80
Java Sea *II-362*, II-363, II-363–4, *II-382*, II-552
— fish landings related to the monsoon II-372
— fishery production II-363
— fragile II-363
— monsoonal climate II-383–4
— pelagic resource base considered heavily exploited II-393
jellyfish I-832
jet drops and film droplets, cause enhancement to particle deposition III-202
Jiangsu Coastal Current II-489
Jinjira *see* St Martin's Island, Bangladesh
John Pennecamp Coral Reef State Park I-412
Johore Strait II-311, II-334–5
— East Strait II-334
— effect of causeway II-334
— increase in sewage and industrial waste discharges II-335
— low wave energy II-334
— poor water quality II-334, II-335
— seagrass beds II-334
— Sungei Buloh Nature Park II-335
— West Strait, problems II-334
Jordan Basin I-309, *I-311*
Juba-Lamu embayment *II-67*
Jutland Current *I-18*, I-23

Kakinada sand spit, Godavari and Krishna deltaic coast II-172, II-272
Kalimantan
— East, SCUBA diving II-377
— Northeastern *II-362*
— coastal habitats II-365
— transmigration programme II-374
Kamchatka *II-464*
— "eastern channel", west coast II-466
Karimata Strait *II-382*
karst I-561
— formation I-417
Kattegatt *I-100*, *I-122*
kelp II-52
— Chile I-707
— Faroes, forests of *Laminaria hyperborea* I-33–5
— Galician coast I-140
— giant, extensive beds, Tasmanian waters II-652
— Gulf of Maine I-312
— Irish Sea I-89
— Norwegian coast I-21
— harvesting causes public concern I-26
— kelp forests denuded I-21
— response to oil spills III-273
— Southern California Bight I-388, I-395
kelp forests III-35
Kelvin wave dynamics III-184
Kelvin waves, and El Niño I-703
Kemp's ridley turtle III-63
— critically endangered III-63
Kent Group ecosystems, Tasmania II-651
Kerch Strait I-287, I-298
Kerguelen Islands II-583
Key Largo *I-406*, I-408
— reefs showing physical or biological stress I-410
Key Largo Formation I-408
— migration of effluent through I-409
Key Largo National Marine Sanctuary I-412
Key West *I-406*, I-408
khawrs
— brackish coastal wetlands II-23, *II-52*
— larger, may host mangroves II-53

— periodic opening to sea II-23, II-52–3
  — roads may interfere with the process II-27
Kilinailau Trench *II-426*
Kishon River, Israel
— carrying nitrogen and phosphorus I-259
— heavy metal contamination in the estuary I-260–1
Kislaya, Bay of, tidal energy project III-316
Kizilirmak delta, Turkey, conservation area I-293
knowledge-based tools in coastal management III-355
Korea, disappearance of tidal marsh zone II-490
Korea Strait II-475
Korean Peninsula, Yellow Sea coast population II-491–2
Kosi Bay system, KwaZulu-Natal, estuarine resource use II-136
krill, Antarctic, food for seabirds III-108
Krishna-Godavari deltaic coast *II-162*, II-163
Kruzenshtern Strait *II-464*, II-465
Kuril Basin *II-464*, II-465
Kuril Islands *II-464*
Kuroshio Current II-489, II-516, II-537
— eastern Taiwan II-501, *II-502*, II-508
Kuwait *II-2*
— coral islands II-7, *II-8*
— coral reef fish fauna II-8
— devastated by Gulf War II-11
— dual-purpose power/desalination plants II-10
— hydrographic changes due to Third River project II-9
— landfill II-9–10
— mariculture II-8–9
— nutrients offshore II-6
— offshore circulation anomaly II-4
Kuwait Action Plan II-29
Kuwait Institute for Scientific Research, research related to pollution by petrochemical industries II-3–4
KwaZulu-Natal
— degraded state of estuaries II-135–6
— from management to co-management: the *Pomatomus saltatrix* fishery II-138
— importance of subsistence and artisanal fishermen II-140
— intertidal harvesting, Maputaland Marine Reserve II-140
— invertebrate fishery/gleaning II-139
  — well managed II-139
— *see also* South Africa
Kyoto Agreement III-304
Kyoto Declaration, and Plan of Action III-139

La Hague, radionuclides transported from *I-14*, I-54–5, I-76
La Niña I-708, I-709
— strong
  — associated coral bleaching III-46
  — bleaching south of typhoon path III-47
La Perouse seamount II-257
Labrador Current *I-352*, I-356
Lac Badana National Park II-80
Laccadives *see* Lakshadweep Islands
lagoonal habitats, important in Mascarene Region II-256
lagoonal shelf, Belize I-506
lagoons
— east coast, Peninsular Malaysia II-349
— Marshall Islands, water residence time II-776
Laguna de Tacarigua, Venezuela I-652–3
— American crocodile I-652
— common birds and fish **I-652**
— eurihaline-mixohaline I-652
— planktonic productivity I-652
Laguna Madre *I-436*
lake waters, airborne spectrometer studies, and chlorophyll content III-297–8
Lakshadweep Islands II-20, *II-190*, II-194–5
— biodiversity II-194
— climate and coastal hydrography II-194
— conservation recommendations II-195
— environmental degradation II-194
— low numbers of seabirds recorded II-204
— population and tourism II-194
Lakshadweep (Laccadive)—Maldives—Chagos Ridge II-201, II-223
— formation of II-223
— spread of coral biodiversity along II-202
*Laminaria*, harvested for alginate I-70
*Laminaria hyperborea* forests I-33–4
land, an increasingly scarce resource, the Comoros II-251
land clearance, for agriculture, impact of silt, nutrients and contaminants I-508
land degradation
— Australia II-585
— Fiji II-758
land mosses, as biomonitors I-116
land reclamation
— for agriculture and spoil disposal I-144
— Coral, Solomon and Bismark Seas region II-438
— effects of, Western Port Bay, Victoria II-667
— Guangdong coast II-557
— and habitat loss, Singapore II-334
— large-scale, Hong Kong II-542–3, II-546, *II-546*
— Malé's artificial breakwater II-208–9
— pros and cons, the Maldives II-208
— small-scale, Tasmania II-655
— Taiwan II-505
  — an earlier policy II-510
  — effects of II-509
— for urban development, western Indonesia II-398, *II-399*
— west coast of Peninsular Malaysia II-339
— west coast Taiwan II-503
land subsidence, caused by artesian wells II-307
land tenure rights
— Marshall Islands II-786
— vs. governance system, Marshall Islands II-785
land use, Colombian Caribbean Coast I-672
land-slips, and associated sediment plumes, show a dynamic landscape II-429
land-use practices, environmental threat, the Comoros II-247
landfill
— American Samoa II-770
— Belize I-510
— biodiversity loss and monetary loss I-145
— Curaçao I-602
— Djibouti, impacting mangroves II-58
— due to urbanization, Arabian Gulf II-9–10
— and habitat loss, major concern, Cambodia II-576
— impacting some lowlands, Gulf of Mexico I-477, I-479
— Kota Kinabalu, Sabah II-375
— the Mascarenes II-263
— New Caledonia II-731
— North and South Korea II-493
— not common in Argentina I-762
— obvious form of estuarine alteration I-448
— a problem throughout the Gulf of Guinea I-787
— Sarawak II-375
Langstone Harbour *I-66*, I-174
Lanzarote, Los Jameos del Agua, endemic invertebrates I-195
Laperuz Strait *II-464*, II-465
Large Marine Ecosystems (LMEs) I-2
— Arabian Gulf II-1–16
— Arabian Sea *II-64*
— Australia *II-580*
— basis for assessing the health of III-230–1
— Bay of Bengal II-287
— Benguela Current LME I-823–37
— East African Marine Ecosystem II-85
— Gulf of Guinea I-773–96
— health impacted by morbidity, mortality and disease events III-218

— Lesser Antilles, Trinidad and Tobago I-627–41
— many becoming stressed III-218
— Mexico, northern Gulf of I-437
— Red Sea II-35–45
— Somali Coastal Current II-63–82
— South Western Pacific Islands Region II-707, II-709
— West Iberian large marine ecosystem I-137
Latin America, coastal management in I-457–66
— coastal zone defined I-458
— integrated coastal zone management defined I-458–9
— the Latin America and Caribbean focus I-460–3
— regional examples I-463–4
— some initiatives on coastal zone management **I-464**
— sustainable development defined I-459–60
Law of the Sea *see* UNCLOS (UN Convention on the Law of the Sea)
lead (Pb)
— Baja California I-495, **I-496**
— in Boston Harbor sediments I-315
— contamination by in Tagus estuary I-158
— decrease in Greenland snows I-14
— from petroleum combustion I-329
— in Greenland marine mammals I-9
— in the North Atlantic III-361
lead pollution
— Greenland I-12
— ships' paint I-37
lead and zinc mining, Chile I-710
leatherback turtles I-780, II-205, II-350, **II-355**, III-61–2
— considered endangered by IUCN III-62
Leeuwin Current II-581, *II-581*, *II-674*, II-676, II-677, II-694, *II-695*
— introduces an Indo-Pacific element II-676
— linked to population dynamics of West and South Australia's commercially important pelagic species II-676, II-679
Leeuwin Under Current *II-674*
Lesser Antilles
— coral bleaching and mortality events III-56
— defined I-629
— economies over reliant on coastal environment I-634
— southern, influence of Amazon and Orinoco rivers I-629–30
Lesser Antilles, Trinidad and Tobago I-627–41
— coastal erosion and landfill I-636
— effects from urban and industrial activities I-636–9
— major shallow water marine and coastal habitats I-631–2
— offshore systems I-632–4
— populations affecting the area I-633–5
— protective measures I-639–40
— rural factors I-635–6
— seasonality, currents, natural environmental variables I-629–31
Levantine Basin *I-254*, I-255
— deep chlorophyll maximum I-256
Levantine Intermediate Water I-255, I-256
— entering the Tyrrhenian Sea I-272
— silicates in I-271
Levantine Sea *I-234*, I-241
Levantine Surface Water I-255
life, beginning on Earth III-394
light availability, and seagrass growth I-358
light-stress-induced mortality, seagrasses III-15
Lighthouse Reef *I-502*, I-504–5
Ligurian Sea *I-268*
limestone, used for cement, Sri Lanka II-181
limestone mountains, Arabian coast II-19, II-21
limestone platforms, Red Sea, foundation for "Little Barrier Reef" II-40
limpets I-71
— harvesting of, the Azores I-206, I-210
lindane, Seine estuary I-75
liquefied natural gas (LNG), Indonesia II-390
Lisbon *I-152*
Lisbon embayment I-153
literacy rate, Gulf of Mannar II-165
litter
— conspicuous source of pollution, Fiji Islands II-761
— dumping of I-512
— and floating wastes, Borneo II-375–6
— and marine debris, western Indonesia II-398–9
— ocean and beach, Australia II-588, II-640, II-684, **II-685**
Little Andamans, reported deforestation II-193
Little Bahama Bank *I-416*, I-421
littoral communities, Faroes, response to wave exposure I-33
littoral transport
— Côte d'Ivoire I-808
— sand, Gulf of Guinea I-778
*Littorina littorea*, lacking in Faroes I-33
live reef fish food trade II-373, II-392–3
— Cambodia II-576
— Coral, Solomon and Bismark Seas region II-440
— Great Barrier Reef region II-623
Liverpool Bay *I-84*, I-85
— discharges to I-95, **I-95**
livestock grazing, and erosion I-601–2
livestock rearing
— changing pattern of, Oman and Yemen II-26
— Gulf of Aden coasts II-57–8
lobster fishery II-80
— Belize I-509
— Lesser Antilles I-635, I-637
— Madagascar, under local control II-126
— Nicaragua I-524
  — Pacific coast I-537
— Yemen II-59
lobster habitats, artificial I-425
local institutions, learning from III-400
loggerhead turtles II-205, II-574, III-63, III-355
— developmental migrations III-63
— Great Barrier Reef II-618
logging
— Alaska, leads to habitat degradation I-381, I-383
— Cambodia II-575
— effects of
  — Andaman and Nicobar Islands II-193, II-195
  — Coral, Solomon and Bismark Seas region II-435–6
— Gulf of Papua II-607
  — probably a threat to coastal and marine environments II-604
Lombok Strait *II-382*, II-383
— tanker traffic II-390
London Dumping Convention II-524, III-338, III-365
— 1996 Protocol III-339
— evolution of III-338–9
— greater prohibition of dumping III-339
— need for more comprehensive perspective on waste management III-339
— permitted amounts of dredged materials **III-365**
Long Bay *I-352*
Long Beach *I-322*
long-distance transport of pollutants
— atmospheric
  — Baltic Sea I-132
  — lead in the southern Baltic I-116, I-214
  — one-hop or multi-hop pathways I-13
— to the Sargasso Sea I-229
Long-Range Transboundary Air Pollution Convention III-340
longline fisheries
— incidental capture of sea turtles a problem for III-146
— Madagascar II-122
— and seabirds III-145–6
— some important by-catch III-125
longshore drift, Belize I-510
longshore sand transport, Israeli coast I-256
Looe Key National Marine Sanctuary I-412
*Lophelia* banks III-379
Lord Howe Island II-635, II-636, II-643

Louisade Archipelago *II-426*
Louisade Plateau *II-426*
Louisiana–Texas Slope and Plateau *I-436*
low energy environments
— biological effects aligned with oil persistence III-275–6
— oil dynamics III-271
lowland evergreen forest, Cambodia II-573–4
Lüderitz upwelling cell *I-822*, I-825
— eddies, filaments and superfilaments I-826
— separates Northern Benguela from Southern Benguela I-823
Lyme Bay *I-66*

Maamorilik, Greenland *I-6*
— sources of lead and zinc pollution identified I-12
Macao *II-550*
mackerel I-49, I-313
— Baltic Sea I-111
Macquarie Harbour, Tasmania, affected by mining pollution II-654
Macquarie Island II-583
macroalgae III-3
— the Bahamas I-429
— Baltic Proper I-109
— benthic I-70
— Baltic, changes in I-109–10
— Carolinas coast
— colonization limited I-357
— offshore I-362
— rich communities occur on rocky outcrops I-357
— a transition zone I-356–7
— Colombian Caribbean Coast I-668–9
— communities, Gulf of Aden II-52
— eutrophication causing community changes I-130–1
— Great Barrier Reef region II-615
— increase in Belize reefs I-505
marine I-490
— northern Patagonia I-755
— Norwegian coast I-21
— red, disappearance from Wadden Sea creeks I-48–9
— reef habitats, Tasmania II-652
— southern Arabia, luxuriant during the Southwest Monsoon *II-22*, II-23
— spring bloom, Gulf of Aden II-50, II-54
— Western Australia II-695, II-701
— habitat alienation and fragmentation II-699
— habitat loss II-699
— western Sumatra II-387
macroalgal blooms I-666
macroalgal mats I-53
macrobenthic assemblages/fauna
— intertidal sedimentary areas I-47
— North Sea I-48
macrobenthos II-277
— southern Yuzhnoprimorsky region, reduction in density of II-481
*Macrocystis pyrifera*, adjacent to *Lessonia* and*Phyllospora* habitats, Tasmania II-652
macrofauna
— intertidal zone, English Channel I-71
— North Sea, northern and southern species I-48
— Portuguese coastal waters I-159
— some limits on life-span and size I-159
macromolluscs, Chagos Islands II-227–8
macrophytes
— free floating, Côte d'Ivoire reservoirs and rivers I-818
— growth restricted, the Bodden III-261
— replaced by nuisance algal species I-53
Madagascar II-101, II-113–31, *II-244*
— coastal erosion and landfill II-124–5
— effects from urban and industrial activities II-125
— islets and islands II-119
— major shallow water marine and coastal habitats II-117–19
— offshore systems II-120
— populations affecting the area II-120–3
— protective measures II-125–9
— Environmental and other legislation II-126–7
— recommendations and prognosis II-129
— rural factors II-123–4
— seasonality, currents, natural environmental variables II-115–17
Madagascar Current *II-114*
Madeira Current I-204, *I-204*
maërl I-69–70
— commercial exploitation I-70
maërl vegetation I-69–70, I-140
Mafia Island, Tanzania *II-84*, II-85
— Multi User Marine Park II-86, II-94
— unsustainable octopus harvesting II-90–1
Magdalena River
— discharges fertilize the Colombian Caribbean I-669
— mangroves in the delta I-671
— rehabilitation project I-671
— sediment movement from I-667
Magdalena River Basin, Colombia I-665–6
Magellan Strait I-753, I-762
magellanic penguin, affected by oil spills I-765–6
Mai Po marshes, Hong Kong II-539–41
— Deep Bay area
— conversion of mangroves to fishponds II-539
— habitat loss a threat II-540
— over-wintering site for cormorants II-539
— shrimp ponds (Gei Wais) II-540
— mudflats II-539
— protective measures II-540
— a RAMSAR Site II-539
— threats to II-540
— Site of Special Scientific Interest II-539
Maine, Gulf of and Georges Bank I-307–20, I-330
— boreal waters I-309
— circulation of the Gulf I-309, I-310, *I-311*
— nontidal surface circulation I-324, *I-324*
— continual addition of new species I-312
— early settlement and development I-313, I-314
— effects from urban and industrial activities I-314–17
— forestland, development of I-313
— growth of environmental regulation I-313–14
— Gulf of Mine Habitat Workshop I-318
— hydrological regions I-310
— major shallow water marine and coastal habitats I-310–12
— natural environmental variables: currents, tides, waves and nutrients I-310
— offshore systems I-312
— population affecting the area I-312–14
— principal basins I-309
— urban areas I-313
— Working Group on Human Induced Biological Change I-318
— workshops and Proceedings I-309
— primary research goals and tasks identified **I-318**
Maine, Gulf of, cetacean–gillnet interaction III-144–5
MAIZCo (ICZM), Nicaragua I-526, I-528, I-534
Makaronesy archipelagoes, physical data **I-187**
Makassar Strait *II-382*, II-383
— tanker traffic II-390
Malacca Strait *II-310*, *II-332*, *II-382*
— diluted by river discharges II-333
— including Singapore and Johore Straits II-331–44
— coastal population II-337
— general environmental setting II-333–4
— impact of human activities II-337–40
— protective measures and sustainable use II-341–2
— water quality II-340–1
— main problems and issues II-341–2
— marine and coastal habitats II-335–7
— organotins II-449
— pre-European trading route II-337

— tanker traffic II-390
Malacca Straits II-309–29
— conclusions and recommendations II-325–7
— critical environmental problems II-317–21
— natural environmental conditions II-311–14
  — climatology and oceanography II-311–12
  — coastal and marine ecosystems II-312–14
  — geography II-311
  — populations affecting the area II-311
  — topography II-311
— ports, trade and navigation II-316–17
— protective measures II-321–5
  — coordination of the management of the Straits II-325
  — international legal regime governing the Straits II-324–5
  — national management systems and legislations II-321–4
— ratification by littoral states of international conventions **II-324**
— resource exploitation, utilization and conflicts II-314–17
— total net economic value, marine and coastal resources **II-326**, II-327
Malacca Straits Demonstration Project (MSDP) II-325
Malacca Straits Strategic Environment Management Plan needed II-327
malaria control I-476
Malaysia II-299
— approaches to prevent and control marine pollution II-323
— coral bleaching and mortality events III-51
— Kuala Selangor Nature Park II-342
— legislation on environmental protection II-323
— mangroves II-312
— Matang managed mangrove forest III-25
— Matang Mangrove Forest Reserve II-342
— move to fisheries management II-341
— National Oil Spill Contingency Plan II-323
— Peninsular, coastal plains and basins on west coast II-311
— regulation
  — of land-based pollution II-323
  — of toxic and hazardous wastes II-323
  — of vessel-related marine pollution II-323
— swine farming source of agricultural waste II-318–19
— water quality monitoring programme II-340–1, **II-341**
Malaysia, Peninsular, East Coast of II-345–59
— coastal erosion and land reclamation II-353
— effects from urban and industrial activities II-353–4
— impact on habitats and communities II-354–5
— major shallow-water marine and coastal habitats II-348–50
— populations affecting the area II-350–1
— protective measures II-355–7
  — 1985 Fisheries Act II-355–6
  — coastal erosion II-356–7
  — prognosis II-357
  — turtle hatcheries II-356
— rural factors II-351–3
— seasonality, currents, natural variables II-347–8
— the shallow seas II-350
Malaysia and the Philippines, coral bleaching and mortality events III-52–3
The Maldives II-199–219
— coastal erosion and landfill II-208–10
— effects from urban and industrial activities II-210–15
— major shallow water marine and coastal habitats II-202–5
— offshore systems II-205
— part of Lakshadweep (Laccadive)—Maldives—Chagos Ridge II-201
— populations affecting the area II-206–7
— protective measures II-215–18
  — carrying capacity, sustainability and future prospects II-217
  — environmental legislation and related measures **II-215**, II-216
  — environmental restoration II-217
  — initiation of Protected Area system II-216–17
  — multidisciplinary Environmental Impact Assessment (EIA) II-216
  — national and regional development plans II-216
— rural factors II-207
— seasonality, currents, natural environmental variables II-201–2
Malé Declaration II-216
Maluan Bay, Xiamen region
— chemical oxygen demand (COD) II-522
— ecosystem changes II-524–5
Malvinas/Falkland Islands *I-750*, I-751, I-752, I-753, I-766–7
— fishing industry I-763
— squid fishery I-763
mammals
— North Sea I-49–50
— as pests, Marshall Islands II-778
man
— and changes in marine mammals and seabirds I-9
— interference with river flows I-365
Man and Biosphere Programme (UNESCO), research project, Moorea and Takapoto reefs II-817
managed coastline retreat I-54, I-79, III-353
management I-3–4
management practices, customary, Madagascan marine and coastal resources II-126
management problems, Xiamen region II-526–8
— conflicting uses in marine waters II-526–7
— lack of knowledge and information II-527
— new management measures II-528–9
— transboundary problems II-527–8
management structures and policy, South Africa
— advisory bodies II-143
— legislation II-143
— principles II-143
Managua, Lake *I-532*, I-533
manatee grass I-601
manatees I-505, I-509, I-591, I-631
mandated science, fisheries management as III-156
mangrove communities, Borneo, mixed, extending up river valleys II-367
mangrove crab culture, sustainable community aquaculture II-375
mangrove deforestation II-438
— Mozambique II-109
— promotes coastal erosion II-108
— Torres Strait and Gulf of Papua II-605
mangrove destruction, West Guangdong II-559
mangrove ecosystem management, sustainable basis for needed, Malacca Strait II-341–2
mangrove ecosystems
— Godavari-Krishna delta II-170
  — impact of deforestation and prawn seed collection II-172
  — indiscriminate exploitation II-171
— interaction with other ecosystems III-18, III-24
— need to improve sustainable use III-29
— Palk Bay–Madras coast II-168–9, II-170
— rehabilitation of III-26–7
  — concern for the human factor III-28
  — criteria and practical considerations III-27
  — goals III-27
  — need for III-26–7
  — replanting programmes III-27
— species poor, but support biodiversity III-24
mangrove forests
— Borneo
  — distribution factors II-366–7
  — successional communities on accreting shores II-367
— Brazilian tropical coast I-722, I-727
— Colombian Pacific Coast I-679–80
  — exploitation of I-680, I-681
  — fisheries in I-682
  — uses of I-682–3
— Gulf of Papua II-597–8
  — lack low salinity species II-598
  — mangrove tree species, Fly River delta II-597
  — pristine II-597
— Indian Sundarbans II-147

— managed, Sundarbans the first II-148
— Palk Bay, areas cleared for salt pans II-167–8
— Palk Bay–Madras coast, reduced by human activity II-169
— the Philippines
— loss of and loss of coastal productivity II-414
— provide nursery grounds II-408
— Sundarbans
— home of the Bengal tiger II-150
— under serious threat II-151
— zoned on the tidal flats II-149–50
— Thailand, use for shrimp farming III-175–6
— western Indonesia, reduced by clearing II-385, II-388
mangrove logging, Cambodia II-575
mangrove palm, Gulf of Guinea, a significant problem I-786
mangrove shrimp ponds III-21–2, III-25
— effects of acid release III-25–6
mangroves III-17–32
— Abrolhos Bank–Cabo Frio, Southern Brazil I-735, I-736
— aerial roots III-18, *III-19*
— American Samoa
— limited occurrence II-768
— loss of II-768
— Andaman and Nicobar Islands II-192
— conservation activities initiated II-195–6
— degraded sites, restoration of II-196
— Australia II-582
— the Bahamas I-421–2
— typical zonation I-421–2
— Bangladesh II-290–1
— land area changes and biodiversity II-290–1
— *see also* Sundarbans
— Bay of Bengal II-277
— depletion in Orissa II-279
— Belize I-506
— some clearance I-510
— biodiversity and human communities III-27–8
— Borneo II-366–70
— clearing for shrimp farms II-368–9
— fauna II-368
— non-conversion uses II-369
— for wood chip industry II-369
— and braided channels *II-71*, II-76–7
— low-energy intertidal environment II-76–7
— British Virgin Islands I-619
— Cabo Frio–Cabo de Santa Marta Grande, southern Brazil
— degraded I-742–3
— estuarine–lagoon complex I-736–7
— lost to urban development I-743
— seriously degraded I-736
— Central America, critical coastal habitat I-460
— clearance for aquaculture III-368
— coastal lagoons, Côte d'Ivoire I-818
— Colombian Caribbean Coast I-669
— rehabilitation important I-669
— Coral, Solomon and Bismark Seas region II-430, II-438
— species diversity II-431
— Côte d'Ivoire I-810–11
— degradation of, the Comoros II-248
— development limited, northwest Arabian Sea and Gulf of Oman II-23
— distribution *III-18*
— patterns of III-20
— Dutch Antilles I-600–1
— east coast, Peninsular Malaysia II-348
— destroyed by development, Pulau Redang II-353
— estuaries, lagoons and mainland II-349
— rate of destruction alarming II-354
— eastern Australian region II-632–3
— loss of II-639
— ecological values III-22–4
— importance to fish populations III-23
— mangrove litter III-23
— productivity III-22–3
— stabilisation of exposed land III-23–4
— El Salvador I-548, I-553
— Fiji Islands II-755
— cleared for development II-760
— the future III-38–9
— Guinea, deforestation I-801
— Gulf of Aden II-53
— Gulf Coast, USA I-440
— Gulf of Guinea I-779–80
— Ghana I-779
— importance of I-779–80
— over-exploited I-786
— Gulf of Mannar II-164
— Gulf of Thailand II-301
— Hawaiian Islands II-797, II-802
— Iran, Saudi Arabia and Bahrain II-7
— Jamaica
— Black River Morass I-565
— Negril Morass I-564
— killed by hypersalinity II-135
— Koh Kong Bay, Cambodia
— may now be seriously degraded II-576
— now cleared for shrimp farming II-577
— pristine II-573
— Lesser Antilles I-631, I-632, I-637
— long term oil retention in sediments III-271
— Madagascar II-118–19
— change in cover II-118–19, II-123
— exploitation of II-125
— harvesting regulations II-126
— majority on the west coast II-118
— Magdalena River Delta I-671
— Malacca Strait II-335–6
— conversion for aquaculture no longer valid II-342
— Malacca Straits II-312
— a natural resource being lost II-314–15
— Malaysia, loss through land reclamation II-312
— The Maldives II-204
— high species richness II-204
— Marshall Islands II-777, II-778, II-779
— Mauritius II-258
— Mayotte Island II-246
— Mexican Pacific coast I-488, I-492
— need for easier availability of existing knowledge III-28
— negative effects of oil spills III-273
— New Caledonia II-726
— Nicaraguan Caribbean coast I-522
— Nicaraguan Pacific coast I-535
— contamination and degradation of the ecosystem I-534
— protected areas I-535
— reduction in, Gulf of Fonseca I-538
— northeastern Australia II-614–15
— reclamation and draining threats II-615
— and other habitats, Tanzania II-87–8
— mangrove harvesting II-89–90
— restoration of, Dar es Salaam II-89
— use of mangrove timber **II-88**
— patterns of use III-24–6
— benefits from mangroves **I-25**
— misuse through international agencies III-26
— pressures for change III-24–5
— recreation and ecotourism III-26
— shrimp aquaculture III-21–2, III-25
— Peru I-691–2
— the Philippines II-409
— source of fishery and forest products II-409
— present extent and loss III-19–22
— areal statistics III-20, **III-20**
— causes of loss **III-21**

— eastern and western groups III-19–20
— loss to shrimp farming III-21–2
— problems of human pressure III-18–19
— Puerto Rico I-582
— clearance of, effects I-584
— the Quirimbas II-102–3
— red, Gulf of Mexico I-472–3
— Red Sea II-40
— hard-bottom (reef) and soft-bottom mangals II-41
— rehabilitation of key system, Colombia I-463
— remote sensing of III-285–6
— detection of change in resources III-285
— future challenges III-286
— mangrove leaf area index (LAI) III-285–6
— role in sediment stabilisation II-414
— Sinai Peninsula II-41, *II-41*
— Somalia II-71, *II-71*, II-77
— South Western Pacific Islands II-710
— spawning and nursery grounds II-335
— Sri Lanka II-179
— damaged by shrimp aquaculture II-179
— Sumatra, zonation depending on tidal regime II-312
— Turks and Caicos Islands I-589–90
— harvesting of mangroves I-592
— Vanuatu, species diversity II-740
— Venezuela
— conversion to shrimp ponds I-655
— mainly in Orinoco Delta and Paria Gulf I-651
— Victoria Province, Australia II-665
— Vietnam II-564, II-565
— effects of destruction of II-566
— west coast Taiwan II-503
— Western Australia II-695
— western Indonesia II-385–8, **II-388**
— western Sumatra II-387
— very productive II-388
— Xiamen region II-518
Mangueira Lagoon, southern Brazil *I-732*, I-737
Mannar, Gulf of II-163–6
— climate and coastal hydrography II-163
— fish and fisheries II-164–5
— human population and environmental degradation II-165
— islands affected by new harbour at Tuticorin II-165
— impact of human activities on the ecosystem II-165–6
— effects of industries and the power station II-165
— main marine ecosystems II-163–4
— National Marine Park and a Marine Biosphere Reserve II-163
Manning Shelf Bioregion, Australia II-634
Manus Basin *II-426*
Manus Trench *II-426*
Maracaibo Lake *I-644*, I-646
— ecosystem under extreme pressure I-656
— population round I-654
Maria Island, Tasmania
— changes in oceanographic climate recorded II-651
— Marine and Estuarine Protected Area II-658–9
mariculture I-52, I-69, I-319, I-362, II-314, II-339
— American Samoa II-770
— culturing of mangrove oysters I-569
— Gulf of Thailand II-306
— Hawaiian Islands, research and development II-803
— Kuwait II-8–9
— northern Gulf of Mexico I-448
— potential for, eastern Russia II-481, II-482
— Taiwan II-505
— Tanzania II-91–2
— Xiamen region II-519, **II-520**
marinas, sources of contaminants I-72
marinculture, Guangdong, main methods II-558
marine antifoulants III-247–56
— biocide-free 'non-stick' coatings, for the future? III-254
— effectiveness of regulations: measuring and monitoring TBT in the environment III-253–4
— leaching during normal operations III-364
— new self-polishing copolmer paints III-248
— organotin-based paints banned on smaller boats III-251
— TBT
— ban on, finding safe alternatives III-254–5
— environmental impacts of III-250–1
— persistence of in the environment III-249–50
— TBT-based, legislative control of III-251, III-253
— regulations have reduced contamination III-253
— *see also* tributyltin (TBT)
marine biota
— Azores I-203
— Chilean I-707
— Peruvian Province and Magellanic Province I-707
— Southern California Bight, effects of anthropogenic activities I-394–9
marine birds
— Patagonian coast, Argentina I-756, I-757
— breeding **I-756**
marine circulation
— Arabian Gulf II-4
— development of two-gyre system
— Arabian Gulf II-4
— Gulf of Aden II-50
— Somali Indian Sea Coast II-66
marine climate data, accessibility of III-216
marine and coastal communities, Côte d'Ivoire, affected by seasonal factors I-809–10
marine coastal and estuarine ecosystems, trophic status categories III-258
marine and coastal habitats
— Baltic Sea I-124–8
— biodiversity of I-124–5
— hard-bottom communities I-125–6
— pelagic communities I-126, I-128
— soft-bottom communities I-126
— Chesapeake Bay I-341–3
— continental seas, western Indonesia II-385–90
— English channel coast I-68–9
— loss of coastal habitats I-74
— protective and remediation measures I-77–9
— French channel coast I-69
— Irish Sea
— intertidal habitats I-87–9
— sub-tidal habitats I-89
— Lesser Antilles, Trinidad and Tobago I-631–2
— Malacca Strait II-335–7
— offshore, English Channel I-69
— and offshore systems, Black Sea I-290–4
— Sargasso Sea and Bermuda Islands I-227–8
— coral reef zones I-227–8
— two inshore nutrient zones I-228
— shallow
— Adriatic Sea I-272–3
— Tyrrhenian Sea I-273–4
— shallow, Dutch Antilles I-598–601
— coastal wilderness I-601
— coral reefs/reefal algal beds I-599–600
— mangroves I-600–1
— salinas I-601
— seagrass beds I-601
— shallow, southern Spain I-170–2
— Eastern sector I-172
— Western sector I-170–2
— Western sector described I-170–2
— shallow water
— Australia II-582
— the Azores I-205–9
— the Bahamas I-419–22

— Bay of Bengal II-149–51
— Belize I-504–6
— Chagos Archipelago II-226–9
— Colombian Pacific Coast I-679–80
— the Comoros II-245–6
— Coral, Solomon and Bismark Seas region II-429–31
— Côte d'Ivoire I-810–13
— east coast, Peninsular Malaysia II-348–50
— eastern Australian region II-632–5
— El Salvador I-548–51
— Fiji Islands II-754–8
— French Polynesia II-817–19
— Great Australian Bight II-678–81
— Great Barrier Reef region II-614–19
— Guinea I-799–800
— Gulf of Aden II-51–4
— Gulf of Alaska I-375
— Gulf of Guinea I-779–81
— Gulf of Maine and Georges Bank I-310–12
— Hawaiian Islands II-795–800
— Hong Kong II-537–8
— Jamaica I-561–5
— Marshall Islands II-777–9
— Mascarene Region II-256–9
— Mozambique II-102–5
— New Caledonia II-725–6
— New York Bight I-325–7
— Nicaraguan Caribbean coast I-520–2
— Nicaraguan Pacific coast I-534–5
— North Sea I-47–50
— northern Gulf of Mexico I-440–2
— Oman 21–4
— the Philippines II-408–10
— Red Sea II-40–2
— Sea of Okhotsk II-167–8
— the Seychelles II-236–7
— Somali Indian Ocean coast II-72–7, *II-78*
— South Western Pacific Islands II-710–12
— southeast South American shelf marine ecosystem I-754–7
— southern Brazil I-735–7
— Southern Gulf of Mexico I-471–4
— southwestern Africa I-828–30
— Sri Lanka II-178–80
— Taiwan Strait II-502–3
— Tanzania II-85–8
— Tasmanian region II-651–3
— The Maldives II-202–5
— Torres Strait and Gulf of Papua II-597–8
— Turks and Caicos Islands I-589–91
— Vanuatu II-740–2
— Venezuela I-648–53
— Victoria Province, Australia II-664–6
— Western Australia II-695–6
— Xiamen region II-517–18
— Spanish north coast I-139–40
— intertidal rocky communities I-140
— soft-bottom communities I-139–40
— subtidal rocky communities I-140
— wetlands and marshes I-139
— Vietnam II-564–5
— littoral habitat II-564–5
marine and coastal protected areas, Madagascar II-127
marine communities
— changes in structure anddiversity III-127
— effects of eutrophication on III-259–60
— Gulf of Alaska, trophic shift in I-376–8
— ocean/climate variability I-375
— major increase in cod and ground fish I-375–6
— population changes of shrimp and forage fish I-375
marine conservation, Vanuatu, traditional and modern practices II-745–6, **II-746, II-747**
marine conservation areas, Borneo II-376–8
marine conservation and resource management, scope of III-376
marine ecosystem health
— concept III-213
— marine epidemiological model III-213
— tracking of HABs III-213
marine ecosystem health as an expression of morbidity, mortality and disease events III-211–34
— basis for assessing the health of large marine ecosystems III-230–1
— categories of disturbance III-217–28
— anoxic/hypoxic disturbances III-219–21
— biotoxins and exposure disturbances III-218–19, *III-220*, **III-220**
— disease disturbances III-225–6
— keystone-endangered and chronic cyclicaldisturbances III-227–8
— mass lethal mortality disturbances III-223–4
— new, novel occurrences and invasive disturbances III-226–7
— physically forced (climate/oceanographic) disturbances III-224–5
— trophic-magnification disturbances III-221–2
— data assimilation methods III-215–17
— disturbance type derivation III-217
— Gulf of Mexico Aquatic Mortality Network III-228
— HEED approach III-214
— network for developing standards and achieving consensus III-229–30
— survey methods III-214–15
marine ecosystems
— Canary Islands I-193–5
— *Cymodocea–Caulerpa* communities I-195
— deepwater species rise at night I-194
— mesolittoral area I-194
— pelagic system I-193–4
— rocky and sandy seabeds I-195
— sublittoral area I-194
— supralittoral area I-194
— and coastal habitats, shallow water, Colombian Caribbean Coast I-666–9
— Gulf of Mannar II-163–4
— long-term concerns about direct and indirect effects of fishing practices III-352
— rocky strata I-194–5
— sustainability of human activities on III-359–73
— exploitation of ecological resources III-366–9
— global commons III-360
— global trends III-369–71
— introduction of hazardous substances and radioisotopes to the marine environment III-360–6
— towards sustainability III-371
— western Sumatra II-386–7
— disturbance and stress II-387
marine environment
— annual input of petroleum hydrocarbons **III-268**
— global legal instruments (at year 2000) III-331–48
— Antarctic regime III-342
— from present to future III-342–7
— Law of the Sea III-332, III-336
— marine organisms III-340–1
— marine pollution III-336–40
— protection from military activities III-342
— "soft law" instruments III-339, III-344
— taking stock III-332, **III-333–5**
— introduction of elevated nutrient loads III-366
— introduction of hazardous substances and radioisotopes to III-360–6
— contminants from shipping III-364–5
— dumping of wastes at sea III-365–6
— land-based sources III-361–3
— operational discharges from the offshore oil and gas sector III-363–4
— protection of from military activities III-342

— bilateral agreements III-342
— regional protection programmes III-363
marine environment strategies, Australia II-590
Marine Environmental Act, Faroes I-38
marine environmental agreements, global
— instruments and their institutional arrangement III-344–5
— move away from binding commitments leads to increased vulnerability III-345
marine environmental protection, legislative power of national governments I-57–9
Marine Environmental Protection Committee (MEPT: IMO), ban on TBT proposed III-365
marine environmental restoration, Hawaiian Islands II-809
marine environmental science, multinational training programmes in III-323–30
— CC:TRAIN III-327–8
— International Ocean Institute Training Programme III-328–9
— synthesis III-329–30
— TRAIN-COAST-SEA III-325–7
— UN training programmes, TRAIN-X strategy III-324–5
marine epidemiological information system, information flows *III-229*
Marine and Estuarine Protected Areas (MPAs), Tasmania II-658
marine fishery reserves I-526
marine habitats, smothered I-383
marine habitats, Brazilian tropical coast I-722–5
— bays I-725
— continental shelf I-722–3
— coral reefs 723–4
marine habitats, shallow
— the Bahamas I-421–2
— coral reefs I-421
— mangroves I-421–2
— seagrass I-421
— Israeli shelf, Red Sea invaders I-256–8
— Tagus and Sado coastal waters I-157–8
— marine biological resources I-157–8
— plankton I-157
— Tagus salt marshes I-157
marine harvesting, attempts to limit impacts of III-378
marine mammals
— abnormalities and perturbations due to toxic chemicals II-456
— accumulate toxins in blubber I-91
— Aegean Sea I-239
— Africa, southwestern I-834
— Alaska, food-limited I-377
— Australian Bight II-680
— Baltic Proper I-109
— Bazaruto Islands II-104–5
— Chesapeake Bay I-343
— and coastal mammals, El Salvador I-549
— contamination by and bioaccumulation of persistent organic organochlorines II-455–6
— Great Barrier Reef II-618–19
— hunted in Greenland I-8–9
— importance of polynyas to I-7
— Jamaica I-561–2
— little affected by radionuclides III-96
— loss of, Java Sea II-364
— Maldives II-205
— Marshall Islands II-779, II-780
— Patagonian coast, Argentina I-756–7
— breeding **I-756**
— the Philippines II-410
— Portuguese coastal waters I-159
— and reptiles, Hawaiian Islands II-800, **II-800**
— small, west coast, Sarawak II-368
— Southern California, effects of anthropogenic activities I-399
— Sri Lanka II-180
— susceptible to chemicals accumulation III-361
— threatened in Hong Kong waters II-542
marine monitoring, Southern California Bight, unable to assist environmental management I-400
Marine National Parks
— Andabar and Nicobar Islands **II-196**
— Madagascar II-127
— Mozambique
— Bazaruto, dugongs and turtles II-110
— Inhaca and Portuguese Islands II-110
— the Quirimbas may be next II-110
— the Seychelles II-239–40, **II-240**
— Curieuse and Sainte Anne II-240, II-241
marine nature reserves
— French Polynesia II-824
— Skomer Island *I-84*, I-96
— Strangford Lough *I-84*, I-96
marine organisms
— common, eastern coast of Taiwan II-508
— harmful effects of increased exposure to UV-B III-370
— legal instrument **III-334–5**
— migration circuit concept III-378
— protection of III-340–1
— TBT, accumulation of III-250
— TBT, toxic to III-250
— Yuzhnoprimorsky region
Marine Park of the North Sporades *I-234*, I-250
Marine Parks
— east coast, Peninsular Malaysia II-356, II-357
— Italy
— Adriatic coast I-280
— Tyrrhenian coast I-280
— Jamaica I-571
— Tanzania II-86, II-94
— further park proposed for Mnazi Bay, Mtwara II-94, II-95
— western Indonesia, zoning schemes II-401
marine plants, Aegean Sea I-239
marine pollution III-336–40
— atmospheric pollution III-340
— Australia II-588–9
— Bay of Bengal II-280
— by persistent organic organochlorines II-455–6
— dismantling of ships III-339
— dumping III-338–9
— from present to future III-3427
— learning-based approach to global codification III-346–7
— from wastes and sewage, Vanuatu II-747–8
— Hong Kong, land-based origin II-542
— Israeli legislation against I-261–2
— Java Sea II-398
— land-based III-339–40
— legal instruments **III-333–4**
— Malacca Strait II-339
— Malacca Straits, sea-based sources II-319–20
— oil and chemical spills II-319–20
— oily discharges II-319
— TBT II-319
— Mozambique II-109
— North Sea, changes in approaching control of I-59
— regulation of land-based sources, Sumatra II-322
— risk of, Lesser Antilles I-638
— sea-based, Indonesian legislation II-322
— seabed activities: peaceful exploration and exploitation III-336, III-338
— shipping III-336
— South China Sea II-354
— Sri Lanka, land-based sources II-184–5
— western Indonesia II-396–8
— Xiamen region
— management of II-532
— monitoring of II-531
marine pollution parameters
— West Guangdong coast II-554–5

— conventional water quality parameters II-555
— pollution sources and waste products II-554–5
— trace toxic organic contaminants in the Pearl River II-555–6
marine populations, dispersal patterns III-380–1
— early life dispersal distance III-381
— larval stage duration III-381
— potential dispersal ranges III-381
marine protected areas
— the Azores I-214, **I-215**
— proposed **I-216**
— the Comoros II-251
— Djibouti II-55
— east coast of Sumatra **II-322**
— insufficient, Strait of Malacca II-342
— Irish Sea, slow process I-96
— a limited success rate, Tanzania II-93–4
— Sweden I-132
Marine Protected Areas (MPAs)
— American Samoa II-771, **II-771**
— Anguilla and British Virgin Islands I-625
— Australia II-590, II-686
— Great Australian Bight Marine Park II-686–8
— Belize **I-513**, I-514
— Borneo, management objectives for II-377
— classification of III-377
— eastern Australian region II-643
— Fiji Islands, proposed **II-761**
— Hawaiian Islands **II-803**, II-808–9
— need to extend coverage II-810
— healthy communities of endangered species at risk from severe storms III-224
— Lesser Antilles I-639–40, **I-639**
— Mexico I-497
— New Caledonia II-734, **II-734**
— Papua New Guinea II-441
— the Philippines II-419
— restrict fishing effort III-161
— South West Pacific Islands II-719–20
— Terminos Lagoon, Southern Gulf of Mexico I-472
— Turks and Caicos Islands I-592–3
— Vanuatu II-746–7
— Victoria Province, Australia II-670
— western Indonesia II-401, **II-402**, II-403
— many only "paper parks" II-401
marine reptiles, Aegean Sea I-240
marine reserves
— assessing the effectiveness of III-384–9
— regional summaries of reserve implementation III-386
— within-reserve effects III-384–5
— demarcate boundaries
— simply and reliably III-383–4
— useful III-390
— design of III-380–4
— dispersal III-380–1
— edge effects III-383–4
— minimum viable biomass III-382–3
— number of III-384
— optimum size III-383
— reserve size and number III-381–2
— effectiveness depends on public acceptance, understanding and compliance III-390
— failures in resource management, reasons for III-378
— fine filter and coarse filter III-380
— framework for design of 'no-take' reserves and networks evolving III-376
— global distribution of *III-376*
— guidelines for the development of III-380
— Madagascar II-127
— Mayotte, Longogori Reserve (S passage) II-250
— need a robust approach to management III-377
— no-take zones III-376, III-377, III-378, III-386
— oceanographic setting crucial for egg and larval stages III-380–1
— placement in current/counter current and gyre systems III-381, *III-381*
— potential benefits of III-384
— potential to meet diverse management objectives III-376
— predicting the effects of III-386–9
— dynamic pool models III-387, III-389
— logistic models III-387
— spatial harvesting models (yield per recruit) III-387, III-388
— and resource management III-375–92
— closed areas and fisheries management III-378
— dealing with uncertainty III-389
— ecosystem effects of fishing III-378–80
— ecosystem management and the reserve concept III-377–8
— social and economic considerations III-389–91
— enforcement and compliance III-390
— user participation in management process III-390–1
— zoning III-389–90
— South Africa, Maputaland Marine Reserve II-140
marine resources
— Australia II-584
— exploitation of, Coral, Solomon and Bismark Seas region II-434
— Greenland
— fishery I-8
— hunting I-8–9
— Guangdong II-556
— utilised in the Bahamas **I-424**
marine scientists I-2
marine species
— and habitats, protection of, duplication or complementary **I-58**
— timing of reproduction and reproductive success II-51
— Xiamen coastal waters **II-517**
marine transgression, Kenya and Somalia II-67
Marine Turtle Specialist Group (IUCN) III-64–5
marine turtles, Fiji Islands II-757
marine zonation scheme, Xiamen region II-529–30, *II-530*, **II-530**
marine zoning III-389–90
— *see also* Marine Protected Areas (MPAs); marine reserves
Marmara Sea *I-234*, I-243
MARPOL II-324
— Gulf of Oman designated a 'Special Area' II-28, II-29
— Malacca Straits, possible designation as a 'Special Area' II-325
— special status for Wider Caribbean area I-640
MARPOL Convention I-38, I-55, I-147, I-248, I-262, I-451, II-658, III-343
— importance of III-337
— operational discharges regulated under III-364
marsh and estuarine systems, central Gulf of Mexico I-438
marsh vegetation, persistent oil effects III-276
Marshall Islands II-773–89
— changes in the country's sociopolitical status II-775
— coastal erosion and landfill II-782–3
— cultural and historical resources II-779
— degrading of Ebeye's natural resources II-783
— effects from urban and industrial activities II-783–4
— ENSOs II-776–7
— geographical gradients in physical features **II-776**
— key environmental and marine regulations **II-786**
— major shallow water marine and coastal habitats II-777–9
— need for greater level of protection II-785
— offshore systems II-779–80
— other environmental concerns II-784–5
— community and individual conflicts linked to land tenure and governance II-785
— conflicts from different user interests II-785
— greater incorporation of environmental concerns in development planning II-785
— limited understanding on cross-sectoral issues II-784
— overall assessment of environmental governance II-788
— populations affecting the area II-780
— protective measures II-786–8

— institutions II-786
— international legislation and regional programmes II-787
— land tenure and customary marine practices II-786
— national legislations II-786
— proposed environmental policies and strategies II-788
— responses to environmental issues: assessment and status II-787–8
— US Army, Kwajalein Atoll (USAKA) procedures and standards II-786–7
— rural factors II-781–2
— seasonality, currents, natural environmental variables II-775–6
— species, habitats and sites of conservational interest II-779
marshes
— coastal, Louisiana, managed by man I-448
— continental, Guadalquivir River I-170
— eastern Russia II-482
— *see also* coastal wetlands; saltmarsh; tidal marshes; wetlands
Martinique, effects of hurricanes I-631
Mascarene Anticyclone II-116
Mascarene Basin II-255
Mascarene Plateau II-255
— an obstacle to deep water flow II-255
— close to a tidal amphidrome II-256
— internal wave generation II-256
Mascarene Region II-116, II-253–68
— coastal erosion and landfill II-262
— effects from urban and industrial activities II-262–5
— artisanal and non-industrial uses of the coast II-262–3
— cities and sewage discharges II-264
— freshwater II-264–5
— light industry II-263
— sand mining and lime production II-263
— shipping, offshore accidents and impacts II-265
— tourism II-264
— major shallow water marine and coastal habitats II-256–9
— offshore systems II-259–60
— populations affecting the area II-260
— protective measures II-265–6
— rural factors II-260–2
— seasonality, currents, natural environmental variables II-255–6
Mascarene Ridge *II-254*
Masirah Island *II-18*
— some coral cover II-22
mass lethal mortality disturbances III-223–4
— many reports coincide with climate extremes III-223–4
mass mortalities, cause ecosystem collapse and reorganization round a new stable state III-223
"maszoperie" I-110
Matang Mangrove Forest Reserve, Malaysia II-342
Mauritius *II-254*, II-255
— agriculture II-260–1
— deforestation causes top soil loss II-260–1
— artisanal fishing II-261
— cities and sewage discharges II-264
— coral reefs II-258
— soft corals II-258
— spur and groove zone II-258
— eutrophication in the lagoons II-261
— Fishing Reserves II-265, **II-265**
— fishponds (barachois) II-262
— habitat diversity II-257–9
— mangroves II-258
— Nature Reserves II-265, **II-265**
— population II-260
— Round Island, free from introduces mammals and plans II-259
— sand mining and lime production, effects of II-263
— Southeast Trades drive most winds II-256
— stone-crushing plants, environmental effects of II-263
— sugar mills and textile plants II-263
— Terre Rouge Bird Sanctuary II-258
— tourism II-264
maximal/maximum sustainable catches/yields (MSY) III-75, III-81, III-118
Mayotte (Maore/Mahore) *II-244*
— coconut, bananas and ylang-ylang grown for export II-246
— Iris Bank II-246
— forest more luxuriant now II-247
— fringing reef II-246
— introduction of sewage treatment II-251
— lagoon a series of hydrologic basins II-248–9
— land-based pollutants and the lagoon II-251
— Marine Reserve of Longogori II-250
— reef threatened by silting II-248
— remained with France II-247
— seagrass beds II-248
— sedimentation in the lagoon affect fishing II-249
Mediterranean
— development of marine aquaculture III-169
— offshore wind energy development slower III-311
— receives water from the Black Sea I-22
Mediterranean Action Plan (MAP) I-249
Mediterranean Climate, Great Australian Bight II-675
Mediterranean fauna
— poor in animal species I-240
— zooplankton I-240–1
Mediterranean Sea III-205
Mediterranean Water (MW) I-139, I-153, I-169, I-205
MEDPOL National Monitoring Programme (Greece) I-249
meiobenthos II-277
Mellish Plateau *II-426*
Menai Strait *I-84*, I-96
mercury contamination
— Colombian Pacific coast I-683
— Haifa Bay I-260, I-263
— Kalimantan rivers II-396
— southern Brazil, east coast I-740
mercury (Hg) I-5, I-95, I-116, I-793, II-390
— accumulation in the Everglades I-444
— in Azorean seabirds, fish and cephalods I-212
— Baja California I-495, **I-496**
— in cinnabar mine tailings, the Philippines II-416
— evidence of increasing atmospheric concentrations I-14
— in Faroese pilot whale meat I-31, I-36, I-39
— in Greenland seabirds and mammals I-9, *I-10*
— low levels in Arabian Gulf II-12
— a multi-hop contaminant I-13
— in seabirds of German North Sea coast I-49
— as seed dressing I-635
— in surficial sediments, Gulf of Thailand II-305, **II-305**
Merka Formation II-68
Merka Red Dune Complex II-68–9, *II-70*
— potential for a national glass industry II-79
*MESA New York Bight Atlas Monograph Series* I-323, I-325
Meso America *see* Central America
Messina Strait I-273, I-281–2
— animal communities I-281
— Atlantic affinity species I-281
— endemic species I-282
— relict species I-281
— intense tidal currents I-281
Mestersvig, Greenland *I-6*
— pollution from lead—zinc mine I-12
metal biomagnification, absent in southern Baltic food chain I-116
metal pollution
— Baltic Sea I-114–16
— atmospheric and riverine fluxes I-115–16
— biota I-116
— Russian Far East II-476
— of surface waters II-477
— *see also* heavy metal pollution
metal sequestration, Tagus salt marshes I-158

metals
— English Channel
— generally low I-74–5
— higher near estuaries and inshore I-75
— entering North Sea from mining and industry I-54
— in land-based pollution III-361
methane
— emissions from the Bodden III-262
— increased in the atmosphere III-188
methane gas hydrates, Blake Plateau I-367
methylmercury I-114, III-95
— in Faroese pilot whale meat I-36
metropolitan areas, located near estuaries, problems associated with climate change and sea level rise III-194
*Metula* oil spill, Chile III-276
— asphalt formation III-270
Mexican Basin, and Sigsbee Deep I-442
Mexican Pacific coast
— case studies I-492–7
— Altata-Ensenda del Pabellón I-493
— Huizache y Caimanero lagoon system I-493–4
— Navachiste—San Ignacio—Macapule bays I-492–3
— Teacapán—Agua Brava lagoon system I-494–5, I-497
Mexican Pacific coastline I-485.**I-486**
Mexico
— coastal zone of Campeche, analysis of environment and its problems I-464
— coasts have ecological and socioeconomic problems I-464
— and Gulf of Mexico, coral bleaching and mortality events III-56
— Pacific coast, coral bleaching and mortality events III-53–4
Mexico, Gulf of
— growth of hypoxic zone III-219
— problem of incidental mortality of juvenile red snappers III-143–4
Mexico, northern Gulf I-435–56
coastal erosion and landfill I-445–8
— defined I-437–8
— eastern sector I-437
— effects from urban and industrial activities I-448–51
— interactions
— with North Atlantic Ocean I-437
— with waters and biota of the Caribbean I-437
— a large marine ecosystem I-437
— major shallow water marine and coastal habitats I-440–2
— offshore systems I-442–3
— populations affecting the area I-443
— rural factors I-443–5
— seasonality, currents, natural environmental variables I-438–40
— western and central sectors I-437
Mexico, Pacific coast I-483–9
— Californian Coastal Province I-486–7
— effects of urban development and industrial activities I-492–7
— case studies I-492–7
— North Pacific Ocean Province I-487–8
— Pacific Center Coastal Province I-488–9
— population I-489
— protective measures I-497–8
— included in international agreements I-497–8
— rural factors and fishing I-489–91
— Sea of Cortes Oceanic Province I-488
— Tropical South Pacific Ocean Province I-489
Mexico, Southern Gulf I-467–82
— coastal erosion and landfill I-476–7
— effects of human activities on natural processes I-478–9
— environmental framework I-469–71
— major shallow water marine and coastal habitats I-471–4
— offshore systems I-474–5
— populations affecting the area I-475–6
— protective measures I-479–80
— international agreements signed by Mexico **I-480**
— National Development Plan (1995–2000) I-479–80
— programs of sustainable regional development (Proders) I-480
— rural factors I-476
— urban and industrial activities I-477–8
micro-organics, sediment contamination by slight in the English Channel I-75
microalgae
— benthic I-357–8
— primary production I-357–8
— toxic I-356
microalgal blooms, increased due to due to seagrass die-off III-15
microbes
— oil degrading, in cyanobacterial mats II-12
— oxydation of organic material III-261
microbial diseases, and prawn mortality II-154
microbiological contamination
— English Channel
— impact on bathing water quality I-76
— impact on shellfish I-76
Micronesia, Marshall Islands a part of II-775
Micronesia, Federated States of, coral bleaching and mortality events III-52
microphytobenthos, North Sea I-48
Middle American Trench I-533
Middle Atlantic Bight I-323
— nontidal surface circulation *I-324*
Midway Atoll National Wildlife Refuge II-798
migration circuit concept, marine organisms III-378
migrations, seasonal, Okhotsk Sea II-466
military activities, protection from **III-335**
military usage
— French Polynesia II-822
— Hawaiian Islands II-804
— Marshall Islands II-783
Mindanao Current II-407
mine tailings
— copper mining, polluting Chilean coast I-713
— construction of large lagoon for tailings I-713
— effect of on the biodiversity of the seabed, Island Copper Mine, Canada III-235–46
— after mine closure sustainable ecological succession soon established III-245
— biodiversity in terms of higher taxa III-241–2
— crab fishery and yields III-244
— data set III-237–8
— habitat change since mine closure III-241
— large infaunal species III-242–3
— species evennness III-240–1
— species richness III-238–9
— tailings deposition levels affecting fauna III-239–40
— time- and cost-effectiveness of the benthos surveys III-243–4
— placement of for minimal and reversible environmental losses III-236
— risk of groundwater contamination I-710
mine tailings deposition, what is the tolerable rate? III-239–40
mineral extraction, areas of interest on Tyrrhenian sea floor I-279
mineral resources, Somalia II-79
mineral springs, west coast, Sea of Japan II-483
minerals
— coastal sands, Sri Lanka II-181
— Colombian Pacific coast I-683
— deep-sea deposits, Mascarene Basin II-260
— Godavari basin II-170
— Malacca Straits II-316
— Palk Bay–Madras coast II-168
— South Africa II-142
Minimata disease II-396, II-398
mining
— Alaska I-381
— Australia II-585
— Chile I-710
— lessons for the future, a case study I-713
— coastal, the Philippines II-415–16

— problems from cinnabar mines, Palawan II-415–16
— wastes and tailings serious threats to the marine environment II-415
— diamonds, Namibia I-835
— Great Barrier Reef hinterland II-622
— Greenland, and heavy metal pollution I-11–12
— Gulf of Papua II-607
— fate of sediments from II-603
— New Caledonia, and its effects II-731–2, *II-732*
— Nicaragua I-525–6, I-541
— Papua New Guinea II-439
— problem of mine discharges II-439
— tailings discharged direct to the sea II-439
— Peru I-695
— runoff from increasing sedimentation and turbidity I-444–5
— small-scale, the Philippines II-414–15
— discharge of untreated mine tailings II-414–15
— southern Spain
— Aznalcóllar disaster I-178
— problems of drainage from I-178
— Sumatra, surface and submarine II-395
— Tasmania, impact on coastal waters II-654
mining pollution I-710
minke whales III-78
— catch limits III-83
— counting of III-79–80
— number estimates III-82
— probably more than one biologically distinct population
Miskito Coast Marine Reserve, Nicaragua I-526
— corals I-520, *I-521*
— recommendations for **I-527**
Mississippi River *I-436*, I-443
— deltaic marshes I-440
— results of 1993 extreme flooding III-219
— silt and clay from I-437
*Mnemiopsis*, effects of introduction to the Black Sea I-290, I-304
Mogadishu Basin *II-67*
Moheli (Moili/Mwali), good soils II-246
molecular markers, used to identify contaminant sources I-392
mollusc aquaculture I-653
— Patagonia I-764
mollusc diversity, Sunda Shelf II-389
mollusc farming III-167, *III-168*
molluscan fauna, larger Norwegian fjords I-20–1
molluscs I-34, I-172, II-586
— carrying bacteria, Sri Lanka II-184
— culturing of II-278
— economically important, Colombian Pacific Coast I-684
— fishing for I-73
— lagoons, Gulf of Mexico I-473
— Nicaraguan Pacific coast I-534
— Spanish north coast I-139
— Tagus and Sado estuaries I-158
— Vietnam II-564
Monin—Obukhov theory 202
monitoring and enforcement, becoming easier on the high seas III-390
monitoring programmes, lacking, Yellow Sea II-496
monk seals
— Hawaiian Islands II-799, II-800
— Mediterranean Sea, decline in III-129
monsoons II-116, II-537
— affecting southeast India II-163, II-166
— affecting Tanzania II-85
— Bay of Bengal II-148
— causes high concentrations of heavy metals in Bay region II-155
— currents and gyres II-273–4, *II-274*
— Borneo II-365
— Cambodia II-571
— Coral, Solomon and Bismark Seas Region II-428
— east coast, Peninsular Malaysia II-347
— intermonsoon changeover period II-347
— winds, waves and currents II-347–8
— Gulf of Thailand, influence of II-300, II-301
— Indian system influences climate of northwestern Arabian Ocean II-19–20
— effects of Southwest Monsoon II-19
— influences in Malacca Straits II-311
— influencing northern Maldivian islands strongly II-201
— and the Intertropical Convergence Zone II-49
— and the Malacca Strait II-333
— the Philippines II-407
— South China Sea coast II-551
— and the Sri Lankan climate II-177
— Vietnam II-564
— winds affect northern and northwestern Madagascar II-117
— winter and summer, Arabian Gulf II-4
Mont Saint Michel Bay
— important bird location I-71
— marshes I-69
Montauk Point *I-322*, I-323
Montego Bay, Jamaica *I-560*
— reef restoration efforts I-546
Monterrey sardine I-488
Montserrat *I-616*, I-617, I-618, I-623
— climate I-617
— coral communities
— before volcanic activity I-618
— effects of volcanic activity I-618–19
— endemic birds I-620
— fishing I-621–2
— regulations not enforced I-624
— hurricane damage I-622
— loss of population due to volcanic activity I-620–1
— mangroves limited I-619
— seagrass beds I-618
— Sustainable Development Plan I-621
— volcanically active *I-616*, I-617
mooring system, British Virgin Islands I-624
moorings, the Bahamas I-431
morbilli virus *see* phocine distemper viruses
Morecambe Bay *I-84*
— shore communities I-88
Moreton Bay, Queensland
— environmental problems II-642
— high biodiversity *II-630*, II-634
— residential marinas II-639
Morondava, Madagascar, problems of land loss II-124–5
morphological abnormalities, indicator of system health III-225
Morrocoy National Park, Venezuela *I-644*
— much damage due to sedimentation I-655
— pollution estimates I-657–8
— subject to man-made disturbances I-655
Morrosquillo, Golfo de *I-664*, I-665, I-666
mother-of-pearl shell II-44
motor vehicles, emissions from I-345
Mozambique II-99–112
— coastal divisions II-101
— coastal erosion and landfill II-108
— effects from urban and industrial activities II-108–9
— high tidal range II-II-101
— major shallow water marine and coastal habitats II-102–5
— Bazaruto Islands II-104–5
— Inhaca and Portuguese Islands Reserves II-105
— Quirimba Archipelago II-102–4, II-245
— offshore systems II-105, II-107
— participation in International Conventions **II-105**
— population II-107
— protected by Madagascar II-101
— protective measures II-109–11
— Framework Environmental Law (1997) II-110
— rural factors II-107–8

— seasonality, currents, natural environmental variables II-101–2
— tropical humid to subhumid climate II-101
Mozambique Channel II-101, *II-114*, II-251
— carried high volume of crude oil traffic II-109
Mozambique Current II-101, *II-101*, *II-114*, II-245
Mozambique Gyre *II-101*, II-105
mucilaginous aggregates, effects of I-278
mud flats *see* soft shores
mud reefs I-583
muddy bottoms, Gulf of Papua II-597
mudflats
— and accreting mangroves, Perak and Selangor II-336
— extensive, Malacca Straits II-312
— Hong Kong II-539
Multiple Marine Ecological Disturbances (MMEDs) III-212
— episodic events and co-occurring anomalies III-214
— HEED approach III-214
— HEED database and GIS III-229–30
— indicators of decline in ecosystem health III-230
— observational reports, additional information for III-215–16
— pooling of co-occurring biological disturbance data, resulting evaluations III-216–17
— use of marine ecosystem health framework III-230–1, *III-230*
Multiple Marine Ecological Disturbances (MMEDs) program
— Health Ecological and Economic Dimensions (HEED) of III-215
— HEED system III-215, III-229
— scale in aggregation of anomaly indicators III-212
multiple-use management model, Great Barrier Reef Marine Park, criticism of II-616
Multivariate ENSO Index (MEI) I-703, *I-704*
munitions, dumping of in Beaufort's Dyke I-96
Murat Bioregion, Australian Bight II-678, II-679
Musandam *II-18*
Muscat *II-18*
— Qurm National Nature Reserve
— at risk from urban development II-30
— mangroves at risk II-28
mussel beds, protect sequestered oil from weathering III-271
mussel cultivation
— Albufeira lagoon I-156
— Galicia I-144
Mussel Watch Project I-452
— and contamination by organotins II-451, **II-453**
mussels I-34–5, I-51, I-290, III-167
— adapted to hypersaline conditions I-442
— Albufeira lagoon, purification needed I-158
— Island Copper Mine dock, bioaccumulation in III-244
— marine, Scope-for-Growth I-56
— metal contamination, Tagus estuary I-158
— response to oil spills III-273
— Venezuela I-649
Muthupet Lagoon, Palk Bay II-167
*Mytilus edulis platensis* I-754
— circalittoral banks, Buenos Aires Province I-755

N/P ratios
— Aegean Sea I-240
— Baltic Sea I-130
— trophic zone, Baltic Proper I-106
Namib Desert I-823
Namib Naukluft National Park I-835
Namibia
— claimed her EEZ at independence I-836
— diamonds I-835
— favourable factors for a sustainable fishery I-837
— the fishing industry I-830
— hake catches I-834
— Kudu gas field I-835
— Namib Naukluft National Park I-835
— other development and commercial activities I-837
— ports, Walvis Bay and Lüderitz I-835
— "sulphur eruptions" I-828
— sustainable harvesting of seals I-834
Nancowrie Biosphere Reserve, Andabar and Nicobar Islands II-196
Nares Strait I-5, *I-6*, I-13
narwhals III-76, III-90
Nassau Grouper fishery I-425
National Coastal Erosion Study, east coast, Peninsular Malaysia II-356
National Estuary Program (USA) I-452
National Marine Fisheries Service (NMFS; USA), publishes bycatch statistics III-159
National Marine, Protection, Research and Sanctuaries Act (USA) I-452
National Parks
— Colombian Pacific I-684
— Fiji Islands II-761–2
— Gulf of Cadiz I-182, **I-182**
— Lac Badana National Park, Somalia II-80
— Masoala National Park, Madagascar II-127
— Southern Gulf of Mexico, "Arrecife Alacranes" I-471
*Natura 2000* network I-52, I-77
natural capital depletion tax III-401
natural events, hazards and uncertainty, Marshall Islands II-776
natural hazards, Xiamen region II-516–17
natural resource depletion I-634
natural resource extraction, Alaska I-381
natural resource regions, Russian East Coast II-480–1
— Severoprimorsky region II-480
— Yuzhnoprimorsky region II-480–1
Nature Reserves, Mascarene region II-265, **II-265**
Nauru Basin *II-426*
Navachiste—San Ignacio—Macapule bays, Mexican Pacific coast I-492–3
— dams I-492
— mangroves I-492
navigation channels, problems created by I-447–8
Nazareth Bank *II-254*, II-255
Nazca Plate I-701
nearshore habitats, Hong Kong II-538
needs, concept of, in Third World I-459–60
Negril Morass, Jamaica I-564
— hummocky swamp I-564
nehrungen *see* sand barriers, southern Baltic coast
nematodes I-48
Netherlands, offshore wind energy production in III-308
neurotoxic shellfish poisoning (NSP) III-221
Neuse River Estuary *I-352*, I-366, III-205
— picoplanktonic cyanophytes I-355
— productivity pulses and algal blooms I-354
— zooplankton abundance and planktonic trophic transfer I-359
New Amsterdam anticyclone *see* Mascarene Anticyclone
New Britain Trench II-427
New Caledonia *II-706*, II-707, *II-708*, II-723–36, II-729
— coastal erosion and landfill II-731
— coastal modifications II-717
— coral reefs II-711
— effects from urban and industrial activities II-731–4
— the lagoon II-715
— major shallow water marine and coastal habitats II-725–6
— offshore systems II-726–7
— Marine Protected Areas (MPAs) II-719
— nickel mining II-729
— case study II-713
— populations affecting the area II-713–14, II-727–9
— activities affecting the sea II-728–9
— protective measures II-734–5
— Marine Protected Areas (MPAs) II-734, **II-734**
— reef-monitoring project II-734
— species-specific regulations II-734
— rural factors II-729–31

— seasonality, currents, natural environmental variables II-725
— subsistence reef fisheries II-715
— ZoNéCo Programme II-727
New Guinea Basin *II-426*
New Guinea Coastal Undercurrent II-428
New Ireland Basin *II-426*
new, novel occurrences and invasive disturbances III-226–7, III-230, III-259
New Providence Island *I-416*
— coastal erosion I-428–9
— perturbations over the last fifty years I-430
New River Estuary *I-352*, I-366
New South Wales
— central, major habitats II-635
— fisheries management system, gives security within adaptive management III-400
— share system conceptual framework relevant to other fisheries III-400–1
New York Bight I-321–33
— Bight restoration plans I-330–1
— dump sites in I-323
— major physical, hydrographic and chemical factors I-323–5
— major shallow water marine and coastal habitats I-325–7
— "new ways forward" I-331
— New York harbor
  — anthropogenic impacts I-327
  — clean up efforts I-327
— offshore systems I-327
— opposing uses I-331
— populations and conditions affecting the Bight I-328–9
— resources at risk I-329–31
— some habitat improvement I-330
New York Metropolitan area, growth of landfill and dumping of wastes I-328
New Zealand, no-take areas III-386
New Zealand Fur Seal II-681
Newfoundland, collapse of Atlantic cod fishery III-120
NGO activities, Jamaica I-572
NGO participation, increased due to "soft law" instruments III-344
Nicaragua
— central highlands, drainage of I-533–4
— coasts *I-533*
— cotton cultivation, after-effects of I-538
— government in I-538
  — laws awaiting approval I-541
— major economic and political trends I-540
— Pacific volcanic chain I-533
Nicaragua, Caribbean coast I-517–42
— coastal resource management need modification of law I-527–8
— effects from urban and industrial activities I-525–6
— geography of I-519
— ICZM I-526–8
— major shallow water marine and coastal habitats I-520–2
— natural reserves I-526
— offshore systems I-522–3
— population I-523
  — eastern slopes of the central highlands I-523
— protective measures I-526
— rural factors I-523–5
— seasonality, currents, natural variables I-519–20
Nicaragua, Lake *I-532*, I-533
Nicaragua, Pacific coast I-531–42
— effects from urban and industrial activities I-539–41
— geography/geology of I-533–4
— major shallow water marine and coastal habitats I-534–5
— marine resources used by Honduras and Salvador I-538
— offshore systems I-535–7
— populations affecting the area I-537–8
— protective measures I-541–2
— rural factors I-538–9
— seasonality, currents geography/geology of I-534
Nicaraguan Center for Hydrobiological Research, estimate of fisheries biomass I-525
nickel mining
— New Caledonia II-713, II-729
  — destructiveness of II-713, II-731
  — environmental effects of processing still largely unknown II-733
Nicobar Islands *II-190*, *II-270*
Niger, river and delta I-776–7
— oil a source of conflict I-788
— onshore oil production in the Delta I-788
— sand movement in the delta I-778
Nigeria, mangroves III-20
Nile Delta, retreat of I-258
nitrate concentrations, Maria Island II-651
nitrate contamination/pollution I-174
nitrate enrichment, euphotic layer, Côte d'Ivoire I-810
nitrate salts, mining of Chile I-710
nitrates I-255–6
— increased load to Baltic Sea I-130
— and seagrass and other aquatic vegetation I-358
nitrification III-262
nitrogen
— anthropogenic
  — in Baltic Proper I-104, *I-104*
  — emissions to the atmosphere III-199
  — load to Chesapeake Bay I-345
— increased load to Baltic Sea I-130
— load to Chesapeake Bay I-344–5
— organic, and phosphorus at depth in Bay of Bengal II-274
nitrogen concentration, Orinoco River I-647
nitrogen cycle, investigations of the impact of eutrophication on III-262–3
nitrogen fixation, important in Baltic waters I-128
nitrogen flux, riverine, increase in III-366
nitrous oxide, increased in the atmosphere III-188
no-take reserves III-376, III-377, III-378, III-386
— for fisheries management, coral reefs III-386
NOAA
— environmental survey baseline investigation, DWD 106 I-327
— National Status and Trends Program I-452
noise pollution, effect on marine mammals III-97
nomadic jellyfish, an immigrant I-157
Nord-Pas de Calais coast I-67
Nordostrundingen *I-6*
Norfolk Island II-635, II-636, II-643
— catastrophic soil erosion II-638
Normandy *I-66*
"nortes" I-489
North Aegean Trough I-235
North Atlantic, Sverdrup transport in I-579
North Atlantic Central Water, South of Finisterre Cape I-139
North Atlantic Current I-5, *I-18*, I-20, I-23, I-204, *I-204*, I-223
North Atlantic Deep Water I-579
North Atlantic Oscillation I-223
— and exceptional North Sea conditions I-45
North Atlantic water, southern Spain I-169
North Brazilian Current I-630, I-722
— retroflection of causes eddies I-630
North Carolina
— demersal marine zooplankton I-360
— eutrophication in some estuaries I-356
— toxic contamination I-360–1
North Channel *I-84*, I-85
North East Monitoring Program I-315
North Equatorial Counter Current I-630, II-407, II-428
— Gulf of Guinea *I-774*, I-778
North Equatorial Current I-223, I-487.I-488, I-489, I-617–18, *I-628*, *II-406*, II-407
North Equatorial Drift *I-222*
North Equatorial Pacific Current II-775, II-779–80, II-795, II-797

North Inlet *I-352*, I-356, I-359
— salt marsh estuary I-353
North Korea, famine due to deforestation, soil erosion and natural disasters II-492
North Loyalty Basin *II-426*
North New Hebrides Trench *II-426*
North Pacific Current *I-374*, *III-180*
North Pacific Fisheries Management Council, observer program on all larger and some smaller fishing vessels III-147
North Pacific Groundfish Observer Program (NPGOP) III-160
North Pacific High I-378
North Pacific Oceanic Province, Mexico I-487–8
North Pacific Pressure Index (NPPI) I-376
North Pacific Subtropical Anticyclonic Gyre II-407, II-795
North Sea I-43–63
— atmospheric nitrogen deposition assessment III-201
— cessation of some polluting inputs I-56
— climate I-45
— coastal erosion and landfill I-53–4
— cyclonic circulation I-45–6
— disposal of dredged material in I-55
— duplication of responsibility for **I-58**
— effects of urban and industrial activities I-54–6
— eutrophication I-53
— general decline in populations of marine mammals and seabirds III-129
— groundfish assemblage, changes in III-128
— major shallow water marine and coastal habitats I-47–50
— partitioning of I-46
— populations affecting the North Sea I-50–3
— protective measures I-56–9
— international arrangements affecting protection of **I-56**
— protection at subregional level **I-58**
— reduction of nutrient input by bordering countries agreed III-220–1
— region defined I-45
— seabirds and fisheries in III-112
— seasonality and natural environmental variables I-45–7
North Sea management
— need to increase management plans I-59
— new system is required I-59
North Sea Task Force I-57
— Monitoring Master Plan I-75, I-77
Northeast Asia Regional Global Observing System (NEARGOOS) II-497
Northeast Atlantic, high by-catch mortality rates III-125
Northeast Pacific shelf, detritus from catch-processor ships III-127
Northeast Providence Channel *I-416*
Northern Adriatic I-269
— communities of I-172
— described I-269
— formation and circulation of water masses I-271
— primary production high in offshore systems I-174
— salinity I-271
northern Benguelan current region *see* Africa, southwestern
northern fur seals, enzyme induction by PCBs II-457, *II-460*
northern Gulf of Mexico shelf, pulses of shrimp, crabs and fish I-438
Northern New South Wales, major habitats II-634
Northwest Arabian Sea and Gulf of Oman II-17–33
— coastal erosion and landfill II-27–8
— georaphy and geology II-19
— major shallow water marine and coastal habitats II-21–4
— offshore systems II-24–5
— populations II-25
— protective measures II-29–31
— rural factors II-35–7
— seasonality, currents and natural environmental stresses II-19–21
— urban and industrial activities II-28–9
Northwest Atlantic Fisheries Organization (NAFO) I-8
Northwest Pacific Action Plan II-496
Northwest Providence Channel *I-416*
Northwestern African Upwelling I-190
Norway
— climate modifies fjord morphology I-19
— coastal wave energy development III-315
— data on contamination in organisms and bottom sediments I-27
— "no discards" policy forces selective fishing III-138
— use of antibiotics in fish cages III-369
— Whaling Act (1929) III-77
— whaling technology III-74–5
— wild fish may contain antibiotic residues III-369
Norwegian Atlantic Current *I-19*
Norwegian coast I-17–30
— anthropogenic influences I-24–6
— environmental setting I-19
— major shallow water marine and coastal habitats I-21–3
— monitoring programmes, environmental quality criteria and protective measures I-26–8
— natural environmental variables I-19–21
— north—south community gradient less than expected I-22–3
— offshore systems I-23–4
Norwegian Coastal Current *I-18*, I-20, I-23
Norwegian Sea, depletion of rorquals III-74
Norwegian Trench I-17
nuclear fuel reprocessing plant, La Hague I-72, I-76
nuclear power, India II-279
nuclear power plants/stations I-54, I-72
— southern Brazil I-738
Nuclear Test Ban Treaty III-342
nuclear waste dump sites, Sea of Japan II-478, *II-480*
nucleation III-200
Nullarbor National Park, Australian Bight II-682, II-684
Nullarbor Plain, Great Australian Bight region II-682
nursery areas, restriction of fishing in III-378
nutricline, Levantine basin I-256
nutrient burden, reduction of III-258–9
nutrient changes, drastic, signs of in the Bodden III-260–1
nutrient concentrations, reef and lagoonal water, French Polynesia II-816
nutrient discharges, to North Sea, changes in I-53
nutrient effects, of seagrasses III-2
nutrient enrichment
— of coastal waters of Florida reef tract I-409
— Cockburn Sound, Western Australia, effects of II-700–1
— from equatorial and open ocean upwelling I-535
— from sewage effluents, The Maldives II-204
— from upwelling, southwestern Africa I-825
— New York Bight I-324–5
— of northern Gulf of Mexico I-449
— problems caused by II-212
— a threat to coral reefs III-36
nutrient fluxes
— and aquatic ecosystem responses III-198–9
— importance of dry deposition processes III-201
nutrient loading, anthropogenic III-366
nutrient loads
— Chesapeake Bay I-344–5
— reduction in I-346
— high
— brought by the Mississippi I-437, I-438, I-443
— introduction of to the marine environment III-366
— increased, North Carolina I-358
— Yellow Sea coastal waters II-495
nutrient ration, effects of alteration III-259
nutrient reduction, Norwegian obligation I-24, I-26
nutrient supply
— Northeast Pacific
— effected by shallow mixed layers III-184
— vulnerable to climatic change III-171
— reduced in Californian coastal wasters III-185
nutrients
— causing eutrophication, Hong Kong II-542

— from fertilisers, Australia II-584
— anthropogenic inputs, to the English Channel I-74
— Arabian Gulf II-5–6, **II-5**
— Argentine Sea I-757, *I-757*
— Baltic Proper, temporal and spatial variability in I-106–8
— calcification and uptake of III-35–6
— concentrations around Tasmania II-651
— concentrations of, Canary Islands I-191–2
  — and nitrite I-191–2, *I-193*
— from fish farming III-369
— from the land, oceans a sink for III-394
— Gulf of Alaska shelf, from deep water I-379, I-380
— high, nearshore, Hawaiian Islands II-795
— horizontal gradients, Tagus and Sado embayments I-155, *I-156*
— input into Haifa Bay I-259–60
— inputs during upwelling events and high algal growth, Côte d'Ivoire I-814
— Irish Sea
  — increasing I-87
  — sources of inputs I-93, **I-93**
— large increase of to Baltic Sea I-130
— low
  — in Australian waters II-582
  — off Central-eastern Australia II-632
— Red Sea II-39
— reduction of input to North Sea by bordering countries agreed III-220–1
— role of in coral reef degradation III-38
— and salinity, off Madagascar II-115–16
— sources of input into Gulf of Guinea I-779
— to Bay of Biscay from Cantabria I-143–4
— transferred from sea to land by birds, Chagos Islands II-229

ocean drift netting, banned II-719
ocean dumping II-322
ocean physics, and foraging seabirds, North Pacific III-113
ocean resources, intergenerational and interspatial effects of use of III-397
ocean species, common, affected by coastal pollution and ocean dumping I-330
Ocean Station Papa, Gulf of Alaska *III-180*
— deviation from normal salinity III-182, *III-182*
— mid-winter mixed layer depth trend III-182, *III-182*, III-185
ocean temperature zones, Australia II-581
ocean temperatures, effects of changes in III-182
Ocean Thermal Energy Conversion (OTEC) III-304
ocean thermal energy plants II-279
— India, impacts of II-279
oceanic islands, associated with Chile I-701
— habitats and faunas I-708
oceanic mixed layer, Gulf of Alaska III-182, III-185, III-814
oceanic productivity, may decrease with climate change III-369
oceanic swell, Chagos Islands, resistance of spur and groove system II-227
oceanographic conditions, as determinants of habitat boundaries, Lesser Antilles I-630
oceanography
— Australia II-583
— of the Bahamas I-419
— Case 1 and Case 2 waters III-294
— eastern Australian region II-631
— Fiji Islands II-754
— in Greece I-242
— Marshall Islands II-779–80
— Vanuatu II-740
oceanography/marine hydrology, Gulf of Guinea I-777–9
— littoral transport and marine sedimentology I-778
— ocean currents littoral transport and marine sedimentology
— productivity and the seasonal cycle I-779
— sea water quality/structure I-777–8
— tides/waves I-778
— upwelling I-778–9
oceans
— biological divisions I-2
— biome scheme I-2
— common property and open access characteristics III-397
— ecological, economic and social importance of III-393–403
  — ecological importance III-394
  — economic importance of III-395
  — social importance III-396–7
  — sustainable governance III-397–402
— human impacts on III-360
— impediments to ecological or scientific divisions I-2–3
— low in nitrates and phosphate, Australia II-583
— missing areas, reasons for I-3
— political divisions I-2
— unique problems III-397
Oceans Policy, Australia II-590–1
OCs *see* organochlorines
octachlorostyrenes (OCSs) I-54
*Oculina* reefs, Florida
— experimental reserve III-390
— mostly destroyed by fishers III-379
Odiel River *I-168*
— estuary receives high metal load I-171, I-174
Odiel saltmarshes I-170–1
ODP oceanographic surveys, of the Somali Basin II-67
off shore wind energy III-304–11
— access issues III-306
— foundations, design concepts III-306
— problems of grid connection III-306
— prospects for the future III-308–11
  — new technology and research needs III-311
— review of current technology III-304–7
— status at the millennium III-3078
— suitable off shore areas constrained III-307
— wind farms offshore III-305–6
  — size of installation III-306
offshore habitat, Cambodia II-574–5
offshore petroleum installations, Norway, monitoring of bottom conditions I-27
Offshore Pollution Liability Agreement (northwest Europe) III-338
offshore systems
— Adriatic Sea I-274
— Aegean Sea I-240–1
  — deep-water fauna I-240–1
— American Samoa II-769
— Australia II-583
  — geomorphology II-583
  — oceanography II-583
  — offshore territories II-583
— the Azores I-209–10
— the Bahamas I-422–3
  — deep water channels and V-shaped canyons I-422
— Belize deep reef habitats I-506
  — pelagic waters I-506
— Borneo II-372–3
— Chagos Archipelago II-229–30
— Chesapeake Bay I-343–4
— Colombian Caribbean Coast, importance of fisheries I-669–71
— Colombian Pacific Coast I-680
— Comoros Archipelago II-246–7
  — Geyser and Zélée Bank II-246–7
— Coral, Solomon and Bismark Seas region II-431–3
  — the environment II-431–2
  — tuna fisheries II-432, *II-433*
— Côte d'Ivoire I-813–15
  — benthic assemblages I-814
  — phytoplankton I-814
  — zooplankton I-814–15
— eastern Australian region
  — Elizabeth and Middleton Reefs II-635

— Lord Howe Island II-635
— Norfolk Island II-635
— El Salvador I-551
— Fiji Islands II-758
— and fisheries resources, Bangladesh II-292–5
— French Polynesia II-817
— Great Australian Bight II-681
— variable abundance of pelagic fish II-681
— Guinea I-800–1
— fishing I-800
— industrial fishing I-800–1
— upwelling I-800
— Gulf of Aden II-54–5
— Gulf of Alaska I-380
— Gulf of Guinea I-781–4
— interannual variability in upwelling I-781–2
— upwellings I-781
— Gulf of Maine and Georges Bank I-312
— Gulf of Papua II-598–9
— Gulf of Thailand II-304–5
— Hawaiian Islands II-800–1
— deep-sea cobalt manganese crusts II-800
— Hong Kong II-538
— Jamaica I-565–6
— deep habitats not well known I-566
— Discovery Bay I-566, *I-567*
— Morant Cays I-565
— Pedro Cays I-566
— Lesser Antilles I-632–3
— low salinity lens of Amazon discharge, Tobago to Barbados I-633
— Madagascar II-120
— The Maldives II-205
— Marshall Islands II-779–80
— oceanography II-779–80
— Mascarene Region II-259–60
— Mozambique II-105, II-107
— New Caledonia II-726–7
— pelagic zone II-736–7
— ZoNéCo programme II-727
— New York Bight I-327
— Nicaraguan Caribbean Coast I-522–3
— Nicaraguan Pacific coast I-535
— oxygen-minimum layer I-535
— northeastern Australia II-619
— northern Gulf of Mexico I-442–3
— neritic province I-442
— oceanic province I-442
— Northwest Arabian Sea and Gulf of Oman II-24–5
— the Philippines II-410–11
— offshore fisheries II-411
— oil and gas II-411
— productivity low II-410
— upwelling and internal waves II-411
— Portuguese coastal waters I-159
— Red Sea II-42–3
— Sea of Okhotsk II-468
— major and permanent zones of vertical intermixing II-468
— mesopelagic layer II-468
— South Western Pacific Islands II-712–13
— pelagic communities II-712–13
— southeast South American shelf marine ecosystem I-757–8
— Southern Gulf of Mexico I-474–5
— fisheries I-475
— upwelling and the Yucatan Current I-474–5
— southern Spain I-172
— southwestern Africa I-830–5
— demersal zone I-833–5
— environmental variability I-832
— epipelagic zone I-830–2
— fishing activity I-832
— mesopelagic zone I-832–3
— Spanish North coast I-140–1
— Tanzania II-88
— Tasmania II-653
— Tyrrhenian Sea I-274–5
— Vanuatu II-743
— Victoria Province
— pelagic system II-666
— slope communities II-666
— West African continental shelf benthic communities I-782, **I-782**
— fish populations I-783
— pelagic variability fish populations
— plankton productivity I-783
— whale migrations I-783–4
— Western Australian region II-696
— Yellow Sea II-490–1
— resident and migratory species II-491
— spawning, nesting and nursery area II-490
offshore wave energy conversion systems III-313–14
— research now limited III-314
oil
— biodegradation increased with clay–oil flocculation III-271
— effects on near shore populations and communities III-271–5
— effects on biological communities III-272–5
— general effects III-271–2
— mutagenic effects of long-term exposure III-278
— penetration affected by viscosity and type III-269
— persistence of on shores III-268–71
— beach wetting: adhesive properties of oil III-270
— dynamics of in low-energy environments III-271
— oil-contaminated sandy beaches III-270–1
— permeability III-269–70
— stickiness of, possible effects on III-270
— total entering northern Gulf of Mexico I-450–1
— trapped in low energy areas I-450
oil contamination
— low levels Gulf of Aden beaches II-59
— southern Brazil I-740
oil deposits, Nicaragua I-520
oil dispersants I-556
— toxic I-76, I-450
— toxicity of III-276
oil exploration I-40
— in Greenland Seas I-14
— Nicaragua, Caribbean coast I-526
oil exploration and production
— Argentina I-766–7
— Gulf of Guinea I-792
— offshore, Australia II-588
— Peru I-696
— southern Brazil I-738
— Western Australia II-702
oil exports
— Colombian Caribbean Coast I-673
— Nigeria and Gabon I-792
oil and gas
— exploration and exploitation, Côte d'Ivoire I-817
— the Philippines II-411
— Trinidad I-638
oil and gas exploration
— drilling waste piles on Norwegian seabed I-17, I-23–4
— Irish Sea I-96
— Mozambique coast II-109
— North Carolina coast I-366–7
— North Sea I-51
— exploitation increasing I-52–3
— and production, Gulf of Thailand II-304–5
— South Africa II-142
oil and gas fields
— Bass Strait II-669
— east coast, Peninsular Malaysia II-354

oil and gas industry
— Australia II-584
— Sakhalin Shelf II-470–1
oil and gas installations
— offshore
— far-field effects possible III-364
— marine pollution from III-360
— operational discharges from III-363–4
— other chemicals in use, possible effects of III-364
oil and gas potential, Somalia II-79
oil and gas production
— Adriatic Sea I-278
— Malacca Straits II-315–16, *II-315*
— North Sea, environmental impacts at all stages I-55–6
— northern Gulf of Mexico I-448
— offshore, pollution by III-338
— transport of crude oil and environmental threat I-450
— Yemen II-59
oil and gas resources, South China shelf II-557
oil industry, offshore, Gulf of Mexico I-478–9
oil installations, offshore, Brunei II-372
oil persistence III-278
— effects may change as oil weathers III-271
— from the Refinería Panama storage tank rupture III-273
oil pipelines II-603–4, II-605
— environmental effect, Gulf of Guinea I-792
— southern Brazil, effects of rupture I-742
oil pollution II-376
— Black Sea I-294
— chronic
— Argentinian coast I-766
— western Indonesia, from oil refineries and production facilities II-390
— Côte d'Ivoire beaches I-817
— Dutch Antilles I-604–5
— False Bay, South Africa II-142
— from shipping III-364
— Gulf of Guinea I-792
— Gulf of Oman
— from routine tanker operations II-28
— worsening II-29
— heavy near Chittagong and Chalna, Bangladesh II-294–5
— increasing, South China Sea II-354
— Jamaica I-570
— and loss of seabirds III-113
— marine ecosystem at Toamasina (Madagascar) threatened II-125
— of marine sediments I-638
— Mexican Pacific coast I-497
— northern Gulf of Mexico I-450–1
— not avoided by small cetaceans III-96
— Sakhalin Shelf II-471
— Saronikos Gulf I-245
— sensitivity of organisms to I-450
— southern Brazil, Sepetiba Bay I-741–2
— Sri Lanka II-185
— Straits of Hormuz II-28
— Sydney Harbour and Botany Bay II-641
— Tierra del Fuego I-762
— Tyrrhenian Sea coasts I-279
— Venezuela
— eastern coastline I-655–6
— western coastline I-656–9
— Xiamen coastal waters II-523
oil production
— Argentina I-761, I-763
— Chile I-712
— Guangdong II-557
— Gulf of Papua, pipelines for delivery of II-603–4
— Mexico I-476
— Niger Delta, onshore I-788
— northern Gulf of Mexico *I-443*
— Northern Sakhalin, causing serious concern II-477
— northwestern Arabian Sea II-28
— onshore, Wytch Farm, Dorset I-72
— western Indonesia II-390
— effects of increase in II-390
oil refineries
— Aden II-59
— Aruba and Curaçao I-604–5
— Fawley I-72
— Malacca Straits *II-315*, II-316
— Mogadishu II-80
— South Korea II-495
— southern Brazil I-738
oil revenues, invested in infrastructure II-28
oil seeps
— Gulf of Alaska I-381
— resulting from salt tectonism, northern Gulf of Mexico I-442, I-451
oil slicks
— Bay of Bengal II-280
— from ballast water I-72
oil spill contingency planning, Coral, Solomon and Bismark Seas region II-441
oil spill response equipment, for the Malacca Straits II-325
oil spills I-512, I-638, II-354, II-417
— accidental, risk of I-14
— acute
— lethal and sublethal effects III-272
— sublethal and chronic effects from III-278
— Aegean Sea I-248
— affect sea turtles III-67
— Alaska, *Exxon Valdez* I-382
— amount spilled *III-268*
— Arabian Gulf II-3
— and the Gulf War II-11–13
— Argentina I-765–7
— "Metula" spill I-765
— "San Jorge" spill I-765
— Australia II-587–8
— Bay of Cadiz I-181
— Black Sea, Nassia disaster I-294
— Chilean coast **I-712**
— differences in lead to different responses III-272–3
— direct causes of mortality III-272
— El Salvador I-556
— endangering western Indonesian coastlines II-388
— English Channel I-72, I-75–6
— Amoco Cadiz, long term effects I-76
— Torrey Canyon, major damage from oil dispersants I-76
— Galician coast I-146
— Great Australian Bight II-686
— Gulf of Aden II-59
— Gulf of Mexico I-473
— high potential for, western Indonesian seas II-390, **II-390**
— Hong Kong, mainly minor II-542
— and illegal discharges, North Sea I-55–6
— increasing, Marshall Islands II-783
— Malacca Strait II-339
— Malacca Straits II-319–20
— serious impact on fragile ecosystems II-319
— Standard Operation Procedures (SOP) II-322–3
— Maracaibo Lake I-656
— Mozambique Channel II-251
— Niger Delta I-788
— northern Gulf of Mexico I-450
— Norway, few I-26
— off Fujeirah II-28
— persistence on beaches III-168–71
— persistent oil effects and their causes III-275–6
— Peru I-696
— a risk for Mozambique II-109
— *Showa Maru* II-334

— small, Sri Lanka II-185
— South China Sea (western) II-559
— southern Brazil
  — Sao Sebastiao City I-742
  — Sepetiba Bay I-741–2
  — southeast coast, and degraded mangroves I-736
— Tasmania II-657
— timing of important III-269, III-270, III-272
— Tobago I-638
— Torres Strait, *Oceanic Grandeur* II-605
— treatment effects III-276–7
  — bioremediation III-277
  — dispersants III-276
  — injuries due to III-275, III-276
  — manual removal of oil III-276–7
  — problems with use of heavy machinery III-277
  — sand-blasting and high/low-pressure-water techniques III-277
— Xiamen coastal waters II-523
— Yellow Sea II-495
oil tanker traffic
— Malacca Strait II-339
— Malacca Straits II-316–17
— Sepetiba Bay, southern Brazil I-741–2
oil terminals, southern Brazil, Sao Sebastiao City I-739, I-742
oil-well fires II-11
okadaic acid III-40, III-219
Okhotsk, Sea of II-463–72
— currents II-465–6
— effects from urban and industrial activities II-469–71
— major shallow-water marine and coastal habitats II-467–8
— offshore systems II-468
— populations affecting the area II-468–9
— protective measures II-471
  — fishery regulations II-471
  — poaching and overfishing II-471
— rural factors II-469
— seasonality, currents, natural environmental variables II-466–7
  — seasonal changes in biota II-466
— winds drive winter water movement II-465
Okhotsk shelf, productive fish area II-467
Old Bahama Channel *I-416*, I-417
Olive Ridley turtle I-780, II-180, II-205, II-289, III-63
— captive breeding programmes, Sundarbans II-158
— critically endangered III-63
— distinctive nesting behaviour III-63
— threatened I-537
Oman *II-2*, *II-18*
— agriculture II-26
— construction of regional fishing harbours II-27
— coral communities
  — damaged by nets *II-30*
  — limiting factors II-22
— increase in beach tar II-28
— industrial diversification, new industrial development, Sohar and Sur II-25, II-27, II-28
— limestone mountains *II-18*
  — unique vegetation II-21
— network of conservation areas proposed II-30
  — national nature reserves II-30
— population II-25
— Ra's Al Hadd National Scenic Reserve II-30
— Ra's Al Junayz National Nature Reserve II-30
— seabirds III-110
— upwellings along coast affect local weather II-50
Oman, Gulf of *II-18*
— current flow *II-18*, II-20
— defined II-19
— fish biodiversity II-24
— low-energy environments II-21
— sea water temperature fluctuations II-21, II-22, *II-22*
— shallow water marine and coastal habitats II-21–4
  — corals, reefs and macroalgae II-22–3
  — seagrasses II-23–4
  — turtles II-24
  — *wadis*, *khawrs* and mangroves II-23
Ontong Java Plateau *II-426*
open ocean habitats I-224–7
— anguilla I-226
— biogeochemistry of the area round Bermuda I-224–6
— humpback whales I-226–7
— *Sargassum* community I-226
open sea banks I-47
Operation Raleigh I-590
optical remote sensing, governing processes involved III-295, *III-296*
orcas ("killer whale") III-76
— high levels of PCBs III-96
— widely distributed III-91
organic carbon, New York Bight, sources of I-324–5
organic matter/material
— Baltic Proper, sinks to soft bottoms I-128
— Baltic Sea I-112–14
— delivered to Baltic Proper **I-103**, I-104
organic pollutants
— Aegean Sea **I-248**
— Baltic Sea I-131
— Faroes I-37–8
— Nervión estuary, northern Spain I-145–6
— Sargasso Sea I-229
organic pollution
— Adriatic Sea I-270
— in the English Channel I-75
— Hong Kong coastal waters II-543
  — trace contaminants II-544
— Sea of Japan *II-479*
— West Guangdong II-559
  — Pearl River mouth II-555
— west Taiwan II-509
— Xiamen region II-522
organisms, sensitivity to oil III-272
organochlorine burdens
— Arctic beluga III-362
— slowing recovery of Baltic Sea seal populations III-361
organochlorine pollution
— fish from Asian waters II-449–51
— oceanwide II-454
organochlorine residues
— in Hoogly (Hugli) river sediments II-279
— in rivers and marine biota, Malaysia II-319, II-338, II-341
organochlorines
— in Australia's marine environment II-588
— Jakarta Bay II-398
— in Liverpool Bay I-95
— southern and western Baltic I-112, *I-113*
  — DDT in herring and perch I-112, *I-114*
— in USA dolphins III-96
— use of in China II-492
organochlorines, persistent
— Asian developing countries *II-448*, II-449, *II-453*
— contamination and bioaccumulation in marine mammals II-455–6
  — toxic effects II-456–8
— contamination in North and South Pacific II-454
— global fate II-455
— Hong Kong II-542
— major pollution sources now II-455
— river and estuarine sediments, Asian developing region II-449, *II-451*, *II-550*
— temporal trend of contamination II-458–60
  — in Antarctic minke whales II-458, *II-460*
— West Guangdong coast II-555–6
organophosphate compounds I-493
organotin pollution, in fish II-451, *II-454*

organotins
— in anti-fouling paints I-146
— Asian developing regions II-449
— in USA dolphins III-96
— *see also* tributyltin
Orinocco effects I-579
Orinoco Basin
— extent of and physiographic units I-646
— geochemical characteristics of rivers I-646–7
Orinoco River
— changed tidal effects after closure of Caño Manamo I-648
— delta *I-644*
  — functions like a wetland I-648
  — main zone I-648
— effects of fluctuation in discharge I-648
— influences of, Venezuelan Atlantic coast and Caribbean Sea I-646–8
Orinoco River Plume I-630
Ortegal Cape *I-136*, I-139
Oslo *I-18*
— contamination of harbour sediments I-27
Oslo and Paris Commissions (OSPARCOM) I-43, III-221
— monitoring contaminants in sediments, biota and waters I-77–8
— monitoring programmes (JMP and JAMP) I-27, I-77–8
OSPAR Convention I-26, I-57, **I-57**, I-147, I-162
— hazardous substance targets III-362–3
— stronger controls on off shore oil and gas III-362–3
Otway Region, Victoria Province II-663
over-exploitation
— of coastal fish resource, Sri Lanka II-182
— littoral fish, Canary Islands I-196
— Mexican artisanal shrimp fisheries I-490, I-491
— Northwest Pacific sardine fishery I-490
— of timber, Vanuatu II-743
— turtle fishery I-491
— Venezuelan coast zone I-645
over-harvesting, of renewable resources, Marshall Islands II-784
overcapacity, world's fishing fleets III-367, III-377
overfishing I-3, I-312, III-119, III-130, III-367
— Adriatic Sea I-278
— Carolinas I-366
— and catch decline, Mozambique II-107
— Colombia
  — Caribbean Coast I-671
  — Pacific Coast, shrimps I-684
— in the Comoros II-248
— Coral, Solomon and Bismark Seas region, boom-and-bust cycles II-437
— Malacca Straits II-320
— 'Malthusian overfishing' III-123
— of marine invertebrates, Fiji Islands II-759
— North Sea I-52
— northern Gulf of Mexico I-445
— Oman and Yemen II-27
  — with industrial methods II-29
— of predators III-128–9
— of salmon and seatrout, in the English Channel I-73–4
— South Western Pacific Islands II-716, II-719
— technique restrictions not effective against recruitment overfishing III-161
— threatens artisanal fishing, Belize I-509
— Vanuatu II-747
overgrazing
— effects of I-444
— Hawaiian Islands II-802
— Oman and Yemen II-26
overgrazing/overcropping eastern Australian region II-637
overharvesting, of fish resources, Palk Bay–Madras coast II-170
Owen Fracture Zone *II-64*, II-67
oxygen deficiency, Baltic Proper, affecting cod spawning I-110
oxygen depletion III-259
— between Mississippi and Sabine rivers I-443
— effects of, North Sea I-53
Oxygen Minimum Layer, Arabian Sea II-24
oyster banks, Chesapeake Bay I-341
oyster beds
— importance of, Carolinas coast I-362
— Texas, disappearing I-447
oyster culture, Venezuela I-653
oyster farms
— Korea II-494
— West Guangdong II-558–9
oyster fishery, small commercial, KwaZulu-Natal II-139
oyster habitat, Chesapeake Bay, destruction of III-126
oyster harvesting
— Chesapeake Bay I-341–2
— northern Gulf of Mexico I-442
oyster reefs, northern Gulf of Mexico I-441–2
oyster shell, for agricultural uses I-447
oysters III-167
— communities destroyed through burial I-447
— decline in production, Tagus estuary I-162
— dragnets and dredging damage seagrass beds III-8–9
— effects of TBT III-250
— El Salvador I-551
  — contamination in I-555
— farming of, Normandy and Brittany coasts I-73
— as indicators of environmental contamination III-253
— rock oysters II-52
— Texas, declines in linked to salinity perturbations I-439–40
— *see also* pearl oyster culture; pearl oyster fishery
ozone, tropospheric, Azores I-204
ozone depletion III-370–1
— a possible impact on small cetaceans III-98

Pacific basin, detection of mid-1970s regime shift III-224–5
Pacific Center Coastal Province, Mexico I-488
— habitats I-488
— important fishing area I-491
— varied marine and coastal fauna I-491
Pacific continental platform I-534
Pacific Deep Water I-679, I-702, II-583
Pacific Ocean
— Central and Eastern, coral bleaching and mortality events III-53
— equatorial current system I-679
— north, surface layer is fresh III-181
— Northwest, coral bleaching and mortality events III-52
— Southwest, coral bleaching and mortality events III-52–3
Pacific Ridley turtle, destroyed by landslides I-539
Pagassitikos Gulf *I-234*, I-236, I-237, I-244
— nutrient rich I-244
— some chlorinated biphenyls I-247–8
Pago Pago harbour, American Samoa
— dredging and filling causing reef loss II-717
— heavy metal pollution II-712, II-771
— industrial activities II-770–1
— loss of habitats II-770
PAH monitoring, Prince William Sound and Cook Inlet I-381
PAHs *see* polycyclic aromatic hydrocarbons (PAHs)
Palk Bay *II-162*, II-163, II-166–8
— climate and coastal hydrography II-166
— coastal ecosystems II-166–7
— fish and fisheries II-167
— geological features II-166
— human population and environmental degradation II-167
— Muthupet Lagoon II-167
— Vedaranyam wildlife sanctuary II-167
Palk Bay–Madras coast II-168–70
— biodiversity II-169
— climate and coastal hydrography II-168
— coastal ecosystems II-168–9
— environmental degradation II-169

— affected by decrease in freshwater II-169
— fish and fisheries biodiversity II-169
— geological features II-168
— impacts of human activities on the ecosystem II-169–70
Pamlico River Estuary *I-352*
— dinoflagellates I-355
— freshwater eelgrass I-359
— productivity pulses and algal blooms I-354
Pamlico Sound *I-352*, I-353
— environmental problems for bottom communities I-360
Panama Bight *I-678*
— seasonal upwelling cycle I-679
— tuna/anchovy/shrimp fishery I-680
Panama Current *I-678*
Panama, Gulf of *I-678*
Panamic Coastal Province, Mexico I-488–9
— habitats and communities I-489
"pantry" reserves, Marshall Islands II-786
Papua, Gulf of *II-426*, *II-594*
— agriculture II-600–1
— fisheries management II-606–7
— low level of development II-599
— water and sediment discharge to II-596, **II-596**
Papua New Guinea (PNG) II-427, **II-427**, II-595
— coral bleaching and mortality events III-53
— coral reefs in good conditions II-431
— effects of logging II-435
— impact of land use on coastal waters II-435–6
— level of social and economic development lo II-434
— mangroves II-430
— population and demography II-433–4, **II-434**
— research providing some information on marine biota II-435
— seagrass species diversity II-431
— sedimentary coast, backed by mountains II-427–8
Papuan Barrier Reef II-430, II-597
paralytic poisoning, and saxitoxin III-40
Paralytic Shellfish Poisoning (PSP) I-69, I-160, I-765, II-320, II-339, II-418, II-495, II-543, III-221
Paria, Gulf of *I-644*
— dry season cirulation of Orinoco waters I-648
Paria Peninsula, cold waters present I-646
Paris Convention, Protection of Marine Pollution from Land-based Sources III-339
particles
— hygroscopic growth of, and deposition velocities III-203
— removal by wet deposition III-200
Pas de Calais, zooplankton in waters off I-70
pastoral nomadism, Somalia II-77, II-80
Patagonian coast, Argentina I-751–2, *I-752*
— eutrophication I-765
— evolution of settlements I-758, I-761
— hydrocarbon concentrations I-761
— intensification of tidal currents I-754
— northern, characteristic sublittoral communities I-754–5
— unique environments I-756–7
— marine birds I-757
— marine mammals I-756–7
patch reefs I-505, II-7–8, II-65, II-191
— Glovers Atoll I-505
— lagoonal
— Bermuda I-227, I-228
— Mauritius II-258
— Madagascar II-118
— northern Kalimantan and eastern Sabah II-371
— Tanzania II-85–6
*Patella*, affected by oil dispersants I-76
Patos Lagoon, southern Brazil *I-732*, I-737
— fishes of I-740
PCBs I-54, I-279, I-392, I-555
— in animals from the Greenland seas I-9, I-11, I-14
— in Asian developing region waters II-449
— in Bergen fish and shellfish I-27
— in blubber II-455
— concentrations in Arctic species III-361, *III-362*
— coplanar PCBs considered more toxic II-457
— estimated loads in the global environment II-455, **II-455**
— in the Gulf of Maine I-317
— high levels in Irish Sea mammals I-95
— Hong Kong coastal waters II-544, **II-545**
— in marine sediments, Vietnam II-565–6
— marine transport of from European waters I-14, *I-14*
— pollution from, Thailand II-449
— reasons for persistence of III-362
— reduction in positively affecting Baltic seal populations I-131
— uniform distribution of in air and surface seawater II-454, *II-458*
— in whale blubber I-36
Peale's dolphins III-91, III-93
Pearl Cays complex, corals I-522
pearl oyster and Chank beds, Gulf of Mannar II-164
pearl oyster culture II-207
— French Polynesia, problems of II-820
— Marshall Islands II-781, II-782
— Solomon Islands II-438
pearl oyster fishery
— Arabian Gulf, in decline II-11
— Fiji, overexploited II-759
— French Polynesia II-820
pearl oysters II-598
Pearl River Delta *II-550*
— phytoplankton II-553
— tributaries II-551
Pearl River mouth
— irregular semi-diurnal tides dominate II-552
— and the Pearl River II-551
pearl shell fishery II-602
peat, Sri Lanka II-181
peat swamp forests, Malacca Strait II-336, II-342
peat swamps
— east Sumatra II-312
— Malacca Strait II-335
pelagic organisms
— Irish Sea I-90–1
— West Guangdong coast II-554
Pemba *II-84*, II-85
Penobscot Bay *I-308*, I-314
Pentland Firth, Scotland, considerable tidal currents III-318
peroxyacetylnitrate (PAN) I-205
Persian Gulf *see* Arabian Gulf
persistent organic compounds, contribute to seal decline I-50
persistent organic pollutants (POPs) I-5, I-39–40, I-381, III-360, III-361
— Asian developing regions II-447–62
— atmospheric transport to colder regions III-361
— decline very slow III-362
— Greenland I-9, I-11, *I-11*
— from Europe and Russia I-13
— may affect human health I-11
— toxicity risk to small cetaceans III-95–6
— transport and deposition in Europe I-56
Peru I-687–97
— biodiversity I-691–3
— Peruvian—Chilean province I-691
— Provincia Panameña I-691
— characteristics of the coast I-689–91
— coastal populations and the main sources of pollution I-693–6
— collapse of anchovy fishery III-119, III-129
— decline in guano birds III-111, III-129
— direct and indirect small cetacean catches III-93
— legislation on environmental protection I-696
— main populated and industrial areas I-695
— National Contingency Plan (oil spills) I-696
Peru (Humboldt) Current I-486, I-487, I-488, I-489, I-543

Peruvian Current
— high productivity I-692
— influences coastal climate I-690
— northward-flowing I-690
pesticide pollution
— and coastal ecosystems, the Seychelles II-239
— from Mozambique's upstream neighbours II-109
— Peru I-694–5
— Southern Gulf of Mexico I-476, I-479
— Xiamen coastal waters II-524
pesticides
— causing environmental stress, Altata-Ensenada del Pabellón I-493
— discharged to Black Sea I-301–2
— El Salvador
  — in oysters I-555
  — in river waters I-552
— golf courses, Hawaiian Islands II-806
— Gulf of Guinea I-784–5
— in land-based pollution III-361
— loss of from shipping III-364
— Malaysian waterways II-341
— organochlorine, Vietnam coastal waters II-565
— and other chemicals used in aquaculture III-369
— in shrimps I-493
— use and abuse, the Philippines II-412–13
— use round Sea of Japan II-475
— use of in Tanzania II-88–9
— used in Greece, concentration in the marine environment I-243–4
Peter the Great Gulf
— changes in bottom communities II-481
— radioactivity in II-470–80
Peter the Great Marine Reserve, endangered II-477
petrochemicals, southern Brazil I-739
petroleum, Venezuela's main export I-655
petroleum hydrocarbons, in Arabian Gulf sediments II-12
petroleum refineries *see* oil refineries
*Pfisteria*, ichthyotoxic I-363–4, I-366
*Phaeocystis* I-69, I-90, I-356
pharmaceuticals
— from marine environments and coral reefs III-39
— natural products and chemicals III-38–9
— natural products from coral reefs III-40
Philippine Tuna Research Project II-410
Philippines, The II-405–23, II-416–17
— the area and its natural environmental variables II-407–8
— coastal erosion and landfill II-413–14
— coral bleaching and mortality events III-51–2
— effects from urban and industrial activities II-414–18
— major shallow-water marine and coastal habitats II-408–10
— mangroves
  — clearance for aquaculture III-368
  — lost to fishponds III-21–2
— offshore systems II-410–11
— populations affecting the area II-411–12
— protective measures II-418–20
  — Coastal Environment Program (CEP) II-419
  — Coastal Resource Management Program II-419
  — community-based coastal resources management (CB-CRM) II-418
  — Fisheries Sector Program II-418–19
  — legalities, utilization, conservation and management of the coastal areas **II-420**
  — marine protected areas II-419
— rural factors II-412–13
— seimically active II-408
phocine distemper virus (PDV) (1988)
— killed seals on English south coast I-71
— reduced seal numbers, North Sea I-50
phocine distemper viruses III-225
phosphate I-255
— increased load to Baltic Sea I-130
— a limiting factor I-240
— release of, the Bodden III-262–3
— seawater, Vietnam II-565
phosphate enrichment, euphotic layer, Côte d'Ivoire I-810
phosphate mining, Makatea, Tuamotu Archipelago II-822
phosphoric acid manufacture I-72
phosphorite ore, Onslow Bay I-366
phosphorus
— from phosphate rock processing I-94
— increased load to Baltic Sea I-130
— as a limiting nutrient I-106
— load to Chesapeake Bay I-344–5
phosphorus level, low, Orinoco River I-647–8
photic layer, Panama Bight I-679
photoinhibition II-42
photosynthesis III-394
— anoxigenic III-263
phthalates, in coastal sediments, west Taiwan II-509
*Phyllophera* meadows, Black Sea, decrease in I-291, I-304
physical environmental anomalies, make entire populations vulnerable III-225
physically forced (climate/oceanographic) disturbances III-224–5
phytohydrographic associations, Arabian Gulf II-6–7, **II-7**
phytoplankton
— Carolinas
  — estuarine I-354–6
  — marine I-356
— Guinea I-800
— Gulf of Guinea I-779
— Gulf of Thailand II-304
— harmful effects of increased exposure to UV-B III-370
— key roles of III-198
— Malacca Strait II-333
— Malacca Straits II-312
— offshore, Côte d'Ivoire I-814
— species diversity, Vietnam II-564
— West Guangdong coast II-553
— wide range in UV-B sensitivity, effects of III-370–1
phytoplankton assemblages, Australia II-583
phytoplankton biomass
— and biodiversity, Pearl River mouth II-553–4
— coastal, Sunda Shelf II-384
— nutrient control of, Tuamoto archipelago II-817
phytoplankton blooms I-244, II-50, II-700
— Baltic Proper
  — changes in composition and dominance I-109
  — spring I-106
— Canary Islands I-192
— Irish Sea I-89
— Norway, spring and summer I-20
— Spanish north coast I-140, I-141
phytoplankton communities, Mascarene Region, nutrient-limited II-259
phytoplankton ecology, Bay of Bengal II-275, *II-276*
phytoplankton growth
— North Spanish coast I-138
— Tagus and Sado coastal waters I-155–6
— western English Channel I-67
phytoplankton processes, Spanish north coast, modified by oceanographic processes I-141
picoplankton, becoming dominant, the Bodden III-261
pilchard fishery
— Australian Bight II-683
  — may affect seabirds II-685
— southwestern Africa I-831
  — failure of I-832
— Victoria Province, Australia II-668
pilchards, killed by herpes virus II-587
pilot whales III-76, III-91
— hunted, Faroe Islands III-92
— long-finned I-31, I-36

Pinatubo, Mount, effects of eruptions II-408
pinnipeds, effects of decline in III-99
*Pinus pinea* forests, southern Spain I-170, I-181
Pitt Bank *II-222*
plaice I-49
— Baltic, stock decreased I-111
plankton
— abundance of, Argentine Sea I-757
— in the Black Sea I-290
— Côte d'Ivoire I-812–13
— eastern Australian region II-634
— North Sea I-48
— Peru, changed during El Niño I-692
— Tagus estuary I-157
— West Guangdong coast **II-554**
plankton assemblages, southeastern Taiwan II-508
plankton productivity, Gulf of Guinea I-783
planktonic communities, French Polynesian lagoons II-816–17
planktonic food web, model for, Takapoto atoll II-817
planktonic systems, Irish Sea I-89–90
Plantagenet Bank I-227
plate boundary, diffuse, through Chagos area II-226
platform reefs II-388
— the Seychelles II-236
plutonium (Pu) I-94
— fall out from nuclear tests I-117
$PO_4$ levels, affected by Tagus and Sado river discharges I-155, *I-157*
Po basin
— agricultural and industrial pollutants from I-275
— evolution of Po delta I-276
poaching, and marine reserves III-390
Poland
— discharged partly treated sewage to River Vistula I-115
— effect of increased standard of living I-112
environmental contamination from mining I-114–15
— pollutants to Baltic Proper I-103–4
"poles of development", Chile I-710
politics, in fisheries management III-154
pollack I-313
pollock fishery
— Alaska III-147
— Bering Sea III-129
pollutant layers, elevated III-200
pollutants
— atmospheric transport of I-13
— baseline studies, Gulf of Alaska I-381, I-383
— brought by river to Black Sea I-287, I-298–9
— from Turkish Black Sea coast *I-200–301*, **I-300**
— Carolinas coast, sources of I-354
— carried to Baltic Proper by rivers I-103
— direct discharge into estuaries, Portugal I-162
— dispersion and deposition processes in the coastal zone III-206
— entering the Bay of Bengal II-278
— environmental, Baltic Sea I-112–17
— global redistribution of *III-361*
— higher sensitivity of Baltic populations I-129
— increase susceptibility to infection III-225
— and the internal boundary layer III-207
— Malacca Strait II-340
— marine current transport of I-13–14, *I-14*
— Palk Bay–Madras coast II-169
— and pathogens, waterborne, spread of in the Caribbean I-503
— reduction of phosphorus and nitrogen to North Sea I-24
— sea ice transport of I-14
— Southern California Bight
— largest reduction from publicly owned treatment works I-389–90
— multiple source discharges I-388–9, *I-389*
— reductions in I-389
— *see also* long-distance transport of pollutants
polluter pays principle II-323, II-327, III-339, III-401
— Jamaica I-570
pollution
— Aegean Sea I-245–9
— affecting eelgrass beds III-7–8
— affecting small cetaceans III-95–6
— airborne, Azores I-204
— atmospheric, from ships III-340
— Bahamas, effects on water quality and near-shore habitat I-429–30
— of beaches, Sri Lanka II-183
— Belize
— control through the Environmental Protection Act I-513
— from industry I-511
— urban I-512
— Black Sea, sources of I-287, I-297, **I-297**, I-298–9, **I-298**, *I-299*
— Cabo Frio–Cabo de Santa Marta Grande, southern Brazil I-738–9
— definition and sources of III-258
— degrading Gulf of Guinea coastal waters I-784
— effects of in the Baltic Sea I-130–1
— eutrophication I-130–1
— heavy metals I-131
— organic pollutants I-131
— pulp mill industry I-131
— effects of climate change and sea level rise III-195
— effects of land-based sources, Israel I-259–61
— Haifa Bay I-259–61
— entering the northern Gulf of Mexico I-449–50
— entering rivers and coastal seas, western Indonesia II-396
— estuarine and marine, South Africa II-142–3
— from fishponds, the Philippines II-414
— from ships, Coral, Solomon and Bismark Seas region II-440–1
— from shrimp farming and agrochemicals, Bangladesh II-294
— in the Great Barrier Reef region II-623–4
— Guinea I-801, I-802
— Gulf of Alaska, mainly from long-distance transport I-381
— Gulf of Mannar II-165
— impacting on seabirds I-49
— industrial
— and domestic, Peru I-694, I-695
— Tanzania II-92
— Kingston Harbour, Jamaica I-570
— land-based
— contributing to marine pollution III-339–40, III-340, III-360, III-361–3
— regulation of III-362–3
— slower treatment of III-339–40
— threat to coral reef biodiversity II-389
— land-based, Turkish Black Sea coast I-296–303
— from city sewerage systems I-300–1, *I-301*
— from rivers I-299–300
— monitoring I-291–302
— pesticides and PCBs I-301–2
— localised, Coral, Solomon and Bismark Seas region II-439
— Malacca Strait
— faecal coliform count II-340
— Indonesian side not systematically monitored II-341
— land-based II-340
— sea-based II-340
— Malacca Straits II-317–20
— agricultural waste II-318–19
— coliform contamination II-319
— land-based II-317–18, **II-318**
— main problem areas II-325–6
— sea-based sources II-319–20
— moderate in Canary Islands I-198
— Nicaraguan Pacific coast I-534
— organic I-835
— Papeete, Tahiti II-817
— point-source, Fiji II-760
— release to marine environment from point and non-point sources III-258
— Saronikos Gulf I-238

— and seabird deaths III-114
— secondary, from Maluan Bay sediments, Xiamen region II-524
— the Seychelles, from habitation and farms II-239
— Taiwan Strait, effects of tidal currents II-502
— through shipping operations and accidents I-146
— Venezuela I-654
— vessel-sourced, reduction of III-343
— Western Australia II-697
pollution abatement, some progress, North Sea countries I-52
pollution hotspots
— identified in Baltic I-132
— the Philippines II-418
pollution prevention programmes
— Yellow Sea
  — economic problems a major impediment to II-496
  — land-based, obstacles to II-495–6
pollution-contamination distinction difficult III-258
polybrominated diphenylethers (PBDEs) I-36
— presence in marine mammals III-362
polychaetes I-157, I-239
— Faroes I-34
— first to colonize mine tailings III-240
— Spanish North coast I-139, I-140
polychlorinated biphenyls *see* PCBs
polycyclic aromatic hydrocarbons (PAHs) I-56, I-392, I-555, II-417, III-361
— Arabian Gulf II-12
— attributed to oil seeps I-381
— in fjords I-24
— Great Barrier Reef region II-623
— Hong Kong coastal waters II-544, **II-545**
  — posing a risk to ecosystems and seafood consumers II-545
— multi-hop contaminants I-13
— in Norwegian shellfish I-25
— Prince William Sound, delayed effects I-382
— in sediments, Baja California I-497
— southern Brazil
  — Guanabara Bay I-741
  — Santos Estuary I-743
— Taiwan Strait II-510
— *see also* PAH monitoring
polynyas
— Baffin Bay I-8
— Greenland Sea I-7
— Sea of Okhotsk II-466
Pomeranian Bay I-109, I-115
Poole Bay *I-66*
POPs *see* persistent organic pollutants
population
— Adriatic and Balkan coastlines I-275
— Aegean, ancient cultures, modern cities I-241–2
— Alaska I-381
— American Samoa II-769
  — population growth rates II-769, *II-769*
— Andaman, Nicobar and Lakshadweep Islands II-193
  — and increasing environmental degradation II-193
  — Lakshadweeps II-194
— Anguilla, British Virgin Islands and Montserrat I-620–1
— Arabian Gulf II-9
— Argentinian coast I-758, I-761–2
— around the Irish Sea I-92–3
— around the North Sea, effects of I-50–3
— around the Sea of Japan II-475
— Australia II-586
  — Aboriginal peoples II-583, II-682
  — general community II-584
  — indigenous communities II-583–4
— the Azores I-210
— the Bahamas I-423
— Baltic catchment I-50
— Bangladesh II-291–2, **II-292**
— Belize I-506–7
— Borneo II-373–4
— Brazil I-733
— Carolinas coast, growing I-354
— Chesapeake Bay area
  — Europeans I-344
  — growth and development I-344
  — Native Americans I-344
  — sprawl development I-345
— coastal
  — Cambodia II-575
  — surrounding the Yellow Sea II-491–2
  — Vietnam II-563
— Colombian Pacific Coast I-680–1
— Comoros Archipelago II-247
— and consumption III-396
— Côte d'Ivoire I-815–16
  — coastal cities I-815
  — indigenous population I-815
— and demography
  — Papua New Guinea II-433–4, **II-434**
  — Solomon Islands II-433–4, **II-434**
— Dutch Antilles I-612
  — leeward group I-597, I-598
— east coast, Peninsular Malaysia II-350–1, *II-351*
  — urban growth rate II-351, **II-351**
— eastern Australian region II-635–6
  — New South Wales II-636
  — south-east Queensland II-636
— El Salvador I-551
— English Channel coasts I-71
— Faroes I-35
— Fiji II-758
— Florida Keys I-408
— Great Australian Bight region II-682
  — European colonisation II-682
— Great Barrier Reef region II-619–20
  — indigenous people II-619–20
  — trends in II-620
— Greenland I-8
— growing beyond sustainable limits II-173
— Guinea coastal zone I-801–2
— Gulf of Aden states, rural and poor II-56, **II-56**
— Gulf of Guinea coast I-784
  — country demographic information **I-785**
— Gulf of Maine I-314
— Hawaiian Islands II-801
  — decline of native population II-801
  — demographic patterns II-801
  — tourism trends II-801
— Hong Kong II-538, II-541
— Huelva and Cadiz, southern Spain I-172
— human
  — growth of III-396
  — total impact of III-396
— and human settlements, Lesser Antilles I-634, **I-634**
— Iranian Gulf of Oman coast II-25
— Israel I-258
— Jamaica I-567
  — development pressure I-561, I-567
— Madagascar II-120–1
  — increase in major coastal towns II-121
— Marshall Islands
  — coastal uses and environmental issues II-780
  — demographic patterns II-780
— Mascarene Region
  — cities and sewage discharges II-264
  — coastal II-260
— Mexican Pacific coast I-489
  — in medium and small communities I-492
— Mozambique, trend to urbanisation II-107

— New Caledonia
  — distribution of II-728
  — structure II-727–8
— of the New York Bight area I-328–9
— Nicaraguan Caribbean coast I-523
— Nicaraguan Pacific coast I-537–8
  — Estero Real, degraded natural resources I-537
  — Gulf of Fonseca I-537–8
  — population-related problems I-537
— northern coast of Spain I-142, *I-143*
— northern Gulf of Mexico I-443, I-448
— Norway I-17, I-24, I-28
— Oman II-25
— Palk Bay coast II-167
— Pearl River delta and West Guangdong II-556–8
— Peruvian coast I-693
  — cities I-695
— the Philippines II-411–12
  — coastal population depends on coastal fisheries II-414
  — growth rate II-412
  — rural–urban migration II-412
— Poland and Lithuania I-103–4
— and politics, French Polynesia II-815
— rates of change, Lesser Antilles I-634
— Red Sea coastline, mainly major ports and cities II-43
— regions round Sea of Okhotsk II-468–**II-469**
  — Sakhalinsky region, unique II-468
— round the Malacca Straits II-311
— rural, Cambodia II-575
— the Seychelles II-237–8
— Singapore II-337
— Somalia 77
  — migration to cities II-77, *II-79*
— South Western Pacific Islands II-713–14
  — population trends II-713
— southeast India, Gulf of Mannar II-165
— southern Black Sea coast, Turkey **I-296**, I-297
— southern Brazil
  — cities growing fast I-738
  — metropolitan areas I-737, I-738
  — Rio de Janeiro and Santos I-738
  — Santos Bay I-743
— Southern California I-388
— Southern Gulf of Mexico I-475
— Southern Portugal I-160–1
— Sri Lanka II-177, II-181
— of the Sundarbans II-153
  — scheduled castes/tribes II-153
— Tanzania II-85, II-88
— Tasmania II-653–4
— The Maldives II-206–7
  — demographic patterns II-206
— Torres Strait and Gulf of Papua II-599–600
— Turks and Caicos Islands I-591
— Tyrrhenian coasts I-275, I-279
— UAE II-25
— Vanuatu II-743
  — dual economic structure II-743
— Venezuela I-654–5
— Victoria, Australia II-666
  — indigenous people II-666
— Vietnam II-566
— west Taiwan coast II-505
— Western Australia II-696–7
— Yellow Sea, high densities inhibit conservation II-495
— Yemen II-25
population biomass, minimum viable in a marine reserve III-382–3
population density, Australia II-586
population growth rates, Oman and Yemen II-25
porpoises III-90
— *see also* Dall's porpoise ; harbo(u)r porpoises
port activities, Colombian Caribbean Coast I-673
port development
— Gulf of Guinea I-789
— Xiamen region II-519, II-521
  — deep harbours II-519
Portland, Maine *I-308*, I-313
Portsmouth Harbour *I-66*
— loss of salt marsh I-74
Portugal, Tagus and Sado estuaries I-151–65
— benthic vegetation I-159–60
— coastal erosion and landfill I-161
— the defined region I-153–4
— dredging I-161–2
— natural environmental variables and seasonality I-153–6
— offshore systems I-159
— populations affecting the area I-160–1
— protective measures I-162, **I-163**
— rural fishing I-161
— shallow marine habitats I-157–8
— upwelling effect on fisheries I-160
— urban and industrial effects I-162
Portuguese coastal waters I-153
*Posidonia oceanica* meadows I-274
power station effluent temperature, Hawaiian Islands II-805–6
power stations, conventional, North Sea coasts I-54
prawn culture
— Grand Bahama I-437
— the Sundarbans II-152–3, II-153–4
  — causing deterioration of coastal water bodies II-154
  — ecological crop loss II-154
prawn fishery
— Borneo, linked to mangroves II-368
— highest discard/catch ratios III-143
— large discard III-125
— northern New South Wales II-638
prawn trawling fishery
— Great Barrier Reef region II-623
  — effects on benthic communities II-623–4
— Torres Strait and Gulf of Papua II-604, II-606
Preah Sihanouk National Park, Cambodia II-571
precautionary action, principle of III-371
precautionary principle
— and equity III-157
— in Straddling Stocks Agreement III-158
precipitation
— Colombian Caribbean Coast I-665
— Gulf of Alaska, large freshwater flux I-378
— increasing III-188
— Norwegian coast I-19
primary production
— Adriatic Sea I-270
— Andaman and Nicobar Islands II-191
— Baltic Proper I-108
— Bay of Bengal coastal waters II-275, *II-276*
— by phytoplankton, usually N and P limited III-198
— Cambodian Sea II-573
— Campeche Sound I-474
— Canary Islands I-192, I-193
— Chagos Islands II-230
— coastal areas, Coral, Solomon and Bismark Seas region II-431–2
— coral reefs III-35
  — the Philippines II-409
— Gulf of Aden II-55
— Gulf of Cadiz I-172
— Gulf of Guinea I-779
— high, northern Gulf of Thailand II-302
— high, Sunda Shelf II-384–5
— important, Tasmania II-654
— inner New York Bight I-324, *I-325*
— Irish Sea, related to stratification of water masses and distribution of fronts I-89–90

— Izmir Bay I-244
— Malacca Strait II-334
— Malacca Straits II-312
— the Maldives II-202
— Norwegian coast, strong seasonality I-20
— and nutrient cycles, Gulf of Alaska I-379–80
— nutrient source probably deep ocean I-379
— pelagic, Baltic Proper I-126
— percentage from the sea III-394
— Red Sea, low II-42
— stimulated by macronutrients III-259
— Tagus estuary I-157
— water column, Lesser Antilles I-633
— western English Channel I-67
primitive earth III-394
Prince William Sound *I-374*, I-375, I-380
— pollution of I-381
— southern, zooplankton community influenced by advection from the Alaskan Shelf I-380
— *see also Exxon Valdez* oil spill
principle components analysis, for grouping of disturbance types III-217
produced water
— containing mercury, Gulf of Thailand II-305
— effects of oil and chemicals in III-363
— from oil and gas production I-24, I-55
productivity
— Arabian Sea
— and the Northeast Monsoon II-20
— and the Southwest Monsoon II-19
— high, Gulf of Paria, Trinidad I-631
— and Nicaraguan Pacific coast fisheries I-536
— primary and secondary, of seagrasses III-2–3
— Somali Current LME II-66
PROKASIH (Clean Rivers Program), western Indonesia II-400
property rights, importance in regulation of the coast III-354
property rights institutions, sophisticated III-400
property rights regimes, and the oceans III-396–7
protected areas
— Borneo, mangroves in II-367–8
— Gulf of Guinea I-793
— Marshall Islands, resistance to II-788
— Turks and Caicos Islands **I-593**
Protected Natural Areas, Mexican Pacific I-497, **I-497**
protected species
— eastern Australian region II-643
— Madagascar II-17
— status and exploitation of, the Comoros II-249–50
— coelacanths II-250
proton secretion III-35
— and nutrient uptake III-36
Providenciales, TCI
— pollution problem I-591
— surrounding waters, fishing and tourism I-591
— tourist destination I-591
*Prymnesium parvum* I-26
public health, Colombian Pacific coast I-680, **I-681**
Puerto Rico I-575–85
— coral bleaching and mortality events III-56
— geology I-577, *I-577*
— physical parameters I-577–81
— population development and land use:: effects from urban and industrial activities I-584
— shallow water and coastal habitats I-582–4
— shelf morphology and sediments I-581–2
— US National Estuarine Sanctuary I-584
Pulicat Lake, Palk Bay–Madras coast II-169, II-278
purse-seine fishery
— Madagascar II-122
— South Africa II-136–7
Puttalam Lagoon, Sri Lanka II-179
Puttalam Lagoon–Dutch Bay–Portugal Bay system, Sri Lanka, seagrass beds II-179
pyrite and evaporites, Laguna de Tacarigua I-652

Qatar *II-2*
— coral bleaching and mortality events III-47
queen conch I-590, I-592
Queen Conch fishery I-425
Queensland Plateau *II-426*, II-619
Queensland Trough *II-426*
Quirimba Archipelago, Mozambique
— fishing techniques II-107–8
— marine/coastal habitats II-102–4
— seagrass fishery, Montepuez Bay II-106
— marema (basket traps), use of II-106
— seagrass preferences II-102–4
— seagrass the most abundant habitat II-104
— source of productivity for South East Africa II-103–4
— net primary productivity calculations II-103
— unspoilt mixture of mangrove, seagrass and coral reefs II-103
quota systems, based on total allowable catch (TAC) III-161–2

radioactive discharges, Irish Sea I-94
radioactive oceanographic tracers, round the Azores I-204
radioactive pollution, Sea of Japan II-478–80
radioactive waste, stored in north Russia, cause for concern I-14–15
radionuclides
— $^{137}Cs$, before and after Chernobyl I-116–17
— Black Sea, from Chernobyl disaster I-294
— from La Hague *I-14*, I-54–5, I-76, I-117
— and Sellafield I-54–5, I-117, III-362
— in Greenland seas I-11
— long-distance marine transport of I-13
— natural III-362
— not affecting small cetaceans III-96
— via Black Sea Water I-248–9
rainbow trout I-35
rainfall
— Coral, Solomon and Bismark Seas Region II-428
— Côte d'Ivoire I-808–9
— Dutch Antilles I-598
— Guinea I-799
— Gulf of Guinea I-776
— Accra dry belt I-776
— Hawaiian Islands II-795
— Marshall Islands II-775
— Pearl River watershed II-551
— South West Pacific Islands II-709
— Torres Strait II-595
— Vanuatu II-739
— within the Fijian Group II-753
— *see also* precipitation
rainy seasons, Tanzania II-85
raised beaches II-349
Raleigh Bay *I-352*
RAMSAR Convention I-147, I-431, I-608, I-793, II-658, **III-334**, III-340
— TCI signed up to I-592
— Walvis Bay and Sandwich Bay wetlands, Namibia I-829
— Western Salt Ponds of Anegada (British Virgin Islands) accepted I-623
RAMSAR sites I-52, I-96, I-148
— Albufeira Lagoon I-162, *I-163*
— Inagua National Park, the Bahamas I-431
Ras al Hadd *II-18*, II-19
Ras Caseyr *II-64*
Ras Muhammed marine park *II-36*, II-44
Raso, Cape *I-152*, I-153
Ratak Submarine Ridge II-779
Recife de Fora Municipal Marine Park, Brazil I-728

recreational angling I-830, **I-830**, II-138–9
— conflict with commercial interests II-138–9
recreational boating, Dutch Antilles I-604
recreational fisheries, management of III-161
recreational fishing
— Australia II-587
— Great Barrier Reef region II-621
— management of II-626
— Hawaiian Islands II-803
— New Caledonia II-729
— Tasmania II-655
— Victoria Province II-668
— Western Australia II-697
recreational industry, New York I-328
Red Sea II-35–45
— biogeographic position II-38
— coral bleaching and mortality events III-47
— endemic species II-38
— extent II-37
— geographical and historical setting II-37–8
— major shallow-water marine and coastal habitats II-40–2
— offshore systems II-42–3
— oil contamination throughout II-43
— population, urban and industrial activities II-43–4
— protective measures II-44
— receives continual supply of larvae for the Indian Ocean *II-36*, II-38
— seasonality, currents, natural environmental variables II-39–40
— turnover time II-40
Red Sea invaders, southeastern Mediterranean I-256–8
Red Sea rift II-37
Red Sea—Gulf of Aden water exchange II-39
red tides I-90, I-244, I-439, I-742, II-376, III-198, III-199, III-298
— Bay of Bengal II-275
— Benguela Current III-218
— and fish kills III-206
— Hong Kong II-542
— impact on fish production II-543
— Izmir Bay I-238
— North Benguela region I-827
— Pearl River mouth II-553
— the Philippines II-418
— Rías Bajas I-141
— Straits of Malacca II-320
— Xiamen coastal waters II-522
— Yellow Sea II-495
Redfield ratios, P:N:Si, Arabian Gulf II-6
reed field, Liaohe Delta II-494
reef ecosystems, decline in abundance from poor fishing practices III-126
reef fish
— Marshall Islands II-778–9
— problem of size-selective fishing III-121
reef flats
— central Red Sea II-40, *II-40*
— Chagos Islands II-227
reef gleaning, Fiji Islands II-759
reef habitats, shallow, Tasmania II-652
reef and lagoonal water, French Polynesia II-816
reef systems, marine reserves, percentage adult population protected III-382
reefs
— artificial I-183
— Belize, buffered from urban pollution I-512
— deep reef habitats I-506
Refinería Panama storage tank rupture III-278
— long-term effects on physical structure of mangrove forest III-273, III-274
— mortality of subtidal corals III-273–4
Reflagging Agreement III-158, III-161
— requirements of III-160
regime shift, Gulf of Alaska shelf I-375
Regional Organisation for the Conservation of the Environment of the Red Sea and the Gulf of Aden (PERGS) II-59
Regional Organisation for the Protection of the Marine Environment (ROPME) II-3, II-29, II-44
Regional Seas Program III-340
— provision for protection against land-based pollution III-339
Relative Penis Size Index (RPSI) III-250
relative sea-level rise, and landform alteration, northern Gulf of Mexico I-446–7
relaxed eddy accumulation (REA) III-204
— application to particle measurement III-205
remineralisation processes, enhanced, the Bodden III-261
remittances, important in South Western Pacific Islands II-714
remote sensing
— applications of III-294
— Brazilian tropical coast *I-725*
— Coastal Zone Color Scanner information
— Chagos Islands II-230
— The Maldives II-202
— could aid seagrass research III-12
— in fisheries management III-160
— modern definition III-294
— provides large-scale synoptic data III-284
— satellite imaging of Chilean coast upwellings *I-705*, I-706
— and sea bottom types III-297
— of tropical coastal resource III-283–91
— coral reef systems III-87–8
— economic considerations III-288–9
— mangroves III-285–6
— sensors relevant to mapping tropical coastal zones III-284
— tropical seagrass ecosystems III-286–7
— types of data achievable III-284
— used to classify and quantify coastal marine habitats, Mauritius II-259
— *see also* AVHRR imagery
renewable energy, from the oceans III-304
reproductive disturbance
— in Baltic biota I-114
— in breeding colonies of seabirds, Southern California I-398
— from *Exxon Valdez* oil spill I-382
reproductive failure
— from organic chemicals III-95
— seabirds, and fish population collapse III-129
reptiles
— El Salvador I-549–50
— Great Barrier Reef II-618
reservoirs
— southern Spain, Guadiana and Guadalquivir basins I-175
— *see also* dams
residence time
— Baltic Sea water I-103
— fjord basin water I-21
residual currents, complex, Irish Sea I-85, *I-86*
resource depletion, the Philippines II-418
resource exploitation, French Polynesia II-819–20
resource management, USA potentially conflicting paradigms III-350
resource utilisation conflicts, Xiamen region **II-518**
resources
— non-renewable, South Africa II-142
— right of access to (South Africa) II-141
restingas
— Abrolhos Bank–Cabo Frio, Southern Brazil I-736
— Brazil I-727
— Santa Catarina State, southern Brazil I-737
retroflection eddies *see* eddies
Reunion *II-254*, II-255
— agriculture II-261
— artisanal fishing II-261–2
— ciguatera poisoning outbreaks II-261–2
— Fishing Reserves **II-265**, II-266

— steep volcanic surfaces support corals II-259
— tourism II-264
Revillagigedo Islands, Mexico I-487
— seasonal surges I-487
— some coral bleaching (1998) III-53–4
Rhine, River I-58–9
— Rhine Action Programme for rehabilitation of I-59
Rhine water, effects on southern North Sea I-46
Rhodos gyre I-236
Rías Bajas, Galicia I-137, I-138, I-141
Riau Archipelago, Sumatra II-311, II-337
— coral reefs II-314
— seagrass beds II-314, II-336
Riga, Gulf of *I-122*
right whales III-74, III-76, III-81
Rim Current, Black Sea I-287
Rincón region, Buenos Aires Province, significant biological activity I-755
ringed seals I-8, III-95
Rio de Janeiro metropolitan area
— industries and port I-738
— Jacarepaguá lagoon systems, receives industrial waste I-742
— pollution in I-741
— Guanabara Bay, sewage and industrial effluents I-741
Rio de la Plata basin system I-751
Rio de la Plata estuary *I-751*, I-759–61
— anthropogenic impacts I-760
— environmental gradients I-759
— formation of I-759
— human impact I-759–60
— impact of urban-industrial zone I-759
— nutrient discharge I-760
— pollutants I-760
— sectors of I-759
— state of knowledge I-760
— system characteristics I-759
Rio Declaration, on transgenerational responsibility III-360
river deltas, New Guinea coastline, support pristine mangrove forests II-597
river dolphins, India, DDT, PCBs and HCHs in II-452
river inflow
— Colombian Caribbean I-665
— Côte d'Ivoire I-807
— and run-off I-809
— Guinea I-799
— Gulf of Guinea I-776–7
river pollution, Russian Far East II-477
river run-off
— Bay of Bengal II-271, **II-272**, II-275
— lessened through irrigation withdrawals II-287
— and river impacts II-273
— Gulf of Thailand, freshwater discharge in addition to II-301
— in Malacca Strait II-333
river and wave material transport, east coast, Peninsular Malaysia II-582
rivers
— Chilean coast, various flow regimes I-704–5, **I-704**
— Gulf of Guinea coast, downstream effects of damming I-789, I-792
— influence on southern Spain I-169–70
— polluted, Venezuelan coastal zone I-645
— Southeast Asia, the most turbid II-306
Rocas atoll, Brazil I-723
— biological reserve I-724
rock lobster fishery II-56
— artisanal, Torres Strait II-602
— Australia II-586
— Gulf of Papua II-604
— South Africa II-137
— Tasmania II-655, II-656
— Victoria Province, Australia II-668
rock and surf angling II-138
rocky coastlines
— Irish Sea I-87–8
— communities of I-88
rocky reefs
— eastern Australian region II-634
— subtidal II-633
— El Salvador I-548–9
— Los Cóbanos the most extensive *I-546*, I-549, I-552
— habitats of, Australian Bight II-679
— Louisiana coast I-421
— Puerto Rico I-582–3
— Victoria Province, subtidal II-664
rocky and sandy seabeds, Canary Islands I-195
rocky shores
— Borneo II-666
— eastern Australian region II-633
— exposed, persistence of oil on III-276
— and headlands, east coast, Peninsular Malaysia II-349
— Malacca Strait II-336
— northern Argentina, macro and megafauna I-754
— Venezuela I-649–50
— algal zone I-649
— barnacle zone I-649
— *Littorina* zone I-649
— microhabitats I-649
— Victoria Province II-664
— western Sumatra II-387
Rodrigues *II-254*, II-255
— agriculture II-261
— artisanal fishing important II-262
— lagoons heavily silted II-259
— mangroves II-259
— population II-260
Rodrigues Bank *II-254*, II-255
Rompido Sand Cliffs *I-168*
Rudnaya River Valley, eastern Russia
— degradation/decay of ecosystems taking place II-485
— health of population requires improvement II-485
— polluted surface and groundwaters II-485
— a pre-crisis situation II-485
runoff
— agricultural, western Indonesia II-396
— annual, mainland Norway I-19–20
— and biota, Papua New Guinea coast II-597
— extensive, Tahiti and Moorea II-821
— increased by impervious surfaces I-445
— polluting Taiwan's coastal environment II-507, **II-510**
— silt-laden, increases turbidity I-446
— soil, Guangdong II-558
— urban, contains pollutants I-450
— *see also* soil runoff; stormwater runoff; surface runoff
Rupat Strait II-311
rural factors
— affecting the Aegean I-242–5
— affecting southern Spanish coastal zones I-174–6
rural land use
— impacts on western Indonesian seas II-395–6
— agriculture II-396
— deforestation II-395
— mining II-395–6
— *see also* land use
rural-urban migration I-567
— the Philippines II-412
— Tanzania II-92–3
Russians, whaling III-75

S:N ratio, Irish Sea I-87
Saba Bank *I-596*, I-598
— fishing by non-Saban fishermen I-603–4
— reefal areas I-600
Saba Island *I-596*, I-598, I-603

— coastal and marine habitats *I-600*
— sediments, due to erosion limiting factor for reef development I-601, *I-602*
Sabah *II-362*
— coastal habitats II-365
— illegal immigrants II-373
— population density II-373–4
— problem of Sipadan Island II-377
— reef destruction II-371, *II-372*
— SCUBA diving II-377
— Semporna Islands Park (proposed) II-377
  — current and potential threats to **II-378**
  — management plan for II-377
— Tunku Abdul Rahman Park, impacts on II-376
Sabellid worm colonies, Yemen II-51
sabkha
— described II-8
— Oman
  — at risk from rising sea levels II-28
  — Bar Al Hickman II-21
*Saccostrea* build-ups *II-76*, II-77
Sado estuary *I-152*, I-154
Saharan low pressure zone I-775
St Bees Head *I-84*
St Brandon Bank *II-254*, II-255
— lagoons with sandy floors II-257
— spur and groove regions with fish II-256–7
St Brandon Islands *II-254*, II-255
— artisanal fishing II-262
— seabirds and turtle nesting sites II-257
— temporary settlements II-260
St Brandon Sea *II-254*, II-257
St. Eustatius *I-596*, I-598, I-603
— coastal erosion I-602
— coastal and marine habitats *I-600*
— deep reef systems I-600
— manatee grass I-601
St. Helena Sound *I-352*, I-354
St. Lucia
— coral bleaching and mortality events III-56
— fuel-wood reforestation I-637
— Marine Islands Nature Reserve I-639
St Lucia system, KwaZulu-Natal, periodic hypersalinity in II-135
St. Maarten *I-586*, I-598, I-603
— coastal and marine habitats *I-600*
— decline in livestock I-601
— filling of saliñas and lagoons I-602
— manatee and turtle grasses I-601
— urbanization and tourist developments I-598
Saint Malo, Golfe de *I-66*
St Martin's Island, Bangladesh, focus for ecotourism II-289–90
saithe I-35, I-49
Sakhalinsky Bay *II-464*
Salalah *II-18*, II-19
saliñas, Dutch Antilles I-601
saline intrusion
— from over abstraction of groundwater II-26
— into aquifers, the Philippines II-414
— Marshall Islands II-781
— some Gulf of Aden coasts II-58
— Taiwan II-507
— to groundwater around Zanzibar II-93
salinisation
— Gulf of Mexico I-476
— of land II-584
salinity
— Aegean Sea I-236
— affecting Baltic biota I-123
— Arabian Gulf II-5
— Baltic Sea
  — and cod reproduction I-128
  — increase in I-111
— Bass Strait II-664
— in Bay of Bengal II-287
— Bien Dong Sea II-563, *II-563*
— Black Sea I-289
— Cambodian Sea II-571
— Campeche Sound I-469
— Canary Islands I-190–1, *I-191*, *I-192*
— changes at the deep salinity minimum, North Pacific III-182
— Chesapeake Bay I-339
— coastal waters, Xiamen region, China II-516
— Colombian Pacific I-679
— dry season, coastal zone of Gulf of Paria and Orinoco delta I-648
— El Salvador estuaries and open water I-547–8
— French Polynesia II-816
— Great Australian Bight II-677
— hypersalinity, Red Sea II-37–8
— impact on fisheries II-117
— Irish Sea I-86–7
— lagoon and open water, Vanuatu II-740
— lower near the Mississippi I-437
— Malacca Straits II-311–12
— Marshall Islands II-776
— near-bottom, Alaskan Gulf I-378–9
— negative anomaly, Alaskan coast III-183
— New York Bight I-323
— North Sea I-45, I-47
— ocean and lagoon, South Western Pacific Islands II-710
— off Guinea coast I-799
— Palk Bay II-166
— of Peruvian coastal waters I-691
— South China Sea II-348, II-552
— surface, Argentine Basin I-753, *I-753*
— surface water, Papua New Guinea coast II-596
— variation in estuaries, correlates with rainfall I-520
— variations, Côte d'Ivoire I-809
— Western Coral Sea II-614
— Yellow Sea II-490
salinity gradients
— Gulf of Suez II-39
— horizontal, Baltic Proper I-102
salinity stress I-364
salinity trend, North Pacific III-182
salinity–$NO_3$ relationship, northern Alaskan Gulf I-379
salmon
— Atlantic I-35, I-313
  — Baltic Sea I-110, I-111, I-128
  — farmed III-168
— Pacific
  — change in salmon survival rate in the open Pacific III-185
  — Okhotsk Sea II-467
  — possible biological effects of climatic change III-185–6
— sockeye, Alaska, accumulating PCBs and DDT I-381, I-383
salt diapirs I-442
salt extraction
— Gulf of Guinea I-786
— Southern Brazil I-738
salt flats, Lac Badana channel, Somalia II-77
salt pans
— Mauritius II-262
— Palk Bay, mangroves cleared for II-167–8
salt ponds
— Anguilla *I-619*, I-620
— British Virgin Islands I-620
— Cambodia II-574
— El Salvador I-553
salt tectonics, effects of I-442
salt wedge, Chesapeake's main tributaries I-337
salt-marsh plants, hybridisation of I-47
saltmarsh 24, I-49, I-358
— decreasing, Chesapeake Bay I-337

— eastern Australian region II-632
— loss of II-639, II-640
— English Channel coasts I-68, I-69, *I-69*
— Gulf Coast, USA I-440
— Gulf of Maine I-311
— impact of dredging I-176
— lost in Cantabria and the Basque country I-144
— lost to infilling I-54
— Mar Chiquita coastal lagoon, Argentina I-755
— North Spanish coast I-139
— southern Spain I-170
— Gulf of Cadiz I-177
— Sri Lanka II-180
— Tagus estuary I-158
— Tasmania II-653
— Victoria Province, Australia II-665
saltwater intrusion
— Belize I-512
— into coastal aquifers I-443, I-447
— changes caused by I-448
Salvage Islands *I-186*, I-187
Salwa, Gulf of *II-2*
Samoa
— coastal modifications, serious impacts II-717
— coral reefs
— cyclone physical destruction II-712
— degraded II-712, II-714
— cyclone damage II-714
— population II-714
Samoa Group *II-706*, II-707, *II-709*
— agriculture and fisheries II-716
— overfishing II-716
— coral reefs II-712
— American Samoa II-712
— Samoa II-712
— Marine Protected Areas (MPAs) II-720
— *see also* American Samoa
San Andrés and Providencia Archipelago
— coral reefs I-668
— strong wave energy I-666
— volcanic origin with reef developments I-667
San Jorge Gulf I-753
— crude oil production I-761
San Matias Gulf, fishing I-763–4
San Salvador Islands, mangroves stressed I-422
sand
— oolithic I-590
— white carbonate, Puerto Rico I-581
sand accumulation, Gulf of Mannar, Palk Strait and Palk Bay II-181
sand barriers, southern Baltic coast III-260
sand dunes I-170
— active, Brazilian tropical coast I-722
— Cabo Frio region, southern Brazil I-736
— embayments behind, nursery grounds I-326
— front edge recession I-429
— Great Australian Bight II-675
— Yalata dunes II-678
— Mexican Pacific coast I-487
— removal of to enlarge beaches I-276
— scrub landscape, mammals in, Doñana National Park, southern Spain I-171–2
— with xerophytic grasses II-51
sand and gravel extraction
— east coast of England I-53
— El Salvador I-554
— English Channel I-74
— impacts from I-77
— New York Harbor I-330
sand loss, from beaches I-176, I-177
sand mining I-604, I-786
— Anguilla I-622
— by suction, east coast, Peninsular Malaysia II-355
— causing beach erosion I-636
— the Comoros, forbidden II-248
— for construction, Israel I-258
— and coral mining, Marshall Islands II-782
— Côte d'Ivoire I-816
— and dredging, in the Bahamas I-428
— from beaches II-27
— Jakarta Bay II-391
— and lime production, Mauritius II-263
— Malacca Straits II-316
— the Maldives II-208
— the Mascarenes II-262
— offshore, Curaçao I-605
— Orissa, for heavy minerals II-279
— the Philippines II-415
— river, east coast, Peninsular Malaysia II-353
— Southern Brazil I-737, I-738
— causing environmental degradation I-740, I-741
— Suva Reefs II-717
— Todos os Santos Bay, Brazil I-725
— Turks and Caicos Islands I-592
— Vanuatu II-744–5
sand spits/sand banks I-170, I-828–9
— east coast, Peninsular Malaysia II-349
— *see also* Kakinada sand spit, Godavari and Krishna deltaic coast
sand vegetation, Doñana National Park, southern Spain I-171
sand-eels I-49
sandstone/beach rock habitats, Sri Lanka II-180
sandy beaches
— Arabian Sea coast II-21
— the Azores I-207
— Borneo II-369–70
— Buenos Aires Province I-754
— Cabo Frio–Cabo de Santa Marta Grande, southern Brazil I-736
— east coast, Peninsular Malaysia II-349
— eastern Australian region II-633
— French channel coast I-69
— Gulf of Aden, deposits of ilmenite and rutile II-58
— Hawaiian Islands, colourful II-797
— high-energy, Yemeni and Somali coasts II-51
— loss of and beach nourishment I-176
— Malacca Strait II-337
— Marshall Islands II-777
— Mauritius II-258
— New York Bight I-325
— and oil spills III-269
— oil-contaminated III-270–1
— Oman II-21
— regeneration of causes serious impacts I-181
— replenished by wadis II-52
— southwestern Africa I-828
— and rocky beaches I-829–30
— Taiwan II-502
— Venezuela I-648–9
— dissipative beaches I-648–9
— high energy beaches, zonation of I-649, *I-649*
— west coast of Malaysia II-311
— western Sumatra II-387
— Xiamen region, polluted II-519
— Yellow Sea II-490
— *see also* beach nourishment/replenishment
sandy bottom habitats I-548
sandy bottoms, southern Somali coast *II-70*, II-72, II-74
Sanganeb Atoll II-40
Santa Maria
— depauperate palagonitic tuff I-206–7
— limpet harvesting I-206
Santa Monica Bay *I-386*
Sarawak
— population II-374

— Pulau Bruit National Park II-368
— well managed coastal National Parks and Protected Areas II-376–7
sardine, horizontal migration of II-466–7
sardine fishery I-141, II-56
— beach seining II-26
— Northwest Pacific I-490
— Pacific I-396–7
— Peru I-693
— Portuguese I-160
— South Africa, collapse of II-137
— Venezuela I-653
sardinella fishery, southern Brazil I-739
Sardinia, 'Smeralda' Coast I-278, I-279
Sargasso Sea, defined I-223
Sargasso Sea and Bermuda I-221–31
— coastal erosion and landfill I-228–9
— effects from urban and industrial activities I-229
— major open ocean habitats I-224–7
— major shallow water marine and coastal habitats I-227–8
— populations I-228
— protective measures I-230
— seasonality, currents, natural environmental variables I-223–4
*Sargassum* community I-226
— displaced benthos I-226
*Sargassum decurrens*, distribution of, Australia II-695
Saronikos Gulf *I-234*, I-235–6, I-247
— industrial pollution I-245–6
— pollution in, affecting the plankton I-238
— seagrass beds and algae I-239
— water masses in I-237
satellite imagery, uses of III-287
satellite remote sensing III-394
— of the coastal ocean: water quality and algal blooms III-293–302
  — areas of application III-295
  — critical bottom depth III-297
  — semianalytical atmospheric radiation transfer models III-295
  — sensor measures upwelling radiance III-297
  — water quality parameters III-295–8
— coasts of III-294
— *see also* AVHRR imagery; remote sensing
satellite-based vessel monitoring systems III-160
Saudi Arabia *II-2*, *II-36*
— coral islands and patch reefs II-7–8
Saya de Malha Bank *II-254*, II-255
— sand, coral and green algae II-256
scallop fishery
— Argentina, collapse of I-764
— New Caledonia II-734
— Scotland I-90
scallop and Puelche oyster culture, Argentina I-764
scavengers, food subsidies from by-catch/discard/ and processing dumping III-127
scheduled castes/tribes
— Lakshadweep Islands II-194
— Sundarbans II-153
science and the seas III-394
scientific realism, in fisheries management III-155, III-156–7
Scilly Isles I-71
Scotian Shelf *I-311*
Scotland
— early attempts to determine effects of fishing on fish populations III-376
— legislation relating to the coastal and marine environment III-353
Scottish scallop fishery, Amnesic Shellfish Toxin contamination I-90
sea breezes, Borneo II-365
sea cliff retreat I-144, I-211
sea colour remote sensing III-294
— need for development III-298
— not as advanced as that for ocean colour III-300
Sea of Cortes Oceanic Province, Mexico I-488
— oceanic habitat I-488
sea cucumber fishery
— the Maldives II-212
— Sri Lanka II-183
sea cucumbers II-90, II-123, II-784
— exploitation of, Mozambique II-108
— Madagascar II-123, II-126
sea horses
— endangered, Palk Bay II-168
— Gulf of Mannar II-165
sea ice transport, of pollutants I-14
sea level
— anomalous rise, Alaskan coast, a propagating Kelvin wave III-84
— high seasonal oscillation, Bay of Bengal II-287
— history of, Brazilian tropical coast I-722
— rising III-188
— *see also* sea-level rise
Sea Moss, harvesting of I-635
sea otters, effected by *Exxon Valdez* oil spill III-274
sea salt extraction, southern Spain I-177
sea surface temperature (SST)
— Aegean Sea I-236
— American Samoa II-767, *II-767*
— Arabian Gulf and Straits of Hormuz II-4–5
— the Bahamas I-419
— Bay of Biscay I-14
— Black Sea I-289
— Brazilian tropical coast I-721
— Cambodian Sea II-572
— Canary Islands, spatial and temporal differences I-190*I-191*, *I-192*
— the Comoros II-245
— continental seas, western Indonesia II-384
— Dutch Antilles I-598
— east coast, Peninsular Malaysia II-348
— eastern Australian region II-631–2
— effect of Southwest Monsoon off Arabian coasts II-19
— French Polynesia II-816
— Great Australian Bight II-677
— Gulf of Guinea I-778
— Gulf of Thailand II-300
— higher, affecting tropical cyclones III-369
— increasing, central Gulf of Alaska III-182
— Indian Ocean and Southeast Asia (1998) *III-45*
  — 'hot spot migration III-44–6
— Irish Sea I-85–6
  — slight rise in I-86, *I-87*
— lagoon and open water, Vanuatu II-740
— long-term increase, Gulf of Guinea, linked to global warming? I-781
— Madagascar II-115
— Malacca Straits II-311
— Maldive Islands II-201
— New Caledonia II-725
— of the Nicaraguan Caribbean coast I-519–20
— off southern Angola *I-824*
— Somali coast II-66
— South China Sea II-407–8
— South China Sea coast II-552
— South Western Pacific Islands II-710
— Spanish north coast I-137
— Sri Lanka II-178
— sub-surface anomaly, Line-P, Northeast Pacific III-183–4
— Tasmania II-650–1
— varies with season, Yellow Sea II-490
— Venezuela
  — AVHRR images *I-647*
  — and upwellings I-646
— Victoria coastal waters II-664
— warmer, effects of III-195
— Western Coral Sea II-614
sea swell, North Spanish coast I-138
sea turtle conservation, Mozambique II-111

sea turtle fibropapilloma disease III-67
sea turtles 59–71, I-522, I-525, I-535, I-591, II-21, II-44, II-250, II-600, II-601, II-797
— Abrolhos National Park I-726
— Bahamas I-426
— breeding grounds, Chagos Islands II-229
— conservation priorities III-67–8
— as flagship species for conservation III-68
— international and regional levels III-68
— national level III-67–8
— Coral, Solomon and Bismark Seas region II-431
— evolutionary history III-60
— Great Barrier Reef II-618
— Gulf of Guinea, status of I-780, **I-780**
— Gulf of Mexico I-473–4
— Hawaiian Islands II-799
— Java Sea II-364
— Lesser Antilles I-620, I-636
— life cycle III-60–1
— delayed maturity III-61
— pelagic migration drifting III-61
— reproduction III-60–1
— living, biological status of 61–4
— and longline fisheries III-146
— Maldives II-205
— migrations, growth and population structure III-61
— modern research needs and tools III-64–5
— information networks III-64–5
— nesting beach studies III-64
— use of PIT tags III-64
— Mozambique II-105
— nesting ground harvests difficult to control III-65
— the Philippines II-410
— protected by Indonesian Law II-400–1
— protected in El Salvador I-550
— Red Sea II-42
— Sri Lanka II-180
— Sunda Shelf II-400–1
— Tanzania II-87
— threats to species survival, historical and modern III-65–7
— damage to nesting and foraging habitat III-66–7
— direct harvest III-65
— fisheries mortality III-65–6
— vulnerable to fisheries III-379
— Yucatan I-471
— *see also* named varieties
sea urchin harvest I-636–7
sea urchins
— effects on reefs, French Polynesia II-819, II-822
— Mauritius II-257–8
sea water temperatures, Turks and Caicos Islands I-589
sea waves, Brazilian tropical coast I-721
sea-bed activities, peaceful exploration and exploitation III-336, III-338
sea-bed current stress II-596
sea-bed exploration, the Philippines II-417
sea-bed and sediments, English Channel I-67, *I-68*, I-69
sea-floor spreading II-66
— Gulf of Aqaba II-37
sea-level rise
— adverse effect on coastal ecosystems III-98
— French Polynesia, effects of II-823
— Gulf of Guinea I-787
— impact on coastal habitats, northern Gulf of Mexico I-446, I-446–7
— Marshall Islands vulnerable to II-777
— significant points II-777
— problem of Bay of, Bengal II-281, II-288, II-295
— puts reclaimed land at risk I-145
— South Western Pacific Islands, effects on II-716–17
— threat taken seriously in the Maldives II-214–15
— *see also* relative sea-level rise
sea-salt production, China II-494
sea-surface microlayer, enrichment of I-391–2
seabirds III-105–15
— Aegean Sea I-239–40
— American Samoa II-769
— Antarctic III-108
— "Arrecife Alacranes" national park, Yucatan I-471
— Australian Bight II-679–80
— the Azores
— mercury in I-212
— roosting/nesting sites I-207
— some protection for I-214
— Chagos Islands, high diversity II-229
— El Salvador
— nesting and migratory I-549
— plentiful, "Colegio de las Aves" I-549
— English Channel, offshore, migrating and breeding species I-71
— evolution and history III-106–7
— derivation III-106
— Tertiary development III-106–7
— exploitation and conservation III-111–14
— loss to growing populations III-111, III-114
— ocean physics-foraging seabirds relationship, North Pacific III-113
— seabirds and fisheries in the North Sea III-112
— value to older cultures 111
— Faroese, concentrations of PCBs and DDT in **I-37**
— Fiji Islands II-758
— threats to II-757
— food and fisheries III-111
— Great Australian Bight, threats to II-684–5
— Great Barrier Reef II-618, II-679–80
— habitat disturbance, Dutch Antilles I-604
— Hawaiian Islands II-798–9, **II-798**
— threatened or endangered **II-799**
— and longline fisheries III-145–6
— development from the Australian zone III-146
— of Madagascar II-120
— breeding sites II-120
— the Maldives II-204–5
— socio-economic importance of II-204
— Marshall Islands II-779
— modern III-107
— main groups **III-106**
— Miocene climax III-106
— shorebirds/waders III-107
— movements and measurements III-107–11
— breeding counts, "apparently occupied nest" III-109
— coastal birds, fluctuating distribution III-109
— complexities of breeding seasons III-109–11
— distributions III-107
— effects of ENSO events III-110
— erratic productivity and mortality III-109
— foraging ranges and feeding areas III-108
— migrations III-107, III-110–11
— seasonal fluctuations at sea III-109
— nesting and migratory, El Salvador I-549
— North Sea I-49
— supported by fishing fleet discards I-51
— plentiful, El Salvador, "Colegio de las Aves" I-549
— reduced numbers of, Greenland I-8–9
— southern Baltic I-109
— Southern California
— dramatic declines in I-398
— effects of major impacts I-398
— and waders, Gulf of Aden coasts II-54
— *see also* fishing mortality; marine birds
seafood
— consumption, Torres Strait II-600, II-601
— health risks from, Southern California I-399–400
seagrass I-71, I-194, I-228, I-239, I-311, II-290

— Arabian Gulf II-7
— Australia II-582
  — major human-induced declines II-700, **II-700**
  — vulnerable to eutrophication and sedimentation II-585
— the Bahamas I-421
— Borneo II-370
— Carolinas I-361–2
  — and other rooted submersed aquatic vegetation I-358–9
  — remapping of I-366
— Coral, Solomon and Bismark Seas region II-430
— damaged by dredging and filling I-569
— distinguished from macroalgae, problem in remote sensing III-286
— east coast, Peninsular Malaysia II-348, II-349–50
  — threats to identified II-354–5, II-357
— eastern Australian region II-633
  — dieback, south-east Queensland II-637
— eastern sector, northern Gulf of Mexico I-438
— English Channel I-70
— epiphytes/epiphyte communities I-357–8
— Fiji Islands II-754–5
— Global Declines and Effects of Climate Change III-10–11
  — some losses documented III-10
— global status of III-1–16
  — declines and effects of climate change III-10–11
  — ecosystems services III-2–3
  — planning, management, policy, goals III-9, III-11
  — research priorities III-12
  — worldwide decline III-7–9
— Great Barrier Reef region II-615
— Guidelines for the Conservation and Restoration of in the USA and adjacent waters, a synopsis III-8
— Gulf of Aden II-53
— Gulf of Mannar, affected by trawling II-165
— Gulf of Thailand II-301
Hawaiian Islands II-796, II-797
— high species diversity, Papua New Guinea and Torres Strait II-431
— impact on of oil and chemical spills I-450
— Jask and Char Bahar, Iran coast II-24
— Koh Kong Bay, Cambodia II-573
— lagoonal, Lakshadweep Islands II-194
— Malacca Strait II-336
— management possibilities, planning, policy and goals III-9, III-11
— Mar Chiquita coastal lagoon, Argentina I-755
— Marshall Islands II-779
— Mexican Pacific coast I-490
— North Sea, 1930s 'wasting disease' I-49
— Oman II-23
— Papua New Guinea coast II-598
— Red Sea II-42
— sediment accumulation and stabilization III-2
— Tasmania II-652
— The Maldives II-204
— Vanuatu II-740–1
— Victoria Province, Australia II-664–5
  — die-back, possibly due to catchment erosion II-667, II-668
  — epiphytic algae II-665
— wasting disease III-9, III-14
— Western Australia II-693, II-696
  — Kimberley coast II-694
— western Sumatra II-387
seagrass beds I-195, I-590
— Anguilla I-618, *I-619*
— Cambodia II-573, II-574
— Chesapeake Bay I-341
— Colombian Caribbean Coast I-668
— damaged by boat propellers III-9
— Dutch Antilles I-601
— El Salvador I-548
— fluctuations in salinity, water temperature and turbidity bad for I-441
— Johore Strait II-334
— in lagoons, Mauritius II-258
— Lesser Antilles, Trinidad and Tobago I-631
— little damaged by hurricanes III-2
— losses from low oxygen III-220
— Madagascar II-119
— Malacca Straits II-312, II-314
— New Caledonia II-726
— Nicaragua, Caribbean coast I-522
— northern Gulf of Mexico, high diversity of I-440–1, *I-441*
— nursery areas I-421, I-440–1
— nursery function III-3
— the Philippines II-410
  — conversion and utilization of II-413–14, **II-413**
— Quirimbas, increasing fishing pressure II-102
— role of, Coral, Solomon and Bismark Seas region II-430
— and sandy bottoms, southern Somali coast *II-70*, II-72, II-74, II-76
— the Seychelles II-236
— shelter function III-3
— South Western Pacific Islands II-711
— Sri Lanka II-179–80
— Tanzania II-8708
— Torres Strait II-598
  — seagrass abundance on reefs II-598
— Venezuela I-650
— West Florida Shelf I-438
— western Indonesia II-388
  — intertidal II-388
seagrass communities
— Australian Bight II-679
— Hong Kong, threatened II-541–2
seagrass dynamics, SPOT XS, Landsat TM and airborne methods all needed III-286–7
seagrass ecosystems
— affected by natural environmental impacts III-8
— divisions of animal community III-3
— nitrogen a rate-limiting factor III-3
— tropical, remote sensing of III-286–7
  — future challenges III-287
seagrass meadows *see* seagrass beds
seagulls, colonies dependent on groundfish trawling discards III-127
sealions I-756
seals I-834, II-797
— common I-109
  — French channel coast I-70
  — north east Irish coast I-91
— in the English Channel I-70–1
— Great Australian Bight II-680–1
— Northwest and Southwest Atlantic, effects of population recovery III-129
seamount fisheries, Emperor Seamounts II-802
seamounts
— the Comoros II-245
— fished and unfished, benthic biomass III-379
— Mascarene Region II-255, II-257
— offshore Tasmania II-653, II-659
— Vening Meinnesz Seamounts II-583
seasonality
— Belize I-503
— El Salvador I-547
— Jamaica I-561
— Nicaraguan Pacific coast I-534
— North Sea, temperature and salinity variability I-47
— Norwegian coastal ecosystem I-20, I-23
seawalls/shore structures, to protect the land, Hawaiian Islands II-805
seawater density, Gulf of Alaska III-181
seawater flooding, western Indonesia, carries contaminants and bacteria II-398
seaweed aquaculture, Fiji Islands II-754
seaweed farming III-167–8
— Tanzania II-91–2
seaweed harvesting, Norway I-26

seaweeds
— alien I-47
— Andaman and Nicobar Islands II-192
— brown and red, diversity decline, North Sea I-47
— Cadiz–Tarifa arc I-172
— cultivated, Yellow Sea II-494
— east coast of Taiwan II-508
— the Philippines, species diversity II-409
Secchi disk depth
— retrieved from satellite data III-296
— and water quality III-296
sediment accumulation, and stabilization, due to seagrasses III-2
sediment contamination, reduced in Gulf of Maine harbors I-317
sediment depletion, an increasing problem in coastal systems III-352
sediment drift I-47
sediment grain size, critical to permeability of oil III-269
sediment load
— high, western Indonesian rivers II-384, II-395
— south-east Queensland II-637
sediment mobility, from loss of aquatic vegetation III-259
sediment sinks, North Sea I-46
sediment supply
— Brazilian tropical coast I-722, I-728
— to Gulf of Guinea I-777
sediment transport
— alteration by coastal development, Western Australia II-699
— Bay of Bengal II-271–2, *II-272*
— into coastal zone from Orinoco River I-648
— patterns altered by harbour and reclamation engineering II-493
sedimentary basins, fault-controlled, cutting Somali coastline *I-67*, II-66
sedimentation
— affecting Andaman Sea coral reefs II-303
— between Rio and Santos I-742
— Borneo
  — a controlling factor in mangrove development II-367
  — high rates of, reducing coral cover and fish abundance II-366, *II-371*
  — a major concern II-375
— Brazilian tropical coast I-722
— can be serious in El Salvador I-552
— Chilka Lake II-279
— coastal, from deforestation, Australia II-666–7
— and contamination, near-shore environment, Belize I-508
— discharge into coastal waters, damage caused, Hawaiian Islands II-801–2
— excessive, Kuwait, and coral bleaching II-10
— Florida Keys, smothering coral I-408
— from logging, threat to Coral, Solomon and Bismark Seas region coral reefs II-436
— Great Barrier Reef region II-620–1
— Gulf of Mexico I-471
— high rates, western Indonesia II-384
— impact on Madagascan coral reefs II-123–4
— lagoonal, from mining, New Caledonia II-731
— Malacca Straits, a growing problem II-317
— Nicaraguan Pacific coast I-539
— Norwegian fjords I-21
— patterns in the Jiulongjiang estuary II-518
— Peruvian deltas I-696
— recent, Mexican coastal plain I-476
— reducing coral cover and fish abundance Borneo II-371, *II-371*
— Vanuatu, threatens some reefs II-742
— Western Australia, increased, carrying pollutants II-701
sedimentology, marine, Gulf of Guinea I-778
sediments
— accumulation of TBT in III-250
— Bay of Bengal II-273
— carbonates, West Florida shelf I-438
— containing agrochemicals, lagoons, French Polynesia II-821
— from rivers systems during the monsoon, Sri Lanka II-178
— Gulf of Aden, settle in deeper water II-53
— Gulf of Cadiz I-170, *I-171*
— high nutrient content I-493
— importance of availability of III-352
— influx to estuaries, El Salvador I-548
— Irish Sea bed I-89
— lagoonal, New Caledonia II-726
— marine
  — colonized by microorganisms III-261
  — trace metals in, Greenland I-9
— pulsed into Chesapeake Bay I-340–1
— shelf, Puerto Rico I-581–2
— size important in benthic community stratification II-665–6
— smothering coral polyps II-193
— Southern California Bight
  — contamination in I-392–4
  — quality assessed by toxicity studies I-396
— surface, Gulf of Maine, metal concentrations in I-315
— transport and deposition of III-352
— trapped by mangroves I-421
sediments loads, Fijian rivers II-758
sei whales III-82, III-82–3
— depletion of III-78
seine netting
— Montepuez Bay, Quirimba Island II-106
— Tanzania, destructive to reefs II-90
seismicity
— Coral, Solomon and Bismark Seas region II-429
— the Philippines II-408
— Vanuatu II-744
— Xiamen region II-516–17
— *see also* earthquake activity
seismology, sea level and island ages, Chagos Islands II-225–6
selenium (Se), detoxifying mercury in Greenland I-9
Sellafield, UK
— radioactive discharges to Irish Sea I-94
— radionuclides transported from I-11, I-13, *I-14*
sensors, satellite and airborne **III-284**
set-back lines, in coastal planning III-353
Setúbal embayment I-153
sewage
— affects coral reefs III-38
— in Bangladeshi rivers II-293
— Borneo
  — island treatment systems II-377–8
  — mainly reaches rivers untreated II-376
— causing problems in the Maldives II-212
— a disposal problem for small islands I-623
— domestic
  — and organic pollution, western Indonesia II-396
  — source of marine contamination, Coral, Solomon and Bismark Seas region II-440
— effluent discharged to deep injection wells I-429
— entering Aden Harbour II-58
— from Mexican towns, carried into Belize by currents I-512
— historically discharged contaminants from I-392
— impact of, Hawaiian Islands II-804, II-809
— inadequate facilities, Fiji Islands II-760–1
— Lesser Antilles I-638
— limited treatment, Gulf of Guinea I-784
— Madras, discharged into the estuaries II-169
— Malacca Strait, from the Sumatran coast II-340
— in Metro Manila II-416
— polluting Turkish Black Sea coast I-300–1, **I-301**
— Rio de la Plata estuary, effects of I-759
— seeps can lead to nutrient enrichment II-135–6
— and sludge discharges, Côte d'Ivoire I-817
— a threat to potable water supplies, Tanzania II-93
— Torres Strait and Gulf of Papua II-605
— treatment demanded in Canary Islands I-197
— untreated, problems of discharge to sea, Marshall Islands II-783

— and water use, the Bahamas I-429–30
sewage discharge
— direct to sea, east coast, Peninsular Malaysia II-351
— and high BOD, Malacca Straits II-317
— to groundwater lens, Bermuda I-228
— and treatment, Dutch Antilles I-606, I-608
— untreated
  — Gulf of Guinea I-789
  — to Irish Sea I-95
sewage disposal
— American Samoa II-770
— and deteriorating water quality, Jamaica I-567
— French Polynesia II-821–2
— into coastal waters, southeast India II-165
— Jamaica I-569
  — and deteriorating water quality I-567
— new scheme implemented, Hong Kong II-543
— a priority, New Caledonia II-733
— proper system needed for Lakshadweep Islands II-195
— Puerto Rico I-584
— Vanuatu, a concern II-745
sewage outfalls
— Southern California
  — contamination of sediments near I-393–4
  — improvement of condition of benthos near I-395–6
sewage pollution
— Cabo Frio–Cabo de Santa Marta Grande, southern Brazil I-741, I-742
— the Comoros II-251
— effects of, Gulf of Aden II-58
— impacting on coral reef building algae, Tanzania II-93
— Maputo Bay II-109
— Morrocoy National Park, Venezuela I-655
— of urban shorelines, Madagascar II-125
— Zanzibar, in nearshore waters II 93
sewage sludge
— disposed of in New York Bight I-327
  — ocean disposal ended I-330
— dumped at sea I-77
— dumping in Irish Sea I-96
sewage treatment
— Tasmania II-657
— Victoria Province, Australia II-669
  — facilities, Port Phillip Bay II-669
sewage treatment facilities
— lack of, Red Sea countries II-43
— Port Phillip Bay, Victoria Province, Australia II-669
sewage treatment plants
— poor design and maintenance I-638, **I-639**
— Tasmania, some improvement in II-657
sewage and waste management
— Australia II-586
— Great Barrier Reef Marine Park II-626
sewage wastes, reaching the Basque coast I-145
The Seychelles II-233–41
— coastal erosion and landfill II-239
— effects from urban and industrial activities II-239
— inner granitic islands II-235
  — Precambrian II-235
— major shallow water marine and coastal habitats II-236–7
— offshore systems II-237
— outer coralline islands and atolls II-235
  — high limestone islands II-235
— populations affecting the area II-237–8
— protective measures II-239–41
  — network of marine protected areas 239–40, **II-240**
  — ratification of international Conventions II-239, **II-240**
— rural factors II-238–9
— seasonality, currents, natural environmental variables II-235–6
— Seychelles National Land Use Plan II-238–9
Seychelles Bank II-235
shallow water ecosystems and biotic communities, Aegean Sea I-237–40
— benthic fauna I-238–9
— fish fauna I-238
— marine mammals I-239
— marine plants I-239
— marine reptiles I-240
— plankton I-237–8
— sea birds I-239–40
— zooplankton communities I-237
shallow water habitats
— Southern China coastal waters II-552–4
  — intertidal zone II-552–3
— Yellow Sea II-490
shark capture, Southern Gulf of Mexico I-475
shark fins, from by-catch, Marshall Island II-784
shark fishery
— Belize I-509
— El Salvador I-550, I-553–4
— Hawaiian Islands II-803
— Mozambique, linked to possible dugong population collapse II-104
— Nicaraguan Pacific coast I-539, I-540
— Yemeni II-57
sharks II-363
— as by-catch III-147
— Chagos Islands, drop in numbers II-228–9
— in the English Channel I-70
Sharm el Sheik *II-36*, II-44
Shatt al Arab *II-2*
— low oxygen saturation II-5
— nitrates, silicates and phosphates II-6
Shaumagin Islands *I-374*
Shebeli alluvial plain *II-67*, *II-70*
Shebeli River, Somalia *II-64*, II-68
shelf benthos
— eastern Australian region II-633–4
— Great Barrier Reef region II-615–16
shelf reefs, Puerto Rico I-582, I-583–4
shelf seas, vulnerable III-295
shelf-edge atolls, Western Australia II-696
Shelikof Strait *I-374*
— lower, interannual variation in copepod biomass I-380
Shelikov Bay *II-464*
shell deposits, Sri Lanka II-181
shellfish, spring mass mortalities, Taiwan II-506
shellfish culture plots, TBT in II-319
shellfisheries
— Albermarle—Pamlico Estuarine System I-360
— Canary Islands I-195
— Carolinas coast
  — barrier islands/coastal rivers region I-361–2
  — Northern Carolina, closed due to high bacterial counts I-364–5
— closure of I-76
— decline in Chesapeake Bay I-346
— Gulf of Cadiz I-175–6
— Irish Sea I-91
— New York Bight I-329
  — overharvested I-327
— northern Spanish coastal communities I-144
— San José Gulf I-764
— shell fish farming, Adriatic Sea I-277
shells, as a form of currency II-436–7
shifting/slash-and-burn cultivation II-88, II-435, II-601
— Gulf of Guinea countries I-786
— Lesser Antilles I-635
shingle areas, Channel coast of England I-68
shipping
— Australia II-587
— eastern Australian region II-641
— in the English Channel I-72–3

— Great Barrier Reef lagoon, inner and outer routes II-622–3
— Gulf of Aden II-55
— impacts of Bay of Bengal II-279–80
— impacts on the Gulf II-13
— intensive activity, South China Sea II-559
— Malacca Strait II-339
— Malacca Straits II-316–17, *II-316*
— and marine pollution III-336
  — inputs of contaminants from III-364–5
  — LOS legal framework III-333–4, III-336
— New Caledonia II-728
— New York Bight I-323
— offshore accidents and impacts
  — Lesser Antilles I-638
  — New Caledonia II-733–4
  — Tasmania II-657–8
— and offshore impacts
  — Belize I-512
  — El Salvador I-556
— potential source of marine pollution, Vietnam II-566
— and seaports, Nicaraguan Pacific coast I-540
— South Western Pacific Islands II-718
— Torres Strait II-605
— Victoria coast II-669
shipping accidents, potential for, the Maldives II-213
ships, dismantling of, potential pollutants III-339
shipwrecks, Madagascan coast II-125
shoalgrass, Carolinas I-358
shore birds
— coastal wetlands of Argentina I-755–6
— Colombian Pacific Coast I-680
— migratory
  — east coast, Peninsular Malaysia II-350
  — Hawaiian Islands II-798
shoreline erosion II-193
Shoreline Management Plans (SMPs) I-59
— England and Wales III-351–2
shorelines, effects of sea level rise III-188
shrimp aquaculture
— Colombian Caribbean Coast I-672
— developing at the expense of mangroves, around Beira II-109
— east coast of India
  — collapse of II-278
  — disease problem II-280
— El Salvador I-553
— environmental consequences cause great concern II-394
— extensive, Bangladesh II-292
— Guinea I-801
— Gulf of Mannar II-166
— Gulf of Thailand II-306
— Malacca Straits II-314
— mangrove clearance for, Rufiji Delta, Tanzania II-89, II-91
— Mexico I-492, I-493
  — extensive aquaculture I-494
— New Caledonia II-732, II-733
— a pollution source, Yellow Sea II-494
— Sri Lanka
  — damaging mangroves II-179
  — major pollution source II-184
— western Indonesia II-394
shrimp farming III-169, *III-170*, III-368
— Asia and Latin America III-368
— Belize I-511
— Cambodia II-576–7
— Colombian Pacific Coast I-683–4
— mangrove clearance for, Borneo II-368–9, *II-368*
— marine, Venezuela I-653–4
— Nicaraguan Caribbean coast I-526
— Nicaraguan Pacific coast I-535, I-538, I-539–40
shrimp fisheries II-27
— Arabian Gulf II-7, II-8, II-11
— Argentina I-763
— Brunei II-372
— Colombian Caribbean Coast I-670
— Côte d'Ivoire I-815
— deep Water, Fiji II-758
— Greenland I-8
— Guinea I-801
— Gulf of Guinea I-787
— Gulf of Thailand II-306
— highest discard/catch ratios III-143
— incidental capture of sea turtles III-66
— large discard III-125
— Madagascar
  — increasingly regulated II-126
  — more productive to the west II-117
  — substantial by-catch II-121
  — variable yields II-121
— Mexican Pacific coast I-490, I-493
  — artisanal I-490–1
— Morecambe Bay I-91
— Nicaraguan Caribbean coast I-524
— Nicaraguan Pacific coast
  — fishing gear not optimally designed I-539
  — penaeid and white shrimp I-534
— northwest Pacific Ocean, high discard rate III-137
— Pamlico Sound I-362–3
— Sofala Bank, Mozambique, changes in II-109
— South Carolina I-363
— southeast coast, Brazil I-739
— Southern Gulf of Mexico I-474
— and water temperature, Alaska I-377
shrimp nurseries I-488
shrimp ponds, western Indonesia II-388
shrimp trawling
— Gulf of Suez II-43
— Norway I-25
shrimp viral diseases III-226
shrimps, amphidromous and diadromous, El Salvador I-550–1
Si:N and Si:P ratios, decline in, northern Gulf of Mexico I-449
Siberian High I-376
— dominant over Black Sea in winter I-287
Siberian High Pressure Core II-465
Sicily, fishing ports I-279
Sierra Leone—Guinea Plateau, continental shelf I-775
Sierra Nevada de Santa Marta, Colombia I-672
Sikhote-Alin, eastern Russia, endemic species II-481
silicates
— Baltic Proper I-106
— indicators of eutrophication in Baltic Proper I-108
sills, associated with fjords I-19
Sinai Peninsula *II-36*
— mangroves II-41, *II-41*
Singapore *II-310*
— coastal modification II-317, II-339
— coral reefs, stressed II-314
— decline in fish catch II-320
— habitat loss severe II-334
— heavy metal pollution II-341
  — Keppel harbour II-318
— improvements in water quality II-341
— industrial centre II-339
— no regulation framework on the environment II-323–4
— Port, provides all major port services II-317
— Prevention of Pollution at Sea Act (1991) II-324
— regulation of land-based pollution II-324
— and sustainable development II-342
Singapore Strait *II-310*, II-311, II-333, II-334
— and South China Sea II-334
Singapore, Thailand, Vietnam, coral bleaching and mortality events III-52
SIORJI growth triangle II-337

Sites of Special Scientific Interest (SSSIs) I-96
Sitka Eddy I-375
Skagerrak *I-100*
Skomer Island *I-84*, I-96
Slovenia and Croatia, special nature reserves I-280
sludge dumping II-543
— New York Bight I-324
— Western Australia II-700
slumping, fjords I-21
small cetaceans III-89–103
— adressing the issues III-99–100
— by-catch in coastal gillnets III-94–5
— classification III-90–2
  — *odontoceti* (dolphins and porpoises) III-90
— and environmental change III-96–8
— human impacts III-92–8
  — direct and indirect catches III-92–5, III-99
  — pollution III-95–6
— populations in a fairly healthy state III-98
— vulnerable to bioaccumulation of toxins III-98
Small Island Developing States
— development problems II-265
— Federal and Islamic Republic of the Comoros II-247
— Sustainable Development of, Conference II-216
Snake Cays, fringing reefs I-505, *I-505*
social choice theory, conventional III-398
Socotra II-49, *II-49*, *II-64*
— artisanal fishery II-56
— coral communities II-52
— effective community management of traditional fisheries II-60
— no regulation on waste disposal II-58
— seabird and raptor nesting site II-54
— seagrass beds II-53
— terrestrial diversity II-49
— use of beach coral debris II-58
Socotra Archipelago II-49
— mangrove forests II-53
— to become a Biosphere Reserve? II-60
Socotra Eddy/Gyre *II-18*, *II-48*, II-50, II-66
sodium cyanide, fish asphyxiant III-123
soft bottom habitats, Malacca Strait II-337
soft corals II-742
soft shores
— eastern Australian region II-633
— Hong Kong II-538
soft-bottom benthos, Western Australia II-695
soft-bottom communities
— Baltic Sea I-126
— Spanish north coast I-139–40
— Tahiti II-819
— Tyrrhenian Sea I-274
soft-bottom habitats
— Southern California Bight I-388
— Taiwan Strait II-502–3
soft-bottom systems
— Carolinas coast I-353
  — diverse communities I-362
soft-sediment communities, Bahamas Archipelago **I-420**
soft-sediment habitats
— Tasmania II-652
— Victoria Province, Australia II-665–6
soil degradation, Lesser Antilles I-635
soil erosion I-142–3, I-174–5
— Australia II-585, II-637
— due to prawn culture II-153
— Fiji II-715, II-758
— Hawaiian Islands II-801–2
— high islands, French Polynesia II-821
— Jamaica I-568
— Mascarenes II-261, II-262
— northern end of Sumatra II-339
— a problem in China II-491
— round Sea of Japan II-475
— sediments raising river beds II-558
— though agriculture, the Philippines II-412
soil runoff, many causes, American Samoa II-770
solar radiation
— and heating of the Baltic Sea I-123–4
— Nicaragua **I-519**
— reduced by Gulf War smoke II-11
— southern Spain I-169
sole I-92, I-162
Solent *I-66*, I-72
solid waste disposal
— American Samoa, improvement in II-770
— inappropriate, Argentina I-765
— Marshall Islands *II-782*, II-783
— not understood, Cambodia II-576
solid wastes
— Côte d'Ivoire I-817
— dumped and burnt II-58, II-262
— dumping of
  — Bangladesh II-293–4
  — municipal, Dutch Antilles I-606–7
— poor disposal of
  — Fiji II-761
  — Vanuatu II-745
— a problem
  — Gulf of Guinea coast I-788–9
  — in the Maldives II-212
  — Metro Manila II-416
solitary waves, coastal erosion and habitat destruction issues III-355–6
Solomon Islands *II-426*, II-427, **II-427**
— concerns about pesticide pollution II-438
— coral reefs in good conditions II-431
— environmental assessment of the only mine II-439
— impact of land use on coastal waters II-435–6
— level of social and economic development low II-434
— mangroves II-430
— plantation agriculture II-435
— population and demography II-433–4, **II-434**
— positive effects of earthquake damage to reefs II-429
— rainforest logging II-435
— small ecosytems knowledge II-435
— tuna fishery II-423
  — "Tuna 2000" policy for sustainability II-432
Solomon Sea *II-426*
Solomon Strait *II-426*
Solway Firth *I-84*
Somali Current, seasonal reversal of II-50, *II-64*, II-66
Somali Natural Resources Management Programme II-60
Somali Plain *II-64*
Somalia II-49
— focus on peace and socio-economic considerations II-60
— Indian Ocean Coast of II-63–82
  — effects from human activities and protective measures II-80
  — Late Pleistocene to present-day event sequence II-72
  — major shallow water marine and coastal habitats 72–7, *II-78*
  — natural environmental parameters II-65–6
  — population and natural resources II-77–80
  — present day features of the coastal zone II-68–72
  — structural framework II-66–7
— location and extent II-65
— modern setting of the coast II-72
— poverty of II-56
— very low level of urbanisation and industrialisation II-58
Songo-Songo Archipelago, Tanzania, reef species richness differences II-86
Soudan Bank *II-254*, II-255
South Aegean volcano arc I-235
South Africa II-133–44

— the coastline II-135
— estuaries II-135–6
— management structures and policy II-143
— pollution and environmental quality II-142–3
— resources use II-136–42
South American Plate I-701
south Atlantic anticyclone I-823
South Atlantic Central Water
— Gulf of Guinea I-778
— South Brazil Bight I-735
South Atlantic Current I-824
South Atlantic high pressure cell I-721
South Atlantic subtropical gyre I-823
South Australia Current *II-674*
South Brazil Bight I-734–5
— bottom thermal front 734
— Inner Shelf, Middle Shelf and Outer shelf water bodies I-734–5
South Caicos, fishing centre I-591
South China Sea *II-310*, II-334, *II-346*, II-357, *II-362*, II-363, *II-382*, II-571
— coastal areas west of Hong Kong II-551
— fish landings related to the monsoon II-372
— marine pollution II-354
— monsoon generation II-365 ·
— presence of thermocline and halocline II-348
— salinity II-348, II-552
— sea surface temperature (SST) II-407–8, II-552
— surface currents II-407
— surface water patterns II-366, *II-366*
— typhoon tracks II-365, *II-366*
— water quality **II-350**
South China Sea Current II-552
South China Sea warm current II-516
South East Anatolia Project(Turkey), may deprive Gulf of river flow and nutrients II-9, II-13
South East Asia, mangrove loss III-21
South East Asian—Great Barrier Reef Region, Indo-Pacific marine species diversity II-753
South East Queensland, major habitats II-634
South Equatorial Current I-598, *II-84*, II-101, *II-101*, *II-254*, II-255, II-428, *II-581*, II-614, II-619, *II-724*, II-767, II-775
— dominant round Madagascar *II-114*, II-115
  — seed stock derives from further East II-117
— the Seychelles lie within II-236
South Florida Slope *I-436*
South Korea
— increase in agrochemical use II-493
— industrialization and urbanization II-492
South Pacific Central Water II-631
south Pacific eastern margin ecosystem I-705–7
— Antarctic ecosystem I-706
— central south Pacific gyre ecosystem
  — islands/archipelago ecosystems I-706
  — pelagic oceanic ecosystem I-706
— coastal ecosystem I-706
— sub-Antarctic ecosystem I-706
South Pacific Forum Fisheries Agency II-442, III-160
South Pacific Marine Region II-739
South Pacific Regional Environmental Program II-442, II-824
South Pacific subtropical gyre II-614
south Tasmanian bioregion, greater number of endemics II-652
South West Pacific Islands Region II-705–22
— aid projects short-lived II-720
— biogeography II-707–9
— coastal erosion and landfill II-716–18
— effects of urban and industrial activities II-718–19
— island groups *II-706*
— major shallow water marine and coastal habitats II-710–12
— new environmental initiatives at community level II-720
— offshore systems II-712–13
— protective measures II-719–20
  — evaluation of marine environmental protection measures II-720
  — marine protected areas II-719–20
  — modern conservation practices II-719
— rural factors II-714–16
— seasonality, currents, natural environmental variables II-709–10
South Western Atlantic burrowing crab I-755
Southampton Water
— refinery discharges I-75
— vegetation loss I-74
Southeast Asia, coral bleaching and mortality events III-49–52
Southeast South American Shelf Marine Ecosystem I-749–71
— coastal erosion and landfill I-762–3
— effects from urban and industrial activities I-763–7
— human populations affecting the area I-758–62
— major shallow water marine and coastal habitats I-754–7
— offshore systems I-757–8
— protective measures I-767
— region defined I-751–3
— rural factors I-762
— seasonality, currents, natural environmental variables I-753–4
Southeast Trade Winds II-116, II-428, II-710, II-753
Southern Adriatic I-269
— beach-rocks habitat I-273
— central Oceanic community I-274
— surface circulation I-271
— water column divisions I-270–1
Southern Bluefin Tuna fishery, Australian Bight II-682, II-683
Southern Brazil Shelf I-735
Southern California I-385–404
— anthropogenic inputs
  — distribution and fate of I-390–4
  — and human contributions I-388–90
— biogeographic provinces and habitats I-387–8
— DDT contamination in I-390
— effects of anthropogenic activities on marine biota I-394–9
— geography and oceanography I-387
— human health concerns I-399–400
— monitoring and management of I-400
Southern California Bight *I-386*
— impaired water bodies I-392
— loss of wetlands affects fish nursery areas and migratory birds I-399
— variations in oceanic environment I-387
Southern China, Vietnam to Hong Kong II-549–60
— effects of urban and industrial activities II-558–9
— major shallow water biota II-552–4
— marine pollution parameters II-554–6
— populations affecting the area II-556–8
— protective measures II-559–60
  — closed fishing seasons II-560
  — local legislation II-559–60
— seasonality, currents, natural variables II-551–2
southern New South Wales, major habitats II-635
Southern Ocean Current *I-700*
Southern Oscillation I-702
Southern Pelagic Province of Australia II-675
southern right whale
— Australian Bight II-676, II-680, II-683
  — disturbance and threats to II-685
Southern Shark fishery, Australian Bight II-682, II-683
southheast Asian archipelago, a zone of megabiodiversity II-389
Southwest Monsoon drift-current II-552
Soya Current II-466
Spain, North Coast I-135–50
— coastal erosion and landfill I-144–5
— effects from urban and industrial activities I-145–7
— lies in Northeast Atlantic Shelf and Eastern Canaries Coastal provinces I-137
— major shallow water marine and coastal habitats I-139–40
— populations affecting the area I-142
— protected coastal areas *I-143*, I-148

— protective measures I-147–9
  — application of Coastal Law I-147–8
  — Law for the Conservation of Natural areas and the Wild Fauna and Flora I-148
  — preservation of marine habitats I-149
  — protected areas being degraded I-148
— region defined I-137–8
— rural factors I-142–4
— seasonality, currents and natural environmental variables I-138–9
Spain, North West, Basques originated whaling III-74
Spain, southern, Atlantic coast I-167–84
— climate I-169
— coastal erosion and landfill I-176–7
— effects from urban and industrial activities I-177–83
— major marine and coastal shallow water habitats I-170–2
— offshore systems I-172
— populations affecting the area I-172–4
— protection measures I-181–3
  — Guiding Plan for Use and Management of La Breña and Barbate saltmarshes I-182–3
  — law on Andalusian Territory Regulation I-182
  — programmes and actions affecting whole coast I-183
— rural factors I-174–6
— seasonality, currents and natural environmental variables I-169–70
*Spartina anglica*
— a hybrid I-47
— spread of I-68
spawning, herring and mackerel, North Sea I-49
spawning stock biomass (SSB)
— Irish Sea, decline in I-92
— percentage preservation III-382
— Southern Bluefin Tuna III-367
— to give maximum sustainable yield (MSS) III-157
Special Areas of Conservation (SACs) I-52, I-77
— marine I-96–7, *I-97*
Special Management Areas, Hawaiian Islands II-808–9
Special Protected Areas (SPAs) I-52, I-77, I-96, I-148, I-162
Specially Protected Natural Territories, eastern Russia II-481–2
— black fir—hardwood forest ecosystem II-482
— "Borisovskoye plato" natural preserve II-482
— "Kedrovaya pad" nature reserve II-482
— Lazovsky Reserve II-482
— Sikhote-Alin Biosphere Reserve II-481–2
species, diversity of functional groups buffers impacts of stressors III-227–8
species abundance, effects of oil spills on III-271
species abundance and distribution, effects of variation in climate, temperature and other factor I-67–8
species diversity
— and distribution, Faroes I-33
— high, northern Australia II-581
— macroalgae, high Carolinas coast I-357
species evenness, mine tailings, Island Copper Mine III-240–1
species extinction, coral reefs III-37
species richness
— of corals in the Chagos Islands II-224–5, II-227
— mine tailings
  — patterns emerging III-238–9
  — under impact and during recovery III-238
— Oman beaches II-21
— Red Sea II-38
species vulnerability, to oil spills III-271
sperm whales/whaling III-74, III-76
— Azores III-74
— numbers indeterminate III-81
spilled oil, persistence of on shores and its effects on biota III-267–81
— effects of oil on nearshore populations and communities III-271–5
— effects of treatments III-276–7
— influences of biota on oil persistence III-271
— oil persistence on shores III-268–71
— recovery of biota and the importance of persistent oil 1275–6
— synthesis III-277–8
  — common themes III-278
Spiny Lobster fishery I-424–5
— Hawaiian Islands II-803
— Nicaragua I-525
— rocky coasts, Arabian Sea II-27
— South Caicos I-591
  — vulnerability to recruitment overfishing I-592
— Turks and Caicos Islands I-592
sponge harvesting I-426, I-511
sponges II-633
— encrusting I-590
— Guinea I-800
— populations, Palk Bay II-167
sport fishing I-621, II-138, II-213, II-260, II-806
— a problem I-197
sprat, Baltic Sea I-110, I-128
Spratly Islands *II-362*, II-364
— disputed claims to II-364
— Layang Layang Island, SCUBA diving II-377
spur and groove structures/systems
— Chagos Island reefs II-227
— Mascarene Region II-256–7, II-258
— outer reef slopes, Tanzania II-85
squid I-90, II-137
squid fishery, Argentina I-763
$^{90}Sr$, levels of fallout decreased in Greenland seas I-11
Sri Lanka II-175–87, *II-270*
— coastal and marine shallow water habitats II-178–80
— coastal resources management II-185–6
— direct and indirect small cetacean catches III-93
— geographical setting II-177
— marine pollution II-184–5
— marine resource use and populations affecting the area II-181–3
— non-living resources II-180–1
— protected marine fish II-183
— regulatory route in coastal planning and management III-353
— rural and urban factors affecting the coastal environment II-183–4
— seasonality, currents, natural environmental variables II-177–8
  — climate and rainfall II-177
  — currents and tides II-178
— shipping activities and fishery harbours II-185
Sri Lanka Turtle Conservation Project II-180
SST *see* sea surface temperature
stagnation periods
— Baltic Proper I-106, I-108
  — and formation of hydrogen sulphide I-105–6, *I-105*
Standard Operation Procedures (SOP), for oil spill response, Malacca Straits II-322–3
State Offshore Island Seabird Sanctuaries (Hawaii) II-797
Steller sea lion, decline in III-98–9, III-99
Stono River Estuary *I-352*, I-354
storm surges I-631
— associated with typhoons III-190, III-193–4, *III-193*, *III-194*
— and estuarine flood risk, effects of climate change and sea level rise 193–4, III-190
— from cyclones, Fiji II-710
— from tsunami waves, Vanuatu II-744
— Mascarene Plateau II-256
— North Sea I-46
— return periods III-194, *III-195*
— Xiamen region II-517
storm waves, Victoria coast II-663–4
storm-water management, problems, Dutch Antilles I-601
stormwater runoff
— contributes nutrients and heavy metals to coastal waters, New South Wales II-642
— increase in I-445
— nutrients in II-586
— polluted
  — Tasmania II-657

— Victoria Province II-669
Straddling Stocks Agreement III-158
— important principles III-158–9
Straits of Malacca *see* Malacca Straits
strandflat I-19
Strangford Lough *I-84*, I-96
stratification
— of Caribbean waters I-579
— near-surface waters, Southern California Bight I-391
— north Pacific Ocean III-181
— North Sea I-46
— seasonal
— New York Bight I-323
— North Spanish coast I-139
— Tagus estuary I-153–4
stratospheric Quasi-Biennial Oscillation (QBO) I-224
stream channelization, impact of I-446
stresses
— causing coral reef deterioration I-563
— favours domination by smaller organisms III-227
— Florida Keys
— global I-409–10
— industrial I-409
— localized, mass mortality of *Diadema* I-410
— to the environment and reefs I-409
striped bass (rockfish) I-348
— juvenile habitat I-342–3
— spawning stock restored I-342
striped dolphin, hunted in Japanese waters III-91, III-94
sturgeon I-342
sub-Antarctic ecosystem I-706
Sub-Antarctic Water I-701–2, I-706, I-735, II-664
sub-cellular damage, in fish, from chlorinated hydrocarbons I-397
sub-Saharan drought, reduced flows of Gulf of Guinea rivers I-777
'sub-tropical underwater' I-666
Subarctic Boundary *III-180*, III-181
— represents abrupt change in stratification III-181
Subarctic Current *III-180*
submarine canyons *I-136*, I-137, I-170, I-667, I-701, I-799
— Australia II-583
— Bay of Bengal II-271
— and shelf valleys I-325–6, *I-326*
— Taiwan Strait II-501
— "Trou sans Fond" canyon, Côte d'Ivoire I-805, *I-806*, I-807, I-808
submarine mining, placer tin, causing high turbidity II-395
subsidence
— coastal
— northern Gulf of Mexico I-446
— Somali coast II-66, II-67
— Colombian Pacific coast, seismic I-682
— marginal Mesozoic, Somali coast II-66
— Sucre coast, Venezuela I-645
— through groundwater extraction, Taiwan's littoral zone II-507
subsidence phenomena, Po delta I-276
subsidies
— political attempt to create legitimacy III-155
— some necessary and justifiable III-155–6
"substituted industrialisation", Chile I-710
Subsurface Equatorial Water I-679
subtidal habitats
— Irish Sea
— hard substrates I-89
— soft substrates I-89
— rocky communities, Spanish north coast I-140
Subtropic Underwater I-579
Subtropical Convergence II-583, II-653
Subtropical Maximum Salinity Water II-255
subtropical mode water, formation of, Sargasso Sea I-223
Subtropical Surface Current *I-688*, I-690
subtropical underwater I-632
Subtropical Water Mass I-701
succession, pattern of altered by oil spills III-275
Sudan *II-36*
Suez Canal *I-254*, *II-36*
— and "Lessepian migration" II-38
— opened the Mediterranean to Red Sea migrants I-257–8
— *see also* Red Sea invaders
Suez, Gulf of *II-36*, II-37, II-39
— coastal pollution by oil II-43
Sulawesi Sea *II-362*, *II-382*
— coastal habitats II-365
sulphate, to the Sargasso I-229
Sulu Sea *II-362*
— circulation II-407
— coastal habitats II-365
— internal waves II-411
Sumatra
— accelerated natural erosion II-338
— central, alluvial coastal plain II-311
— east, peat swamps II-312
— maintenance of a mangrove buffer belt II-342
— mangroves II-312
— converted to aquaculture ponds II-314–15
— massive deforestation II-338
— North, traditional fisheries management II-341
— sources of industrial pollution II-317–18
— western, marine ecosystems of II-386–7
sunbelt development, northern New South Wales II-636
Sunda Shelf II-312, II-333
Sunda Shelf region II-383
— Asian floral and faunal realm II-389
— exposed during Pleistocene lowstands II-383
— river discharge on Indonesian portion of II-384, **II-385**
Sunda Strait *II-382*
Sundarban Biosphere Reserve II-147, II-156–7
— Project Tiger II-156, II-157
Sundarban ecosystem II-148, II-290
Sundarbans
— Bangladeshi *II-146*, II-147
— British occupation II-151
— cultural convergence II-152
— early settlers II-151
— a hostile environment for settlement II-151–2
— Indian *II-146*, II-147
— name derivation II-290
— natural protection by II-290
— a necessity for Bangladesh II-291
— salinity intrusion close to II-287
— spread of colonization II-152
— *see also* Deltaic Sundarbans
sunlight penetration, reduced by nuisance micro- and macro-algae III-219
superficial reefs, Brazil I-723
surf, high energy I-487.I-486
surf zone, southwestern Africa I-830
Surface Equatorial Current I-690
surface runoff
— and river flow, Texas I-437
— source of pollutants to the Southern Californian Bight I-390, **I-391**
— stormwater discharges, little regulation and control of I-390
surface temperature, mean global, rising III-188
surface waves, direction and velocity, Bay of Bengal II-148
surfaces, rainfall-impervious, effects on flood frequency/intensity I-341
surrogate surfaces, particle collection on III-204–5
surveillance, in fisheries management III-159–60
suspended particulate matter, Haifa Bay I-261
suspended sediments/solids
— estuarine waters, Perak and Johore II-338
— flowing into the Aegean I-242–3
— off southern Spanish coast I-175
sustainability
— adaptive management III-400

— in coastal management III-350, III-356
— Coral, Solomon and Bismark Seas region, threats to II-437–8
— of current and future practices III-360
— gauged against certain criteria III-360
— of human activities on marine ecosystems III-359–73
— requires precautionary approach to ecosystems management III-400
sustainable development I-465
— application to society III-350–1
— in coastal zones III-351, III-356
— definitions I-459–60
— elements of **I-459**
— Xiamen coastal waters, measures for II-531
sustainable energy, from the oceans III-304
Sustainable Fisheries Act (USA) I-451
sustainable governance, of the oceans III-397–402
— the deliberative process in governance III-397–8
  — moving from public opinion to public judgement III-397
  — use of "visions" III-397–8
— the deliberative process in governance two-tier social decision structure III-398, *III-398*
— integrated ecological–economic modelling III-398–9
— property rights regimes III-397
  — conflict-resolution mechanisms III-399
  — design principles III-399
  — new III-399–401
— taxes and other economic incentives III-401–2
  — ecological tax reform III-401
  — shifting tax burden to ecological damage and consumption of non-renewables III-410–12
sustainable regional development, programs for, Mexico I-480
Suva Harbour, Fiji, pollution in II-719
Svalbard I-13, I-27
— fjords I-19
swamp forests
— Cambodia II-573
— Coral, Solomon and Bismark Seas region II-430–1
— Gulf of Mexico I-472
swamp lands, filled and canalised, no longer trap sediments II-136
swamps
— freshwater
  — Coral, Solomon and Bismark Seas region II-430–1
  — Gulf of Papua II-598
— Gulf of Mexico, Centla I-472
— Peru, 'Reserved Zone of Villa Swamps' I-692
Swatch of No Ground II-289
Sweden, offshore wind farm development III-309–10
swell waves I-190, I-618
— the Bahamas I-419
— Gulf of Aden II-50
— reaching Puerto Rico I-578
swine industry, Carolinas, a significant environmental threat I-366–7
synthetic drilling muds, impacting on marine benthic communities III-363
synthetic organic chemicals, in land-based pollution III-361–2
Szczecin Lagoon (Oder Haff) I-109
— fishery I-111
— metal pollution I-115

Tadjora, Golfe de II-49
— mangroves II-53
Tagus embayment, frontal boundary I-155
Tagus estuary *I-152*, I-153–4
— disposal of wastes in I-162, **I-162**
— effects of fertilizers and pesticides I-162
— flow patterns at the mouth I-154
— two distinct regions I-153
Tagus and Sado coastal waters I-155–6
Tagus and Sado coastal zone I-153
— protected areas I-162, *I-163*, **I-163**
Taiwan
— coral bleaching and mortality events III-50
— eastern II-508
Taiwan Current II-537, II-538
Taiwan Shoal *II-500*, II-501
Taiwan Strait II-499–512
— coastal erosion and landfill II-506–7
— effects from urban and industrial activities II-507–10
— major shallow water marine and coastal habitat II-502–3
— offshore systems II-503–5
— populations affecting the areas II-505
— protective measures II-510–11
  — coastal zone management II-510–11
  — instability of coastal policy II-510
  — protection of coastal resources II-511
— rural factors II-506
— seasonality, currents, natural environmental variables II-501–2
Tangier Sound *I-336*
— decline of aquatic vegetation I-345, I-347
Tanker Safety and Pollution Prevention Conference III-337
Tanzania II-83–98
— coastal erosion II-93
— effects from urban and industrial activities II-92–3
— offshore systems II-88
— population II-88
— protected areas and integrated coastal management II-93–6
— rural factors II-88–92
— seasonality, currents, natural environmental variables II-85
— shallow water marine and coastal habitats II-85–8
— signatory to international conventions supporting ICZM **II-94**, II-95
— Southern, local community training and education II-95
Tanzanian Coastal Management Partnership II-95
tar
— on Aegean coasts I-248
— Bay of Bengal II-280
— contamination, Dutch Antilles beaches I-605
— on Israeli beaches I-262–3
— in the Sargasso Sea I-229
tar balls
— Belize I-512
— Gulf of Guinea beaches I-792
— Jamaica I-570
— Malacca Straits II-319
— Sri Lankan beaches II-185
Taranto, Gulf of *I-268*
Tasman Bay, destruction of coralline grounds III-126
Tasmania, catchment management policy II-655
Tasmanian Province II-649
Tasmanian region II-647–60
— coastal erosion and landfill II-655
— effects from urban and industrial activities II-655–8
— major shallow water marine and coastal habitats II-651–2
— offshore systems II-653
— populations affecting the area II-653–4
— protective measures II-658–9
  — international conventions II-658
  — Marine and Estuarine Protected Areas II-658–9
  — state legislation II-658
— rural factors II-654–5
— seasonality, currents, natural variables II-649–51
Tasmanian Wilderness World Heritage Area II-653
Tatarsky Strait *II-464*, II-465, II-475
Taura Syndrome, spread by shrimp-eating birds? III-226
TBT *see* tributyltin (TBT)
TCI *see* Turks and Caicos Islands
Teacapán—Agua Brava lagoon system I-494–5, I-497
— adjacent agricultural area uses N and P fertilisers I-495
— estuarine system I-494
— mangroves I-494–5
$^{99}$Tc, increase in due in Greenland Seas from Sellafield I-11

technetium (Tc), conservative behaviour I-13
tectonic activity
— the Azores I-203, I-211
— extensional, Aegean Sea I-235
— northwestern Arabian Sea II-19, II-27
— off the Mexican Pacific coast I-487, I-488
Tehuantepec, Gulf of, fishery resources I-491
Tehuantepec winds I-488
'teleconnection', Pacific and Equatorial Atlantic, maybe? I-827
temperature, and the water column, Gulf of Alaska III-181
Terceira (Azores), destruction and creation of marshes I-207–9
Terminos Lagoon *I-468*
terraces, Somali coast II-72
terrestrial vegetation, Marshall Islands II-777–8
Texas Louisiana Shelf *I-436*
Thailand II-299
— butyltin compounds in sediments **II-452**
— mangrove loss, uncontrolled conversion to shrimp ponds III-21
— "shifting aquaculture" III-368
— shrimp farming III-175–6
  — move to more intensive methods III-176
  — regulations III-177, **III-177**
  — source of land for III-175–6
— species used in aquaculture III-171
Thailand, Gulf of II-297–308, *II-310*
— coastal erosion, land subsidence and sea-level rise II-307
— coastal habitats II-301–4
— defined II-299
— depletion through human intervention III-120
— effects from urban and industrial activities II-307
— geological description II-299
— offshore systems II-304–5
— physical oceanography II-299–300
— populations affecting the area II-306
  — rural factors II-306–7
— protective measures II-307
— ray and shark species reduced III-120
— seasonality, currents, natural environmental variables II-301
Thermaikos Gulf *I-234*, I-235, I-237, I-244
— increased phytoplankton abundance I-238
— organophosphorus pesticides present I-244
— river input I-244
— sewage pollution I-246
thermal pollution
— Curaçao I-605
— damaging to seagrass beds III-9
— discharges from desalination plants II-10, II-28
— Indonesian power plants II-390
thermal stratification
— around Bermuda I-225
— English Channel I-67
— summer
  — Irish Sea I-85
  — Spanish north coast I-140
thermal vents, deep sea, South Western Pacific Islands region II-712
thermal winds, Gulf of Aden II-50
thermohaline circulation, and climate change III-369–70, *III-370*
Thessaloniki *I-234*, I-244
— summer anoxia I-241
Third River (Main Outfall Drainage), Iraq *II-6*
— may seriously impact Gulf ecosystem II-9, II-13
— purpose of II-9
threshold hypothesis, for economic growth and welfare III-395
tidal amplitude
— Faroes I-33
— La Rance I-72
tidal barrages III-316–18
— construction will change the environment III-318
— double basin systems *318*, III-317
— electricity generation
  — ebb generation III-316–17, III-318
  — flood generation III-317
  — two-way generation III-317
— high costs of, but 21st century development likely III-319
— single basin schemes *317*, III-316
tidal bars *II-71*, II-76
tidal current generation III-318–19
— current and future prospects III-319
— SeaFlow Project III-319
— vertical axis and horizontal axis turbines III-318
  — problem of fixing III-319
tidal currents
— coastal waters, Xiamen region, China II-516
— Guinea coast I-799
— Puerto Rico I-580
  — Guayanilla Bay I-580
  — Mayagüez-Añasco Bay I-580
— strong
  — Irish Sea I-85
  — Yellow Sea II-490–1
— Taiwan Strait II-502
tidal energy III-304, III-315–19
— affects height of oil deposition III-269
— current and future prospects III-319
— harnessing the energy in tides III-316–19
— public perception of III-318
tidal flats
— clay–oil flocculation, reduces oil retention in fine sediments III-271
— vulnerable to oil spills III-271
— Xiamen region II-517–18
— Yellow Sea II-490
tidal marshes
— Cabo de Santa Marta Grande-Chui, southern Brazil I-737
— northwest Florida, impounded for mosquito control I-445–6
tidal power, electricity generation I-54
tidal range
— Baltic Sea I-124
— Galicia I-138–9
tidal regimes, effects of, Madagascar II-117
tidal residual, in homogeneous and stratified water III-188
tidal residual circulation, magnitude of III-188–90
tidal surges I-487, I-488, I-489
tidepools, with algae I-649
tides
— Chagos Islands II-225
  — water accumulates oxygen II-225
— Colombian Pacific Coast I-679
— and currents, affected by monsoons, Sri Lanka II-178
— diurnal and semi-diurnal, Puerto Rico I-580, I-581
— east coast, Peninsular Malaysia II-347
— extracting energy from III-315–16
  — Spring and Neap tides III-316
— harnessing the energy in III-316–19
  — tidal barrage methods III-316–18
  — tidal current generation III-318–19
— Indian Ocean II-66
— influences on III-316
— local tidal currents, Tanzania II-85
— Red Sea, annual, importance of II-39
— and tidal currents, effects of climate change and sea level rise III-188–90
Tierra del Fuego Island I-752–3, *I-753*
— whales I-756–7
Tierra del Fuego Province, coastal economy I-761–2
tiger reserves, India II-156, II-157
Tigris, River II-3
*Tilapia*, in aquaculture, Bahamas I-427
timber exploitation/harvesting
— Colombian Pacific Coast I-681
— problems from I-444
timber industry, Sakhalin, profitable but overharvesting II-469
TINRO Basin *II-464*, II-465

Tiran, Strait of *II-36*, II-37
titanium dioxide processing I-72
titanium—magnesium deposits, Yuzhnoprimorsky region II-480
*Tivela mactroides*, dissipative beaches I-648
Tobago *I-628*, *I-644*
Todos os Santos Bay, Brazil *I-720*
— mining of calcareous sand I-725
— Pinaunas Reef Environmental Protected Area I-728
Toliara region, Madagascar
— barrier reef II-116, II-118
— signs of over-exploitation II-123
— changes to the Grand Récif, 1964–1996 II-123, II-124
— long swell II-116
— types of fishing and target species II-123
Tonga *II-706*, II-707, *II-709*
— agriculture and fisheries II-716
— at risk from sea-level rise II-716
— coastal modifications II-717
— coral reefs II-711–12
— Marine Protected Areas (MPAs) II-720
— population II-714
Tongatapu
— coral reefs II-711
— degraded II-712
— Fanga'uta lagoon II-712
Tongue of the Ocean *I-416*, I-417, I-428
— V-shaped canyons I-422
Tordesillas, Treaty of (1494) I-2
Torres Strait *II-426*
— commercial use of marine resources II-600
— coral reefs II-597
— defined II-595
— environmental management II-608
— fisheries
— artisanal II-602
— management of II-606
— problems of over-exploitation II-602
— reef II-601
— sedentary resources II-602
— subsistence II-601
— high indigenous population II-600
— high seagrass species diversity II-431
— indigenous fishing rights protected II-584
— low level of development II-600
— population mainly on "Inner islands" II-599
— tidal circulation II-596
Torres Strait Baseline Study II-600
Torres Strait and the Gulf of Papua II-593–610
— coastal erosion and landfill II-603
— effects from urban and industrial activities II-603–5
— major shallow water marine and coastal habitats II-597–8
— offshore systems II-598–9
— populations affecting the area II-599–600
— protective measures II-605–8
— environmental management II-607–8
— fisheries management II-605–7
— international agreements II-608
— rural factors II-600–3
— seasonality, currents and natural environmental variables II-595–7
Torres Strait Islands II-583
Torres Strait Treaty (Australia–Papua New Guinea) II-595, II-605–6
— Torres Strait Protected Zone (TSPZ) II-606
*Torrey Canyon* oil spill, increase in *F. vesiculosus* III-272, III-274, III-275
Total Allowable Catch (TAC)
— for exploited Baltic fish stocks I-112
— southwestern Africa I-836
— under CAP I-92
total suspended matter (TSM) III-297
tourism I-215, II-43–4
— Adriatic coasts I-277
— Alaska I-381, I-383
— Anguilla I-620, I-622
— Argentine coast I-758
— around the North Sea I-50, I-51
— Australia II-584
— marine and coastal II-586
— awareness of, Côte d'Ivoire I-816
— the Bahamas, effects on the population I-423
— Belize I-511, I-514
— demands on the coastal zone I-507
— regulation of I-513
— Bermuda I-228
— British Virgin Islands I-620
— Canary Islands I-188–9, I-197
— coastal
— development for unlikely to meet high standards, Mozambique II-108
— Malacca Straits II-316, II-326
— Xiamen region II-519, **II-520**, **II-521**
— and coastal pollution, Sri Lanka II-185
— Colombian Caribbean Coast I-673
— the Comoros II-251
— and coral reefs II-315
— coral reefs important for II-398
— east coast, Peninsular Malaysia II-353–4
— and economic value of coral reefs III-38, III-41
— English Channel coasts I-72
— Fiji Islands II-714, II-758
— impacting on turtles II-757
— Florida Keys I-408
— French Polynesia, importance of II-823
— Godavari–Krishna delta II-172
— Great Australian Bight II-683–4
— Great Barrier Reef II-622
— Gulf of Maine and Georges Bank I 314
— Gulf of Mannar, affects coastal water quality II-166
— Hawaiian Islands II-797, II-806–7
— boating, surfing and submarines II-807
— development of Waikiki and shoreline erosion II-805
— diving and snorkelling II-806
— golf courses II-806
— infrastructure II-806
— sport fishing II-806
— impact of, Great Barrier Reef Marine Park II-626
— Irish Sea coast I-93
— Jamaica, demands on the environment I-567
— Lakshadweep Islands II-194
— and landscapes, west coast, Sea of Japan II-482–3
— Lesser Antilles I-637–8, **I-637**
— the Maldives
— boating, fishing and other effects II-213
— consideration of carrying capacity II-217
— impacts from diving and snorkelling II-213
— impacts from infrastructure II-212–13
— many positive effects II-217
— Marshall Islands II-780
— infrastructures and activities II-783
— Mauritius and Reunion, already well-developed II-264
— Mexican Pacific coast
— development plans I-491
— potential for I-491
— Namibia I-830
— New Caledonia, a developing sector II-732
— Nicaragua, Caribbean coast I-523
— North and South Carolina I-354
— Palk Bay area II-168
— potential, Nicaraguan Pacific coast I-535
— and recreation, Latin America I-462–3
— returning to Montserrat I-620
— round the Tyrrhenian Sea I-278–9
— the Seychelles II-238, II-240
— Sainte Anne Marine National Park II-241

— small industry, Coral, Solomon and Bismark Seas region II-439
— southern Brazil I-737, I-738, I-739
  — destructive side of I-740
  — indiscriminate I-742
— Southern Gulf of Mexico, poorly developed I-475
— Southern Portugal I-161
  — pressure of, Gulf of Cadiz I-179, I-181
— southern Spain I-173, I-177
— southwestern Africa I-835
— Taiwan Strait coral reefs II-504
— Tanzania II-92
— Tonga II-714
— Turks and Caicos Islands I-591
— Tyrrhenian coast I-275
— upward trends in The Maldives II-206
— Venezuela I-654
tourism trends, Hawaiian Islands II-801
Townsville Trench *II-426*
toxic materials, accumulation in parts of Chesapeake Bay I-348
toxic residues, Peruvian coastal waters I-695
toxic waste
— deliberate dumping of II-59
— Indonesia, cradle-to-grave approach II-322
toxic waste trade, Nigeria I-788
toxicity
— in Argentine coast molluscs I-764–5
— of harmful algal blooms III-98
toxin bioaccumulation, from algal blooms III-199
toxins
— defensive III-39
— dinoflagellate III-40
trace elements
— atmospheric input into North Sea I-56
— Bay of Bengal II-274
trace metal pollution, the Philippines II-416
trace metals
— Faroes I-36–7
— Gulf of Maine I-314
— Hong Kong coastal waters II-544
— oysters and sediments, El Salvador I-555
— in sediments I-26
trade, Somalia II-80
trade agreement, affect fisheries management III-146
Trade Winds I-188, I-591, III-369
— Marshall Islands II-775
— Mexican Pacific coast I-488
— and the Northwestern African Upwelling I-190
— and the Puerto Rican wave climate I-577–8
— tropical coast of Brazil I-721
tragedy of the commons I-462
— and fishing management III-154
TRAIN-SEA-COAST III-325–7
— integrated management of coastal and marine areas III-325
— network III-326
  — capabilities of III-326
  — central unit responsibilities III-326
  — development units for specific training priorities III-326–7
  — range of training courses under development III-327
— UN/DOALOS (UN Division for Ocean Affairs and the Law of the Sea) III-325
  — programme of action III-325
TRAIN-X strategy (UN)
— main elements of III-324
— methodology III-325, III-329
— training networks III-324–5
— *see also* CC:TRAIN; TRAIN-SEA-COAST
trans-oceanic floating debris, even in Chagos Islands II-230
transboundary pollution
— affecting Mozambique II-109
— air pollution, Sea of Japan II-476
— oil spills in Malacca Straits II-325
— Xiamen region II-527–8
— Yellow Sea II-496
transboundary straddling stocks/species, the Philippines II-410
transgressive–regressive cycles, Holocene, Argentine coastlines I-762
Transkei coast
— artisanal and subsistence collectors II-140
— overexploitation of mussels II-141
transport, the Maldives, environmental concerns II-213–14
trap fisheries I-397
trawl fisheries
— eastern Bass Strait II-668
— restricted, Western Australia II-697
— southern Brazil I-739
— Venezuela I-653
trawl nets, by-catch excluders II-587
trawling
— affecting Black Sea biota I-290
— banned, by Indonesia II-320, II-338, II-341
— beam trawls, North Sea III-368
— bottom trawling
  — unselective III-124
  — very destructive III-379
— causing damage to the Norwegian continental shelf I-25
— destructive III-122–3
  — Gulf of Aden II-59
— detrimental effect on the English Channel I-76, I-77
— ecological impact of I-178
— and habitat destruction III-379–80
— North Sea I-51, *I-51*, I-52
— pelagic trawls more selective III-124
— restricted in southern Spain I-183
— Southern California Bight I-397
— trawls and dredges scour bottoms I-445
trawling exclusion zones, Gulf of Alaska I-383
treaties, traditional, substituted by global plans and programs of action III-344
Treaty of the Rio de la Plata and its Maritime Front I-463
tributyltin (TBT) I-72, I-95, II-12, II-339, II-449
— in Australia II-588, II-669, II-701
— in the Baltic I-114, I-131
— currently only from larger vessels III-364–5
— degradation of III-249–50
— and endocrine disrupter III-249
— environmental impacts of III-250–1
— high concentrations, Malacca Straits 319
— Hong Kong II-542
— and imposex I-25, I-36–7, I-55, I-75, I-210–11, III-364
— major antifoulant III-248
— a moiety III-248
— as part of free association paints III-248
— persistence of in the environment III-249–50
— in Prince William Sound I-381
— Suva Harbour, Fiji II-719, II-745
— Tagus and Sado estuaries I-158
Trinidad *I-628*
Trinidad and Tobago, estuarine conditions and turbid waters I-630
Triste, Gulf of *I-644*
— coastal retreat I-645
— contamination by oil, domestic and rural wastes I-645
— great industrial impact I-659
— high heavy metal levels I-659
Trobriand Trough *II-426*
trochus fishery
— Fiji, overexploited II-759
— New Caledonia II-730
— Torres Strait II-602
*Trochus* shell II-784, II-820
Tromelin seamount II-255, II-257, II-260
Trondheim *I-18*
— contaminated harbour I-27–8
trophic cascade model III-127

— cases of top-down controls in community structure III-128–9
trophic-magnification disturbances III-221–3
trophodynamic disturbances
— affect habitat supporting organisms III-221–2
— and habitat lost, significant portion of lost GDP III-222–3
Tropical Atlantic Central Water I-579
Tropical Cyclones I-223, I-224
— impact on the Sargasso Sea and Bermuda I-224
tropical habitats, Puerto Rico I-582–4
tropical lows, Puerto Rico I-578
Tropical South Pacific Oceanic Province, Mexico I-489
— oceanic habitat I-489
tropical storms I-488, I-489, I-617, I-631
— "Agnes" (1972), effects of I-341, I-346
— impacts on Chesapeake Bay I-340–1
— Xiamen region II-517
Tropical Surface Waters I-777
tropicalization, of Peruvian coastal waters I-692
troposphere, thermal stratification of, Canary Islands I-188
'trottoir', Tyrrhenian Sea I-273
"Trou sans Fond" canyon, Côte d'Ivoire I-805, *I-806*, I-807, I-808
trout, farming of III-168
*Tsesis* oil spill III-272
tsunamis II-744
— Colombian Pacific Coast I-679
— Coral, Solomon and Bismark Seas region II-429
— Fiji Islands II-753
Tubataha Reef *II-406*
tuna fisheries
— the Azores I-209
— Borneo II-373
— Canary Islands I-196
— Coral, Solomon and Bismark Seas region II-432, *II-433*
— French Polynesia II-819
— Gulf of Guinea I-787
— long line and purse-seine, Colombian Caribbean Coast I-670–1
— longline, Bay of Bengal II-277
— Madagascar II-121–2
— the Maldives II-210
— Marshall Islands II-784
— Mexico I-490
— Mozambique II-105, II-109
— Nicaraguan Pacific coast I-536
— the Philippines II-411
— recreational, Chagos Islands II-230
— South Western Pacific Islands II-718–19
— Southern Gulf of Mexico I-475
tuna–dolphin problem III-141–3
turbidity I-548
— affects kelp beds I-395
— changes in alter biological processes II-193
— coastal, Cameroon I-790
— Côte d'Ivoire I-809–10
— due to seagrass die-off III-14–15
— in fjords I-21
— harmful effects
— in estuaries I-365
— on habitats I-445
— high, inshore, northern Gulf of Mexico I-437
— Kuwait, caused by landfill II-9–10
— offshore, Guinea I-799
— a problem in Bermuda I-228–9
— Tagus estuary I-154
— of water, related to maximum depth of living reef II-397
— of water above seagrasses, Western Australia II-700
turbidity currents, responsible for Bahamian V-shaped canyons I-422
*Turbo* shell II-820
turbot, Baltic predator I-111
turbulence, thermally-induced or mechanically generated III-207
turbulent transport 202
Turkey
— Black Sea fishing industry I-292
— the southern Black Sea I-294–6
— characteristics and population of the coast I-295–6
— development of the coastal areas I-296
— land-based pollution I-296–303
— rural factors affecting the coast I-296
Turks Bank *I-588*
— reef areas I-590
Turks and Caicos Islands I-587–94
— artisanal and non-industrial uses of the coasts I-592
— coastal erosion and landfill I-592
— coastal habitat map III-289
— defined I-589
— major shallow water marine and coastal habitats I-589–91
— offshore systems I-591
— populations affecting the area I-591
— protective measures I-592–3
— regulations to protect local fisheries I-593
— rural areas I-591–2
— seasonality, currents, natural variables I-589
Turneffe Atoll (Island) *I-502*, I-504–5
turtle eggs
— collection now banned, east coast, Peninsular Malaysia II-355
— eaten, Socotra II-57
Turtle Excluder Devices (TEDs) II-127, III-66, III-141, III-143
turtle fisheries I-620
— Mexican Pacific coast I-491
turtle grass I-562, I-601, I-632
— die-off in Florida Bay, a trend for coastal waters in the new Millennium? III-14–16
turtle hatcheries, east coast, Peninsular Malaysia II-356
Turtle Island *II-406*
turtle nesting beaches II-24, II-54, II-350, II-377
— the Comoros II-250
— Mozambique II-105
— St Brandon Islands II-257
— St Martin's Island, Bangladesh II-289
turtle nesting sites I-550
— West Africa I-780
turtlegrass II-598
Tweed-Moreton Bioregion, Australia II-634
Twofold Region, Victoria Province II-663
typhoon shelters, Hong Kong, polluted *II-543*, II-544–5
typhoons II-483
— Cambodia II-571
— Marshall Islands II-776
— present and future effects III-190, III-193, *III-194*
— South China Sea II-552, *II-552*
— Xiamen region II-516, II-517, **II-517**
— Yellow Sea II-489, II-490
Tyrrhenian Sea *I-268*, I-271–2
— algal species of tropical origin I-282
— Atlantic Water in I-272
— Central-Southern, abyssal fauna I-274–5
— coastal erosion I-276–7
— coastline I-270
— deep waters I-272
— effects of urban and industrial activities I-278–9
— effects of winds I-272
— limits I-270
— nutrient availability I-271
— shipping activity intense I-275, I-279
— typical Mediterranean biocenoses I-273

UAE *see* United Arab Emirates
UK
— commitment to offshore wind energy III-310
— 'insensitive structures', potential for legal action III-354
— many laws relating to the coast III-353

— NERC models., southern North Sea I-50
— nutrient input into the English Channel I-74
UK government, White Paper, biodiversity issues in overseas territories I-623–4
UK Overseas Territories in the Northeast Caribbean: Anguilla, British Virgin Islands, Montserrat I-615–26
— Darwin Initiative funds I-623
— legislation and protective measures I-623–5
— prospects and prognoses I-625
UN Conference on Environment and Development (UNCED) III-343
— Agenda 21 III-158
— and the London Dumping Convention reviews III-338–9
— Preparatory Committee, process-oriented criteria III-346
UN Convention on Biodiversity I-147, I-162
UN Convention on the Law of the Sea (LOS: UNCLOS) *see* UNCLOS (UN Convention on the Law of the Sea)
UN Framework Convention on Climate Change *see* Climate Change Convention
UN/DOALOS (UN Division for Ocean Affairs and the Law of the Sea), and the TRAIN-SEA-COAST programme III-325–6
UNCHE III-343
UNCLOS (UN Convention on the Law of the Sea) I-56, I-147, III-158, III-332, **III-333**, III-336
— adopted by the Philippines II-408–10
— authorises littoral states to undertake enforcement measures II-324–5
— basis for international fisheries law, little detail on application III-158
— requirements of ships transitting Malacca Straits II-324
— *see also* Reflagging Agreement; Straddling Stocks Agreement
UNEP
— global conference on sewage III-345
— need to accurately assess program effectiveness III-346
— progress with global 'POPs' Convention III-363
— Regional Seas Convention for East Africa (Nairobi Convention) II-127
— Regional Seas Programme I-2, III-339, III-340
— Kuwait Action Plan I-2
— Mediterranean Action Plan (MAP) I-249
— Red Sea Action Plan II-44
— South Pacific Region I-2
Unguja Island II-85
unique environments, Patagonian coast, Argentina I-756–7
UNITAR (UN Institute for Training and Research)
— approach of I-327
— country team approach I-327
— Climate Change Programme training packages III-328
— development of CC:TRAIN III-327
— regional partners III-327
United Arab Emirates *II-2, II-19*
— coral bleaching and mortality events III-47
— special protection proposed for Khawr Kalba II-31
— *see also* Fujeirah
United Joint Group of Experts on the Scientific Aspects of Marine Pollution (GESAMP), definition of pollution III-258
uplift, Venezuelan coast I-645
upper water column characteristics, the Maldives II-201–2
upwelling index, Alaskan coast I-378, I-380
upwellings II-614
— act as barrier to gene flow and marine organism distribution II-51
— Arabian Sea II-19, II-66
— cool and nutrient rich II-19
— open-ocean and coastal II-19
— stimulate phytoplankton II-24
— Arabian Sea system II-50
— attractive to seabirds III-113
— Bay of Bengal II-274
— Benguela ecosystem II-135
— Cambodia II-571, II-572, II-574
— Chagos Islands II-230
— Colombian Caribbean Coast I-666
— causes special environmental conditions I-666
— recedes in rainy season I-666
— restrict coral formations I-668
— Colombian Pacific Coast I-679
— Côte d'Ivoire I-808, I-809
— equatorial and open ocean I-535
— Galician and Cantabrian coasts I-139, I-141
— Great Australian Bight, provide nutrients to surface waters II-676, II-677
— Guinea coast I-800
— Gulf of Aden II-50–1, II-54
— Gulf of Alaska I-376
— from Ekman pumping I-380, I-381
— Gulf of Guinea I-781
— central subsystem I-778–9
— intensification of winter upwelling I-781
— interannual variability in I-781–2
— intensity may reduce with global warming III-369
— Long Island and New Jersey coastlines I-324
— Madagascan coast II-115
— the Maldives II-202
— North Sea I-46
— Northern Benguela I-823
— and nutrient enrichment, Messina Strait I-281
— off Luzon, South China Sea II-411
— off Peru I-690
— importance of in El Niño events I-691
— off West Greenland coast I-7
— Oman coast II-19, II-50, II-54
— Sea of Okhotsk II-468
— shelf-break area, Argentinian continental shelf I-754
— shelf-edge, Carolinas coast I-353
— Somali coast II-50, II-51
— southeast coast, southern Brazil I-733
— southern Baltic I-111
— Southern Gulf of Mexico I-475
— southern Portuguese coast I-153, I-155
— effects on fisheries I-160
— southern Spanish coast I-172
— southwestern Africa
— central Namibian region I-825–6
— effects of remote forcing from the equatorial Atlantic I-825
— interannual variability I-827
— low-oxygenated bottom water, northern Namibia I-827
— Lüderitz upwelling cell *I-822*, I-823, I-825, I-826–7
— northern Namibian Region I-825
— wind-driven I-824
— upwelling ecosystem, Chilean coast I-706
— Venezuelan coast I-646
— Vietnam II-564
— Yemeni coast II-54
uranium, in the Hoogly (Hugli) river II-278–9
urban development
— coastal, Dutch Antilles I-606
— Colombian Caribbean Coast I-673
— Côte d'Ivoire, encroaching on lagoons and estuaries I-817
— Hong Kong
— pressure on local marine environment II-543
— rapid habitat loss II-541–2
— poorly planned, impact of, Coral, Solomon and Bismark Seas region II-439–40
— Tasmania II-654
— pressure on the marine environment II-657
— Torres Strait and Gulf of Papua II-605
— unplanned, southern Brazil I-740, I-742
— Victoria Province, Australia II-669, II-670
urban, environmental and health problems, the Maldives II-212
— sewage-related problems II-212
urban impacts
— Australia II-586
— Noumea, New Caledonia II-733

urban and industrial activities
— Gulf of Mexico I-477–8
 — artisanal and non-industrial uses I-477
 — industrial uses I-477
 — shipping and offshore accidents I-477–8
— Yellow Sea II-494–5
 — aquaculture and coastal industries II-494
 — oil and oil spills II-495
 — wastewater and solid waste discharges II-494–5
urban and industrial activities, effects of
— American Samoa
 — habitat loss II-770–1
 — industrial activities and impacts II-770–1
 — other urban activities and impacts II-770
— Argentine coast I-763–7
 — commercial fisheries I-763
— Bay of Bengal II-279–80
— British Virgin Islands I-623
— Cambodia II-576–7
— Colombian Caribbean Coast I-673
— Colombian Pacific Coast I-682–4
 — artisanal and non-industrial uses of the coast I-682–3
 — cities I-683
 — industrial uses of the coast I-683–4
— the Comoros II-251
— Côte d'Ivoire I-817–18
 — depletion and degradation of coastal habitats I-817–18
 — oil and gas exploration I-817
 — waste disposal I-817–18
— Dutch Antilles I-502–7
— east coast, Peninsular Malaysia II-353–4
— eastern Australian region II-639–42
 — cities II-640
 — coastal settlements II-640
 — commercial fisheries II-640–1
 — industrial uses II-640
 — mining and dredging II-641
 — ports and shipping II-641
 — regional issues II-641–2
— Fiji Islands II-760–1
— Great Australian Bight II-684–6
— Great Barrier Reef region II-622–4
— Gulf of Aden II-58–9
— Gulf of Guinea I-787–93
 — artisanal and non-artisanal coastal use I-787
 — cities I-788–9
 — dams I-789, I-792
 — industrial fishing I-787–8
 — onshore oil production I-788
 — shipping and offshore I-792–3
 — toxic waste trade I-788
— Gulf of Thailand II-307
— Hong Kong II-543–6
 — anthropogenic contamination widespread II-543
— Lesser Antilles, Trinidad and Tobago I-636–9
— Madagascar II-125
— the Maldives II-210–15
— Marshall Islands II-783–4
— the Mascarene Region II-262–5
— Mozambique II-108–9
— New Caledonia II-731–4
 — artisanal and industrial uses of the coast II-731–2
 — cities II-732–3
 — shipping and offshore accidents and impacts II-733–4
— the Philippines II-414–18
 — artisanal and non-industrial uses of the coast II-414–15
 — cities II-416–17
 — industrial uses of the coast II-415–16
— Sea of Okhotsk II-469–71
— the Seychelles II-239
— shipping and offshore accidents and impacts II-417–18
 — pollution hot spots II-418
— South Western Pacific Islands II-718–19
— southwestern Africa I-835
— Taiwan Strait II-507–10
— Tanzania II-92–3
— Tasmania
 — artisanal and non-industrial coastal uses II-655
 — commercial usage of coastal resources II-655–6
— Torres Strait and Gulf of Papua II-603–5
 — artisanal and non-industrial coastal uses II-603
 — cities II-605
 — fisheries II-604–5
 — industrial uses II-603–4
 — shipping and offshore accidents, impacts II-605
— Vanuatu II-745
— Venezuela I-655–9
 — eastern coastline I-655–6
 — western coastline I-656–9
— Victorian Province, Australia II-667–9
 — commercial fisheries II-667–8
 — industrial uses of the coast II-669
 — recreational fisheries II-668
 — shipping and offshore accidents and impacts II-669
 — urban use (cities) II-669
— Western Guangdong coast II-558–9
 — artisanal and non-industrial uses II-558–9
 — cities II-559
 — industrial uses II-559
 — shipping and offshore accidents and impacts II-559
— western Indonesian II-396–9
urban, industrial and other activities, effects of, Australia II-586–9
urban pollution
— Borneo II-375
— Patagonia, and eutrophication I-765
urban sewage, a pollutant I-24
urban waste, a problem along entire Gulf of Guinea coast I-788–9
urbanisation
— adjoining an estuary, creates problems II-135–6
— of Carolinas coast I-365
— coastal and waste disposal, Tanzania II-92–3
— effects of, Hawaiian Islands II-804–7
— expansion allows debris to enter northern Gulf of Mexico I-450
— Florida Keys I-408
— Great Barrier Reef catchment II-622
— Pearl River delta II-554–5, II-558, II-559
— the Philippines
 — major industrial regions II-417
 — rapid, and informal settlements II-416
— rapid
 — east coast of India II-278
 — effects of I-449
— South Western Pacific Islands II-718
— Southern California I-388
— Vietnam, impact on the marine environment II-566
Uruguay, EcoPlata I-463
USA
— and adjacent waters, Guidelines for the Conservation and Restoration of seagrasses, a synopsis III-8
— Coastal Zone Management Act
 — CZM evaluation model III-347
 — independent evaluation of program effectiveness III-346–7
— eastern, atmospheric nitrogen pathway to watersheds III-201
— Endangered Species Act II-798
— marine reserves III-386
— property rights important in regulation of the coast III-354
— stringent environmental standards for uses of Kwajalein atoll II-786–7
user conflict
— conflict resolution over use of Aliwal Shoal, South Africa II-141, II-142
— from increased resource use II-142

Ushant *I-66*
USSR
— catch reports false, collusion with Japan III-77, III-86
— natives of eastern Siberia, need for whale meat III-83
UV radiation, increase in III-370

Vanuatu *II-706*, II-707, *II-708*, II-717, II-737–49
— agriculture II-714–15
— biogeography II-739
— coastal erosion and landfill II-744–5
— coral reefs II-711
  — cyclone damage II-711
— customary law and reef ownership II-741
— cyclone-prone II-710
— effects of urban and industrial activities II-745
— island groups II-739
— major shallow water marine and coastal habitats II-740–2
— Marine Protected Areas (MPAs) II-719
— offshore systems II-743
— population II-713
— population affecting the area II-743
— protective measures II-745–7
  — marine conservation II-745–6, **II-746**, **II-747**
  — Marine Protected Areas (MPAs) II-746–7
— rural factors II-743–4
— seagrasses II-711
— seasonality, currents, natural environmental variables II-739–40
— unsatisfied demand for fresh fish II-744
vaquita, Gulf of California
— endangered III-90
— incidental catch III-94
Vas Deferens Sequence Index (VDSI) III-250
Vedaranyam Wildlife Sanctuary, Palk Bay II-167
Venezuela I-643–61
— coastal erosion and landfill I-655
— the continental coastline I-645–6
— effects from urban and industrial activities I-655–9
— fisheries I-653–4
— major shallow water marine and coastal habitats I-648–53
— populations affecting the area I-654–5
Venezuela, Gulf of *I-644*, I-645, I-646
Venice, Gulf of *I-268*
— salinity I-271
Vening Meinesz Seamounts II-583
vermetid reefs *I-66*, I-255, I-256–7
vermin and insect infestations, Tanzanian coast II-88–9
vertebrates
— accumulation of TBT in III-250
— endangered, southern Brazil I-737
— Gulf of Aden II-54
Victoria Province, Australia II-661–71
— coastal erosion and landfill II-667
— effects from urban and industrial activities II-667–9
— effects of removal of native vegetation from catchments II-666–7
— Fisheries Management Plans II-670
— Gippsland Lakes, adjusting to changed environment II-667
— human populations affecting the area II-666
— major shallow water marine and coastal habitats II-664–6
— marine species, distribution patterns II-663
— offshore systems II-666
— problems in marine environment II-670
— protective measures II-670
  — endangered and vulnerable species II-670
— rural factors II-666–7
— seasonality, currents, natural variables II-663–4
Victorian Biodiversity Strategy II-670
Vietnam II-299
— coral bleaching and mortality events III-52
— mangroves lost to fishponds III-22, III-25
Vietnam and adjacent Bien Dong (South China Sea) II-561–8
— biodiversity II-564
— impacts from development II-566
— legislation II-566–7
  — Environmental Protection Law II-566–7
  — species in need of protection II-567, **II-567**
— marine and coastal habitats II-564–5
— physical parameters II-563–4
— regional setting II-546
— river systems and estuaries II-566
— water quality II-565–6
Virginian Sea *see* Middle Atlantic Bight
Vistula Lagoon I-109
— fishery I-111
volcanic islands, Mascarene region *see* Mauritius; Reunion; Rodrigues
volcanic pinnacle, off Tasmania II-583
volcanicity
— Coral, Solomon and Bismark Seas region II-429
— latent, American Samoa II-767
— the Philippines II-408
Volterra—Hjort formulation III-79
Vulnerability Index *see* Environmental Sensitivity Index (ESI)

Wadden Sea I-45, I-53
— black and white spots III-263
— decline in eelgrass beds III-7–8
— disappearance of red macroalgae I-48–9
— long-term changes from bottom trawling noted III-126
— suspended sediment I-46
— and *Zostera marina* III-6–7
  — decrease in area suitable for re-establishment III-7
  — large-scale decline III-6
wadi systems, show past erosional processes II-37
wadis
— Al Batinah II-27
— and development of khawrs II-23
— Gulf of Aden coast, flow from percolates into groundwater II-52
Wake Island II-775
— extinction of Wake Rail II-778
— war-time activities II-782
war-time effects, Marshall Islands II-782–3
warming events, El Niño, Northeast Pacific Ocean III-183–4, *III-183*
waste disposal I-634
— an environmental problem, Jamaica I-569
— Côte d'Ivoire I-817–18
— Faroes I-38–9
waste dumping
— English Channel I-77
  — regional, global and European regulation I-77, **I-78**
waste management
— difficulties of, Guinea I-801
— Vanuatu, by industry, minimal II-745
waste plastic pellets, dangers of, Arabian Gulf II-12
wastes
— disposal of, Bangladesh II-293–4
— domestic, dumpsite for, New Caledonia II-733
— dumping of
  — at sea III-365–6
  — Suva Harbour, Fiji II-719
  — in Xiamen coastal waters II-523–4
— Guangdong Province **II-555**
— Halong City, Vietnam II-566
— Turkish industry, input to Black Sea **I-298**, I-302
— urban and industrial, disposal of, Tasmania II-655
— *see also* chemical waste; hazardous waste; industrial waste; radioactive waste; solid waste; toxic waste; urban waste
wastewater
— direct discharge into coastal waters
  — Chinese Yellow sea II-494–5
  — West Guangdong coast II-555
— domestic and industrial, New York Bight I-323
— El Salvador

— domestic, untreated, discharged to marine environment I-555–6
— industrial discharges I-554–5
— entering Amursky Bay II-483
— Hong Kong, arrives in coastal waters II-538
— industrial, Chile I-712–13
— municipal, Chile I-713
— final discharge to the sea I-713
— submarine disposal solution I-715
— problems of discharge to sea, Marshall Islands II-783
— released to the Yellow Sea, Korea II-495
— reuse of, Oman and UAE II-29
— Tasmania, polluting II-657
— treatment seen as a priority, Mauritius II-263, II-264
— urban, causing contamination, southern Portugal I-179
— used for irrigation, Curaçao I-606
wastewater discharge
— major issue in Noumea, New Caledonia II-733
— Peru I-695
— to sea, Gulf of Aden II-58
*Wastewater Management for Coastal Cities* (C. Gunnerson and J. French) I-331
wastewater pollution, Arabian Gulf II-10
wastewater treatment
— Chesapeake Bay
— biological nutrient removal I-347
— improving I-347
— Norway I-27
water
— clean, needed for aquaculture III-170
— Colombian Pacific, from artisanal wells I-680
— potable, demand for, Gulf coast, USA I-448
— resources in Somalia II-79
water bodies and their circulations, around Puerto Rico I-579–81
water clarity/transparency
— Andaman Sea II-302
— decreasing, Baltic Sea I-130
water column habitats, Hawaiian Islands II-796
water column stability, Gulf of Alaska III-181
water masses
— Bien Dong Sea II-563–4, *II-564*
— transformation and spreading of II-564
— Chilean coast I-701–2
— Gulf of Mexico *II-470*
water pollution
— local, Turkish Black Sea coast I-302
— Pearl River delta II-554–5
— the Philippines II-416–17
— Sea of Japan II-477
— sugar and rum industry, Jamaica I-569–70
— threatens Mai Po marshes RAMSAR site II-540
water quality
— American Samoa, a concern II-770
— and aquaculture III-172
— declining, Australia's inland waterways and lakes II-585
— degraded, areas of Victoria coast, Australia II-669
— Great Barrier Reef Marine Park II-626
— impaired
— Adriatic coasts I-278
— northern Gulf of Mexico I-449, I-449–50, I-450
— and industry, Nicaraguan Caribbean coast I-525
— Jakarta Bay and Pulau Seribu, influence of Java on II-397
— Jamaican beaches I-569
— Jiulongjiang estuary II-518
— Malacca Strait II-340–1
— marine, often exceeding standards, east coast, Peninsular Malaysia II-351–2, **II-351, II-352**
— New York Bight I-328–9
— off Carolinas coast I-353
— Papeete Harbour, Tahiti II-817
— parameters, optical sensing III-295–8
— bottom depth and reflectance III-297
— poor
— central New South Wales metropolitan areas II-642
— creeks and groundwater, some areas of French Polynesia II-822
— northern New South Wales rivers II-637
— some improvement, Cubatao, southern Brazil I-743
— South China Sea **II-350**
— Southern Californian shoreline I-399
— Vietnam II-565
— west coast of Taiwan II-509–10
— West Guangdong coast II-555
— Western Australia II-699
— declining II-700–1
water quality parameters, optical sensing
— chlorophyll pigment, coloured dissolved organic matter and total suspended matter III-297–8
— retrieval of chlorophyll content from spectral radiance III-297
— secchi disk depth and diffusive attenuation coefficient III-296–7
water residence time, lagoonal waters, French Polynesia II-816
water scarcity, and dam building, side effects II-306
water surfaces, dry deposition to III-201–2
water temperature, Baltic Proper, seasonal variations I-102–3
water transit time, English Channel I-67
water weeds, invading mangrove canals, Côte d'Ivoire I-818
water-column nitrate inhibition I-358
water-leaving radiance, satellite data of, measurement of diffuse attenuation coefficient III-296
water-purification technology, Tyrrhenian coasts, Arno valley I-280
watershed management planning II-496
watershed management units, Jamaica I-565
wave climate
— deep wave heights I-578
— deep-water, off Fiji II-758
— Puerto Rico
— generated by Trade Winds I-577–8
— swell waves I-578
wave energy III-311–15
— current and future prospects III-315
— extraction of from waves III-311–12
— and persistence of oil III-269
— Puerto Rico I-578
— the resource III-313–15
wave energy programme, Chennai, India II-279
wave exposure, a key role in Belize atolls I-505
wave height, increase in I-46, I-68
wave intensity, and community structure I-22
wave power III-304
wave (swell) surges
— Marshall Islands II-776
— post-hurricane **I-519**
wave-power devices
— absorber mode III-313, *III-314*
— attenuator mode III-314
— design criteria III-313
— enclosed water column devices III-313
— flexible membrane devices III-313
— relative motion devices III-313
— tethered buoyant structures III-313
waves
— energy in is kinetic III-311, III-312
— from Antarctic winter storms, Hawaii II-795
— generation of at sea III-311–12, *III-312*
— Great Australian Bight, west coast swell environment II-675–6
— growth of through differential pressure distribution III-311, *III-312*
— Gulf of Guinea I-778
— interception of by an energy converter 312
— Levantine basin I-256
— modified by atolls, Belize I-503
— permanent surf, Côte d'Ivoire I-808
— the Philippines, generated by North Pacific storms II-407
— Significant Height III-311

— wind-induced, east coast, Peninsular Malaysia II-347
weather effects, the Comoros II-245
West African flyway, Gulf of Guinea a part of I-780
West African manatee I-780
West Bengal
— felling of mangrove forests II-151
— structure of Coastal Regulation Zone (CRZ) II-157–8, **II-158**
West Florida Shelf *I-436*
— seagrass beds I-438
West Greenland
— changes in marine climate I-7
— decline in marine mammal and seabird stocks I-8–9
— lower level of POPs than East Greenland I-13
West Greenland Current I-7
West Guangdong Province coast II-551
— intertidal habitats II-552–3
— phytoplankton and red tide organisms II-553–4
West Iberian large marine ecosystem I-137
West Kamchatka Current II-466
West Kamchatka Shelf, productive fish area II-467
West Wind cold water mass II-676
West Wind Drift Current (WWDC) I-701
Western Atlantic Ocean Experiment (WATOX) I-224, I-229
Western Australian region II-691–704
— coastal erosion, landfill and effects from urban and industrial activities II-700–2
— geomorphology of the coast II-693–4
— Kimberley coast, ria system II-694
— major shallow water marine and coastal habitats II-695–6
— offshore systems II-696
— populations affecting the area II-696–7
— protective measures II-702
  — Western Australian Marine Parks and Reserves Authority II-702
— rural factors II-697–700
— seasonality, currents, natural environmental variables II-694–5
Western Central South Pacific Current II-619
Western Indian Ocean Marine Scientists Association (WIOMSA) II-95–6
Western Somali Basin II-67
wet deposition I-13, I-56, I-223, III-200
wetland forest, Cambodia II-574
wetlands
— Bangladeshi coast II-289
— drained, Tasmania II-655
— forested, Gulf Coast, USA I-440
— Gulf of Guinea, some protected areas I-793
— Jamaica
  — Black River Morass I-565
  — Negril Morass I-564
— loss of
  — American Samoa II-770
  — northern Gulf of Mexico I-445–6, I-448
  — to agriculture, Peru I-695
  — to urban development *II-718*
— Mai Po marshes, Hong Kong II-539–41
— and marshes, North Spanish coast I-139
— miniature, unique, the Azores I-207, *I-208*, I-215
  — destruction and creation of I-207–9, I-211
— and protected areas, Peru I-691–2
  — National Sanctuary of the Mangroves of Tumbes I-691
  — National Sanctuary of Mejia Lagoons I-692
— reclamation of, Sri Lanka II-183–4
— tropical Brazil I-727
— Victoria Province, Australia II-666
Whale Research Programme under Special Permit (JARPA), Japan III-82
Whale Sanctuaries III-84–5
— entire Southern Ocean a sanctuary III-84–5
— further suggestions III-85
— in part of Antarctic Pacific sector III-77, III-84
— proposed for Indian Ocean III-78
whale watching II-251, III-85
— Great Australian Bight II-683
— south east Queensland II-636
whales I-343, I-591, I-834, II-24, II-205, II-582
— in the Azores I-213–14
— baleen (rorqual) whales III-76
— bottle-nosed III-76, III-383
— Colombia I-680
— Great Australian Bight II-680
— Great Barrier Reef II-618–19
— individual, tags and radio tags III-80
— Marshall Islands II-780
— Mexican Pacific coast I-487, I-488
— migrations of II-116
  — Gulf of Guinea I-783–4
— Mozambique waters II-105
— in the Southern California Bight I-399
— Sri Lanka II-180
— *see also* individual types
whales and whaling III-73–88
— assumption about whales as food competitors not sound III-86
— in a changing world III-86–7
— competition III-85–6
— conservation III-75–7
  — Mørch's memorandum III-75
  — quotas controversial III-77
— historical perspective III-74–5
  — expansion of whaling fleets III-75
  — factory ships III-75
  — modern whaling III-74
  — sale of whaling technology to Japan III-74
— how many? what is happening? III-80–3
— just lookin' III-85
— nations withdrawing from whaling III-78
— 'the orderly development of the whaling industry' III-77–8
  — installation of on-ship freezers III-77
— sanctuary III-84–5
— 'scientific' whaling III-78
— sharing resources III-84
— subsistence, and indigenous rights III-83–4
— visual counting and acoustic listening III-79–80
— whales in a changing world III-86–7
whaling I-3, I-36, I-636
— regulation of III-341
whiting I-49
Wider Caribbean I-460–2
— coastal crises I-461–2, **I-461**
— coastline activity **I-462**
— lack of port reception facilities for garbage III-345
— pollution along heavily urbanized/industrialized coasts I-461
widgeon-grass, Carolinas I-358
wildfowl and waders (shorebirds) III-107, III-110
Wildlife Management Areas
— Gulf of Papua II-608
— Papua New Guinea II-441
Wildlife Reserve, Kure Atoll, Hawaiian Islands II-798
Wilkinson Basin I-309, *I-311*
wind power III-304
wind shear III-305
wind speed, varies with height over different roughnesses III-305, *III-305*
wind stress, and residual flows in the English Channel I-67
wind turbines
— lifetimes of III-306
— modified to operate in sea conditions III-305
— new technology for installation in deeper water III-311
— power output of III-304–5, *III-305*
wind waves I-190
winds
— Aegean, summer Etesians I-236

— affecting the Bahama I-418, *I-418*
— hurricanes I-418–19
— Arabian Gulf, Shamal and Kaus II-4
— Azores I-203–4
— Coral, Solomon and Bismark Seas Region II-428, *II-428*
— Côte d'Ivoire
— monsoon I-808
— Northeast Trade/Harmattan I-808
— Gulf of Aden, controlled by monsoons II-49–50
— Gulf of Alaska I-378
— Gulf of Guinea I-776
— Levantine basin I-256
— Mexican Pacific coast I-486, I-487
— monsoonal
— Gulf of Thailand II-300
— may cause damage, Palk Bay II-167
— Northern Benguela I-823
— northern Gulf of Mexico I-439
— northwest Arabian Sea II-19
— prevailing, Red Sea, and extreme sea breezes II-39
— Puerto Rico I-578–9
— *see also* hurricanes; Trade Winds; tropical storms
Winyah Bay *I-352*, I-353
women, trained to restore degraded mangrove forest, Dar es Salaam II-89
World Commission on Environment and Development I-459
World Conference on Fisheries Management and Development III-161
World Heritage Convention, TCI signatories to I-592
World Trade Organization (WTO), hindrance to environmental protection III-159

xerophytic shrub vegetation I-170
Xiamen region, China II-513–33
— clean up of Yuan Dang Lagoon II-531
— coastal waters of II-515
— critical environmental problems II-521–4
— ecosystem changes II-524–5
— geography of II-515
— local agencies with marine waters mandates II-526
— major shallow water marine and coastal habitats II-517–18
— protective measures II-525–32
— a functional marine zonation scheme II-529–30
— institutional arrangements for marine environment management II-526
— land—sea integration II-531–2
— laws and regulations II-525–6, II-529
— major management problems II-526–8
— Marine Management Co-ordination Committee (MMCC) II-528–9
— marine pollution management II-532
— monitoring, surveillance and emergency preparedness II-531
— preparedness and response systems II-532
— present limitations and recommendations II-531
— resource exploitation, utilisation and conflicts II-518–21
— seasonality, currents, natural environmental variables II-515–17
— a Special Economic Zone II-519
— towards integrated coastal management II-528
— new management issues II-528–9
— strategic management plan (SMP) and its implementation II-528

Yellow River, modern source of sediment to the Yellow Sea II-493
Yellow Sea *II-474*, II-487–98
— coastal erosion and landfill II-493–4
— main rivers entering II-489
— offshore systems II-490–1
— physical parameters and environmental variables II-489–90
— population II-491–2
— protection and conservation measures II-495–7
— current conservation and marine protection measures II-496–7
— factors inhibiting conservation II-495–6
— rural factors II-492–3
— shallow water habitats II-490
— shift away from demersal fish III-128
— urban and industrial activities II-494–5
Yellow Sea Large Marine Ecosystem (YSLME) programme II-496–7
Yellow Sea Warm Current II-489
Yellowfin fishery II-56
Yemen *II-18*, *II-36*, II-43, II-49
— beaches and mudflats important for seabirds and waders II-54
— dense and flourishing coastal algal community II-51
— fishing subsidies II-43
— limestone cliffs II-19
— new licensing round for oil and gas exploration II-59
— population II-25
— seasonal rainfall on southern mountains II-50

Zagros Mountains *II-18*, II-19
zährte, in southern Baltic rivers I-111
Zanzibar *II-84*, II-88
— problems of tourism expansion II-92
Zeehan Current II-649, *II-649*, II-650, *II-674*
Zhejiang—Fujian Coastal Current II-489, II-516
zinc pollution, Greenland I-12
zinc tolerance I-75
zinc (Zn) I-162
— Azorean amphipods I-213
zoogeographic provinces, South African shore II-135
zooplankton
— Antarctic I-707
— Bay of Bengal II-275–7
— Black Sea I-290–1
— Carolinas
— estuarine I-359
— marine I-360
— English Channel I-70
— Guinea upwellings, Senegal—Mauritanian, cold waters from I-800
— Gulf of Alaska I-379–80
— Gulf of Guinea I-779, I-783
— Irish Sea I-89–90
— Malacca Straits II-312
— the Maldives II-202
— North Spanish coast I-138, I-141
— offshore, Côte d'Ivoire I-814–15
— Peru, affected by El Niño I-692–3
— Southern Gulf of Mexico I-474
— southwestern Africa I-830, I-832
— Tagus estuary I-157
— Tyrrhenian Sea I-274
— Vietnam II-564
— West Guangdong coast II-554
zooplankton community, Deltaic Sundarbans II-150
*Zostera* I-70
— Amursky Bay, degradation of II-481
— *Z. marina*
— propagation of from seed III-4–5
— recovery of I-70
— and the Wadden Sea III-6–7